PHYSICAL CONSTANTS

Acceleration of gravity at sea level	$g = 9.807\ \text{m/s}^2$
	$= 32.174\ \text{ft/s}^2$
Avogadro's number	$N_A = 6.024 \times 10^{26}$ atoms/kg·mol·K
Boltzmann's constant	$k = 1.380 \times 10^{-16}$ erg/K·molecule
	$= 1.380 \times 10^{-23}$ J/K·molecule
Planck's constant	$h = 6.625 \times 10^{-27}$ erg·s/molecule
	$= 6.625 \times 10^{-34}$ J
Speed of light in a vacuum	$c = 2.998 \times 10^8$ m/s
Universal gas constant	$\overline{R} = 8.314\ \text{kPa}\cdot\text{m}^3/\text{kg}\cdot\text{mol}\cdot\text{K}$
	$= 8.314$ kJ/kg·mol·K
	$= 8314$ N·m/kg·mol·K
	$= 0.08205$ liter·atm/g·mol·K
	$= 1545\ \text{ft}\cdot\text{lb}_f/\text{lb}_m\cdot\text{mol}\cdot{}^\circ\text{R}$
	$= 1.986\ \text{Btu}/\text{lb}_m\cdot\text{mol}\cdot{}^\circ\text{R}$
	$= 1.986$ cal/g·mol·K
	$= 10.73\ \text{psia}\cdot\text{ft}^3/\text{lb}_m\cdot\text{mol}\cdot{}^\circ\text{R}$

$g_c = \frac{32.2\ \text{lbm}\cdot\text{ft}}{1\ \text{lbf}\cdot\text{s}^2}$

SECOND EDITION

Thermodynamics

English/SI Version

WILLIAM Z. BLACK
Georgia Institute of Technology

JAMES G. HARTLEY
Georgia Institute of Technology

HarperCollins*Publishers*

Sponsoring Editor: Don Childress
Project Editor: Kristin Syverson/Andrea B. Coens
Art Direction and Text Design: Julie Anderson
Cover Coordinator: Julie Anderson
Cover Design: Tom Kosak Design
Cover Illustration/Photo: © Morocco Flowers
Photo Research: Karen Koblik
Production Administrator: Beth Maglione
Compositor: Waldman Graphics, Inc.
Printer and Binder: R. R. Donnelley & Sons Company
Cover Printer: The Lehigh Press, Inc.

Thermodynamics, Second Edition, English/SI Version

Library of Congress Cataloging-in-Publication Data
Black, William Z.
Thermodynamics / William Z. Black, James G. Hartley.—2nd ed., English/SI version.
p. cm.
Includes bibliographical references and index.
ISBN 0-06-040734-4
1. Thermodynamics. I. Hartley, James G. II. Title.
QC311.B64 1991
536'.7—dc20 90-49800
CIP

90 91 92 93 9 8 7 6 5 4 3 2 1

Contents

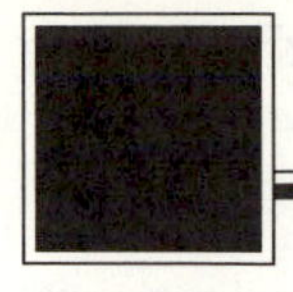

Preface

This text presents an introductory treatment of thermodynamics designed for use in undergraduate engineering courses. The classical approach to thermodynamics is emphasized, and a background in elementary physics and calculus is assumed. The material is designed to be covered in three quarters or two semesters. Subject material is presented in a simple, straightforward manner, using examples to familiarize the student with those subjects that are known to be difficult to comprehend.

Throughout the text we emphasize a physical understanding of the thermodynamic processes. Each chapter contains numerous examples, which are often used to expand on and clarify text material. The examples draw on familiar experiences, and most are derived from practical problems. Each example problem is worked in detail, and particular attention has been given to the proper use of units and unit conversions in the solutions. Many of the end-of-chapter problems are also formulated so that they model practical engineering situations. The problems are arranged so that the first few that deal with a given topic require routine calculations, but later problems are more sophisticated, requiring the student to exercise engineering judgment and make reasonable assumptions in order to obtain a solution.

After introductory definitions are established, properties of substances are discussed in detail, and emphasis is placed on the use and understanding of tabulated property data. The concepts of the ideal gas and the incompressible substance are then introduced as approximations to the actual behavior of pure substances. In particular, the range of applicability of the ideal-gas equation of state is explained and emphasized.

Properties of ideal gases, incompressible substances, and pure substances are discussed prior to the introduction of the basic conservation principles so that the applications of these concepts are not limited to a particular substance. As the conservation princples are introduced, examples and problems can, therefore, be applied to a complete spectrum of substances including solids, liquids, gases, and mixtures.

The developments of the conservation-of-mass and conservation-of-energy equations are carefully designed to closely parallel each other, and each is discussed in a separate chapter. The equivalent treatment allows the student to become familiar with the development of general conservation equations for any conserved quantity. By introducing the conservation of mass first, we can build on concepts discussed in introductory courses such as chemistry and physics.

The conservation equations are first developed in a general form, which is later simplified for the special cases of open or closed systems, uniform flow, and steady-state conditions. Emphasis is placed on a thorough understanding of the physical meaning of each term in the conservation equations. Detailed explanations accompany the simplifications when the general equations are reduced to the forms that apply to the special cases. With this approach we believe that the student will gain a better understanding of the conservation principles as well as greater awareness and appreciation of these powerful analytical tools. Examples specifically address the questions of which form of the conservation laws should be used and why certain assumptions can be applied to simplify the solutions.

Early in the text the student is introduced to an organizational framework that can be used to advantage in the solution of all thermodynamic problems. The recommended procedure includes a rational method of problem formulation, criteria for simplifying the governing equations, and a systematic approach that can be used to solve for the desired quantities. The purpose of the organizational procedure is to establish a logical approach to the application of the basic thermodynamic principles so that the student can begin to master the confusing aspects of thermodynamic analyses. The recommended procedure is used consistently in the illustrative examples.

The development of the second law of thermodynamics is similar to that used for the introduction of the conservation of mass and energy. The coverage of the second law is more extensive than that found in most other undergraduate thermodynamics texts. We believe that the second-law analysis is becoming increasingly important and that it should play a major role in the analysis of engineering problems. The results of the second law, therefore, are carried over into subsequent chapters where they are applied to thermodynamic systems such as power and refrigeration cycles as well as air-conditioning processes.

A short summary is placed at the end of each chapter as an aid to chapter review. The more important material introduced in the chapter is summarized in a brief and concise fashion.

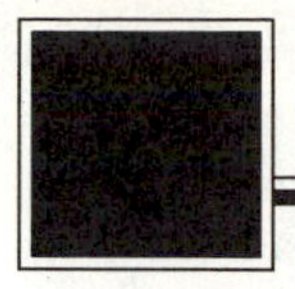

New to this Edition

This edition of *Thermodynamics* features a number of significant revisions and additions that will improve and enhance the teaching and understanding of classical thermodynamics. Although many changes have been made throughout the text, the major revisions and additions include the following:

- The text and problems have been revised to include both English and SI units.
- A thermodynamic property software package has been included with the text.
- A computerized tutorial program for solving thermodynamics problems has been provided.
- Nearly 40 percent new problems have been added.
- The treatment of the second law has been improved and clarified.
- Photographs of common thermodynamic equipment have been included.

The decision to include both SI and English units in this edition was motivated by interactions with and comments from faculty members throughout the United States. Many faculty members prefer to use a text that employs SI units exclusively, and a separate edition of this text is available to fulfill their needs. Many other faculty members, however, prefer to have a mixture of SI units and English units, and we have provided equal treatment of those two systems of units with this edition in the example problems and in the end-of-chapter problems. A complete set of tables for each system of units is included in the Appendixes.

New formulations have been used to generate the thermodynamic properties for steam, refrigerant-12, and common gases at low pressures. The steam tables are based upon the 1984 formulation from the NBS/NRC Steam Tables. The formulation used for the properties of refrigerant-12 is that presented in the 1986 *ASHRAE Thermodynamic Properties of Refrigerants*. The properties for gases at low pressure are derived from data contained in the 1983 *Gas Tables* by Keenan, Chao, and Kaye and from data in the 1986 JANAF Tables. Since the SI-and-English-unit tables for each substance were generated from the same formulation, the values are entirely consistent from one table to the other for each substance. We have expanded the superheat tables for steam and the saturation and superheat tables for refrigerant-12. We also have presented all of the gas tables in Appendix D on a mass basis rather than having only the properties of air on a mass basis. In addition, we have included values for the reduced pressure, the reduced specific volume and the constant-pressure specific heat in each of the gas tables.

The programs used to generate the values contained in the tables for properties of steam, refrigerant-12, and air plus eight other low-pressure gases are available on a diskette that is included as part of this textbook. The diskette contains property software that is menu-driven and user-friendly. The software can be used to evaluate properties in either SI or English units, and it greatly simplifies the often tedious task of determining values for thermodynamic properties.

In conjunction with this edition of the text, HarperCollins has entered into an agreement with S.A. Klein and W.A. Beckman at the University of Wisconsin-Madison to provide their excellent tutorial program, CP/Thermo, to schools that adopt the text. CP/Thermo includes more than 100 problems covering most of the subjects appearing in the text. Each problem is designed to teach specific concepts. The CP/Thermo problems are included with the end-of-chapter problems and are identified with a computer icon placed in the margin of the text. They may be integrated into a thermodynamics course along with conventional homework assignments.

CP/Thermo encourages problem solving, which is essential to learning thermodynamics. Tutorial program assignments offer significant advantages over conventional homework. They assist the student in mastering conceptual difficulties, and they encourage the student to persevere in the problem-solving process. Furthermore, CP/Thermo solutions require little algebraic manipulation. Unhampered by mundane tasks, the student is free to concentrate on learning the underlying principles of basic thermodynamics and the methodology of problem solving.

In this edition we have increased the number of end-of-chapter problems by nearly forty percent. We have not only increased the quantity of problems, but we have also increased the quality of the problems by including more problems that are design-oriented, more problems that investigate parametric analyses with the aid of computer solutions, and more problems that are open-ended in nature. We have also reordered the problems at the end of each chapter so that they more closely reflect the order of presentation of concepts and principles in the text.

The treatment of the second law of thermodynamics has been improved in this edition. The proof of the Clausius inequality has been made clearer, and the introduction of entropy in Chapter 5 has been streamlined. In addition, the derivation of the general equation for the rate of change of total entropy in Chapter 6 has been made more logical. These new developments of the second law are more straightforward, and they will be easier for the student to follow and understand. A thorough comprehension of those concepts and their meaning is extremely important for the student, and the new approaches provide more clarity and better physical interpretations of those concepts.

Throughout the text, we have improved the clarity of presentations. To complement this effort, we have included photographs that will give the student a greater appreciation of components and systems that are integral parts of actual thermodynamic systems.

Although we have made many improvements to this second edition, the primary goals set out in our first edition remain unchanged. We develop the fundamental concepts in a pedagogically logical order and lead students through the difficult areas in thermodynamics, especially those that are encountered in a first course in thermodynamics. We have retained those elements of the first edition that students have found useful in the learning process. We introduce the concepts of conservation of mass and energy in a parallel fashion, building a firm foundation in the student's mind with the conservation of mass before turning to the more difficult concept of energy. We first derive the conservation equations in a general form and then simplify them for specific applications such as closed systems, open steady systems, and open transient systems.

We also have retained our emphasis on thorough explanations of thermodynamic principles and processes and on a logical, organized approach to the analysis and solution of thermodynamics problems. Each example problem in the text has been worked out in detail, and particular attention has been given to making appropriate assumptions, using units and conversions properly, and reflecting upon the physical implications of results obtained.

Acknowledgments

We would like to thank the following associates and colleagues who have provided valuable contributions and suggestions during the revision process: Keith Herold at the University of Maryland, William Moses at Texas A & M University, Peter A. Liley at Purdue University, Muhammad Metghalchi at Northeastern University, and John Henry at Penn State University. We also acknowledge suggestions from Georgia Tech students who used the text while it was in its early stages of development. Their comments and questions have played a major role in the evolution of the material in this edition.

We especially wish to thank our good friend and colleague Professor Alan Larson at Georgia Tech for his careful review of the first edition, his insightful comments, and his penetrating questions. His thoughtful suggestions and contributions, which have led to many improvements in our approach to teaching thermodynamics, are too numerous to list. We express our appreciation to Rosie Atkins for her exceptional typing skills, her attention to detail, and her patience in dealing with the many revisions to the manuscript. Finally, we wish to renew our expressions of gratitude to our wives, Linda and Julene, who continue to offer the encouragement, patience, and understanding, without which we would not have completed this text.

William Z. Black
James G. Hartley

Nomenclature

Symbol	*Definition*	*SI units*	*English units*
a	Acceleration	m/s^2	ft/s^2
a	Specific Helmholtz function	kJ/kg	Btu/lb_m
A	Helmholtz function	kJ	Btu
A	Area	m^2	ft^2
AFR	Air-fuel ratio		
c	Specific heat	kJ/kg·K	$Btu/lb_m \cdot °R$
C	Arbitrary constant		
c_p	Constant-pressure specific heat	kJ/kg·K	$Btu/lb_m \cdot °R$
c_v	Constant-volume specific heat	kJ/kg·K	$Btu/lb_m \cdot °R$
d	Differential of a point function		
d,D	Diameter	m	ft
e	Specific total energy	kJ/kg	Btu/lb_m
e_k	Specific kinetic energy	kJ/kg	Btu/lb_m
e_p	Specific potential energy	kJ/kg	Btu/lb_m
E	Total energy	kJ	Btu
E_k	Total kinetic energy	kJ	Btu
E_p	Total potential energy	kJ	Btu
f	Functional relation		
F	Force	N	lb_f
g	Acceleration of gravity	m/s^2	ft/s^2
g	Specific Gibbs function	kJ/kg	Btu/lb_m
g_c		$kg \cdot m/N \cdot s^2$	$lb_m \cdot ft/lb_f \cdot s^2$
G	Gibbs function	kJ	Btu
h	Vertical height	m	ft
h	Specific enthalpy	kJ/kg	Btu/lb_m
H	Enthalpy	kJ	Btu
HHV	Higher heating value	kJ/kg·mol	$Btu/lb_m \cdot mol$
i	Specific irreversibility	kJ/kg	Btu/lb_m

Symbol	*Definition*	*SI units*	*English units*
I	Irreversibility	kJ	Btu
$\dot{I}$	Irreversibility rate	W	Btu/h
k	Ratio of specific heats c_p/c_v		
K_p	Equilibrium constant		
L	Length	m	ft
LHV	Lower heating value	kJ/kg·mol	$\text{Btu/lb}_\text{m}\cdot\text{mol}$
m	Mass	kg	lb_m
$\dot{m}$	Mass flow rate	kg/s	$\text{lb}_\text{m}\text{/s}$
M	Molecular weight	kg/kg·mol	$\text{lb}_\text{m}\text{/lb}_\text{m}\cdot\text{mol}$
MEP	Mean effective pressure	kPa	$\text{lb}_\text{f}\text{/in}^2$
mf	Mass fraction		
n	Polytropic exponent		
N	Number of moles		
$p;p_i$	Partial pressure; component pressure	kPa	$\text{lb}_\text{f}\text{/in}^2$
P	Pressure, partial pressure	kPa	$\text{lb}_\text{f}\text{/in}^2$
P	Product		
q	Specific heat transfer	kJ/kg	Btu/lb_m
Q	Heat transfer	kJ	Btu
$\dot{Q}$	Heat-transfer rate	W	Btu/h
r	Compression ratio		
R	Gas constant	kJ/kg·K	$\text{Btu/lb}_\text{m}\cdot{}^\circ\text{R}$
R	Reactant		
r_c	Cutoff ratio		
r_p	Pressure ratio		
R	Universal gas constant	kJ/kg·mol·K	$\text{Btu/lb}_\text{m}\cdot\text{mol}\cdot{}^\circ\text{R}$
s	Displacement	m	ft
s	Specific entropy	kJ/kg·K	$\text{Btu/lb}_\text{m}\cdot{}^\circ\text{R}$
S	Entropy	kJ/K	Btu/°R
Sp.Gr.	Specific gravity		
t	Time	s	s
T	Temperature	°C or K	°F or °R
T	Torque	N·m	$\text{ft}\cdot\text{lb}_\text{f}$
u	Specific internal energy	kJ/kg	Btu/lb_m
U	Internal energy	kJ	Btu
v	Specific volume	$\text{m}^3\text{/kg}$	$\text{ft}^3\text{/lb}_\text{m}$
V	Velocity	m/s	ft/s
$\mathbf{V}$	Volume	m^3	ft^3
w	Specific work	kJ/kg	Btu/lb_m
W	Weight	N	lb_f
W	Work	kJ	$\text{ft}\cdot\text{lb}_\text{f}$
$\dot{W}$	Power	W	hp

Symbol	Definition	SI units	English units
x	Quality		
X	Arbitrary point function		
y	Mole fraction		
Y	Arbitrary path function, extensive property		
z	Elevation	m	ft
Z	Compressibility factor		
Greek symbols			
β	Coefficient of performance		
β	Volume expansivity	1/K	1/°R
γ	Specific weight	N/m^3	lb_f/ft^3
Δ	Finite change in a quantity		
δ	Differential of a path function		
ϵ	Degree of chemical reaction		
ϵ	Electric potential, voltage	V	V
ϵ	Second-law efficiency		
η	Efficiency		
θ	Angle	° or rad	° or rad
κ	Isothermal compressibility	m^2/N	ft^2/lb_f
μ	Joule-Thompson coefficient	$m^2 \cdot K/N$	$ft^2 \cdot °R/lb_f$
ν	Coefficients in stoichiometric equation		
ρ	Density	kg/m^3	lb_m/ft^3
Σ	Summation		
Φ	Closed-system availability	kJ	Btu
ϕ	Relative humidity		
ϕ	Closed-system availability on a unit-mass basis	kJ/kg	Btu/lb_m
ϕ	Stream or open-system availability	kJ/kg	Btu/lb_m
ω	Humidity ratio		
ω	Angular velocity	rad/s	rpm
Subscripts			
a	Actual, air		
A	Based on Amagat's law		
abs	Absolute		
act	Actual		
atm	Atmospheric		
avg	Average		
b	Boiler		
B	Based on Bartlett's law		

Symbol	*Definition*	*SI units*	*English units*
Brayton	Brayton cycle		
c	Condenser, compressor, cutoff, cooling, combustion		
c	Critical value		
cd	Condenser		
cp	Compressor		
Car	Carnot		
cyc	Cycle		
CL	Compressed liquid		
D	Based on Dalton's law		
diesel	Diesel cycle		
dp	Dew point		
db	Dry bulb		
elec	Electric		
e	Exit conditions		
e	Evaporator		
E	Heat engine		
env	Environment		
Ericsson	Ericsson cycle		
f	Saturated liquid; final; formation		
fg	Change in property between saturated-liquid and saturated-vapor states		
g	Saturated vapor		
H	Heat pump, high temperature		
h	Heating		
i	Inlet condition; initial		
int	Internal		
i	ith component		
int rev	Internally reversible		
irr	Irreversible		
k	Property of the surroundings; kinetic energy		
K	Based on Kay's rule		
L	Low temperature		
m	Mixture		
max	Maximum		
min	Minimum		
n	Normal component		
N	Nozzle		
p	Pump, propulsive; potential energy; pressure		

Symbol	*Definition*	*SI units*	*English units*
P	Products		
prod	Production, product		
r	Relative value, reheat		
R	Reduced value, reservoir, refrigerator, reactants, reversible		
ref	Reference state		
regen	Regeneration		
res	Reservoir		
rev	Reversible		
s	Isentropic path, steam, component in the direction s		
sat	Saturated		
stm	Steam		
surr	Surroundings		
sys	System		
Stirling	Stirling cycle		
t	Tangential component		
t, T	Turbine		
th	Thermal		
tot	Total value		
v	Water vapor		
wb	Wet bulb		
0	Dead state, environment, reference state		
1, 2	Time 1, time 2, or state 1 state 2		
Superscripts			
dot	Rate quantity or quantity per unit time		
bar	Molar quantity or quantity per unit mole		
prime	Pseudo value		
$^\circ$	Used for entropy function, s°		
^{0}	Standard reference state h^0, g^0, etc.		

1 Concepts of Thermodynamics

1.1 INTRODUCTION

Thermodynamics is a physical science; that is, the principles that form the framework of thermodynamics are all based on observations of physical phenomena. Following the observation of a phenomenon, experimental evidence is collected to verify that the observation is indeed a correct one. Finally, once the principle has been accepted, the physical observation can be recast into a mathematical formulation that will provide a mechanism by which the principle can be applied to engineering problems.

A large portion of the subject matter of thermodynamics deals with a study of energy. In fact, many people define thermodynamics as a study of energy and its relationship with the properties of matter. While most people are familiar with the concept of energy, few are able to give a rigorous definition of energy. On a very simplified level, *energy* could be defined as a capacity to produce change. The energy output of an automobile engine provides the capacity to move from one location to another. The energy output of a power plant provides the capacity to produce a wide variety of changes—to operate motors, television sets, and lights, to name only a few possibilities. Energy derived from petroleum products can be used to power many different devices. Solar energy provides a capacity for change by heating water and air for comfort purposes.

The basic principles that are the starting point for the study of thermodynamics are the conservation of mass, the conservation of energy, and the second law of thermodynamics. The conservation of mass and energy are usually discussed in some detail in introductory courses in physics, so most students are somewhat familiar with these basic principles. The second law of thermodynamics, however, is usually unique to a course in thermodynamics, and it is a basic principle that is developed from the physical observation that without external sources of energy, heat transfer always occurs in a preferred direction; that is, heat transfer is always from a region of high temperature to a region of lower temperature. From this observation the concept of entropy can be formulated and used to predict whether a particular process can occur and to what extent the process will occur.

Thermodynamics provides important relationships among heat transfer, work interactions, kinetic and potential energy, and quantities that are called *properties*, which describe the condition of any substance. In fact, a major contribution of thermodynamics is the mathematical relationship between the amount of energy that is transferred to a substance and the change in the properties of that substance. This relationship is used to study the operation of devices that utilize and transform the various forms of energy. Thermodynamics is therefore particularly important in an era of dwindling supplies of readily available energy and increased interest in energy conservation.

In the preceding paragraph terms such as *work interaction, heat transfer, kinetic energy*, and *potential energy* were used without definition, but most students should be familiar with these terms since they are used in physics, statics, and dynamics courses. Rigorous definitions of these terms are included later in this chapter.

Traditionally, the study of thermodynamics has emphasized applications to devices such as turbines, pumps, engines, compressors, air conditioners, and so on. The association of thermodynamics with predominantly mechanical devices is extremely restrictive and somewhat unfortunate because it gives a narrow view of areas where thermodynamics can be applied. Actually, the principles of thermodynamics apply equally well to other devices of contemporary interest such as solar collectors, MHD generators, rocket engines, fuel cells, wind and wave energy systems, and other systems that transform energy from one form to another. As the basic concepts of thermodynamics unfold, devices used to illustrate the basic principles will be seen to stem from an extremely broad cross section of disciplines. A firm command of thermodynamics, therefore, is essential to practically every phase of an engineer's or scientist's career.

Because thermodynamics deals with such a broad and diverse subject as energy, it is traditionally introduced early in a student's formal education. The principles are expanded in a course in thermodynamics and carry over to courses in fluid mechanics and heat transfer, two other disciplines that, along with thermodynamics, are an integral part of a broader area referred to as the *thermal sciences*. Thermodynamics also has an important impact on the design of engineering systems, and it plays a major role in the selection of materials as well as in the design methodology of practically all engineering systems.

Before you proceed with the study of thermodynamics, a few words of caution are appropriate. Studying thermodynamics can be compared to constructing a building. The structural integrity of the building can be guaranteed only if the foundation is sound. Similarly, a thorough understanding of thermodynamics can be ensured only if the knowledge of a few underlying principles is sound. The analogy is true of practically all courses in engineering and science, but achieving the results of the analogy in thermodynamics is often complicated by the facts that the introductory material appears to be introduced rather slowly and that the accompanying mathematics is on a very fundamental level. Students often overlook the subtle implications of this introductory material, and they frequently achieve a false sense of security early in the course. They are often tempted to race through the first few chapters before acquiring a firm grasp of the basic concepts. This approach may be successful for a short time, but weaknesses will soon develop as a result of an incomplete understanding of the fundamental and underlying principles.

This chapter begins with definitions of several terms, such as *state, process, system*, and *property*, that will be used repeatedly throughout the text. The section on definitions is followed by a brief discussion of the two systems of units that are most commonly encountered in technical fields. The Système Internationale d'Unités (SI), or *International system*, is used in many of the examples and end-of-chapter problems, while the *English system* is used in the remainder of the problems. The next two sections are devoted to a discussion of pressure and temperature, and the chapter concludes with a discussion of two distinctly different forms of energy transfer: heat transfer and work interactions.

1.2 DEFINITIONS

1.2.1 Systems: Closed, Open, and Isolated

Common engineering terms such as *system, property, process*, and *path* have subtle but distinct differences in meaning in various engineering disciplines. In thermodynamics these terms have particular significance, and an understanding of their definitions in the proper thermodynamic context is essential. A *thermodynamic system*, or simply a *system*, is a region enclosed by an imaginary boundary that may be rigid or flexible. The imaginary boundary often coincides with a physical boundary. The concept of a system is essential in analyzing practically all thermodynamic problems.

Systems may be classified as being closed, open, or isolated. A *closed system* is one for which no mass crosses the boundary of the system. Figure 1.1(a) shows an example of a closed system. The imaginary system boundary, indicated by a dashed line, encloses the gas within the piston-cylinder assembly. Since the piston is assumed to fit tightly within the cylinder, no mass is permitted to cross the imaginary system boundary shown in Figure 1.1(a). The fact that a system is closed does not eliminate the possibilities of energy crossing the system boundary or of the system changing its shape. If a bunsen burner is placed below the piston and cylinder shown in Figure 1.1(a), energy will cross the boundary of the system and the temperature of the gas will increase. The gas will therefore expand and cause the piston to move upward. *Energy* clearly crosses the boundary of the system in this example, and yet the system is classified as being closed because no *mass* crosses the boundary.

Systems that permit the transfer of both mass and energy across their boundaries are called *open systems*.* Figure 1.1(b) shows an example of an open system consisting of a solar collector that uses the sun's energy to heat water. The dashed line in the figure

*Some authors refer to an open system as a *control volume*. An open system is equivalent in every respect to a control volume, but the term *open system* is used throughout this text because it specifically implies that the system can have mass and energy crossing the system boundary.

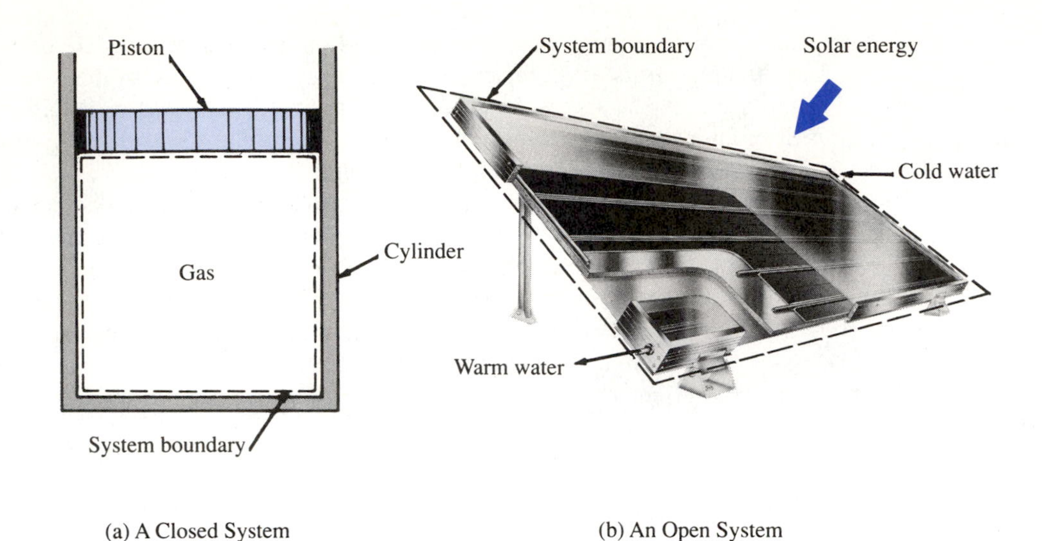

Figure 1.1 Examples of thermodynamic systems. (a) A closed system. (b) An open system. [(b) Courtesy of American Energy Technologies, Inc.]

represents the boundary of the open system. Since water crosses the boundary at two locations, the water and collector constitute an open system.

A third type of system, called an *isolated system*, has neither mass nor energy crossing its boundary. While practical examples of isolated systems are rare, the concept of an isolated system is particularly useful in formulating the principles derived from the second law of thermodynamics that are introduced in Chapter 5.

A system can consist of a single substance, as in the piston-cylinder example illustrated in Figure 1.1(a), or it can consist of several substances, as in the solar collector example, where the system boundary surrounds the collector material as well as the water inside the collector tubes. Most often the choice of a system will be limited to a single substance and a single device, but occasionally a combination of devices such as an entire power plant might be chosen as a system for convenience. Through an analysis of a complex, multicomponent system such as a power plant, we can draw general conclusions regarding the overall operation of the system without becoming involved in analyzing the detailed operation of any one of its individual components.

All thermodynamic systems consist of three basic elements: the imaginary surface that bounds the system, called the *system boundary*; the volume within the imaginary surface, called the *system volume*; and the surroundings. The *surroundings* are defined as everything external to the system. The piston-cylinder discussed earlier can be used to illustrate the elements of a system, as shown in Figure 1.2. The system volume is composed of the space occupied by the gas inside the piston and cylinder. The imaginary system boundary that encloses the gas coincides with the interior surfaces of the piston and cylinder. The surroundings for this example consist of everything outside the system boundary, including the piston and the cylinder.

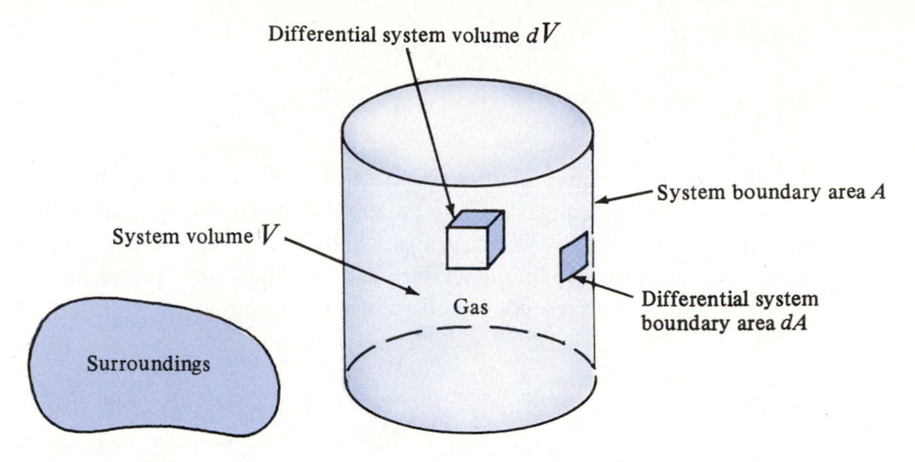

Figure 1.2 Three elements of all thermodynamic systems.

All energy or mass that enters or leaves a system must cross the surface area of the system boundary. When it does so, the properties inside the volume of the system may change. One of the objectives of our initial study of thermodynamics is to relate the amount of energy and mass that enter and leave the system to the changes in properties experienced by the mass within the volume of the system. For example, as our study of thermodynamics proceeds, we will be able to relate the amount of solar energy that is absorbed by the collector in Figure 1.1(b) to the mass flow rate and temperature rise of the water as it passes through the open system.

1.2.2 Property, Equilibrium, State, and Process

A *property* is any measurable characteristic of a system. Familiar examples of properties are pressure, temperature, volume, and mass. Still other properties might come to mind, such as viscosity, modulus of elasticity, thermal expansion coefficient, coefficient of friction, and electrical resistivity. Some properties are defined in terms of other properties. An example is the *density* ρ of a substance, which is defined* as the mass of a substance per unit volume, or

$$\rho \equiv \frac{m}{V} \tag{1.1}$$

Another property used frequently in thermodynamics is the *specific volume*, which is defined as the volume per unit mass. The specific volume v of a substance is therefore

*In this text an equality is designated by the familiar equal sign and a defined quantity is specified by the symbol $\equiv$.

the reciprocal of its density, or

$$v \equiv \frac{V}{m} = \frac{1}{\rho} \tag{1.2}$$

An important aspect of thermodynamics is the derivation of relationships among thermodynamic properties. Various equations that relate properties will be discussed as the study of thermodynamics develops. Some of these equations are based on experimental measurements, while others are derived from a theoretical analysis. Regardless of origin, a relationship among properties y_i of the form

$$f(y_1, y_2, \ldots, y_n) = 0 \tag{1.3}$$

is called an *equation of state*.

A *state* is the condition of a system as specified by its properties. As an example, consider the gas inside the piston-cylinder arrangement shown in Figure 1.1(a). The state of the gas can be specified by the properties of the gas such as the pressure, density, and temperature. The minimum number of properties that are necessary to fix the state of the system and to specify the remaining properties of the gas will be considered in the next chapter.

The transformation of a system from one state to another is called a *process*. To illustrate a simple process, consider once again the system of the gas inside a piston-cylinder arrangement and assume that a flame is placed beneath the cylinder so that the gas is heated slowly. Further, assume that the piston is restrained by a constant force such that the pressure of the gas remains constant during the heating process. As the gas is heated by the flame, its temperature and volume increase while the pressure remains constant during the process. Suppose that the temperature, pressure, and volume of the gas are all periodically recorded during the heating process. The gas temperature, pressure, and volume at the initial state are recorded as T_1, P_1, and V_1, and the same properties at the end of the process at state 2 are T_2, P_2, and V_2. The recorded property values can be plotted on *process diagrams*, which are graphical representations of the changes in properties that occur between the initial and final states.

The P-V and T-V process diagrams for this example are shown in Figure 1.3. Process diagrams are valuable aids in the analysis of thermodynamic systems because they provide a convenient visualization of the means by which a change in state occurs during a process. Inherent in the use of process diagrams to depict a process is the assumption that a system actually passes through the series of equilibrium states indicated by the process curve. In other words, the system would remain infinitesimally close to a condition of equilibrium throughout the process. Such a process is called a *quasi-equilibrium* or *internally reversible process*. The latter term is used to indicate that the process could be reversed in direction and the system would retrace the same series of equilibrium states.

The internally reversible process is an ideal process because it would have to occur at an infinitesimally slow rate in order for the system to remain infinitesimally close to equilibrium conditions. Real processes are not internally reversible because they occur at a finite rate, and factors such as friction, fluid shear, and temperature gradients within the system cause the system to depart from equilibrium. Process that are not internally

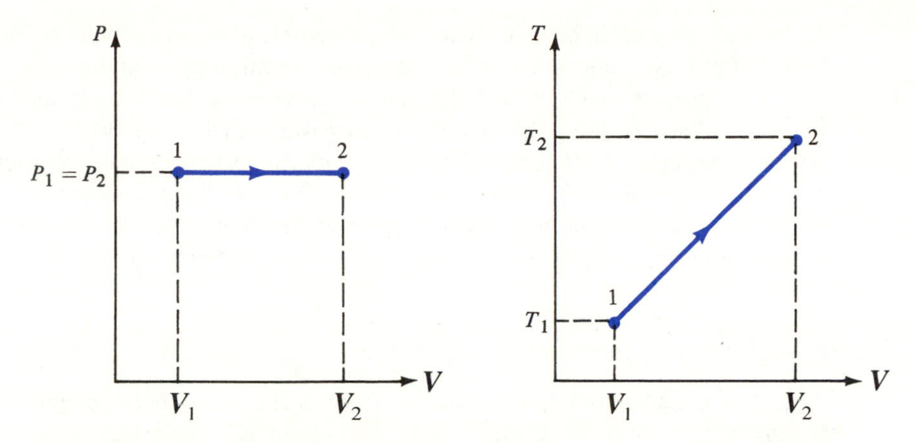

Figure 1.3 Process diagrams for a constant-pressure process of a gas heated in a piston and cylinder.

reversible are irreversible. Even though an internally reversible process would have to proceed at an infinitesimally slow rate, the concept of the internally reversible process is often useful in describing or analyzing actual processes. Reversible and irreversible processes are discussed more fully in Chapters 5 and 6.

In thermodynamics the prefix *iso-* is frequently used to designate a process for which a particular property remains constant. For example, a process in which the temperature remains constant is called an *isothermal process.* A constant-pressure process, such as the one illustrated by the diagram in Figure 1.3, is called an *isobaric process.* Several other constant-property processes are possible, and they are discussed as they are encountered in the text.

Properties of a system have meaning only when the system is in equilibrium. If a system is isolated from its surroundings and the properties of the system do not change with respect to time, the system is in *thermodynamic equilibrium.* If a system exists in thermodynamic equilibrium, the properties of the system can change only if there is a change in the properties of the surroundings. There are other kinds of equilibrium as well; thermal and mechanical equilibrium are particularly important in thermodynamics. *Thermal equilibrium* refers to a condition in which the temperature of a system will remain unchanged when the system is isolated from its surroundings. *Mechanical equilibrium* implies that the system forces, which for common systems are usually caused solely by pressures within the system, will remain unchanged when the system is isolated from its surroundings.

A system can proceed from one condition of equilibrium to another only if the system is disturbed from its equilibrium state. For example, if energy is transferred from the surroundings into a system consisting of a gas, the temperature of the gas will begin to increase. If the gas is confined, other properties such as the pressure will also begin to change. At the end of the energy-transfer process, a condition of equilibrium can be established by isolating the system from its surroundings. At this time the properties again become representative of the system. If the energy-transfer process occurs at a finite rate, which is always the case in real processes, deviations from equilibrium conditions can become significant. Since the properties of the system change at finite rates for a real process, indicated properties may vary significantly from the values they would have if the process were slowed and the properties were allowed to reach their equilibrium values at all states during the time that energy was added to the system. Requiring that processes proceed at infinitesimal rates may appear to be extremely restrictive, but as the study of thermodynamics develops, we will discover that properties at the end states of processes are often of more interest than how the process occurs or what the property values are during the process when nonequilibrium conditions may exist.

1.2.3 Point and Path Functions, Cycles

Properties exhibit some important characteristics that deserve special emphasis. The value for any property of a system at any state is independent of the path or process used to reach that state. For example, the temperature and pressure of the gas depicted in Figure 1.3 at state 2 are always T_2 and P_2, respectively, regardless of which path is followed in reaching state 2. Because of this characteristic, properties are referred to as *point functions*. If the differential change in any arbitrary point function X is integrated between states 1 and 2, the result is

$$\int_1^2 dX = X_2 - X_1$$

regardless of the path used to connect the two states. Since all thermodynamic properties are point functions, the integral of the differential of any property is simply the difference between the values of the property evaluated at the final and initial states. If the volume (a property) is used as an example, then

$$\int_1^2 d\mathbf{V} = \mathbf{V}_2 - \mathbf{V}_1$$

and similar expressions can be written for the integral of all other thermodynamic properties.

Other quantities of importance in thermodynamics display characteristics vastly different from point functions. A quantity whose values do depend on the path followed during a particular change in state is called a *path function*. The path must be specified before the value of a path function can be determined. Properties cannot be path functions because they are measurable characteristics of the system at a given state. Suppose that the symbol Y represents a path function and its differential is designated by δY. The Greek symbol δ is used to denote the differential of a path function to distinguish it from the differential of a point function, which is identified by the letter d. If the integration of the quantity δY between two arbitrary states 1 and 2 is attempted, the following statement can be made:

$$\int_1^2 \delta Y \neq Y_2 - Y_1 \tag{1.4}$$

because the integral of the path function Y cannot be evaluated solely from a knowledge of states 1 and 2. In fact, the symbols Y_1 and Y_2 have no meaning because Y is a path

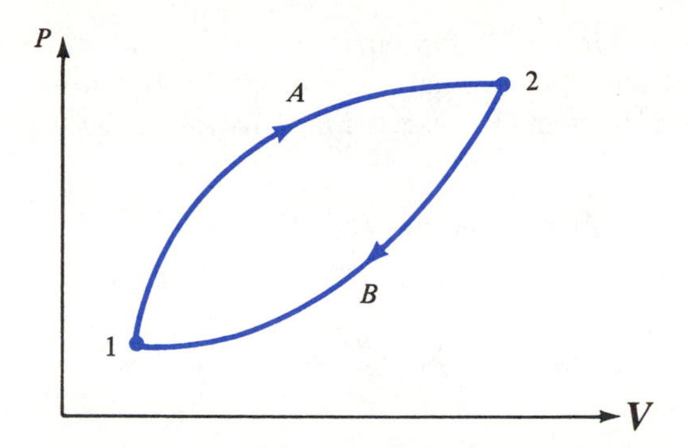

Figure 1.4 An example of a thermodynamic cycle.

function and the integral in Equation 1.4 can be evaluated only if the path followed between the two states is completely specified.

To help distinguish between the integrals of point and path functions, a different set of nomenclature is used. The integral of an arbitrary path function Y is designated as

$$\int_1^2 \delta Y = Y_{12} \tag{1.5}$$

where the double subscript is used to show that the value of Y for the process between states 1 and 2 can be determined only when the path followed between those states is specified.

A thermodynamic *cycle* is a process or series of processes whose initial and final states are identical. Figure 1.4 shows an example of a cycle on a P-V process diagram. In this cycle the system proceeds from state 1 to 2 along path A and returns to state 1 along path B, completing the cycle. The pressure and volume of the system vary continually throughout the cycle, but the initial and final values are the same. The path followed during any cycle on a process diagram is a closed path. To indicate integration over a cycle, a circle is superimposed over the integral sign. Since the initial and final states of a cycle are identical, the change in any property X (a point function) must always be zero for a complete cycle:

$$\oint dX = 0 \tag{1.6}$$

The converse of this statement is also true; that is, if a quantity dX is integrated over any arbitrary cycle and the result is zero, then the quantity X is a property.

The integral of any nonproperty or path function Y around a cycle is not necessarily zero, or

$$\oint \delta Y \neq 0 \tag{1.7}$$

because the value of the integral depends on the path followed during the cycle. In fact, the integral in Equation 1.7 would be different for each cycle composed of a different series of processes.

The behavior of a point function is the same as that of exact differentials, often discussed in beginning calculus courses, and a test or condition can be established to determine whether or not a differential is exact. Mathematically, the first-order differential

$$M(x, y)\,dx + N(x, y)\,dy \tag{1.8}$$

is said to be *exact* if it is the differential du of a continuous function $u(x, y)$ given by

$$du = \frac{\partial u}{\partial x}\,dx + \frac{\partial u}{\partial y}\,dy$$

That is,

$$\frac{\partial u}{\partial x} = M$$

and

$$\frac{\partial u}{\partial y} = N$$

If M and N are defined and have continuous first partial derivatives, then the order of differentiation is immaterial, so that

$$\frac{\partial M}{\partial y} = \frac{\partial^2 u}{\partial y\,\partial x} \qquad \frac{\partial N}{\partial x} = \frac{\partial^2 u}{\partial x\,\partial y}$$

and

$$\frac{\partial M}{\partial y} = \frac{\partial N}{\partial x} \tag{1.9}$$

The condition expressed by Equation 1.9 is necessary and sufficient for $M\,dx + N\,dy$ to be an exact differential.

1.3 UNITS AND DIMENSIONS

A *dimension* is a name given to any measurable quantity. For example, the name used to describe the distance between two points is the dimension called length. Other examples of dimensions are force, mass, time, temperature, and pressure. *Units* are measures for each of the dimensions. For example, some of the more common units for the dimension of length are the meter, millimeter, foot, yard, and mile.

The *SI* (*Système Internationale d'Unités*) system of units is the most widely used system throughout the entire world with the single major exception of the United States. The SI system of units is the most widely adopted system because it is based on a decimal relationship among the various units. For example, each unit of length is related to other length units by multiples of 10;* the meter (m) is equivalent to 100 centimeters (cm)

*A listing of commonly used SI unit prefixes is given in Table A.1 in the Appendix.

and a centimeter is equal to 10 millimeters (mm). The decimal feature of the SI system has made it well suited for use by the engineering and scientific community.

At the present time in the United States the *U.S. Customary System of Units* (or, more simply, the *English System of Units*) is still the system most commonly used by those in the nonscientific community. As a result of the widespread use of the English system in the United States, engineers must remain familiar with and conversant in the English system of units while still being able to apply the SI system. In contrast to the decimal nature of the SI system, the units in the English system are related to each other in a rather arbitrary fashion; 1 mile (mi) is equal to 1760 yards (yd), to 5280 feet (ft) and to 63,360 inches (in). Also 1 pound (lb) contains 16 ounces (oz) and is equal to 1/2000 ton. This seemingly haphazard relationship among units is one of the major drawbacks of the English system of units.

Calibrated standards of measurement do not need to be maintained for all units, because not all units are independent of each other. Those units for which reproducible standards are maintained are called *primary units*. Those units that are related to the primary units through defining equations, and therefore require no standard, are called *secondary units*. For example, a secondary unit of volume called the liter is defined in terms of the meter, which is a primary unit of length. The equation that relates the volume of a cube, for example, to the length L of its sides is

$$V = L^3$$

A liter is the volume occupied by a cube whose sides are 10 cm in length. Therefore, the liter and the meter are related to each other by

$$1 \text{ liter} = (10 \text{ cm})^3 = 10^{-3} \text{ m}^3$$

The SI system of units is based on seven primary units, which are listed in Table A.2 in the Appendix. Some of the SI secondary units along with their definitions in terms of the primary units are given in Table A.3.

The English system of units is based upon the primary units listed in Table A.2E. Several secondary units in the English system along with their definitions in terms of the primary units are shown in Table A.3E.

The important primary units in the SI system are the units of mass, length, and time. Therefore, force becomes a secondary unit whose definition must be consistent with Newton's second law, which for a constant-mass system is

$$F = ma \tag{1.10}$$

The unit of mass in the SI system is the kilogram (kg) and the unit of force is the newton (N), defined as the force necessary to accelerate a mass of 1 kg at a rate of 1 m/s^2, or

$$1 \text{ N} \equiv 1 \text{ kg·m/s}^2 \tag{1.11}$$

Newton's law can be also used to relate the mass of a body to its weight. The *weight* of a body is simply the force exerted on the body by the local acceleration of gravity. If the local acceleration of gravity is g and the weight of a body is W, then according to Equation 1.10, the weight and mass of a body in the SI system are related by

$$W = mg \tag{1.12}$$

The acceleration of gravity is a function of location. The acceleration of gravity is greater at sea level than on a mountaintop, and the acceleration of gravity on the moon is roughly one-sixth of the sea-level value on earth. The mass of an object, on the other hand, does not vary with location; its value remains the same regardless of its location. Even though the mass of an object is constant, Equation 1.12 shows that the weight of the object may vary with position because the local acceleration of gravity may change.

In specifying the weight of an object, the sea-level value is most frequently assumed. The mean sea-level value for the acceleration of gravity in the SI system of units is

$$g = 9.807 \text{ m/s}^2 \tag{1.13}$$

For this value of g, a body with a mass of 1 kg weighs 9.807 N at sea level.

The English system of units differs slightly from the SI system in terms of primary and secondary units. The important primary units in the English system are length, time, and *both* mass and force. When both force and mass are selected as primary units, Newton's law in the form of Equation 1.10 will not be dimensionally consistent, because a defined force will not have units consistent with the units of a mass times an acceleration (length/time^2). Therefore, a dimensional constant, denoted by g_c, must be used in Newton's second law to ensure dimensional homogeneity:

$$F = \frac{ma}{g_c}$$

In the English system with units of mass, force, length, and time selected as pounds mass (lb_m), pounds force (lb_f), foot (ft), and second (s) the value for g_c is

$$g_c = 32.174 \frac{\text{lb}_\text{m}\cdot\text{ft}}{\text{lb}_\text{f}\cdot\text{s}^2}$$

Note that g_c is not a dimensionless quantity, although it has a constant value. When working in the English system of units it will frequently be necessary to insert g_c into equations or terms relating force units to mass units to ensure dimensional consistency.

The equation equivalent to Equation 1.12 in the English system is

$$W = \frac{mg}{g_c}$$

Notice that a force of 1 lb_f will accelerate a 1 lb_m mass at a rate of 32.174 ft/s^2. The value for the sea-level acceleration of gravity is 32.174 ft/s^2, which means that a mass of 1 lb_m weighs 1 lb_f at sea level. At other locations where the acceleration of gravity is not equal to 32.174 ft/s^2, however, the mass of an object is not numerically equal to its weight.

EXAMPLE 1.1

An object has a mass of 10 kg. Calculate the following quantities:

(a) The weight, in newtons, of the object at sea level
(b) The weight, in newtons, of the object at a location where $g = 9.4$ m/s^2
(c) The weight, in lb_f, of the object where $g = 31.8$ ft/s^2

Solution.

(a) Using Equation 1.12 and the acceleration of gravity at sea level, we have

$$W = mg = (10\ \text{kg})(9.807\ \text{m/s}^2) = 98.07\ \text{kg·m/s}^2$$

Using the conversion factor in Equation 1.11 gives the sea-level weight:

$$W = \frac{98.07\ \text{kg·m/s}^2}{1\ \text{kg·m/N·s}^2} = \underline{\underline{98.07\ \text{N}}}$$

(b) At a location where $g = 9.4\ \text{m/s}^2$,

$$W = \frac{(10\ \text{kg})(9.4\ \text{m/s}^2)}{1\ \text{kg·m/N·s}^2} = \underline{\underline{94.0\ \text{N}}}$$

(c) First converting the mass into English units,

$$m = 10\ \text{kg} = 22.05\ \text{lb}_\text{m}$$

Then using Newton's law to compute the weight

$$W = \frac{mg}{g_c} = (22.05\ \text{lb}_\text{m})\left[31.8\ \frac{\text{ft}}{\text{s}^2}\right]\left[\frac{1}{32.174}\ \frac{\text{lb}_\text{f}\text{·s}^2}{\text{ft·lb}_\text{m}}\right] = \underline{\underline{21.8\ \text{lb}_\text{f}}}$$

Notice that even though the mass of the object remains constant, its weight changes with location. ■

The SI system is convenient to use and is widely accepted in the scientific community. Therefore, a majority of the examples and problems at the end of the chapters will be formulated in the SI system. Units used in both the SI and the English systems for several dimensions are given in Tables A.4 and A.4E. Conversion factors that can be used to change from one system to another are listed in Appendix A.5 and Appendix A.5E. A copy of this table is placed on the inside cover of the text for convenience.

1.4 PRESSURE

The *pressure* is defined as a normal force per unit area acting on the surface of a system. For fluid systems (either liquids or gases), the pressure on the inside surface of the container that holds the fluid is due to the cumulative effect of individual molecules striking the walls of the container, causing a normal force on the surface. For a fluid in equilibrium the pressure is defined by the equation

$$P \equiv \frac{dF_n}{dA} \tag{1.14}$$

where the differential area dA is the smallest surface area for which the effects of the fluid are the same as those for a continuous medium. The symbol dF_n represents the total normal force caused by the fluid on the area dA. The normal force per unit area within a solid is usually called a normal stress rather than a pressure. According to Equation

1.14, the units of pressure are those of force per unit area. In the SI system of units pressures are normally expressed in terms of N/m^2 or *pascals* (abbreviated as Pa) where

$$1\ \text{Pa} \equiv 1\ \text{N/m}^2$$

In the English system pressures are usually measured in pounds per square inch (psi or lb_f/in^2).

Fluid pressures can be measured with a variety of electrical or mechanical devices. *Bourdon gauges* are simple mechanical devices calibrated to read pressure directly by detecting the movement of a needle attached to a hollow tube connected to a pressurized container. Pressures can also be measured with *pressure transducers*, which convert the deflection of a flexible diaphragm to an electric output by means of a calibrated strain gauge. The height of a fluid column is also frequently used for pressure measurement. A pressure gauge based on this principle is called a *manometer* and is illustrated in Figure 1.5. When a manometer is used to measure the pressure of the atmosphere, it is called a *barometer*. A manometer employs a tube partially filled with a fluid of density ρ, as shown in Figure 1.5. The tube is connected to a container that encloses a gas at a pressure P_1. The pressure difference between two surfaces in the manometer fluid separated by a differential height dy can be related to the density of the manometer fluid by applying Newton's law for static equilibrium:

$$\sum F_y = 0$$

The y component of forces on the elemental fluid volume is a result of pressure forces on the surface and gravitational forces in the volume, or

$$PA - (P + dP)A + \rho g A\ dy = 0$$

$$dP = \rho g\ dy$$

Integrating this expression over the height h of the fluid column gives

$$\int_{P_2}^{P_1} dP = \int_0^h \rho g\ dy$$

or

$$P_1 - P_2 = \rho g h \qquad (1.15)$$

for a fluid of constant density. Notice that the pressure difference depends only on the net vertical height between the fluid levels in a single, constant-density manometer fluid and not on the shape of the tube. Thus, the pressure at points R and S in Figure 1.5 must be the same.

The property ρg in Equation 1.15 is the weight of the substance per unit volume and is called the *specific weight* of the fluid, γ, or*

$$\gamma \equiv \rho g \qquad (1.16)$$

---

*Since Equation 1.16 relates force and mass units when the English system is used, it must be written as

$$\gamma \equiv \frac{\rho g}{g_c}$$

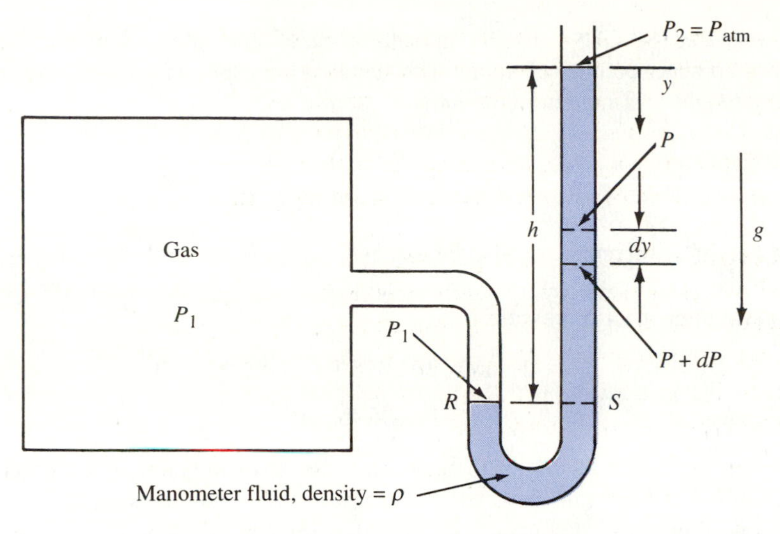

Figure 1.5 A simple manometer for measuring fluid pressure.

Therefore, the pressure difference between the two layers of fluid with a constant specific weight of γ separated by a vertical distance h can be expressed as

$$P_1 - P_2 = \gamma h \tag{1.17}$$

The specific gravity is another property frequently used to fix the density of a substance. The *specific gravity* is defined as the ratio of the specific weight of the substance to the specific weight of water, or

$$\text{Sp. Gr.} \equiv \frac{\gamma}{\gamma_{H_2O}} \tag{1.18}$$

The expression for the pressure difference in a static fluid can now be expressed in terms of the specific gravity of the fluid and the specific weight of water, or

$$P_1 - P_2 = (\text{Sp. Gr.})\gamma_{H_2O} h \tag{1.19}$$

The density of liquid water at 1 atmosphere pressure and 24° C (or 75.2° F) is

$$\rho_{H_2O} = 1.0\,\text{g/cm}^3 = 10^6\,\text{g/m}^3 = 1000\ \text{kg/m}^3$$
$$= 62.43\ \text{lb}_\text{m}/\text{ft}^3$$

Thus, the specific weight of water at these conditions and at sea level where g = 9.807 m/s² (32.174 ft/s²) is

$$\gamma_{H_2O} = \rho g = \frac{(1000\ \text{kg/m}^3)(9.807\ \text{m/s}^2)}{1\ \text{kg}\cdot\text{m/N}\cdot\text{s}^2} = 9.807 \times 10^3\,\text{N/m}^3$$
$$= 62.43\ \text{lb}_\text{f}/\text{ft}^3$$

Barometers, used to measure atmospheric pressure, usually use mercury as a manometer fluid. Since mercury has a specific weight that is approximately 13.59 times that of water (Sp.Gr. = 13.59), measurements of atmospheric pressure can be achieved with a compact column of liquid. The standard atmospheric pressure measured in terms of the height of a mercury column is

$$1 \text{ atm} = 760 \text{ mm Hg}$$
$$= 29.92 \text{ in Hg}$$

The height of mercury can also be expressed in pressure units by using Equation 1.17. Since the specific weight of mercury is $\gamma = 1.333 \times 10^5$ N/m^3 (848.4 lb$_f$/ft^3), the standard atmospheric pressure is

$$1 \text{ atm} = 101{,}325 \text{ Pa} = 101.325 \text{ kPa}$$
$$= 14.696 \text{ lb}_f/\text{in}^2$$

If water were used as the manometer fluid instead of mercury, a fluid column of about 10.3 m (33.0 ft) would be needed to measure the standard atmospheric pressure. This example illustrates why a water barometer is impractical. Occasionally pressures in the SI system are expressed in *bars*, where the definition of a bar is

$$1 \text{ bar} \equiv 10^5 \text{ Pa}$$

This unit of pressure is particularly convenient because standard atmospheric pressure is approximately equal to one bar.

Two different pressures are common in engineering practice: gauge pressures and absolute pressures. Pressures measured relative to the local atmospheric pressure are called *gauge pressures*, while pressures measured relative to absolute zero pressure are called *absolute pressures*. The absolute pressure is therefore the sum of the gauge pressure and the atmospheric pressure, or

$$P_{abs} = P_{gauge} + P_{atm} \tag{1.20}$$

The relationship among absolute, gauge, and atmospheric pressure is shown schematically in Figure 1.6.

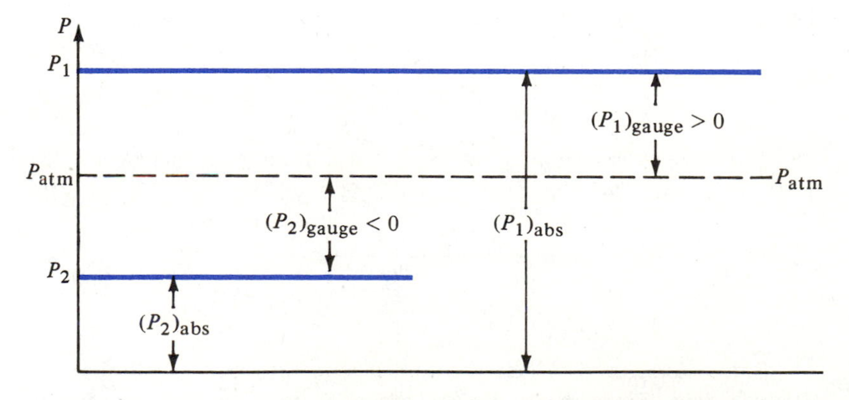

Figure 1.6 Relationship among absolute, gauge, and atmospheric pressure.

Pressure gauges such as Bourdon gauges record gauge pressures because they are calibrated to read zero pressure when they are open to the atmosphere. Therefore, if a Bourdon gauge attached to a pressurized system indicates a pressure of 1.0 MPa, the pressure of the system is actually 1.0 MPa higher than the local atmospheric pressure, or 1.0 MPa gauge pressure. The absolute pressure is determined by adding the gauge pressure and the local atmospheric pressure.

Absolute pressures are always positive, while gauge pressures can be either positive or negative. Positive gauge pressures indicate pressures above atmospheric pressure, while negative gauge pressures indicate pressures below local atmospheric pressure. Pressures below atmospheric pressure are often called *vacuum pressures*. That is, a pressure of 50 kPa vacuum is equivalent to a gauge pressure of -50 kPa. Similarly, a gauge pressure of -5 psi is equal to a pressure of 5 psi below the local atmospheric pressure.

In this text if a pressure is not explicitly stated as being either gauge or absolute pressure, the implication is that the value is an absolute pressure.

EXAMPLE 1.2

A manometer is attached to a pressurized container as shown in Figure 1.5. One end of the manometer is open to the atmosphere, and the local atmospheric pressure is 760 mm Hg. Calculate the absolute pressure on the inside surface of the container under the following conditions:

(a) The height of the manometer fluid is 420 mm and the fluid has a specific gravity of 1.6.
(b) The height of the manometer fluid is 850 mm and the fluid has a density of 1100 kg/m^3.
(c) The manometer fluid is water and it has a height of 25 in.

Solution. The manometer measures gauge pressure, and atmospheric pressure must be added to the manometer pressure to give the absolute pressure. Equation 1.19 is used to convert fluid heights to pressures.

(a)
$$P_{gauge} = (\text{Sp. Gr.})\gamma_{H_2O}h = (1.6)(9.807 \times 10^3\ N/m^3)(0.42\ m)$$
$$= 6590\ N/m^2$$
$$P_{abs} = P_{gauge} + P_{atm}$$
$$= 6590\ N/m^2 + 101{,}300\ N/m^2$$
$$= 107{,}890\ N/m^2 = \underline{\underline{107.9\ kPa}}$$

(b) Using Equation 1.15, we have
$$P_{gauge} = \rho g h = (1100\ kg/m^3)(9.807\ m/s^2)(0.85\ m)$$
$$= 9170\ N/m^2$$
$$P_{abs} = 9170\ N/m^2 + 101{,}300\ N/m^2$$
$$= 110{,}470\ N/m^2 = \underline{\underline{110.5\ kPa}}$$

(c) Using Equation 1.17 and the specific weight of water in English units gives

$$P_{gauge} = \gamma_{H_2O} h = (62.43\ lb_f/ft^3)(25\ in)(1\ ft/12\ in)$$

$$= 130.1\ lb_f/ft^2 = 0.90\ lb_f/in^2$$

$$P_{abs} = 0.90\ lb_f/in^2 + 14.7\ lb_f/in^2 = \underline{\underline{15.6\ lb_f/in^2}}$$

1.5 TEMPERATURE

Temperature is often thought of as being a measure of the "hotness" or "coldness" of a substance, because a body with a higher temperature than another is said to be hotter. This statement is a rather poor definition of temperature because the words *hot* and *cold* are subjective rather than quantitative terms.

One means of attaching some physical significance to the meaning of the temperature is to relate the temperature of a system to the movement of the molecules that comprise the system. As the temperature increases, the molecular activity also increases. In fact, the mean or average velocity of the molecules can be shown to increase as the temperature increases. As a result of this qualitative observation, one would expect, for example, that water-vapor (steam) molecules at a high temperature would have a relatively high velocity. As the temperature of the water vapor decreases, the mean molecular velocity also decreases. Further cooling of the water vapor could result in condensation into a liquid phase and eventual freezing into solid water or ice. Throughout the entire cooling process the average molecular velocity decreases.

Another way to attach physical significance to the temperature of a body is to perform a simple experiment and to formulate a fundamental law that summarizes the results of the experiment. Suppose that a thermometer is brought into contact with a body and is allowed to reach thermal equilibrium with that body. The reading of the thermometer (for example, the length of the mercury column in a mercury-in-glass thermometer) is recorded. Then the thermometer is placed in contact with a second body and after thermal equilibrium is attained, the reading of the thermometer is again recorded. If the two thermometer readings are the same, we conclude that the temperatures of the two bodies are equal. This conclusion may seem quite obvious, but it cannot be derived from other fundamental principles. It was formulated after the first and second laws of thermodynamics had been established and is called the *zeroth law of thermodynamics*:

When two bodies have equality of temperature with a third body, they have equality of temperature with each other.

The zeroth law addresses only equality of temperature and does not attempt to quantify or assign numerical values of temperature. The temperature measured by the thermometer in the experiment described above can be quantified by establishing a set of reproducible standard temperatures with which the thermal state of other bodies may be compared. Once the standard scale of temperatures has been established, the temperature of various substances can be determined.

Thermometers are not the only temperature-measurement devices. *Thermistors* and *resistance thermometers* are devices that are calibrated so that the electric resistance of a semiconducting element or a wire is related to the temperature of a body. *Thermocouples* can be used to measure temperatures by relating a voltage generated by two dissimilar metals to the temperature of the junction of the dissimilar metals. Temperatures can also be determined by measuring the pressure of a gas in a constant-volume container, as is done with a *gas thermometer*.

Two absolute temperature scales are defined such that a temperature of zero corresponds to a theoretical state of no molecular movement of the substance; they are the *Kelvin scale*, where the temperatures are designated in *kelvin* (K), and the *Rankine scale*, in which temperatures are measured in degrees Rankine (° R). The kelvin is the unit of absolute temperature in the SI system, and the degree Rankine is the unit of absolute temperature in the English system. In the Kelvin and Rankine scales negative temperatures are impossible, but a rigorous proof of this statement is possible only after the thermodynamic property called entropy has been introduced in Chapter 5.

Historically, several temperature scales were introduced before thermodynamic principles were well established and certainly well before all substances were known to consist of molecules. Early temperature scales were proposed by arbitrarily selecting reference temperatures corresponding to easily reproducible state points. Two of the most widely used reference temperatures are the boiling point of water at a pressure of 1 atm and the triple point of water where solid, liquid, and vapor phases of water exist in equilibrium.

The SI temperature scale based on these two reference states is the *Celsius scale*, for which the unit of temperature is the degree Celsius (° C). The scale is named after Anders Celsius, a Swedish astronomer, who first proposed it in the 1740s. The Celsius temperature selected for the boiling point of water was 100° C and the triple-point temperature was chosen as 0° C. As a result of these selected reference temperatures, the Celsius scale is subdivided into 100 equal divisions between the triple point and the boiling point of water. The Celsius scale is related to the thermodynamic or absolute temperatures measured in kelvin degrees by the relationship

$$K = {}^{\circ}C + 273.15^{\circ} \tag{1.21}$$

For most calculations the constant in Equation 1.21 can be rounded off to 273° without significant loss in accuracy, and in this text the rounded figure will usually be used when converting between the Celsius and Kelvin temperature scales. Equation 1.21 indicates that the absolute zero temperature is -273.15° C. Negative temperatures exist for the Celsius scale, but temperatures in the absolute temperature scale are always positive.

The temperature scale adopted in the English system of units was proposed in the early 1700s by Daniel Gabriel Fahrenheit, a Dutch instrument maker. He arbitrarily selected the value of zero on his temperature scale as the temperature of a mixture of water, salt, and ice at atmospheric pressure. The other reference temperature on the Fahrenheit scale was chosen to be the body temperature of a healthy human. Later the freezing and boiling temperatures of fresh water at one atmosphere were found to be 32 and 212 degrees, respectively, and these two fixed temperatures were defined to be exact standard values on the Fahrenheit scale. As a result of the choice of these two temperatures, the temperature difference between the two established reference temperatures is

180° F and the Celsius degree is 1.8 times the *size* of the Fahrenheit degree, or

$$1^\circ\ C = 1.8^\circ\ F$$

and the two temperature scales are related by

$$^\circ F = (9/5)^\circ C + 32^\circ$$

or

$$^\circ C = (5/9)(^\circ F - 32^\circ)$$

The Fahrenheit temperature scale is related to the Rankine scale, which is the absolute temperature scale based on thermodynamic principles, by the relationship

$$^\circ R = ^\circ F + 459.67^\circ$$

The constant in this equation is usually rounded to 460 because the error produced by this approximation is negligible in most engineering calculations. As a result of this relationship, the absolute zero temperature is -459.67° F.

1.6 HEAT TRANSFER

Even in the absence of mass flow across the boundary of a system, energy can be transported across the boundary by two distinct mechanisms: heat transfer and work interactions. Energy transfer across a boundary as a result of a temperature difference between a system and its surroundings is called *heat transfer*. An in-depth study of the field of heat transfer is beyond the scope of this text, and most engineering disciplines teach heat transfer as a separate subject.

Even though a quantitative study of heat transfer is quite involved, several general observations regarding heat transfer are important to an understanding of thermodynamics. Heat transfer may occur by three distinct modes: conduction, convection, and radiation. *Conduction* occurs primarily through solids, *convection* occurs in fluids, while *radiation* is an electromagnetic wave phenomenon in which energy can be transported through transparent substances and even through a vacuum. While the three modes are quite different, they have one factor in common: All three modes occur across the boundary of a system because of a temperature difference between the system and the surroundings.

In general, the heat-transfer rate increases as the temperature difference between the system and its surroundings increases and approaches zero as the temperature difference approaches zero. Factors other than the temperature difference affect the rate of heat transfer. One such factor is the *thermal resistance* at the boundary. Just as the electric resistance is a measure of how well a material will resist the flow of an electric current when a voltage is maintained across the material, the thermal resistance relates the flow of heat to the temperature difference across the material. Substances that have a high value for thermal resistance are classified as *thermal insulators*, while materials that have low values of thermal resistance are termed good *conductors* of heat. As the thermal resistance is increased for a given temperature difference between the system and surroundings, the rate of heat transfer across the boundary is reduced. Placing an insulating material over the boundary of a system is one means of increasing the thermal resistance and thereby reducing the heat-transfer rate across the bounding surface of a system.

In many instances the resistance to heat flow is so large or the temperature difference is so small that the heat-transfer rate across a system boundary is negligible. When there is no heat transfer across the boundary of a system, the system is said to undergo an *adiabatic* process. An adiabatic process should not be confused with an isothermal process during which the temperature of the system remains constant. An adiabatic process does not imply that a process is isothermal, nor does an isothermal process imply that the process is adiabatic.

Heat transfer occurs from a region of high temperature to a region of lower temperature. Therefore, the direction of the heat transfer is to a system that has a temperature lower than its surroundings, and from a system that has a temperature higher than its surroundings. Because heat transfer is directional in nature, the establishment of a sign convention is necessary to designate its direction. Throughout this text heat transfer *to* a system is assigned a positive value. Heat transfer *from* the system is negative. The sign convention on heat transfer is illustrated in Figure 1.7.

Because heat transfer is a transfer of energy, it has units of energy. In the SI system the unit of energy is the *joule* (J), which is also equal to 1 W·s and 1 N·m. In the English system the unit of energy most commonly used for heat transfer is the *British thermal unit*, abbreviated as Btu. One Btu is defined as the amount of heat transfer required to raise the temperature of 1 lb_m of water 1° F at an average temperature of 68° F and at atmospheric pressure. The symbol for heat transfer is Q, and heat transfer per unit mass

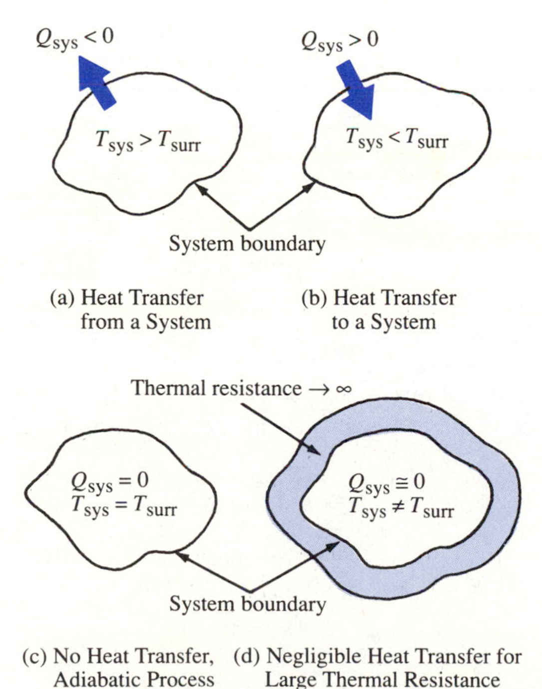

Figure 1.7 Heat transfer between system and surroundings.

is designated by the symbol q. Common units for q are kilojoules per kilogram (kJ/kg) in the SI system and Btu per pound mass (Btu/lb_m) in the English system. The *heat-transfer rate* across a boundary is given by the symbol $\dot{Q}$, where the dot superscript signifies a quantity per unit time. SI units for $\dot{Q}$ are watts (W) and English units for $\dot{Q}$ are Btu/s or Btu/h.

Heat transfer is a mechanism for energy transfer and is not a property, and therefore a system does not contain heat at any state. Heat transfer can be associated with a particular process only between one state and another, and it can be identified as an energy transfer only across the boundary of the system. Furthermore, since heat transfer is not a property, the heat transfer during a particular process is known only if the process is specified. The heat transfer during a specific process from state 1 to state 2 is denoted by Q_{12}:

$$\int_1^2 \delta Q = Q_{12} \tag{1.22}$$

In the absence of other forms of energy transfer across the system boundary, heat transfer to a system produces an increase in the energy level of the system. In a similar fashion, heat transfer from the system results in an overall reduction in the total energy of the system. Once heat transfer across the system boundary has occurred, its influence on the state of the system can be detected through the change in properties of the system. These observations are very qualitative in nature, and equations that permit a precise evaluation of the change in properties resulting from energy transfer across the system boundary are developed in Chapter 4.

1.7 WORK INTERACTIONS

In thermodynamics a *work interaction* is defined as energy transfer across the boundary of a system that is equivalent to a force acting through a distance. Work is assigned the symbol W, and the sign convention used in this text considers work done *by* the system as positive and work done *on* the system as negative. Notice that the sign convention used for work is *opposite* to that used for heat transfer in that a positive heat transfer indicates energy *entering* the system while a positive value for work indicates energy *leaving* the system. This sign convention for work is chosen so that the work output of a system such as an engine is positive.

If a force F, as shown in Figure 1.8, is used to move a body through a differential distance ds, the magnitude of the differential work performed during the differential displacement is

$$\delta W = F_s \, ds = F \cos \theta \, ds \tag{1.23}$$

where θ is the angle between the force F and the displacement ds. If the body in Figure 1.8 is displaced from an initial position s_1 to a final position s_2, the work performed during the process is

$$W_{12} = \int_1^2 \delta W = \int_1^2 F_s \, ds = \int_1^2 F \cos \theta \, ds \tag{1.24}$$

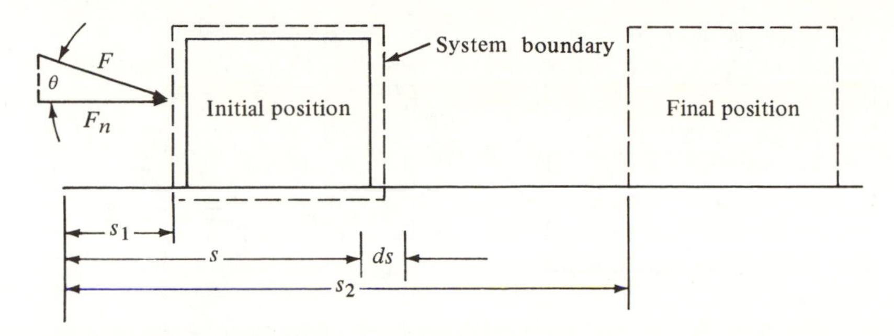

Figure 1.8 Work performed by a force acting through a displacement.

Notice that Equations 1.23 and 1.24 only provide a magnitude for the work of a force through a displacement and that the sign associated with the work can only be determined once the system is selected and the direction of the force relative to the displacement is known. For example, the force in Figure 1.8 is shown acting in the direction of the displacement of the system. The force therefore performs work on the system, which, according to the sign convention for work, is negative. In a similar fashion, if the force acts in a direction opposite to the displacement, the work is done by the system on the surroundings, and the work for the system is positive.

Recognizing that the work equations above (and those that follow in the remainder of this section) merely give the magnitude of the work, the user must account for the sign of the work once specifics of the process are known.

Work, like heat transfer, is a mechanism for energy transfer and is not a thermodynamic property; therefore, it is a path function, and its value depends on the particular path followed during the process. For example, if the path followed during the process illustrated in Figure 1.8 is one for which the force and the angle θ are constant over the entire displacement, then the magnitude of the work performed on the body by the external force is

$$W_{12} = \int_1^2 F \cos \theta \, ds = F \cos \theta \, (s_2 - s_1) \tag{1.25}$$

If the magnitude of the force, the angle θ, or the path vary during the process, then the expression for the work will be different from the expression in Equation 1.25. This behavior is characteristic of a path function. Since work is not a property and can be identified only at the boundary of the system, referring to a particular system as containing work is incorrect. Once a work interaction has occurred at the boundary of a system, its influence is reflected by changes in other forms of energy.

Work has units of energy—joules in the SI system. Work per unit mass is designated by the symbol w, and it is measured in units of kilojoules per kilogram in the SI system. In the English system of units work is measured in $\text{ft}\cdot\text{lb}_\text{f}$ and work on a unit-mass basis is expressed in $\text{ft}\cdot\text{lb}_\text{f}/\text{lb}_\text{m}$.

There are other examples of work interactions at the boundary of a system without a readily identifiable force acting through a distance. One typical example is shown in Figure 1.9(a), in which a voltage supply external to a thermodynamic system is connected

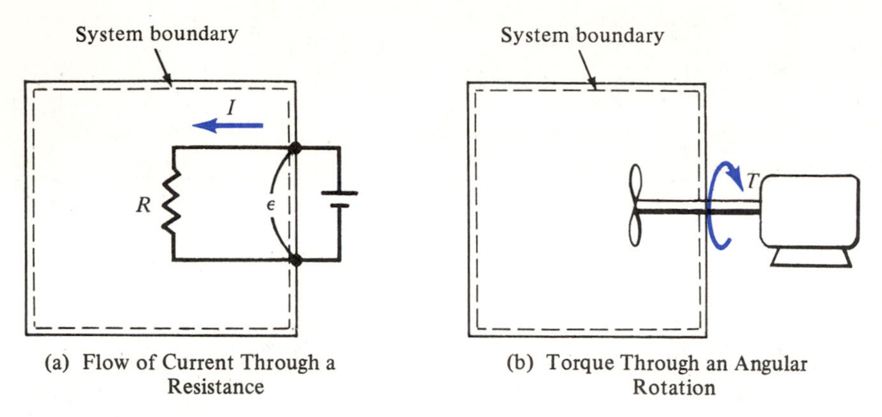

Figure 1.9 Examples of (a) electric work and (b) shaft work.

to an electric resistance inside the system. In this example the voltage supply creates a current flow through the resistance, and while there is no identifiable force acting through a distance, an electric force is necessary to displace the electrons in the wire. Therefore, an electric current across the boundary of a system is thermodynamically equivalent to work.

The expression for the magnitude of the *electric work* performed on the system when the current is I, the electric potential is ε, and the process occurs between an initial time t_1 and a final time t_2 is

$$W_{12} = \int_1^2 \varepsilon I \, dt \tag{1.26}$$

If the process followed is one for which the current and voltage are constant, then Equation 1.26 can be integrated, resulting in

$$W_{12} = \varepsilon I(t_2 - t_1) \tag{1.27}$$

Once again the form of Equation 1.26 for the electric work shows that the path followed during a process must be specified before the amount of electric work can be determined.

Another common example of work is illustrated in Figure 1.9(b), where a motor rotates a shaft that extends into the thermodynamic system. The rotating shaft is equivalent to a force acting through a distance, since the shaft could be used to raise a weight in the presence of the earth's gravitational field. If the torque required to rotate the shaft is T and the angle through which the shaft rotates is $d\theta$, then the magnitude of the *shaft work* performed on the system by the rotating shaft is

$$W_{12} = \int_1^2 T \, d\theta \tag{1.28}$$

The shaft work can be calculated only after the relationship between torque and angular displacement is known. For the case of rotation under a constant torque, the expression for shaft work becomes

$$W_{12} = T \int_1^2 d\theta = T(\theta_2 - \theta_1) \tag{1.29}$$

Rotating shafts are often present in thermodynamics systems, because many mechanical systems transmit energy by means of rotating shafts. Motors, pumps, compressors, turbines, and many other devices transmit useful work by means of rotating shafts.

As a final example of work frequently encountered in thermodynamics, consider a gas enclosed in a piston and cylinder, as shown in Figure 1.10. The system boundary surrounds the gas, and the boundary expands or contracts with the gas as the piston moves. The pressure of the gas is P, and the volume of the gas is $\boldsymbol{V}$. If the piston is displaced by a distance ds in a quasi-equilibrium or internally reversible process, the work performed by the gas during the process is

$$\delta W = F_s\, ds = PA\, ds = P\, d\boldsymbol{V} \tag{1.30}$$

and the work for a finite displacement of the piston, called *P dV work*, from state 1 to state 2 is

$$W_{12} = \int_1^2 \delta W = \int_1^2 P\, d\boldsymbol{V} \tag{1.31}$$

Since work is a path function, Equation 1.31 cannot be integrated until the pressure-volume path is known.

The $P\, d\boldsymbol{V}$ work performed during the expansion or compression of a system against a resisting pressure is graphically equal to the area under the P-$\boldsymbol{V}$ curve, as illustrated in Figure 1.11. If the system boundary expands, then work is performed by the system on the surroundings, and the $P\, d\boldsymbol{V}$ work of the system is positive. If the volume decreases, work is performed on the system, and the value for work of the system is negative.

For a system to be in equilibrium, there must be no unbalanced forces between the system and its surroundings. This condition would necessarily require that the process that the system undergoes be executed at an infinitesimally slow rate. If these conditions are met, the process is *quasistatic* or internally reversible, and any work that occurs

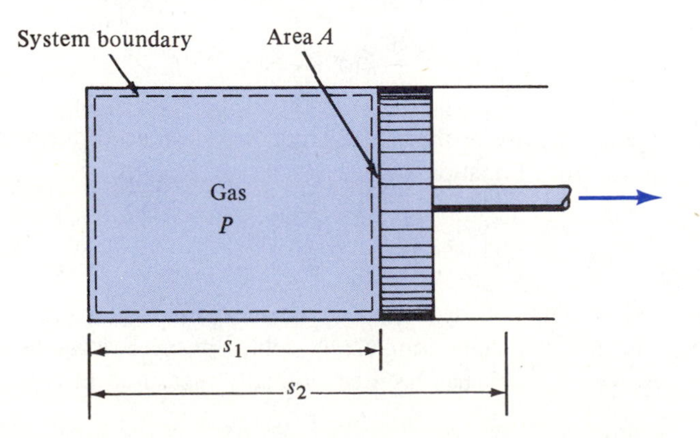

Figure 1.10 Work involving the motion of a system boundary.

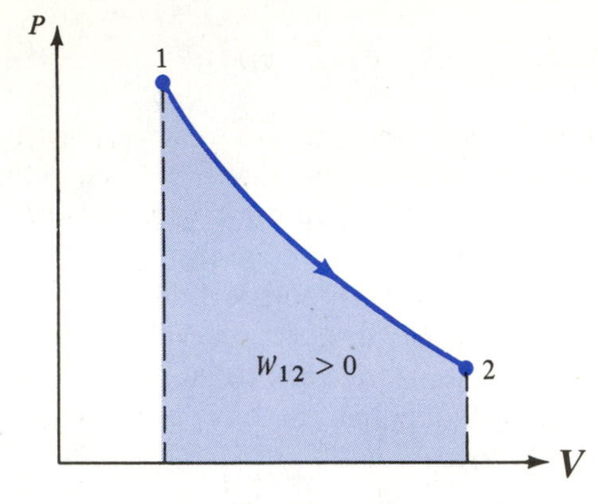

Figure 1.11 Graphical interpretation of $P\,d\mathbf{V}$ work on a P-$\mathbf{V}$ process diagram.

during the process is called *reversible work*. An example of this work is the $P\,d\mathbf{V}$ work performed during slow expansion, as discussed in the preceding paragraphs. Other forms of work can be termed reversible work provided that they are the result of internally reversible processes.

The *power* is defined as the work per unit time crossing the boundary. The symbol for power is $\dot{W}$, where the dot superscript is used to signify a rate quantity:

$$\dot{W} \equiv \frac{\delta W}{dt} \tag{1.32}$$

The power generated when a force acts through a distance is then

$$\dot{W} = \frac{\delta W}{dt} = \frac{F_s\,ds}{dt} = F_s V \tag{1.33}$$

where V (or ds/dt) is the velocity* of the boundary. The power associated with electric work, from Equation 1.26, is

$$\dot{W} = \frac{\delta W}{dt} = \varepsilon I \tag{1.34}$$

The shaft power involved when a shaft rotates through an angle is

$$\dot{W} = \frac{\delta W}{dt} = T\frac{d\theta}{dt} = T\omega \tag{1.35}$$

where ω is the angular velocity of the shaft. The power produced during the expansion of a system boundary, from Equation 1.30, is

$$\dot{W} = \frac{\delta W}{dt} = PA\frac{ds}{dt} = PAV \tag{1.36}$$

Since work interactions and heat transfer are mechanisms for energy transfer, the units of heat transfer and work are equivalent—joules in the SI system and Btu in the English system. However, work has been traditionally measured in foot·pounds (ft·lb_f)

---

*In this text the symbol for velocity is V to distinguish it from the volume, which is designated by the larger symbol $\mathbf{V}$.

in the English system instead of the unit of heat transfer, which is the Btu. The two energy units are related by

$$1 \text{ Btu} = 778.169 \text{ ft}\cdot\text{lb}_f$$

The units of power or energy per unit time are joules per second or watts. In the English system the most frequently used unit of power is the *horsepower*, which is defined as

$$1 \text{ hp} \equiv 550 \text{ ft}\cdot\text{lb}_f/\text{s}$$

EXAMPLE 1.3

A gas inside a flexible container expands such that the product of its pressure and volume remains constant during the process; that is,

$$PV = C$$

where C is a constant.

(a) Calculate the work done by the gas on the surroundings during an expansion from P_1, V_1 to P_2, V_2.

(b) Compare the work calculated for the path in part (a) with a path that has the same initial and final states but consists of a constant-volume process that continues until the pressure drops to P_2 followed by a constant-pressure process that terminates at state 2.

Solution.

(a) A sketch of the process on a P-V diagram helps in visualizing the work performed. Since the product of P and V is a constant, the process appears as a hyperbola in Figure 1.12(a), and according to Equation 1.31, the work can be interpreted graphically as the area under the process curve:

$$W_{12} = \int_1^2 P\, dV$$

Substituting the path equation results in

$$W_{12} = \int_1^2 \frac{C\, dV}{V}$$

which can be integrated to produce

$$W_{12} = C \int_1^2 \frac{dV}{V} = C \ln\left(\frac{V_2}{V_1}\right)$$

Finally, the value for the constant C can be determined in terms of the pressure and volume at any single state of the process, so that

$$\underline{\underline{W_{12} = P_1 V_1 \ln\left(\frac{V_2}{V_1}\right)}}$$

Since $V_2 > V_1$, the value for W_{12} is positive, indicating that during this process work is done by the system on the surroundings.

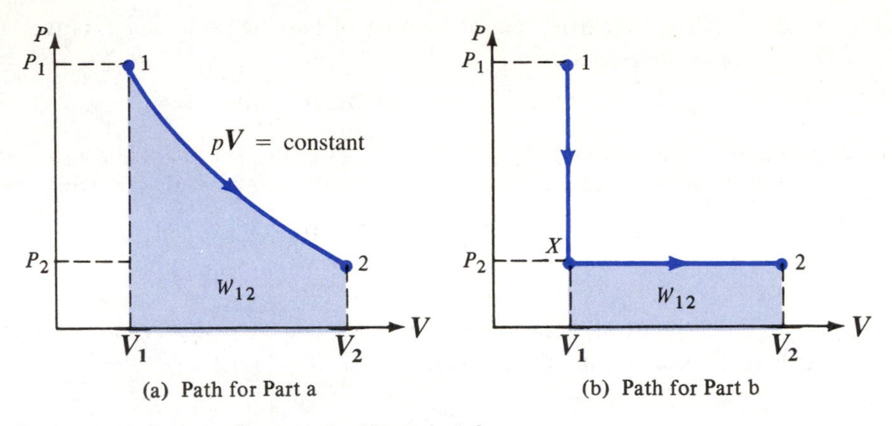

Figure 1.12 Process diagrams for Example 1.3.

(b) The process for part (b) follows the path 1-*X*-2, as illustrated in Figure 1.12(b), where the value for the work is once again equal to the area under the process curve. Obviously, the work during the processes for parts (a) and (b) is different. During the process from state 1 to state X there is no $P\,d\mathbf{V}$ work crossing the system boundary because the boundary does not expand or contract during a constant-volume process:

$$W_{1X} = 0$$

During the process from state X to state 2 the work done by the system on the surroundings is

$$W_{X2} = \int_X^2 P\,d\mathbf{V} = P_2 \int_X^2 d\mathbf{V} = P_2(\mathbf{V}_2 - \mathbf{V}_X) = P_2\,(\mathbf{V}_2 - \mathbf{V}_1)$$

The total work during the process from 1 to 2 is the sum of W_{1X} and W_{X2}, so

$$\underline{\underline{W_{12} = P_2(\mathbf{V}_2 - \mathbf{V}_1)}}$$

Clearly, the work for this path is not equal to the work done during the path in part (a) of this problem, even though both processes have the same initial and final states. ■

■ EXAMPLE 1.4

A shaft such as the one shown in Figure 1.9(b) rotates at a rate of 100 revolutions per minute (rev/min) against a constant torque of 10^3 N·m. Calculate the power required to rotate the shaft. Also, calculate the work required to rotate the shaft through 50 revolutions.

Solution. The power associated with a shaft rotating against a constant torque is

$$\dot{W} = T\omega = (10^3\ \text{N}\cdot\text{m})(100\ \text{rev/min})(2\pi\ \text{rad/rev})\left[\frac{1\ \text{min}}{60\ \text{s}}\right]$$

$$= 1.047 \times 10^4\ \text{N}\cdot\text{m/s} = \underline{\underline{10.5\ \text{kW}}}$$

The power crosses the boundary of the system in Figure 1.9(b), and from the standpoint of the system, the power is negative. If the system for this problem had been the electric motor, the power would still have been numerically equal to 10.5 kW, but its sign would have been positive because energy would leave this system (work done by the electric motor). The work for the case of a shaft rotated against a constant torque is given by Equation 1.29:

$$W = T(\theta_2 - \theta_1) = (10^3\ \text{N}\cdot\text{m})(50\ \text{rev})(2\pi\ \text{rad/rev}) = 314{,}000\ \text{N}\cdot\text{m} = \underline{\underline{314\ \text{kJ}}}$$ ■

■ EXAMPLE 1.5

The combustion gases within an enclosed piston and cylinder expand such that the gas pressure-cylinder volume relationship follows the path

$$PV^{1.5} = C$$

where C is a constant. The gas pressure at the beginning of the power stroke is 450 lb_f/in^2 when the volume within the cylinder is 2 in^3. At the end of the stroke the cylinder volume is 50 in^3. Calculate the work developed by the gas during a single stroke of the piston and the average power developed by the gas if there are 20 power strokes per second.

Solution. The work of the expanding gas is determined from Equation 1.31. Substituting the P-V path relationship and integrating between the given end states gives

$$W_{12} = \int_1^2 P\,dV = \int_1^2 \frac{C}{V^{1.5}}\,dV = -1.5\,C\left[V_2^{-0.5} - V_1^{-0.5}\right]$$

$$= -1.5\,P_1V_1^{1.5}\left[V_2^{-0.5} - V_1^{-0.5}\right]$$

$$= -1.5(450\ \text{lb}_f/\text{in}^2)(2\ \text{in}^3)^{1.5}\left[(50\ \text{in}^3)^{-0.5} - (2\ \text{in}^3)^{-0.5}\right]$$

$$= 1080\ \text{in}\cdot\text{lb}_f = \underline{\underline{90\ \text{ft}\cdot\text{lb}_f}}$$

The power developed by the gas is

$$\dot{W} = \left[90\ \frac{\text{ft}\cdot\text{lb}_f}{\text{stroke}}\right]\left[20\ \frac{\text{stroke}}{\text{s}}\right] = \frac{1800\ \text{ft}\cdot\text{lb}_f/\text{s}}{550\ \text{ft}\cdot\text{lb}_f/\text{s}\cdot\text{hp}}$$

$$= \underline{\underline{3.27\ \text{hp}}}$$

The work can be determined once the path and end states are specified, but the power can be determined only after the engine speed is known. In this problem the work is performed by the gas on the surroundings and therefore is positive when the gas is the system. ■

1.8 SUMMARY

Thermodynamics is a study of energy. It involves applying basic principles such as conservation of mass and energy to thermodynamic systems. In a closed system, energy can cross the boundary of the system but mass cannot. An open system can have both mass and energy crossing the boundary. An isolated system can have neither mass nor energy crossing its boundary. Properties of a system are measurable characteristics of the system, such as pressure, temperature, and specific volume. The state of the system is the condition described by the properties of the system.

All properties are point functions whose changes in value between two states do not depend on the path followed during the process. The cyclic integral of any property is always zero, since the initial and final states of a cycle are identical. Energy-transfer mechanisms such as heat transfer and work, which are path-dependent functions, are not properties. Values for path-dependent functions depend on the path followed between the end states, and the cyclic integral of path functions is not necessarily zero.

The pressure is the normal force exerted per unit area on the surface of a system. The units of pressure are pascals (Pa) in the SI system and psi in the English system. Pressures can be either gauge or absolute pressures, and the two are related by

$$P_{abs} = P_{gauge} + P_{atm} \tag{1.20}$$

Pressure differences can also be expressed in terms of the height h of a fluid with specific weight γ by the equation

$$P_1 - P_2 = \gamma h \tag{1.17}$$

The temperature of a substance can be interpreted as a measure of the mean molecular velocity of the substance. As the molecular velocity increases, the temperature of the substance increases. Temperatures are usually measured with thermometers, thermistors, thermocouples, or other devices that relate the temperature to an easily measured property. The units of temperature in the SI system are the Kelvin degree or the degree Celsius. The Kelvin temperature scale is the absolute temperature scale, and it is related to the Celsius scale by

$$\text{K} = {}^\circ\text{C} + 273.15^\circ \tag{1.21}$$

The units of temperature in the English system of units are degrees Rankine or degrees Fahrenheit. The Rankine scale is the absolute temperature scale, and it is related to the Fahrenheit scale by

$${}^\circ\text{R} = {}^\circ\text{F} + 459.67^\circ$$

Heat transfer and work are both mechanisms for the transfer of energy across the boundary of a system. The sign conventions for heat transfer and work are: heat transfer to a system is positive and work done by a system is positive. Heat transfer is caused by a temperature difference between a system and its surroundings, while work is equiv-

TABLE 1.1 SUMMARY OF EXPRESSIONS FOR WORK AND POWER

Forms of Work	General Expression for Work		Power	
Force-displacement	$W_{12} = \int_1^2 F_s ds$	(1.24)	$\dot{W} = F_s V$	(1.33)
Electric	$W_{12} = \int_1^2 \varepsilon I \, dt$	(1.26)	$\dot{W} = \varepsilon I$	(1.34)
Shaft	$W_{12} = \int_1^2 T \, d\theta$	(1.28)	$\dot{W} = T\omega$	(1.35)
$P \, d\mathbf{V}$	$W_{12} = \int_1^2 P \, d\mathbf{V}$	(1.31)	$\dot{W} = PAV$	(1.36)

alent to a force acting through a distance. Heat transfer increases as the temperature difference between the system and surroundings increases, and it decreases as the thermal resistance at the boundary increases. A process during which no heat transfer occurs at the boundary of the system is called an adiabatic process. The equations for work and power associated with the various work modes are summarized in Table 1.1.

PROBLEMS

1.1 Identify the system and the surroundings for the systems shown in Figures 1.9 and 1.10.

1.2 Explain the difference between the terms in each of the following sets:
- **(a)** dimension; unit
- **(b)** point function; path function
- **(c)** open system; closed system
- **(d)** primary units; secondary units
- **(e)** gauge pressure; absolute pressure
- **(f)** work; power
- **(g)** adiabatic; isothermal

1.3 Briefly define the following terms:
- **(a)** property
- **(b)** process
- **(c)** cycle
- **(d)** thermodynamic system
- **(e)** surroundings
- **(f)** isolated system
- **(g)** equilibrium
- **(h)** work interaction
- **(i)** heat transfer
- **(j)** temperature
- **(k)** adiabatic
- **(l)** isothermal

1.4 An object weighs 150 N at sea level and it occupies a volume of 0.60 m^3. Calculate the density of the object in kg/m^3 and lb_m/ft^3 and its specific volume in m^3/kg and ft^3/lb_m.

1.5 An individual weighs 150 lb_f at sea level on earth. Calculate the mass of the individual in lb_m on earth. Calculate this person's weight, in N and lb_f, on the surface of the moon where the acceleration of gravity is one-sixth of that on the earth. Determine the mass of the individual, in kg and lb_m, on the moon.

1.6 What is the weight of a 5-kg mass at a location where the acceleration of gravity is 9.4 m/s^2?

1.7 Convert the following units:

3 lb_m to kg	1 kW to hp
25 lb_f to N	3×10^3 Btu/h to kW
5×10^5 Pa to psi	45 in Hg to psi
560 kg to lb_m	850 mm Hg to Pa
495 Btu to J	750 J to ft·lb_f

1.8 Calculate the weight, in both N and lb_f, of a 25-kg mass where the local acceleration of gravity is as follows:
(a) 9.0 m/s^2
(b) 8.5 m/s^2

1.9 Calculate the sea-level weight in lb_f and N of an object that has a mass of 5 kg.

1.10 Convert the following gauge pressures to absolute pressures in Pa and psi if the atmospheric pressure is 101 kPa:
(a) 76.2 mm Hg
(b) 13.8 kPa
(c) −41.4 kPa
(d) −29 mm Hg

1.11 A 200-kg mass has a uniform density of 4.5×10^3 kg/m^3. Determine the weight of the object where the local acceleration of gravity is (a) 27 ft/s^2 and (b) 32 ft/s^2 in newtons and lb_f. Calculate the volume of the mass in ft^3 and m^3. Calculate the specific volume of the mass in ft^3/lb_m and m^3/kg.

1.12 A liquid has a density of 1350 kg/m^3. Calculate the specific weight and specific volume of the liquid at sea level.

1.13 A mercury barometer is used to measure atmospheric pressure. The pressure indicated by the barometer is 749 mm Hg. Calculate the atmospheric pressure in (a) meters of water, (b) bars, and (c) Pa.

1.14 A gas is enclosed by a vertical, frictionless piston and cylinder. The surface area of the piston is 20 cm^2, and its mass is 5 kg. Calculate the gauge pressure of the gas, in Pa.

1.15 The gauge pressure of a gas inside a container is 200 kPa. Calculate the vertical height of manometer fluid that can be supported by this gas pressure if the fluid is (a) mercury, (b) water, and (c) an oil with specific gravity of 0.95.

1.16 An inclined manometer is filled with a fluid with a specific gravity of 2.0. The manometer is used to measure the pressure of a gas in a container as shown in the figure. If the local atmospheric pressure is 29.90 in of mercury, calculate the gauge and absolute pressure of the gas in lb_f/in^2 and ft of water.

Problem 1.16

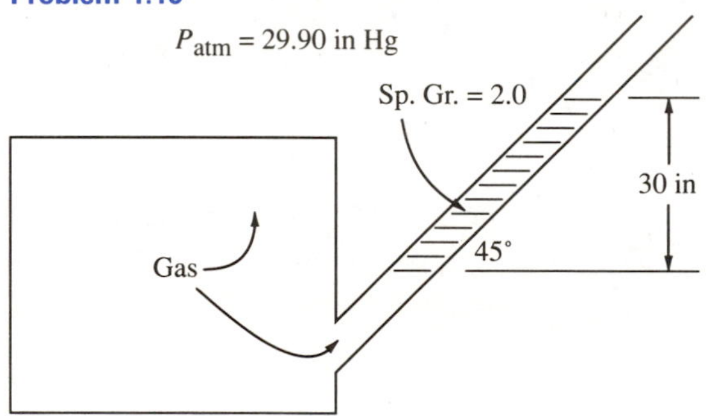

1.17 A manometer is used to measure the gas pressure in the vessel shown in the figure. The manometer contains two immiscible fluids, one with a density of 40 lb_m/ft^3 and a second one with a density of 50 lb_m/ft^3. Determine the gauge and absolute pressure of the gas for conditions shown in the figure.

Problem 1.17

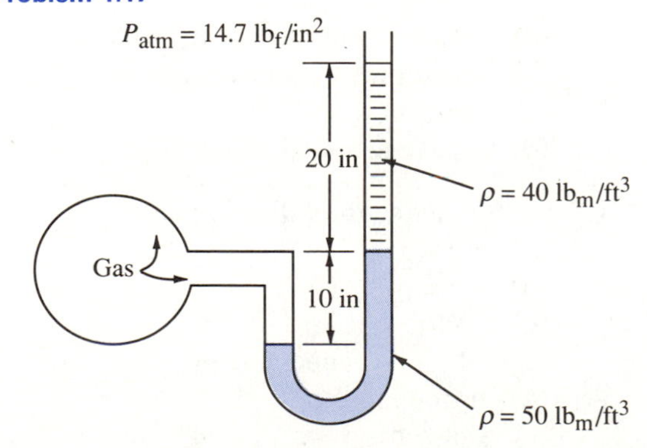

1.18 What is the force exerted by the water on the bottom of a cylindrical tank of water with a diameter of 10 m and a depth of 5 m?

1.19 The gauge pressure within a chamber is equivalent to a height of 356 mm of a fluid with a specific gravity of 0.75. The barometric pressure is 749 mm Hg. Compute the absolute pressure within the chamber, in Pa and psia.

1.20 [CDA111] A manometer is used to measure the pressure of air in a tank. The manometer fluid has a specific volume of 0.0012 m^3/kg. The difference in height of the two columns is 0.5 m. The barometric pressure is 99.3 kPa. Determine the following:

(a) the density of the manometer fluid in kg/m^3

(b) the gauge pressure measured by the manometer in kPa

(c) the absolute pressure measured by the manometer in kPa

1.21 [CDA211] A gas inside a container has a pressure of 1.5 psig. Calculate the following:

(a) the corresponding height difference in feet of a mercury manometer

(b) the absolute pressure in psia

(c) the force that the gas exerts on the container which has a total surface area of 0.75 ft^2

1.22 The gauge pressure of a system is 140 mm of a fluid with a specific gravity of 1.35. The atmospheric pressure is 750 mmHg. What is the absolute pressure of the system in Pa and psia?

1.23 A submarine is cruising in the ocean at a depth of 320 m. Salt water has a specific gravity of 1.03. Determine the pressure on the hull of the submarine in Pa and psia.

1.24 Assume that the atmosphere may be assumed to be a static, isothermal, gas. Further assume that the pressure and density of the gas are related by $P/\rho = RT_0$. Derive an expression for the variation in atmospheric pressure with elevation from sea level in terms of pressure at sea level, P_o, temperature, T_o, acceleration of gravity, g, and gas constant R.

1.25 A cylindrical container is filled with water and the valve is closed. The container is inverted as shown in the figure and the valve is opened, allowing some water to drain from the container. Assuming that the valve prevents air from entering the container, determine the equilibrium depth of the vapor layer, x, above the liquid. Suppose that the container is then relocated to a place where the atmospheric pressure is reduced to 14.0 lb_f/in^2. Determine the additional amount of water (volume and mass) that

Problem 1.25

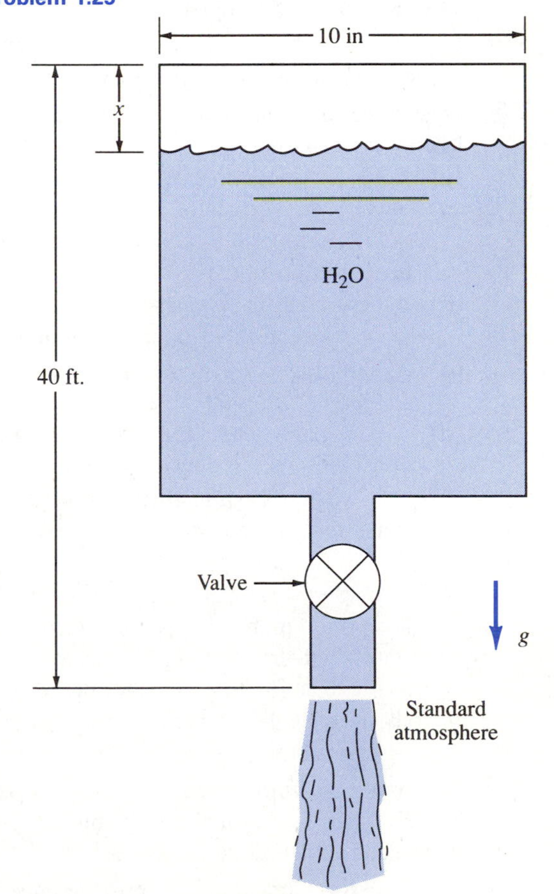

drains from the container during the relocation process.

1.26 Convert the following temperatures:
- **(a)** 20° C to kelvin
- **(b)** 70 K to degrees Celsius
- **(c)** 400 K to degrees Celsius
- **(d)** 250° F to degrees C
- **(e)** 100° F to degrees R
- **(f)** 800° F to kelvin
- **(g)** −50° F to degrees C

1.27 What is the temperature of an object when Fahrenheit and Celsius thermometers indicate the same magnitude of temperature of the object?

1.28 [CDA411] A thermometer reads a temperature of 168 on the Fahrenheit scale. What are the corresponding readings on the Celsius, Kelvin, and Rankine scales?

1.29 Explain the difference between an adiabatic process and an isothermal process.

1.30 Determine what is incorrect about the following statements:
- **(a)** A TV news commentator states that the heat on the surface of a manned spacecraft reaches 1400° C as it reenters the earth's atmosphere.
- **(b)** A basketball is stamped with the statement "Inflate to 15 pounds pressure."
- **(c)** A shipping crate is stamped with the statement "Weight = 250 kg."
- **(d)** A newspaper photograph shows workers erecting a building during a summer heat wave. The title on the photograph states the workers are "suffering in the 110° F heat."
- **(e)** A new pasteurization process for milk is reported in the newspaper. The article states that the "raw milk is heated for a few seconds to about 280 degrees—about twice the heat required for normal pasteurization."

1.31 Determine the work required to compress an elastic spring a distance of 85 mm if the spring constant is 5×10^4 N/m.

1.32 Calculate the work required to move an object horizontally a distance of 30 ft with a constant force of 850 lb_f inclined at an angle of 30° above the horizontal.

1.33 Determine the work required to move an object horizontally a distance of 30 m with a force inclined at an angle of 45° above the horizontal. The force F changes with displacement x of the object according to the equation

$$F = 200x + 0.1x^2$$

where F is in newtons and x is in meters.

1.34 Determine the work required to move an object horizontally a distance of 50 ft with a constant force of 500 lb_f inclined at an angle of θ to the horizontal. The angle θ varies with the displacement according to the equation

$$\cos \theta = 0.1 + 0.015x$$

where θ is in degrees and x is in feet. Compare this work with the work that is required assuming the same force is applied at a constant angle of $\theta = 50°$. Determine the average value for θ that will produce the same work for the force of 500 lb_f when θ varies according to the above expression.

1.35 Denver's Mile High football stadium incorporates a system by which a section of stands can be moved to accommodate different sports on the field. The grandstand, which has a mass of 4×10^6 kg, can be moved a distance of 45 m on a thin film of water. The movable grandstand rests on 46 bearing pads that are each 1.22 m in diameter. Water at high pressure is pumped into each of the pads until the stands are lifted a vertical distance of 3.8 cm. Excess water forms a lubricating film over which the grandstand is moved. The force required to move the stands is approximately 4.45 N per 450 kg of stadium mass. Calculate the following:
- **(a)** the pressure of the water under each bearing pad

(b) the power of the motor required to move the grandstands over the distance of 45 m if the job takes 1 h

(c) the work required to raise the grandstand 3.8 cm

1.36 As a garage door is lowered, it stretches a spring. The spring has a spring constant of 30 lb_f/ft, and it is stretched 4 ft as the door is lowered. Calculate the work done by the door on the spring as it is lowered.

1.37 Calculate the amount of work required to compress a spring 25 in from its uncompressed position if the force-displacement relation is given by

$$F = 20x^3$$

where F is in lb_f and x is in inches.

1.38 A jet engine produces a thrust of 80,000 lb_f while the aircraft is moving with a velocity of 500 ft/s. Calculate the power developed by the engine and the work produced by the engine in 1 h.

1.39 A garden sprayer is pressurized by a hand-operated pump. The total length of the stroke of the pump is 35 cm. The force required to move the pump increases linearly with displacement according to the equation

$$F = M + 0.12x$$

where x is in centimeters and F is in newtons. The value of M is 1.0 N for the first stroke, 5 N for the second stroke, 9 N for the third stroke, 13 N for the fourth stroke, and so on. Determine the work required to pump the handle 10 times.

1.40 Determine the size of a motor, in kilowatts and horsepower, necessary to pull a 2-mm wire through a die at a velocity of 15 m/s if the tension in the wire is 250 N.

1.41 A stapling gun drives staples by means of compressed air. The force required to drive the staple is known to be equal to $F = kx^2$, where $k = 300\ lb_f/in^2$, F is measured in lb_f, and x is the distance the staple is driven into the substance in inches. The device that provides the force to drive the staple is a 2-in diameter piston pressurized by 100-psia air. The travel of the piston is limited to 0.5 in. Calculate the depth that a staple can be driven under this design if all the energy of the piston is available to seat the staple.

1.42 A garden hose (shown in the figure) supplies water at a pressure of 100 lb_f/in^2 at a rate of 4 gal/min. Determine the maximum amount of power that can be derived from the garden hose.

Problem 1.42

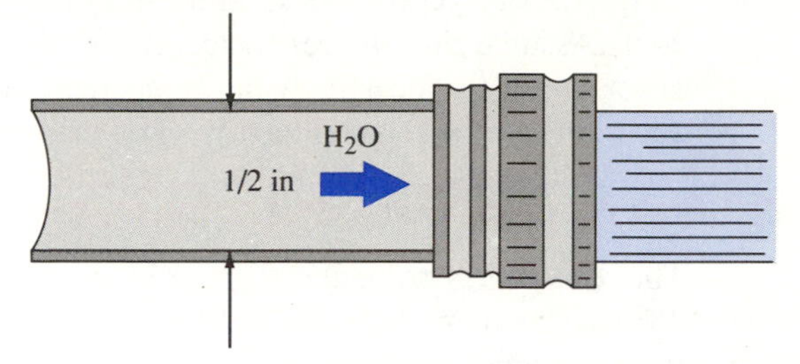

1.43 A wire is pulled through a die at a constant velocity of 10 ft/s with a constant force of 250 lb_f. Determine the power required to pull the wire at this velocity and the work required to pull the wire a distance of 100 ft. The force on the wire is reduced linearly

Problem 1.43

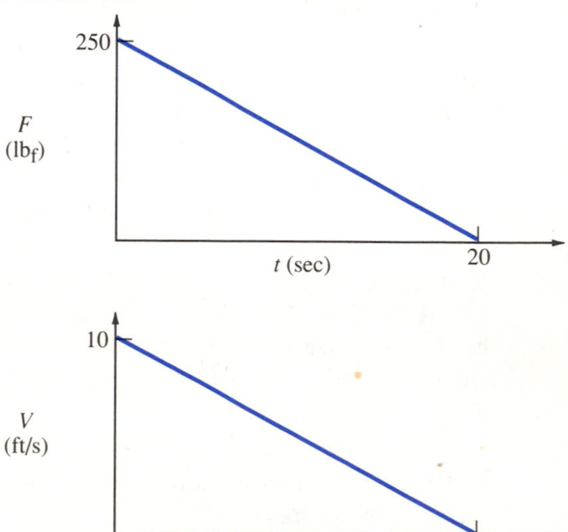

from 250 lb_f until the wire stops moving when the force is zero. During the time that the force is reduced to zero, the velocity of the wire reduces linearly to zero from 10 ft/s. Both the force-time and the velocity-time curves are shown in the figure. Calculate the work required for the process that occurs during the 20-s period indicated in the figure.

1.44 A constant current of 20 A is circulated through a resistor. As a result of heating in the resistor, its temperature increases linearly from 30° C to 500° C in a 3-min period. Assume that the resistance of the resistor varies linearly with temperature according to the relationship

$$R(T) = R_o[1 + \alpha(T - T_o)]$$

The resistor has properties $R_o = 5\ \Omega$, $T_o = 20°$ C, and $\alpha = 5 \times 10^{-4}$ 1/° C. Determine the amount of work required to circulate the constant current for the 3-min time interval. Also calculate the power required at the beginning of the process when the resistor temperature is 30° C and at the end of the process when the temperature is 500° C.

1.45 An integrated-circuit package is powered by a 5-V supply. The package is turned on and it initially draws a current of 5 mA. As the temperature of the circuit increases, the current drops linearly until it reaches 4 mA after 1 min. Determine the total work performed by the power supply on the package during the 1-min period. Calculate the power delivered by the power supply at the beginning and end of the process.

1.46 A battery charger delivers a steady current of 2 A at 12 V. Determine the power that the battery charger can deliver and the amount of work it can perform over a period of 1 h.

1.47 A transistor draws a current of 2 mA at a voltage of 6 V. Determine the power consumption of the transistor.

1.48 An electronic device operates at a constant voltage of 120 V. The device is energized, and the current drawn by the device varies with time according to the equation

$$I = 10e^{-t/60}$$

where I is in amperes and t is in seconds. Calculate the work performed on the device during the first 2 min of operation. Also, determine the power delivered to the device at $t = 0$ and $t = 2$ min.

1.49 A 6-V battery causes a constant current of 2 A to pass through a resistor for 1 min. Calculate the total work performed by the battery and the instantaneous power delivered to the resistor.

1.50 A 12-V battery provides a current through a resistor. The initial current is 1 A, but because of the heating of the resistor, the current drops linearly with time until it reaches a value of 0.8 A after 5 min. Determine the total work performed by the battery on the resistor during the 5-min period. Calculate the power delivered by the battery at the beginning and at the end of the process.

1.51 As a weight falls, its energy is stored in a coiled spring. The relationship between the torque in ft·lb_f in the spring and the angular displacement in radians is given by

$$T(\theta) = 500\ \theta^{1.5}$$

Determine the work required to rotate the spring through:

(a) 10 revolutions
(b) 20 revolutions

1.52 For the conditions given in Problem 1.51, determine the average power necessary to coil the spring at a constant angular velocity if the process takes place over a period of:

(a) 1 min
(b) 5 min

1.53 The torque, in newton-meters, required to rotate a shaft through an angle θ is given by the equation

$$T = 500(1 + \sin\theta)$$

What is the average size motor, in kilowatts, required to turn the crank at an average speed of 1000 rev/min? Determine the work required to rotate the crank through one revolution.

1.54 A shaft is rotated at constant angular speed of 100 rpm. During the rotation the torque-angular displacement path is as shown in the figure. Calculate the work required to rotate the shaft through 100 revolutions and the power required to rotate the shaft for angular locations of $\theta = 0°$ and $\theta = 180°$ if $T_0 = 500$ ft·lb$_f$.

Problem 1.54

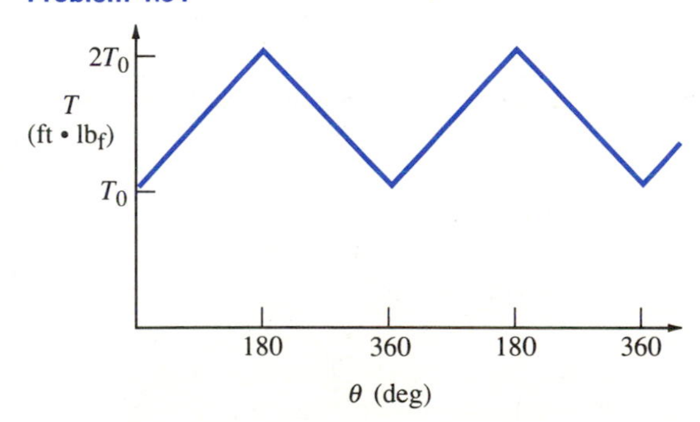

1.55 Determine the power delivered by a shaft rotating at 1500 rev/min against a constant torque of 5.0 N·m.

1.56 A scissor jack requires a constant torque 350 ft·lb$_f$ and 36 complete revolutions to raise a weight. Calculate the work and power required to raise the weight in (a) 1 min and (b) 10 min.

1.57 A turbine in a power plant turns a generator against a constant torque of 7×10^8 N·m at 3600 rev/min. Determine the power generated by the turbine.

1.58 An internal-combustion engine is placed on a dynamometer and the output torque of the engine is measured as a function of engine speed. The torque, in newton-meters, is found to vary with ω, in revolutions per minute, according to the equation

$$T = 400 \sin\left(\frac{\pi}{2} \cdot \frac{\omega}{3000}\right)$$

Calculate the power delivered by the engine at speeds of 1000 and 2000 rev/min, and calculate the work delivered by the engine if it accelerates linearly from rest to 1000 rev/min in 1 min.

1.59 A foot pump shown in the sketch in the fully open position is used to inflate an air mattress. The force-angle relationship required to activate the pump is

$$F(\theta) = 5.0 + 0.01\,\theta$$

where F is measured in lb$_f$ and θ is measured in degrees ($0° \le \theta \le 60°$). The pump delivers 0.1 ft^3 of air when pumping against no resistance and for each successive stroke after the first, it delivers 1×10^{-3} ft^3 less of air. Assuming the air mattress requires 3 ft^3 of air to be fully inflated, calculate the number of complete strokes ($\theta = 60°$) to fully inflate the mattress. Also calculate the work required to inflate the mattress and the power required to inflate the mattress in 5 and 10 min.

Problem 1.59

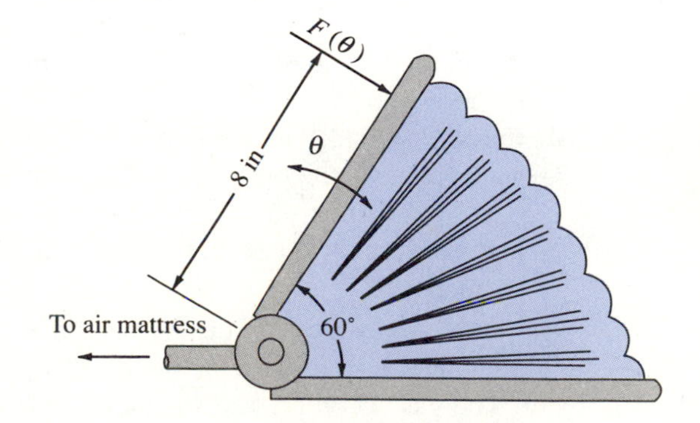

1.60 A coiled spring with a mean diameter of 16 in is rotated and the stored energy in the spring is released and used to start a small internal-combustion engine. A spring with a spring constant equal to 2×10^{-1} lb_f per degree of rotation is proposed for the job. The minimum power necessary to start the engine is estimated to be 2.0 hp. The time required to uncoil the spring is directly proportional to the angle through which the spring has been rotated and the time required to uncoil the spring is 1 s when it uncoils from two full rotations. Estimate the number of rotations that the spring must be coiled in order to provide enough power plus a 50 percent surplus to start the engine.

1.61 For each process and each system listed, indicate whether work is positive, negative, or zero; do the same for heat transfer, observing the appropriate sign convention.

(a) A perfectly elastic spring stands on the end of a table. The spring is compressed by a book, which is placed on top of the spring. Consider as the system (1) the book, (2) the spring, and (3) the table.

(b) A paddle wheel turned by a motor stirs a liquid in an insulated vessel. Consider as the system (1) the liquid, and (2) the paddle wheel.

(c) A gas in an insulated cylinder is compressed so that its pressure and temperature both increase. The system is the gas.

(d) A steel wire is bent back and forth until it becomes hot to the touch. The system is the wire.

(e) Carbon dioxide is compressed in a water-cooled compressor. The system is (1) the carbon dioxide, (2) the cooling water, and (3) a section of the compressor wall.

1.62 A gas at the temperature of the surroundings is placed in a piston and cylinder arrangement. The gas is compressed so that its volume decreases. Assuming that the gas is considered to be the system, determine the sign of the work and heat transfer for the system during the process.

1.63 An electric resistor is placed in room air. Current is passed through the resistor until it becomes hot. The resistor is considered to be the system. Determine the sign of the work and heat transfer for the system.

1.64 An electric drill is used to drill a hole in a piece of wood. Indicate whether the work and heat transfer are positive, negative, or zero if the system is (1) the electric drill or (2) the drill bit.

1.65 A piston and cylinder contains air at a pressure of 200 psia and a volume of 0.015 ft^3. The air expands until its volume doubles. Calculate the work done by the air during the process if the path for the process is $PV = C$.

1.66 A gas enclosed in a flexible container undergoes a thermodynamic cycle consisting of three processes. The properties at the original state are P_1, T_1, and V_1. The gas first expands in a constant-pressure process until its volume doubles. The gas then is cooled in a constant-volume process until the temperature reaches T_1. The third process is isothermal compression until the gas returns to the original state. During the isothermal process the path followed is $PV =$ constant.

(a) Sketch the cycle on a P-V diagram.

(b) Determine the work done by the gas during each of the three processes in terms of P_1, T_1, and V_1.

(c) Determine the net work done by the gas during the cycle.

1.67 A gas with a constant absolute pressure of 3.5 MPa acts on a piston having a diameter of 7.5 cm. Calculate the work done by the gas when the piston moves frictionlessly through a displacement of 65 cm.

1.68 A pressure of 300 psia acts on a piston having a diameter of 7.5 in. Calculate the work done when this piston moves through a displacement of 15 in.

1.69 The pressure inside a system varies with the volume of the system according to the equation

$$PV^n = C$$

where n and C are constants and $n > 1$. Show that the work performed by the system during a change from state 1 to state 2 is given by

$$W_{12} = \frac{P_2V_2 - P_1V_1}{1 - n}$$

1.70 The pressure inside a system varies inversely with the square of the system volume. Determine the work required to compress the system from a volume of 2 m^3 and a pressure of 200 kPa to a pressure of 600 kPa.

1.71 Matthew Boulton designed a rotating steam engine in 1788 which was frequently used to drive laps for grinding and polishing metal objects. His lap engine, as it was frequently called, was one of the first engines to free industry from the reliance on wind, water, and animal power, and it had a considerable impact on industrial expansion in the eighteenth and nineteenth centuries. A single-cylinder engine developed by Boulton had a bore of 18 in, a stroke of 4 ft, and developed 13.75 hp. Compare the horsepower output per cubic inch of displacement of this engine with a modern gasoline engine, which has a displacement of 350 in^3 and a power output of 220 hp.

1.72 A common method used to install cables in a conduit involves placing a piston in the conduit and attaching a small lightweight cord to the piston. The conduit is pressurized with air and the piston pulls the cord through the conduit. Suppose the piston is held in place until the air pressure behind it reaches a value of 50 lb_f/in^2. It is then released and allowed to drag the cord behind it while no additional air is introduced into the conduit. Assume the conduit in front of the piston is vented sufficiently to maintain a constant pressure of 1 atm. The drag force of the piston on the conduit wall is a constant value of 2 lb_f and the drag provided by the cord is proportional to the length of cord that has been pulled through the conduit. The drag force of the cord is known to be 3 lb_f when 100 ft of cord is extended into the conduit. Assume the air during the expansion process follows the path PV = constant and the original volume of air behind the piston is 1000 in^3. Calculate the length of conduit that can be serviced in this manner for the given conditions. The conduit is 1 inch in diameter.

Problem 1.72

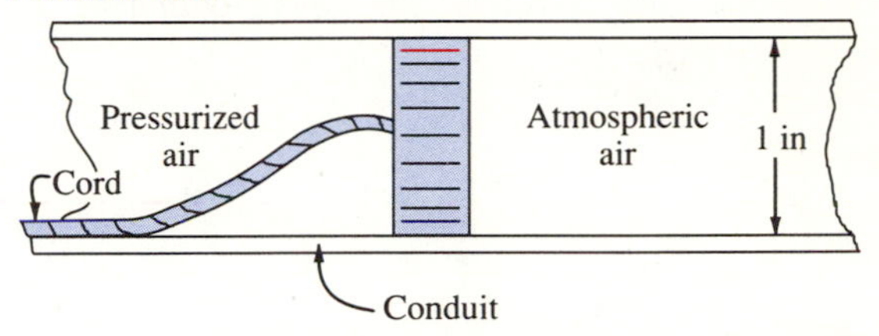

1.73 In 1796, Matthew Boulton and James Watt first introduced *indicator diagrams* as an aid to obtaining the correct settings and timing of valves in their steam engines. An indicator diagram is a record of the cylinder pressure plotted as a function of displacement. The figure shown below is an indicator diagram of a steam engine that operates at a steam supply pressure of 110 psig and a speed of 200 rpm. Estimate the horsepower developed by this particular engine.

Problem 1.73

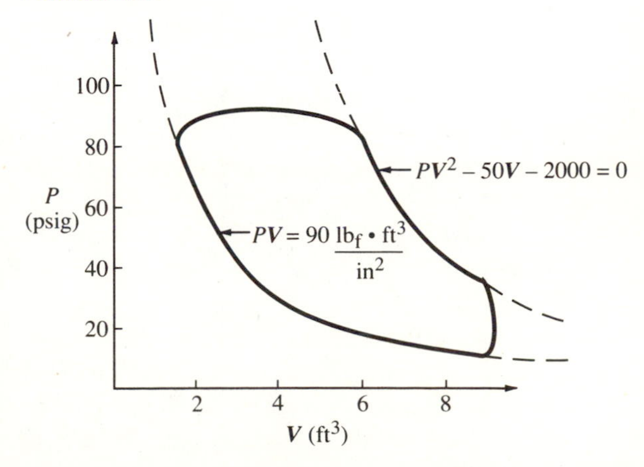

1.74 Compound steam engines were introduced to improve upon the efficiency of early steam engine designs. In these engines the expansion process occurs in two cylinders rather than a single cylinder. The compound expansion process reduces the losses that result from leakage past valves and pistons because the steam that leaks in the high-pressure cylinder can be used in the second, low-pressure cylinder. The expansion process for a compound steam engine is shown in the figure along with its indicator diagram. If the particular engine operates at 80 rpm, estimate the power output of the engine.

Problem 1.74

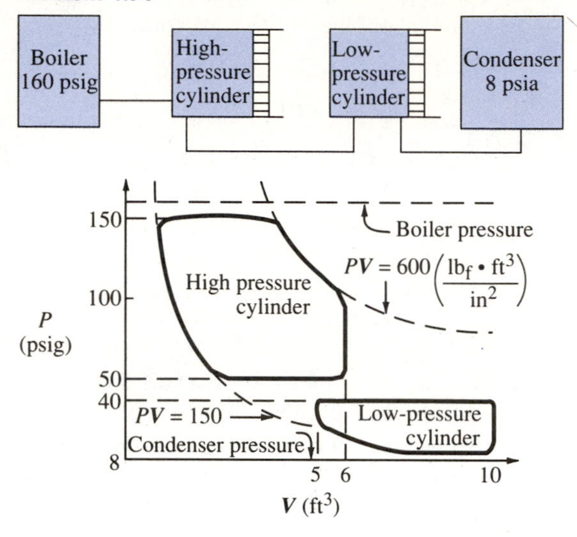

2 Properties of Substances

2.1 INTRODUCTION

An introductory course in thermodynamics is devoted primarily to an in-depth study of a few basic physical principles. These principles, such as the conservation of mass developed in Chapter 3 and the conservation of energy developed in Chapter 4, are applied to a variety of thermodynamic systems that can be composed of different classes of substances. The conservation of energy relates energy transfer across the boundary of a system due to work interactions and heat transfer to the change in a property of the system called the total energy. The total energy, in turn, is related to other properties of the system. Therefore, a detailed study of properties and property behavior is also of paramount importance in thermodynamics.

This chapter has been divided into two major parts. The first part begins with a discussion of several classifications of properties and the relationship between properties and the thermodynamic state of a substance. Next, a postulate is presented that prescribes the number of properties required to specify the state of a substance. The forms of energy associated with a substance are identified, and the properties called the internal energy and enthalpy are introduced. These properties are then related to other measurable properties so that values of internal energy and enthalpy can be determined.

The remainder of the chapter deals with property determination and property behavior for different classes of substances. The qualitative behavior of pure substances is considered, and the determination of the thermodynamic properties of pure substances from tabular data is studied and illustrated through the use of examples. The concepts of an ideal gas and an incompressible substance are introduced as simple models for the behavior of pure substances for certain restricted ranges of temperature and pressure. Finally, the property relationships for these substances are developed and discussed in detail.

2.2 DEFINITIONS

The thermodynamic behavior of a substance is characterized by its properties. Since a property is independent of path, unlike quantities such as heat transfer and work, its value depends only on the state of the substance and not on how the substance achieved the state. To merely distinguish between properties and nonproperties is not sufficient, however. A number of important and useful classifications of properties exist, and these will be discussed in this section.

2.2.1 Extensive and Intensive Properties

Properties can be classified as being either extensive or intensive. An *extensive property* is one that depends on the extent of a system, or the amount of mass present. If a system is subdivided into n (possibly unequal) parts, the value of the extensive property for the system is equal to the sum of the contributions from each of the n individual parts. For an arbitrary extensive property Y,

$$Y_{\text{sys}} = \sum_{j=1}^{n} Y_j$$

Obviously, then, the volume V is an extensive property. If the volume of a system is divided into three parts, for example, the total volume of the system is the sum of the volumes of each of the three parts. Generally, uppercase letters, such as V for volume, are used to denote extensive properties.

An *intensive property*, on the other hand, is independent of the extent of the system or the amount of mass present. If a single-phase system in thermodynamic equilibrium is subdivided into n parts, the value of any intensive property is the same for each of the n subdivisions provided that the size of the subdivisions is large compared with the molecular mean free path, that is, the average distance between molecules. Temperature is a typical example of an intensive property because the temperature of each of the n subdivisions of a larger system is identical.

In most cases the value of an intensive property y can be obtained by dividing the corresponding extensive property Y by the mass of the system:

$$y = \frac{Y}{m}$$

For example, the volume (an extensive property) can be converted to an intensive property (the specific volume) by dividing it by the mass:

$$v = \mathbf{V}/m$$

Lowercase letters are usually used to denote intensive properties, while capital letters are usually reserved for extensive properties. Three exceptions to this convention are the temperature T and the pressure P, which are intensive properties, and the mass m, which is an extensive property.

2.2.2 Physical and Thermodynamic Properties

Properties can also be classified as being either physical properties or thermodynamic properties. A *physical property* is one that requires the specification of an external, coordinate frame to define a reference value for the property. Physical properties include velocity, kinetic energy, elevation, and potential energy. The velocity and elevation of a system might be measured relative to a coordinate system fixed at the earth's surface or some other convenient location, whereas the velocity of a fluid is often measured relative to a reference frame fixed with respect to the thermodynamic system.

Unlike physical properties, *thermodynamic properties* do not require an external coordinate frame to define a reference value. Examples of thermodynamic properties are temperature, pressure, volume, and specific volume.

2.2.3 Homogeneous Substances and Phases

A substance that has a uniform physical structure and chemical composition is said to be *homogeneous*. The chemical composition of water is denoted by H_2O, but water may exist in various forms. At ordinary room temperature and atmospheric pressure, water exists as a *liquid*. If the water is heated sufficiently, it can become a *vapor*; and if cooled, it can be transformed into ice, or a *solid*. Water can also exist as a mixture of liquid and vapor or liquid and solid. These and other mixtures of water have the same chemical composition throughout, but the physical structure of the mixtures is not uniform. Therefore, mixtures such as these are not homogeneous.

Each of the three forms of water (liquid, vapor, and solid) is called a *phase*, a quantity of matter that is homogeneous throughout. The vapor phase is also often called the *gaseous phase*. In thermodynamics, however, the term *vapor* is usually used to indicate that the substance can be easily condensed to the liquid phase.

In a two- or three-phase mixture the phases are separated by phase boundaries. The phase boundaries can be very distinct, as is the case with an ice cube in a glass of water, or they may be difficult to perceive, as in a mist of water vapor and droplets of liquid water.

2.3 THE STATE POSTULATE

Even though the thermodynamic state of a substance is determined by its properties, the question of how many properties are required to completely specify the state of a substance has not been considered. This question can be resolved with the *state postulate*:

> **The number of independent, intensive thermodynamic properties required to completely and uniquely specify the thermodynamic state of a homogeneous substance is one more than the number of relevant, reversible modes of work.**

Notice that the state postulate refers to homogeneous substances and, therefore, is limited to single-phase substances. Also, thermodynamic properties specified by the state postulate must be independent properties. In succeeding sections the fact that various intensive properties are not independent of each other in some instances will become evident. Furthermore, only those reversible modes of work that might possibly exert significant influence on the substance need be considered when applying the state postulate.

Although there are many reversible modes of work, some of which were discussed in Chapter 1, PdV work is often the only significant reversible work mode in many engineering systems. Such systems are composed of substances referred to as *simple compressible substances*. For a simple compressible substance, then, the state postulate leads to the conclusion that only two independent, intensive thermodynamic properties are required to completely specify the state of the substance.

An important implication of the state postulate is that any two independent, intensive thermodynamic properties associated with the state of a simple compressible substance can, in principle at least, determine all other properties associated with that state. Thus all other properties associated with the state are dependent properties. Suppose that the temperature and specific volume are selected to be the independent properties that fix the state of a simple compressible substance. The fact that any other thermodynamic property y depends only on these two independent properties can be indicated mathematically as

$$y = y(v,T)$$

For example, the relation between pressure, volume, and temperature can be expressed as $P = P(v,T)$. Similar equations of state can be written for other thermodynamic properties. The state postulate can then be used as the basis for relationships between properties that are difficult to measure and properties that are much easier to measure such as pressure, temperature, and specific volume. This notion will be discussed further in Section 2.7 and in much more detail in Chapter 9.

2.4 THE TOTAL ENERGY

In Chapter 1 work interactions and heat transfer were described as being mechanisms for energy transfer across the boundary of a system. Heat transfer was defined as energy transfer across the boundary arising from a temperature difference between the system and the surroundings, while a work interaction was defined as energy transfer across the boundary of the system arising from an effect equivalent to a force acting through a distance. Neither heat transfer nor work is a property, and they cannot be represented by exact differentials since their values depend on the path followed during a change of state.

Many forms of energy are important in the study of thermodynamics. Kinetic, potential, chemical, electrical, surface tension, and magnetic energies are familiar examples. In a classification of all the different forms of energy that play a role in thermodynamic systems, distinguishing between microscopic and macroscopic energy forms is helpful. *Microscopic forms* of energy are related to the energy possessed by the individual molecules and to the interaction between the molecules that comprise the system under consideration. *Macroscopic forms* of energy, on the other hand, are related to the gross characteristics of a substance on a scale that is large compared with the mean free path of the molecules. These forms of energy can be identified without considering the fact that the substance consists of molecules, but rather by considering the system as an equivalent mass concentrated at the center of gravity of the system.

To clarify the distinction between microscopic and macroscopic forms of energy, we consider a simple example. Suppose that the center of mass of a container of water has a velocity with respect to some reference frame such as the earth's surface. By virtue of its mass and velocity, the water possesses kinetic energy on a macroscopic scale. The kinetic energy identified here is independent of any energy possessed by the water on a molecular scale. A similar argument could be used to show that the water possesses potential energy on a macroscopic scale because the mass of the system is above some horizontal reference plane in a gravitational field. On a molecular, or microscopic, scale the water is composed of molecules that move about randomly, colliding with one another. On this scale the individual molecules possess kinetic energy and other forms of energy that are independent of the kinetic energy identified on a macroscopic scale.

The *total energy* E is a property of a system and is defined as the sum of all macroscopic forms of energy plus the total of the microscopic forms of energy:

$$E \equiv E_{\text{macroscopic}} + E_{\text{microscopic}} \tag{2.1}$$

A thermodynamic analysis usually includes a determination of the change in the total energy of a system during a process or series of processes. However, only rarely does a system experience significant changes in more than a few of the many different forms of energy that sum to the total energy of the system. For example, if an elevator is considered to be a thermodynamic system, then changes in kinetic and potential energy are significant, while changes in electrical, chemical, and magnetic energies are negligible. On the other hand, when a chemical reaction occurs, as in a lead-acid storage battery, changes in chemical energy may completely overshadow changes in potential and kinetic energy. Such a system is called a *chemically reacting* system.

If the chemical composition of a system is not altered, then changes in the chemical energy of the system are negligible, and the system is called a *nonreacting* system, an example of which is water vapor passing through a turbine. The chemical composition of the water leaving and entering the turbine is identical, so the chemical energy level of the water does not change as the water flows through the turbine. The application of thermodynamic principles will be limited to nonreacting systems until chemically reacting systems are discussed in Chapters 12 and 13.

Processes involving significant changes in magnetic, electrical, and surface tension energy levels, for example, are somewhat rare in an introductory course in thermodynamics. Most often $P\, d\mathbf{V}$ work is the only significant reversible work mode, and therefore attention is focused primarily on simple compressible systems. While the thermodynamic principles that are developed here are not restricted to simple compressible systems, most of the applications of those principles are limited to such systems throughout the remainder of this text.

The microscopic and macroscopic forms of energy, which sum to the total energy of a system, are considered in further detail in the following sections. Kinetic and potential energies on a macroscopic scale are discussed in Sections 2.4.1 and 2.4.2., while microscopic energy forms are discussed in Section 2.4.3.

2.4.1 Kinetic Energy

The *kinetic energy* E_k of a quantity of mass m with velocity V is defined by the expression

$$E_k \equiv mV^2/2 \tag{2.2}$$

The kinetic energy and the velocity of the center of mass are physical properties and, as such, must be measured with respect to some external coordinate frame. Often the most convenient reference frame is one that is stationary relative to the earth. For this frame of reference a quantity of mass with no motion relative to the earth has a relative velocity of zero, and its kinetic energy is zero. Other choices for the reference frame are possible, but once the reference frame has been selected, it should remain unchanged throughout the analysis of a system.

Thermodynamic analyses are most often concerned with determining the change in kinetic energy of a quantity of mass as it proceeds through a process from one state to another. Since the kinetic energy is a property, the change in kinetic energy of a system is independent of the path followed between the two end states of the process. The magnitude of the change is solely dependent on the mass and velocity of the system at the end states.

The kinetic energy per unit mass, an intensive physical property, is

$$e_k = V^2/2 \tag{2.3}$$

This expression is also used to evaluate the kinetic energy associated with mass flow across the boundary of open systems.

2.4.2 Potential Energy

A quantity of mass m possesses *potential energy* in a gravitational field with acceleration of gravity g by virtue of its elevation z above some arbitrary coordinate frame. The potential energy is defined by the expression

$$E_p \equiv mgz \tag{2.4}$$

The potential energy and the elevation of the center of mass, like the kinetic energy and velocity, are physical properties that require a physical, external reference. Usually, potential energy is assigned a value of zero at an arbitrary reference elevation, which might be the earth's surface or any other convenient elevation.

Since thermodynamic analyses are most often concerned with *changes* in properties, the choice of the reference elevation for potential energy is completely arbitrary. For example, the change in a mountain climber's potential energy from the base of a mountain to its summit is the same regardless of whether the zero elevation point is selected at the base of the mountain, at sea level, or at any other fixed location. Furthermore, the change in the climber's potential energy between the base and the peak of the mountain is identical regardless of the path the climber takes during ascent. In other words, the potential energy is a property, and the magnitude of the change in potential energy of a system depends only on the mass, the elevation, and the local acceleration of gravity at the end states of a process.

The potential energy per unit mass, an intensive property, is

$$e_p = gz \tag{2.5}$$

Equation 2.5 can also be used to evaluate the potential energy associated with mass flow across the boundary of open systems.

EXAMPLE 2.1

Two identical automobiles each have a mass of 1500 kg. Both autmobiles start from rest at the same location, indicated by point 1 (elevation of 1000 m) in Figure 2.1. Automobile

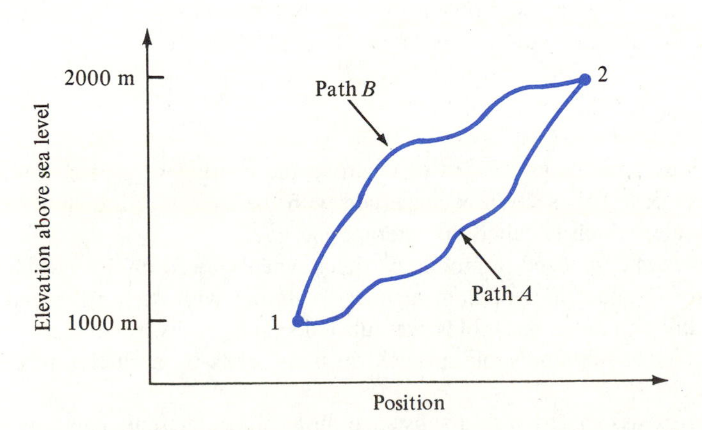

Figure 2.1 Paths for Example 2.1.

A follows path A and passes through point 2 (elevation of 2000 m) with a velocity of 15 m/s, while automobile B follows path B and passes through point 2 with a velocity of 20 m/s. Determine the change in potential energy and kinetic energy of each automobile between points 1 and 2.

Solution. Reference states are first selected for the potential and kinetic energies. Suppose that velocities are measured with respect to the surface of the earth and sea level is used as the point for zero potential energy. The changes in kinetic energy and potential energy for automobile A are

$$E_{k_2} - E_{k_1} = \frac{m}{2}(V_2^2 - V_1^2)$$

$$= \frac{(1500 \text{ kg})[(15 \text{ m/s})^2 - 0]}{2(1 \text{ kg} \cdot \text{m/N} \cdot \text{s}^2)} = \underline{\underline{1.69 \times 10^5 \text{ N} \cdot \text{m}}}$$

$$E_{p_2} - E_{p_1} = mg(z_2 - z_1)$$

$$= \frac{(1500 \text{ kg})(9.81 \text{ m/s}^2)(2000 - 1000) \text{ m}}{(1 \text{ kg} \cdot \text{m/N} \cdot \text{s}^2)}$$

$$= \underline{\underline{1.47 \times 10^7 \text{ N} \cdot \text{m}}}$$

The changes in kinetic energy and potential energy for automobile B are

$$E_{k_2} - E_{k_1} = \frac{(1500 \text{ kg})[(20 \text{ m/s})^2 - 0]}{2(1 \text{ kg} \cdot \text{m/N} \cdot \text{s}^2)} = \underline{\underline{3.00 \times 10^5 \text{ N} \cdot \text{m}}}$$

$$E_{p_2} - E_{p_1} = \frac{(1500 \text{ kg})(9.81 \text{ m/s}^2)(2000 - 1000) \text{ m}}{(1 \text{ kg} \cdot \text{m/N} \cdot \text{s}^2)} = \underline{\underline{1.47 \times 10^7 \text{ N} \cdot \text{m}}}$$

The potential energy change for each automobile is identical even though the paths followed are different. The change in potential energy depends only on the change in elevation and not on the path followed in achieving the change in elevation. The change in kinetic energy is also independent of path, but it is different for each auto because the velocities of the two automobiles are different at point 2. ■

2.4.3 Internal Energy

In the previous sections two forms of macroscopic energy were discussed, kinetic and potential energies. This section is concerned with the energy of a quantity of mass on a molecular scale, which is called the internal energy.

All matter is composed of atomic particles when viewed on a molecular scale. The molecules move about in a random manner, colliding with each other and striking the surface of any container that holds the substance. On an even smaller scale electrons orbit the nuclei of the atoms and are held in their orbits by attractive forces within the nuclei.

The energy associated with a substance on a molecular scale can consist of several forms. Molecules possess kinetic energy resulting from their individual mass and velocity as they move about along a linear path. The molecules also possess vibrational and rotational energies as they rotate and vibrate as a consequence of their random motion, and yet another form of energy is associated with the intermolecular forces between molecules. The sum of all these molecular or microscopic energies is called the *internal energy* of the substance.

At this point several general observations regarding the magnitude of the internal energy and its relationship to other measurable properties can be made. These observations will help portray a property that is associated with submicroscopic particles. For instance, the average molecular velocity of a substance is known to be proportional to its temperature, and a substance with a high temperature contains molecules with higher velocities than those of molecules in the same material with a lower temperature. Because the internal energy is, in part, a measure of the microscopic kinetic energy of the molecules, the internal energy increases as the temperature of the substance increases. Furthermore, the contribution of intermolecular forces to the internal energy of a substance will increase as the intermolecular forces increase. These forces are strongest for solids that have small molecular spacing, moderate for liquids whose molecules are spaced farther apart, and weakest for gases whose intermolecular forces are relatively small. To change a substance from a solid phase to a liquid phase would require an increase in energy to overcome the strong intermolecular forces of the solid. Therefore, an increase in the internal energy as a substance changes phase from solid to liquid to vapor should be expected.

The internal energy is a thermodynamic property, and therefore its change during a process from one equilibrium state to another depends only on the end states of the process and is independent of the path connecting the state points. The symbol for the internal energy of a quantity of mass is U. The internal energy is an extensive property, and the corresponding intensive property, or *internal energy per unit mass*, is designated by u, where

$$u \equiv \frac{U}{m} \tag{2.6}$$

The value for the internal energy cannot be measured, but changes in the internal energy can be related to changes in other measurable properties such as temperature, pressure, and the specific volume. (Thermodynamic relationships between the internal energy and measurable properties are presented in Chapter 9.) Since the total energy of the system consists of both macroscopic and microscopic energy forms, Equation 2.1 can be written as

$$E = E_k + E_p + E_{\text{electric}} + E_{\text{magnetic}} + E_{\text{chemical}} + \cdots + U \tag{2.7}$$

For simple compressible systems in which the changes in electrical, magnetic, chemical, and other macroscopic forms of energy are small compared with the kinetic, potential, and internal energy, the total energy can be expressed as

$$E = E_k + E_p + U \tag{2.8}$$

Substituting Equations 2.2 and 2.4 for the kinetic and potential energy, respectively, the total energy for a simple compressible system becomes

$$E = mV^2/2 + mgz + mu \tag{2.9}$$

The total energy per unit mass of simple compressible systems, an intensive property, is therefore

$$e = \frac{E}{m} = e_k + e_p + u$$

or

$$e = \frac{1}{2} V^2 + gz + u \tag{2.10}$$

EXAMPLE 2.2

A mass of 20 lb_m has an internal energy of 50 Btu/lb_m, an elevation of 150 ft, and a velocity of 225 ft/s, both measured relative to the surface of the earth. Determine the kinetic, potential, and total energies of the mass relative to the surface of the earth assuming that the local acceleration of gravity is 32.2 ft/s^2.

Solution. The kinetic energy is given by Equation 2.2,

$$E_k = \frac{1}{2} mV^2 = \frac{1}{2} \frac{(20\ lb_m)(225\ ft/s)^2}{(32.2\ lb_m \cdot ft/lb_f \cdot s^2)}$$

$$= \underline{\underline{15{,}700\ ft \cdot lb_f}}$$

The potential energy is determined from Equation 2.4,

$$E_p = mgz = \frac{(20\ lb_m)(32.2\ ft/s^2)(150\ ft)}{(32.2\ lb_m \cdot ft/lb_f \cdot s^2)}$$

$$= \underline{\underline{3000\ ft \cdot lb_f}}$$

The total energy of the mass from Equation 2.8 is

$$E = E_k + E_p + U = 15{,}700\ ft \cdot lb_f + 3000\ ft \cdot lb_f + (20\ lb_m)\left(50\ \frac{Btu}{lb_m}\right)\left(778\ \frac{ft \cdot lb_f}{Btu}\right)$$

$$= \underline{\underline{797{,}000\ ft \cdot lb_f}}$$

This problem illustrates the important point that g_c must be inserted into the denominator of both the expressions for kinetic and potential energy when the English system of units is used so that the result has the familiar units of energy rather than a mixture of force and mass units. ■

2.4.4 Enthalpy

In the analysis of open systems the combination of properties $U + PV$ is frequently encountered. For the sake of convenience this combination of properties is defined as the *enthalpy* (en′-thal-pē) and is assigned the symbol H:

$$H \equiv U + PV \tag{2.11}$$

Following the convention of using the capital letter to represent the extensive property and the lowercase letter for the intensive form of the property, the *enthalpy per unit mass* is defined by the equation

$$h \equiv u + Pv \tag{2.12}$$

The enthalpy is a continuous function of other properties, and therefore it is also a property. Because it is a property, its value can be determined for a simple compressible substance once two independent, intensive thermodynamic properties of the substance are known, and the change in enthalpy is independent of the path followed between two equilibrium states. Like the internal energy, the enthalpy cannot be measured directly but must be related to other measurable properties. The development of the appropriate relationships is presented in Chapter 9.

2.5 EQUILIBRIUM DIAGRAMS

All known substances can exist in several phases: the solid phase, the liquid phase, and the vapor or gaseous phase. A mixture of more than one phase of a substance in equilibrium is also possible. Because of this complex behavior, a single equation of state in the form of Equation 1.3 between the pressure, temperature, and specific volume that is valid for all possible states of a substance has not yet been developed. However, the qualitative aspects of these substances can be discussed in order to gain insight into their P-v-T behavior.

In this section attention is focused primarily on the behavior of the liquid and vapor phases and mixtures of these phases. Although water is used as a specific example, its general behavior is typical of most substances.

Consider the following simple experiment devised to monitor the changes in temperature and specific volume (or density) of a substance such as water as it is heated at constant pressure. For this purpose imagine a cylinder initially filled with water at 20°C (68°F) and maintained at atmospheric pressure or 101.32 kPa (14.7 lb_f/in^2) by means of a piston, as shown in Figure 2.2. Under these conditions water exists in the liquid phase.

As heat transfer to the water in the cylinder takes place, the temperature of the water will begin to increase and at the same time the water will expand. In other words, the specific volume of the contents of the cylinder will increase as the energy of the water is increased at constant pressure. Since the cylinder is closed, the expansion will also cause the piston to move.

If the heat transfer to the water occurs continuously, then a plot of the temperature of the water as a function of specific volume could be constructed for this process. A sketch of the results is indicated by the constant-pressure line 1-2-3-4-5 in Figure 2.3. Notice that the temperature of the water continues to increase until state 2 is reached. During subsequent heating the specific volume continues to increase, while the temperature remains constant. State 2 is the point at which the temperature of the water reaches 100°C (212°F). The fact that water at atmospheric pressure begins to vaporize at 100°C or 212°F is common knowledge, and state 2 indicates the beginning of the vaporization process. At this point the water is still completely a single phase (the liquid phase), but

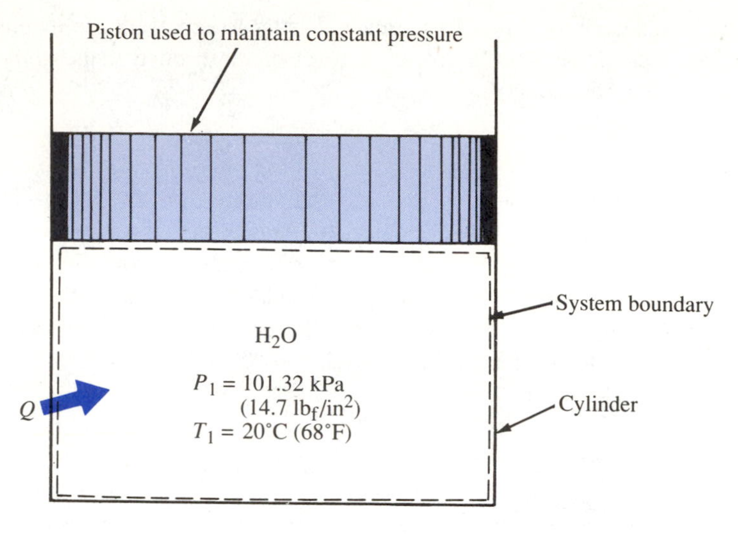

Figure 2.2 System consisting of water inside a piston-cylinder.

any further heat transfer to the water will cause vaporization to begin and a mixture of liquid and vapor would be present. State 2 is called a *saturated-liquid* state.

If the heat-transfer process is continued, then vaporization will commence, and the temperature of the two-phase mixture of liquid and vapor will not increase. At state 3 a point is reached where approximately one-half of the mass of water in the cylinder is in the liquid phase and the other half is in the vapor phase. Eventually, at state 4 all of the water in the cylinder is in the vapor phase, and further heating will cause the temperature to increase once more. State 4, where the water has again become a single phase (the vapor phase), is called a *saturated-vapor* state.

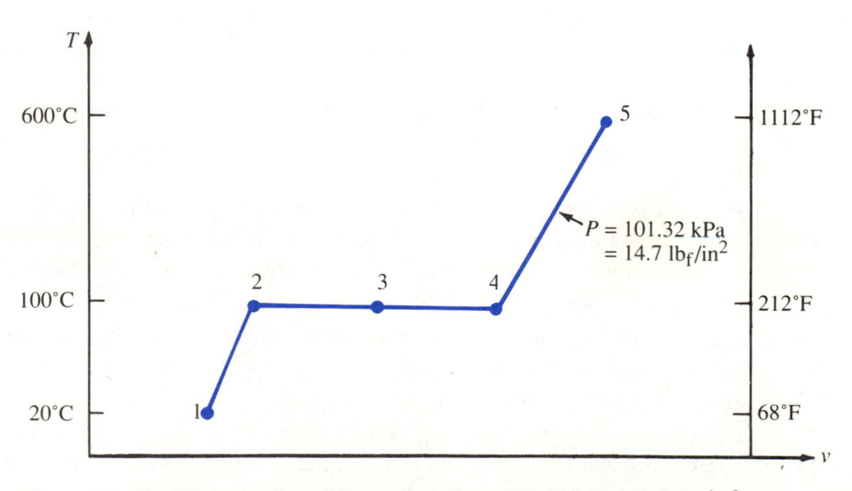

Figure 2.3 Equilibrium states for water at P = 101.32 kPa (14.7 lb_f/in^2).

Further heating causes the temperature of the water to increase once again. Any state to the right of state 4 on the constant-pressure line is called a *superheated-vapor* state. The vapor is called superheated because the temperature of the vapor is higher than the temperature at which it is a saturated vapor for this pressure. An example of a superheated-vapor state is state 5 at 600°C (1112°F) in Figure 2.3.

The temperature of 100°C (212°F) at states 2, 3, and 4 is called the *saturation temperature* for the pressure of 101.32 kPa (14.7 lb_f/in^2) since this temperature is the one at which the saturation states occur at this pressure. Conversely, this pressure is called the *saturation pressure* for a temperature of 100°C or 212°F.

State 1 (or other liquid states to the left of state 2) is called a *subcooled-liquid* state because its temperature is lower than the temperature at which it is a saturated liquid for this pressure; the term *compressed-liquid* state is more frequently used to describe this state. This term will be explained further in the next section.

If the process just described is reversed by cooling the water while maintaining the pressure at the same value, the water would begin to retrace the same path indicated in the figure. At state 4 further cooling would cause some of the vapor to begin to condense, and eventually all of the vapor will have condensed back to the liquid phase at state 2. During heating from state 2 to state 4 the process is called *vaporization*, while during cooling from state 4 to state 2 the process is termed *condensation*.

Suppose now that the pressure of the liquid is increased to 10 MPa (1450 lb_f/in^2) by adding weight to the piston. If the initial temperature remains at 20°C (68°F), the level of the piston will drop slightly as the water is compressed. Thus the specific volume of liquid water at 20°C (68°F) decreases somewhat as indicated by state 1′ in Figure 2.4. As heat transfer to the water occurs at this higher pressure, the relationship between temperature and specific volume change is indicated by the line segment 1′-2′-3′-4′-5′.

Notice that the sketch of the process looks very much like the first process, although there are some important differences. First of all, the temperature at which vaporization (or condensation) occurs is much higher (311.03°C or 591.9°F) at this pressure. Second, the specific volume of the saturated liquid (point 2′) is larger and the specific volume of the saturated vapor (point 4′) is smaller than the corresponding values at 101.32 kPa or 14.7 lb_f/in^2. In other words, the horizontal line that joins the saturated-liquid state and the saturated-vapor state is shorter in length. As the pressure is increased further, this line will become shorter and shorter until the saturated-liquid state and the saturated-vapor state coincide. This state is called the *critical state* (also called *critical point*). Tables H.1 and H.1E contain values for the critical-state properties of common substances.

Above the critical state there is no clear division between the superheated-vapor region and the compressed-liquid region. Furthermore, a phase change from liquid to vapor (or vice versa) cannot occur at pressures or temperatures that are higher than the corresponding values at the critical state for the substance.

If constant-pressure lines are drawn for a number of other pressures, behavior similar to that shown in Figure 2.4 would be observed. When the points that represent the saturated-liquid and saturated-vapor states are connected, a saturation region is constructed, as shown in Figure 2.5. The figure that emerges is called an *equilibrium* diagram in which each point on the diagram represents equilibrium values for T and v for water.

The sketch of the T-v equilibrium diagram in Figure 2.5 has been drawn grossly out of proportion in order to provide clarity of presentation. Even though this practice is

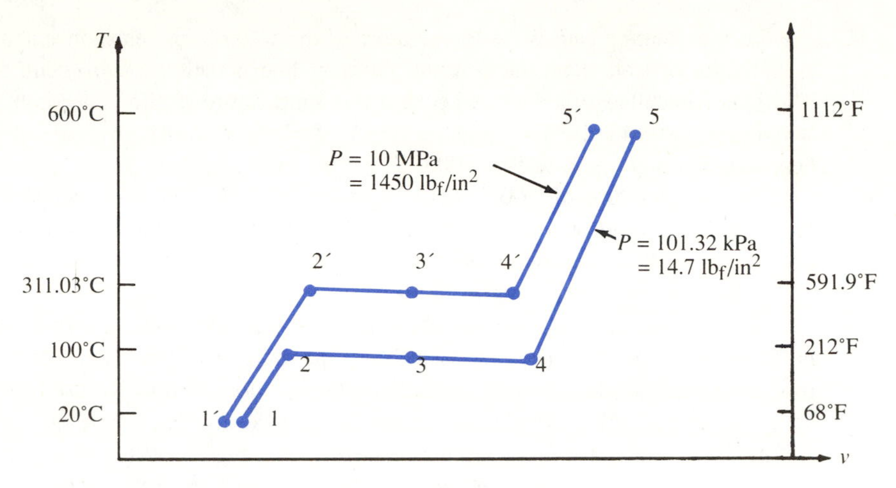

Figure 2.4 Equilibrium states for water at $P = 10$ MPa and 101.32 kPa (1450 and 14.7 lb_f/in^2).

common, you should be aware that such figures are grossly distorted, especially in the area to the left of the critical state, as can be easily illustrated. Suppose the *T-v* diagram in Figure 2.4 is to be drawn to scale, and the distance along the horizontal specific volume axis between states 2 and 2′ is taken to be 1 cm. Then to maintain this same scale, we would have to have the horizontal distance from state 2 to state 4 be nearly 42 m, and the horizontal distance between state 2 and state 5 would have to be nearly 100 m—approximately the length of a football field.

The dome-shaped region in Figure 2.5 consists of the *saturated-liquid line* to the left of the critical state and the *saturated-vapor line* to the right of the critical state. The

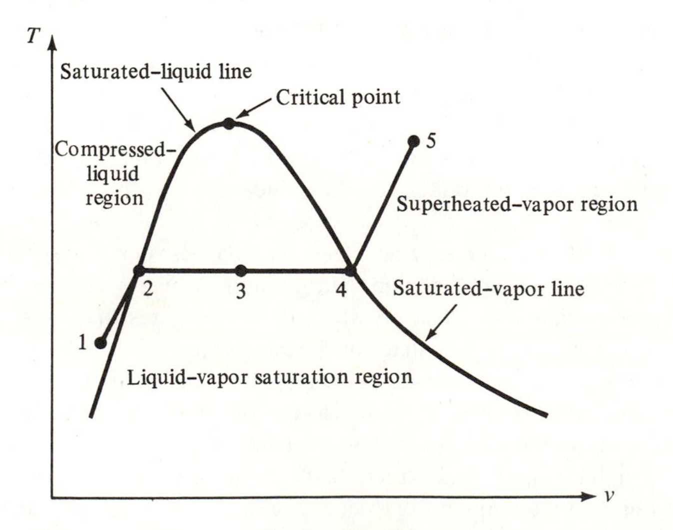

Figure 2.5 The *T-v* equilibrium states for water.

single-phase region immediately to the left of the saturated-liquid line is known as the *compressed-liquid region*. The region to the right of the saturated-vapor line is called the *superheated-vapor region*. A state within the saturation dome, such as state 3, is called a liquid-vapor *saturation state* and is composed of a two-phase mixture of liquid and vapor.

Above the critical state there is no clear division between the superheated-vapor region and the compressed-liquid region. A customary, although arbitrary, practice is to include as a part of the superheated-vapor region all states for which the temperature is above the critical temperature.

Since a state in the saturation region can be composed of any combination of liquid and vapor from pure liquid to pure vapor, the definition of a property that describes the relative amount of vapor in the mixture is useful. This property, known as the *quality* and given the symbol x, is defined as the mass fraction of vapor present in a liquid-vapor mixture, or

$$x \equiv \frac{m_g}{m_{\text{mixture}}} = \frac{m_g}{m_f + m_g} \tag{2.13}$$

The subscript f is used to denote properties of the saturated-liquid phase, and the subscript g denotes properties of the saturated-vapor phase. (The use of the subscripts f and g is standard in the thermodynamic literature and has its origin in the German words for liquid and gas.) The quality has meaning only in the saturation region, and its value is always between zero and one. A quality of zero denotes a saturated-liquid state, while a quality of one (or 100 percent) denotes a saturated-vapor state.

The equilibrium diagram developed for water in Figure 2.5 does not represent all possible equilibrium states. The solid phase and the solid-liquid and solid-vapor saturation regions, which were not included in Figure 2.5, are added in Figure 2.6. This diagram

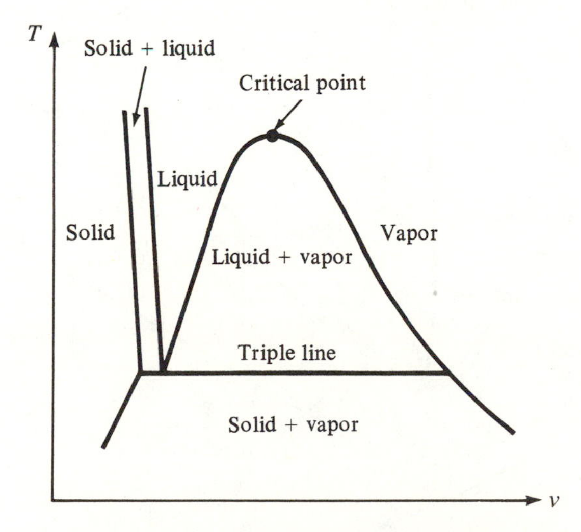

Figure 2.6 The T-v projection of an equilibrium diagram for a substance that contracts upon freezing.

is typical of substances that contract upon freezing. Water is an unusual substance because it expands as it freezes. Therefore, the same diagram for water differs primarily in the representation of the solid-liquid region, since the specific volume of ice is greater than the specific volume of liquid water. This behavior accounts for the fact that ice floats in liquid water.

With one exception the solid lines in Figure 2.6 separate a single-phase region from a two-phase region. The exception is called the *triple line*. Along the triple line all three phases coexist in equilibrium. The pressure and temperature corresponding to the triple line are called *triple-point properties*, and values for several substances are presented in Tables H.4 and H.4E.

If the equilibrium values for pressure, temperature, and specific volume are plotted on a three-dimensional set of coordinates, a surface results for which the locus of all points on the surface represents the equilibrium values for P, T, and v of the substance. The projection of the three-dimensional equilibrium surface shown in Figure 2.7 when projected parallel to the pressure axis becomes the T-v equilibrium diagram shown in

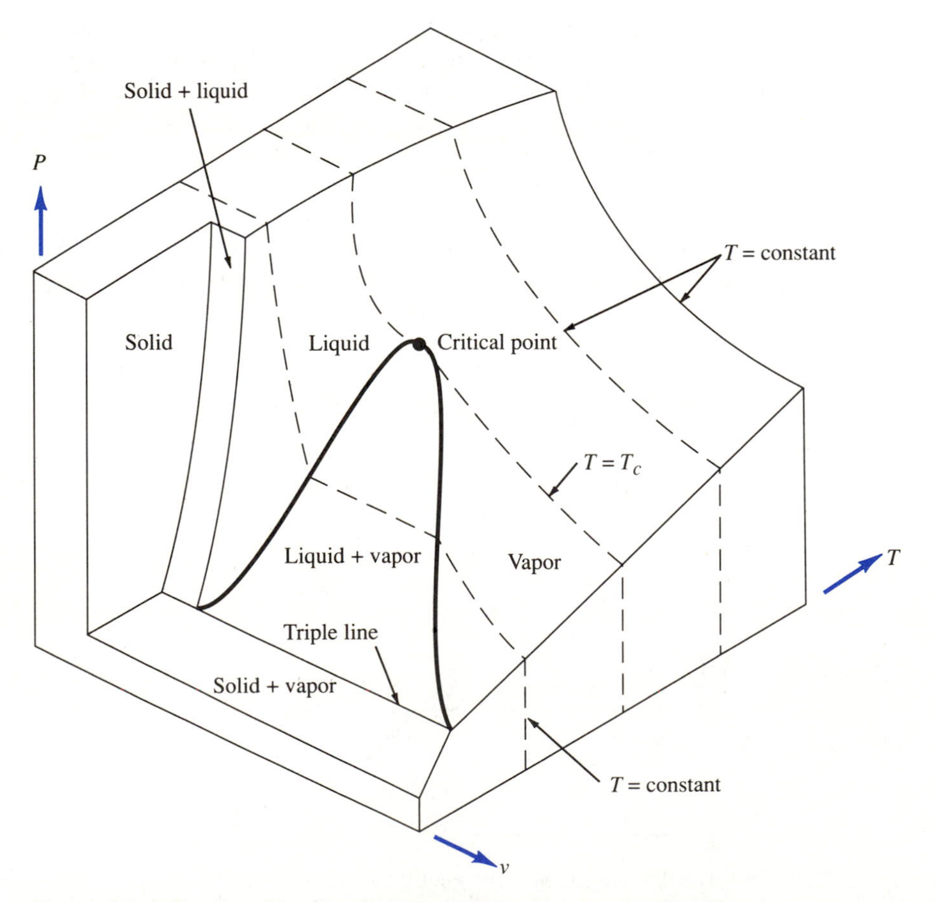

Figure 2.7 A P-v-T surface for a substance that contracts upon freezing.

Figures 2.5 and 2.6. The complexity of the surfaces that comprise the three-dimensional surface in Figure 2.7 should make it evident that it is unlikely that a single equation of state could be formulated to represent *P-v-T* data in all regions of the diagram. Instead, tabulated data will be used to provide properties for all phases of the substance. Much of the remainder of this chapter will be devoted to formulating methods of determining thermodynamic properties from tabulated values. Furthermore, you should be aware that the increase in popularity of personal computers has stimulated the development of software that can be used to calculate properties once the state of the substance has been specified.

A software package capable of providing thermodynamic properties of many common working substances is enclosed in the back cover of this text. The program calculates properties of water, refrigerant-12, air and eight other gases. The program is capable of calculating properties in both SI and English Units and it can provide properties for the gases on both a molar and mass basis. The program greatly simplifies the process of determining property values and it also eliminates the tedious task of interpolation when the given properties fall between tabular values. Furthermore, it saves a significant amount of time and effort when working problems that require properties at multiple states. The program was used to generate the property values for most of the substances that appear in the appendices. Even though the program and tabular values are identical, we have assumed the use of tabular values when working the example problems in the text and we point out that interpolation is sometimes required to work the problems.

A *P-v-T* surface for a substance that contracts on freezing is shown in Figure 2.7. If this equilibrium diagram is projected onto the *P-v* plane, the *P-v* diagram of Figure 2.8 results. An isotherm (a line of constant temperature) is indicated by the dashed line in the figure. Notice that discontinuities are present in the slope of an isotherm each time a saturation curve is crossed; since isotherms are horizontal on this diagram in any two-phase region, pressure and temperature are not independent properties in those regions.

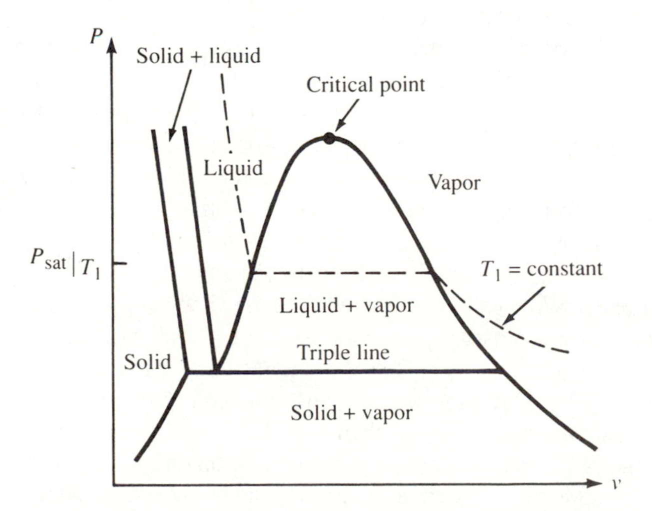

Figure 2.8 A *P-v* projection of an equilibrium diagram for a substance that contracts upon freezing.

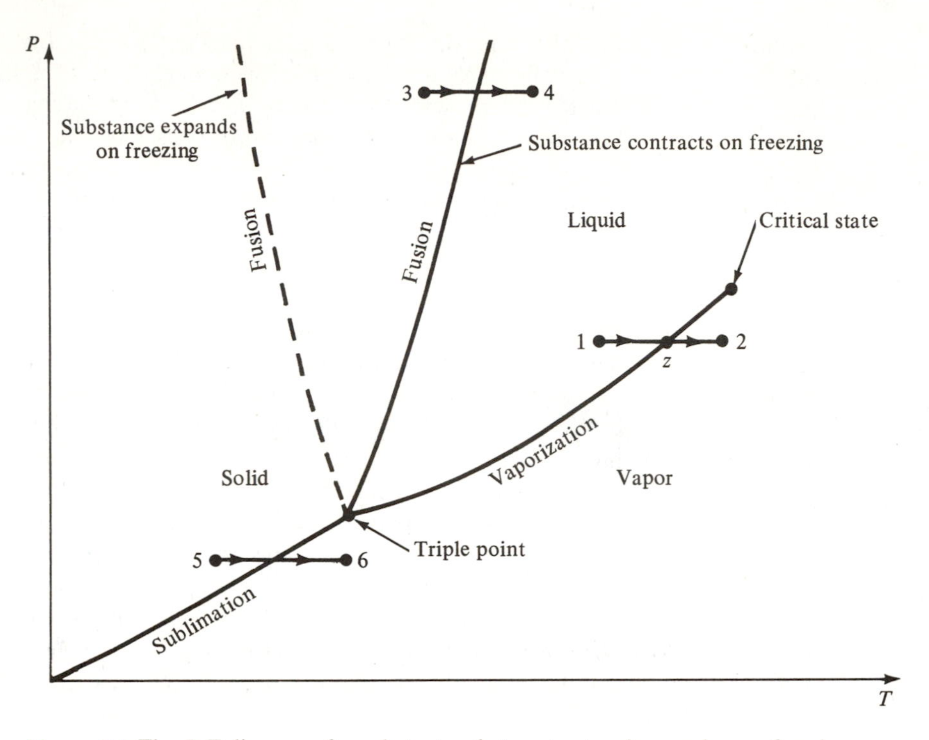

Figure 2.9 The P-T diagrams for substances that contract and expand upon freezing.

The relationship between the saturation pressure and saturation temperature in the two-phase regions can best be seen on a P-T diagram, the projection of the P-v-T surface on the P-T plane, as illustrated in Figure 2.9. Notice that the triple line appears as a single point on the P-T diagram. This point is often called the *triple point*. The triple point is the only point on the P-T diagram at which solid, liquid, and vapor can coexist in equilibrium. This state is not, however, a single state but rather a point that represents an infinite number of possible states at which the solid, liquid, and vapor phases coexist.

The vaporization line represents the projection of the saturated-liquid and saturated-vapor lines on the P-T plane. This line terminates at the critical state, as indicated in the figure. The transition from liquid to solid (freezing) is called *fusion*. For a substance that contracts on freezing, the fusion line has a positive slope on the P-T diagram; whereas, for a substance that expands on freezing, the fusion line has a negative slope (Figure 2.9). The transition from the solid phase directly to the vapor phase is called *sublimation*.

Examples of constant-pressure processes that are accompanied by phase changes are shown on the P-T diagram in Figure 2.9: the liquid-to-vapor transition (1–2), the solid-to-liquid transition (3–4), and the solid-to-vapor transition (5–6). Notice that a constant-temperature or constant-pressure phase-change process, which apppears as a line on the T-v and the P-v diagrams, respectively, is a single point on the P-T diagram. For example, constant-pressure vaporization from a saturated-liquid state to a saturated-vapor state would be represented by point z in Figure 2.9.

While fusion (melting) and vaporization (boiling) are familiar phenomena, sublimation is not as commonplace. Examples of substances that sublime at atmospheric pressure are solid CO_2 (dry ice) and paradichlorobenzene (mothballs). For both of these substances the triple-point pressure is much higher than atmospheric pressure. Thus at atmospheric pressure these solids cannot reach the liquid phase. When these solids are at atmospheric pressure, they undergo a process represented by the line segment 5–6 in Figure 2.9.

Another interesting fact can be illustrated with the aid of a P-T diagram. Since water expands on freezing, ice can be melted by increasing the pressure on the solid phase. When the weight of an ice skater is concentrated on the thin blade of a skate, the pressure on the ice is increased greatly, and the ice melts, forming a thin liquid layer between the blade and the ice. The skate therefore actually glides on a film of liquid water rather than experiencing solid-to-solid contact.

2.6 PROPERTIES OF PURE SUBSTANCES

A substance that has a uniform chemical composition throughout is called a *pure substance*. The thermodynamic behavior of pure substances cannot be conveniently described with a single equation of state, since each substance can exist in the solid, liquid, or vapor phase. Consequently, the relationships between pressure, volume, and temperature and other thermodynamic properties are most often presented in tabular or graphical form and more recently in the form of computer software.

In this section discussion is concentrated on the use of the tables of properties for pure substances for states in the compressed-liquid, liquid-vapor saturation, and superheated-vapor regions. Even though the specific applications are restricted to water and refrigerant-12 (dichlorodifluoromethane),* the principles involved apply equally well to any other pure substance. In most instances separate tables are provided for each of the three regions, and the region that contains the state of interest must first be identified. For this reason each region is discussed separately in the following paragraphs.

2.6.1 The Superheated-Vapor Region

Consider first the superheated-vapor region. By referring to a typical P-v diagram (Figure 2.10), some of the characteristics of a superheated vapor can be visualized. An arbitrary point in the superheated-vapor region is denoted by state 1, where the pressure, temperature, and specific volume are P_1, T_1, and v_1, respectively. From the figure the following observations can be made concerning states in the superheated-vapor region.

1. If the saturation temperature corresponding to the pressure P_1 is denoted by T_{sat}, then state 1 is a superheated-vapor state when T_1 is greater than T_{sat}. The word *superheated* arises from the fact that the temperature of the substance is greater than the saturation temperature corresponding to its pressure.

---

*In this text refrigerant-12 will often be abbreviated as R-12.

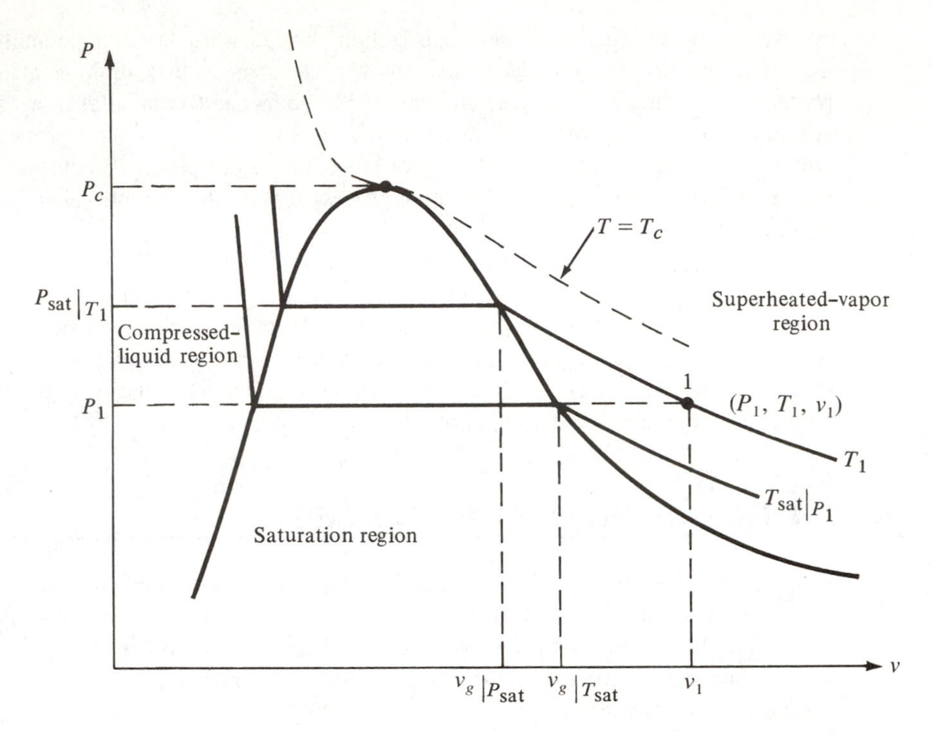

Figure 2.10 A *P-v* diagram illustrating the equilibrium state in the superheated-vapor region.

2. If the saturation pressure corresponding to the temperature T_1 is denoted by P_{sat}, then state 1 is a superheated-vapor state when P_1 is less than P_{sat}.
3. Furthermore, for a superheated-vapor state the specific volume v_1 is greater than either of the saturated-vapor specific volumes that correspond to P_{sat} or T_{sat}; that is,

$$v_1 > v_g|_{P_{sat}} \quad \text{and} \quad v_1 > v_g|_{T_{sat}}$$

Although not readily apparent from the *P-v* diagram in Figure 2.10, the same conclusion holds for the internal energy and enthalpy of a superheated vapor; for example,

$$u_1 > u_g|_{P_{sat}} \quad \text{and} \quad u_1 > u_g|_{T_{sat}}$$

4. If T_1 is greater than the critical temperature, the state is also referred to as a superheated-vapor state.
5. Any two intensive properties—for example, P and T or T and u—are independent and are therefore sufficient to specify the state of a simple compressible substance in the superheated-vapor region.

2.6.2 The Compressed- or Subcooled-Liquid Region

Point 1 in Figure 2.11 denotes an arbitrary compressed-liquid state on a *P-v* diagram. In this case the following observations can be made:

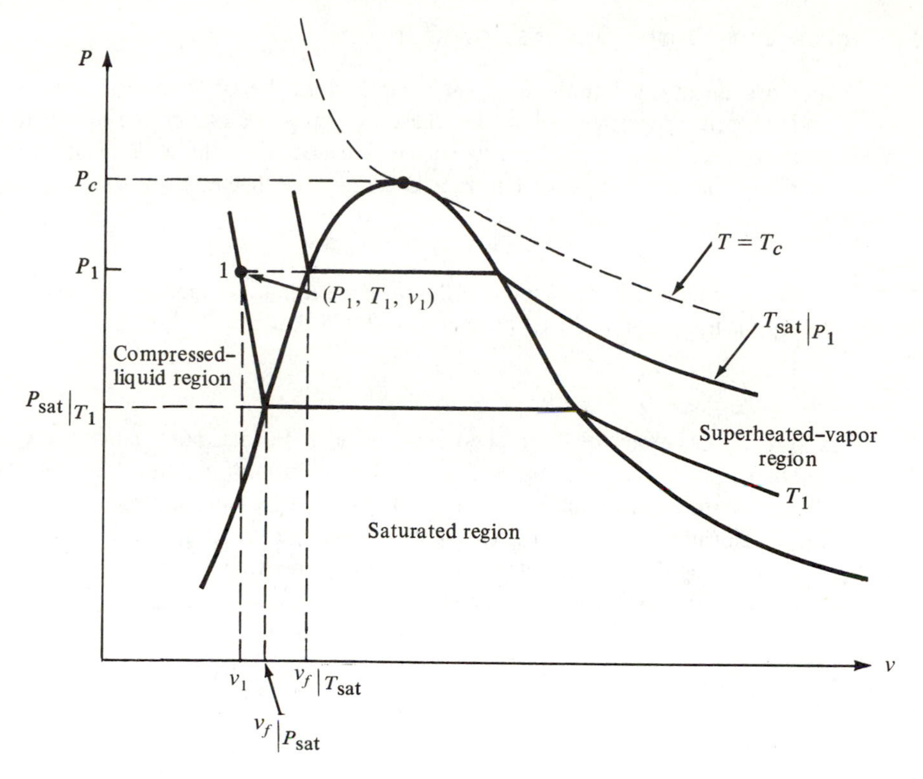

Figure 2.11 A P-v diagram illustrating the equilibrium state in the compressed-liquid region.

1. If T_{sat} is the saturation temperature corresponding to the pressure P_1, then state 1 is a compressed-liquid state when T_1 is less than T_{sat} but greater than the solidification temperature for pressure P_1. For this reason a compressed-liquid state is also referred to as a subcooled-liquid state.
2. If P_{sat} is the saturation pressure corresponding to the temperature T_1, then state 1 is a compressed-liquid state when P_1 is greater than P_{sat}. This observation is the motivation for the term *compressed liquid.*
3. For a compressed-liquid state the specific volume v_1 is less than either of the saturated-liquid specific volumes that correspond to P_{sat} or T_{sat}; that is,

$$v_1 < v_f|_{P_{sat}} \quad \text{and} \quad v_1 < v_f|_{T_{sat}}$$

   Likewise, the internal energy and enthalpy of a compressed-liquid state are less than the values of these properties at the corresponding saturated-liquid states.
4. If T_1 is less than the critical temperature and greater than the solidification temperature corresponding to P_1, the state is referred to as a compressed-liquid state.
5. Any two intensive properties are independent and are therefore sufficient to specify the state of a simple compressible substance in the compressed-liquid region.

2.6.3 The Liquid-Vapor Saturation Region

For a state in the liquid-vapor saturation region, the pressure and temperature are not independent properties. Therefore, these two properties alone are not sufficient to completely specify the state of the substance. The specific volume at a state in the saturation region—for example, state 1 in Figure 2.12—is such that

$$v_f \leq v_1 \leq v_g$$

where v_f and v_g are the saturated-liquid and saturated-vapor specific volumes, respectively, at the pressure P_1 (or temperature T_1). Likewise,

$$y_f \leq y_1 \leq y_g$$

where y can be any of the properties v, u, or h. Furthermore, since the quality x is an intensive property, it can be used in the specification of a state in the saturation region.

A mixture within the liquid-vapor saturation region is composed of a saturated liquid in equilibrium with a saturated vapor. Any extensive property Y of a mixture in the saturation region can be obtained by summing the values for the extensive property of the saturated liquid and the saturated vapor:

$$Y = Y_f + Y_g$$

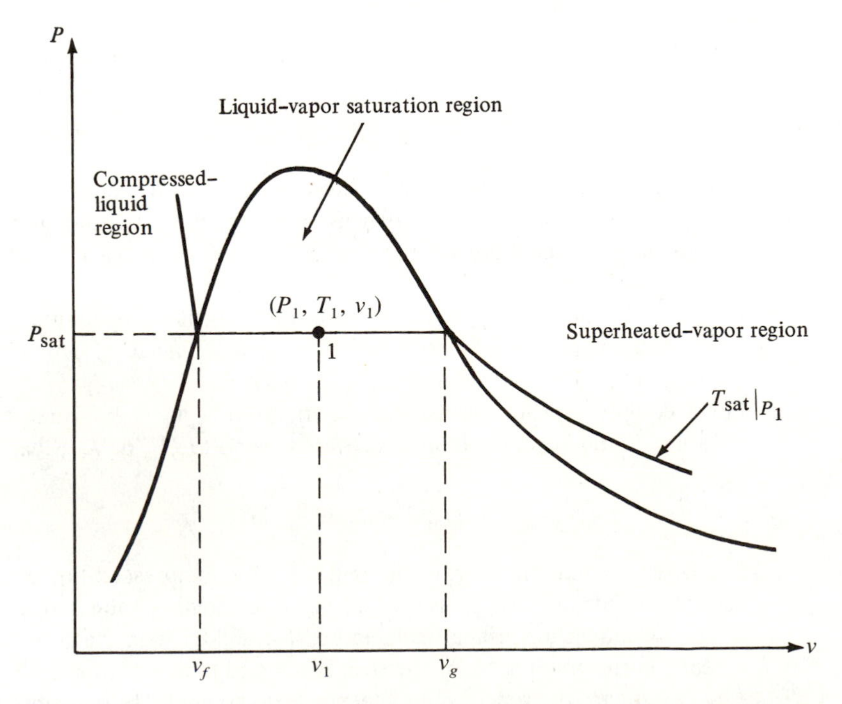

Figure 2.12 A P-v diagram illustrating the equilibrium state in the liquid-vapor saturation region.

The extensive property for each component is the product of the mass and the corresponding intensive property; therefore

$$Y = m_f y_f + m_g y_g$$

With some rearrangement this equation can be expressed in terms of the quality of the mixture as

$$Y = m(1 - x)y_f + mxy_g$$
$$= m(y_f + xy_{fg})$$

where m is the total mass of the mixture and y_{fg} is a shorthand notation that represents the difference between the intensive property of the saturated vapor and the intensive property of the saturated liquid,

$$y_{fg} \equiv y_g - y_f$$

Thus the *average* intensive property y of the mixture can be written as

$$y = \frac{Y}{m} = y_f + xy_{fg} \tag{2.14}$$

EXAMPLE 2.3

The enthalpy of refrigerant-12 is 213.06 kJ/kg at 40°C. Determine the pressure of the refrigerant.

Solution. From the saturation table, Table C.1, at T_{sat} = 40°C, we find that h_g = 204.8 kJ/kg. Since $h > h_g$, the state is a superheated-vapor state, and from Table C.3 the pressure is 0.4 MPa.

EXAMPLE 2.4

Suppose 2 kg of H_2O at 200°C and 300 kPa are contained in a weighted piston-cylinder assembly. As a result of heating at constant pressure, the temperature of the H_2O increases to 400°C. Determine the change in volume, the change in internal energy, and the change in enthalpy of the H_2O for this process.

Solution. The saturation pressure corresponding to the initial temperature of the H_2O is 1.5536 MPa from Table B.1. Since $P_1 < P_{sat}$, the initial state is in the superheated-vapor region. Figure 2.13 indicates that state 2 is also a superheated-vapor state since $P_2 = P_1$ and $T_2 > T_1$.

From the superheated-vapor table, Table B.3, the properties at states 1 and 2 are

$$v_1 = 0.7163 \text{ m}^3/\text{kg} \qquad u_1 = 2650.2 \text{ kJ/kg} \qquad h_1 = 2865.1 \text{ kJ/kg}$$

and

$$v_2 = 1.0315 \text{ m}^3/\text{kg} \qquad u_2 = 2965.4 \text{ kJ/kg} \qquad h_2 = 3274.9 \text{ kJ/kg}$$

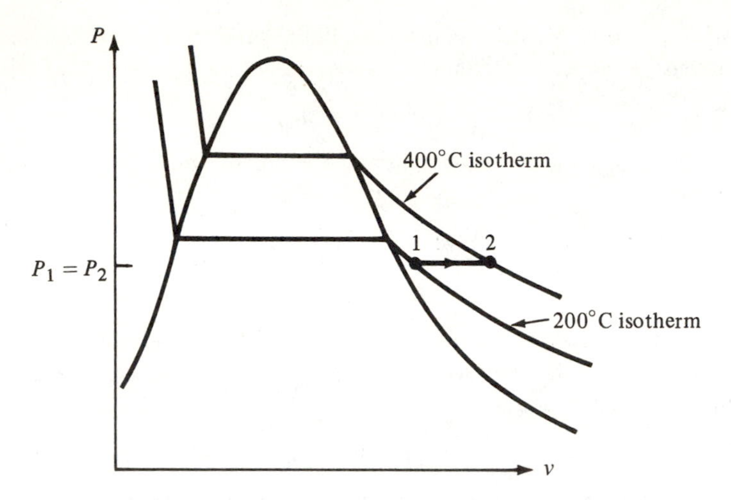

Figure 2.13 Process diagram for Example 2.4.

Thus for this closed system

$$\mathbf{V}_2 - \mathbf{V}_1 = m(v_2 - v_1) = (2 \text{ kg})(1.0315 - 0.7163)\text{m}^3/\text{kg} = \underline{\underline{0.6304 \text{ m}^3}}$$

$$U_2 - U_1 = m(u_2 - u_1) = (2 \text{ kg})(2965.4 - 2650.2)\text{kJ/kg} = \underline{\underline{630.4 \text{ kJ}}}$$

$$H_2 - H_1 = m(h_2 - h_1) = (2 \text{ kg})(3274.9 - 2865.1)\text{kJ/kg} = \underline{\underline{819.6 \text{ kJ}}}$$

As a matter of practical interest, consider the relationship between the property changes determined above. Recall that the enthalpy is defined as

$$h = u + Pv$$

Therefore, $dh = du + P\,dv + v\,dP$. Since the process under consideration is one of constant pressure, this expression can be integrated, with the result

$$(h_2 - h_1) = (u_2 - u_1) + P(v_2 - v_1)$$

and thus

$$H_2 - H_1 = U_2 - U_1 + P(\mathbf{V}_2 - \mathbf{V}_1)$$

This result can be verified for this problem:

$$H_2 - H_1 = 630.4 \text{ kJ} + (300 \text{ kPa})\left(\frac{1 \text{ kN}}{\text{m}^2 \cdot \text{kPa}}\right)(0.6304 \text{ m}^3) = 819.5 \text{ kJ}$$ ■

■ EXAMPLE 2.5

A closed, rigid vessel contains 1.57 lb_m of liquid water and 1.28 lb_m of water vapor at a pressure of 100 psia. Heat transfer to the water causes the water to reach a pressure of 300 psia. Determine the internal energy change during the process.

Solution. The initial state is in the saturated region because the liquid and vapor phases coexist. Properties are given in Table B.2E (pressure table) and the quality must be calculated before the initial internal energy can be determined

$$x_1 = \frac{m_g}{m_f + m_g} = \frac{1.28 \text{ lb}_\text{m}}{(1.57 + 1.28) \text{ lb}_\text{m}} = 0.449$$

and the initial internal energy is

$$u_1 = u_f + x_1 u_{fg} = 298.34 \frac{\text{Btu}}{\text{lb}_\text{m}} + (0.449)\left(807.4 \frac{\text{Btu}}{\text{lb}_\text{m}}\right)$$

$$= 660.9 \frac{\text{Btu}}{\text{lb}_\text{m}}$$

At first glance the final internal energy may seem indeterminant because only the pressure is specified at the final state and the state postulate requires that two properties must be specified before the state of a simple compressible substance is known. However, a second property at the final state is actually given in the problem statement. The water in the vessel proceeds along a constant-mass (closed system) and constant-volume (rigid system) path, resulting in a constant-specific-volume path or

$$v_1 = v_2$$

The initial specific volume can be determined from the initial quality and the saturation properties in Table B.2E

$$v_1 = v_2 = v_f + x_1 v_{fg} = 0.01774 \frac{\text{ft}^3}{\text{lb}_\text{m}} + (0.449)\left(4.415 \frac{\text{ft}^3}{\text{lb}_\text{m}}\right)$$

$$= 2.00 \text{ ft}^3/\text{lb}_\text{m}$$

At this value for specific volume and the given pressure at the final state, the final internal energy is located in the superheated vapor table, Table B.3E,

$$u_2 = 1202.3 \frac{\text{Btu}}{\text{lb}_\text{m}}$$

The change in internal energy is then

$$u_2 - u_1 = 1202.3 \frac{\text{Btu}}{\text{lb}_\text{m}} - 660.9 \frac{\text{Btu}}{\text{lb}_\text{m}} = \underline{\underline{541.4 \frac{\text{Btu}}{\text{lb}_\text{m}}}}$$

As this problem illustrates, it is always important to be aware of the path that a substance takes during a process. Not only is the path information essential in determining path-dependent quantities such as work and heat transfer, it is also an important ingredient in determining property changes that occur during a process. ■

■ EXAMPLE 2.6

Determine the pressure, specific volume, and internal energy of H_2O at 25°C and a quality of 70 percent.

Solution. Since a quality is associated with the H_2O, this state is necessarily in the saturation region. Thus the pressure is equal to the saturation pressure corresponding to 25°C. From Table B.1

$$P = P_{sat}|_{25°C} = \underline{\underline{3.169 \text{ kPa}}}$$

The other properties can be determined from the quality and the tabulated properties for the saturated liquid and saturated vapor:

$$v = v_f + xv_{fg} = 0.001003 \text{ m}^3/\text{kg} + 0.7(43.36)\text{m}^3/\text{kg}$$

$$= \underline{\underline{30.353 \text{ m}^3/\text{kg}}}$$

$$u = u_f + xu_{fg} = 104.75 \text{ kJ/kg} + 0.7(2304.1)\text{kJ/kg} = \underline{\underline{1717.6 \text{ kJ/kg}}}$$

■

■ EXAMPLE 2.7

Determine the pressure and specific volume of H_2O at 80°F that has an internal energy of 500 Btu/lb$_m$.

Solution. The state of the H_2O in this example can be determined by referring to the saturation table (Table B.1E) at 80°F. The internal energy of the saturated liquid and saturated vapor are u_f = 48.03 Btu/lb$_m$ and u_g = 1036.6 Btu/lb$_m$. Since $u_f < u < u_g$, the state is a saturation state, and the pressure must be the corresponding saturation pressure:

$$P = \underline{\underline{0.5073 \text{ psia}}}$$

The specific volume can be determined once the quality is known. Solving for the quality, we have

$$x = \frac{u - u_f}{u_{fg}} = \frac{(500 - 48.03)\text{Btu/lb}_m}{988.6 \text{ Btu/lb}_m} = 0.457$$

Therefore, the specific volume of the H_2O is

$$v = v_f + xv_{fg} = 0.01607 \text{ ft}^3/\text{lb}_m + 0.457(632.71)\text{ft}^3/\text{lb}_m$$

$$= \underline{\underline{289.2 \text{ ft}^3/\text{lb}_m}}$$

This value is an average specific volume for the mixture of liquid and vapor, that is, the total volume of the mixture divided by the total mass of the mixture. ■

■ EXAMPLE 2.8

Refrigerant-12 is contained in a rigid storage tank initially at 400 kPa and 105.4°C. The refrigerant is cooled until the pressure reaches 300 kPa. Determine the final temperature, the change in specific volume, and the change in enthalpy of the refrigerant-12.

Solution. Since the tank is rigid, the volume does not change during the process. Because this system is closed, the mass remains constant also. Thus the process is one of constant specific volume.

From Table C.1 at 105°C, the saturation pressure for R-12 is 3.6538 MPa. The saturation pressure at 105.4°C is slightly higher than this value, so $P_1 < P_{sat}$ and the initial state is in the superheated-vapor region.

Property values are not tabulated for 105.4°C in the superheat table, but linear interpolation can be used to approximate the properties of state 1. Linear interpolation is merely a first-order approximation for values that lie between tabulated values. Geometrically, linear interpolation is equivalent to joining the tabulated values by a straight-line segment and estimating the unknown quantity by locating the point of intersection with the straight line. This technique is illustrated in Figure 2.14. Since linear interpolation is only an approximation, some error can be expected, but the error is usually small if the increments in the tabulated data are small.

Using this procedure and values at 400 kPa in Table C.3, we find

$$\frac{105.4°\text{C} - 100°\text{C}}{110°\text{C} - 100°\text{C}} = \frac{v_1 - v_{100°\text{C}}}{v_{110°\text{C}} - v_{100°\text{C}}} = \frac{v_1 - 0.061761 \text{ m}^3/\text{kg}}{(0.063621 - 0.061761)\text{m}^3/\text{kg}}$$

or

$$v_1 = 0.06277 \text{ m}^3/\text{kg}$$

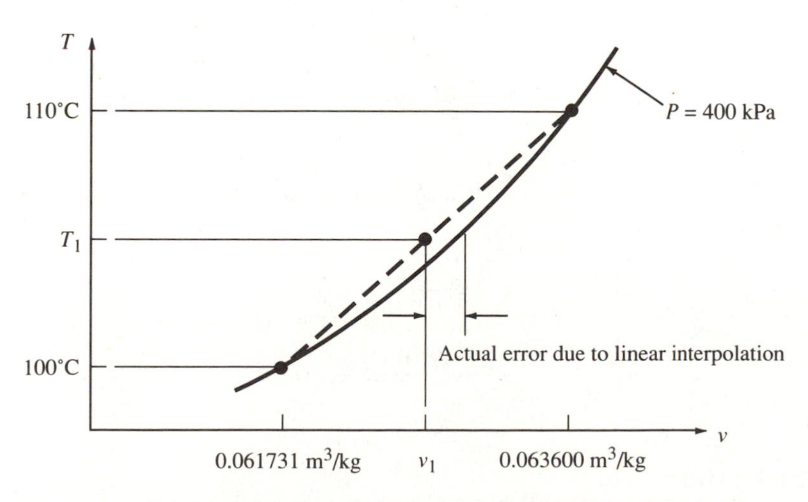

Figure 2.14 Errors produced by linear interpolation of tabulated data.

Similarly, we find

$$h_1 = 256.75 \text{ kJ/kg}$$

State 2 is now specified by the pressure (300 kPa) and the specific volume ($v_2 = v_1 = 0.06277 \text{ m}^3\text{/kg}$). From Table C.3 this value corresponds to the superheat state at which approximately

$$T_2 = \underline{\underline{19.2°\text{C}}}$$

For this state

$$h_2 = 200.98 \text{ kJ/kg}$$

Therefore

$$h_2 - h_1 = (200.98 - 256.75)\text{kJ/kg} = \underline{\underline{-55.77 \text{ kJ/kg}}}$$

The enthalpy of the R-12 decreases during the process. ■

■ EXAMPLE 2.9

Refrigerant-12 enters a compressor of an air-conditioning unit at 50 psia and 50°F and leaves at 200 psia, 180°F. Determine the phase(s), density, and enthalpy of the refrigerant-12 as it enters and leaves the compressor.

Solution. At the inlet temperature of 50°F, the saturation pressure of R-12 is 61.43 psia (see Table C.1E). Since the inlet pressure is less than the saturation pressure, the refrigerant is a superheated vapor. This same conclusion could be reached by referring to the properties in Table C.2E and concluding that $T_1 > T_{sat}$ for the inlet pressure of 50 psia.

The inlet properties from Table C.3E are (interpolation is necessary)

$$\rho = \frac{1}{v} = \frac{1}{0.8297 \text{ ft}^3/\text{lb}_m} = \underline{\underline{1.205 \frac{\text{lb}_m}{\text{ft}^3}}}$$

$$h = \underline{\underline{83.61 \text{ Btu/lb}_m}}$$

At the outlet conditions $P < P_{sat}$ for a temperature of 180°F (see Table C.1E), so the R-12 is once again in the superheated-vapor region. From Table C.3E, the properties at the outlet are

$$\rho = \frac{1}{v} = \frac{1}{0.2361 \text{ ft}^3/\text{lb}_m} = \underline{\underline{4.235 \frac{\text{lb}_m}{\text{ft}^3}}}$$

$$h = \underline{\underline{99.40 \text{ Btu/lb}_m}}$$

■

■ EXAMPLE 2.10

A closed system holds a mixture of 1 kg of liquid water and 1 kg of water vapor in equilibrium at 700 kPa.

(a) Determine the initial temperature.
(b) Heat transfer to the contents occurs until the temperature reaches 350°C. The pressure is maintained constant during the process. Determine the change in volume of the system.

Solution.
(a) The initial state must be in the saturation region since an equilibrium mixture of liquid and vapor exists. Thus from Table B.2

$$T_1 = T_{sat}\big|_{700\ \text{kPa}} = \underline{\underline{164.98°\text{C}}}$$

(b) The initial volume can be determined as the sum of the volume of liquid present and the volume of vapor present; that is,

$$\begin{aligned} \mathbf{V}_1 &= \mathbf{V}_{1f} + \mathbf{V}_{1g} = m_f v_f + m_g v_g \\ &= (1\ \text{kg})(0.001108\ \text{m}^3/\text{kg}) + (1\ \text{kg})(0.2728\ \text{m}^3/\text{kg}) = 0.274\ \text{m}^3 \end{aligned}$$

We could also have determined the initial volume by first calculating the quality of the mixture:

$$x_1 = \frac{m_g}{m_{\text{total}}} = \frac{1\ \text{kg}}{2\ \text{kg}} = 0.5$$

Then

$$\begin{aligned} v_1 &= v_f + x_1 v_{fg} = 0.001108\ \text{m}^3/\text{kg} + 0.5(0.2717)\text{m}^3/\text{kg} \\ &= 0.137\ \text{m}^3/\text{kg} \end{aligned}$$

and

$$\mathbf{V}_1 = m_1 v_1 = (2\ \text{kg})(0.137\ \text{m}^3/\text{kg}) = 0.274\ \text{m}^3$$

State 2 is a superheated-vapor state since P_{sat} at 350°C is 16.521 MPa, and this value is greater than P_2. Thus from Table B.3 at 350°C, the specific volume at state 2 is found by interpolation between 600 and 800 kPa:

$$\frac{700\ \text{kPa} - 600\ \text{kPa}}{800\ \text{kPa} - 600\ \text{kPa}} = \frac{v_2 - v_{600\ \text{kPa}}}{v_{800\ \text{kPa}} - v_{600\ \text{kPa}}} = \frac{v_2 - 0.4742\ \text{m}^3/\text{kg}}{(0.3544 - 0.4742)\text{m}^3/\text{kg}}$$

from which

$$v_2 = 0.4143\ \text{m}^3/\text{kg}$$

and

$$\mathbf{V}_2 = m_2 v_2 = (2\ \text{kg})(0.4143\ \text{m}^3/\text{kg}) = 0.8286\ \text{m}^3$$

so that the volume of the system increases by

$$\mathbf{V}_2 - \mathbf{V}_1 = (0.8286 - 0.274)\text{m}^3 = \underline{\underline{0.5546\ \text{m}^3}}$$

■

2.7 SPECIFIC HEATS AND LATENT HEATS

2.7.1 Specific Heats

The state postulate discussed in Section 2.3 was used to conclude that the state of a simple compressible substance is determined by values of two independent, intensive thermodynamic properties. The values for the remaining properties associated with the state are uniquely determined by specifying the two independent properties. As a consequence of the state postulate, the internal energy of a simple compressible, homogeneous substance could be considered a function of only the temperature and the specific volume of the substance. This idea can be expressed mathematically as

$$u = u(T,v) \tag{2.15}$$

The choice of T and v as independent variables in Equation 2.15 is entirely arbitrary, but they are logical choices because they can be used to express the internal energy in terms of two independent properties that are easily measured. A complete set of values of internal energy for a substance as a function of temperature and specific volume could be plotted as a three-dimensional surface, as shown in Figure 2.15. Each point on the surface represents an equilibrium state for the substance in accordance with the functional relationship of Equation 2.15.

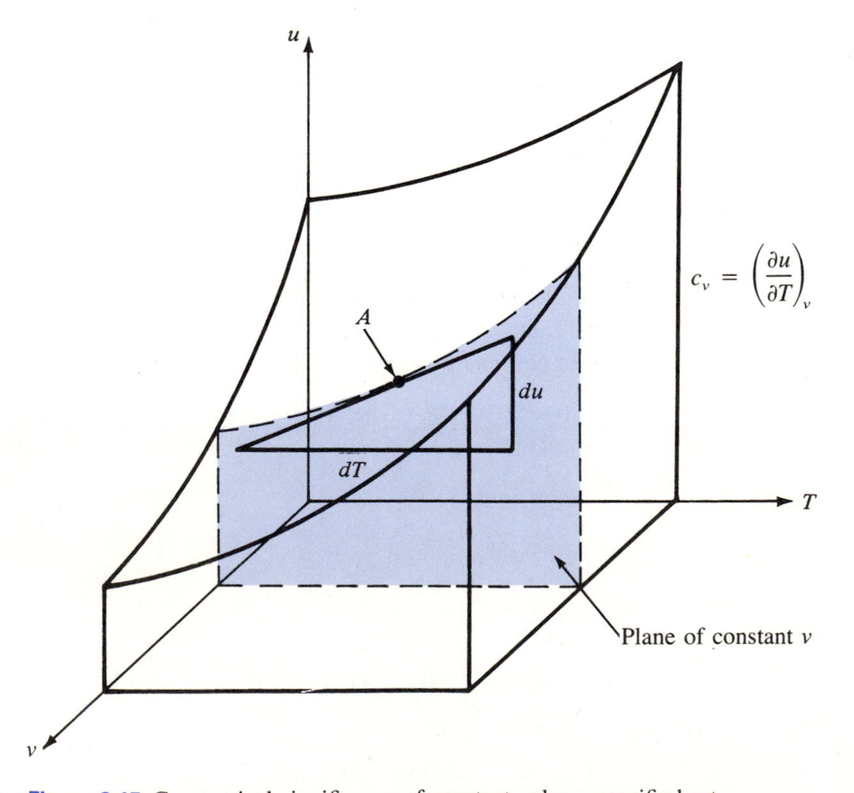

Figure 2.15 Geometrical significance of constant-volume specific heat.

A general expression for the change in internal energy of any simple compressible substance can be derived by differentiating the functional expression in Equation 2.15. The result is

$$du = \left(\frac{\partial u}{\partial T}\right)_v dT + \left(\frac{\partial u}{\partial v}\right)_T dv \tag{2.16}$$

The first partial derivative in Equation 2.16 is a thermodynamic property called the *constant-volume specific heat*,

$$c_v \equiv \left(\frac{\partial u}{\partial T}\right)_v \tag{2.17}$$

Physically, the constant-volume specific heat is the ratio of the incremental change in internal energy to the corresponding incremental change in temperature as the specific volume is maintained constant. Geometrically, the value for c_v at a particular state A shown in Figure 2.15 can be interpreted as the slope of the curve produced when a plane of constant v intersects the surface $u = u(T,v)$. Substituting the definition of c_v in Equation 2.16 gives

$$du = c_v\, dT + \left(\frac{\partial u}{\partial v}\right)_T dv \tag{2.18}$$

for the differential change in internal energy for any simple compressible substance.

The enthalpy of any simple compressible, homogeneous substance can be assumed to be a function of the independent properties T and P, or

$$h = h(T, P) \tag{2.19}$$

Using a procedure similar to that used in the development of the expression for the differential change in internal energy, the differential change in enthalpy can be written as

$$dh = \left(\frac{\partial h}{\partial T}\right)_P dT + \left(\frac{\partial h}{\partial P}\right)_T dP \tag{2.20}$$

The first partial derivative in Equation 2.20 is a thermodynamic property of a substance similar to c_v and is called the *constant-pressure specific heat*:

$$c_p \equiv \left(\frac{\partial h}{\partial T}\right)_P \tag{2.21}$$

The constant-pressure specific heat is the ratio of the incremental change in enthalpy to the corresponding incremental change in temperature as the pressure is maintained constant. It can also be interpreted as the slope of the equilibrium surface of $h = h(T, P)$ when cut by a plane of constant pressure. Substituting Equation 2.21 in Equation 2.20 gives

$$dh = c_p\, dT + \left(\frac{\partial h}{\partial P}\right)_T dP \tag{2.22}$$

for the differential change in enthalpy for any simple compressible substance.

The values of the constant-pressure specific heat and the constant-volume specific heat are generally much more affected by changes in temperature than by changes in pressure. For instance, as the pressure of a gas is reduced to near zero, the values for c_p and c_v are essentially independent of pressure and depend primarily on temperature. In the SI system the units of c_p and c_v are kJ/kg · K, and the corresponding molar specific heats, $\bar{c}_p$ and $\bar{c}_v$, have units* of kJ/kg · mol · K. In the English system of units c_p and c_v are expressed in Btu/lb$_m$ · °R and the molar quantities $\bar{c}_p$ and $\bar{c}_v$ have units of Btu/lb$_m$ · mol · °R.

Another quantity that is frequently used in thermodynamics is the *specific heat ratio*, a dimensionless quantity defined as

$$k \equiv \frac{c_p}{c_v} \tag{2.23}$$

Since k is the ratio of two thermodynamic properties, it is also a thermodynamic property. The specific heat ratio will be used later when the second law of thermodynamics is developed.

2.7.2 Latent Heats

Whenever the phase of a substance changes at constant pressure and temperature, energy must either be added to or removed from the substance. The change in enthalpy that the substance experiences during the phase-change process is called the *latent heat* of the substance. Later, after the development of the conservation of energy is discussed in Chapter 4, we will be able to show that the latent heat is defined such that it is equal to the amount of energy required to change the phase of a unit mass of the substance at a given pressure and temperature. For example, the enthalpy of a saturated vapor is greater than the enthalpy of a saturated liquid by an amount equal to h_{fg}. This quantity is defined as the *latent heat of vaporization* of the substance at a given pressure and temperature. In a similar fashion, the *latent heat of fusion* is defined as the difference between the enthalpy of a saturated liquid and the enthalpy of a saturated solid, or h_{if}. The third latent heat of a substance is the *latent heat of sublimation,* and it is defined as the difference between the enthalpy of a saturated vapor and the enthalpy of a saturated solid, or h_{ig}.

The latent heats of a substance are thermodynamic properties that are functions of pressure and temperature. The latent heat of vaporization of substances like water and R-12 can easily be determined by inspecting the h_{fg} column in the property tables that appear in the appendixes.

■ EXAMPLE 2.11

Suppose that the internal energy and enthalpy of a particular substance are functions of temperature only and that the specific heats of the substance are constant. Determine expressions for the change in internal energy and enthalpy of this substance as it undergoes a process during which the temperature increases from T_1 to T_2.

---

*The overbar here and elsewhere throughout the text is used to indicate a quantity on a molar basis to distinguish it from the same quantity on a mass basis.

Solution. Since the internal energy is a function of temperature only for this substance, $u = u(T)$ and

$$\left(\frac{\partial u}{\partial v}\right)_T = 0$$

and Equation 2.18 reduces to

$$du = c_v\, dT$$

And since c_v is constant,

$$u_2 - u_1 = \underline{\underline{c_v(T_2 - T_1)}}$$

Similarly, the enthalpy is only a function of temperature, or $h = h(T)$, so that

$$\left(\frac{\partial h}{\partial P}\right)_T = 0$$

and Equation 2.22 reduces to

$$dh = c_p\, dT$$

and

$$h_2 - h_1 = \underline{\underline{c_p(T_2 - T_1)}}$$

Notice that these results are true for this particular substance regardless of the process. That is, $du = c_v\, dT$ and $dh = c_p\, dT$ even though the process might be something other than a constant-volume or a constant-pressure process. The subscripts on the specific heats simply designate the definitions of the specific heats and do not limit their use to a particular process. ■

2.8 IDEAL GASES

An *ideal gas* is defined as a gas whose absolute pressure, absolute temperature, and specific volume obey the equation of state

$$Pv = RT \tag{2.24}$$

Equation 2.24 is commonly called the *ideal-gas equation of state*. As a result of Equation 2.24, the internal energy of an ideal gas is a function of temperature alone.

$$u = u(T) \tag{2.25}$$

The fact that the internal energy of an ideal gas depends on temperature alone will be verified in Chapter 9. The symbol R in the ideal-gas equation of state is called the *gas constant*; its value depends on the particular gas being considered. The value of R for each gas is determined from the equation

$$R = \frac{\overline{R}}{M} \tag{2.26}$$

where $\overline{R}$ is a physical constant called the *universal gas constant*:

$$
\begin{aligned}
\overline{R} &= 8.314 \text{ kPa} \cdot \text{m}^3/\text{kg} \cdot \text{mol} \cdot \text{K} \\
&= 8.314 \text{ kJ/kg} \cdot \text{mol} \cdot \text{K} \\
&= 8314 \text{ N} \cdot \text{m/kg} \cdot \text{mol} \cdot \text{K} \qquad (2.27) \\
&= 1545 \text{ ft} \cdot \text{lb}_f/\text{lb}_m \cdot \text{mol} \cdot {}^\circ\text{R} \\
&= 1.986 \text{ Btu/lb}_m \cdot \text{mol} \cdot {}^\circ\text{R}
\end{aligned}
$$

Values for the molecular weight M and the gas constant R in SI units for several common gases are given in Table D.10 in the Appendix. Equivalent values in English units are given in Table D.10E.

The concept of an ideal gas is only a model for the behavior of *real gases*, that is, pure substances in the vapor phase, at relatively low pressures. Consequently, the ideal-gas equation of state does not accurately predict the behavior of real gases under all conditions. Equations 2.24 and 2.25 are reasonably accurate, however, for a real gas as the pressure of the gas is reduced to near zero. Furthermore, the accuracy of the ideal-gas equation of state improves as the temperature of the gas increases and as the molecular weight of the gas decreases. In general, the behavior of a real gas is more nearly ideal when the pressure of the gas is less than the critical-state pressure and the temperature of the gas is greater than the critical-state temperature. (Values for the critical pressure and critical temperature for several gases are given in Tables H.1 and H.1E.)

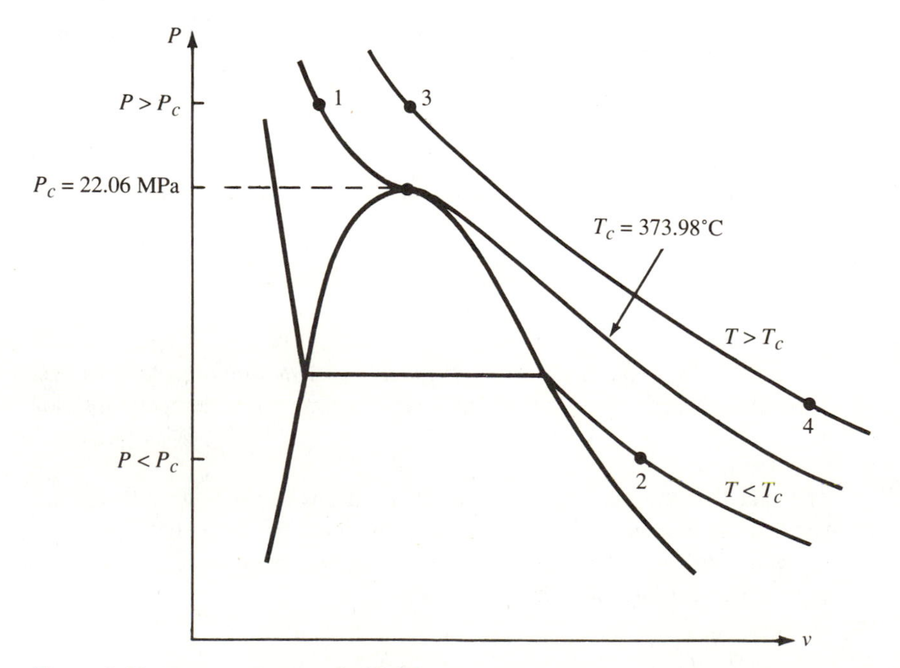

Figure 2.16 The *P*-*v* diagram for H_2O.

These observations concerning ideal-gas behavior can be illustrated by examining the accuracy of Equations 2.24 and 2.25 for water at several states. Four vapor states that represent various combinations of temperature and pressure relative to the critical-state values for water are depicted in Figure 2.16: state 1, $T \simeq T_c$, $P > P_c$; state 2, $T < T_c$, $P < P_c$; state 3, $T > T_c$, $P > P_c$; state 4, $T > T_c$, $P < P_c$. Table 2.1 summarizes the values for specific volume and internal energy from Table B.3 for water at several of these representative states. This table also shows the percentage change in internal energy owing to a change in pressure while the temperature is maintained constant, as well as the value of Pv/RT for each state. For ideal-gas behavior the value of Pv/RT at each state should be unity, and according to Equation 2.25, the internal-energy change should be zero for a change of state along an isothermal path. The results shown in Table 2.1 indicate that ideal-gas conditions are best approximated by superheated-vapor states for which the temperature is greater than T_c and the pressure is less than P_c.

Air at normal room conditions of 300 K and 101 kPa is highly superheated; the pressure is much less than the critical pressure of 3.74 MPa, and the temperature is much greater than the critical temperature of 133 K. The ideal-gas equation of state (Equation 2.24), therefore, will accurately predict the properties of air at room conditions. In fact, the error in the ideal-gas equation of state is less than 1 percent for air at 25°C for pressures as high as 2.7 MPa. The error is also less than 1 percent when the temperature of air is as low as -130°C for a pressure of 101 kPa.

For a gas such as hydrogen that has a low molecular weight, the errors in the properties predicted by the ideal-gas equation of state are small for even larger ranges of pressure and temperature. For example, the error in the specific volume when calculated from $Pv = RT$ is less than 1 percent from the true value for hydrogen at 101 kPa pressure, even when the temperature is as low as -220°C.

In essence, the ideal-gas equation of state neglects the intermolecular forces in a gas. Thus the behavior of a real gas should be more nearly ideal when intermolecular forces are weakest. As the pressure is reduced, the temperature is increased, and the molecular weight is reduced, the mean free path between gas molecules is increased,

TABLE 2.1 EVALUATION OF IDEAL-GAS BEHAVIOR FOR WATER

State	T,°C	P, MPa	v, m^3/kg	u, kJ/kg	% Change in u	Pv/RT
1a	375	35	0.0017007	1702.4		0.1991
1b	375	40	0.0016405	1676.5	-1.54	0.2194
2a	300	1	0.2579	2792.7		0.9753
2b	300	5	0.04530	2697.0	-3.5	0.8565
3a	1300	35	0.02089	4626.4		1.007
3b	1300	40	0.018304	4617.9	-0.18	1.009
4a	1300	1	0.7261	4685.1		1.000
4b	1300	5	0.14529	4678.1	-0.15	1.001
Critical state	373.98	22.06				

producing lower intermolecular interactions. Under these conditions the assumption of ideal-gas behavior is quite accurate.

While the ideal-gas equation of state will not accurately represent the characteristics of real gases for all states, it does provide sufficient accuracy for engineering calculations over large pressure and temperature ranges for many gases. Other more complex equations of state have been proposed, and a few of these are discussed in Chapter 10.

Several different forms of the ideal-gas equation of state are possible. Equation 2.24 can be expressed in terms of the mass and volume of the gas by substituting the definition of the specific volume:

$$v = \frac{\mathbf{V}}{m}$$

It can also be written in terms of the number of moles N of a gas, since

$$N = \frac{m}{M}$$

where M is the molecular weight of the gas. Verification of five different forms of the ideal-gas equation of state is left as an exercise in the problems at the end of this chapter.

EXAMPLE 2.12

The volume of the passenger compartment of an aircraft is 75,000 ft^3. Automatic equipment maintains the air inside the plane at a pressure of 14.3 psia and a temperature of 75°F. Calculate the mass of the air inside the plane. Determine the percent increase in the mass of the air if the pressure is increased to 14.7 psia and the temperature drops to 65°F.

Solution. For the given pressure and temperature the air inside the plane behaves as an ideal gas, which obeys the equation

$$Pv = RT$$

Since the volume of the air is given rather than its specific volume, a more convenient form of the idea-gas equation of state for this problem is

$$P\mathbf{V} = mRT$$

Solving for the mass of air, we have

$$m_1 = \frac{P_1 \mathbf{V}_1}{RT_1} = \frac{(14.3\ \text{lb}_f/\text{in}^2)(144\ \text{in}^2/\text{ft}^2)(75{,}000\ \text{ft}^3)}{\left(\dfrac{53.34\ \text{ft} \cdot \text{lb}_f}{\text{lb}_m \cdot {}^\circ\text{R}}\right)(535^\circ\text{R})}$$

$$= \underline{\underline{5410\ \text{lb}_m}}$$

If the pressure of the air is increased and the temperature is decreased, the mass of the air must increase because the volume of air remains constant. Denoting the initial conditions with a subscript 1 and the final conditions by a subscript 2, the ratio of the initial mass of air in the plane to the final mass is

$$\frac{m_2}{m_1} = \frac{P_2}{P_1}\frac{T_1}{T_2} = \left(\frac{14.7 \text{ psia}}{14.3 \text{ psia}}\right)\left(\frac{535°\text{R}}{525°\text{R}}\right) = 1.048$$

The percent increase in the mass of the air inside the plane is

$$\text{Percent increase} = \left(\frac{m_2 - m_1}{m_1}\right) 100\% = \left(\frac{m_2}{m_1} - 1\right) 100\% = \underline{\underline{4.8\%}}$$

■ EXAMPLE 2.13

A 4 ft³ rigid cylinder contains helium initially at 50°F and 80 psia (state 1). More helium is added to the container until conditions reach 95°F and 250 psia (state 2) at which time the inlet valve is closed and the contents cool to the original temperature of 50°F (state 3). Calculate the mass of helium added to the tank and the final pressure in the cylinder.

Solution. For the pressure and temperature values given in the problem, helium behaves as an ideal gas. The initial mass of helium in the tank is

$$m_1 = \frac{P_1 V}{RT_1} = \frac{(80 \text{ lb}_f/\text{in}^2)(144 \text{ in}^2/\text{ft}^2)(4 \text{ ft}^3)(4 \text{ lb}_m/\text{lb}_m \cdot \text{mol})}{(1545 \text{ ft} \cdot \text{lb}_f/\text{lb}_m \cdot \text{mol} \cdot °\text{R})(510°\text{R})} = 0.234 \text{ lb}_m$$

After helium is added to the tank the mass is

$$m_2 = \frac{P_2 V}{RT_2} = \frac{(250 \text{ lb}_f/\text{in}^2)(144 \text{ in}^2/\text{ft}^2)(4 \text{ ft}^3)(4 \text{ lb}_m/\text{lb}_m \cdot \text{mol})}{(1545 \text{ ft} \cdot \text{lb}_f/\text{lb}_m \cdot \text{mol} \cdot °\text{R})(555°\text{R})} = 0.672 \text{ lb}_m$$

The amount of mass added is

$$m_2 - m_1 = 0.672 \text{ lb}_m - 0.234 \text{ lb}_m = \underline{\underline{0.438 \text{ lb}_m}}$$

The pressure in the cylinder after the contents cool at constant mass is

$$P_3 = \frac{m_2 R T_3}{V} = \frac{(0.672 \text{ lb}_m)(1545 \text{ ft} \cdot \text{lb}_f/\text{lb}_m \cdot \text{mol} \cdot °\text{R})(510°\text{R})}{(4 \text{ ft}^3)(4 \text{ lb}_m/\text{lb}_m \cdot \text{mol})}$$

$$= 33{,}100 \frac{\text{lb}_f}{\text{ft}^2} = \underline{\underline{230 \frac{\text{lb}_f}{\text{in}^2}}}$$

■ EXAMPLE 2.14

A worker pressurizes a rigid pipe (1 in inside diameter, 20 ft long) with dry air to check for leaks. The temperature and gauge pressure of the air in the pipe are 110°F and 85 psi. The worker returns 24 h later and the gauge pressure has dropped to 75 psi, while the air temperature inside the pipe has decreased to 80°F. Has the pipe leaked? If so, calculate the mass of air that has leaked through the fittings.

Solution. A decrease in pressure does not always mean that the pipe has leaked. The pressure could decrease because of cooling of the air.

The given pressure and temperature are such that the air behaves as an ideal gas. The ratio of the mass of gas initially in the pipe to that after 24 h can be determined by forming the ratio of the ideal-gas equation of state at the initial and final states. Recognizing that the volume of gas does not change during the process, we find

$$\frac{m_1}{m_2} = \left(\frac{P_1}{T_1}\right)\left(\frac{T_2}{P_2}\right) = \left(\frac{99.7 \text{ psia}}{570°\text{R}}\right)\left(\frac{540°\text{R}}{89.7 \text{ psia}}\right) = 1.053$$

Since $m_1/m_2 > 1.0$, the pipe has leaked during the process. Notice that absolute pressures and temperatures must be used in this equation because it is derived from the ideal-gas equation of state.

The ideal-gas equation of state can be applied to determine the mass of air initially in the pipe:

$$P_1\mathbf{V} = m_1RT_1$$

The volume occupied by the air is

$$\mathbf{V} = \frac{\pi d^2 L}{4} = \frac{\pi(1/12)^2\text{ft}^2(20 \text{ ft})}{4} = 0.1091 \text{ ft}^3$$

and the initial mass of air is

$$m_1 = \frac{P_1\mathbf{V}}{RT_1} = \frac{(99.7 \text{ lb}_f/\text{in}^2)(144 \text{ in}^2/\text{ft}^2)(0.1091 \text{ ft}^3)}{(53.34 \text{ ft}\cdot\text{lb}_f/\text{lb}_m\cdot°\text{R})(570°\text{R})} = 0.0515 \text{ lb}_m$$

The final mass of air in the pipe is therefore

$$m_2 = \frac{m_1}{1.053} = \frac{0.0515 \text{ lb}_m}{1.053} = 0.0489 \text{ lb}_m$$

The mass of air leaked is

$$m_1 - m_2 = \underline{\underline{2.6 \times 10^{-3} \text{ lb}_m}}$$

■

The concept of the ideal gas not only provides for a simple equation of state, but it also simplifies the determination of properties other than P, v, and T. For example, if a substance is an ideal gas, then u is a function of temperature alone, and therefore

$$\left(\frac{\partial u}{\partial v}\right)_T = 0$$

Consequently, the general expression for the differential change in internal energy of a simple compressible substance (Equation 2.18) can be simplified considerably for an ideal gas. The result is

$$du = c_v \, dT \tag{2.28}$$

Furthermore, since the internal energy of an ideal gas depends only on temperature, an important implication of Equation 2.28 is that the constant-volume specific heat also depends, at most, on temperature.

The change in internal energy for an ideal gas corresponding to a change of state from state 1 to state 2 can therefore be determined by integrating Equation 2.28:

$$u_2 - u_1 = \int_{T_1}^{T_2} c_v(T)\, dT \tag{2.29}$$

A convenient expression for the enthalpy of an ideal gas can be obtained by substituting Equation 2.24 into Equation 2.12, with the result

$$h = u + RT \tag{2.30}$$

Both terms on the right side of Equation 2.30 are functions of temperature alone, so the enthalpy of an ideal gas is also a function of temperature only or

$$h = h(T)$$

For an ideal gas, therefore

$$\left(\frac{\partial h}{\partial P}\right)_T = 0$$

and Equation 2.22 can be reduced to

$$dh = c_p dT \tag{2.31}$$

Examination of Equation 2.31 also reveals that the constant-pressure specific heat of an ideal gas depends, at most, on temperature.

The change in enthalpy for an ideal gas corresponding to a change of state can be determined by integrating Equation 2.31:

$$h_2 - h_1 = \int_{T_1}^{T_2} c_p(T)\, dT \tag{2.32}$$

Although real gases are not ideal gases, the behavior of a real gas at low pressures and at temperatures above its critical temperature is very nearly ideal. Under these conditions the specific heats of a gas are, to a good approximation, functions of temperature alone and can be measured quite accurately. The results of such experimental measurements are usually presented as functions of temperature and are denoted by $\bar{c}_{p0}$ or $\bar{c}_{v0}$. These terms are referred to as *molar zero-pressure specific heats*. The zero subscript indicates that these are specific heats for the "ideal" state of a real gas, or as the pressure of the gas is reduced to zero. The bar superscript denotes that the specific heats are molar quantities rather than mass quantities. Several curves of the molar zero-pressure specific heat, $\bar{c}_{p0}$, are plotted in Figure 2.17 as functions of temperature. Equations for $\bar{c}_{p0}$ in SI units for several gases are included in Table D.11, and the corresponding expressions in English units appear in Table D.11E.

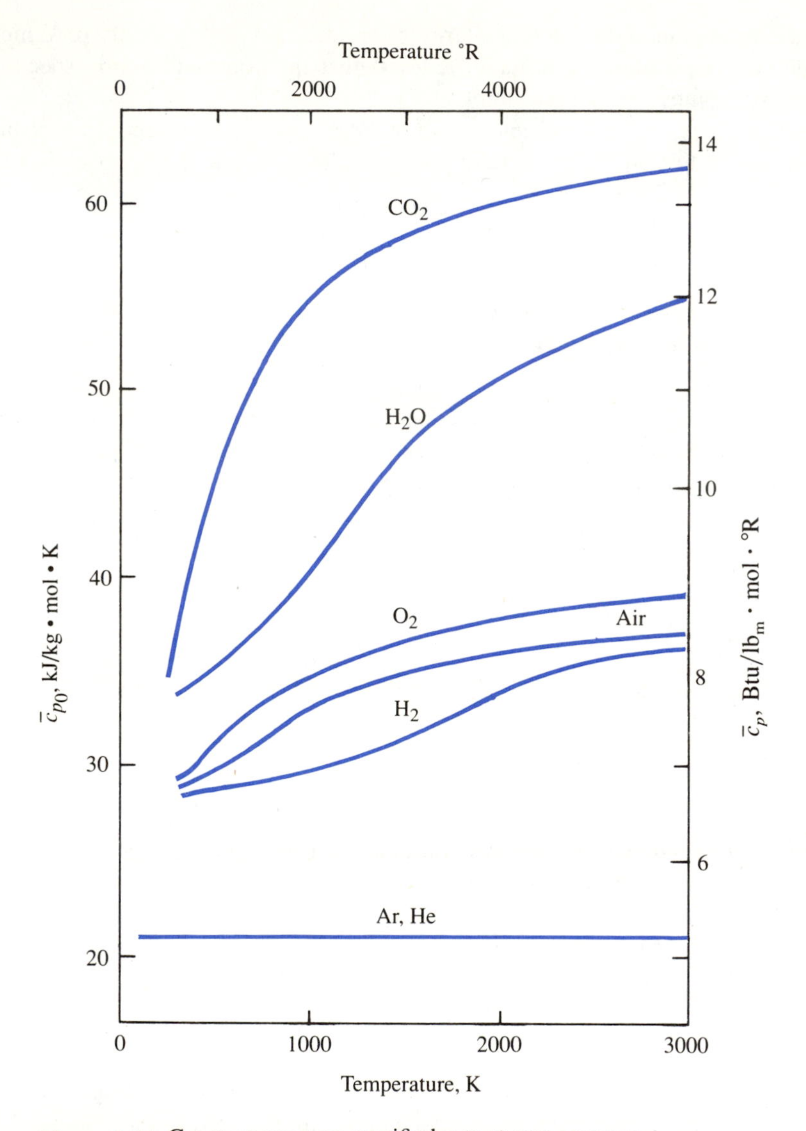

Figure 2.17 Constant-pressure specific heats at zero pressure.

Equations for $\bar{c}_{p0}(T)$ given in Table D.11 and Table D.11E may be used in Equation 2.32 to determine the change in enthalpy, and this procedure is illustrated in Example 2.15. While direct integration of Equations 2.29 and 2.32 is straightforward, the procedure is laborious. The results of the integration for many common gases are tabulated in Appendix D. In these tables a reference value of zero has been assigned to the enthalpy and internal energy of the ideal gas at zero absolute temperature. The choice of the reference temperature is arbitrary since only changes in internal energy and enthalpy are of primary interest.

The assumption of ideal-gas behavior affords an additional simplification of properties. Differentiating Equation 2.30 term by term, while recognizing that each term is only a function of temperature for an ideal gas, yields

$$\frac{dh}{dT} = \frac{du}{dT} + R$$

Substituting the definitions of the specific heats for an ideal gas gives

$$c_p - c_v = R \tag{2.33}$$

Therefore, even though the specific heats are temperature-dependent, the difference between the constant-pressure and constant-volume specific heats for an ideal gas is equal to the gas constant R. If either specific heat is known, the other may be calculated from Equation 2.33. For this reason it is customary to give values of only the constant-pressure specific heat knowing that the constant-volume specific heat can then be calculated from Equation 2.33.

The specific heats for an ideal gas can also be expressed in terms of k and R by combining Equation 2.33 with the definition of the specific heat ratio k from Equation 2.23:

$$c_p = \frac{kR}{k - 1} \tag{2.34}$$

and

$$c_v = \frac{R}{k - 1} \tag{2.35}$$

These results indicate that the gas constant R and the specific heat ratio are sufficient to determine both c_p and c_v for an ideal gas.

EXAMPLE 2.15

Carbon dioxide at a pressure of 101 kPa is heated from 300 to 800 K. Calculate the change in enthalpy and internal energy of the CO_2 for this process.

Solution. The low pressure and high temperatures during the process relative to the critical-state values of 7.39 MPa and 304.2 K ensure that the CO_2 behaves as an ideal gas. The constant-pressure specific heat for CO_2 on a molar basis, $\bar{c}_{p0}$, is given as a function of temperature in Table D.11 in the Appendix:

$$\frac{\bar{c}_{p0}}{\bar{R}} = a + bT + cT^2 + dT^3 + eT^4$$

where T is in kelvins. The enthalpy change given by Equation 2.32, written on a mass basis, is

$$h_2 - h_1 = \frac{\bar{R}}{M} \int_{300\text{ K}}^{800\text{ K}} (a + bT + cT^2 + dT^3 + eT^4)dT$$

Substituting values for the constants from Table D.11 and carrying out the integration, we find that the change of enthalpy is

$$h_2 - h_1 = \underline{\underline{516.9 \text{ kJ/kg}}}$$

Using Equation 2.30 and the value of R from Table D.9, we find the internal-energy change to be

$$\begin{aligned} u_2 - u_1 &= h_2 - h_1 - R(T_2 - T_1) \\ &= 516.9 \text{ kJ/kg} - (0.1889 \text{ kJ/kg} \cdot \text{K})(800 - 300)\text{K} \\ &= \underline{\underline{422.5 \text{ kJ/kg}}} \end{aligned}$$

Notice that the final pressure of the CO_2 was not given in the problem statement. The change in enthalpy and internal energy of an ideal gas is dependent only on the temperatures at the end states of the process. The final pressure therefore does not affect the solution.

These same answers could be obtained more easily by using tabular values for properties of CO_2. Table D.2 in the Appendix gives the values on a mass basis:

$$h_2 - h_1 = (731.0 - 214.3)\text{kJ/kg} = 516.7 \text{ kJ/kg}$$

$$u_2 - u_1 = (579.8 - 157.7)\text{kJ/kg} = 422.1 \text{ kJ/kg}$$

The tabular values, when they are available, are preferred because of their improved accuracy and ease of use.

2.8.1 Ideal Gases with Linearly Varying Specific Heats

The specific heats of an ideal gas depend only on temperature, and there are many instances when they may be assumed to be constant or to vary linearly with temperature. While these assumptions are not always strictly valid, they can often be made without appreciable error. The specific heats of diatomic gases such as oxygen and hydrogen, as well as air at low pressure, are approximately linear functions of temperature above about 1000 K (1800°R), as shown in Figure 2.17. The variation of the specific heats for these gases at lower temperatures and for other gases, such as H_2O and CO_2, is nearly linear for small temperature changes (on the order of 100 to 200 K, or about 200 to 300°R).

When c_p is a linear function of temperature ($c_p = a + bT$), the enthalpy change of an ideal gas corresponding to a temperature change from T_1 to T_2 can be determined from Equation 2.32 as

$$\begin{aligned} h_2 - h_1 &= \int_{T_1}^{T_2} (a + bT)\, dT \\ &= a(T_2 - T_1) + \frac{b}{2}(T_2^2 - T_1^2) \\ &= \left[a + b\left(\frac{T_2 + T_1}{2}\right)\right](T_2 - T_1) \end{aligned}$$

$$h_2 - h_1 = c_{p,\text{avg}}(T_2 - T_1) \tag{2.36}$$

where

$$c_{p,\text{avg}} = a + b\left(\frac{T_1 + T_2}{2}\right)$$

represents the average value of the specific heat (i.e., the specific heat evaluated at the average temperature). A similar result holds for the change in internal energy, obtained by integrating Equation 2.29:

$$u_2 - u_1 = c_{v,\text{avg}}(T_2 - T_1) \tag{2.37}$$

Zero-pressure specific heats for several common ideal gases are listed in Table D.9 for a temperature of 27°C. Table D.8 also contains SI values for c_p, c_v, and k for a number of gases at various temperatures. The corresponding values in English units appear in Tables D.8E and D.9E.

The constant-volume and constant-pressure specific heats (defined by Equations 2.17 and 2.21) each have the units of kJ/kg · K. The denominator represents a *temperature difference*, and therefore the unit of temperature used in Equations 2.36 and 2.37 can be either degrees Celsius or kelvins in the SI system of units. The magnitude of the temperature *difference* is the same regardless of whether temperatures are measured in degrees Celsius or kelvins. In the English system of units c_p and c_v are expressed in Btu/lb_m · °R or in Btu/lb_m · °F.

EXAMPLE 2.16

Compressed air is used to power an air wrench. Air enters the wrench at a pressure of 150 psia and a temperature of 140°F. Air exhausts from the wrench at atmospheric pressure and a temperature of 40°F. Calculate the internal-energy change and the enthalpy change of the air per lb_m between the inlet and exit of the air wrench.

Solution. The air wrench does work by virtue of a decrease in the energy of the air. Equations are derived in Chapter 4 that relate the change in energy level of the air to the power output of the wrench.

For the given pressures and temperatures the ideal-gas assumption is valid for air. The specific heats for air can be assumed linear for the small temperature difference experienced in the expansion process. The average temperature during the process is 550°R, and the average values of the constant-pressure and constant-volume specific heats can be determined by linear interpolation in Table D.8E:

$$c_{p,\text{avg}} = 0.2395\ \text{Btu/lb}_\text{m}\cdot{}^\circ\text{R}$$

$$c_{v,\text{avg}} = 0.1705\ \text{Btu/lb}_\text{m}\cdot{}^\circ\text{R}$$

Using these average values of the specific heats, the change in internal energy is

$$u_2 - u_1 = c_{v,\text{avg}}(T_2 - T_1) = (0.1705\ \text{Btu/lb}_\text{m}\cdot{}^\circ\text{R})(500^\circ\text{R} - 600^\circ\text{R})$$

$$= \underline{\underline{-17.1\ \text{Btu/lb}_\text{m}}}$$

and the change in enthalpy is

$$h_2 - h_1 = c_{p,\text{avg}}(T_2 - T_1) = (0.2395 \text{ Btu/lb}_\text{m} \cdot {}^\circ\text{R})(500^\circ\text{R} - 600^\circ\text{R})$$
$$= \underline{\underline{-24.0 \text{ Btu/lb}_\text{m}}}$$

The minus signs indicate that the internal energy and enthalpy of the air decrease during the process.

Notice that the pressure does not affect the answers because the values for enthalpy and internal energy are independent of pressure for an ideal gas.

A more accurate result for the change in internal energy and enthalpy could be obtained by using the tabulated values for properties of air. From Table D.1E,

$$h_2 - h_1 = (119.6 - 143.6)\text{Btu/lb}_\text{m} = -24.0 \text{ Btu/lb}_\text{m}$$
$$u_2 - u_1 = (85.3 - 102.4)\text{Btu/lb}_\text{m} = -17.1 \text{ Btu/lb}_\text{m}$$

These values are identical to the values obtained by assuming a linear variation in specific heats. The linear approximation produces accurate results in this instance because the temperature change for the process is only 100°F. ■

2.8.2 Ideal Gases with Constant Specific Heats

At low pressure the constant-pressure specific heat of monatomic gases, such as argon and helium, is essentially independent of temperature, as shown in Figure 2.17. From the kinetic theory of gases the magnitude of the constant-pressure specific heat of an ideal monatomic gas is given by

$$c_p = \frac{5}{2} R \tag{2.38}$$

With this result and Equation 2.33 the constant-volume specific heat of an ideal monatomic gas is found to be

$$c_v = \frac{3}{2} R \tag{2.39}$$

Since the specific heats of an ideal monatomic gas are constant, Equations 2.29 and 2.32 can be integrated with the result

$$u_2 - u_1 = c_v(T_2 - T_1) \tag{2.40}$$

and

$$h_2 - h_1 = c_p(T_2 - T_1) \tag{2.41}$$

These equations are also valid for other instances in which the specific heats of an ideal gas can be assumed constant.

2.8.3 Polytropic Processes for Ideal Gases

The particular path followed by a substance during a process is often given in the form of a relationship between the pressure and specific volume during the process:

$$P = P(v)$$

One particularly useful form is expressed by the equation

$$Pv^n = \text{constant} = c \tag{2.42}$$

where n is a constant. A process described by Equation 2.42 is called a *polytropic process*, n is the *polytropic exponent*, and Equation 2.42 is the *polytropic equation*.

The polytropic process as described by the P-v path is a convenient expression, because it can be used in determining path-dependent quantities such as the work. For internally reversible polytropic processes the $P\,dV$ work for a closed system, on a unit-mass basis, is obtained by integrating

$$w_{12} = \int_1^2 P\,dv$$

Substituting Equation 2.42 and performing the integration yields

$$\begin{aligned} w_{12} &= \int_1^2 P\,dv = \int_1^2 cv^{-n}\,dv \\ &= \left.\frac{cv^{1-n}}{1-n}\right|_1^2 = \left.\frac{Pv}{1-n}\right|_1^2 \end{aligned}$$

or

$$w_{12} = \frac{P_2v_2 - P_1v_1}{1-n} \tag{2.43}$$

Equation 2.43 is valid for all values of n except $n = 1$. When $n = 1$, the work for the internally reversible polytropic process is

$$\begin{aligned} w_{12} &= \int_1^2 P\,dv = \int_1^2 c\frac{dv}{v} = c\ln\left(\frac{v_2}{v_1}\right) \\ &= P_1v_1\ln\left(\frac{v_2}{v_1}\right) = P_2v_2\ln\left(\frac{v_2}{v_1}\right) \end{aligned} \tag{2.44}$$

The use of the polytropic equation, Equation 2.42, and the expressions for $P\,dV$ work given in Equations 2.43 and 2.44 are not restricted to ideal gases.

If Equation 2.42 is applied to ideal gases, additional relationships can be derived among the pressure, temperature, and specific volume at the end states of the polytropic process. For example, if v is eliminated from Equation 2.42 by using the ideal-gas equation of state,

$$Pv = RT$$

the result is

$$P\left(\frac{T}{P}\right)^{n} = \text{constant}$$

or

$$P = \text{constant}(T)^{n/(n-1)}$$

Thus the relationship between the temperatures and pressures at the end states of a polytropic process for an ideal gas can be expressed as

$$\frac{P_2}{P_1} = \left(\frac{T_2}{T_1}\right)^{n/(n-1)} \tag{2.45}$$

Similarly, P may be eliminated from Equation 2.42 by using the ideal-gas equation of state, with the result

$$\left(\frac{T}{v}\right)v^{n} = \text{constant}$$

or

$$Tv^{n-1} = \text{constant}$$

and the temperatures and specific volumes at the end states of a polytropic process for an ideal gas are related by

$$\frac{v_2}{v_1} = \left(\frac{T_1}{T_2}\right)^{1/(n-1)} \tag{2.46}$$

Several important processes can be described by the polytropic equation. For example, the value of $n = 0$ represents a constant-pressure process; the value of $n = 1$ represents a constant-temperature process for an ideal gas; the value of $n = k$ (the specific heat ratio) represents an internally reversible, adiabatic process for an ideal gas with constant specific heats (this result is established in Chapter 5); the value of $n \to \infty$ represents a constant-specific-volume process. These processes are summarized in Table 2.2 and illustrated in Figure 2.18.

TABLE 2.2 POLYTROPIC EXPONENTS FOR VARIOUS PROCESSES

Process	Polytropic exponent
Constant pressure	0
Constant temperature, ideal gas	1
Constant volume	∞
Reversible, adiabatic, ideal gas, constant specific heats	k

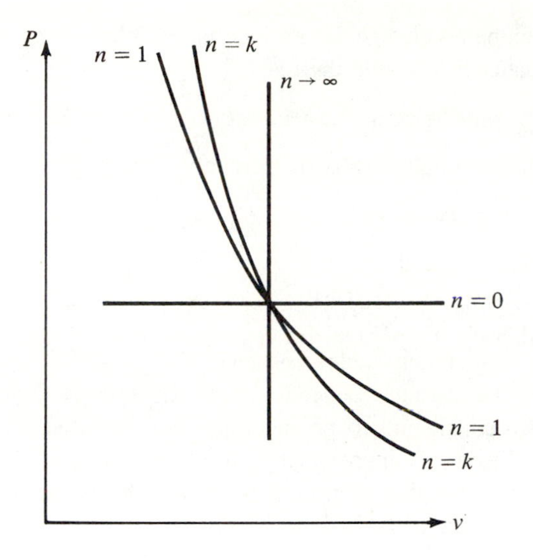

Figure 2.18 Polytropic processes for ideal gases.

2.9 INCOMPRESSIBLE SUBSTANCES

An *incompressible substance* is one whose density (or specific volume) remains constant regardless of changes in other properties. Many solids and most liquids can be assumed to be incompressible without much loss in accuracy. For example, liquid water at 100°C and a pressure of 5 MPa has a specific volume of 0.001041 m^3/kg. If the temperature is maintained constant and the pressure is decreased to 101 kPa, the specific volume of the water will be 0.001043 m^3/kg, an increase of less than 0.3 percent. On the other hand, if the pressure is maintained constant at 5 MPa and the temperature is reduced from 100 to 20°C, the specific volume decreases by about 4.0 percent, to 0.0009995 m^3/kg.

An expression for the internal-energy change of an incompressible substance can be obtained from Equation 2.18. The change in specific volume is zero ($dv = 0$), so Equation 2.18 reduces to

$$du = c_v \, dT \tag{2.47}$$

for an incompressible substance. The change in internal energy for an incompressible substance from state 1 at temperature T_1 to state 2 at temperature T_2 is, then

$$u_2 - u_1 = \int_{T_1}^{T_2} c_v \, dT \tag{2.48}$$

The difference between the constant-pressure specific heat and the constant-volume specific heat of an incompressible substance is zero (this result is verified in Chapter 10). Therefore, for substances such as solids and many liquids, which are essentially incompressible, the subscripts are often dropped and the specific heat of an incompressible substance is simply designated by c:

$$c_p = c_v = c \tag{2.49}$$

An expression for the enthalpy change for an incompressible substance can be derived by differentiating Equation 2.12, with the result

$$dh = du + P\,dv + v\,dP$$

Since v is constant for an incompressible substance, then

$$dh = du + v\,dP = c_v\,dT + v\,dP \tag{2.50}$$

This equation can be integrated to produce

$$h_2 - h_1 = u_2 - u_1 + v(P_2 - P_1) \tag{2.51}$$

for the enthalpy change for an incompressible substance.

Figure 2.19 depicts the temperature dependence of the specific heat for several substances that are normally considered to be incompressible. For small temperature changes the specific heat can be closely approximated by a linear function of temperature. Therefore, as discussed in the preceding section, Equation 2.48 for the internal-energy change of an incompressible substance can be integrated, with the result

$$u_2 - u_1 = c_{avg}(T_2 - T_1) \tag{2.52}$$

where c_{avg} is the specific heat evaluated at the average temperature $(T_1 + T_2)/2$. Values of the specific heats at various temperatures for several common liquids and solids are tabulated in Table H.3 for SI units and in Table H.3E for English units.

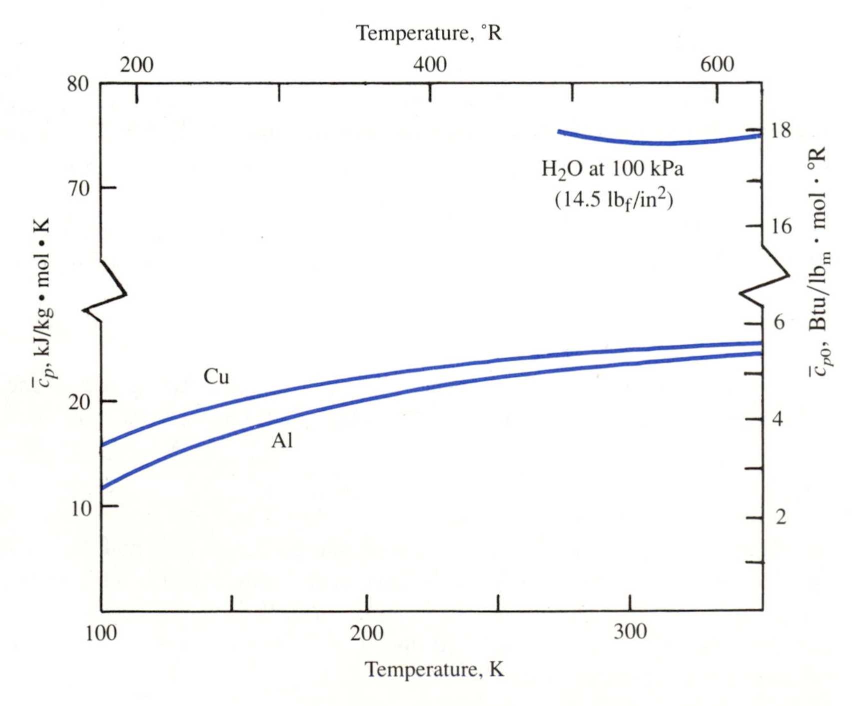

Figure 2.19 Typical constant-pressure specific heats of substances that are nearly incompressible.

EXAMPLE 2.17

At the beginning of a certain industrial process a 10-kg block of copper (ρ = 8930 kg/m^3) is at a temperature and pressure of 300 K and 100 kPa. At the end of the process the temperature and pressure of the copper are 600 K and 1 MPa. Calculate the change in internal energy and enthalpy of the copper during this process.

Solution. Copper is nearly incompressible, and the change in internal energy is given by Equation 2.52. Using the average specific heat of copper from Table H.3, the change in internal energy is

$$U_2 - U_1 = mc_{avg}(T_2 - T_1)$$

$$= (10\,\text{kg})(0.401\ \text{kJ/kg}\cdot\text{K})(600 - 300)\ \text{K} = \underline{\underline{1.20 \times 10^3\ \text{kJ}}}$$

The change in enthalpy is

$$H_2 - H_1 = m(u_2 - u_1) + mv(P_2 - P_1)$$

$$= 1.20 \times 10^3\ \text{kJ} + \frac{(10\,\text{kg})(1000 - 100)10^3\ \text{N/m}^2}{(8930\ \text{kg/m}^3)(10^3\ \text{N}\cdot\text{m/kJ})}$$

$$= 1.20 \times 10^3\ \text{kJ} + 1.0\ \text{kJ} \simeq \underline{\underline{1.20 \times 10^3\ \text{kJ}}}$$

■

Notice that in the preceding example the increase in enthalpy due to the pressure change is insignificant compared with the increase due to the temperature change. This result is typical of incompressible substances. The internal energy is independent of pressure, and a substantial pressure change is required to produce a relatively small change in enthalpy. To quantify this result, we calculate the temperature change and the pressure change required to produce an enthalpy change of only 1 kJ/kg for copper at an average temperature of 450 K:

$$c(T_2 - T_1) = 1\ \text{kJ/kg}$$

or

$$(T_2 - T_1) = \frac{1\ \text{kJ/kg}}{0.401\ \text{kJ/kg}\cdot\text{K}} = 2.5\ \text{K}$$

and

$$v(P_2 - P_1) = 1\ \text{kJ/kg}$$

or

$$P_2 - P_1 = (8930\ \text{kg/m}^3)(1\ \text{kJ/kg})(10^3\ \text{N}\cdot\text{m/kJ})$$

$$= 8.93 \times 10^6\,\text{N/m}^2 = 8.93\ \text{MPa}$$

Thus while a temperature increase of only 2.5 K will cause a change in enthalpy of 1 kJ/kg, the pressure on the copper would have to increase by 8.93 MPa to cause the same 1-kJ/kg increase in enthalpy. The corresponding values for water at 300 K that would result in a 1-kJ/kg increase in enthalpy are an increase of 0.24 K in temperature or an increase in pressure of 960 kPa.

2.10 APPROXIMATION OF PROPERTIES FOR COMPRESSED-LIQUID STATES

Often compressed-liquid property values are not tabulated, or only brief tables are available. In such cases approximate property values can be obtained by assuming that the liquid is essentially incompressible. Recall from Equation 2.52 that the internal-energy change for an incompressible substance is

$$u_2 - u_1 = c(T_2 - T_1)$$

That is, the internal energy is a function of temperature only. If state 2 is a compressed-liquid state at the same temperature as a saturated liquid at state 1, the implication of this equation is that the internal energy for a compressed-liquid state is approximately the same as the internal energy of the saturated liquid at the same temperature,

$$u_{CL}|_{T_1} \simeq u_f|_{T_1}$$

where the subscript *CL* is used to signify the compressed liquid. The specific volume at the compressed-liquid state is also very nearly equal to the specific volume of a saturated liquid at the same temperature. This condition was illustrated in the preceding section.

Unlike the internal energy, the enthalpy of an incompressible substance does depend somewhat on pressure, because Equation 2.51 states that

$$h_2 - h_1 = (u_2 - u_1) + v(P_2 - P_1)$$

for incompressible substances. Thus, considering an isothermal change of state from a saturated-liquid state to a compressed-liquid state, we have

$$h_{CL} - h_f = (u_{CL} - u_f) + v_f(P_{CL} - P_{\text{sat}})$$

or

$$h_{CL}|_{T_1} \simeq h_f|_{T_1} + (P_{CL} - P_{\text{sat}})v_f|_{T_1} \tag{2.53}$$

Unless the pressure of the compressed-liquid state is significantly greater than the saturation pressure corresponding to the temperature of the compressed liquid, the correction due to the pressure difference is usually quite small. This result is illustrated in Example 2.18.

To summarize the conclusions of this section, the properties of a compressed-liquid state can be approximated by using the values for the *saturated liquid at the same temperature* as the compressed-liquid state. For large pressure differences the enthalpy value should be corrected for the pressure dependence.

■ EXAMPLE 2.18

Compare values of v, u, and h for H_2O at 40°C and 10 MPa obtained from the compressed-liquid table with those obtained by assuming that the H_2O is incompressible.

Solution. The property values from the water tables are summarized as follows:

Property	Values from Table B-4 at 10 MPa and 40°C	Values of saturated liquid at 40°C	% error
v	0.0010035 m^3/kg	0.001008 m^3/kg	+0.45
u	166.30 kJ/kg	167.50 kJ/kg	+0.72
h	176.33 kJ/kg	167.50 kJ/kg	−5.0

Notice that the largest error involved in the approximate values is associated with the enthalpy. As indicated previously, this result is due to the fact that the enthalpy is much more pressure-dependent than the other properties. If a correction is applied for the pressure difference, as suggested in Equation 2.53, the approximation can be improved:

$$h \simeq h_f + v_f(P - P_{sat})$$

The saturation pressure at 40°C is 7.381 kPa, so

$$\begin{aligned} h &\simeq 167.50 \text{ kJ/kg} + (0.001008 \text{ m}^3\text{/kg})[(10^4 - 7.381)\text{kPa}] \\ &\quad \times (1 \text{ kN/m}^2 \cdot \text{kPa})(1 \text{ kJ/kN} \cdot \text{m}) \\ &\simeq 177.57 \text{ kJ/kg} \end{aligned}$$

With the pressure correction applied, the percent error in the approximate value of the enthalpy compared with the tabulated value becomes + 0.70 percent. As a general rule, the enthalpy of liquid water increases by about 1 kJ/kg for each 1-MPa increase in pressure (or about 3 Btu/lb_m for a pressure increase of 1000 psia). In this example, for instance, the enthalpy of the compressed liquid state is about 10 kJ/kg higher than that of the saturated liquid because the pressure is nearly 10 MPa higher than the saturation pressure. ■

2.11 SUMMARY

The first portion of this chapter dealt with definitions used to classify the properties of substances.

The state postulate was introduced, and it was used to establish the fact that the thermodynamic state of a simple compressible substance is completely specified by two independent, intensive thermodynamic properties.

Two physical properties, the kinetic energy and potential energy, were introduced as

$$E_k = mV^2/2 \tag{2.2}$$

and

$$E_p \equiv mgz \tag{2.4}$$

Then two thermodynamic properties were introduced: the internal energy and the enthalpy. The internal energy u was defined as the sum of all the microscopic forms of energy, and the enthalpy was defined as

$$h \equiv u + Pv \tag{2.12}$$

Equilibrium diagrams are useful in visualizing regions where single phases and various combinations of phases can exist in equilibrium. The triple line and the critical point are well-defined conditions that can be identified on the equilibrium diagram. The quality is an intensive thermodynamic property defined as the mass fraction of vapor in a liquid-vapor mixture, or

$$x \equiv \frac{m_g}{m_f + m_g} \tag{2.13}$$

The quality is used in the liquid-vapor saturation region to specify the composition of the mixture and to determine the mixture properties. Using Equation 2.14, for example, we have

$$u = u_f + xu_{fg}$$

and

$$h = h_f + xh_{fg}$$

The superheated-vapor states and compressed-liquid states were described, and property evaluations for these states were considered in connection with the use of tabulated thermodynamic properties. The constant-volume specific heat, defined as

$$c_v \equiv \left(\frac{\partial u}{\partial T}\right)_v \tag{2.17}$$

and the constant-pressure specific heat, defined as

$$c_p \equiv \left(\frac{\partial h}{\partial T}\right)_P \tag{2.21}$$

were shown to be helpful in relating the changes in internal energy and enthalpy to measurable thermodynamic properties.

An ideal gas was defined as a gas that obeys the equation of state

$$Pv = RT \tag{2.24}$$

and whose internal energy depends on temperature alone:

$$u = u(T) \tag{2.25}$$

The behavior of all gases approaches ideal-gas behavior when the pressure on the gas is much less than the critical pressure and the temperature of the gas is much greater than the critical temperature.

The internal energy, enthalpy, constant-volume specific heat, and constant-pressure specific heat of an ideal gas are each functions of temperature only. The internal-energy and the enthalpy changes of an ideal gas between state 1 and state 2 are given by

$$u_2 - u_1 = \int_{T_1}^{T_2} c_v(T)\, dT \tag{2.29}$$

and

$$h_2 - h_1 = \int_{T_1}^{T_2} c_p(T)\, dT \tag{2.32}$$

If the specific heats of the ideal gas vary linearly with temperature, these expressions reduce to

$$u_2 - u_1 = c_{v,\text{avg}}(T_2 - T_1) \tag{2.37}$$

and

$$h_2 - h_1 = c_{p,\text{avg}}(T_2 - T_1) \tag{2.36}$$

where the specific heats are evaluated at the average temperature during the process $(T_1 + T_2)/2$.

The assumption of constant specific heats can be made with little error for monatomic gases such as argon, helium, and neon. Tabular values of internal energy and enthalpy for several gases are included in the appendixes, and these values account for the variation in the specific heats with temperature.

If the specific heats of an ideal gas are constant, then the changes in internal energy and enthalpy are

$$u_2 - u_1 = c_v(T_2 - T_1) \tag{2.40}$$

$$h_2 - h_1 = c_p(T_2 - T_1) \tag{2.41}$$

Incompressible substances are those substances whose density (or specific volume) remains constant regardless of changes in other properties. Changes in internal energy and enthalpy for an incompressible substance are determined from

$$u_2 - u_1 = c_{\text{avg}}(T_2 - T_1) \tag{2.52}$$

and

$$h_2 - h_1 = u_2 - u_1 + v(P_2 - P_1) \tag{2.51}$$

PROBLEMS

2.1 Briefly define the following terms:
- **(a)** simple compressible substance
- **(b)** critical point
- **(c)** saturated conditions
- **(d)** triple point
- **(e)** compressed liquid
- **(f)** ideal gas
- **(g)** superheated vapor
- **(h)** phase

2.2 Describe the difference between the terms in each of the following sets
- **(a)** macroscopic and microscopic
- **(b)** extensive and intensive
- **(c)** physical properties and thermodynamic properties

2.3 State whether the following properties are intensive or extensive: volume, mass, density, weight, kinetic energy, specific volume, and potential energy.

2.4 Calculate the kinetic energy of a bullet with a mass of 50 g traveling at a velocity of 350 m/s.

2.5 Calculate the kinetic energy of a 2000-lb_m automobile traveling at a velocity of 60 mph.

2.6 Calculate the kinetic energy, in joules, of a jet aircraft with a mass of 80,000 kg and a velocity of 250 m/s. Compare this kinetic energy with the potential energy change of the plane as it climbs from sea level to an altitude of 12,000 m.

2.7 A substance contained in a closed system undergoes a change of state from P_1, v_1 to P_2, v_2 where $P_2 < P_1$ and $v_2 > v_1$ by two different paths consisting of internally reversible processes. Path *A* consists of a constant-pressure process followed by a constant-volume process. Path *B* consists of a constant-volume process followed by a constant-pressure process.

(a) The work for path *A* is (greater than, equal to, less than) zero.
(b) The work for path *B* is (greater than, equal to, less than) zero.
(c) The magnitude of the work for path *A* is (greater than, equal to, less than) the magnitude of the work for path B.
(d) The internal energy change for path *A* is (greater than, equal to, less than) that for path *B*.

2.8 Determine the missing properties in the accompanying table. Identify any state for which insufficient properties are given to determine the properties of the system.

System	*P*, MPa	*T*, °C	*v*, m³/kg	*u*, kJ/kg
H_2O	4.0		0.030	
H_2O	0.225	124		
Refrigerant-12	0.14	−20		

2.9 Determine the following properties for water:

(a) v in ft^3/lb_m at $x = 1.00$, $P = 200$ psia
(b) v in ft^3/lb_m at $T = 200°F$, $P = 100$ psia
(c) u in $ft \cdot lb_f/lb_m$ at $T = 350°F$, $x = 0.5$
(d) h in Btu/lb_m at $T = 600°F$, $P = 1000$ psia
(e) v in ft^3/lb_m at $P = 100$ psia, $u = 800$ Btu/lb_m
(f) h in Btu/lb_m at $P = 300$ psia, saturated liquid

2.10 State the phase or phases of water that may exist at the given states:

(a) $T = 215°C$, $P = 2.0$ MPa
(b) $T = 215°C$, $P = 2.2$ MPa
(c) $T = 240°C$, $x = 0.4$
(d) $T = 260°C$, $v = 0.40$ m³/kg
(e) $P = 30$ kPa, $T = 75°C$
(f) $P = 50$ kPa, $v = 2$ m³/kg

2.11 A tank contains 1 lb_m of liquid water and 0.1 lb_m of water vapor at 400°F. Determine the following properties:

(a) the quality of the water
(b) the total volume of the container
(c) the volume occupied by the liquid
(d) the pressure of the water

2.12 A container with a volume of 4 ft³ is filled with water at a pressure of 200 psia and a temperature of 800°F. Determine the mass of the water in the container.

2.13 [CDC311] A rigid tank with a volume of 0.0705 m³ contains 8 kg of water at 40°C. The tank and the water are heated until only water vapor remains. Determine the following:

(a) the final temperature in °C
(b) the final pressure in kPa
(c) the final density in kg/m³

2.14 Determine the phase(s) and specific volume of H_2O under the following conditions:

(a) $P = 1$ MPa, $T = 200°C$
(b) $x = 0.2$, $P = 100$ kPa
(c) $P = 2$ MPa, $T = 400°C$

2.15 Calculate the mass of water contained in a 0.6-ft^3 vessel under the following conditions:
(a) 15 psia, 300°F
(b) 100 psia, 500°F
(c) 20 psia, 210°F
(d) 60 psia, saturated vapor
(e) 60 psia, saturated liquid
(f) at the critical state

2.16 Determine the internal energy of water at 10 MPa and 450°C. Determine the enthalpy of water at a quality of 0.3 and a temperature of 300°C. Determine the specific volume of water at 5 MPa and 1000°C.

2.17 Determine the following properties of water for the given state points. Indicate any state for which insufficient information is given to find the requested property.
(a) h at the critical point
(b) u at $P = 250$ psia, $T = 401.04°F$
(c) v at saturated vapor, $T = 400°F$
(d) u at $T = 1200°F$, $P = 40$ psia
(e) h at $x = 0.5$, $T = 400°F$

2.18 Calculate the mass, in kilograms, of water contained in a 3-m^3 tank under the following conditions:
(a) saturated vapor, $T = 160°C$
(b) saturated liquid, $T = 160°C$
(c) $T = 600°C$, $P = 2$ MPa
(d) $T = 50°C$, $P = 101$ kPa
(e) $x = 0.5$, $P = 500$ kPa

2.19 A tank with a volume of 1 ft^3 is half-filled with liquid water; the remainder of the tank is occupied by water vapor. The pressure of the water is 200 psia. Calculate (a) the quality of the water, (b) the mass of the vapor and the mass of the liquid water, and (c) the specific volume and internal energy of the water.

2.20 Calculate the change in internal energy as water vapor follows a constant-pressure path at 30 psia from 600 to 1200°F.

2.21 Determine the specific volume of water at 1000 psia and 200°F.

2.22 Calculate the density and specific volume of water at 20°C and 100 kPa.

2.23 Quite often, liquid water is assumed to be an incompressible substance. Assume that water at 50°F and 20 psia is compressed along an isothermal path. Determine the percent change in density as the water is compressed to (a) 1000 psia, (b) 2000 psia, and (c) 5000 psia.

2.24 Determine the change in the enthalpy of water as it changes state from 20 MPa, 100°C to 30 MPa, 200°C.

2.25 Calculate the quality of refrigerant-12 for the following states:
(a) 40°F, 0.5 ft^3/lb$_m$
(b) 50 psia, $h = 40$ Btu/lb$_m$
(c) 0°F, 1.0 ft^3/lb$_m$

2.26 [CDC211] A rigid refrigeration tank with a 0.025-m^3 volume is evacuated and then slowly charged with refrigerant-12. During this process, the temperature of the refrigerant-12 remains constant at the ambient temperature of 20°C. Determine the following:
(a) the mass of refrigerant-12 in the system when the pressure reaches 250 kPa (state 1)
(b) the mass of refrigerant-12 in the system when the system is filled with saturated vapor (state 2)
(c) The fraction of the refrigerant-12 that will be vapor when 1.148 kg of refrigerant-12 has been placed in the system (state 3)

2.27 [CDC411] A sealed glass tube contains refrigerant-12 at 80°F (state 1). It is desired to measure the pressure without breaking the seal. If the tube is cooled to 50°F (state 2), liquid droplets begin to appear on the walls of the glass tube. Determine the pressure in psia when the temperature is 80°F.

2.28 Calculate the specific volume for refrigerant-12 at 400 kPa and a quality of 0.5.

2.29 Refrigerant-12 is placed in a rigid, closed container at a pressure of 120 psia and a temperature of 100°F. Heat transfer from the refrigerant to the surroundings occurs until the temperature drops to 60°F.

(a) Sketch the process on a P-v diagram.

(b) Calculate the work done during the process.

(c) Determine the initial and final values of specific volume and internal energy of the refrigerant-12.

(d) Calculate the quality of the refrigerant-12 at the final state.

2.30 A rigid tank has a volume of 1 m^3 and contains refrigerant-12 at 20°C. Originally, the tank is filled with one-third liquid and two-thirds vapor, by volume. The contents of the tank are heated until the temperature of the R-12 rises to 80°C. Determine the original pressure within the tank, the original mass of vapor, the original quality, and the final pressure within the tank.

2.31 Determine the temperature of water at the following states:

(a) $P = 0.50$ MPa, $v = 0.6173$ m^3/kg

(b) $P = 0.20$ MPa, $v = 0.6173$ m^3/kg

(c) $P = 0.85$ MPa, $v = 0.0011114$ m^3/kg

(d) $P = 1.00$ MPa, $v = 0.2453$ m^3/kg

2.32 For water at each of the following states, determine the saturation temperature for the given pressure. Also determine whether the temperature of the state specified is greater than, less than, or equal to that saturation temperature:

(a) $P = 0.4$ MPa, $v = 0.5$ m^3/kg

(b) $P = 10.0$ kPa, $v = 0.5$ m^3/kg

(c) $P = 0.1$ MPa, $u = 2600$ kJ/kg

2.33 A closed, rigid tank contains 10 lb_m water at a pressure of 60 psia and a quality of 0.92. The water is heated until its pressure has increased to 120 psia. Calculate the following quantities:

(a) the mass of liquid water in the tank at the initial state

(b) the mass of the water vapor in the tank at the initial state

(c) the initial temperature of the water in the tank

(d) the volume of the tank

(e) the enthalpy change of the water for the process

(f) the final temperature of the water in the tank

2.34 Refrigerant-12 is contained in a closed piston and cylinder assembly with the arrangement shown in the figure. There is no friction between the piston and cylinder walls. The piston has a mass of 4080 kg and the cross-sectional area of the piston is 0.4 m^2. The initial temperature and volume of the R-12 are 50°C and 2 m^3. The refrigerant is heated until its temperature reaches 80°C. Calculate the following quantities:

(a) the initial and final pressure of the R-12

(b) the mass of the R-12 in the cylinder

(c) the internal-energy change of the R-12 during the process

Problem 2.34

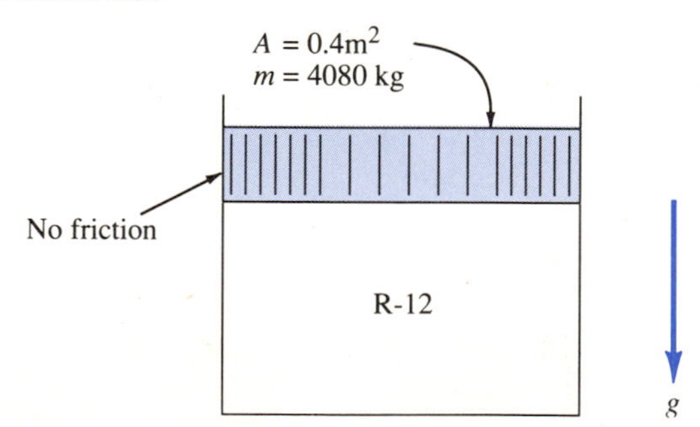

2.35 Water exists at a pressure of 20 MPa and a temperature of 80°C. Use the compressed-liquid table (Table B.4) and determine the specific volume, internal energy, and enthalpy of the water. Compare these answers with the ones calculated by assuming that the water has properties of a saturated liquid at 80°C.

2.36 A closed, flexible container initially contains 2 lb_m of saturated liquid water at a temperature of 450°F. The water expands in an internally reversible, constant-temperature process until the volume of the container reaches 2.0 ft^3. Sketch the P-v diagram for this process, determine the final pressure of the water, and determine the work for this process.

2.37 [CDC111] Water is contained in a cylinder fitted with a frictionless piston. The outside surface of the piston is exposed to an atmospheric pressure of 100 kPa. The mass of the water is 2 kg and the area of the piston is 0.2 m^2. At state 1, the water is at 110°C with a quality of 0.9 and a spring just touches the piston, but exerts no force on it. Then heat transfer to the water causes the piston to begin to rise. During the heating process, the resisting force of the spring is proportional to the distance moved, with a spring constant of 10.4 kN/m. Determine the following:

(a) the pressure at state 2 in kPa in the cylinder when the temperature reaches 250°C and the spring has been compressed 1.09 m

(b) the specific volume of the steam at states 1 and 2

2.38 Dry saturated steam at 400°F is heated in a constant-pressure process. Determine the amount of work per lb_m the steam performs on the surroundings if the final temperature of the steam is 900°F.

2.39 A piston-cylinder arrangement containing refrigerant-12 at 400 kPa and 80°C is compressed. While the compression process is taking place, the properties of the refrigerant-12 are known to obey the relation

$$Pv^{1.2} = \text{constant}$$

The compression process stops when the pressure on the refrigerant-12 reaches 1 MPa. Determine the amount of work required to compress 5 kg of refrigerant-12 in this manner.

2.40 Water is contained with a piston-cylinder arrangement. The water proceeds through a quasi-equilibrium, constant-pressure process from 80 psia, 900°F to 1200°F. Calculate the work per unit mass done by the water during the process.

2.41 Refrigerant-12 expands in a quasi-equilibrium process along a path such that

$$Pv = \text{constant}$$

The initial state is P = 200 psia, T = 260°F. The final pressure is 100 psia. Calculate the work done per lb_m of refrigerant.

2.42 Ten lb_m of water is contained in a closed piston and cylinder arrangement. The water is initially a saturated liquid at 320°F. Determine the amount of work done by the water when it changes frictionlessly and isothermally to a state of saturated vapor.

2.43 Saturated water vapor at a pressure of 1 MPa expands in a closed container along a constant-pressure path until the temperature reaches 400°C. Determine the work done per kilogram of water if the expansion process occurs frictionlessly.

2.44 One lb_m of water is compressed in a closed, frictionless piston and cylinder arrangement. The original volume and temperature of the water are 0.5 ft^3 and 520°F. The water is compressed isothermally until it becomes a saturated liquid. Draw the process on a P-v diagram. Calculate the work done on the water during the process.

2.45 Refrigerant-12 is initially at a quality of 0.50 and a temperature of 20°C. The refrigerant is compressed frictionlessly and isothermally in a closed system until it becomes a saturated liquid. Calculate the work done on a unit mass of the refrigerant.

2.46 Twenty kilograms of refrigerant-12 are enclosed in a flexible container such that the initial temperature is −30°C and the initial quality is 0.80. The container expands frictionlessly so that the refrigerant follows a constant-pressure path until the temperature of the refrigerant at the final state is 40°C. Determine the work required during the process. Draw the P-v diagram for the process.

2.47 Determine the work required per lb_m to isothermally and frictionlessly compress refrigerant-12 in a closed system from a saturated vapor to a saturated liquid at 20°F.

2.48 Determine the work provided by expanding 10 lb_m of water frictionlessly and isothermally in a closed system from a quality of 0.20 to a quality of 0.80 at 520°F.

2.49 A closed, flexible container initially encloses saturated liquid refrigerant-12 at a pressure of 120 psia and a volume of 3 ft^3. The refrigerant expands frictionlessly along a constant-pressure path until the volume of the container is 12 ft^3. Determine the work required for the given change of state.

2.50 A closed, flexible container initially contains 1 kg of saturated liquid water at a temperature of 160°C. The water expands along a constant-pressure path until the volume of the container reaches 1 m^3. Draw the process on a P-v diagram, calculate the final temperature of the water, and determine the work done by the water during the process.

2.51 Write a computer program to evaluate the work required to reduce the volume of 1 kg of water in a closed system initially at 350°C and 5 MPa to 15 percent of its initial volume in a quasi-equilibrium, isothermal process.

2.52 A closed system contains 0.15 kg of saturated water vapor initially at 205°C. The water undergoes the following sequence of internally reversible processes. The water is first cooled at constant volume until the temperature reaches 150°C. Then the water is further cooled at constant temperature (150°C) until the volume is reduced to one-half of the initial volume. Sketch this sequence of processes on a P-v diagram. Determine the final pressure of the water, the work for the first process, and the work for the second process.

2.53 A closed, flexible vessel contains 20 lb_m of water initially at 20 psia, 300°F. The water is compressed reversibly at constant pressure until the temperature reaches 500°F. Determine the initial and final volume of the container and the work done on the water during the process.

2.54 Water in a closed system proceeds through a reversible, constant-pressure path from $x_1 = 0.2$, $T_1 = 300°C$ to $v_2 = 0.06\ m^3/kg$. Determine the work per kg performed by the water during the process.

2.55 Starting with the equation $Pv = RT$, show that the ideal gas equation of state can be written in the following forms:

(a) $P\mathbf{V} = N\overline{R}T$
(b) $P\mathbf{V} = mRT$
(c) $Pv = \overline{R}T/M$
(d) $Pv = N\overline{R}T/m$

2.56 Ten moles of an ideal gas with a molecular weight of 85 and $k = 1.35$ exist at 300°F and 50 psia in a closed, rigid container. As a result of heat transfer with the surroundings the gas temperature drops to 130°F. Calculate the following quantities:

(a) the volume of the container
(b) the mass of the gas in the container
(c) the change in the enthalpy of the gas during the process
(d) the change in internal energy of the gas during the process

2.57 *Boyle's law* states that the volume occupied by a given mass of a gas varies inversely with the absolute pressure if the temperature of the gas is held constant. *Charles' law* states that the volume of a given mass of gas is directly proportional to the absolute temperature if the pressure is held constant. Show that both Boyle's and Charles' law are special cases of the general ideal gas equation of state.

2.58 A pressurized container shown in the figure is designed to prolong the life of tennis balls. Tennis balls are placed in the container and the lid is screwed on and the air surrounding the balls is thereby pressurized. One rotation of the lid advances the lid by 0.3 in. Significant aging of a tennis ball ceases if the ball is maintained at an external pressure of 10 lb_f/in^2 above atmospheric pressure. Estimate the number of turns through which the lid of the container must be rotated in order to preserve the life of tennis balls stored inside the container.

Problem 2.58

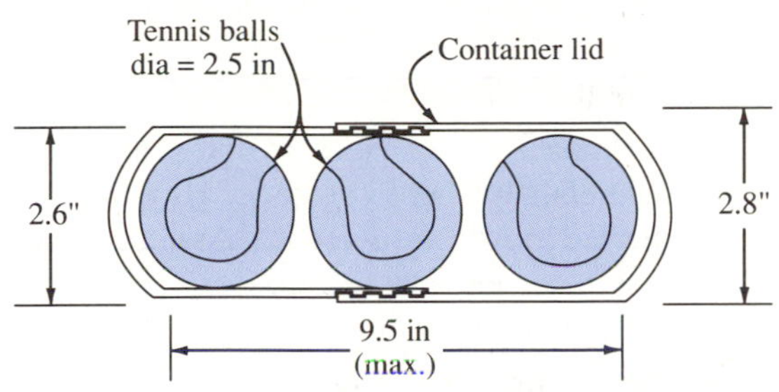

2.59 Air is compressed in a closed system from a pressure of 100 kPa and a temperature of 25°C to a pressure of 600 kPa and a temperature of 150°C. The initial volume of the air is 2 m^3. Calculate the final volume of the air.

2.60 Nitrogen gas is stored in a cylindrical tank having an internal diameter of 45 cm and a length of 1.6 m. The maximum allowable pressure and temperature are 2 MPa and 75°C. How many kilogram moles and kilograms of nitrogen can be safely stored in the tank?

2.61 Suppose 4 $lb_m \cdot$ mol of CO is contained in a vessel with a volume of 30 ft^3. What is the gauge pressure of the CO if its temperature is 80°F? Atmospheric pressure is 14.7 psi.

2.62 An automobile tire is inflated to a pressure of 30 psi according to a tire gauge. Determine the mass of the air in the tire if it has a volume of 1.4 ft^3 and a temperature of 110°F.

2.63 A fixed mass of air is compressed from a volume of 1 m^3 at 30°C, 101 kPa pressure, to 500 kPa and 100°C. Calculate the mass of the air and its final volume.

2.64 A rigid pressure vessel contains 0.6 kg of an ideal gas at a pressure of 70 kPa and a temperature of 15°C. After more of the same gas has been added to the vessel, the pressure and temperature are 200 kPa and 35°C. Determine the mass of gas added to the vessel.

2.65 Archimedes' principle states that the buoyancy force of a submerged object is equal to the weight of the displaced fluid. Suppose that a spherical balloon 35 ft in diameter is filled with helium at a gauge pressure of 10 psi and a temperature of 80°F. Calculate the net lifting capacity of the balloon when immersed in air at 14.7 psia, 70°F. The balloon material weighs 70 lb_f.

2.66 A balloon contains 6000 m^3 of helium at 28°C and 760 mmHg just before launching. Calculate the mass of helium in the balloon. What is the volume of the balloon if it rises to an altitude where the helium pressure is 14 kPa and the temperature is −15°C?

2.67 [CDB211] A rigid vessel is connected by a closed valve to a spherical elastic balloon. Both contain dry air at ambient temperature, 300 K. The volume of the rigid vessel is 0.03 m^3 and the initial air pressure is 275 kPa (state 1). The initial radius of the balloon is 0.15 m and its pressure is 100 kPa (state 2). The valve is opened and air flows from the vessel to the balloon until, after some time, the pressure and temperature are uniform in the vessel and the balloon. The final radius of the balloon is 0.21 m. Denote this final state of the air in the balloon and vessel as state 3. Assume that the pressure in the balloon is directly proportional to its radius. Neglect the air contained in the valve and in the lines.

Heat transfer to the vessel and/or balloon may occur. Determine the following:

(a) the final volume of the balloon in m^3

(b) the final temperature in K and pressure in kPa of the air at state 3

2.68 A rigid tank with a volume of 2.3 ft^3 contains an ideal gas having a molecular weight of 27. The tank contains 0.2 $lb_m \cdot$ mol of the gas at a temperature of 140°F. Calculate the pressure of the gas. Heat transfer from the gas to the surroundings causes the temperature of the gas to drop to 40°F. Calculate the pressure of the gas at this temperature.

2.69 A 35-m^3 tank contains 5 kg · mol of an ideal gas with a molecular weight of 19. The pressure of the gas is 360 kPa. Determine the temperature of the gas. Determine the mass of the gas, in kilograms, that must be removed from the tank so that the pressure is reduced to 100 kPa at the same temperature that the gas had when its pressure was 360 kPa.

2.70 The average adult breathes in 30 in^3 of air with each breath. Estimate the mass of air that each hour enters the lungs of an adult located at sea level (14.7 psia, 72°F), assuming the person takes 25 breaths per minute. How rapidly does a person have to breathe on top of a 15,000-ft mountain, where the pressure is 11 psia and the temperature is −20°F, if the person is to take in the same mass of air as was taken in at sea level?

2.71 Some automobiles are equipped with collapsible spare tires and rigid pressurized containers for inflating the tire. The rigid container is filled with air at the factory to a pressure of 80 psia at a temperature of 80°F. The burst strength of the container is 400 psia. The container is placed in the trunk of a car where the air temperature reaches 130°F during the summer. Is there any danger of the container exploding, assuming that a pressure safety factor of 5 must be applied to the burst pressure of the container to ensure safe pressure levels?

2.72 Determine the mass of air in a pressure vessel that has a volume of 2.6 m^3, a pressure of 6 MPa, and a temperature of 200°C.

2.73 Calculate the change in internal energy of CO_2 as it changes state from 20 to 700°C, assuming that the specific heats of CO_2 vary with temperature. Compare your answer with the change in internal energy calculated when c_v is assumed constant and evaluated at the average temperature. Assume ideal-gas behavior.

2.74 Calculate the changes in internal energy and enthalpy of an ideal gas when it changes state from 40 to 100°F and properties of the gas are $k = 1.4$, $R = 0.05$ Btu/$lb_m \cdot$ °R.

2.75 Calculate the changes in internal energy and enthalpy per kilogram of H_2 when it is heated from 20 to 100°C. Assume ideal-gas behavior, but account for variation in specific heats with temperature.

2.76 Assuming constant specific heats and ideal-gas behavior, calculate the changes in internal energy and enthalpy per lb_m when helium changes from 100 to 1200°F.

2.77 Assuming constant specific heats and ideal-gas behavior, calculate the changes in enthalpy and internal energy per lb_m as a gas ($R = 500$ ft · $lb_f/lb_m \cdot$ °R) changes state from 700 to 1200°F.

2.78 Calculate the enthalpy and the internal-energy changes for CO_2 and N_2 between 400 and 800 K, assuming ideal-gas behavior and (a) constant specific heats, (b) specific heats as functions of temperature.

2.79 Suppose 6 lb_m of CO_2 at 60 psia changes state from 420 to 560°F. Determine the changes in enthalpy and internal energy for the process.

2.80 Calculate the change in enthalpy per unit mass of O_2 as it changes from 200 kPa, 17°C to 400 kPa, 37°C. Assume constant specific heats.

2.81 Water exists at a pressure of 450 psia and a temperature of 600°F. Determine the specific volume of the water (a) by assuming the water is an ideal gas and (b) by using tabulated data. Which answer is more accurate?

2.82 Determine the percent error in the calculated value for the specific volume that results if water at a pressure of 200 kPa and a temperature 130°C is assumed to be an ideal gas.

2.83 Water vapor at 500 kPa and 500°C follows a constant-pressure process until its temperature increases to 800°C. Determine the enthalpy and internal-energy changes that occur during the process, using two different sources of property data: (a) use tabular property data; (b) assume ideal-gas behavior and use property values from Table D.7. Which answer is more accurate?

2.84 Calculate the specific volume for the following substances at the given states:

(a) water at a quality of 0.5, 400°F
(b) refrigerant-12 at 30 psia, 120°F
(c) nitrogen at 30 psia, 600°F
(d) air at 5 psig, 400°F
(e) CO_2 at -3 psig, 250°F

2.85 Fill in the properties listed in the accompanying table.

Substance	*P*, kPa	*T*, °C	*v*, cm³/g
Air	100	300	
CO_2	50	200	
H_2O	300	240	
H_2O		320	85
N_2	50		2000

2.86 Determine the following properties at the given states:

(a) x for refrigerant-12; $T = 0°C$, $v = 0.03\ m^3/kg$
(b) v for refrigerant-12 at $T = 30°C$, $P = 100$ kPa
(c) u for oxygen at $T = 400$ K, $P = 100$ kPa
(d) v for nitrogen at $P = 100$ kPa, $T = 300°C$
(e) v for carbon dioxide at $P = 100$ kPa, $T = 400°C$
(f) h for hydrogen at $P = 300$ kPa, $T = 600°C$

2.87 Determine the internal energy, enthalpy, and specific volume of hydrogen at a pressure of 80 psia and a temperature of 360°F.

2.88 Carbon monoxide at a low pressure changes temperature from 300 to 600 K. Determine the changes in internal energy and enthalpy for this process.

2.89 Determine the specified properties at the given states:

(a) v for air at 450 psia and 1000°F
(b) u for N_2 at a low pressure and 1000°R
(c) h for CO_2 at 20 psia and 750°R
(d) u for air at 30 psia and 750°R

2.90 Suppose you wish to investigate the errors resulting from the assumption of constant specific heats when calculating the enthalpy change of an ideal gas. Write a computer program that will calculate the enthalpy change of CO_2 at a low pressure between 300 K and 3000 K in equal intervals of 100 K. Determine the enthalpy change per kg for three different cases:

(a) constant specific heat evaluated at the original temperature
(b) constant specific heat evaluated at the average temperature during the process
(c) variable specific heats using the $c_p(T)$ expression in Table D.11

2.91 A closed, rigid tank contains 2 lb_m of liquid water and 0.5 lb_m of water vapor at 350°F. Determine the following properties of the water at this state:

(a) quality
(b) volume of the tank
(c) volume occupied by the liquid
(d) pressure of the water
(e) enthalpy of the water

2.92 Determine the most accurate values for the properties for the following substances at the given states:

(a) H_2O, $P = 1.0$ MPa, $h = 2000$ kJ/kg: T, u, v

(b) refrigerant-12, $T = 50°C$, $P = 0.25$ MPa: h, u, v

(c) Air, $P = 200$ kPa, $T = 527°C$: u, h, v

(d) H_2, $T = 407°C$, $v = 8$ m^3/kg: P, h, u

2.93 Determine the specific enthalpy and specific internal energy changes as accurately as possible when a substance changes from 600°F to 800°F along a constant-pressure path at 20 psia, if the substance is:

(a) water

(b) air

(c) argon

2.94 A plastic bag contains frozen foods. The bag is removed from the freezer at 20°F and placed in a microwave oven. At that point the bag is fully expanded and the vapor space surrounding the frozen foods has a volume of 1 in^3 and a pressure of 14.7 psia. The burst strength of the bag is estimated to be 20 psia. Approximate the temperature of the vapor in the bag which will break the bag, assuming the properties of the vapor in the bag are similar to those of air.

2.95 Canned goods are cooked and placed in a tin can as shown in the figure. The lid is placed on the can at a point in the processing line where the canned goods have a temperature of approximately 160°F. The contents of the can cool until they reach a temperature of 70°F. At that time estimate the mass of vapor and the pressure in the vapor space at the top of the can if the vapor in the can has properties similar to air.

Problem 2.95

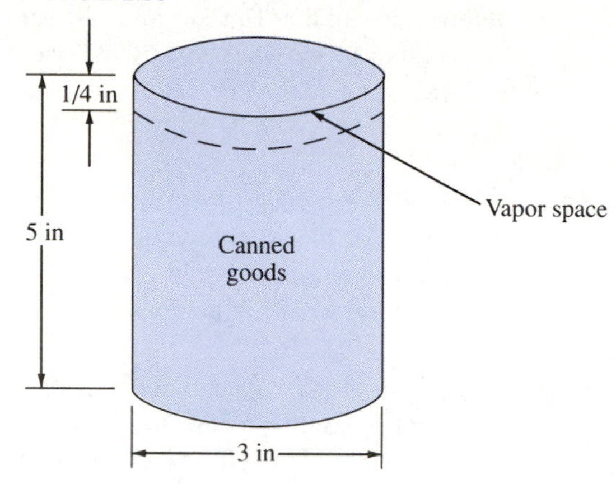

2.96 A closed, flexible vessel contains nitrogen at 150 kPa and 90°C. The nitrogen is compressed along the path Pv^2 = constant until the volume of the nitrogen is reduced to one-half of its initial value. Determine the work done on the nitrogen during the process per kg of N_2 and the change in specific volume of the nitrogen.

2.97 Air in a closed system proceeds through a reversible, constant-temperature process from $v_1 = 0.25$ m^3/kg, $P_1 = 350$ kPa to $P_2 = 35$ kPa. Determine the work per kg performed by the air during the process.

2.98 [CDB111] Air is contained in a vertical cylinder fitted with a frictionless piston and a set of stops. The cross-sectional area of the piston is 0.5 ft^2. At state 1, the initial state, the air is at 800°F and 30 psia and the piston is 4 ft above the cylinder bottom. The stops are positioned 2 ft above the cylinder bottom. The air is cooled by heat transfer to the surroundings. Determine the following:

(a) the mass of air in the cylinder in lb_m

(b) the volume of the air at state 1 in ft^3

(c) the air temperature in °F at state 2 when the piston just reaches the stops

(d) the pressure at state 3 in psia when the air temperature is 70°F

(e) the work for process 1-2

Problem 2.98

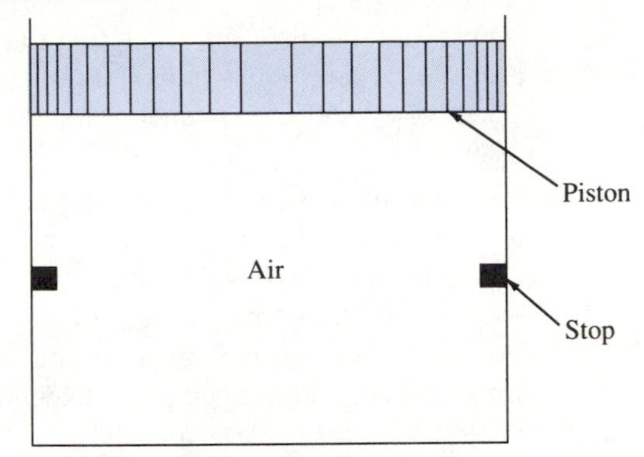

2.99 A closed system contains a substance whose volume is reduced to one-half of its initial volume during an isothermal process.

(a) The substance is water initially in a saturated vapor state at 200°C.

(1) Sketch the P-v diagram for this process.

(2) The work for this process is (greater than, equal to, less than) zero.

(3) The internal energy change of the water for this process is (greater than, equal to, less than) zero.

(b) The substance is air initially at 30°C and 400 kPa

(1) Sketch the P-v diagram for this process.

(2) The work for this process is (greater than, equal to, less than) zero.

(3) The internal energy change of the air for this process is (greater than, equal to, less than) zero.

2.100 An ideal gas initially in a piston-cylinder at 150 kPa, 50°C, and 0.03 m^3 is first heated at constant pressure until the volume doubles. It is then allowed to expand isothermally until the volume doubles again.

(a) Determine the total work done by the gas, in newton-meters.

(b) Determine the pressure of the gas at the final state, in kilopascals.

(c) Determine the final temperature of the gas, in degrees Celsius.

2.101 A piston and cylinder contains 5 lb_m of gaseous methane (CH_4) at 370°F and 15 psia. The methane is compressed slowly so that its temperature remains constant. Calculate the amount of work required to compress the methane to a pressure of 180 psia.

2.102 Air is enclosed in a flexible container at a pressure of 200 kPa and a temperature of 50°C. The air expands isothermally along a quasi-equilibrium path until the pressure reaches 100 kPa. Calculate the work performed per kilogram of air during the process.

2.103 A balloon contains 2 lb_m of methane at 200°F and 15 psia. The balloon is compressed slowly so that the temperature of the methane remains constant. Determine the amount of work required to compress the methane to a pressure of 65 psia.

2.104 Two kilograms of an ideal gas having constant specific heats begin a process at 200 kPa and 60°C. The gas is first expanded at constant pressure until its volume doubles. Then it is heated at constant volume until its pressure doubles. Properties of the gas are $R = 0.6$ kJ/kg · K and $c_p = 2.0$ kJ/kg · K. Calculate the work done by the gas during the entire process and the change in internal energy of the gas during the entire process.

2.105 One gram of air undergoes a process at constant pressure from 3 kPa, 200°C in a closed system. The final temperature of the air is 435°C. Calculate the following quantities:

(a) the initial and final volumes of the air

(b) the work done by the air during the process, assuming an equilibrium expansion

(c) the change in internal energy of the air, assuming constant specific heats

2.106 Helium is compressed frictionlessly and isothermally in a closed system from 50°C, 200 kPa to 500 kPa. Determine the work per unit mass required to compress the helium between these two states.

2.107 Carbon monoxide is compressed frictionlessly and isothermally in a closed system from 280°F, 60 psia to a pressure of 190 psia. The original volume of the carbon monoxide is 1.3 ft^3. Calculate the work required for the compression process.

2.108 Air is contained in a flexible, closed container. The original volume of the container is 500 mm^3 and the original state of the air is 100°C, 300 kPa. The air is compressed along a frictionless, constant-pressure path until its volume is reduced to one-half the original volume. Determine the final temperature of the air and the work required during the process.

2.109 Air is compressed polytropically in a closed system from 25 psia, 60°F to 90 psia. The polytropic exponent for the process is 1.28. Determine the work per unit mass of air required for the compression process.

2.110 Compare the work required to compress air in a closed system from a pressure of 45 psia and a temperature of 100°F with a final pressure of 320 psia for the following processes:

(a) a polytropic process with $n = 1.4$
(b) a polytropic process with $n = 1.2$
(c) an isothermal process

2.111 Write a computer program to analyze the following cycle: Process 1-2, Pv^n = constant; process 2-3, v = constant; process 3-4, Pv^n = constant; process 4-1, v = constant. The working fluid is air, the system is closed, and the polytropic exponent is (a) $n = 1$ for cycle A and (b) $n = 1.4$ for cycle B. Sketch the cycle processes and plot the net work for a maximum cycle temperature of 900°C and a range of compression ratios (v_1/v_2) from 2.0 to 12.0 for cycles A and B. At state 1, $P_1 = 110$ kPa and $T_1 = 50$°C. Assume ideal-gas behavior.

2.112 Write a computer program to analyze the following cycle: Process 1-2, Pv^n = constant; process 2-3, P = constant; process 3-4, Pv^n = constant; process 4-1, P = constant. The working fluid is air, the system is closed, and the polytropic exponent is (a) $n = 1$ for cycle A and (b) $n = 1.4$ for cycle B. Sketch the cycle processes and plot the net work for a maximum cycle temperature of 1200°C and a range of pressure ratios (P_2/P_1) from 2.0 to 20.0 for cycles A and B. At state 1, P_1 = 100 kPa and T_1 = 37°C. Assume ideal-gas behavior.

2.113 Five lb$_m$ of air is placed in a closed, frictionless piston and cylinder arrangement. It expands along a polytropic path for which the polytropic exponent is 1.8. The initial state of the air is given by $P_1 = 160$ psia, $T_1 = 1400$°F. After the expansion process the pressure has dropped to 20 psia. For this process calculate the following information:

(a) the initial volume of the air
(b) the final volume of the air
(c) the work performed by the air during the process
(d) Sketch the path of the process on a Pv diagram.

2.114 Calculate the work performed by an ideal gas (R = 30 kJ/kg · K) when it expands reversibly and isothermally in a closed system from state 1 to state 2. The following properties are known: P_1 = 20 MPa, V_1 = 10 m^3, T_1 = 300°C, P_2 = 0.2 MPa. Also calculate the specific volume and temperature of the gas at the final state.

2.115 Calculate the work done by an ideal gas with molecular weight of 50 that expands in a closed system along the polytropic path Pv^3 = constant from a temperature of 350°F, a volume of 3 ft^3, and a pressure of 500 psia to a volume of 9 ft^3. Express the work in units of Btu and ft · lb$_f$. Calculate the temperature of the gas at the final state.

2.116 Determine the work required to compress 2 lb$_m$ of hydrogen in a closed system from 150°F, 10 psia to 50 psia along a frictionless isothermal path.

2.117 Three kilograms of CO_2 is placed within a closed system at 300 kPa and 50°C. The carbon dioxide is compressed along a polytropic path until the pressure reaches 1 MPa. The polytropic exponent is 1.45. Calculate the final temperature and volume of the carbon dioxide and the work during the process.

2.118 Air expands in a closed piston-cylinder arrangement along a polytropic process with n = 1.25. Determine the work performed on a unit mass of air if the initial pressure and temperature are 35 psia and 185°F and the final pressure is 195 psia.

2.119 Helium is placed in a flexible, closed container at 200 kPa and 40°C. The helium is compressed so that it follows a polytropic process with n = 1.2 until the temperature reaches 135°C. The initial volume of the container is 0.45 m^3. Determine the final pressure of the helium and the work performed during the process.

2.120 Hydrogen is placed in a closed, flexible container at 20 psia and 60°F. It is compressed along a polytropic path with a polytropic exponent equal to 1.38. The container can withstand a maximum pressure of 100 psia. Determine the maximum amount of work per lb$_m$ that can be performed on the hydrogen during the compression process.

2.121 A closed piston-cylinder arrangement has an initial internal volume of 0.09 ft^3. The initial state of the air inside the cylinder is 300 psia, 1400°R. The air expands along a polytropic process (n = 1.45) until the pressure reaches 50 psia. Calculate the final temperature of the air and the work done by the air during the process.

2.122 Three kilograms of nitrogen is compressed in a closed system along a polytropic path. The initial state of the gas is 100 kPa, 300 K and the gas is compressed until its volume is reduced to one-third its original value. Write a computer program that will calculate the final gas pressure, temperature, and volume as well as the work required to compress the nitrogen during the process. Calculate those values for a polytropic exponent between 0.2 and 1.4 in increments of 0.1.

2.123 Two lb$_m$ of CO_2 is placed within a closed system at 120 psia and 180°F. The gas is compressed in a polytropic process until the temperature reaches 390°F. The polytropic exponent is 1.45. Calculate the final pressure and final volume of the gas and the work for the process.

2.124 Nitrogen is contained in a closed system at an initial temperature of 125°C and an initial pressure of 2 MPa. The nitrogen undergoes an internally reversible, polytropic process until the pressure reaches 0.4 MPa. The polytropic exponent is 1.25. Sketch this process on a P-v diagram. Determine the work per unit mass for this process and the change in internal energy per unit mass of the nitrogen.

2.125 Solid copper has a density of 8930 kg/m^3. The state of copper changes from 100 kPa, 20°C to 3000 kPa, 200°C. Calculate the changes in internal energy and enthalpy per kilogram of copper.

2.126 Water at 20 psia and 70°F has its state changed to 150 psia and 120°F. Calculate the changes in internal energy and enthalpy per lb$_m$ of water, assuming that water is an incompressible substance.

2.127 Solid aluminum with a density of 169 lb$_m$/ft^3 is compressed from a state of 15 psia, 50°F to 1000 psia, 200°F. Calculate the changes in internal energy and enthalpy per lb_m of aluminum.

2.128 Suppose 7 kg of liquid H_2O is compressed from 100 kPa, 20°C to 30 MPa, 80°C. Calculate the changes in internal energy and enthalpy of the H_2O, assuming the H_2O is incompressible.

2.129 Solid copper with a density of 557 lb_m/ft^3 is compressed from 25 psia, 70°F to 5000 psia, 300°F. Calculate the internal-energy and enthalpy changes per lb_m for this process.

2.130 Calculate the internal energy and enthalpy change of 10 kg of a metal that has a density of 2000 kg/m^3 and a specific heat of 2 kJ/kg · K when its state changes from $P_1 = 1$ MPa, $T_1 = 200$°C to $P_2 = 100$ MPa, $T_2 = 500$°C.

2.131 Calculate the change in enthalpy and internal energy on a per unit mass basis for the following substances and the given changes in state.

(a) water
state 1: $x = 0.5$, $P = 200$ kPa
state 2: $T = 50$°C, $P = 100$ kPa

(b) carbon dioxide
state 1: $P = 200$ kPa, $T = 927$°C
state 2: $P = 100$ kPa, $T = 27$°C

(c) an ideal gas with $R = 35$ N · m/kg · K, $k = 1.35$
state 1: $P = 200$ kPa, $T = 500$°C
state 2: $P = 100$ kPa, $T = 100$°C

(d) steel with $\rho = 500$ lb_m/ft^3, $c = 0.09$ Btu/lb_m · °F
state 1: $P = 5000$ lb_f/in^2, $T = 50$°F
state 2: $P = 100$ lb_f/in^2, $T = 1000$°F

3 Conservation of Mass

3.1 INTRODUCTION

Two of the most important principles developed in an introductory course in thermodynamics are the conservation of mass and the conservation of energy. A thorough knowledge of both is essential to an understanding of fundamental thermodynamics, because the application of the conservation-of-mass and -energy equations is necessary for practically all thermodynamic analyses.

A common set of concepts and techniques is used to develop mathematical statements of these two basic conservation principles. In this chapter a general equation for the conservation of mass is developed first. Its application to various thermodynamic systems is discussed and illustrated with examples. A similar treatment of the conservation-of-energy principle follows in the next chapter.

3.2 GENERAL CONSERVATION-OF-MASS EQUATION

The general principle of conservation of mass is quite simple. It states the following:

> **Mass is a conserved property. It can be neither created nor destroyed; only its composition can be altered from one form to another.**

This statement of the conservation of mass is technically correct for practically all problems encountered in engineering thermodynamics. However, there is an equivalence between mass and energy given by the famous equation proposed by Einstein, $E = mc^2$. For all energy reactions, with the exception of nuclear reactions, the amount of mass converted to energy is extremely small, and it can be neglected for all practical purposes.

In this book discussion is limited to processes in which the conversion of mass to energy can be neglected, and the statement of conservation of mass given above can be used without any significant loss in accuracy.

The statement of conservation of mass simply means that the mass of a system must always be accounted for, or conserved. This principle is taught early in an engineering education. Students in chemistry, for example, learn that whenever a chemical reaction occurs, the mass of products and reactants must be equal in order to properly balance a chemical equation.

Our simple statement of the conservation-of-mass principle alludes to the fact that the composition of the mass may be altered during a process. In a chemical reaction, the chemical composition prior to the reaction will differ from the composition of the mass after the reaction occurs. Even though there is a change in the chemical composition of the system, the mass of the system still must be conserved. For example, suppose that a closed system initially contains a mixture of hydrogen and oxygen, and that a spark ignites the mixture so that it reacts chemically, producing water. Since the system is closed, the mass of the water present after the reaction equals the sum of the mass of the hydrogen and the oxygen before the reaction. Furthermore, the mass of each chemical species is also conserved. That is, the mass of hydrogen and the mass of oxygen initially in the system are equal, respectively, to the mass of hydrogen and the mass of oxygen chemically combined in the water at the final state. The conservation of mass applied to chemical reactions is discussed in further detail in Chapter 12.

In the next several chapters only processes for which there are no chemical reactions are considered. If no chemical reactions occur during a process, the composition of the mass at the end of the process will be chemically identical to the composition at the beginning of the process. Thus, for nonreacting systems the conservation-of-mass principle can be simplified because changes in chemical composition do not occur.

The conservation-of-mass equation derived in this section is expressed in a very general form, and it can be applied to a wide variety of problems. In subsequent sections of the chapter special limiting cases are considered that lead to simplifications of the general form of the conservation-of-mass equation. This same approach is followed in the derivation and application of the conservation-of-energy equation in the next chapter.

The previous statement of the conservation-of-mass principle can be developed into a word equation that is consistent with that statement. Applying the conservation of mass to a thermodynamic system results in the following expression:

$$\begin{bmatrix}\text{Rate at which mass enters}\\ \text{at the boundary of the system}\end{bmatrix} - \begin{bmatrix}\text{rate at which mass leaves at}\\ \text{the boundary of the system}\end{bmatrix} = \begin{bmatrix}\text{time rate of change of the}\\ \text{mass within the system}\end{bmatrix} \tag{3.1}$$

Equation 3.1 is simply another way of stating that mass is conserved, since any difference between the amount of mass entering and leaving a system must cause a corresponding change in the amount of mass inside the system.

To illustrate the application of the conservation of mass, consider the operation of an air compressor. Suppose air enters the compressor at a mass flow rate of 1 kg/s. Assuming the operation of the compressor is steady (i.e., air is not accumulated or stored

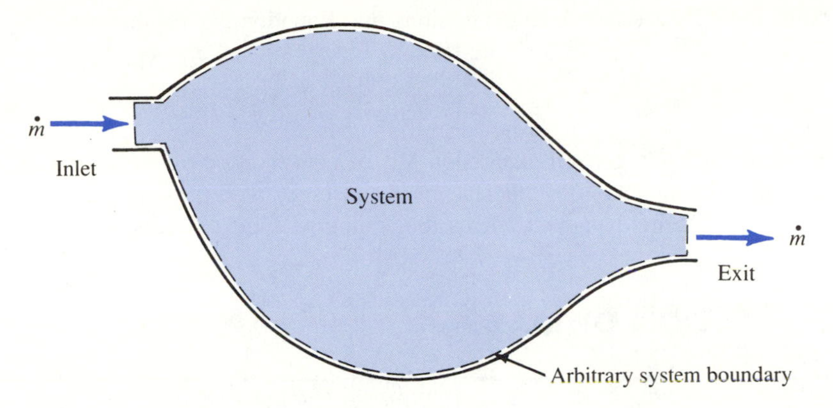

Figure 3.1 General system used for development of conservation-of-mass equation.

in the compressor), then conservation of mass requires that the rate at which the mass of air leaves the compressor must also be 1 kg/s.

Systems that exhibit unsteady, or *transient*, behavior will also be encountered. A simple example of transient conditions is the flow of water into a container, such as a bucket, where the water in the bucket is assumed to be the system. If the container is originally empty when the water flow is first started, the rate of mass accumulation within the system equals the mass flow rate of water entering the system. Eventually, the container will fill, and it will begin to overflow. Under these conditions the principle of mass conservation states that the rate of mass flowing over the sides of the container must equal the rate of mass flow into the container. Also, the rate of accumulation of mass inside the container must equal zero. At this time the rates of mass flow into and out of the system are equal.

Equation 3.1 can be easily converted into a mathematical expression for the conservation of mass for a general system such as the one shown in Figure 3.1. The time rate of change of the mass within the system is

$$\frac{dm_{\text{sys}}}{dt}$$

The total *mass flow rate* entering at the boundary of the system is

$$\sum_{\text{inlet}} \dot{m}$$

where the summation sign indicates that the mass flow rate $\dot{m}$ at each inlet area must be included.* Similarly, the total mass flow rate leaving at the boundary of the system is

$$\sum_{\text{exit}} \dot{m}$$

*The dot over a symbol, as in $\dot{m}$, is used throughout the text to indicate a quantity per unit time. In this case $\dot{m}$ represents mass per unit time or mass flow rate.

Substituting these last three expressions into Equation 3.1 results in

$$\sum_{\text{inlet}} \dot{m} - \sum_{\text{exit}} \dot{m} = \frac{dm_{\text{sys}}}{dt} \tag{3.2}$$

Equation 3.2 is a general expression for the conservation of mass for a thermodynamic system. The remainder of this chapter is concerned with the application of Equation 3.2 to specific thermodynamic systems, beginning with the closed system.

3.3 CONSERVATION OF MASS FOR CLOSED SYSTEMS

Many problems of engineering importance involve a system that has no mass crossing its surface area or boundary. Such a system was defined in Chapter 1 to be a closed system. Thus for a closed system the conservation-of-mass equation, Equation 3.2, simplifies to

$$\frac{dm_{\text{sys}}}{dt} = 0 \tag{3.3}$$

because no mass crosses the surface area of a closed system. Equation 3.3 states that the amount of mass within a closed system does not change with respect to time, or

$$m_{\text{sys}} = \text{constant} \tag{3.4}$$

The following three examples illustrate the conservation of mass applied to closed systems.

■ EXAMPLE 3.1

An automobile tire is inflated to a pressure of 30 psig at a temperature of 70°F. After a trip the temperature of the air in the tire increases to 120°F, while the volume of the tire increases by 2 percent owing to stretching of the tire material. Calculate the air pressure in the tire after the trip.

Solution. To begin the solution we draw a sketch of the tire and identify the system, as shown in Figure 3.2. Assuming that the air inside the tire is the thermodynamic system, as shown in the figure, no mass crosses the surface area of the system. Notice that the pressure of the air is much lower than its critical pressure, and its temperature is higher than the critical temperature. Thus the air can be assumed to be an ideal gas for the given conditions, and the equation of state for the system is

$$PV = mRT$$

If subscripts 1 and 2 are used to denote initial and final states, then the equation of state at both states can be written as

$$P_1 V_1 = m_1 R T_1$$
$$P_2 V_2 = m_2 R T_2$$

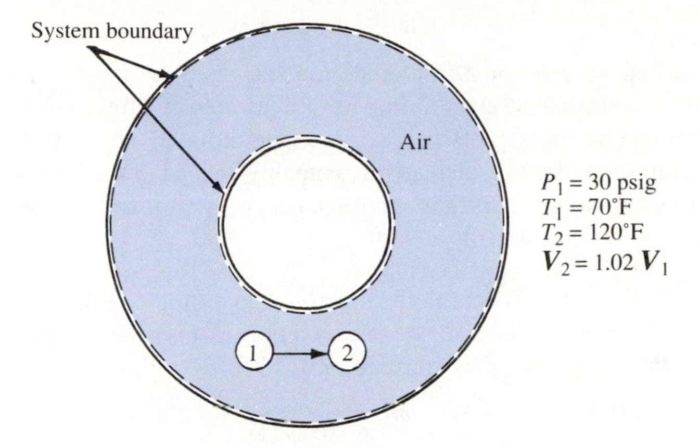

Figure 3.2 System for Example 3.1.

Applying the closed-system form of the conservation-of-mass equation (Equation 3.4), then

$$m_1 = m_2$$

and therefore

$$\frac{P_1 \mathbf{V}_1}{T_1} = \frac{P_2 \mathbf{V}_2}{T_2}$$

Notice that when the equations are combined in the form of a ratio, the need to solve for intermediate quantities that are unnecessary for the ultimate solution is eliminated. In addition, manipulating equations in the form of ratios helps to minimize problems that arise from the use of inconsistent units.

From the given information

$$\mathbf{V}_2 = 1.02\ \mathbf{V}_1$$

so solving for P_2 gives

$$P_2 = \frac{P_1}{1.02}\left(\frac{T_2}{T_1}\right)$$

If we assume that the atmospheric pressure is 14.7 $\text{lb}_\text{f}/\text{in}^2$, then the initial absolute pressure is 44.7 $\text{lb}_\text{f}/\text{in}^2$, and

$$P_2 = \frac{44.7\ \text{lb}_\text{f}/\text{in}^2}{1.02}\left(\frac{580°\text{R}}{530°\text{R}}\right) = \underline{\underline{48.0\ \text{lb}_\text{f}/\text{in}^2}}$$

Notice that absolute pressures and absolute temperatures must be used in this expression.

The increase in absolute air pressure is due to heating of the air caused by frictional effects between the tire and the road. The absolute pressure increases by 7.4 percent, so if a tire is to be inflated to a reproducible value, the temperature at the time of filling must also be specified. ■

■ EXAMPLE 3.2

Nitrogen is enclosed by a piston and cylinder as shown in the figure. The nitrogen is initially at 40°F. The piston weighs 50 lb_f, has a diameter of 5 in, and is initially 4 in above the bottom of the cylinder. Heat transfer to the nitrogen occurs until the nitrogen reaches a temperature of 190°F. Calculate the enthalpy change of the nitrogen during the process and the work done by the nitrogen against the piston and the surrounding atmosphere if the process is internally reversible.

Solution. In Figure 3.3 the system is identified as the nitrogen. The boundaries of the system contain no inlets or exits, so the conservation of mass (Equation 3.4) requires that the mass of the system remains constant or

$$m_1 = m_2$$

At the specified conditions, the nitrogen behaves as an ideal gas and the change in enthalpy is only a function of the change in temperature regardless of path followed by the nitrogen during the process. However, the work cannot be evaluated until the path is specified. The path is determined by the fact that the nitrogen must support a piston of constant weight as well as constant-pressure atmosphere. This means that the process is one of constant pressure. The cylinder walls do not exert any frictional force on the piston because the process is internally reversible. The work performed during the process is then

$$W_{12} = \int_1^2 P\, d\mathbf{V} = P(\mathbf{V}_2 - \mathbf{V}_1)$$

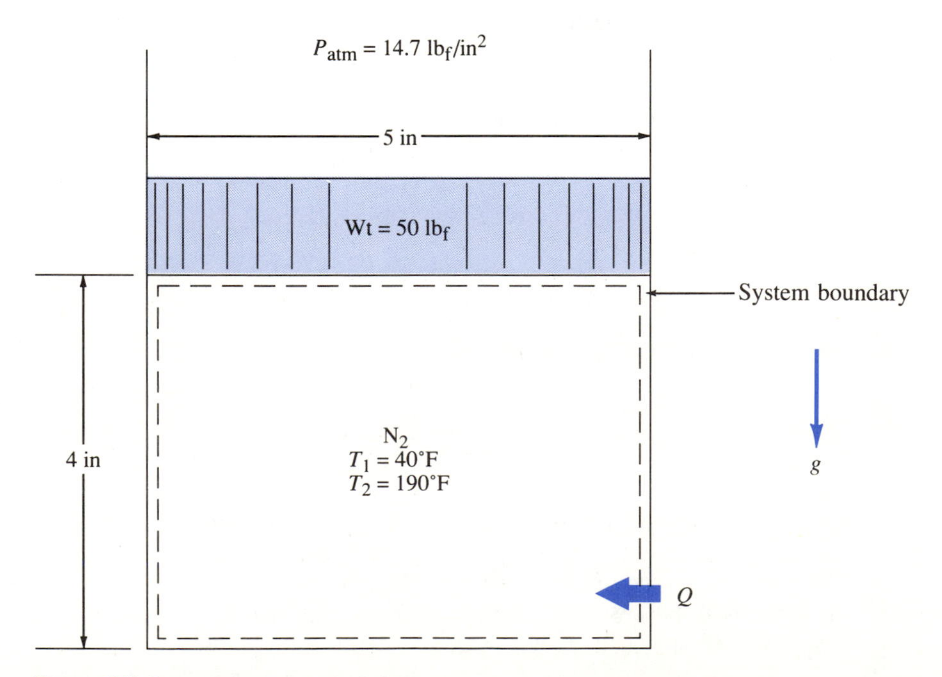

Figure 3.3 System for Example 3.2.

Substituting the ideal-gas equation of state, we may express the work in terms of the temperature change during the process

$$W_{12} = mR(T_2 - T_1)$$

The pressure during the process is

$$P = \frac{W_{\text{piston}}}{A_{\text{piston}}} + P_{\text{atm}} = \frac{50 \text{ lb}_f}{\left(\frac{\pi 5^2}{4}\right) \text{in}^2} + 14.7 \frac{\text{lb}_f}{\text{in}^2} = 17.2 \frac{\text{lb}_f}{\text{in}^2}$$

The initial volume of the nitrogen is

$$V_1 = \frac{\pi}{4}\left(\frac{5}{12} \text{ ft}\right)^2 \left(\frac{4}{12} \text{ ft}\right) = 0.0455 \text{ ft}^3$$

and the mass of the nitrogen is

$$m_1 = \frac{P_1 V_1}{RT_1} = \frac{(17.2 \text{ lb}_f/\text{in}^2)(144 \text{ in}^2/\text{ft}^2)(0.0455 \text{ ft}^3)}{(55.17 \text{ ft} \cdot \text{lb}_f/\text{lb}_m \cdot {}^\circ\text{R})(500^\circ\text{R})}$$

$$= 4.09 \times 10^{-3} \text{lb}_m$$

The work during the process is

$$W_{12} = mR(T_2 - T_1) = (4.09 \times 10^{-3} \text{lb}_m)\left(55.17 \frac{\text{ft} \cdot \text{lb}_f}{\text{lb}_m \cdot {}^\circ\text{R}}\right)(650 - 500)^\circ\text{R}$$

$$= \underline{\underline{33.8 \text{ ft} \cdot \text{lb}_f}}$$

The enthalpy values from Table D.5E are

$$h_2 - h_1 = (161.2 - 124.0)\text{Btu/lb}_m = 37.2 \frac{\text{Btu}}{\text{lb}_m}$$

and the total enthalpy change is

$$H_2 - H_1 = m(h_2 - h_1) = (4.09 \times 10^{-3} \text{ lb}_m)(37.2 \text{ Btu/lb}_m)$$

$$= \underline{\underline{0.152 \text{ Btu}}}$$

■

■ EXAMPLE 3.3

A closed, rigid pressure vessel is filled with a saturated mixture of 1.78 kg of liquid water and 0.22 kg of water vapor at an absolute pressure of 700 kPa. The vessel is heated until the water pressure reaches 8 MPa. Determine the following properties of the water at the final state:

(a) Temperature
(b) Enthalpy
(c) Internal energy

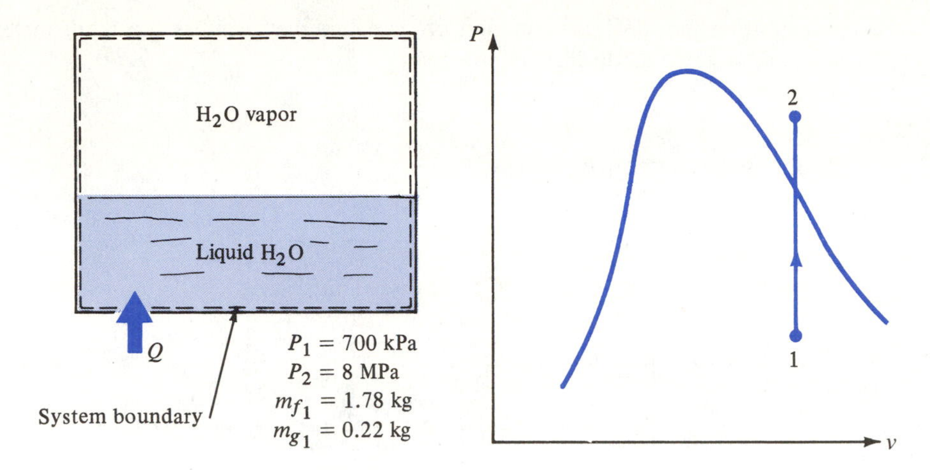

Figure 3.4 System and P-v diagram for Example 3.3.

Solution. The water contained within the vessel is chosen to be the system, as shown in Figure 3.4. This system is a closed system, and conservation of mass requires that the mass of the system remain constant, or

$$m_1 = m_2$$

The total volume of water is also constant as it is heated, because the vessel is rigid. Coupling this fact with the fact that the mass is constant means that the path followed during the heating process is one of constant specific volume, or

$$v_1 = v_2$$

The P-v diagram for this process is shown in the figure. The properties known at the initial state are

$$P_1 = 700 \text{ kPa}$$

and

$$x_1 = \frac{m_{g1}}{m_1} = \frac{m_{g1}}{m_{g1} + m_{f1}} = \frac{0.22 \text{ kg}}{(0.22 + 1.78)\text{kg}} = 0.11$$

(The quality is much lower than is indicated by the sketch in Figure 3.4. This result is attributed to the fact that the left side of the P-v diagram is distorted considerably for clarity of presentation.) From Table B.2 the specific volume of the mixture at this state is calculated as

$$v_1 = v_f + x_1 v_{fg} = 0.001108 \text{ m}^3/\text{kg} + (0.11)(0.2717)\text{m}^3/\text{kg}$$
$$= 0.031 \text{ m}^3/\text{kg}$$

At the final state two properties are now known:

$$v_2 = 0.031 \text{ m}^3/\text{kg}$$
$$P_2 = 8 \text{ MPa}$$

Thus the final state is a superheated-vapor state. The remaining properties at the final state are found by interpolation in Table B.3.

(a) $T_2 = \underline{\underline{362°C}}$
(b) $h_2 = \underline{\underline{3023\,kJ/kg}}$
(c) $u_2 = \underline{\underline{2775\,kJ/kg}}$ ■

In this problem the energy transferred to the water by heat transfer simultaneously causes an increase in pressure and changes the phase of the water from a mixture of liquid and vapor to a superheated vapor.

3.4 CONSERVATION OF MASS FOR OPEN SYSTEMS

Far more problems of engineering interest concern systems that involve mass flow across the surface area of the system. Compressors, pumps, internal-combustion engines, turbines, and heat exchangers are only a few examples of devices that operate with mass transfer occurring across their boundaries. Systems such as these were defined in Chapter 1 as open systems. For open systems the quantities appearing in the general conservation-of-mass equation (Equation 3.2) can be expressed in terms of characteristic parameters such as the fluid properties and system geometry.

An expression for the mass flow rate $\dot{m}$ of fluid entering or leaving a system can be derived by considering a fluid with a velocity V flowing across a differential area dA, as shown in Figure 3.5. The velocity vector can be resolved into components normal and tangential to the plane containing dA. The normal component V_n accounts for mass that is transported across dA. The tangential component V_t does not account for any mass

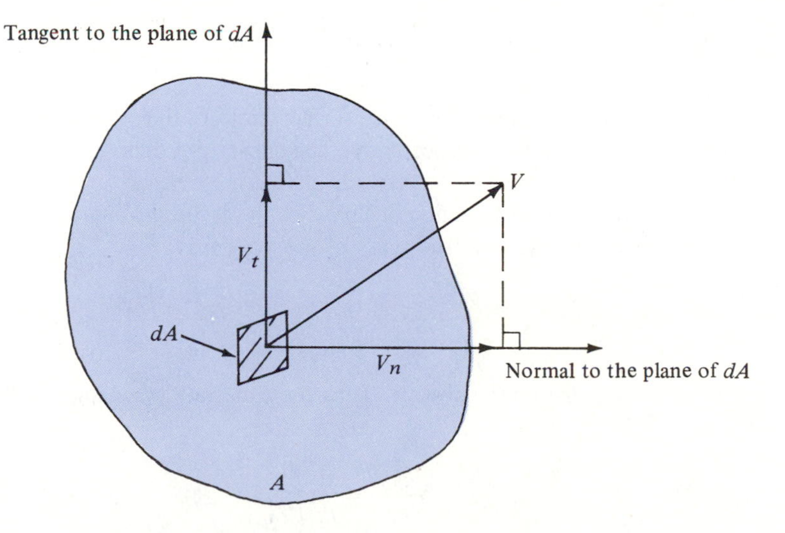

Figure 3.5 Mass flow rate at the surface area of a system.

crossing the area dA because it simply moves the mass in the plane of the area dA but never across dA. Therefore, to calculate the mass flow rate of a substance crossing an area when the velocity vector is oblique to the area, we must use the component of the velocity vector that is perpendicular, or normal, to the plane of the area. If the system boundary is moving, the correct value of V_n to use in calculating the mass flow rate across the surface area is the normal component of the fluid velocity *relative to* the velocity of the surface area.

The distance that a fluid particle will move normal to the area dA during a time interval dt is $V_n dt$. The differential mass of fluid crossing the differential area is, then, the product of the mass per unit volume (ρ) and the differential volume, or

$$\rho V_n \, dA \, dt \tag{3.5}$$

The mass flow rate, or mass per unit time crossing the differential surface area dA, is therefore

$$\rho V_n \, dA \tag{3.6}$$

The mass flow rate of fluid crossing the finite area A, obtained by integrating Equation 3.6, is

$$\dot{m} = \int_A \rho V_n \, dA \tag{3.7}$$

With the aid of Equation 3.7 the two mass-flow-rate terms in Equation 3.2 become

$$\sum_{\text{inlet}} \dot{m} = \int_{A_i} \rho V_n \, dA \tag{3.8}$$

and

$$\sum_{\text{exit}} \dot{m} = \int_{A_e} \rho V_n \, dA \tag{3.9}$$

The symbols A_i and A_e represent the *total* surface areas of the system through which flow enters and leaves the system, respectively. The subscript i denotes inlet areas, while the subscript e denotes exit areas.

The remaining term in Equation 3.2 physically represents any accumulation or storage of mass inside the system. The mass of the system is

$$\int_{\mathbf{V}} \rho \, d\mathbf{V} \tag{3.10}$$

where $\mathbf{V}$ is the volume of the system. Thus the time rate of change of the mass of the system is

$$\frac{dm_{\text{sys}}}{dt} = \frac{d}{dt} \int_{\mathbf{V}} \rho \, d\mathbf{V} \tag{3.11}$$

Substituting Equations 3.8, 3.9, and 3.11 into Equation 3.2 gives an alternative form of the equation that represents the principle of conservation of mass:

$$\int_{A_i} \rho V_n \, dA - \int_{A_e} \rho V_n \, dA = \frac{d}{dt} \int_{V} \rho \, d\mathbf{V} \tag{3.12}$$

Several important characteristics of Equation 3.12 should be noted before attempting to apply it to problems. First of all, the equation involves only four measurable quantities: time, density, velocity, and geometry. Also, the equation requires a knowledge of the fluid velocities at all inlet and exit surface areas of the system, but a knowledge of the velocity of the particles inside the system is not necessary in order to apply the conservation-of-mass equation. Finally, Equation 3.12 contains both the integral and differential operations. The equation is therefore rather difficult to solve in the form presented in Equation 3.12. Fortunately, there are a number of assumptions that can be applied in most practical engineering problems that permit simplification of Equation 3.12.

Equations 3.2 and 3.12 are equivalent in the sense that they both express the conservation of mass in a mathematical form. However, Equation 3.2 is written in a somewhat more convenient form than Equation 3.12, and it is usually preferred because the mass flow rate is often known from measurements, while the velocity distribution across a cross section is generally unknown. Equation 3.12 is a more suitable form of the conservation of mass when the velocity and density distributions are known over the inlet and exit areas of the system under consideration.

3.4.1 Uniform Flow

One assumption that will permit simplification of Equation 3.12 is a condition referred to as *uniform flow*. Uniform flow exists at an inlet or an exit area of a system when all measurable properties are uniform throughout the cross-sectional area.

The condition of uniform flow does not preclude the possibility of differences in fluid properties from one inlet or exit to another, nor does it prohibit changes in properties in the direction of flow. To illustrate these points, consider the simple pipe flow shown in Figure 3.6. The fluid enters the pipe with properties ρ_i, P_i, T_i, which have uniform values over the entire inlet area. However, the fluid properties experience changes in the

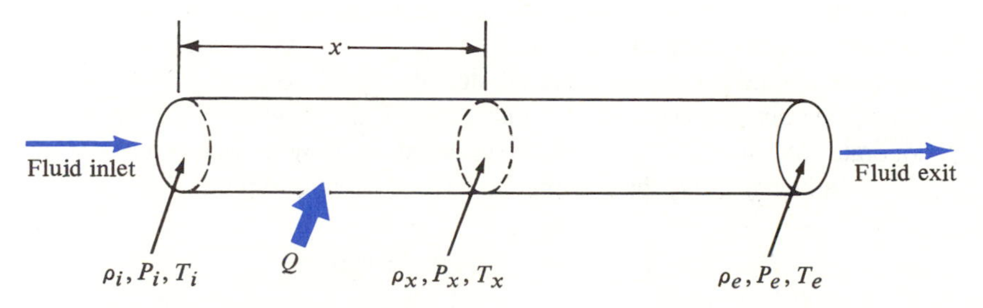

Figure 3.6 Uniform flow in a pipe.

direction of flow because of energy transfer across the boundary of the system by heat transfer, for example. At a position x the properties have changed to ρ_x, P_x, T_x, but these values are still uniform over the cross-sectional area of the pipe at that position. Finally, the flow properties as the fluid leaves the pipe are uniform over the exit area, although they are numerically different from the property values at the inlet to the pipe and at the location x.

The assumption of uniform flow should be applied only when flow conditions warrant its application. Fortunately, in many cases the flow of a fluid through a device is completely turbulent so that the fluid particles are thoroughly mixed in a random fashion. When these conditions occur, the fluid properties at an inlet or an exit area are indeed relatively uniform, and the assumption of uniform flow does not lead to significant errors. Most problems in this book give a single value for the properties at the inlet and exit areas to the thermodynamic system. Those properties are assumed to be suitably averaged over their respective areas so that conditions of uniform flow apply.

If the flow is uniform as it enters and leaves a system, then the area integrals in Equations 3.8 and 3.9 can be replaced by simple algebraic summations. The mass flow rate at an area A when the flow is uniform over the area is

$$\dot{m} = \int_A \rho V_n \, dA = \rho V_n A \tag{3.13}$$

since the density and velocity are uniform throughout the cross-sectional area A. If Equation 3.13 is used to express the mass flow rate in uniform flow, Equation 3.12 can be rewritten as

$$\sum_{\text{inlet}} (\rho V_n A) - \sum_{\text{exit}} (\rho V_n A) = \frac{d}{dt} \int_V \rho \, d\mathbf{V} \tag{3.14}$$

Equation 3.2 is still valid under the assumption of uniform flow, and Equation 3.14 merely indicates that for uniform flow the mass flow rate can be evaluated as the product of the density, the normal component of the velocity, and the cross-sectional area at each inlet and exit of the system.

3.4.2 Steady State

Perhaps the most common simplifying assumption that can be made in engineering problems is the condition of *steady state* or *steady flow*. When none of the extensive properties associated with a system vary with time, the system is said to operate steadily or to exist under the conditions of steady state. Two examples of steady-state behavior were mentioned previously in this chapter; the first involved an air compressor that operates steadily and the second concerned the flow of water into a filled container.

Mathematically, the assumption of steady state implies

$$\frac{dY_{\text{sys}}}{dt} = 0 \tag{3.15}$$

where Y_{sys} is any average extensive property of the system. For a system to operate steadily, each extensive property of the system must satisfy Equation 3.15.

The assumptions of steady state and uniform flow should not be confused. Steady state implies no change in properties with respect to *time*, while uniform flow implies no change in properties at a particular cross-sectional *area*. A system can operate both uniformly and steadily. For example, the flow of a fluid through the pipe shown in Figure 3.6 was previously said to illustrate uniform flow. If none of the flow properties change with respect to time, then the flow is also classified as steady. When conditions of steady state are applied to Equation 3.2, the equation for conservation of mass reduces to

$$\sum_{\text{exit}} \dot{m} = \sum_{\text{inlet}} \dot{m} \tag{3.16}$$

since the mass of the system (an extensive property) cannot change with time. If, in addition to steady-state conditions, the flows into and out of the system are uniform, then the conservation-of-mass equation further simplifies to

$$\sum_{\text{exit}} (\rho V_n A) = \sum_{\text{inlet}} (\rho V_n A) \tag{3.17}$$

Physically, the steady-state form of the conservation-of-mass equation states that the total mass flow rate into and out of the system must be equal, because the system cannot store or accumulate mass as time progresses.

If the flow of a substance through a system is steady and uniform and the substance is also incompressible, then the steady form of the conservation-of-mass equation can be simplified even further. An incompressible substance is one that has a constant density, so Equation 3.17 reduces to

$$\sum_{\text{exit}} (V_n A) = \sum_{\text{inlet}} (V_n A) \tag{3.18}$$

The product of the normal component of the velocity and the area through which the flow occurs is the *volume flow rate* of the substance. Therefore, the conservation of mass under the assumptions of steady state and incompressible flow states that the total volume flow rate into and out of a system must be equal.

The next two examples illustrate the application of conservation of mass to steady-state, uniform-flow problems.

EXAMPLE 3.4

Steam enters a turbine through a pipe with a 15-cm diameter. The inlet steam velocity is 90 m/s, and the inlet pressure and temperature are 20 MPa and 600° C. The exit pipe of the turbine has a diameter of 60 cm, and the exit pressure and temperature of the steam are 300 kPa and 150° C. Assuming that the steam flows steadily through the turbine, calculate the following quantities:

(a) Inlet steam density
(b) Mass flow rate of steam through the turbine
(c) Exit steam velocity

Solution. We first sketch the turbine and identify the steam as the system, as shown in Figure 3.7. The flow of steam entering and leaving the turbine is assumed to be uniform.

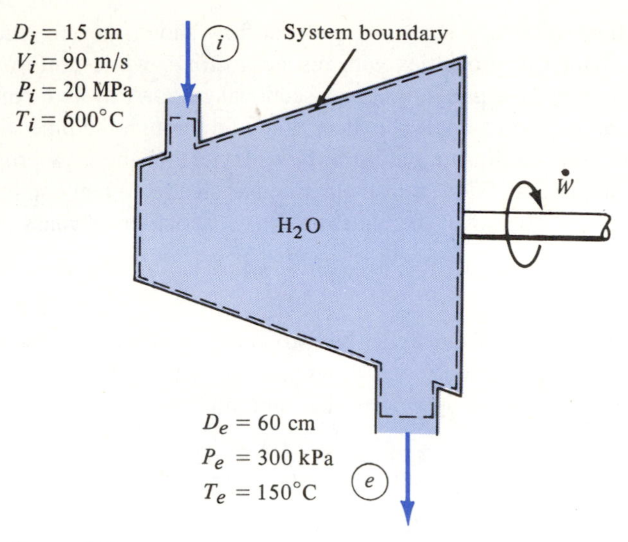

Figure 3.7 System for Example 3.4.

(a) Since the inlet pressure and temperature are given, the density of the steam at the inlet to the turbine can be determined from Table B.3:

$$\rho_i = \frac{1}{v_i} = \frac{1}{0.018169\,\text{m}^3/\text{kg}} = \underline{\underline{55.04\,\text{kg/m}^3}}$$

(b) The mass flow rate of steam at the inlet is

$$\dot{m}_i = \rho_i V_i A_i = (55.04\,\text{kg/m}^3)(90\,\text{m/s})\left[\frac{\pi(0.15)^2}{4}\right]\text{m}^2 = \underline{\underline{87.5\,\text{kg/s}}}$$

(c) The exit density can be determined at the given exit pressure and temperature from Table B.3:

$$\rho_e = \frac{1}{v_e} = \frac{1}{0.6339\,\text{m}^3/\text{kg}} = 1.578\,\text{kg/m}^3$$

Since the steam flows steadily through the turbine, the conservation-of-mass equation states that

$$\dot{m}_i = \dot{m}_e$$

or

$$\dot{m}_e = \rho_e V_e A_e = 87.5\,\text{kg/s}$$

Solving for the exit velocity gives

$$V_e = \frac{\dot{m}_e}{\rho_e A_e} = \frac{87.5\,\text{kg/s}}{(1.578\,\text{kg/m}^3)\left[\dfrac{\pi(0.60)^2}{4}\right]\text{m}^2} = \underline{\underline{196\,\text{m/s}}}$$

■

EXAMPLE 3.5

A steady-flow pump takes in 1500 gpm (gallons per minute) of liquid water through an inlet pipe that has a diameter of 12 in. The pump supplies water to two pipes having diameters of 4 and 7 in, respectively. The volume flow rate in the smaller of the two exit pipes is 300 gpm. Determine the velocities in the inlet pipe and the two exit pipes.

Solution. A sketch of the system is shown in Figure 3.8. The inlet is designated by i and the two exits are referred to by the symbols 1 and 2.

The flow is steady and uniform, and the water can be assumed to be an incompressible substance because its density does not change significantly as a result of the pumping process.

The velocity at the inlet can be calculated from the volume flow rate

$$V_i = \left(\frac{V_i A_i}{A_i}\right) = \frac{\left(1500\ \dfrac{\text{gal}}{\text{min}}\right)(4)}{\pi(1\ \text{ft})^2(7.48\ \text{gal/ft}^3)} = \underline{\underline{255\ \text{ft/min}}}$$

For steady, uniform flow of an incompressible fluid the conservation-of-mass equation (Equation 3.18) may be written as

$$(VA)_i = (VA)_1 + (VA)_2$$

$$(1500\ \text{gal/min}) = (300\ \text{gal/min}) + V_2\left[\frac{\pi(7/12)^2}{4}\ \text{ft}^2\right](7.48\ \text{gal/ft}^3)$$

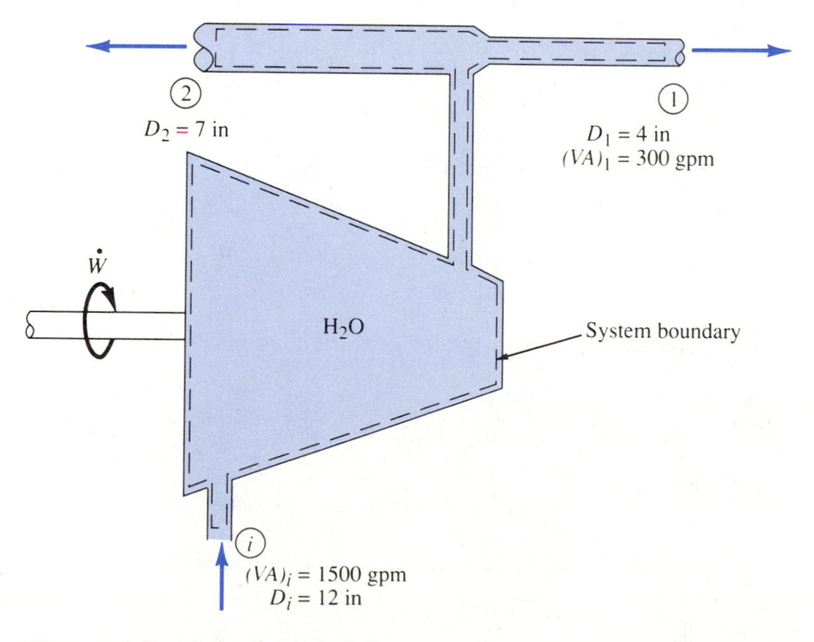

Figure 3.8 Sketch for Example 3.5.

Solving for the exit velocity in the larger pipe gives

$$V_2 = \underline{\underline{600\ \text{ft/min}}}$$

The velocity of water in the smaller exit pipe is

$$V_1 = \frac{\left(300\ \dfrac{\text{gal}}{\text{min}}\right)(4)}{\left(\pi(4/12)^2\text{ft}^2\right)(7.48\ \text{gal/ft}^3)} = \underline{\underline{460\ \text{ft/min}}}$$

■

3.4.3 Transient Analysis

Some problems of engineering interest involve a transient, or time-varying, analysis. Start-up or shutdown of equipment and detailed analyses of fluid flow in reciprocating equipment during an individual cycle are typical examples of problems requiring a transient analysis.

Whenever flow conditions dictate transient conditions, the mass of the system can change with respect to time. The change in the mass of the system is considered in the conservation-of-mass equation by the last term in Equation 3.2.

Most transient problems involve evaluation of properties at the beginning or the end of a specified time interval Δt that starts at the initial time t_1 and ends at the final time t_2, or

$$\Delta t = t_2 - t_1 \tag{3.19}$$

Since transient conditions are involved, the general conservation-of-mass equation, Equation 3.2, must be used:

$$\sum_{\text{inlet}} \dot{m} - \sum_{\text{exit}} \dot{m} = \frac{dm_{\text{sys}}}{dt} \tag{3.20}$$

The mass flow rates can be replaced by equivalent time rates of change of mass at the surface area of the system, so that

$$\sum_{\text{inlet}} \left(\frac{dm}{dt}\right) - \sum_{\text{exit}} \left(\frac{dm}{dt}\right) = \frac{dm_{\text{sys}}}{dt} \tag{3.21}$$

When Equation 3.21 is integrated over the time interval from t_1 to t_2, the result is

$$\sum_{\text{inlet}} \int_{t_1}^{t_2} \left(\frac{dm}{dt}\right) dt - \sum_{\text{exit}} \int_{t_1}^{t_2} \left(\frac{dm}{dt}\right) dt = \int_{t_1}^{t_2} \left(\frac{dm_{\text{sys}}}{dt}\right) dt \tag{3.22}$$

which can be reduced to

$$\sum_{\text{inlet}} m - \sum_{\text{exit}} m = (m_2 - m_1)_{\text{sys}} \tag{3.23}$$

where the symbol $(m_2 - m_1)_{\text{sys}}$ represents the change of mass of the system during the time interval,

$$(m_2 - m_1)_{sys} = m_{sys}(t_2) - m_{sys}(t_1) \tag{3.24}$$

and $\sum m$ represents the amount of mass that enters (or leaves) the system during the interval Δt.

If no inlets or exits are available for mass flow across the system boundary, then Equation 3.23 reduces to the conservation-of-mass equation for a closed system:

$$(m_2 - m_1)_{sys} = 0$$

or

$$m_{sys} = \text{constant}$$

Before turning to numerical examples, we should consider one final implication of the transient form of the conservation-of-mass equation. Equation 3.11 indicates that during a transient process, the mass of the system can change with time if either the size of the system changes or the volume-averaged density of the fluid within the system changes with time. Therefore, if the system boundary is rigid and the fluid is incompressible, the mass of the system cannot change with time.

■ EXAMPLE 3.6

An air compressor is used to supply air to a rigid tank that has a volume of 4 m^3. Initially, the pressure and temperature of the air in the tank are 101 kPa and 35° C, respectively. The supply pipe to the tank is 7 cm in diameter, and the velocity of the air in the inlet pipe remains constant at 12 m/s. The pressure and temperature of the air in the inlet pipe are constant at 600 kPa and 35° C. Calculate the following quantities:

(a) The time rate of change of mass inside the tank
(b) The mass of air added to the tank if the compressor stops operating when the tank reaches 400 kPa and 55° C.
(c) The time that the compressor must be operated to produce a tank pressure of 400 kPa and a temperature of 55° C.

Solution. If the air inside the tank is considered to be the system shown in Figure 3.9, the problem is a transient one. Properties at the inlet are designated by the subscript i, the initial and final properties of the air have subscripts 1 and 2, respectively, and the flow is assumed to be uniform at the inlet.

(a) The time rate of change of the mass of air inside the tank can be determined by applying the conservation-of-mass equation (Equation 3.2):

$$\frac{dm_{sys}}{dt} = \dot{m}_i = \rho_i V_i A_i$$

Assuming ideal-gas behavior, the density of the air in the inlet pipe is

$$\rho_i = \frac{P_i}{RT_i} = \frac{600\,\text{kPa}}{(0.287\,\text{kPa}\cdot\text{m}^3/\text{kg}\cdot\text{K})(308\,\text{K})} = 6.79\,\text{kg/m}^3$$

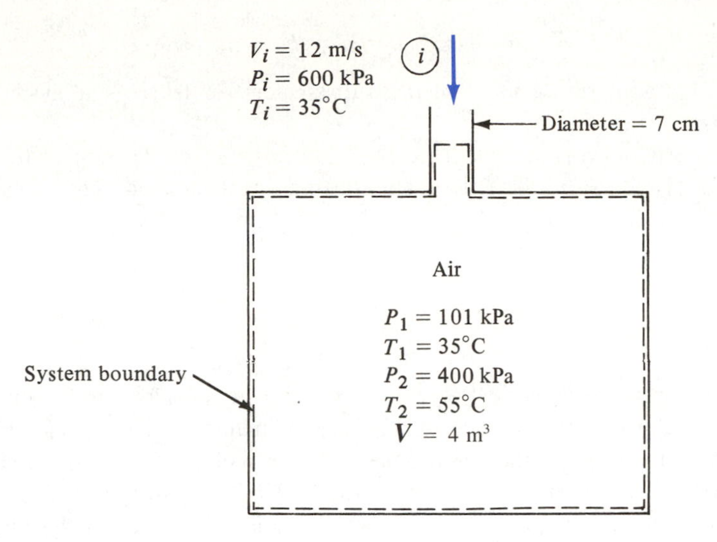

Figure 3.9 System for Example 3.6.

Substituting this value in the preceding equation results in

$$\frac{dm_{sys}}{dt} = (6.79\,\text{kg/m}^3)(12\,\text{m/s})\left[\frac{\pi(0.07)^2\text{m}^2}{4}\right] = \underline{\underline{0.314\,\text{kg/s}}}$$

(b) The mass in the tank when the air reaches 400 kPa, 55° C is

$$m_2 = \frac{P_2V_2}{RT_2} = \frac{(400\,\text{kPa})(4\,\text{m}^3)}{(0.287\,\text{kPa}\cdot\text{m}^3/\text{kg}\cdot\text{K})(328\,\text{K})} = 17.0\,\text{kg}$$

The initial mass in the tank before the compressor is started is

$$m_1 = \frac{P_1V_1}{RT_1} = \frac{(101\,\text{kPa})(4\,\text{m}^3)}{(0.287\,\text{kPa}\cdot\text{m}^3/\text{kg}\cdot\text{K})(308\,\text{K})} = 4.57\text{ kg}$$

The mass of air added to the tank during the compression process is

$$m_2 - m_1 = 17.0\,\text{kg} - 4.57\,\text{kg} = \underline{\underline{12.43\,\text{kg}}}$$

(c) The time required to reach this pressure is

$$\Delta t = \left(\frac{m_2 - m_1}{dm_{sys}/dt}\right) = \left(\frac{12.43\,\text{kg}}{0.314\,\text{kg/s}}\right) = \underline{\underline{39.6\,\text{s}}}$$

assuming that the air is added to the tank at a constant rate. ■

3.5 SUMMARY

In this chapter a general mathematical expression for the principle of conservation of mass has been developed, which can be written as

$$\sum_{\text{inlet}} \dot{m} - \sum_{\text{exit}} \dot{m} = \frac{dm_{\text{sys}}}{dt} \tag{3.2}$$

This one equation was applied to the following two types of thermodynamic systems, which are frequently encountered when solving engineering problems.

Closed System

In a closed system there are no inlets or exits, so there can be no mass flow into or out of the system. For the closed system, therefore, Equation 3.2 reduces to

$$m_{\text{sys}} = \text{constant} \tag{3.4}$$

Open System

Uniform Flow Whenever fluid properties are uniform across the cross-sectional area at each inlet and exit of a system, the flow is said to be uniform, and the conservation-of-mass equation reduces to

$$\sum_{\text{inlet}} (\rho V_n A) - \sum_{\text{exit}} (\rho V_n A) = \frac{d}{dt}\int_{V} \rho \, d\mathbf{V} \tag{3.14}$$

Steady State For a system operating at steady state, the time rate of change of all extensive properties of the system is zero. Since the mass of the system is an extensive property, there can be no change in the mass of the system with time. Therefore, Equation 3.2 reduces to

$$\sum_{\text{exit}} \dot{m} = \sum_{\text{inlet}} \dot{m} \tag{3.16}$$

If the flow is both steady and uniform, the conservation-of-mass equation becomes

$$\sum_{\text{exit}} (\rho V_n A) = \sum_{\text{inlet}} (\rho V_n A) \tag{3.17}$$

Transient Analysis Integration of Equation 3.2 over a finite time interval provides a form of the conservation-of-mass equation that applies to an open thermodynamic system that undergoes transient changes. The result is

$$\sum_{\text{inlet}} m - \sum_{\text{exit}} m = (m_2 - m_1)_{\text{sys}} \tag{3.23}$$

PROBLEMS

3.1 A rain gauge is designed with a small funnel-shaped attachment at the top of the gauge to collect the moisture as shown in the figure. Determine the spacing between ¼-in marks on the gauge body so that they accurately reflect the true depth of rainfall.

Problem 3.1

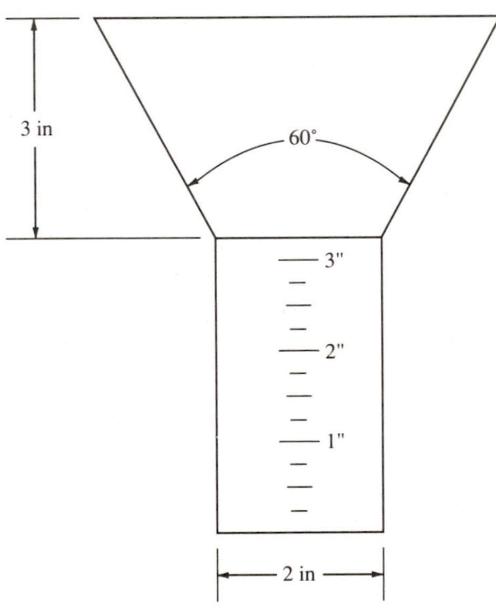

3.2 Nitrogen at 200 kPa pressure and a temperature of 25°C flows through a 35-mm-diameter pipe with a velocity of 20 m/s. Determine the mass flow rate of nitrogen through the pipe.

3.3 Water with a pressure of 60 psia and a temperature of 80°F flows through a pipe with a diameter of 6 in. Calculate the mass flow rate of water through the pipe if the velocity of water is 20 ft/s.

3.4 Water at 100 kPa, 20°C flows through a pipe with a cross-sectional area of 5 cm^2. The volume flow rate of the water is 5 liter/s. Calculate the average velocity of the water through the pipe and the mass flow rate.

3.5 Liquid water enters the square duct shown in the accompanying figure with an average velocity of 10 m/s. Determine the average velocity and mass flow rate of the water as it leaves the duct.

Problem 3.5

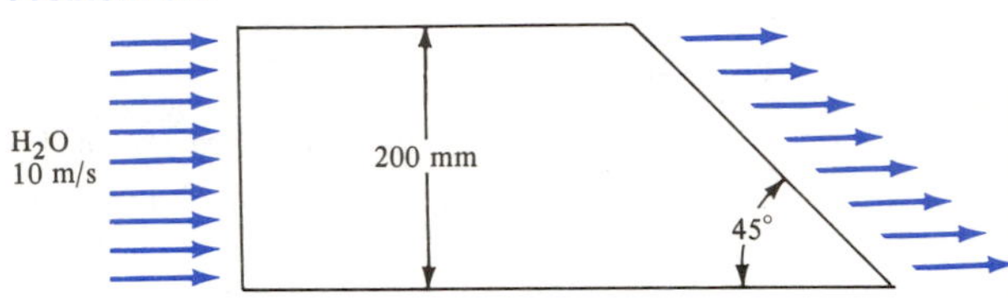

3.6 A pipe is to be selected so that it will carry liquid water at 20°C with a mass flow rate of 150 kg/s and with an average velocity that does not exceed 2.3 m/s. Determine the minimum inside diameter of the pipe.

3.7 Air leaves a compressor at 160 psia and 120°F with a mass flow rate of 5.0 lb_m/s. Determine the inside diameter of the pipe leaving the compressor such that the velocity of air in the pipe will not exceed 300 ft/s.

3.8 Water at 5 MPa and 400°C and a velocity of 30 m/s enters a device through a pipe with a cross-sectional area of 0.35 m^2. Calculate the mass flow rate of water in the pipe.

3.9 Refrigerant-12 enters a compressor at 40 psia and 70°F with a mass flow rate of 0.1 lb_m/s. Determine the smallest inlet diameter tubing that can be used if the average inlet velocity of the refrigerant should not exceed 15 ft/s.

3.10 A circular pipe with inside radius R contains an incompressible fluid with a parabolic velocity profile; see the accompanying figure. Determine an expression for the average fluid velocity, in terms of the centerline velocity V_{max}, that could replace the profile shown in the figure and produce the same mass flow rate through the pipe.

Problem 3.10

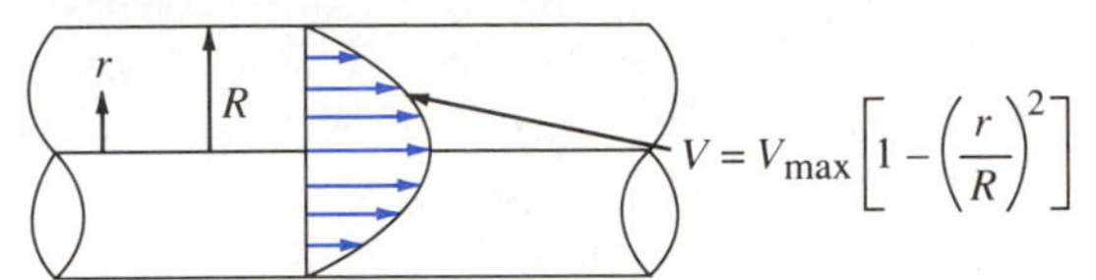

3.11 Determine the mass flow rate of refrigerant-12 at a pressure of 40 psia and a temperature of 100°F flowing through a pipe with an inside diameter of 8 in and a parabolic velocity profile with centerline velocity of 40 ft/s.

3.12 A circular pipe with inside radius R contains an incompressible fluid with a linear velocity profile, as shown in the accompanying figure. Determine an expression for the average fluid velocity, in terms of the centerline velocity V_{max}, that could replace the profile shown in the figure and produce the same mass flow rate through the pipe.

Problem 3.12

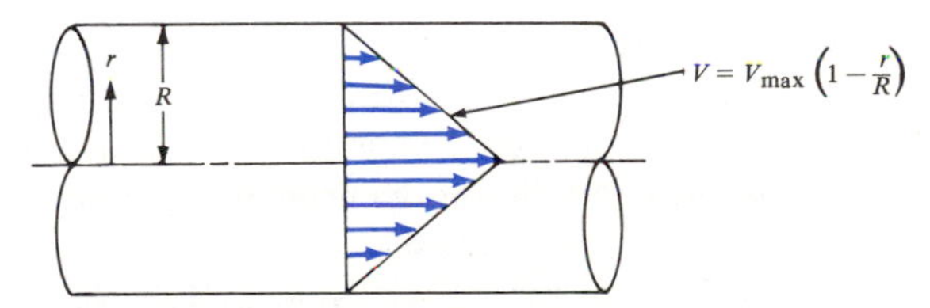

3.13 Water flows through an opening with a rectangular cross section as shown in the figure. Determine the mass flow rate of water across that cross section. The depth of the cross section into the plane of the page is 2 ft. The centerline velocity of the water is 25 ft/s and the velocity profile is linear with y.

Problem 3.13

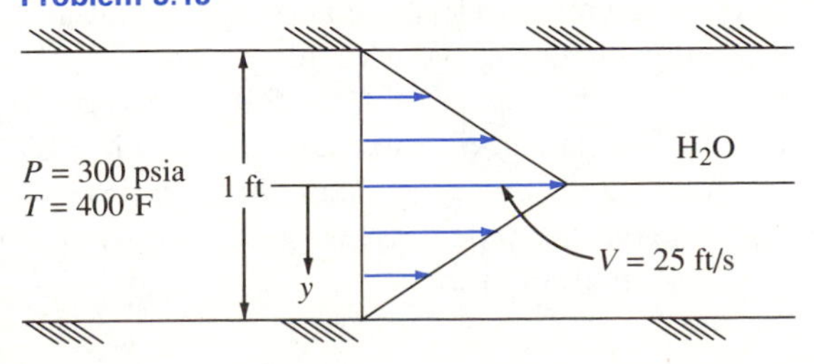

3.14 Calculate the mass flow rate of air through a circular pipe with a diameter of 70 mm if the velocity and density of the air are known to vary parabolically with radius r according to the equations.

$$V = 25\left[1 - \left(\frac{r}{35}\right)^2\right]$$

$$\rho = 300\left[4 - \left(\frac{r}{35}\right)^2\right]$$

where V is in meters per second, ρ is in grams per cubic meter, and r is in millimeters.

3.15 A 40-mm-diameter pipe is used to carry a fluid whose velocity is 15 m/s. The temperature and pressure of the fluid are 25°C and 1 MPa, respectively. Calculate the mass flow rate when the fluid is (a) nitrogen and (b) refrigerant-12.

3.16 A refrigeration system requires a steady flow of refrigerant-12 at 80 psia, 140°F at a mass flow rate of 0.5 lb_m/s. If the velocity of the refrigerant is to be limited to 5 ft/s, what is the minimum allowable pipe diameter?

3.17 Liquid water flows steadily through the Y connector shown in the accompanying figure. Determine the volume flow rate through the connector and the minimum diameter of the upper exit pipe required to prevent the average flow velocity in that branch from exceeding 10 m/s.

Problem 3.17

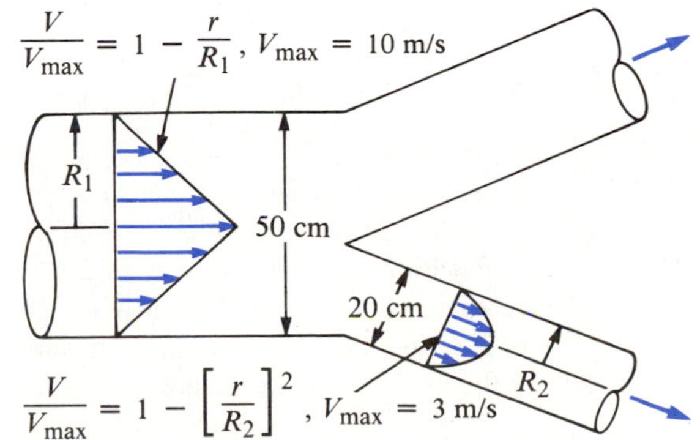

3.18 Air under steady-state conditions enters a constant-diameter pipe at 90°F and 60 psia. The inlet velocity of the air is uniform at 35 ft/s. The average exit velocity from the pipe is 70 ft/s, and the air at the exit has a pressure of 50 psia. Determine the exit temperature of the air.

3.19 Hydrogen enters a steady-flow device with an inlet-to-exit-area ratio of 2 to 1. The inlet conditions are 50 psia, 100°F, and 60 ft/s. The exit conditions are 45 psia and 90°F. Calculate the mass flow rate of the hydrogen and the velocity of the H_2 at the exit conditions if the exit area is 1 ft^2.

3.20 **[CDE111]** Two streams of air are mixed in a steady-flow mixing chamber that has two inlets and a single exit; 1 m^3/s of air at 120 kPa, 40°C is mixed with 2 m^3/s of air at 120 kPa, 90°C, resulting in a mixture at 120 kPa and 70°C. The area of the pipe that contains the mixture has a cross-sectional area of 0.1 m^2. Determine the mass flow rate of each of the two entering airstreams and the exit velocity of the air mixture.

3.21 Water flows steadily through the circular pipe shown in the figure with a mass flow rate of 1.0 kg/s. Properties are given at both the inlet and exit to the pipe. Calculate the following:

(a) the inlet velocity of the water
(b) the exit velocity of the water

Problem 3.21

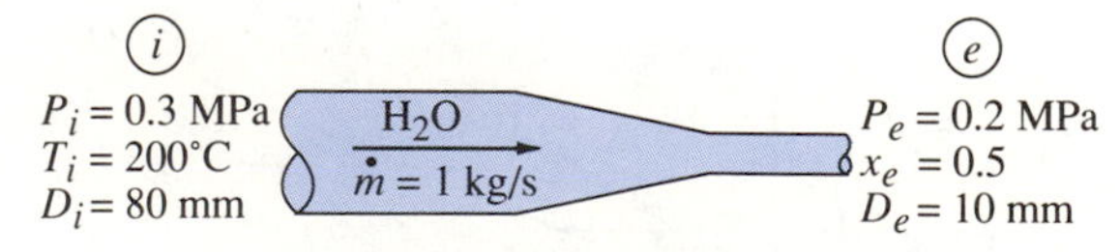

3.22 Air flows steadily through a constant-diameter pipe. The air enters the pipe at 60°F, 50 psia, and with an average velocity of 2 ft/s. The average velocity of the air leaving the pipe is 2.2 ft/s. Calculate the exit temperature of the air if the exit pressure is 40 psia.

3.23 A pipe with a constant diameter of 10 in transports nitrogen pressurized to 85 psia. At one location along the pipe the nitrogen has a temperature of 120°F and an average velocity of 60 ft/s. At a second location downstream of the first, the nitrogen has a temperature of 110°F and an average velocity of 55 ft/s. Somewhere between these two locations there is a leak in the pipe. Assume steady flow and that the pressure is essentially constant. Determine the mass flow rate of the leak.

3.24 Nitrogen flows steadily through a pipe with a constant cross-sectional area. The nitrogen enters the pipe at a pressure of 500 kPa, a temperature of 50°C, an average velocity of 15 m/s, and a mass flow rate of 5.0×10^{-2} kg/s. At the exit of the pipe the pressure is 200 kPa and the temperature is 25°C.

(a) Calculate the inside diameter of the pipe.
(b) Calculate the velocity of the nitrogen at the exit of the pipe.

3.25 A jet aircraft is flying steadily at a 35,000-ft altitude and at a velocity of 500 mph into air that has a temperature of -40°F and a pressure of 8 lb$_f$/in^2. Each engine is burning fuel at a steady rate of 1.8 lb$_m$/s. Determine the average exhaust-gas velocity relative to the engine if it leaves the engine at a pressure of 8 lb$_f$/in^2 and a temperature of 850°F. You may assume that the air and the exhaust gas are both ideal gases. The inlet area of the engine is 2 ft^2 and the exit area is 1.1 ft^2. For the exhaust gases assume that R = 0.069 Btu/lb$_m$ · °R.

3.26 Water at a pressure of 200 psia and a temperature of 1400°F enters a constant-area pipe with an average velocity of 275 ft/s. The water leaves the pipe at a pressure of 190 psia and a temperature of 1000°F. Calculate the average velocity of the water as it leaves the pipe, assuming that steady flow exists in the pipe.

3.27 Dry air enters an industrial dryer with a pressure of 101 kPa, a temperature of 250°C, and at a mass flow rate of 2.6 kg/h through a duct with a cross-sectional area of 0.85 m^2. A wet material is placed inside the dryer, and its rate of drying is 5.5 g/min. Determine the mass flow rate of moist air leaving the dryer and the average velocity of air entering the dryer, assuming steady operation.

3.28 A caulking cartridge is moved steadily at a velocity of 30 mm/s and discharges a circular bead of caulking compound with a diameter of 9 mm, as shown in the accompanying figure. The volume of caulking compound in a new cartridge is 600 cm^3, and the density of the caulking compound is 0.96 g/cm^3. Calculate the following quantities:

(a) the velocity of the piston V_p necessary to steadily discharge the compound

(b) the time rate of change of the mass of the compound inside the cartridge

(c) the length of bead that can be caulked with a single cartridge

(d) the time that a new cartridge can be used before it is empty

Problem 3.28

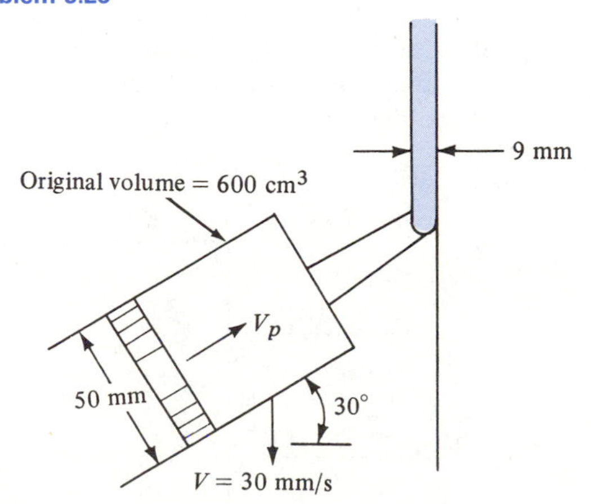

3.29 A section of a constant-diameter pipe carries a steady flow of nitrogen. The nitrogen enters at 120°F and 80 psia with a velocity of 12 ft/s. At the exit the pressure and velocity are 70 psia and 19 ft/s, respectively. Determine the exit temperature of the nitrogen.

3.30 At one section in a 30-cm-diameter pipe methane flows at an absolute pressure of 800 kPa, a temperature of 40°C, and an average velocity of 25 m/s.

(a) What is the mass flow rate in kg/s?

(b) What is the pressure farther downstream where the average velocity is 250 m/s and the temperature is 20°C if the flow is steady?

3.31 In a certain steady-flow industrial mixing chamber, 50 lb_m/min of H_2O at 80°F and 60 psia is mixed with H_2O at 500°F and 60 psia to produce a stream of saturated vapor at 60 psia that flows at a rate of 600 ft^3/min.

(a) Determine the mass flow rate of the 500°F H_2O entering the chamber.

(b) What inlet area should be used if the velocity of the 80°F H_2O entering the chamber is to be limited to 0.2 ft/s?

3.32 Water at 50 kPa and 150°C enters a constant-area pipe ($A = 0.08\ m^2$) with a velocity of 11 m/s. At some point farther downstream in the pipe, the pressure and temperature are 10 kPa and 200°C, respectively. The flow is steady. Determine the mass flow rate of water entering the pipe and the mass flow rate and velocity at the downstream section.

3.33 Liquid water at 25°C flowing in a 75-mm-diameter pipe discharges into a rectangular tank 1.5 m by 2.5 m in horizontal dimensions. Water rises in the tank at a rate of 0.15 m in 40 s. What is the average flow velocity in the pipe?

3.34 A rigid, high-pressure tank has a volume of 3.8 ft^3 and contains nitrogen at 185 psia and at 290°F. A small leak develops in a valve connected to the tank, and the mass flow rate through the leak is 2×10^{-5} lb_m/s. The area of the hole causing the leak is 0.01 in^2. Calculate the velocity of the nitrogen through the hole, assuming that the state of the nitrogen at the location of the hole is the same as the state of the nitrogen inside the tank. Determine the mass of the nitrogen in the tank if the leak continues for 24 h, assuming that the leak rate is constant.

3.35 [CDE211] A tank containing liquid water, at 20°C as shown in the accompanying figure, drains through a whole near the bottom of the tank. The hole is circular with a diameter of 150 mm, and the average velocity of water through the hole is 20 m/s. Determine the time rate of change of the mass of water in the tank and the velocity of water at the free surface.

Problem 3.35

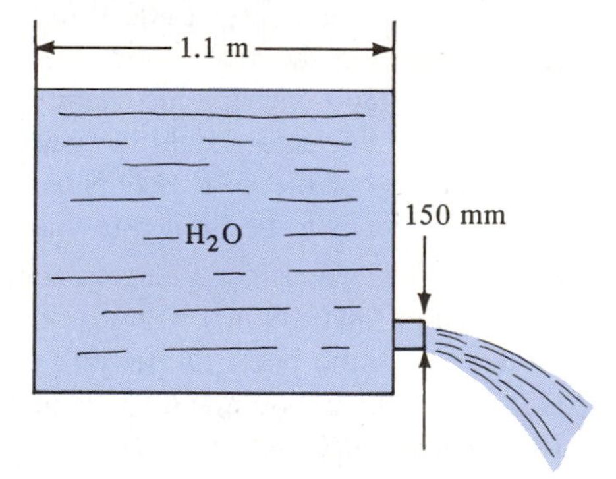

3.36 [CDE211] A tank containing liquid water at 70°F drains through a hole near the bottom of the tank. The diameter of the tank is 3.6 ft. The hole is circular with a diameter of 0.5 ft, and the average velocity of water through the hole is 66 ft/s. Determine the following:

(a) the velocity of the free water surface in ft/s

(b) the time rate of change of the mass of water in the tank in lb_m/s

3.37 In a solid-fuel rocket, gases leave the rocket with a temperature of 1150°C at a pressure of 135 kPa through a nozzle area of 1.45 m^2. The velocity of the gases through the nozzle is 565 m/s. Assuming that the gases behave as an ideal gas with a molecular weight of 28, calculate the time rate of change of the mass of the rocket due to the gases leaving through the nozzle.

3.38 A rigid tank with volume of 1.2 ft^3 has one inlet and one exit. A compressible substance with an average velocity of 6 ft/s and an average density of 0.05 lb_m/ft^3 flows through the 4-in-diameter inlet pipe. The substance in the 1-in-diameter exit pipe has an average velocity and density of 3 ft/s and 0.12 lb_m/ft^3, respectively. Calculate the time rate of change of the average density of the substance inside the rigid tank.

3.39 A rigid container holds pressurized CO_2. A hole 2 mm in diameter is drilled through the container, and the CO_2 escapes through the hole with an average density of 1.1 kg/m^3. Calculate the time rate of change of mass inside the container when the average velocity of the CO_2 through the hole is (a) 150 m/s and (b) 50 m/s.

3.40 A rigid tank with a volume of 30 m^3 is connected to a 30-cm-diameter pipe containing compressed air. The velocity profile across the inlet line is parabolic, as shown in the accompanying figure, and it is assumed to remain independent of time. The average density of the air entering the tank is constant at a value of 1.06 kg/m^3. Determine the time rate of change of the mass of air inside the tank.

Problem 3.40

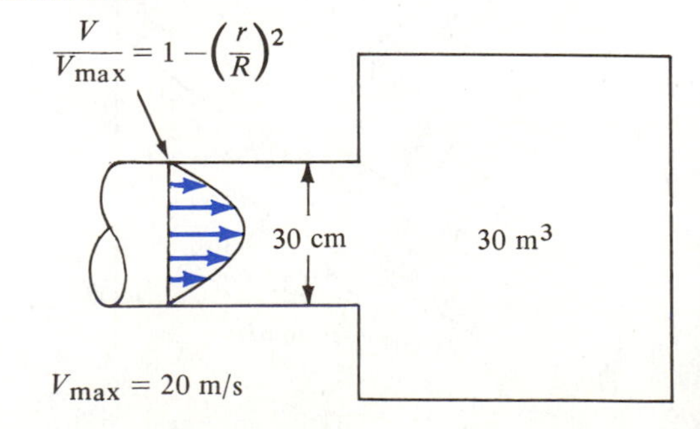

3.41 Assume that a compressible substance flows through a rigid container, as shown in the accompanying figure. The velocity profiles as shown do not vary with depth into the page. Determine an expression for the time rate of change of mass within the container per unit depth.

Problem 3.41

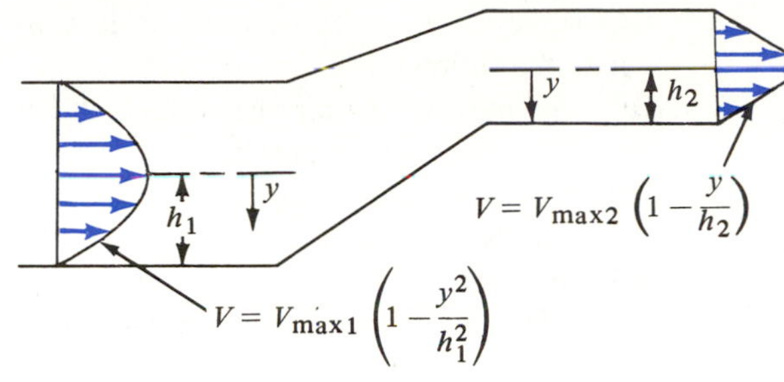

3.42 A piston and cylinder arrangement shown in the figure has one circular inlet and one circular exit. Water enters and leaves the system uniformly at the states shown in the figure. Assume that the given inlet and exit conditions are independent of time. Calculate the following quantities:

(a) the mass flow rate of the water entering the system

(b) the mass flow rate of the water leaving the system

(c) the time rate of change of the mass inside the system

(d) the velocity of the piston, and state whether the piston is moving upward or downward

Problem 3.42

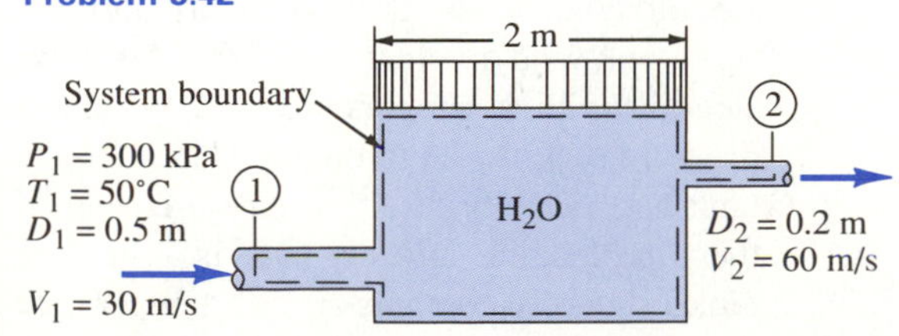

3.43 A rigid air cylinder with a volume of 10^5 mm^3 is punctured with a hole having a cross-sectional area of 0.3 mm^2. See the accompanying figure. The original pressure and temperature of the air inside the cylinder are 800 kPa and 35°C. As the air leaves the hole in the cylinder, it reaches a pressure of 100 kPa and a temperature of 5°C. The velocity of the air as it escapes through the hole is 100 m/s. Calculate the original mass of air inside the tank and the mass in the tank 5 s after it is punctured, assuming that the exit conditions of the air remain independent of time.

Problem 3.43

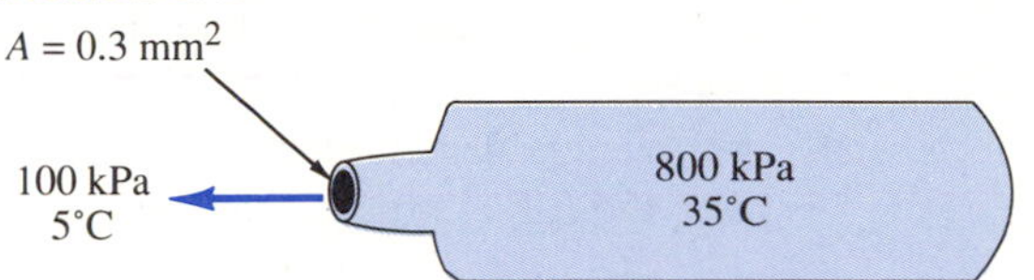

3.44 A 1-ft-diameter pipe carries helium into a closed tank. The average velocity of the helium in the pipe is 1 ft/s and the pressure and temperature of the helium in the pipe as it enters the tank are 15 psia, 400°F. Calculate the mass flow rate of the helium as it enters the tank and determine the time rate of change of the mass of helium inside the tank.

3.45 [CDE311] A rigid air cylinder has a volume of 3.5 ft^3 and is punctured with a circular hole having a cross-sectional area of 0.003 ft^2. The original pressure and temperature of the air inside the cylinder are 120 psia and 65°F. Air escapes from the cylinder for 5 s with an average density of 0.0794 lb$_m$/ft^3 and an average exit velocity of 300 ft/s. The pressure in the cylinder after 5 s is 98.2 psia. Assume that the exit conditions of the air are independent of time during the 5-s period. Calculate the following:

(a) the original mass of air in the cylinder in lb$_m$

(b) the mass of air in the tank after 5 s

3.46 A compressible substance flows through a pipe into a rigid (V = 2 m^3) tank through a pipe as shown in the figure. The velocity profile at the entrance to the tank is parabolic and the density is constant at the entrance. All conditions at the entrance are independent of time. The average density inside the tank varies with time and the tank originally contains 4.0 kg before mass begins to enter through the pipe. Calculate the following information:

(a) the time rate of change of mass inside the tank

(b) the mass in the tank 2 min after the filling process begins

(c) the average density of the substance in the tank before the filling process begins

(d) the average density in the tank 2 min after the filling process begins

Problem 3.46

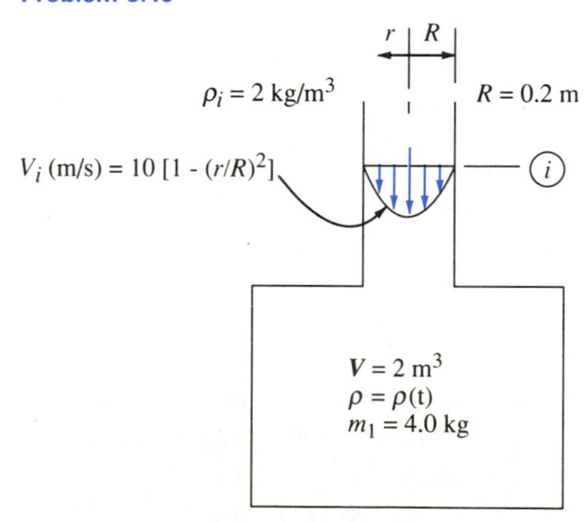

3.47 Exhaust gases with a molecular weight of 30 leave a rocket nozzle with an average velocity of 1200 ft/s. The cross-sectional area of the nozzle exit is 3.5 ft^2 and the gases at that area are at 1400°F and 25 psia. The gases behave as an ideal gas. Calculate the time rate of change of the rocket mass due to gases leaving through the nozzle. If the volume of the rocket is 375 ft^3, calculate the time rate of change of the average gas density inside the rocket.

3.48 A rigid tank contains 33 ft^3 of air at 165 psia and 80°F. A valve connected to the tank is opened slightly, and air escapes from the tank with an average velocity of 100 ft/s through an opening with an area of 0.8 in^2. The pressure and temperature of the air as it leaves the valve are 15 psia and 60°F. Calculate the following:

(a) the mass of the air inside the tank before the valve is opened

(b) the mass flow rate of air escaping through the valve

(c) the mass of air that escapes in 1 min

(d) the mass of air inside the tank after 1 min

3.49 Helium enters and leaves the system shown in the figure. The properties at the inlet and exit of the system do not vary over the circular cross-sectional areas. Determine the rate at which the mass of the helium is changing inside the system and state whether the mass inside the system is increasing or decreasing with time.

Problem 3.49

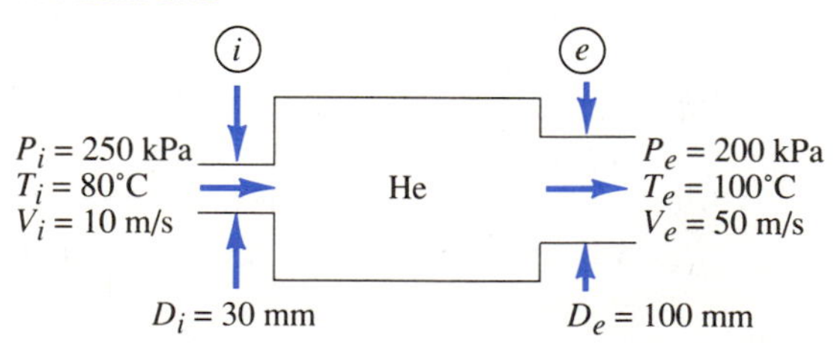

3.50 An air compressor is used to pressurize a rigid tank with a volume of 15 ft^3. Air flows steadily with a temperature of 90°F and a pressure of 120 psia through a valve into the tank at a velocity of 4 ft/s. The diameter of the pipe leading into the tank is 2 in. The tank originally contains air at a temperature of 70°F and a pressure of 30 psia. Calculate the following:

(a) the original mass of air in the tank
(b) the mass flow rate of air into the tank
(c) the time rate of change of average density inside the tank while the compressor is running
(d) the mass of air in the tank after 1 min of running the compressor

3.51 Write a computer program to determine the time required to lower the liquid level (at atmospheric pressure) in a cylindrical water tank from 10 m to 2 m. The diameter of the tank is 3 m and water leaves the tank at the bottom through an orifice having a diameter, d, of 30 mm. The volumetric flow rate through the orifice varies with the liquid level, h, according to the following relation:

$$(AV)_{\text{orifice}} = 0.15\ \pi d^2 \sqrt{2gh}$$

Solve numerically the first-order differential equation arising from application of the conservation of mass.

3.52 Write a computer program to determine the time required to drain 75 percent of the liquid water from a spherical tank through an orifice at the bottom having a diameter, d, of 40 mm. The diameter of the tank is 8 m and the water level (at atmospheric pressure) is initially 7 m above the bottom of the tank. Use the expression for the volumetric flow rate given in Problem 3.51 and solve the first-order differential equation numerically.

Problem 3.52

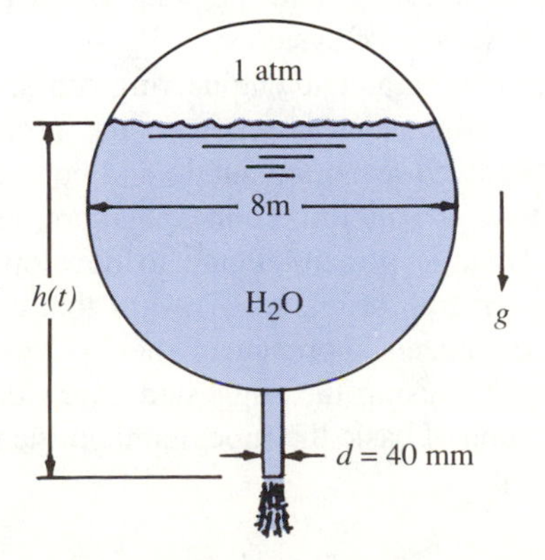

3.53 An empty flask with the shape shown in the sketch is filled with 40°F water at a constant flow rate of 10^{-2} lb_m/s. Determine expressions for the height of water $h(t)$ as a function of time and the velocity of the free surface dh/dt. At what time after the initiation of the filling process will the flask begin to overflow?

Problem 3.53

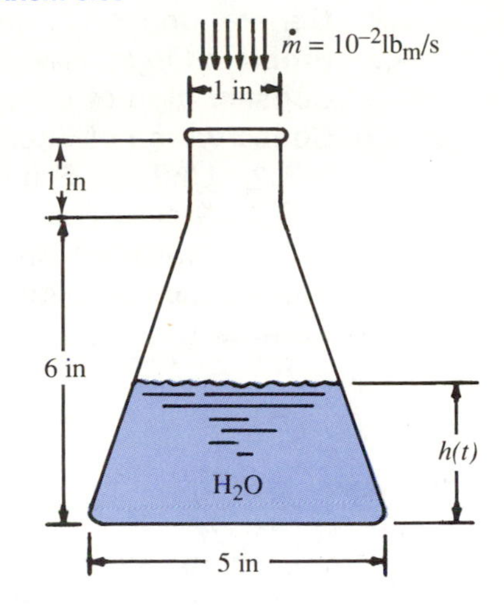

4 Conservation of Energy

4.1 INTRODUCTION

The principle of conservation of energy is a cornerstone for the analysis of thermodynamic systems. This principle provides the necessary framework required to study the relationships among the various forms of energy and energy transformations. Historically, this principle has been the foundation of the study of thermodynamics, and it is, therefore most often referred to as the *first law of thermodynamics*. The fact that the name *the first law* is traditionally preferred is, perhaps, unfortunate, because this name conceals the nature of the principle it describes.

The principle of conservation of energy enables the engineer to study the relationships among work, heat transfer, and various forms of energy. For example, one can determine the power produced by a turbine from the properties of the working fluid entering and leaving the turbine and the heat transfer from the turbine casing. This principle is a very valuable tool of thermodynamic analysis, and the student must understand it completely and be able to apply it to a variety of systems.

Engineering thermodynamics, like most other undergraduate engineering courses, is a problem-oriented subject. While the level of mathematics required in the problem solutions is relatively elementary, the concepts involved are often subtle and confusing to a person studying thermodynamics for the first time. To minimize conceptual problems, the student should strive to understand thermodynamic principles and to develop the ability to analyze and successfully solve thermodynamic problems. Toward this end a separate section in this chapter emphasizes an organized approach to thermodynamic problems. All examples in this chapter are solved by using the suggested approach so that the student can appreciate a systematic application of basic thermodynamic principles and concepts.

In this chapter the general mathematical form of the conservation-of-energy principle is developed first. The method of development closely parallels that used to derive the general conservation-of-mass equation in Chapter 3.

4.2 GENERAL CONSERVATION-OF-ENERGY EQUATION

The general principle of conservation of energy can be stated in simple terms, as follows:

Energy is a conserved property. It can be neither created nor destroyed; only its form can be altered from one form of energy to another.

In other words, the energy of a thermodynamic system must always be accounted for, or conserved. For example, suppose that a rock at an elevation z_1 is released from rest and allowed to fall to the earth, as shown in Figure 4.1. If the resistance due to the air is neglected, then as the elevation of the rock decreases, the velocity of the rock must increase. This fact follows directly from the conservation-of-energy principle. A decrease in the elevation z of the rock results in a decrease in its potential energy. Thus the conservation-of-energy principle requires that the kinetic energy of the rock increase by an equal amount, or

$$E_2 - E_1 = (E_{k2} - E_{k1}) + (E_{p2} - E_{p1}) = 0$$

The initial velocity of the rock is zero, and the velocity V_2 at the instant of time when the elevation has decreased to z_2 can be calculated from

$$E_{k2} - E_{k1} = -(E_{p2} - E_{p1})$$

or

$$\frac{m}{2}(V_2^2 - V_1^2) = mg(z_1 - z_2)$$

Solving for the velocity at this elevation yields

$$V_2 = \sqrt{2g(z_1 - z_2)}$$

This result should be a familiar one from earlier courses in physics.

Suppose the rock has an initial elevation of 30 m and a mass of 1 kg. The potential energy (relative to the earth's surface, where $z = 0$) and kinetic energy at several elevations can be calculated and the results summarized as follows:

Elevation, m	E_p, J	Velocity, m/s	E_k, J	$E = E_k + E_p$, J
30	294	0	0	294
20	196	14	98	294
10	98	19.8	196	294
0 (prior to impact)	0	24.2	294	294

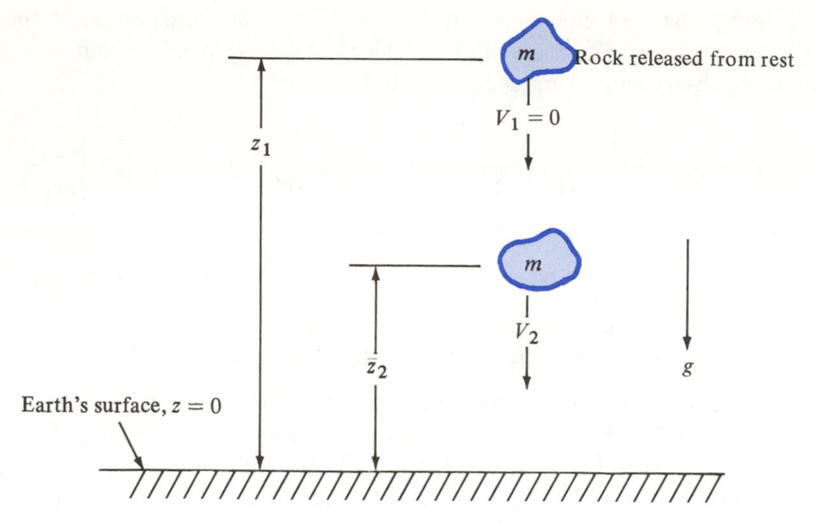

Figure 4.1 Conservation of energy applied to a falling rock.

Note that during the fall the total energy of the rock remains constant at 294 J, but the form of the energy changes. The simple example of the falling rock is used to illustrate the application of the conservation of energy at an elementary level without introducing unnecessary complications. Problems considered later include more and different forms of energy, and they require more complex solutions. Regardless of the complexity of the problem, the principle of the conservation of energy remains unchanged.

The mathematical statement of the conservation of energy used above in considering the falling rock is a special form of a much more general equation that is developed in this section. However, for the more general case the conservation-of-energy equation must include all forms of energy and energy transfers. These are as follows:

1. Energy transfer associated with differences in temperature, pressure, and so on. Heat transfer and work, the mechanisms responsible for such energy transfer, are identifiable only at the boundaries of a thermodynamic system.
2. Energy directly related to the mass of a substance. In Chapter 2 this energy was called the total energy E (or e, on a unit-mass basis) of the mass, and it is a property. Since the total energy is associated with the mass, a system will contain energy simply because it contains a quantity of mass. Furthermore, energy may be transported into or out of an open system by virtue of mass flow across the boundaries of the system.

The derivation of the general conservation-of-energy equation begins with a word equation, similar to the derivation of the conservation-of-mass equation (Equation 3.1):

$$\begin{bmatrix} \text{Rate at which} \\ \text{energy enters} \\ \text{at the boundary} \\ \text{of the system} \end{bmatrix} - \begin{bmatrix} \text{rate at which} \\ \text{energy leaves} \\ \text{at the boundary} \\ \text{of the system} \end{bmatrix} = \begin{bmatrix} \text{time rate of} \\ \text{change of the} \\ \text{energy within} \\ \text{the system} \end{bmatrix} \tag{4.1}$$

Simply stated, Equation 4.1 says that whatever energy enters the system must either leave the system or cause a change in the energy within the system. Notice the similarity between Equation 4.1 and Equation 3.1. When the word *mass* in Equation 3.1 is replaced by the word *energy*, the statement of the conservation of energy expressed by Equation 4.1 is obtained. Similarly, the form of all statements of conservation principles are identical whether the conserved quantity is mass, energy, linear or angular momentum, and so forth.

The first two terms of Equation 4.1 refer to the boundary of the system and therefore relate to energy transfer by work interactions and heat transfer and energy transport due to mass flow across the boundary. The last term, on the other hand, relates only to the energy associated with the mass within the system at any instant of time.

To transform this word equation into a mathematical equation, we consider the system shown in Figure 4.2, whose volume is free to change during the process. The rate at which energy enters or leaves at the boundary of the system includes the heat-transfer rate $\dot{Q}$, the rate at which work is performed by the system due to all work interactions, $\dot{W}_{tot}$, as well as the rate of energy transport due to mass crossing the system boundary. Since the shape of the boundary is not fixed, the term $\dot{W}_{tot}$ includes PdV work.

The properties at an inlet or exit to the system usually vary over the cross-sectional area, and for this reason the rate of energy transport is evaluated by determining the rate at which energy crosses a differential area dA and then adding each of these contributions, or integrating, over the entire cross-sectional area. The rate at which mass enters an elemental area dA was developed in Chapter 3 (Equation 3.6) and is given by

$$\begin{bmatrix} \text{Rate at which mass flows} \\ \text{through the elemental area } dA \end{bmatrix} = \rho V_n \, dA \tag{4.2}$$

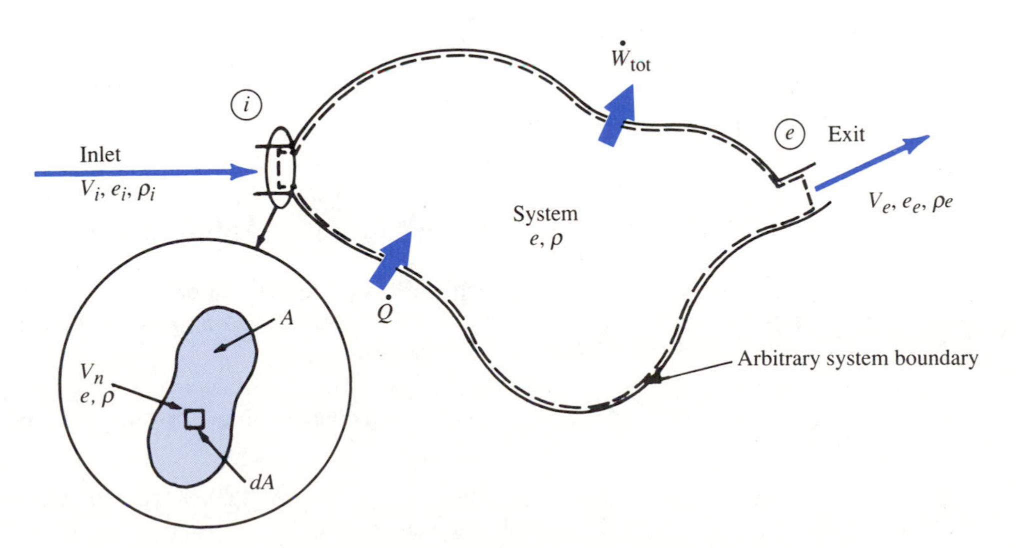

Figure 4.2 General system used for development of conservation-of-energy equation.

The rate at which energy is transported through this same elemental area is simply the product of the energy per unit mass of the fluid, e, and the mass flow rate:

$$\begin{bmatrix}\text{Rate of energy transport due to mass} \\ \text{flowing through the elemental area } dA\end{bmatrix} = e(\rho V_n\, dA) \tag{4.3}$$

Integrating this expression over all cross-sectional areas available for flow into the system (A_i) results in

$$\begin{bmatrix}\text{Rate of energy transport due to} \\ \text{mass entering the system}\end{bmatrix} = \int_{A_i} e(\rho V_n\, dA) \tag{4.4}$$

Therefore, the first term of the word equation, Equation 4.1, can be written as

$$\begin{bmatrix}\text{Rate at which energy enters at} \\ \text{the boundary of the system}\end{bmatrix} = \dot{Q} - \dot{W}_{\text{tot}} + \int_{A_i} e(\rho V_n\, dA) \tag{4.5}$$

Similarly, the second term of Equation 4.1 consists of the energy transport due to mass leaving at the boundary of the system,

$$\begin{bmatrix}\text{Rate at which energy leaves at} \\ \text{the boundary of the system}\end{bmatrix} = \int_{A_e} e(\rho V_n\, dA) \tag{4.6}$$

where A_e is the total area through which mass leaves the system. Note that $\dot{Q}$ and $\dot{W}_{\text{tot}}$ do not appear in Equation 4.6, because all heat transfer and work interactions are accounted for in Equation 4.5. For example, if the heat transfer is to the system, then $\dot{Q} > 0$ in Equation 4.5. On the other hand, if the heat transfer is from the system, then $\dot{Q} < 0$ in Equation 4.5. A similar argument holds for the rate of work performed on or by the system.

Finally, the rate of change of energy within the system is determined by evaluating the energy associated with an elemental volume, integrating over the entire volume of the system, and taking the time derivative of the result:

$$\begin{bmatrix}\text{Time rate of change of} \\ \text{the energy within the system}\end{bmatrix} = \frac{d}{dt}[E_{\text{sys}}] = \frac{d}{dt}\int_V e\rho\, d\mathbf{V} \tag{4.7}$$

Substituting Equations 4.5, 4.6, and 4.7 into the word equation, Equation 4.1, results in

$$\dot{Q} - \dot{W}_{\text{tot}} + \int_{A_i} e(\rho V_n\, dA) - \int_{A_e} e(\rho V_n\, dA) = \frac{d}{dt}\int_V e\rho\, d\mathbf{V} \tag{4.8}$$

where $\dot{W}_{\text{tot}}$ signifies the rate at which work is performed on or by the system due to all forms of work. This result is the general mathematical expression for the conservation-of-energy principle.

For successful application of Equation 4.8 to a particular thermodynamic system, the physical significance of each of the terms it contains should be thoroughly understood. These are summarized below:

$\dot{Q} \equiv \dfrac{\delta Q}{dt}$ Rate of energy entering the system due to heat transfer. (Note that $\dot{Q} > 0$ indicates heat transfer to the system, while $\dot{Q} < 0$ indicates heat transfer from the system).

$\dot{W}_{\text{tot}} \equiv \dfrac{\delta W_{\text{tot}}}{dt}$	Rate of energy leaving the system due to all work interactions, including *PdV* work, shaft work, electric work, and any other form of work that may be significant. (Note that $\dot{W}_{\text{tot}} > 0$ indicates work done by the system, while $\dot{W}_{\text{tot}} < 0$ indicates work done on the system.)
$\int_{A_e} e(\rho V_n\, dA)$	Total rate of energy leaving the system by virture of mass flow across the boundary through the surface area A_e
$\int_{A_i} e(\rho V_n\, dA)$	Total rate of energy entering the system by virtue of mass flow across the boundary through the surface area A_i
$\dfrac{d}{dt}\int_{V} e\rho\, d\mathbf{V}$	Time rate of change of the total energy within the system

If mass enters or leaves on open system, work is required to cause the mass to flow across the boundaries of the system. This work is commonly referred to as *flow work*. In the analysis of such systems this work term is customarily separated from the other work interactions that appear in the general conservation-of-energy equation, Equation 4.8.

Consider an element of mass entering a thermodynamic system, as indicated in Figure 4.3. The rate of work performed on the system in pushing the small element of mass through the elemental area dA is given by

$$\bar{F} \cdot \frac{d\bar{s}}{dt} = \bar{F} \cdot \bar{V} = F_n V_n \tag{4.9}$$

where the subscripts n denote the components of the force and velocity that are normal to the area dA. The normal force F_n can be expressed in terms of the pressure acting on

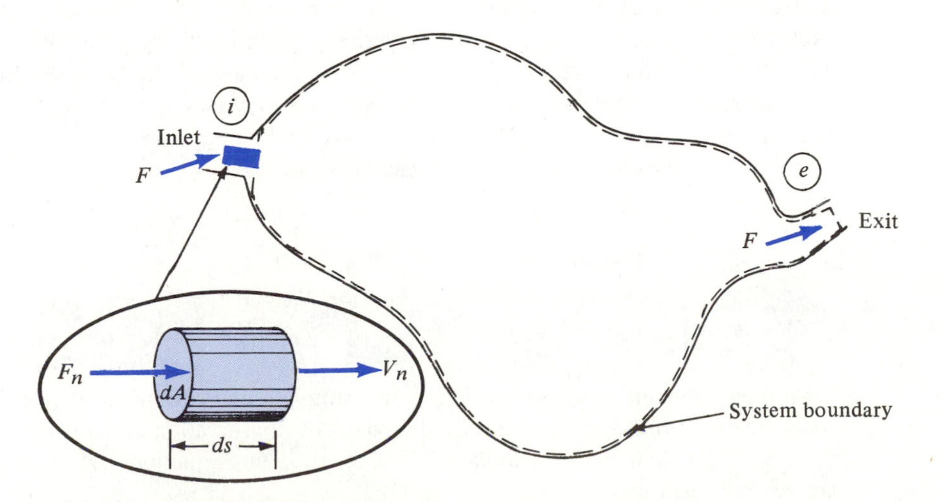

Figure 4.3 System illustrating flow work.

the element of mass. Therefore

$$F_n V_n = (PdA)V_n \tag{4.10}$$

The total rate of work associated with mass entering the system is obtained by integrating Equation 4.10 over all cross-sectional areas available for flow into the system (A_i). At the same time Equation 4.10 can be multiplied by $v\rho = 1$ for convenience:

$$\left(\frac{\delta W}{dt}\right)_i = -\int_{A_i} PV_n \, dA = -\int_{A_i} Pv(\rho V_n \, dA) \tag{4.11}$$

The negative sign in Equation 4.11 is a result of the sign convention for work; that is, work done on the system is negative. Similarly, the total rate of work associated with mass leaving the system through all cross-sectional areas available for flow out of the system (A_e) is

$$\left(\frac{\delta W}{dt}\right)_e = \int_{A_e} Pv(\rho V_n \, dA) \tag{4.12}$$

Combining Equations 4.11 and 4.12 results in an expression for the total flow work rate associated with an arbitrary system:

$$\dot{W}_{\text{flow}} = \int_{A_e} Pv(\rho V_n \, dA) - \int_{A_i} Pv(\rho V_n \, dA) \tag{4.13}$$

The flow work rate is separated from $\dot{W}_{\text{tot}}$ by defining

$$\dot{W}_{\text{tot}} \equiv \dot{W} + \dot{W}_{\text{flow}} \tag{4.14}$$

and substituting Equation 4.13 into Equation 4.8, with the result

$$\dot{Q} - \dot{W} + \int_{A_i} (e + Pv)\rho V_n \, dA - \int_{A_e} (e + Pv)\rho V_n \, dA = \frac{d}{dt}\int_{\mathbf{V}} e\rho \, d\mathbf{V} \tag{4.15}$$

where $\dot{W}$ now includes contributions from all reversible and irreversible work modes except the work associated with mass entering or leaving the system (i.e., flow work). This $\dot{W}$ is sometimes referred to as *shaft work* in the case of steady-state systems.

For simple compressible systems the energy e may be expressed as the sum of the internal energy, the kinetic energy, and the potential energy (see Equation 2.10), so the terms in parentheses in Equation 4.15 can be written as

$$e + Pv = u + Pv + e_k + e_p = h + e_k + e_p \tag{4.16}$$

and Equation 4.15 becomes

$$\dot{Q} - \dot{W} + \int_{A_i} (h + e_k + e_p)\rho V_n \, dA - \int_{A_e} (h + e_k + e_p)\rho V_n \, dA = \frac{d}{dt}\int_{\mathbf{V}} e\rho \, d\mathbf{V} \tag{4.17}$$

for a thermodynamic system consisting of a simple compressible substance. The occurrence of the sum $u + Pv$ in the formulation of the conservation of energy is the primary motivation *for defining the enthalpy*, $h = u + Pv$. Thus Equation 4.17 is merely a more convenient form of Equation 4.8.

4.3 PROBLEM ORGANIZATION FOR ANALYSIS OF THERMODYNAMIC SYSTEMS

Before we proceed with the application of the conservation-of-energy equation to specific systems, we will discuss the basic approach to the analysis of thermodynamic systems in some detail.

Much of the confusion and difficulty arising in the solution to thermodynamic problems can be attributed to the lack of organization of problem solutions. Recognition of the common ingredients in the solution to all thermodynamic problems is an important aspect of the analysis, and such problems should be approached in a logical, straightforward manner rather than in a haphazard fashion. Listed below are several suggestions that should be considered in solving thermodynamic problems. By observing and following these suggestions, the student will avoid many of the common pitfalls that accompany the problem solution.

STEP 1: IDENTIFY THE SYSTEM BOUNDARY

Draw a sketch of the thermodynamic system and clearly identify the system boundary. The sketch does not have to be elaborate, but it should resemble the system being analyzed. Be sure to identify and label all energy interactions with the surroundings. Identify all heat transfer and work interactions at the boundary of the system. Show all inlets and exits, identify mass leaving and entering the system, and identify the working substance.

STEP 2: LIST GIVEN INFORMATION

Identify the material that is inside the system boundaries and indicate it on the sketch. List the given information on the sketch of the system. Listing numerical values for the given quantities is not necessary, but you may wish to simply place symbols on the figure to designate known quantities.

STEP 3: LIST ALL ASSUMPTIONS

Most assumptions that are made while solving a problem will simplify a solution. For example, if the given conditions are such that the system can be assumed to operate under steady-state conditions, the general form of the conservation principles can be greatly simplified. Be particularly aware of other simplifying assumptions. Are conditions of a working fluid such that it behaves as an ideal gas? Can potential- and kinetic-energy changes be neglected? Is the system boundary sufficiently insulated so that the heat transfer can be assumed negligible? All of the assumptions have an important bearing on whether a simplified solution can be obtained for the problem. Remember, however, that assumptions made during the course of a solution should be fully justified or checked for validity.

STEP 4: APPLY BASIC PRINCIPLES

Once the system is identified and the given information and assumptions are listed, the actual solution to the problem can proceed. Rather than attempt to apply equations in a random fashion, use a more organized and logical approach. While all problems do not require the application of an identical sequence of equations, most problems do require application of the following concepts and principles:

1. Conservation of mass
2. Conservation of energy
3. Property relationships

This list is not intended to be all-inclusive, and other concepts and basic principles can be added to it as they are developed in the study of thermodynamics. The property relationships frequently are in the form of tabular or graphical values. If the assumption of ideal-gas behavior is appropriate for the working fluid, use the ideal-gas equation of state and other property equations that apply when the behavior of the substance is ideal.

Additional property information can be gained if the process is one of constant pressure, temperature, or volume. A sketch of the process on a P-v, T-v, or other process diagram is often helpful. A process diagram can assist in identifying phases present during a process and can aid in visualizing the relationship between properties as the process takes place.

STEP 5: SIMPLIFY EQUATIONS

Using the equations identified in step 4, apply appropriate assumptions to simplify them. Reduce the equations until there is enough information to solve for the required quantities. If there is insufficient information for a solution, consider further assumptions and be sure that all equations that apply to the problem have been considered.

Avoid premature substitution of numerical values into the equations. Manipulate the equations in symbolic form until the unknown quantities can be written in terms of known quantities. Ensure that the units in the equations are consistent throughout. Remember that the ideal-gas equation of state and any relationship derived from it require the use of absolute pressures and absolute temperatures.

STEP 6: ANALYZE THE RESULTS

Once you have completed the problem solution, take time to analyze the results. Is the magnitude of the numerical value reasonable? Is the algebraic sign correct? Are the units associated with the numerical value correct? Try to interpret the results and draw conclusions.

Obviously all engineering problems are not solved by using the same series of steps. Nevertheless, these six steps relate to common characteristics of most thermodynamic problems. In simpler problems some of these steps may be unnecessary, while more complex problems may require additional steps. As you gain more experience, you may wish to modify the list to include hints that you find particularly helpful.

The solutions to example problems in this chapter follow these six steps. The intent here is not to establish a rigid pattern for problem solutions but, rather, to emphasize the value of a logical approach to thermodynamic analyses.

4.4 CONSERVATION OF ENERGY FOR CLOSED SYSTEMS

One particular thermodynamic system of interest to engineers is the closed system. Such a system is characterized by the fact that mass cannot enter or leave the system. That is, the surface area of the closed system contains no inlets or exits. Therefore, the conservation of mass, Equation 3.2, requires that the amount of mass within a closed system remain constant:

$$\frac{dm_{sys}}{dt} = 0 \tag{4.18}$$

or

$$m_{sys} = \text{constant}$$

Similarly, since flow into or out of the system is excluded, the conservation-of-energy principle for the closed system, Equation 4.17, reduces to

$$\dot{Q} - \dot{W} = \frac{d}{dt}\int_V e\rho \, dV = \frac{dE_{sys}}{dt} \tag{4.19}$$

Inserting the definitions of $\dot{Q}$ and $\dot{W}$ into Equation 4.19 results in

$$\frac{\delta Q}{dt} - \frac{\delta W}{dt} = \frac{dE_{sys}}{dt} \tag{4.20}$$

which can be written as

$$\delta Q - \delta W = dE_{sys} \tag{4.21}$$

Equations 4.20 and 4.21 are the differential forms of the conservation of energy for a closed system.

The thermodynamic analysis of a closed system is usually concerned with the effects of a finite change of state as the system proceeds from some initial state to some final state, and integration of Equation 4.20 with respect to time is appropriate. Designating the total energy of the system at state 1 by E_1 at the initial time t_1 and by E_2 at state 2 after an interval of time $\Delta t = t_2 - t_1$, the integrated form of Equation 4.20 can be written as

$$Q_{12} - W_{12} = (E_2 - E_1)_{sys} \tag{4.22}$$

This equation is the convservation-of-energy equation as it applies to a closed system that undergoes a process from state 1 to state 2. The work term in Equation 4.22 includes both reversible and irreversible work modes.

Recall that both heat transfer and work are path functions and that the magnitudes of the heat transfer and work interactions that occur during a change of state depend on how the system achieves the change of state. The notations Q_{12} and W_{12} are used to represent the heat transfer and work, respectively, for the particular process that the system undergoes during a finite change of state from state 1 to state 2.

If both $\dot{Q}$ and $\dot{W}$ are constant over the time interval Δt, Equation 4.19 can be integrated with respect to time, with the result

$$\dot{Q}\,\Delta t - \dot{W}\,\Delta t = (E_2 - E_1)_{\text{sys}} \tag{4.23}$$

When a closed system undergoes a cycle, an energy analysis of the cycle can be performed by integrating Equation 4.21 throughout the cycle. For this purpose the cyclic integral $\oint$ is used to indicate integration over the entire cycle. Performing the cyclic integral of Equation 4.21 results in

$$\oint \delta Q - \oint \delta W = \oint dE = 0 \tag{4.24}$$

Since the energy of the system is a property and since the end states of a cycle are identical, the cyclic integral of the energy change is identically zero.

Equation 4.24 is a mathematical statement of a conclusion obtained experimentally by Joule in the 1840s. In a series of experiments the temperature of a fluid was increased by performing work on the fluid by means of a paddle wheel, and Joule found that the amount of heat transfer required to return the fluid to its original temperature (a complete cycle) was equal to the amount of work done on the fluid. Thus, as Equation 4.24 states, the algebraic sum of the net heat transfer for the cycle is equal to the algebraic sum of the net work for the cycle.

For a cycle composed of a number of individual processes, evaluation of the cyclic integral is a simple matter. For example, suppose that a cycle is composed of three processes, A, B, and C, as shown in Figure 4.4. Process A begins at state 1 and terminates at state 2. Process B begins at state 2 and terminates at state 3. Process C begins at state

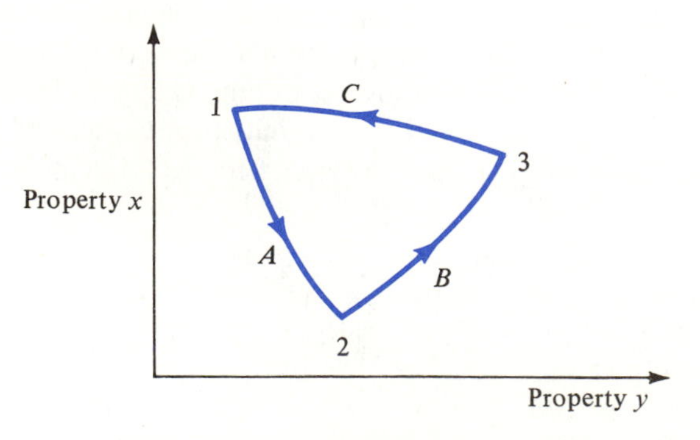

Figure 4.4 Example of a cyclic process.

3 and terminates at state 1 in order to complete the cycle. For this cycle the cyclic integral of the heat transfer is evaluated as

$$\oint \delta Q = \int_1^2 \delta Q + \int_2^3 \delta Q + \int_3^1 \delta Q = Q_{12} + Q_{23} + Q_{31}$$

Thus the value of the cyclic integral of a quantity such as heat transfer is determined by simply summing the various heat-transfer contributions for each process of the cycle. For this cycle Equation 4.24 requires that

$$(Q_{12} + Q_{23} + Q_{31}) = (W_{12} + W_{23} + W_{31})$$

The fact that the cyclic integral of the energy change is zero can also be verified, since

$$\oint dE = \int_1^2 dE + \int_2^3 dE + \int_3^1 dE$$
$$= (E_2 - E_1) + (E_3 - E_2) + (E_1 - E_3) = 0$$

Closed systems are not as frequently encountered in actual thermodynamic analyses as are open systems. Nevertheless, the analysis of closed systems is an important aspect of engineering thermodynamics. Such systems are very often quite simple and straightforward, so the problem solution is not overshadowed by unnecessary complexities. The experience gained in closed-system analysis, then, should serve to strengthen the student's ability to analyze more difficult problems. In addition, the concept of a closed system can often be used to advantage in the analysis of the gross behavior of large systems. For example, the simple steam power plant shown in Figure 4.5 consists of four devices (boiler, turbine, condenser, and pump), which could be analyzed separately as open-system devices as described in the next section. On the other hand, if the system

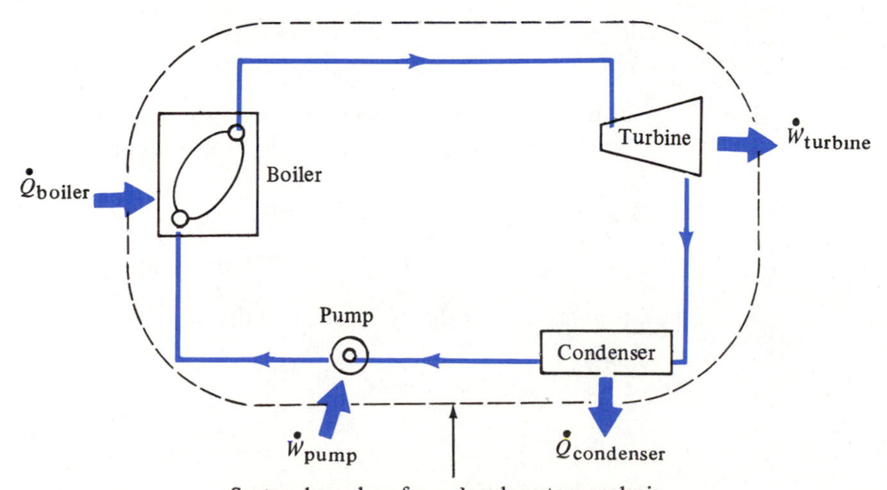

Figure 4.5 Schematic for a simple steam power plant.

boundary includes all four devices, the entire power plant can be analyzed as a closed system since no mass crosses the system boundary.

The following examples illustrate the thermodynamic analysis of closed systems.

EXAMPLE 4.1

A closed system executes a cycle composed of three separate processes. During the first process 8 kJ of heat transfer to the system occurs while the system performs 2 kJ of work. The second process is adiabatic. During the third process 3 kJ of work is performed on the system, and the total energy of the system decreases by 2 kJ. Determine the change in total energy of the system during each of the first two processes, the heat transfer for the last process, and the work for the second process.

Solution. The identity of the system is unknown in this problem and there is no need to attempt to sketch it. We do know, however, that the system involved is a closed system and that it undergoes a complete cycle. Therefore, Equation 4.22 applies to each individual process and Equation 4.24 applies to the entire cycle.

Applying Equation 4.22 to the first process, we have

$$Q_{12} - W_{12} = E_2 - E_1$$

or

$$E_2 - E_1 = 8 \text{ kJ} - 2 \text{ kJ}$$

$$E_2 - E_1 = \underline{\underline{6 \text{ kJ}}} \qquad \text{(increase)}$$

Applying Equation 4.22 to the last process yields

$$Q_{31} - W_{31} = E_1 - E_3$$

or

$$Q_{31} = -3 \text{ kJ} - 2 \text{ kJ} = \underline{\underline{-5 \text{ kJ}}} \qquad \text{(from the system)}$$

The heat transfer for the second process is zero since the process is adiabatic, and therefore the cyclic integral of the heat transfer can be evaluated as

$$\oint \delta Q = \int_1^2 \delta Q + \int_2^3 \delta Q + \int_3^1 \delta Q$$
$$= Q_{12} + Q_{23} + Q_{31} = +8 \text{ kJ} + 0 - 5 \text{ kJ} = +3 \text{ kJ}$$

Applying Equation 4.24 for the cycle, we find that

$$\oint \delta Q = \oint \delta W$$

or

$$3 \text{ kJ} = 2 \text{ kJ} + W_{23} + (-3 \text{ kJ})$$

$$W_{23} = \underline{\underline{4 \text{ kJ}}} \qquad \text{(work performed on the surroundings)}$$

Equation 4.22 can be used to find the total energy change during the second process:

$$Q_{23} - W_{23} = E_3 - E_2$$

$$E_3 - E_2 = 0 - 4 \text{ kJ} = \underline{\underline{-4 \text{ kJ}}} \qquad \text{(decrease)}$$

Finally, note that the cyclic integral of the total energy change should sum to zero. This condition provides a check on the previous calculations:

$$\oint dE = (E_2 - E_1) + (E_3 - E_2) + (E_1 - E_3)$$

$$= +6 \text{ kJ} - 4 \text{ kJ} - 2 \text{ kJ} = 0 \qquad \text{(check)}$$

■

There are many examples of closed thermodynamic systems; some are real and others are merely idealizations of actual devices. For instance, an incandescent light bulb, a television picture tube, a storage battery, an inflated automobile tire, storage tanks, and blocks of solid materials are all examples of closed systems. On the other hand, even though the cylinder of an internal-combustion engine is actually an open system because fuel and air enter the cylinder and combustion products leave, the piston-cylinder assembly is often idealized as a closed system for the purpose of simple thermodynamic analysis.

■ EXAMPLE 4.2

A football official inflates a football to the required gauge pressure of 13 lb_f/in^2 prior to a game. The football has an internal volume of 160 in^3 and the air is at a temperature of 75°F when the football is first inflated. The ball is taken onto the field, and by the time it is put into play, the air temperature inside the football has dropped to 30°F. Assuming that the volume of the football does not change significantly during the cooling process, calculate the following quantities:

(a) The mass of air in the ball
(b) The pressure of the air in the ball when play begins
(c) The amount of heat transfer from the air in the ball during the process
(d) The initial pressure to which the ball must be inflated so that it will be at the required 13 lb_f/in^2 gauge pressure when the temperature reaches 30°F

Solution. We consider the air to be the system, as shown in Figure 4.6. Since no mass crosses the boundary of the system, the air inside the football is a closed system. The air is at low pressure and high temperature relative to the critical-state values, so the assumption of ideal-gas behavior is appropriate.

(a) The mass of air in the ball can be determined from the ideal-gas equation of state at the given initial conditions:

$$m_1 = \frac{P_1 V_1}{RT_1} = \frac{[(13 + 14.7)\text{lb}_f/\text{in}^2](160\,\text{in}^3)(1\text{ ft}/12\text{ in})}{53.34\,\dfrac{\text{ft}\cdot\text{lb}_f}{\text{lb}_m\cdot{}^\circ\text{R}}\,(75 + 460)^\circ\text{R}}$$

$$= \underline{\underline{1.29 \times 10^{-2}\ \text{lb}_m}}$$

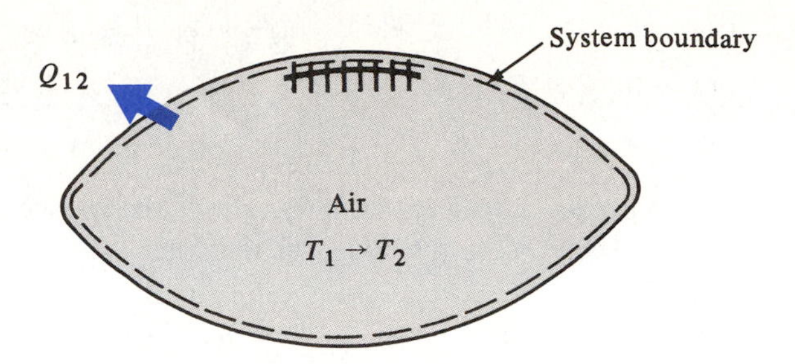

Figure 4.6 System for Example 4.2.

(b) Applying the conservation-of-mass equation, Equation 3.4, for a closed system results in

$$m_1 = m_2$$

and the pressure when the temperature drops to 30°F can be calculated by applying the ideal-gas equation of state at the final state, or

$$P_2 = \frac{m_2 R T_2}{\mathbf{V}_2}$$

$$= \frac{(1.29 \times 10^{-2}\ \text{lb}_\text{m})(53.34\ \text{ft}\cdot\text{lb}_\text{f}/\text{lb}_\text{m}\cdot°\text{R})(30 + 460)°\text{R}(1728\ \text{in}^3/\text{ft}^3)}{(160\ \text{in}^3)}$$

$$= 3641\ \text{lb}_\text{f}/\text{ft}^2 = 25.3\ \text{lb}_\text{f}/\text{in}^2 \qquad \text{(absolute pressure)}$$

$$= \underline{\underline{10.6\ \text{lb}_\text{f}/\text{in}^2}} \qquad \text{(gauge pressure)}$$

(c) The amount of heat transfer from the air during the process can be determined by applying the conservation-of-energy equation for a closed system, Equation 4.22:

$$Q_{12} - W_{12} = E_2 - E_1$$

If the football is assumed to be rigid, there is no $Pd\mathbf{V}$ work, and irreversible-work modes are absent. Also, potential- and kinetic-energy changes can be neglected, so the conservation-of-energy equation reduces to

$$Q_{12} = U_2 - U_1 = m(u_2 - u_1)$$

The air has been assumed to be an ideal gas and values for internal energy can be determined from Table D.1E:

$$u_2 = 83.6\ \text{Btu}/\text{lb}_\text{m}$$

$$u_1 = 91.3\ \text{Btu}/\text{lb}_\text{m}$$

Thus,

$$Q_{12} = (1.29 \times 10^{-2}\ \text{lb}_\text{m})(83.6 - 91.3)\text{Btu/lb}_\text{m}$$

$$= \underline{\underline{-0.0993\ \text{Btu}}}$$

The heat transfer Q_{12} is negative, indicating that the air in the football is cooled.

(d) Since the volume and mass of air are constant, the ratio of the ideal-gas equation of state evaluated at the initial and final states gives

$$\frac{P_1}{P_2} = \frac{T_1}{T_2}$$

or

$$P_1 = \left(\frac{P_2 T_1}{T_2}\right) = \frac{[(13 + 14.7)\text{lb}_\text{f}/\text{in}^2](75 + 460)°\text{R}}{(30 + 460)°\text{R}}$$

$$= 30.2\ \text{lb}_\text{f}/\text{in}^2 \quad \text{(absolute)} \qquad = \underline{\underline{15.5\ \text{lb}_\text{f}/\text{in}^2}} \quad \text{(gauge)}$$

Note that since this equation was developed from the equation of state for an ideal gas, absolute pressures and absolute temperatures must be used. ■

■ EXAMPLE 4.3

A refrigerant-12 tank is located outdoors. The tank has a volume of 20 ft^3 and it is filled with 75 lb$_\text{m}$ of R-12 at a pressure of 100 psia. During the daytime the tank is exposed to the sun, and heat transfer to the R-12 from the sun causes the R-12 to reach a saturated-vapor state. Calculate the following quantities:

(a) The initial temperature and state of the R-12 prior to heating
(b) The final temperature and pressure of the R-12 after heating
(c) The amount of heat transfer to the R-12

Solution. A sketch of the tank is shown in Figure 4.7(a). The R-12 is a closed system, so the conservation of mass requires that

$$m_1 = m_2$$

The initial specific volume of the R-12 can be calculated from the given volume and mass:

$$v_1 = \frac{V}{m_1} = \frac{20\ \text{ft}^3}{75\ \text{lb}_\text{m}} = 0.2667\ \text{ft}^3/\text{lb}_\text{m}$$

From Table C.2E at a pressure of 100 psia, the specific volume is between the value of v_f and v_g, indicating that the original state of the R-12 is in the saturation (two-phase) region.

The process on a P-v diagram is shown in Figure 4.7(b). Notice that the path followed during the heating process is one of constant specific volume because both the mass and volume of the system remain constant, so

$$v_1 = v_2$$

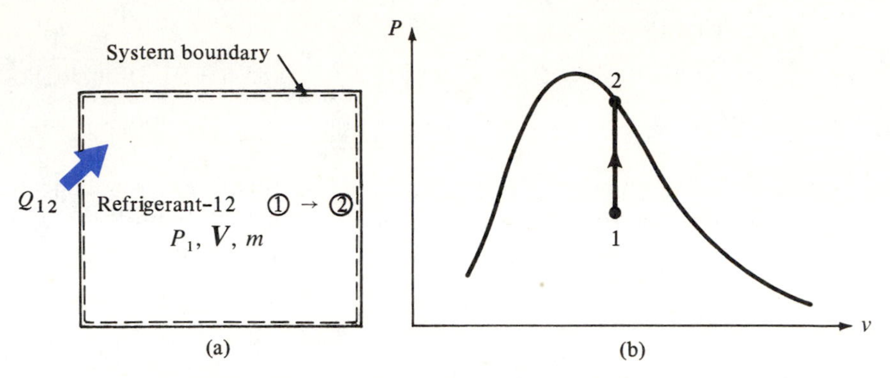

Figure 4.7 (a) System consisting of refrigerant-12 for Example 4.3; (b) process diagram for refrigerant-12.

Note also that the final state 2 is on the saturated-vapor curve. During the heating process the pressure and temperature increase, while the specific volume remains constant.

(a) Since state 1 is in the saturation region, the initial temperature and quality of the R-12 can be determined from the properties in Table C.2E. These two properties completely specify the state of the R-12 at the initial conditions:

$$T_1 = T_{\text{sat}} = \underline{\underline{80.8°\text{F}}}$$

$$x_1 = \frac{v_1 - v_f}{v_{fg}} = \frac{(0.2667 - 0.01229)\text{ft}^3/\text{lb}_\text{m}}{0.3979\ \text{ft}^3/\text{lb}_\text{m}} = \underline{\underline{0.639}}$$

(b) The final temperature and pressure can be determined once two independent properties at the final state are known. These two properties are the quality and the specific volume:

$$x_2 = 1.0$$

$$v_2 = v_1 = 0.2667\ \text{ft}^3/\text{lb}_\text{m}$$

From Table C.2E the pressure and temperature at the final state are

$$P_2 = \underline{\underline{153.4\ \text{psia}}}$$

$$T_2 = \underline{\underline{111.2°\text{F}}}$$

(c) The heat transfer to the R-12 during the process can be determined by applying the conservation of energy for a closed system:

$$Q_{12} - W_{12} = E_2 - E_1$$

No work is performed during the process, and the system is stationary, so kinetic- and potential-energy changes are zero. The conservation-of-energy equation reduces to

$$Q_{12} = m(u_2 - u_1)$$

The initial and final internal energies can be determined from Table C.2E because both state points are known:

$$u_1 = u_f + x_1 u_{fg} = 26.59 \text{ Btu/lb}_m + 0.639(51.84 \text{ Btu/lb}_m) = 59.72 \frac{\text{Btu}}{\text{lb}_m}$$

$$u_2 = u_g = 81.15 \text{ Btu/lb}_m$$

Substituting these values into the conservation-of-energy equation yields

$$Q_{12} = m(u_2 - u_1) = (75 \text{ lb}_m)(81.15 - 59.72)\text{Btu/lb}_m$$

$$= \underline{\underline{1607 \text{ Btu}}} \qquad \text{(to the system)}$$

As a result of solar heating during the day, the pressure in the tank has increased by more than 50 percent. If the possibility of heating is overlooked during the design of the pressure vessel, a catastrophic failure can result. Pressure-vessel design codes require some means of limiting the pressure, such as pressure-relief valves, in an attempt to eliminate failures due to overpressurization.

In this example, a P-v diagram was used to sketch the path followed by the system during the process. The use of such a sketch often helps to increase the understanding of a process while providing additional insight or information that might otherwise be overlooked. ■

■ EXAMPLE 4.4

A piston-cylinder assembly contains 1 kg of a substance at 100 kPa. The initial volume is 0.5 m^3. Heat transfer to the substance causes a slow expansion at constant temperature. This process is terminated when the final volume is twice the initial volume. Determine the magnitude of the heat transfer required if the substance is

(a) Nitrogen
(b) Water

Solution. The substance within the cylinder is chosen to be the system, as shown in Figure 4.8(a). The system is closed, so the conservation of mass reduces to $m_1 = m_2$. Assuming that the system is stationary, the kinetic and potential energies of the system remain unchanged, and the conservation-of-energy equation, Equation 4.22, reduces to

$$Q_{12} - W_{12} = U_2 - U_1 = m(u_2 - u_1)$$

The process proceeds slowly and the work consists only of $Pd\boldsymbol{V}$ work;

$$W_{12} = \int_1^2 Pd\boldsymbol{V}$$

(a) Nitrogen can be assumed to behave as an ideal gas since its pressure is much lower than the critical pressure and its specific volume is much greater than the critical-state specific volume. A sketch of the isothermal process for nitrogen is shown in Figure 4.8(b) on a P-$\boldsymbol{V}$ diagram.

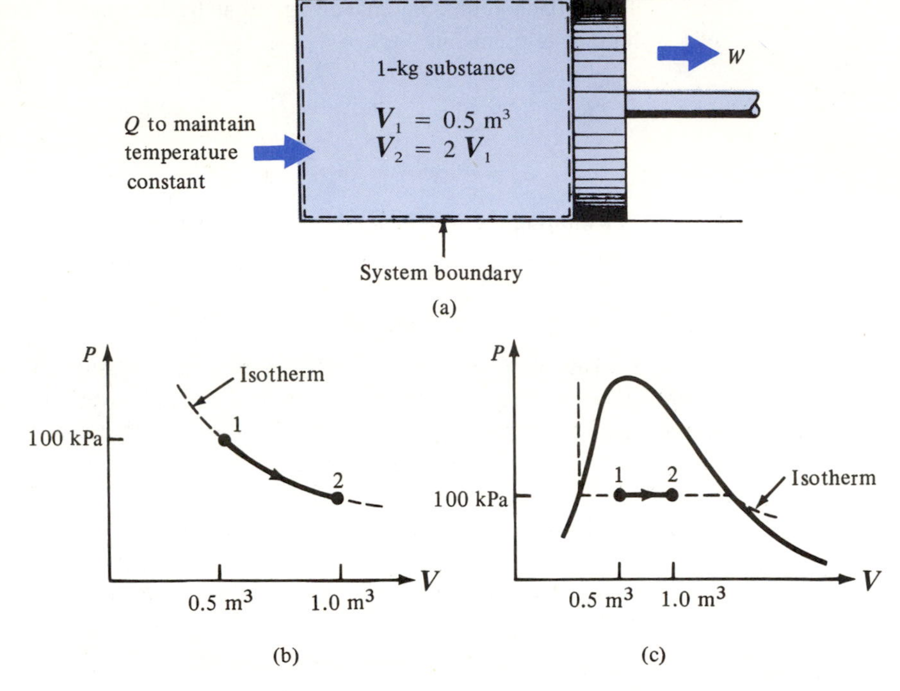

Figure 4.8 (a) Piston-cylinder assembly for Example 4.4; (b) process diagram for nitrogen; (c) process diagram for water.

Since the internal energy of an ideal gas is a function of the absolute temperature only, the change in internal energy of the nitrogen during an isothermal process is zero, or

$$u_2 = u_1$$

Therefore, the heat transfer is

$$Q_{12} = W_{12} = \int_1^2 P dV$$

The integration can be performed once the functional relation between P and V is known. This relationship can be determined for this process by employing the equation of state for an ideal gas:

$$PV = mRT = \text{constant}$$

Solving for P yields

$$P = \frac{mRT}{V}$$

and

$$Q_{12} = W_{12} = mRT \int_1^2 \frac{dV}{V} = mRT \ln\left[\frac{V_2}{V_1}\right]$$

or, alternatively,

$$Q_{12} = W_{12} = P_1 \boldsymbol{V}_1 \ln\left(\frac{\boldsymbol{V}_2}{\boldsymbol{V}_1}\right)$$

$$= \frac{(100 \text{ kPa})(0.5 \text{ m}^3)}{(1 \text{ kPa}\cdot\text{m}^2/\text{kN})(1 \text{ kN}\cdot\text{m/kJ})} (\ln 2)$$

$$= \underline{\underline{34.7 \text{ kJ}}} \qquad \text{(into the system)}$$

(b) To sketch the P-$\boldsymbol{V}$ process diagram for water, we first determine the state of the water.

The initial state is fixed by the pressure and the specific volume

$$P_1 = 100 \text{ kPa}$$

$$v_1 = \frac{\boldsymbol{V}_1}{m} = \frac{0.5 \text{ m}^3}{1 \text{ kg}} = 0.5 \text{ m}^3/\text{kg}$$

From the saturation-pressure table for water at 100 kPa, we find that

$$v_f < v_1 < v_g$$

so that the initial state is in the saturation region where $T_1 = T_{sat} = 99.63°\text{C}$. The final state is fixed by the temperature and the specific volume:

$$T_2 = T_1 = 99.63°\text{C}$$

$$v_2 = \frac{\boldsymbol{V}_2}{m} = \frac{2\boldsymbol{V}_1}{m} = 2v_1 = 1.0 \text{ m}^3/\text{kg}$$

From the saturation table at T_2 this state is also in the saturation region since

$$v_f < v_2 < v_g$$

Because the water remains in the saturation region during the process while the temperature remains constant at 99.63°C, the pressure also remains constant at the saturation pressure of 100 kPa. A sketch of this process for water is shown in Figure 4.8(c).

From the conservation of energy

$$Q_{12} - W_{12} = U_2 - U_1$$

For the constant-pressure process the work is simply

$$W_{12} = \int_1^2 P d\boldsymbol{V} = P(\boldsymbol{V}_2 - \boldsymbol{V}_1)$$

Thus the heat transfer is found to be

$$Q_{12} = U_2 - U_1 + P(\boldsymbol{V}_2 - \boldsymbol{V}_1)$$

which is equivalent to

$$Q_{12} = H_2 - H_1 \qquad \text{since} \qquad P = \text{constant}$$

The change in enthalpy during this isothermal process is not zero, however, because water in the saturation region does not behave as an ideal gas. The enthalpy of the water at the end states can be found by using the quality. From Table B.2 at 100 kPa,

$$v_f = 0.001043 \text{ m}^3/\text{kg}$$

$$v_{fg} = 1.693 \text{ m}^3/\text{kg}$$

And the qualities are determined from the relation $v = v_f + xv_{fg}$:

$$x_1 = \frac{v_1 - v_f}{v_{fg}} = \frac{(0.5 - 0.001043)\text{m}^3/\text{kg}}{1.693 \text{ m}^3/\text{kg}} = 0.295$$

$$x_2 = \frac{v_2 - v_f}{v_{fg}} = \frac{(1.0 - 0.001043)\text{m}^3/\text{kg}}{1.693 \text{ m}^3/\text{kg}} = 0.590$$

The enthalpies are determined from the relation $h = h_f + xh_{fg}$ and Table B.2 at 100 kPa:

$$h_1 = 417.51 \text{ kJ/kg} + 0.295(2257.6 \text{ kJ/kg}) = 1083.5 \text{ kJ/kg}$$

$$h_2 = 417.51 \text{ kJ/kg} + 0.590(2257.6 \text{ kJ/kg}) = 1749.5 \text{ kJ/kg}$$

Finally, the heat transfer is

$$Q_{12} = m(h_2 - h_1) = (1 \text{ kg})(1749.5 - 1083.5)\text{kJ/kg}$$

$$= \underline{\underline{666 \text{ kJ}}} \qquad \text{(into the system)}$$

■

■ EXAMPLE 4.5

A well-insulated frictionless piston-cylinder assembly contains 2 lb_m of air initially at 50°F and 150 psia. An electric-resistance heating element inside the cylinder is energized and causes the air temperature to reach 400°F. The pressure of the air is maintained constant throughout the process. Determine the work for the process and the amount of electrical work.

Solution. A sketch of the system and a P-v process diagram are shown in Figure 4.9. Since the system is closed, the mass remains constant:

$$m_1 = m_2 = m$$

Furthermore, the air pressure is low and the temperatures are high relative to the critical-state values, so the assumption of ideal-gas behavior is appropriate:

$$PV = mRT$$

Applying the conservation-of-energy equation for a closed system, we have

$$Q_{12} - W_{12} = m(u_2 - u_1)$$

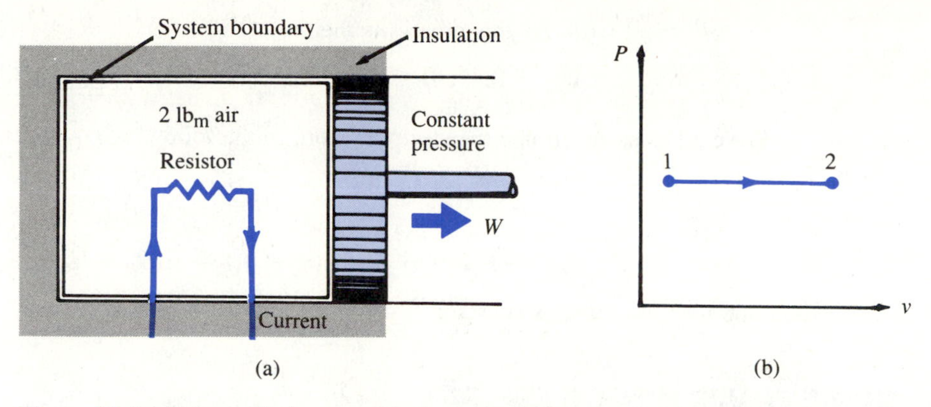

Figure 4.9 (a) Piston-cylinder assembly for Example 4.5; (b) process diagram for air.

For the system boundary chosen in Figure 4.9 the heat transfer is zero because the cylinder is well-insulated. Thus

$$W_{12} = -m(u_2 - u_1)$$

The temperature change of the air is small, so we can assume an average value of c_v and use Equation 2.40 for the internal-energy change:

$$\begin{aligned} W_{12} &= -m(u_2 - u_1) = -mc_v(T_2 - T_1) \\ &= -(2 \text{ lb}_m)(0.173 \text{ Btu/lb}_m \cdot {}^\circ\text{R})(860 - 510){}^\circ\text{R} \\ &= \underline{\underline{-121 \text{ Btu}}} \qquad \text{(on the system)} \end{aligned}$$

Thus the net work for this process is negative. The net work W_{12} includes all forms of work that are present during the process. In this example it is the sum of the electrical work and the $Pd\mathbf{V}$ work:

$$W_{12} = W_{12,\text{elec}} + W_{12,Pd\mathbf{V}}$$

The $Pd\mathbf{V}$ work can be determined from

$$W_{12,Pd\mathbf{V}} = \int_1^2 Pd\mathbf{V} = P(\mathbf{V}_2 - \mathbf{V}_1)$$

since the pressure remains constant. Ideal-gas behavior is assumed, so the preceding equation can be written as

$$\begin{aligned} W_{12,Pd\mathbf{V}} &= mR(T_2 - T_1) \\ &= (2 \text{ lb}_m)\left(\frac{1.986 \text{ Btu/lb}_m\cdot\text{mol}\cdot{}^\circ\text{R}}{28.97 \text{ lb}_m/\text{lb}_m\cdot\text{mol}}\right)(860 - 510){}^\circ\text{R} \\ &= 48.0 \text{ Btu} \qquad \text{(by the system)} \end{aligned}$$

The electrical work for this process is therefore

$$W_{12,\text{elec}} = W_{12} - W_{12,P\,d\mathbf{V}} = -121\ \text{Btu} - 48.0\ \text{Btu} = \underline{\underline{-169\ \text{Btu}}}$$

We could also determine the electrical work in the following equivalent manner:

$$W_{12} = W_{12,\text{elec}} + W_{12,P\,d\mathbf{V}} = -m(u_2 - u_1)$$

or

$$W_{12,\text{elec}} = -P(\mathbf{V}_2 - \mathbf{V}_1) - m(u_2 - u_1) = -(H_2 - H_1)$$

since the pressure remains constant. ■

4.5 CONSERVATION OF ENERGY FOR OPEN SYSTEMS

The open system, characterized by the fact that both mass and energy can cross the boundary of the system, is frequently encountered in engineering applications. Open systems include a large number of practical devices that can be conveniently subdivided into systems that undergo either steady-state or transient processes.

In this section the application of the conservation-of-energy principle and problem-solving techniques to open systems is discussed and demonstrated. Particular attention is given to the characteristics of the more common types of steady-state devices operating with uniform flow conditions.

4.5.1 Uniform Flow

In most engineering applications the assumption of uniform flow at a cross section can be used without appreciable error. The validity of the uniform-flow assumption and its application to the conservation-of-mass principle were discussed in detail in Chapter 3. The conservation-of-energy principle for uniform flow is developed in the following paragraphs.

Recall that the general form of the conservation-of-energy equation, Equation 4.17, involves integrals that must be evaluated at each inlet and exit on the bounding surface of a system:

$$\dot{Q} - \dot{W} + \int_{A_i} (h + e_k + e_p)\rho V_n\, dA - \int_{A_e} (h + e_k + e_p)\rho V_n\, dA = \frac{d}{dt}\int_{\mathbf{V}} e\rho\, d\mathbf{V} \tag{4.25}$$

Under the conditions of uniform flow, the fluid properties at each inlet and exit are uniform over each cross-sectional area. Thus the fluid properties can be removed from the integrals with the following results:

$$\int_{A_e} (h + e_k + e_p)\rho V_n\, dA = \sum_{\text{exit}} \int_{A} (h + e_k + e_p)\rho V_n\, dA = \sum_{\text{exit}} (h + e_k + e_p) \int_{A} \rho V_n\, dA$$

Thus

$$\int_{A_e} (h + e_k + e_p)\rho V_n\, dA = \sum_{\text{exit}} \dot{m}(h + e_k + e_p) \tag{4.26}$$

Similarly

$$\int_{A_i} (h + e_k + e_p)\rho V_n \, dA = \sum_{\text{inlet}} \dot{m}(h + e_k + e_p) \tag{4.27}$$

The summation signs in Equations 4.26 and 4.27 indicate that the quantity $\dot{m}(h + e_k + e_p)$ is to be evaluated at each inlet and each exit at the boundary of the system. Furthermore, the volume integral in Equation 4.25 is the total energy of the mass within the system, so that

$$\frac{d}{dt}\int_V e\rho \, dV = \frac{dE_{\text{sys}}}{dt} \tag{4.28}$$

With the substitution of Equations 4.26 through 4.28 into Equation 4.25, the conservation-of-energy equation for uniform flow can be written as

$$\dot{Q} - \dot{W} + \sum_{\text{inlet}} \dot{m}(h + e_k + e_p) - \sum_{\text{exit}} \dot{m}(h + e_k + e_p) = \frac{dE_{\text{sys}}}{dt} \tag{4.29}$$

This particular form of the conservation-of-energy principle will be used throughout the remainder of this chapter for the analysis of open systems undergoing steady-state and transient processes. Before we proceed with the discussion, some general comments concerning simplifying assumptions other than uniform flow are in order.

In working with the conservation-of-energy equation, a comparison of the magnitudes of the changes in kinetic energy and potential energy and the changes in enthalpy or internal energy is often valuable. Those energy changes that are found to be comparatively small can be neglected without significant loss of accuracy. The energy change associated with elevation changes and velocity changes can easily be determined. For example, the elevation change required to produce a potential-energy change of 1 kJ/kg of a substance is

$$g \, \Delta z = \Delta e_p = 1 \, \text{kJ/kg}$$

or

$$\Delta z = \frac{1 \text{ kJ/kg}}{9.81 \text{ m/s}^2} = 0.102 \text{ km} = 102 \, \text{m}$$

Thus an elevation change of over 100 m is required to produce an energy change of just 1 kJ/kg. This same energy change in liquid water, for instance, could be produced by a temperature change of about 1/4°C. Similarly, an elevation change of 778 ft is required to produce an energy change of just 1 Btu/lb_m. Therefore, in an analysis of many steady-flow devices, the potential-energy change can often be neglected relative to the enthalpy change of the working fluid because the difference in elevation between inlet and exit in many devices is less than several meters.

A similar analysis can be used to determine the magnitude of the velocity change required to produce this same energy change of 1 kJ/kg. This problem, however, is not as straightforward because the kinetic-energy change depends on the magnitudes of both of the velocities involved rather than simply the velocity difference. The same energy change can be produced with a small velocity change if the velocities are large, but if the velocities are small the velocity change must be much greater. To illustrate, suppose that $V_e > V_i$ and that an energy change of only 1 kJ/kg is considered. Then

$$\frac{V_e^2 - V_i^2}{2} = \Delta e_k = 1 \text{ kJ/kg}$$

An expression for the velocity ratio V_e/V_i, which depends on the magnitude of the inlet velocity, can be written in terms of the kinetic energy change as

$$\frac{V_e}{V_i} = \left(1 + \frac{2\Delta e_k}{V_i^2}\right)^{1/2} = \left(1 + \frac{2 \times 10^3}{V_i^2}\right)^{1/2}$$

where V_i is in meters per second. This relationship between the velocity ratio V_e/V_i and the inlet velocity is shown in Figure 4.10. Notice that for an inlet velocity of 10 m/s (32.8 ft/s) an exit velocity of about 46 m/s (151 ft/s), that is, $V_e/V_i = 4.6$, a 360 percent increase is required to produce an energy change of 1 kJ/kg (0.430 Btu/lb$_m$).

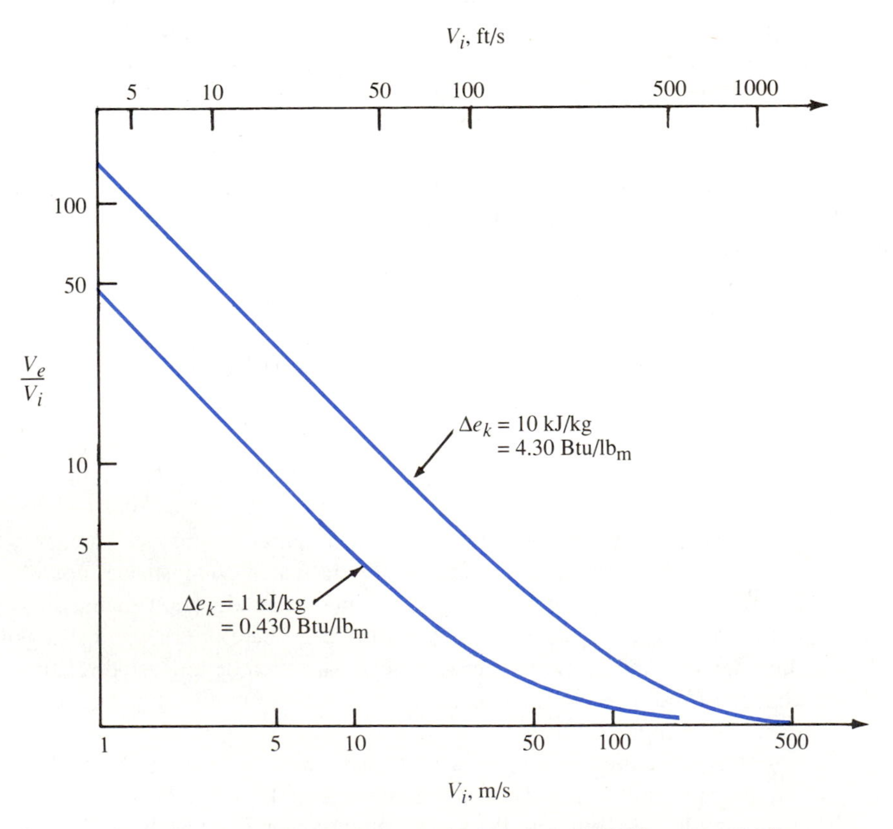

Figure 4.10 Exit-to-inlet velocity ratio required to produce 1 kJ/kg and 10 kJ/kg energy changes.

On the other hand, for an inlet velocity of 100 m/s (328 ft/s) an exit velocity of about 110 m/s (361 ft/s)(V_e/V_i = 1.1), an increase of only 10 percent is required.

An order-of-magnitude analysis (comparing the relative magnitude of terms appearing in an equation) can be quite useful not only in thermodynamics but in other engineering disciplines as well. It is one of the best ways to verify the validity of assumptions made during a problem solution.

4.5.2 Steady State

A system is said to be operating in steady state if the time rate of change of all the extensive properties of the system is zero; that is, from Equation 3.15

$$\frac{dY_{\text{sys}}}{dt} = 0 \tag{4.30}$$

where Y_{sys} is any extensive property of the system. Since the volume of the thermodynamic system is an extensive property, Equation 4.30 implies that the size or shape of an open, steady-state system does not change with time. As a result, there will be no work due to expansion or contraction of the system, and the *PdV* work is zero.

The total energy of the system E_{sys} is also an extensive property. Therefore, another implication of Equation 4.30 is that the total energy associated with the mass within the system does not change with time in the steady state. As a result, the conservation-of-energy equation for open, steady-state systems reduces to

$$\dot{Q} - \dot{W} + \sum_{\text{inlet}} \dot{m}(h + e_k + e_p) - \sum_{\text{exit}} \dot{m}(h + e_k + e_p) = 0 \tag{4.31}$$

The work associated with expansion or contraction of the system boundary vanishes for a steady-state system because the volume of the system is not permitted to change with time. Thus only other work modes, typically associated with rotating-machinery shafts in steady-state devices, remain in Equation 4.31. For this reason the work for steady-state systems is commonly called *shaft work*. Several examples of open, steady-state systems will now be considered.

■ EXAMPLE 4.6

Refrigerant-12 flows through a 40-mm-diameter horizontal pipe. At a point where the velocity is 40 m/s the temperature and pressure of the refrigerant are 40°C and 300 kPa, respectively. As a result of heat transfer from the surroundings, the temperature at a point downstream reaches 50°C. Assuming a negligible pressure drop, determine the heat-transfer rate to the refrigerant-12.

Solution. A sketch of the system is shown in Figure 4.11. This example of pipe flow can be represented by a steady-state system, and the solution to the problem is governed by the conservation of mass (Equation 3.16) and the conservation of energy (Equation 4.31) for an open system in steady state. From the conservation of mass

$$\sum_{\text{inlet}} \dot{m} = \sum_{\text{exit}} \dot{m}$$

or

$$\dot{m}_i = \dot{m}_e = \dot{m} = \text{constant}$$

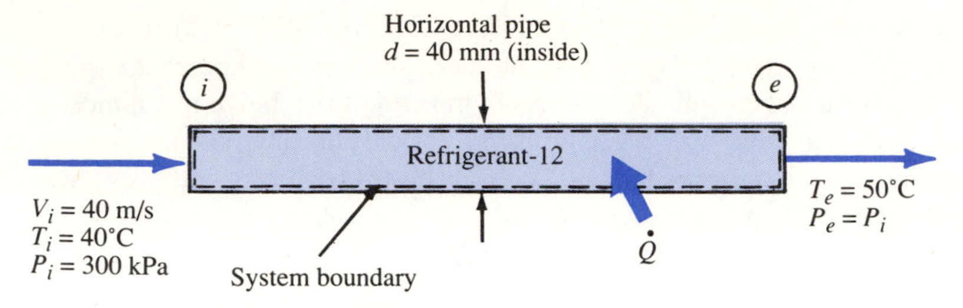

Figure 4.11 Sketch for Example 4.6.

From the conservation of energy

$$\dot{Q} - \dot{W} + \sum_{\text{inlet}} \dot{m}(h + e_k + e_p) - \sum_{\text{exit}} \dot{m}(h + e_k + e_p) = 0$$

There is no work and there is only one inlet and one exit; therefore,

$$\dot{Q} = \dot{m}_e(h + e_k + e_p)_e - \dot{m}_i(h + e_k + e_p)_i$$

$$= \dot{m}\left[(h_e - h_i) + \left(\frac{V_e^2 - V_i^2}{2}\right) + g(z_e - z_i)\right]$$

The potential-energy change is zero since the pipe is horizontal. The kinetic-energy change is likely to be negligible, but the velocities should be compared to be sure. Since $\dot{m} = AV/v$ is constant and the diameter of the pipe is constant, then

$$\frac{V_e}{v_e} = \frac{V_i}{v_i}$$

or

$$V_e = V_i\left(\frac{v_e}{v_i}\right)$$

Substituting values of the specific volume from Table C.3 gives

$$V_e = \left(\frac{0.07077 \text{ m}^3/\text{kg}}{0.06821 \text{ m}^3/\text{kg}}\right) V_i = 1.04\, V_i$$

The velocity increases by only 4 percent, and since the inlet velocity is only 40 m/s, the kinetic-energy change can be neglected (see Figure 4.10). Thus the heat-transfer rate can be calculated as

$$\dot{Q} = \dot{m}(h_e - h_i)$$

The mass flow rate of the refrigerant can be calculated on the basis of the conditions at the inlet:

$$\dot{m} = \frac{A_i V_i}{v_i} = \left(\frac{\pi d^2}{4}\right)\left(\frac{V_i}{v_i}\right) = \left(\frac{\pi}{4}\right)\left[\frac{(0.04 \text{ m})^2(40 \text{ m/s})}{0.06821 \text{ m}^3/\text{kg}}\right] = 0.737 \text{ kg/s}$$

Using enthalpy values at 300 kPa from Table C.3, we find the heat-transfer rate to be

$$\dot{Q} = \dot{m}(h_e - h_i) = (0.737 \text{ kg/s})(220.77 - 214.31)\text{kJ/kg} = 4.76 \text{ kJ/s} = \underline{\underline{4.76 \text{ kW}}}$$ ■

In the preceding example the pressure drop was assumed to be negligible. However, fluid flow in a pipe may be accompanied by a decrease in pressure due to the presence of friction. To evaluate the significance of the pressure drop in the preceding calculations, suppose that the pressure of the refrigerant at the downstream section has been reduced to 250 kPa. The effect of the pressure drop will be to increase the specific volume of the refrigerant, thereby increasing the downstream velocity. With a specific volume of 0.08560 m^3/kg at the downstream section, the fluid will have a velocity of 50 m/s, or 25 percent greater than the inlet velocity. This velocity change will produce a corresponding change in kinetic energy of only about 0.5 kJ/kg, but the enthalpy change of the fluid is also small, so the kinetic-energy term should not be neglected in the evaluation of the heat-transfer rate.

The enthalpy of R-12 at 250 kPa and 50°C is found from Table C.3, and the heat-transfer rate in this instance is

$$\dot{Q} = \dot{m}\left[h_e - h_i + \frac{V_e^2 - V_i^2}{2}\right]$$

$$= (0.737 \text{ kg/s})\left[(221.33 - 214.31)\text{kJ/kg} + \frac{(50 \text{ m/s})^2 - (40 \text{ m/s})^2}{2(1 \text{ kg}\cdot\text{m/N}\cdot\text{s}^2)(10^3 \text{ N}\cdot\text{m/kJ})}\right]$$

$$= 5.51 \text{ kJ/s} = 5.51 \text{ kW}$$

The heat-transfer rate is increased by about 16 percent above the value obtained when the drop in pressure which results from fluid friction is neglected.

■ EXAMPLE 4.7

A stream of liquid water at 50 psia and 70°F is mixed in an adiabatic mixing chamber with steam at 50 psia and 500°F, which is entering at the rate of 200 $\text{lb}_\text{m}/\text{s}$. The mixture leaves as a saturated vapor at 50 psia. Determine the mass flow rate of the liquid water entering the chamber and the mass flow rate of saturated vapor leaving the chamber.

Solution. The operation of the mixing chamber is assumed to be steady state; the chamber is shown schematically in Figure 4.12. The system boundary is drawn around the water inside the mixing chamber.

Applying the conservation of mass for an open system in steady state, Equation 3.16, results in

$$\sum_{\text{inlet}} \dot{m} = \sum_{\text{exit}} \dot{m}$$

or

$$\dot{m}_1 + \dot{m}_2 = \dot{m}_e$$

Applying the conservation-of-energy equation, Equation 4.31, and noting that no work is present and that the chamber is adiabatic, we have

$$\sum_{\text{inlet}} \dot{m}(h + e_k + e_p) - \sum_{\text{exit}} \dot{m}(h + e_k + e_p) = 0$$

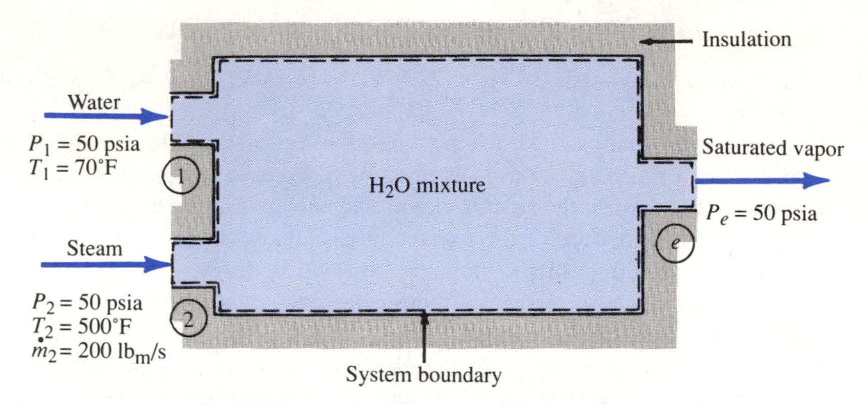

Figure 4.12 Schematic of mixing chamber for Example 4.7.

Since the changes in elevation between the inlets and exit are typically less than several meters for a device such as the mixing chamber, the potential-energy terms can be neglected relative to the much larger changes in enthalpy. Furthermore, the velocities would likely be low to reduce pressure losses, so the changes in kinetic energy can also be neglected in this device. The conservation-of-energy equation therefore reduces to

$$\dot{m}_1 h_1 + \dot{m}_2 h_2 - \dot{m}_e h_e = 0$$

The enthalpies at the inlets and exit represent known quantities because two independent, intensive properties are known at each location. All properties can therefore be found from the tables. Thus there are two unknowns, $\dot{m}_1$ and $\dot{m}_e$, and two equations available for the solution. Substituting the conservation-of-mass equation into the conservation-of-energy equation in order to eliminate $\dot{m}_e$ gives

$$\dot{m}_1 h_1 + \dot{m}_2 h_2 = (\dot{m}_1 + \dot{m}_2) h_e$$

Solving for $\dot{m}_1$ gives

$$\dot{m}_1 = \dot{m}_2 \left(\frac{h_e - h_2}{h_1 - h_e} \right)$$

Water at 50 psia, 70°F is in a slightly compressed liquid state and the enthalpy can be approximated by using Equation 2.53.

$$h_1 \simeq h_f|_{70°\text{F}} + (P_1 - P_{\text{sat}})v_f = 38.04\ \text{Btu/lb}_\text{m}$$

$$+ \frac{(50 - 0.3633)\text{lb}_\text{f}/\text{in}^2(144\ \text{in}^2/\text{ft}^2)(0.01605\ \text{ft}^3/\text{lb}_\text{m})}{778\ \text{ft}\cdot\text{lb}_\text{f}/\text{Btu}}$$

$$= 38.2\ \text{Btu/lb}_\text{m}$$

The other two enthalpies are obtained from Table B.3E and B.2E:

$$h_2 = 1283.8\ \text{Btu/lb}_\text{m}$$

$$h_e = h_g|_{50\ \text{psia}} = 1174.4\ \text{Btu/lb}_\text{m}$$

(a)

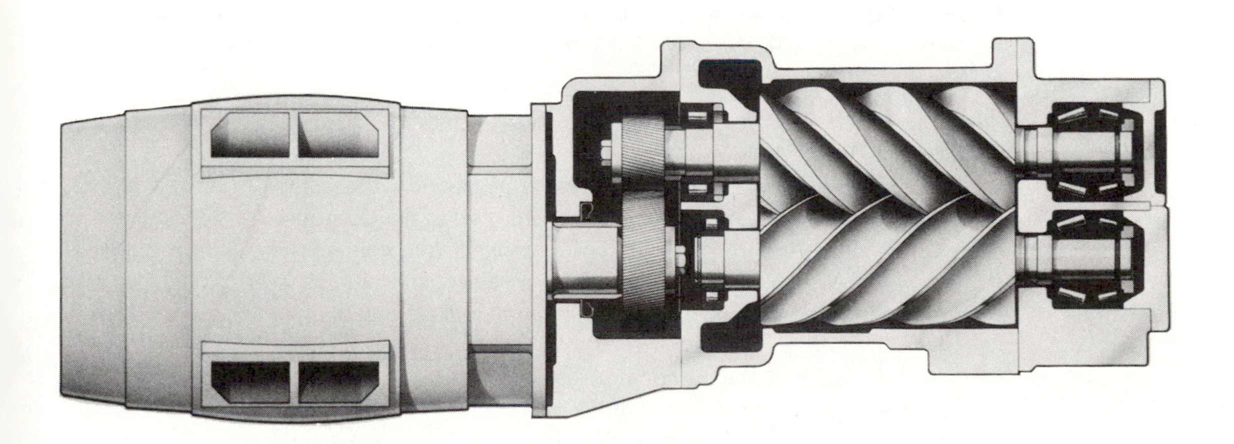

(b)

Figure 4.13 Two examples of positive-displacement compressors. (a) A three-cylinder reciprocating compressor; (b) a rotary-screw compressor driven by an integral electric motor. (Courtesy of Ingersoll-Rand.)

The mass flow rate of water entering is then

$$\dot{m}_1 = (200 \text{ lb}_m/\text{s}) \left[\frac{(1174.4 - 1283.8)\text{Btu/lb}_m}{(38.2 - 1174.4)\text{Btu/lb}_m} \right] = \underline{\underline{19.3 \text{ lb}_m/\text{s}}}$$

The mixture mass flow rate at the exit can now be determined from the conservation of mass:

$$\dot{m}_e = \dot{m}_1 + \dot{m}_2 = 19.3 \text{ lb}_m/\text{s} + 200 \text{ lb}_m/\text{s} = \underline{\underline{219.3 \text{ lb}_m/\text{s}}}$$

■

An important class of mechanical devices that can be analyzed and designed with the aid of thermodynamic principles are pumps and compressors. Pumps and compressors are used to increase the pressure of a working fluid. If the working fluid is a liquid, then a pump is used to increase the pressure of the liquid. If the working fluid is a gas, then a compressor is used to increase the pressure of the gas.

Compressors are usually distinguished as being either positive-displacement compressors or centrifugal compressors. In *positive-displacement compressors*, such as those shown in Figure 4.13, the motion of a solid member creates the force required to compress a fixed volume of gas during each cycle (i.e., one shaft revolution). The *reciprocating compressor* employs a piston and cylinder to compress a gas. The *rotary-screw compressor* is driven by an integral gear drive attached directly to an electric motor. The rotating, mating screw surfaces trap a volume of gas at the intake port and force it into a smaller volume before discharging it at a higher pressure.

In a *centrifugal compressor* an impeller rotates on a shaft within a rigid casing. The gas to be compressed is introduced toward the center of the rotating impeller where it is accelerated by vanes. Centrifugal compressors are often arranged to achieve *multistage compression* where the discharge from one set of vanes is ducted into the center of the impeller of the following stage. With multistage compression very high pressures can be obtained. A multistage refrigeration compressor is one component of the air-conditioning unit shown in Figure 4.26. *Axial-flow compressors* are discussed in Chapter 7.

■ EXAMPLE 4.8

An air compressor is designed to compress atmospheric air (assumed to be at 100 kPa, 20°C) to a pressure of 1 MPa. The heat-transfer rate to the environment is anticipated to be about equal to 10 percent of the power input to the compressor. The air enters at 50 m/s where the inlet area is 9×10^{-3} m^2 and leaves at 120 m/s through an area 5×10^{-4} m^2. Determine the exit-air temperature and the power input to the compressor.

Solution. A sketch of the air compressor, which operates as an open system in steady state, is shown in Fig. 4.14.

The conservation of mass, Equation 3.16, for this steady-state, open system is

$$\sum_{\text{inlet}} \dot{m} = \sum_{\text{exit}} \dot{m}$$

or

$$\dot{m}_i = \dot{m}_e = \dot{m}$$

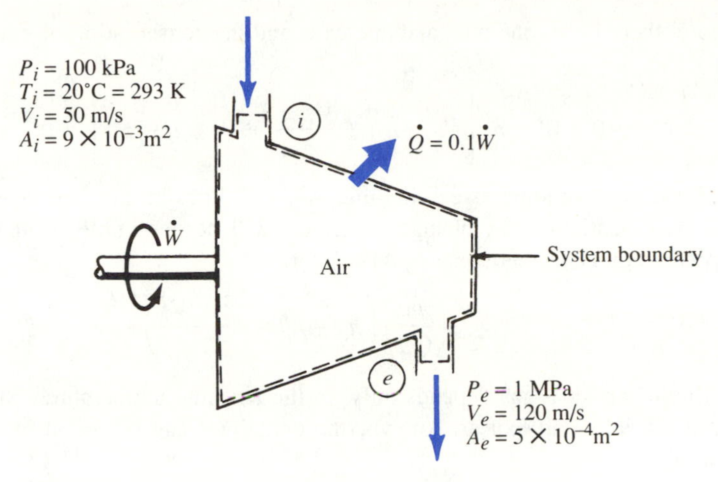

Figure 4.14 Sketch of air compressor for Example 4.8.

Under the conditions specified in the problem statement, air can be assumed to behave ideally. The mass flow rate of air through the compressor can be determined from the inlet conditions and the equation of state for an ideal gas ($Pv = RT$):

$$\dot{m}_i = \frac{A_i V_i}{v_i} = \frac{A_i V_i P_i}{RT_i}$$

$$= \frac{(9 \times 10^{-3}\ \text{m}^2)(50\ \text{m/s})(100\ \text{kPa})(28.97\ \text{kg/kg·mol})}{(8.314\ \text{kPa·m}^3/\text{kg·mol·K})(293\ \text{K})}$$

$$= 0.535\ \text{kg/s}$$

The exit temperature of the air can now be calculated by using the conservation of mass and the ideal-gas equation of state:

$$\dot{m}_i = \dot{m}_e = \frac{A_i V_i P_i}{RT_i} = \frac{A_e V_e P_e}{RT_e}$$

Solving for T_e, we have

$$T_e = T_i\left(\frac{A_e V_e P_e}{A_i V_i P_i}\right) = (293\ \text{K})\left(\frac{5 \times 10^{-4}\text{m}^2}{9 \times 10^{-3}\text{m}^2}\right)\left(\frac{120\ \text{m/s}}{50\ \text{m/s}}\right)\left(\frac{1000\ \text{kPa}}{100\ \text{kPa}}\right)$$

$$= \underline{\underline{391\ \text{K} = 118°\text{C}}}$$

The power input to the compressor is determined by applying the conservation of energy, Equation 4.31:

$$\dot{Q} - \dot{W} + \sum_{\text{inlet}} \dot{m}(h + e_k + e_p) - \sum_{\text{exit}} \dot{m}(h + e_k + e_p) = 0$$

Since the compressor requires a power input ($\dot{W} < 0$) and heat transfer occurs from the compressor to the surroundings ($\dot{Q} < 0$), the heat-transfer rate is

$$\dot{Q} = +0.1\ \dot{W}$$

Furthermore, there is but one inlet and one exit, and the conservation-of-energy equation reduces to

$$-0.9\,\dot{W} + \dot{m}\left[(h_i - h_e) + \frac{V_i^2 - V_e^2}{2} + g(z_i - z_e)\right] = 0$$

The difference in elevation between the inlet and exit lines of a compressor is generally small, so the potential-energy change of the air will be negligible compared with the change in enthalpy and kinetic energy. Therefore

$$\dot{W} = -\left(\frac{\dot{m}}{0.9}\right)\left(h_e - h_i + \frac{V_e^2 - V_i^2}{2}\right)$$

The enthalpy of an ideal gas depends only on the absolute temperature. Since the temperatures at the inlet and exit are known, the enthalpies can be found by interpolation from Table D.1:

$$h_e|_{391\text{ K}} = 392.1 \text{ kJ/kg}$$

$$h_i|_{293\text{ K}} = 293.3 \text{ kJ/kg}$$

The power input to the compressor is therefore

$$\dot{W} = -\left(\frac{0.535 \text{ kg/s}}{0.9}\right)$$

$$\times\left[(392.1 - 293.3)\text{kJ/kg} + \frac{(120^2 - 50^2)\text{m}^2/\text{s}^2}{(2)(1\text{ kg}\cdot\text{m/N}\cdot\text{s}^2)(10^3\text{N}\cdot\text{m/kJ})}\right]$$

$$= -(0.5944 \text{ kg/s})(98.8 + 5.95)\text{kJ/kg} = \underline{\underline{-62 \text{ kW}}} \qquad \text{(into the system)}$$

Note that even though the velocity of the air at the exit has increased by nearly 140 percent, the kinetic-energy change is only about 6 percent as large as the enthalpy change experienced by the air during the compression process. This small kinetic-energy change is typical of the operation of many gas compressors, and therefore the kinetic-energy change can often be neglected in engineering calculations involving such devices. ■

Nozzles and *diffusers* are devices that are used to alter the characteristics of a flow and are often merely ducts whose area either decreases or increases in the direction of the flow. A nozzle is used to increase the velocity or kinetic energy of a fluid at the expense of a decrease in pressure, whereas a diffuser is used to increase the pressure of a fluid at the expense of a decrease in its kinetic energy.

■ EXAMPLE 4.9

Air flows steadily through a diverging diffuser. The inlet area of the diffuser is 0.01 m^2, and the mass flow rate of air into the diffuser is 3.0 kg/s. The pressure and temperature of the air as it enters the diffuser are 300 kPa and 100°C. The air leaves the diffuser with a velocity that is very low compared with the inlet velocity. Determine the temperature of the air leaving the diffuser, assuming the diffuser to be well-insulated.

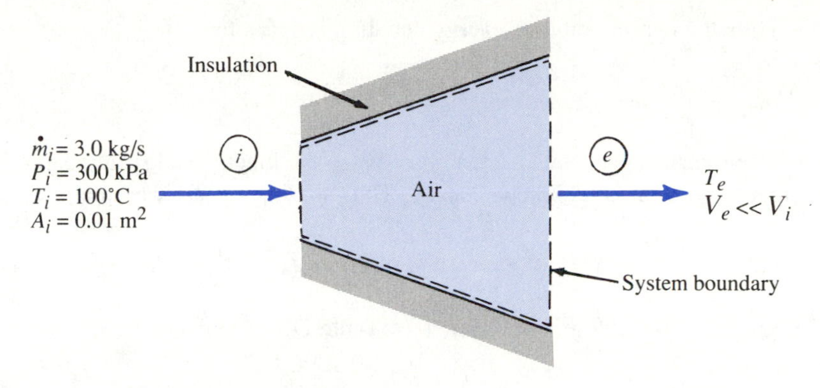

Figure 4.15 System for Example 4.9.

Solution. The system boundary is drawn around the air, as shown in Figure 4.15. The conservation-of-mass equation for this steady-flow device reduces to

$$\dot{m}_i = (\rho VA)_i = \dot{m}_e = \dot{m}$$

Since the air is at low pressure and high temperature relative to the critical pressure and temperature, ideal-gas behavior can be assumed. Then the density at the inlet can be determined from

$$\rho_i = \frac{P_i}{RT_i} = \frac{(300 \text{ kPa})(28.97 \text{ kg/kg·mol})}{(8.314 \text{ kPa·m}^3/\text{kg·mol·K})(373 \text{ K})}$$
$$= 2.8 \text{ kg/m}^3$$

Next, the inlet velocity can be calculated from the definition of the mass flow rate:

$$V_i = \frac{\dot{m}_i}{\rho_i A_i} = \frac{3.0 \text{ kg/s}}{(2.8 \text{ kg/m}^3)(0.01 \text{ m}^2)} = 107 \text{ m/s}$$

The air does not perform any work as it passes through the diffuser, and there is negligible heat transfer across the boundary of the system because the diffuser is well-insulated. Furthermore, the inlet and exit areas are at the same elevation, so the change in potential energy is zero. Under these conditions the conservation-of-energy equation simplifies to

$$\dot{m}\left(h_i - h_e + \frac{V_i^2 - V_e^2}{2}\right) = 0$$

Since the air is an ideal gas and the temperature change of the air is small enough that the specific heats are nearly constant, the enthalpy change is approximately

$$h_i - h_e = c_p(T_i - T_e)$$

The assumption of constant specific heats should be verified after the exit temperature has been determined.

Thus the conservation-of-energy equation reduces to

$$c_p(T_i - T_e) + \frac{V_i^2 - V_e^2}{2} = 0$$

From the problem statement, $V_e << V_i$, so the kinetic energy at the exit is negligible compared with the inlet kinetic energy. Thus solving for T_e, we obtain

$$T_e = \frac{V_i^2}{2c_p} + T_i$$

Using $c_p = 1.01$ kJ/kg·K at 373 K (see Table D.8), we have

$$T_e = \frac{(107 \text{ m/s})^2}{2(1.01 \text{ kJ/kg·K})(1 \text{ kg·m/N·s}^2)(10^3 \text{ N·m/kJ})} + 373 \text{ K} = \underline{\underline{378.7 \text{ K}}}$$

Note that since the air temperature increases by only about 6 K, the assumption of a constant specific heat is justified. ■

A *throttling valve* is simply a flow-restricting device. It can take the form of a thin capillary tube or an adjustable valve. Regardless of design, the purpose of a throttling valve is to produce a sizable pressure drop in the fluid, which results in an accompanying temperature drop. A throttling valve is usually a relatively small device so that even though there is a temperature difference between it and the surroundings, the rate of heat transfer is usually negligible because the surface area is small. When potential- and kinetic-energy changes and heat transfer can be neglected, the inlet and exit values for the enthalpy of the fluid are equal. For this reason a throttling valve is frequently called a constant-enthalpy device.

■ EXAMPLE 4.10

A steady flow of refrigerant-12 enters a throttling valve in a refrigeration unit as a saturated liquid with a pressure of 150 psia. The refrigerant is discharged with a pressure of 40 psia. Determine the inlet and exit temperatures and the exit quality, assuming that the heat transfer and the kinetic-energy change of the refrigerant are negligible.

Solution. Applying the conservation-of-mass equation to the system shown in Figure 4.16 results in

$$\dot{m}_i = \dot{m}_e = \dot{m}$$

The conservation of energy for this system is

$$\dot{Q} - \dot{W} + \sum_{\text{inlet}} \dot{m}(h + e_k + e_p) - \sum_{\text{exit}} \dot{m}(h + e_k + e_p) = 0$$

The refrigerant does not change elevation, so the potential-energy change is zero. The refrigerant does no work, and since the heat transfer and the kinetic-energy change are negligible, the conservation-of-energy equation reduces to

$$\dot{m}_i h_i - \dot{m}_e h_e = 0$$

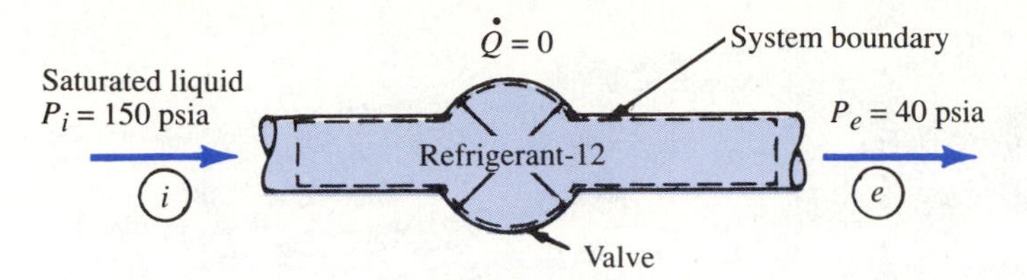

Figure 4.16 System for Example 4.10.

Substituting the results of the conservation of mass gives

$$h_i = h_e$$

which shows that the throttling valve for the assumptions in this problem operates with equal inlet and exit enthalpies.

The enthalpy and temperature of the refrigerant at the inlet conditions are found in Table C.2E.

$$T_i = \underline{\underline{109.6°\text{F}}}$$

$$h_i = 33.83\ \text{Btu/lb}_\text{m}$$

Thus two independent properties are known at the exit state:

$$h_e = 33.83\ \text{Btu/lb}_\text{m}$$

$$P_e = 40\ \text{psia}$$

These two properties are sufficient to determine all other properties at the exit state. From Table C.2E notice that at the exit pressure of 40 psia the enthalpy is between h_f and h_g, so the exit state of the refrigerant is in the saturation region. The exit temperature and quality are

$$T_e = \underline{\underline{25.9°\text{F}}}$$

$$x_e = \frac{h_e - h_f}{h_{fg}}$$

$$= \frac{33.83\ \text{Btu/lb}_\text{m} - 14.18\ \text{Btu/lb}_\text{m}}{66.3\ \text{Btu/lb}_\text{m}}$$

$$= \underline{\underline{0.296}}$$

In this problem the throttling valve caused the pressure to drop by a factor of nearly 4, and the temperature dropped by nearly 85°F. In a typical refrigeration or air-conditioning system the cold refrigerant leaving the throttling valve would be routed through a heat exchanger, where it would be used to cool and dehumidify air. ■

Another piece of equipment frequently encountered by engineers is a heat exchanger. The primary purpose of a heat exchanger is to provide heat transfer between the fluids. Typical designs include *finned-tube heat exchangers* like those shown in Figure 4.17 and

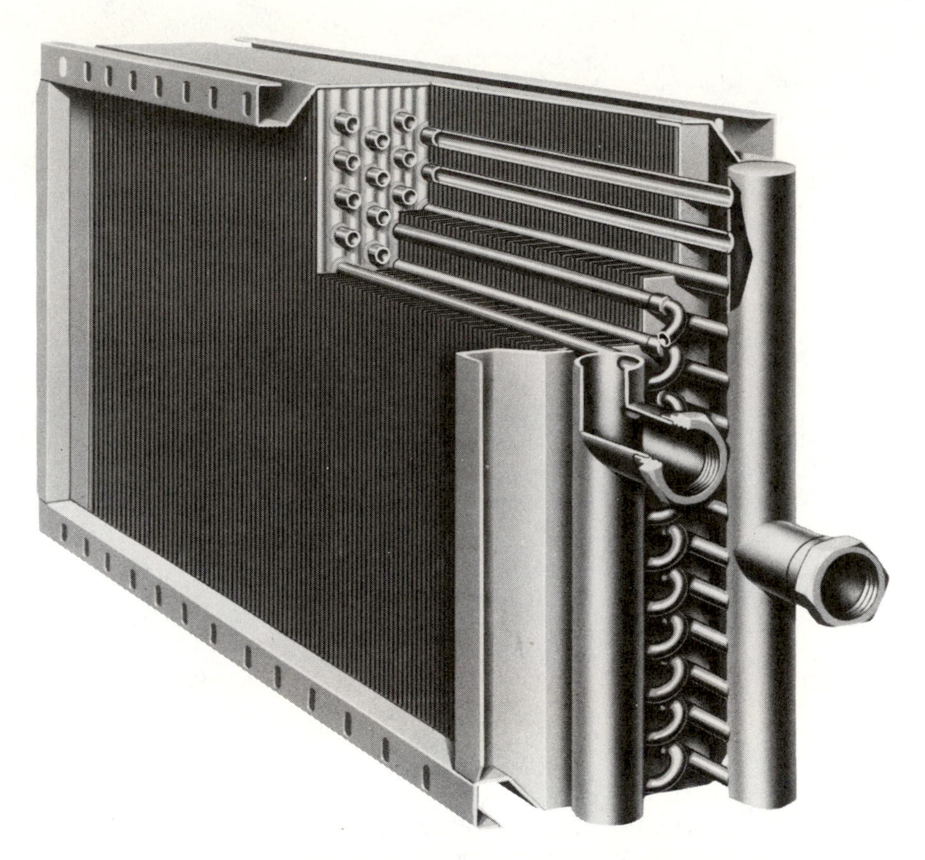

Figure 4.17 A cross-flow finned-tube heat exchanger. This unit is used in an air conditioner in which the cool refrigerant is circulated through the tubes that are fed by two manifolds shown in section. Warm air is drawn over the tubes by a fan and flows perpendicular to the tubes. The air circulates over the extended surfaces, or fins, that increase the surface area of the tubes, thereby increasing the heat transfer between the refrigerant and the air. (Drawing courtesy of The Trane Company.)

shell-and-tube heat exchangers like the evaporator and condenser designs shown in Figure 4.27. Since the heat-transfer rate from a body to the surrounding fluid is proportional to the surface area of the body, heat exchangers can be quite large when high heat transfer rates are required. *Compact heat exchangers* attempt to reduce the size of the heat exchangers by using innovative designs involving extended surfaces and turbulence producing devices which help augment heat-transfer rates.

■ EXAMPLE 4.11

A heat exchanger is designed to use exhaust steam from a turbine to heat air in a manufacturing plant. Steam enters the heat exchanger with a flow rate of 1.2 kg/s, a pressure of 200 kPa, and a temperature of 200°C. The steam leaves the heat exchanger as a saturated vapor at 200 kPa. The air enters the heat exchanger at 20°C and 101 kPa

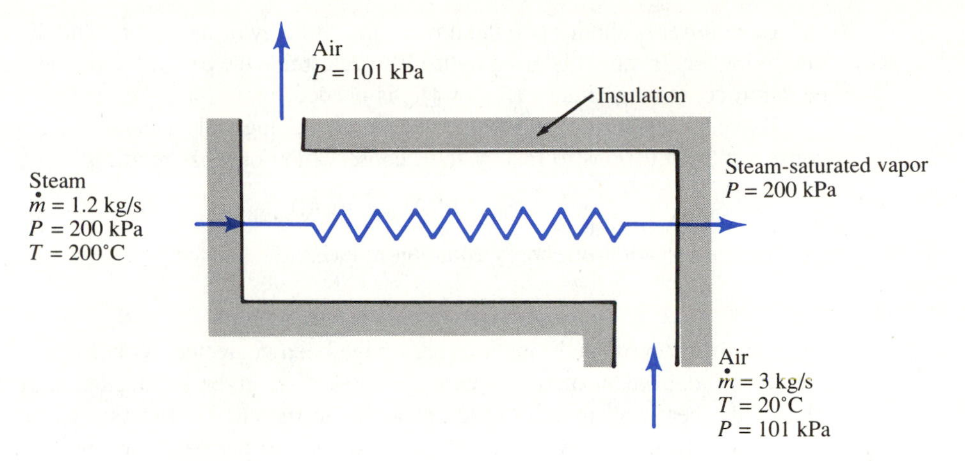

Figure 4.18 Heat exchanger for Example 4.11.

and leaves at a pressure of 101 kPa. The design flow rate of air is 3 kg/s. Assuming that the outer shell of the heat exchanger is well-insulated, calculate the temperature of the air as it leaves the heat exchanger.

Solution. A simplified cross section of the heat exchanger is shown in Figure 4.18. There are several different possible choices for the boundary of the system in this problem since the two fluid streams do not mix. Suppose we start the analysis by using a system that contains only the air. Assuming a steady flow of air, the conservation-of-mass equation for this choice of system states that

$$(\dot{m}_i)_{\text{air}} = (\dot{m}_e)_{\text{air}}$$

In a heat exchanger the differences in elevation between the inlets and exits are usually less than several meters, and therefore potential-energy changes are negligible compared with the change in enthalpy of the working fluids. The kinetic-energy changes can be assumed to be small also, since fluid velocities are generally low in order to keep pressure losses at a minimum. In addition, the air performs no work during the process, so the conservation-of-energy equation written for the air reduces to

$$\dot{Q}_{\text{air}} = (\dot{m}_e h_e - \dot{m}_i h_i)_{\text{air}}$$

Substituting the results of the conservation-of-mass equation results in

$$\dot{Q}_{\text{air}} = [\dot{m}(h_e - h_i)]_{\text{air}} \tag{a}$$

Since the air pressure is much lower than the critical pressure and its temperature is well above the critical value, the ideal-gas assumption is appropriate. Thus the air enthalpies depend only on the temperature and can be determined from Table D.1. The inlet-air temperature is known, so the inlet enthalpy is

$$(h_i)_{\text{air}} = 293.3 \text{ kJ/kg}$$

Equation (a) contains two unknowns, the enthalpy of the exit air and the heat-transfer rate to the air. It cannot be used by itself to determine the exit-air temperature, so another equation containing either $(h_e)_{air}$ or $\dot{Q}_{air}$ is needed.

If we now consider a second system containing only steam and apply the basic conservation equations to that system, conservation of mass requires

$$(\dot{m}_i)_{steam} = (\dot{m}_e)_{steam}$$

and the conservation-of-energy equation reduces to

$$\dot{Q}_{steam} = [\dot{m}(h_e - h_i)]_{steam} \tag{b}$$

Changes in potential and kinetic energy have been neglected, as before.

Two independent thermodynamic properties are given for the steam at the inlet and exit of the heat exchanger, so the enthalpies in the energy equation can be determined from Table B.3. The energy equation for the steam permits determination of the heat-transfer rate from the steam, but the heat-transfer rate *from* the steam is also equal in magnitude (but opposite in sign) to the heat-transfer rate *to* the air, or

$$\dot{Q}_{steam} = -\dot{Q}_{air} \tag{c}$$

Substituting Equations (b) and (c) into Equation (a) gives an expression for the exit-air enthalpy in terms of known quantities:

$$[\dot{m}(h_i - h_e)]_{steam} = [\dot{m}(h_e - h_i)]_{air} \tag{d}$$

From Table B.3 the steam enthalpies are

$$(h_i)_{steam} = 2870 \text{ kJ/kg}$$

$$(h_e)_{steam} = 2707 \text{ kJ/kg}$$

Substituting values, we have

$$(1.2 \text{ kg/s})(2870 \text{ kJ/kg} - 2707 \text{ kJ/kg}) = (3 \text{ kg/s})[(h_e)_{air} - 293.3 \text{ kJ/kg}]$$

or

$$(h_e)_{air} = 358.5 \text{ kJ/kg}$$

Using Table D.1 at this value of air enthalpy, the exit-air temperature is

$$(T_e)_{air} = 358 \text{ K} = \underline{\underline{85°\text{C}}}$$

At this point another way to solve this problem might be apparent. If a system that contains *both* the steam and the air is chosen, it will have two inlets and two exits. The conservation of mass for this new system states that

$$(\dot{m}_i)_{air} + (\dot{m}_i)_{steam} = (\dot{m}_e)_{air} + (\dot{m}_e)_{steam}$$

For this system the heat-transfer rate is zero because the boundary is insulated. As before, the steam and air perform no work, potential- and kinetic-energy changes can be neglected, and the fluid streams do not mix. The conservation-of-energy equation therefore reduces to

$$[\dot{m}(h_i - h_e)]_{\text{air}} + [\dot{m}(h_i - h_e)]_{\text{steam}} = 0$$

This expression is the same equation as Equation (d) that resulted from a slightly more complicated analysis using two systems instead of one.

This problem illustrates an important point. The basic conservation principles are valid regardless of the choice of system. In most problems the choice of the system that will provide the required information is reasonably obvious. However, in some problems, like this example, the choice may not be so clear-cut. Regardless of what is eventually selected for a system, the correct solution will be obtained provided that the governing equations are applied properly. Recognize, however, that a prudent choice of system will often save much work by eliminating the determination of intermediate quantities. ■

■ EXAMPLE 4.12

A pump is used to raise water steadily at a volume flow rate of 800 gal/min, as shown in Figure 4.19. The diameter of the inlet pipe is 6 in and the diameter of the exit pipe is 7 in. The power input to the pump is 80 hp. The water is drawn from a lake at atmospheric pressure and 70°F, and it is discharged 300 ft above the lake at atmospheric pressure. The pump is well-insulated and heating of the water due to frictional effects is small enough that it may be neglected. Calculate the temperature of the water at the pipe exit.

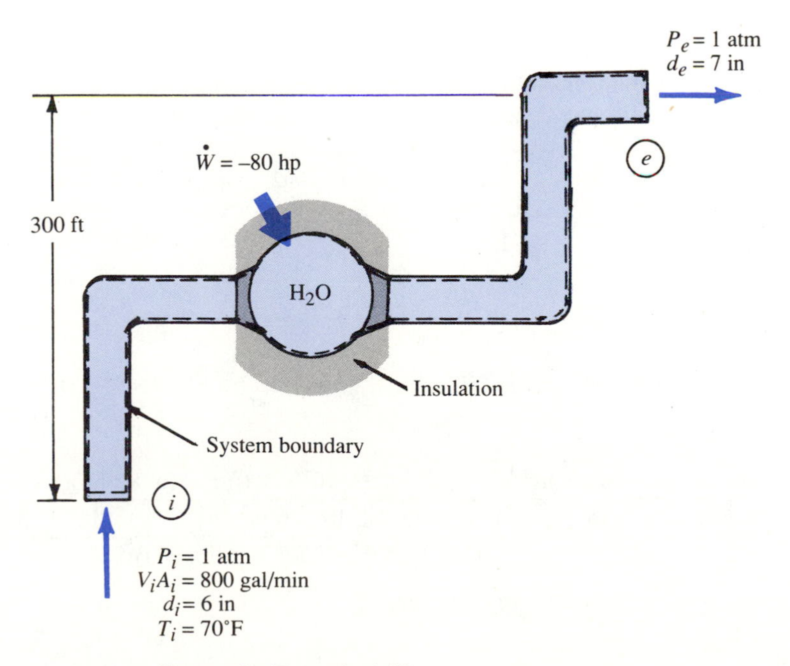

Figure 4.19 System for Example 4.12.

Solution. A system boundary that encloses the water from the inlet to exit pipes will be used to determine the exit temperature of the water. The conservation-of-mass equation for this steady, incompressible flow requires that

$$V_i A_i = V_e A_e$$

The volume flow rate at the inlet is $V_i A_i = 800$ gal/min. Thus

$$V_i = \frac{(800 \text{ gal/min})(1 \text{ ft}^3/7.48 \text{ gal})}{\pi(0.5 \text{ ft})^2/4} = 545 \text{ ft/min}$$

$$= 9.08 \text{ ft/s}$$

The exit velocity is

$$V_e = \frac{V_i A_i}{A_e} = (9.08 \text{ ft/s})\left(\frac{6 \text{ in}}{7 \text{ in}}\right)^2 = 6.67 \text{ ft/s}$$

The mass flow rate can be determined from

$$\dot{m} = \rho_i (AV)_i = (62.3 \text{ lb}_\text{m}/\text{ft}^3)(800 \text{ gal/min})(1 \text{ ft}^3/7.48 \text{ gal})$$

$$= 6663 \frac{\text{lb}_\text{m}}{\text{min}} = 111.1 \frac{\text{lb}_\text{m}}{\text{s}}$$

The temperature of the water at the exit pipe can be determined from a conservation-of-energy analysis. The conservation of energy for steady flow is

$$\dot{Q} - \dot{W} + \sum_{\text{inlet}} \dot{m}(h + e_k + e_p) - \sum_{\text{exit}} \dot{m}(h + e_k + e_p) = 0$$

Neglecting heat transfer and replacing the enthalpy-change term by $c(T_i - T_e)$ for an incompressible fluid, the conservation-of-energy equation can be reduced to

$$-\dot{W} + \dot{m}\left[c(T_i - T_e) + \frac{V_i^2 - V_e^2}{2} + g(z_i - z_e)\right] = 0$$

Therefore,

$$T_e = T_i + \frac{(V_i^2 - V_e^2)/2 + g(z_i - z_e) - \dot{W}/\dot{m}}{c}$$

Using the specific heat c from Table H.3E gives

$$T_e = 530°\text{R}$$

$$+ \left[\frac{\dfrac{(9.08^2 - 6.67^2)\text{ft}^2/\text{s}^2}{2(32.2 \text{ ft}\cdot\text{lb}_\text{m}/\text{lb}_\text{f}\cdot\text{s}^2)} + \dfrac{(32.2 \text{ ft/s}^2)(0 - 300)\text{ft}}{32.2 \text{ ft}\cdot\text{lb}_\text{m}/\text{lb}_\text{f}\cdot\text{s}^2} - \dfrac{(-80 \text{ hp})(550 \text{ ft}\cdot\text{lb}_\text{f}/\text{s}\cdot\text{hp})}{111.1 \text{ lb}_\text{m}/\text{s}}}{(1.0 \text{ Btu/lb}_\text{m}\cdot°\text{R})(778 \text{ ft}\cdot\text{lb}_\text{f}/\text{Btu})}\right]$$

$$= 530°\text{R} + (0.0008 - 0.386 + 0.509)°\text{R}$$

or

$$T_e = 530.12°\text{R} = \underline{\underline{70.12°\text{F}}}$$

Now that the problem has been analyzed, the fact that the change in kinetic energy of the water is negligible compared with the change in potential energy is evident. Also, even though the temperature rise of the water is only 0.12°F the enthalpy rise of the water accounts for 24 percent of the 80 hp of energy transferred into the water by the pump. The remaining 76 percent of the energy is transformed into increased potential energy of the water. ■

■ EXAMPLE 4.13

Steam enters a turbine with a pressure and temperature of 2000 psia and 1100°F and leaves at 10 psia as a saturated vapor. The flow area at the turbine inlet is 0.5 ft^2 and at the exit it is 3.5 ft^2. The steam flows steadily through the turbine at a mass flow rate of 60 lb$_m$/s. Calculate the power that can be produced by the turbine, assuming negligible heat transfer from the system.

Solution. The system along with the given information is shown in Figure 4.20. The power produced by the turbine will be calculated by applying the conservation-of-energy principle. Since the conservation-of-energy equation contains the velocities at the inlet and exit, they will be calculated first. Notice also that the state of the water is known at the inlet and exit of the turbine because two independent thermodynamic properties are given at both locations.

The velocities can be determined by applying the conservation of mass, which, for steady flow with one inlet and exit, reduces to

$$\dot{m}_i = \dot{m}_e = \dot{m}$$

or

$$\frac{V_i A_i}{v_i} = \frac{V_e A_e}{v_e} = \dot{m}$$

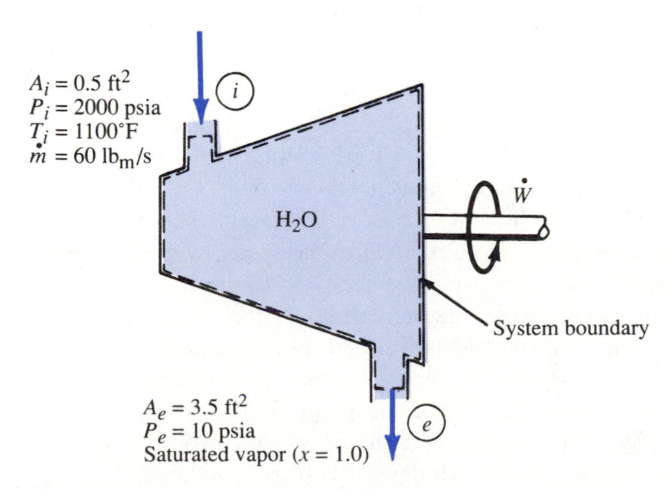

Figure 4.20 System for Example 4.13.

The specific volumes from Tables B.3E and B.2E at the given states are

$$v_i = 0.4324 \text{ ft}^3/\text{lb}_\text{m}$$

$$v_e = 38.4 \text{ ft}^3/\text{lb}_\text{m}$$

Thus the inlet and exit velocities are

$$V_i = \frac{\dot{m}v_i}{A_i} = \frac{(60 \text{ lb}_\text{m}/\text{s})(0.4324 \text{ ft}^3/\text{lb}_\text{m})}{0.5 \text{ ft}^2} = 51.9 \text{ ft/s}$$

$$V_e = \frac{\dot{m}v_e}{A_e} = \frac{(60 \text{ lb}_\text{m}/\text{s})(38.4 \text{ ft}^3/\text{lb}_\text{m})}{3.5 \text{ ft}^2} = 658 \text{ ft/s}$$

For a steady, open system with negligible heat transfer, the conservation-of-energy equation (Equation 4.31) is

$$-\dot{W} + \sum_{\text{inlet}} \dot{m}(h + e_k + e_p) - \sum_{\text{exit}} \dot{m}(h + e_k + e_p) = 0$$

Typical elevation changes for a large turbine are on the order of only several feet, and therefore potential-energy changes should be very small compared with the rather large change in enthalpy of the fluid stream. In addition, the turbine has only one inlet and one exit, so

$$\dot{W} = \dot{m}\left(h_i - h_e + \frac{V_i^2 - V_e^2}{2}\right)$$

The enthalpy values at the given inlet and exit states, from Tables B.3E and B.2E, are

$$h_i = 1536.9 \text{ Btu/lb}_\text{m}$$

$$h_e = 1143.1 \text{ Btu/lb}_\text{m}$$

Substituting into the conservation-of-energy equation yields

$$\dot{W} = (60 \text{ lb}_\text{m}/\text{s})\left[(1536.9 - 1143.1)\text{Btu/lb}_\text{m} + \frac{(51.9^2 - 658^2)\text{ft}^2/\text{s}^2}{2(778 \text{ ft·lb}_\text{f}/\text{Btu})(32.2 \text{ ft·lb}_\text{m}/\text{lb}_\text{f}\text{·s}^2)}\right]$$

$$= 23{,}100 \text{ Btu/s} = \underline{\underline{32{,}700 \text{ hp}}}$$

The positive value for $\dot{W}$ implies that the work is done by the steam on the surroundings; that is, the turbine produces useful power.

Finally, we should try to reflect on the problem to see what conclusions can be reached from the analysis. The steam produces the power by a decrease in its own energy level or its enthalpy. The decrease in enthalpy of the steam also occurs at the expense of an increase in kinetic energy of the steam. Of the total rate of decrease in steam enthalpy, 98 percent is converted into usable power, and the remaining 2 percent is converted into an increase in kinetic energy of the steam. Even though the steam velocity increases by nearly a factor of 13, the change in kinetic energy is small compared with the enthalpy change. These figures indicate that neglecting kinetic-energy changes when dealing with turbines and other such devices is often justifiable. ■

4.5.3 Transient Analysis

In the preceding section attention was focused on the analysis of open thermodynamic systems in steady state. In this section the thermodynamic analysis of open systems that exhibit transient behavior is considered.

Transient thermodynamic analysis is important when the time rate of change of the properties of the system is significant. In such cases, changes that occur over some interval of time Δt are of interest. In this respect the analysis is similar to that for a closed system. The most significant difference, is, however, that the amount of mass within the system does not remain constant.

In transient problems particular attention should be given to the condition of the fluid streams entering and leaving the system. The usual assumption when dealing with transient problems is to consider the flow of the fluid as it enters or leaves the system to be uniform. Although not always essential, assuming perfect mixing of the mass *within* the system is often necessary. In other words, the state of the substance *inside* the system is the same throughout the system even though the thermodynamic state may change with time. Note that this assumption is quite different from the steady-state assumption for which the thermodynamic state may vary with *location* throughout the system but the state of the substance at a particular point in the system does not change with *time*.

Since the changes that occur over an interval of time Δt are of interest, the conservation-of-energy equation, Equation 4.31, is integrated with respect to time. Since

$$\dot{m} = \frac{dm}{dt} \qquad \dot{Q} = \frac{\delta Q}{dt} \qquad \dot{W} = \frac{\delta W}{dt}$$

Equation 4.31 can be written as

$$\frac{\delta Q}{dt} - \frac{\delta W}{dt} + \sum_{\text{inlet}} (h + e_k + e_p) \frac{dm}{dt} - \sum_{\text{exit}} (h + e_k + e_p) \frac{dm}{dt} = \frac{dE_{\text{sys}}}{dt} \tag{4.32}$$

Integrating this expression with respect to time, then, results in the conservation-of-energy equation as it will be used for transient analyses:

$$\begin{aligned} \int_{t_1}^{t_2} \frac{\delta Q}{dt} dt - \int_{t_1}^{t_2} \frac{\delta W}{dt} dt + \sum_{\text{inlet}} \int_{t_1}^{t_2} (h + e_k + e_p) \frac{dm}{dt} dt \\ - \sum_{\text{exit}} \int_{t_1}^{t_2} (h + e_k + e_p) \frac{dm}{dt} = \int_{t_1}^{t_2} \frac{dE_{\text{sys}}}{dt} dt \end{aligned} \tag{4.33}$$

or

$$\begin{aligned} Q_{12} - W_{12} + \sum_{\text{inlet}} \int_1^2 (h + e_k + e_p)\, dm - \sum_{\text{exit}} \int_1^2 (h + e_k + e_p)\, dm \\ = (E_2 - E_1)_{\text{sys}} \end{aligned} \tag{4.34}$$

Note that if the surface area of the system contains neither inlets nor exits, Equation 4.34 is the same as Equation 4.22 for the closed system. The work term in Equation 4.34 includes both reversible and irreversible work modes since the size of the system is allowed to change.

■ EXAMPLE 4.14

A rigid tank initially contains 0.5 kg of steam at 800 kPa and 300°C and is connected through an insulated valve to a steam supply line that is capable of supplying steam at a constant condition of 1.4 MPa, 300°C. The valve is opened so that the supply steam flows slowly into the tank until the pressure and temperature inside are 1.2 MPa and 300°C. Determine the final mass of steam in the tank and the heat transfer to (or from) the steam in the tank during the process.

Solution. The charging process for a rigid tank is a classic example of an open system in transient flow. A sketch of the system is shown in Figure 4.21. Selecting a system boundary that cuts across the supply pipe on the supply-line side of the valve greatly simplifies the solution to the problem. The inlet conditions to the system remain constant at the state of the steam in the supply line because the quantity of mass withdrawn from the line is small enough to leave conditions in the supply line unchanged. If the system boundary were drawn on the other side of the valve, the state of the steam would change with time and the solution to the problem could be more involved. Notice that the valve behaves like the throttling valve discussed in Example 4.10. The valve restricts the flow of steam so that a large pressure drop occurs.

The final mass of steam in the tank can be determined very simply. The volume of the tank remains constant, so that

$$\boldsymbol{V} = m_1 v_1 = m_2 v_2$$

and

$$m_2 = m_1 \frac{v_1}{v_2}$$

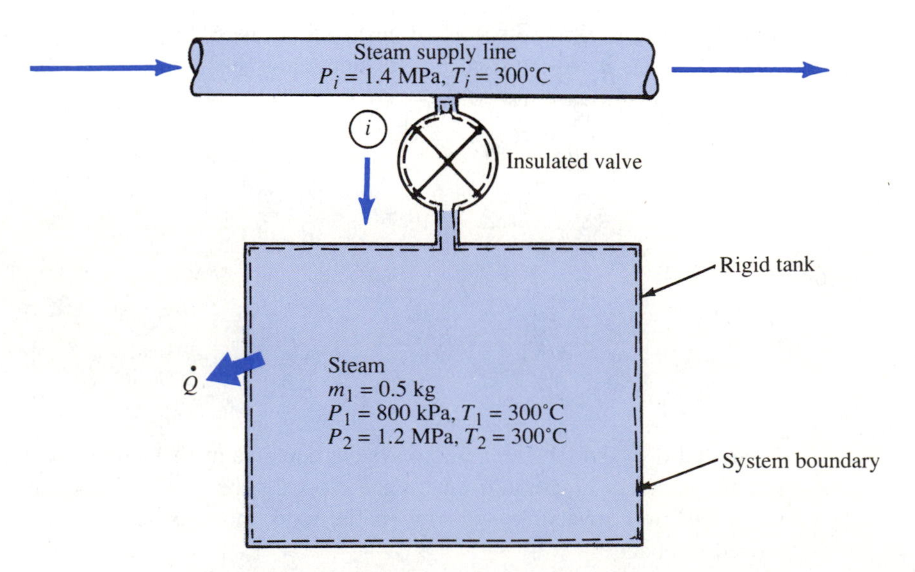

Figure 4.21 Sketch for Example 4.14.

The specific volumes at the initial and final states are determined from Table B.3:

$$v_1 = 0.3241 \text{ m}^3/\text{kg}$$

$$v_2 = 0.2138 \text{ m}^3/\text{kg}$$

and

$$m_2 = (0.5 \text{ kg})\left(\frac{0.3241 \text{ m}^3/\text{kg}}{0.2138 \text{ m}^3/\text{kg}}\right) = \underline{\underline{0.76 \text{ kg}}}$$

The conservation of mass for transient flow is given by Equation 3.23:

$$\sum_{\text{inlet}} m - \sum_{\text{exit}} m = (m_2 - m_1)_{\text{sys}}$$

which, for this example, reduces to

$$m_i = (m_2 - m_1)_{\text{sys}}$$

Therefore

$$m_i = m_2 - m_1 = 0.76 \text{ kg} - 0.5 \text{ kg} = 0.26 \text{ kg}$$

The conservation-of-energy equation, Equation 4.34, is used to calculate the heat transfer for the process. The work is zero since the volume of the tank is constant and there are no other identifiable work modes. Therefore

$$Q_{12} + \left[\int_1^2 (h + e_k + e_p)\, dm\right]_i = (E_2 - E_1)_{\text{sys}}$$

The velocity and elevation of the center of mass of the system do not change, so

$$(E_2 - E_1)_{\text{sys}} = (U_2 - U_1)_{\text{sys}} = (m_2 u_2 - m_1 u_1)_{\text{sys}}$$

When the elevation of the tank is selected as the reference, the potential energy of the inlet fluid is negligible compared with its enthalpy. The enthalpy of the steam entering will be several thousand kilojoules per kilogram, and the fluid velocities should be small enough that the kinetic energy of the steam will also be very small compared with the enthalpy. Therefore, the kinetic and potential energies at the inlet can be neglected, and the conservation-of-energy equation reduces to

$$Q_{12} + \int_1^2 (h\, dm)_i = (m_2 u_2 - m_1 u_1)_{\text{sys}}$$

The state of the steam as it enters the system remains constant, and therefore h_i = constant, and

$$\int_1^2 (h\ dm)_i = h_i \int_1^2 dm_i = h_i m_i$$

where m_i represents the amount of mass that enters the tank during the interval of time $t_2 - t_1$. Hence

$$Q_{12} = -h_i m_i + (m_2 u_2 - m_1 u_1)_{\text{sys}}$$

The enthalpy of the supply steam and the initial and final internal energies are found from Table B.3:

$$h_i = 3039.7 \text{ kJ/kg}$$

$$u_1 = 2796.6 \text{ kJ/kg}$$

$$u_2 = 2788.6 \text{ kJ/kg}$$

The heat transfer for the process is, then

$$\begin{aligned} Q_{12} &= (-3039.7 \text{ kJ/kg})(0.26 \text{ kg}) + (0.76 \text{ kg})(2788.6 \text{ kJ/kg}) \\ &\quad - (0.5 \text{ kg})(2796.6 \text{ kJ/kg}) \\ &= \underline{\underline{-69.3 \text{ kJ}}} \qquad \text{(heat transfer from the system)} \end{aligned}$$

■

■ EXAMPLE 4.15

An insulated piston-cylinder assembly has an initial volume of 1.2 ft^3 and contains air at 100 psia and 400°F. Air is supplied to the cylinder at 200 psia and 300°F through a valve fitted into the cylinder. The piston is restrained in such a manner that the pressure of the air in the cylinder remains constant at 100 psia during the process of filling. The filling process is terminated when the final volume is twice the initial volume. Determine the final temperature of the air in the cylinder and the mass of air added through the valve, assuming that the piston is frictionless.

Solution. The piston-cylinder arrangement is shown in Figure 4.22. As in the previous example, the system boundary is drawn so that it cuts across the inlet where the state of the air entering the system is uniform and remains constant with time. Applying the conservation-of-mass equation, Equation 3.23, for this transient system, we have

$$\sum_{\text{inlet}} m - \sum_{\text{exit}} m = (m_2 - m_1)_{\text{sys}}$$

which reduces to

$$m_i = (m_2 - m_1)_{\text{sys}}$$

Since the system is adiabatic, the conservation-of-energy equation, Equation 4.34, with the assumption that the kinetic- and potential-energy terms are negligible (see Example 4.14), reduces to

$$-W_{12} + \int_1^2 (h\,dm)_i = (U_2 - U_1)_{\text{sys}}$$

Since irreversible-work modes are absent and the pressure is constant, the work is given by

$$W_{12} = \int_1^2 P d\mathbf{V} = P(\mathbf{V}_2 - \mathbf{V}_1)_{\text{sys}}$$

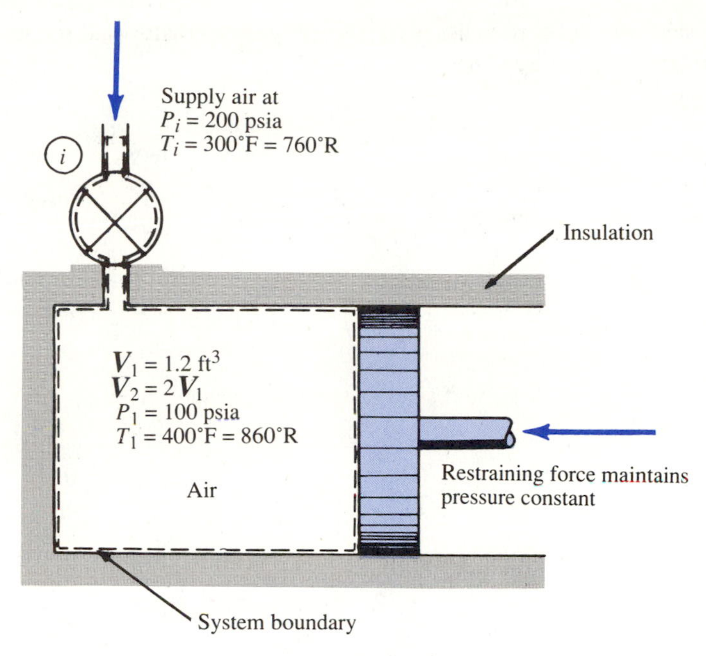

Figure 4.22 Insulated piston-cylinder assembly for Example 4.15.

The enthalpy of the entering air remains constant because the supply conditions are constant. Therefore

$$\int_1^2 (h\,dm)_i = h_i \int_1^2 (dm)_i = h_i m_i = h_i(m_2 - m_1)_{\text{sys}}$$

where the results of the conservation-of-mass equation have been incorporated in the last step. With these results the conservation-of-energy equation can be written as

$$P(V_2 - V_1)_{\text{sys}} - h_i(m_2 - m_1)_{\text{sys}} + (m_2u_2 - m_1u_1)_{\text{sys}} = 0 \qquad \text{(a)}$$

Since there are two unknowns, m_2 and u_2, with only one equation, a solution does not appear possible. However, if ideal-gas behavior is assumed, there is actually only one unknown (the temperature T_2), since T_2, together with other given data, is sufficient to uniquely determine both m_2 and u_2. Ideal-gas behavior is a reasonable assumption because the air pressure is low and the air temperature is high relative to the critical-state values. An interative solution could be used to arrive at the correct value of T_2, but an approximate solution can be obtained directly if the specific heats of the air are assumed to remain constant. This assumption must be verified after T_2 has been estimated. For an ideal gas

$$dh = c_p\,dT \quad \text{and} \quad du = c_v\,dT$$

and a reference value of zero is usually assigned to the enthalpy and the internal energy at zero absolute temperature. Then

$$h = c_p T \quad \text{and} \quad u = c_v T$$

where T is in absolute units.

Substituting these equations and the ideal-gas equation of state into Equation (a) results in

$$P(V_2 - V_1) - c_p T_i \left(\frac{P_2 V_2}{RT_2} - \frac{P_1 V_1}{RT_1} \right) + \frac{P_2 V_2}{RT_2} (c_v T_2) - \frac{P_1 V_1}{RT_1} (c_v T_1) = 0$$

But the pressure can be eliminated since it remains constant. Thus,

$$(V_2 - V_1) - \left(\frac{c_p T_i}{R} \right) \left(\frac{V_2}{T_2} - \frac{V_1}{T_1} \right) + \frac{c_v}{R} (V_2 - V_1) = 0$$

or

$$(V_2 - V_1) \left(1 + \frac{c_v}{R} \right) - \left(\frac{c_p T_i}{R} \right) \left(\frac{V_2}{T_2} - \frac{V_1}{T_1} \right) = 0$$

Since $c_p - c_v = R$, this equation can be written as

$$\frac{c_p}{R} (V_2 - V_1) - \left(\frac{c_p T_i}{R} \right) \left(\frac{V_2}{T_2} - \frac{V_1}{T_1} \right) = 0$$

Solving for T_2 yields

$$T_2 = \frac{V_2}{(V_1/T_1) + [(V_2 - V_1)/T_i]}$$

Since the final volume is twice the initial volume, this equation reduces to

$$T_2 = \frac{2V_1}{(V_1/T_1) + (V_1/T_i)} = \frac{2}{(1/T_1) + (1/T_i)}$$

or

$$T_2 = \frac{2}{(1/860°\text{R}) + (1/760°\text{R})} = \underline{\underline{807°\text{R} = 347°\text{F}}}$$

The equation above indicates that the final temperature of the contents of the cylinder for this problem is independent of the gas. That is, provided the behavior of the gas is ideal, the final temperature would be the same if, for example, nitrogen were used instead of air. Note also that the maximum and minimum temperatures of the air vary by only 100°R, so the assumption of constant specific heats is justified in this case.

The mass of air that enters is

$$m_i = m_2 - m_1 = \frac{PV_2}{RT_2} - \frac{PV_1}{RT_1}$$

$$= \left(\frac{P}{R}\right)\left(\frac{2V_1}{T_2} - \frac{V_1}{T_1}\right) = \frac{PV_1}{R}\left(\frac{2}{T_2} - \frac{1}{T_1}\right)$$

$$= \frac{(100\ \text{lb}_f/\text{in}^2)(144\ \text{in}^2/\text{ft}^2)(1.2\ \text{ft}^3)}{\left(53.34\ \dfrac{\text{ft}\cdot\text{lb}_f}{\text{lb}_m\cdot{}^\circ\text{R}}\right)}\left(\frac{2}{807^\circ\text{R}} - \frac{1}{860^\circ\text{R}}\right) = \underline{\underline{0.426\ \text{lb}_m}}$$

EXAMPLE 4.16

A home pressure cooker has an internal volume of 0.004 m^3 and employs a pressure regulator to limit the internal gauge pressure to 99 kPa. A sufficient quantity of water is added to the pressure cooker to ensure that saturation conditions prevail while cooking is in progress. This feature allows cooking with water at a controlled temperature, which is higher than the usual 100°C, and cooking time is consequently reduced.

A medium-heat setting is used until the pressure cooker reaches the operating pressure. The heat-transfer rate is then reduced to eliminate excessive loss of moisture through the pressure regulator during cooking. Suppose that when the operating pressure is first reached, the pressure cooker contains 0.25 kg of a mixture of liquid water and water vapor. Determine the following:

(a) The temperature at which cooking occurs

(b) The mass of liquid and the mass of vapor present when the operating pressure is first reached

(c) The maximum allowable heat-transfer rate to the cooker if the final mass of liquid in the cooker at the end of 20 min operation is required to be equal to one-half of the initial mass of liquid

Solution.

(a) The design of the pressure cooker ensures that saturation conditions are maintained in the cooker by regulating the internal pressure and limiting the amount of vapor escaping. Therefore, cooking occurs at the saturation temperature corresponding to the pressure of 200 kPa (99 kPa gauge). From Table B.2 this temperature is $\underline{\underline{120.24^\circ\text{C}}}$.

(b) A sketch of the system for this problem is shown in Figure 4.23. The mass of liquid and the mass of vapor present when the operating pressure is first reached can be determined from property relations and given information. The initial state is fixed by the pressure (200 kPa) and the specific volume of the mixture. Therefore, the initial quality can be calculated. The initial proportions of liquid and vapor present can be determined from these properties.

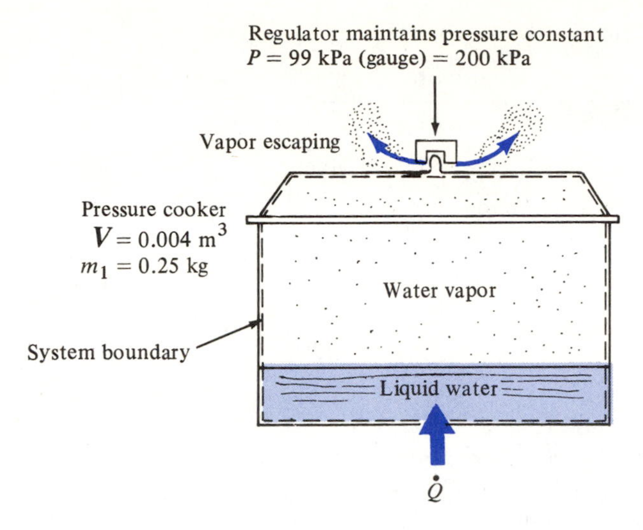

Figure 4.23 Sketch for Example 4.16.

The total initial mass and total volume are known, so the initial mixture specific volume is

$$v_1 = \frac{\mathbf{V}}{m_1} = \frac{0.004 \text{ m}^3}{0.25 \text{ kg}} = 0.016 \text{ m}^3/\text{kg}$$

and the initial quality is determined from

$$x_1 = \frac{v_1 - v_f}{v_{fg}}$$

Using Table B.2, we have

$$x_1 = \frac{(0.016 - 0.001061)\text{m}^3/\text{kg}}{0.8848 \text{ m}^3/\text{kg}} = 0.0169$$

The initial mass of vapor can now be calculated from the definition of the quality:

$$m_{g1} = x_1 m_1 = 0.0169(0.25 \text{ kg}) = \underline{\underline{0.00423 \text{ kg}}}$$

Therefore, the initial mass of liquid present is

$$m_{f1} = m_1 - m_{g1} = (0.25 - 0.00423)\text{kg} = \underline{\underline{0.246 \text{ kg}}}$$

(c) Applying the conservation-of-mass equation, Equation 3.23, to this transient system yields

$$\sum_{\text{inlet}} m - \sum_{\text{exit}} m = (m_2 - m_1)_{\text{sys}}$$

Since there is one exit and no inlet,

$$-m_e = (m_2 - m_1)_{sys}$$

The conservation-of-energy equation, Equation 4.34, with the assumption that the kinetic- and potential-energy terms are negligible (see Example 4.14), reduces to

$$Q_{12} - W_{12} + \int_1^2 (h\ dm)_i - \int_1^2 (h\ dm)_e = (U_2 - U_1)_{sys}$$

Irreversible-work modes are absent, and the system boundary is rigid. Therefore, the work for this process is zero. Only saturated vapor leaves the cooker through the pressure regulator. Since the saturation pressure remains constant, the enthalpy of the saturated vapor escaping from the cooker is also constant. This result allows the evaluation of the integral in the conservation-of-energy equation as

$$\int_1^2 (h\ dm)_e = h_e \int_1^2 (dm)_e = h_e m_e = -h_e(m_2 - m_1)_{sys}$$

where the result of the conservation-of-mass equation has been substituted in the last step.

The conservation-of-energy equation can now be written as

$$Q_{12} = -h_e(m_2 - m_1)_{sys} + (m_2u_2 - m_1u_1)_{sys}$$

The heat transfer can be determined once the property values are known. From Table B.2

$$h_e = h_g|_{200\ \text{kPa}} = 2706.5\ \text{kJ/kg}$$

$$u_1 = u_f + x_1u_{fg} = 504.59\ \text{kJ/kg} + 0.0169(2024.8\ \text{kJ/kg}) = 538.8\ \text{kJ/kg}$$

Also,

$$m_{f2} = \frac{1}{2}m_{f1} = \frac{1}{2}(0.246\ \text{kg}) = 0.123\ \text{kg}$$

Since the volume of the cooker remains constant,

$$V = m_{f2}v_{f2} + m_{g2}v_{g2}$$

and the final mass of the vapor is

$$m_{g2} = \frac{V - m_{f2}v_{f2}}{v_{g2}}$$

$$= \frac{0.004\ \text{m}^3 - (0.123\ \text{kg})(0.001061\ \text{m}^3/\text{kg})}{0.8859\ \text{m}^3/\text{kg}}$$

$$= 0.00437\ \text{kg}$$

Therefore, the final mass of water and the final quality are

$$m_2 = m_{f2} + m_{g2} = (0.123 + 0.00437)\text{kg} = 0.127 \text{ kg}$$

$$x_2 = \frac{m_{g2}}{m_2} = \frac{0.00437 \text{ kg}}{0.127 \text{ kg}} = 0.0344$$

The final internal energy can now be calculated as

$$u_2 = u_f + x_2 u_{fg} = 504.59 \text{ kJ/kg} + 0.0344(2024.8 \text{ kJ/kg}) = 574.2 \text{ kJ/kg}$$

The heat transfer required is

$$\begin{aligned} Q_{12} &= (-2706.5 \text{ kJ/kg})(0.127 - 0.25)\text{kg} + (0.127 \text{ kg})(574.2 \text{ kJ/kg}) \\ &\quad - (0.25 \text{ kg})(538.8 \text{ kJ/kg}) \\ &= 271 \text{ kJ} \end{aligned}$$

The maximum allowable heat-transfer rate to the water is

$$\dot{Q} = \frac{Q_{12}}{\Delta t} = \frac{271 \text{ kJ}}{(20 \text{ min})(60 \text{ s/min})} = \underline{\underline{226 \text{ W}}}$$

Here the heat-transfer rate has been assumed to be constant throughout the 20-min period.

The heat-transfer rate of 226 W does not represent the electrical power consumption by the appliance. It is only the rate at which energy must be added to the water. The appliance would have to supply this energy, and in addition, it would have to supply the energy that is transferred to the surroundings as a result of unavoidable heat losses. ■

4.6 INTRODUCTION TO SIMPLE THERMODYNAMIC CYCLES

Thus far the basic principles of thermodynamics and the equations derived from them have been applied largely to systems consisting of a single device. A powerful method of analyzing problems is to apply these same equations to larger systems composed of several pieces of equipment.

A typical large-scale system is a conventional *power plant*. A schematic of a simple steam power plant is shown in Figure 4.24. Regardless of energy source (nuclear fuel, oil, coal, or natural gas), steam power plants in their simplest form consist of four basic components: a boiler, a turbine, a condenser, and a pump. In the boiler the energy from the fuel is transferred to the water, changing its state from a compressed liquid to a superheated vapor. The pressure change of the water in the boiler is quite small, and it is often neglected when a simplified analysis is applied to the operation of a boiler.

In the turbine the energy extracted from the steam is used to produce useful work. The rotating shaft of the turbine turns a generator to produce electrical energy. The steam, as it leaves the turbine, is usually slightly superheated or at least has a very high

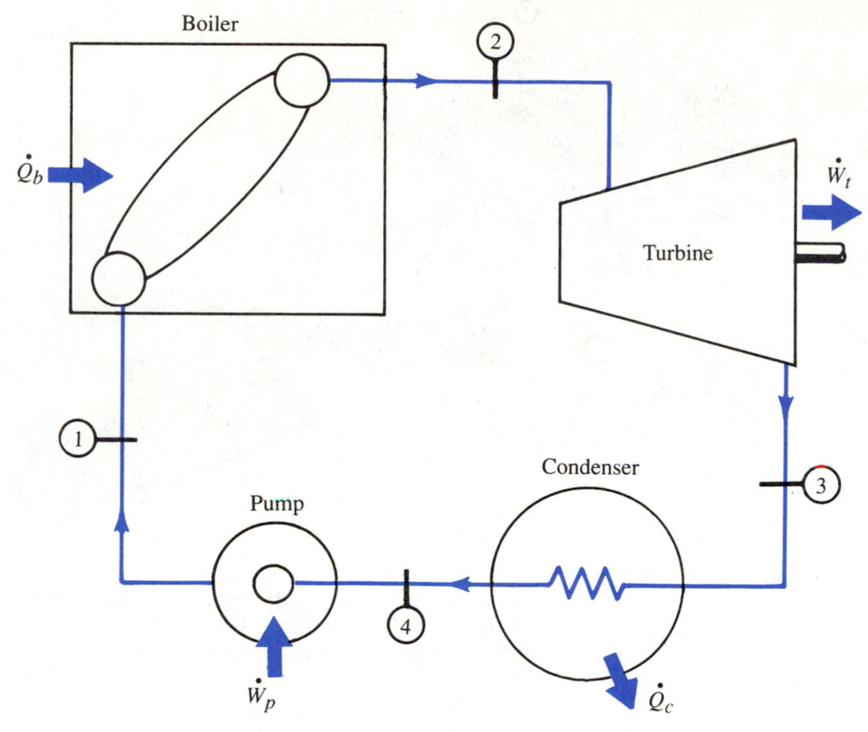

Figure 4.24 A simple steam-power-plant cycle.

quality. The exhaust pressure of the steam is typically below atmospheric pressure because the condenser creates a partial vacuum by greatly reducing the volume of steam leaving the turbine.

In the condenser, there is heat transfer from the water vapor, causing it to condense. Heat transfer often occurs from the steam to cooling water taken from a river or lake, although cooling towers are also used to promote heat transfer to the air surrounding a power plant. Pressure changes of the steam in the condenser are small, and they are usually neglected in a simplified thermodynamic analysis.

The fourth component in a simple steam power plant is a pump. The purpose of the pump is to increase the pressure of the condensate that leaves the condenser so that the condensate can be forced into the boiler at high pressure. The increase in energy level of the water in the pump is provided by work input from some external source, usually an electric motor.

The high-pressure, high-temperature steam entering the turbine impacts on rotating turbine blades, and the enthalpy of the steam decreases as it passes through the turbine. The steam eventually exhausts from the turbine and enters the condenser. The turbine shaft is attached to an electric generator which provides an electrical power output. The turbine is normally well-insulated so that most of the energy extracted from the steam is used to produce work by rotating the turbine shaft. The photograph in Figure 4.25 shows the internal mechanism of a large steam turbine.

Figure 4.25 A 45-MW steam turbine with the top casing removed. The entering high-pressure steam impacts on the small turbine blades shown at the right in the photograph. The steam reaches the large, low-pressure blades at the left where it enters the void area in the casing on its way to the condenser. (Photograph courtesy of General Electric Co.)

The water that flows through the four components of the power plant proceeds through a thermodynamic cycle because the water continually retraces the same series of states. If the water is considered to be a thermodynamic system that proceeds through a cycle, changes in any property of the water will be zero for a complete cycle. Therefore Equation 4.24 written for the water as the system yields

$$\oint \delta Q = \oint \delta W \tag{4.35}$$

That is, the net heat transfer to the water for a complete cycle must be equal to the net work performed by the water. From Figure 4.24 the net work of the cycle is the algebraic sum of the work performed by the water in the turbine and the work performed on the water in the pump. The net heat transfer for the cycle is the algebraic sum of the heat transfer to the water in the boiler and the heat transfer from the water in the condenser, or

$$\oint \delta W = W_p + W_t$$

and

$$\oint \delta Q = Q_b + Q_c$$

Substituting these results into Equation 4.35 gives

$$Q_b + Q_c = W_P + W_t \tag{4.36}$$

Since each component in the steam power plant is assumed to operate in steady state, this equation can be written on a rate basis as

$$\dot{Q}_b + \dot{Q}_c = \dot{W}_P + \dot{W}_t \tag{4.37}$$

One means of measuring the performance of a thermodynamic cycle is to calculate the ratio of the desired effect of the cycle to the energy input required to produce the desired effect. That is,

$$\text{Measure of performance} = \frac{\text{desired effect}}{\text{energy input}}$$

A measure of the performance of a power cycle is the *thermal efficiency*, defined as the ratio of the net power output of the cycle (the desired effect) to the total rate of heat transfer to the working fluid:

$$\eta_{\text{th}} \equiv \frac{\dot{W}_{\text{net}}}{\dot{Q}_{\text{input}}} \tag{4.38}$$

The net power output in the simple steam-power cycle is the algebraic sum of the power output of the turbine and the power input of the pump. The input heat-transfer rate during the cycle is the rate of heat transfer to the water in the boiler. Therefore, the thermal efficiency of the simple steam-power cycle is

$$\eta_{\text{th}} \equiv \frac{\dot{W}_t + \dot{W}_p}{\dot{Q}_b} \tag{4.39}$$

Factors that produce increases in the thermal efficiency are desirable, and the thermal efficiency is used by engineers as a measure of how efficiently a cycle uses the heat transfer to perform useful work. Parametric studies can be used to determine the effect of the thermodynamic states on the cycle's thermal efficiency. Limitations on the maximum thermal efficiency that can be achieved with a power cycle are considered in Chapter 5. The power-plant cycle, as well as other thermodynamic cycles, are discussed in further detail in Chapters 7 and 8.

■ EXAMPLE 4.17

Properties of water at various locations throughout the steam-power-plant cycle shown in Figure 4.24 are given in the following table:

State	T, °F	P, psia
1	170	3000
2	1200	3000
3	(x_3 = 0.96)	10
4	140	10

The turbine is well-insulated and the velocity of the water throughout the cycle is much less than the velocity at location 3, which is 450 ft/s. The water flows steadily through each component at a rate of 20 lb_m/s. Determine the following quantities:

(a) The heat-transfer rate to the water in the boiler, $\dot{Q}_b$
(b) The rate at which work is performed by the water in the turbine, $\dot{W}_t$
(c) The heat-transfer rate from the water in the condenser, $\dot{Q}_c$
(d) The rate at which work is performed on the water in the pump, $\dot{W}_p$
(e) The thermal efficiency of the cycle, η_{th}

Solution. The water is the system for this problem, and the flow is steady and can be assumed to be uniform for each of the four devices that compose the cycle. Potential-energy changes can be neglected because elevation changes are at most only several feet for each of the devices. Kinetic energies can be neglected except at the exit from the turbine because all velocities are much less than that at location 3.

The conservation-of-mass equation for steady flow requires that the mass flow rates through each of the four components of the cycle are equal, or

$$\dot{m} = 20 \text{ lb}_m/\text{s} = \text{constant}$$

Applying the conservation-of-energy equation to the water in the boiler yields

$$\dot{Q}_b = \dot{m}(h_2 - h_1)$$

because no work interactions occur at the boundary of the boiler. Applying the conservation-of-energy equation to the water in the turbine gives

$$\dot{W}_t = \dot{m}\left(h_2 - h_3 - \frac{V_3^2}{2}\right)$$

because the heat transfer from the turbine can be neglected when compared with the work output. For the condenser, the conservation-of-energy equation reduces to

$$\dot{Q}_c = \dot{m}\left(h_4 - h_3 - \frac{V_3^2}{2}\right)$$

and for the pump the result is

$$\dot{W}_p = \dot{m}(h_4 - h_1)$$

because the heat transfer from the pump is small compared with $\dot{W}_p$.

The four values for the enthalpies in the above equations can be determined from the water tables. The state of the water as it enters the boiler is a compressed-liquid state, and the enthalpy can be determined from Table B.4E:

$$h_1 = 145.1 \text{ Btu/lb}_m$$

From Table B.3E (superheated vapor) at 3000 psia and 1200°F

$$h_2 = 1575.5 \text{ Btu/lb}_m$$

From Table B.2E at 10 psia and $x_3 = 0.96$,

$$h_3 = h_f + x_3 h_{fg} = 161.25 \text{ Btu/lb}_\text{m} + (0.96)(981.9 \text{ Btu/lb}_\text{m})$$

$$= 1103.9 \text{ Btu/lb}_\text{m}$$

State 4 is in the compressed-liquid region, and the enthalpy can be approximated by the enthalpy of a saturated liquid at 140°F since the pressure is not substantially higher than the saturation pressure at 140°F:

$$h_4 \simeq h_f|_{140°\text{F}} = 107.98 \text{ Btu/lb}_\text{m}$$

(a) Substituting these enthalpy values into the equation for $\dot{Q}_b$ results in

$$\dot{Q}_b = \dot{m}(h_2 - h_1) = (20 \text{ lb}_\text{m}/\text{s})(1575.5 - 145.1)\text{Btu/lb}_\text{m} = \underline{\underline{28{,}600 \text{ Btu/s}}}$$

(b) The power output of the turbine is

$$\dot{W}_t = \dot{m}\left(h_2 - h_3 - \frac{V_3^2}{2}\right)$$

$$= (20 \text{ lb}_\text{m}/\text{s})\left[(1575.5 - 1103.9)\text{Btu/lb}_\text{m} - \frac{(450 \text{ ft/s})^2}{2(32.2 \text{ ft}\cdot\text{lb}_\text{m}/\text{lb}_\text{f}\cdot\text{s}^2)(778 \text{ ft}\cdot\text{lb}_\text{f}/\text{Btu})}\right]$$

$$\dot{W}_t = \underline{\underline{9{,}350 \text{ Btu/s} = 13{,}230 \text{ hp}}}$$

The kinetic-energy change reduces the power output of the turine by about 1 percent. The positive sign on $\dot{W}_t$ indicates that the turbine produces a power output.

(c) The heat-transfer rate in the condenser is

$$\dot{Q}_c = \dot{m}\left(h_4 - h_3 - \frac{V_3^2}{2}\right)$$

$$= (20 \text{ lb}_\text{m}/\text{s})\left[(107.98 - 1103.9)\text{Btu/lb}_\text{m} - \frac{(450 \text{ ft/s})^2}{2(32.2 \text{ ft}\cdot\text{lb}_\text{m}/\text{lb}_\text{f}\cdot\text{s}^2)(778 \text{ ft}\cdot\text{lb}_\text{f}/\text{Btu})}\right]$$

$$= \underline{\underline{-20{,}000 \text{ Btu/s}}}$$

(d) The power input to the pump is

$$\dot{W}_p = \dot{m}(h_4 - h_1) = (20 \text{ lb}_\text{m}/\text{s})(107.98 - 145.1)\text{Btu/lb}_\text{m}$$

$$\dot{W}_p = \underline{\underline{-742 \text{ Btu/s} = -1050 \text{ hp}}}$$

(e) The thermal efficiency of the cycle is given by Equation 4.39:

$$\eta_{\text{th}} = \frac{\dot{W}_t + \dot{W}_p}{\dot{Q}_b} = \frac{(9350 - 742)\text{Btu/s}}{28{,}600\ \text{Btu/s}} = \underline{\underline{0.30}}$$

Therefore, only 30 percent of the energy available by burning the fuel used to heat the water in the boiler is actually available to produce net usable work. Seventy percent of the energy from the fuel is rejected to the surroundings in the condenser. The 30 percent efficiency value may seem low, but it is actually a reasonable value for simple steam-power plants. Ways to improve the thermal efficiency of the cycle are discussed in Chapters 7 and 8.

As a check of the answers, the net heat-transfer rate of the cycle must equal the net work rate (power) of the cycle, or

$$\dot{Q}_b + \dot{Q}_c = \dot{W}_p + \dot{W}_t$$

Substituting values gives

$$(28{,}600 - 20{,}000)\text{Btu/s} = (-742 + 9350)\text{Btu/s}$$

which verifies that the energy of the cycle is conserved. ■

Another cycle of considerable importance is the *vapor-compression refrigeration cycle*, the cycle most frequently used in air-conditioning and refrigeration applications. It consists of four major components: a compressor, a condenser, a throttling valve, and an evaporator, as shown in Figure 4.26. The refrigerant is compressed in the compressor so that its temperature is greater than the surroundings. The refrigerant leaves the compressor as a superheated vapor and enters the condenser, where heat transfer from the refrigerant causes the refrigerant to condense to the liquid phase. The refrigerant then enters a throttling valve, where the pressure and temperature of the refrigerant are greatly reduced. The cold, low-pressure refrigerant then enters the evaporator section, where heat transfer occurs from the refrigerated space to the cold refrigerant. The refrigerant is then routed back into the compressor, where it enters as a superheated vapor, and the cycle is repeated.

Pressure changes in both the condenser and evaporator are usually neglected because both of these components are simply finned-tube heat exchangers designed to minimize pressure losses and maximize heat-transfer rates. Heat-transfer rates in the throttling valve can be neglected because the small size of the valve inhibits heat flow. The heat-transfer rate from the compressor is quite small compared with the power input to the compressor, so it is usually neglected in simplified analyses.

The vapor-compression refrigeration cycle is similar to a reversed power-plant cycle with the exception that the pump is replaced with a throttling valve. The purpose of a power-plant cycle is to produce work at the expense of burning a fuel, that is, heat transfer to the working fluid. On the other hand, the purpose of the vapor-compression refrigeration cycle is to provide heat transfer from a cool region or to a warm region at the expense of work done on the working fluid in the compressor. Refrigerators and

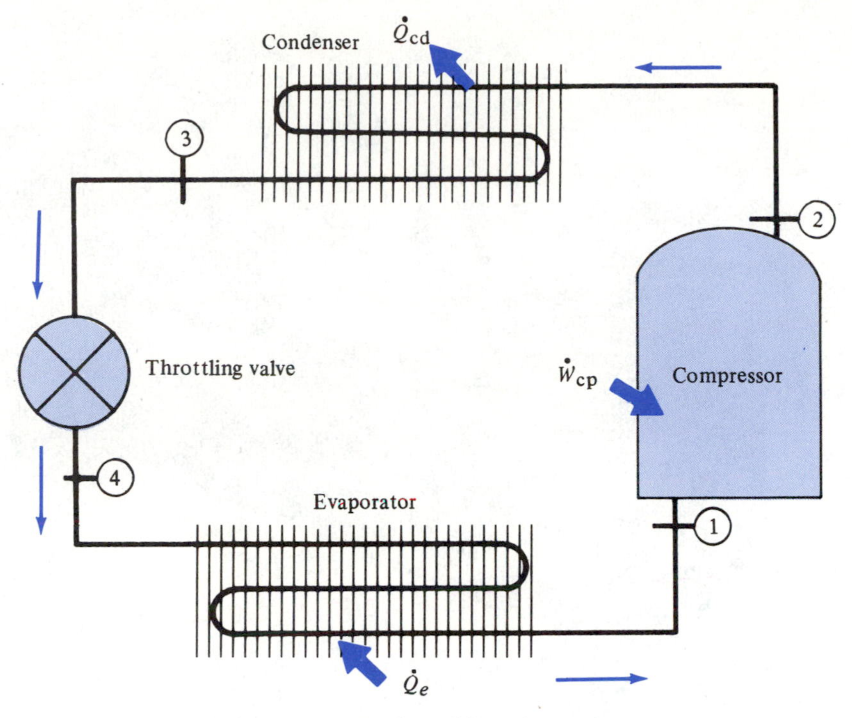

Figure 4.26 A simple vapor-compression refrigeration cycle.

heat pumps both usually operate on a vapor-compression refrigeration cycle. An air-conditioning unit designed for large commercial loads is shown in Figure 4.27.

A *refrigerator* is used to provide heat transfer from a low-temperature region at the expense of a work input. A measure of the performance of a refrigerator is called the *coefficient of performance* (COP) and is designated by the symbol β_R. The COP is the heat-transfer rate to the evaporator (the desired effect) from a refrigerated space divided by the power input required by the compressor. The definition of the COP of a refrigeration cycle can therefore be expressed as

$$\beta_R \equiv -\frac{\dot{Q}_e}{\dot{W}_{cp}} \tag{4.40}$$

For a refrigerator $\dot{Q}_e$ is positive and $\dot{W}_{cp}$ is negative. The negative sign is used in the definition in Equation 4.40 to ensure that the coefficient of performance is positive. Large values for the COP are desirable since they indicate that more cooling capacity is provided per unit of power input.

Applying the conservation of energy, Equation 4.24, on a rate basis to the refrigeration cycle shown in Figure 4.26 results in

$$\dot{Q}_e + \dot{Q}_{cd} = \dot{W}_{cp} \tag{4.41}$$

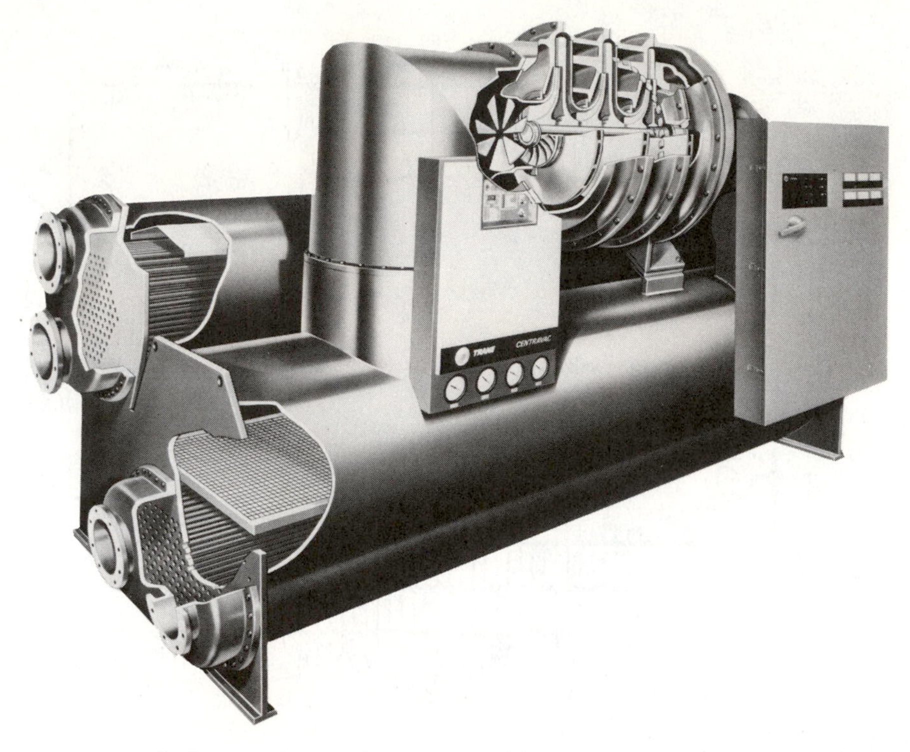

Figure 4.27 An integrated water chiller, air-conditioning unit. The three-stage centrifugal compressor shown in the upper right is the type typically used in large-capacity commercial units. The condenser and evaporator are both shell-and-tube design heat exchangers. In the condenser shown in the rear, the refrigerant flows through the shell and exchanges heat with water circulating through the tubes. In the evaporator shown in the foreground, the water in the tubes is cooled and is pumped to a remote location where cooling is desired. (Courtesy of The Trane Company.)

Therefore, an alternative form for β_R is

$$\beta_R = -\frac{\dot{Q}_e}{\dot{Q}_{cd} + \dot{Q}_e}$$

or

$$\beta_R = -\frac{1}{(\dot{Q}_{cd}/\dot{Q}_e) + 1} \tag{4.42}$$

A *heat pump* is used to provide heat transfer to a high-temperature region at the expense of a work input. The measure of performance of a heat pump is also called the *coefficient of performance* and is designated by the symbol β_H. The COP of a heat pump is the heat transfer rate from the condenser of the system to a warm region (the desired effect) divided by the power input required to drive the compressor:

$$\beta_H \equiv \frac{\dot{Q}_{cd}}{\dot{W}_{cp}} \tag{4.43}$$

By substituting Equation 4.41, we can write this expression as

$$\beta_H = \frac{1}{(\dot{Q}_e/\dot{Q}_{cd}) + 1} \tag{4.44}$$

EXAMPLE 4.18

A residential air-conditioning unit uses refrigerant-12 that circulates at a steady rate of 0.07 kg/s. Properties of the refrigerant at various points in the cycle shown in Figure 4.26 are

$$P_1 = 300 \text{ kPa} \qquad P_2 = 1.40 \text{ MPa} \qquad T_3 = 40^\circ\text{C}$$

$$T_1 = 10^\circ\text{C} \qquad T_2 = 100^\circ\text{C} \qquad P_3 = P_2$$

Determine the power required by the compressor, the heat transfer rate to the refrigerant in the evaporator, and the coefficient of performance of the cycle.

Solution. Kinetic-energy and potential-energy changes across each of the components can be neglected. The heat transfer from the compressor can be neglected, and the conservation of energy applied to the refrigerant-12 in the compressor gives

$$\dot{W}_{cp} = \dot{m}(h_1 - h_2)$$

The enthalpies at states 1 and 2 can be determined from Table C.3 (superheated vapor):

$$h_1 = 195.12 \text{ kJ/kg}$$

$$h_2 = 244.27 \text{ kJ/kg}$$

and

$$\dot{W}_{cp} = (0.07 \text{ kg/s})(195.12 - 244.27)\text{kJ/kg} = \underline{\underline{-3.44 \text{ kW}}}$$

Applying the conservation of energy to the refrigerant in the evaporator yields

$$\dot{Q}_e = \dot{m}(h_1 - h_4)$$

because there is no work performed on the evaporator. The state 4 is unknown, so h_4 cannot be determined directly from the given information. However, the change in enthalpy through the throttling valve is zero (see Example 4.10), so

$$h_3 = h_4$$

and the state 3 is known because

$$T_3 = 40^\circ\text{C}$$

$$P_3 = P_2 = 1.40 \text{ MPa}$$

These conditions are in the compressed-liquid region, and the enthalpy at state 3 is approximately equal to the enthalpy of saturated-liquid R-12 at 40°C since the pressure is not substantially higher than the saturation pressure at 40°C. Thus

$$h_4 = h_3 \simeq h_f|_{40°C} = 75.43 \text{ kJ/kg}$$

The heat-transfer rate to the evaporator becomes

$$\dot{Q}_e = (0.07 \text{ kg/s})(195.12 - 75.43)\text{kJ/kg} = \underline{\underline{8.38\,\text{kW}}}$$

The coefficient of performance of the cycle is calculated by using Equation 4.40:

$$\beta_R = -\frac{\dot{Q}_e}{\dot{W}_{cp}} = -\left(\frac{8.38\,\text{kW}}{-3.44\text{ kW}}\right) = \underline{\underline{2.44}}$$

For every kilowatt of power input to the compressor, the heat-transfer rate to the refrigeration unit from the refrigerated space is 2.44 kW. This result may appear to be a violation of the conservation of energy, but energy is conserved because Equation 4.41 is satisfied. The upper limit of the coefficient of performance is discussed in Chapter 5.

4.7 SUMMARY

In this chapter a very general mathematical expression for the principle of conservation of energy has been developed:

$$\dot{Q} - \dot{W} + \int_{A_i} (h + e_k + e_p)\rho V_n dA - \int_{A_e} (h + e_k + e_p)\rho V_n\, dA = \frac{d}{dt}\int_V e\rho\, d\mathbf{V} \quad (4.17)$$

This single equation was applied to the following two types of thermodynamic systems, which are frequently encountered in engineering problems.

Closed System

In a closed system the system boundary contains no inlets or exits, so there is no mass flow into or out of the system. For the closed system, therefore, Equation 4.17 reduces to

$$\dot{Q} - \dot{W} = \frac{dE_{\text{sys}}}{dt} \quad (4.19)$$

Integration of Equation 4.19 with respect to time results in the conservation-of-energy equation for a closed system that undergoes a finite change of state:

$$Q_{12} - W_{12} = (E_2 - E_1)_{\text{sys}} \quad (4.22)$$

For a closed system that undergoes a cycle, Equation 4.19 was evaluated throughout the cycle by using the cyclic integral, or

$$\oint \delta Q - \oint \delta W = \oint dE = 0 \quad (4.24)$$

Open Systems

Uniform Flow For uniform flow, fluid properties do not vary across inlet or exit cross-sectional areas. This assumption permits evaluation of the integrals in Equation 4.17, so that for uniform flow

$$\dot{Q} - \dot{W} + \sum_{\text{inlet}} \dot{m}(h + e_k + e_p) - \sum_{\text{exit}} \dot{m}(h + e_k + e_p) = \frac{dE_{\text{sys}}}{dt} \tag{4.29}$$

Steady State For a system operating in steady state the time rate of change of any extensive property of the system is zero. Since the total energy E_{sys} of the mass within the system is an extensive property of the system, there can be no change in the total energy in the system with time. Therefore, Equation 4.29 reduces to

$$\dot{Q} - \dot{W} + \sum_{\text{inlet}} \dot{m}(h + e_k + e_p) - \sum_{\text{exit}} \dot{m}(h + e_k + e_p) = 0 \tag{4.31}$$

for an open system operating in steady, uniform flow.

Transient Analysis The analysis of an open thermodynamic system that undergoes transient changes requires the application of the general equation for the conservation of energy, Equation 4.29. This equation was integrated with respect to time, with the result

$$Q_{12} - W_{12} + \sum_{\text{inlet}} \int_1^2 (h + e_k + e_p)\, dm - \sum_{\text{exit}} \int_1^2 (h + e_k + e_p)\, dm = (E_2 - E_1)_{\text{sys}} \tag{4.34}$$

PROBLEMS

4.1 The orbiting portion of the American space shuttle has a mass of 72,600 kg. As it reenters the outer edge of the earth's atmosphere at an altitude of approximately 37,000 m, the shuttle travels at a velocity of 7600 m/s. Calculate the kinetic and potential energy of the shuttle vehicle relative to the surface of the earth as it reenters the atmosphere. After the shuttle lands, it comes to rest and its kinetic energy and potential energy are zero. Considering the shuttle as a thermodynamic system, discuss the significant terms in the conservation-of-energy equation for the vehicle as it reenters the atmosphere.

4.2 An automobile engine has a rated output of 100 hp. Calculate the percent of the total output of the engine required for the following functions:

(a) operate the headlights, which have a total power of 400 W

(b) operate the air conditioner, which has a cooling capacity of 10,000 Btu/h and a coefficient of performance (ratio of cooling capacity to power input) of 2.2

(c) accelerate a 2000-lb_m car from rest to a speed of 60 mph in 12 s

(d) climb vertically 300 ft up a hill in 60 s in a 2000-lb_m car at a constant velocity

(e) climb a vertical rise of 200 ft in 2 min at a constant velocity in a 2000-lb_m car

(f) overcome an air-drag force of 150 lb_f at a velocity of 30 mph

4.3 A slingshot, shown in the accompanying figure, is used to propel a projectile of mass m. The unstretched length of the elastic is x_0. The elastic is stretched a length L, and the force in the elastic is proportional to its displacement. The projectile is released from rest. Determine an expression for the velocity of the projectile when it passes through the plane $x = 0$. Suppose the projectile is shot vertically upward. Determine an expression for the maximum height the projectile will reach if the air friction is neglected.

Problem 4.3

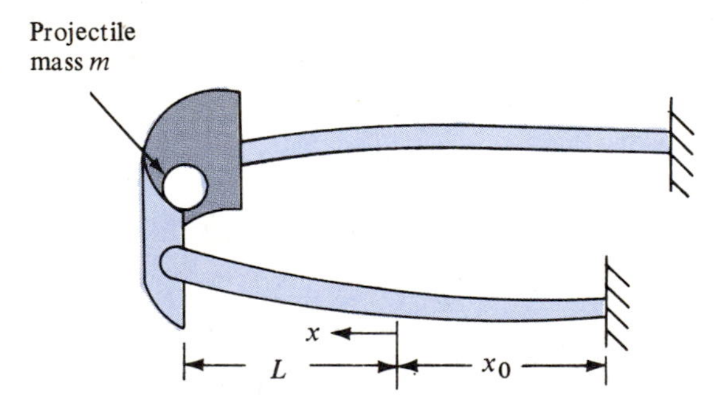

4.4 A fireworks shell, shown in the accompanying figure, has a mass of 12 kg and a height of 20 cm. It is placed vertically in a tube. The height of the tube is 70 cm and its diameter is 27 cm. An explosive charge is set off under the shell, and the pressure reaches 5 MPa before the shell begins to move. After the shell moves, the pressure of the gases drops linearly at a rate of 15 kPa for each centimeter the shell moves in the tube. Calculate the velocity of the shell as it completely leaves the tube. Determine the maximum height the shell can reach in the air.

Problem 4.4

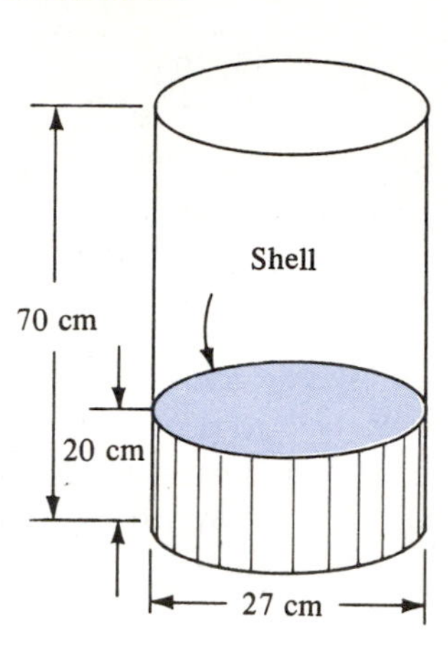

4.5 The world record for a 100-m run is 9.8 s. Assuming that a pole-vaulter can only run 90 percent of the average velocity for the world-record 100-m run, calculate the maximum height that a pole-vaulter can jump regardless of the construction of the pole. Neglect the height that the pole-vaulter could attain from the strength of his legs and any added height that he could achieve by moving his hands up the pole. The world record for the pole vault is 5.5 m. Why is the world record greater than the apparent ability of the pole-vaulter on the basis of a simplified conservation-of-energy analysis?

4.6 Suppose you wish to estimate the armor-piercing capability of an artillery shell. As a simplification, assume that the shell simply shears out a cylindrical hole in the armor without any deformation of the metal. Assume that a shell with mass of 40 kg, diameter of 105 mm, and velocity of 700 m/s impacts on an armor plate that has an ultimate shear stress of 3.5×10^{10} N/m^2; see the accompanying figure. Estimate the maximum thickness x of armor plate that the shell could pierce.

Problem 4.6

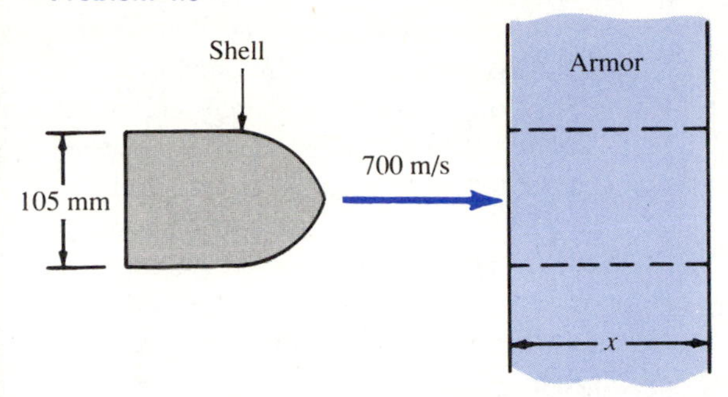

4.7 A homeowner has a lot situated on a small stream that has an average flow rate of 1500 gal/min. By damming up the stream, a small lake with a water elevation of 12 ft above the stream can be achieved. Estimate the maximum power that can be generated by the water in the lake. If the homeowner expects his power requirements to always be less than 20 kW, can the water in the stream be expected to provide sufficient power for his needs?

4.8 The Itaipu hydroelectric power-generating station located on the Parana river at the Brazil-Paraguay border is the world's largest. The plant has nine turbine-generator sets that operate at 60 Hz and supply power to Brazil. Nine other sets provide energy to Paraguay and operate at 50 Hz. Other statistics relating to the plant are shown below:

Parana river average daily flow	8500 m^3/s
Reservoir normal water level above turbine discharge	220 m
Rated turbine output (each)	715 MW
Total plant electric output at switchyard	12,600 MW

Calculate the overall plant efficiency (ratio of actual switchyard power output to maximum power that can be obtained from river flow). Calculate the turbine efficiency (ratio of actual turbine power output to maximum power that can be obtained from the river flow).

4.9 An exercise machine is designed to dissipate the energy of the person who is exercising by a device such as the one illustrated in the figure. The tension in the stationary rope is maintained at 20 lb_f and the exerciser rotates the wheel at a constant angular velocity of 120 rpm. The coefficient of friction between the surface of the notched wheel and the rope is 0.4. Assuming that the wheel has a specific heat of 0.3 $Btu/lb_m{\cdot}°R$ and a mass of 14 lb_m calculate the maximum average temperature rise of the wheel if the person exercises for 1 h under these conditions.

Problem 4.9

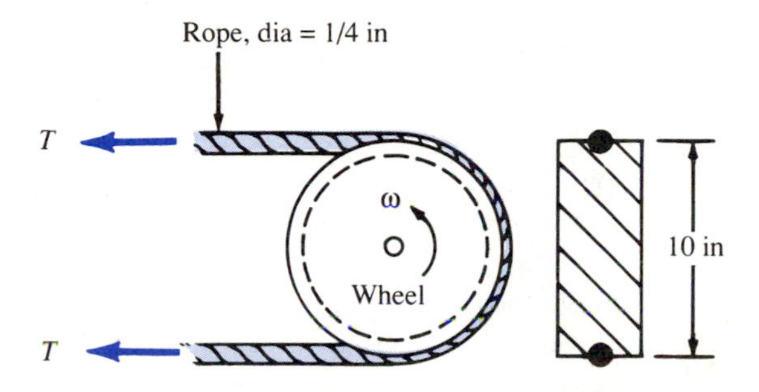

4.10 A fully loaded plane weighs 200,000 lb_f as it lands at a speed of 125 mph. The mass of the brakes on the plane is 200 lb_m and the brake assembly has an equivalent specific heat of 0.20 $Btu/lb_m{\cdot}°R$. Approximately 20 percent of the energy of the plane is dissipated by means of the brakes while the remainder is dissipated by means of reversing the thrust of the engines and viscous drag to the surrounding air. Estimate the maximum temperature rise that the brake assembly will reach when bringing the plane to rest. Recalculate the maximum brake temperature rise if the engines cannot be used to help stop the plane and 80 percent of the energy of the plane must be dissipated via the brakes.

4.11 Explain how spraying grapes with a fine mist of water will protect them from potential frost damage on a cold night. Suppose that a single grape (d = 0.5 in) is covered with a layer of water 0.05 in thick. Determine the maximum heat transfer to the grape if the initial temperature of the water is 50°F and its final state is ice at 32°F. The latent heat of fusion of water is 144 Btu/lb_m.

4.12 A sledge hammer has a weight of 70 N and it is used with a wedge to split logs. Suppose the sledge hammer strikes the wedge 20 times with an impact velocity of 10 m/s. Calculate the maximum energy that can be transferred to the log via the wedge. The wedge has a specific heat of 0.4 kJ/kg·K. Estimate the maximum temperature rise of the wedge as a result of being struck by the sledge hammer under these conditions if the wedge weighs 30 N.

4.13 A *pumped storage hydroelectric plant* consists of a large reservoir and convertible pump-turbine units that can function either as pumps or as turbines. Water is pumped from a lake into the reservoir during off-peak hours such as nights or weekends when the demand for electricity is low. The water is stored in the reservoir until the daytime hours of peak electrical demand when it is released through the pump-turbines that now act as turbines, generating electricity on the way back to the lake. Even though losses are incurred during both the pumping and the turbine processes, the pumped storage plant is economically feasible because of the differential charges to the customer during off-peak and peak hours. A large pumped storage facility has the following statistics: 6 pump-turbines, maximum electrical output of each turbine is 312 MW, maximum flow rate of water through all turbines is 3.3×10^7 gal/min, maximum reservoir level above the lake is 360 ft, maximum usable reservoir capacity is 1.8×10^{10} gal of water, pump efficiency of 0.87, and turbine efficiency of 0.90. Calculate the overall plant efficiency which is defined as the ratio of the actual plant electrical power output to the maximum possible power output. Also determine the minimum ratio of peak to off-peak power charges before the plant is economically practical if the plant is expected to return a 50 percent profit on each kWh of energy generated. Assume that the power company must pay for power just like its customers at the off-peak rate when it pumps the water up to the reservoir.

4.14 When fishing line is first removed from the spool it contains a "set" that can be removed by heating the line. The set is removed by pulling the line (ρ = 50 lb_m/ft^3) through a piece of leather as illustrated in the figure. The specific heat of the line is 0.85 Btu/lb_m°R and the line is pulled with a velocity of 2 ft/s. The coefficient of friction between the leather and line is 0.8 and the normal force exerted on the line is 5 lb_f. Estimate the temperature increase of the line that can be achieved by this process.

Problem 4.14

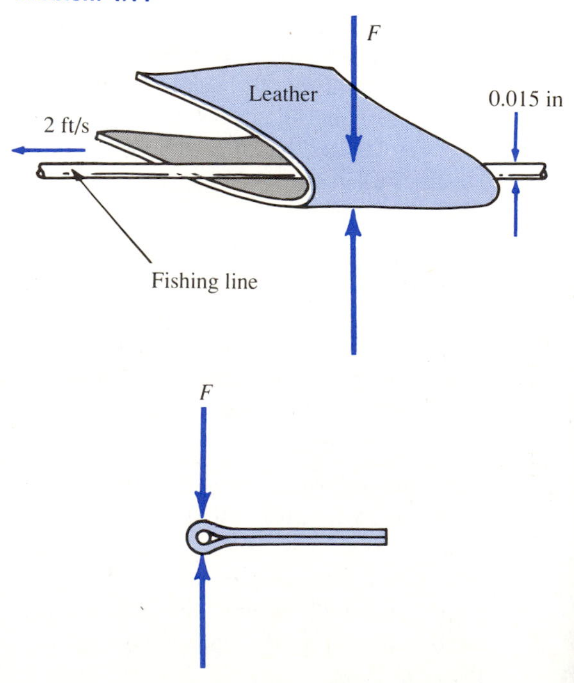

4.15 A ski lift has a vertical rise of 250 m and a length of 800 m. The capacity of each chair is two people, and the chairs are spaced 10 m apart. At steady operating conditions the velocity of the chairs is 3 m/s. The average mass of each chair and its two passengers is 190 kg. Estimate the power of a motor used to operate the ski lift, neglecting friction and wind drag. Recalculate the power of the motor if it is to be large enough to accelerate a fully loaded lift from rest to operational speed in 10 s.

4.16 A Cornish steam-operated pump was a steady-flow device frequently used in the nineteenth century to pump water out of mines. One particular model built in 1840 operated at the unusually high pressure of 40 psig. Working at its maximum capacity and a speed of 7.5 rpm, it developed 222 hp. The single-cylinder engine had a stroke of 11 ft and a cylinder diameter of 85 in. This engine worked a number of pumps via a belt system, raising water in stages through a total height of 938 ft. Assuming that the cylinder was exhausted to a condenser pressure of −10 psig, calculate the maximum flow rate of water that could be removed from the mine.

4.17 A vertical-span drawbridge has a mass of 10,000 kg, and it has a vertical lift of 7 m. Calculate the minimum power of a motor required to open the bridge to its maximum height in 4 min.

4.18 Determine the size of a motor, in kilowatts and horsepower, necessary to pull a 2-mm wire through a die at a velocity of 15 m/s if the tension in the wire is 250 N.

4.19 A scissor jack is used to raise a car. Thirty turns of the handle of the jack causes the car to rise vertically a distance of 1 m. Calculate the average torque necessary to raise the car, with a mass of 900 kg, a distance of 1 m. Determine the average power delivered by the jack if the crank is turned at an angular speed of 15 rev/min.

4.20 What average power is needed to accelerate a 1800-lb_m dragster from a standing start through a distance of ¼ mile in 7 s to a final velocity of 200 mph? Assume negligible wind drag. How much larger would the engine have to be to achieve the same performance if the track were inclined at an angle of 10° with the horizon? Assume that the engine performs steadily throughout the entire distance. Determine the size of the engine necessary for the same performance if the wind-drag force increases linearly from 0 to 1000 lb_f through the ¼-mile inclined track.

4.21 An automobile engine has a thermal efficiency of 20 percent. The rated mechanical-power output of the engine is 150 hp. Determine the fuel-consumption rate in gallons per hour of the engine if the fuel has a heat content of 20.6×10^3 Btu/lb_m and a density of 56 lb_m/ft^3.

4.22 The U.S. energy demand is estimated to be 3600 GW·yr of energy each year by 2015. Estimate the length of time that the U.S. coal reserves of 2.5×10^{15} kg will last if they are used exclusively to provide 100 percent of the U.S. energy demand at a thermal efficiency of 35 percent. Assume an average heating value of 26,600 kJ/kg for coal.

4.23 A rigid, closed, insulated container with a volume of 45 ft^3 is filled with carbon dioxide at a pressure of 14.7 psia and a temperature of 90°F. A 100-W electric-resistance heater inside the tank is energized for 10 min. Calculate the mass of CO_2 in the tank and the temperature and pressure in the tank at the end of the 10-min period.

4.24 Determine the heat transfer required to change the temperature of 50 g of solid lead by 15°C.

4.25 [FLA411] In a measurements lab, one of the experiments involves using a bomb calorimeter to measure the energy released by a chemical reaction. The bomb is a closed vessel placed in a large tank of water. When the chemicals react, there is heat transfer from the bomb to the water. A small stirring device circulates the water to maintain uniform temperature. In an experiment, the heat transfer from the bomb to the water is 1266.4 kJ, the heat transfer from the water to the surroundings is 63.3 kJ, and 52.7 kJ of work is required to operate the stirring device. The calorimeter contains 100 kg of liquid water initially at 20°C. Assume that no water evaporates during the process. Determine the following:

(a) the change in internal energy of the water in kJ

(b) the final temperature of the water in °C

4.26 Water in a closed, rigid, 3-ft^3 container exists at 80 psia and 500°F. The water is cooled by heat transfer to the surroundings until its temperature reaches 50°F. Calculate the following quantities:

(a) the pressure at the end of the process

(b) the mass of liquid water in the container at the end of the process

(c) the mass of the water vapor in the container at the end of the process

(d) the heat transfer from the water during the process

4.27 Refrigerant-12 is heated in a closed, rigid tank that has a volume of 1 m^3. Liquid refrigerant-12 at 20°C occupies one-tenth of the tank volume, and the remainder of the tank contains R-12 vapor. Determine the initial quality of the refrigerant and the heat transfer to the refrigerant required to vaporize all of the liquid.

4.28 Suppose that 0.5 ft^3 of air at 80°F and 20 psia is compressed in a closed system without friction until the volume is one-tenth of the initial value. During the compression process the air follows the path $Pv^{1.2}$ = constant. Calculate the temperature and pressure of the air at the end of the process and the heat transfer during the process.

4.29 Hydrogen in a closed, rigid container with a volume of 2.3 m^3 has a temperature of 45°C and a pressure of 125 kPa. How much heat transfer to the hydrogen is required to increase the gas temperature to 100°C? Determine the hydrogen pressure at this temperature.

4.30 A 100-W light bulb is placed in a closed room that has a volume of 55 m^3. The air in the room is at atmospheric pressure and at 25°C before the light is turned on. Calculate the temperature of the air 2 h after the light is turned on, assuming that the room is well-insulated.

4.31 A closed, rigid tank has a volume of 40 ft^3 and contains refrigerant-12 at 50°F. Originally, the tank is filled with one-third liquid and two-thirds vapor, by volume. The contents of the tank are heated until the temperature rises to 200°F. Determine the following:

(a) the original pressure within the tank

(b) the original mass of liquid and vapor

(c) the original quality

(d) the final pressure within the tank if the refrigerant-12 is superheated, or the final quality if it is saturated

4.32 A gas company stores its home-heating gas in tanks that are designed to move vertically so that they deliver gas to homes under constant pressure regardless of tank capacity. A schematic diagram of the tank is shown in the accompanying sketch. During a summer day the demand for gas is rather low, so the tank is completely closed for 12 h. During this time the tank is observed to move upward 1 m, and the sun adds 5.2×10^7 J of energy to the gas. Determine the change in internal energy of the gas during this period.

Problem 4.32

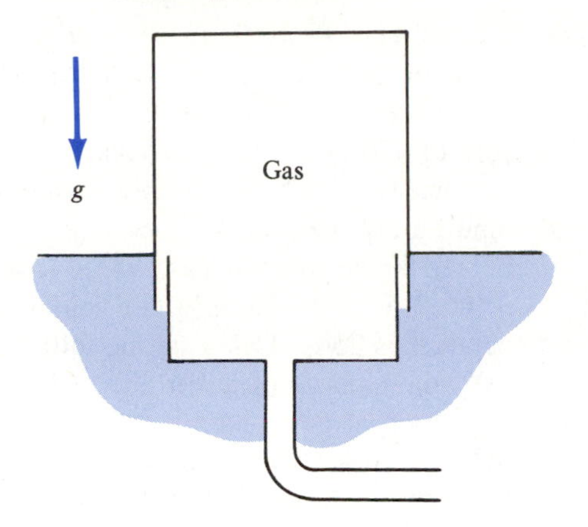

4.33 Saturated water vapor is contained in a rigid, closed tank at 90 psia. Calculate the heat transfer per lb_m required to reduce the quality to 50 percent. Also calculate the final pressure for this process.

4.34 A closed system contains 5 kg of an ideal gas having constant specific heats (R = 0.35 kJ/kg·K, and c_p = 0.75 kJ/kg·K) initially at 75 kPa and 50°C. The gas is first expanded at constant pressure until its volume doubles. Then it is heated at constant volume until its pressure doubles. Determine the following:

(a) the amount of work done by the gas during the entire process

(b) the heat transfer to the gas during the entire process

(c) the change in the internal energy of the gas during the entire process

4.35 A closed, rigid container with a volume of 20 ft^3 contains water at the critical point. Determine the heat transfer from the water required to reduce the water pressure to 100 psia.

4.36 A 1-kW resistance heater is placed in a 10-m^3, closed, rigid container filled with air at 101 kPa, 25°C. The heater is allowed to operate for 10 min and the container is insulated. Calculate the temperature and pressure of the air at the end of the time period.

4.37 Determine the work done by 30 lb_m of H_2O in expanding slowly in a closed system from P = 60 psia, x = 0.5 to T = 400°F if the pressure is maintained constant. Draw the process on a P-v diagram. Calculate the heat transfer.

4.38 You are asked to design a shock absorber, in the form of a piston and cylinder containing an ideal gas with constant specific heats, that will safely stop a falling elevator in the event of cable breakage. Your first concern is to make sure you have allowed enough length of stroke of the piston to stop the elevator under given conditions. Assume that the elevator has mass m_e and velocity V_e just before impact and that it transfers all of its energy reversibly to the gas inside the cylinder. The process occurs so quickly there is negligible heat transfer from the gas to the cylinder walls. Determine an expression for x, the distance the piston travels during the time that it brings the elevator to rest, in terms of thermodynamic properties of the gas and m_e, m_g, V_e, A, H, P_1 (pressure of the gas before impact), and P_2 (final gas pressure). See the accompanying figure.

Problem 4.38

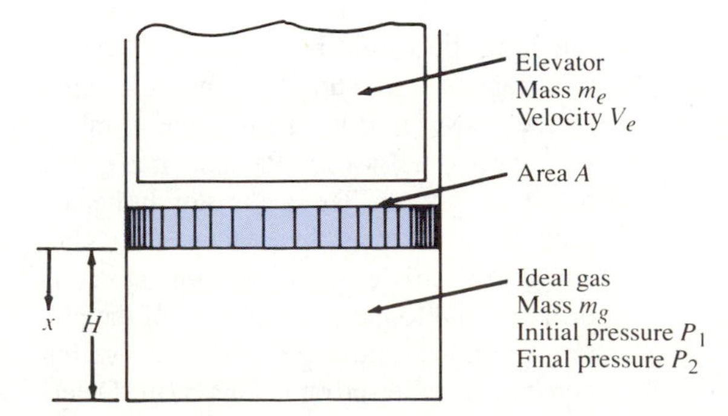

4.39 Water at 60 psia and 800°F is confined to a closed cylinder by a piston, as shown in the accompanying figure. The initial volume is twice as great as that which exists within the cylinder when the piston rests on the stop. The water is cooled until the piston rests on the stops.

(a) Determine the temperature (if superheated) or quality (if saturated) at the end of the process.

(b) At what temperature is the water in the saturated-vapor state?

(c) Determine the work per lb_m done during this process.

(d) Determine the heat transfer per lb_m during the process.

Problem 4.39

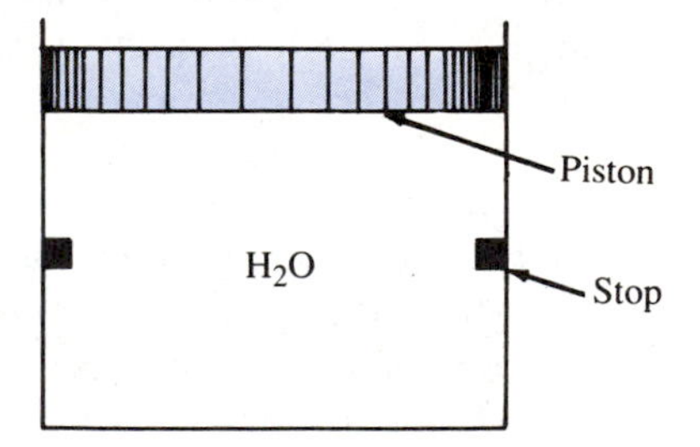

4.40 [FLA112] Air with a mass of 1 kg is contained inside a closed piston-cylinder with an initial volume of 0.82 m^3. At this state the temperature is 300 K and the air pressure inside the cylinder is 105 kPa, which just balances the atmospheric pressure outside plus the piston weight. A spring is touching the piston but exerts no force at this state. Assume that the initial elevation of the piston is 0 m at this state. Atmospheric pressure is 100 kPa. The area of the piston is 0.1 m^2. The air is slowly heated until the final pressure reaches 200 kPa, and during this process the spring force is proportional to the displacement of the piston from the initial position. The spring constant for the spring is 10 kN/m. Determine the following:

(a) the final temperature of the air in K

(b) the final volume of the air in m^3

(c) the total work done by the air in kJ

4.41 [FLA211] An ideal gas is compressed quasistatically in a closed system from an initial pressure of 100 kPa to a final pressure of 420 kPa. The final volume is 0.03 m^3. The relation between pressure and volume during the process is closely approximated by the equation $P = aV + b$ where $a = -36{,}514$ kPa/m^3. The initial temperature is 25°C. Calculate the following:

(a) the amount of gas in the cylinder in kg·mol

(b) the initial volume in m^3

(c) the work required for this process in kJ

4.42 Suppose that 80 ft^3 of air at 100°F is enclosed in a closed, frictionless, vertical-cylinder arrangement. The weight of the piston is 300 lb_f. The cylinder has a cross-sectional area of 1.0 ft^2. The air is heated until the volume of the air doubles.

(a) Determine the mass of air in the cylinder.

(b) Calculate the final air temperature.

(c) Determine the work done by the air on both the atmosphere and the piston.

4.43 Steam is commonly used to clean the inside of railroad tank cars that are used to transport edible materials. After the workers finish cleaning a 15,000-ft^3 car with 240°F, 14.7 psia steam, they tightly close the car in preparation for filling. The steam inside the car eventually cools until its pressure reaches 12 psia, at which time the tank car is ready to be filled with salad oil. Determine the following quantities:

(a) the heat transfer from the steam during the time that it cools

(b) the amount of liquid water in the bottom of the tank car that will dilute the salad oil

4.44 An adiabatic container holds 10 lb_m of water at the critical point. The closed container expands until the water reaches the state of saturated vapor at 120 psia. Determine the following:

(a) the original temperature and pressure of the water

(b) the amount of work during the process

(c) the water temperature at the final state

4.45 A 200-W fan is mounted inside a 30-m^3, rigid, closed box filled with air at 101 kPa and 25°C. The outside of the box is perfectly insulated, and the fan is allowed to run for 1 h. Calculate the final pressure and temperature of the air.

4.46 A spherical elastic membrane confines 5 lb_m of water and supports an internal pressure proportional to its diameter. The initial condition of the water is saturated vapor at 240°F. The water is heated until the pressure reaches 30 psia. Determine (a) the final temperature of the water and (b) the heat transfer to the water during the process.

4.47 A rigid, closed container holds 1.5 kg of liquid water and 0.05 kg of water vapor at 450 kPa. Calculate the heat transfer to the water required to transform it into a saturated vapor. Calculate the volume of the container.

4.48 A closed, insulated tank holds 10 kg of water at 1 MPa in the saturated-vapor state. How much work must be done on the water in order to convert it into a superheated vapor at 4 MPa, 500°C?

4.49 Saturated water vapor is contained in a closed, rigid tank at 100 psia. Calculate the heat transfer per lb_m required to reduce the quality to 0.50. Also, calculate the final pressure for this process.

4.50 The state of 5 kg of CO_2 is changed without friction in a closed system along a constant-pressure path from 150 kPa, 45°C to 90°C. Determine the change in internal energy, the change in enthalpy, the heat transfer, and the work performed on the CO_2 during the process.

4.51 Helium is contained in a closed cylinder fitted tightly with a piston. The initial conditions of the helium are 7 ft^3, 80 psia, and 100°F. The helium expands reversibly according to the process

$$PV^{1.5} = \text{constant}$$

until the pressure reaches 15 psia. Calculate the work and the heat transfer during the process. Assume that helium is an ideal gas.

4.52 A closed, rigid vessel with a volume of 0.2 m^3 is filled with helium at 200 kPa and 65°C. The helium is cooled by heat transfer to the surroundings until the temperature of the helium drops to 20°C. Calculate the heat transfer during the process.

4.53 A rigid, insulated, closed tank consists of two compartments of equal volume separated by a valve. Initially, one compartment contains air at a pressure of 145 psia and a temperature of 130°F, while the second compartment is completely empty. The valve is opened and it remains open until the two compartments reach pressure and temperature equilibrium. Determine the final pressure and temperature of the air, assuming that each compartment has a volume of 20 ft^3.

4.54 Air in a rigid, insulated, closed tank with a volume of 15 m^3 is initially at a pressure of 70 kPa and a temperature of 0°C. An electric component inside the tank draws 1 A at 120 V. Determine the pressure and temperature of the air inside the tank if the component remains energized for 25 min.

4.55 Compressed air in a 2-m^3 pressure vessel is originally at 600 kPa, 20°C. A valve separating the pressure vessel is opened, and the air slowly leaks into a second tank that contains a piston supported by a spring, as shown in the accompanying figure. The spring is elastic with a spring constant of 2×10^6 N/m, and the surface area of the piston is 0.5 m². The piston moves slowly without friction during the expansion process, and it originally is at the top of the cylinder so that there is no air in the cylinder prior to the opening of the valve. The valve remains open until the air reaches pressure and temperature equilibrium. Assuming that the entire expansion process is adiabatic, determine the final pressure and temperature of the air.

Problem 4.55

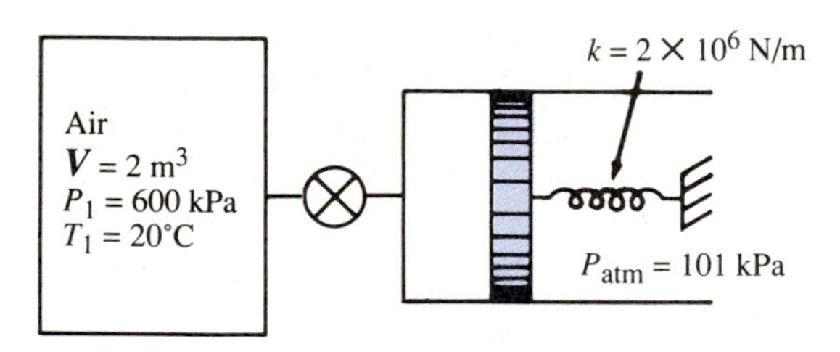

4.56 A closed system undergoes a process from state 1 to state 3 along path 1-2-3 during which the heat transfer to the system is 47.5 Btu and 30 Btu of work is done by the system.

(a) Determine the magnitude and direction of the heat transfer for process 1-4-3 if 15 Btu of work is done by the system during this process.

(b) If the system is returned to state 1 from state 3 by means of process 1-3, as shown in the accompanying figure, the work done on the system is 6 Btu. Determine the magnitude and direction of the heat transfer for this process.

(c) If $E_2 = 175$ Btu and $E_3 = 87.5$ Btu, determine the magnitude and direction of the heat transfer for process 2-3 and the value of E_1.

Problem 4.56

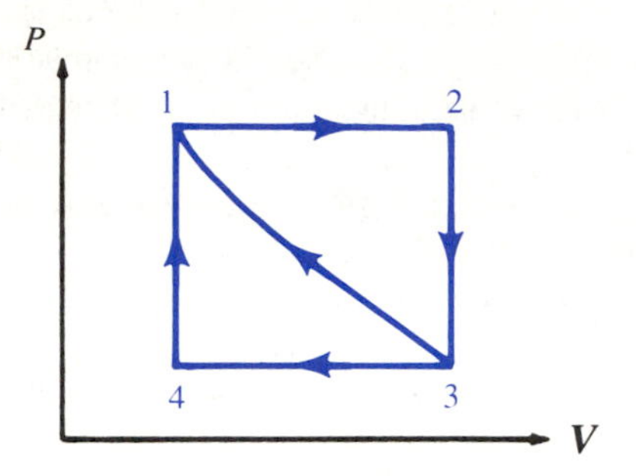

4.57 Two rigid storage tanks containing air are interconnected through a valve. Tank A has a volume of 0.03 m³ and is initially at 150°C and 4 MPa. Tank B is well-insulated, has a volume of 0.6 m³, and contains air initially at 40°C and 200 kPa. The valve is opened allowing air to flow from tank A into tank B. At the same time the air in tank A is cooled so that the temperature of the air in this tank remains constant. When the valve is finally closed, the heat transfer to tank A is found to be 78 kJ. Determine the final pressure in each tank and the final temperature of the air in tank B.

4.58 A closed system undergoes a cycle composed of two quasi-static processes. During process 1-2 the energy of the system increases by 30 Btu. During process 2-1, the heat transfer from the system is 40 Btu. The net work done by the system during the cycle is 10 Btu. Determine the magnitude and direction of Q_{12}, W_{12}, and W_{21}.

4.59 A closed system undergoes a cycle composed of four processes. Complete the accompanying table if the net work done by the system during this cycle is 200 kJ.

Process	Q, kJ	W, kJ	$(E_f - E_i)$, kJ
1-2	180	70	
2-3	90		130
3-4		120	
4-1	0		

4.60 At the moment when the valves are closed on the radiator of a steam-heating system, the radiator contains saturated vapor at 9.3 psig. The radiator has an internal volume of 1.3 ft^3, and the pressure within the radiator eventually drops to -0.7 psig as a result of heat transfer to the surroundings.

(a) Sketch the process described above on a *P*-*v* diagram relative to the saturation region.

(b) Determine the final temperature of the steam.

(c) Calculate the volume and the mass of liquid in the radiator at the end of the process.

(d) Calculate the heat transfer from the steam.

4.61 A closed system containing 2 kg of an ideal gas whose specific heats are constant undergoes the following series of internally reversible processes: The gas is first heated at constant volume from an initial temperature of 40°C to a temperature of 200°C. It is then cooled at constant pressure until its temperature has returned to 40°C. During the first process the heat transfer to the gas is 220 kJ, and during the second process the heat transfer from the gas is 340 kJ.

(a) Sketch these processes on a *P*-*v* diagram.

(b) Calculate c_p and c_v for the gas.

(c) Determine the molecular weight.

(d) Determine the magnitude and direction of the work for each process.

4.62 A closed, vertical cylinder containing 1.2 lb_m of nitrogen at 175°F is fitted with a weighted, frictionless piston so that a constant pressure of 80 psia is maintained on the gas. The nitrogen is stirred by a paddle wheel inserted through the cylinder wall until the absolute temperature of the gas is doubled. During the process the heat transfer from nitrogen to the surroundings is 20 Btu. Determine the amount of paddle-wheel work required for this process.

4.63 The *P*-*v* relationship for the expansion process that occurs during the power stroke in the cylinder for an internal-combustion engine is usually approximated by an expression of the form $Pv^n = C$, where n and C are constants. Assume that the cylinder is closed, that the combustion gases have the thermodynamic properties of air, and that the pressure and temperature of the gases at the beginning of the power stroke are 9 MPa and 1500 K. The pressure at the end of the power stroke is 600 kPa, and $n = 1.29$. Calculate the work and heat transfer per unit mass of air for the power stroke.

4.64 A certain closed system containing an ideal gas initially at P_1 and v_1 expands to P_2 and v_2. The expansion could be accomplished by either of the two following quasistatic processes.

Process *A* is an isothermal expansion to state 2.

Process *B* consists of a constant-pressure expansion to specific volume v_2, followed by a constant-volume expansion to pressure P_2.

Sketch those two processes on a single *P*-*v* diagram and complete the following questions. Explain your responses.

(a) The internal energy change for process *A* is (greater than, equal to, less than) the internal energy change for process *B*.

(b) The work for process *A* is (greater than, equal to, less than) the work for process *B*.

(c) The heat transfer for process *A* is (greater than, equal to, less than) the heat transfer for process *B*.

4.65 A sealed storage tank for refrigerant-12 is fitted with a sight glass to indicate the liquid level, as shown in the accompanying figure. The cross-sectional area of the tank is 1.2 ft^2, and the height of the tank is 1.5 ft. The refrigerant is initially at a temperature of 80°F, and the level of the liquid is

8 in. Some time later it is noticed that the final temperature has decreased by 50°F. Determine the liquid level of the R-12 and the magnitude and direction of the heat transfer that has taken place.

Problem 4.65

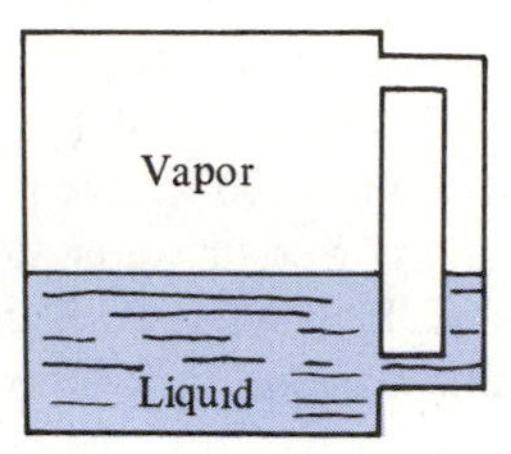

4.66 A closed, rigid tank contains 0.5 kg of carbon monoxide, which is heated until the gas pressure is doubled. The initial temperature of the gas is 35°C. Determine the heat transfer, in kilojoules, to the carbon monoxide.

4.67 [FLB311] A rigid, open container having a volume of 0.28 m^3 is placed in a furnace that contains air at 460°C and 100 kPa. The container is sealed, removed from the furnace, and then allowed to cool to 25°C. Determine the following:

(a) the final pressure of the air assuming it is an ideal gas

(b) the heat transfer from the air using a constant-volume specific heat of 0.718 kJ/kg·K

4.68 A closed, frictionless, piston-cylinder assembly contains 4 lb_m of steam. Beginning with an initial volume of 10 ft^3, the steam is compressed slowly until the volume has been reduced to 10 percent of the initial volume. During the process heat transfer from the steam occurs at such a rate as to keep the temperature constant at 300°F. Sketch the process on a *P*-*v* diagram relative to the saturation region. Calculate the work input and the heat transfer during the process. What fraction of the final volume is occupied by liquid?

4.69 [FLB211] An insulated vessel has an evacuated compartment separated by a membrane from a second compartment which contains 0.983 kg of water at 65°C and 600 kPa. The membrane then ruptures and the water fills the entire volume. The final equilibrium pressure is 10 kPa. Determine the following:

(a) the final temperature of the water in °C

(b) the volume of the vessel in m^3

4.70 Consider a closed system containing refrigerant-12 that undergoes a quasi-static process under the following circumstances:

(a) The refrigerant is initially superheated at pressure P_1 and temperature T_1. At the end of the process the specific volume is twice the initial value. For these conditions the work done during a constant-pressure process at pressure P_1 is (greater than, less than, or equal to) the work done during an isothermal process at temperature T_1. Justify your answer.

(b) The refrigerant is initially saturated liquid at pressure P_1 and temperature T_1. At the end of the process the specific volume is twice the initial value (and $v_2 < v_g$ at pressure P_1). For these conditions the work done during a constant-pressure process at pressure P_1 is (greater than, less than, or equal to) the work done during an isothermal process at temperature T_1. Justify your answer.

4.71 A closed piston-cylinder assembly contains 1 kg of air and has an initial volume of 0.015 m^3. The air expands in a quasi-equilibrium process until the volume reaches 0.03 m^3. During the process the air is cooled so that the temperature remains constant at 315°C. Sketch the *P*-*v* diagram and determine the heat transfer during the process.

4.72 A closed, insulated tank holds 25 lb_m of water at 30 psia in a saturated-vapor state. Determine the amount of work that must be done on the water in order to convert it into a superheated vapor at 3000 psia, 2000°F.

4.73 [FLB611] A closed piston-cylinder device is fitted with a set of stops upon which the piston initially rests. The 0.3-m^3 cylinder is initially evacuated except for a 0.006-m^3 capsule which contains water at 200°C and 7000 kPa. The piston mass is such that a pressure of 1400 kPa is required to lift the piston. The capsule suddenly ruptures. Heat transfer maintains the cylinder contents at 200°C. Determine the heat transfer and the work in kJ for this process.

4.74 Oxygen is compressed in a closed piston-cylinder arrangement from 50 psia, 500°R to 500 psia. Calculate the work done on the gas and the heat transfer per lb_m of O_2 assuming that the oxygen is an ideal gas. The path followed by the O_2 during the process is given by $PV^{1.4}$ = constant.

4.75 A rigid storage container is divided into two parts by an uninsulated partition that is held in a fixed position. One section of the container contains 0.22 lb_m of nitrogen initially at 450 psia and 930°F. The other section contains 0.022 lb_m of water initially at 930°F and 900 psia. The contents of the container are cooled, and the equilibrium temperature of the contents of the container is found to be 445°F. Determine the final pressure of the water, the final pressure of the nitrogen, and the heat transfer during the process.

4.76 Write a computer program to determine the heat transfer required for a quasi-equilibrium, isothermal process in which 3 kg of water in a closed system initially at 300°C and 4 MPa is reduced to 30 percent of its initial volume.

4.77 A 20-W heater is placed in a well-insulated, rigid, closed room (V = 36 ft^3) that contains air at 62°F and atmospheric pressure. The heater is turned on for 1 h. Calculate the temperature and pressure of the air in the room at the end of the process.

4.78 A closed, rigid 10-m^3 tank contains water at 120°C and a quality of 0.5. The tank is cooled to 10°C. Calculate the following information:

(a) the initial pressure in the tank
(b) initial internal energy of the water in the tank
(c) final internal energy of the water in the tank
(d) final pressure in the tank
(e) the heat transfer during the process

4.79 A rigid, insulated container filled with water is initially at 400°F, 30 psia. A paddle wheel inserted into the water adds energy to the water until the pressure reaches 80 psia. Determine the work done on the water by the paddle wheel and the change in the internal energy experienced by the water during the process.

4.80 Air expands reversibly in a well-insulated, closed piston and cylinder arrangement. The initial conditions are 180 kPa, 77°C and the final pressure is 100 kPa. Determine the heat transfer and work per kg of air during the process. Calculate the final temperature of the air.

4.81 Twenty lb_m of water is compressed reversibly in a closed system along a path given by Pv = constant. The initial pressure and temperature of the water are P_i = 20 psia and T_i = 300°F, and the final pressure is 50 psia. Determine the heat transfer and work that occurs during the process and the temperature of the water at the end of the process.

4.82 In 1844, Joule used the apparatus shown to prove that "no change of temperature occurs when air is allowed to expand in such a manner as not to develop mechanical power.'' Two copper vessels were placed in a water bath whose temperature was accurately known. The first vessel contained air at an elevated pressure while the second vessel was initially evacuated. A valve connecting the two vessels was opened so that the air expanded from the first to the second vessel and the process terminated with pressure equilibrium between the two vessels. During the process no change in temperature of the water bath was recorded. Show that Joule's experiment is a verification of the conservation of energy.

Problem 4.82

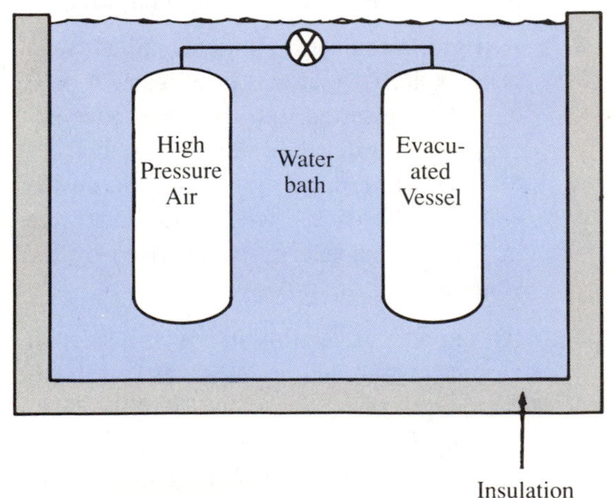

4.83 Carbon dioxide is contained in a rigid, closed vessel ($V = 50\ ft^3$). The CO_2 is originally at a pressure of 50 psia and a temperature of 60°F. The heat transfer to the contents of the vessel is 50 Btu. Determine the pressure and temperature of the CO_2 at the end of the process.

4.84 A rigid, closed container holds 2.684 kg of liquid refrigerant-12 and 0.3158 kg of refrigerant-12 vapor at a pressure of 100 kPa. The R-12 is heated until all of the liquid in the container is vaporized. Determine the following information:

(a) the initial temperature of the R-12

(b) the volume of the container

(c) the heat transfer to the R-12 during the process

(d) the final temperature and pressure of the R-12

4.85 Water in a closed system is initially at 250°C and 300 kPa. A total of 700 kJ/kg of work is done on the water in order to isothermally reduce its volume to 1/20 of its initial volume. Determine the final pressure of the water and the heat transfer for this process.

4.86 A closed system contains 5 lb_m of refrigerant-12 initially at 60 psia with an initial volume of 2.5 ft^3. The R-12 undergoes an internally reversible, isothermal process until the volume has decreased by 50 percent. Sketch this process on a *P-v* diagram. Determine the final temperature and pressure of the R-12, the internal energy change for the R-12, and the work for this process.

4.87 A closed system contains 1.5 kg of CO_2 initially at a temperature of 427°C and an initial volume of 0.3 m^3. The system undergoes a cycle consisting of a series of internally reversible processes as depicted in the accompanying figure. During process 1-2 the volume doubles. At the end of process 2-3 the temperature of the gas is the same as the initial temperature.

Problem 4.87

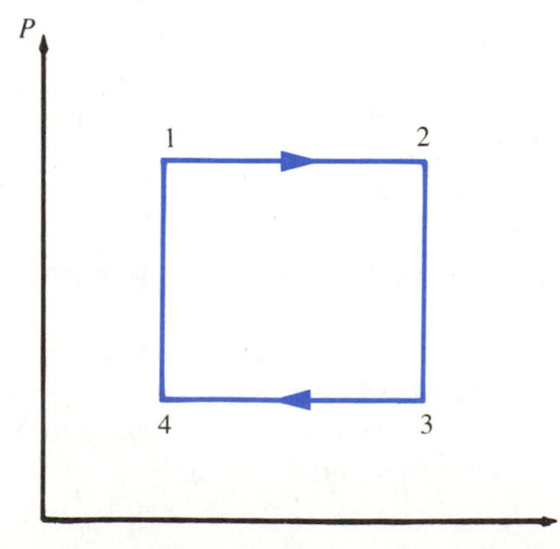

(a) Determine P, T, v, and u at states 1 through 4.

(b) Calculate the work for each process and the net work for the cycle.

4.88 A closed piston-cylinder assembly contains 0.44 lb_m of helium at an initial temperature of 172°F and an initial pressure of 25 psia. The helium undergoes a process during which there is 30.7 Btu of heat transfer to the helium. After the process has been completed, the final equilibrium pressure of the helium is found to be the same as the initial pressure, and the final volume is twice the initial volume. Determine the final temperature of the helium and the work for this process.

4.89 A rigid storage container is divided into two parts by an uninsulated partition that is held in a fixed position. One section of the container contains 0.1 kg of nitrogen initially at 3 MPa and 500°C. The other section contains 0.01 kg of water initially at 500°C and 6 MPa. Heat transfer from the contents occurs and the equilibrium temperature of the contents of the container is found to be 230°C. Determine the final pressure of the water, the final pressure of the nitrogen, and the heat transfer for this process.

4.90 A closed system contains air and it performs a thermodynamic cycle. The heat transfer from the air during the cycle is 350 kJ. Determine:

(a) the work during the cycle

(b) the change in the internal energy of the air during the cycle

4.91 Which terms (e_k, e_p, h, q, w) are generally more significant when the conservation-of-energy equation is applied to the following steady-flow devices? Briefly explain your responses: (a) turbine; (b) pump; (c) boiler; (d) nozzle; (e) diffuser.

4.92 Water is heated in a constant-pressure process at 75 psia from a state of saturated vapor to a state where the temperature is 800°F. Calculate the heat transfer per lb_m in (a) a nonflow process in which the steam is contained behind a piston in a cylinder and (b) a steady-flow process for which there is no external work.

4.93 Refrigerant-12 vapor enters a steady-flow compressor as a saturated vapor at 20°F. The outlet conditions are 70 psia and 110°F, and the process is assumed to be adiabatic. Calculate the power, in hp, required if the refrigerant flow rate is 1.0 lb_m/s. Determine the diameter of the inlet tubing to the compressor if the inlet velocity is not to exceed 10 ft/s.

4.94 Steam enters a turbine at 20 MPa, 600°C, with negligible velocity, and exhausts at 7 kPa with a velocity of 180 m/s. The flow rate is 5.1 kg/s, and the turbine power output is 1500 kW. The flow is steady and adiabatic. Determine the quality (if saturated) or temperature (if superheated) of the exhaust steam.

4.95 Estimate the size of a pump in hp you would need to pump 40 gal/min of liquid water at 70°F through a rise in elevation of 400 ft. Assume that the water is pumped adiabatically in steady flow and with negligible changes in temperature and kinetic energy.

4.96 A steady flow of steam enters a well-insulated nozzle at 6.8 MPa and 600°C with a velocity of 100 m/s. The area of the inlet is 2.8×10^3 mm^2, and the steam exits the nozzle at 4.5 MPa and 500°C. Determine the mass flow rate of steam and the area of the nozzle exit.

4.97 A steady-flow, adiabatic diffuser is employed to decrease the velocity of an airstream from 600 to 50 ft/s. The air enters the diffuser at a rate of 15 lb_m/s with a temperature of 400°F and a pressure of 18 psia. Determine the exit area of the diffuser if the exit pressure is 25 psia.

4.98 Nitrogen flows steadily through a constant-area pipe. The N_2 enters the pipe at 3.5 MPa, 60°C with a velocity of 10 m/s and a mass flow rate of 160 kg/h. At the exit of the pipe the N_2 pressure is 0.2 MPa and the temperature is 50°C. Determine (a) the cross-sectional area of the pipe, (b) the exit velocity of the nitrogen, and (c) the heat-transfer rate to the nitrogen in the pipe.

4.99 Air enters an air compressor at a steady volume flow rate of 45 m^3/min. The air pressure increases from 100 to 700 kPa while the air is cooled at a rate of 20 kW. The air temperature at the inlet to the compressor is 25°C, and the exit temperature is 150°C. The air leaves the compressor through a pipe with an area of 0.03 m^2. Determine the power necessary to operate the compressor under these conditions.

4.100 A centrifugal compressor is steadily supplied with saturated water vapor at 5 psia; 1200 lb_m of water is compressed per hour to 100 psia, 400°F. During the process the heat-transfer rate from the water is 3000 Btu/h. Determine the power required to compress the water.

4.101 Steam enters a steady-flow, adiabatic turbine, shown in the accompanying sketch, at a rate of 10 kg/s at 3.5 MPa and 350°C. At a point where the steam is at 1 MPa and 250°C, 15 percent of the total mass flow is extracted and used to preheat water entering a boiler. The rest of the steam expands further and is exhausted from the turbine at 30 kPa with a quality of 90 percent. Determine the turbine's power output.

Problem 4.101

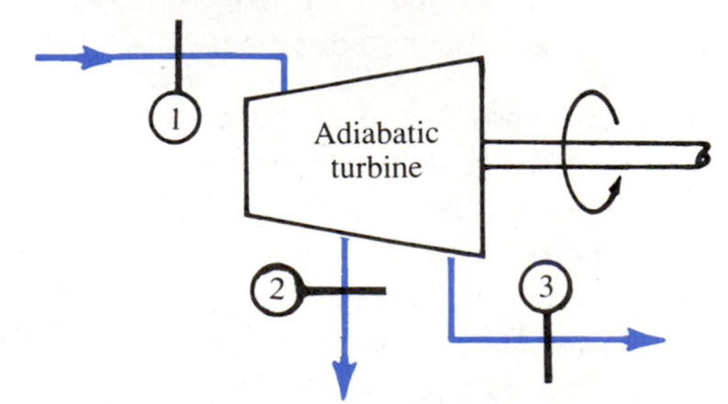

4.102 Saturated steam at 80°F enters a steady-flow, centrifugal compressor at a rate of 30 lb_m/s. The vapor is heated during the compression process at a rate of 6.5×10^5 Btu/h, and the vapor exits at 300 psia and 600°F. Calculate the power input required to drive the compressor.

4.103 Steam enters a steady-flow turbine with a low velocity at 8 MPa and 600°C and leaves at 30 kPa and a quality of 0.95. The steam leaves the turbine through an area of 0.3 m^2 with a velocity of 200 m/s. Calculate the power generated by the turbine, assuming that it is well-insulated.

4.104 A steady-flow heat exchanger is designed to take in cool air and circulate it over tubes containing steam. The air is thus heated, and the steam leaves the heat exchanger at a lower temperature than that at which it entered. Steam enters at 150 kPa and 200°C, while the air enters at 35°C and leaves at 45°C. Calculate the mass flow rate of steam required to heat 50 kg/min of air when the exit conditions of the steam are $x = 0.75$ and $P = 140$ kPa.

4.105 Steam is heated in a constant-pressure process at 150 psia from saturated-vapor state to a state where the temperature is 680°F. Calculate the heat transfer per kilogram in (a) a nonflow process in which the steam is contained behind a piston in a cylinder and (b) a steady-flow process for which there is no external work.

4.106 Refrigerant-12 vapor enters a steady-flow compressor as saturated vapor at 4°C. The outlet conditions are 1.2 MPa and 100°C, and the process is assumed to be adiabatic. Calculate the power required to drive the compressor if the refrigerant flow rate is 5 kg/min. What diameter of inlet tubing to the compressor is required if the inlet velocity cannot exceed 5 m/s?

4-107 [FLC611] A steady-flow steam turbine is used to drive a refrigerant-12 compressor. Steam enters the turbine with a mass flow rate of 60 lb_m/h at 800°F, 500 psia and exits at 10 psia with a quality of 0.90. R-12 enters the compressor at a mass flow rate of 540 lb_m/h at 100°F, 60 psia. The R-12 then flows into an aftercooler from which it exits at 140°F, 200 psia. The steam turbine is adiabatic and delivers 6600 Btu/h of power to the compressor. The balance of the power produced by the steam turbine is used to drive an electric generator. Determine the following:

(a) the power available for driving the electric generator in Btu/h

(b) the heat-transfer rate in Btu/h from the R-12 as it flows through the compressor and aftercooler

4.108 Refrigerant-12 at 30°C, 400 kPa enters a steady-flow heat exchanger at a rate of 6 kg/min. The inlet to the heat exchanger is a tube with a 30-mm diameter. Calculate the average velocity into the heat exchanger. The exit tube from the heat exchanger has an inner diameter that is twice the diameter of the inlet tube, and the average velocity of refrigerant at the exit is 0.6 m/s. Calculate the specific volume of the exiting refrigerant.

4.109 Air enters an adiabatic, converging nozzle at 60 psia, 120°F, and a velocity of 230 ft/s. The air exits the nozzle at 35 psia and a velocity of 540 ft/s. Assuming that the inlet area to the nozzle is 40 in^2 and that the flow is steady, calculate the exit area and the exit-air temperature.

4.110 Air at 50°C, 200 kPa enters a nozzle through a cross-sectional area of 0.3 m^2 with a velocity of 200 m/s. The air leaves the nozzle with a velocity of 250 m/s and a pressure of 100 kPa. Assuming that the nozzle is well-insulated and that the flow is steady, calculate (a) the mass flow rate of air through the nozzle, (b) the exit area of the nozzle, and (c) the air temperature at the nozzle exit.

4.111 Calculate the power required to drive a steady-flow compressor if air flowing at a rate of 1 kg/s enters at 100 kPa, 20°C with a velocity of 60 m/s and leaves at 240 kPa, 70°C with a velocity of 120 m/s. Heat transfer from the air to the cooling water circulating through the compressor casing amounts to 19 kJ/kg of air.

4.112 A steady-flow air turbine produces 10^4 $ft{\cdot}lb_f/lb_m$ of shaft work. The conditions of the air entering and leaving the turbine are as follows: inlet, 100 psia, 120°F, 165 ft/s; exit, 15 psia, 40°F, 30 ft/s. Calculate the magnitude and direction of heat transfer per lb_m of the air in the turbine.

4.113 Steam at a pressure of 1.0 MPa and temperature of 200°C enters a 300-mm-diameter diffuser with a velocity of 250 m/s. The steam leaves the diffuser at 75 m/s and a pressure of 300 kPa as a saturated vapor. Assuming steady flow, calculate the mass flow rate of the steam, the ratio of the inlet and exit areas of the diffuser, and the heat-transfer rate from the steam.

4.114 [FLC311] Steam enters a steady-flow, adiabatic diffuser as a saturated vapor at 90°C with a velocity of 330 m/s. At the exit the pressure is atmospheric and the temperature is 110°C. The exit area is 1.779 × 10^3 mm^2. Determine the following:

(a) the exit velocity in m/s

(b) the mass flow rate in kg/s

(c) the entrance area in m^2

4.115 A well-insulated valve is used to throttle high-pressure steam to a lower pressure that is suitable for a steady-flow industrial process. The steam enters the valve at 520°F with a quality of 98 percent and is throttled to a pressure of 100 psia. Determine the temperature (if superheated) or the quality of the steam (if saturated) at the exit of the valve, and calculate the change in internal energy of the steam.

4.116 Water passes through a well-insulated nozzle that has a ratio of inlet to exit area of 2:1. The water enters the nozzle at 1200°F, 300 psia with a velocity of 375 ft/s. The water leaves the nozzle with a velocity of 820 ft/s. Assuming steady flow, calculate the exit pressure and temperature.

4.117 Refrigerant-12 is throttled adiabatically in a steady-flow nozzle from a saturated liquid at 219 kPa to a temperature of -20°C. Calculate the exit pressure and density of the refrigerant, assuming negligible changes in kinetic energy.

4.118 [FLC111] Refrigerant-12 with a mass flow rate of 0.1 kg/s, a temperature of 90°C, and a pressure of 700 kPa is throttled to 200 kPa. The R-12 is then mixed with 1 kg/s of R-12 at -40°C and 200 kPa. The heat-transfer rate to the mixer is 2.7 kW from the surroundings. Determine the temperature in °C of the R-12 at the outlet of the throttling valve assuming that the flow is steady. Also, determine the temperature and quality (as applicable) of the R-12 at the mixer outlet if the pressure is 200 kPa.

4.119 [FLC411] Refrigerant-12 is throttled by a valve from a saturated liquid at 90°F to a temperature of -10°F. The mass flow into the valve is 720 lb_m/h. The inlet and exit areas of the valve are equal. The velocity of the refrigerant at the inlet is 0.1 ft/s. For steady-flow conditions, determine the following:

(a) the exit pressure is psia
(b) the density of the refrigerant at the outlet in lb_m/ft^3
(c) if the refrigerant is a two-phase mixture at the exit, determine the quality
(d) the velocity of the refrigerant leaving the valve in ft/s

4.120 Air is throttled steadily and adiabatically from 101 kPa, 50°C, and a velocity of 160 m/s through a nozzle that has an exit-to-inlet area ratio of 1:1.2. The exit-air temperature is 35°C. Calculate the exit-air velocity and pressure.

4.121 The cooling coil of a small air-conditioning system is to be designed for a heat-transfer rate of 48,000 Btu/h from air at 14.7 psia and 90°F, which flows steadily at a rate of 1500 ft^3/min. The cooling coil is a finned-tube heat exchanger, as shown in the accompanying figure. The air flows over the finned surfaces, and heat transfer from the air causes the refrigerant inside the tubes to evaporate. Refrigerant-12 flows inside the tubes and evaporates at a constant pressure of 50 psia from saturated-liquid conditions to an exit temperature of 50°F. Calculate the mass flow rate of liquid required and the exit temperature of the air.

Problem 4.121

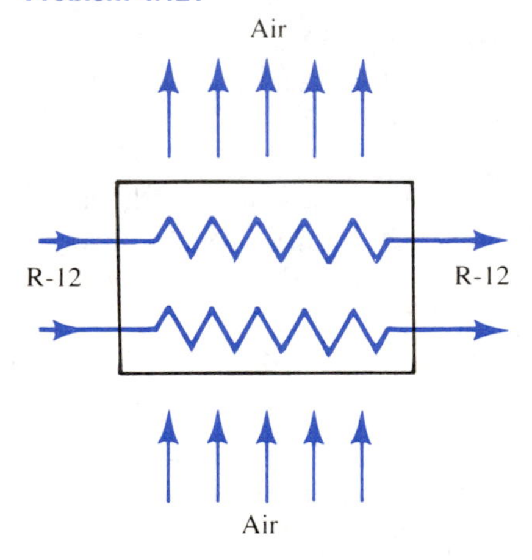

4.122 A supply of liquid water at 80°C flowing at a rate of 180 kg/min is required for an industrial cleaning process. The plant supervisor suggests that this requirement can be met by mixing water that is available at 200 kPa and 15°C with steam at 200 kPa and 150°C from a nearby supply line. The mixing is done in an insulated, steady-flow mixer, with the mixture leaving at 80°C and 200 kPa. At what rates, in kilograms per minute, should the cold water and steam be fed to the mixer?

4.123 Steam at 600 kPa and 250°C flowing at a rate of 2.5 kg/min is allowed to mix freely with 5 kg/min of steam at 600 kPa and a quality of 50 percent in a well-insulated, steady-flow device. The exit pressure of the mixture is 600 kPa. Determine the temperature (if superheated) or quality (if saturated) of the mixture at the exit of the device.

4.124 [FLC211] A steady-flow feedwater heater for preheating water before entering a boiler mixes water at pressure 100 psia and temperature of 90°F with steam at the same pressure and a quality of 0.9. Water leaves the heater at 95 psia and 300°F and a mass flow rate of 200 lb_m/h. The feedwater heater is well-insulated. Determine the following:

(a) the temperature of the entering steam in °F

(b) the mass flow rate of the entering steam in lb_m/h

(c) the mass flow rate of the entering water in lb_m/h

(d) the ratio of mass flow rate of water to steam entering the heater

4.125 The exhaust gases from the boiler in a power plant are used to preheat the air before it enters the boiler. Exhaust gases leave the boiler and enter an air preheater (a steady-flow heat exchanger) at a pressure of 100 kPa and a temperature of 500°C, with a mass flow rate of 75 kg/min. Outside air enters the preheater with a flow rate of 70 kg/min at a pressure of 101 kPa and a temperature of 15°C. The exhaust gases leave the preheater at 250°C. Assuming that the properties of the exhaust gases can be approximated by those of air, calculate the outside-air temperature as it leaves the preheater and the heat-transfer rate between the two gas streams.

4.126 [FLC511] In the boiler of a power plant, water is steadily heated at a rate of 100 kg/s from 260°C, 8 MPa to 500°C, 8 MPa. Combustion gas is used to heat the water and it enters the boiler at a rate of 750 kg/s at 1260°C, 200 kPa and leaves the boiler at 900°C, 100 kPa. The combustion gas may be considered to have the same properties as air, with a constant-pressure specific heat of 1.005 kJ/kg·K. Determine the following:

(a) the heat-transfer rate from the boiler jacket in kW

(b) the rate of energy transfer to the water in kW

4.127 In the evaporator section of an air-conditioning unit, refrigerant-12 is used to cool room air. Air enters the evaporator with a mass flow rate of 0.32 lb_m/s, a pressure of 14.7 psia, and a temperature of 95°F. The refrigerant enters the evaporator with a mass flow rate of 0.038 lb_m/s as a saturated liquid at a temperature of 35°F and leaves with a temperature of 60°F. Neglect changes in kinetic energy and neglect the pressure drop of the refrigerant in the evaporator. Assume steady flow. Calculate the heat-transfer rate in the evaporator and the temperature of the air as it leaves the evaporator.

4.128 The cooling coil in an air-conditioning system is designed to remove 20 kW from 50 m^3/min of air at 30°C and 101 kPa. The cooling is accomplished in a steady-flow heat exchanger in which the refrigerant-12 is evaporated at constant pressure from saturated-liquid conditions at 5°C to a superheated state at 20°C. Calculate the mass flow rate of refrigerant required and the exit-air temperature.

4.129 An assembly line in a plant is to be expanded by adding 15 new stations where workers use wrenches run by compressed air. Specifications of the wrenches call for an operating pressure of 150 psia, a minimum flow rate of 0.03 lb_m/s, and a cross-sectional area of the supply air hose of 0.25 in^2. You are asked to purchase the compressor that will supply air for the wrenches, assuming all 15 wrenches operate steadily, adiabatically, and simultaneously. After checking compressor catalogs, you find that the heat transfer from the compressor for every lb_m of air supplied is approximately 4 Btu. The catalog also lists the supply-air temperature of 150°F when the compressor operates at a pressure of 150 psia. The inlet pressure and temperature to the compressor are 14.7 psia and 85°F. Neglecting any changes in potential and kinetic energy, what size compressor, in hp, should you buy? What is the velocity of air entering one of the air wrenches?

4.130 Air enters a jet engine at 40°F, 14.7 psia through an intake with a cross-sectional area of 8 ft^2 and a velocity of 150 ft/s. The air leaves the engine with a velocity of 1500 ft/s and a pressure of 12 psia. The flow of air through the engine is steady and the heat-transfer rate to the air as a result of combustion of a fuel is 1.5×10^8 Btu/h. Calculate the mass flow rate of air through the engine, the exit area of the engine, and the air temperature as it leaves the engine.

4.131 A heat exchanger in a residential air-conditioning unit operates steadily and utilizes R-12 as the refrigerant. Air enters the heat exchanger at 40°C, 100 kPa with a mass flow rate of 1.0 kg/s. The refrigerant enters with a mass flow rate of 0.15 kg/s as a saturated liquid at 0°C and leaves at 25°C. Pressure drops of both fluids may be neglected and changes in potential energy and kinetic energy may also be neglected. Calculate the following information:

(a) the cooling capacity of the heat exchanger, in kW

(b) the temperature of the air as it leaves the heat exchanger

4.132 Nitrogen at a volume flow rate of 350 ft^3/s enters a steady-flow heater at 14.7 psia, 40°F and exits at 25 psia, 1440°F, Calculate:

(a) the mass flow rate of nitrogen through the heater

(b) the heat-transfer rate to the nitrogen in the heater

(c) the volume flow rate of nitrogen leaving the heater

4.133 Steam enters an insulated, horizontal, steady-flow nozzle at 3 MPa, 350°C with a low velocity and exits the nozzle at 1.6 MPa at a velocity of 700 m/s. The mass flow rate of the steam is 0.7 kg/s. Calculate:

(a) the exit steam quality (if saturated) or temperature (if superheated)

(b) the exit area of the nozzle

4.134 What mass flow rate of water is necessary to produce a power output of 500 hp from a steady-flow, adiabatic steam turbine if the inlet and exit conditions to the turbine are:

$P_i = 400$ psia $\quad T_i = 600°F \quad V_i = 500$ ft/s
$P_e = 10$ psia $\quad x_e = 0.95 \quad V_e = 100$ ft/s

4.135 Suppose that steam is mixed with liquid water in a steady-flow device as shown in the figure. Determine the exit temperature of the water if the device is well-insulated.

Problem 4.135

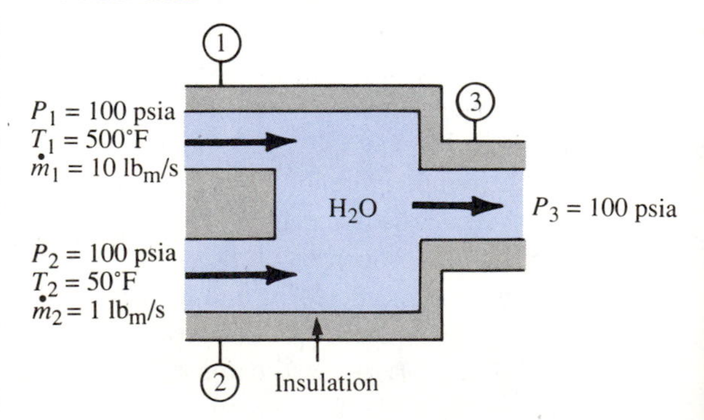

4.136 A centrifugal pump operates steadily compressing water from 10 psia, 20°F to 500 psia, 100°F and delivers water at a volume flow rate of 125 gal/min. Calculate the power required to operate the pump if the heat transfer from the pump and the changes of kinetic and potential energies of the water in the pump are negligible.

4.137 Water steadily enters a long, insulated, constant-diameter (D = 100 mm), horizontal pipe at 20 MPa and 600°C. At the exit of the pipe the average velocity is 20 m/s, and the pressure of the water has dropped to 17.5 MPa. Calculate:

(a) the mass flow rate of water through the pipe

(b) the exit temperature of the water

(c) the water velocity at the inlet of the pipe

4.138 Steam enters a well-insulated, steady-flow turbine at 1500 psia, 1300°F and leaves as a saturated vapor at 100°F. The inlet velocity of the steam is 2000 ft/s and it leaves the turbine at a velocity 600 ft/s. The mass flow rate of steam through the turbine is 45 lb_m/s. Calculate the shaft power delivered by the turbine.

4.139 Air enters a steady-flow, converging nozzle at a low velocity at 227°C, 500 kPa and leaves the nozzle at 27°C, 100 kPa and a velocity of 100 m/s. The exit area of the nozzle is 0.1 m^2. Determine the magnitude and direction of the heat-transfer rate during the process.

4.140 A water heater has an input rating of 100,000 Btu/h and an estimated efficiency of 80 percent (i.e., the ratio of the energy input into the water to the energy input to the heater). Water is supplied to the heater at 1 atm, 50°F, through a pipe with an internal diameter of 0.5 in. Suppose you wish to operate the water heater under a continuous mode. Calculate the velocity and volume flow rate that can be continually supplied if the temperature of the water is not expected to drop below 170°F.

4.141 Consider the information given in Problem 4.153. Write a computer program that will calculate the instantaneous temperature, pressure, and mass of air inside the tank for 10-s intervals up to a total time span of 2 min if the radius of the pipe into the tank is (a) 15 cm and (b) 30 cm.

4.142 An energy-storage scheme is proposed to generate electric energy during the day when electric rates are high by forcing compressed air into an abandoned mine during the night when electric rates are low. Air is compressed in a mine that has an average depth below the surface of the earth of 5000 ft and has a volume of 2.4 $\times$ 10^6 ft^3. When the compression process ends the air in the mine has reached 250 psia, 100°F. During the day the air is routed through a turbine that exhausts to the atmosphere. The expansion process takes place over a period of 3.5 h and the operation of the turbine is regulated so that the exhaust conditions are constant at 15 psia, 40°F. During the process the pressure in the mine decreases linearly at a rate of 60 psia per hour of turbine operation. The temperature of the air in the mine remains at 100°F and 1 $\times$ 10^8 Btu of heat is added to the air during the entire expansion process. Determine the mass of air that passes through the turbine and the maximum energy that can be generated by the turbine during the 3.5 h of operation. Calculate the average power output of the turbine.

4.143 [FLB111] A rigid tank with a volume of 0.14 m^3 is initially filled with dry saturated steam at a pressure of 200 kPa. The tank transfers heat with the surroundings until the temperature of the steam in the tank drops to 40°C. Calculate the following information:

(a) the mass of water in kg contained in the tank

(b) the heat transfer from the tank in kJ

(c) the pressure in kPa and the quality of the water at the end of the charging process

4.144 A rigid, insulated tank is initially evacuated. Atmospheric air at 14.7 psia and 70°F is allowed to leak into the tank until the pressure reaches 14.7 psia. What is the final temperature of the air within the tank?

4.145 Write a computer program to solve the following problem. A cubical tank having a volume of 1 m^3 initially contains air at 150 kPa and 27°C and is connected to an air supply line by a valve. The air in the supply line is at 750 kPa and 80°C. The valve is opened and air enters the tank until the pressure in the tank has reached 500 kPa, at which time the valve is closed. The mass flow rate of air into the tank varies directly with the pressure difference between the supply air and the air in the tank and is initially 0.2 kg/s. The instantaneous heat-transfer rate from the air in the tank at temperature T to the tank wall at temperature T_w is

$$\dot{Q}_A = h_{ci}A(T - T_w)$$

and the heat-transfer rate from the tank wall to the ambient air at 27°C is

$$\dot{Q}_w = h_{cw}A(T_w - T_{\text{amb}})$$

where $h_{ci} = h_{cw} = 40\ \text{W/m}^2\cdot\text{K}$, A is the surface area of the tank, and $T_{\text{amb}} = 27°\text{C}$. The tank wall is steel having a total mass of 240 kg and a specific heat of 0.434 kJ/kg·K. Assume that the air in the tank is well-mixed and that the wall temperature is uniform at any instant of time. Determine the time required for air in the tank to reach a pressure of 500 kPa and the air temperature at this time.

4.146 Refrigerant-12 is contained in a 1.2-ft^3 tank at 80°F and 14.7 psia. It is desired to fill the tank 80 percent full of liquid (by volume) at this temperature. The tank is connected to a line with refrigerant-12 at 160 psia and 100°F, and the valve is opened slightly. Calculate the final mass in the tank at 80°F. Determine the required heat transfer during the filling process if the temperature is to remain at 80°F.

4.147 How much time is required for the air in Problem 4.145 to reach 90 percent of its final equilibrium temperature after the valve is closed?

4.148 Water flows in a pipe at a pressure of 3.5 MPa and at a temperature of 350°C. Attached to the pipe is an evacuated enclosure separated from the pipe by a valve, as shown in the accompanying figure. An impeller is used to thoroughly mix the contents of the enclosure. The valve is opened and 10 kg of steam is allowed to enter the enclosure. At the end of the process the pressure inside the enclosure is found to be 3.5 MPa, and during the process 700 kJ of impeller work is required to mix the contents. The enclosure has a volume of 0.15 m^3. Determine the final temperature (if superheated) or quality (if saturated) of the steam inside the enclosure. Also, calculate the heat transfer from the enclosure during the process.

Problem 4.148

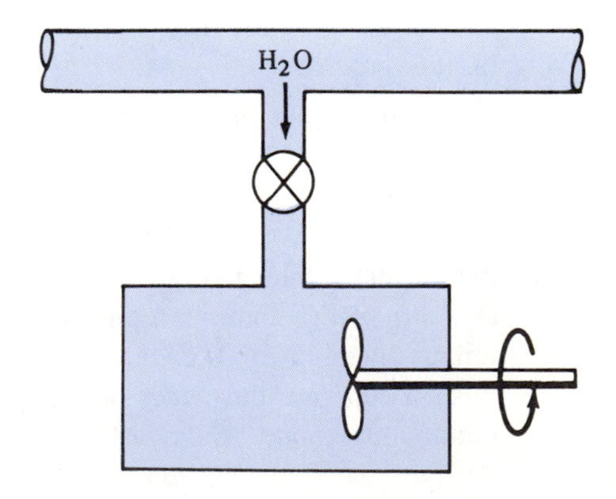

4.149 [FLD311] A cylinder fitted with a frictionless piston initially contains 0.9 lb_m of steam at 25 psia and 350°F. The cylinder is connected by a short pipe fitted with a valve to a steam line containing steam at 500°F and 100 psia. The valve is opened in a constant pressure process and 1.8 lb_m of steam enters the cylinder. At this point, the temperature of the steam in the cylinder is 400°F. The valve is then closed. Assuming that the tank has negligible heat capacity, determine the following:

(a) the work, in Btu, done on the piston

(b) the heat transfer to the system in Btu

4.150 A tank of 20-ft^3 volume is half-filled with liquid water, and the remainder is filled by vapor at 550 psia. The water is heated until one-half of the liquid (by volume) is evaporated, while an automatic valve lets saturated vapor escape at such a rate that the pressure is held constant at 550 psia. Determine the heat transfer during the process.

4.151 An empty, well-insulated, rigid tank is attached to a large line containing pressurized nitrogen. A valve leading to the tank is opened allowing nitrogen to enter the tank. The state of the N_2 in the line is 1.8 MPa and 300°C. The valve remains open until the nitrogen inside the tank reaches pressure equilibrium with the nitrogen in the line. Determine the temperature of the nitrogen inside the tank at the end of the process.

4.152 [FLD111] Air flows in a pipeline at a pressure of 200 psia and a temperature of 80°F. Connected to the pipeline is a rigid tank that initially contains 1.4 lb_m of air at 70°F and 14.7 psia. A valve is opened allowing air to flow into the tank until the pressure is 100 psia. During this process, 7.7 lb_m of air enters the tank. Assume that the change in the internal energy of the tank is negligible. Also, assume air to be an ideal gas with c_p = 0.24 Btu/lb_m·°R. Determine the following:

(a) the volume of the tank in ft^3

(b) the final temperature in °F of the air in the tank

(c) the heat transfer in Btu between the environment and the tank

4.153 A rigid, insulated tank with a volume of 30 m^3 is connected to a 30-cm-diameter pipe containing compressed air. The velocity profile across the inlet line is parabolic, as shown in the accompanying figure, and it is assumed to remain independent of time. The average density of the air entering the tank is constant at a value of 1.06 kg/m^3. The air is known to enter the tank at a temperature of 80°C. The tank is initially empty, and the process is allowed to continue for 2 min. Calculate the mass, pressure, and temperature of the air in the tank at the end of the 2-min period. Assume that the air is an ideal gas with constant specific heats.

Problem 4.153

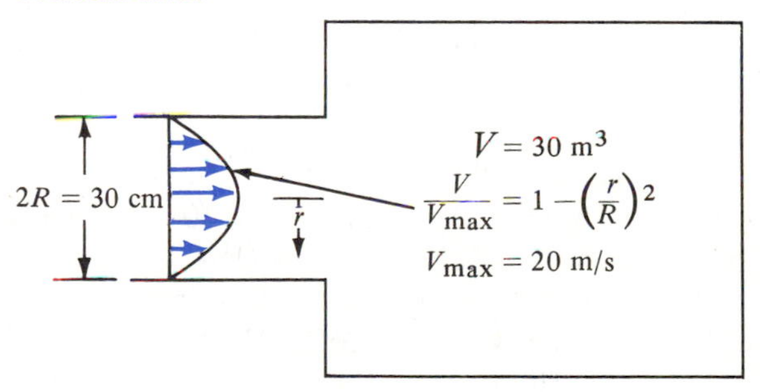

4.154 A rigid tank with a volume of 15 ft^3 contains one-half liquid water and one-half water vapor by volume at 400°F. The water is heated until one-half of the volume of liquid water is evaporated, while an automatic valve lets saturated vapor escape at such a rate that the pressure is held constant at its original value. Determine the heat transfer to the water during the process.

4.155 So that helium is supplied at a constant temperature of 140°F for a laboratory experiment, a helium supply tank is heated by an electric-resistance heater. The volume of the tank is 5 ft^3, and the tank initially contains helium at 140°F and 40 psia. A valve connected to the tank is opened, allowing helium to escape until the pressure in the tank has been reduced by 40 percent. How much heat transfer to the helium is required during this process to maintain the temperature of the helium at 140°F?

4.156 An empty, rigid tank has a volume of 5 m^3. A valve connected to the tank is opened and room air at 120 kPa, 30°C is allowed to enter the tank until the pressure in the tank reaches 50 kPa, at which time the valve is closed. The filling process occurs rapidly and therefore it can be assumed to be adiabatic. The tank then remains in the room with the valve closed until the air reaches a temperature of 30°C. Calculate:

(a) the temperature of the air in the tank immediately after the valve is closed
(b) the final pressure of the air in the tank
(c) the heat transfer from the air during the entire process

4.157 A rigid tank has a volume 10 m^3. Initially one-half of the tank's volume is filled with liquid water and one-half is filled with water vapor at 2 MPa. The water is heated until one-half of the total mass of water is evaporated and the vapor leaves the tank through a valve in the top of the tank. The heat transfer is just sufficient to maintain the pressure in the tank constant. Determine:

(a) the initial mass of water in the tank
(b) the initial temperature of the water in the tank
(c) the final mass of water in the tank
(d) the final quality of the water in the tank
(e) the heat transfer to the water during the process

4.158 An empty, rigid container (V = 10 ft^3) is connected through a valve to a line containing water at a pressure of 120 psia and a temperature of 400°F. The valve into the container is opened slowly and water enters the container until there is 2.3 lb_m of water at 15 psia inside. Calculate the temperature of the water at the end of the process and the heat transfer during the process.

4.159 A rigid tank is attached to a large supply line containing steam at 200°C, 200 kPa. The tank initially contains 2 kg of liquid water and 0.5 kg of water vapor at 50°C. A valve separating the line from the tank is opened and water enters the tank. The filling process continues until the tank contains 4 kg of water and the tank is in pressure equilibrium with the steam in the line. Determine the following information:

(a) the volume of the tank
(b) the original quality of the water in the tank
(c) the final temperature of the water in the tank
(d) the heat transfer from the water in the tank during the filling process

4.160 An industrial operation requires a supply of argon gas at a constant temperature of 93°C. To satisfy this requirement a large supply tank having a volume of 1.2 m^3 containing argon at 25°C and 2 MPa is heated by an electric-resistance heater until the temperature of the argon reaches 93°C. Then a valve on the tank is opened so that argon from the tank can be supplied for the industrial process. While the valve is open, the heater is also energized and regulated so that the temperature of the argon in the tank remains at 93°C. When the valve is closed, the mass of argon in the tank has been reduced by 80 percent. Determine the final pressure of the argon in the tank and the heat transfer required for the total process.

4.161 A tank is connected to a pipe by a valve. The tank has a volume of 1 m^3, and it initially contains air at 150 kPa and 270°C. The air in the pipe has a pressure of 750 kPa and a temperature of 27°C. The tank is uninsulated. The valve is opened and left open until the air in the tank reaches pressure equilbrium with the air in the pipe and the air in the tank reaches 27°C. Determine the mass of air that has entered the tank and the heat transfer during the process.

4.162 [FLD211] A tank having a volume of 5 ft^3 initially contains nitrogen gas at 11 atm and 70°F. The tank develops a leak and, after a long time, the pressure drops to 5 atm. Assume that nitrogen is an ideal gas with c_p = 0.248 Btu/lb_m·°R and neglect the heat capacity of the tank and piping. Assume an isothermal expansion.

(a) determine the mass of nitrogen that escapes from the tank

(b) determine the heat transfer, in Btu

4.163 A rigid, insulated tank is initially evacuated. Atmospheric air at 100 kPa and 20°C is allowed to leak into the tank until the pressure reaches 100 kPa. Calculate the final temperature of the air within the tank.

4.164 A simple engine uses an ideal gas with constant specific heats as the working fluid in a closed, piston-cylinder system. The gas is first heated at constant pressure from state 1 to state 2, then cooled at constant volume to state 3, where $T_3 = T_1$, and then compressed at constant temperature, thereby returning to state 1. Derive expressions for the work and heat transfer per unit mass of gas for each process in terms of the temperatures and pressures at each state and the thermodynamic properties c_p, c_v, and R.

4.165 [FLA311] Air is contained inside a closed piston-cylinder with an initial volume of 3 ft^3. At this state the temperature is 1040°F and the pressure inside is 15 psia. The air then undergoes four reversible processes during which it passes from its initial state 1 through states 2, 3, 4 and then back to state 1 to complete a cycle along the path shown in the figure. At state 2, the pressure is 15 psia and the volume is 1 ft^3. State 3 is at 45 psia with a volume of 1 ft^3 and state 4 is at 45 psia with a volume of 3 ft^3. Determine the following:

(a) the mass of air in lb_m in the cylinder

(b) the work in ft·lb_f done on (or by) the air in process 1-2

(c) the work in ft·lb_f done on (or by) the air in process 2-3

(d) the total work done by the air in ft·lb_f during a cycle

(e) the maximum temperature of the air in °F during the cycle

Problem 4.165 and 4.166

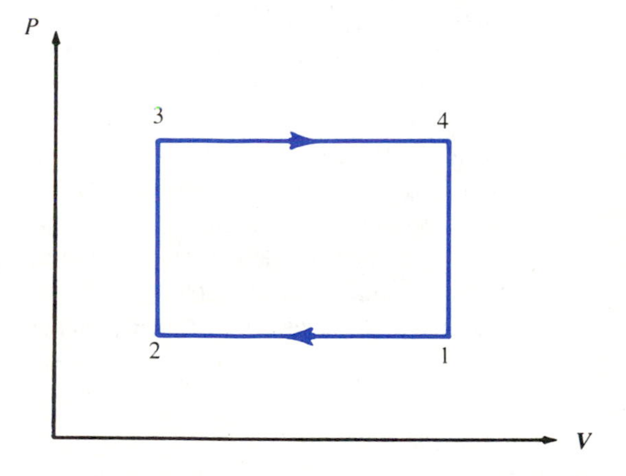

4.166 [FLA312] Air is contained inside a piston-cylinder with an initial volume of 0.1 m^3. At this state the temperature is 300°C and the pressure is 105 kPa. The air undergoes four reversible processes during which it passes from its initial state 1 through states 2, 3, 4 and then back to state 1 to complete a cycle along the paths shown in the figure.

At state 2, the pressure is 105 kPa and the volume is 0.03 m^3. The air at state 3 is at a pressure of 400 kPa with a volume of 0.03 m^3. At state 4 the air is at 400 kPa with a volume of 0.1 m^3. Calculate the following quantities:

(a) the mass of air in kg in the cylinder
(b) the work in kJ done on (or by) the air in process 1-2
(c) the work in kJ done on (or by) the air in process 2-3
(d) the total work done by the air in kJ during a cycle
(e) the maximum temperature of the air in °C during the cycle

4.167 The operating data for the simple, steady-flow, stationary, gas-turbine power plant shown in the accompanying figure are summarized in the table below. Atmospheric air enters the compressor at a rate of 10 kg/s. The power required to drive the compressor is supplied by the turbine and is transmitted through a shaft connecting the turbine and compressor. The net power output from the plant is used to drive an electric generator. Assuming that the kinetic and potential energy changes through each component are negligible, determine the power input to the compressor, the net power produced by the power plant, and the heat-transfer rate in the heat exchanger.

Problem 4.167

Location	Pressure, kPa	Temperature, °C
1	100	27
2	550	262
3		
4	170	477

Other data:
Mass flow rate of air = 10 kg/s
Power input to compressor is 60% of the total power output of the turbine
Negligible pressure drop through heat exchanger

4.168 The operating data from the simple, steady-flow, steam-power-plant cycle shown in the accompanying figure are summarized in the table below. Determine the power output of the turbine, the heat-transfer rate in the steam generator, the power input to the pump, and the thermal efficiency of the cycle.

Problem 4.168

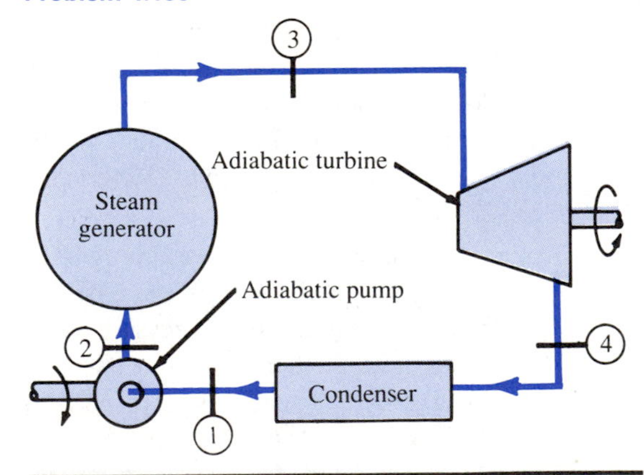

Location	Pressure, psia	Quality or temperature
1	2	0.0
2		140°F
3	1500	1600°F
4		0.95

Other data:
Mass flow rate of steam = 50 lb_m/s
Negligible pressure drops through steam generator and condenser

5

Entropy and the Second Law of Thermodynamics

5.1 INTRODUCTION

While the conservation-of-mass and conservation-of-energy principles might, at first, appear to provide a means of developing a complete thermodynamic analysis of systems, finding examples of processes from everyday experiences that cannot be explained with these principles alone is not difficult. For example, the fact that a hot cup of coffee, left undisturbed, will be cooled as a result of heat transfer until its temperature reaches the temperature of the surroundings is easy to accept. Experience is convincing ''proof'' that heat transfer is a directional process; that is, heat transfer alone occurs from a high temperature to a lower temperature, but never in the opposite direction. Similarly, if a well-insulated tank of fluid is stirred by a rotating paddle wheel or mixer, the energy of the fluid is increased. If the mixer is stopped, however, the energy of the fluid is not expected to suddenly decrease and cause the mixer to rotate in the opposite direction.

Even though the fact that each of the two processes described above will proceed in one direction only is intuitively evident, the possibility of the processes proceeding in the opposite direction is not excluded by the conservation-of-energy principle. For instance, a process during which heat transfer occurs from the cool surroundings to the hot cup of coffee could be proposed. Application of the conservation-of-energy principle to this proposed process would indicate that the energy of the cup of coffee must increase by an amount equal to the heat transfer to the coffee. The conservation-of-energy principle can be satisfied in this thought experiment even though the process itself is impossible. In many instances deciding strictly on the basis of intuition or experience whether a proposed process can take place is difficult, if not impossible. However, the second law of thermodynamics that is developed in this and the next chapter can be used to describe the directional nature of processes.

Since the 1973 oil embargo, much has been said about the need to conserve energy, the desirability of practicing energy-conservation measures, and the idea that the world's energy reserves are dwindling. However, energy conservation is not the major concern because energy conservation is unavoidable—the conservation-of-energy principle dictates that energy can be neither created nor destroyed. What is of concern is the preservation of the *quality* of energy. To illustrate this point, suppose that 10 kJ of work is to be produced by causing a gas at 300 K to expand isothermally in a piston-cylinder assembly. In accordance with the conservation-of-energy principle, this work could be accomplished by transferring 10 kJ to the gas by means of heat transfer. That heat transfer occurs from a high temperature to a lower temperature can be accepted from experience; therefore, the 10 kJ of energy must be transferred to the gas from a body at a temperature higher than 300 K if work is to be produced as described. In this instance energy at temperatures above 300 K would be valuable for this application (or have a high quality), while energy at temperatures equal to or below 300 K would not be valuable. Furthermore, the quality of energy could be completely degraded so that the energy is no longer valuable. For instance, 10 kJ of energy at 400 K could be used to produce work with the piston-cylinder assembly or 10 kJ of heat transfer could occur to, say, a lake at a temperature below 300 K. In the latter case the quality of the energy would have been degraded to the point that the energy could not be used to produce work with the piston-cylinder.

This example serves to illustrate the concepts of the quality of energy and the degradation of energy. While expressing concern for what is called energy "conservation," the world community is actually expressing the need to conserve the quality of energy or to reduce the degradation of energy. These concepts are the province of the second law of thermodynamics.

With the principles developed from the second law, engineers can evaluate the performance of actual systems by comparing them with ideal systems. They can also decide how best to optimize individual processes from a thermodynamic standpoint. However, engineering most often involves the practical application of scientific principles and economic realities. For this reason a complete thermodynamic analysis alone is usually not sufficient. However, in addition to being essential, the need to use our energy resources to the best advantage is clearly recognized and appreciated.

In this chapter reversible and irreversible processes are discussed, classical statements of the second law of thermodynamics are introduced, corollaries and implications of the second law are examined, the property *entropy* is defined, and the means of evaluating entropy changes of various substances are developed. The application of the second law of thermodynamics to the analysis of thermodynamic systems is discussed in Chapter 6.

5.2 REVERSIBLE AND IRREVERSIBLE PROCESSES

Two processes that can proceed in only one direction were discussed in the preceding section: heat transfer through a finite temperature difference and mixing. These are examples of processes that are described as being irreversible. To clarify the distinction between reversible and irreversible processes, we must examine the characteristics of reversible processes in more detail.

Reversible processes can be classified as being either totally reversible or internally reversible. During an *internally reversible process* a system proceeds through a series of equilibrium states, and if the direction is reversed at any point in the process, the system can be returned to its initial equilibrium state without leaving any permanent change in the system. Thus the quasi-equilibrium process introduced in Chapter 1 is an internally reversible process. Notice that internal reversibility is a characteristic ascribed to a process from the viewpoint of the system of interest and does not place requirements on the surroundings.

The internally reversible process is an idealization of an actual process. It is purely conceptual and may never be observed in nature. Common idealized processes that are internally reversible include the slow expansion of a gas in a frictionless-piston-cylinder assembly and a frictionless pendulum swinging in a perfect vacuum. Notice that in each of these processes dissipative effects (friction, air resistance) were removed and quasi-equilibrium (slow expansion) was specified in order to achieve internal reversibility. Internal reversibility, therefore, precludes the presence of dissipative effects in any form within the system. If a gas is allowed to expand rapidly, it does not proceed through a series of equilibrium states and therefore the expansion is irreversible. If the pendulum is not frictionless, energy is dissipated by frictional heating and the effects of the frictional heating cannot be reversed by reversing the pendulum's direction of motion.

Another example of a process that is internally reversible is a quasi-equilibrium heat-transfer process. If a closed system is simply heated slowly to a state of higher energy, the system can be returned to its original state by reversing the direction of the heat transfer. While this experiment requires that the surroundings be changed, the process that the system undergoes is internally reversible from the point of view of the system.

For a process to be *totally reversible* it must be internally reversible, and, in addition, the interactions between the system and its surroundings must also be reversible. That is, if the process is reversed, both the system and its surroundings must be capable of being returned to their initial equilibrium states without leaving any permanent changes in either. In this text the term *reversible process* is used to indicate the totally reversible process. Furthermore, if a process is not internallly reversible, it is *irreversible*.

The heat-transfer process described above as being internally reversible is not totally reversible. While the substance is being heated, energy must be transferred to the system from an external energy source that must be at a higher temperature than that of the system. If this process is reversed in direction, the system can only be returned to its initial equilibrium state by transferring energy to an external source that is at a lower temperature than that of the system. Thus the system can be returned to its original state by reversing the process, but permanent changes are present in the surroundings; as a result of the heat-transfer processes, the higher-temperature source is at a lower-energy level and the lower-temperature source is at a higher-energy level.

A totally reversible heat-transfer process is not physically possible. Since heat transfer can only occur when a temperature difference exists between a system and its surroundings, reversing the direction of heat transfer necessitates a net change in the surroundings, as described in the preceding paragraph. However, a theoretical heat-transfer process that is totally reversible is conceptually possible. This imaginary process involves heat transfer between a system and its surroundings whose temperature differs from the temperature of the system by only an infinitesimal amount dT. As dT approaches zero, the heat-transfer process can theoretically be reversed in direction, leaving no net change in the surroundings. This process of heat transfer across a vanishingly small temperature difference is *totally reversible heat transfer*.

Totally reversible heat transfer is a conceptual idealization and would be impossible to achieve with real processes. The heat-transfer rate is proportional to the temperature difference between the system and its surroundings as well as the surface area available for heat transfer. Therefore, if a finite quantity of heat transfer is caused by an infinitesimally small temperature difference, then an infinitely large surface area would be required. While approaching totally reversible heat transfer might be desirable for thermodynamic reasons, attempting to do so would certainly not be economically feasible.

Even in the absence of heat transfer, reversible processes or even internally reversible processes are not possible. Friction renders a process irreversible and can be caused by the movement of mechanical components, the motion of all real fluids, fluid resistance, bending, or the presence of motion of any kind. This friction could theoretically be made vanishingly small only for very slow motion; and while an engine operating under such conditions would have superior performance characteristics on a thermodynamic basis, it would hardly be practical, because it could not deliver power in finite amounts.

5.3 THERMAL-ENERGY RESERVOIRS

Throughout the discussion of the second law of thermodynamics and the development of its corollaries, the concept of a *thermal-energy reservoir*, a body that remains at a constant temperature regardless of the amount of heat transfer to or from it, is quite valuable. While the thermal-energy reservoir is an idealization, many instances arise when this idealization is a fairly accurate approximation. These situations occur when the magnitude of the heat transfer to or from a body is small compared with the *thermal mass* of the body (the product of the mass and the constant-volume specific heat). For example, in most interactions between a thermodynamic system and the earth's atmosphere, the magnitude of the heat transfer during a process is not large enough to change the temperature of the atmosphere by more than an infinitesimal amount. The atmosphere, therefore, can be assumed to be a thermal-energy reservoir. For the same reason the oceans and large lakes and rivers are also examples of thermal-energy reservoirs.

A body does not necessarily have to possess a very large absolute thermal mass, however, for it to be considered a thermal-energy reservoir. The thermal mass of the body must only be large compared with the magnitude of the heat transfer that occurs. For instance, if a small, hot ball bearing is quenched in a large container of cool water or oil, the temperature of the liquid will increase only slightly. In this instance the liquid in the container could be considered to be a thermal-energy reservoir. Thermal-energy reservoirs are also often called *heat sinks* and *heat sources* depending on whether the direction of heat transfer is to or from the reservoir.

Heat transfer is an integral part of our existence, and the effect of heat transfer on our surroundings should be of concern to an engineer. While *thermal pollution*, an increase in the temperature of portions of the environment created by heat transfer from commercial or residential sources, is unavoidable when heat transfer to the surroundings occurs, the extent of the thermal pollution is not necessarily significant. For instance, heat transfer from a home central-air-conditioning system causes the temperature of the air in the vicinity of the condenser to increase somewhat. This localized thermal pollution is practically insignificant since the effects essentially vanish as the air is dispersed throughout the environment. Thermal pollution can have lasting effects, however. The temperature of large bodies of water, for example, can be increased measurably owing to heat transfer from industrial processes if adequate precautions are not taken to prevent such occurrences.

EXAMPLE 5.1

An industrial process uses nearby river water to completely condense dry saturated steam at 125 kPa. The steam mass flow rate is 20,000 kg/h. At a location *A* upstream of the heat-exchanger loop, the river water flows at the rate of 2×10^5 m^3/h, and the temperature of the water is 15°C. Determine the heat-transfer rate to the river water and the temperature rise of the river water at a location *B* downstream from the industrial site where the river water and heat-exchanger effluent have been completely mixed.

Solution. A sketch of the situation described in the problem statement is shown in Figure 5.1.

The heat-transfer rate to the river water can be determined by applying the conservation-of-mass and -energy equations to the steam side of the heat exchanger, which is chosen as the system (S_1 in Figure 5.1), or

$$\dot{m}_3 = \dot{m}_4 = \dot{m}_s$$

And since the heat exchanger is a steady-flow device and kinetic- and potential-energy changes of the steam are expected to be negligible, we have

$$\dot{Q}_s = \dot{m}_s(h_4 - h_3)$$

The enthalpies of the steam at the inlet and exit are obtained directly from Table B.2 at 125 kPa:

$$x_3 = 1.0 \qquad h_3 = h_g = 2685.1 \text{ kJ/kg}$$

$$x_4 = 0.0 \qquad h_4 = h_f = 444.4 \text{ kJ/kg}$$

Thus the heat-transfer rate from the steam is

$$\dot{Q}_s = (20{,}000 \text{ kg/h})(444.4 - 2685.1)\text{kJ/kg} = -4.48 \times 10^7 \text{ kJ/h}$$

The heat-transfer rate to the river water is equal in magnitude but opposite in sign to the heat-transfer rate from the steam:

$$\dot{Q}_R = -\dot{Q}_s = \underline{\underline{4.48 \times 10^7 \text{ kJ/h}}}$$

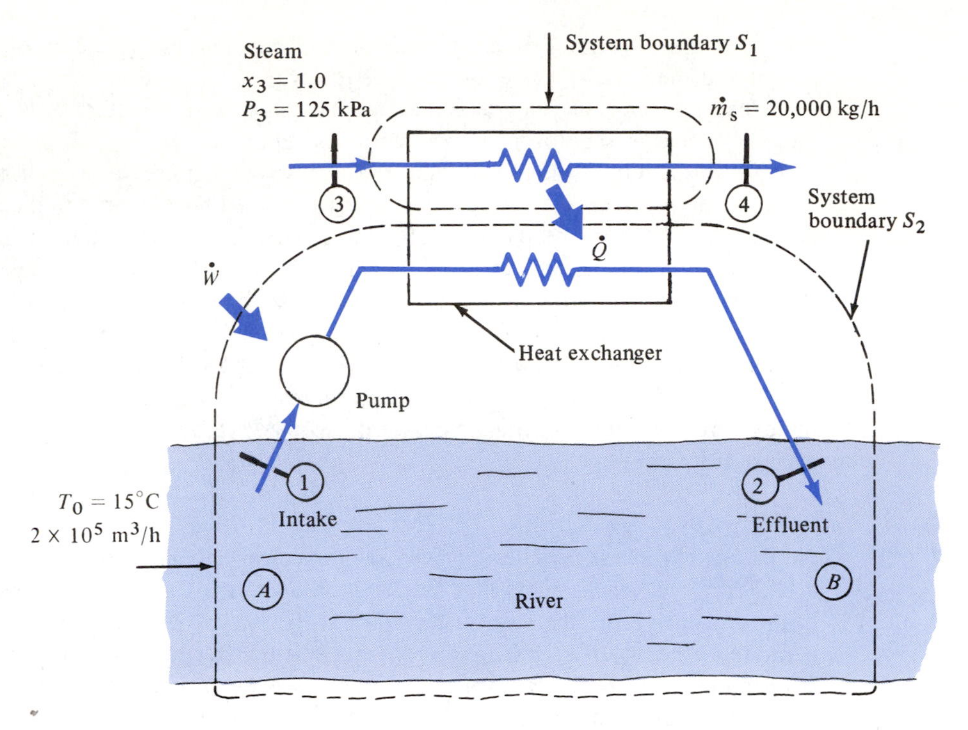

Figure 5.1 Sketch for Example 5.1.

By applying the conservation of mass and energy to the river as the system from location A to location B (S_2 in Figure 5.1), we can determine the state of the water downstream from the industrial process:

$$\dot{m}_A = \dot{m}_B = \dot{m}_R$$

$$\dot{Q}_R = \dot{m}_R(h_B - h_A)$$

The changes in kinetic energy and potential energy of the river water have been assumed negligible, and the power input required for the pump has also been assumed to be negligible compared with the heat-transfer rate. Since the river water is nearly incompressible, the change in enthalpy can be determined from Equation 2.51:

$$h_B - h_A = c_R(T_B - T_A) + v(P_B - P_A)$$

The pressure of the river water is assumed to remain essentially constant, so

$$\dot{Q}_R \simeq \dot{m}_R c_R(T_B - T_A)$$

The mass flow rate of river water is related to the volumetric flow rate (AV) by $\dot{m} = (AV)/v$, and the temperature rise of the water is

$$T_B - T_A = \frac{v\dot{Q}_R}{c_R(AV)}$$

Using values of v from Table B.1 and c_R from Table H.3 yields

$$T_B - T_A = \frac{(0.001\ \text{m}^3/\text{kg})(4.48 \times 10^7\ \text{kJ/h})}{(4.19\ \text{kJ/kg·K})(2 \times 10^5\ \text{m}^3/\text{h})} = \underline{\underline{0.05\ \text{K}}}$$

Since the river-water temperature increased by only 0.05 K, the river can be considered to be a thermal-energy reservoir. Notice that in this problem the temperature of the effluent from the heat exchanger was not specified in the problem statement, nor can it be evaluated since the mass flow rate of river water through the heat exchanger is not given. In the design of the heat exchanger consideration must, of course, be given to the temperature rise of the river water downstream from the industrial process; furthermore, the conditions of the effluent (location 2 in Figure 5.1) must be examined to ensure that localized thermal pollution is not significant. This consideration would limit the allowable exit temperature of the water at location 2 and fix the minimum allowable mass flow rate of water through the heat exchanger as well. ■

5.4 THE CLAUSIUS STATEMENT OF THE SECOND LAW OF THERMODYNAMICS; REFRIGERATORS AND HEAT PUMPS

Like the conservation-of-mass and conservation-of-energy principles, the second law of thermodynamics is a statement that expresses an observation relating to the behavior of physical processes. The validity of these principles cannot be proved; however, experimental evidence has always supported their validity, and for this reason they are accepted as being accurate statements of physical laws.

On the basis of the observation that heat transfer alone always occurs from a high temperature to a low temperature, one begins to intuitively accept this observation as being a ''law of nature.'' A more formal statement of this observation was formulated by R. Clausius in 1850 and is now called the *Clausius statement of the second law of thermodynamics*:

> **A device that operates in a cycle and has no effect on the surroundings other than heat transfer from a lower-temperature body to a body at a higher temperature is impossible to construct.**

The Clausius statement has a much broader implication than the mere fact that heat transfer does not, of its own, proceed from a low temperature to a higher temperature. The Clausius statement goes beyond this fact and precludes the possibility, by whatever means might be devised, of causing heat transfer to proceed from a low temperature to a higher temperature with no other effect on the surroundings.

This statement does not mean that a cyclic device that will result in heat transfer from a low-temperature body to a higher-temperature body is impossible to construct. In fact, a common household refrigerator performs precisely this task. The refrigeration system maintains a refrigerated space (the low-temperature body) at a low temperature by heat transfer from the refrigerated space to the refrigerator (the cyclic device) followed by heat transfer from the refrigerator to the kitchen (the high-temperature body), as shown

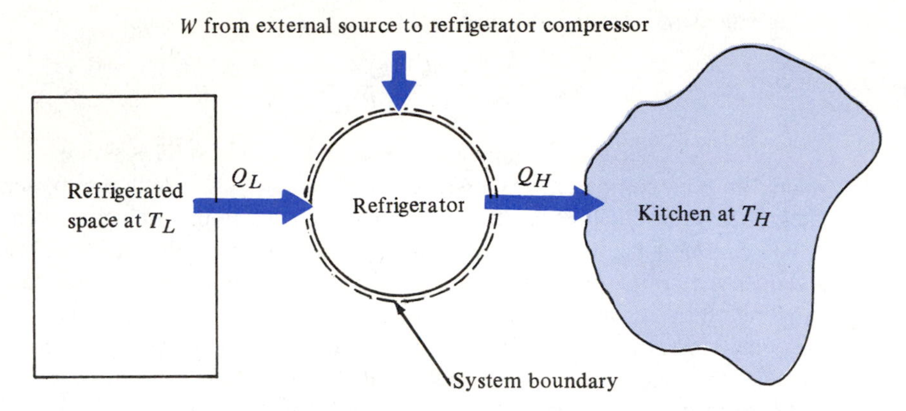

Figure 5.2 Schematic of refrigerator operation.

in Figure 5.2. In order for the refrigeration system to operate, however, a compressor must be driven by an external source of power, such as an electric motor. Thus the net effect on the surroundings is not only heat transfer from a low-temperature body and to a higher-temperature body but also an input of work to the cyclic device. This operation clearly does not violate the Clausius statement of the second law of thermodynamics. A second-law violation would occur, however, if the refrigerator could perform its task without being powered by an external source.

Devices that operate in a cycle and achieve the objective of heat transfer from a low-temperature body to a body at a higher temperature are called *refrigerators* and *heat pumps*. The operation of both devices requires an input of work or an energy input from another body. While refrigerators and heat pumps achieve the same overall objective, the basic purpose of each is quite different. A *refrigerator* is used to maintain the temperature of a refrigerated space at a temperature lower than the temperature of the environment. A *heat pump*, on the other hand, is used to maintain the temperature of a heated space at a higher temperature than that of the environment. Examples of these two devices are illustrated in Figure 5.3.

The measure of performance used for refrigerators and heat pumps, first introduced in Chapter 4, is called the coefficient of performance β. The coefficient of performance is defined as the ratio of the desired heat transfer to the net energy input required to produce the desired effect. For the simple refrigeration cycle shown in Figure 5.3, the coefficient of performance is*

$$\beta_R \equiv -\frac{Q_L}{W_{\text{net}}} \tag{5.1}$$

Since the heat transfer is from the cool space to the refrigerator (the system) ($Q_L > 0$) and the system must be powered by an external source ($W_{\text{net}} < 0$), the negative sign is

*Here and elsewhere in this chapter, the algebraic signs of terms designated Q_L and Q_H are determined relative to the system under consideration.

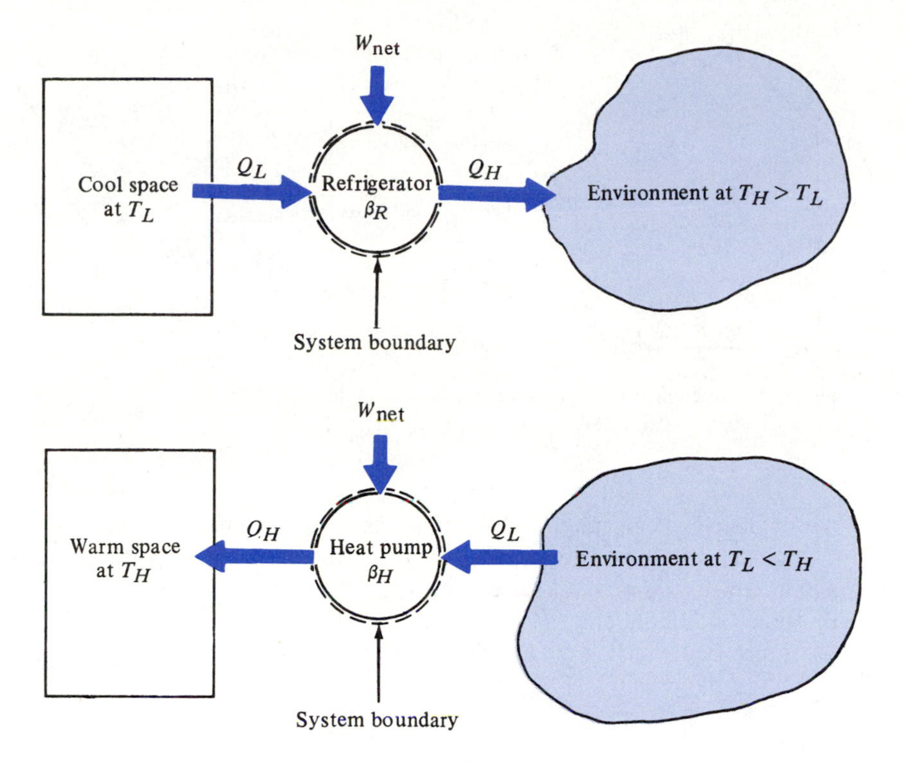

Figure 5.3 Schematics of a refrigerator and a heat pump.

introduced in Equation 5.1 to ensure positive values for β_R. For the simple heat-pump cycle shown in Figure 5.3, the coefficient of performance is

$$\beta_H \equiv \frac{Q_H}{W_{\text{net}}} \tag{5.2}$$

EXAMPLE 5.2

A residential air-source heat pump is used to provide heating during the winter season. A house is to be maintained at 70°F, and on a typical day the heat transfer from the house amounts to 70,000 Btu/h when the outdoor-air temperature is 24°F. The heat pump has a coefficient of performance of 3.7 under these conditions. Determine the power input required for the heat pump and the heat-transfer rate to the heat pump from the outdoor air.

Solution. The air-source heat pump extracts energy from the outdoor air, and then heat transfer from the heat pump to the living areas of the house maintains the temperature of the heated space at 70°F. Because the house cannot be perfectly insulated, there is continuous heat transfer from the house to the surrounding air, and the rate of heat transfer is proportional to the difference between the temperature of the heated space and

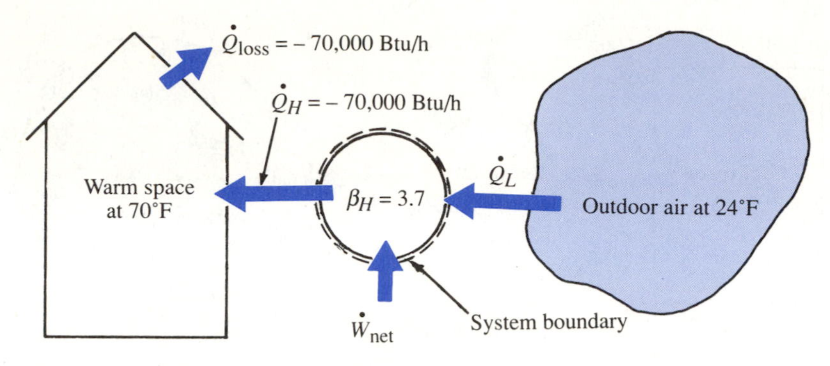

Figure 5.4 Sketch of heat-pump operation for Example 5.2.

the outdoor-air temperature. To maintain the temperature of the rooms in the house at 70°F under steady conditions, the heat pump must provide heat transfer to the house at a rate equal to the heat-transfer rate from the house, 70,000 Btu/h. This system is illustrated in Figure 5.4.

From Equation 5.2 the coefficient of performance of the heat pump is

$$\beta_H \equiv \frac{\dot{Q}_H}{\dot{W}_{net}}$$

and the power input to the heat pump can be determined as

$$\dot{W}_{net} = \frac{\dot{Q}_H}{\beta_H} = \left(\frac{-70{,}000 \text{ Btu/h}}{3.7}\right) = \underline{\underline{-18{,}920 \text{ Btu/h}}}$$

The heat-transfer rate to the heat pump from the air, $\dot{Q}_L$, can be determined by applying the conservation-of-energy equation, Equation 4.24, to the heat pump operating in a cycle. On a rate basis the result is

$$\dot{Q}_H + \dot{Q}_L = \dot{W}_{net}$$

and therefore

$$\dot{Q}_L = \dot{W}_{net} - \dot{Q}_H = -18{,}920 \text{ Btu/h} - (-70{,}000 \text{ Btu/h}) = \underline{\underline{+51{,}080 \text{ Btu/h}}}$$

Notice that, in effect, the heat-transfer rate from the heat pump to the house must be equal to the sum of the power input plus the heat-transfer rate to the heat pump from the outdoor air. For this reason the heat pump is superior to a simple electric-resistance heater, which can only provide a heat-transfer rate equal in magnitude to the electric power input. ■

5.5 THE KELVIN-PLANCK STATEMENT OF THE SECOND LAW OF THERMODYNAMICS; HEAT ENGINES

In the preceding section the Clausius statement of the second law of thermodynamics was introduced. This statement places a limitation on the operation of refrigerators and heat pumps; neither can operate without an input of work. Similarly, the second law can also be stated with reference to devices called heat engines. Specifically, a *heat engine* is a device that operates in a cycle and produces net positive work while heat transfer occurs across the boundaries of the device. An example of a heat engine is a simple steam power plant. Heat transfer to the system occurs in the boiler, work is produced by the turbine, heat transfer occurs from the system in the condenser, and a work input is required to compress the liquid to the operating pressure of the boiler.

The following statement, known as the *Kelvin-Planck statement of the second law of thermodynamics*, places a limitation on the operation of heat engines:

> **A device that operates in a cycle and has no effect on the surroundings other than the conversion of heat transfer to an equivalent amount of net positive work is impossible to construct.**

This statement is not, perhaps, as easy to accept as the Clausius statement since conceiving of an ideal device that would produce an amount of net positive work that is equivalent to the heat transfer is relatively easy. One example would be a frictionless-piston-cylinder assembly containing an ideal gas that is caused to expand isothermally as it is heated. The heat transfer would be equal to work of expansion of the ideal gas since the internal energy of the gas would not change during an isothermal process. This process does not contradict the Kelvin-Planck statement since the gas does not complete a cycle. The Kelvin-Planck statement does say, however, that the gas could not undergo a complete cycle in this or any other device and at the same time produce an amount of net positive work that is equivalent to the heat transfer and have no other effect on the surroundings. Such a device is impossible to construct even if it could be made to operate without friction.

The measure of performance for a heat engine is called the *thermal efficiency* η_{th}, which was first defined in Chapter 4 as the ratio of the desired effect (the net work output) to the energy input required to produce the desired effect:

$$\eta_{th} \equiv \frac{W_{net}}{Q_{input}} \tag{5.3}$$

One direct implication of the Kelvin-Planck statement of the second law of thermodynamics is that a heat engine, even an ideal heat engine, cannot attain a thermal efficiency of 100 percent. Thus a heat engine must reject some energy by means of heat transfer to a body at a lower temperature, as shown schematically in Figure 5.5.

The Kelvin-Planck statement and the Clausius statement are equivalent, and either statement can be used as a statement of the second law of thermodynamics. The other could then be considered to be a corollary of the second law. Demonstrating that a

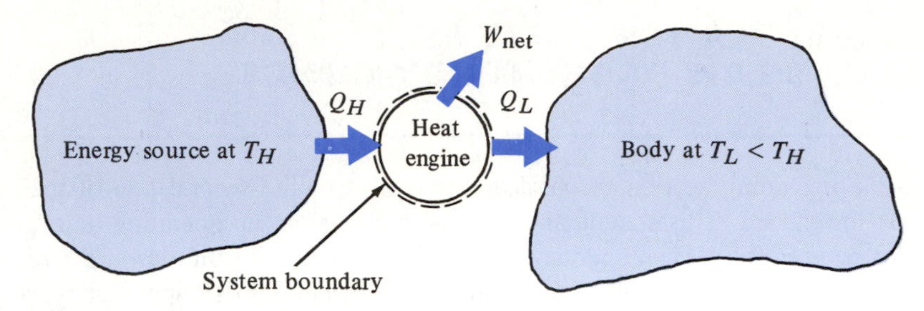

Figure 5.5 Schematic of a heat engine.

violation of one statement necessarily leads to a violation of the other is sufficient to prove that this result is true.

For example, suppose that a heat engine that has a thermal efficiency of 100 percent could be constructed so that there would be heat transfer Q_{H1} to the heat engine from a thermal-energy reservoir at temperature T_H, and that the heat engine produced an equivalent amount of net positive work W_{net} (a violation of the Kelvin-Planck statement). The work output of the heat engine could then be used to drive a refrigerator (or heat pump) that had heat transfer Q_L to the refrigerator from a thermal-energy reservoir at temperature T_L and heat transfer Q_{H2} from the refrigerator to the thermal-energy reservoir at temperature T_H. This arrangement is shown schematically in Figure 5.6(a). If the combination of the heat engine and refrigerator is considered as a single system, as shown in Figure 5.6(b), the result is a device that operates in a cycle and has no effect on the surroundings

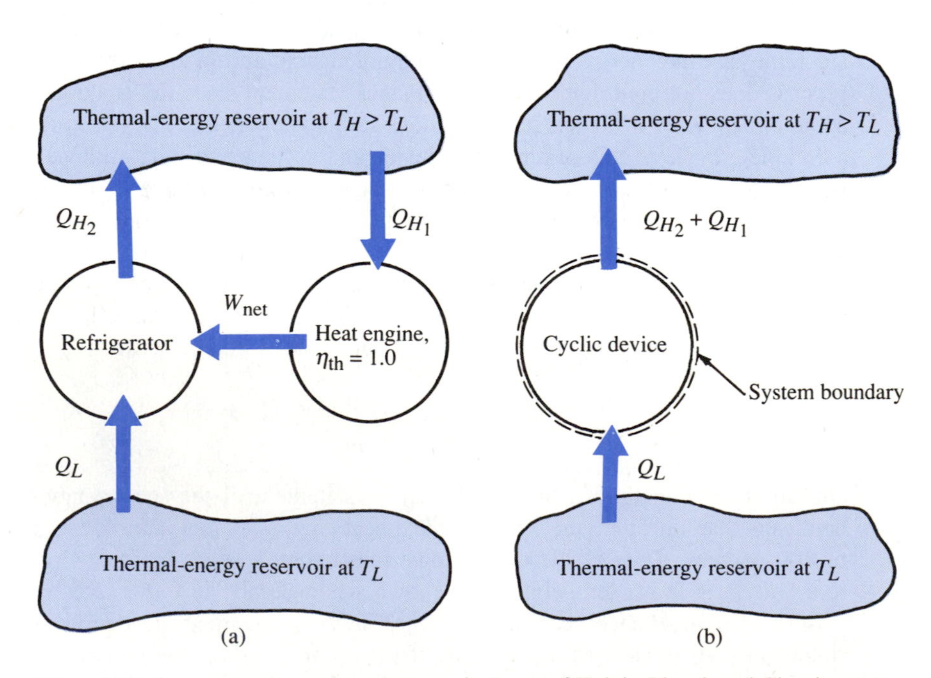

Figure 5.6 Arrangements used to show equivalence of Kelvin-Planck and Clausius statements of the second law.

other than heat transfer Q_L from a low-temperature reservoir to a reservoir at a higher temperature. Thus a violation of the Kelvin-Planck statement leads to a violation of the Clausius statement of the second law. The proof that a violation of the Clausius statement leads to a violation of the Kelvin-Planck statement is similar and is presented as a problem for the reader at the end of this chapter.

Notice that if the magnitudes of Q_{H1} and Q_{H2} were the same for the arrangement described in the preceding paragraph, then only a portion of the net work of the heat engine would be required to operate the refrigerator. Thus, the equivalent cyclic device would then produce excess positive work and provide cooling at the same time. Such a device would violate the second law of thermodynamics and would be called a *perpetual motion machine of the second kind*. Similarly, a heat engine that produced net positive work in excess of the amount of energy added during a cycle would violate the first law of thermodynamics and would be called a *perpetual motion machine of the first kind*.

■ EXAMPLE 5.3

A heat engine with a thermal efficiency of 35 percent produces 750 kJ of work. Heat transfer to the engine is from a reservoir at 550 K, and the heat transfer from the engine is to the surrounding air, which is at 300 K. Determine the heat transfer to the heat engine and the heat transfer from the heat engine to the air.

Solution. A sketch of the heat engine is shown in Figure 5.7, and the heat transfer to the engine can be determined directly by using the definition of the thermal efficiency, Equation 5.3;

$$\eta_{th} = \frac{W_{net}}{Q_H}$$

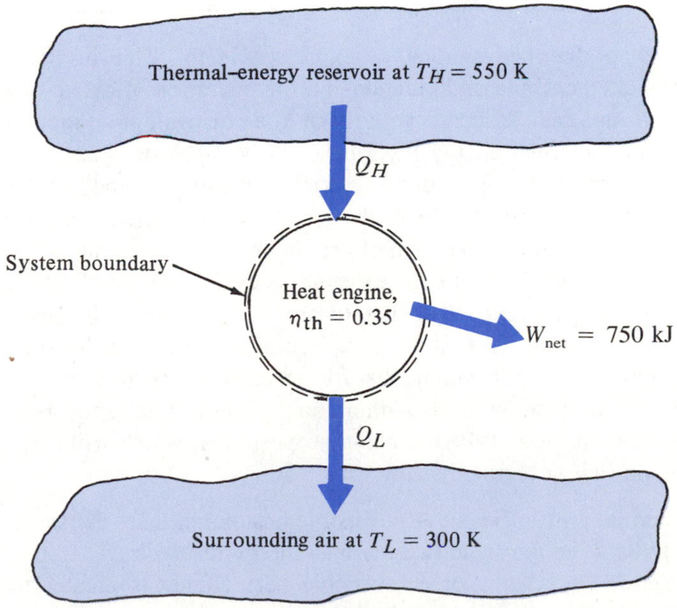

Figure 5.7 Sketch of heat engine for Example 5.3.

or

$$Q_H = \frac{W_{\text{net}}}{\eta_{\text{th}}} = \frac{750 \text{ kJ}}{0.35} = \underline{\underline{2140 \text{ kJ}}}$$

From the conservation of energy for a cycle, Equation 4.24,

$$Q_H + Q_L = W_{\text{net}}$$

thus

$$Q_L = W_{\text{net}} - Q_H = 750 \text{ kJ} - 2140 \text{ kJ} = \underline{\underline{-1390 \text{ kJ}}}$$

The magnitude of the heat transfer from the heat engine to the surrounding air is the difference between the heat transfer to the heat engine and the net work produced by the heat engine. Notice that these results are independent of the reservoir temperatures. ■

If the heat transfer from the engine to the air in the preceding example could somehow be reduced or eliminated, the heat engine could produce more work, but the Kelvin-Planck statement of the second law implies that the heat-rejection process cannot be eliminated completely. Furthermore, as the principles of the second law are developed further, the fact that there is a limit to how much the heat transfer from the heat engine can be reduced will become evident. This limit is related to the reservoir temperatures.

5.6 CARNOT'S PRINCIPLE AND THE THERMODYNAMIC TEMPERATURE SCALE

The second law of thermodynamics, as expressed in the Clausius statement and the Kelvin-Planck statement, places limitations on the operation of cyclic devices; refrigerators and heat pumps cannot operate without an input of work or an input of energy from a high-temperature thermal energy reservoir, and heat engines cannot operate without heat transfer from the heat engine during a cycle. However, some very important questions naturally arise in connection with these statements of the second law: How much work input is required for the operation of heat pumps and refrigerators? How much heat transfer must occur from the heat engine during its cyclic operation? Can the second law be applied to a single process rather than to a complete cycle? The last question is the subject of the next section. The first two questions are addressed in this section, beginning with a discussion of two propositions that form Carnot's principle.

Carnot's principle, which deals with a comparison of totally reversible and irreversible heat engines, consists of the following propositions, which will be proved with the aid of the second law of thermodynamics.

1. The thermal efficiencies of all reversible heat engines are the same if the engines operate between the same two thermal-energy reservoirs.
2. The thermal efficiency of a reversible heat engine is greater than that of an irreversible heat engine when both heat engines operate between the same two thermal-energy reservoirs.

The reversible heat engine mentioned in Carnot's principle is only theoretically possible. For the heat engine to be reversible, all processes that make up the heat-engine cycle must be totally reversible. As mentioned previously, this condition would require that the heat transfer to the reversible heat engine from the high-temperature reservoir and the heat transfer from the reversible heat engine to the low-temperature reservoir take place through infinitesimally small temperature differences. One such reversible heat engine is called a *Carnot heat engine*, and it operates in a cycle that is called a *Carnot cycle*. The specific processes that make up the Carnot cycle are discussed later in this chapter.

The method of proof for each of the statements of Carnot's principle consists of showing that the opposite statements violate the second law of thermodynamics, and therefore these propositions must be correct. To prove the first statement, we consider the two reversible heat engines, A and B, shown schematically in Figure 5.8(a). Both heat engines operate between the same thermal-energy reservoirs, one at a high temperature T_H and the other at a lower temperature T_L. In contradiction with the first statement in Carnot's principle, both heat engines are proposed to produce the same amount of work, with heat transfer $Q_{H,B}$ to reversible heat engine B from the high-temperature reservoir that is less than the heat transfer $Q_{H,A}$ to reversible heat engine A. This proposed arrangement would require that the thermal efficiency of heat engine B be greater than the thermal efficiency of heat engine A:

$$\frac{W}{Q_{H,B}} = \eta_{\text{th},B} > \eta_{\text{th},A} = \frac{W}{Q_{H,A}} \tag{5.4}$$

since $Q_{H,B} < Q_{H,A}$.

Since heat engine A is totally reversible, its operation can be reversed. The result is a refrigerator requiring an input of work W that can be supplied by reversible heat engine B, as indicated in Figure 5.8(b). If the combination of these two devices is considered as a single system, as shown in Figure 5.8(c), the result is a device that operates in a cycle and produces no effect on the surroundings other than heat transfer from a low-temperature reservoir to a high-temperature reservoir—a violation of the Clausius statement of the second law. Thus the original proposition was in error, and the opposite must be true; the two reversible heat engines must each have the same thermal efficiency. The only restriction placed on the engines is that they be reversible, and therefore the result applies for any and all reversible engines operating between the same two thermal-energy reservoirs. That is, all reversible engines operating between the same two thermal-energy reservoirs have the same thermal efficiency.

The second part of Carnot's principle can be proved in much the same manner by replacing the reversible heat engine B in Figure 5.8(a) by an irreversible heat engine that is proposed to have a thermal efficiency equal to or greater than the thermal efficiency of engine A. If the efficiency of the irreversible engine were greater than that of engine A, a violation of the second law of thermodynamics would result. If the efficiency of the engine were the same as that of engine A, then according to the first statement of Carnot's principle, the engine would be reversible. Therefore, the conclusion is that the thermal efficiency of a reversible engine must be higher than the thermal efficiency of an irreversible engine if both engines operate between the same two thermal-energy reservoirs.

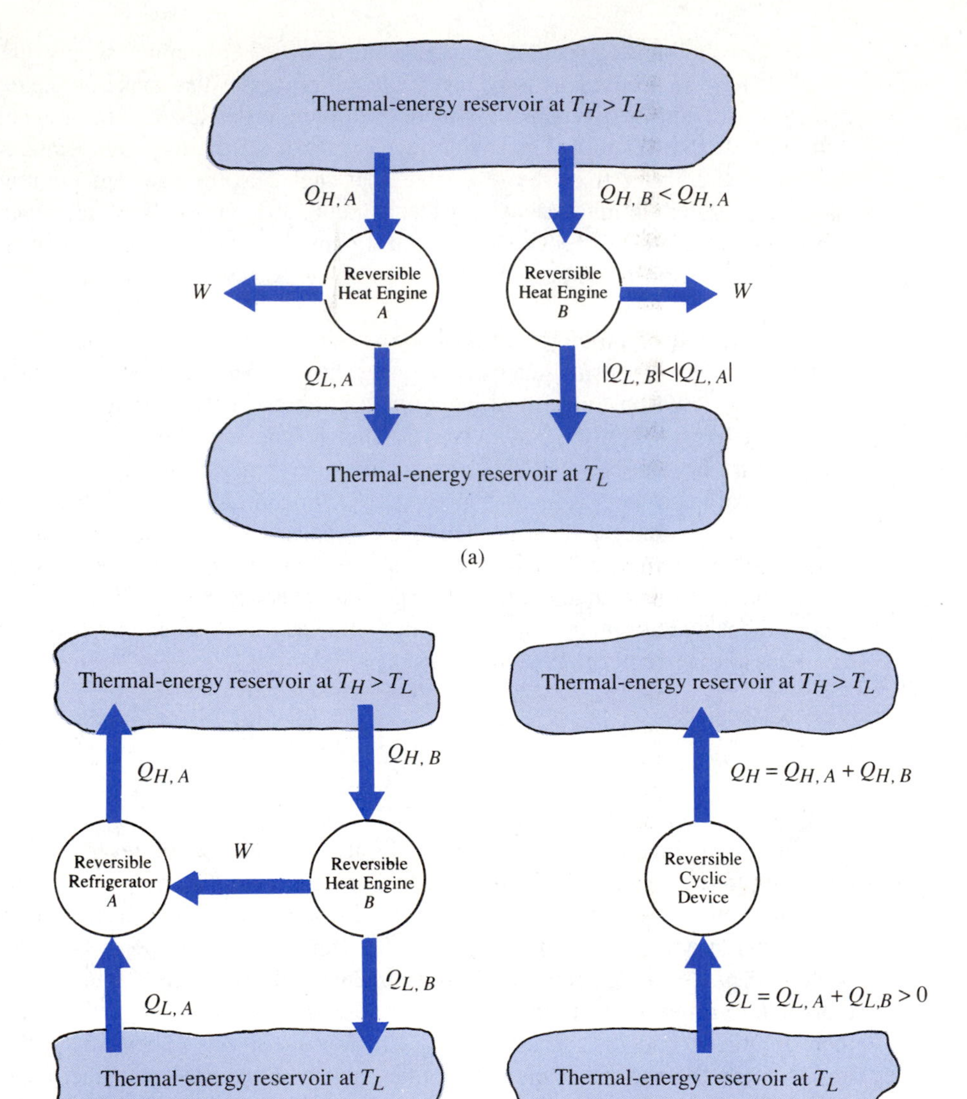

Figure 5.8 Proof of Carnot's principle.

The Kelvin-Planck statement of the second law limits the thermal efficiency of a heat engine to a value less than unity, but it does not give a direct indication of the maximum theoretical thermal efficiency. Having proved Carnot's principle, we can now evaluate the maximum theoretical thermal efficiency. For a given pair of high- and low-temperature thermal-energy reservoirs, the maximum theoretical thermal efficiency must be achieved with a reversible heat engine. Furthermore, this same maximum thermal efficiency will be achieved with all reversible heat engines operating between the two reservoirs, even though the heat transfer and work, as well as the working fluid, may be different for various reversible heat engines. The conclusion is that the maximum thermal efficiency, or the thermal efficiency of a reversible heat engine, must depend only on the temperatures of the two thermal-energy reservoirs involved. This result can be expressed mathematically as

$$\eta_{\text{th,rev}} = f(T_L, T_H)$$

From the definition of the thermal efficiency of a heat engine, Equation 5.3, however,

$$\eta_{\text{th}} = \frac{W_{\text{net}}}{Q_H}$$

Furthermore, the net work can be expressed in terms of the heat transfer to and from the heat engine through the conservation of energy for the heat-engine cycle, Equation 4.24:

$$Q_H + Q_L = W_{\text{net}}$$

Substituting into Equation 5.3 and simplifying leads to the result

$$\eta_{\text{th,rev}} = 1 + \frac{Q_L}{Q_H} = f(T_L, T_H)$$

In other words, the heat transfer from the reversible heat engine divided by the heat transfer to the reversible heat engine for a cycle must be solely a function of the absolute temperatures of the thermal-energy reservoirs:

$$\left(\frac{Q_L}{Q_H}\right)_{\text{rev}} = g(T_L, T_H)$$

One such function, proposed by Lord Kelvin, that satisfies this and other required mathematical criteria is simply the negative of the ratio of the absolute temperatures:

$$\left(\frac{Q_L}{Q_H}\right)_{\text{rev}} = -\frac{T_L}{T_H} \tag{5.5}$$

This relationship defines a temperature scale known as the *absolute thermodynamic temperature scale*.

With this relationship the maximum theoretical thermal efficiency of the heat engine is the thermal efficiency of the reversible heat engine, given by

$$\eta_{\text{th,rev}} = 1 - \frac{T_L}{T_H} \tag{5.6}$$

The efficiency given in Equation 5.6 is also referred to as the *Carnot efficiency*.

The maximum theoretical thermal efficiency cannot be achieved by actual heat engines since all such engines have sources of irreversibility, such as friction associated with moving parts. Moreover, even if the friction could be eliminated and an internally reversible heat engine could be constructed, the heat engine could still not achieve the maximum theoretical thermal efficiency. The actual heat-transfer processes, which must occur through finite temperature differences between the engine and the reservoirs under practical circumstances, would cause the operation of the engine to be irreversible.

The coefficient of performance for a reversible heat pump or reversible refrigerator can also be expressed in terms of the thermal-energy-reservoir temperatures with the aid of Equation 5.5. For the heat pump

$$\beta_H = \frac{Q_H}{W_{\text{net}}} = \frac{Q_H}{Q_H + Q_L} = \frac{1}{1 + (Q_L/Q_H)} \tag{5.7}$$

Therefore, the coefficient of performance of the reversible heat pump is

$$\beta_{H,\text{rev}} = \frac{1}{1 - (T_L/T_H)} \tag{5.8}$$

For the refrigerator

$$\beta_R = -\frac{Q_L}{W_{\text{net}}} = -\frac{Q_L}{Q_H + Q_L} = -\frac{1}{(Q_H/Q_L) + 1} \tag{5.9}$$

and the coefficient of performance of the reversible refrigerator is

$$\beta_{R,\text{rev}} = \frac{1}{(T_H/T_L) - 1} \tag{5.10}$$

EXAMPLE 5.4

Suppose that the heat pump in Example 5.2 is replaced with a Carnot heat pump. Determine the coefficient of performance and the power input for the conditions stated.

Solution. Since the Carnot heat pump is totally reversible, its coefficient of performance is determined solely by the temperatures of the heated space and the outdoor air. From Equation 5.8

$$\beta_{H,\text{rev}} = \frac{1}{1 - (T_L/T_H)}$$

$$= \frac{1}{1 - (24 + 460)°\text{R}/(70 + 460)°\text{R}} = \underline{\underline{11.5}}$$

If the heat-transfer rate from the Carnot heat pump to the house is the same as that given in Example 5.2, then the power input can be calculated from Equation 5.2 written on a rate basis:

$$\beta_{H,\text{rev}} = \frac{\dot{Q}_H}{\dot{W}_{\text{rev}}}$$

or

$$\dot{W}_{\text{rev}} = \frac{\dot{Q}_H}{\beta_{H,\text{rev}}} = \frac{-70{,}000 \text{ Btu/h}}{11.5} = \underline{\underline{-6090 \text{ Btu/h}}}$$

A Carnot heat pump would require 68 percent less power input than the actual heat pump for the same conditions. Remember, however, that for a Carnot heat pump (a totally reversible heat pump), infinitely large heat exchangers would be required to achieve totally reversible heat transfer. ■

EXAMPLE 5.5

Suppose that the heat engine in Example 5.3 is replaced by a Carnot heat engine that produces the same work output. Determine its thermal efficiency, the heat transfer to this heat engine, and the heat transfer from the heat engine to the low-temperature reservoir.

Solution. The thermal efficiency of a Carnot heat engine (a totally reversible heat engine) is given by Equation 5.6:

$$\eta_{\text{th,rev}} = 1 - \frac{T_L}{T_H}$$

For reservoir temperatures of 300 K and 550 K the thermal efficiency is

$$\eta_{\text{th,rev}} = 1 - \frac{300 \text{ K}}{550 \text{ K}} = \underline{\underline{0.455 \text{ or } 45.5 \text{ percent}}}$$

This value is higher than the thermal efficiency of the actual heat engine, and it is the highest thermal efficiency that could possibly be achieved for these temperature limits.

The heat transfer to the heat engine is related to the work output through the definition of the thermal efficiency, Equation 5.3:

$$\eta_{\text{th}} = \frac{W_{\text{net}}}{Q_H}$$

Thus

$$Q_{H,\text{rev}} = \frac{W_{\text{net,rev}}}{\eta_{\text{th,rev}}} = \frac{750 \text{ kJ}}{0.455} = \underline{\underline{1650 \text{ kJ}}}$$

The heat transfer from the heat engine can be determined from the conservation of energy for a cycle, Equation 4.24:

$$Q_H + Q_L = W_{\text{net}}$$

or for this reversible cycle

$$Q_{L,\text{rev}} = W_{\text{net,rev}} - Q_{H,\text{rev}} = 750 \text{ kJ} - 1650 \text{ kJ} = \underline{\underline{-900 \text{ kJ}}}$$

Since the Carnot heat engine is reversible, we could also have determined the heat transfer from the heat engine by using Equation 5.5:

$$\left(\frac{Q_L}{Q_H}\right)_{\text{rev}} = -\frac{T_L}{T_H}$$

or

$$Q_{L,\text{rev}} = -Q_{H,\text{rev}}\left(\frac{T_L}{T_H}\right) = (-1650 \text{ kJ})\left(\frac{300 \text{ K}}{550 \text{ K}}\right) = \underline{\underline{-900 \text{ kJ}}}$$

Since the maximum thermal efficiency for the given temperature limits is 45.5 percent, in order to produce 750 kJ of work, a heat engine must have heat transfer of at least 900 kJ to the low-temperature reservoir. Otherwise, a violation of the second law would result. The actual thermal efficiency of 35 percent given in Example 5.3 is not as low as one might think at first. The actual thermal efficiency is more than 76 percent of the maximum theoretical thermal efficiency. ■

5.7 THE CLAUSIUS INEQUALITY AND ENTROPY

The Clausius inequality is an important thermodynamic relationship that can be derived with the aid of the second law of thermodynamics. Its importance is twofold since it provides a means of applying second-law principles to cycles composed of arbitrary processes, and it will be used to develop the classical definition of the thermodynamic property called entropy.

To establish the Clausius inequality, we consider an arbitrary cyclic device that is illustrated in Figure 5.9 and is composed of two reversible heat engines, engine A and engine B, and an arbitrary closed system that executes a cycle. The temperature, T, of the closed system may change as the cycle is executed. For simplicity, the cyle for the closed system is considered to be composed of two processes, A and B. The closed system undergoes a change of state from state 1 to state 2 by means of process A and then returns to state 1 by process B to complete the cycle.

As the closed system executes process A, reversible heat engine A operates in a cycle while receiving thermal energy from the high-temperature thermal-energy reservoir and rejecting thermal energy to the closed system. Furthermore, as reversible heat engine A executes an integral number of cycles, there is heat transfer δQ_H to the heat engine and heat transfer $\delta Q_{\text{sys},A}$ to the closed system. The algebraic sign of δQ_H is taken relative to heat engine A and the subscript sys indicates that the algebraic sign of this heat transfer it taken relative to the closed system. Because heat engine A is reversible and operates in a cycle, then from Equation 5.5,

$$\frac{\delta Q_H}{T_H} = \frac{\delta Q_{\text{sys},A}}{T}$$

For the complete process A, between states 1 and 2, we have

$$\int_{1,A}^{2} \frac{\delta Q}{T}\bigg|_{\text{sys}} = \oint \frac{\delta Q_H}{T_H} = \frac{Q_H}{T_H}$$

where Q_H represents the total heat transfer to engine A as the closed system completes process A.

Similarly, during process B, there is heat transfer $\delta Q_{\text{sys},B}$ from the closed system to reversible, cyclic heat engine B. This heat engine executes an integral number of cycles in order to make this heat transfer possible, and the engine rejects thermal energy by means of heat transfer δQ_L to the low-temperature thermal-energy reservoir. Because heat engine B is reversible and operates in a cycle then Equation 5.5 applies to its operation also.

$$\frac{\delta Q_L}{T_L} = \frac{\delta Q_{\text{sys},B}}{T}$$

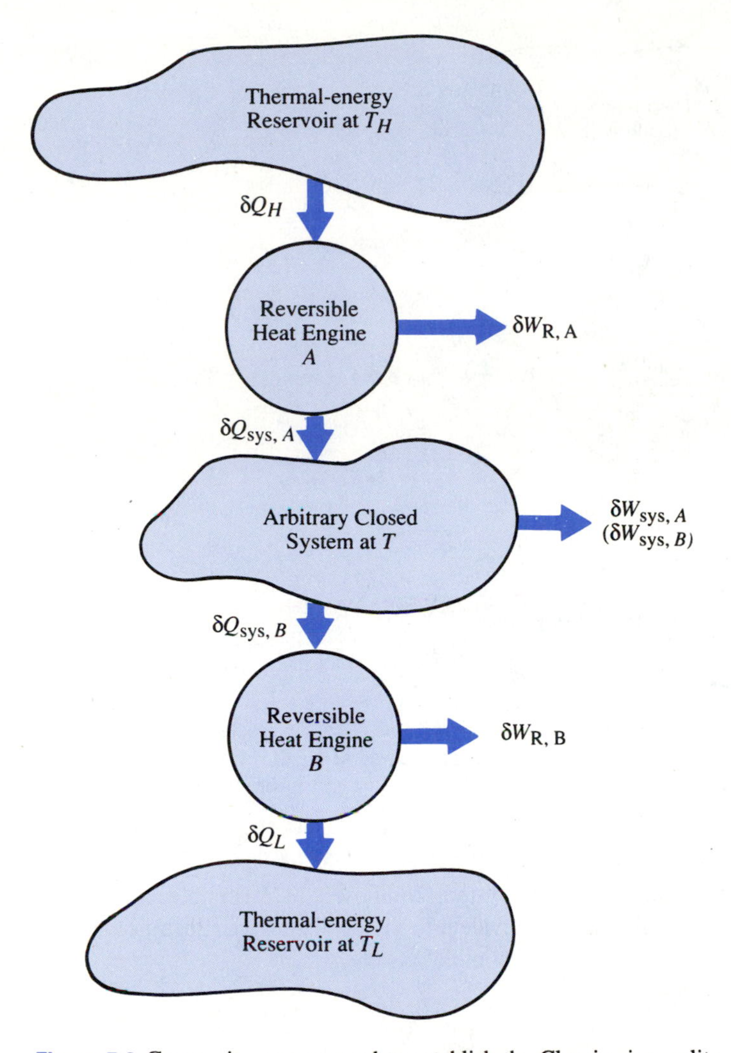

Figure 5.9 Composite system used to establish the Clausius inequality.

Then, for the complete process B, between states 2 and 1,

$$\int_{2,B}^{1} \frac{\delta Q}{T}\bigg|_{\text{sys}} = \oint \frac{\delta Q_L}{T_L} = \frac{Q_L}{T_L}$$

where Q_L represents the total heat transfer from engine B as the closed system completes process B.

When the closed system has executed processes A and B, thereby completing one cycle, then

$$\int_{1,A}^{2} \frac{\delta Q}{T}\bigg|_{\text{sys}} + \int_{2,B}^{1} \frac{\delta Q}{T}\bigg|_{\text{sys}} = \oint \frac{\delta Q}{T}\bigg|_{\text{sys}} = \frac{Q_H}{T_H} + \frac{Q_L}{T_L}$$

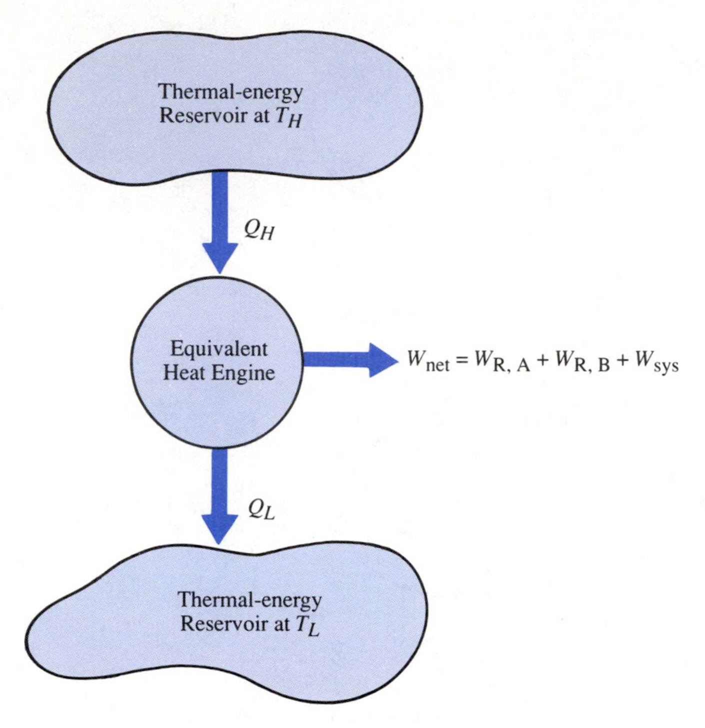

Figure 5.10 The cyclic device that is equivalent to the arrangement of Figure 5.9 after one complete cycle of the arbitrary closed system.

Furthermore, the composite system composed of reversible engines A and B and the arbitrary closed system is equivalent to the cyclic device illustrated in Figure 5.10. Thus, from Equation 4.24, for one complete cycle,

$$Q_H + Q_L = W_{net}$$

or

$$Q_L = W_{net} - Q_H$$

and therefore

$$\oint \frac{\delta Q}{T}\bigg|_{sys} = \frac{Q_H}{T_H} + \frac{W_{net} - Q_H}{T_L} = \frac{W_{net}}{T_L} - \frac{Q_H(1 - T_L/T_H)}{T_L}$$

The numerator of the last term on the right side of this equation represents the work, W_{rev}, that would be produced by the equivalent cyclic device if the operation were totally reversible. Thus,

$$\oint \frac{\delta Q}{T} = \frac{W_{net} - W_{rev}}{T_L}$$

The work W_{net} represents the actual work produced by the equivalent cyclic device. Since the closed system depicted in Figure 5.9 is an arbitrary system, its operation may be irreversible or it may be internally reversible. If it is internally reversible, the equivalent cycle of Figure 5.10 is totally reversible, and the first statement of Carnot's principle implies that W_{net} is equal to W_{rev}. If, however, the arbitrary system is irreversible, then the equivalent cycle is also irreversible, and the second statement of Carnot's principle implies that W_{net} is less than W_{rev}. Therefore, from these facts and the preceding equation, we conclude that for any arbitrary closed system

$$\oint \frac{\delta Q}{T} \leq 0 \tag{5.11}$$

since T_L is always positive. This result, that the cyclic integral of $\delta Q/T$ for any closed system must be less than or equal to zero, is known as the *Clausius inequality*.

Since the equality in Equation 5.11 applies when the closed system is internally reversible, we may also state that

$$\oint \frac{\delta Q}{T}\bigg|_{\text{int rev}} = 0 \tag{5.12}$$

Equation 5.12 is a significant result because of the fact that if

$$\oint dY = 0$$

then dY represents the differential of a property Y, as was established in Chapter 1. The fact that the cyclic intergral of $\delta Q/T$ for an internally reversible process is equal to zero, therefore, means that $(\delta Q/T)_{\text{int rev}}$ represents the differential change of a property. This property, called the *entropy*, is assigned the symbol S and is defined in the following manner:

$$dS \equiv \left(\frac{\delta Q}{T}\right)_{\text{int rev}} \tag{5.13}$$

As can be verified from Equation 5.13, the units of entropy are those of energy per unit of absolute temperature (kJ/K or Btu/°R). On a unit-mass basis the entropy change can be written as

$$ds = \left(\frac{\delta q}{T}\right)_{\text{int rev}} \tag{5.14}$$

Equation 5.13 has a convenient geometrical interpretation when T-S coordinates are used for a process diagram for internally reversible processes. For an internally reversible process,

$$\delta Q = T\, dS$$

and δQ may be represented by the differential area shown in Figure 5.11. Therefore, the total heat transfer during an internally reversible process is given by

$$Q_{12,\text{int rev}} = \int_1^2 T\, dS$$

and may be represented by the area under the process curve on the T-S diagram.

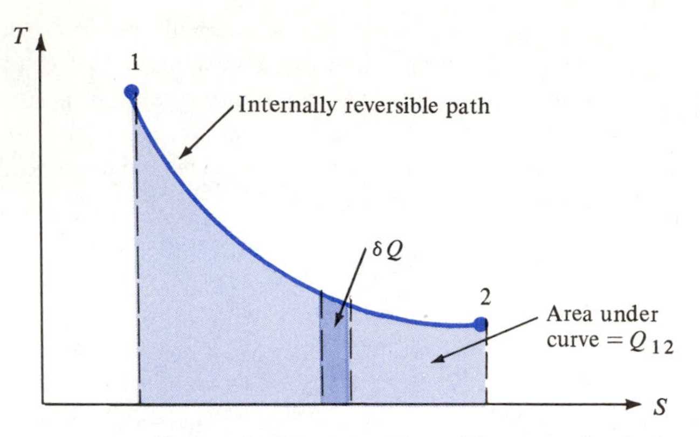

Figure 5.11 Geometrical interpretation of heat transfer during an internally reversible process.

The relationship between the entropy change and heat-transfer interactions during irreversible processes can be illustrated with a simple cycle composed of two processes, one of which is internally reversible and the other of which is irreversible, as shown in Figure 5.12. The Clausius inequality applied to this irreversible cycle can be written as

$$\oint \frac{\delta Q}{T} = \int_{1,A}^{2} \left(\frac{\delta Q}{T}\right)_{\text{irr}} + \int_{2,B}^{1} \left(\frac{\delta Q}{T}\right)_{\text{int rev}} < 0$$

Since process B is internally reversible, this process can be reversed, and therefore

$$\int_{1,A}^{2} \left(\frac{\delta Q}{T}\right)_{\text{irr}} - \int_{1,B}^{2} \left(\frac{\delta Q}{T}\right)_{\text{int rev}} < 0$$

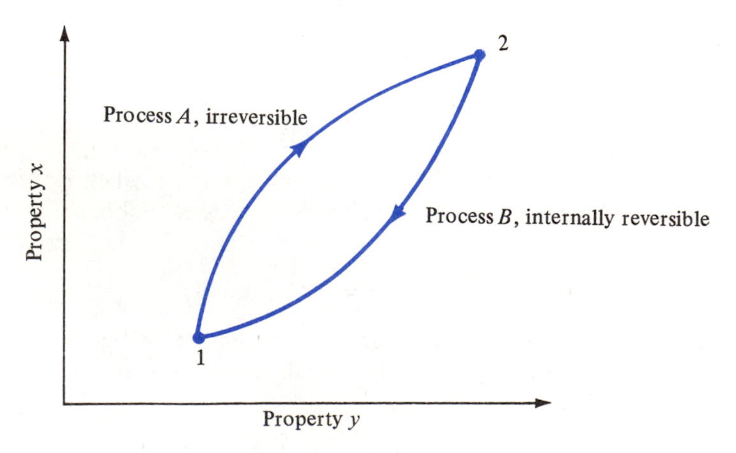

Figure 5.12 Simple cycle composed of an irreversible process and an internally reversible process.

or

$$\int_{1,B}^{2}\left(\frac{\delta Q}{T}\right)_{\text{int rev}} > \int_{1,A}^{2}\left(\frac{\delta Q}{T}\right)_{\text{irr}}$$

and use of the definition of the entropy (Equation 5.13) to replace the integral on the left side yields

$$S_2 - S_1 > \int_{1,A}^{2}\left(\frac{\delta Q}{T}\right)_{\text{irr}} \tag{5.15}$$

In other words, the change in entropy associated with a change of state is greater than the integral of $\delta Q/T$ evaluated for an irreversible process between the same end states. Equation 5.15 can also be written in differential form as

$$dS > \left(\frac{\delta Q}{T}\right)_{\text{irr}} \tag{5.16}$$

Combining Equations 5.13 and 5.16, we can express the entropy change for any process as

$$dS \geq \frac{\delta Q}{T} \tag{5.17}$$

or in integrated form,

$$S_2 - S_1 \geq \int_{1}^{2}\frac{\delta Q}{T}$$

where the equality holds for all internally reversible processes and the inequality holds for all irreversible processes. An important conclusion follows from Equation 5.17. That is, if a system undergoes an adiabatic process, then $dS \geq 0$. In other words, the entropy change for an adiabatic process must be either positive (for an irreversible process) or zero (for an internally reversible process).

The implications of Equation 5.15 may not be obvious at first. Both of the processes depicted in Figure 5.12 have the same end states, and therefore both processes also have the same entropy change $S_2 - S_1$, even though one process is irreversible. This must be so because entropy is a property, and as such, its value depends on the state of a substance alone. The entropy change $S_2 - S_1$ during process B, however, is also equal to $\int \delta Q/T$ because this process is internally reversible. While the entropy change during process A is equal to $S_2 - S_1$, this value is greater than $\int \delta Q/T$ for process A because that process is irreversible.

The preceding comments might suggest that one way of evaluating the entropy change of an irreversible process is to imagine an internally reversible process between the same two end states and to calculate the entropy change for the internally reversible process from Equation 5.13. Since the irreversible process has the same end states, the entropy change for the irreversible process must be the same as the entropy change of the imaginary reversible process. This observation is illustrated in the following example.

■ EXAMPLE 5.6

Nitrogen at 75 psia and 720°R is contained in a closed piston-cylinder assembly that has an initial volume of 45 in^3. The nitrogen is heated isothermally and expands until its pressure is reduced to 15 psia. During this process the work done by the nitrogen amounts to 400 ft·lb_f. Determine whether this process is internally reversible or irreversible, and calculate the entropy change.

Solution. The piston-cylinder assembly is depicted in Figure 5.13(a). The nitrogen is at a low pressure and a high temperature relative to the critical-state values, and therefore ideal-gas behavior is a reasonable assumption. Since the work for the process is given, we can determine whether or not the process is internally reversible by comparing the amount of work done in an internally reversible process with the actual work done. If the process were internally reversible, only $P d\mathbf{V}$ *work would occur since we are dealing with a simple compressible substance. Thus*

$$W_{12,\text{int rev}} = \int_1^2 P d\mathbf{V}$$

The integration can be performed in this instance by eliminating P, using the ideal-gas equation of state, and noting that the process is isothermal:

$$W_{12,\text{int rev}} = \int_1^2 mRT \frac{d\mathbf{V}}{\mathbf{V}} = mRT \int_1^2 \frac{d\mathbf{V}}{\mathbf{V}} = mRT \ln\left(\frac{\mathbf{V}_2}{\mathbf{V}_1}\right)$$

Furthermore, since the temperature is constant,

$$P_1\mathbf{V}_1 = mRT$$

and

$$P_1\mathbf{V}_1 = P_2\mathbf{V}_2$$

The work for the internally reversible process can therefore be expressed as

$$W_{12,\text{int rev}} = P_1\mathbf{V}_1 \ln\left(\frac{P_1}{P_2}\right)$$

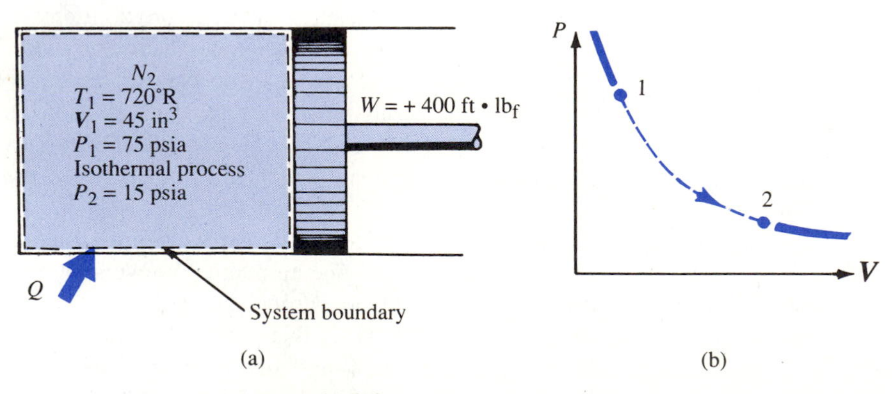

Figure 5.13 Sketch for Example 5.6.

Thus

$$W_{12,\text{int rev}} = \frac{(75\ \text{lb}_f/\text{in}^2)(45\ \text{in}^3)}{(12\ \text{in/ft})} \ln\left(\frac{75}{15}\right) = 453\ \text{ft}\cdot\text{lb}_f\ \text{(done by the system)}$$

Since the work for the actual isothermal process is not equal to the work that would have been done in an internally reversible isothermal process, the actual process is irreversible.

At this point in our development of the second law of thermodynamics, entropy changes can be calculated only for internally reversible processes, but all actual processes are irreversible. This difficulty can be overcome, however, because entropy is a property, and the change in entropy depends only on the end states of the process. Thus the entropy change that would occur during any internally reversible process between the actual end states can be calculated, and that value will be the same entropy change that occurs during the actual process.

From Equation 5.13

$$S_2 - S_1 = \int_1^2 \left(\frac{\delta Q}{T}\right)_{\text{int rev}}$$

If we choose the internally reversible isothermal process, the integration is straightforward:

$$S_2 - S_1 = \frac{1}{T}\int_1^2 (\delta Q)_{\text{int rev}} = \frac{Q_{12,\text{int rev}}}{T}$$

The heat transfer can be determined from the conservation-of-energy equation for the closed stationary system:

$$Q_{12,\text{int rev}} - W_{12,\text{int rev}} = (U_2 - U_1)$$

The internal-energy change of the nitrogen during the isothermal process is zero because the internal energy of an ideal gas depends only on temperature. Therefore

$$Q_{12,\text{int rev}} = W_{12,\text{int rev}} = \frac{453\ \text{ft}\cdot\text{lb}_f}{778\ \text{ft}\cdot\text{lb}_f/\text{Btu}} = 0.582\ \text{Btu}$$

and

$$S_2 - S_1 = \frac{0.582\ \text{Btu}}{720°\text{R}} = \underline{\underline{0.00081\ \text{Btu/°R}}}$$

Thus the entropy of the nitrogen increases by 0.00081 Btu/°R during the actual process between state 1 and state 2.

According to Equation 5.15, the entropy change must be greater than the integral of $\delta Q/T$ for the actual process. The actual heat transfer is equal in magnitude to the actual work since the internal-energy change of the nitrogen is zero and the system is closed:

$$Q_{12,\text{act}} = W_{12,\text{act}} = \frac{400\ \text{ft}\cdot\text{lb}_f}{(778\ \text{ft}\cdot\text{lb}_f/\text{Btu})} = 0.514\ \text{Btu}$$

Since the process is isothermal,

$$\int_1^2 \left(\frac{\delta Q}{T}\right)_{\text{act}} = \frac{Q_{12,\text{act}}}{T}$$

Therefore

$$\int_1^2 \left(\frac{\delta Q}{T}\right)_{\text{act}} = \frac{0.514\ \text{Btu}}{720°\text{R}} = 0.00071\ \text{Btu/°R}$$

and for this reason

$$S_2 - S_1 > \int_1^2 \frac{\delta Q}{T}$$

in accordance with the second law.

As a final comment on this example, suppose the work for the process was specified as being 500 ft·lb$_f$ (or any amount greater than $W_{12,\text{int rev}}$). Application of the conservation-of-energy equation would lead to the conclusion that

$$Q_{12,\text{act}} = W_{12,\text{act}} = \frac{500\ \text{ft·lb}_f}{(778\ \text{ft·lb}_f/\text{Btu})} = 0.643\ \text{Btu}$$

but since the process is isothermal, we would find that

$$\int_1^2 \left(\frac{\delta Q}{T}\right)_{\text{act}} = \frac{Q_{12,\text{act}}}{T} = \frac{0.643\ \text{Btu}}{720°\text{R}} = 0.00089\ \text{Btu/°R}$$

This result would mean that

$$S_2 - S_1 < \int_1^2 \frac{\delta Q}{T}$$

However, the second law of thermodynamics requires that the integrated form of Equation 5.17 be satisfied; that is,

$$S_2 - S_1 \geq \int_1^2 \frac{\delta Q}{T}$$

We must conclude that this latter process is impossible since it would lead to a violation of the second law. In the next chapter we will formalize and generalize this conclusion. The methods of second-law analysis will be used to show that the maximum amount of work that can be produced with a given change of state is the work associated with a reversible process. Furthermore, irreversibilities always cause the actual work produced to be less than this maximum amount. ■

5.8 THE *T ds* EQUATIONS

In the previous discussion of entropy, the entropy change corresponding to a change of state was evaluated by conceiving of an internally reversible process connecting the actual end states. This procedure can be used because entropy is a property, and as such, the change in entropy associated with a change of state depends only on the initial and final states and not on the path or process followed during the change of state. Any process between the desired end states could conceivably be chosen, but an internally reversible process is a logical choice, because the entropy change for such a process is related to the heat transfer through Equation 5.13.

If the conservation of energy is applied to a closed stationary system containing a simple compressible substance that undergoes an internally reversible process, the result on a unit-mass basis is

$$\delta q - \delta w = du$$

For these conditions the work per unit mass is given by Equation 1.31:

$$\delta w = P dv$$

and the heat transfer per unit mass is related to the entropy change through Equation 5.14:

$$\delta q = T\,ds$$

Substitution of these expressions for work and heat transfer into the conservation-of-energy equation and rearrangement yields an expression usually referred to as the first of the *T ds (or Gibbs) equations*:

$$T\,ds = du + P dv \tag{5.18}$$

Another useful relationship is obtained when the internal-energy change is eliminated by using the definition of the enthalpy:

$$du = d(h - Pv) = dh - P dv - v dP \tag{5.19}$$

If Equation 5.19 is substituted into Equation 5.18, then the second $T\,ds$ equation is obtained:

$$T\,ds = dh - v dP \tag{5.20}$$

The $T\,ds$ equations, Equations 5.18 and 5.20, can be used to evaluate the entropy change associated with the change of state of any simple compressible substance. Even though these equations were developed from an analysis of a closed system undergoing an internally reversible process, the $T\,ds$ equations are valid for both closed and open systems as well as reversible and irreversible processes. This conclusion follows from the fact that entropy is a property, and the change in entropy associated with a change of state is independent of the process that the system undergoes. Solving for ds in Equations 5.18 and 5.20 yields

$$ds = \frac{du}{T} + \frac{P dv}{T} \tag{5.21}$$

and

$$ds = \frac{dh}{T} - \frac{v dP}{T} \tag{5.22}$$

Thus the entropy change can be determined by integrating either of these two equations. To perform the integration, we must know two things: the relationship between temperature and internal energy or enthalpy as well as the *P-v-T* behavior of the substance. This information, developed in Chapter 2, will be used to evaluate the entropy change for ideal gases, incompressible substances, and pure substances such as water and refrigerant-12.

In conjunction with the discussion of entropy changes, a process of considerable importance for the second-law analysis of systems, the *internally reversible adiabatic process*, will also be discussed. According to Equation 5.13, the entropy change during such a process is zero, and the process is therefore referred to as an *isentropic* (constant-entropy) process. The isentropic process is often used as an ideal model for actual processes. Many devices or systems of interest to the engineer are very nearly adiabatic in their operation. Therefore, the internally reversible adiabatic (or isentropic) process is an appropriate model for use in comparing ideal performance with actual performance. At this point an important distinction concerning the isentropic process can be emphasized. While an internally reversible adiabatic process is isentropic, an isentropic process is not necessarily reversible and adiabatic, although the use of the term *isentropic* is often meant to imply internally reversible and adiabatic. On the basis of Equation 5.13, if any two of the three terms *internally reversible, adiabatic*, and *isentropic* apply, then the third must also apply. As is customary thermodynamic usage, however, the term *isentropic process* will be used in this text to imply an internally reversible, adiabatic process.

5.9 THE ENTROPY CHANGE FOR IDEAL GASES

5.9.1 Arbitrary Processes for Ideal Gases

An expression for the entropy change of an ideal gas undergoing any arbitrary process can be obtained from Equation 5.21 or 5.22 by employing the ideal-gas equation of state. By substituting $du = c_v\, dT$ from Equation 2.28 into Equation 5.21 and using the ideal-gas equation of state, we can express the differential entropy change of an ideal gas as

$$ds = c_v \frac{dT}{T} + R\frac{dv}{v} \tag{5.23}$$

Thus for a finite change of state the entropy change is

$$s_2 - s_1 = \int_1^2 c_v \frac{dT}{T} + R \ln \left(\frac{v_2}{v_1}\right) \tag{5.24}$$

The integral in Equation 5.24 can be evaluated only if the temperature dependence of the constant-volume specific heat is known. However, unlike internal-energy and enthalpy changes, the entropy change of an ideal gas is not a function of temperature alone.

If Equation 5.22 is used instead of Equation 5.21, while substituting $dh = c_p\, dT$ from Equation 2.31 and using the ideal-gas equation of state, the result is

$$ds = c_p \frac{dT}{T} - R\frac{dP}{P} \tag{5.25}$$

or

$$s_2 - s_1 = \int_1^2 c_p \frac{dT}{T} - R \ln\left(\frac{P_2}{P_1}\right) \tag{5.26}$$

Since the entropy change does not depend on temperature alone, the entropy cannot be tabulated as a function of temperature for the ideal gas, as was done for both the internal energy and the enthalpy. However, the temperature-dependent part of the entropy change can be tabulated since it can be represented by either of the integrals appearing in Equations 5.24 and 5.26. Most tables of thermodynamic properties of ideal gases include values that are related to the integral in Equation 5.26 and are defined in the following manner: Since the constant-pressure specific heat of an ideal gas depends on temperature alone, a function $s°$, which also depends on temperature alone, can be defined such that

$$s° \equiv \int_0^T c_p \frac{dT}{T} \tag{5.27}$$

In this definition a reference temperature of zero degrees absolute has been chosen so that the value of $s°$ is zero at an absolute temperature of zero. The quantity $s°$ has the same units as entropy and it is a function of temperature alone. Values for $s°$ are tabulated for many common gases in the tables in Appendix D. With this definition Equation 5.26 can be expressed as

$$s_2 - s_1 = \int_0^{T_2} c_p \frac{dT}{T} - \int_0^{T_1} c_p \frac{dT}{T} - R \ln\left(\frac{P_2}{P_1}\right)$$

or

$$s_2 - s_1 = s_2^\circ - s_1^\circ - R \ln\left(\frac{P_2}{P_1}\right) \tag{5.28}$$

Arbitrary Processes for Ideal Gases with Constant Specific Heats

As was discussed in Chapter 2, to assume that the specific heats of an ideal gas are constant during a process is often appropriate. For ideal monatomic gases, such as argon and helium, the specific heats are independent of the temperature. For most other common ideal gases the specific heats vary almost linearly with temperature if the temperature change during a process is not too large. In this instance the assumption of constant specific heats is appropriate provided that average values based on the average temperature for the change of state are used.

The entropy change of an ideal gas can be obtained from Equation 5.24 by using an average value of the constant-volume specific heat over the temperature interval corresponding to the change of state:

$$s_2 - s_1 = c_{v,\text{avg}} \ln\left(\frac{T_2}{T_1}\right) + R \ln\left(\frac{v_2}{v_1}\right) \tag{5.29}$$

By use of an average value of the constant-pressure specific heat in Equation 5.26, the approximate value of the entropy change for an ideal gas could also be expressed as

$$s_2 - s_1 = c_{p,\text{avg}} \ln\left(\frac{T_2}{T_1}\right) - R \ln\left(\frac{P_2}{P_1}\right) \tag{5.30}$$

The equations for the entropy change of an ideal gas given by either Equation 5.29 or Equation 5.30 are special cases of the more general expression given by Equation 5.28. Equation 5.28 was derived with the assumption that the specific heat of the ideal gas varies with temperature. If the temperature change during the process is greater than about 100 to 200 K (or about 200 to 400°R) for most ideal gases, or if the specific heats of the gas vary considerably within the temperature range of the process, the variation in specific heat is important when calculating the entropy change. In these situations the use of Equation 5.28 is preferable to the use of Equation 5.29 or 5.30 because the latter two equations do not account for the temperature dependence of the specific heat.

EXAMPLE 5.7

Carbon dioxide gas undergoes a change of state from an initial temperature and pressure of 45°C and 190 kPa to a final state of 80°C and 375 kPa. Determine the entropy change of the gas by using (a) the tables of properties for carbon dioxide and (b) average values for the specific heats of the gas.

Solution. Since the carbon dioxide is at low pressure and high temperature relative to its critical-state values, the ideal-gas assumption is appropriate. The entropy change can therefore be determined by using Equation 5.28 and interpolating in the CO_2 table, Table D.2:

$$\begin{aligned} s_2 - s_1 &= s_2^\circ - s_1^\circ - R \ln\left(\frac{P_2}{P_1}\right) \\ &= (5.0047 - 4.9114)\,\text{kJ/kg·K} \\ &\quad - \frac{(8.314\ \text{kJ/kg·mol·K}) \ln (375\ \text{kPa}/190\ \text{kPa})}{44\,\text{kg/kg·mol}} \\ &= \underline{\underline{-0.0352\ \text{kJ/kg·K}}} \end{aligned}$$

If an average value of the constant-pressure specific heat of the gas is used, an approximate value for the entropy change can be obtained by using Equation 5.30:

$$s_2 - s_1 = c_{p,\text{avg}} \ln\left(\frac{T_2}{T_1}\right) - R \ln\left(\frac{P_2}{P_1}\right)$$

With $c_{p,\text{avg}} = 0.883$ kJ/kg·K at 63°C from Table D.8, this expression gives

$$\begin{aligned} s_2 - s_1 &= (0.883\ \text{kJ/kg·K}) \ln\left(\frac{353\ \text{K}}{318\,\text{K}}\right) \\ &\quad - \left(\frac{8.314\ \text{kJ/kg·mol·K}}{44\ \text{kg/kg·mol}}\right) \ln\left(\frac{375\ \text{kPa}}{190\ \text{kPa}}\right) \\ &= \underline{\underline{-0.0363\,\text{kJ/kg·K}}} \end{aligned}$$

In this example the assumption of a constant specific heat for CO_2 results in an error of about 3 percent in the calculated entropy change. When larger temperature changes are involved, however, large errors will result because the specific heat of CO_2 varies considerably with temperature, more so than many of the common ideal gases. ■

The fact that the entropy change for this process is negative does not violate the second law. It does mean, however, that during the process heat transfer from the gas takes place since

$$s_2 - s_1 \geq \int_1^2 \frac{\delta q}{T}$$

and the absolute temperature is always positive.

5.9.2 Isentropic Processes for Ideal Gases

When an ideal gas undergoes an isentropic (or constant-entropy) process from the reference state of 0 K or 0°R and pressure P_0, then by Equation 5.28

$$0 = s^\circ - R \ln \left(\frac{P}{P_0} \right)$$

or

$$\ln \left(\frac{P}{P_0} \right)_s = \frac{s^\circ}{R}$$

where the subscript s indicates that the results are valid only for an isentropic process. The ratio of the pressure P at the end of an isentropic process to the pressure P_0 at the initial reference state is called the *relative pressure* P_r

$$P_r \equiv \left(\frac{P}{P_0} \right)_s \tag{5.31}$$

Therefore, for an isentropic process with an ideal gas,

$$P_r = \exp \left(\frac{s^\circ}{R} \right) \tag{5.32}$$

Equation 5.32 shows that since s° is solely a function of temperature, the relative pressure is also a function of temperature alone. Thus, values of P_r can be tabulated as is done in Appendix D. Notice that P_r is a ratio of pressures and is therefore a dimensionless quantity.

For an isentropic process between two states, states 1 and 2, Equation 5.28 can be expressed in the form

$$\left(\frac{P_2}{P_1} \right)_s = \exp \left(\frac{s_2^\circ - s_1^\circ}{R} \right) = \frac{\exp (s_2^\circ / R)}{\exp (s_1^\circ / R)} \tag{5.33}$$

or

$$\left(\frac{P_2}{P_1}\right)_s = \frac{P_{r2}}{P_{r1}} \tag{5.34}$$

The ratio (P_{r2}/P_{r1}) is called the *isentropic-pressure ratio*. The values of P_r calculated from Equation 5.32 are very large in magnitude, and for this reason values of P_r divided by a large constant are tabulated instead. This manipulation is performed merely for convenience, since the relative-pressure functions are always used to form ratios for isentropic-process calculations, as indicated in Equation 5.34.

A *relative specific volume*, denoted by v_r, is also defined for the isentropic process of an ideal gas. One suggested form for the definition of v_r is

$$v_r = \left(\frac{v}{v_0}\right)_s$$

where v_0 is the specific volume at the reference state of zero absolute temperature and pressure P_0. With the ideal-gas equation of state this expression would yield

$$v_r = \left(\frac{TP_0}{PT_0}\right)_s = \left(\frac{1}{P_r}\right)\left(\frac{T}{T_0}\right)_s$$

But since T_0 is 0 K or 0°R, this expression would become infinite. An alternative is to define the relative specific volume as

$$v_r \equiv \left(\frac{v}{v_{\text{ref}}}\right)_s \tag{5.35}$$

so that

$$v_r = \left(\frac{TP_{\text{ref}}}{PT_{\text{ref}}}\right)_s = \left(\frac{TP_{\text{ref}}P_0}{T_{\text{ref}}P_0P}\right)_s$$

or

$$v_r = \frac{CT}{P_r} \tag{5.36}$$

The choice of the constant C is arbitrary. The relative specific volume, like the relative pressure, is a dimensionless quantity. It is a function of temperature alone and is used only for isentropic processes involving ideal gases.

For an isentropic process between two states, states 1 and 2, the ratio of the temperatures and pressures for the process can be expressed as

$$\left(\frac{T_2P_1}{T_1P_2}\right)_s = \left(\frac{T_2P_{r1}}{T_1P_{r2}}\right)_s \tag{5.37}$$

or substituting Equation 5.36 and the ideal-gas equation of state,

$$\left(\frac{v_2}{v_1}\right)_s = \frac{v_{r2}}{v_{r1}} \tag{5.38}$$

Values for the relative specific volume are tabulated for air and other gases and can be found in Appendix D.

Isentropic Processes for Ideal Gases with Constant Specific Heats

In those instances when the specific heats are constant, or when the use of average values of the specific heats is appropriate, simple algebraic expressions relating the temperatures, pressures, and specific volumes at the end states of an isentropic process can be derived. From Equation 5.29 for an isentropic process of an ideal gas with constant specific heats,

$$\ln\left(\frac{v_2}{v_1}\right)_s = -\frac{c_v}{R}\ln\left(\frac{T_2}{T_1}\right)_s$$

or

$$\left(\frac{v_2}{v_1}\right)_s = \left(\frac{T_1}{T_2}\right)_s^{c_v/R}$$

and since $R = c_p - c_v$ and $k = c_p/c_v$, this equation can be expressed as

$$\left(\frac{v_2}{v_1}\right)_s = \left(\frac{T_1}{T_2}\right)_s^{1/(k-1)} \tag{5.39}$$

Similarly, from Equation 5.30 for an ideal gas undergoing an isentropic process,

$$\ln\left(\frac{P_2}{P_1}\right)_s = \frac{c_p}{R}\ln\left(\frac{T_2}{T_1}\right)_s$$

or

$$\left(\frac{P_2}{P_1}\right)_s = \left(\frac{T_2}{T_1}\right)_s^{c_p/R}$$

Therefore

$$\left(\frac{P_2}{P_1}\right)_s = \left(\frac{T_2}{T_1}\right)_s^{k/(k-1)} \tag{5.40}$$

Finally, a relationship between the pressures and specific volumes of the ideal gas at the end states of an isentropic process results when Equation 5.39 is substituted into Equation 5.40:

$$\left(\frac{P_2}{P_1}\right)_s = \left(\frac{v_1}{v_2}\right)_s^{k} \tag{5.41}$$

The equations for isentropic processes involving ideal gases presented in the preceding paragraphs were developed with regard to the temperature dependence of the specific heats of the gas. The most general equation, which accounts for the variation of specific heat with temperature, is Equation 5.33. The relative pressure and the relative specific volume were developed from this expression, and therefore they also account for the variation of specific heat with temperature. When the use of constant values for the specific heat of the gas is appropriate, Equations 5.39 through 5.41 are valid. In these equations k is the specific heat ratio.

■ EXAMPLE 5.8

Nitrogen at 750°R and 210 psia is expanded reversibly and adiabatically in a nozzle to an exit pressure of 105 psia as shown in Figure 5.14. Determine the temperature of the nitrogen at the exit of the nozzle.

Solution. Since the process is reversible and adiabatic, it is also isentropic. That is, the entropy change of the nitrogen is zero for this process.

The exit temperature for this isentropic process can be determined with the aid of Equation 5.33, since the nitrogen can be assumed ideal for the stated conditions:

$$\left(\frac{P_2}{P_1}\right)_s = \exp\left(\frac{s_2^\circ - s_1^\circ}{R}\right)$$

The s° functions depend on temperature alone, and solving for s_2° allows us to find the corresponding exit temperature from Table D.5E.

$$\begin{aligned} s_2^\circ &= s_1^\circ + R \ln\left(\frac{P_2}{P_1}\right) \\ &= 1.717\ \text{Btu/lb}_\text{m}\cdot{}^\circ\text{R} + \left(\frac{1.986\ \text{Btu/lb}_\text{m}\cdot\text{mol}\cdot{}^\circ\text{R}}{28.01\ \text{lb}_\text{m}/\text{lb}_\text{m}\cdot\text{mol}}\right) \ln\left(\frac{105\ \text{psia}}{210\ \text{psia}}\right) \\ &= 1.6679\ \text{Btu/lb}_\text{m}\cdot{}^\circ\text{R} \end{aligned}$$

From Table D.5E the exit temperature is found to be

$$T_2 = \underline{\underline{616^\circ\text{R}}}$$

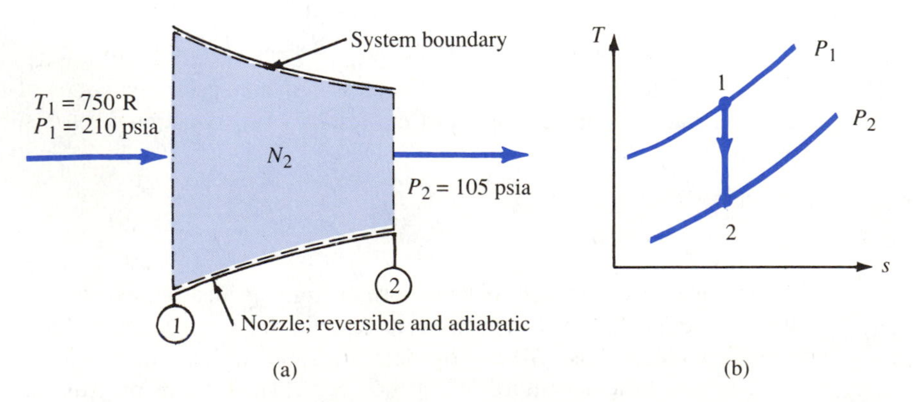

Figure 5.14 Sketch for Example 5.8.

The exit temperature could also be approximated by assuming appropriate average values for the specific heats of the nitrogen and using Equation 5.40:

$$\left(\frac{T_2}{T_1}\right)_s = \left(\frac{P_2}{P_1}\right)_s^{(k-1)/k}$$

An average value of k is obtained from Table D.8E, and the exit temperature is approximately

$$T_2 = (750°\text{R})\left(\frac{105 \text{ psia}}{210 \text{ psia}}\right)^{(1.4-1)/1.4} = \underline{\underline{615°\text{R}}}$$

Even though the temperature change in this example was more than 130°R, the constant-specific-heat assumption yields very accurate results. This result should be expected, since the specific heat of nitrogen varies almost linearly with temperature over this temperature range, and an average specific heat is therefore quite suitable. ■

■ EXAMPLE 5.9

Air is compressed in an isentropic process from an initial state of 300 K and 101 kPa to a final temperature of 870 K. Determine the pressure of the air at the final state.

Solution. At these temperatures and with low pressure, the assumption of ideal-gas behavior is appropriate. Since the process is isentropic, the final pressure can be determined by using Equation 5.34 and the relative pressure functions:

$$\left(\frac{P_2}{P_1}\right)_s = \frac{P_{r2}}{P_{r1}}$$

The relative pressures are functions of temperature alone and are tabulated in Table D.1:

$$P_{r2} = 65.74$$

$$P_{r1} = 1.380$$

Thus the exit pressure for the isentropic process is

$$P_2 = P_1\left(\frac{P_{r2}}{P_{r1}}\right) = (101 \text{ kPa})\left(\frac{65.74}{1.380}\right) = \underline{\underline{4.81 \text{ MPa}}}$$

In the absence of tabulated values for the relative pressures, this result could also be obtained by using Equation 5.33 for an isentropic process:

$$\left(\frac{P_2}{P_1}\right)_s = \exp\left(\frac{s_2^\circ - s_1^\circ}{R}\right)$$

$$P_2 = (101 \text{ kPa}) \exp\left[\frac{(6.8103 - 5.7016)\text{kJ/kg·K}}{(8.314 \text{ kJ/kg·mol·K})(1 \text{ kg·mol}/28.97 \text{ kg})}\right] = 4.81 \text{ MPa}$$

Instead of using the table values of P_r or $s°$, which account for the variation of the specific heats with temperature, we could assume constant specific heats for the air and find an approximate value for P_2 from Equation 5.40:

$$\left(\frac{P_2}{P_1}\right)_s = \left(\frac{T_2}{T_1}\right)_s^{k/(k-1)}$$

Using $k = 1.38$ from Table D.8 at the average air temperature, we obtain

$$P_2 = (101 \text{ kPa}) \left(\frac{870 \text{ K}}{300 \text{ K}}\right)^{1.38/(1.38-1)} = 4.83 \text{ MPa}$$

The constant-specific-heat assumption results in an error of less than 1 percent for this example. The specific heat of air changes gradually with temperature, and even though the temperature change of 570 K is quite large, the use of the constant-specific-heat assumption results in an accurate calculation for the final pressure. ■

5.10 THE ENTROPY CHANGE FOR INCOMPRESSIBLE SUBSTANCES

An incompressible substance is characterized by the fact that the specific volume of the substance remains constant. Thus the change in specific volume is zero for an incompressible substance during any process and Equation 5.21 reduces to

$$ds = \frac{du}{T} \tag{5.42}$$

The internal-energy change of an incompressible substance is given by Equation 2.47, and $c_p = c_v = c$ (Equation 2.49), so that

$$ds = c\,\frac{dT}{T}$$

and

$$s_2 - s_1 = \int_1^2 c\,\frac{dT}{T} \tag{5.43}$$

With an appropriate average value of the specific heat, this expression can be integrated to obtain

$$s_2 - s_1 = c_{\text{avg}} \ln\left(\frac{T_2}{T_1}\right) \tag{5.44}$$

The entropy change of an incompressible substance depends only on the temperature change during a change of state. Thus an isothermal process for a truly incompressible substance would also be an isentropic process.

EXAMPLE 5.10

A 0.5-kg block of copper initially at 80°C is cooled by immersion in an insulated tank containing 5 kg of liquid water at 23°C. The heat-transfer process continues until the copper and the water reach thermal equilibrium. Determine the entropy change for the copper and the entropy change for the water during this process.

Solution. To calculate the entropy change for the copper and for the water, we must first determine the final equilibrium temperature by applying the conservation of energy to the system, which is taken to be composed of the copper and the water (a closed system):

$$Q_{12} - W_{12} = (U_2 - U_1)_{\text{sys}}$$

The work and heat transfer for this system are both zero. Thus

$$(U_2 - U_1)_{\text{sys}} = (U_2 - U_1)_{\text{Cu}} + (U_2 - U_1)_{\text{H}_2\text{O}} = 0$$

Both substances can be assumed to be essentially incompressible, so Equation 2.47 applies for the internal-energy change:

$$m_{\text{Cu}}c_{\text{Cu}}(T_2 - T_1)_{\text{Cu}} + m_{\text{H}_2\text{O}}c_{\text{H}_2\text{O}}(T_2 - T_1)_{\text{H}_2\text{O}} = 0$$

and solving for the equilibrium temperature T_2 yields

$$T_2 = \frac{m_{\text{Cu}}c_{\text{Cu}}T_{1,\text{Cu}} + m_{H_2\text{O}}c_{H_2\text{O}}T_{1,\text{H}_2\text{O}}}{m_{\text{Cu}}c_{\text{Cu}} + m_{H_2\text{O}}c_{H_2\text{O}}}$$

Using estimated average values for the specific heats of copper and water from Table H.3, we have

$$T_2 = \frac{(0.5\text{ kg})(0.39\text{ kJ/kg·K})(353\text{ K}) + (5.0\text{ kg})(4.18\text{ kJ/kg·K})(296\text{ K})}{(0.5\text{ kg})(0.39\text{ kJ/kg·K}) + (5.0\text{ kg})(4.18\text{ kJ/kg·K})} = 296.5\text{ K}$$

The entropy change of each substance can be calculated by using Equation 5.44:

$$(S_2 - S_1)_{\text{Cu}} = m_{\text{Cu}}(s_2 - s_1)_{\text{Cu}} = m_{\text{Cu}}c_{\text{Cu}} \ln\left(\frac{T_2}{T_{1,\text{Cu}}}\right)$$

$$= (0.5\text{ kg})(0.39\text{ kJ/kg·K}) \ln\left(\frac{296.5\text{ K}}{353\text{ K}}\right)$$

$$= \underline{\underline{-0.0340\text{ kJ/K}}}$$

$$(S_2 - S_1)_{\text{H}_2\text{O}} = m_{H_2\text{O}}(s_2 - s_1)_{\text{H}_2\text{O}} = m_{\text{H}_2\text{O}}c_{\text{H}_2\text{O}} \ln\left(\frac{T_2}{T_{1,\text{H}_2\text{O}}}\right)$$

$$= (5.0\text{ kg})(4.18\text{ kJ/kg·K}) \ln\left(\frac{296.5\text{ K}}{296\text{ K}}\right)$$

$$= \underline{\underline{0.0353\text{ kJ/K}}}$$

The entropy change of the copper is less than zero and the entropy change of the water is greater than zero because heat transfer occurs from the copper and to the water. ■

5.11 THE ENTROPY CHANGE FOR PURE SUBSTANCES

The entropy change for pure substances such as water and refrigerant-12 is determined in much the same manner as is done for the specific volume, internal energy, and enthalpy. Specifically, in the saturation region the entropy is related to the quality by

$$s = s_f + x s_{fg} \tag{5.45}$$

and in the compressed-liquid and superheated-vapor regions, the entropy can be obtained from the tables if two independent properties associated with the state of the substance are specified.

In the absence of adequate compressed-liquid data, the entropy of the compressed liquid can usually be approximated quite well by using the value of the entropy of the saturated liquid at the same temperature as that of the compressed-liquid state.

EXAMPLE 5.11

Steam enters a reversible, adiabatic turbine at 4 MPa and 520°C with a velocity of 60 m/s. The steam exhausts from the turbine at a pressure of 75 kPa with a velocity of 140 m/s. Determine the work output of the turbine per unit mass of steam flowing through the turbine.

Solution. A sketch of the turbine and a *T-s* diagram for the process are shown in Figure 5.15. The work for the steady-flow turbine can be determined from the conservation of energy, Equation 4.31:

$$\dot{Q} - \dot{W} + \dot{m}\left[(h_i - h_e) + \left(\frac{V_i^2 - V_e^2}{2}\right) + g(z_i - z_e)\right] = 0$$

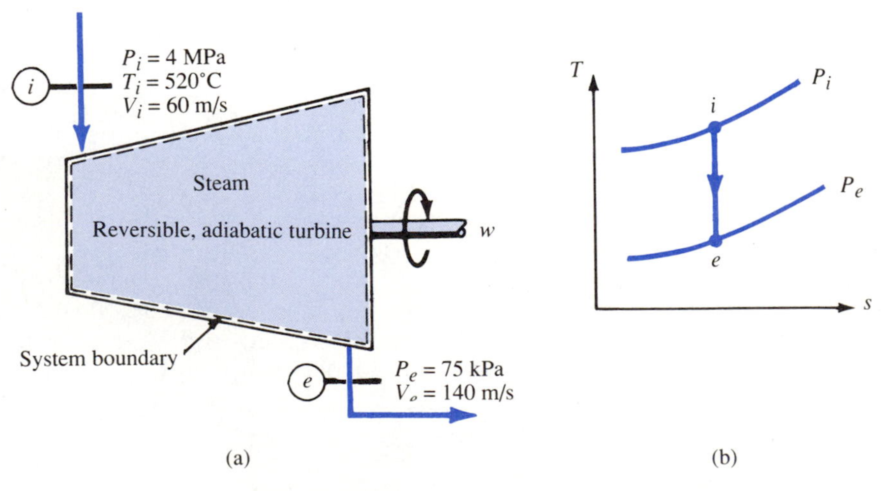

Figure 5.15 Sketch for Example 5.11.

The mass flow rates at the inlet and exit must be the same since the turbine operates steadily. The heat transfer is zero since the turbine is adiabatic, and the potential-energy change is likely to be negligible. Thus the work output per unit mass of steam is given by

$$w = (h_i - h_e) + \left(\frac{V_i^2 - V_e^2}{2}\right)$$

The enthalpy of the steam at the inlet can be determined by interpolation from Table B.3 at 4 MPa and 520°C:

$$h_i = 3491.1 \text{ kJ/kg}$$

However, at the exit state only the pressure is known. Another independent property at this state can be determined by noting that the process is isentropic because the steam undergoes a reversible adiabatic expansion in the turbine. Thus

$$s_e = s_i = 7.1475 \text{ kJ/kg·K}$$

from Table B.3.

The quality corresponding to the exit conditions, $P_e = 75$ kPa and $s_e = 7.1475$ kJ/kg·K, is obtained by interpolation in Table B.2:

$$x_e = \frac{s_e - s_f}{s_{fg}} = \frac{(7.1475 - 1.2131)\text{kJ/kg·K}}{6.2425 \text{ kJ/kg·K}} = 0.951$$

and therefore the exit enthalpy is

$$h_e = h_f + x_e h_{fg} = 384.43 \text{ kJ/kg} + 0.951(2278.1)\text{kJ/kg}$$
$$= 2551 \text{ kJ/kg}$$

Finally, the work output of the turbine is

$$w = (3491.1 - 2551)\text{kJ/kg} + \frac{(60^2 - 140^2)\text{m}^2/\text{s}^2}{2(1 \text{ kg·m/N·s}^2)(10^3 \text{ N·m/kJ})}$$
$$= \underline{\underline{+932 \text{ kJ/kg}}}$$

■

5.12 THE INCREASE-IN-ENTROPY PRINCIPLE

The isolated system has received little attention in the discussion of thermodynamic principles thus far. The real utility of the concept of the isolated system cannot be fully appreciated until the second law of thermodynamics is discussed. Since neither energy nor mass can cross the boundary of an isolated system, an isolated system can be considered to be the composite of any open or closed system and everything external to the open or closed system. In other words, any system plus its surroundings constitutes an isolated system. The entropy change of the isolated system therefore includes the net entropy change of the closed or open system plus the net entropy change of the surroundings. This entropy change is called the *total entropy change* associated with the process or series of processes that the closed or open system undergoes. That is,

$$dS_{\text{tot}} = dS_{\text{net,sys}} + dS_{\text{net,surr}} = dS_{\text{isolated}} \tag{5.46}$$

The second law of thermodynamics places a restriction on the total entropy change, as can be seen by applying Equation 5.17 to an isolated system:

$$dS_{\text{tot}} = dS_{\text{isolated}} \geq \left(\frac{\delta Q}{T}\right)_{\text{isolated}}$$

Since heat transfer cannot occur at the boundary of an isolated system, the total entropy change of an isolated system is restricted in the following manner:

$$dS_{\text{tot}} \geq 0 \tag{5.47}$$

Equation 5.47 is a mathematical statement that the total entropy change, that is, the net entropy change of the system plus the net entropy change of the surroundings associated with any process, must be greater than or equal to zero. The equality in Equation 5.47 holds only if all processes within the isolated system are reversible so that the isolated system would be internally reversible.

Equation 5.47 is called the *increase-in-entropy principle*. This equation does not imply that the entropy change for all processes is positive or zero. The entropy can decrease during a process, but according to Equation 5.47, the sum of the entropy change of the system and the entropy change of its surroundings cannot be negative.

■ EXAMPLE 5.12

A block of copper with a mass of 3.3 lb_m is initially at 1260°R. It is allowed to cool by means of heat transfer to the surrounding air at 540°R. Determine the change in entropy of the copper and the total change in entropy of the copper and surrounding air after the copper reaches thermal equilibrium. Assume that the specific heat of copper is constant at 0.093 Btu/lb_m·°R.

Solution. The copper is essentially incompressible, and its mass remains constant during the process. The conservation-of-energy equation for the copper, a closed, stationary system, reduces to

$$\delta Q - \delta W = dU$$

No work is associated with the process; therefore

$$\delta Q = dU = m\, du = mc\, dT$$

where we have used the fact that the internal-energy change of an incompressible substance depends only on the temperature change. The surrounding air represents a thermal-energy reservoir, so that the equilibrium temperature of the copper will be 540°R, and the heat transfer is

$$Q_{12,\text{Cu}} = \int_1^2 mc\, dT = mc \int_1^2 dT$$

$$= mc(T_2 - T_1) = (3.3\ \text{lb}_\text{m})(0.093\ \text{Btu/lb}_\text{m}\cdot°\text{R})(540 - 1260)°\text{R}$$

$$= -221\ \text{Btu}$$

The entropy change of the copper can be calculated from Equation 5.44:

$$(S_2 - S_1)_{\text{Cu}} = mc \ln\left(\frac{T_2}{T_1}\right)$$

$$= (3.3\ \text{lb}_\text{m})(0.093\ \text{Btu/lb}_\text{m}\cdot°\text{R}) \ln(540°\text{R}/1260°\text{R})$$

$$= \underline{\underline{-0.26\ \text{Btu}/°\text{R}}}$$

The entropy of the copper decreases during the process because the direction of the heat transfer is from the copper.

To evaluate the total entropy change, we must first determine the entropy change of the surrounding air. The air is a thermal-energy reservoir, so the air temperature remains essentially constant, and the heat-transfer process for the reservoir is internally reversible. Thus from Equation 5.13

$$(S_2 - S_1)_{\text{air}} = \int_1^2 \left(\frac{\delta Q}{T}\right)_{\text{int rev}} = \frac{Q_{12,\text{air}}}{T_{\text{air}}}$$

The heat transfer to the air is equal in magnitude but opposite in direction to the heat transfer from the copper:

$$(S_2 - S_1)_{\text{air}} = -\frac{Q_{12,\text{Cu}}}{T_{\text{air}}} = -\frac{(-221\ \text{Btu})}{(540°\text{R})} = 0.409\ \text{Btu}/°\text{R}$$

The total entropy change is, then,

$$(S_2 - S_1)_{\text{tot}} = (S_2 - S_1)_{\text{Cu}} + (S_2 - S_1)_{\text{air}}$$

$$= -0.26\ \text{Btu}/°\text{R} + 0.409\ \text{Btu}/°\text{R} = \underline{\underline{0.149\ \text{Btu}/°\text{R}}}$$

The entropy of the copper decreases during the process, and the entropy of the surroundings increases. The change in total entropy is positive, in accordance with the increase-in-entropy principle, indicating that the process is irreversible. The irreversibility is a result of heat transfer through a finite temperature difference. ■

■ EXAMPLE 5.13

Using the increase-in-entropy principle, show that the direction of heat transfer must be from a higher-temperature body to a body at a lower temperature.

Solution. Consider two bodies, one at temperature T_A and the other at temperature T_B, as shown in Figure 5.16. Suppose that the only interaction that occurs between the two

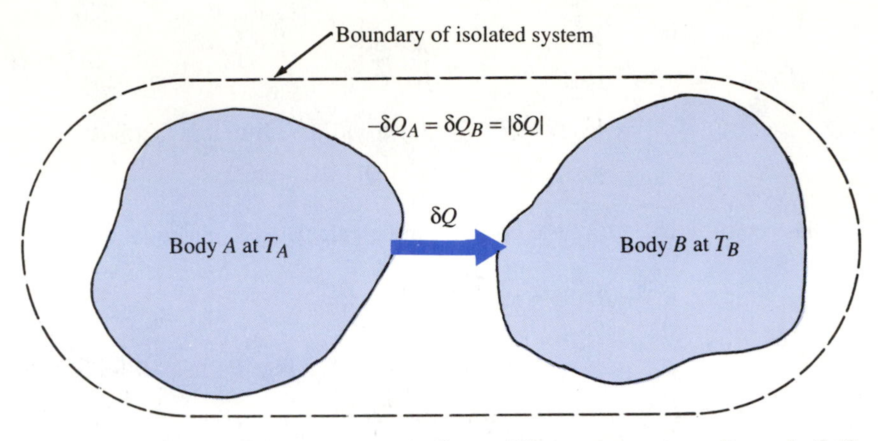

Figure 5.16 Heat transfer between two bodies at different temperatures; Example 5.13.

bodies is heat transfer, so that the combination of the two bodies forms an isolated system. The total entropy change associated with the process is the sum of the entropy change of body A and the entropy change of body B:

$$dS_{\text{tot}} = dS_A + dS_B$$

Heat transfer alone to or from a body is considered to be an internally reversible process, and therefore the entropy change for both A and B can be obtained from Equation 5.13:

$$dS_{\text{tot}} = \left(\frac{\delta Q}{T}\right)_{A,\text{int rev}} + \left(\frac{\delta Q}{T}\right)_{B,\text{int rev}}$$

$$= \frac{\delta Q_A}{T_A} + \frac{\delta Q_B}{T_B}$$

As indicated in Figure 5.16, heat transfer occurs from body A and to body B; therefore

$$|\delta Q_A| = -|\delta Q|$$

and

$$|\delta Q_B| = |\delta Q|$$

and therefore

$$dS_{\text{tot}} = |\delta Q|\left(\frac{1}{T_B} - \frac{1}{T_A}\right)$$

From Equation 5.47, the increase-in-entropy principle requires that the total entropy change be greater than or equal to zero. From this and the preceding equation we conclude that

$$T_A \geq T_B$$

Thus if heat transfer is to take place from body A to body B, then T_A must be greater than or equal to T_B. If $T_A > T_B$, a finite temperature difference exists between A and B and the overall process is irreversible ($dS_{tot} > 0$), even though the individual processes are internally reversible. For a fixed value of T_A, as the temperature difference between the two bodies becomes larger, the heat-transfer process becomes more irreversible. This result is reflected in the fact that the total entropy change also increases.

If $T_A = T_B$, heat transfer does not occur since there is no difference in temperature between A and B. If T_A is only infinitesimally higher than T_B, $T_A = T_B + dT$, then the total entropy change approaches zero (a reversible process). This result is precisely what is required for the idealization of the totally reversible heat-transfer process, which was used, for example, in connection with the totally reversible heat-engine cycle. ■

The total entropy change associated with a process is zero only if the process is totally reversible. As irreversibilities of any kind are introduced into the process, the total entropy change is increased to a positive value. The more irreversible the process is, the larger will be the total entropy change. This result was illustrated in the preceding example, in which heat transfer between two bodies was irreversible owing to the finite temperature difference between them. The total entropy change, therefore, can be used as a direct measure of the extent of the irreversibility of a process.

5.13 THE CARNOT CYCLE

The Carnot cycle was introduced briefly in Section 5.6 in the discussion of the Carnot principle and heat engines. In this section the Carnot cycle is examined in more detail.

A totally reversible cycle such as the Carnot cycle cannot be achieved in practice, because irreversibility accompanies the motion of all fluids and mechanical components. Furthermore, heat transfer to or from a system cannot take place reversibly since a finite temperature difference is necessary to have a finite amount of heat transfer. However, the study of reversible cycles is instructive because these cycles provide upper limits on the performance of real cycles. The performance of actual heat engines and refrigerators can best be evaluated by comparison with the performance of their reversible counterparts. In addition, improvements in the actual cycle are often deliberate attempts to cause the actual cycles to more nearly approximate the reversible cycle.

In order for a cycle to be totally reversible, each of the individual processes that make up the cycle must be internally reversible and all heat-transfer interactions with the surroundings must occur in a reversible manner. The Carnot cycle operates between two constant-temperature reservoirs and is composed of the following four reversible processes:

1-2: a reversible isothermal expansion during which heat transfer occurs from the high-temperature reservoir to the working fluid

2-3: a reversible adiabatic expansion that continues until the working fluid reaches the temperature of the low-temperature reservoir

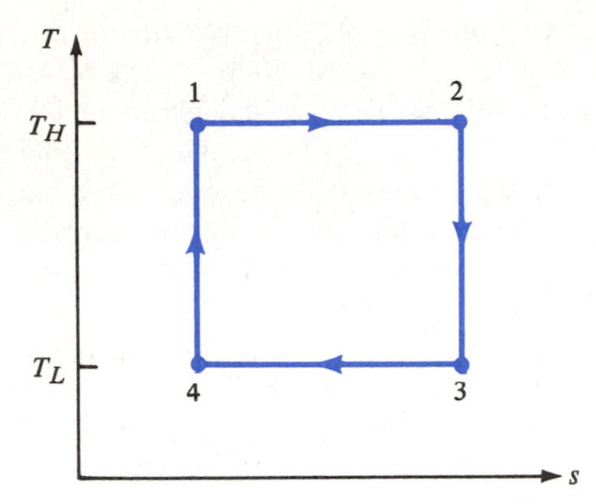

Figure 5.17 Temperature-entropy diagram for a Carnot cycle.

3-4: a reversible isothermal compression during which heat transfer occurs from the working fluid to the low-temperature reservoir

4-1: a reversible adiabatic compression that continues until the working fluid reaches the temperature of the high-temperature reservoir

These processes are illustrated with a temperature-entropy diagram in Figure 5.17. Notice that since processes 2-3 and 4-1 are both reversible and adiabatic, they are isentropic processes and appear as vertical lines on the *T-s* diagram. The Carnot cycle has a rectangular shape on a *T-s* diagram regardless of the working fluid. However, if the cycle is represented on a *P-v* diagram, it can have many different shapes depending on the working fluid and the state of the working fluid in the various parts of the cycle. Two possibilities are illustrated in Figure 5.18(a) for an ideal gas and Figure 5.18(b) for a fluid that remains in the saturation region throughout the cycle.

Since the processes that make up the Carnot cycle are reversible, the entropy change associated with each process is given by Equation 5.14:

$$ds = \left(\frac{\delta q}{T}\right)_{\text{int rev}}$$

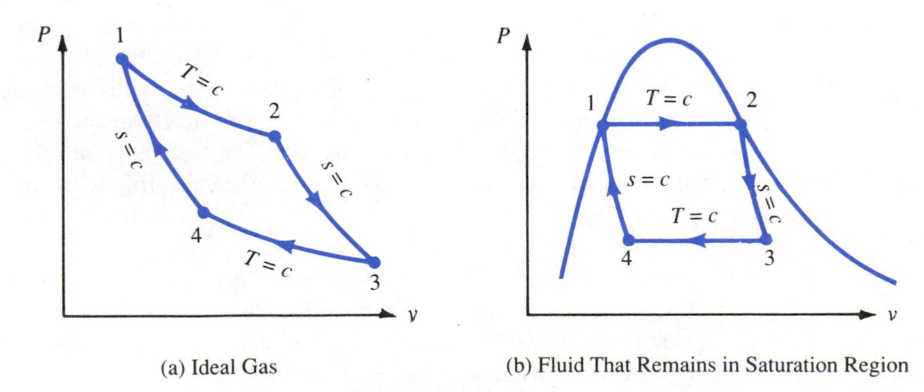

Figure 5.18 Typical *P-v* diagrams for Carnot cycles.

or

$$\delta q_{\text{int rev}} = T\,ds$$

and

$$q_{\text{if,int rev}} = \int_i^f T\,ds$$

The physical interpretation of this equation is that the heat transfer for each internally reversible process is represented by the area under the process curve on the T-s diagram. Therefore, for process 1-2 in Figure 5.17,

$$q_{H,\text{int rev}} = T_H(s_2 - s_1)$$

and for process 3-4

$$q_{L,\text{int rev}} = T_L(s_4 - s_3)$$

and since

$$(s_4 - s_3) = -(s_2 - s_1)$$

the ratio of the heat-transfer quantities can be written as

$$\left(\frac{q_H}{q_L}\right)_{\text{int rev}} = -\frac{T_H}{T_L}$$

The thermal efficiency of the Carnot heat engine is therefore given by

$$\eta_{\text{th,Carnot}} = 1 - \frac{T_L}{T_H}$$

These are the same results (see Equations 5.5 and 5.6) obtained in previous discussions of reversible heat engines. The area enclosed by the cycle on the T-s diagram represents the net heat transfer for the cycle, since the cycle is internally reversible. The net heat transfer is also equal to the net work of the cycle according to the conservation of energy.

For closed systems the area under the P-v diagram represents the $P\,dv$ work done by the system during an internally reversible process. Notice, therefore, that a closed system executing a Carnot cycle has work output during the expansion processes 1-2 and 2-3 (see Figure 5.18) and work input during the compression processes 3-4 and 4-1.

For open, steady systems $P\,dv$ work is not present, since the boundary of the system cannot expand or contract. Therefore, the areas under the process curves on the P-v diagram do not represent the work done for the open, steady system. However, in these systems work is done by the system during the reversible adiabatic expansion (process 2-3), and work is done on the system during the reversible adiabatic compression (process 4-1).

If the direction of each process in the Carnot cycle is reversed, the cycle that results is called a *reversed Carnot cycle* or a *Carnot refrigerator* cycle or a *Carnot heat-pump* cycle. The cycles for the Carnot heat engine and the Carnot refrigerator are used frequently for comparison with actual heat-engine and refrigerator cycles.

EXAMPLE 5.14

Suppose that a closed, frictionless piston-cylinder that contains water executes a Carnot cycle. The water is initially at 250°C and has a quality of 80 percent. The water is expanded isothermally until its pressure reaches 2 MPa. This process is followed by an adiabatic expansion to a temperature of 175°C. Determine the thermal efficiency of the cycle, the heat transfer during the isothermal expansion, the work associated with the adiabatic expansion, the heat transfer during the isothermal compression, and the net work of the cycle.

Solution. A sketch of the system is shown in Figure 5.19(a). The T-s and P-v diagrams for this cycle are illustrated in Figure 5.19(b) and (c).

Since the cycle is totally reversible, the thermal efficiency depends only on the temperature limits. From Equation 5.6

$$\eta_{\text{th,rev}} = 1 - \frac{T_L}{T_H} = 1 - \frac{448\text{ K}}{523\text{ K}} = \underline{\underline{0.143}}$$

The properties at the four states of the cycle are summarized in the following table:

State	T, °C	P, MPa	u, kJ/kg	s, kJ/kg·K	Note
1	250	3.974	2297.4	5.4159	$x_1 = 0.8$
2	250	2.0	2678.8	6.5438	$T_2 = T_1$
3	175	0.892	2546.9	6.5438	$s_3 = s_2$
4	175	0.892	2089.2	5.4159	$s_4 = s_1$

Each process is reversible; therefore, the heat transfer can be determined from Equation 5.14:

$$\delta q_{\text{int rev}} = T\,ds$$

For the isothermal expansion (1-2),

$$q_{12} = \int_1^2 T\,ds = T_1(s_2 - s_1)$$

$$= (523\text{ K})(6.5438 - 5.4159)\text{kJ/kg·K} = \underline{\underline{589.9\text{ kJ/kg}}}$$

Similarly, for the isothermal compression (3-4),

$$q_{34} = T_3(s_4 - s_3)$$

$$= (448\text{ K})(5.4159 - 6.5438)\text{kJ/kg·K} = \underline{\underline{-505.3\text{ kJ/kg}}}$$

The heat transfer during each of the adiabatic processes is zero:

$$q_{23} = 0 \qquad q_{41} = 0$$

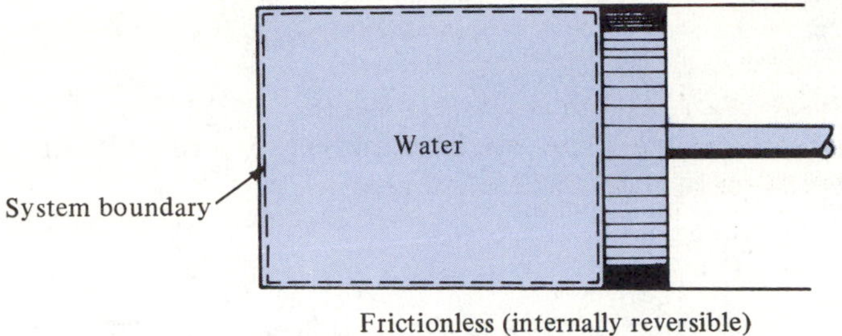

(a) Sketch for Example 5-14

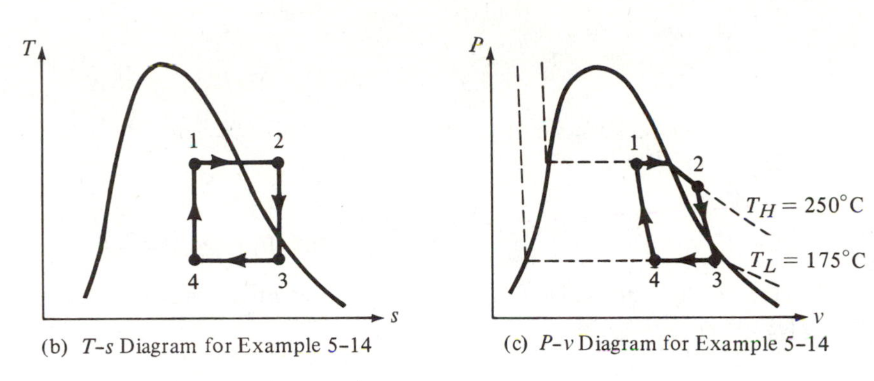

(b) T-s Diagram for Example 5-14 (c) P-v Diagram for Example 5-14

Figure 5.19 A Carnot cycle example.

The work for each process can be determined from the conservation-of-energy equation for a closed system:

$$q_{if} - w_{if} = (u_f - u_i)$$

Thus

$$\begin{aligned} w_{12} &= q_{12} - (u_2 - u_1) \\ &= 589.9\,\text{kJ/kg} - (2678.8 - 2297.4)\text{kJ/kg} = 208.5\ \text{kJ/kg} \end{aligned}$$

For process 2-3,

$$\begin{aligned} w_{23} &= q_{23} - (u_3 - u_2) \\ &= 0 - (2546.9 - 2678.8)\text{kJ/kg} = \underline{\underline{131.9\ \text{kJ/kg}}} \end{aligned}$$

For process 3-4,

$$\begin{aligned} w_{34} &= q_{34} - (u_4 - u_3) \\ &= -505.3\,\text{kJ/kg} - (2089.2 - 2546.9)\text{kJ/kg} = -47.6\ \text{kJ/kg} \end{aligned}$$

and for process 4-1,

$$w_{41} = q_{41} - (u_1 - u_4)$$
$$= 0 - (2297.4 - 2089.2)\text{kJ/kg} = -208.2 \text{ kJ/kg}$$

The net work of the cycle is therefore

$$w_{net} = w_{12} + w_{23} + w_{34} + w_{41}$$
$$= (208.5 + 131.9 - 47.6 - 208.2)\text{kJ/kg} = \underline{\underline{84.6\,\text{kJ/kg}}}$$

We can check this result since for a cycle the net heat transfer is equal to the net work:

$$q_{net} = q_{12} + q_{23} + q_{34} + q_{41}$$
$$= (589.9 + 0 - 505.3 + 0)\text{kJ/kg} = 84.6\,\text{kJ/kg}$$

The thermal efficiency of the cycle could also be determined from the definition of the thermal efficiency:

$$\eta_{th} = \frac{w_{net}}{q_{in}} = \frac{w_{net}}{q_{12}} = \frac{84.6 \text{ kJ/kg}}{589.9 \text{ kJ/kg}} = \underline{\underline{0.143}}$$

This result is the same as that obtained using Equation 5.6, since the cycle is totally reversible. ■

5.14 SUMMARY

The second law of thermodynamics, through the property called entropy, permits a qualitative evaluation of thermodynamic systems and predicts the directional nature of physical processes. It provides information that complements an analysis based on the conservation of mass and energy.

The development and explanation of the second law of thermodynamics rely heavily on the concept of the reversible process, which may be either totally reversible or internally reversible. During an internally reversible process, a system proceeds through a series of equilibrium states, and if the direction is reversed at any point in the process, then the system can be returned to its initial equilibrium state without leaving any permanent change in the system. For a process to be totally reversible it must be internally reversible, and the interaction between the system and its surroundings must also be reversible. Common factors present in practically all processes that render a process irreversible are heat transfer across a finite temperature difference and friction.

The second law of thermodynamcis can be expressed by either of two equivalent statements that have evolved from observations of practical experiences. The Clausius statement of the second law places restrictions on the operation of heat pumps and refrigerators. Practically speaking, the Clausius statement says that if cyclic heat pumps or refrigerators are to be used to provide heat transfer from a low-temperature body to a high-temperature body, then the cyclic devices must be powered by an external energy supply. The Kelvin-Planck statement of the second law places restrictions on the operation of heat engines. It states that a cyclic heat engine must have some heat transfer from the heat engine to a low-temperature reservoir. This statement is tantamount to restricting the efficiency of a heat engine to less than 100 percent.

The Carnot cycle is an example of a totally reversible cycle that consists of two reversible isothermal heat-exchange processes between two energy reservoirs, a reversible adiabatic compression and a reversible adiabatic expansion. Corollaries of the second law stipulate that the efficiency of a Carnot cycle (or any other cycle based totally on reversible processes) is greater than the efficiency of any irreversible cycle operating between the same two thermal-energy reservoirs. Furthermore, the thermal efficiencies of all reversible cycles operating between the same thermal-energy reservoirs are indentical. These corollaries lead to a definition of the absolute thermodynamic temperature scale:

$$\left(\frac{Q_L}{Q_H}\right)_{\text{rev}} = -\frac{T_L}{T_H} \tag{5.5}$$

With this definition the thermal efficiency of all reversible heat engines operating between thermal-energy reservoirs that have temperatures of T_H and T_L is

$$\eta_{\text{th,rev}} = 1 - \frac{T_L}{T_H} \tag{5.6}$$

Similarly, the coefficients of performance of all reversible heat pumps and refrigerators are

$$\beta_{H,\text{rev}} = \frac{1}{1 - (T_L/T_H)} \tag{5.8}$$

and

$$\beta_{R,\text{rev}} = \frac{1}{(T_H/T_L) - 1} \tag{5.10}$$

The Clausius inequality is an important stepping-stone to the definition of entropy. It states that

$$\oint \frac{\delta Q}{T} \leq 0 \tag{5.11}$$

where the inequality applies to all irreversible processes and the equality sign applies to all internally reversible processes. The Clausius inequality was used to define a thermodynamic property, called the entropy, as

$$dS \equiv \left(\frac{\delta Q}{T}\right)_{\text{int rev}} \tag{5.13}$$

As a result of the definition of the entropy in Equation 5.13, the fact that

$$dS \geq \frac{\delta Q}{T} \tag{5.17}$$

was established, where the inequality applies to all irreversible processes and the equality applies to all internally reversible processes.

Even though Equation 5.13 is restricted to internally reversible processes, we know that the entropy is a property and its change from one state to another is independent of the path connecting the initial and final states. That fact was exploited to develop two expressions, the $T\,ds$ equations, that are written entirely in terms of thermodynamic properties and that can be used to calculate the entropy change of simple compressible substances.

$$T\,ds = du + P\,dv \tag{5.18}$$

and

$$T\,ds = dh - v\,dP \tag{5.20}$$

The $T\,ds$ equations can be used to determine the entropy change once the specific heat as a function of temperature and the equation of state of the substance are known. For an ideal gas with variable specific heats, the entropy change becomes

$$s_2 - s_1 = s_2^\circ - s_1^\circ - R \ln\left(\frac{P_2}{P_1}\right) \tag{5.28}$$

where the quantity s° is solely a function of temperature. If the ideal gas is assumed to have constant specific heats, the entropy change can be written as

$$s_2 - s_1 = c_{v,\text{avg}} \ln\left(\frac{T_2}{T_1}\right) + R \ln\left(\frac{v_2}{v_1}\right) \tag{5.29}$$

or

$$s_2 - s_1 = c_{p,\text{avg}} \ln\left(\frac{T_2}{T_1}\right) - R \ln\left(\frac{P_2}{P_1}\right) \tag{5.30}$$

For the special case of an ideal gas with variable specific heats undergoing an isentropic (constant-entropy) process, the initial and final states are related by the relative pressures (P_r):

$$\left(\frac{P_2}{P_1}\right)_s = \frac{P_{r2}}{P_{r1}} \tag{5.34}$$

The states are also related by the relative specific volume (v_r);

$$\left(\frac{v_2}{v_1}\right)_s = \frac{v_{r2}}{v_{r1}} \tag{5.38}$$

If an ideal gas with constant specific heats proceeds along an isentropic path, the properties are related by

$$\left(\frac{P_2}{P_1}\right)_s = \left(\frac{T_2}{T_1}\right)_s^{k/(k-1)} = \left(\frac{v_1}{v_2}\right)_s^k \qquad (5.40)\ (5.41)$$

The entropy change of an incompressible substance is

$$s_2 - s_1 = c_{avg} \ln\left(\frac{T_2}{T_1}\right) \tag{5.44}$$

The chapter concluded with the increase-in-entropy principle, which states that

$$dS_{tot} \geq 0 \tag{5.47}$$

where dS_{tot} is the sum of the net entropy change of the system and the net entropy change of the surroundings. The equality sign applies to all reversible processes and the inequality applies to all irreversible processes. The increase-in-entropy principle provides a means for analyzing individual processes on the basis of the second law. Those processes that result in a total entropy change that is positive are possible according to the second law. Those processes that suggest a decrease in the total entropy are impossible, and only totally reversible processes could produce no change in the total entropy. The implications of these observations are investigated and discussed in further detail in the following chapter.

PROBLEMS

5.1 Explain the difference between a totally reversible process and an internally reversible process. Give an example of each.

5.2 Comment on the following observation: An irreversible process always leaves a history that it has occurred, while a reversible process leaves no history of its occurrence.

5.3 List the factors, such as heat transfer across a finite temperature difference, that can render a process irreversible.

5.4 Using a kitchen refrigerator as an example of a cyclic device, shown that it does not violate the Clausius statement of the second law.

5.5 A heat pump maintains a heated space at 70°F on a day when the outdoor air temperature is −10°F. The heating requirements of the heat pump are 100,000 Btu/h, and the power input to the heat pump is 10 hp. Calculate the heat-transfer rate from the outside air, and determine the coefficient of performance of the heat pump.

5.6 Prove that a violation of the Clausius statement of the second law leads to a violation of the Kelvin-Planck statement.

5.7 Using an automobile engine as an example of a cyclic heat engine, show that it does not violate the Kelvin-Planck statement of the second law.

5.8 [SLA111] A heat pump is to be used to heat a house in the winter. The interior temperature is to be maintained at 20°C. Heat transfer through the walls and roof is estimated to be 0.6 kW per °C temperature difference between the interior and outdoors. Determine the following:

(a) the minimum power in kW required to drive the heat pump if the outdoor temperature is 4°C

(b) the coefficient of performance for the conditions of part a

5.9 A heat pump is used to maintain the temperature of a heated space at 25°C. The heat-transfer rate from the heat pump to the heated space must be 20 kW when the outdoor air temperature is 3°C. The coefficient of performance of the heat pump is 5.4. Determine the power required to operate the heat pump and the rate of change of the total entropy. If the heat pump could be replaced by a totally reversible heat pump, what would be the power required and the rate of change of the total entropy?

5.10 Two heat engines, one reversible and the other irreversible, operate between the same two energy reservoirs. The heat transfer to the engine from the high-temperature reservoir is the same for both engines. Prove the following statements:

(a) The reversible engine produces more work than the irreversible engine.

(b) The heat transfer to the low-temperature reservoir from the reversible engine is larger in magnitude than the heat transfer for the reversible engine.

5.11 A heat pump is used to maintain a heated space at 70°F. Heat transfer is to the heat pump from outdoor air on a day when the air temperature is 32°F. Determine the maximum coefficient of performance that the heat pump can achieve under these circumstances.

5.12 Plot the coefficient of performance of a Carnot heat pump for a fixed-high-temperature reservoir of 25°C as the low-temperature reservoir ranges between −20 and +20°C. Does the decrease in the coefficient of performance at low temperatures suggest that the heat pump may have to be supplemented by resistance heating during periods of unusually cold weather? Why is the use of heat pumps limited to those geographical areas where winter temperatures are relatively moderate?

5.13 What is the maximum thermal efficiency of a heat engine that operates between temperature reservoirs of 2000 and 0°F?

5.14 A refrigerator maintains a cooled space at 2°C when the ambient air around the refrigerator is 25°C. The refrigerator has a coefficient of performance of 2.5. The rate of cooling in the refrigerated space is 8000 kJ/h. Determine the power consumption of the refrigerator, the heat-transfer rate to the surrounding air, and the maximum coefficient of performance of any cyclic device operating between these two reservoirs. Why is the coefficient of performance of this device less than the maximum coefficient of performance?

5.15 Suppose that the refrigerator in Problem 5.14 is replaced by an energy-conserving unit that has a coefficient of performance of 4.0. Calculate the power input for this unit and the heat-transfer rate to the surrounding air.

5.16 A manufacturer claims that an automobile engine has a thermal efficiency of 18 percent. The temperature of the fuel when it burns is 1700°F and the surrounding air is at 75°F. Would the operation of this engine violate the second law of thermodynamics?

5.17 Show that $\beta_{H,\text{rev}} = 1 + \beta_{R,\text{rev}}$.

5.18 For the three situations described below state whether the cycles do or do not violate the second law of thermodynamics and support your answers with calculations:

(a) A heat pump operates between two reservoirs at −10°C and 50°C while requiring a power input of 5 kW. The heat-transfer rate from the environment to the heat pump is 10 kW.

(b) A refrigerator operates between the same two temperature reservoirs as given in part a. The refrigerator provides a cooling rate of 4 kW and a heat-transfer rate to the surroundings of 6 kW.

(c) A heat engine has an efficiency of 50 percent while operating between two reservoirs at 100°C and 300°C.

5.19 Calculate the maximum coefficient of performance for the heat pump specified in Problem 5.5. Would the heat pump described in Problem 5.5 violate the second law of thermodynamics?

5.20 Calculate the maximum coefficient of performance for the refrigerator specified in Problem 5.14. Would the refrigerator described in Problem 5.14 violate the second law of thermodynamics?

5.21 A heat engine operates between two thermal-energy reservoirs with temperatures of 2000 and 200°F, respectively. The heat-transfer rate to the heat engine from the high-temperature reservoir is 10^6 Btu/h. Calculate the maximum power, in hp, that this engine can produce. Determine the minimum heat-transfer rate with the low-temperature reservoir.

5.22 [SLA211] A heat engine operates on a Carnot cycle between temperatures of 1900°R and 520°R. The power output is 50 Btu/h. Determine the following:

(a) the heat-transfer rate to and from the engine in Btu/h

(b) the thermal efficiency

5.23 [SLA311] A Carnot heat engine is operating between thermal reservoir *A* and the environment. The engine powers a Carnot heat pump operating between the environment and thermal reservoir *B*. The temperature of reservoir *A* is 550 K and the temperature of reservoir *B* is 1200 K. The temperature of the environment is 293 K. The heat-transfer rate to the engine is 2 kW. All of the power produced by the engine is utilized to drive the heat pump. Determine the following:

(a) the power developed by the engine in kW

(b) the heating coefficient of performance of the heat pump

(c) the heat-transfer rate from the heat pump to reservoir *B* in kW

5.24 A Carnot heat pump supplies 4.7×10^4 Btu/h to a large room in order to maintain the temperature at 70°F. The low-temperature reservoir is the environment that is at -10°F. Determine the power input necessary to drive the heat pump.

5.25 A heat pump is used to heat a residential structure in the winter and to cool it in the summer. Assume that the heat pump is required to maintain the inside air temperature at 23°C during both summer and winter. On a winter day when the outside air temperature is 5°C the heating requirements for the structure are 15 kW. On a summer day the outside air temperature is 40°C and the cooling requirements are known to be 8 kW. Assuming that the heat pump has a coefficient of performance only one-half of the maximum possible value, calculate the following information:

(a) the maximum coefficient of performance that the heat pump could achieve during the winter day described in the problem

(b) the maximum coefficient of performance that the heat pump could achieve during the summer day described in the problem

(c) the actual power requirements of the heat pump during the winter day described in the problem

(d) the power requirements of the heat pump during the summer day described in the problem

5.26 The world's known coal reserves are estimated to be 7.6×10^{12} tons. The world's requirements for energy are equivalent to 1.1×10^{18} Btu per year by 2050. Assuming that the coal is to be converted to electric energy in devices having an average thermal efficiency of 35 percent and that coal has an average heating value of 11,500 Btu/lb_m, determine how many years this coal will last if it is used exclusively to meet all energy requirements. Suppose that the coal is burned and produces an equivalent high-temperature reservoir at 2200°F and that heat is rejected to a low-temperature reservoir at 50°F. Determine the maximum number of years this amount of coal could be used to satisfy the energy requirements of the world.

5.27 You wish to run an experiment on a device that is to be maintained at a constant temperature of -40°C in a laboratory. At this temperature the device is known to require 3 kW of cooling. You select a refrigeration unit with a coefficient of performance of

3.5 for the job. Determine the power required to operate the refrigeration unit and the minimum power input of any refrigeration unit if the laboratory temperature is 20°C.

5.28 An inventor claims to have perfected a heat engine that receives energy by means of heat transfer from a fuel at 900°F and rejects energy to the surroundings at 100°F while achieving a thermal efficiency of 60 percent. How would you evaluate the inventor's claims?

5.29 The work output from a Carnot engine is to be used to drive a Carnot refrigerator. The heat transfer to the engine from a thermal-energy reservoir that has a temperature of 600°C is 75 kJ, and the heat transfer from the engine is to the surrounding air, which is at 30°C. The refrigerator is to be used to maintain the temperature of a refrigerated space at −25°C. The refrigerator also has heat transfer to the surrounding air. Determine the cooling load that the refrigerator is capable of achieving.

5.30 Heat transfer to a heat pump is from a reservoir at 20°F, and the heat-transfer rate from the heat pump to a reservoir at 80°F is 3×10^4 Btu/h. Determine the coefficient of performance if a power input of 6 hp is required to drive the heat pump. Compare this value with the coefficient of performance of a Carnot heat pump operating between the same two reservoirs.

5.31 A promising method of generating power involves operating a turbine between the warm water near the surface of the ocean and the deep layers of cold water. The surface water heated by the sun provides a heat reservoir at 30°C of practically infinite extent. Cold water at depths of as little as 700 m has a temperature of 3°C. Calculate the maximum thermal efficiency of a cyclic device operating between these two reservoirs.

5.32 Studies suggest that a significant amount of energy can be generated by using ocean thermal gradients as a potential energy source. The warm surface water is used to evaporate a working fluid such as ammonia. The ammonia vapor is circulated through a low-pressure turbine where energy is generated much like a traditional steam turbine. The ammonia is condensed as a result of heat transfer to the cool seawater that exists at lower levels in the ocean. One potential location for an ocean thermal-energy-conversion power plant is near the Hawaiian Islands where surface temperatures average 82°F and temperatures at depths near 3000 ft average 36°F on a yearly basis. Calculate the maximum thermal efficiency that can be expected from a power plant that utilizes this type of conversion process near the Hawaiian Islands.

5.33 The heat transfer from a thermal-energy reservoir at 500 K to the environment at 15°C is 50 kJ. Calculate the change in entropy of the reservoir and the change in entropy of the environment. Show that the total entropy change satisfies the second law. If the heat transfer remains unchanged, but the reservoir temperature is lower, what effect does this have on the total entropy change? Explain.

5.34 Fusion reactions using deuterium and tritium in a small Tokamak reactor have produced temperatures of 60×10^6 degrees Celsius for a fraction of a second. If this temperature could be sustained indefinitely, determine the efficiency of a Carnot cycle operating between a thermal reservoir at this temperature and the atmosphere. Suppose that a typical hydrocarbon fuel is burned in a traditional combustion process producing a reservoir temperature of 2500°C. Compare the Carnot efficiency for this situation with the Carnot efficiency when the reservoir is provided by the fusion reaction.

5.35 An ideal gas is put in two different piston-cylinder assemblies of equal volume at the same temperature and pressure. One of these is frictionless and the other is not. The gas in each is heated until the absolute temperature is doubled while the final pressure is the same as the initial pressure. The entropy change of the gas in the piston-cylinder assembly having no friction is (greater than, less than, equal to) the entropy change of the gas in the other piston-cylinder assembly. Justify your response.

5.36 A Carnot engine operates between two thermal-energy reservoirs whose temperatures are 1000 and 400°R, respectively. The entropy of the low-temperature reservoir increases by 0.7 Btu/°R. Determine the heat transfer to the engine and the net work.

5.37 For the same change of state, the entropy change of a substance during an irreversible process is (greater than, equal to, less than) the entropy change of the same substance during a reversible process. Explain.

5.38 Nitrogen in a closed, piston-cylinder assembly undergoes an internally reversible, isentropic expansion from an initial pressure of 800 kPa to a final state where the temperature and pressure are 407°C and 250 kPa, respectively. Determine the heat transfer and work for this process.

5.39 Five lb_m of air is heated at constant pressure from 14.7 psia, 140°F to 560°F. Determine the entropy, enthalpy, and internal-energy changes of the air during the process.

5.40 Air at an initial state described by 7°C, 300 kPa changes to a final state of 107°C, 100 kPa. Calculate the entropy change per kg of the air that occurs during each of the following processes:

(a) an internally reversible process
(b) an irreversible process
(c) a totally reversible process

5.41 Air proceeds along a constant-pressure path at 100 kPa from 20 to 100°C. Determine the entropy change during the process assuming the following:

(a) Air is an ideal gas with constant specific heats.
(b) Air is an ideal gas with variable specific heats.

5.42 Carbon dioxide changes state from 40 psia, 190°F to 80 psia, 540°F. Calculate the entropy change of the CO_2 during the process, assuming the following:

(a) The CO_2 is an ideal gas with constant specific heats.
(b) The CO_2 is an ideal gas with variable specific heats.

5.43 Nitrogen proceeds along an irreversible path between 50 psia, 600°R and 75 psia, 1000°R. Calculate the entropy change of the nitrogen on a per-unit-mass basis.

5.44 Carbon dixoide initially at 50 kPa and 420 K changes state until its pressure and temperature are 2 MPa and 800 K. Compute the entropy change, assuming the following:

(a) ideal-gas behavior with temperature-dependent specific heats
(b) ideal-gas behavior with constant specific heats

5.45 Determine the entropy change per kilogram of air as it changes from 150 kPa, 300 K to 2 MPa, 900 K, assuming the following:

(a) The air is an ideal gas with variable specific heats.
(b) The air is an ideal gas with constant specific heats.

5.46 Carbon monoxide is initially at a temperature of 80°F and a specific volume of 17.6 ft^3/lb_m. The final state of the carbon monoxide is 900°R and 12 ft^3/lb_m. Calculate the entropy change per lb_m of CO for this change in state, assuming the following:

(a) ideal-gas behavior with variable specific heats
(b) ideal-gas behavior with constant specific heats

5.47 Five lb_m of air is initially at a temperature of 60°F and a pressure of 14.7 psia. The state of the air is changed until the pressure and temperature are 100 psia and 520°F, respectively. Calculate the entropy change during the process, assuming ideal-gas behavior and (a) constant specific heats and (b) variable specific heats.

5.48 Air proceeds along an irreversible path between a state at which $T_1 = 127°C$ and $v_1 = 0.6\ m^3/kg$, and a state at which $T_2 = 327°C$ and $P_2 = 200$ kPa. Calculate the entropy change per kilogram of air during the process, assuming that air has constant specific heats.

5.49 Three lb_m of hydrogen changes state from 40°F, 20 psia to 540°F, 220 psia. Calculate the entropy change during the process, assuming ideal-gas behavior and (a) variable specific heats and (b) constant specific heats.

5.50 Carbon dioxide changes state from 300 kPa, 10°C to 800 kPa, 500°C along an irreversible path. Determine the entropy change for 1 kg of CO_2 during this process for (a) constant specific heats and (b) variable specific heats.

5.51 Two kilograms of nitrogen change state from 400 K, 200 kPa to 10 MPa, 900 K. Calculate the entropy change during the process, assuming ideal-gas behavior and (a) variable specific heats and (b) constant specific heats.

5.52 Air at an initial state of 600°R, 50 psia follows a process until it reaches a final state of 1200°R, 350 psia. Assume that the air is an ideal gas with constant specific heats. Calculate the entropy change per unit mass during the process, assuming the following:

(a) The process is internally reversible.
(b) The process is irreversible.
(c) The process is totally reversible.

5.53 Air initially has a specific volume of 0.6 m^3/kg and a temperature of 147°C. The final state of the air is specified by a pressure of 1.4 MPa and a temperature of 367°C. Determine the entropy change per kilogram of air during the process, assuming that air is an ideal gas with constant specific heats.

5.54 Air undergoes an internally reversible adiabatic process from 60 psia, 80°F to a final pressure of 400 psia. Assuming ideal-gas behavior with constant specific heats, calculate the temperature and the specific volume at the final state.

5.55 Nitrogen at 800 K, 2 MPa proceeds along an isentropic path until its temperature is reduced to 300 K. Assuming ideal-gas behavior, calculate the pressure at the final state for the following conditions:

(a) The nitrogen has variable specific heats.
(b) The nitrogen has constant specific heats.

5.56 Three lb_m of air in a closed system proceeds along an internally reversible, isothermal path from 120°F and 250 psia to 100 psia. Assume that air has constant specific heats. For this process calculate the following quantities of the air:

(a) the entropy change
(b) the heat transfer
(c) the work
(d) the enthalpy change

5.57 Carbon dioxide proceeds along an internally reversible adiabatic path from 357°C, 0.7 MPa to a final temperature of 47°C. Assuming that the CO_2 is an ideal gas, calculate the specific volume and pressure at the final state.

5.58 Hydrogen proceeds along an isentropic path from 130 psia, 500°R to 15 psia. Calculate the temperature at the final state, assuming that hydrogen is an ideal gas with constant specific heats.

5.59 Air proceeds along an internally reversible adiabatic path from a specific volume of 0.7 m^3/kg and a temperature of 125°C to a temperature of 310°C. Assuming that air is an ideal gas, calculate the pressure and specific volume of the air at the final state.

5.60 A well-insulated, steady-flow nozzle receives air at 800 kPa, 7°C and discharges it at 100 kPa. The velocity of the air as it leaves the nozzle is much greater than the velocity at the inlet and the exit area of the nozzle is 0.002 m^2. The process in the nozzle is reversible. Calculate the exit-air velocity, the exit-air temperature, and the mass flow rate of the air through the nozzle.

5.61 Air enters a steady-flow air compressor at 80°F, 1 atm pressure. The air follows a reversible adiabatic process in the compressor and leaves at 300 psia. The mass flow rate of the air through the compressor is 5 lb_m/s. Neglecting changes in kinetic energy, calculate the power required by the compressor and the temperature of the air as it leaves the compressor.

5.62 An ideal gas in a closed system undergoes an internally reversible, constant-pressure process during which the temperature of the gas decreases as a result of heat transfer to the environment. Determine the correct response to each of the following statements and briefly justify your responses:

(a) The entropy change of the gas is (greater than, equal to, less than) zero.

(b) The entropy change of the environment is (greater than, equal to, less than) zero.

(c) The total entropy change is (greater than, equal to, less than) zero.

5.63 An ideal gas contained in a piston-cylinder assembly is compressed in an internally reversible isothermal process. For this process the entropy change of the gas is (greater than, equal to, less than) zero. Explain.

5.64 Air flows steadily and irreversibly through a turbine and produces a shaft work output of 50 kJ/kg. The inlet air conditions are: 400 kPa, 47°C, 25 m/s and the exit state is 50 kPa, 10°C, 120 m/s. Calculate the following data:

(a) the magnitude and direction of the heat transfer per kg of air for the process

(b) the entropy change of the air per kg during this process

5.65 Determine whether or not it is possible to compress air adiabatically from 140 psia and 600°R to 360 psia and 700°R.

5.66 Determine whether or not it is possible to compress air adiabatically from 70 kPa and 310 K to 140 kPa and 400 K.

5.67 It is proposed that an ideal gas with an initial pressure P_1 undergo an isothermal process during which the volume of the gas is reduced by one-half.

(a) Show that $S_2 - S_1$ is independent of the gas present, whereas $s_2 - s_1$ does depend on the identity of the gas.

(b) Could the proposed process take place without violating the second law? If so, is it possible to determine whether the process is internally reversible or irreversible? Explain your answer.

5.68 Air enters a steady-flow adiabatic turbine with a pressure of 260 psia and a temperature of 1040°F and exhausts at 35 psia and 380°F. Assuming negligible changes in kinetic and potential energies, determine the turbine work per unit mass of air flowing. Is this process reversible, irreversible, or impossible?

5.69 Air is expanded reversibly in a steady-flow adiabatic nozzle. The diameter of the nozzle exit is 25 mm, and the air enters at 810 kPa and 250 K. The velocity at the nozzle exit is much larger than the inlet velocity. Find the exit-air temperature, the exit velocity, and the mass flow rate of air for an exit pressure of 160 kPa.

5.70 Compressed air at 900 psia and 1600°R enters a steady-flow air turbine and exhausts at 1 atm pressure, while within the turbine the air proceeds along an internally reversible adiabatic path. Calculate the exhaust temperature of the air and the work developed by the turbine for each lb_m of air that passes through the turbine.

5.71 Air enters a steady-flow nozzle at 40°C and 5 MPa and leaves at 3.0 MPa. Determine the minimum temperature that the air can attain at the exit of the nozzle.

5.72 A piston and cylinder arrangement contains 12 lb_m of an ideal gas that has a molecular weight of 25. Initially the pressure and temperature of the gas are 45 psia and 200°F. The gas follows a polytropic process where the polytropic exponent is 1.2 until a pressure of 30 psia is reached. The ratio of specific heats for the gas is 1.3. Assume that the gas has constant specific heats and that the process is reversible. Calculate the following quantities:

(a) the original and final volumes of the gas
(b) the change in internal energy of the gas
(c) the change in enthalpy of the gas
(d) the entropy change of the gas
(e) the heat transfer to or from the gas
(f) the work done by or on the gas
(g) the temperature change of the gas

5.73 Nitrogen in a rigid, insulated tank with a volume of 15 m^3 is initially at a pressure of 70 kPa and a temperature of 0°C. An electric component inside the tank draws 1 A at 120 V. The component is energized for 25 min and then shut off, and the nitrogen reaches equilibrium. Determine the final temperature and the final pressure of the nitrogen, the entropy change of the nitrogen, the entropy change of the surrounding ($T_0 = 25$°C), and the total entropy change.

5.74 A simple heat engine uses an ideal gas with constant specific heats as the working fluid in a frictionless piston-cylinder assembly. The gas is first heated at constant pressure from state 1 to state 2; then it is cooled at constant volume to state 3, where $T_3 = T_1$. The gas is then compressed at constant temperature, thereby returning to state 1. Determine the work and the heat transfer for each of the three processes, the thermal efficiency of the heat engine, and the thermal efficiency of a totally reversible heat engine operating between the maximum and minimum temperatures of this cycle. Express all answers in terms of variables from the following list: m, c_v, c_p, R, T_1, T_2, T_3, P_1, P_2, P_3.

5.75 Ten lb_m of solid lead initially at −50°F has its temperature increased to 500°F along an irreversible path. Calculate the entropy change during the process.

5.76 A 20-lb_m piece of solid copper changes temperature from 0 to 200°F. Calculate the entropy change of the copper during the process.

5.77 Solid aluminum at −200°C has its temperature increased to 100°C. Calculate the entropy change per kilogram of aluminum during the process.

5.78 Calculate the entropy change of water when it changes state from 15 psia, 500°F to 90 psia, 800°F.

5.79 Calculate the entropy change of water when it changes state from saturated vapor conditions at 10 MPa to 25 MPa, 400°C along (a) a reversible path and (b) an irreversible path.

5.80 Water at 1 atm pressure and a temperature of 50°F undergoes a process until a final pressure of 30 psia and a final temperature of 250°F are reached. Calculate the entropy change per lb_m of water during the process.

5.81 [SLA511] A steady flow of steam at a temperature of 175°C and a quality of 0.5 is throttled to a temperature of 138°C. Determine the following:

(a) the final pressure in kPa
(b) the final quality
(c) the entropy change per unit mass

5.82 Refrigerant-12 proceeds from 40 psia, 60°F to 120°F along an irreversible constant-pressure path. Calculate the entropy change of the refrigerant.

5.83 Water at 10 psia and 1000°F proceeds along a constant-pressure path until it reaches 400°F. Calculate the entropy change of the water during the process.

5.84 Refrigerant-12 proceeds along an internally reversible, isothermal path from a state of saturated vapor at 60°F to a pressure of 60 psia. Calculate the following:

(a) the entropy change during the process

(b) the heat transfer per unit mass during the process

5.85 Saturated water vapor at 140 kPa is compressed reversibly and adiabatically at a flow rate of 9 kg/s in a steady-flow compressor to a pressure of 800 kPa. Calculate the exit temperature of the water and the power input required for the compressor.

5.86 Water proceeds along a constant-pressure path from 50 kPa and 400°C to 200°C. Determine the entropy change per unit mass of the water if the process is:

(a) internally reversible

(b) irreversible

Does this process violate the second law of thermodynamics?

5.87 Determine the work per lb_m produced when R-12 is heated in a closed system along an internally reversible isothermal path from a saturated-liquid state at 60°F. During the process the heat transfer to the refrigerant is 100 Btu/lb_m.

5.88 [SLA411] Two kg of saturated water vapor is heated in a tank at 100°C. The heat source is a thermal-energy reservoir at a temperature of 773 K. Assume that at its final state, the water is in thermal equilibrium with the reservoir. Determine the following:

(a) the entropy change of the water in kJ/K

(b) the entropy change of the thermal-energy reservoir in kJ/K

(c) the entropy change of the universe in kJ/K

5.89 Water with a quality of 50 percent and a temperature of 280°F changes state to 500°F, 80 psia. Determine the entropy change during the process on a per-unit-mass basis.

5.90 Water enters an adiabatic, steady-flow turbine at 10 MPa, 500°C and exhausts at 100 kPa, 150°C. The mass flow rate of the steam through the turbine is 20 kg/s. Neglect changes in kinetic and potential energy. Calculate the power developed by the turbine and determine if the process is reversible, irreversible, or impossible.

5.91 Refrigerant-12 is compressed steadily and adiabatically in a reversible compressor. The volume flow rate entering the compressor is 800 cfm. The R-12 enters the compressor at 120 psia, 100°F and leaves with its pressure being twice the inlet value. Neglect kinetic-energy changes, and calculate the power required to operate the compressor.

5.92 Dry saturated steam enters a steady-flow system with a flow rate of 5 lb_m/s and a pressure of 80 psia. It proceeds along a reversible isothermal path and leaves the device at 20 psia. Calculate the heat-transfer rate and power developed during the process.

5.93 An aluminum container having a mass of 4 kg is heated to 80°C and then quickly filled with 20 kg of a liquid that has a specific heat of 0.8 kJ/kg·°C and an initial temperature of 50°C. The liquid-filled container is exposed to outdoor air at 25°C for an unspecified period of time, and then the container is completely covered with a very good thermal insulation. Thereafter, the liquid and the container are allowed to reach thermal equilibrium, and the equilibrium temperature is 52°C. Determine the heat transfer from the liquid-filled container to the air, the entropy change of the liquid, the entropy change of the aluminum container, the entropy change of the air, and the total entropy change. Discuss your results.

5.94 Saturated steam at 180°F enters a steady-flow compressor at a flow rate of 300 lb_m/s. During the compression process the heat-transfer rate from the water is 5×10^6 Btu/h and the water leaves the compressor at 20 psia, 400°F. Calculate the power of the motor required to drive the compressor and show that this process does not violate the second law of thermodynamics if the temperature of the surroundings is constant at 60°F.

5.95 A closed piston-cylinder assembly contains 0.5 kg of water initially at 800 kPa and 435 K. As a result of heat transfer during an internally reversible isothermal process, the water reaches a final pressure of 150 kPa. Determine the heat transfer and work, in kilojoules, for this process.

5.96 Refrigerant-12 in a closed system is heated in an internally reversible isothermal process from saturated-liquid conditions at 80°F. The heat transfer to the R-12 during this process is 75 Btu/lb_m. Sketch the T-s and P-v diagrams for this process, and determine the magnitude and the direction of the work.

5.97 As a result of heat transfer to the surroundings during a constant-pressure process, 0.3 kg of steam in a frictionless-piston-cylinder assembly experiences a decrease in entropy of 2.9 kJ/kg·K. The initial temperature and pressure of the steam are 600°C and 1.4 MPa, respectively, and the process is internally reversible. Calculate the work and heat transfer for this process and the total entropy change ($T_0 = 27$°C).

5.98 Refrigerant-12 is to be accelerated in a steady-flow nozzle to a velocity of 840 ft/s. At the nozzle inlet the pressure and temperature of the refrigerant are 90 psia and 220°F, and the inlet velocity is negligible. The pressure at the nozzle exit is 15 psia. Calculate the heat transfer per unit mass during the expansion if the exit temperature of the refrigerant is 80°F. Determine the exit temperature assuming that the nozzle is well-insulated.

5.99 A steady-flow centrifugal compressor is supplied with saturated water vapor at 15 kPa and 500 kg of the water is compressed per hour to 250 kPa, 400°C. During the process there is heat transfer to the surroundings ($T_0 = 30$°C) at a rate of 0.8 kW. Determine the power required to compress the water, the rate of change of entropy of the water, and the rate of change of entropy of the surroundings.

5.100 Water at a pressure of 2000 psia and a temperature of 1500°F enters a steady-flow, internally reversible, adiabatic turbine. The water leaves the turbine at 20 psia. Neglecting changes in kinetic energy, calculate the power developed by the turbine if the mass flow rate of water through the turbine is 50 lb_m/s.

5.101 A turbine salesman makes the following claim for one of his products: Steam enters a steady-flow, adiabatic turbine at 4 MPa, 600°C with negligible velocity and exhausts at 200 kPa with a velocity of 180 m/s. The flow rate is 2.2 kg/s, and the turbine power output is 2 MW.

(a) What would be the temperature of the exhaust steam for the turbine as described by the salesman?

(b) Could the salesman's claim be valid? Explain and justify your response.

5.102 Steam enters a steady-flow nozzle with a pressure of 1500 psia and a temperature of 1200°F. The steam leaves the nozzle at a pressure of 100 psia. Determine the exit temperature of the steam and the change in kinetic energy if the process in the nozzle is isentropic.

5.103 Ten kilograms of refrigerant-12 is contained in a closed system at 20°C, and the initial volume is 0.2 m^3. The R-12 undergoes a process during which the heat transfer to the system is 750 kJ from a reservoir at a temperature of 100°C. The final state of the R-12 is 20°C and 0.15 MPa. Determine the work for this process, the entropy change of the reservoir, the entropy change of the R-12, and the total entropy change.

5.104 Water in a closed system is initially at saturated-liquid conditions. Consider two internally reversible processes by which the water may reach a superheated vapor state where the specific volume is v_2. Process *A* is a constant-pressure process, and process *B* is an isothermal process. Sketch these processes on the same *P-v* diagram and respond to the following:

(a) The work for process *A* is (greater than, less than, equal to) the work for process *B*.

(b) The internal-energy change for process *A* is (greater than, less than, equal to) that for process *B*.

(c) The heat transfer for process *B* is (positive, negative, zero).

(d) The entropy change for process *B* is (positive, negative, zero).

5.105 A closed, rigid vessel having a volume of 28 ft^3 contains 50 lb_m of R-12 initially at 100°F. The vessel is placed outside where the air temperature is 50°F, and after some period of time the temperature of the R-12 is found to be 60°F. Determine the heat transfer required for this process, the entropy change of the R-12, and the entropy change of the outdoor air. Is this process totally reversible, irreversible, or impossible? Explain your response.

5.106 A totally reversible heat engine operating in a Carnot cycle uses water as the working fluid. At the beginning of the isothermal expansion, the water is a saturated liquid at 480 K. It is expanded during this process until saturated-vapor conditions are achieved. The water is then further expanded in a reversible adiabatic process until the pressure is 40 kPa. Sketch the cycle on a *T-s* diagram and on a *P-v* diagram relative to the saturation region. Determine the thermal efficiency of the engine, the heat transfer to the cycle, in kilojoules per kilogram, and the net work of the engine, in kilojoules per kilogram.

5.107 A Carnot heat engine uses 2.92 lb_m of water as a working substance. At the beginning of the isothermal expansion process the water is at 80 psia and occupies a volume of 25 ft^3. The pressure and volume of the water at the end of the adiabatic expansion process are 14.7 psia and 120 ft^3. Calculate the following information:

(a) the temperature of the high-temperature reservoir

(b) the temperature of the low-temperature reservoir

(c) the pressure of the water at the end of the isothermal expansion process

(d) the pressure of the water at the end of the isothermal compression process

(e) the thermal efficiency of the cycle.

5.108 Suppose a closed system containing 2 kg of steam operates in a Carnot cycle. At the beginning of the isothermal expansion, the temperature is 160°C and the volume is 0.434 m^3. The isothermal expansion continues until the volume of the steam has tripled. The temperature at the conclusion of the adiabatic expansion is 50°C. Determine the thermal efficiency of the cycle, the heat transfer to the cycle, and the net work for a single cycle. If the working fluid is changed to argon gas, what effect would this have on the thermal efficiency?

5.109 A Carnot heat engine uses air as a working fluid. The temperatures of the high- and low-temperature reservoirs are 570 and 200°F, respectively. The pressure of the air at the beginning of the isothermal expansion process is 150 psia, and the pressure of the air at the beginning of the isothermal compression process is 15 psia. Calculate the following quantities:

(a) the heat transfer from the high-temperature reservoir per lb_m of air
(b) the heat transfer to the low-temperature reservoir per lb_m of air
(c) the net work produced by the cycle per lb_m of air
(d) the thermal efficiency of the cycle
(e) the pressure of the air at the end of the isothermal expansion process
(f) the pressure of the air at the end of the isothermal compression process

5.110 A closed Carnot heat-engine cycle utilizes 10 lb_m water as a working fluid. Refer to Figure 5.17; the following information is known:

$$P_1 = 1000 \text{ psia} \qquad T_1 = 1400°\text{F}$$
$$T_4 = 500°\text{F} \qquad P_2 = 500 \text{ psia}$$

Calculate the following quantities:

(a) the heat transfer from the high-temperature reservoir
(b) the heat transfer to the low-temperature reservoir
(c) the net work developed by the cycle
(d) the thermal efficiency of the cycle
(e) the temperature at state 2
(f) the pressure at state 3
(g) the pressure at state 4

6 Second-Law Analysis of Thermodynamic Systems

6.1 INTRODUCTION

The growing awareness of the importance of using the world's energy resources wisely has stimulated interest in second-law analysis. Such analyses provide insight into the best thermodynamic performance of a device or system from a thermodynamic viewpoint, and an energy analysis alone cannot achieve this objective. Keep in mind, however, that the best thermodynamic performance is not necessarily a desirable or practical objective since thermodynamic criteria are seldom the only considerations of importance in the design of devices or systems. Instead, an acceptable design or problem solution is likely to be based on many different criteria, which might include thermodynamic performance, state-of-the-art technology, material selection, and economics.

Consider, for example, the problem of deciding whether or not to buy a new car. The criteria for the decision can be vastly different from one individual to another. If the criteria are purely aesthetic, the decision might be made without regard to price or fuel economy. If fuel economy is the primary consideration, one might be willing to sacrifice appearance, size, and even price to own a car with better fuel economy. Most often the decision is ultimately made after consideration has been given to the initial cost, fuel economy, projected maintenance costs, appearance, size, resale value, and so forth.

Design of equipment and processes must also reflect the compromises that arise from attempts to satisfy a variety of criteria. Thermodynamic performance, while often very important, is rarely the sole consideration. Second-law analysis, therefore, is not a panacea for the design process, but it does provide a means of evaluating the thermodynamic performance of systems and processes so that technically sound design decisions can be made.

In this chapter the thermodynamic aspects of second-law analysis are discussed and developed. The concepts and applications of irreversibility, maximum work, and availability are illustrated. The methods of analysis introduced in this chapter are used extensively in much of the remainder of the text to complement the use of the conservation-of-mass and conservation-of-energy equations.

6.2 A GENERAL EXPRESSION FOR THE RATE OF CHANGE OF TOTAL ENTROPY

The goal of this section is to develop an equation that may be used to evaluate the total rate of entropy change for arbitrary systems and processes. The first step in the formulation of second-law-analysis methods is the development of an entropy-balance equation that is similar in many respects to the conservation-of-mass and -energy equations. However, unlike mass and energy, entropy is not a conserved quantity, and for this reason a conservation-of-entropy equation cannot be formulated. The fact that entropy is not a conserved quantity is evident from the increase-in-entropy principle as expressed by Equation 5.47:

$$dS_{\text{tot}} \geq 0$$

The total entropy change, that is, the entropy change of an isolated system, must be greater than or equal to zero, and therefore entropy is not conserved. On the other hand, if the conservation-of-energy equation is written for an isolated system, the result is

$$dE_{\text{tot}} = 0$$

since energy is a conserved quantity.

The rate at which the total entropy change occurs is simply called the *rate of change of total entropy*. One conclusion from the second law is that the rate of change of total entropy must be greater than or equal to zero. This result can be seen by writing Equation 5.47 on a rate basis:

$$\left(\frac{dS}{dt}\right)_{\text{tot}} \geq 0 \tag{6.1}$$

In a totally reversible process, then, the total entropy does not change, while in an irreversible process, the total entropy increases. Furthermore, in accordance with Equation 5.46, the rate of change of total entropy must be the sum of the net rate of change of entropy for the system and the net rate of change of entropy for the surroundings:

$$\left(\frac{dS}{dt}\right)_{\text{tot}} = \left(\frac{dS}{dt}\right)_{\text{net,sys}} + \left(\frac{dS}{dt}\right)_{\text{net,surr}} \geq 0 \tag{6.2}$$

To develop the techniques for a second-law analysis of systems, we focus attention on the process that the system undergoes and those components of the surroundings that exchange heat with the system of interest. Thus the rate of change of total entropy is composed of the net rate of change of entropy of the system and the net rate of change of entropy of the surroundings brought about solely by heat-transfer interactions with the system.

The net rate of change of entropy for the system is the difference between the time rate of change of entropy associated with the mass inside the system and the net rate of change of entropy due to mass flow into the system across the boundary. A mathematical expression for the net rate of change of entropy for the system can be developed in much the same manner as was employed in the development of the conservation-of-mass and conservation-of-energy equations. The development begins with a word equation:

$$\begin{bmatrix}\text{Net rate of}\\ \text{change of}\\ \text{entropy for}\\ \text{the system}\end{bmatrix} = \begin{bmatrix}\text{time rate of}\\ \text{change of the}\\ \text{entropy of the}\\ \text{mass within}\\ \text{the system}\end{bmatrix} + \begin{bmatrix}\text{rate at which}\\ \text{entropy leaves}\\ \text{at the boundary}\\ \text{of the system}\end{bmatrix} - \begin{bmatrix}\text{rate at which}\\ \text{entropy enters}\\ \text{at the boundary}\\ \text{of the system}\end{bmatrix} \tag{6.3}$$

A comparison of Equation 6.3 with Equations 3.1 and 4.1 reveals that the terms on the right side of Equation 6.3 are similar to terms present in the mass- and energy-conservation statements. However, the term on the left side of Equation 6.3 is not zero because entropy is not a conserved quantity. Furthermore, the mathematical forms of these terms are identical to the corresponding terms developed in Chapters 3 and 4:

$$\begin{bmatrix}\text{Rate at which}\\ \text{entropy enters}\\ \text{at the boundary}\\ \text{of the system}\end{bmatrix} = \int_{A_i} s(\rho V_n\, dA) \tag{6.4}$$

$$\begin{bmatrix}\text{Rate at which}\\ \text{entropy leaves}\\ \text{at the boundary}\\ \text{of the system}\end{bmatrix} = \int_{A_e} s(\rho V_n\, dA) \tag{6.5}$$

and

$$\begin{bmatrix}\text{Time rate of}\\ \text{change of the}\\ \text{entropy of the}\\ \text{mass within}\\ \text{the system}\end{bmatrix} = \frac{d}{dt}[S_{\text{sys}}] = \frac{d}{dt}\int_{V} s\rho\, d\mathbf{V} \tag{6.6}$$

Substituting Equations 6.4, 6.5, and 6.6 into the word equation, Equation 6.3, and rearranging, we obtain

$$\left(\frac{dS}{dt}\right)_{\text{net,sys}} = \int_{A_e} s(\rho V_n\, dA) - \int_{A_i} s(\rho V_n\, dA) + \frac{d}{dt}\int_{V} s\rho\, d\mathbf{V} \tag{6.7}$$

As a consequence of the second law, Equation 5.17, the net rate of change of entropy of a system must be greater than or equal to the rate at which heat transfer occurs at the boundary of the system divided by the local absolute temperature at the bounding surface. In general,

$$\left(\frac{dS}{dt}\right)_{\text{net,sys}} = \int_{A_e} s(\rho V_n\, dA) - \int_{A_i} s(\rho V_n\, dA) + \frac{d}{dt}\int_V s\rho\, d\mathbf{V} \geq \sum_j \frac{\dot{Q}_j}{T_j}$$

The summation sign indicates that heat transfer could possibly occur at a number of locations at the boundary of the system where the absolute temperature of the system, T_j, might vary from one location to another.

As was explained in Chapter 4, if uniform-flow conditions are assumed to exist at all inlets and exits of the system, the area integrals in Equation 6.7 can be replaced by summations that involve the mass flow rates and suitable averaged values of entropy at each inlet and exit:

$$\left(\frac{dS}{dt}\right)_{\text{net,sys}} = \sum_{\text{exit}} \dot{m}s - \sum_{\text{inlet}} \dot{m}s + \left(\frac{dS}{dt}\right)_{\text{sys}} \geq \sum_j \frac{\dot{Q}_j}{T_j} \tag{6.8}$$

Notice that for a closed system the net change of entropy for the system, given by Equation 6.8, is equal to the change in entropy of the mass inside the system since mass does not flow across the boundary of a closed system. For an open, steady system the net rate of change of entropy is equal to the *net* flow of entropy out of the system due to mass flow. For more general cases (e.g., an open, transient system) the net rate of change of entropy of the system is the sum of the time rate of change of entropy associated with the mass inside the system and the net rate of change of entropy due to mass flow out of the system across the boundary.

With the aid of Equation 6.8 the rate of change of total entropy in Equation 6.2 can be written for uniform flow conditions as

$$\left(\frac{dS}{dt}\right)_{\text{tot}} = \sum_{\text{exit}} \dot{m}s - \sum_{\text{inlet}} \dot{m}s + \left(\frac{dS}{dt}\right)_{\text{sys}} + \left(\frac{dS}{dt}\right)_{\text{net,surr}} \geq 0 \tag{6.9}$$

The net rate of change of entropy for the surroundings could be determined by applying Equation 6.8 to each subsystem of the surroundings that experiences heat transfer with the system being analyzed. For example, if the surroundings are composed of a single thermal-energy reservoir, then the net rate of change of entropy for the surroundings would reduce to

$$\left(\frac{dS}{dt}\right)_{\text{net,surr}} = \left(\frac{dS}{dt}\right)_{\text{res}} = \frac{\dot{Q}_{\text{res}}}{T_{\text{res}}}$$

In this manner systems of arbitrary nature could be included as components of the surroundings. However, constructing the boundaries of the system to be analyzed such that the surroundings are composed entirely of thermal-energy reservoirs is usually more convenient because heat-transfer interactions with thermal-energy reservoirs represent internally reversible processes from the viewpoint of the surroundings. (This result does not limit the applicability of the equations that follow, as will be illustrated in Example

6.1.) With this convention, therefore, the net rate of change of entropy of the surroundings can be related to the heat transfer to or from the surroundings, as indicated above, by

$$\left(\frac{dS}{dt}\right)_{\text{net,surr}} = \sum_k \frac{\dot{Q}_k}{T_k} \tag{6.10}$$

The summation sign in Equation 6.10 indicates that the surroundings could possibly be composed of any number of reservoirs, each having a different temperature T_k and experiencing heat transfer with the system of interest. Furthermore, both $\dot{Q}_k$ and T_k in Equation 6.10 are associated with the surroundings. That is, T_k is the absolute temperature of the surroundings and $\dot{Q}_k$ has an algebraic sign that is determined from the direction of heat transfer relative to the surroundings (e.g., $\dot{Q}_k > 0$ for heat transfer to the surroundings).

Substitution of Equation 6.10 into Equation 6.9, therefore, results in a general expression for the rate of change of total entropy:

$$\left(\frac{dS}{dt}\right)_{\text{tot}} = \sum_{\text{exit}} \dot{m}s - \sum_{\text{inlet}} \dot{m}s + \left(\frac{dS}{dt}\right)_{\text{sys}} + \sum_k \frac{\dot{Q}_k}{T_k} \geq 0 \tag{6.11}$$

The application of this equation is illustrated in the following examples.

EXAMPLE 6.1

Water enters a heat exchanger at a rate of 60 kg/s as a saturated liquid at 50 kPa and leaves at 250°C. The heating of the water is accomplished by heat transfer from a hot stream of air that enters the heat exchanger at 1000°C and leaves at 450°C. Assume ideal-gas behavior for the air, neglect pressure drops of the fluids in the heat exchanger, and neglect changes in kinetic and potential energies. Determine the rate of change of total entropy associated with this process.

Solution. A sketch of the heat exchanger is shown in Figure 6.1. The heat exchanger is a steady-flow device and heat transfer to the environment is assumed negligible. If we construct the system boundaries as shown in Figure 6.1 so that the system consists of both the water and the air, the conservation-of-mass and conservation-of-energy equations, Equation 3.16 and Equation 4.31, reduce to

$$\dot{m}_1 + \dot{m}_3 = \dot{m}_2 + \dot{m}_4$$

and

$$0 = (\dot{m}_2 h_2 + \dot{m}_4 h_4) - (\dot{m}_1 h_1 + \dot{m}_3 h_3)$$

Since the two fluid streams do not mix,

$$\dot{m}_1 = \dot{m}_2 \quad \text{and} \quad \dot{m}_3 = \dot{m}_4$$

Thus

$$0 = \dot{m}_1(h_2 - h_1) + \dot{m}_3(h_4 - h_3)$$

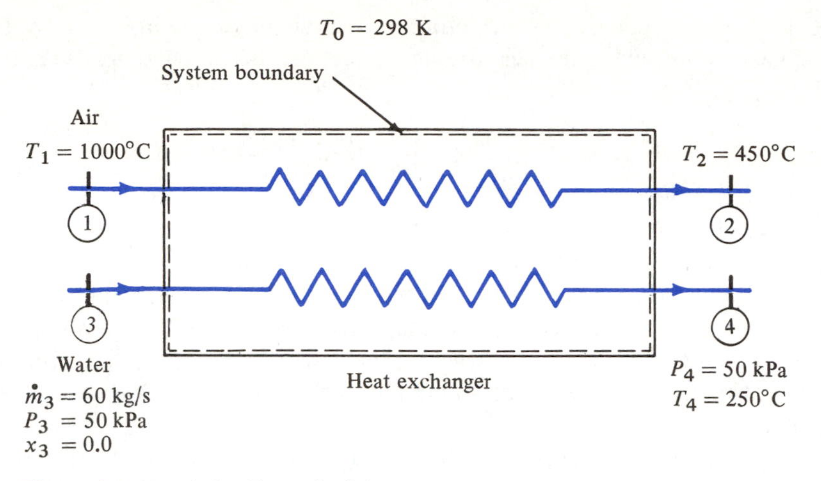

Figure 6.1 Sketch for Example 6.1.

The mass flow rate of air through the heat exchanger can therefore be determined from

$$\dot{m}_1 = -\dot{m}_3 \left(\frac{h_4 - h_3}{h_2 - h_1}\right)$$

Using Table D.1 for air properties and Tables B.2 and B.3 for water properties, we obtain

$$\dot{m}_1 = -(60 \text{ kg/s}) \frac{(2975.6 - 340.54)\text{kJ/kg}}{(738.3 - 1363.8)\text{kJ/kg}}$$

$$= 252.8 \text{ kg/s}$$

The rate of change of total entropy associated with this process can be determined by using Equation 6.11. The system being analyzed is a steady-flow system. Therefore, the entropy of the system, an extensive property, cannot change with time. Thus Equation 6.11 reduces to

$$\left(\frac{dS}{dt}\right)_{\text{tot}} = \sum_{\text{exit}} \dot{m}s - \sum_{\text{inlet}} \dot{m}s + \sum_k \frac{\dot{Q}_k}{T_k}$$

The last term is zero in this instance because heat transfer from the heat exchanger to the surroundings is assumed to be negligible. Therefore

$$\left(\frac{dS}{dt}\right)_{\text{tot}} = (\dot{m}_2 s_2 + \dot{m}_4 s_4) - (\dot{m}_1 s_1 + \dot{m}_3 s_3)$$

$$= \dot{m}_1(s_2 - s_1) + \dot{m}_3(s_4 - s_3)$$

The entropy change of the air, using Table D.1, is

$$s_2 - s_1 = s_2^\circ - s_1^\circ - R \ln \frac{P_2}{P_1} = (6.6073 - 7.2477)\text{kJ/kg·K} + 0 = -0.6404 \text{ kJ/kg·K}$$

and for the water, from Tables B.2 and B.3,

$$s_4 - s_3 = (8.3548 - 1.0912)\text{kJ/kg·K} = 7.2636 \text{ kJ/kg·K}$$

Therefore, the rate of change of total entropy associated with the process is

$$\left(\frac{dS}{dt}\right)_{\text{tot}} = (252.8 \text{ kg/s})(-0.6404 \text{ kJ/kg·K}) + (60 \text{ kg/s})(7.2636 \text{ kJ/kg·K})$$

$$= \underline{\underline{274 \text{ kJ/K·s}}} = \underline{\underline{274 \text{ kW/K}}}$$

In order to demonstrate that Equation 6.11 and Equation 6.9 are equally applicable, we also determine the rate of change of total entropy by using Equation 6.9. Let us consider the air side of the heat exchanger to be the system in this instance. Equation 6.9 reduces to

$$\left(\frac{dS}{dt}\right)_{\text{tot}} = \sum_{\text{exit}} \dot{m}s - \sum_{\text{inlet}} \dot{m}s + \left(\frac{dS}{dt}\right)_{\text{net,surr}}$$

$$= \dot{m}_1(s_2 - s_1) + \left(\frac{dS}{dt}\right)_{\text{net,surr}}$$

For this choice for the system, the surroundings include the water side of the heat exchanger as well as the environment:

$$\left(\frac{dS}{dt}\right)_{\text{net,surr}} = \left(\frac{dS}{dt}\right)_{\text{net,water}} + \left(\frac{dS}{dt}\right)_{\text{net,env}}$$

Applying Equation 6.8 for steady flow, we find

$$\left(\frac{dS}{dt}\right)_{\text{net,water}} = \dot{m}_3(s_4 - s_3)$$

Since interactions with the environment (a reservoir) would be internally reversible and heat transfer to the environment is absent, then

$$\left(\frac{dS}{dt}\right)_{\text{net,env}} = \left(\frac{dS}{dt}\right)_{\text{env}} = 0$$

Therefore, the rate of change of total entropy becomes

$$\left(\frac{dS}{dt}\right)_{\text{tot}} = \dot{m}_1(s_2 - s_1) + \dot{m}_3(s_4 - s_3)$$

This result is precisely the same expression developed previously by using Equation 6.11 and considering the entire heat exchanger as the system.

The rate of change of total entropy for this process is greater than zero, indicating that the process is irreversible. The source of the irreversibility is the heat transfer between the air and the water, which occurs as a result of a finite temperature difference between the two fluid streams. ■

■ EXAMPLE 6.2

Helium is expanded in a well-insulated, steady-flow nozzle from 45 psia and 810°R to 25 psia. The velocity of the helium at the nozzle inlet is small compared with the exit velocity.

(a) Determine the maximum exit velocity that could be attained by the helium.
(b) The temperature of the helium is measured to be 670°R at the nozzle exit. Determine the actual exit velocity.
(c) Determine the increase in total entropy per unit mass of helium for the actual process and explain the source of the irreversibility.

Solution.

(a) As the problems become more complex, approaching the solutions in a systematic fashion is increasingly important. From the problem organization outlined in Section 4.3, a sketch of the system is shown in Figure 6.2(a). Since the nozzle operates in steady flow and has a single inlet and exit, the conservation-of-mass equation, Equation 3.16, reduces to

$$\dot{m}_i = \dot{m}_e$$

Furthermore, the nozzle is adiabatic, and no work is done on or by the helium, so that the conservation-of-energy equation, Equation 4.31, reduces to

$$\dot{m}_i(h + e_k + e_p)_i - \dot{m}_e(h + e_k + e_p)_e = 0$$

or

$$(h_i - h_e) + \left(\frac{V_i^2 - V_e^2}{2}\right) + g(z_i - z_e) = 0$$

The potential-energy change of the helium is negligible in the nozzle, and the inlet velocity is small compared with the exit velocity. Thus

$$V_e = \sqrt{2(h_i - h_e)}$$

For calculation of the exit velocity of the helium, the change in enthalpy of the helium as it flows through the nozzle must be determined. The inlet state is fixed, but in part (a) of the problem the exit state is constrained by the exit pressure alone. The application of the second law is necessary in order to determine a second property at the exit state. From Equation 6.11 we know that the rate of change of total entropy must be greater than or equal to zero. For steady flow and an adiabatic nozzle, this equation reduces to

$$\dot{m}_e s_e - \dot{m}_i s_i \geq 0$$

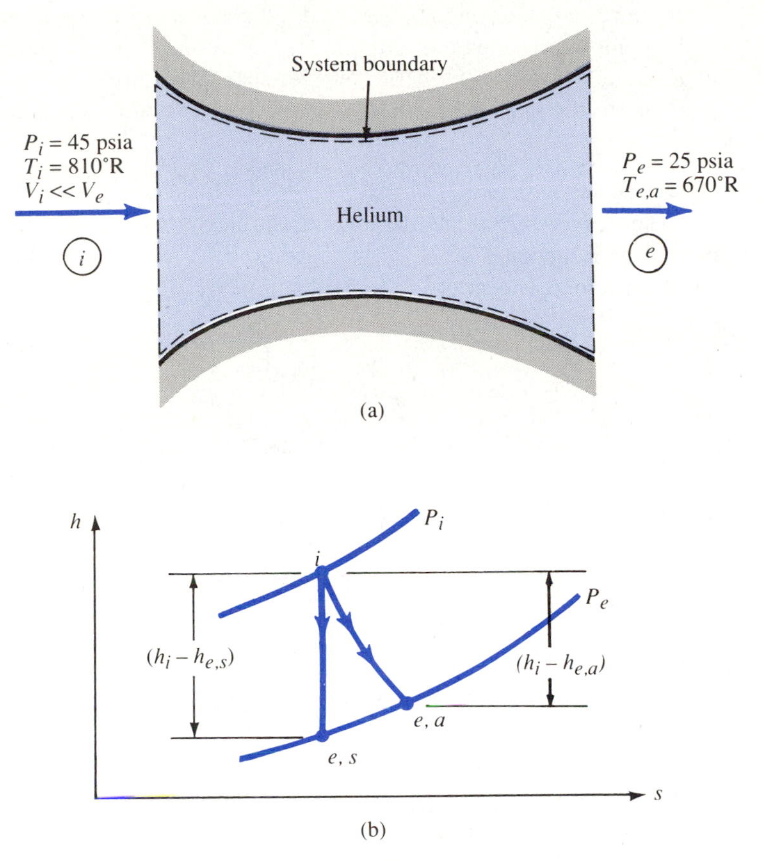

Figure 6.2 (a) Sketch of adiabatic, steady-flow nozzle; (b) enthalpy-entropy diagram for Example 6.2.

since $(dS/dt)_{sys} = 0$ for steady flow and $\dot{Q}_k = 0$ for a perfectly insulated system. Therefore, we can conclude that

$$s_e \geq s_i$$

This result can be used to determine the exit state that would produce the maximum velocity. For this purpose an *h-s* diagram is used, as shown in Figure 6.2(b). First, we must recognize that the second law requires that the exit state for this process be either directly below or to the right of the inlet state on the constant-pressure line representing the exit pressure P_e. Second, the exit velocity is proportional to the square root of the enthalpy change of the helium, and therefore the greater the vertical distance between the inlet and exit states on the *h-s* diagram, the greater will be the exit velocity. With these constraints the maximum exit velocity will be achieved if the exit state of the helium corresponds to the point e, s in Figure 6.2(b). Under these conditions the entropy at the exit state is equal to the entropy at the inlet state. Therefore, the exit state corresponding to the maximum exit velocity is fixed by the two conditions

$$P_e = 25\,\text{psia} \quad \text{and} \quad s_{e,s} = s_i$$

The temperature at the exit for an isentropic process can be determined with the aid of Equation 5.40 since the specific heats of a monatomic gas such as helium are independent of temperature:

$$\frac{T_{e,s}}{T_i} = \left(\frac{P_e}{P_i}\right)_s^{(k-1)/k}$$

Using the value of $k = 1.667$ from Table D.9E, we have

$$T_{e,s} = (810°\text{R})\left(\frac{25\ \text{psia}}{45\ \text{psia}}\right)^{(1.667-1)/1.667} = 640°\text{R}$$

and the maximum exit velocity is

$$\begin{aligned} V_{e,s} &= \sqrt{2(h_i - h_{e,s})} = \sqrt{2c_p(T_i - T_{e,s})} \\ &= [2(1.24\,\text{Btu/lb}_\text{m}\cdot°\text{R})(810 - 640)°\text{R} \\ &\quad \times (778\,\text{ft}\cdot\text{lb}_\text{f}/\text{Btu})(32.2\,\text{lb}_\text{m}\cdot\text{ft/lb}_\text{f}\cdot s^2)]^{1/2} \\ &= \underline{\underline{3250\,\text{ft/s}}} \end{aligned}$$

(b) If the actual exit temperature is 670°R the actual exit velocity will be somewhat less than the maximum velocity:

$$\begin{aligned} V_{e,a} &= \sqrt{2c_p(T_i - T_{e,a})} \\ &= [2(1.24\ \text{Btu/lb}_\text{m}\cdot°\text{R})(810 - 670)°\text{R} \\ &\quad \times (778\ \text{ft}\cdot\text{lb}_\text{f}/\text{Btu})(32.2\ \text{lb}_\text{m}\cdot\text{ft/lb}_\text{f}\cdot s^2)]^{1/2} \\ &= \underline{\underline{2950\ \text{ft/s}}} \end{aligned}$$

(c) The increase in total entropy per unit mass of helium can be determined from the previously simplified form of Equation 6.11,

$$(ds)_\text{tot} = s_e - s_i$$

Notice that the exit state corresponding to the maximum velocity ($s_{e,s} = s_i$) results in a zero value for the total entropy change. Thus this process is reversible. Since the actual process is irreversible, the total entropy change is greater than zero. This entropy change can be evaluated by using Equation 5.30 since the specific heats of helium are constant:

$$(ds)_{\text{tot}} = s_{e,a} - s_i = c_p \ln\left(\frac{T_{e,a}}{T_i}\right) - R \ln\left(\frac{P_e}{P_i}\right)$$

$$= (1.24 \text{ Btu/lb}_\text{m}\cdot°\text{R}) \ln\left(\frac{670°\text{R}}{810°\text{R}}\right)$$

$$- \left(\frac{1.986 \text{ Btu/lb}_\text{m}\cdot\text{mol}\cdot°\text{R}}{4 \text{ lb}_\text{m}/\text{lb}_\text{m}\cdot\text{mol}}\right) \ln\left(\frac{25 \text{ psia}}{45 \text{ psia}}\right)$$

$$= \underline{\underline{0.057 \text{ Btu/lb}_\text{m}\cdot°\text{R}}}$$

Examination of Figure 6.2(b) reveals that any allowable exit state other than the state corresponding to the maximum velocity must have an exit enthalpy that is larger than $h_{e,s}$ for this adiabatic nozzle. For an ideal gas, then, the actual exit temperature will be greater than the exit temperature corresponding to the reversible process because the enthalpy of an ideal gas depends on temperature alone. The source of the irreversibility in the actual process is friction associated with the fluid flow. The dissipation of energy resulting from frictional effects causes the helium to leave the nozzle at a higher temperature and prevents the maximum velocity from being achieved. ■

6.3 REVERSIBLE WORK AND IRREVERSIBILITY

The rate of change of total entropy, as expressed by Equation 6.11, gives an indication of the extent of irreversibility associated with a process as well as those interactions that occur between a system and its surroundings during that process. However, a physical interpretation of the terms appearing in this equation is not particularly easy because entropy is more abstract than, say, energy or mass. A physical interpretation, important for an understanding of the concepts and principles involved, can be obtained by applying the equations of conservation of mass and energy together with the equation for the rate of change of total entropy, Equation 6.11, to an arbitrary thermodynamic system.

Consider the thermodynamic system shown in Figure 6.3. The system can have any number of inlets and exits, work can be done on or by the system, and heat-transfer interactions with the surroundings can occur. The surroundings could possibly be composed of any number of subsystems—for example, several reservoirs at different temperatures. The general form of the conservation-of-energy equation, Equation 4.29, applies to this system:

$$\dot{Q} - \dot{W} + \sum_{\text{inlet}} \dot{m}(h + e_k + e_p) - \sum_{\text{exit}} \dot{m}(h + e_k + e_p) = \frac{dE_{\text{sys}}}{dt} \tag{6.12}$$

This system uses some available energy resources (e.g., the initial energy of the system and the energy in the fluid entering the system as well as heat transfer to or from the surroundings) and produces a change in these resources. Now we wish to consider

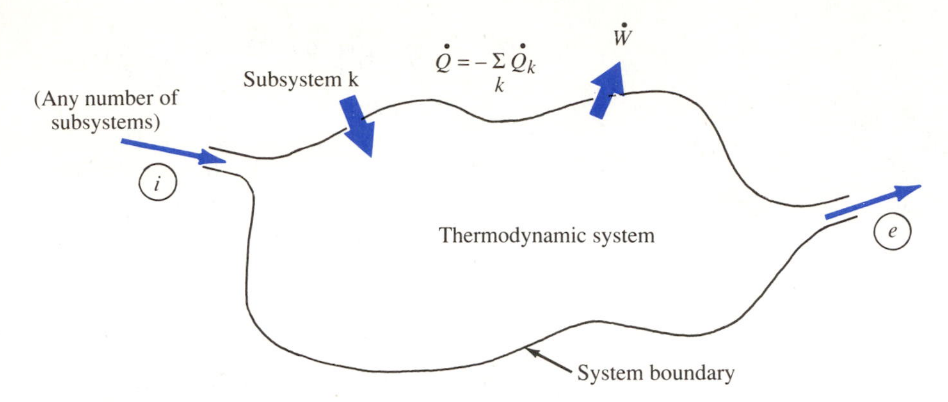

Figure 6.3 Arbitrary thermodynamic system.

the following question: For given changes in these available resources, what is the maximum power output (or minimum power input) that may result from those changes? To answer this question we stipulate that the changes in the resources are fixed:

1. The initial and final states of the system are given, but the process or processes by which the change of state is achieved are not fixed.
2. The change in the energy of each external resource is the same as that which occurs with the actual system. Each thermal-energy reservoir, except the environment because it is not a resource, has the same net heat transfer to or from it. Each external subsystem supplies (or takes in) the same amount of mass at the same rate and with the same fluid properties as that which enters and leaves the actual system.

Notice that these requirements place no restrictions on the process that a system undergoes (or the devices used during the process) while using these resources. Thus, to use these same resources to produce the maximum power output (or minimum power input) we select internally reversible devices and components, and we require that all processes be totally reversible. Since this restriction means that any heat transfer must occur reversibly, reversible heat engines and reversible heat pumps (or refrigerators) may be required to achieve total reversibility.

One of the simplest arrangements that could be used to produce the maximum power output is illustrated in Figure 6.4. Reversible heat engines and reversible heat pumps are inserted between the reservoirs and the environment so that the same heat transfer occurs for each resource. The power, $\dot{W}_k$, associated with each of these reversible, cyclic devices is, from Equation 5.6

$$\dot{W}_k = -\dot{Q}_k\left(1 - \frac{T_0}{T_k}\right) \tag{6.13}$$

and the power associated with reversible device R_0 is

$$\dot{W}_0 = -\dot{Q}_{\text{rev,sys}}\left(1 - \frac{T_0}{T}\right) \tag{6.14}$$

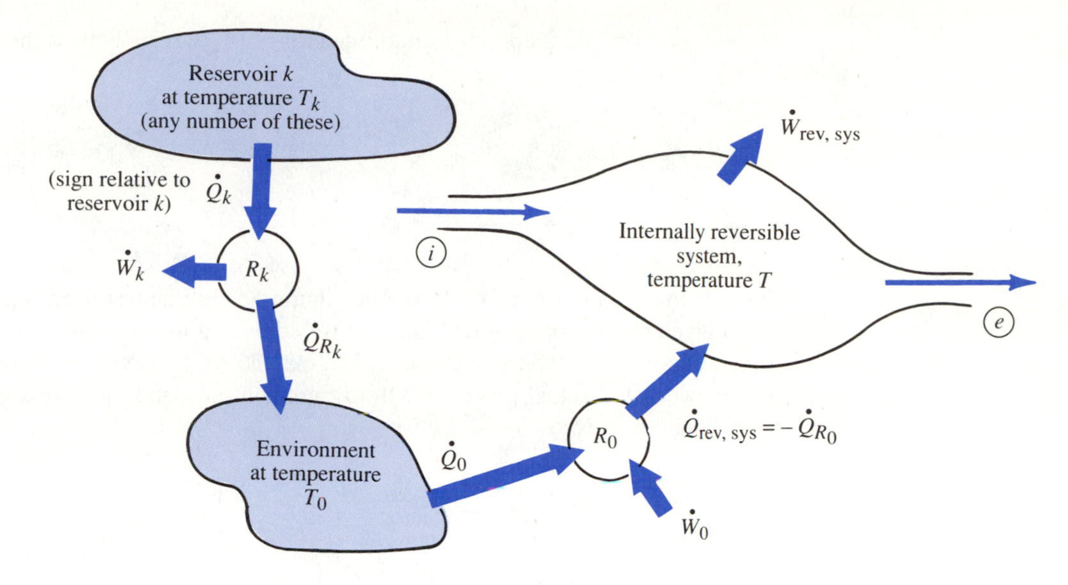

Figure 6.4 One arrangement for producing the maximum power for a given change of resources.

The power produced by the internally reversible system is, from Equation 4.29,

$$\dot{W}_{\text{rev,sys}} = \dot{Q}_{\text{rev,sys}} - \left[\sum_{\text{exit}} \dot{m}(h + e_k + e_p) - \sum_{\text{inlet}} \dot{m}(h + e_k + e_p)\right] - \frac{dE_{\text{sys}}}{dt} \quad (6.15)$$

Note that $\dot{W}_{\text{rev,sys}}$ and $\dot{Q}_{\text{rev,sys}}$ are, in general, different from the actual work and heat transfer, $\dot{W}$ and $\dot{Q}$, associated with the actual system.

Now the net reversible power produced, $\dot{W}_{\text{rev}}$, with this given change of resources is simply the algebraic sum of the power associated with the internally reversible system and each reversible cyclic device (i.e., the sum of Equations 6.13 through 6.15):

$$\dot{W}_{\text{rev}} = \dot{W}_{\text{rev,sys}} + \sum_k \dot{W}_k + \dot{W}_0$$

$$\dot{W}_{\text{rev}} = \dot{Q}_{\text{rev,sys}}\left(\frac{T_0}{T}\right) - \left[\sum_{\text{exit}} \dot{m}(h + e_k + e_p) - \sum_{\text{inlet}} \dot{m}(h + e_k + e_p)\right] - \frac{dE_{\text{sys}}}{dt} - \sum_k \dot{Q}_k\left(1 - \frac{T_0}{T_k}\right) \quad (6.16)$$

Since the system undergoes an internally reversible process, then, from Equation 6.8,

$$\dot{Q}_{\text{rev,sys}} = T\left(\frac{dS}{dt}\right)_{\text{net,sys}} = T\left\{\sum_{\text{exit}} \dot{m}s - \sum_{\text{inlet}} \dot{m}s + \left(\frac{dS}{dt}\right)_{\text{sys}}\right\} \quad (6.17)$$

After substitution of Equation 6.17 into Equation 6.16, we can express the reversible power as

$$\dot{W}_{rev} = -\left[\sum_{exit} \dot{m}(h + e_k + e_p - T_0 s) - \sum_{inlet} \dot{m}(h + e_k + e_p - T_0 s)\right] - \frac{d(E - T_0 S)_{sys}}{dt} - \sum_k \dot{Q}_k \left(1 - \frac{T_0}{T_k}\right) \tag{6.18}$$

While we may suspect that $\dot{W}_{rev}$ is the maximum power output (or minimum power input) for a given change of resources because we stipulated that all changes must occur reversibly, we have not established that fact. We may do so, however, by examining the difference between the actual power, $\dot{W}$, from Equation 6.12 and the reversible power, $\dot{W}_{rev}$, from Equation 6.18:

$$\dot{W}_{rev} - \dot{W} = T_0 \left[\sum_{exit} \dot{m}s - \sum_{inlet} \dot{m}s + \left(\frac{dS}{dt}\right)_{sys}\right] - \dot{Q} - \sum_k \dot{Q}_k \left(1 - \frac{T_0}{T_k}\right) \tag{6.19}$$

For the actual system illustrated in Figure 6.3,

$$\dot{Q} = -\sum_k \dot{Q}_k$$

and therefore Equation 6.19 may be written as

$$\dot{W}_{rev} - \dot{W} = T_0 \left[\sum_{exit} \dot{m}s - \sum_{inlet} \dot{m}s + \left(\frac{dS}{dt}\right)_{sys} + \sum_k \frac{\dot{Q}_k}{T_k}\right]$$

or

$$\dot{W}_{rev} - \dot{W} = T_0 \left(\frac{dS}{dt}\right)_{tot} \geq 0 \tag{6.20}$$

where we have used Equation 6.11 in arriving at the final result in Equation 6.20.

Thus the reversible power is always greater than or equal to the actual power associated with given changes in resources. We may say, then, that the reversible power represents the maximum power output (or the minimum power input) for given changes in available resources. Real systems produce less power or work output (or require more power or work input) than could be produced if the changes could be carried out in a totally reversible manner.

$$\dot{W} \leq \dot{W}_{rev} \tag{6.21}$$

The difference between $\dot{W}_{rev}$ and $\dot{W}$, which can be interpreted as a lost opportunity to produce work or power, is called the *irreversibility rate*, $\dot{I}$:

$$\dot{I} \equiv T_0 \left(\frac{dS}{dt}\right)_{tot} = \dot{W}_{rev} - \dot{W} \geq 0 \tag{6.22}$$

Substitution of Equation 6.11 into Equation 6.22 yields the following expression for the irreversibility rate:

$$\dot{I} = T_0 \left(\frac{dS}{dt}\right)_{\text{tot}} = T_0 \left(\sum_{\text{exit}} \dot{m}s - \sum_{\text{inlet}} \dot{m}s + \frac{dS_{\text{sys}}}{dt} + \sum_k \frac{\dot{Q}_k}{T_k}\right) \tag{6.23}$$

where the k subscripts denote thermal-energy reservoir quantities. Thus the irreversibility rate must be greater than or equal to zero. If a process is totally reversible, the irreversibility is zero and the actual work produced is equal to the reversible work. Irreversibilities, however, cause the actual work to be less than the reversible work.

Equation 6.22 may also be written in differential form as

$$\delta I = T_0 \, dS_{\text{tot}} \tag{6.24}$$

and on a unit-mass basis as

$$\delta i = T_0 \, ds_{\text{tot}}$$

Equations 6.22 through 6.24 reveal that the irreversibility is directly proportional to the rate of change of total entropy. This result is consistent with the interpretation of the increase-in-entropy principle discussed in Chapter 5. That is, the irreversibility is zero for the totally reversible process, and the irreversibility increases as the total entropy change increases.

A quantity of more practical significance than the work actually done on or by a system is the useful work of the system. The *useful work* (or *useful power*) is defined as the difference between the work (or power) actually done on or by a system and the work (or power) done on or by the environment at pressure P_0 as the boundary of the system expands or contracts:

$$\dot{W}_{\text{useful}} \equiv \dot{W} - P_0 \frac{d\mathbf{V}_{\text{sys}}}{dt} \tag{6.25}$$

Since the difference between work and useful work is merely the work done on or by the environment at pressure P_0, the same is also true for the reversible work and the reversible useful work; the difference between these two quantities is the work, $P_0(d\mathbf{V}_{\text{sys}}/dt)$, done on or by the environment at pressure P_0. Thus an expression similar to Equation 6.25 may also be written to relate the reversible useful power (or reversible useful work) to the reversible power (or reversible work).

$$\dot{W}_{\text{rev,useful}} = \dot{W}_{\text{rev}} - P_0 \frac{d\mathbf{V}_{\text{sys}}}{dt} \tag{6.26}$$

To clarify the distinction between work and useful work, consider the compression of a fluid contained in a piston-cylinder assembly. If the fluid is compressed without friction at a constant pressure P from an initial volume $\mathbf{V}_1$ to a final volume $\mathbf{V}_2$, the work done on the fluid is $P(\mathbf{V}_2 - \mathbf{V}_1)$. This work is the actual amount of work that must be done on the fluid in order to compress it from state 1 to state 2. This work must be supplied from external sources, that is, the surroundings. However, part of the work is done by the environment, which exerts a pressure P_0 on the piston, and this work done by the environment amounts to $P_0(\mathbf{V}_2 - \mathbf{V}_1)$. In this example, therefore, the amount of additional work (useful work) that must be supplied to compress the fluid is $P(V_2 - V_1) - P_0(V_2 - V_1)$; that is, the magnitude of the useful work is less than the work actually done on the system by an amount equal to the work done by the environment. Only the useful work would have to be produced by a device external to the system.

Upon substituting Equation 6.18 into 6.26, we obtain an expression for the *reversible useful power*:

$$\dot{W}_{\text{rev,useful}} = -\left[\sum_{\text{exit}} \dot{m}(h + e_k + e_p - T_0 s) - \sum_{\text{inlet}} \dot{m}(h + e_k + e_p - T_0 s)\right] - \frac{d(E + P_0 V - T_0 S)_{\text{sys}}}{dt} - \sum_k \dot{Q}_k \left(1 - \frac{T_0}{T_k}\right) \quad (6.27)$$

If we subtract Equation 6.25 from Equation 6.26, then we find, from Equation 6.22, that the difference between the reversible power and the power actually produced is equal to the difference between the reversible useful work and the useful work. This difference is just the *irreversibility rate* $\dot{I}$,

$$\dot{I} \equiv \dot{W}_{\text{rev}} - \dot{W} = \dot{W}_{\text{rev,useful}} - \dot{W}_{\text{useful}} \geq 0 \quad (6.28)$$

EXAMPLE 6.3

A storage vessel with a volume of 0.15 m^3 contains air initially at 600 K and 300 kPa. The temperature of the air is allowed to decrease to 400 K by means of heat transfer to the environment, which is at 298 K and 100 kPa. Determine the reversible useful work and the irreversibility associated with this process. Explain the significance of the reversible useful work.

Solution. A sketch of the storage vessel is shown in Figure 6.5(a). Since the system is closed, the conservation of mass requires that

$$m_1 = m_2$$

At the stated pressure and temperature the air can be assumed to behave as an ideal gas.

The reversible useful work can be determined with the aid of Equation 6.27. Since the system is closed, this equation reduces to

$$\dot{W}_{\text{rev,useful}} = -\frac{d(E + P_0 V - T_0 S)_{\text{sys}}}{dt} - \sum_k \dot{Q}_k \left(1 - \frac{T_0}{T_k}\right)$$

The energy E of the system includes the kinetic and potential energy of the center of mass of the system. For a stationary system the time rate of change of these quantities is zero. Furthermore, the storage vessel is rigid, so dV_{sys}/dt is also zero. Thus

$$\dot{W}_{\text{rev,useful}} = -\frac{d(U - T_0 S)_{\text{sys}}}{dt} - \sum_k \dot{Q}_k \left(1 - \frac{T_0}{T_k}\right)$$

The surroundings in this example are composed entirely of the environment at temperature T_0 (i.e., $T_k = T_0$). Therefore, even though heat transfer to the environment occurs, the

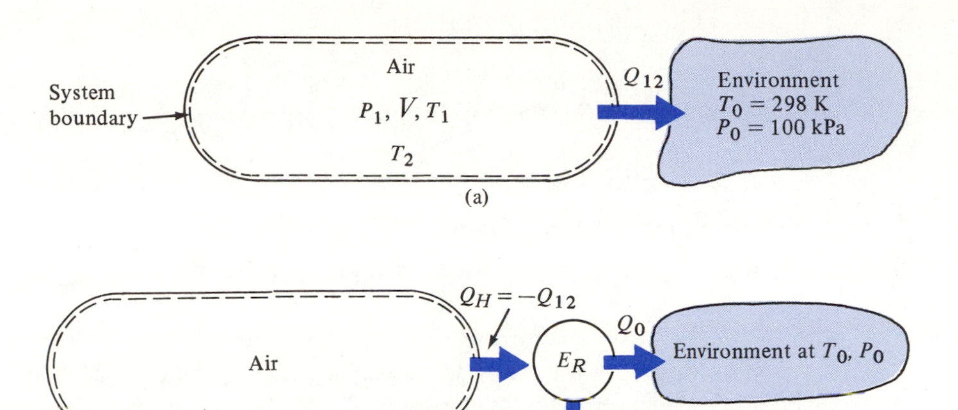

Figure 6.5 (a) Sketch for Example 6.3; (b) production of reversible work, Example 6.3.

last term in the previous equation is zero. Thus

$$\dot{W}_{\text{rev,useful}} = -\frac{d}{dt}(U - T_0 S)_{\text{sys}}$$

Integrating this expression with respect to time for the initial state 1 to the final state 2, we obtain

$$\begin{aligned} W_{12,\text{rev,useful}} &= -[(U_2 - U_1) - T_0(S_2 - S_1)]_{\text{sys}} \\ &= -m[(u_2 - u_1) - T_0(s_2 - s_1)]_{\text{sys}} \end{aligned}$$

The mass of air in the storage vessel is determined from the initial temperature, pressure, and volume by using the ideal-gas equation of state:

$$m = \frac{P_1 \mathbf{V}_1}{RT_1} = \frac{(3 \times 10^5\,\text{N/m}^2)(0.15\ \text{m}^3)(28.97\ \text{kg/kg·mol})}{(8.314\ \text{kJ/kg·mol·K})(600\text{K})(10^3\ \text{N·m/kJ})} = 0.261\ \text{kg}$$

Since ideal-gas behavior has been assumed and the end-state temperatures are known, the change in specific internal energy can be determined directly from Table D.1:

$$u_2 - u_1 = (286.4 - 435.0)\text{kJ/kg} = -148.6\,\text{kJ/kg}$$

The entropy change of the air can also be calculated with the aid of Table D.1, using Equation 5.28:

$$s_2 - s_1 = s_2^\circ - s_1^\circ - R \ln \frac{P_2}{P_1}$$

The pressure ratio can be related to the end-state temperatures by using the ideal-gas equation of state since the specific volume remains constant:

$$\frac{P_2}{P_1} = \frac{T_2}{T_1}$$

Thus

$$s_2 - s_1 = (5.9916 - 6.4087)\text{kJ/kg·K} - \left(\frac{8.314\ \text{kJ/kg·mol·K}}{28.97\ \text{kg/kg·mol}}\right) \ln\left(\frac{400\ \text{K}}{600\ \text{K}}\right)$$

$$= -0.301\ \text{kJ/kg·K}$$

The reversible useful work is therefore

$$W_{12,\text{rev,useful}} = -(0.261\ \text{kg})[-148.6\ \text{kJ/kg} - (298\ \text{K})(-0.310\ \text{kJ/kg·K})]$$

$$= \underline{\underline{+15.4\ \text{kJ}}}$$

The irreversibility can easily be determined from Equation 6.28 since there is no useful work associated with this process. (The volume of the closed system does not change.)

$$I_{12} = W_{12,\text{rev,useful}} - W_{12,\text{useful}}$$

Thus

$$I_{12} = W_{12,\text{rev,useful}} = \underline{\underline{+15.4\ \text{kJ}}}$$

Notice that the irreversibility for this process is greater than zero, indicating that the process is not totally reversible. The process that the air undergoes is internally reversible since an equal amount of heat transfer to the air would be sufficient to cause the air to regain its initial state. Similarly, the environment also undergoes an internally reversible process. Even though these processes are each internally reversible, the overall process is irreversible because the heat transfer between the system and the environment takes place through a finite temperature difference.

Although the work (and the useful work as well) is zero because the system boundary is rigid, the reversible useful work is not zero. If the system does no work, the fact that the reversible useful work could be nonzero might seem implausible. Recall, however, that the concept of reversible work is based on a totally reversible process. The reversible work could be produced only if a totally reversible process could be conceived for the conditions stated in the problem.

For the overall process to be totally reversible, the heat transfer from the system must occur with an infinitesimally small temperature difference. This effect could be achieved with a reversible heat engine that has heat transfer to the heat engine from the air in the storage vessel and from the heat engine to the environment. This arrangement is shown schematically in Figure 6.5(b). The work produced by the reversible heat engine can be shown to be equal to the reversible work calculated above in this instance, as follows.

For the heat engine, which operates in a cycle, the conservation-of-energy equation states that

$$\delta W_E = \delta Q_H + \delta Q_0$$

Since the heat engine is totally reversible, the heat-transfer terms are related to the absolute temperatures by Equation 5.5, or

$$\frac{\delta Q_H}{\delta Q_0} = -\frac{T_H}{T_0}$$

The heat transfer from the heat engine to the environment is therefore

$$Q_0 = -T_0 \oint \frac{\delta Q_H}{T_H}$$

But

$$\delta Q_H = -\delta Q_{12}$$

and for the closed system with no work present,

$$\delta Q_{12} = dU_{\text{sys}}$$

Furthermore, T_H is the temperature of the air, so the expression for Q_0 can be written as

$$Q_0 = T_0 \int_1^2 \left(\frac{dU}{T}\right)_{\text{sys}}$$

But this integral is equal to the entropy change of the air because the volume remains constant (see Equation 5.18):

$$(S_2 - S_1) = \int_1^2 \frac{dU}{T} + R \ln \frac{V_2}{V_1}$$

Therefore

$$Q_0 = T_0(S_2 - S_1)_{\text{sys}}$$

and the work produced by the reversible engine can be expressed as

$$\begin{aligned} W_E = Q_H + Q_0 &= -Q_{12} + T_0(S_2 - S_1)_{\text{sys}} \\ &= -[(U_2 - U_1) - T_0(S_2 - S_1)]_{\text{sys}} \end{aligned}$$

This result is precisely the same expression developed for the reversible useful work. ■

The significance of the reversible useful work should now be evident. The reversible useful work represents the amount of work that *could* be produced if all processes associated with the given change in resources (in this case the given change of state) occurred in a reversible manner. Because of this interpretation, the irreversibility can also be thought of as representing a loss of work or, more properly, a loss of the opportunity to do work. That is, an additional amount of work, equal in magnitude to the irreversibility, could be produced if all processes occurred in a totally reversible manner.

6.4 MAXIMUM WORK AND AVAILABILITY

While the reversible work represents the maximum amount of work output for a given *change of resources*, determining the *maximum* amount of reversible useful work that could theroretically be obtained from a substance in a given *intital state* is also of interest.

That is, considering all of the processes that could be conceived for the substance to undergo from given initial conditions, what is the maximum amount of reversible useful work that could be achieved? In addressing this question, keep in mind that a substance in a given initial state is capable of doing work only until it is in thermodynamic equilibrium with the environment. Once thermodynamic equilibrium with the environment is reached, the substance is said to be in the *dead state* and is not capable of further producing work. The pressure and temperature of the environment are denoted by P_0 and T_0, respectively, and the properties of the substance at this *dead state* are also denoted by the subscript 0. Furthermore, in the dead state the kinetic and potential energies of the substance are at a minimum; the velocity is zero relative to the environment, and the elevation is the same as that of the environment. Thus the *maximum useful work* is obtained when the substance in a given initial state proceeds reversibility to the dead state while heat transfer to the environment at T_0, the lowest naturally occurring reservoir temperature, occurs reversibly. For reversible heat transfer to occur, a reversible heat engine or reversible heat pump might have to be employed. The processes required to produce the maximum useful work with an arbitrary system are illustrated in Figure 6.6.

The maximum useful power for the system shown in Figure 6.6 can be determined from Equation 6.27 provided that the fluid leaving the system exits at the dead state, only reversible heat transfer with the environment at T_0 occurs, and the final state of the mass within the system is at a state of thermodynamic equilibrium with the environment. For these conditions Equation 6.27 gives

$$\dot{W}_{\text{max,useful}} = -\left[\sum_{\text{exit}} \dot{m}(h_0 + e_{k_0} + e_{p_0} - T_0 s_0) - \sum_{\text{inlet}} \dot{m}(h + e_k + e_p - T_0 s)\right] - \frac{d(E + P_0 V - T_0 S)_{\text{sys}}}{dt} \tag{6.29}$$

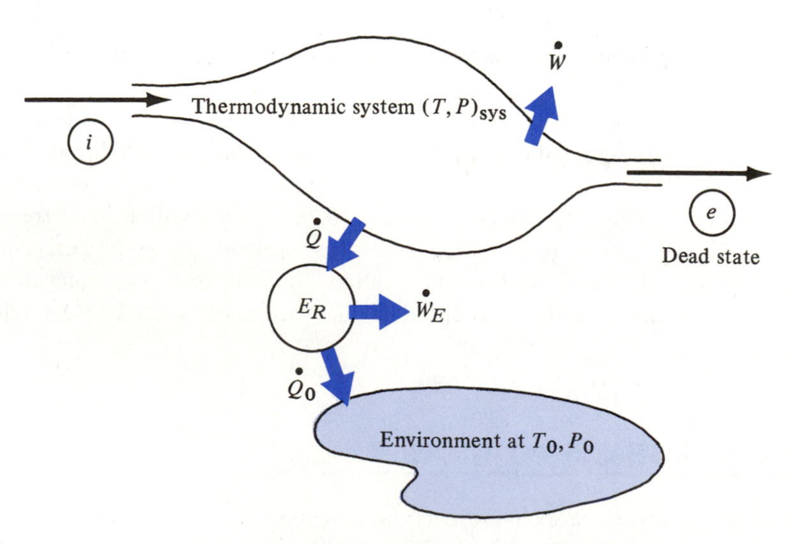

Figure 6.6 A method for producing the maximum useful work.

The first term in brackets in Equation 6.29 can be written in terms of the mass flow into the system and the rate of change of mass of the system by using the conservation-of-mass equation, Equation 3.2, and noting that all properties at the dead state are constant:

$$\sum_{\text{exit}} \dot{m}(h_0 + e_{k_0} + e_{p_0} - T_0 s_0)$$

$$= \sum_{\text{inlet}} \dot{m}(h_0 + e_{k_0} + e_{p_0} - T_0 s_0) - \frac{d[m(h_0 + e_{k_0} + e_{p_0} - T_0 s_0)]_{\text{sys}}}{dt}$$

and since $h_0 = u_0 + P_0 v_0$, then

$$\sum_{\text{exit}} \dot{m}(h_0 + e_{k_0} + e_{p_0} - T_0 s_0)$$

$$= \sum_{\text{inlet}} \dot{m}(h_0 + e_{k_0} + e_{p_0} - T_0 s_0) - \frac{d(E_0 + P_0 \mathbf{V}_0 - T_0 S_0)_{\text{sys}}}{dt} \tag{6.30}$$

Substituting this expression into Equation 6.29 gives

$$\dot{W}_{\text{max,useful}} = \sum_{\text{inlet}} \dot{m}[(h + e_k + e_p - T_0 s) - (h_0 + e_{k_0} + e_{p_0} - T_0 s_0)] - \frac{d[(E + P_0 \mathbf{V} - T_0 S) - (E_0 + P_0 \mathbf{V}_0 - T_0 S_0)]_{\text{sys}}}{dt} \tag{6.31}$$

Just as the definition of enthalpy was motivated by the sum $u + Pv$ appearing in the conservation-of-energy equation, the sums appearing in Equation 6.31 motivate the definitions of two *availability functions*. The quantity ψ, commonly called the *stream availability* or *open-system availability* since it is normally associated with fluid streams, is defined as

$$\psi \equiv (h + e_k + e_p - T_0 s) - (h_0 + e_{k_0} + e_{p_0} - T_0 s_0) \tag{6.32}$$

The remaining sum in Equation 6.31 is usually called the *closed-system availability*, Φ, and is defined as

$$\Phi \equiv (E + P_0 \mathbf{V} - T_0 S) - (E_0 + P_0 \mathbf{V}_0 - T_0 S_0) \tag{6.33}$$

On a unit-mass basis the closed-system availability is

$$\phi \equiv (e + P_0 v - T_0 s) - (e_0 + P_0 v_0 - T_0 s_0) \tag{6.34}$$

The open- and closed-system availabilities are defined in terms of sums of properties and they are therefore thermodynamic properties. Once P_0 and T_0 are specified, the values for the open- and closed-system availabilities are determined by specifying the state of the system.

In the remainder of this text the term *availability* alone will be used to denote both ψ and Φ. The context in which the term is used is sufficient to determine which of these quantities is intended.

With the definitions in Equations 6.32 and 6.33, Equation 6.31 can be written in a more compact form as

$$\dot{W}_{\text{max,useful}} = \sum_{\text{inlet}} \dot{m}\psi - \frac{d\Phi_{\text{sys}}}{dt} \tag{6.35}$$

The term *availability* is used because, as can be seen from Equation 6.35, the functions in Equations 6.32 and 6.33 are related to the maximum amount of energy that could be converted to useful work by means of reversible processes during which heat transfer occurs only with the environment. For a closed system (no inlets or exits) the maximum useful work is obtained by integrating Equation 6.35 with respect to time and noting that the availability in the dead state is zero ($\Phi_0 = 0$),

$$W_{\text{max,useful}} = \Phi_{\text{initial}}$$

In other words, the maximum useful work for a closed system is equal to the availability in the initial state. For an open, steady system Equation 6.35 reduces to

$$\dot{W}_{\text{max,useful}} = \sum_{\text{inlet}} \dot{m}\psi$$

Therefore, the maximum useful power for a steady-flow system is equal to the net inflow of availability into the system.

The reversible useful power from Equation 6.27 can also be expressed in terms of the availability functions by adding Equation 6.30 to Equation 6.27 and rearranging:

$$\dot{W}_{\text{rev,useful}} = -\left(\sum_{\text{exit}} \dot{m}\psi - \sum_{\text{inlet}} \dot{m}\psi\right) - \frac{d\Phi_{\text{sys}}}{dt} - \sum_k \dot{Q}_k\left(1 - \frac{T_0}{T_k}\right) \qquad (6.36)$$

Equation 6.36 indicates that if heat transfer occurs only with a reservoir at the lowest naturally occurring sink temperature T_0, the reversible useful power is equal to the net rate of decrease of the availability.

Substituting the expression for the reversible useful power from Equation 6.36 into Equation 6.28 yields an expression for the irreversibility rate in terms of the availability functions,

$$\dot{I} = -\left(\sum_{\text{exit}} \dot{m}\psi - \sum_{\text{inlet}} \dot{m}\psi\right) - \frac{d\Phi_{\text{sys}}}{dt} - \sum_k \dot{Q}_k\left(1 - \frac{T_0}{T_k}\right) - \dot{W}_{\text{useful}} \geq 0 \qquad (6.37)$$

Equation 6.23 is usually the most convenient expression to use for determining the irreversibility associated with a process. In the examples in this and succeeding chapters, it is used extensively as the basis for a second-law analysis.

At this point a summary of the preceding development is worthwhile. The important terms introduced thus far are summarized in Table 6.1. A general expression for the rate of change of total entropy was developed first (Equation 6.9). Next we established an upper bound on the power (or work) that could be produced for a given change in available resources, the reversible power (Equation 6.18). The concept of useful power or useful work was introduced, and an upper bound for the useful power that could be achieved for a given change of resources, the reversible useful power, was also established (Equation 6.27). The irreversibility rate was defined as the difference between the reversible power and the power actually produced (Equation 6.22). A physical interpretation of the irreversibility rate is that it represents an additional amount of power that could have been produced if a change of resources had been achieved by means of reversible processes rather than the actual process. The irreversibility rate was also found to be directly proportional to the rate of change of total entropy. Finally, the maximum

TABLE 6.1 SUMMARY OF IMPORTANT DEFINITIONS DEVELOPED FOR SECOND-LAW ANALYSIS

Term	Symbol	Definition	Equation/comments
Power (or work)	$\dot{W}$	The actual amount of power output (or input) associated with a *given change of resources*	This term is the power term that appears in the conservation-of-energy equation, Equation 4.29
Reversible power	$\dot{W}_{rev}$	The maximum amount of power output (or minimum input) associated with a *given change of resources*; this term is the power that would be produced if the change in the resources could be achieved with a totally reversible process	Defined in Equation 6.18; note that $\dot{W} \leq \dot{W}_{rev}$ from Equation 6.21
Useful power	$\dot{W}_{useful}$	The difference between the actual power output (or input) of the system and the power output (or input) of the system on (or by) the environment as the system boundary expands (or contracts) against the pressure P_0 of the environment, for a *given change of resources*	Defined in Equation 6.25
Reversible useful power	$\dot{W}_{rev,useful}$	The maximum amount of useful power output (or minimum useful input) associated with a *given change of resources*; this term is the useful power that would be produced if the change in the resources could be achieved with a totally reversible process	Defined in Equation 6.27; note that $\dot{W}_{useful} \leq \dot{W}_{rev,useful}$ from Equation 6.28, and $\dot{W}_{rev} = \dot{W}_{rev,useful} + P_0(dV_{sys}/dt)$
Irreversibility rate	$\dot{I}$	The difference between the reversible power and the power actually produced; also equal to the difference between the reversible useful power and the useful power	Defined in Equation 6.22; note that $\dot{I} \geq 0$; also expressed in terms of availability in Equation 6.37; the irreversibility rate can also be expressed in terms of the rate of change of total entropy, Equation 6.23
Maximum useful power	$\dot{W}_{max,useful}$	The maximum useful power is obtained when the system, *in a given initial state*, proceeds reversibly to the *dead state* with only reversible heat transfer to the environment at T_0	Defined in Equation 6.35
Open-system availability	ψ	Equation 6.32	Defined in Equation 6.32; a physical interpretation can be attached to ψ since the maximum useful power for a steady-flow system is equal to the net inflow of availability into the system
Closed-system availability	Φ	Equation 6.33	Defined in Equation 6.33; a physical interpretation for Φ is that the maximum useful work for a closed system is equal to the availability in the initial state

useful work was defined as the amount of work that could be produced by a substance in a given initial state if the substance were allowed to proceed to the dead state in a reversible manner while heat transfer occurs only with the environment at temperature T_0. An expression for the maximum useful power, in terms of the availability function introduced during the development, is given in Equation 6.35.

These concepts and definitions are simplified somewhat in succeeding sections wherein closed systems; open, steady systems; and transient systems are discussed.

EXAMPLE 6.4

A piston-cylinder assembly contains 10 lb_m of water initially at 300°F and 145 psia. During an internally reversible, isothermal-expansion process, the heat transfer to the water from a thermal-energy reservoir (whose temperature is 1100°F) is 10,000 Btu. The temperature and pressure of the environment are 537°R and 14.7 psia, respectively. Determine the work for this process, the useful work, the reversible useful work, the irreversibility, and the change in availability of the water. Explain the significance of the irreversibility and the change in availability.

Solution. A sketch of the arrangement described in this example is shown in Figure 6.7(a). Since the system is closed, the conservation-of-mass equation reduces to

$$m_1 = m_2$$

The work for this process is obtained by application of the conservation-of-energy equation, Equation 4.22, for a closed, stationary system:

$$Q_{12} - W_{12} = (U_2 - U_1)_{sys}$$

Although the heat transfer is known, the final state of the water is undetermined, so the work cannot yet be calculated. Since the process is internally reversible, the heat transfer is related to the entropy change of the water through Equation 5.13:

$$dS = \left(\frac{\delta Q}{T}\right)_{\text{int rev}}$$

Integrating this equation for the isothermal process permits us to evaluate the entropy of the water in the final state:

$$S_2 - S_1 = \frac{1}{T}\int_1^2 \delta Q = \frac{Q_{12}}{T}$$

or

$$s_2 = s_1 + \frac{Q_{12}}{mT}$$

The initial state of the water is a compressed-liquid state, and s_1 is therefore approximately equal to the entropy of the saturated liquid at the same temperature

$$s_1 \simeq s_{f@300°F} = 0.4373 \text{ Btu/lb}_m\cdot°R$$

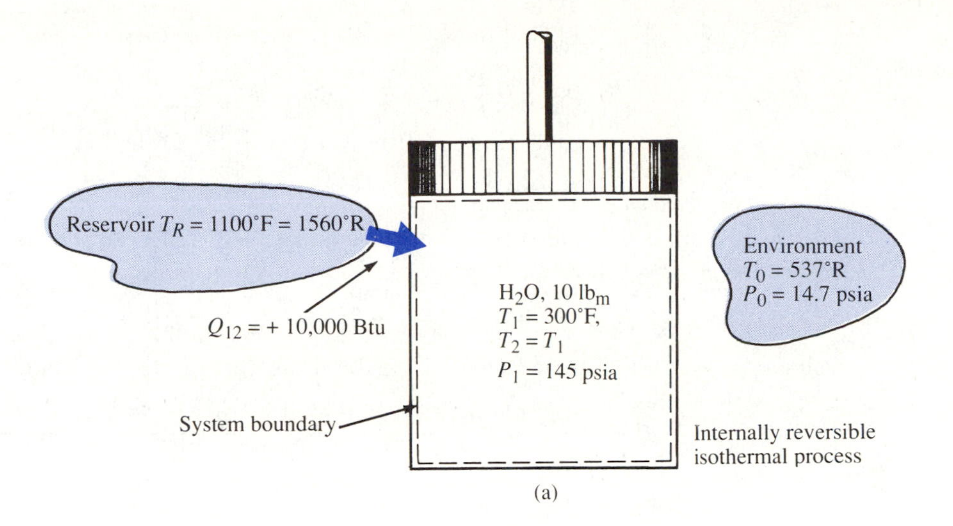

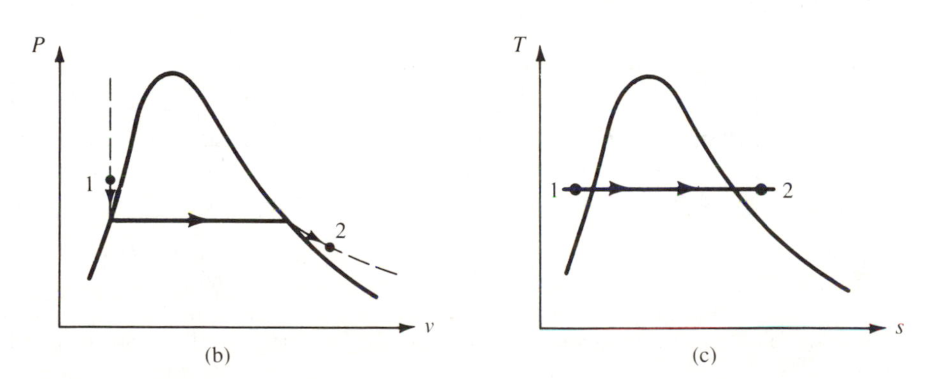

Figure 6.7 (a) Sketch for Example 6.4; (b) *P*-*v* diagram for Example 6.4; (c) *T*-*s* diagram for Example 6.4.

Therefore

$$s_2 = 0.4373 \text{ Btu/lb}_m\cdot°\text{R} + \frac{10{,}000 \text{ Btu}}{(10 \text{ lb}_m)(300 + 460)°\text{R}}$$
$$= 1.753 \text{ Btu/lb}_m\cdot°\text{R}$$

This value of entropy at 300°F indicates that the final state of the water is a state of superheated vapor. The process is therefore represented on the *P*-*v* and *T*-*s* diagrams as shown in Figures 6.7(b) and (c).

The internal energy in state 2 can be determined by interpolation in Table B.3E:

$$u_2 = 1107.4 \text{ Btu/lb}_m \quad \text{at} \quad P_2 = 26.8 \text{ psia}$$

and the work for this process is

$$W_{12} = Q_{12} - (U_2 - U_1)_{sys} = Q_{12} - m(u_2 - u_1)_{sys}$$
$$= 10{,}000\,\text{Btu} - (10\,\text{lb}_m)(1107.4 - 269.6)\text{Btu/lb}_m$$
$$= \underline{\underline{+1620\,\text{Btu}}}$$

The useful work is the difference between the work actually produced and the work done against the environment at P_0:

$$W_{12,\text{useful}} = W_{12} - P_0(\mathbf{V}_2 - \mathbf{V}_1) = W_{12} - mP_0(v_2 - v_1)$$

Using $v_1 \simeq v_f$ at 300°F from Table B.1E and v_2 from Table B.3E, we have

$$W_{12,\text{useful}} = 1620\text{ Btu} - \frac{(10\text{ lb}_m)(14.7\,\text{lb}_f/\text{in}^2)(144\text{ in}^2/\text{ft}^2)(18.53 - 0.01745)\text{ft}^3/\text{lb}_m}{(778\,\text{ft}\cdot\text{lb}_f/\text{Btu})}$$
$$= \underline{\underline{1120\text{ Btu}}}$$

The reversible useful work for a closed system can be determined from Equation 6.36:

$$W_{12,\text{rev,useful}} = -(\Phi_2 - \Phi_1)_{sys} - \sum_k \int_1^2 \delta Q_k \left(1 - \frac{T_0}{T_k}\right)$$

The last term in this expression reduces to a single integral since heat transfer occurs only with the reservoir at temperature $T_R = 1560°\text{R}$. The reservoir temperature is constant, and therefore the evaluation of the integral is quite simple:

$$W_{12,\text{rev,useful}} = -m(\phi_2 - \phi_1)_{sys} - Q_R\left(1 - \frac{T_0}{T_R}\right)$$

The change in availability is evaluated by using the definition of ϕ from Equation 6.34,

$$\phi_2 - \phi_1 = (u_2 - u_1) + P_0(v_2 - v_1) - T_0(s_2 - s_1)$$

where the changes in kinetic and potential energies have been neglected since the system is stationary. Thus

$$\phi_2 - \phi_1 = (1107.4 - 269.6)\text{Btu/lb}_m$$
$$+ \frac{(14.7\text{ lb}_f/\text{in}^2)(144\text{ in}^2/\text{ft}^2)(18.53 - 0.01745)\text{ft}^3/\text{lb}_m}{778\text{ ft}\cdot\text{lb}_f/\text{Btu}}$$
$$- (537°\text{R})(1.753 - 0.4373)\text{Btu/lb}_m\cdot°\text{R}$$
$$= 182\text{ Btu/lb}_m$$

and

$$\Phi_2 - \Phi_1 = m(\phi_2 - \phi_1) = 10\text{ lb}_m(182\,\text{Btu/lb}_m)$$
$$= \underline{\underline{1820\text{ Btu}}}$$

The heat transfer is from the reservoir so $Q_R = -10{,}000$ Btu; thus the reversible useful work is

$$W_{12,\text{rev,useful}} = -(1820 \text{ Btu}) - (-10{,}000 \text{ Btu})\left(1 - \frac{537°\text{R}}{1560°\text{R}}\right) = \underline{\underline{4740 \text{ Btu}}}$$

Since the useful work and the reversible useful work have both been calculated, the irreversibility for this process can be determined by using Equation 6.22:

$$I_{12} = W_{12,\text{rev,useful}} - W_{12,\text{useful}} = 4740 \text{ Btu} - 1120 \text{ Btu} = \underline{\underline{3620 \text{ Btu}}}$$

The irreversibility is greater than zero because the process is not totally reversible. Although the piston is frictionless and the water undergoes an internally reversible process, the heat transfer between the reservoir and the water occurs through a finite temperature difference and renders the process irreversible.

The increase in availability is a measure of the ability of the water to produce useful work. Since the availability has increased, the water is capable of producing more useful work in the final state as a result of this process than could have been produced by the water in the initial state.

The irreversibility can be interpreted as being a lost opportunity to produce work. If the process had been totally reversible, an additional amount of work, equal in magnitude to the irreversibility, could have been produced with the same change of resources, that is, the same change of state of the water and the same amount of heat transfer from the high-temperature reservoir and to the water.

The ratio of the actual useful work produced to the reversible useful work that could have been produced can be used to measure the second-law efficiency for this process. If we call this ratio the *effectiveness* ϵ of the process, then

$$\epsilon = \frac{W_{12,\text{useful}}}{W_{12,\text{rev,useful}}} = \frac{1120 \text{ Btu}}{4740 \text{ Btu}} = 0.24$$

A meaningful measure of performance for such a process is difficult to achieve without using a second-law analysis. ■

■ EXAMPLE 6.5

Five kilograms of air is contained in a rigid storage vessel at an initial pressure and temperature of 200 kPa and 550°C, respectively. A paddle wheel inserted through the side of the vessel is used to agitate the air. The paddle wheel is turned until a total of 70 kJ of work has been performed on the air. During the process the temperature of the air is maintained constant by heat transfer from the air to a thermal-energy reservoir whose temperature is 400°C. Determine the irreversibility associated with this process when (a) the air is considered to be the system, (b) the reservoir is considered to be the system, and (c) the combination of the air plus the reservoir is considered to be the system. Assume $T_0 = 298$ K and $P_0 = 100$ kPa.

Solution. A sketch of the arrangement described above is shown in Figure 6.8. The pressure and temperature of the air are such that the assumption of ideal-gas behavior is appropriate. During the process described, the air does not undergo a change of state;

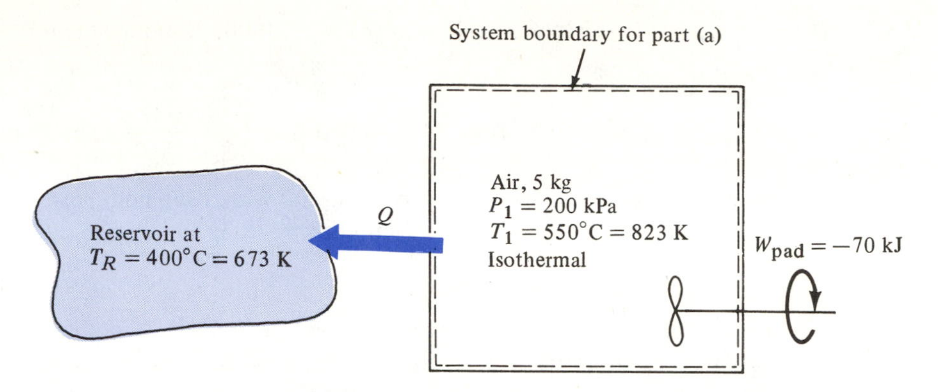

Figure 6.8 Sketch for Example 6.5.

the volume and mass of the air remain unchanged, and the temperature of the air remains constant.

The conservation-of-energy equation for the air reduces to

$$Q_{12,a} - W_{12,a} = (U_2 - U_1)_a$$

where the subscript a denotes the air in the system. And since the internal energy of the air does not change, the heat transfer from the air is equal in magnitude to the amount of work done on the air by the paddle wheel:

$$Q_{12,a} = W_{12,a} = -70 \text{ kJ}$$

(a) The irreversibility associated with this process can be determined by using Equations 6.11 and 6.22 if the air is considered to be the system. These equations reduce to

$$I_{12} = T_0(S_2 - S_1)_a + T_0 \int_1^2 \frac{\delta Q_R}{T_R}$$

The reservoir temperature is constant, and $Q_R = -Q_{12,a}$; therefore,

$$I_{12} = T_0(S_2 - S_1)_a - T_0\left(\frac{Q_{12,a}}{T_R}\right)$$

The entropy of the air does not change during this process because the state of the air does not change. Thus

$$I_{12} = -T_0\left(\frac{Q_{12,a}}{T_R}\right) = (-298 \text{ K})\left(\frac{-70 \text{ kJ}}{673 \text{ K}}\right)$$

$$= \underline{\underline{+31 \text{ kJ}}}$$

(b) If the reservoir is considered to be the system, Equation 6.22 can be used to determine the irreversibility, but Equation 6.9 should be used to determine the total entropy change since the surroundings do not consist entirely of thermal-energy reservoirs in this instance. For a closed system the integrated form of these equations gives

$$I_{12} = T_0[(S_2 - S_1)_{sys} + (S_2 - S_1)_{net,surr}]$$

The system is the reservoir, so

$$(S_2 - S_1)_{sys} = (S_2 - S_1)_{res} = \frac{Q_R}{T_R} = -\frac{Q_{12,a}}{T_R}$$

The surroundings consist of the air in the cylinder, and using Equation 6.8, we have

$$(S_2 - S_1)_{net,surr} = (S_2 - S_1)_a = 0$$

since the state of the air does not change. Hence the irreversibility is given by

$$I_{12} = -T_0\left(\frac{Q_{12,a}}{T_R}\right) = \underline{\underline{31 \text{ kJ}}}$$

(c) Finally, if the combination of the air plus the reservoir is considered to be the system, an isolated system results. The entropy change of the surroundings is therefore zero, and the irreversibility is simply

$$I_{12} = T_0(S_2 - S_1)_{sys} = T_0\left[(S_2 - S_1)_a + \frac{Q_R}{T_R}\right]$$

Again this expression becomes

$$I_{12} = -T_0\left(\frac{Q_{12,a}}{T_R}\right) = \underline{\underline{31 \text{ kJ}}}$$

The conclusion from the preceding results is that the irreversibility associated with a process is the same regardless of the choice of the system provided that the equations for irreversibility are applied properly. The irreversibility therefore accounts for internal as well as external irreversibilities.

Also worth noting at this point is that a quantity called the *entropy production*, usually defined as the difference between the net rate of change of entropy for the system and the rate of heat transfer for the system divided by the absolute temperature, is often encountered in thermodynamics. Using this definition, the entropy production is:

$$\left(\frac{dS}{dt}\right)_{prod,sys} = \left(\frac{dS}{dt}\right)_{net,sys} - \sum_j \frac{\dot{Q}_j}{T_j} \geq 0$$

Substituting the results of Equation 6.8 the rate of entropy production would be

$$\left(\frac{dS}{dt}\right)_{prod,sys} = \sum_{exit} \dot{m}s - \sum_{inlet} \dot{m}s + \left(\frac{dS}{dt}\right)_{sys} - \sum_j \frac{\dot{Q}_j}{T_j} \geq 0$$

An *internal irreversibility* is then usually defined as the product of T_0 and the rate of entropy production,

$$\dot{I}_{\text{int}} \equiv T_0 \left(\frac{dS}{dt}\right)_{\text{prod,sys}}$$

These definitions are much more limited in scope than the irreversibility based on the rate of change of total entropy. This limitation can be illustrated with this example. Notice that from the above definition the internal irreversibility is zero if a process is internally reversible. For the present example the internal irreversbility for the air is given by

$$I_{\text{int},a} = T_0(S_2 - S_1)_{\text{prod},a} = T_0 \left[(S_2 - S_1)_a - \frac{Q_{12,a}}{T_a} \right]$$

$$= -T_0 \left(\frac{Q_{12,a}}{T_a}\right) = -(298\text{ K}) \left(\frac{-70\text{ kJ}}{823\text{ K}}\right)$$

$$= 25.3\text{ kJ}$$

The fact that this value is greater than zero indicates that the process that the air undergoes is not internally reversible. The internal irreversibility of the process is due to the friction that results from the operation of the paddle wheel.

The internal irreversibility for the surroundings, that is, the reservoir, becomes

$$I_{\text{int,res}} = T_0(S_2 - S_1)_{\text{prod,res}}$$

$$= T_0 \left[(S_2 - S_1)_{\text{res}} - \frac{Q_R}{T_R} \right] = 0$$

The internal irreversibility associated with the reservoir is zero because the process that the reservoir undergoes is internally reversible.

The sum of the internal irreversibilities, however, is *not* equal to the irreversibility associated with the overall process. The reason is that the concept of internal irreversibility cannot account for the external irreversibility that arises from the heat transfer between the air and the reservoir through a finite temperature difference. ■

6.5 SECOND-LAW ANALYSIS OF CLOSED SYSTEMS

An examination of the equations governing second-law analysis presented in the preceding sections reveals that there are a number of approaches that could be used to determine the irreversibility and the reversible work. Each of these, as they apply to closed systems, is discussed in this section.

Consider first Equation 6.23 for the irreversibility rate. Since flow into and out of a closed system is not permitted, the first two terms on the right side of this equation are zero, so that for closed systems

$$\dot{I} = T_0\left(\frac{dS_{sys}}{dt}\right) + \sum_k T_0\left(\frac{\dot{Q}_k}{T_k}\right) \geq 0 \tag{6.38}$$

This equation simply represents the product of T_0 and the rate of change of total entropy. The rate of change of entropy of the system is equivalent to the net rate of entropy change for the closed system, and this equivalence can be easily verified by examination of Equation 6.8. The last term in Equation 6.38 is equal to the product of T_0 and the net rate of entropy change for the surroundings, which are assumed to be composed entirely of thermal-energy reservoirs.

For the analysis of closed systems Equation 6.38 can be integrated with respect to time to obtain the irreversibility associated with a change of state,

$$I_{12} = T_0(S_2 - S_1)_{sys} + \sum_k T_0 \int_1^2 \frac{\delta Q_k}{T_k}$$

and since the reservoir temperatures are constant,

$$I_{12} = T_0(S_2 - S_1)_{sys} + \sum_k T_0\left(\frac{Q_k}{T_k}\right) \tag{6.39}$$

The irreversibility can also be expressed in terms of the availability, as in Equation 6.37. For closed systems this equation reduces to

$$\dot{I} = -\frac{d\Phi_{sys}}{dt} - \sum_k \dot{Q}_k\left(1 - \frac{T_0}{T_k}\right) - \dot{W}_{useful} \tag{6.40}$$

After integration with respect to time for a specific change of state, this equation becomes

$$I_{12} = -(\Phi_2 - \Phi_1)_{sys} - \sum_k\left[Q_k\left(1 - \frac{T_0}{T_k}\right)\right] - W_{12,useful} \tag{6.41}$$

The useful work for the closed system is related to the work by integrating Equation 6.25:

$$W_{12,useful} = W_{12} - P_0(\mathbf{V}_2 - \mathbf{V}_1)_{sys} \tag{6.42}$$

That is, the useful work is the difference between the work actually done and the work done by (or on) the system on the surroundings at pressure P_0. Finally, the irreversibility can also be expressed as the difference between the reversible useful work and the useful work from Equation 6.28. Thus

$$I_{12} = W_{12,rev,useful} - W_{12,useful} \tag{6.43}$$

where the reversible useful work of a closed system is, by integration of Equation 6.36,

$$W_{12,rev,useful} = -(\Phi_2 - \Phi_1)_{sys} - \sum_k\left[Q_k\left(1 - \frac{T_0}{T_k}\right)\right] \tag{6.44}$$

If the work done on (or by) the surroundings is added to each of these useful-work terms, the irreversibility can also be written as the difference between the reversible work and the work actually produced:

$$I_{12} = W_{12,rev} - W_{12} \tag{6.45}$$

The application of these concepts and equations to closed systems is illustrated in the following examples.

EXAMPLE 6.6

A frictionless piston-cylinder assembly contains 0.2 kg of steam at a quality of 40 percent. The piston is weighted to maintain the pressure of the steam at 900 kPa. Heat transfer to the steam from a thermal-energy reservoir at 600 K continues until the steam reaches saturated-vapor conditions. Assume that the environment is at 298 K and 100 kPa. Determine the heat transfer required, the work produced by the steam, the useful work, the reversible useful work, and the irreversibility for this process.

Solution. A sketch of this system is shown in Figure 6.9(a). For the closed system the conservation-of-mass equation requires that

$$m_1 = m_2 = m$$

This system is also assumed to be stationary, so the conservation-of-energy equation reduces to

$$Q_{12} - W_{12} = m(u_2 - u_1)_{\text{sys}}$$

The piston is frictionless, and other irreversible-work modes are absent. Thus the work is simply the $Pd\mathbf{V}$ work that occurs at constant pressure during this process:

$$W_{12} = \int_1^2 Pd\mathbf{V} = P(\mathbf{V}_2 - \mathbf{V}_1)_{\text{sys}} = mP(v_2 - v_1)_{\text{sys}}$$

The specific volumes, from Table B.2, are

$$v_1 = v_f + xv_{fg} = 0.001121 \text{ m}^3/\text{kg} + 0.4(0.2138)\text{m}^3/\text{kg}$$

$$= 0.0866 \text{ m}^3/\text{kg}$$

$$v_2 = v_g = 0.215 \text{ m}^3/\text{kg}$$

The internal-energy values can also be determined from Table B.2;

$$u_1 = u_f + xu_{fg} = 741.92 \text{ kJ/kg} + 0.4(1838.3 \text{ kJ/kg}) = 1477.2 \text{ kJ/kg}$$

$$u_2 = u_g = 2580.2 \text{ kJ/kg}$$

Thus the work is

$$W_{12} = mP(v_2 - v_1)_{\text{sys}} = \frac{0.2 \text{ kg}(900 \text{ kN/m}^2)(0.215 - 0.0866)\text{m}^3/\text{kg}}{1 \text{ kN·m/kJ}}$$

$$= \underline{\underline{+23.1 \text{ kJ}}}$$

and the heat transfer, therefore, amounts to

$$Q_{12} = W_{12} + m(u_2 - u_1)_{\text{sys}} = 23.1 \text{ kJ} + (0.2 \text{ kg})(2580.2 - 1477.2)\text{kJ/kg}$$

$$= \underline{\underline{244 \text{ kJ}}}$$

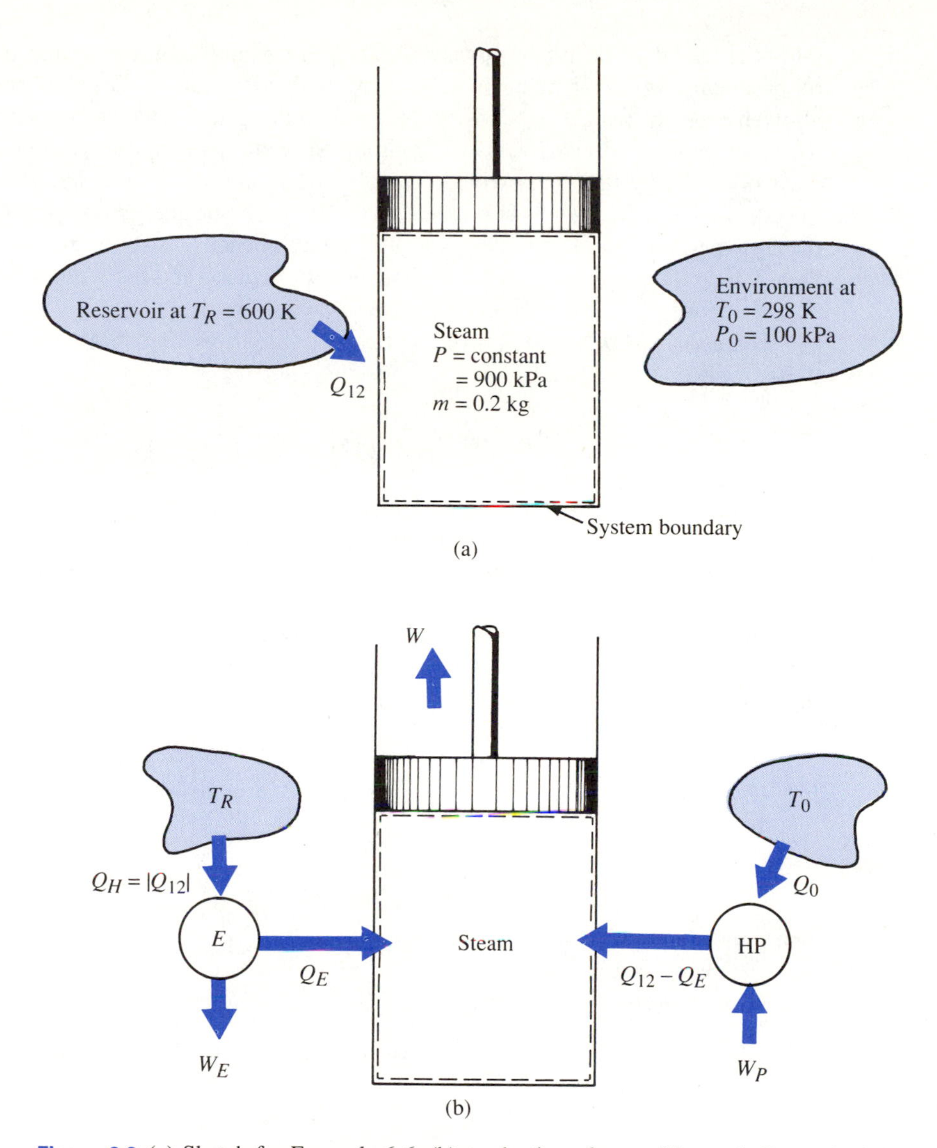

Figure 6.9 (a) Sketch for Example 6.6; (b) production of reversible work, Example 6.6.

The useful work is the difference between the work actually produced and the work performed by the system in expanding against the environment at pressure P_0:

$$W_{12,\text{useful}} = W_{12} - P_0(\mathbf{V}_2 - \mathbf{V}_1)_{\text{sys}}$$

$$= 23.1 \text{ kJ} - \frac{(0.2 \text{ kg})(100 \text{ kN/m}^2)(0.215 - 0.0866)\text{m}^3/\text{kg}}{1 \text{ kN}\cdot\text{m/kJ}}$$

$$= \underline{\underline{20.5 \text{ kJ}}}$$

Only the useful work has any practical value. For example, in this instance the purpose of producing work might be to raise the weighted piston. If so, then the maximum allowable piston weight depends on the area of the piston and the pressure exerted by the environment. Obviously, if the environment were a vacuum at zero pressure, all of the work done by the steam would be useful work and a more massive piston could be raised. If the environment is at pressure P_0, however, some energy is expended in moving the piston against the force exerted by the environmental pressure, and the useful work then is the difference between the work actually produced and the work done against the environment.

Equation 6.44 will be employed here to illustrate its use in determining the reversible useful work:

$$W_{12,\text{rev,useful}} = -(\Phi_2 - \Phi_1)_{\text{sys}} - \left[Q\left(1 - \frac{T_0}{T}\right)\right]_{\text{res}} - \left[Q\left(1 - \frac{T_0}{T}\right)\right]_{\text{env}}$$

Notice that two different subsystems make up the surroundings in this problem—the environment at T_0 and the thermal-energy reservoir at T_R. The last bracketed term in the expression for the reversible useful work is zero because there is no heat transfer to the environment. Even if heat transfer to the environment did occur, this term would not contribute to the reversible useful work since in this instance the temperature of the environment corresponds to the lowest naturally occurring reservoir temperature T_0. Since the heat transfer to the system is from the reservoir,

$$Q_R = -Q_{12}$$

The change in availability is evaluated by using the definition in Equation 6.33:

$$(\Phi_2 - \Phi_1)_{\text{sys}} = m(\phi_2 - \phi_1)_{\text{sys}}$$
$$= m[(u_2 - u_1) + P_0(v_2 - v_1) - T_0(s_2 - s_1)]$$

where the changes in kinetic energy and potential energy of the system are zero because the system is stationary.

Evaluating the entropies from Table B.2, we have

$$s_1 = s_f + xs_{fg} = 2.0948 \text{ kJ/kg·K} + 0.4(4.5274 \text{ kJ/kg·K})$$
$$= 3.9058 \text{ kJ/kg·K}$$
$$s_2 = s_g = 6.6222 \text{ kJ/kg·K}$$

The change in availability is found to be

$$(\Phi_2 - \Phi_1)_{\text{sys}} = (0.2 \text{ kg})\left[(2580.2 - 1477.2)\text{kJ/kg} + \frac{(100 \text{ kN/m}^2)(0.215 - 0.0866)\text{m}^3\text{/kg}}{1 \text{ kN·m/kJ}} - (298 \text{ K})(6.6222 - 3.9058)\text{kJ/kg·K}\right]$$
$$= 61.3 \text{ kJ}$$

and the reversible useful work is

$$\begin{aligned} W_{12,\text{rev,useful}} &= -(\Phi_2 - \Phi_1)_{\text{sys}} - Q_R\left(1 - \frac{T_0}{T_R}\right) \\ &= -(\Phi_2 - \Phi_1)_{\text{sys}} + Q_{12}\left(1 - \frac{T_0}{T_R}\right) \\ &= -61.3 \text{ kJ} + (244 \text{ kJ})\left(1 - \frac{298 \text{ K}}{600 \text{ K}}\right) \\ &= \underline{\underline{61.5 \text{ kJ}}} \end{aligned}$$

Notice that the last term in the expression for the reversible useful work represents the amount of work that could be produced by a totally reversible heat engine operating between the high-temperature reservoir and the environment and having the same heat transfer that was actually experienced by the steam. This work is the maximum amount of work that could be produced with the 244 kJ of heat transfer if that heat transfer from the high-temperature reservoir went instead to a reversible heat engine that had heat rejection to the environment at T_0.

The irreversibility of the process is the difference between the reversible useful work and the useful work:

$$\begin{aligned} I_{12} &= W_{12,\text{rev,useful}} - W_{12,\text{useful}} \\ &= 61.5 \text{ kJ} - 20.5 \text{ kJ} = \underline{\underline{41 \text{ kJ}}} \end{aligned}$$

The reservoir and the steam undergo internally reversible processes, but the heat transfer between the two through a finite temperature difference renders the overall process irreversible.

One means of producing the reversible work for the same change of state of the steam would be to operate a totally reversible heat engine between the reservoir and the piston-cylinder assembly. However, we must require that the amount of heat transfer for the reservoir and the sytem be the same as exists in the actual problem. For satisfaction of this requirement, a heat pump can be operated between the system and the environment as shown in Figure 6.9(b). For this arrangement the reversible useful work is the sum of the work produced by the heat engine, the work input to the heat pump, and the useful work done by the steam:

$$W_{12,\text{rev,useful}} = W_E + W_P + W_{12,\text{useful}}$$

The work produced by the reversible engine operating between the constant-temperature reservoir and the constant-temperature steam at a saturation temperature of 448.5 K is simply

$$W_E = Q_H \eta_{\text{Car}} = Q_H\left(1 - \frac{T_{\text{stm}}}{T_R}\right)$$

and since

$$Q_H = |Q_{12}|$$

then

$$W_E = Q_{12}\left(1 - \frac{T_{\text{stm}}}{T_R}\right) = (244 \text{ kJ})\left(1 - \frac{448.5 \text{ K}}{600 \text{ K}}\right) = 61.6 \text{ kJ}$$

The work input to the reversible heat pump is

$$W_P = -\frac{|Q_{12} - Q_E|}{\beta_{H,\text{rev}}} = -|Q_{12} - Q_E|\left(1 - \frac{T_0}{T_{\text{stm}}}\right)$$

But since the heat engine is reversible,

$$\left|\frac{Q_E}{Q_{12}}\right| = \frac{T_{\text{stm}}}{T_R}$$

Therefore

$$\begin{aligned} W_P &= -Q_{12}\left(1 - \frac{T_{\text{stm}}}{T_R}\right)\left(1 - \frac{T_0}{T_{\text{stm}}}\right) \\ &= (-244 \text{ kJ})\left(1 - \frac{448.5 \text{ K}}{600 \text{ K}}\right)\left(1 - \frac{298 \text{ K}}{448.5 \text{ K}}\right) \\ &= -20.7 \text{ kJ} \end{aligned}$$

and the reversible useful work is

$$\begin{aligned} W_{\text{rev,useful}} &= W_E + W_P + W_{12,\text{useful}} \\ &= 61.6 \text{ kJ} + (-20.7 \text{ kJ}) + 20.5 \text{ kJ} = 61.4 \text{ kJ} \end{aligned}$$

This result is the same value obtained by using the equations developed for second-law analysis. ■

■ EXAMPLE 6.7

The heat transfer from a thermal-energy reservoir at a temperature of 990°R to a heat engine is 50 Btu. The heat engine produces 10 Btu of work and the heat transfer from the heat engine is to (a) a thermal-energy reservoir at 700°R and (b) the environment at 537°R. For each case, determine the thermal efficiency of the engine and compare this result to the maximum theoretical thermal efficiency. Calculate the irreversibility and discuss its significance.

Solution. A schematic representation of the heat engine is shown in Figure 6.10. The temperature of the low-temperature reservoir is denoted by T_0 in both cases since the heat transfer from the heat engine is to the lowest available reservoir temperature in each case.

The heat engine operates in a cycle, and the conservation of energy for the cycle states that

$$\oint \delta Q = \oint \delta W$$

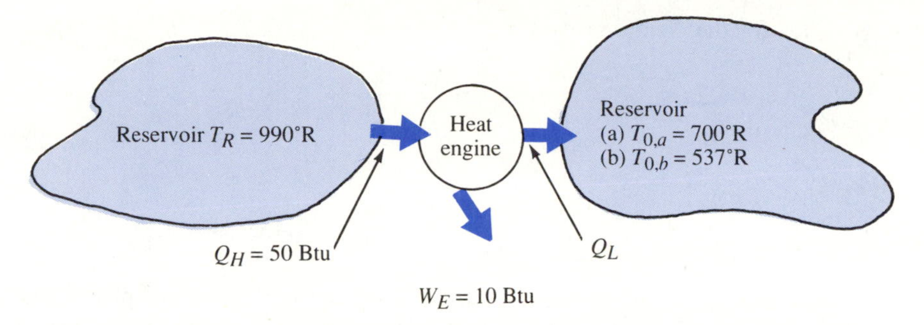

Figure 6.10 Sketch for Example 6.7.

Therefore, for the heat engine

$$Q_H + Q_L = W_E$$

Since the engine produces the same amount of work in both instances, the thermal efficiency of this engine does not depend on the temperature of the low-temperature reservoir:

$$\eta_{\text{th}} = \frac{W_{\text{net}}}{Q_{\text{added}}} = \frac{W_E}{Q_H} = \frac{10 \text{ Btu}}{50 \text{ Btu}} = \underline{\underline{0.2}}$$

The maximum theoretical efficiency does, however, depend on the sink temperature. This is the efficiency that would be achieved by a totally reversible heat engine. From Equation 5.6

$$\eta_{\text{th,rev}} = 1 - \frac{T_L}{T_H}$$

Therefore, the maximum theoretical efficiencies are

$$\eta_{\text{th,rev},a} = 1 - \frac{T_{0,a}}{T_H} = 1 - \frac{700°\text{R}}{990°\text{R}} = \underline{\underline{0.293}}$$

and

$$\eta_{\text{th,rev},b} = 1 - \frac{T_{0,b}}{T_H} = 1 - \frac{537°\text{R}}{990°\text{R}} = \underline{\underline{0.458}}$$

The best theoretical efficiency is achieved with engine (b) since it operates in the presence of the lowest sink temperature for a given source temperature.

The fact that the actual thermal efficiency is the same in both instances is deceptive. This so-called first-law efficiency gives no indication of how well the actual device performs relative to the best possible performance. To overcome this disadvantage, we could define a *second-law-effectiveness* as the ratio of the thermal efficiency of the actual engine to the thermal efficiency of a reversible heat engine operating between the same temperature limits. We denote the second-law effectiveness by ϵ,

$$\epsilon \equiv \frac{\eta_{\text{th,act}}}{\eta_{\text{th,rev}}}$$

Thus for the conditions stated

$$\epsilon_a = \frac{0.2}{0.293} = 0.683$$

and

$$\epsilon_b = \frac{0.2}{0.458} = 0.437$$

Two observations can be made concerning the second-law effectiveness. First, the second-law effectiveness provides a more meaningful measure of the performance of the device. A thermal efficiency of only 20 percent naturally seems very low because the tendency is to compare it with a value of 100 percent. If, however, it is compared with the maximum possible thermal efficiency, we find that the actual performance is much better than might have been perceived originally. Second, the performance of the heat engine in part (a) is much better than in part (b) since it is much closer to the performance of a reversible engine. This conclusion could not be reached on the basis of an energy analysis alone.

The irreversibility associated with the engine described in this problem can be determined from Equation 6.43:

$$I = W_{\text{rev,useful}} - W_{\text{useful}}$$

The useful work is equal to the work produced by the heat engine since it operates in a cycle. The reversible useful work is found from the product of the thermal efficiency of the reversible engine and the heat transfer to the engine:

$$W_{\text{rev,useful},a} = Q_H \eta_{\text{th,rev},a} = (50\,\text{Btu})(0.293) = 14.7 \text{ Btu}$$

and

$$W_{\text{rev,useful},b} = Q_H \eta_{\text{th,rev},b} = (50\,\text{Btu})(0.458) = 22.9 \text{ Btu}$$

Thus

$$I_a = W_{\text{rev,useful},a} - W_{\text{useful},a} = (14.7 - 10)\text{Btu} = \underline{\underline{4.7\,\text{Btu}}}$$

$$I_b = W_{\text{rev,useful},b} - W_{\text{useful},b} = (22.9 - 10)\text{Btu} = \underline{\underline{12.9\,\text{Btu}}}$$

The irreversibility can be thought of as being a loss of work. In other words, rather than produce 10 Btu of work with an irreversible engine, we could produce a greater amount of work if we could use a reversible engine. The irreversibility of the actual engine might possibly be reduced through better design, but the possible increase in second-law efficiency must be evaluated from an economic viewpoint to determine if the increase is worthwhile. Since the cost of energy, equipment, and so forth is constantly changing, the economic evaluation might be unfavorable this year but favorable at some later time. ■

■ EXAMPLE 6.8

Two methods are proposed to increase the temperature of a rigid tank of nitrogen from 550 to 700°R. Method *A* would be to insulate the tank and turn a paddle wheel to agitate the contents until the desired temperature is reached. In method *B* the paddle wheel and insulation would be removed and the temperature of the nitrogen would be increased by means of heat transfer to the nitrogen from a reservoir at 800°R. The initial pressure of the nitrogen is 30 psia, and the environment is at a temperature of 537°R. Explain which proposal would be preferred from a thermodynamic standpoint.

Solution. The two proposals are illustrated in Figure 6.11. The nitrogen forms a closed, stationary system and behaves as an ideal gas at the temperatures and pressure stated. Therefore, the conservation-of-mass and conservation-of-energy equations reduce to

$$m_1 = m_2$$

and

$$Q_{12} - W_{12} = (U_2 - U_1)_{\text{sys}}$$

In method A the heat transfer is zero and the work consists entirely of paddle-wheel work since the tank is insulated and rigid:

$$W_{12,A} = (U_1 - U_2)_{\text{sys}}$$

On a unit-mass basis

$$w_{12,A} = (u_1 - u_2)_{\text{sys}}$$

In method B the work is zero, and the heat transfer per unit mass of nitrogen is

$$q_{12,B} = (u_2 - u_1)_{\text{sys}}$$

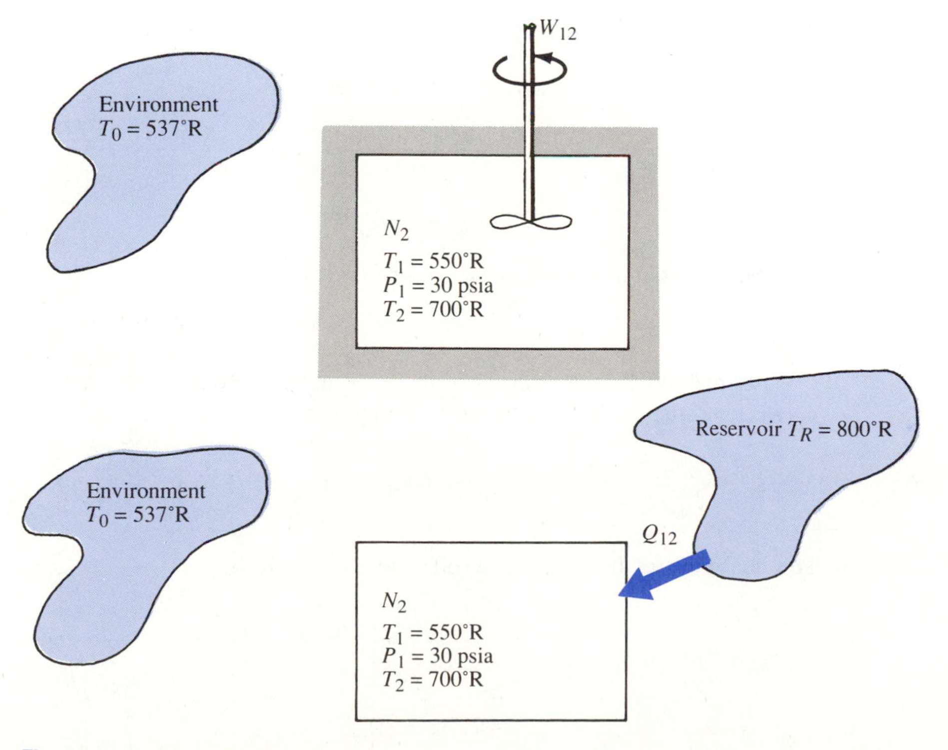

Figure 6.11 Sketch for Example 6.8.

The end-state temperatures of the nitrogen remain unchanged from one method to the other, and therefore the amount of energy required to achieve the change of state is the same for both methods. Thus both methods are equally acceptable if an energy analysis alone is used for evaluation.

Let us next examine the irreversibility associated with each method. From Equation 6.39

$$I_{12} = T_0(S_2 - S_1)_{sys} + \sum_k T_0\left(\frac{Q_k}{T_k}\right)$$

In method A there is no heat transfer, so the irreversibility is equal to the product of T_0 and the entropy change of the nitrogen. On a unit-mass basis

$$i_{12,A} = T_0(s_2 - s_1)_{sys}$$

and the entropy change of the nitrogen can be evaluated from Equation 5.28:

$$s_2 - s_1 = s_2^\circ - s_1^\circ - R \ln\left(\frac{P_2}{P_1}\right)$$

But since the specific volume of the nitrogen remains constant,

$$\frac{P_2}{P_1} = \frac{T_2}{T_1}$$

From Table D.5E the entropy change of the nitrogen is calculated to be

$$\begin{aligned} s_2 - s_1 &= (1.6998 - 1.6398)\text{Btu/lb}_\text{m}\cdot{}^\circ\text{R} \\ &\quad - \frac{(1.986\ \text{Btu/lb}_\text{m}\cdot\text{mol}\cdot{}^\circ\text{R}) \ln (700^\circ\text{R}/550^\circ\text{R})}{28\ \text{lb}_\text{m}/\text{lb}_\text{m}\cdot\text{mol}} \\ &= 0.0429\ \text{Btu/lb}_\text{m}\cdot{}^\circ\text{R} \end{aligned}$$

and the irreversibility associated with method A is

$$i_{12,A} = (537^\circ\text{R})(0.0429\ \text{Btu/lb}_\text{m}\cdot{}^\circ\text{R}) = 23.0\ \text{Btu/lb}_\text{m}$$

For method B, heat transfer from a constant-temperature reservoir at temperature T_R occurs, and the irreversibility for process 1-2 is

$$i_{12,B} = T_0(s_2 - s_1)_{sys} + T_0\left(\frac{q_R}{T_R}\right)$$

The heat transfer from the reservoir is to the system, so

$$\begin{aligned} i_{12,B} &= T_0(s_2 - s_1)_{sys} - q_{12}\left(\frac{T_0}{T_R}\right) \\ &= T_0(s_2 - s_1)_{sys} - (u_2 - u_1)_{sys}\left(\frac{T_0}{T_R}\right) \end{aligned}$$

and the irreversibility per unit mass is

$$i_{12,B} = (537°\text{R})(0.0429\ \text{Btu/lb}_\text{m}\cdot°\text{R}) - (124.1 - 97.4)\text{Btu/lb}_\text{m}\left(\frac{537°\text{R}}{800°\text{R}}\right)$$

$$= 5.1\ \text{Btu/lb}_\text{m}$$

A clear choice exists between the two proposed methods when the second-law analysis is considered. Thermodynamically, method B is preferred since the irreversibility associated with this method is much less than that of method A. This conclusion holds regardless of the allowable reservoir temperature. The irreversibility for method B would increase if the reservoir temperature were increased because a larger temperature difference would exist between the reservoir and the system. But for all finite, possible (i.e., greater than 700°R) reservoir temperatures, the irreversibility would remain smaller in magnitude than the irreversibility of method A.

From a thermodynamic standpoint the conversion of work to heat transfer is not a wise use of energy resources. To illustrate this point, consider that a Carnot engine operating between the 800°R reservoir and the environment at 537°R ($\eta_{\text{Car}} = 0.33$) must have heat transfer equivalent to 81.2 Btu/lb$_\text{m}$ of nitrogen in order to produce the 26.7 Btu/lb$_\text{m}$ of work proposed in method A. If the heat transfer occurs directly to the nitrogen from the reservoir as proposed in method B, only 26.7 Btu/lb$_\text{m}$ is required. Recognize, however, that the two proposed methods have been assumed to be viable alternatives and the cost of the energy in each instance has not been considered. Economics would ordinarily influence the ultimate decision, and if one option is not available, the choice is made for us. ■

6.6 SECOND-LAW ANALYSIS OF OPEN SYSTEMS

Considerable emphasis is placed on the analysis of open systems in engineering thermodynamics because of the prevalence and importance of these systems in engineering applications. While an energy analysis is indispensable, a thermodynamic assessment of the performance of open systems is incomplete without a second-law analysis. In this section the methods of second-law analysis will be applied to both steady-state and transient open systems.

6.6.1 Steady State

Open systems operating in steady state are characterized by the time invariance of all extensive properties of the system. From Equation 3.15

$$\frac{dY_{\text{sys}}}{dt} = 0 \tag{6.46}$$

The entropy S_{sys}, being an extensive property of the system, cannot change with time when steady-state conditions prevail, so that the irreversibility rate, Equation 6.23, for an open, steady system becomes

$$\dot{I} = T_0\left(\sum_{\text{exit}} \dot{m}s - \sum_{\text{inlet}} \dot{m}s + \sum_k \frac{\dot{Q}_k}{T_k}\right) \geq 0 \tag{6.47}$$

Notice that this equation represents the product of T_0 and the rate of change of total entropy. The first two terms on the right side are equivalent to the product of T_0 and the net rate of change of entropy of the system; for conditions of steady flow these terms are equal to the net flow of entropy associated with mass flow across the boundary of the system. The last term is proportional to the net rate of change of entropy of the surroundings, which are assumed to be composed entirely of thermal-energy reservoirs.

With Equation 6.37 the irreversibility rate can also be expressed in terms of the open-system availability ψ:

$$\dot{I} = -\left(\sum_{\text{exit}} \dot{m}\psi - \sum_{\text{inlet}} \dot{m}\psi\right) - \sum_k \dot{Q}_k\left(1 - \frac{T_0}{T_k}\right) - \dot{W}_{\text{useful}} \tag{6.48}$$

since the rate of change of Φ_{sys} must be zero for the steady state. Another important point is that the power for an open, steady system is identical to the useful power because the volume of the system does not expand or contract under steady-state conditions. Therefore, no $Pd\mathbf{V}$ work is done against the environment. All power, therefore, is useful power, as can be seen from Equation 6.25.

In terms of the reversible useful power, the irreversibility rate can be determined from Equation 6.28:

$$\dot{I} = \dot{W}_{\text{rev,useful}} - \dot{W}_{\text{useful}} \geq 0 \tag{6.49}$$

where the reversible useful power for the open, steady system is determined from Equation 6.36:

$$\dot{W}_{\text{rev}} = \dot{W}_{\text{rev,useful}} = -\left(\sum_{\text{exit}} \dot{m}\psi - \sum_{\text{inlet}} \dot{m}\psi\right) - \sum_k \dot{Q}_k\left(1 - \frac{T_0}{T_k}\right) \tag{6.50}$$

EXAMPLE 6.9

Consider the heat exchanger described in Example 6.1. Determine the irreversibility rate associated with the heat exchanger by using first Equation 6.47 and then Equation 6.48. Assume $T_0 = 298$ K.

Solution. The sketch of the heat exchanger from Example 6.1 is reproduced in Figure 6.12. Equation 6.47,

$$\dot{I} = T_0\left(\sum_{\text{exit}} \dot{m}s - \sum_{\text{inlet}} \dot{m}s + \sum_k \frac{\dot{Q}_k}{T_k}\right)$$

simply represents the product of T_0 and the rate of change of total entropy associated with this steady-flow process. In Example 6.1 the rate of change of total entropy was determined to be 274 kJ/K·s, and therefore the irreversibility rate is

$$\dot{I} = T_0\left(\frac{dS}{dt}\right)_{\text{tot}} = (298\text{ K})(274\text{ kJ/K·s}) = \underline{\underline{81.7\text{ MW}}}$$

Equation 6.48, which expresses the irreversibility rate in terms of the availability, can also be used to arrive at the same result:

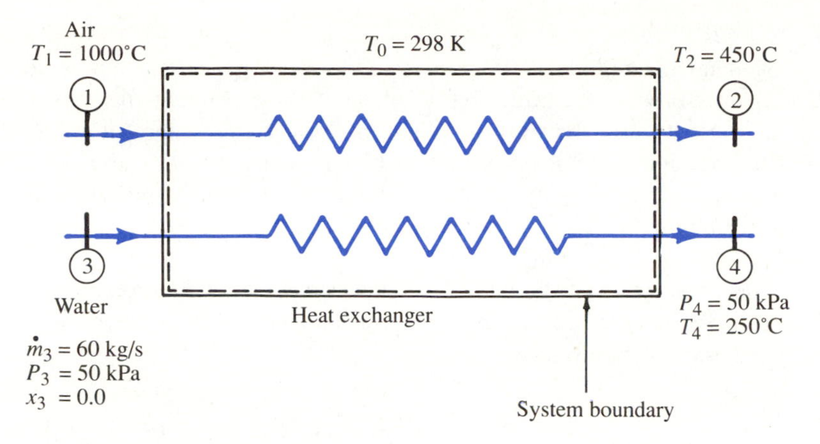

Figure 6.12 Sketch for Example 6.9.

$$\dot{I} = -\left(\sum_{\text{exit}} \dot{m}\psi - \sum_{\text{inlet}} \dot{m}\psi\right) - \sum_k \dot{Q}_k\left(1 - \frac{T_0}{T_k}\right) - \dot{W}_{\text{useful}}$$

The choice of which equation should be used is primarily a matter of convenience, since both equations produce the same results.

When the heat exchanger is the system, no work is done and no heat transfer occurs at the boundary of the system. Therefore

$$\dot{I} = -\left(\sum_{\text{exit}} \dot{m}\psi - \sum_{\text{inlet}} \dot{m}\psi\right) = -[\dot{m}_1\psi_2 + \dot{m}_3\psi_4 - \dot{m}_1\psi_1 - \dot{m}_3\psi_3]$$

$$= -[\dot{m}_1(\psi_2 - \psi_1) + \dot{m}_3(\psi_4 - \psi_3)]$$

Using the definition of ψ from Equation 6.32 and neglecting changes in kinetic and potential energy of the fluid streams as they pass through the heat exchanger, we find that this equation becomes

$$\dot{I} = -\{\dot{m}_1[(h_2 - h_1) - T_0(s_2 - s_1)] + \dot{m}_3[(h_4 - h_3) - T_0(s_4 - s_3)]\}$$

With values of $\dot{m}$, h, and entropy changes from Example 6.1, we obtain

$$\dot{I} = -\{(252.8 \text{ kg/s})[(738.3 - 1363.8)\text{kJ/kg} - (298 \text{ K})(-0.6404 \text{ kJ/kg·K})]$$
$$+ (60 \text{ kg/s})[(2975.6 - 340.54)\text{kJ/kg} - (298 \text{ K})(7.2636 \text{ kJ/kg·K})]\}$$
$$= 81{,}650 \text{ kJ/s} = \underline{\underline{81.7 \text{ MW}}}$$

Notice that since the actual power produced is zero, the reversible power is equal to the irreversibility rate. That is, if the process described in this example could be achieved in a reversible manner (using reversible heat engines and heat pumps), approximately 82 MW of power could be produced while the air and water experienced the changes of state described in the problem statement. ■

EXAMPLE 6.10

Nitrogen enters a turbine at 100 psia and 1800°R and exhausts at a pressure of 14.7 psia. During this process the work output of the turbine per unit mass of nitrogen is 165 Btu/lb$_m$ and the heat transfer from the nitrogen to the environment at 537°R amounts to 20 Btu/lb$_m$. For negligible changes in kinetic and potential energies of the nitrogen, determine the irreversibility (in Btu/lb$_m$) associated with the process and the reversible work per unit mass.

Solution. A sketch of the turbine is shown in Figure 6.13. For the conditions stated, nitrogen can be assumed to behave as an ideal gas, and the irreversibility can be determined from Equation 6.47 for steady flow:

$$\dot{I} = T_0 \left(\sum_{\text{exit}} \dot{m}s - \sum_{\text{inlet}} \dot{m}s + \sum_k \frac{\dot{Q}_k}{T_k} \right)$$

Since there is a single inlet and a single exit, the conservation-of-mass equation reduces to

$$\dot{m}_i = \dot{m}_e = \dot{m}$$

Furthermore, heat transfer occurs only with the environment at temperature T_0. Therefore, the irreversibility rate can be expressed as

$$\dot{I} = T_0 \left[\dot{m}(s_e - s_i) + \frac{\dot{Q}_0}{T_0} \right]$$

or

$$\frac{\dot{I}}{\dot{m}} = T_0 \left[(s_e - s_i) + \frac{q_0}{T_0} \right]$$

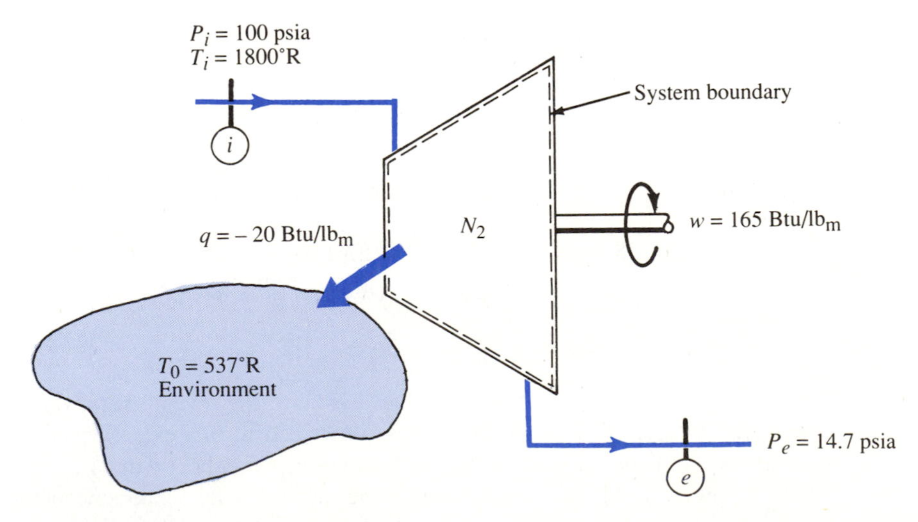

Figure 6.13 Sketch for Example 6.10.

The heat transfer to the environment is from the nitrogen, so that

$$q_0 = -q = -(-20 \text{ Btu/lb}_m) = +20 \text{ Btu/lb}_m$$

The entropy change of the nitrogen is determined from Equation 5.28 and Table D.5E:

$$s_e - s_i = s_e^\circ - s_i^\circ - R \ln\left(\frac{P_e}{P_i}\right)$$

However, the exit temperature of the nitrogen is unknown and must be determined by application of the conservation-of-energy equation. For negligible changes in kinetic and potential energies, this equation reduces to

$$\dot{Q} - \dot{W} + \dot{m}(h_i - h_e) = 0$$

Thus

$$h_e = q - w + h_i$$

Using Table D.5E for enthalpy yields

$$h_e = -20 \text{ Btu/lb}_m - 165 \text{ Btu/lb}_m + 462.5 \text{ Btu/lb}_m$$
$$= 277.5 \text{ Btu/lb}_m$$

For this value of enthalpy the exit temperature can be determined by interpolation in Table D.5E:

$$T_e = 1111°\text{R}$$

and therefore

$$s_e^\circ = 1.8162 \text{ Btu/lb}_m\cdot°\text{R}$$

The entropy change of the nitrogen can now be evaluated as

$$s_e - s_i = s_e^\circ - s_i^\circ - R \ln\left(\frac{P_e}{P_i}\right)$$
$$= (1.8162 - 1.9454)\text{Btu/lb}_m\cdot°\text{R}$$
$$- \frac{1.986 \text{ Btu/lb}_m\cdot\text{mol}\cdot°\text{R}}{28 \text{ lb}_m/\text{lb}_m\cdot\text{mol}} \ln\left(\frac{14.7 \text{ psia}}{100 \text{ psia}}\right)$$
$$= 0.00679 \text{ Btu/lb}_m\cdot°\text{R}$$

and the irreversibility per unit mass is

$$\frac{\dot{I}}{\dot{m}} = i = (537°\text{R})\left(0.00679 \text{ Btu/lb}_m\cdot°\text{R} + \frac{20 \text{ Btu/lb}_m}{537°\text{R}}\right)$$
$$= \underline{\underline{23.6 \text{ Btu/lb}_m}}$$

The reversible work per unit mass can be determined with the aid of Equation 6.49:

$$\dot{I} = \dot{W}_{\text{rev,useful}} - \dot{W}_{\text{useful}}$$

thus

$$i = w_{\text{rev,useful}} - w_{\text{useful}}$$

For steady flow the work input or output is useful work because no PdV work is associated with steady-flow systems. Thus

$$w_{\text{rev}} = i + w = 23.6\ \text{Btu/lb}_{\text{m}} + 165\ \text{Btu/lb}_{\text{m}}$$

$$= \underline{\underline{188.6\ \text{Btu/lb}_{\text{m}}}}$$

■

Adiabatic Efficiencies of Some Steady-Flow Devices

From a thermodynamic standpoint minimizing the irreversibility of a device or system is desirable. A closer examination of Equation 6.47 reveals that heat transfer from a system to its surroundings always tends to increase the irreversibility, and therefore the best thermodynamic performance can be achieved if heat transfer from the system is eliminated. With equipment such as power-plant and refrigeration-system condensers, heat transfer to the surroundings cannot be avoided, but in other applications heat transfer can often be minimized or reduced with a subsequent decrease in irreversibility. When heat transfer can be eliminated, the process is adiabatic and has an irreversibility rate that is proportional to the net flow of entropy across the boundary of the open, steady system. Furthermore, the net flow of entropy during the adiabatic process must be greater than or equal to zero, as dictated by the second law. Thus the irreversibility can be minimized by minimizing the entropy change of the fluid streams. In the limit the irreversibility approaches zero for the reversible adiabatic (or isentropic) process. For these reasons the thermodynamic performance of many open, steady systems, such as turbines, compressors, pumps, nozzles, and diffusers, is often compared with the performance that could be achieved with a reversible adiabatic or isentropic process. For many of these devices adiabatic efficiencies are used to compare the actual performance of an adiabatic device with the performance that would be achieved with an isentropic device.

Adiabatic-Nozzle Efficiency

A simple nozzle is used to accelerate a fluid stream. As the fluid flows through the nozzle, the kinetic energy of the fluid is increased while its pressure is decreased. Work interactions are absent, and the potential-energy change is negligible. Thus for adiabatic nozzles or nozzles that have negligible heat transfer, the conservation-of-energy equation, Equation 4.31, for a single fluid stream reduces to

$$h_i - h_e + \frac{V_i^2 - V_e^2}{2} = 0$$

Thus the kinetic energy of the fluid is increased at the expense of a decrease in the enthalpy of the fluid.

The *adiabatic-nozzle efficiency* is defined as the ratio of the actual kinetic energy at the exit of an adiabatic nozzle to the kinetic energy that could be achieved at the exit of

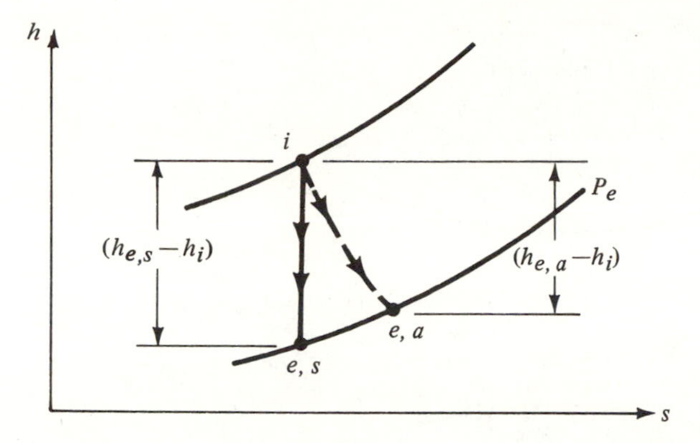

Figure 6.14 Enthalpy-entropy diagram for an adiabatic nozzle.

an isentropic nozzle for the same inlet conditions and the same exit pressure as exist in the actual nozzle:

$$\eta_N \equiv \frac{V_{e,a}^2/2}{V_{e,s}^2/2} \tag{6.51}$$

where the subscripts, e,a and e,s denote the actual exit state and the isentropic exit state, respectively. Using the conservation-of-energy equation from above, we can write this equation as

$$\eta_N = \frac{h_i - h_{e,a} + (V_i^2/2)}{h_i - h_{e,s} + (V_i^2/2)} \tag{6.52}$$

If, as is often the case, the inlet kinetic energy is small compared with the exit kinetic energy and the enthalpy change, the nozzle efficiency can be approximated by

$$\eta_N \simeq \frac{h_i - h_{e,a}}{h_i - h_{e,s}} \tag{6.53}$$

The relationship between the actual process and the isentropic process for the adiabatic nozzle is shown in Figure 6.14 with an enthalpy-entropy, or h-s, diagram. Since the nozzle process is assumed adiabatic, the second law (see Equation 5.17) dictates that the entropy change of the fluid must be greater than or equal to zero. Thus exit states to the left of state e,s in the diagram in Figure 6.14 are impossible to achieve with an adiabatic nozzle. Therefore, for fixed pressure limits P_i and P_e, the maximum decrease in enthalpy (and hence the maximum increase in kinetic energy) is achieved with the isentropic process. Thus the adiabatic-nozzle efficiency is less than unity.

Adiabatic-nozzle efficiencies are generally very high. For instance, nozzles used in aircraft jet engines typically have adiabatic efficiencies of 90 to 95 percent.

EXAMPLE 6.11

A converging-diverging nozzle is used to accelerate a fluid from subsonic velocity to supersonic velocity. The nozzle is adiabatic, and argon enters the inlet of the nozzle at

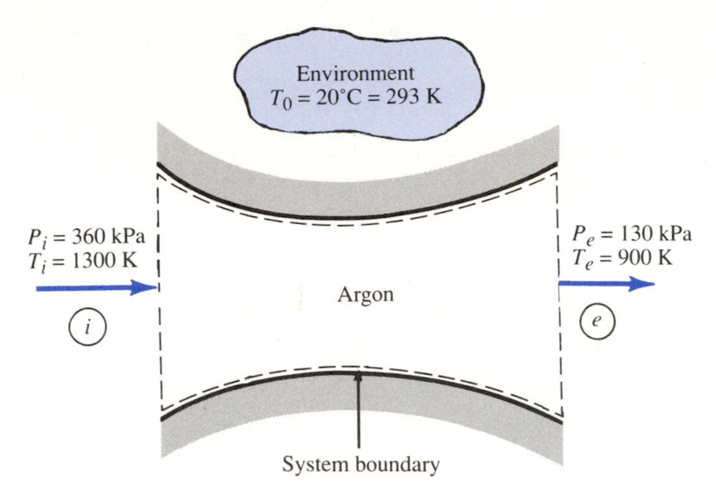

Figure 6.15 Sketch for Example 6.11.

low velocity with a pressure of 360 kPa and a temperature of 1300 K. At the nozzle exit the pressure and temperature of the argon are 130 kPa and 900 K, respectively. Determine the actual exit velocity of the argon, the adiabatic-nozzle efficiency, and the irreversibility per unit mass of argon associated with the process. Assume that $T_0 = 20°C$ and that the flow through the nozzle is steady.

Solution. A sketch of the nozzle is shown in Figure 6.15. At the stated temperatures and pressures the assumption of ideal-gas behavior for argon is appropriate. The actual exit velocity of the argon can be determined by using the conservation-of-energy equation for steady flow, Equation 4.31:

$$\dot{Q} - \dot{W} + \sum_{\text{inlet}} \dot{m}(h + e_k + e_p) - \sum_{\text{exit}} \dot{m}(h + e_k + e_p) = 0$$

The nozzle is adiabatic, no work is done on or by the argon, the potential-energy change is negligible, and the mass flow rate is the same at the inlet and exit. Thus

$$h_i - h_e + \frac{V_i^2 - V_e^2}{2} = 0$$

The inlet velocity is much less than the exit velocity, so that

$$V_e \simeq \sqrt{2(h_i - h_e)}$$

Since argon is a monatomic gas, the specific heats are independent of temperature, and the enthalpy change can be expressed in terms of the constant-pressure specific heat and the temperature change:

$$V_e \simeq \sqrt{2c_p(T_i - T_e)}$$

Using $\bar{c}_p = 20.8$ kJ/kg·mol·K from Table D.8, we have

$$V_e = \left[\frac{2(20.8 \text{ kJ/kg·mol·K})}{40 \text{ kg/kg·mol}}(1300 - 900) \text{ K } (10^3 \text{ N·m/kJ})(1 \text{ kg·m/N·s}^2)\right]^{1/2}$$

$$= \underline{\underline{645 \text{ m/s}}}$$

The adiabatic-nozzle efficiency is determined from Equation 6.53. However, the enthalpy $h_{e,s}$ that would exist at the exit if the nozzle process were isentropic must first be determined. For the isentropic process of an ideal gas with constant specific heats, the end-state temperatures are related to the pressure ratio by Equation 5.40:

$$\left(\frac{T_e}{T_i}\right)_s = \left(\frac{P_e}{P_i}\right)_s^{(k-1)/k}$$

or

$$T_{e,s} = T_i\left(\frac{P_e}{P_i}\right)_s^{(k-1)/k}$$

Using $k = 1.666$ from Table D.9, we find

$$T_{e,s} = (1300 \text{ K})\left(\frac{130 \text{ kPa}}{360 \text{ kPa}}\right)^{0.666/1.666} = 865 \text{ K}$$

Thus from Equation 6.53

$$\eta_N = \frac{h_i - h_{e,a}}{h_i - h_{e,s}} = \frac{c_p(T_i - T_{e,a})}{c_p(T_i - T_{e,s})}$$

$$= \frac{(1300 - 900)\text{K}}{(1300 - 865)\text{K}} = \underline{\underline{0.92}}$$

The irreversibility for this process can be determined from Equation 6.47:

$$\dot{I} = T_0\left(\sum_{\text{exit}} \dot{m}s - \sum_{\text{inlet}} \dot{m}s + \sum_k \frac{\dot{Q}_k}{T_k}\right)$$

Since the nozzle is adiabatic and has a single exit and inlet, this expression can be simplified to

$$i = T_0(s_e - s_i)$$

From Equation 5.30 this expression can be written as

$$i = T_0\left(c_p \ln\frac{T_e}{T_i} - R \ln\frac{P_e}{P_i}\right)$$

$$= (293 \text{ K})\left[\frac{(20.8 \text{ kJ/kg·mol·K}) \ln(900 \text{ K}/1300 \text{ K})}{40 \text{ kg/kg·mol}} - \frac{(8.314 \text{ kJ/kg·mol·K}) \ln(130 \text{ kPa}/360 \text{ kPa})}{40 \text{ kg/kg·mol}}\right]$$

$$= \underline{\underline{6.0 \text{ kJ/kg}}}$$

The irreversibility for this process can be attributed to friction associated with the flow of the fluid. ■

Adiabatic-Turbine Efficiency

The *adiabatic-turbine efficiency* is defined as the ratio of the actual power output of an adiabatic turbine to the power output that would be achieved with a reversible, adiabatic turbine that has exhaust pressure(s) identical to the actual turbine:

$$\eta_T \equiv \frac{\dot{W}}{\dot{W}_s} \tag{6.54}$$

The power of the isentropic turbine is denoted by $\dot{W}_s$.

For simple turbines that have a single inlet and a single exit or exhaust, the turbine efficiency can be expressed in terms of the work output per unit mass of fluid flowing through the turbine:

$$\eta_T = \frac{w}{w_s}$$

The potential-energy change of the fluid is usually negligible for turbine processes, and the kinetic-energy change is often small compared with the change in enthalpy of the fluid stream. Furthermore, since the turbine is adiabatic, the preceding expression for the adiabatic-turbine efficiency becomes

$$\eta_T \simeq \frac{h_i - h_{e,a}}{h_i - h_{e,s}}$$

The enthalpy-entropy diagram for a simple adiabatic gas turbine is identical to the sketch shown in Figure 6.14 for an adiabatic nozzle.

■ EXAMPLE 6.12

Steam at 200 psia and 660°F enters an adiabatic turbine at a rate of 990 lb_m/h. The turbine has an adiabatic efficiency of 80 percent, and the exhaust pressure is 14.7 psia. Assuming that the kinetic-energy change of the steam is negligible, determine the temperature of the exhaust steam, the power produced by the turbine, and the irreversibility rate associated with the turbine. Assume T_0 = 77°F and that the flow of steam is steady.

Solution. A sketch of the turbine and a *T-s* diagram for the process described above are shown in Figure 6.16. Since the turbine operates in steady flow and there is a single inlet and exit, the conservation-of-mass equation reduces to

$$\dot{m}_i = \dot{m}_e = \dot{m}$$

and the conservation-of-energy equation becomes

$$\dot{Q} - \dot{W} + \dot{m}[(h_i - h_e) + (e_{k,i} - e_{k,e}) + (e_{p,i} - e_{p,e})] = 0$$

The turbine is adiabatic, and changes in kinetic energy and potential energy of the steam are negligible. Thus the power that would be produced by the turbine if it were isentropic is given by

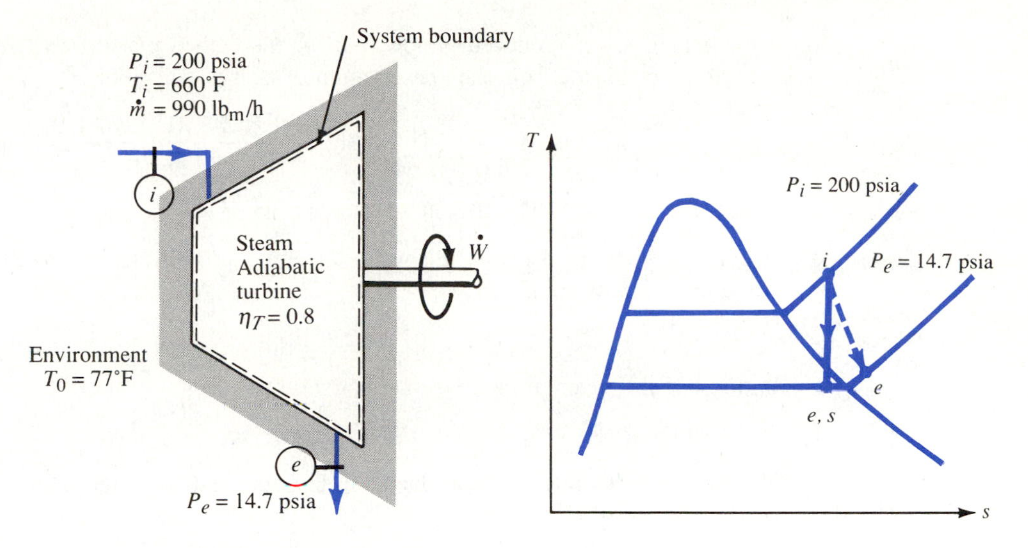

Figure 6.16 Sketches for Example 6.12.

$$\dot{W}_s = \dot{m}(h_i - h_{e,s})$$

To determine the enthalpy at the exit of the isentropic process, we note that

$$s_{e,s} = s_i = 1.705 \text{ Btu/lb}_\text{m}\cdot{}^\circ\text{R}$$

With an exit pressure of 14.7 psia, this value of entropy corresponds to a saturation state, and the quality is

$$x_{e,s} = \frac{s_{e,s} - s_f}{s_{fg}} = \frac{(1.705 - 0.3121)\text{Btu/lb}_\text{m}\cdot{}^\circ\text{R}}{1.4444 \text{ Btu/lb}_\text{m}\cdot{}^\circ\text{R}}$$
$$= 0.964$$

The exit enthalpy for the isentropic process is therefore

$$h_{e,s} = h_f + x_{e,s}h_{fg} = 180.17 \text{ Btu/lb}_\text{m} + 0.964(970.2 \text{ Btu/lb}_\text{m})$$
$$= 1115.4 \text{ Btu/lb}_\text{m}$$

The enthalpy at the inlet is $h_i = 1353 \text{ Btu/lb}_\text{m}$ and the power produced by the turbine for isentropic operation is

$$\dot{W}_s = \dot{m}(h_i - h_{e,s}) = (990 \text{ lb}_\text{m}\text{/h})(1353 - 1115.4)\text{Btu/lb}_\text{m}$$
$$= 2.352 \times 10^5 \text{ Btu/h} = 92.4 \text{ hp}$$

The power produced by the actual turbine can now be determined by using Equation 6.54:

$$\dot{W} = \eta_T\dot{W}_s = 0.8(92.4 \text{ hp}) = \underline{\underline{73.9 \text{ hp or } 1.882 \times 10^5 \text{Btu/h}}}$$

Since the actual power produced is now known, the enthalpy at the exit of the actual process can be evaluated from the conservation-of-energy equation:

$$h_e = h_i - \frac{\dot{W}}{\dot{m}} = 1353\,\text{Btu/lb}_\text{m} - \frac{1.882 \times 10^5\ \text{Btu/h}}{990\ \text{lb}_\text{m}\text{/h}}$$

$$= 1162.9\,\text{Btu/lb}_\text{m}$$

At an exit pressure of 14.7 psia this value of enthalpy indicates a superheated-vapor state, and the temperature is

$$T_e \simeq \underline{\underline{238°\text{F}}}$$

The entropy of the steam at the exit state is therefore

$$s_e = 1.774\ \text{Btu/lb}_\text{m}\cdot°\text{R}$$

The irreversibility rate can be evaluated from Equation 6.47, which, for this example, reduces to

$$\dot{I} = T_0\dot{m}(s_e - s_i)$$

Using values of entropy from Table B.2E, we obtain

$$\dot{I} = (537°\text{R})(990\ \text{lb}_\text{m}\text{/h})(1.774 - 1.705)\text{Btu/lb}_\text{m}\cdot°\text{R}$$

$$= 3.67 \times 10^4\ \text{Btu/h} = \underline{\underline{14.4\,\text{hp}}}$$

The irreversibility can be attributed to frictional effects present in the actual turbine operation.

The reversible useful power can easily be determined from Equation 6.49:

$$\dot{W}_\text{rev,useful} = \dot{I} + \dot{W}_\text{useful} = 14.4\ \text{hp} + 73.9\ \text{hp} = 88.3\,\text{hp}$$

Notice that the magnitudes of the power output can be ranked in the following order:

$$\dot{W}_s > \dot{W}_\text{rev,useful} > \dot{W}_\text{useful}$$

The value of $\dot{W}_s$ represents the maximum amount of power that could be produced for the given inlet state and the given exit *pressure*. The value of $\dot{W}_\text{rev,useful}$ represents the maximum amount of power that could be produced for the given inlet state and given exit *state*. ■

Adiabatic-Compressor Efficiency

The *adiabatic-compressor efficiency* is defined as the ratio of the power input required to compress a fluid to a given exit pressure in a reversible adiabatic process to the actual power input required to achieve adiabatic compression to the same pressure:

$$\eta_C \equiv \frac{\dot{W}_s}{\dot{W}} \qquad (6.55)$$

Here the power input for the reversible, adiabatic compressor is denoted by $\dot{W}_s$. Equation 6.55 is also used to define the adiabatic efficiency for pumps. For a compressor with a

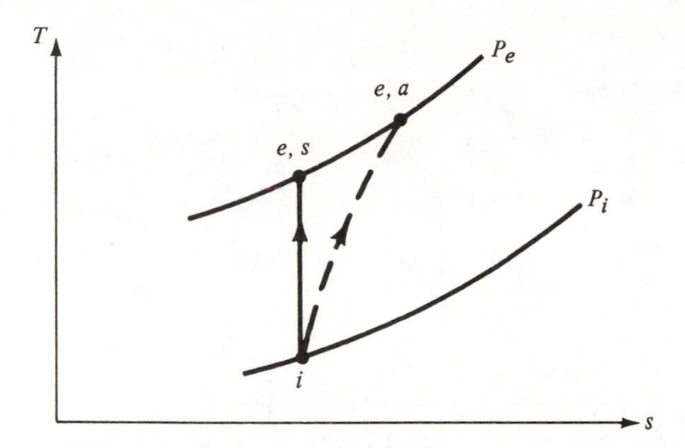

Figure 6.17 Temperature-entropy diagram for a gas compressor.

single inlet and exit, the adiabatic-compressor efficiency can be written as

$$\eta_C = \frac{w_s}{w} \tag{6.56}$$

and for negligible changes in kinetic and potential energies of the fluid as it passes through an adiabatic compressor, this expression is approximately

$$\eta_C \simeq \frac{h_{e,s} - h_i}{h_{e,a} - h_i} \tag{6.57}$$

A temperature-entropy diagram for a simple adiabatic gas compressor is shown in Figure 6.17. Notice that the inlet state and the exit pressure for both the actual compressor and the isentropic compressor are the same. Furthermore, since the compressor is assumed to be adiabatic, the entropy at the actual exit state must be greater than the entropy at the inlet state. Thus exit states to the left of state e,s in Figure 6.17 are impossible to achieve with an adiabatic compressor.

EXAMPLE 6.13

An adiabatic compressor in a refrigeration system compresses 90 kg/h of refrigerant-12 from 200 kPa and 0°C to 1.2 MPa and 80°C. The refrigerant-12 experiences a negligible change in kinetic energy. Determine the adiabatic-compressor efficiency and the irreversibility rate associated with the compressor. Assume $T_0 = 25°C$ and a steady process.

Solution. The compressor is illustrated in Figure 6.18. Since the compressor operates in steady flow and has a single inlet and exit, the conservation-of-mass and -energy equations reduce to

$$\dot{m}_i = \dot{m}_e = \dot{m}$$

and

$$\dot{Q} - \dot{W} + \dot{m}[(h_i - h_e) + (e_{k,i} - e_{k,e}) + (e_{p,i} - e_{p,e})] = 0$$

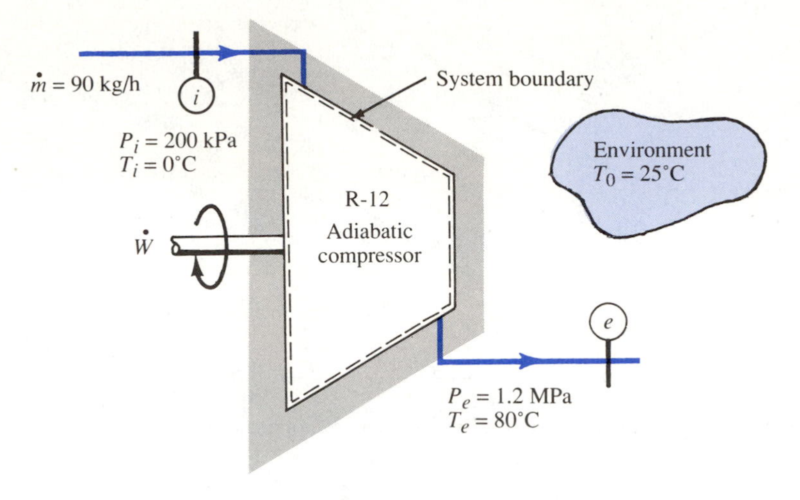

Figure 6.18 Sketch for Example 6.13.

For an adiabatic compressor with negligible changes in kinetic energy and potential energy of the R-12, this equation becomes

$$\dot{W} = \dot{m}(h_i - h_e)$$

With values of enthalpy for R-12 from Table C.3, the power input to the compressor is

$$\dot{W} = \dot{m}(h_i - h_e) = \frac{(90\,\text{kg/h})(190.62 - 231.47)\text{kJ/kg}}{3600\ \text{s/h}}$$

$$= -1.02\,\text{kW}$$

The adiabatic efficiency can be evaluated once the power required by an isentropic compressor operating between the same pressure limits is determined. The exit state of the isentropic compressor is found by noting that

$$s_{e,s} = s_i = 0.7357\ \text{kJ/kg·K}$$

Interpolation in Table C.3 at 1.2 MPa yields

$$h_{e,s} = 224.72\ \text{kJ/kg}$$

and the power required for an isentropic compressor is

$$\dot{W}_s = \dot{m}(h_i - h_{e,s}) = \frac{(90\,\text{kg/h})(190.62 - 224.72)\text{kJ/kg}}{3600\,\text{s/h}}$$

$$= -0.85\,\text{kW}$$

Thus the adiabatic compressor efficiency, from Equation 6.55, is

$$\eta_C = \frac{\dot{W}_s}{\dot{W}} = \frac{-0.85\ \text{kW}}{-1.02\,\text{kW}} = \underline{\underline{0.833}}$$

The irreversibility rate associated with the actual compressor operation can be determined from Equation 6.47, which reduces to

$$\dot{I} = T_0 \dot{m}(s_e - s_i)$$

Using entropy values from Table C.3, we find

$$\dot{I} = \frac{(298\ \text{K})(90\ \text{kg/h})(0.7551 - 0.7357)\text{kJ/kg·K}}{3600\ \text{s/h}} = \underline{\underline{0.14\ \text{kW}}}$$

■

6.6.2 Transient Systems

Transient systems are most often associated with the start-up or shutdown of equipment and the many charging and discharging processes. Pressurizing and depressurizing of storage vessels, the intake and exhaust processes of reciprocating equipment, and the initial start-up of a power plant are all examples of transient processes.

In the analysis of transient systems the usual approach for a thermodynamic analysis is to integrate the appropriate governing equations with respect to time. This procedure was used in Chapter 3 and in Chapter 4 in discussing the application of the equations of conservation of mass and energy to transient systems.

The appropriate integral forms can be obtained directly from the general equations for second-law analysis presented in earlier sections of this chapter. In each case the equation is integrated with respect to time from time t_1 when the system is at state 1 and time t_2 when the system is at state 2. Following the procedure used in Chapter 4, we obtain the irreversibility by integrating Equation 6.23, with the result

$$\begin{aligned} I_{12} &= T_0(S_2 - S_1)_{\text{tot}} \\ &= T_0\left[\sum_{\text{exit}} \int_1^2 s\, dm - \sum_{\text{inlet}} \int_1^2 s\, dm + (S_2 - S_1)_{\text{sys}} + \sum_k \frac{Q_k}{T_k}\right] \end{aligned} \tag{6.58}$$

And the reversible useful work is determined from Equation 6.36:

$$\begin{aligned} W_{12,\text{rev,useful}} = &-\left(\sum_{\text{exit}} \int_1^2 \psi\, dm - \sum_{\text{inlet}} \int_1^2 \psi\, dm\right) \\ &- (\Phi_2 - \Phi_1)_{\text{sys}} - \sum_k Q_k\left(1 - \frac{T_0}{T_k}\right) \end{aligned} \tag{6.59}$$

Furthermore, the irreversibility is also related to the useful work by Equation 6.22, which in integrated form is

$$I_{12} = W_{12,\text{rev,useful}} - W_{12,\text{useful}} \tag{6.60}$$

The use of these equations is illustrated in the following example.

■ EXAMPLE 6.14

A well-insulated air storage tank initially contains 0.065 lb_m of air at 580°R and 20 psia. The tank is connected to a valve in order to charge the tank. The valve is opened, allowing supply air at 290 psia and 1220°R to enter the tank. When the pressure of the

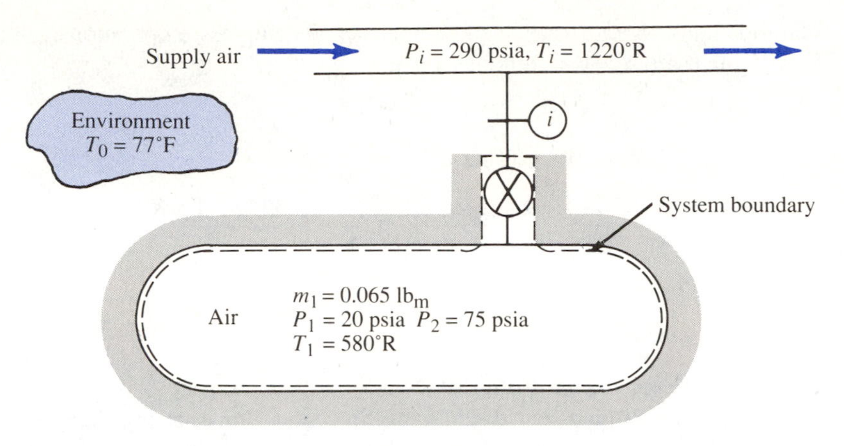

Figure 6.19 Sketch for Example 6.14.

air in the tank has reached 75 psia, the valve is closed. Determine the irreversibility associated with this process. Assume T_0 is 77°F.

Solution. A sketch of the storage tank is shown in Figure 6.19. The transient system in this example has a single inlet and no exits. Thus the conservation-of-mass equation, Equation 3.23, reduces to

$$m_i = m_2 - m_1$$

and the conservation-of-energy equation, Equation 4.34, reduces to

$$Q_{12} - W_{12} + \int_1^2 h_i dm_i = (U_2 - U_1)_{\text{sys}}$$

The storage tank is adiabatic, and no work is done during the process. Furthermore, the enthalpy of the air at the inlet to the tank is constant, enabling us to evaluate the integral term above. Therefore, the conservation-of-energy equation can be written as

$$h_i m_i = (m_2 u_2 - m_1 u_1)$$

and substituting the results of the conservation-of-mass equation, we have

$$h_i(m_2 - m_1) = (m_2 u_2 - m_1 u_1)$$

or

$$0 = m_2(u_2 - h_i) + m_1(h_i - u_1)$$

This expression can be used to determine the temperature of the air in the tank at the end of the process. To facilitate this determination, we assume that the specific heats of the air remain constant during the process. This assumption is appropriate because the temperature change of the air is not expected to be more than a few hundred degrees. The validity of the assumed values, however, should be checked when the final temperature has been determined.

With the assumption of constant specific heats, the internal energy and enthalpy of air, assumed to be ideal, can be expressed as

$$u = c_v T \quad \text{and} \quad h = c_p T$$

where T is the absolute temperature and we have assumed zero reference values for h and u at zero degrees absolute temperature. Using these expressions and the ideal-gas equation of state, we can write the simplifed conservation-of-energy equation as

$$0 = \left(\frac{P_2 \mathbf{V}_2}{RT_2}\right)(c_v T_2 - c_p T_i) + m_1(c_p T_i - c_v T_1)$$

With some algebraic manipulation this equation can be simplified and solved for the final temperature:

$$T_2 = \frac{T_i}{(c_v/c_p) + (P_1/P_2)[(T_i/T_1) - (c_v/c_p)]}$$

Using a value of $c_v/c_p = 1/k$ from Table D.8E at an estimated average temperature of 900°R, we find

$$T_2 = \frac{1220°\text{R}}{(1/1.390) + (20\text{ psia}/75\text{ psia})[(1220°\text{R}/580°\text{R}) - (1/1.390)]}$$

$$= 1120°\text{R}$$

For this value of T_2 the average temperature would be approximately 850°R, and therefore the assumed value of k is appropriate.

The mass in the tank at the end of the process can now be calculated by using the ideal-gas equation of state. Since the volume is constant.

$$m_2 = m_1\left(\frac{T_1}{T_2}\right)\left(\frac{P_2}{P_1}\right)$$

$$= (0.065\text{ lb}_\text{m})\left(\frac{580°\text{R}}{1120°\text{R}}\right)\left(\frac{75\text{ psia}}{20\text{ psia}}\right) = 0.126\text{ lb}_\text{m}$$

The irreversibility associated with this process is found by using Equation 6.58, which for this example reduces to

$$I_{12} = T_0\left[-\int_1^2 s_i\, dm_i + (S_2 - S_1)_{\text{sys}}\right]$$

Since the properties of the air at the inlet to the tank remain constant, the integral can be evaluated, yielding

$$I_{12} = T_0[-s_i m_i + (S_2 - S_1)_{\text{sys}}]$$

or

$$I_{12} = T_0[-s_i(m_2 - m_1) + (m_2 s_2 - m_1 s_1)]$$

$$= T_0[m_2(s_2 - s_i) + m_1(s_i - s_1)]$$

Using Equation 5.30 for the entropy change of an ideal gas with constant specific heats, we have

$$I_{12} = T_0\left[m_2\left(c_p \ln\frac{T_2}{T_i} - R\ln\frac{P_2}{P_i}\right) + m_1\left(c_p \ln\frac{T_i}{T_1} - R\ln\frac{P_i}{P_1}\right)\right]$$

$$= (537°\text{R})\left\{(0.126\ \text{lb}_\text{m})\left[(0.246\ \text{Btu/lb}_\text{m}\cdot°\text{R})\ln\left(\frac{1120°\text{R}}{1220°\text{R}}\right) - \left(\frac{1.986\ \text{Btu/lb}_\text{m}\cdot\text{mol}\cdot°\text{R}}{28.97\ \text{lb}_\text{m}/\text{lb}_\text{m}\cdot\text{mol}}\right)\ln\left(\frac{75\ \text{psia}}{290\ \text{psia}}\right)\right]\right.$$

$$\left. + (0.065\ \text{lb}_\text{m})\left[(0.246\ \text{Btu/lb}_\text{m}\cdot°\text{R})\ln\left(\frac{1220°\text{R}}{580°\text{R}}\right) - \left(\frac{1.986\ \text{Btu/lb}_\text{m}\cdot\text{mol}\cdot°\text{R}}{28.97\ \text{lb}_\text{m}/\text{lb}_\text{m}\cdot\text{mol}}\right)\ln\left(\frac{290\ \text{psia}}{20\ \text{psia}}\right)\right]\right\}$$

$$= \underline{4.84\ \text{Btu}}$$

Since no work is performed during this process, the irreversibility is equal in magnitude to the reversible useful work that would have been produced if the process were reversible. ■

6.7 SUMMARY

In this chapter the methods of second-law analysis were introduced, and general mathematical expressions for reversible power (or work) and irreversibility rate were developed. These concepts evolved from an equation for the rate of change of total entropy,

$$\left(\frac{dS}{dt}\right)_{\text{tot}} = \sum_{\text{exit}} \dot{m}s - \sum_{\text{inlet}} \dot{m}s + \left(\frac{dS}{dt}\right)_{\text{sys}} + \left(\frac{dS}{dt}\right)_{\text{net,surr}} \geq 0 \tag{6.9}$$

If the boundaries of the system to be analyzed are constructed so that the surroundings consist entirely of thermal-energy reservoirs, Equation 6.9 can be written as

$$\left(\frac{dS}{dt}\right)_{\text{tot}} = \sum_{\text{exit}} \dot{m}s - \sum_{\text{inlet}} \dot{m}s + \left(\frac{dS}{dt}\right)_{\text{sys}} + \sum_k \frac{\dot{Q}_k}{T_k} \geq 0 \tag{6.11}$$

where the subscript k refers to reservoir k. This convention for the boundaries of the system was used throughout for the development of succeeding equations.

Next an expression for the reversible power, which represents the maximum amount of power that could be produced with given changes in available resources was developed:

$$\dot{W}_{\text{rev}} = -\left[\sum_{\text{exit}} \dot{m}(h + e_k + e_p - T_0 s) - \sum_{\text{inlet}} \dot{m}(h + e_k + e_p - T_0 s)\right] - \frac{d(E - T_0 S)_{\text{sys}}}{dt} - \sum_k \dot{Q}_k\left(1 - \frac{T_0}{T_k}\right) \tag{6.18}$$

The power actually produced for the same conditions is always less than the reversible power,

$$\dot{W} \le \dot{W}_{\text{rev}} \tag{6.21}$$

The irreversibility rate associated with a process was defined as the difference between the reversible power and the power actually produced. This rate is also equal to the product of T_0 and the rate of change of total entropy:

$$\dot{I} \equiv T_0 \left(\frac{dS}{dt}\right)_{\text{tot}} = \dot{W}_{\text{rev}} - \dot{W} \ge 0 \tag{6.22}$$

and

$$\dot{I} = T_0 \left(\frac{dS}{dt}\right)_{\text{tot}} = T_0 \left(\sum_{\text{exit}} \dot{m}s - \sum_{\text{inlet}} \dot{m}s + \frac{dS_{\text{sys}}}{dt} + \sum_k \frac{\dot{Q}_k}{T_k}\right) \tag{6.23}$$

The useful power was defined as the difference between the power actually produced and the power done on or by the surroundings at pressure P_0. Thus the useful power is

$$\dot{W}_{\text{useful}} = \dot{W} - P_0 \frac{d\mathbf{V}_{\text{sys}}}{dt} \tag{6.25}$$

From Equations 6.18 and 6.26 the reversible useful power is given by

$$\begin{aligned} \dot{W}_{\text{rev,useful}} = & -\left[\sum_{\text{exit}} \dot{m}(h + e_k + e_p - T_0 s) - \sum_{\text{inlet}} \dot{m}(h + e_k + e_p - T_0 s)\right] \\ & - \frac{d(E + P_0\mathbf{V} - T_0 S)_{\text{sys}}}{dt} - \sum_k \dot{Q}_k \left(1 - \frac{T_0}{T_k}\right) \end{aligned} \tag{6.27}$$

and it follows from Equation 6.28 that

$$\dot{W}_{\text{useful}} \le \dot{W}_{\text{rev,useful}}$$

Through a consideration of the maximum amount of work that could be produced with a given initial state or a given initial condition, two availability functions were defined. The stream availability ψ was defined as

$$\psi \equiv (h + e_k + e_p - T_0 s) - (h_0 + e_{k_0} + e_{p_0} - T_0 s_0) \tag{6.32}$$

and the closed-system availability Φ was defined as

$$\Phi \equiv (E + P_0\mathbf{V} - T_0 S) - (E_0 + P_0\mathbf{V}_0 - T_0 S_0) \tag{6.33}$$

With these definitions Equation 6.27 was written in a more compact form as

$$\dot{W}_{\text{rev,useful}} = -\left(\sum_{\text{exit}} \dot{m}\psi - \sum_{\text{inlet}} \dot{m}\psi\right) - \frac{d\Phi_{\text{sys}}}{dt} - \sum_k \dot{Q}_k \left(1 - \frac{T_0}{T_k}\right) \tag{6.36}$$

Closed System

In a closed system the system boundary contains no inlets or exits, so there is no mass flow into or out of the system. For closed systems the general equations were simplified and integrated with respect to time for a change of state from state 1 to state 2.

Thus from Equation 6.25 the work and the useful work are related by

$$W_{12,\text{useful}} = W_{12} - P_0(V_2 - V_1)_{\text{sys}} \tag{6.42}$$

The reversible useful work expressed in terms of the availability, Equation 6.33, reduced to

$$W_{12,\text{rev,useful}} = -(\Phi_2 - \Phi_1)_{\text{sys}} - \sum_k \left[Q_k \left(1 - \frac{T_0}{T_k}\right)\right] \tag{6.44}$$

and the irreversibility for closed-system processes, from Equations 6.22 and 6.23, was simplified to

$$I_{12} = W_{12,\text{rev,useful}} - W_{12,\text{useful}} \tag{6.43}$$

and

$$I_{12} = T_0(S_2 - S_1)_{\text{sys}} + \sum_k T_0 \left(\frac{Q_k}{T_k}\right) \tag{6.39}$$

Open Systems (Uniform Flow)

Steady State For a system operating in steady state, the time rate of change of any extensive property of the system is zero. Therefore, from Equation 6.25 the power and the useful power are equal. The reversible useful power expressed in terms of the availability, Equation 6.36, reduced to

$$\dot{W}_{\text{rev}} = \dot{W}_{\text{rev,useful}} = -\left(\sum_{\text{exit}} \dot{m}\psi - \sum_{\text{inlet}} \dot{m}\psi\right) - \sum_k \dot{Q}_k \left(1 - \frac{T_0}{T_k}\right) \tag{6.50}$$

and the irreversibility rate for steady-state processes, from Equations 6.22 and 6.23, was simplified to

$$\dot{I} = \dot{W}_{\text{rev,useful}} - \dot{W}_{\text{useful}} \geq 0 \tag{6.49}$$

and

$$\dot{I} = T_0\left(\sum_{\text{exit}} \dot{m}s - \sum_{\text{inlet}} \dot{m}s + \sum_k \frac{\dot{Q}_k}{T_k}\right) \geq 0 \tag{6.47}$$

Adiabatic efficiencies were defined in terms of actual adiabatic processes and internally reversible adiabatic (or isentropic) processes:

$$\text{adiabatic-nozzle efficiency} \qquad \eta_N \equiv \frac{V_{e,a}^2/2}{V_{e,s}^2/2} \tag{6.51}$$

$$\text{adiabatic-turbine efficiency} \qquad \eta_T \equiv \frac{\dot{W}}{\dot{W}_s} \tag{6.54}$$

$$\text{adiabatic-compressor efficiency} \qquad \eta_C \equiv \frac{\dot{W}_s}{\dot{W}} \tag{6.55}$$

The subscript s in these equations indicates the isentropic process.

Transient Analysis The analysis of an open thermodynamic system that undergoes transient changes requires the application of the governing equations in their most general form. However, the equations were integrated with respect to time, and the result for the reversible useful work, from Equation 6.36, was

$$W_{12,\text{rev,useful}} = -\left(\sum_{\text{exit}} \int_1^2 \psi\, dm - \sum_{\text{inlet}} \int_1^2 \psi\, dm\right) - (\Phi_2 - \Phi_1)_{\text{sys}} - \sum_k Q_k \left(1 - \frac{T_0}{T_k}\right) \tag{6.59}$$

and the irreversibility, from Equations 6.22 and 6.23, was written as

$$I_{12} = W_{12,\text{rev,useful}} - W_{12,\text{useful}} \tag{6.60}$$

and

$$I_{12} = T_0(S_2 - S_1)_{\text{tot}} = T_0\left[\sum_{\text{exit}} \int_1^2 s\, dm - \sum_{\text{inlet}} \int_1^2 s\, dm + (S_2 - S_1)_{\text{sys}} + \sum_k \frac{Q_k}{T_k}\right] \tag{6.58}$$

PROBLEMS

6.1 It is proposed that nitrogen gas be compressed adiabatically from 15 psia and 100°F to 45 psia and 250°F. The proposed process is (internally reversible, irreversible, impossible). Explain.

6.2 A closed system containing nitrogen executes the following internally reversible cycle: The nitrogen is compressed from 100 kPa and 100°C to a pressure of 350 kPa in a constant-volume process. It is then expanded adiabatically to the initial pressure. The cycle is completed by a constant-pressure process that returns the nitrogen to its initial state. Determine the thermal efficiency for the cycle. Determine the thermal efficiency of a Carnot engine that operates between thermal-energy reservoirs whose temperatures correspond to the maximum and minimum temperatures of the cycle described above. The actual cycle is composed of internally reversible processes but the thermal efficiency is different from that of the Carnot efficiency. Why?

6.3 Consider two energy sources: equal amounts of steam at 250 psia and 2400°F and steam at 3000 psia and 800°F each in a closed system. Which of these has the highest quality of energy? Explain. (Assume T_0 = 77°F, P_0 = 14.7 psia.)

6.4 Consider two energy sources: equal amounts of refrigerant-12 at 0.7 MPa and 140°C and refrigerant-12 at 1.6 MPa and 110°C each in a closed system. Which of these has the highest quality of energy? Explain. (Assume T_0 = 25°C, P_0 = 100 kPa.)

6.5 Three lb_m of refrigerant-12 exists at 300 psia and 200°F in a closed system. What is the maximum useful work that could be obtained by using the refrigerant? (Assume T_0 = 77°F, P_0 = 14.7 psia.)

6.6 Air is contained in a closed system at 7.5 MPa and 475°C. Determine the maximum useful work that could be obtained by using this air. (Assume T_0 = 25°C, P_0 = 100 kPa.)

6.7 Two kilograms of steam exists at 6 MPa and 450°C in a closed system. What is the maximum useful work that could be obtained by using this steam? (Assume T_0 = 25°C, P_0 = 100 kPa.)

6.8 A closed, rigid pressure vessel has a volume of 30 ft^3 and contains 6.0 lb_m nitrogen at 220 psia. The nitrogen in the vessel is cooled to 550°R by means of heat transfer to the surroundings at 500°R and 14.7 psia. Calculate the heat transfer (in Btu), the change in availability of the nitrogen (in Btu), and the irreversibility of the process (in Btu). Explain the source of the irreversibility.

6.9 Two kilograms of steam expands isentropically in a closed, adiabatic system from 2 MPa and 350°C to a final temperature of 90°C. Sketch the *T-s* and *P-v* diagrams, and calculate the work and the irreversibility for this process. (Assume T_0 = 25°C, P_0 = 100 kPa.)

6.10 A rigid pressure vessel with a volume of 70 ft^3 initially contains 1.8 ft^3 of liquid water and 68.2 ft^3 of water vapor in equilibrium at 20 psia. Heat transfer to the contents of the vessel is from a reservoir whose temperature is 600°F until the pressure in the vessel reaches 600 psia. The contents of the vessel are then cooled by heat transfer to the environment until the initial state is reached once again. Determine the heat transfer from the high-temperature reservoir and the irreversibility of the overall process. (Assume T_0 = 77°F, P_0 = 14.7 psia.)

6.11 A small storage tank with a volume of 0.12 m^3 initially contains 0.9 kg of argon at 40°C. Heat transfer to the argon from a reservoir whose temperature is 1300°C continues until the entropy of the argon has increased by 0.46 kJ/kg·K. Determine the heat transfer, the final pressure of the argon, the reversible work, and the irreversibility of the process. (Assume T_0 = 25°C, P_0 = 100 kPa.)

6.12 [SLC111] A constant-volume tank with a volume of 5 ft^3 contains water at a saturated vapor state at 200°F. Heat transfer from a thermal reservoir at a temperature of 1200°R to the water in the tank continues until the water pressure is 20 psia. The ambient conditions are 70°F and 14.7 psia. Determine the following:

(a) the heat transfer to the tank in Btu
(b) the increase in the availability of the water in the tank in Btu
(c) the irreversibility of this process in Btu

6.13 A 2-liter container of cold liquid water at 1.5°C is allowed to warm by exposing it to air at 25°C and 100 kPa. The container is constructed from 0.2 kg of aluminum. Determine the total heat transfer that has occurred from the air when the container and its contents have reached equilibrium with the surrounding air. Determine the entropy change of the aluminum, the entropy change of the water, the entropy change of the air, and the irreversibility.

6.14 A rigid storage tank with a volume of 20 ft^3 is connected by a valve to a piston-cylinder assembly whose volume is initially zero.

The tank and cylinder are well-insulated, and the tank initially contains nitrogen at 30 psia and 80°F. The valve is opened and nitrogen is allowed to escape slowly into the piston-cylinder assembly. During the process, which continues until flow ceases naturally, the pressure in the cylinder is maintained constant at 15 psia. Determine the equilibrium temperature of the nitrogen, the work for the process, the maximum useful work, and the irreversibility. (Assume $T_0 = 70°\text{F}$, $P_0 = 15$ psia.)

6.15 A frictionless piston-cylinder assembly contains 1 kg of steam initially at 1.6 MPa and 500°C. The steam is compressed isothermally until the pressure reaches 2.5 MPa, while heat transfer takes place from the steam to the environment, which is at 25°C and 100 kPa. Sketch this process on T-s and P-v diagrams. Calculate the heat transfer, the work, and the irreversibility for this process.

6.16 A piston-cylinder assembly initially contains 5.0 lb_m of water at 300°F and 150 psia. During an internally reversible, isothermal-expansion process, heat transfer to the water from a thermal-energy reservoir whose temperature is 2200°R amounts to 2500 Btu. The temperature and pressure of the environment are 500°R and 15 psia, respectively. Determine the final pressure of the water, the work for this process, the useful work, the irreversibility, and the change in availability of the water. Explain the significance of the change in availability and the irreversibility.

6.17 Two metallic spheres, each having a mass of 2.5 kg, are to be cooled by placing them in an uninsulated tank of water. In an attempt to increase the cooling rate, the water in the tank is agitated by means of an impeller. One sphere is made of copper, and its initial temperature is 90°C. The other sphere is made of lead, and its initial temperature is 70°C. The water in the tank has a volume of 0.028 m^3 and is initially at 20°C and 100 kPa. During the process the impeller work amounts to 330 kJ, and the heat transfer to the environment (at 20°C and 100 kPa) amounts to 150 kJ. Determine the final equilibrium temperature of the contents of the tank, the change in entropy of each metallic sphere, the total entropy change, and the irreversibility of the process.

6.18 Heat transfer occurs from a reservoir at 600°R to air contained in a piston-cylinder assembly. The air is initially at 60 psia and 500°R and occupies a volume of 22 ft^3. The air expands in an internally reversible isothermal process until its pressure reaches 14.0 psia. The environment is at 540°R and 14.0 psia. Determine the heat transfer, the change in entropy of the air, the change in availability of the air, the useful work, the reversible useful work, and the irreversibility for this process.

6.19 [SLC311] A frictionless piston-cylinder device contains 3 kg·mol of an ideal gas that is initially at 370 K and 150 kPa. The gas is heated by heat transfer from a thermal reservoir at 1100 K. During the process the gas pressure remains constant until the temperature of the gas is 510 K. The gas is then expanded reversibly and adiabatically resulting in 5580 kJ of work. For this gas, $c_p = 29.3$ kJ/kg·mol·K. Assume that the environment (state 0) is at 298 K and 100 kPa. Determine the following:

(a) the heat transfer to the gas in kJ
(b) the irreversibility in kJ
(c) the final temperature (K) and pressure (kPa)
(d) the availability changes of the gas during the heating process and during the adiabatic-expansion process

6.20 A well-insulated tank contains 5.0 lb_m of nitrogen initially at 18 psia and 520°R. An impeller inside the tank is rotated by an external mechanism until the pressure of the nitrogen reaches 25 psia. Determine the work for the process, the reversible useful work associated with this change of state, and the irreversibility of the process. (Assume $T_0 = 530°R$, $P_0 = 14.7$ psia.)

6.21 [SLB311] Five kg of air, initially at 700 kPa and 340 K, is contained in a closed piston-cylinder device that maintains the air at a constant pressure. Heat transfer from the air in the cylinder to the atmosphere, which is at 273 K, amounts to 114 kJ and 25 kJ of work is performed on the air by the piston. Use the air tables to evaluate properties. Calculate the following:

(a) the final air temperature in K
(b) the entropy of the air at the final state
(c) the total entropy change for this process

6.22 [SLE411] An insulated piston-cylinder apparatus is fitted with an electrical heating element through which there is a steady power dissipation. The cylinder initially contains 6.6 lb_m of ice and 2.2 lb_m of liquid water, both at 32°F. A thermocouple on the surface of the heating element registers a steady temperature of 625°F. After 300 s of operation, there are 4.6 lb_m of ice remaining in the cylinder. The cylinder contents are maintained at constant pressure during this process and the temperature of the surroundings is 68°F. Data for ice at 32°F: $u = -143.3$ Btu/lb_m; $h = -143.3$ Btu/lb_m; $s = -0.292$ Btu/lb_m·°R. Determine the following:

(a) the energy-transfer rate provided by the heater in Btu/s
(b) the entropy change of the heater during the 300s of operation in Btu/°R
(c) the entropy change of the water (ice and liquid) during the 300s of operation in Btu/°R
(d) the total change in entropy for this process

6.23 [SLC211] A small container with a volume of 0.001416 m^3 contains an ideal gas at 6893 kPa and 300 K. This container is placed in a large, rigid, insulated chamber with a volume of 0.2832 m^3 which is initially evacuated. The small container is punctured so that the ideal gas is allowed to fill the larger chamber. The specific heats of this gas are $c_p = 14.36$ kJ/kg·K and $c_v = 10.21$ kJ/kg·K. The "dead" state is $T_0 = 294.4$ K and $P_0 = 101.3$ kPa. Determine the following:

(a) the final temperature of the gas in K
(b) the change of entropy in kJ/K
(c) the change in the availability in kJ
(d) the minimum work necessary to compress the gas back to its original state
(e) the final pressure of the gas in kPa

6.24 Nitrogen is confined in a closed piston-cylinder assembly initially at 25 psia and 600°F. The gas undergoes an internally reversible polytropic process to a final pressure of 130 psia. The polytropic exponent for this process is 1.45, and the cylinder contains 1.2 lb_m of nitrogen. Heat transfer to the gas during the process is from a thermal-energy reservoir whose temperature is 2200°F. The environment is at 77°F and 14.7 psia. Determine the work and heat transfer for the process, the reversible work, and the irreversibility. Explain why the reversible work is different from the work produced during this internally reversible process.

6.25 Three kilograms of H_2O in a piston-cylinder assembly undergoes an internally reversible, constant-pressure process. During the process there is heat transfer from a thermal-energy reservoir at 400°C to the H_2O, causing the entropy of the H_2O to increase by 9.04 kJ/K. The H_2O is initially at 1 MPa and occupies a volume of 0.27 m^3. The environment is at 27°C and 100 kPa. Determine the following:

(a) the magnitude and direction of the work for this process (in kilojoules)
(b) the magnitude of the heat transfer (in kilojoules)

(c) the irreversibility associated with the process (in kilojoules)

(d) the reversible work (in kilojoules)

6.26 Helium is contained in a closed system initially at 20 psia and 85°F and undergoes a constant-pressure process until the volume is doubled. The mass of helium present is 2.6 lb_m, and heat transfer to the helium during the process amounts to 1660 Btu. Determine whether this process is internally reversible, internally irreversible, or impossible.

6.27 Steam initially at 500 kPa and 553 K is contained in a cylinder fitted with a frictionless piston. The initial volume of the steam is 0.057 m^3. The steam undergoes an internally reversible isothermal process until it reaches a quality of 70 percent. During the process heat transfer occurs between the steam and the environment, which is at 300 K and 100 kPa.

(a) Sketch the T-s and P-v diagrams for this process.

(b) Determine the magnitude and direction of the heat transfer and the work (in kilojoules).

(c) Determine the irreversibility and the reversible work for the process (in kilojoules).

6.28 Energy from a thermal-energy reservoir at 600°F is used to heat hydrogen, which is contained in a closed, rigid pressure vessel having a volume of 80 ft^3. The hydrogen, initially at 100°F and 35 psia, is heated until its temperature reaches 320°F. Calculate the irreversibility of this process and the change in availability of the hydrogen. Discuss the significance of the difference between the irreversibility and the change in availability. (Assume T_0 = 77°F, P_0 = 15 psia.)

6.29 One kilogram of air in a frictionless piston-cylinder assembly undergoes the following series of processes: From an initial state of 130 kPa and 95°C, it is first expanded at constant pressure until its volume doubles as a result of heat transfer to the air from a thermal-energy reservoir at 800°C. Then heat transfer from a thermal-energy reservoir at 1200°C to the air causes the air pressure to double while the volume remains constant. Calculate the irreversibility for this series of processes, and determine the maximum amount of work that could be produced by the air for the same initial state and final state of this series of processes and the same heat transfer with thermal-energy reservoirs other than the environment. (Assume T_0 = 25°C, P_0 = 100 kPa.)

6.30 A rigid, insulated tank consists of two compartments each having a volume of 25 ft^3, which are separated by a valve. One compartment initially contains nitrogen at 90 psia and 180°F. The second compartment is initially evacuated. The valve is opened and the contents of the two compartments are allowed to reach equilibrium. Determine the equilibrium temperature and pressure and the irreversibility of the process. Explain why the process is irreversible. (Assume T_0 = 80°F, P_0 = 15 psia.)

6.31 A 5.5-kg turkey initially at 25°C is placed in an oven whose temperature is 175°C. The turkey is cooked until its average temperature reaches 75°C. Assume that the oven temperature remains uniform at 175°C, that the turkey is incompressible, that c_{avg} for the turkey is 3.3 kJ/kg·K, and that T_0 = 25°C. How much irreversibility is associated with cooking the turkey?

6.32 Air enters a steady-flow, adiabatic diffuser at 70°F and 15 psia with a velocity of 680 ft/s. Determine the maximum possible exit pressure.

6.33 An inventor claims that a steady-flow, adiabatic air compressor that compresses air at a rate of 6 kg/s from an inlet state of 100 kPa and 310 K to a discharge pressure of 350 kPa with a power input of 785 kJ/s has been developed. Evaluate this claim.

6.34 Steam at 3000 psia and 1800°F enters a steady-flow, adiabatic turbine at a rate of 20 lb_m/s. The discharge pressure is 30 psia. Calculate the maximum possible power output of the turbine.

6.35 Oxygen is to enter a steady-flow nozzle with a negligible velocity at 3.8 MPa and 385°C. A manufacturer claims that with his nozzle the oxygen can be expanded to 150 kPa and 35°C and that the exit velocity will be 750 m/s. Evaluate this claim. (Assume T_0 = 25°C.)

6.36 A supply line carries a steady flow of nitrogen at 1000 psia and 800°R with a velocity of 650 ft/s and a mass flow rate of 1.5 lb_m/s. Determine the maximum useful power that could be produced with this nitrogen supply. (Assume T_0 = 77°F, P_0 = 15.0 psia.)

6.37 Steam at 30 MPa and 800°C is flowing at a rate of 5 kg/s with a velocity of 135 m/s. Determine the maximum useful power that could be produced in a steady-flow device with this steam supply. (Assume T_0 = 25°C, P_0 = 100 kPa.)

6.38 Refrigerant-12 at 160 psia and 180°F is flowing at a rate of 8 lb_m/s with a low velocity. Determine the maximum useful power that could be produced in a steady-flow device with this R-12 supply. (Assume T_0 = 80°F, P_0 = 15 psia.)

6.39 An engineer in an industrial plant instruments a steady-flow mixing chamber and reports the following measurements: inlet steam, 200 kPa, 150°C, 1.5 kg/s; inlet water, saturated liquid, 15°C, 0.8 kg/s; exit mixture, 400 kPa, 2.3 kg/s; heat transfer to environment, 175 kW; environment, 25°C, 100 kPa. Determine whether it is possible for these data to be correct.

6.40 Liquid water is used to condense steam in a heat exchanger. The water enters at 16 psia and 80°F, and it leaves at 120°F. Steam enters at 12 psia with a quality of 96 percent and a mass flow rate of 18 lb_m/s and it leaves as a saturated liquid. Neglect pressure drops and assume T_0 = 80°F. Determine the mass flow rate of cooling water required and the irreversibility of the process.

6.41 Refrigerant-12 enters a cooling coil (an evaporator) with an enthalpy of 74 kJ/kg. The pressure drop in the coil is negligible, and the refrigerant is to leave as a saturated vapor. The coil is used to cool air, which enters at 100 kPa and 23°C and leaves at 100 kPa and 8°C at a rate of 4.1 kg/s. Two evaporator temperatures are being considered. In one case the refrigerant would enter at −20°C, and in the other the refrigerant would enter at −25°C. Which of these evaporator temperatures would be preferred from a thermodynamic viewpoint if only the evaporator operation is considered? Justify your answer.

6.42 [SLB211] One type of feedwater heater for preheating water before entering a boiler mixes 150 lb_m/h of water at pressure 100 psia and temperature 90°F with 50 lb_m/h of steam at the same pressure and a quality of 0.9. Water leaves the heater at pressure 95 psia and temperature 300°F. Assuming steady flow and that the environment is at 530°R and 14.7 psia, determine the following:

(a) the temperature of the entering steam in °F

(b) the heat-transfer rate between the heater and the surroundings in Btu/h

(c) the rate of change of total entropy in Btu/h·°R

6.43 Two alternatives are being considered for heating 500 m^3/min of air from 15°C and 100 kPa at 28°C. One method would use a heat exchanger with steam entering at 200 kPa and 200°C and leaving at 150°C. The other method would use an electric-resistance heater to heat the air. Which method is superior from a thermodynamic viewpoint? Assume negligible pressure drops and T_0 = 15°C.

6.44 Calculate the minimum power input required to drive a steady-flow, adiabatic air compressor that compresses 3.1 lb_m/s of air from 15 psia and 80°F to a pressure of 80 psia.

6.45 Calculate the minimum power input required to compress 2.3 kg/s of air from 110 kPa and 20°C to 240 kPa and 70°C if only heat transfer with the environment occurs. (Assume $T_0 = 20°C$.)

6.46 A well-insulated air preheater is to be installed to preheat combustion air for a furnace. The preheating is achieved by cooling the products of combustion from the furnace, as shown in the accompanying sketch. The combustion air, which flows at a rate of 50 kg/s, is initially at 40°C and is to be preheated to 170°C. The products of combustion leave the furnace at 315°C at a rate of 54 kg/s. The average value of the constant-pressure specific heat for the products of combustion is 1.09 kJ/kg·K, and the apparent molecular weight of the products is 28.72. Assume ideal-gas behavior, negligible changes in kinetic and potential energies, negligible pressure drops, and environmental conditions of 298 K and 100 kPa.

(a) Determine the temperature of the products at the exit of the preheater.

(b) Determine the irreversibility associated with the preheater. Explain the source of this irreversibility.

(c) Determine the reversible power associated with the given conditions.

Problem 6.46

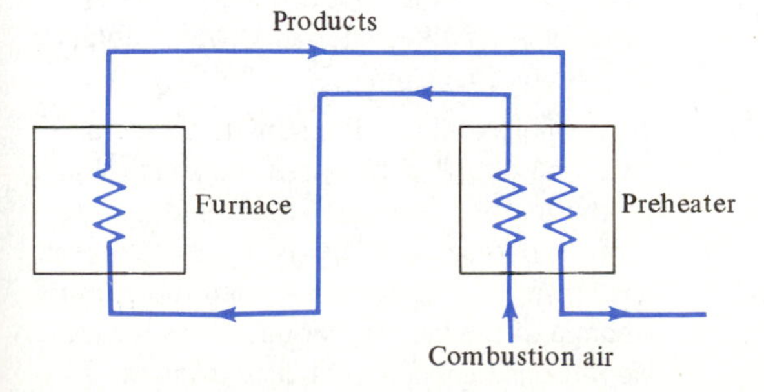

(d) Adding the preheater to the system introduces an additional source of irreversibility. What effect will this have on the total irreversibility of the system (i.e., compare *qualitatively* the irreversibility of the furnace alone to the irreversibility of the furnace-preheater combination)?

6.47 At one inlet to a certain steady-flow device, 0.4 lb_m/s of steam enters at 1500 psia and 1300°F. At the other inlet to the device steam enters at 1250 psia and 1000°F. The mass flow rate at this second inlet is twice as large as the flow rate at the first inlet. At the exit of the device steam leaves as a saturated vapor at 30 psia. Heat transfer to the environment (at 15 psia and 530°R) occurs during this process in the amount of 30 Btu/lb_m steam leaving the device. Kinetic- and potential-energy changes are negligible. Determine the power associated with this device, the irreversibility rate associated with the device, and the reversible power that could be produced for the given conditions.

6.48 A steady-flow, internally reversible compressor is used to compress air isothermally from 400 kPa to 1.8 MPa. The air enters the compressor at the rate of 30 m^3/min at a temperature of 35°C. The kinetic-energy change of the air is assumed negligible, and $T_0 = 25°C$. Calculate the power input to the compressor and the irreversibility rate for the process.

6.49 A heat exchanger is designed to condense steam from the saturated-vapor state to the saturated-liquid state at a constant pressure of 15 psia. The coolant is atmospheric air, which enters the heat exchanger at 70°F and exits at 150°F and experiences a negligible pressure drop. The heat exchanger is well-insulated and operates in steady flow. The environment is at 530°R and 14.7 psia. For each lb_m of steam condensed in the heat exchanger, determine (a) the mass flow of air required, (b) the change in availability of the steam, (c) the change in availability of the air, and (d) the irreversibility of the process.

6.50 [SLD111] In the boiler of a power plant, water is continuously heated at a rate of 1 kg/s from 260°C, 8000 kPa to 500°C, 8000 kPa with combustion gas. The combustion gas enters the boiler at a rate of 10 kg/s at 1267°C, 200 kPa and exits at 927°C, 100 kPa. The combustion gas may be considered to have the same properties as air, as given in the air tables. The surroundings are at 300 K and 100 kPa. Determine the following:

- **(a)** the heat-transfer rate from the boiler jacket
- **(b)** the entropy per unit mass of the combustion gas leaving the boiler
- **(c)** the availability per unit mass of the water leaving the boiler
- **(d)** the irreversibility rate for the overall process

6.51 [SLD211] In a power plant steam is transported from a boiler to a turbine through high-pressure pipes. The steam enters at 1000°F and 700 psia, exits at 700°F and 650 psia. Assuming that the environment is at a state of 70°F and 14.7 psia, determine on a per unit mass basis:

- **(a)** the total change in entropy
- **(b)** the decrease in the availability of the steam as it flows through the pipe

6.52 Steam enters one inlet of an adiabatic mixing chamber at 1.4 MPa and 300°C. At the second inlet steam enters at 1.4 MPa with a quality of 92 percent. The mass flow rates of these two streams are equal. Determine the temperature (or quality, if saturated) of the mixture as it leaves the chamber at a pressure of 1.4 MPa and the irreversibility of the process per unit mass of mixture. Explain the source of the irreversibility. (Assume $T_0 = 25°C$.)

6.53 A steady-flow, water-cooled air compressor is used to compress air from 15 psia and 80°F to 150 psia and 260°F. The air enters the compressor at a rate of 350 ft^3/min. Cooling water enters the cooling passages of the compressor at a rate of 20 lb_m/min at a temperature of 60°F and a pressure of 30 psia. The water experiences an increase in temperature of 20°F and a negligible pressure drop. Determine the power input required to drive the compressor, the irreversibility rate, and the reversible power. (Assume $T_0 = 70°F$.)

6.54 Refrigerant-12 flowing at a rate of 7 lb_m/s is to be accelerated in a steady-flow nozzle to a velocity of 850 ft/s. At the nozzle inlet the pressure and temperature of the refrigerant are 100 psia and 260°F, respectively, and the inlet velocity is negligible. It is claimed that under these conditions there will be heat transfer to the surroundings, which are at 77°F and 15 psia, and that the refrigerant will leave the nozzle at 25 psia and 90°F.

- **(a)** Determine the magnitude and direction of the heat transfer that would occur for the process (in Btu/h).
- **(b)** Determine whether it is possible for the refrigerant to reach the exit state specified in the claim.

6.55 Steam at 150 kPa and 200°C enters a steady-flow diffuser with a high velocity and exits with a velocity that is negligible compared with the inlet velocity. If the diffuser operation were isentropic, the pressure of the steam at the exit would attain 600 kPa. However, during the actual process the heat transfer to the surroundings amounts of 85.9 kJ/kg, and the steam exits at 350°C. For these conditions, determine the inlet-steam velocity and the actual exit pressure. Determine also whether the actual process is internally reversible, internally irreversible, or impossible. (Assume $T_0 = 25°C$.)

6.56 It is reported that helium enters a pipe at 30 psia and 480°R. The gas flows steadily and isothermally through the pipe, and the pressure at one section downstream is 10 psia. Determine the heat transfer and the entropy change of the helium per unit mass for these conditions. This process, as described (is internally reversible, is internally irreversible, violates the second law, violates the conservation of energy). (Assume T_0 = 80°F.)

6.57 The exhaust from the turbine of a turbojet engine enters a nozzle at 180 kPa and 975 K with a velocity of 70 m/s. In the nozzle the gas expands adiabatically to a pressure of 70 kPa. Assume that the exhaust gases have approximately the same properties as those of air. Determine the exit velocity for a nozzle efficiency of 90 percent, and calculate the irreversibility per unit mass. (Assume T_0 = 25°C.)

6.58 Refrigerant-12 enters a steady-flow, adiabatic compressor at 30 psia and 40°F and is compressed to 150 psia. The adiabatic-compressor efficiency is 90 percent. Determine the exit temperature of the refrigerant, the work per unit mass required to operate the compressor, and the irreversibility per unit mass. Calculate the change in availability of the refrigerant for this process and discuss its significance. (Assume T_0 = 530°R.)

6.59 Steam at 1.8 MPa and 400°C is required for use in an industrial process, and steam at 2.5 MPa and 700°C is available in the plant. An engineer in the plant suggests that a system could be devised to deliver the steam at the desired state and at the same time produce extra power. The steam mass flow rate is 3 kg/s, and T_0 = 25°C. What is the maximum amount of power that could be produced for the stated conditions? Illustrate the process(es) involved on a *T-s* diagram.

6.60 Steam at 200 psia and 500°F enters a steady-flow, adiabatic nozzle with negligible velocity and exits at a pressure of 25 psia and a quality of 96 percent. Calculate the exit velocity, the adiabatic-nozzle efficiency, and the irreversibility per unit mass. (Assume T_0 = 77°F.)

6.61 A steady-flow, adiabatic nozzle with a nozzle efficiency of 96 percent delivers air at 500 m/s, 110 kPa, and 10°C. Assume that the inlet velocity is negligible and that the specific heats of the air are constant. Determine the inlet temperature and pressure of the air and the irreversibility of the process per unit mass of air. (Assume T_0 = 25°C.)

6.62 Air at 50 psia and 580°R enters an internally reversible, isothermal turbine and exhausts at 15 psia. Heat transfer to the air in the turbine is from a reservoir whose temperature is 900°R, and T_0 = 500°R. Determine the work per lb_m of air and turbine effectiveness, assuming that changes in kinetic and potential energies are negligible.

6.63 Refrigerant-12 enters a steady-flow, adiabatic compressor in a refrigeration system as a saturated vapor at 200 kPa and exits at 1.2 MPa. The mass flow rate of refrigerant is 85 kg/h. The environment is at 25°C and 100 kPa. Determine the motor size (in kilowatts) required to drive the compressor if the compressor efficiency is 82 percent. Determine the minimum power input associated with the actual change of state.

6.64 Steam enters a steady-flow, adiabatic nozzle at 1000 psia and 1000°F with a velocity of 300 ft/s through an area of 4.0 in^2. The steam exits at 500 psia and 800°F. Determine the adiabatic-nozzle efficiency, the irreversibility rate, and the maximum possible exit velocity for an exit pressure of 500 psia and the given inlet conditions. (Assume T_0 = 77°F.)

6.65 A steady-flow air compressor inducts 2000 ft^3/min of air at 15 psia and 75°F and compresses it to 120 psia and 240°F. During this process the heat transfer from the air to the environment is 2.2×10^5 Btu/h. The inlet velocity is negligible, and the exit area is 60 in^2. Determine the power input.

6.66 Steam at 4 MPa and 450°C enters a steady-flow, adiabatic turbine at a rate of 12 kg/s. Ten percent of the mass flow is extracted at a point where the temperature and pressure of the steam are 300°C and 1.2 MPa, respectively, for use in an industrial process.

The remainder of the steam is further expanded and exhausts from the turbine at 75 kPa with a quality of 98 percent. Determine the adiabatic efficiency of the turbine, the irreversibility rate, and the reversible power. (Assume $T_0 = 25°C$.)

6.67 Refrigerant-12 enters a steady-flow, adiabatic compressor as a saturated vapor at $-10°F$ at a rate of 46 lb_m/min. At the exit of the compressor the temperature and pressure of the refrigerant are 80°F and 60 psia. Calculate the power input, the compressor efficiency, and the irreversibility rate if $T_0 = 80°F$.

6.68 Steam enters a steady-flow, adiabatic turbine at 5 MPa and 550°C with a negligible velocity and exhausts at 100 kPa and with a quality of 99 percent and a velocity of 140 m/s. Calculate the turbine efficiency, the irreversibility, and the reversible work per unit mass of steam flowing through the turbine. (Assume $T_0 = 25°C$.)

6.69 [SLB411] The mass rate of flow into a steady-flow steam turbine is 10,000 lb_m/h, and the heat-transfer rate from the turbine to the environment is 30,000 Btu/h. Steam enters the turbine at 700°F and 300 psia, with a velocity of 200 ft/s at 16 ft above a reference plane. Saturated vapor leaves the turbine 10 ft above the reference plane at 15 psia and 600 ft/s. The ambient temperature is 70°F. Determine the following:

(a) the power output of the turbine in Btu/h
(b) the rate of change of total entropy in Btu/h·°R
(c) the turbine effectiveness

6.70 Steam enters a three-stage, steady-flow, adiabatic turbine at 12.5 MPa and 500°C. Before the steam is finally exhausted at 25 kPa, steam is extracted at the end of the first two stages at pressures of 3 MPa and 200 kPa, respectively. The amount of steam extracted at the end of each stage is equal to 10 percent of the mass flow through the stage. The turbine produces 8 MW of power. Assume that the adiabatic efficiency of each stage is 93 percent and neglect changes in kinetic energy. Determine the mass flow rate of steam entering the turbine and the irreversibility rate of the process. (Assume $T_0 = 25°C$.)

6.71 Consider two steady-flow, adiabatic nozzles, each having an efficiency of 95 percent. Both nozzles take in nitrogen at 300 psia with a negligible velocity and expand the gas to 30 psia. The mass flow rate of nitrogen is also the same for each nozzle. However, for nozzle *A* the entering temperature of the nitrogen is 800°F, while for nozzle *B* the entering temperature is 1000°F. Which of the nozzle processes is superior from a thermodynamic viewpoint? Justify and explain your response. (Assume $T_0 = 80°F$.)

6.72 Under full-load conditions steam enters a steady-flow, adiabatic turbine at 5 MPa and 450°C and exhausts at 20 kPa. However, for part-load operation the steam is throttled from 5 MPa and 450°C to 2.5 MPa before it enters the turbine. Assume that the adiabatic efficiency is 85 percent for all operating conditions and that $T_0 = 25°C$. Calculate the work and irreversibility per unit mass of steam flowing for both full-load and part-load conditions.

6.73 Argon enters a steady-flow, adiabatic turbine at 1840°F and 350 psia and exhausts at 75 psia. The turbine produces 175 hp of power when the mass flow rate of argon is 1.0 lb_m/s. Determine the exit temperature of the argon, the efficiency of the turbine, and the irreversibility rate. (Assume $T_0 = 70°F$.)

6.74 An internally reversible, isothermal gas turbine is suggested for use to power an adiabatic gas compressor. Air would enter the turbine at 300 kPa and 45°C and exhaust at 110 kPa. Heat transfer to the air in the turbine would be from a reservoir whose temperature is 100°C. The gas compressor, which has an adiabatic efficiency of 80 percent, would be used to compress 0.42 kg/s of nitrogen from 150 kPa and 28°C to 500 kPa. Assume steady flow. Determine the mass flow rate of air required, the heat-transfer rate to the air in the turbine, and the irreversibility rate for the overall arrangement. How much of the irreversibility rate would be associated with the turbine process? (Assume $T_0 = 25°C$.)

6.75 [SLB111] A steady-flow, adiabatic air turbine has been chosen to produce power at a factory for minor tasks. Air enters the turbine at 870°C and 500 kPa and is expanded to 101 kPa. Assume that the turbine operation is internally reversible and that the air behaves as an ideal gas with $c_p = 1.005$ kJ/kg·K. Determine the following:

(a) the exit temperature in °C
(b) the specific volumes of the air at the inlet and outlet
(c) the output work per unit mass

6.76 [SLE211] A tank having a volume of 0.14 m^3 initially contains nitrogen gas at 1 MPa and 20°C, the temperature of the surroundings. The tank develops a leak and, over a long time, the pressure drops to 500 kPa. The atmospheric pressure of the air surrounding the tank is 100 kPa. Assume nitrogen is an ideal gas with $c_p = 1.038$ kJ/kg·K. Neglect the heat capacity of the tank and piping and assume that the nitrogen remains at 20°C. Calculate the following:

(a) the mass of nitrogen that escapes the tank
(b) the heat transfer in kJ to the tank
(c) the total entropy change during the process in kJ/K
(d) the irreversibility in kJ

6.77 A pressure vessel with a volume of 30 ft^3 contains 1 percent liquid water by volume in equilibrium with water vapor at 120°F. Heat transfer to the water is from a reservoir with a temperature of 575°F, and a relief valve at the top of the vessel maintains the pressure constant by allowing vapor to escape. The process continues until one-half of the mass of the liquid has vaporized. Determine the amount of heat transfer required and the irreversibility. The overall process is irreversible. Is the process that the water and vapor undergoes internally reversible? Justify your response quantitatively. (Assume $T_0 = 80°F$, $P_0 = 15$ psia.)

6.78 Steam at 800 kPa and 300°C is bled off from a large supply line, passed through a valve, and allowed to enter an initially evacuated, uninsulated tank with a volume of 0.3 m^3. When the pressure in the tank reaches 100 kPa, the valve is closed. The temperature in the tank when the valve is closed is reported to be 150°C. Determine whether this process is possible. (Assume $T_0 = 25°C$.)

6.79 An insulated nitrogen storage tank with a volume of 6.0 ft^3, which is initially evacuated, is to be filled with nitrogen at a pressure of 200 psia. The tank is to be filled by bleeding nitrogen from a supply line connected to the tank by a valve. The supply line carries nitrogen at 300 psia and 90°F. Determine the mass and temperature of nitrogen at the end of the process and the irreversibility. (Assume $T_0 = 80°F$, $P_0 = 15$ psia.)

6.80 [SLE111] Air flows in a pipeline at a pressure of 1380 kPa and a temperature of 27°C. Connected to the pipeline is a tank that initially contains 0.63 kg of air at 20°C and 100 kPa. A valve is opened allowing air to flow into the tank until the pressure is 700 kPa. During this process, 3.5 kg of air enters the tank. Assume that the tank has negligible heat capacity. Also, assume air to be an ideal gas with $c_p = 1.005$ kJ/kg·K. The surroundings (state 0) are at 20°C and 100 kPa. Calculate the following:

(a) the final temperature T_2 in °C of the air in the tank
(b) the heat transfer in kJ between the environment and the tank
(c) the total entropy change in kJ/K during this process

6.81 [SLE311] A cylinder fitted with a frictionless piston initially contains 0.9 lb_m of steam at 25 psia and 350°F. The cylinder is connected to a steam line containing steam at 500°F and 100 psia with a short pipe fitted with a valve. The valve is opened and 1.8 lb_m of steam enters the cylinder. At this time, the temperature of the steam in the cylinder is 400°F. The valve is then closed. Assume that the tank has negligible heat capacity. Also assume that the pressure in the cylinder remains constant. The environment surrounding the cylinder is at 530°R and 14.7 psia. Determine the following:

(a) the work in Btu done on the piston
(b) the heat transfer to the system in Btu
(c) the total change in entropy in Btu/°R during this process
(d) the irreversibility for the process

6.82 A leak occurs in a rigid, well-insulated, evacuated tank that has a volume of 5 m^3, and as a result, atmospheric air at 100 kPa and 20°C enters the tank. When the pressure in the tank reaches 100 kPa, what is the mass of air in the tank? Calculate the irreversibility for this process.

7 Gas Cycles

7.1 INTRODUCTION

The study of cycles is an important aspect of thermodynamics that includes the analysis of systems such as power plants, gasoline engines, jet engines, diesel engines, gas-turbine power plants, and refrigeration and air-conditioning systems. The analysis of thermodynamic cycles was introduced in a cursory fashion in Chapter 4 prior to the introduction of the second law of thermodynamics. Application of the second-law principles permits a more in-depth study of thermodynamic cycles. Before considering the details of the analysis, though, we discuss general concepts that apply to all cycles.

7.2 BASIC CONSIDERATIONS

Thermodynamic cycles can be divided into two broad categories, power cycles and refrigeration cycles, depending on the purpose of the cycle. *Power cycles* are designed to produce a net positive work output, and the devices or systems that execute power cycles are often simply called *engines*. For example, Carnot engines and diesel engines would execute Carnot cycles and diesel cycles, respectively. An automobile engine and a steam power plant are other examples of systems that operate on a power cycle. *Refrigeration cycles*, on the other hand, are designed to provide cooling or heating. A *refrigerator* is a device that executes a refrigeration cycle and whose purpose is to provide cooling. A home air-conditioning unit is a typical example of a device that executes a refrigeration cycle. A *heat pump*, commonly used in residential heating, also operates on a refrigeration cycle, although its purpose is to provide heating.

The substance that circulates through the cyclic device is called a *working fluid*. The working fluid in a refrigerator or air-conditioning unit is a refrigerant such as refrigerant-12, while the working fluid in an automobile engine is a mixture of gasoline vapor and air. Depending on the phase of the working fluid, thermodynamic cycles can be categorized as gas cycles or vapor cycles. In a *gas cycle* the working fluid remains in the gaseous phase. By contrast, the working fluid in a *vapor cycle* exists in the vapor phase and the liquid phase at various points throughout the cycle since the working fluid alternately vaporizes and condenses. This chapter deals with gas power cycles and gas refrigeration cycles; vapor cycles are discussed in the following chapter.

A frequent assumption in the analysis of thermodynamic cycles is that the changes in the potential and kinetic energies of the working fluid are relatively small. In turbines, pumps, and compressors, for example, the magnitude of the work is usually much larger than the changes in the potential and kinetic energies that may occur. In heat exchangers, such as boilers, condensers, and evaporators, the fluid velocities are kept relatively low to minimize pressure drops; velocity changes are also small, tending to minimize the changes in the kinetic energy of the fluid. For these reasons kinetic- and potential-energy changes can usually be neglected in the analysis of many thermodynamic cycles. However, when one is dealing with diffusers and nozzles and other devices designed specifically to create large changes in the velocity of the working fluid, the changes in the kinetic energy should always be considered when applying the conservation of energy to such devices.

The measure of performance for power cycles and refrigeration cycles can be defined in general terms as the ratio of the desired effect to the energy input required to produce the desired effect. For a power cycle the measure of performance is called the *thermal efficiency* and is defined as the ratio of the net work output during the cycle (the desired effect) to the heat transfer to the working fluid from the high-temperature reservoir during the cycle, or

$$\eta_{\text{th}} = \frac{W_{\text{net}}}{Q_H} \tag{7.1}$$

To evaluate the performance of a refrigeration cycle, the *coefficient of performance* (COP), designated by the greek symbol β, is used. Since the desired effect for a refrigerator is to provide cooling sufficient to maintain a reservoir at a low temperature, the COP of a refrigeration cycle is defined as the heat transfer to the system from the low-temperature reservoir divided by the work required to operate the refrigerator:

$$\beta_R = -\frac{Q_L}{W_{\text{net}}} \tag{7.2}$$

The negative sign appears in Equation 7.2 because net work is performed on the system and it is a negative quantity according to the sign convention established for work while the COP and Q_L are both positive quantities. Similarly, the desired effect for a heat pump is to provide heating sufficient to maintain a reservoir at a high temperature. Thus the

COP of a heat pump is defined as the heat transfer from the system to the high-temperature reservoir divided by the work required to operate the heat pump:

$$\beta_H = \frac{Q_H}{W_{\text{net}}} \tag{7.3}$$

The thermal efficiency of a power cycle and the coefficient of performance of a refrigeration cycle each provide a single parameter that can be used to evaluate the effect that design changes and modifications to the basic cycles have on cycle performance. Throughout the study of cycles, one of the major objectives is to select modifications to the processes and states throughout the cycles such that the performance of the cycle is maximized. For engines used in automobiles, trucks, ships, and airplanes, for example, this desire would translate into designing the engine so that it consumes a minimum amount of fuel for a given trip. In the operation of power plants any change that increases the cycle efficiency will produce more electricity for the same consumption of fuel. Similarly, any modification to a refrigeration unit or a heat pump that results in an increase in the COP is a desirable change. An increase in the coefficient of performance of a refrigeration cycle would produce an air-conditioning or refrigeration system that would require less work to provide the same amount of cooling or a heat pump that would provide more heating for the same work input.

Second-law analysis is also important in the study of thermodynamic cycles. For example, it can be used as an aid to improve designs by focusing attention on those components or processes in the cycle that have large irreversibilities. The insight provided by a second-law analysis of individual components generally cannot be achieved with an energy analysis alone. Perhaps one of the most important contributions of second-law analysis is that it provides a methodology that can be used to choose among alternatives that might otherwise appear to have equal merit. This feature is illustrated in a number of examples in this and the next chapter.

7.3 IDEAL AND ACTUAL CYCLES

The study of thermodynamic cycles is often approached by first considering *ideal cycles*, which consist entirely of internally reversible processes. By considering ideal cycles first, we need not consider complicating effects such as friction and other sources of internal irreversibilities. Devices or engines that operate on ideal cycles are impossible to construct, because all real devices have some irreversibility. Therefore, an ideal cycle is merely a conceptual cycle that is easy to analyze from a thermodynamic standpoint. *Actual cycles*, on the other hand, are ones that do contain irreversibilities, and the analysis of engines that operate on actual cycles takes into consideration the irreversibilities that accompany all real processes.

Factors such as friction result in irreversibilities that are detrimental to the thermal efficiency of a cycle. For this reason the ideal cycle, void of all internal irreversibilities, is used to determine the maximum efficiency that could be achieved for a given cycle. Although ideal cycles are internally reversible, they are not necessarily totally reversible because they may include heat transfer between the system and its surroundings across finite temperature differences. The thermal efficiency of an ideal cycle, therefore, will be less than the thermal efficiency of a Carnot cycle operating between the same temperature limits as those of the ideal cycle.

Although the ideal-cycle analysis may seem to be somewhat limited, the general trends that emerge from an analysis of ideal cycles often apply to the actual irreversible cycle. Once the trends in design are established from the analysis of the ideal cycle, modifications that will increase the thermal efficiency or coefficient of performance of actual cycles can be proposed.

In Chapter 5 the *T-s* diagram was introduced to facilitate the analysis of thermodynamic processes. A *T-s* diagram has particular significance in cycle analysis because it aids in the visualization of the processes that compose the cycle. The area under the *T-s* process curve represents the heat transfer during an internally reversible process. Also, the net work produced during a cycle is numerically equal to the net heat transfer during the cycle. Therefore, the area enclosed by the process curves of an internally reversible or ideal cycle on a *T-s* diagram is equal to both the net heat transfer and the net work for the cycle. This fact helps in understanding the impact of changes in the cycle on the thermodynamic performance. For instance, the thermal efficiency of an ideal power cycle can be interpreted geometrically as being the ratio of the net area enclosed by the process curve on the *T-s* diagram to the area under the *T-s* curve that represents the heat transfer to the system. Any modification to the cycle that will increase the former or decrease the latter will necessarily produce an increase in the thermal efficiency of the cycle.

Certain physical limitations impact on the design of cyclic devices and should be considered before starting a detailed discussion of gas cycles. Heat exchangers, designed for high heat-transfer rates, offer a good example of these limitations. Often heat exchangers are simply tubes covered with fins that promote heat transfer by increasing the tube surface area. Heat transfer occurs between the fluid in the tubes and a fluid at a different temperature that circulates over the tubes. In a well-designed heat exchanger the restriction to the motion of the working fluid within the tubes should be minimized in order to reduce fluid pressure drops and to reduce the energy required to circulate the fluid. For this reason the pressure drop in the ideal heat exchanger is assumed to be zero and the working fluid is assumed to undergo an internally reversible, constant-pressure process.

With these concepts pertaining to ideal and actual cycles, general conclusions can be drawn regarding the expected ranking of the thermal efficiencies of cycles. Suppose that the temperatures of two reservoirs are fixed, and that the efficiencies of several heat engines operating between these two reservoirs are to be compared. If one of the engines operates on a Carnot cycle, it will possess the maximum thermal efficiency of all engines operating between these two reservoirs because each of the processes in the cycle will be totally reversible. Suppose that two other heat engines are available that also operate between the same temperature limits as those of the Carnot cycle. One engine operates on an ideal cycle consisting entirely of internally reversible processes, while the second engine operates on an actual cycle with the same set of processes but the processes are irreversible. Even though the ideal cycle is internally reversible, external irreversibilities cause the cycle's thermal efficiency to be less than that of a Carnot cycle. However, the efficiency of the ideal cycle exceeds that of the actual cycle since additional irreversibilities are present in the actual cycle. Thus, if a Carnot engine, an ideal engine, and an actual engine operate between the same temperature limits, then the general conclusion regarding the ranking of the thermal efficiencies would be

$$\eta_{Car} > \eta_{ideal} > \eta_{act}$$

This rank order is intimately related to the amount of irreversibility associated with each of the engines.

The concept of irreversibility introduced in Chapter 6 takes on added significance now that the notion of ideal and actual cycles has been introduced. Consider an actual (or even ideal) power cycle that has heat transfer Q_H to the cycle from a reservoir at a temperature of T_H and heat transfer Q_L from the cycle to a reservoir at T_L. Furthermore, assume that T_L is equal to the temperature of the surroundings, T_0, because the heat transfer from the cycle is to the lowest available energy sink, which, under normal circumstances, is the environment. The irreversibility associated with the cycle can be determined by integrating Equation 6.38 for the cycle:*

$$\begin{aligned} I_{cyc} &= T_0 \oint dS_{sys} + \sum_k T_0 \oint \frac{\delta Q_k}{T_k} \\ &= 0 + T_0 \left(-\frac{Q_H}{T_H} - \frac{Q_L}{T_0} \right) \\ &= Q_H \left(-\frac{T_0}{T_H} - \frac{Q_L}{Q_H} \right) \end{aligned}$$

Notice that the entropy change of the system over an entire cycle is zero because entropy is a property and the cyclic integral of a property is zero.

The thermal efficiency of the actual cycle can be expressed as

$$\eta_{act} = 1 + \frac{Q_L}{Q_H}$$

Combining this expression with the preceding equation gives

$$I_{cyc} = Q_H \left(1 - \frac{T_0}{T_H} - \eta_{act} \right)$$

The sum of the first two terms within the parentheses is equal to the thermal efficiency of a totally reversible cycle, such as a Carnot cycle:

$$\eta_{Car} = 1 - \frac{T_0}{T_H}$$

Therefore, the expression for the irreversibility of the actual cycle can be written as

$$I_{cyc} = Q_H(\eta_{Car} - \eta_{act}) \tag{7.4}$$

Equation 7.4 shows that the difference between the thermal efficiency of a Carnot cycle and the thermal efficiency of an actual cycle (or an ideal cycle) is proportional to the irreversibility of the cycle. For an ideal cycle the irreversibility results entirely from heat

*Recall that the signs of Q_L and Q_H are taken relative to the heat engine as the system.

transfer between the system and its surroundings through a finite temperature difference. For the actual cycle, internal irreversibilities as well as irreversibility associated with heat transfer cause the thermal efficiency of the cycle to be less than the Carnot cycle efficiency.

7.4 AIR-STANDARD ASSUMPTIONS

The *air-standard assumptions*, often used when considering gas cycles, consist of the following:

1. The working fluid has properties that are the same as those of air.
2. The working fluid is an ideal gas.
3. The exhaust process is replaced by heat transfer from the cycle to a low-temperature reservoir and the combustion process is replaced by heat transfer to the cycle from a high-temperature reservoir.

The air-standard assumptions simplify the analysis of gas cycles without imposing unrealistic restrictions on the analysis. The working fluid for gas cycles is usually a mixture of fuel and air, but since the percentage of air in the mixture for most gas cycles is relatively high, the presence of a small amount of fuel does not significantly alter the properties of the mixture from the properties of pure air. Assuming that the working fluid is composed entirely of air usually results in rather small errors in the analysis.

Heat engines such as gasoline or diesel engines are open systems that draw in fresh air at ambient conditions and expel high-temperature combustion products to the environment. With use of the air-standard assumptions, the combustion process is replaced by an equivalent process consisting of heat transfer with a high-temperature reservoir. Likewise, the exhaust process is replaced with a process of heat transfer with a low-temperature reservoir. For the air-standard assumptions the working fluid is always recirculated and the heat-rejection process simply restores the properties of the working fluid to the values that correspond to the initial state so that the cycle can be repeated.

The air-standard assumptions apply only to gas cycles; they should not be applied to vapor cycles in which the working fluid does not remain in the gaseous phase throughout the cycle.

While the air-standard assumptions simplify the analysis of gas cycles, the additional assumption that the specific heats of air are constant provides a further simplification of the analysis. The justification of the assumption of constant specific heats is sometimes questionable because the specific heats are known to be functions of temperature and the temperature of the working fluid can vary considerably throughout the cycle. The term *cold air-standard assumption* is used to imply that the air-standard assumption is used in addition to assuming that the specific heats of the working fluid are constant. Several examples in this chapter illustrate the magnitude of the errors that can result when the cold air-standard assumption is applied. For reduction of the size of the error associated with the cold air-standard assumption, the specific heats should be evaluated at an average temperature during a particular process, or if a single value of specific heat is needed for the cycle, then it should be evaluated at an average temperature for the cycle.

7.5 GAS CARNOT CYCLE

Before we study individual gas cycles, a few general conclusions resulting from the application of the second law of thermodynamics and the Carnot principle should be considered. Because the Carnot cycle is a totally reversible cycle, the thermal efficiency of a *Carnot power cycle* is solely a function of the temperatures of the high- and low-temperature reservoirs, or

$$\eta_{\text{Car}} = 1 - \frac{T_L}{T_H} \tag{7.5}$$

Just as the Carnot power cycle is a theoretical standard for all other power cycles, the *totally reversible* (or Carnot) *refrigeration cycle* is a theoretical standard for all refrigeration cycles. The COP of a Carnot refrigeration cycle operating between the temperature limits T_L and T_H is, from Equation 5.10,

$$\beta_{R,\text{Car}} = \frac{1}{(T_H/T_L) - 1} \tag{7.6}$$

Similarly, the maximum COP of a heat pump operating between temperature limits T_L and T_H is that of a heat pump operating on a Carnot refrigeration cycle, which, from Equation 5.8, is

$$\beta_{H,\text{Car}} = \frac{1}{1 - (T_L/T_H)} \tag{7.7}$$

The Carnot cycle, discussed in some detail in Section 5.13, consists of four totally reversible processes: two reversible, isothermal heat-transfer processes, one at a temperature of T_H and a second at a temperature of T_L; a reversible, adiabatic- (or isentropic-) compression process; and a reversible, adiabatic-expansion process. Because each process is totally reversible, the Carnot cycle is also totally reversible. The combination of isothermal and isentropic processes results in a T-s process diagram with a rectangular shape for the Carnot cycle, as shown in Figure 7.1. Areas under the T-s diagram represent heat transfer during a reversible process. No heat transfer occurs during processes 4-1 and 2-3. The heat transfer to the system from the high-temperature reservoir is equal to the area under curve 1-2 on the T-s diagram, and the heat transfer from the system to the low-temperature reservoir is equal to the area under curve 3-4. Therefore, the net heat transfer during the cycle is equal to the area inside the closed curve 1-2-3-4-1, and from the conservation-of-energy equation for a cycle, this area is also equal to the net work for the Carnot cycle.

The working fluid in a Carnot engine could be any number of substances. However, this chapter considers only working fluids that remain in the gaseous phase, and the cycle is therefore called a *gas Carnot cycle*. A typical P-v diagram for a gas Carnot cycle is shown in Figure 7.1. If a closed system executes a Carnot cycle, the area under each process curve on the P-v diagram represents the work during the reversible process. Notice that work output occurs during both expansion processes from 1-2 and 2-3 and that work is required to compress the gas during both reversible processes 3-4 and 4-1.

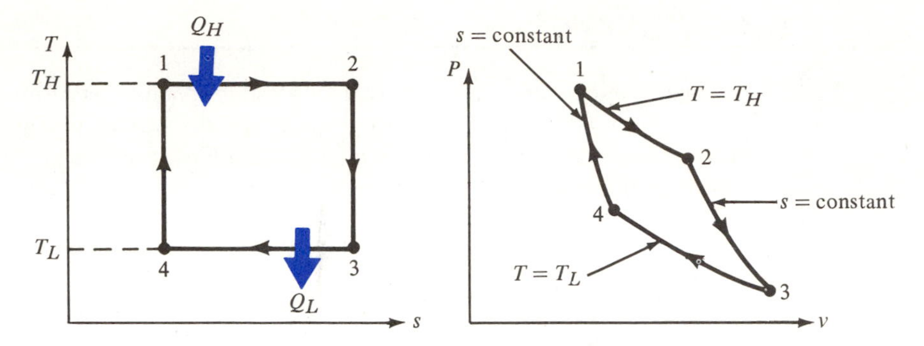

Figure 7.1 Process diagrams for a totally reversible, Carnot, gas power cycle.

Throughout this chapter the performance of several devices that operate on ideal cycles are compared with the theoretical performance of a Carnot cycle in order to determine how closely the cycle's maximum possible thermal efficiency or coefficient of performance is approached. Even though the Carnot cycle is used as a standard of comparison, attempting to construct an actual cycle that operates on the Carnot cycle would not be practical even if friction could be eliminated, because reversible heat transfer is extremely difficult to approach in reality. In fact, to design and fabricate a heat exchanger with finite heat-transfer rates while maintaining the temperature of the working fluid infinitesimally close to the temperature of a thermal-energy reservoir would be very difficult, if not impossible.

Examination of Equation 7.5 reveals that increasing the temperature of the high-temperature reservoir or reducing the temperature of the low-temperature reservoir would have the effect of increasing the thermal efficiency of the Carnot cycle. The same general conclusion holds for all power cycles. However, technical barriers exist that restrict the reservoir temperatures. In most power cycles the high-temperature reservoir is created by burning a fuel. For example, combustion of a mixture of gasoline and air occurs in an automobile engine, and the combustion temperature is restricted by the maximum temperature that the cylinder walls, piston, and other components can withstand. In fact, significant heat transfer from the cylinder to cooling water circulating through the engine block near the combustion zone is required to keep the engine temperature within a safe operating limit.

The temperature of the low-temperature reservoir is likewise restricted. According to the second law, all heat engines must have heat transfer to the surroundings. But the limits of the environmental temperature are small indeed, and while reducing the temperature of the environment in an attempt to increase the power cycle's thermal efficiency might be desirable, using a refrigeration system for this purpose is neither technically nor economically practical. From Equations 7.6 and 7.7 the fact that the maximum COP for refrigerators and heat pumps occurs as the temperatures of the high- and low-temperature reservoirs approach each other should also be evident. This theoretical limit of $T_H \rightarrow T_L$ certainly does not represent a reasonable design limit for refrigeration systems. In few, if any, situations would refrigerating or air-conditioning a space to achieve a temperature that is the same temperature as the surroundings be desirable or necessary, even though the refrigeration process would be very efficient. In practice, the need for refrigeration usually occurs in situations that tend to lower the coefficient of performance. The need for air conditioning is greatest when the outdoor temperature is the highest or when the difference between the temperatures of the high- and low-temperature reservoirs is the greatest. Furthermore, the demands placed on the operation of a heat pump are greatest when the outdoor temperatures reach seasonal lows. Thus, refrigeration cycles would operate most efficiently when they are not needed, and the COP is lowest when the demand on the cycle is the greatest. Although engineers must frequently design systems that operate as efficiently as possible, the system performance is often restricted by conditions that severely hamper achievement of this goal. This situation is one that often faces an engineer who is responsible for designing energy systems.

The concepts involved in a gas Carnot cycle are illustrated in the following example.

EXAMPLE 7.1

A closed piston and cylinder operate on a cold air-standard Carnot cycle. The air at the beginning of the isentropic-compression process is at a pressure of 100 kPa and a temperature of 25°C, and it is compressed to a pressure of 1 MPa. The heat transfer to the air per unit mass from the high-temperature reservoir is 150 kJ/kg. Determine the following quantities:

(a) The heat transfer from the system to the low-temperature reservoir per unit mass of air
(b) The net work produced by the cycle per unit mass of air
(c) The thermal efficiency of the cycle
(d) The irreversibility of the cycle per unit mass of air

Solution. Referring to Figure 7.1, we see that the given properties are

$$P_4 = 100\,\text{kPa} \qquad T_4 = 25°\text{C} \qquad P_1 = 1\,\text{MPa}$$

Process 4-1 is isentropic, and for the cold air-standard analysis the working fluid (air) is an ideal gas with constant specific heats. Thus Equation 5.40 can be used to relate the pressures and temperatures at the end states:

$$T_1 = T_4\left(\frac{P_1}{P_4}\right)^{(k-1)/k} = (298\,\text{K})\left(\frac{1.0\ \text{MPa}}{0.1\ \text{MPa}}\right)^{(1.4-1)/1.4} = 575\,\text{K}$$

Process 1-2 is isothermal, so that

$$T_2 = T_1 = 575\ \text{K}$$

The conservation-of-energy equation applied to the closed system (assumed stationary) during process 1-2 gives

$$q_{12} - w_{12} = u_2 - u_1$$

But for an internally reversible isothermal process of an ideal gas,

$$w_{12} = \int_1^2 P\,dv = RT_1\int_1^2 \frac{dv}{v} = RT_1 \ln\left(\frac{v_2}{v_1}\right)$$

and

$$u_2 - u_1 = 0$$

Furthermore, since process 1-2 is isothermal and the gas is ideal,

$$\frac{v_2}{v_1} = \frac{P_1}{P_2}$$

and the conservation-of-energy equation can therefore be written as

$$q_H = q_{12} = RT_1 \ln\left(\frac{P_1}{P_2}\right)$$

Thus

$$P_2 = P_1 \exp\left(-\frac{q_H}{RT_1}\right)$$

Substituting known values into this equation, we can calculate the pressure at state 2:

$$P_2 = (1 \text{ MPa}) \exp\left(\frac{-1.5 \times 10^5 \text{ J/kg}}{(287 \text{ J/kg·K})(575 \text{ K})}\right)$$

$$P_2 = 0.403 \text{ MPa}$$

Finally, the properties at state 3 can be determined by noting that path 2-3 is also an isentropic process for which

$$P_3 = P_2\left(\frac{T_3}{T_2}\right)^{k/(k-1)}$$

Therefore, the pressure at state 3 is

$$P_3 = (0.403 \text{ MPa})\left(\frac{298 \text{ K}}{575 \text{ K}}\right)^{1.4/0.4} = 0.0404 \text{ MPa} = 40.4 \text{ kPa}$$

(a) The heat transfer from the system to the low-temperature reservoir can be calculated by applying the conservation of energy to process 3-4 in a manner analogous to that used for process 1-2:

$$q_L = q_{34} = u_4 - u_3 + w_{34} = \int_3^4 P\,dv = RT_3 \ln\left(\frac{v_4}{v_3}\right)$$

$$= RT_3 \ln\left(\frac{P_3}{P_4}\right) = (287 \text{ J/kg·K})(298 \text{ K}) \ln\left(\frac{40.4 \text{ kPa}}{100 \text{ kPa}}\right)$$

$$= -77{,}500 \text{ J/kg} = \underline{\underline{-77.5 \text{ kJ/kg}}}$$

(b) Applying the conservation-of-energy equation to the entire cycle yields

$$w_{net} = q_{net} = q_H + q_L = 150 \text{ kJ/kg} - 77.5 \text{ kJ/kg} = \underline{\underline{72.5 \text{ kJ/kg}}}$$

(c) The thermal efficiency of the cycle is

$$\eta_{th} = \frac{w_{net}}{q_H} = \frac{72.5 \text{ kJ/kg}}{150 \text{ kJ/kg}} = 0.483 = \underline{\underline{48.3 \text{ percent}}}$$

The thermal efficiency could also be obtained by using Equation 7.5:

$$\eta_{th} = 1 - \frac{T_L}{T_H} = 1 - \frac{298 \text{ K}}{575 \text{ K}} = 48.2 \text{ percent}$$

(d) The Carnot cycle is composed of four totally reversible processes, and the irreversibility of each process is zero. Therefore, the irreversibility for the entire cycle is zero:

$$I = \underline{\underline{0}}$$

This result can be verified by applying Equation 6.24:

$$I = T_0(\Delta S_{tot}) = T_0(\Delta S_{sys} + \Delta S_{surr})$$

The system in this case undergoes a thermodynamic cycle; therefore

$$\Delta S_{sys} = 0$$

and heat transfer occurs only during reversible isothermal processes, so that the entropy change of the surroundings is

$$\Delta S_{surr} = \Delta S_{H,res} + \Delta S_{L,res} = \frac{Q_{H,res}}{T_H} + \frac{Q_{L,res}}{T_L}$$

If Equation 5.5 is now applied, then the entropy change of the surroundings is found to be zero:

$$\Delta S_{surr} = 0$$

And the irreversibility of the Carnot cycle is therefore zero:

$$I = 0$$

7.6 STIRLING AND ERICSSON CYCLES

If a cycle is to be totally reversible and if heat transfer occurs with two constant-temperature reservoirs, then the working fluid must be isothermal and at the same temperature as the reservoir during the heat-exchange processes to eliminate any temperature differences between the energy reservoirs and the working fluid within the cycle. Such is the case with the Carnot cycle. However, this cycle is not the only reversible cycle. Others have been proposed, such as the *Stirling* and *Ericsson cycles*, which have nonisothermal heat-transfer processes. Both of these cycles employ *regeneration processes* during which heat transfer to an energy-storage device is used to cool the working fluid during one portion of the cycle and then during another portion of the cycle the stored energy is used to heat the working fluid. The ideal regeneration processes are not isothermal, but the heat transfer occurs through an infinitesimally small temperature difference.

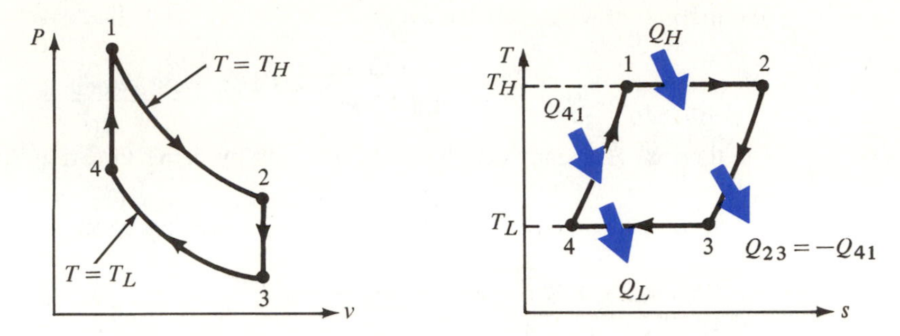

Figure 7.2 Process diagrams for the totally reversible Stirling cycle.

The *Stirling cycle* consists of two reversible, isothermal, heat-transfer processes and two reversible, constant-volume processes. Regeneration occurs during the constant-volume processes. The *P-v* and *T-s* diagrams for the Stirling cycle are shown in Figure 7.2. During process 1-2 the working fluid expands isothermally at T_H, as it is heated with energy from the high-temperature reservoir. During process 2-3 the volume remains constant, while heat transfer to the regenerator occurs so that the energy may be stored for use later in the cycle. During process 2-3 the temperature of the working fluid decreases from T_H to T_L. Heat transfer to the low-temperature reservoir occurs during the isothermal process between states 3 and 4 while the working fluid is at the temperature of T_L. Finally, the working fluid is returned to its original state by following the constant-volume path from state 4 to state 1, while the energy stored in the regenerator is returned to the fluid by means of heat transfer. Since there is no heat transfer from the regenerator to the surroundings, the heat transfer to the regenerator during process 2-3 is exactly equal in magnitude to the heat transfer to the working fluid during process 4-1.

The *Ericsson cycle* is a totally reversible cycle consisting of two isothermal processes and two constant-pressure processes. The *T-s* and *P-v* diagrams for the Ericsson cycle are shown in Figure 7.3. During the isothermal process 1-2, energy is added to the working fluid by means of heat transfer from the high-temperature reservoir at T_H. The working fluid is then compressed from state 2 to state 3, while heat transfer from the working fluid to the regenerator occurs until the temperature of the working fluid decreases to T_L. The working fluid is then compressed isothermally at T_L from state 3 to

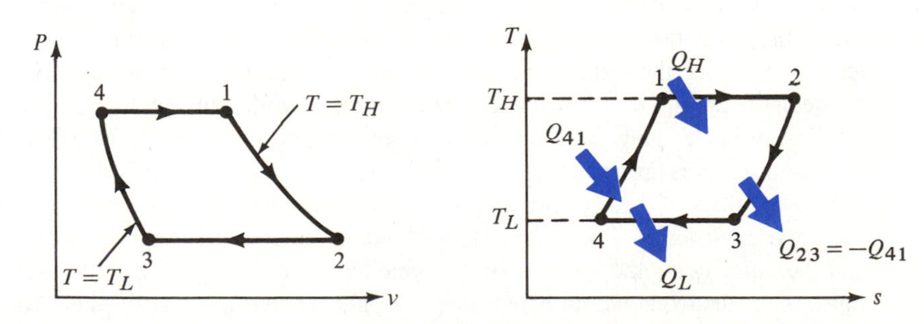

Figure 7.3 Process diagrams for the totally reversible Ericsson cycle.

state 4 as heat transfer between the working fluid and the low-temperature reservoir takes place. Finally, the working fluid is returned to its original state by expanding along a constant-pressure path 4-1, while the energy stored in the regenerator during process 2-3 is returned to the fluid. The cycle is totally reversible, and the heat transfer to the regenerator during process 2-3 is equal in magnitude to the heat transfer to the working fluid during process 4-1.

The Carnot, Stirling, and Ericsson cycles are all totally reversible cycles. Therefore, according to Carnot's principle discussed in Section 5.6, each of these cycles would have an identical thermal efficiency if they operated between the same temperature reservoirs:

$$\eta_{\text{Car}} = \eta_{\text{Stirling}} = \eta_{\text{Ericsson}} = 1 - \frac{T_L}{T_H} \tag{7.8}$$

Even though the thermal efficiencies of the Stirling and Ericsson cycles are equal to that of the Carnot cycle for the same values of T_H and T_L, the regeneration process presents practical limitations on the operation of these cycles that do not exist in the Carnot cycle. From a practical standpoint a regenerator that can transfer heat across an infinitesimal temperature difference while the working fluid continually changes temperature would be difficult to construct. Furthermore, a regenerator that operates at 100 percent efficiency, while having heat transfer only with the working fluid within the system, would also be impractical.

EXAMPLE 7.2

A Stirling cycle operates as a closed cycle with 1.0 lb_m of hydrogen as a working fluid between reservoirs at $T_H = 1800°\text{R}$ and $T_L = 540°\text{R}$. The highest pressure during the cycle is 450 psia, and the lowest pressure is 75 psia. Calculate the following:

(a) The heat transfer for the two energy reservoirs per cycle
(b) The heat transfer to the regenerator per cycle
(c) The net work per cycle
(d) The thermal efficiency of the cycle

Solution. Referring to Figure 7.2, we see that the given information is

$$T_1 = T_2 = T_H = 1800°\text{R}$$

$$T_3 = T_4 = T_L = 540°\text{R}$$

$$P_1 = 450\ \text{psia}$$

$$P_3 = 75\ \text{psia}$$

For the given states the hydrogen behaves as an ideal gas and properties are found in Table D.4E.

For a closed system the conservation-of-energy equation for process 1-2 is

$$Q_{12} - W_{12} = m(u_2 - u_1) = 0$$

The internal-energy change is zero since the gas is ideal and the process is isothermal. For a reversible isothermal process of an ideal gas,

$$W_{12} = \int_1^2 P d\mathbf{V} = mRT_1 \ln\left(\frac{\mathbf{V}_2}{\mathbf{V}_1}\right)$$

For process 2-3

$$Q_{23} - W_{23} = m(u_3 - u_2)$$

The work for this process is zero since the volume does not change. Similarly, for process 3-4

$$Q_{34} - W_{34} = m(u_4 - u_3) = 0$$

$$W_{34} = mRT_3 \ln\left(\frac{\mathbf{V}_4}{\mathbf{V}_3}\right)$$

and for process 4-1

$$Q_{41} - W_{41} = m(u_1 - u_4)$$

and W_{41} is zero. Internal-energy values from Table D.4E are

$$u_1 = u_2 = 4443 \text{ Btu/lb}_m$$

$$u_3 = u_4 = 1285 \text{ Btu/lb}_m$$

The volume ratios in the work terms can be determined by applying the ideal-gas equation of state:

$$\mathbf{V}_1 = \mathbf{V}_4 = \frac{mRT_1}{P_1}$$

$$\mathbf{V}_3 = \mathbf{V}_2 = \frac{mRT_3}{P_3}$$

or

$$\frac{\mathbf{V}_4}{\mathbf{V}_3} = \frac{\mathbf{V}_1}{\mathbf{V}_2} = \frac{P_3 T_1}{P_1 T_3}$$

(a) The heat transfer during process 1-2 is therefore

$$Q_{12} = W_{12} = mRT_1 \ln\left(\frac{P_1 T_3}{P_3 T_1}\right)$$

$$= (1.0 \text{ lb}_m)(0.9851 \text{ Btu/lb}_m\cdot°\text{R})(1800°\text{R}) \ln\left[\frac{(450 \text{ psia})(540°\text{R})}{(75 \text{ psia})(1800°R)}\right]$$

$$= Q_H = 1040 \text{ Btu} \qquad \text{(to the system)}$$

The heat transfer for the high-temperature reservoir is equal in magnitude and opposite in sign:

$$Q_{H,\text{res}} = \underline{\underline{-1040 \text{ Btu}}} \qquad \text{(from the reservoir)}$$

The heat transfer during process 3-4 is

$$Q_{34} = W_{34} = mRT_3 \ln\left(\frac{P_3T_1}{P_1T_3}\right)$$

$$= (1.0\ \text{lb}_\text{m})(0.9851\ \text{Btu/lb}_\text{m}\cdot{}^\circ\text{R})(540^\circ\text{R}) \ln\left[\frac{(75\ \text{psia})(1800^\circ\text{R})}{(450\ \text{psia})(540^\circ\text{R})}\right]$$

$$= Q_L = -313\ \text{Btu} \qquad \text{(from the system)}$$

The heat transfer for the low-temperature reservoir is equal in magnitude and opposite in sign:

$$Q_{L,\text{res}} = \underline{\underline{313\ \text{Btu}}} \qquad \text{(to the reservoir)}$$

(b) Since the regenerator performs no work, the heat transfer from the hydrogen to the regenerator during process 2-3 is

$$Q_{23} = m(u_3 - u_2) = (1.0\ \text{lb}_\text{m})(1285 - 4443)\text{Btu/lb}_\text{m}$$

$$= \underline{\underline{-3160\ \text{Btu}}}$$

Similarly, the heat transfer to the hydrogen from the regenerator during process 4-1 is

$$Q_{41} = m(u_1 - u_4) = (1.0\ \text{lb}_\text{m})(4443 - 1285)\text{Btu/lb}_\text{m}$$

$$= \underline{\underline{3160\ \text{Btu}}}$$

(c) The net work of the cycle is

$$W_\text{net} = W_{12} + W_{34} = 1040\ \text{Btu} - 313\ \text{Btu} = \underline{\underline{727\ \text{Btu}}}$$

(d) The thermal efficiency of the cycle is

$$\eta = \frac{W_\text{net}}{Q_H} = \frac{727\ \text{Btu}}{1040\ \text{Btu}} = \underline{\underline{0.70}}$$

This same result could be obtained by using Equation 7.8, since the Stirling cycle is totally reversible:

$$\eta_\text{rev} = 1 - \frac{T_L}{T_H} = 1 - \frac{540^\circ\text{R}}{1800^\circ\text{R}} = 0.70$$

■

7.7 IDEAL OTTO CYCLE

The *internal-combustion* (IC) *engine* is one of the most reliable and most widely used power sources in the world. It is used to power most automobiles, light trucks, and light aircraft as well as pumps, compressors, generators, conveyors, and a variety of other devices.

An internal-combustion engine can be classified as a *spark-ignition* (SI) engine or a *compression-ignition* (CI) engine, depending on the source of ignition for the mixture of fuel and air. In a spark-ignition engine the mixture is ignited by an external source such as a spark plug. In a compression-ignition engine the mixture is compressed to a high pressure and a temperature that is higher than the ignition temperature so that the mixture ignites spontaneously as the fuel is injected into the cylinder.

Two- and four-stroke engines as well as gasoline engines operating on a *Wankel cycle* are usually classified as spark-ignition IC engines and are the subject of this section. Diesel engines, discussed in the next section, are categorized as compression-ignition engines, because the fuel ignites in the cylinder without the presence of an external spark.

In a *four-stroke engine* the piston executes four complete strokes within the cylinder as the crankshaft completes two revolutions per cycle. The salient features of a four-stroke engine (sometimes called a *four-cycle engine*) are shown in Figure 7.4. On the *intake stroke* the intake valve is open and the piston moves downward in the cylinder, drawing in a premixed charge of gasoline and air until the piston reaches its lowest point

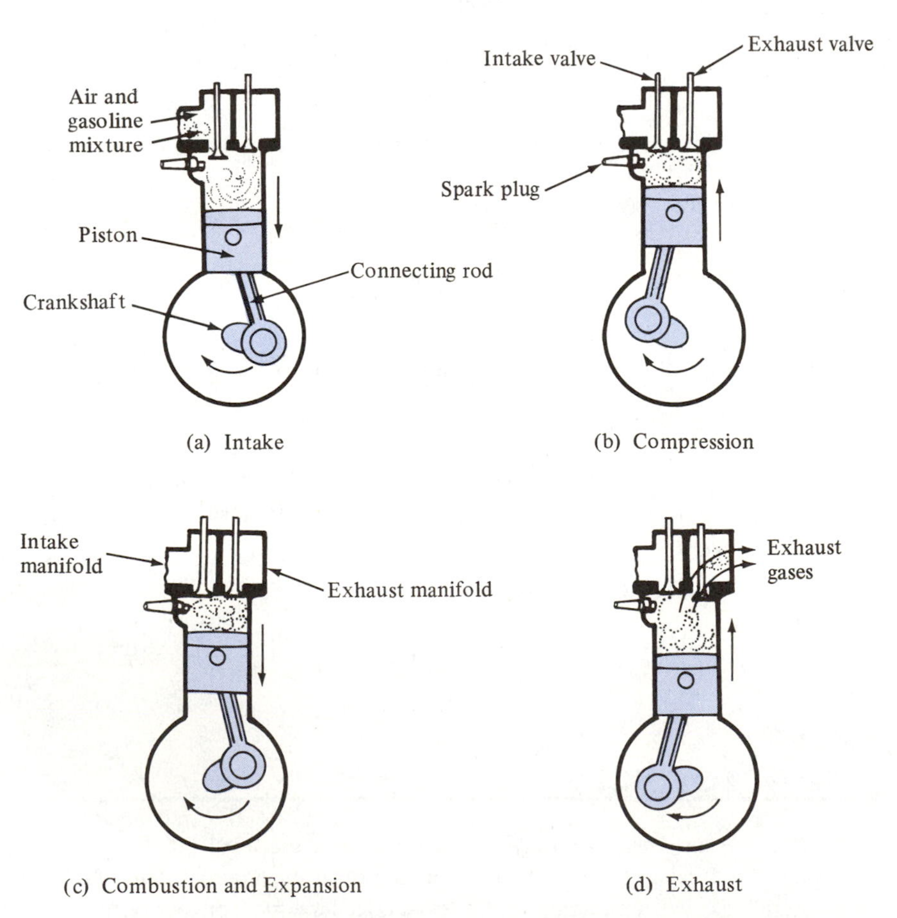

Figure 7.4 Four basic components of a cycle of a four-stroke internal-combustion engine. (Source: "The four-stroke spark-ignition (SI) cycle" (p. 3), from *Internal Combustion Engines* by Edward F. Obert. Copyright 1944, 1950, (c) 1968, 1973 by Harper & Row, Publishers, Inc. Reprinted by permission of the publisher.)

of the stroke (a position called *bottom dead center*, or BDC). During the *compression stroke* that follows, the intake valve closes and the piston moves toward the top of the cylinder, compressing the gas-air mixture. As the piston approaches the top of the cylinder (a position called *top dead center*, or TDC), the spark plug is energized and the mixture ignites, creating an increase in the temperature and pressure of the gas. During the *expansion stroke* the piston is forced down by the high-pressure gases, producing a useful work output by rotating the crankshaft against a resisting torque. The cycle is then completed when the exhaust valve opens and the piston travels toward the top of the cylinder, expelling the products of combustion.

Many modern, high-performance internal-combustion engines use *turbochargers* to increase power output. The turbocharger uses a compressor to force the air and fuel into the cylinder and it thereby overcomes some of the fluid resistance encountered in the intake manifold. In a turbocharger the exhaust gases leaving the engine pass through a small turbine that is connected by a shaft to the compressor. Power derived from the pressurized exhaust gases drives the compressor which forces the air into the cylinder during the intake stroke. The increase in exhaust pressure that results from the presence of the turbine in the exhaust stream decreases the engine efficiency somewhat, but the overall influence of the turbocharger is to improve the power output for a given engine displacement. The performance of the turbocharger can be further enhanced by compressing in stages and employing *intercooling* between each stage to minimize the work required during the compression process. The process of *intercooling* is discussed in more detail in Section 7.12. A section view of a turbocharger used on a diesel engine is shown in Figure 7.5.

Modern internal-combustion-engine designs are affected by environmental constraints as well as desires to increase gas mileage. Recent design improvements include the use of four valves per cylinder, fuel injectors to replace traditional carburetors, and turbochargers to increase the fuel and air flow to each cylinder. An example of an improved six-cylinder engine is shown in Figure 7.6.

Unlike a four-stroke engine, which requires two revolutions of the crankshaft to complete each cycle, a *two-stroke IC engine* requires only a single revolution of the crankshaft per cycle. In the two-stroke engine both the intake and exhaust valves are replaced by openings, or ports, in the lower portion of the cylinder wall. During the intake process the piston uncovers the intake ports, and air and fuel are drawn into the cylinder. The mixture is subsequently compressed as the piston moves toward the top of the cylinder. The spark plug ignites the mixture, the piston moves downward, and the work causes the crankshaft to rotate. Toward the end of the power stroke, the exhaust ports are uncovered, and the exhaust gases are partially expelled from the cylinder. At the same time a fresh charge of fuel and air is drawn into the cylinder, and the process is repeated.

Because of the relatively inefficient purging of the exhaust gases from the cylinder and expulsion of part of the fresh air-fuel mixture with the exhaust gases, the fuel economy of the two-stroke engine is not as good as that of a four-stroke engine of comparable size. Therefore, most two-stroke engines produced for today's market are for applications requiring small engines (e.g., lawn mowers, small boat engines, chain saws), where fuel consumption is not a primary consideration. One advantage of the small two-stroke engines is that they operate at higher engine speeds and provide rela-

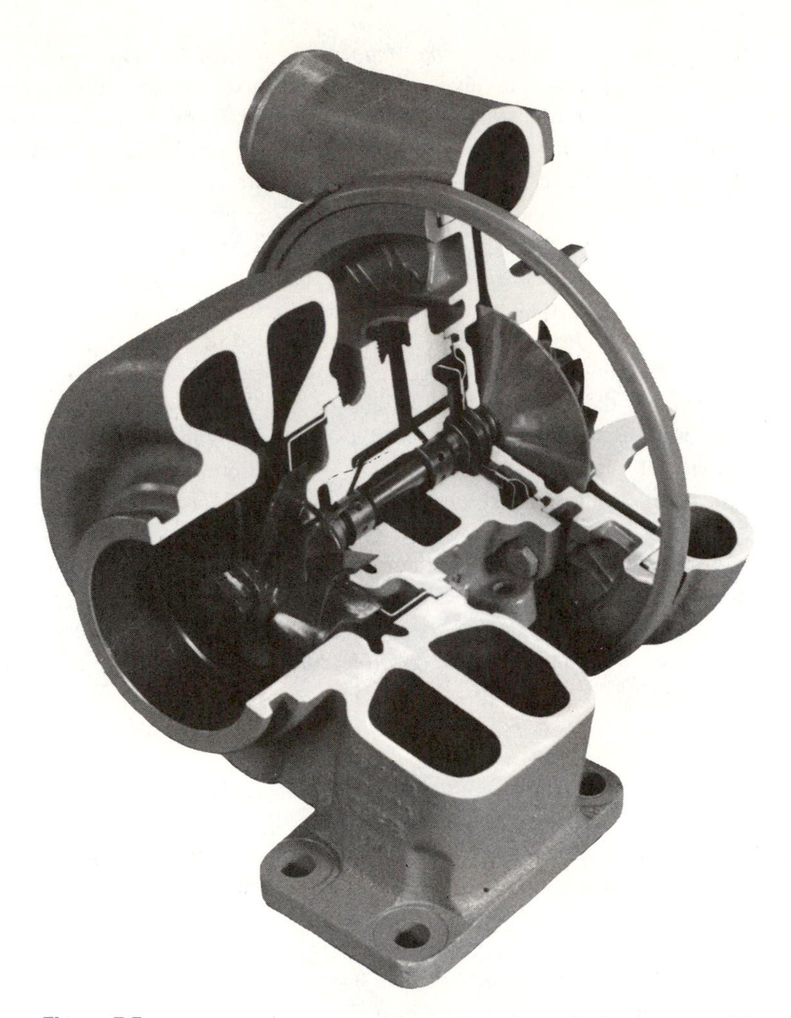

Figure 7.5 A diesel turbocharger. The turbine shown in the foreground is connected to the centrifugal compressor in the background. The exhaust gases are routed to the turbine through dual intake ports while the compressor supplies air to the engine through a single port. (Courtesy of Cummins Engine Company.)

tively high torques for their size, because the smaller moving parts and the absence of valves produce lower inertia forces within the engine.

The thermodynamic analysis of an actual IC engine is somewhat complicated and involved. To simplify the analysis, we consider an ideal cycle composed entirely of internally reversible processes. This cycle, called the ideal *Otto cycle*, closely resembles the operation of a spark-ignition IC engine. Even though the analysis of an Otto cycle does not consider the irreversibilities that are present in real internal-combustion engines, the conclusions drawn from the analysis help to predict the expected behavior of an actual IC engine. A brief treatment of a more realistic spark-ignition IC engine including irreversibilities is included later in this chapter.

Figure 7.6 A modern 3.0-liter, naturally aspirated 60° V-6 fuel-injected engine. The engine features a dual overhead cam 24 valve design, a compression ratio of 9.8, and when it is coupled with a 5-speed manual transmission it produces 220 hp at 6000 rpm and 200 ft·lb$_f$ of torque at 4800 rpm. The multivalve overhead cam design improves torque at high rpm's by reducing the flow restriction on the air as it enters and leaves the cylinder. (Courtesy of Ford Motor Company.)

Nicholas Otto, a German engineer, built a four-stroke IC engine in 1876, and the thermodynamic cycle patterned after his design is called the Otto cycle. In the ideal Otto cycle a closed piston-cylinder assembly is used as a model. The compression and the expansion processes are approximated by internally reversible, adiabatic or isentropic processes; the combustion process is assumed to be an internally reversible, constant-volume process during which heat transfer occurs to the working fluid, and the exhaust process is imagined to be an internally reversible, constant-volume process during which the working fluid is cooled. The *T-s* and *P-v* process diagrams for the ideal Otto cycle are illustrated in Figure 7.7, where 1-2 is the path followed during the isentropic-compression process, 2-3 is the combustion process, 3-4 is the isentropic-expansion process, and the cycle is completed by the exhaust process 4-1.

To determine how effectively a spark-ignition IC engine converts the energy in the fuel into useful work output, we use the thermal efficiency of the cycle:

$$\eta_{\text{th}} = \frac{W_{\text{net}}}{Q_H} = 1 + \frac{Q_L}{Q_H} \tag{7.9}$$

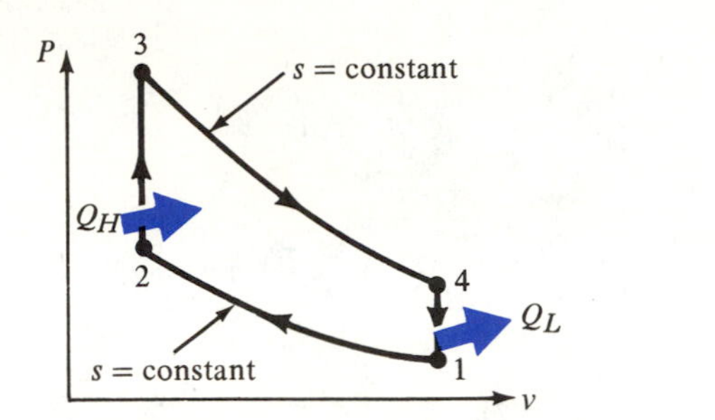

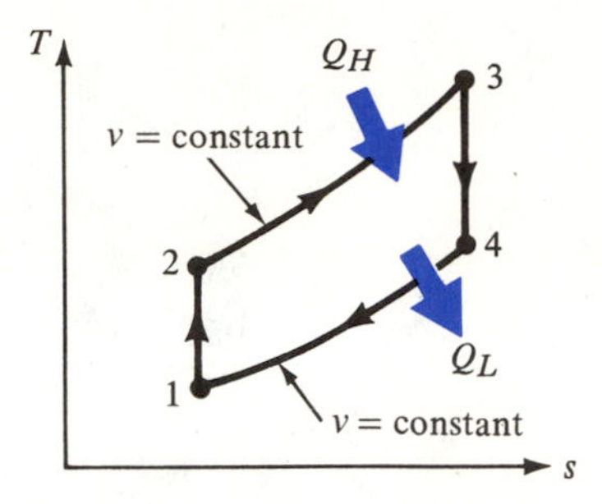

Figure 7.7 The P-v and T-s process diagrams for the ideal Otto cycle.

The heat transfer to the working fluid, which occurs during process 2-3, is equal to the change in internal energy of the working fluid, because the system is closed and the volume remains constant during the process:

$$Q_H = m(u_3 - u_2) \tag{7.10}$$

Similarly, the heat transfer from the working fluid during constant-volume process 4-1 is

$$Q_L = m(u_1 - u_4) \tag{7.11}$$

After substitution of Equations 7.10 and 7.11 into Equation 7.9, the thermal efficiency of the ideal Otto cycle becomes

$$\eta_{\text{th,Otto}} = 1 + \left(\frac{u_1 - u_4}{u_3 - u_2}\right) \tag{7.12}$$

This expression for the thermal efficiency of an ideal Otto cycle can be simplified if the cold air-standard assumptions are used. The working fluid is then assumed to be air, which behaves as an ideal gas with constant specific heats, and Equation 7.12 reduces to

$$\eta_{\text{th,Otto}} = 1 + \left(\frac{T_1 - T_4}{T_3 - T_2}\right) \tag{7.13}$$

The relation in Equation 7.13 can be expressed in terms of the volume ratio by using the isentropic-process relationship (Equation 5.39) for an ideal gas with constant specific heats. Since processes 1-2 and 3-4 are isentropic,

$$\frac{T_1}{T_2} = \left(\frac{v_2}{v_1}\right)^{k-1} \qquad \frac{T_3}{T_4} = \left(\frac{v_4}{v_3}\right)^{k-1} \tag{7.14}$$

Substituting these two equations into Equation 7.13 and simplifying (since $v_2 = v_3$ and $v_1 = v_4$) gives the following expression for the thermal efficiency of an ideal, cold air-standard Otto cycle:

$$\eta_{\text{th,Otto}} = 1 - (r)^{1-k} \tag{7.15}$$

where r is the *compression ratio* for the engine defined by the equation

$$r \equiv \frac{v_1}{v_2} \tag{7.16}$$

The compression ratio is the ratio of the cylinder volume at the beginning of the compression process (BDC) to the volume at the end of the compression process (TDC).

Equation 7.15 shows that the thermal efficiency of an ideal, cold air-standard Otto cycle is only a function of the compression ratio of the engine and the specific heat ratio of the working fluid, k. A plot of Equation 7.15 is shown in Figure 7.8 for a value of $k = 1.4$. Since an ideal Otto cycle is internally reversible and an actual IC engine is irreversible, the thermal efficiency given by Equation 7-15 is the maximum possible efficiency for a spark-ignition IC engine operating at a given compression ratio.

The conclusion that the thermal efficiency of an ideal, spark-ignition IC engine is solely a function of the compression ratio is only theoretically correct, because the analysis is based on the ideal, cold air-standard Otto cycle. A real IC engine is far more complex, and factors such as cam design, valve design, carburization, and bearing friction have a significant impact on the engine efficiency. Nevertheless, any engine design that increases the compression ratio should result in an increased engine efficiency and increased gas mileage, both very desirable goals in an era of dramatically changing energy prices and supplies.

Even though the compression ratio has a dominant influence on the thermal efficiency of an IC engine, it cannot be increased indefinitely. A practical limit on the compression

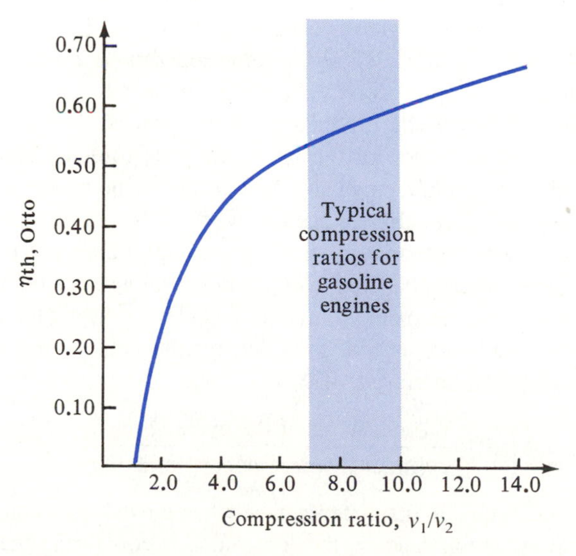

Figure 7.8 Thermal efficiency of the ideal, cold air-standard Otto cycle as a function of compression ratio ($k = 1.4$).

ratio exists because of two overriding factors, one relating to technical aspects of engine design and one relating to environmental factors. As compression ratios are increased, the temperature of the working fluid (an air-fuel mixture) is also increased during the compression process. Eventually, a temperature is reached that is sufficiently high to ignite the mixture prematurely without the presence of a spark. This condition, called *preignition*, causes the engine to produce an audible noise, often called ''knock'' or ''ping.'' The presence of engine knock places a barrier on the upper limit of engine compression ratios. For reduction of engine knock most gasoline blends were once formulated with tetraethyl lead (TEL) to increase the *octane rating* of the fuel. During the 1950s, 1960s, and early 1970s, spark-ignition IC engines having compression ratios in the range 9 to 12 were not uncommon, and practically all gasoline contained some lead compounds.

During the late 1960s and throughout the 1970s, environmentalists advocated the removal of TEL from gasoline on the grounds that exhaust emissions containing lead compounds were health hazards. As a result, lead compounds have been largely removed from gasoline, and the octane ratings have correspondingly decreased. For counteraction of the reduction in octane rating and for reduction of the likelihood of engine knock, the compression ratios of engines produced for today's market have been reduced to the range 7 to 10. The trend toward lower compression ratios brought on by changes in fuel formulations has, unfortunately, produced slight reductions in engine efficiencies. Other factors, however, such as the reduction in overall automobile weight, the reduction in wind drag by improved aerodynamic design of the automobile profile, and the emphasis on smaller engine capacities, have resulted in increased fuel economy even though lower compression ratios have produced a slight decrease in overall engine efficiency.

The pressure during the compression and expansion processes varies continually during the operation of the ideal IC engine cycle, as shown in the P-v diagram of Figure 7.7. Because the ideal cycle assumes a closed system undergoing a series of internally reversible processes, the net work provided by the cycle can be calculated by determining the difference between the Pdv work during the expansion (process 3-4) and compression (process 1-2) processes. Determination of the net work, therefore, usually requires integration of the P-v curves during the expansion and compression processes.

One way to simplify the calculation of the net work of the cycle and to provide a comparative measure of the performance of a reciprocating engine is to introduce the concept of the mean effective pressure. The *mean effective pressure* is a fictitious pressure that, when multiplied by the displacement volume of the cylinder during the power stroke, would produce the same net work as that provided by the actual cycle. That is, the definition of the mean effective pressure (MEP) is

$$\text{MEP} \equiv \frac{\text{net work of the cycle}}{\text{cylinder displacement volume}}$$

The *displacement volume* is equal to the difference between the cylinder volume at bottom dead center and the volume at top dead center. When one is comparing the performance of two engines of equal piston displacement, the engine with the larger value of the MEP would be the engine with superior performance, because it would produce a greater net work output.

■ EXAMPLE 7.3

An engine operates on an ideal, cold air-standard Otto cycle with a compression ratio of 8. Heat transfer to the engine is from a high-temperature reservoir at 1000°C, and heat transfer from the engine is to the surroundings at 20°C and 100 kPa. At the beginning of the isentropic-compression process the pressure and the temperature of the working fluid are 110 kPa and 50°C. At the end of the combustion process the temperature is 900°C. Assume that the specific heat ratio of the working fluid is 1.4 and that c_v = 0.718 kJ/kg·K. Calculate the following quantities:

(a) The pressure and the temperature at the remaining three states of the Otto cycle
(b) The thermal efficiency of the cycle
(c) The irreversibility per unit mass for each process and for the cycle (explain the source of the irreversibilities)
(d) The mean effective pressure
(e) The thermal efficiency of a Carnot cycle operating between the same high- and low-temperature reservoirs.

Solution.

(a) Referring to the states in Figure 7.7, we see that compression process 1-2 is isentropic. For an ideal gas with constant specific heats, the isentropic relation from Equation 5.39 applies, so

$$\frac{T_2}{T_1} = \left(\frac{v_1}{v_2}\right)^{k-1} = (r)^{k-1} = (8)^{0.4} = 2.3$$

or

$$T_2 = 2.3\ T_1 = 2.3(323\ \text{K}) = 743\ \text{K} = \underline{\underline{470°\text{C}}}$$

Furthermore, the isentropic relation from Equation 5.41 can be used to relate the pressure and specific volume during the process 1-2:

$$\frac{P_2}{P_1} = \left(\frac{v_1}{v_2}\right)^{k} = (8)^{1.4} = 18.4$$

The combustion process (2-3) occurs at constant volume, and

$$P_2 = 18.4(110\ \text{kPa}) = \underline{\underline{2.02\ \text{MPa}}}$$

The gas is ideal and $v_2 = v_3$, so that

$$\frac{P_3}{P_2} = \frac{T_3}{T_2}$$

or

$$P_3 = P_2\left(\frac{T_3}{T_2}\right) = (2.02\ \text{MPa})\left(\frac{1173\ \text{K}}{743\ \text{K}}\right) = \underline{\underline{3.19\ \text{MPa}}}$$

The expansion process 3-4 is isentropic, and therefore

$$T_4 = T_3 \left(\frac{v_3}{v_4}\right)^{k-1} = T_3 \left(\frac{v_2}{v_1}\right)^{k-1} = (1173 \text{ K}) \left(\frac{1}{8}\right)^{0.4}$$

$$= 511 \text{ K} = \underline{\underline{238°\text{C}}}$$

$$P_4 = P_3 \left(\frac{v_3}{v_4}\right)^{k} = P_3 \left(\frac{v_2}{v_1}\right)^{k} = (3.19 \text{ MPa}) \left(\frac{1}{8}\right)^{1.4}$$

$$= \underline{\underline{174 \text{ kPa}}}$$

(b) The thermal efficiency of the cold air-standard Otto cycle is determined from Equation 7.15:

$$\eta_{\text{th}} = 1 - r^{1-k} = 1 - 8^{-0.4} = \underline{\underline{0.565}}$$

Since the volume remains constant during process 2-3, the heat transfer to the cycle from the high-temperature reservoir on a unit-mass basis is

$$q_{23} = q_H = u_3 - u_2 = c_v(T_3 - T_2)$$

$$= (0.718 \text{ kJ/kg·K})(1173 - 743)\text{K} = 308.7 \text{ kJ/kg}$$

The heat transfer from the cycle can be determined from the conservation-of-energy equation written for process 4-1. Since the volume is constant,

$$q_{41} = q_L = u_1 - u_4 = c_v(T_1 - T_4)$$

$$= (0.718 \text{ kJ/kg·K})(323 - 511)\text{K} = -135 \text{ kJ/kg}$$

The net work of the cycle per unit mass is therefore

$$w_{\text{net}} = q_H + q_L = (308.7 - 135)\text{kJ/kg} = 173.7 \text{ kJ/kg}$$

(c) The irreversibility for each process executed by the closed system can be determined from Equation 6.39:

$$I_{if} = T_0(S_f - S_i)_{\text{sys}} + \sum_k T_0 \left(\frac{Q_k}{T_k}\right)$$

where the subscripts i and f are used to indicate the initial and final states of a process, respectively. None of the processes involves heat transfer with more than a single reservoir. Therefore, the irreversibility can be written as

$$I_{if} = T_0 \left[(S_f - S_i)_{\text{sys}} + \frac{Q_{\text{res}}}{T_{\text{res}}} \right]$$

For each of the reversible adiabatic processes, 1-2 and 3-4, the irreversibility is identically zero since the processes are isentropic:

$$I_{12} = \underline{\underline{0}} \quad \text{and} \quad I_{34} = \underline{\underline{0}}$$

For processes 2-3 and 4-1 the entropy change can be determined from Equation 5.29 since the gas is assumed to be ideal with constant specific heats:

$$s_f - s_i = c_v \ln\left(\frac{T_f}{T_i}\right) + R \ln\left(\frac{v_f}{v_i}\right)$$

Furthermore, the specific volume is constant during each of these processes. Thus

$$s_3 - s_2 = c_v \ln\left(\frac{T_3}{T_2}\right) = (0.718 \text{ kJ/kg·K}) \ln\left(\frac{1173 \text{ K}}{743 \text{ K}}\right)$$
$$= 0.328 \text{ kJ/kg·K}$$

and

$$s_1 - s_4 = c_v \ln\left(\frac{T_1}{T_4}\right) = (0.718 \text{ kJ/kg·K}) \ln\left(\frac{323 \text{ K}}{511 \text{ K}}\right)$$
$$= -0.329 \text{ kJ/kg·K}$$

Notice that within the accuracy of the calculations, $(s_3 - s_2) = (s_4 - s_1)$. This fact is evident also from the T-s diagram for the ideal cycle in Figure 7.7. Therefore, the irreversibilities for these processes, on a unit-mass basis, are

$$i_{23} = T_0\left(s_3 - s_2 + \frac{q_H}{T_H}\right) = T_0\left(s_3 - s_2 - \frac{q_{23}}{T_H}\right)$$
$$= (293 \text{ K})\left(0.328 \text{ kJ/kg·K} - \frac{308.7 \text{ kJ/kg}}{1273 \text{ K}}\right) = \underline{\underline{25.1 \text{ kJ/kg}}}$$

and

$$i_{41} = T_0\left(s_1 - s_4 + \frac{q_L}{T_L}\right) = T_0\left(s_1 - s_4 - \frac{q_{41}}{T_0}\right)$$
$$= (293 \text{ K})\left(-0.329 \text{ kJ/kg·K} - \frac{-135 \text{ kJ/kg}}{293 \text{ K}}\right) = \underline{\underline{38.6 \text{ kJ/kg}}}$$

The irreversibility for the cycle is the sum of the irreversibilities of the individual processes:

$$i_{\text{cyc}} = i_{12} + i_{23} + i_{34} + i_{41}$$
$$= 0 + 25.1 \text{ kJ/kg} + 0 + 38.6 \text{ kJ/kg} = \underline{\underline{63.7 \text{ kJ/kg}}}$$

Each process in the cycle is internally reversible, but the irreversibility for the cycle is not zero because all processes are not totally reversible. The isentropic processes (1-2 and 3-4) are both internally reversible and adiabatic and therefore are totally reversible. Hence the irreversibility for each of these two processes is zero. The two heat-transfer processes, however, are irreversible because each involves heat transfer between the system and its surroundings (the reservoirs) through a finite temperature difference.

In this cycle the irreversibility associated with process 4-1 is the largest contribution to the irreversibility of the cycle. If the performance of the cycle is to be improved, consideration might first be given to this process.

The pertinent characteristics of this cycle are summarized in Table 7.1. The work associated with each process is determined from Equation 4.22:

$$w_{if} = q_{if} - (u_f - u_i)$$

and could also be calculated from

$$w_{if} = \int_i^f P\,dv$$

The internal-energy change for the ideal gas is calculated from

$$u_f - u_i = c_v(T_f - T_i)$$

The change in the closed-system availability of the air for each process is also included and is calculated, based on Equation 6.34, from

$$\phi_f - \phi_i = (u_f - u_i) + P_0(v_f - v_i) - T_0(s_f - s_i)$$

Several observations can be made concerning the data in the table. First, the net heat transfer and the net work of the cycle are equal. This observation provides a check of our calculations. Second, the net change for each of the properties u, s, and ϕ is zero since the series of processes makes up a cycle. Notice in particular that the net entropy change for the cycle is zero. This result must be so for *any* cycle, even if the individual processes are not internally reversible, because entropy is a property. Third, the irreversibility for a process is zero only if the process is totally reversible. Fourth, since process 1-2 and process 3-4 are totally reversible, the actual work for each of these processes is equal to the reversible work. This result is also reflected in the fact that the irreversibility is zero (the irreversibility is the difference between the reversible work and the actual work). Finally, an energy input in the form of a work

TABLE 7.1 IDEAL OTTO CYCLE

T_H = 1273 K, $T_L = T_0$ = 293 K, P_0 = 100 kPa

Process	q_{if}, kJ/kg	w_{if}, kJ/kg	$u_f - u_i$, kJ/kg	$s_f - s_i$, kJ/kg·K	$\phi_f - \phi_i$, kJ/kg	i_{if}, kJ/kg
1-2	0.0	−301.6	301.6	0.0	220.5	0.0
2-3	308.7	0.0	308.7	0.329	212.3	25.1
3-4	0.0	475.3	−475.3	0.0	−394.2	0.0
4-1	−135.0	0.0	−135.0	−0.329	−38.6	38.6
Cycle	173.7	173.7	0.0	0.0	0.0	63.7

interaction or heat transfer causes an increase in the availability of the working fluid; that is, the capacity of the fluid to do work is increased. The availability decreases when work is done by the fluid or there is heat transfer from the fluid. Also, for the reversible adiabatic processes the change in the availability is equal in magnitude but opposite in sign to the reversible useful work (see Equation 6.44). Therefore, the sum of the change in the availability and the work for each of the isentropic processes is equal to the work done on or by the surroundings, $P_0(v_f - v_i)$, during the process.

(d) The definition of the MEP is

$$\text{MEP} = \frac{w_{\text{net}}}{v_4 - v_3} = \frac{w_{\text{net}}}{R[(T_4/P_4 - T_3/P_3)]}$$

$$= \frac{173.7 \text{ kJ/kg}}{(0.287 \text{ kPa}\cdot\text{m}^3/\text{kg}\cdot\text{K})[(511/174) - (1173/3190)] \text{ K/kPa}}$$

$$\times [1 \text{ kPa}/(\text{kJ/m}^3)] = \underline{\underline{235.6 \text{ kPa}}}$$

Thus, a constant pressure of 235.6 kPa acting during the entire expansion stroke would provide a work output equal to the net work output of the engine.

(e) The thermal efficiency of a Carnot cycle operating between the given temperature reservoirs is

$$\eta_{\text{Car}} = \eta_{\text{th,rev}} = 1 - \frac{T_L}{T_H} = 1 - \frac{293 \text{ K}}{1273 \text{ K}} = \underline{\underline{0.77}}$$

The irreversibility per unit mass that occurs during the cycle can be checked by using Equation 7.4:

$$i = q_H(\eta_{\text{Car}} - \eta_{\text{Otto}})$$
$$= (308.7 \text{ kJ/kg})(0.770 - 0.565) = 63.3 \text{ kJ/kg}$$

The only way in which the thermal efficiency of the ideal Otto cycle could approach the efficiency of the Carnot cycle would be to eliminate the irreversibilities that occur during the two heat-transfer processes. However, this is an impossible task because, with the ideal Otto cycle, heat transfer occurs between the working fluid, which has a continually varying temperature, and a reservoir that has a constant temperature. This restriction automatically produces irreversibilities that cause the thermal efficiency of the ideal Otto cycle to be less than that for the Carnot cycle. The thermal efficiency of 56.5 percent for the ideal Otto cycle in this example represents the maximum efficiency of any spark-ignition IC engine for the given compression ratio and specific heat ratio. Furthermore, it is 73 percent of the maximum theoretical efficiency corresponding to the given temperature limits. Any actual IC engine with a compression ratio of 8 should be expected to be less efficient because the actual engine contains irreversibilities that are absent in an ideal-Otto-cycle engine. ■

7.8 IDEAL DIESEL CYCLE

The *diesel cycle* is somewhat similar to the Otto cycle, except that ignition of the fuel-air mixture is caused by spontaneous combustion owing to the high temperature that results from compressing the mixture to a pressure significantly higher than the pressure that exists in an SI engine. The basic components of the diesel engine are the same as those of the spark-ignition engine shown in Figure 7.4 except that the spark plug is replaced by a fuel injector and the stroke of the piston is lengthened to provide a greater compression ratio.

The elements of the diesel cycle were first proposed by Rudolf Diesel in the 1890s for an engine that was to operate on coal dust. Diesel soon discovered that coal dust was an unsatisfactory fuel and decided to use more convenient liquid fuels instead. Like the ideal Otto cycle, the ideal diesel cycle consists of internally reversible, adiabatic-compression and -expansion processes and an internally reversible, constant-volume cooling process. In the ideal diesel cycle, however, the duration of the ignition process is extended because combustion occurs continuously while the fuel is injected into the cylinder, and the fuel continues to burn as the piston is moving away from top dead center. In the ideal cycle this process is modeled by an internally reversible, constant-pressure heating process. The *P-v* and *T-s* process diagrams for an ideal diesel cycle are illustrated in Figure 7.9.

The measure of performance of the ideal diesel cycle is the thermal efficiency.

$$\eta_{th} = \frac{W_{net}}{Q_H} = 1 + \frac{Q_L}{Q_H} \tag{7.17}$$

Applying the conservation-of-energy equation for a closed system, we find that the heat transfer to the diesel cycle during the constant-pressure process 2-3 is

$$Q_{23} - W_{23} = m(u_3 - u_2)$$

Thus,

$$\begin{aligned} Q_H = Q_{23} &= mP_3(v_3 - v_2) + m(u_3 - u_2) \\ &= m(h_3 - h_2) \end{aligned} \tag{7.18}$$

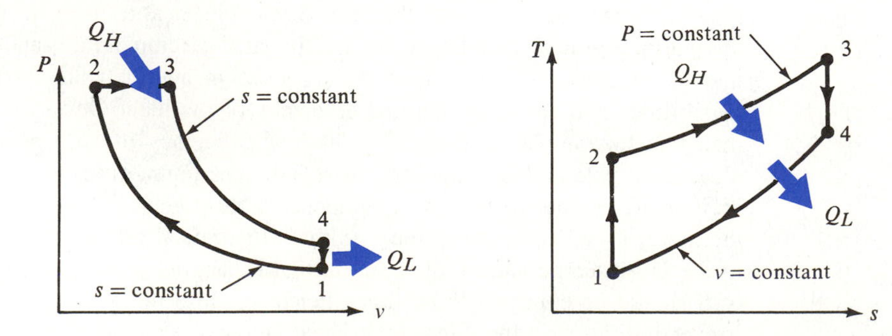

Figure 7.9 The *P-v* and *T-s* process diagrams for the ideal diesel cycle.

Similarly, the heat transfer from the diesel cycle during the constant-volume process 4-1 is

$$Q_L = Q_{41} = m(u_1 - u_4) \tag{7.19}$$

Therefore, the expression for the thermal efficiency of an ideal diesel cycle is

$$\eta_{\text{th,diesel}} = 1 + \left(\frac{u_1 - u_4}{h_3 - h_2}\right) \tag{7.20}$$

This expression can be simplified if the cold air-standard assumptions are used:

$$\begin{aligned} u_1 - u_4 &= c_v(T_1 - T_4) \\ h_3 - h_2 &= c_p(T_3 - T_2) \end{aligned} \tag{7.21}$$

Thus the thermal efficiency for the ideal, cold air-standard diesel cycle is

$$\eta_{\text{th,diesel}} = 1 + \frac{(T_1 - T_4)}{k(T_3 - T_2)} \tag{7.22}$$

Introducing the *cutoff ratio* r_c as the ratio of the cylinder volume V_3 after the combustion process to the volume V_2 after the compression process, or

$$r_c \equiv \frac{V_3}{V_2} \tag{7.23}$$

and recalling the definition of the compression ratio r given in Equation 7.16, we can express the thermal efficiency of the ideal, cold air-standard diesel cycle as

$$\eta_{\text{th,diesel}} = 1 - r^{(1-k)}\left[\frac{r_c^k - 1}{k(r_c - 1)}\right] \tag{7.24}$$

A comparison of Equation 7.24 and Equation 7.15 for the efficiency of an ideal, cold air-standard Otto cycle shows that the two expressions differ by the quantity in the brackets. This term is always greater than 1, and as a result, the thermal efficiency of an ideal, cold air-standard diesel cycle is always less than the efficiency of an ideal, cold air-standard Otto cycle operating at the same compression ratio. As indicated by the graph of Equation 7.24 in Figure 7.10, higher diesel-cycle thermal efficiencies result from lower cutoff ratios. Finally, the expression for the diesel-cycle thermal efficiency reduces to that for the Otto cycle when the cutoff ratio is equal to unity.

One factor that greatly enhances the desirability of a diesel engine is that the fuel is not injected into the cylinder until after the air has been completely compressed, so that there is no possibility of preignition. Once injected into the cylinder, the fuel burns immediately upon contact with the high-temperature air. Eliminating the possibility of preignition permits the use of less refined fuel with a higher ignition temperature, and the diesel engine can be designed to operate at much higher compression ratios than are practical for spark-ignition engines that use more volatile fuels. In fact, diesel engines typically operate at compression ratios between 12 and 20, values that are as much as double the values for spark-ignition engines. As a result of their higher compression ratios, diesel engines are typically slightly more efficient than spark-ignition IC engines.

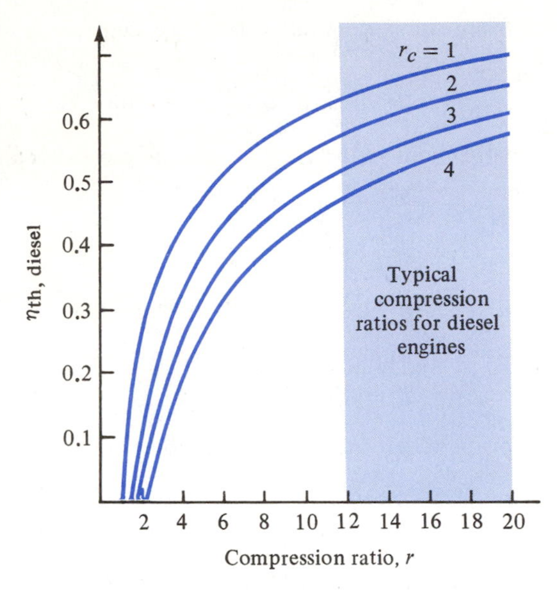

Figure 7.10 Thermal efficiency of the ideal, cold air-standard diesel cycle ($k = 1.4$).

Diesel engines are popular where relatively large amounts of power are required, and they are used almost exclusively in locomotives, large ships, standby power-generation units, and heavy trucks and buses. Diesel engines are usually more robust than SI engines of equivalent power output, because they must be built to withstand higher operating pressures. Diesel engines usually require less maintenance, a factor that is particularly important when selecting an engine for heavy trucks and buses. Recently, price differentials between gasoline and diesel fuel and greater fuel economy of diesel power plants have made diesel engines more competitive with spark-ignition engines for small automobile engines, even though the spark-ignition engine is somewhat lighter and less expensive to manufacture. A modern diesel engine is shown in Figure 7.11. Note the placement of the turbocharger in the foreground and the fuel injector directly above the center of the piston.

EXAMPLE 7.4

An ideal diesel cycle with a compression ratio of 17 and a cutoff ratio of 2 has a temperature of 105°F and a pressure of 15 psia at the beginning of the isentropic-compression process. During the constant-pressure heating process, heat transfer to the working fluid occurs from a reservoir that has a temperature of 3270°F. During the constant-volume cooling process, heat transfer is from the working fluid to the environment, which is at 77°F and 15 psia. Use the cold air-standard assumptions, assume that $k = 1.4$, and calculate the following quantities:

(a) The temperature and pressure of the gas at the end of the isentropic-compression process and at the end of the combustion process

Figure 7.11 A 6-cylinder turbocharged diesel engine that has a compression ratio of 15.5 and develops 450 hp at 2100 rpm while weighing 3600 lb. (Courtesy of Cummins Engine Company.)

(b) The thermal efficiency of the cycle

(c) The maximum thermal efficiency of any cycle operating between the given reservoir temperatures

(d) The irreversibility for each process and the cycle irreversibility

Solution.

(a) The process diagram for the ideal diesel cycle is shown in Figure 7.9. For this problem

$$r_c = \frac{V_3}{V_2} = 2 \qquad r = \frac{V_1}{V_2} = 17$$

The compression process is isentropic, so

$$T_2 = T_1 \left(\frac{V_1}{V_2}\right)^{k-1} = (565°\text{R})(17)^{0.4} = 1755°\text{R} = \underline{\underline{1295°\text{F}}}$$

$$P_2 = P_1 \left(\frac{V_1}{V_2}\right)^{k} = (15 \text{ psia})(17)^{1.4} = \underline{\underline{792 \text{ psia}}}$$

The combustion process is one of constant pressure, so that

$$P_3 = P_2 = \underline{\underline{792 \text{ psia}}}$$

Using the ideal-gas equation of state for a constant-pressure process yields

$$T_3 = T_2 \left(\frac{V_3}{V_2}\right) = (1755°\text{R})(2) = 3510°\text{R} = \underline{\underline{3050°\text{F}}}$$

(b) From Equation 7.24 the thermal efficiency of the ideal, cold air-standard diesel cycle is

$$\eta_{\text{th,diesel}} = 1 - (17^{-0.4}) \left[\frac{2^{1.4} - 1}{1.4(2 - 1)}\right] = \underline{\underline{0.623}}$$

(c) The maximum thermal efficiency of any cycle operating between the given reservoir temperatures of 77°F and 3270°F is the thermal efficiency of a Carnot cycle:

$$\eta_{\text{Car}} = 1 - \frac{T_L}{T_H} = 1 - \frac{537°\text{R}}{3730°\text{R}} = \underline{\underline{0.856}}$$

Notice that the efficiency of the ideal, cold air-standard diesel cycle is only about 73 percent of the efficiency of the Carnot cycle. This trend should be expected because the efficiency of the Carnot cycle is greater than that of all irreversible cycles operating between the same temperature reservoirs.

(d) The irreversibility for each process executed by the closed system is determined from Equation 6.39. Since the heat transfer for each process occurs with at most one reservoir, Equation 6.39 can be written on a unit-mass basis as

$$i_{if} = T_0 \left[(s_f - s_i)_{\text{sys}} + \frac{q_{\text{res}}}{T_{\text{res}}}\right]$$

For each of the reversible adiabatic (isentropic) processes, 1-2 and 3-4, the irreversibility is zero:

$$i_{12} = \underline{\underline{0}} \quad \text{and} \quad i_{34} = \underline{\underline{0}}$$

For the constant-pressure process 2-3 the entropy change can be determined most conveniently from Equation 5.30. Since constant specific heats have been assumed:

$$s_3 - s_2 = c_p \ln \left(\frac{T_3}{T_2}\right) - R \ln \left(\frac{P_3}{P_2}\right) = c_p \ln \left(\frac{T_3}{T_2}\right)$$

and c_p can be evaluated from Equation 2.34. For air this value is

$$c_p = \frac{kR}{k-1} = \frac{1.4(1.986\ \text{Btu/lb}_\text{m}\cdot\text{mol}\cdot°\text{R})}{(1.4-1)(28.97\ \text{lb}_\text{m}/\text{lb}_\text{m}\cdot\text{mol})} = 0.24\ \text{Btu/lb}_\text{m}\cdot°\text{R}$$

Therefore

$$s_3 - s_2 = (0.24\ \text{Btu/lb}_\text{m}\cdot°\text{R})\ \ln\left(\frac{3510°\text{R}}{1755°\text{R}}\right) = 0.166\ \text{Btu/lb}_\text{m}\cdot°\text{R}$$

From Figure 7.9 we find that

$$s_4 = s_3 \quad \text{and} \quad s_1 = s_2$$

Therefore, the entropy change for process 4-1 is

$$s_1 - s_4 = s_2 - s_3 = -0.166\ \text{Btu/lb}_\text{m}\cdot°\text{R}$$

The heat transfer for each process can be determined from the conservation-of-energy equation for a closed system:

$$q_{if} - w_{if} = (u_f - u_i)$$

For the constant-pressure process 2-3,

$$w_{23} = \int_2^3 P\,dv = P_2(v_3 - v_2)$$

and therefore

$$q_{23} = P_2(v_3 - v_2) + (u_3 - u_2) = h_3 - h_2$$

since the pressure is constant. Thus

$$q_{23} = h_3 - h_2 = c_p(T_3 - T_2)$$
$$= (0.24\ \text{Btu/lb}_\text{m}\cdot°\text{R})(3510 - 1755)°\text{R} = 421.2\ \text{Btu/lb}_\text{m}$$

The irreversibility for process 2-3 is therefore

$$i_{23} = T_0\left[(s_3 - s_2) + \frac{q_{H,\text{res}}}{T_H}\right] = T_0\left[(s_3 - s_2) - \frac{q_{23}}{T_H}\right]$$

$$= (537°\text{R})\left[0.166\ \text{Btu/lb}_\text{m}\cdot°\text{R} - \frac{421.2\ \text{Btu/lb}_\text{m}}{3730°\text{R}}\right] = \underline{\underline{28.5\ \text{Btu/lb}_\text{m}}}$$

For process 4-1

$$q_{41} - w_{41} = (u_1 - u_4) = c_v(T_1 - T_4)$$

since the volume remains constant. The constant-volume specific heat can be calculated from Equation 2.35:

$$c_v = \frac{R}{k-1} = \frac{1.986\ \text{Btu/lb}_\text{m}\cdot\text{mol}\cdot°\text{R}}{(1.4-1)(28.97\ \text{lb}_\text{m}/\text{lb}_\text{m}\cdot\text{mol})} = 0.171\ \text{Btu/lb}_\text{m}\cdot°\text{R}$$

The temperature at state 4 can be determined by combining the expression

$$\frac{P_4}{T_4} = \frac{P_1}{T_1}$$

which is valid for the constant-volume process, and the expression

$$\frac{T_3}{T_4} = \left(\frac{P_3}{P_4}\right)^{(k-1)/k}$$

for the isentropic process 3-4. Solving for T_4, we obtain

$$T_4 = 1491°\text{R}$$

Thus

$$q_{41} = (0.171\ \text{Btu/lb}_\text{m}\cdot°\text{R})(565 - 1491)°\text{R} = -158.3\ \text{Btu/lb}_\text{m}$$

and the irreversibility for process 4-1 is

$$i_{41} = T_0\left[(s_1 - s_4) + \frac{q_{L,\text{res}}}{T_L}\right] = T_0\left[(s_1 - s_4) - \frac{q_{41}}{T_0}\right]$$

$$= (537°\text{R})\left[-0.166\ \text{Btu/lb}_\text{m}\cdot°\text{R} - \frac{(-158.3\ \text{Btu/lb}_\text{m})}{537°\text{R}}\right] = \underline{\underline{69.2\ \text{Btu/lb}_\text{m}}}$$

The irreversibility for the cycle is the sum of the irreversibilities of each of the individual processes, or

$$i_\text{cyc} = i_{12} + i_{23} + i_{34} + i_{41}$$

$$= 0 + 28.5\ \text{Btu/lb}_\text{m} + 0 + 69.2\ \text{Btu/lb}_\text{m} = \underline{\underline{97.7\ \text{Btu/lb}_\text{m}}}$$

If the irreversibility of the cycle alone was desired, it could be calculated more quickly by evaluating the cyclic integral of Equation 6.38:

$$i_\text{cyc} = T_0 \oint ds_\text{sys} + \sum_k T_0 \oint \frac{\delta q_k}{T_k}$$

$$= T_0\left(\frac{q_{H,\text{res}}}{T_H} + \frac{q_{L,\text{res}}}{T_L}\right) = -T_0\left(\frac{q_{23}}{T_H} + \frac{q_{41}}{T_0}\right)$$

$$= (-537°\text{R})\left[\frac{421.2\ \text{Btu/lb}_\text{m}}{3730°\text{R}} + \frac{(-158.3\ \text{Btu/lb}_\text{m})}{537°\text{R}}\right] = \underline{\underline{97.7\ \text{Btu/lb}_\text{m}}}$$

As was done in Example 7.3, a summary of the pertinent characteristics of the cycle is presented in Table 7.2. The details of the calculations for the work and internal-energy change of each process are omitted.

Notice that the constant-volume cooling process accounts for about 70 percent of the total irreversibility for this cycle. The irreversibility of this process would be reduced by decreasing the average temperature at which heat transfer occurs in process 4-1 or by using some of that energy in a recovery scheme

TABLE 7.2 IDEAL DIESEL CYCLE
T_H = 3730°R, T_L = T_0 = 537°R, P_0 = 15 psia

Process	q_{if}, Btu/lb$_m$	w_{if}, Btu/lb$_m$	$u_f - u_i$, Btu/lb$_m$	$s_f - s_i$, Btu/lb$_m$·°R	i_{if}, Btu/lb$_m$
1-2	0.0	−203.5	203.5	0.0	0.0
2-3	421.2	121.1	300.0	0.166	28.5
3-4	0.0	345.2	−345.2	0.0	0.0
4-1	−158.3	0.0	−158.3	−0.166	69.2
Cycle	262.9	262.8	0.0	0.0	97.7

rather than allowing heat transfer to the surroundings where the availability of the energy is reduced to zero. Notice that if process 4-1 is restricted to a constant-volume process, the minimum temperature in the cycle (state 1) could not be less than T_0(537°R), and therefore there is a limit to the amount of reduction in the irreversibility that could be achieved for the process if heat transfer occurs directly to the surroundings. ■

7.9 IDEAL BRAYTON CYCLE

The gas-turbine cycle is named after an American engineer, George Brayton, who first proposed the basic elements for a reciprocating oil-burning engine in the 1870s. The ideal *Brayton cycle* consists of four internally reversible processes; an isentropic-compression process, a constant-pressure combustion process, an isentropic-expansion and a constant-pressure cooling process. The *T*-*s* and *P*-*v* process diagrams for an ideal Brayton cycle are illustrated in Figure 7.12.

A schematic of a gas-turbine engine, which is assumed to operate steadily as an open system, is shown in Figure 7.13. Air is drawn into the compressor section, where the pressure and temperature of the air are increased (process 1-2 in Figure 7.12). After the air leaves the compressor section, the fuel is added to it and the mixture is ignited in the combustion section. This causes energy to be added at constant pressure (process

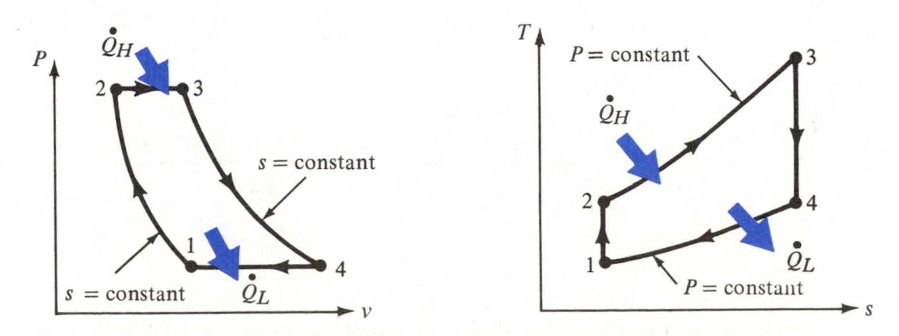

Figure 7.12 The *T*-*s* and *P*-*v* process diagrams for the ideal Brayton cycle.

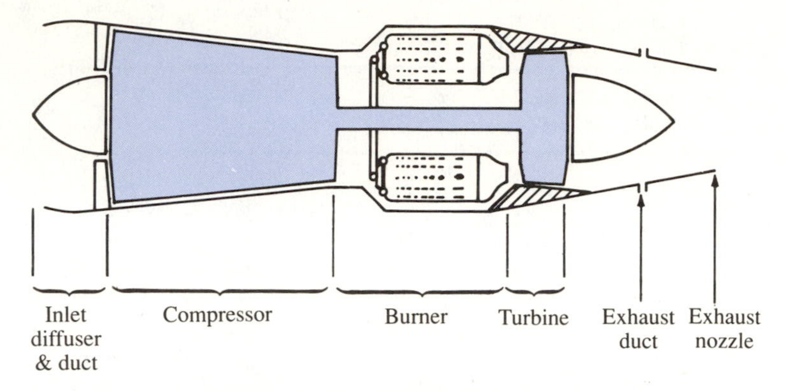

Figure 7.13 Basic elements of an open-system gas turbine. [Source: *The Aircraft Gas Turbine Engine and Its Operation*. © United Aircraft Corporation (now United Technologies Corporation), 1951, 1974.]

2-3). The resulting high-pressure, high-temperature gases then pass through the turbine section, where energy is extracted from the fluid (process 3-4), causing the turbine shaft to rotate. In the open cycle the gases leave the turbine section and are exhausted to the atmosphere as shown in Figure 7.13. In the closed Brayton cycle illustrated in Figure 7.14 the exhaust process is replaced by a constant-pressure cooling process (path 4-1 in Figure 7.12). Some of the work produced in the turbine section is used to operate the compressor, and the remainder can be used to power an external device. When a gas turbine is used to power a jet aircraft, the entire work output from the turbine is used to operate the compressor and auxiliary systems such as electric generators and hydraulic

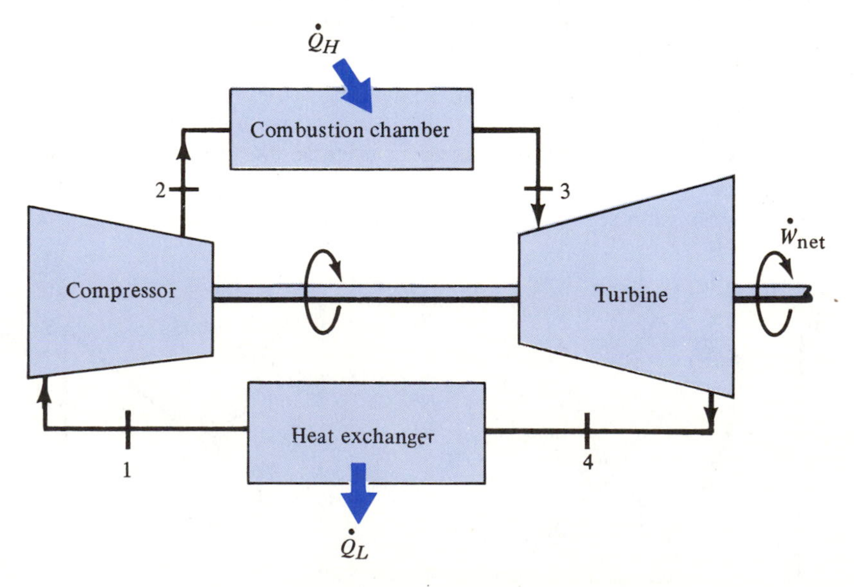

Figure 7.14 Components of the ideal, closed-system Brayton cycle.

systems, whereas the thrust for propelling the aircraft is produced by the high-velocity exhaust gases. When a gas turbine is used as a stationary power plant, the power produced by the turbine in excess of that necessary to power the compressor is used to rotate an electric generator.

Even though a gas-turbine power plant operates in an open cycle, drawing in air and exhausting combustion products, an analysis based on a closed ideal cycle wherein the working fluid executes a complete thermodynamic cycle will lead to fruitful results. The ideal Brayton cycle operates with individual components that are open, steady-flow devices. However, including a heat exchanger between the turbine exhaust and compressor inlet allows a gas-turbine power plant to be transformed into a closed-system device. The ideal, closed-system Brayton cycle is illustrated in Figure 7.14.

The thermal efficiency of an ideal Brayton cycle is

$$\eta_{\text{th}} = 1 + \frac{\dot{Q}_L}{\dot{Q}_H} \tag{7.25}$$

No work interactions occur during heat-transfer processes 2-3 and 4-1, so that the heat transfer rate to the working fluid from the high-temperature reservoir and the heat transfer rate from the working fluid to the low-temperature reservoir are given by

$$\dot{Q}_H = \dot{Q}_{23} = \dot{m}(h_3 - h_2) \tag{7.26}$$

and

$$\dot{Q}_L = \dot{Q}_{41} = \dot{m}(h_1 - h_4) \tag{7.27}$$

Substituting these results into Equation 7.25 and rearranging gives the following expression for the thermal efficiency of an ideal Brayton cycle:

$$\eta_{\text{th,Brayton}} = 1 + \left(\frac{h_1 - h_4}{h_3 - h_2}\right) \tag{7.28}$$

If the cold air-standard assumptions are applied, then the expression for the thermal efficiency of an ideal, air-standard Brayton cycle becomes

$$\eta_{\text{th,Brayton}} = 1 + \left(\frac{T_1 - T_4}{T_3 - T_2}\right) \tag{7.29}$$

Finally, if the isentropic relationships for ideal gases with constant specific heats are applied,

$$\frac{P_2}{P_1} = \left(\frac{T_2}{T_1}\right)^{k/(k-1)} = \frac{P_3}{P_4} = \left(\frac{T_3}{T_4}\right)^{k/(k-1)} \tag{7.30}$$

the temperatures in Equation 7.29 can be expressed in terms of P_2/P_1. The *pressure ratio* of the gas turbine cycle, P_2/P_1, is the ratio of the outlet pressure to inlet pressure of the compressor and is designated by the symbol r_p:

$$r_p \equiv \frac{P_2}{P_1}$$

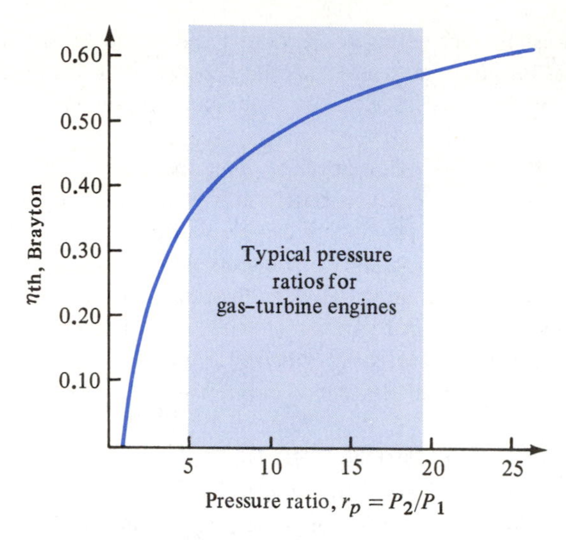

Figure 7.15 Thermal efficiency of the ideal, cold air-standard Brayton cycle ($k = 1.4$).

The expression for the thermal efficiency of the Brayton cycle in terms of the pressure ratio is

$$\eta_{\text{th,Brayton}} = 1 - (r_p)^{(1-k)/k} \tag{7.31}$$

Equation 7.31 shows that the efficiency of an ideal, cold air-standard, gas-turbine cycle is solely a function of the pressure ratio in the compressor section and the specific heat ratio for the working fluid. The thermal efficiency of an ideal, cold air-standard Brayton cycle for a gas with $k = 1.4$ is shown in Figure 7.15.

EXAMPLE 7.5

An ideal Brayton cycle uses air as a working fluid. The air enters the compressor at 101 kPa and 37°C. The pressure ratio of the compressor is 12:1, and the temperature of the air as it leaves the turbine is 497°C. The temperature and the pressure of the surroundings are 37°C and 100 kPa, respectively. Use the air-standard assumptions and determine the following quantities:

(a) The work per unit mass required to operate the compressor
(b) The work per unit mass produced by the turbine
(c) The heat transfer per unit mass during the combustion process and during the exhaust process
(d) The irreversibility of the cycle on a unit-mass basis, assuming that the temperatures of the low- and high-temperature reservoirs are 37°C and 1300°C, respectively.
(e) The thermal efficiency of the cycle

Solution. Referring to the process diagram in Figure 7.12, we see that the given information is

$$\frac{P_2}{P_1} = 12 \qquad T_4 = 497°\text{C} = 770\,\text{K}$$

$$T_1 = 37°\text{C} = 310\text{ K} \qquad P_1 = 101\text{ kPa}$$

(a) The steady-flow compressor process is internally reversible and adiabatic, so that the work of compression per unit mass is

$$w_{12} = h_1 - h_2$$

The temperature at the end of the isentropic-compression process can be determined by applying Equation 5.28:

$$s_2^\circ = s_1^\circ + R \ln \frac{P_2}{P_1}$$

Using properties from Table D.1, we have

$$s_2^\circ = 5.7335\text{ kJ/kg·K} + (0.287\text{ kJ/kg·K}) \ln (12) = 6.447\text{ kJ/kg·K}$$

Again using Table D.1 to determine T_2 at this value of s° gives

$$T_2 = 622\text{ K} = 349°\text{C}$$

and

$$h_1 = 310.5\text{ kJ/kg}$$

$$h_2 = 630.7\,\text{kJ/kg}$$

Therefore

$$w_{12} = (310.5 - 630.7)\text{kJ/kg} = \underline{\underline{-320.2\,\text{kJ/kg}}}$$

(b) The work produced by the turbine is

$$w_{34} = h_3 - h_4$$

and the enthalpies at states 3 and 4 can be evaluated once the temperatures at states 3 and 4 are known. At state 3

$$P_3 = P_2 = 12P_1 = 1.212\,\text{MPa}$$

Another property at state 3 can be determined by first determining the conditions at state 4 and realizing states 3 and 4 are on an isentropic path:

$$P_4 = P_1 = 101\text{ kPa}$$

$$T_4 = 770\,\text{K and } s_3 = s_4$$

thus

$$s_3^\circ = s_4^\circ - R \ln \left(\frac{P_4}{P_3}\right)$$

$$= 6.6757 \text{ kJ/kg·K} - (0.287 \text{ kJ/kg·K}) \ln \left(\frac{101 \text{ kPa}}{1212 \text{ kPa}}\right)$$

$$= 7.3889 \text{ kJ/kg·K}$$

Therefore, the properties at states 3 and 4 are

$$T_3 = 1433 \text{ K} = 1160°\text{C}$$

$$h_3 = 1554 \text{ kJ/kg}$$

$$h_4 = 789.3 \text{ kJ/kg}$$

and

$$w_{34} = h_3 - h_4 = \underline{\underline{764.7 \text{ kJ/kg}}}$$

(c) The heat transfer per unit mass during the steady-flow, constant-pressure combustion process is

$$q_{23} = h_3 - h_2 = (1554 - 630.7)\text{kJ/kg} = \underline{\underline{923.3 \text{ kJ/kg}}}$$

and the heat transfer during the constant-pressure exhaust process is

$$q_{41} = h_1 - h_4 = (310.5 - 789.3)\text{kJ/kg} = \underline{\underline{-478.8 \text{ kJ/kg}}}$$

(d) From Equation 6.39 the irreversibility per unit mass produced during the closed cycle is

$$i = T_0 \left[\frac{q_H}{T_H} + \frac{q_L}{T_L}\right] = -T_0 \left[\frac{q_{23}}{T_H} + \frac{q_{41}}{T_L}\right]$$

since the entropy change for the cycle is zero. The heat-transfer values were determined in part c. Substitution gives

$$i = -(310 \text{ K}) \left(\frac{923.3}{1573} + \frac{-478.8}{310}\right) \text{kJ/kg·K} = \underline{\underline{296.8 \text{ kJ/kg}}}$$

Notice that the magnitude of the irreversibility for this cycle is significantly larger than that calculated for the Otto cycle in Example 7.3. The size of the irreversibility results from the large temperature differences that exist during the two heat-transfer processes.

To illustrate the aspects of a second-law analysis of the cycle, we also determine the irreversibility as well as the change in availability associated with each process. Since the ideal Brayton cycle is composed of four steady-flow processes, the irreversibility rate for each process is given by Equation 6.47,

$$\dot{I} = T_0 \left(\sum_{\text{exit}} \dot{m}s - \sum_{\text{inlet}} \dot{m}s + \sum_k \frac{\dot{Q}_k}{T_k}\right)$$

This expression can be simplified considerably since each component in the cycle has a single inlet and a single exit and heat transfer occurs with, at most, a single reservoir. Therefore, the irreversibility can be written on a unit-mass basis as

$$i = T_0 \left(s_e - s_i + \frac{q_{\text{res}}}{T_{\text{res}}} \right)$$

Thus the irreversibility for each of the internally reversible adiabatic (isentropic) processes 1-2 and 3-4 is zero:

$$i_{12} = 0 \quad \text{and} \quad i_{34} = 0$$

The entropy change for each of the constant-pressure processes is given by

$$s_e - s_i = s_e^\circ - s_i^\circ - R \ln \left(\frac{P_e}{P_i} \right) = s_e^\circ - s_i^\circ$$

Therefore

$$\begin{aligned} s_3 - s_2 = s_3^\circ - s_2^\circ &= (7.3889 - 6.447)\text{kJ/kg·K} \\ &= 0.942 \text{ kJ/kg·K} \end{aligned}$$

and

$$\begin{aligned} s_1 - s_4 = s_1^\circ - s_4^\circ &= (5.7335 - 6.6757)\text{kJ/kg·K} \\ &= -0.942 \text{ kJ/kg·K} \end{aligned}$$

and the irreversibility for each of these processes is

$$\begin{aligned} i_{23} &= T_0 \left(s_3 - s_2 + \frac{q_{H,\text{res}}}{T_H} \right) = T_0 \left(s_3 - s_2 - \frac{q_{23}}{T_H} \right) \\ &= (310\,\text{K}) \left(0.942 \text{ kJ/kg·K} - \frac{923.3\,\text{kJ/kg}}{1573 \text{ K}} \right) = 110 \text{ kJ/kg} \end{aligned}$$

and

$$\begin{aligned} i_{41} &= T_0 \left(s_1 - s_4 + \frac{q_{L,\text{res}}}{T_L} \right) = T_0 \left(s_1 - s_4 - \frac{q_{41}}{T_0} \right) \\ &= (310 \text{ K}) \left(-0.942 \text{ kJ/kg·K} - \frac{(-478.8 \text{ kJ/kg})}{310 \text{ K}} \right) = 186.8 \text{ kJ/kg} \end{aligned}$$

The change in the availability of the air for each process can be determined with Equation 6.32 as

$$\psi_e - \psi_i = (h_e - h_i) - T_0(s_e - s_i)$$

The results of such calculations as well as other characteristics of this cycle are summarized in Table 7.3.

TABLE 7.3 IDEAL BRAYTON CYCLE
T_H = 1573 K, T_L = T_0 = 310 K

Process	q, kJ/kg	w, kJ/kg	$h_e - h_i$, kJ/kg	$s_e - s_i$, kJ/kg·K	$\psi_e - \psi_i$, kJ/kg	i, kJ/kg
1-2	0.0	−320.2	320.2	0.0	320.2	0.0
2-3	923.3	0.0	923.3	0.942	631.3	110.0
3-4	0.0	764.7	−764.7	0.0	−764.7	0.0
4-1	−478.8	0.0	−478.8	−0.942	−186.8	186.8
Cycle	444.5	444.5	0.0	0.0	0.0	296.8

Notice that the change in each of the properties h, s and ψ is zero for the cycle and that the net heat transfer is equal to the net work for the cycle. Furthermore, since all processes are internally reversible, irreversibilities arise only in those processes that are not totally reversible due to heat transfer to or from the system through a finite temperature difference.

From Equation 6.50, written on a unit-mass basis for each process, other interesting facts emerge:

$$w_{\text{rev}} = w_{\text{rev,useful}} = -(\psi_e - \psi_i) - q_{\text{res}}\left(1 - \frac{T_0}{T_{\text{res}}}\right)$$

If no heat transfer occurs during a process or if heat transfer is only to the environment at temperature T_0, then the reversible work (and the reversible useful work, since the system boundary is rigid) is equal in magnitude but opposite in sign to the change in the availability of the fluid. Such is the case for processes 1-2, 3-4, and 4-1. Furthermore, the actual work is equal to the reversible work for the totally reversible processes 1-2 and 3-4.

The irreversibility of the cycle is due entirely to heat transfer that occurs between the system and its surroundings through a finite temperature difference. In this example, most of the irreversibility is attributed to the cooling process. Furthermore, the lowest temperature during this process (T_1) is the same as the temperature of the environment (T_0). Therefore, the minimum temperature of the cycle could not be reduced further. Some other means of reducing the average temperature for the heat-rejection process is necessary if the irreversibility is to be reduced. This idea is explored further as we discuss common modifications to the ideal Brayton cycle.

(e) The thermal efficiency of the cycle is

$$\eta_{\text{th}} = \frac{w_{\text{net}}}{q_{23}} = \frac{w_{34} + w_{12}}{q_{23}}$$

$$= \frac{(764.7 - 320.2)\text{kJ/kg}}{923.3 \text{ kJ/kg}} = \underline{\underline{0.481}}$$

Notice that a large portion of the work output of the turbine is required to operate the compressor. In this example 42 percent of the useful work output of the turbine is necessary to compress the air. This result is characteristic of a Brayton cycle and contributes to the relatively low thermal efficiency of the gas-turbine cycle.

If the cold air-standard analysis is used with $k = 1.4$ and $c_p = 1.0\ \text{kJ/kg·K}$, this problem could be reworked, giving the following results:

$T_1 = 310\ \text{K}$	$q_{23} = 935.6\ \text{kJ/kg}$
$T_2 = 630.5\ \text{K}$	$w_{34} = 796.1\ \text{kJ/kg}$
$T_3 = 1566.1\ \text{K}$	$q_{41} = -460\ \text{kJ/kg}$
$T_4 = 770\ \text{K}$	$\eta_{\text{th}} = 0.508$
$w_{12} = -320.5\ \text{kJ/kg}$	

Comparing these answers with the ones calculated above shows that the variations in the specific heats due to the large temperature changes have a significant influence on the calculated thermal efficiency. ■

7.10 IDEAL BRAYTON CYCLE WITH REGENERATION

Most modern gas-turbine units operate with a pressure ratio that ranges from about 5 for small, single-stage compressors to approximately 20 for high-performance military jet engines. The efficiency of an ideal-gas turbine cycle for these pressure ratios, as given in Figure 7.15, ranges between 40 and 55 percent. The relatively low efficiency of a Brayton cycle, particularly when compared with the other gas power cycles such as the Otto and diesel cycles (see Figures 7.8 and 7.10), suggests that modifications to the Brayton cycle that would increase the thermal efficiency would be very beneficial.

One common modification to the Brayton cycle that can produce an increase in efficiency is called *regeneration*. Regeneration involves using the high-temperature exhaust gases from the turbine to heat the gas as it leaves the compressor. A sketch of the regeneration process along with the T-s diagram for an ideal Brayton cycle modified with regeneration is shown in Figure 7.16.

The regeneration process produces an increase in the thermal efficiency of the Brayton cycle because some of the energy that is normally rejected to the surroundings by the turbine exhaust gases is used to preheat the air entering the combustion chamber. Therefore, in those cases where the temperature of the air leaving the compressor is less than the air leaving the turbine (that is, $T_2 < T_5$), increased cycle efficiency will result if the regeneration process is utilized. The feasibility of regeneration becomes more apparent when the temperature of the gases at the turbine outlet is considered. As an illustration, the temperature of the gases leaving the turbine in the previous example is approximately 500°C.

The turbine and compressor exhaust gases that enter the regenerator are assumed to follow internally reversible, constant-pressure processes since the heat-exchanger process

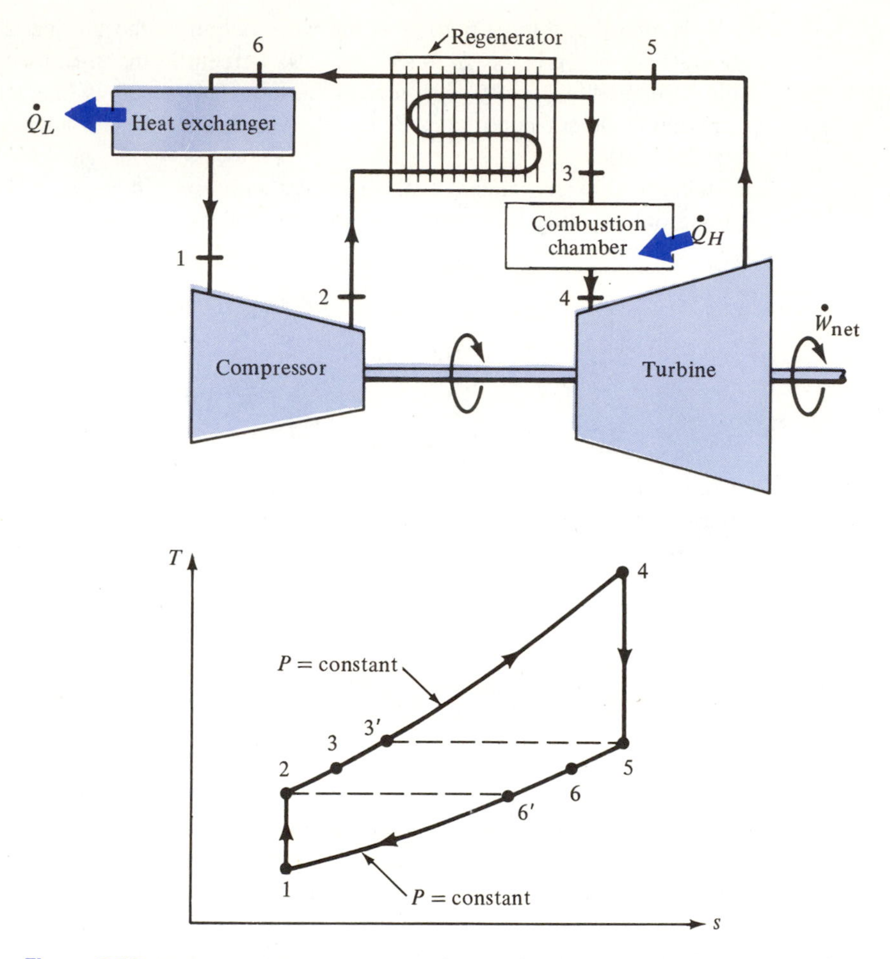

Figure 7.16 Schematic and *T-s* process diagram for an ideal Brayton cycle modified with regeneration.

is assumed to be ideal. As the *T-s* process diagram in Figure 7.16 indicates, the gas leaves the compressor at state 2 and follows a constant-pressure path to state 3. Even under the most ideal conditions, the temperature of the gas leaving the regenerator could never exceed T_5, the temperature of the exhaust gas from the turbine. Therefore, under ideal conditions the compressor exhaust gases could leave the regenerator at state 3′. Any air temperature at the exit of the regenerator in excess of $T_{3'}$ would result in a violation of the second law since heat transfer must occur from a higher to a lower temperature. The turbine exhaust gas enters the regenerator at state 5 and follows a constant-pressure path to state 6. As the turbine gases pass through the regenerator, the temperature could never drop below T_2, the lowest temperature that occurs within the regenerator. State point 6′, therefore, represents the temperature of the turbine gases leaving the regenerator under the most ideal conditions.

The *efficiency of the regenerator* is defined as the actual heat-transfer rate from the turbine exhaust gases to the compressor exhaust gases in the regenerator divided by the maximum heat-transfer rate that could occur in the regenerator, or

$$\eta_{\text{regen}} \equiv \frac{\dot{Q}_{\text{act}}}{\dot{Q}_{\text{max}}} \tag{7.32}$$

The regenerator is a steady-flow, constant-pressure device, so the regenerator efficiency can be expressed in terms of the enthalpies of the gas as

$$\eta_{\text{regen}} = \frac{h_3 - h_2}{h_5 - h_{6'}} \tag{7.33}$$

Applying the cold air-standard assumptions, we can simplify the expression for the efficiency of the regenerator to

$$\eta_{\text{regen}} = \frac{T_3 - T_2}{T_5 - T_2} \tag{7.34}$$

since $T_2 = T_{6'}$. Notice that the regenerator efficiency is equal to 1.0 when $T_3 = T_5$. Under this condition ($\eta_{\text{regen}} = 1.0$) the regeneration process is called *ideal* regeneration.

The thermal efficiency of an ideal Brayton cycle modified with regeneration can be shown to be

$$\eta_{\text{th}} = \frac{\dot{W}_{\text{net}}}{\dot{Q}_H} = \frac{(h_4 - h_5) + (h_1 - h_2)}{h_4 - h_3} \tag{7.35}$$

Using cold air-standard assumptions, we can write the thermal efficiency of the ideal, air-standard Brayton cycle with ideal regeneration as

$$\eta_{\text{th}} = 1 - \left(\frac{T_1}{T_4}\right)(r_p)^{(k-1)/k} \tag{7.36}$$

This expression reveals that the thermal efficiency of an ideal, cold air-standard Brayton cycle modified with ideal regeneration is a function of the pressure ratio of the compressor and the ratio of the maximum to minimum temperatures of the cycle. The thermal efficiency of the ideal, cold air-standard Brayton with and without regeneration is illustrated in Figure 7.17. The figure shows the large improvement in efficiency for the regeneration process at the lower pressure ratios. The addition of ideal regeneration would always result in an increase in thermal efficiency over the corresponding ideal Brayton cycle without regeneration. That is, the regeneration curves in Figure 7.17 terminate at the intersection with the curve labeled ''without regeneration.''

■ EXAMPLE 7.6

Rework Example 7.5 assuming that a regenerator with an efficiency of 70 percent is inserted into the cycle. Use the cold air-standard assumptions, and assume $k = 1.4$ and $c_p = 1.0$ kJ/kg·K.

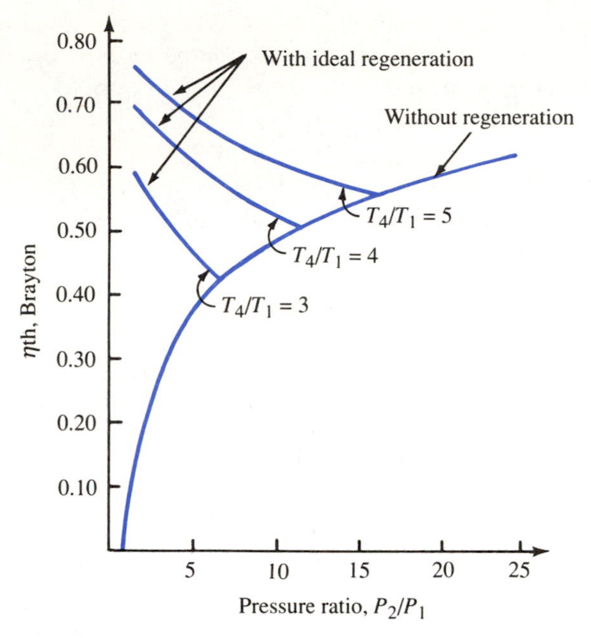

Figure 7.17 Thermal efficiency of the ideal, cold air-standard Brayton cycle with and without regeneration ($k = 1.4$).

Solution. With reference to Figure 7.16 the given information is

$$\frac{P_2}{P_1} = 12 \qquad T_1 = 310 \text{ K} \qquad P_1 = 101 \text{ kPa} \qquad T_5 = 770 \text{ K}$$

The temperature leaving the isentropic compressor is

$$T_2 = T_1 \left(\frac{P_2}{P_1}\right)^{(k-1)/k} = (310 \text{ K})(12)^{0.2857} = 630.5 \text{ K}$$

The temperature entering the isentropic turbine is

$$T_4 = T_5 \left(\frac{P_4}{P_5}\right)^{(k-1)/k} = T_5 \left(\frac{P_2}{P_1}\right)^{(k-1)/k}$$

$$= (770 \text{ K})(12)^{0.2857} = 1566.1 \text{ K}$$

The temperature T_3 can be determined by using Equation 7.34:

$$T_3 = T_2 + \eta_{\text{regen}}(T_5 - T_2)$$

$$= 630.5 \text{ K} + (0.7)(770 - 630.5) \text{ K} = 728.2 \text{ K}$$

(a) The work per unit mass of fluid in the compressor is

$$w_{12} = h_1 - h_2 = c_p(T_1 - T_2)$$

$$= (1.0 \text{ kJ/kg·K})(310 - 630.5) \text{ K} = \underline{\underline{-320.5 \text{ kJ/kg}}}$$

(b) The work output of the turbine is

$$w_{45} = h_4 - h_5 = c_p(T_4 - T_5)$$
$$= (1.0 \text{ kJ/kg·K})(1566.1 - 770) \text{ K} = \underline{\underline{796.1 \text{ kJ/kg}}}$$

The work required by the compressor and the work output of the turbine are not influenced by the presence of regeneration. Regeneration does, however, reduce the heat transfer in the combustion process. Notice that during the regeneration process,

$$q_{23} = -q_{56} = h_5 - h_6 = c_p(T_5 - T_6) = c_p(T_3 - T_2)$$

or

$$T_6 = T_5 - T_3 + T_2 = 770 \text{ K} - 728.2 \text{ K} + 630.5 \text{ K} = 672.3 \text{ K}$$

(c) The heat transfer from the working fluid in the exhaust process is

$$q_{61} = h_1 - h_6 = c_p(T_1 - T_6)$$
$$= (1.0 \text{ kJ/kg·K})(310 - 672.3) \text{ K} = \underline{\underline{-362.3 \text{ kJ/kg}}}$$

The heat transfer during the combustion process is

$$q_{34} = h_4 - h_3 = c_p(T_4 - T_3)$$
$$= (1.0 \text{ kJ/kg·K})(1566.1 - 728.2) \text{ K} = \underline{\underline{837.9 \text{ kJ/kg}}}$$

(d) The irreversibility for the cycle can be calculated from Equation 6.38. The only processes that involve heat transfer with the surroundings are processes 3-4 and 6-1. There is no heat transfer between the regenerator and the surroundings. Heat transfer only occurs internally between the two gas streams. Therefore, the expression for the irreversibility of the cycle on a per-unit-mass basis reduces to

$$i = \sum_k T_0 \oint \frac{\delta q_k}{T_k} = T_0 \left(\frac{q_{H,\text{res}}}{T_H} + \frac{q_{L,\text{res}}}{T_L} \right)$$

or

$$i = -T_0 \left(\frac{q_{34}}{T_H} + \frac{q_{61}}{T_L} \right)$$
$$= -(310 \text{ K}) \left(\frac{837.9 \text{ kJ/kg}}{1573 \text{ K}} + \frac{-362.3 \text{ kJ/kg}}{310 \text{ K}} \right) = \underline{\underline{197.2 \text{ kJ/kg}}}$$

With a second-law analysis we can examine the performance of each component in the cycle to determine the full effect of regeneration. The procedure used is identical to that employed in Example 7.5 except for the fact that the specific heats of the air are assumed constant in this example. In addition, the net entropy change for the regenerator process is determined from Equation 6.47, which reduces to

$$\dot{I}_{\text{regen}} = T_0(\dot{m}_3 s_3 + \dot{m}_6 s_6 - \dot{m}_2 s_2 - \dot{m}_5 s_5)$$

since the heat transfer from the regenerator to the surroundings is assumed to be zero. Furthermore, each of the mass flow rates is identical, and this equation can be reduced to

$$i_{\text{regen}} = T_0[(s_3 - s_2) + (s_6 - s_5)]$$

Each fluid stream remains at constant pressure and the specific heat is assumed constant. Therefore

$$\begin{aligned} i_{\text{regen}} &= T_0\left(c_p \ln\frac{T_3}{T_2} + c_p \ln\frac{T_6}{T_5}\right) \\ &= (310\text{ K})\left[(1.0\text{ kJ/kg·K}) \ln\left(\frac{728.2\text{ K}}{630.5\text{ K}}\right)\right. \\ &\qquad \left. + (1.0\text{ kJ/kg·K}) \ln\left(\frac{672.3\text{ K}}{770\text{ K}}\right)\right] \\ &= (310\text{ K})(0.0084\text{ kJ/kg·K}) = 2.6\text{ kJ/kg} \end{aligned}$$

The net change in the availability of the fluid streams passing through the regenerator can be evaluated as

$$\dot{m}_3\psi_3 + \dot{m}_6\psi_6 - \dot{m}_2\psi_2 - \dot{m}_5\psi_5$$

or on a unit-mass basis,

$$(\psi_3 - \psi_2) + (\psi_6 - \psi_5)$$

With the definition of the availability from Equation 6.32, we have

$$\begin{aligned} (\psi_3 - \psi_2) + (\psi_6 - \psi_5) &= (h_3 - h_2) - T_0(s_3 - s_2) + (h_6 - h_5) - T_0(s_6 - s_5) \\ &= -T_0[(s_3 - s_2) + (s_6 - s_5)] \\ &= (-310\text{ K})(0.0084\text{ kJ/kg·K}) = -2.6\text{ kJ/kg} \end{aligned}$$

Processes 1-2 and 4-5 are totally reversible; therefore

$$i_{12} = 0 \quad \text{and} \quad i_{45} = 0$$

The entropy change and irreversibility for each of the constant-pressure heat-transfer processes 3-4 and 6-1 are determined from Equation 5.30 and Equation 6.39, respectively:

$$\begin{aligned} s_4 - s_3 &= c_p \ln\frac{T_4}{T_3} - R \ln\frac{P_4}{P_3} = c_p \ln\frac{T_4}{T_3} \\ &= (1.0\text{ kJ/kg·K}) \ln\left(\frac{1566.1\text{ K}}{728.2\text{ K}}\right) = 0.7658\text{ kJ/kg·K} \end{aligned}$$

$$\begin{aligned} s_1 - s_6 &= c_p \ln\frac{T_1}{T_6} - R \ln\frac{P_1}{P_6} = c_p \ln\frac{T_1}{T_6} \\ &= (1.0\text{ kJ/kg·K}) \ln\left(\frac{310\text{ K}}{672.3\text{ K}}\right) = -0.7741\text{ kJ/kg·K} \end{aligned}$$

and

$$i_{34} = T_0\left[(s_4 - s_3) + \frac{q_{H,\text{res}}}{T_H}\right] = T_0\left[(s_4 - s_3) - \frac{q_{34}}{T_H}\right]$$

$$= (310\text{ K})\left[0.7658\text{ kJ/kg·K} - \frac{837.9\text{ kJ/kg}}{1573\text{ K}}\right] = 72.3\text{ kJ/kg}$$

$$i_{61} = T_0\left[(s_1 - s_6) + \frac{q_{L,\text{res}}}{T_L}\right] = T_0\left[(s_1 - s_6) - \frac{q_{61}}{T_0}\right]$$

$$= (310\text{ K})\left[-0.7741\text{ kJ/kg·K} - \frac{(-362.3\text{ kJ/kg})}{310\text{ K}}\right]$$

$$= 122.3\text{ kJ/kg}$$

The change in the availability for each process (other than the regenerator) is determined by using Equation 6.32:

$$\psi_e - \psi_i = (h_e - h_i) - T_0(s_e - s_i)$$

The results of the thermodynamic analysis are summarized in Table 7.4.

From the data in this table we see that adding the regenerator to the cycle also adds another source of irreversibility. The irreversibility associated with the regenerator is very small in magnitude and is due to the energy transfer between the two fluid streams, which are at different temperatures. One benefit derived from the regenerator is that the average temperature of the fluid in the heating process is increased, resulting in a decrease in the irreversibility for this process. Another benefit is that the average temperature of the fluid in the cooling process is decreased, resulting in a decrease in the irreversibility for this process. The net effect is a rather substantial decrease in the irreversibility for the ideal Brayton cycle with regeneration compared with the irreversibility of the simple cycle considered in Example 7.5.

TABLE 7.4 IDEAL BRAYTON CYCLE WITH REGENERATION
$T_H = 1573$ K, $T_L = T_0 = 310$ K

Process	q, kJ/kg	w, kJ/kg	$h_e - h_i$, kJ/kg	$s_e - s_i$, kJ/kg·K	$\psi_e - \psi_i$ kJ/kg	i, kJ/kg
1-2	0.0	−320.5	320.5	0.0	320.5	0.0
3-4	837.9	0.0	837.9	0.7658	600.5	72.3
4-5	0.0	796.1	−796.1	0.0	−796.1	0.0
Regenerator	0.0	0.0	0.0	0.0084	−2.6	2.6
6-1	−362.3	0.0	−362.3	−0.7741	−122.3	122.3
Cycle	475.6	475.6	0.0	0.0	0.0	197.2

(e) The thermal efficiency of the ideal Brayton cycle with regeneration is

$$\eta_{th} = \frac{w_{net}}{q_H} = \frac{w_{45} + w_{12}}{q_{34}}$$

$$= \frac{796.1 \text{ kJ/kg} - 320.5 \text{ kJ/kg}}{837.9 \text{ kJ/kg}} = \underline{\underline{0.568}}$$

The efficiency of the Brayton cycle modified with regeneration is 8.7 percentage points higher than the value obtained without regeneration.

If Equation 7.36 is used to determine the maximum cycle efficiency that can be obtained for the given conditions with regeneration, that is, ideal regeneration, the result is

$$\eta_{\text{ideal regen}} = 0.597$$

The lower efficiency obtained for the cycle having a 70 percent efficiency regeneration process specified in this problem results from the irreversibilities that occur in the regenerator. The efficiencies can now be ranked as follows

$$\eta_{\text{ideal regen}} > \eta_{\text{70\% eff. regen}} > \eta_{\text{no regen}}$$

■

7.11 IDEAL JET-PROPULSION CYCLES

Modified gas turbines are frequently used to power aircraft because they are capable of providing a large amount of power per unit engine weight. This consideration is an important one in the design of an engine to be used to power an aircraft. An open Brayton cycle can be adapted for use in a *jet-propulsion engine*. For this application the turbine-exhaust conditions are such that the turbine power output equals the power requirements of the compressor and other auxiliary systems such as small generators and hydraulic pumps. The turbine exhaust is then expanded in a nozzle, providing the thrust to propel the aircraft.

An ideal Brayton cycle used for jet propulsion (frequently called a *turbojet engine*) is shown schematically in Figure 7.18 along with the T-s diagram for the cycle. In the diffuser section of the engine upstream of the compressor, the pressure of the air is increased isentropically from state 1 to 2. In the compressor the pressure of the air is increased further along an isentropic path to state 3. In the combustion chamber the fuel is added to the air and the mixture is burned at constant pressure. The working fluid enters the turbine at state 4, where it expands isentropically and exhausts at state 6.

According to Newton's second law, the thrust produced by the jet engine is equal to the rate of change of the momentum of the fluid flowing through the engine. When the inlet and exit pressures of the engine are identical, the *thrust* produced by the engine reduces to a simple expression,

$$T = \dot{m}(V_{out} - V_{in}) \tag{7.37}$$

The efficiency of a turbojet engine can be defined in various ways, but the most frequently used measure of performance is the *propulsive efficiency*, defined as the ratio

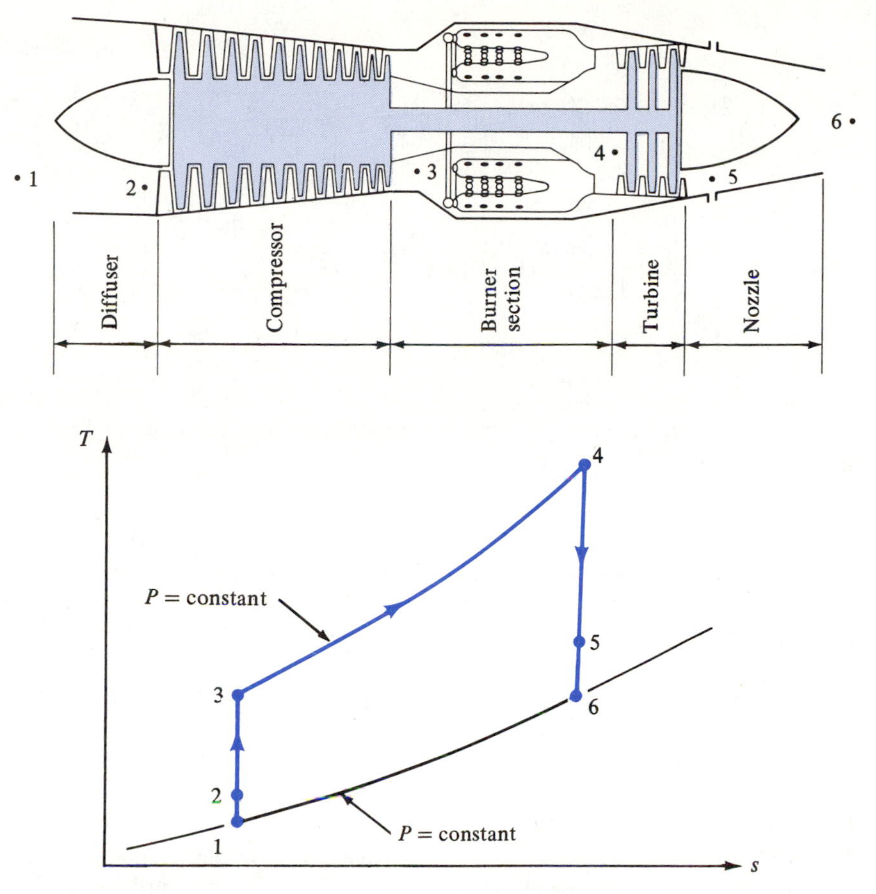

Figure 7.18 Components of a simple turbojet engine and the *T-s* diagram for the ideal turbojet cycle. [Source: *The Aircraft Gas Turbine Engine and Its Operation.* © United Aircraft Corporation (now United Technologies Corporation), 1951, 1974.]

of the power developed from the thrust of the engine (product of engine thrust and engine velocity) to the heat-transfer rate derived from the fuel, or

$$\eta_p = \frac{TV_{\text{engine}}}{\dot{Q}_H} = \frac{\dot{m}(V_{\text{out}} - V_{\text{in}})V_{\text{engine}}}{\dot{Q}_H} \tag{7.38}$$

To define a turboject efficiency in terms of the net power developed by the engine would be inappropriate. In a turbojet engine no net power is developed by the engine because all of the turbine output is used to power the compressor and auxiliary systems. Therefore, the power developed owing to the motion of the engine rather than the net power output is used to define the propulsive efficiency.

Several modifications to the basic jet-engine design shown in Figure 7.18 are commonly used in aircraft applications. A *turboprop* (or *propjet*) engine is one in which the exhaust gases from the turbine are used to rotate a set of turbine blades, which then rotate a propeller through a gear-reduction system, as shown in Figure 7.19. The majority

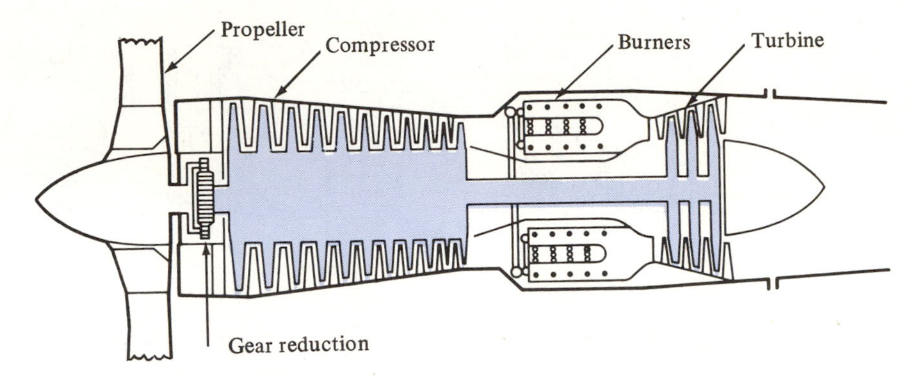

Figure 7.19 Direct-propeller-drive turboprop engine. [Source: *The Aircraft Gas Turbine Engine and Its Operation*. © United Aircraft Corporation (now United Technologies Corporation), 1951, 1974.]

of the thrust provided by the turboprop engine is therefore produced by the propeller rather than the exhaust gases from the engine. The turboprop is more complicated and heavier than a turbojet engine of equivalent power; however, the turboprop engine is capable of providing greater thrust at low speeds. As the speed of the aircraft increases, the propulsive efficiency of a turboprop decreases, until a speed is reached for which a turbojet is more efficient than a turboprop. As a result, turboprop engines are limited in their operation to speeds significantly less than the speed of sound.

A *turbofan* (or *fan-jet*) engine consists of an axial-flow fan with rotating blades and stationary vanes placed in front of the compressor. After the air passes through the fan section it bypasses the burner and turbine sections. The bypass air may be either exhausted to the ambient or passed through ducts that surround the engine, exhausting at the rear of the engine, as shown in Figure 7.20. The fan exhaust contributes a substantial amount to the total thrust of the engine, often between 30 and 70 percent. The turbofan engine combines the relatively high efficiency and the excellent thrust capabilities of a turboprop engine with the superior high-speed characteristics of a turbojet engine. The fan-jet engine

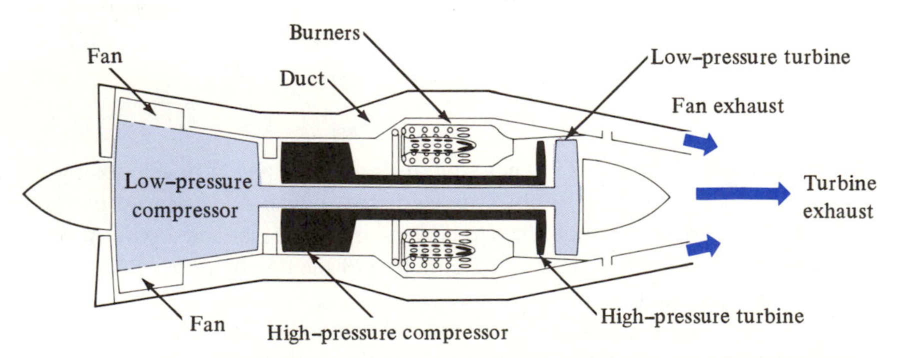

Figure 7.20 Simplified diagram of a turbofan engine. [Source: *The Aircraft Gas Turbine Engine and Its Operation*. © United Aircraft Corporation (now United Technologies Corporation), 1951, 1974.]

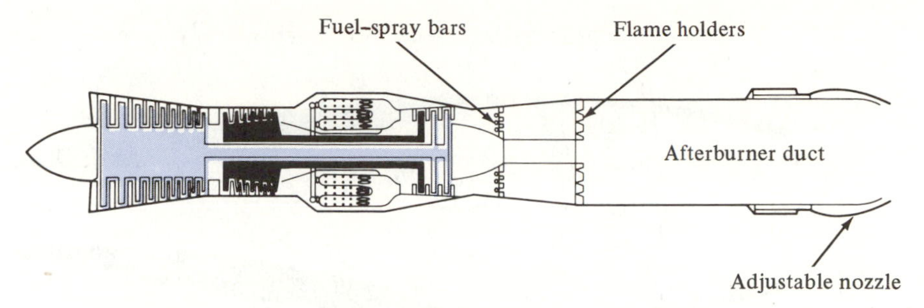

Figure 7.21 Simplified diagram of a turbojet engine with afterburner. [Source: *The Aircraft Gas Turbine Engine and Its Operation*. © United Aircraft Corporation (now United Technologies Corporation), 1951, 1974.]

has become the most widely used power plant for both military and commercial jet aircraft.

One cycle modification that is often used for thrust augmentation is the use of an *afterburner*. Essentially, an afterburner is a segment of ductwork attached to the exhaust section of the turbine. In a forward section of the afterburner a series of fuel-spray bars are placed so that fuel can be injected and burned in the turbine exhaust. The spray bars are followed by flameholders, which prevent the flame from being blown out of the afterburner section. The process in the afterburner more than doubles the fuel consumption, but it can produce up to 50 percent additional thrust for situations that require short periods of additional speed. Military aircraft use an afterburner to produce extra power for short takeoffs and during combat conditions. A typical jet engine adapted with an afterburner is shown in Figure 7.21.

A section view of a turbofan engine used in military aircraft is shown in Figure 7.22. This particular engine has three fan stages, ten compressor stages, and four turbine stages. It provides a maximum thrust of 23,800 lb_f when the afterburner is in use although the engine weighs only 3200 lb_f. The engine has an overall pressure ratio of 25, and 38 percent of the air entering the engine bypasses the burner and turbine sections.

A *ramjet engine* is occasionally used to power missiles and aircraft. Simply stated, a ramjet is an open-ended engine without a compressor or turbine. A fuel-injecting and

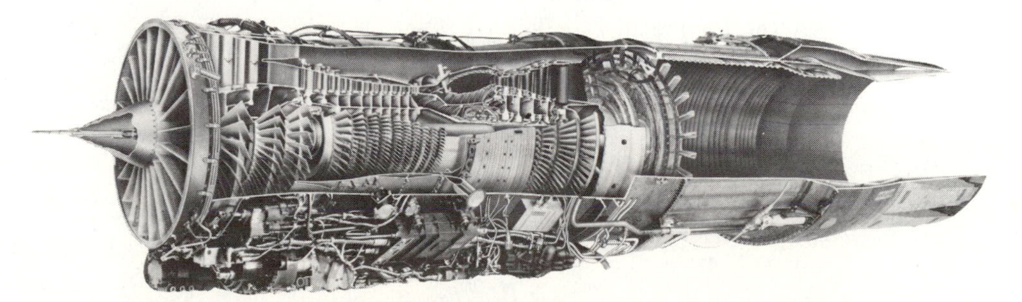

Figure 7.22 Section view of a Pratt-Whitney F100-Turbofan engine. This particular engine is used in the F-15 Eagle and F-16 Falcon fighter aircraft. (Courtesy of United Technologies, Pratt & Whitney.)

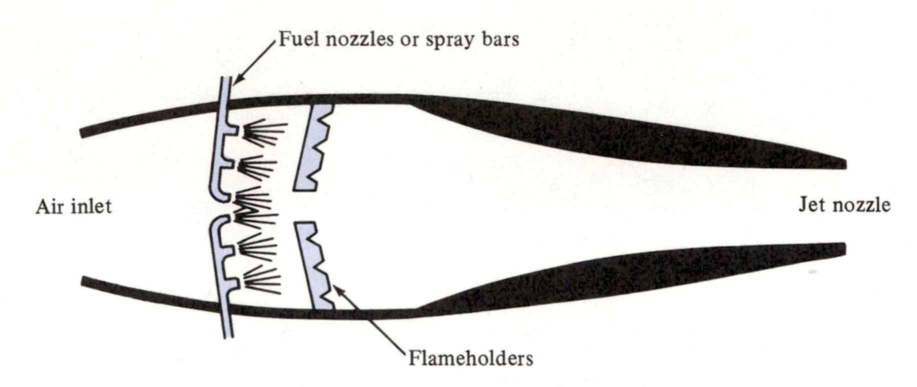

Figure 7.23 Schematic of a ramjet engine. [Source: *The Aircraft Gas Turbine Engine and Its Operation*. © United Aircraft Corporation (now United Technologies Corporation), 1951, 1974.]

-metering device is located inside the skin of the engine, as shown in Figure 7.23. The pressure required to operate the ramjet results solely from the "ram" effect of the entering air. The kinetic energy of the air entering the ramjet is converted to a sufficient pressure to operate the engine. Consequently, the ramjet cannot be operated until it has been accelerated by an external power supply. Often ramjets are carried aloft by aircraft or missiles and ignited once they reach operating speeds.

EXAMPLE 7.7

A jet aircraft flies with a velocity of 650 ft/s at an altitude where the air temperature is −40°F. The inlet and exit areas of the turbojet engine are 6.5 ft^2 and 3.0 ft^2, respectively. The pressure of the air at the inlet and exit areas of the engine is 10.0 psia. The compressor has a pressure ratio of 10.0, and the inlet temperature to the turbine is 1240°F. The pressure increase in the diffuser is 5.0 psia. Assume that the engine operates on an ideal cycle, use air-standard assumptions, and calculate the following information:

(a) The temperature of the air leaving the compressor
(b) The temperature of the air leaving the engine
(c) The velocity of the air leaving the engine
(d) The thrust produced by the engine
(e) The propulsive efficiency of the cycle

Solution. With reference to the states identified in Figure 7.18, the given information is

$$P_1 = 10.0 \text{ psia} \qquad T_1 = 420°\text{R}$$

$$P_2 = P_1 + 5.0 \text{ psia} = 15.0 \text{ psia}$$

$$P_3 = 10P_2 = 150 \text{ psia}$$

$$P_4 = P_3$$

$$T_4 = 1240°\text{F} = 1700°\text{R}$$

$$P_6 = 10.0 \text{ psia}$$

(a) States 1 and 3 are connected by an isentropic path, so

$$\frac{P_{r3}}{P_{r1}} = \frac{P_3}{P_1} = \frac{150\,\text{psia}}{10\ \text{psia}} = 15.0$$

The temperature at state 3 may be determined by using the reduced-pressure values from Table D.1E:

$$P_{r1} = 0.5739$$

$$P_{r3} = 15.0(0.5739) = 8.609$$

$$h_3 = 218.0\ \text{Btu/lb}_\text{m}$$

$$T_3 = \underline{\underline{907°\text{R}}}$$

(b) States 4 and 6 are connected by an isentropic path, so the reduced-pressure values from Table D.1E can be used to determine the temperature at state 6:

$$T_4 = 1700°\text{R} \qquad h_4 = 422.7\ \text{Btu/lb}_\text{m}$$

$$P_{r4} = 90.54$$

$$P_{r6} = P_{r4}\left(\frac{P_6}{P_4}\right) = 90.54\left(\frac{10\ \text{psia}}{150\ \text{psia}}\right) = 6.036$$

$$h_6 = 197.0\,\text{Btu/lb}_\text{m}$$

$$T_6 = \underline{\underline{821°\text{R}}}$$

(c) The mass flow rate of air entering the engine is

$$\dot{m}_1 = \rho_1 V_1 A_1 = \left(\frac{P_1}{RT_1}\right) V_1 A_1$$

$$= \frac{(10.0\,\text{lb}_\text{f}/\text{in}^2)(650\ \text{ft/s})(6.5\ \text{ft}^2)(144\ \text{in}^2/\text{ft}^2)}{(53.34\,\text{ft}\cdot\text{lb}_\text{f}/\text{lb}_\text{m}\cdot°\text{R})(420°\text{R})}$$

$$= 271.6\ \text{lb}_\text{m}/\text{s}$$

For steady flow through the engine and a small fuel mass flow rate in the burner section, the mass flow rate at the engine exit is approximately equal to $\dot{m}_1$. The velocity of the air leaving the engine is therefore

$$V_6 = \frac{\dot{m}_1 RT_6}{P_6 A_6} = \frac{(271.6\ \text{lb}_\text{m}/\text{s})(53.34\,\text{ft}\cdot\text{lb}_\text{f}/\text{lb}_\text{m}\cdot°\text{R})(821°\text{R})}{(10.0\ \text{lb}_\text{f}/\text{in}^2)(144\ \text{in}^2/\text{ft}^2)(3.0\,\text{ft}^2)}$$

$$= \underline{\underline{2753\ \text{ft/s}}}$$

(d) The thrust produced by the engine is

$$T = \dot{m}(V_6 - V_1) = \frac{(271.6\,\text{lb}_\text{m}/\text{s})(2753 - 650)\text{ft/s}}{(32.2\ \text{lb}_\text{m}\cdot\text{ft}/\text{lb}_\text{f}\cdot\text{s}^2)} = \underline{\underline{17{,}740\ \text{lb}_\text{f}}}$$

(e) The power developed by the engine is

$$TV_{\text{engine}} = (17{,}740 \text{ lb}_f)(650 \text{ ft/s}) = 11.5 \times 10^6 \text{ft}\cdot\text{lb}_f/\text{s} = 20{,}910 \text{ hp}$$

The heat-transfer rate from the fuel may be determined by applying the conservation-of-energy equation to the air as it flows through the combustion zone neglecting potential-energy changes:

$$\dot{Q}_H = \dot{m}q_{34} = \dot{m}(h_4 - h_3)$$

$$= (271.6 \text{ lb}_m/\text{s})(422.7 - 218.0)\text{Btu/lb}_m$$

$$= 55{,}600 \text{ Btu/s} = 78{,}660 \text{ hp}$$

The propulsive efficiency of the engine is

$$\eta_P = \frac{TV_{\text{engine}}}{\dot{Q}_H} = \frac{20{,}910 \text{ hp}}{78{,}660 \text{ hp}} = \underline{\underline{26.6 \text{ percent}}}$$

7.12 IDEAL BRAYTON CYCLE WITH INTERCOOLING AND REHEATING

In Chapter 1 the expression for the work required to compress a substance in a closed system while following a reversible path was shown to be

$$w_{12} = \int_1^2 P dv \tag{7.39}$$

An expression similar to Equation 7.39 can be developed for an internally reversible process that occurs in a steady-flow, open system containing a single inlet and exit. For this situation the conservation-of-energy equation written in differential form is

$$\delta q - \delta w - dh - de_k - de_p = 0 \tag{7.40}$$

When changes in kinetic and potential energy can be neglected, Equation 7.40 can be further simplified and rearranged to

$$\delta w = \delta q - dh \tag{7.41}$$

From Equation 5.13 the heat transfer for an internally reversible process can be expressed in terms of the entropy change, or

$$\delta q_{\text{int rev}} = T\, ds \tag{7.42}$$

Substituting the second $T\,ds$ equation (Equation 5.20) for dh results in

$$\delta w = T\, ds - (T\, ds + v\, dP) = -v dP \tag{7.43}$$

Integrating this expression between the state of the fluid at the exit and inlet for the steady-flow system gives

$$w_{ie} = -\int_i^e v dP \tag{7.44}$$

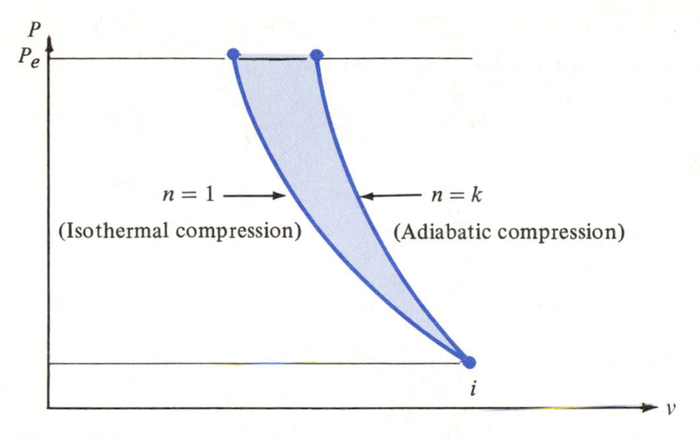

Figure 7.24 Compression of a gas along isothermal and isentropic paths.

The restrictions placed on Equation 7.44 can be summarized as follows: It applies to an internally reversible, steady-flow process in an open system with a single inlet and exit in which changes in kinetic and potential energy are negligible.

Geometrically, Equation 7.44 states that the area between the P-v process curve and the P axis numerically represents the work per unit mass during the internally reversible, steady-flow process.

The concept of a polytropic process in which the path of an ideal gas is given by Equation 2.42 now takes on added importance. Suppose that an ideal gas is compressed steadily in an internally reversible process and changes in kinetic and potential energies can be neglected. The inlet state is fixed by the state i shown on the P-v diagram in Figure 7.24, and the exit pressure of the gas is P_e. Two different paths are to be examined for the proposed compression process. The first path is an adiabatic process for which the polytropic exponent is equal to k, and the second path is isothermal for which $n = 1$. Application of Equation 7.44 shows that the choice of the isothermal process requires less work to achieve the compression process than the adiabatic process. Furthermore, the difference between the amount of work required to compress the gas isothermally and the work for adiabatic compression is equal to the shaded area in Figure 7.24. These concepts are illustrated in the next two example problems.

EXAMPLE 7.8

Air enters a steady-flow compressor at 101 kPa and 25°C, where it undergoes an internally reversible compression to 800 kPa. Assume constant specific heats and neglect changes in kinetic and potential energy during the compression process. Calculate the work per unit mass required to compress the air and the temperature of the air as it leaves the compressor for the following:

(a) An isentropic-compression process
(b) An isothermal-compression process

Solution.

(a) For the conditions stated in the problem, Equation 7.44 can be used to determine the work per unit mass of air:

$$w_{ie} = -\int_i^e v\,dP$$

The path followed during an isentropic process of an ideal gas with constant specific heats is

$$Pv^k = \text{constant}$$

Substitution followed by integration gives

$$\begin{aligned} w_{ie} &= -\int_i^e \left(\frac{\text{constant}}{P}\right)^{1/k} dP \\ &= -(\text{constant})^{1/k}\left[\frac{P_e^{(k-1)/k} - P_i^{(k-1)/k}}{(k-1)/k}\right] \\ &= -\frac{k(P_e v_e - P_i v_i)}{k-1} = -\frac{kR(T_e - T_i)}{k-1} \end{aligned}$$

The exit temperature can be determined from the isentropic relations for an ideal gas with constant specific heats. With $k = 1.4$ for air,

$$T_e = T_i\left(\frac{P_e}{P_i}\right)^{(k-1)/k} = (298\text{ K})\left(\frac{800\text{ kPa}}{101\text{ kPa}}\right)^{0.2857} = 538\text{ K} = \underline{\underline{265°\text{C}}}$$

The work per unit mass of the air is then

$$w_{ie} = -\frac{1.4(0.287\text{ kJ/kg·K})(538 - 298)\text{K}}{0.4} = \underline{\underline{-241.1\text{ kJ/kg}}}$$

(b) For an isothermal process of an ideal gas, the pressure and specific volume obey the path equation:

$$Pv = \text{constant}$$

The work per unit mass is

$$\begin{aligned} w_{ie} &= -\int_i^e v\,dP = -\int_i^e \frac{(\text{constant})\,dP}{P} = -(\text{constant})\ln\left(\frac{P_e}{P_i}\right) \\ &= -Pv\ln\left(\frac{P_e}{P_i}\right) = -RT\ln\left(\frac{P_e}{P_i}\right) \\ &= -(0.287\text{ kJ/kg·K})(298\text{ K})\ln\left(\frac{800\text{ kPa}}{101\text{ kPa}}\right) = \underline{\underline{-177\text{ kJ/kg}}} \end{aligned}$$

The exit-air temperature is

$$T_e = T_i = 298\text{ K} = \underline{\underline{25°\text{C}}}$$

The work required to compress the air isothermally is 27 percent less than that for an isentropic-compression process. ■

■ EXAMPLE 7.9

Hydrogen is compressed reversibly in a steady-flow system along a polytropic path for which the polytropic exponent is equal to 1.25. The inlet state of the hydrogen is 30 psia, 85°F and the exit pressure is 135 psia. Calculate the work per unit mass during the process and the exit temperature of the hydrogen.

Solution. The work required per unit mass to compress an ideal gas in an open system along a polytropic path is given by Equation 7.44:

$$w_{ie} = -\int_i^e v\, dP = \frac{-n(P_e v_e - P_i v_i)}{n-1} = -\frac{nR(T_e - T_i)}{n-1}$$

The exit temperature is determined from Equation 2.45:

$$T_e = T_i \left(\frac{P_e}{P_i}\right)^{(n-1)/n} = (545°\text{R}) \left(\frac{135\ \text{psia}}{30\ \text{psia}}\right)^{(1.25-1)/1.25} = 736.3°\text{R}$$

The work per unit mass during the process is therefore

$$w_{ie} = \frac{-1.25(0.9851\ \text{Btu/lb}_\text{m}\cdot°\text{R})(736.3 - 545)°\text{R}}{1.25 - 1} = \underline{\underline{-942\ \text{Btu/lb}_\text{m}}}$$ ■

Equation 7.44 coupled with the results shown in Figure 7.24 illustrate the fact that an isothermal compression of an ideal gas requires less work than an adiabatic-compression process, assuming both processes are steady-flow, internally reversible processes. This result suggests that cooling of the gas as it is compressed will reduce the energy requirements of the compression process. While cooling during compression is desirable, cooling in an amount necessary to maintain isothermal conditions is difficult to achieve in practice. Adequate cooling is particularly difficult to achieve if the gas remains in the compressor during the cooling process. However, if the gas is compressed in stages and between each stage the gas is circulated through a heat exchanger called an *intercooler*, then sufficient heat transfer can take place from the gas to reduce the power required to compress the gas.

The path followed by a gas that is compressed in a two-stage compression process separated by an intercooler is shown in Figure 7.25. The first-stage compression of the gas occurs along the polytropic path 1-2. The gas is then cooled at constant pressure in an intercooler along path 2-3. The temperature at state 3 is assumed to be equal to the initial temperature at state 1. The gas is then compressed in a second stage of the compressor from state 3 to 4 along a polytropic path. For comparison purposes the single-stage polytropic compression is illustrated as path 1-2-6 and the single-stage isothermal compression is shown as path 1-3-5. The shaded area on the P-v diagram represents the work saved by means of the two-stage compression separated with intercooling.

The areas on the T-s diagram in Figure 7.25 graphically illustrate the heat transfer that must occur from the gas during the internally reversible compression process. For compression with intercooling the heat transfer must be equal to the area under curve 1-

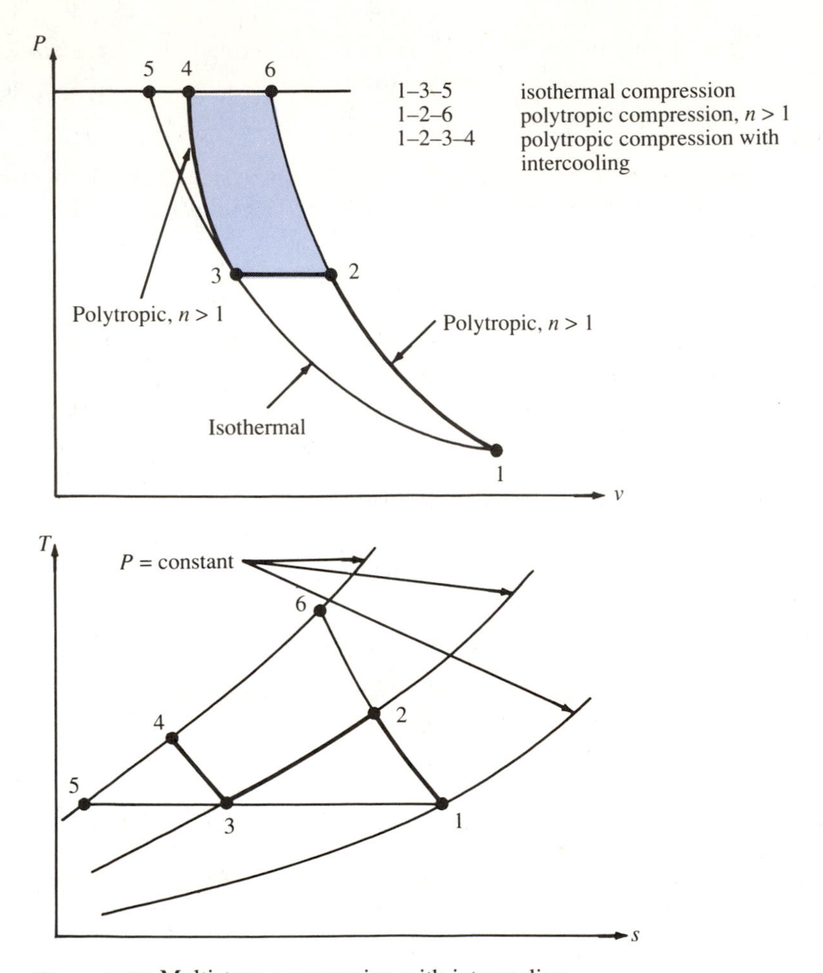

Figure 7.25 Multistage compression with intercooling.

2-3-4. If the compression process is assumed to occur isothermally, the heat transfer must be equal to the area under path 1-3-5 on the *T-s* diagram in Figure 7.25.

One way to increase the efficiency of a Brayton cycle is to reduce the work required for the compressor. This reduction can be achieved by the use of multistage compression with intercooling. Another way to increase the thermal efficiency of a Brayton cycle would be to increase the output of the turbine. The latter goal can be reached by expanding the gas in separate turbine stages with a heating process called *reheat* occurring between each expansion stage. A simplified, ideal, two-stage expansion process with a single reheat process is shown in Figure 7.26. The gas enters the first-stage turbine at state 3 after leaving the combustion chamber and thereafter expands isentropically to state 4. The gas then is reheated in a second combustor at constant pressure, reaching state 5 before expanding isentropically to state 6 in the second-stage turbine.

The use of intercooling, reheating, and regeneration will complement each other in improving the thermal efficiency of a Brayton cycle. Intercooling decreases the temper-

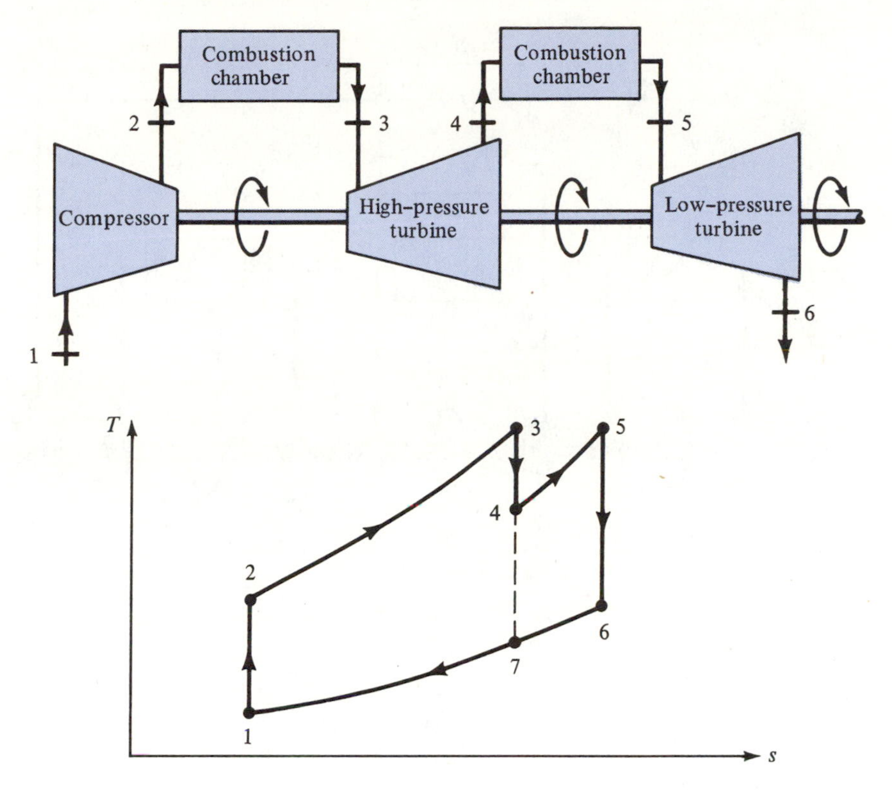

Figure 7.26 Ideal Brayton cycle modified with the reheat process.

ature of the gas that leaves the compressor. This feature can be seen in Figure 7.25, where T_4, the temperature of the gas after intercooling, is less than T_6, the temperature without intercooling. Following similar reasoning, the temperature of the gas leaving the turbine is greater with reheating than without, as shown in Figure 7.26, where $T_6 > T_7$. These trends also make the regeneration process more attractive because reducing the temperature of the compressor exhaust and increasing the temperature of the turbine exhaust provide greater potential for heat transfer in the regenerator.

A simple example of an ideal Brayton cycle modified with intercooling, reheating, and regeneration is shown in Figure 7.27, where two stages of compression and expansion are indicated. If the number of intercooling and reheat stages is very large, the ideal cycle approaches the Ericsson cycle discussed in Section 7.6, and the thermal efficiency would approach the theoretical maximum value. However, the use of many intercoolers and reheaters is economically impractical, and more than two or three are rarely used.

■ EXAMPLE 7.10

An ideal Brayton cycle is modified with intercooling, reheating, and regeneration. Air enters the first-stage compressor at 27°C and 100 kPa. Both stages of the compressor and the turbine have a pressure ratio of 3. The temperature of the air as it enters the turbine is 727°C. The efficiency of the regenerator is 100 percent. Determine the thermal efficiency of the cycle and compare it with the efficiency of a simple, ideal Brayton cycle

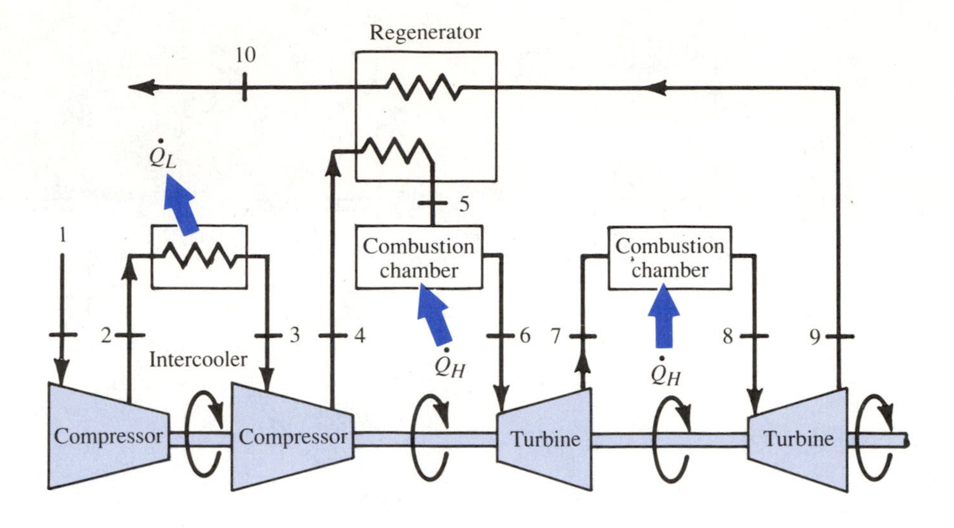

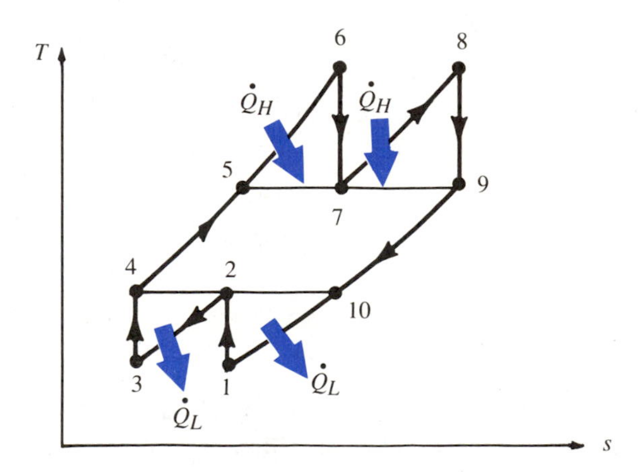

Figure 7.27 Ideal Brayton cycle with reheat, regeneration, and intercooling.

with a single-stage compressor and turbine, both with a pressure ratio of 9 and no regeneration. Use the air-standard assumptions. Assume that heat transfer to each cycle is from a reservoir that has a temperature of 1100 K and that heat transfer from each cycle is to the surroundings at 20°C. Compare the irreversibilities of the processes for both cycles.

Solution. For a simple cycle with no intercooling, reheating, or regeneration and with reference to Figure 7.12, the given information is

$$T_1 = 300 \text{ K} \qquad P_1 = 100 \text{ kPa}$$

$$P_2 = 900 \text{ kPa} \qquad T_3 = 1000 \text{ K}$$

$$\frac{P_2}{P_1} = \frac{P_3}{P_4} = 9$$

The compression process 1-2 is isentropic, so the temperature at state 2 can be determined.

$$P_{r2} = P_{r1}\left(\frac{P_2}{P_1}\right) = 1.38(9) = 12.4$$

$$T_2 = 557 \text{ K}$$

The enthalpies from Table D.1 are

$$h_1 = 300.4 \text{ kJ/kg}$$

$$h_2 = 562.3 \text{ kJ/kg}$$

$$h_3 = 1046.2 \text{ kJ/kg}$$

The turbine process is isentropic, so the temperature and enthalpy at state 4 are

$$P_{r4} = P_{r3}\left(\frac{P_4}{P_3}\right) = \frac{113.5}{9} = 12.6$$

$$T_4 = 560 \text{ K}$$

$$h_4 = 565.4 \text{ kJ/kg}$$

The thermal efficiency of the simple cycle, which is composed of steady-flow components, is

$$\eta = \frac{w_t + w_c}{q_H} = \frac{(h_3 - h_4) + (h_1 - h_2)}{h_3 - h_2}$$

$$= \frac{(1046.2 - 565.4 + 300.4 - 562.6)\text{kJ/kg}}{(1046.2 - 562.6)\text{kJ/kg}} = \underline{\underline{0.452}}$$

A second-law analysis of the simple cycle can be performed with the same procedure used in Example 7.5 to examine the performance of each component. The irreversibility of each component is given by

$$i = T_0\left[(s_e - s_i) + \frac{q_k}{T_k}\right]$$

and the entropy change of each process is calculated from

$$s_e - s_i = s_e^\circ - s_i^\circ - R \ln\frac{P_e}{P_i}$$

The results of the complete analysis are summarized in Table 7.5.

For an ideal cycle including intercooling, reheating, and regeneration and with reference to Figure 7.27, the given information is

$$T_1 = T_3 = 300 \text{ K}$$

$$T_6 = T_8 = 1000 \text{ K}$$

$$\frac{P_2}{P_1} = \frac{P_4}{P_3} = 3$$

$$\frac{P_6}{P_7} = \frac{P_8}{P_9} = 3$$

TABLE 7.5 IDEAL BRAYTON CYCLE
T_H = 1100 K, T_L = T_0 = 293 K

Process	q, kJ/kg	w, kJ/kg	$h_e - h_i$, kJ/kg	$s_e - s_i$, kJ/kg·K	i, kJ/kg
1-2	0.0	−262.2	262.2	0.0	0.0
2-3	483.6	0.0	483.6	0.636	57.5
3-4	0.0	480.8	−480.8	0.0	0.0
4-1	−265.0	0.0	−265.0	−0.636	78.6
Cycle	218.6	218.6	0.0	0.0	136.1

Enthalpy values from Table D.1 are

$$h_1 = h_3 = 300.4 \text{ kJ/kg}$$

$$h_6 = h_8 = 1046.2 \text{ kJ/kg}$$

Properties at state 2 are

$$P_{r2} = P_{r1}\left(\frac{P_2}{P_1}\right) = 1.38(3) = 4.14$$

$$T_2 = 410 \text{ K}$$

$$h_2 = h_4 = 411.5 \text{ kJ/kg}$$

The turbine processes are isentropic, so that

$$P_{r7} = P_{r6}\left(\frac{P_7}{P_6}\right) = \frac{113.5}{3} = 37.8$$

$$T_7 = T_9 = 753 \text{ K}$$

$$h_7 = h_9 = 771 \text{ kJ/kg}$$

For a regenerator with an efficiency of 100 percent,

$$T_4 = T_{10} \qquad T_5 = T_9$$

The thermal efficiency of the cycle is

$$\eta = \frac{w_t + w_c}{q_H} = \frac{(h_6 - h_7) + (h_8 - h_9) + (h_1 - h_2) + (h_3 - h_4)}{(h_6 - h_5) + (h_8 - h_7)}$$

$$= \frac{[2(1046.2 - 771) + 2(300.4 - 411.5)]\text{kJ/kg}}{[1046.2 - 771) + (1046.2 - 771)]\text{kJ/kg}} = 0.596$$

$$= \underline{\underline{59.6 \text{ percent}}}$$

The thermal efficiency of the Brayton cycle with intercooling, reheating, and regeneration is 14.4 percentage points higher than the thermal efficiency of the simple Brayton cycle.

A second-law analysis of the modified cycle can be used to more fully explain the impact of the modifications on the ideal Brayton cycle. The analysis is identical in form to that used for the simple cycle except as it is applied to the regenerator. The details of the regenerator analysis are included here where we consider the more general case of a regenerator with an efficiency denoted by η_{regen}. Applying the conservation-of-energy equation for steady flow to the adiabatic regenerator, we find

$$0 = \dot{m}_5 h_5 + \dot{m}_{10} h_{10} - \dot{m}_4 h_4 - \dot{m}_9 h_9$$

where the changes in kinetic and potential energies have been neglected. Since the fluid streams do not mix,

$$\dot{m}_5 = \dot{m}_4 = \dot{m} \quad \text{and} \quad \dot{m}_9 = \dot{m}_{10} = \dot{m}$$

Therefore

$$0 = (h_5 - h_4) + (h_{10} - h_9)$$

The irreversibility rate associated with the regenerator is evaluated by using Equation 6.47, which reduces to

$$\dot{I}_{\text{regen}} = T_0(\dot{m}_5 s_5 + \dot{m}_{10} s_{10} - \dot{m}_4 s_4 - \dot{m}_9 s_9)$$

or

$$i_{\text{regen}} = T_0[(s_5 - s_4) + (s_{10} - s_9)]$$

If pressure drops in the regenerator are zero, as they are for the ideal cycle, then from Equation 5.28

$$s_5 - s_4 = s_5^\circ - s_4^\circ \quad \text{and} \quad s_{10} - s_9 = s_{10}^\circ - s_9^\circ$$

Therefore

$$i_{\text{regen}} = T_0[(s_5^\circ - s_4^\circ) + (s_{10}^\circ - s_9^\circ)]$$

Recall that the s° values for an ideal gas depend only on temperature.

The general effect of the regenerator efficiency on the irreversibility can best be examined if we consider a differential length of the regenerator. For this purpose we use the subscript a to denote the properties of the high-pressure fluid stream that follows path 4-5 and the subscript b to denote the properties of the low-pressure fluid stream that follows path 9-10. If we write an equation for the irreversibility for the differential length, we obtain

$$\delta i = T_0[(ds)_a + (ds)_b]$$

This equation can be written in terms of the enthalpy change by substituting the second $T\,ds$ equation.

$$T\,ds = dh - v\,dP$$

Since the processes are assumed to occur at constant pressure, this equation reduces to

$$T\,ds = dh$$

and the irreversibiity equation can be written as

$$\delta i = T_0 \left[\left(\frac{dh}{T} \right)_a + \left(\frac{dh}{T} \right)_b \right]$$

Next, the conservation-of-energy equation is applied to the differential length, with the result

$$0 = (dh)_a + (dh)_b$$

or

$$(dh)_a = -(dh)_b$$

Thus the irreversibility can be expressed as

$$\delta i = T_0 (dh)_a \left(\frac{1}{T_a} - \frac{1}{T_b} \right)$$

For the regenerator the temperature of fluid stream a increases, so that $(dh)_a > 0$. Furthermore, the temperature of fluid stream b is everywhere higher than the temperature of stream a if the regenerator efficiency is less than 100 percent. Therefore

$$\frac{1}{T_a} - \frac{1}{T_b} > 0$$

and we conclude that the irreversibility is positive for regenerators having efficiencies less than 100 percent.

If the regenerator efficiency is 100 percent, then the temperatures of the two fluid streams are everywhere infinitesimally close to each other,

$$\left(\frac{1}{T_a} - \frac{1}{T_b} \right) \rightarrow 0$$

and the irreversibility in the regenerator is zero.

In other words, a regenerator with an efficiency of 100 percent is totally reversible. Efficiencies less than 100 percent result in irreversibilities. For a regenerator to have an efficiency of 100 percent, the fluid streams flow in opposite directions (counterflow) and, in addition, the fluid temperatures would be infinitesimally close to one another throughout the regenerator. Therefore, the heat transfer between the two fluid streams would be totally reversible. For lesser efficiencies a finite temperature difference would exist between the two fluid streams, rendering the process irreversible.

The regenerator in this example is assumed to have an efficiency of 100 percent. Therefore, the net entropy change for the adiabatic-regenerator process is zero, and the regenerator has a zero irreversibility. The characteristics of the cycle are summarized in Table 7.6.

Comparing the operation of the modified cycle and the simple cycle, we find that the irreversibility per unit mass has been reduced by about one-half. The most significant improvement is associated with the cooling process. The effect of regeneration and intercooling is to provide an overall decrease in the average temperature during the

TABLE 7.6 IDEAL BRAYTON CYCLE WITH INTERCOOLING AND REGENERATION

T_H = 1100 K, T_L = T_0 = 293 K

Process	q, kJ/kg	w, kJ/kg	$h_e - h_i$, kJ/kg	$s_e - s_i$, kJ/kg·K	i, kJ/kg
1-2	0.0	−111.1	111.1	0.0	0.0
2-3	−111.1	0.0	−111.1	−0.3153	18.7
3-4	0.0	−111.1	111.1	0.0	0.0
5-6	275.2	0.0	275.2	0.3153	19.1
6-7	0.0	275.2	−275.2	0.0	0.0
7-8	275.2	0.0	275.2	0.3153	19.1
8-9	0.0	275.2	−275.2	0.0	0.0
Regenerator	0.0	0.0	0.0	0.0	0.0
10-1	−111.1	0.0	−111.1	−0.3153	18.7
Cycle	328.2	328.2	0.0	0.0	75.6

cooling process, resulting in a decrease in the irreversibility of the process. The effect of regeneration and reheating is to increase the average temperature of the fluid during the heating process, thereby decreasing the irreversibility. ■

7.13 IDEAL GAS REFRIGERATION CYCLE

A gas power cycle such as the Brayton cycle can be reversed to obtain a *gas refrigeration cycle*. A reversed Brayton power cycle can be used as a refrigeration cycle in which the compressor produces high-temperature air that is cooled in a heat exchanger by heat transfer to the surroundings. The gas is further cooled during expansion through a turbine. It is then passed through a heat exchanger, where its temperature is increased by heat transfer from the refrigerated space.

In the ideal Brayton refrigeration cycle the compression and expansion processes are assumed to be isentropic, while the heat exchangers are assumed to be internally reversible, constant-pressure devices. A schematic of the cycle components and process diagrams for an ideal, Brayton refrigeration cycle is shown in Figure 7.28.

The coefficient of performance for the Brayton refrigeration cycle shown in Figure 7.28 is

$$\beta_R = -\frac{\dot{Q}_L}{\dot{W}_{\text{net}}} = -\frac{\dot{Q}_L}{\dot{W}_T + \dot{W}_c} \tag{7.45}$$

If changes in potential and kinetic energies are neglected, the heat-transfer rates in the heat exchangers, the power output of the turbine, and the power input to the compressor can be expressed in terms of the enthalpy changes of the working fluid, so that the expression for β_R is

$$\beta_R = -\frac{h_1 - h_4}{(h_3 - h_4) + (h_1 - h_2)} = \frac{1}{[(h_2 - h_3)/(h_1 - h_4)] - 1} \tag{7.46}$$

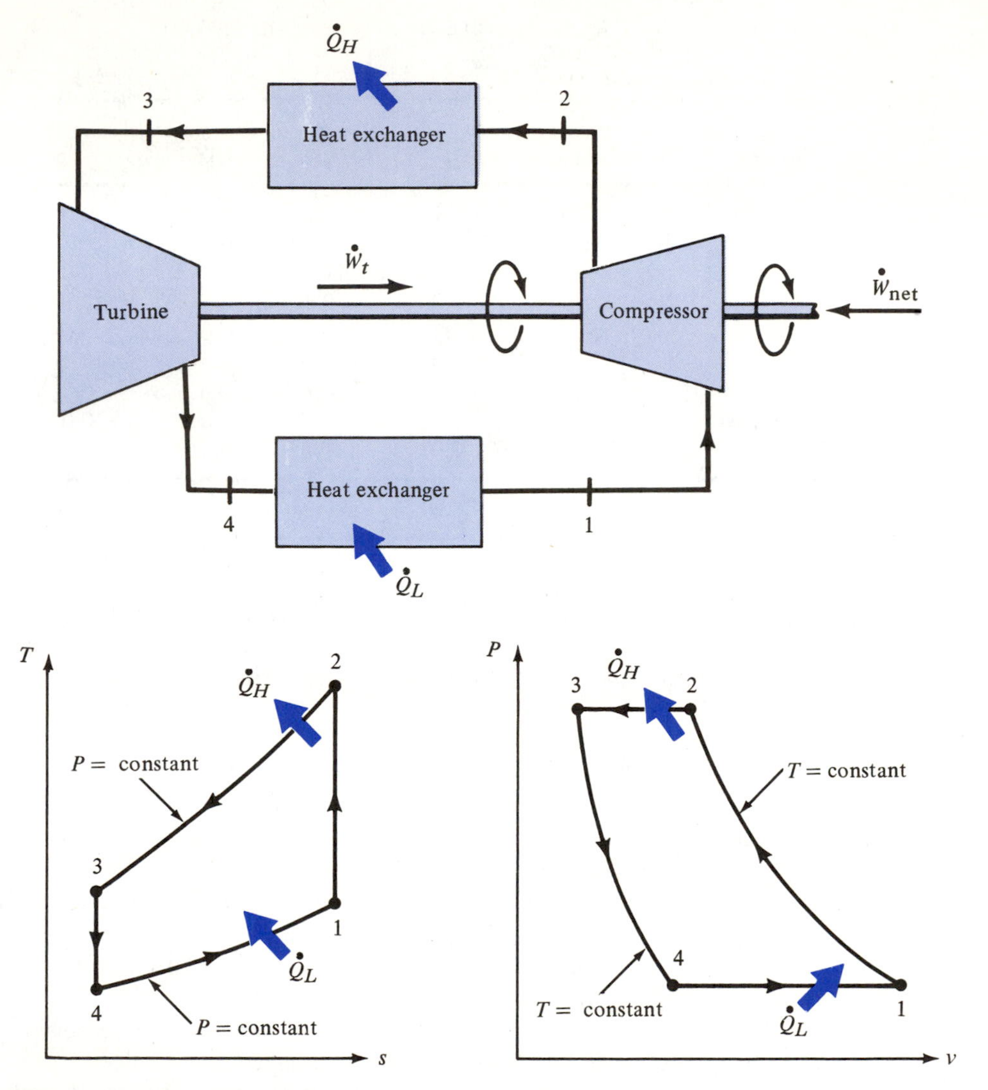

Figure 7.28 Schematic and process diagrams for an ideal Brayton refrigeration cycle.

With the cold air-standard assumptions the COP of an ideal Brayton refrigeration cycle used as a refrigerator is

$$\beta_R = \frac{1}{[(T_2 - T_3)/(T_1 - T_4)] - 1} \tag{7.47}$$

When a Brayton refrigeration cycle is used as a heat pump, the desired effect is the heat transfer from the cycle, or $\dot{Q}_H$. The COP of the cycle is then

$$\beta_H = \frac{\dot{Q}_H}{\dot{W}_{\text{net}}} = \frac{\dot{Q}_{23}}{\dot{Q}_{\text{net}}} = \frac{\dot{Q}_{23}}{\dot{Q}_{23} + \dot{Q}_{41}} \tag{7.48}$$

If changes in kinetic and potential energies are neglected, the expression for β_H can be reduced to

$$\beta_H = \frac{h_3 - h_2}{(h_3 - h_2) + (h_1 - h_4)} \tag{7.49}$$

With the cold air-standard assumptions the coefficient of performance of an ideal Brayton refrigeration cycle used as a heat pump becomes simply a function of the temperatures of the four states shown in Figure 7.28, or

$$\beta_H = \frac{1}{1 + [(T_1 - T_4)/(T_3 - T_2)]} \tag{7.50}$$

One of the major disadvantages of the Brayton refrigeration cycle is the fact that the gas temperature changes significantly in both heat exchangers, as can be seen in the process diagram in Figure 7.28. The temperature variations necessitate a deviation from the Carnot refrigeration cycle for which heat transfer with a reservoir occurs at a constant temperature. As a result, the coefficient of performance of the Brayton refrigeration cycle is relatively low when compared with the vapor-compression refrigeration cycle, which will be discussed in the next chapter. Because of its inherently low COP, the Brayton refrigeration cycle sees limited applications, although it does possess an advantage of having lighter components than other refrigeration cycles. Gas refrigeration devices have the ability to produce greater cooling capacities per unit mass of the components. As a result, the Brayton refrigeration cycle is commonly used when weight is a primary consideration, as in aircraft air-conditioning systems.

EXAMPLE 7.11

An ideal Brayton refrigeration cycle uses air as a working fluid. The pressure ratio of the compressor is 4:1. The inlet conditions to the compressor are $P_1 = 14.7$ psia and $T_1 = 100°F$. The temperature at the outlet of the turbine is 25°F. Use the air-standard assumptions and calculate the following:

(a) The heat transfer per unit mass in both heat exchangers
(b) The work per unit mass in both the compressor and the turbine
(c) The cycle coefficient of performance

Solution. The heat transfer and work can be calculated once the enthalpy at all four states in Figure 7.28 are determined. Using properties in Table D.1E, we have

$$h_1 = 134.0 \text{ Btu/lb}_m \qquad P_{r1} = 1.563$$

Process 1-2 is an isentropic path, so

$$P_{r2} = P_{r1}\left(\frac{P_2}{P_1}\right) = 1.563(4) = 6.252$$

$$T_2 = 829°\text{R} = 369°\text{F} \qquad h_2 = 199 \text{ Btu/lb}_m$$

Using properties from Table D.1E at the known value of T_4, we have

$$T_4 = 485°\text{R} \qquad h_4 = 116 \text{ Btu/lb}_m \qquad P_{r4} = 0.953$$

Process 3-4 is also an isentropic process, so

$$P_{r3} = P_{r4}\left(\frac{P_3}{P_4}\right) = P_{r4}\left(\frac{P_2}{P_1}\right) = 0.953(4) = 3.812$$

$$T_3 = 721°\text{R} = 261°\text{F} \qquad h_3 = 172.8\ \text{Btu/lb}_\text{m}$$

(a) The heat transfer from the air to the surroundings is

$$q_H = h_3 - h_2 = (172.8 - 199)\text{Btu/lb}_\text{m} = \underline{\underline{-26.2\ \text{Btu/lb}_\text{m}}}$$

The heat transfer to the air from the low-temperature reservoir is

$$q_L = h_1 - h_4 = (134 - 116)\text{Btu/lb}_\text{m} = \underline{\underline{18\ \text{Btu/lb}_\text{m}}}$$

(b) The work in the compressor and turbine is

$$w_c = h_1 - h_2 = (134 - 199)\text{Btu/lb}_\text{m} = \underline{\underline{-65\ \text{Btu/lb}_\text{m}}}$$

and

$$w_t = h_3 - h_4 = (172.8 - 116)\text{Btu/lb}_\text{m} = \underline{\underline{56.8\ \text{Btu/lb}_\text{m}}}$$

(c) The coefficient of performance for the cycle is

$$\beta = -\frac{q_L}{w_\text{net}} = -\frac{q_L}{w_t + w_c}$$

$$= \frac{-(18.0\ \text{Btu/lb}_\text{m})}{(56.8 - 65)\text{Btu/lb}_\text{m}}$$

$$= \underline{\underline{2.2}}$$

■

7.14 ACTUAL GAS CYCLES

The ideal gas cycles considered in the preceding sections consist of internally reversible processes. The behavior of actual cycles, composed of components with varying degrees of internal irreversibilities, can only approach the behavior of the ideal cycles. This section investigates actual cycles in an attempt to determine how component irreversibilities influence the performance of gas cycles.

There are two major sources of irreversibilities present in systems that execute thermodynamic cycles. The first results from friction, primarily friction caused by motion of the working fluid. The second major source of irreversibility is associated with heat transfer through a finite temperature difference. Fluid friction always accompanies motion of the working fluid. As the fluid is circulated through a pipe, heat exchanger, boiler, turbine, or any other system component, fluid viscosity tends to retard the motion, thereby producing a pressure drop. To maintain the motion of the fluid, an external source of energy such as a pump or compressor must be used to overcome the overall pressure drop produced by the fluid friction.

The friction represents a source of irreversibility and, by itself, will always produce an increase in the entropy of the fluid. Furthermore, fluid friction causes the availability of the fluid to decrease, and therefore design steps should be taken to reduce possible sources of friction. Fluid-flow components should be designed with gradual changes in cross-sectional area, while sharp corners and rough surfaces should be eliminated. Heat-exchanger tubes should be designed with large areas so that the detrimental effects of fluid friction are minimized. As with any design, however, there are trade-offs that must be considered. From the standpoint of minimizing irreversibilities, a heat exchanger having a large heat-transfer surface area would provide thermodynamic benefits by reducing the temperature difference between the two fluids exchanging energy. On the other hand, the size of a heat exchanger is not unlimited because component size and cost must be considered in the overall design. The final design will evolve after a number of factors, including initial investment, operating costs, size, weight, and thermal efficiency, have been considered.

While fluid friction always produces an increase in entropy, heat transfer can cause the entropy of a fluid to either increase or decrease, depending on the direction of the heat transfer. If heat transfer is the only source of irreversibility, heat transfer from a system to the surroundings will cause the entropy of a system to decrease. Heat transfer to a system from the surroundings will produce an increase in the entropy of the system. When both fluid friction and heat transfer occur, the entropy of the working fluid could either increase or decrease. In a steam turbine, for example, the temperature of the working fluid exceeds that of the surroundings, and heat transfer from the steam to the surroundings occurs. The heat transfer represents a lost opportunity for producing useful work with the steam. In an effort to reduce irreversibilities, the turbine casing is usually insulated to minimize the heat transfer to the surroundings. The presence of the insulation on the turbine can effectively reduce the heat transfer from the turbine to a negligible amount compared with the enthalphy change experienced by the steam between the inlet and exit to the turbine. The frictional effects that occur within the turbine cannot be entirely eliminated, and their contribution to the irreversibility of the turbine process is often more significant than that caused by heat transfer to the surroundings.

Irreversibilities that occur in pumps, compressors, and other fluid-handling devices are similar to those described for the turbine. The main source of irreversibilities is often fluid friction. Heat transfer with the surroundings usually is a minor loss, unless the temperature difference between the system and surroundings becomes large and the thermal resistance at the boundary becomes very small.

Some of the sources of irreversibilities can be accounted for by using the concepts of component efficiencies introduced in Section 6.6. For example, consider a Brayton cycle consisting of a nonideal compressor and a nonideal turbine. Friction in the compressor and turbine causes the entropy of the working fluid to increase. The amount of entropy change in each of these devices can be determined with the aid of the adiabatic-compressor and turbine efficiencies.

The adiabatic-compressor efficiency for a simple adiabatic compressor, given in Equation 6.57, is

$$\eta_c = \frac{\dot{W}_s}{\dot{W}_{\text{act}}} \simeq \frac{h_{e,s} - h_i}{h_{e,a} - h_i} \tag{7.51}$$

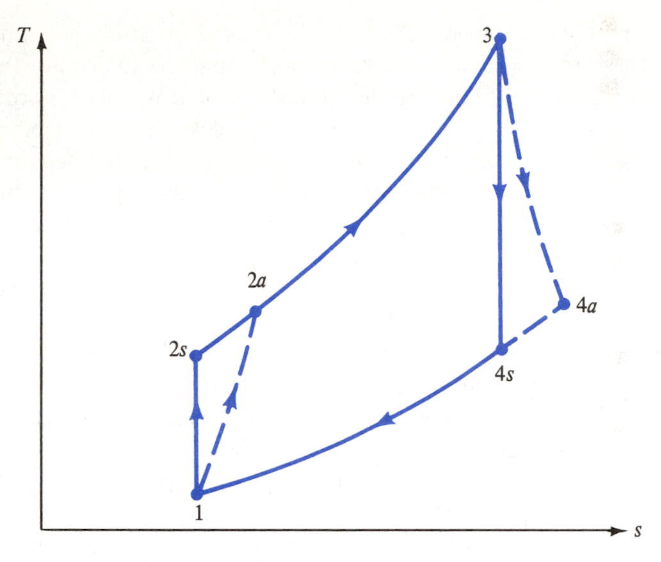

Figure 7.29 Influence of compressor and turbine irreversibilities on the Brayton cycle.

where the subscript s denotes the isentropic process and the subscript a denotes the actual, irreversible process. The adiabatic-turbine efficiency for a simple adiabatic turbine is given by Equation 6.54,

$$\eta_T = \frac{\dot{W}_{\text{act}}}{\dot{W}_s} \simeq \frac{h_i - h_{e,a}}{h_i - h_{e,s}} \tag{7.52}$$

The T-s diagram for an ideal Brayton cycle, which consists of internally reversible processes, is shown in Figure 7.29 as a solid line. Deviations from the ideal cycle caused by irreversibilities in the compressor and turbine are shown in the same figure as dashed lines. The next example illustrates the analysis of cycles consisting of nonideal or actual processes.

■ EXAMPLE 7.12

The Brayton cycle described in Example 7.5 has a compressor with an adiabatic efficiency of 84 percent and a turbine with an adiabatic efficiency of 87 percent. Neglect heat transfer in both the compressor and turbine. Apply the air-standard assumptions and calculate the following quantities:

(a) The work per unit mass required by the compressor
(b) The work per unit mass produced by the turbine
(c) The thermal efficiency of the cycle
(d) The irreversibility of each process and the irreversibility of the cycle

Solution. The T-s diagrams for both the ideal and the actual Brayton cycles are shown in Figure 7.29. The enthalpies at the four states of the ideal cycle were determined in Example 7.5:

$$h_1 = 310.5 \text{ kJ/kg} \qquad h_3 = 1554 \text{ kJ/kg}$$

$$h_{2s} = 630.7 \text{ kJ/kg} \qquad h_{4s} = 789.3 \text{ kJ/kg}$$

(a) The work per unit mass required to compress the air along an isentropic path, as calculated in Example 7.5, is

$$w_s = -320.2 \text{ kJ/kg}$$

The actual work required by the compressor can be determined by using the definition of the adiabatic-compressor efficiency, Equation 7.51:

$$w_{c,\text{act}} = \frac{w_s}{\eta_c} = \frac{-320.2 \text{ kJ/kg}}{0.84} = \underline{\underline{-381.2 \text{ kJ/kg}}}$$

(b) The work per unit mass developed by the isentropic turbine was calculated in Example 7.5:

$$w_s = 764.7 \text{ kJ/kg}$$

and the actual work developed by the turbine can be calculated by using Equation 7.52:

$$w_{T,\text{act}} = \eta_T w_s = 0.87(764.7 \text{ kJ/kg}) = \underline{\underline{665.3 \text{ kJ/kg}}}$$

The enthalpy at state $2a$ can be determined from the definition of the adiabatic-compressor efficiency:

$$\eta_c = \frac{h_{2s} - h_1}{h_{2a} - h_1}$$

or

$$0.84 = \frac{(630.7 - 310.5)\text{kJ/kg}}{(h_{2a} - 310.5)\text{kJ/kg}}$$

Solving for h_{2a} gives

$$h_{2a} = 691.7 \text{ kJ/kg}$$

The heat transfer per unit mass in the combustion process is

$$q_{2a-3} = h_3 - h_{2a} = 1554 \text{ kJ/kg} - 691.7 \text{ kJ/kg} = 862.3 \text{ kJ/kg}$$

(c) The thermal efficiency of the cycle is

$$\eta_{\text{th}} = \frac{w_{\text{net}}}{q_{2a-3}} = \frac{(665.3 - 381.2)\text{kJ/kg}}{862.3 \text{ kJ/kg}} = 0.33 = \underline{\underline{33 \text{ percent}}}$$

The thermal efficiency of the ideal Brayton cycle considered in Example 7.5 was 48.1 percent. The thermal efficiency in this problem is significantly less because irreversibilities are present in the compressor and turbine. Further reductions in the thermal efficiency would result if pressure drops in the combustion and exhaust processes were included.

(d) The component and cycle irreversibilities are obtained from a second-law analysis. The irreversibility for each process is determined from Equation 6.47, which reduces to

$$i = T_0\left[(s_e - s_i) + \frac{q_k}{T_k}\right]$$

and the entropy change during each process is calculated from Equation 5.28:

$$s_e - s_i = s_e^\circ - s_i^\circ - R\ln\frac{P_e}{P_i}$$

Furthermore, the change in availability for each process, from Equation 6.32, is given by

$$\psi_e - \psi_i = (h_e - h_i) - T_0(s_e - s_i)$$

The requisite properties for the actual states are summarized in the table below.

State	T, K	h, kJ/kg	s°, kJ/kg·K	Note
1	310	310.5	5.7338	$P_{2a}/P_1 = 12$
$2a$	679.7	691.7	6.5409	
3	1433	1554	7.3889	$P_{4a}/P_3 = 1/12$
$4a$	860.2	888.7	6.7978	

These property values were used to obtain the data from the cycle analysis, which are summarized in Table 7.7.

If we compare the results of this analysis with the analysis of the ideal cycle in Example 7.5, then a number of interesting conclusions can be drawn. First, the turbine and compressor processes for the actual cycle are irreversible because

TABLE 7.7 IDEAL BRAYTON CYCLE
T_H = 1573 K, T_L = T_0 = 310 K

Process	q, kJ/kg	w, kJ/kg	$h_e - h_i$, kJ/kg	$s_e - s_i$, kJ/kg·K	$\psi_e - \psi_i$, kJ/kg	i, kJ/kg
1-$2a$	0.0	−381.2	381.2	0.094	352.1	29.1
$2a$-3	862.3	0.0	862.3	0.848	599.4	93.1
3-$4a$	0.0	665.3	−665.3	0.122	−703.1	37.8
$4a$-1	−578.2	0.0	−578.2	−1.064	−248.4	248.4
Cycle	284.1	284.1	0.0	0.0	0.0	408.4

of friction in these devices. Notice, however, that the irreversibility per unit mass for the actual combustion process is less than that for the ideal cycle. The reason for this decrease is that the temperature of the air leaving the actual compressor is higher, resulting in a lower irreversibility per unit mass. On the other hand, the irreversibility is higher for the actual cooling process. Irreversibilities in the turbine cause the temperature of the air at the turbine exit to be higher than it would be for an isentropic turbine. Therefore, the average temperature of the air during the cooling process is higher, resulting in an increase in the irreversibility per unit mass.

The situation is actually much worse if we consider what would happen if both cycles were required to produce the same power output,

$$\dot{W} = \dot{m}_{\text{act}} w_{\text{act}} = \dot{m}_{\text{ideal}} w_{\text{ideal}}$$

We find that the mass flow rate would have to be increased substantially in the actual cycle:

$$\dot{m}_{\text{act}} = \dot{m}_{\text{ideal}} \left(\frac{w_{\text{ideal}}}{w_{\text{act}}} \right) = \dot{m}_{\text{ideal}} \left(\frac{444.5 \text{ kJ/kg}}{284.1 \text{ kJ/kg}} \right) = 1.56\, \dot{m}_{\text{ideal}}$$

An increase of more than 50 percent in the mass flow rate would be required. Not only would this result mean that much larger components would be necessary, but also the irreversibility rate for the actual cycle would be significantly higher than the ideal cycle that produces the same net power output:

$$\begin{aligned} \dot{I}_{\text{act}} &= \dot{m}_{\text{act}} i_{\text{act}} = 1.56\, \dot{m}_{\text{ideal}} i_{\text{act}} \\ &= 1.56(408.4)\, \dot{m}_{\text{ideal}} = 637.1\, \dot{m}_{\text{ideal}} \end{aligned}$$

and

$$\dot{I}_{\text{ideal}} = \dot{m}_{\text{ideal}} i_{\text{ideal}} = 296.8\, \dot{m}_{\text{ideal}}$$

Thus the percent increase in the irreversibility rate would be

$$\begin{aligned} \text{Percent increase} &= \left(\frac{\dot{I}_{\text{act}} - \dot{I}_{\text{ideal}}}{\dot{I}_{\text{ideal}}} \right) (100) \\ &= \left(\frac{637.1 - 296.8}{296.8} \right) (100) = \underline{\underline{115 \text{ percent}}} \end{aligned}$$

■

As a final example of how component irreversibilites can influence the behavior of thermodynamic cycles, we may compare the behavior of an actual Otto-cycle engine with that of an engine operating on an ideal Otto cycle. The ideal Otto cycle approximates the behavior of the more complicated internal-combustion engine. Friction and other factors that produce irreversibilities cause an actual IC engine to deviate somewhat from the behavior of the ideal Otto cycle. Figure 7.30 shows a comparison of the P-v diagrams of an actual spark-ignition IC engine and the corresponding ideal Otto cycle for the purpose of illustrating the deviations that can occur between the actual and ideal cycles.

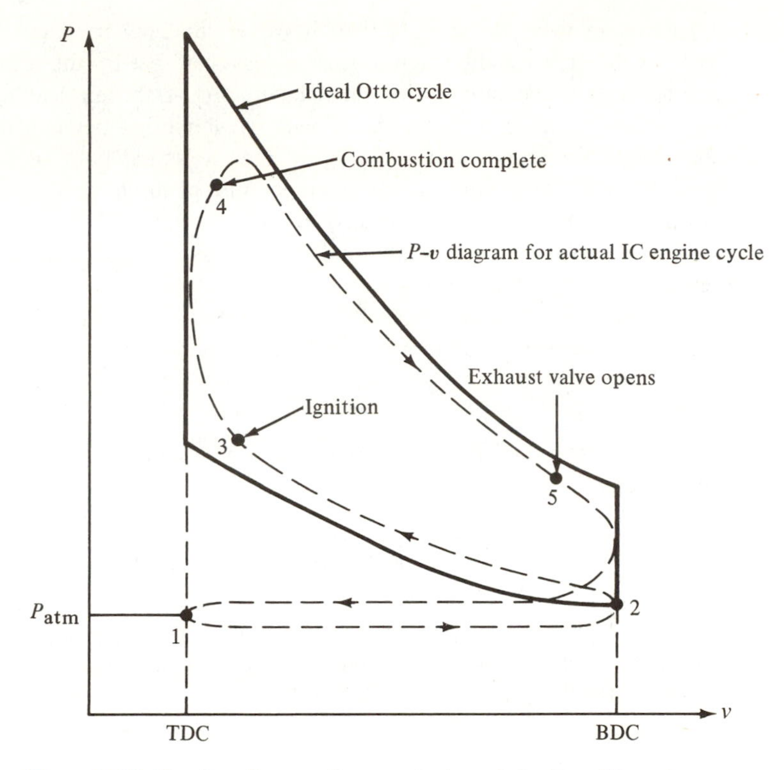

Figure 7.30 The *P-v* diagram for an actual spark-ignition IC engine.

During the intake stroke of an actual engine, the pressure inside the cylinder is slightly below the atmospheric pressure so that a charge of vaporized fuel and air can be drawn into the cylinder. The ideal Otto cycle does not consider frictional effects of the air entering the cylinder through the valves, carburetor, and air cleaner. At approximately state 3 during the compression stroke, the spark plug is energized and the fuel mixture is ignited. The fuel burns until approximately state 4, and the power stroke continues until approximately state 5, where the exhaust valve opens. The exhaust process occurs from state 5 to 1, with the pressure in the cylinder being slightly above the atmospheric pressure so that the exhaust gases can be expelled from the cylinder, overcoming frictional pressure drops resulting from the exhaust valves, manifold, muffler, and exhaust system. When state 1 is reached, the cycle can be repeated, beginning with the intake stroke.

7.15 SUMMARY

Gas cycles are those thermodynamic cycles in which the working fluid remains in the gaseous phase. Examples of heat engines that operate on a gas cycle include both the spark-ignited and compression-ignited internal-combustion engine and the gas-turbine engine. An example of a gas refrigeration cycle is the Brayton refrigeration cycle, which can be used as either a refrigerator or a heat pump.

Because the Carnot cycle is a totally reversible cycle, the thermal efficiency of a Carnot cycle that operates between thermal-energy reservoirs at temperatures T_H and T_L is

$$\eta_{\text{Car}} = 1 - \frac{T_L}{T_H} \tag{7.5}$$

Two other examples of a totally reversible cycle are the Stirling and Ericsson cycles, both of which use the concept of regeneration.

The air-standard assumptions are used to simplify calculations in gas-cycle analysis and consist of the following three conditions:

1. The working fluid has properties identical to those of air.
2. The working fluid is an ideal gas.
3. The system that executes the cycle is a closed system.

While the air-standard assumptions greatly simplify the calculations, they approximate the conditions that occur in many devices and therefore provide general conclusions that are not clouded by unnecessary complications. The cold air-standard assumptions are identical to the air-standard assumptions, but in addition the specific heats of the working fluid are assumed constant.

To maximize the efficiency of a device that operates on a thermodynamic cycle, internal irreversibilities should be minimized. Furthermore, in the design of a heat engine, attempts should be made to modify the cycle to increase the average temperature in the cycle where heat transfer to the working fluid occurs and to decrease the average temperature in the cycle where heat transfer from the working fluid occurs.

The ideal Otto cycle is a model for the spark-ignited IC engine. The thermal efficiency of a cold air-standard, ideal Otto cycle is a function of the engine-compression ratio only, or

$$\eta_{\text{th,Otto}} = 1 - (r)^{1-k} \tag{7.15}$$

The ideal diesel cycle is the ideal, compression-ignited, IC engine cycle. The thermal efficiency of a cold air-standard, ideal diesel cycle that has a compression ratio of r and a cutoff ratio of r_c is

$$\eta_{\text{th,diesel}} = 1 - (r)^{1-k}\left[\frac{r_c^k - 1}{k(r_c - 1)}\right] \tag{7.24}$$

For typical compression ratios and cutoff ratios, engines operating on a diesel cycle are slightly more efficient than engines operating on the Otto cycle.

The work required during an internally reversible, steady-flow process for which changes in kinetic and potential energies are negligible is

$$w_{ie} = -\int_i^e v\, dP \tag{7.44}$$

The concept of a polytropic process is valuable because it provides a relationship between P and v that can be used in Equation 7.44. The polytropic process is one that follows the path

$$Pv^n = \text{constant} \tag{2.42}$$

where n is the polytropic exponent.

The gas-turbine cycle is called the Brayton cycle. The thermal efficiency of an ideal, cold air-standard Brayton cycle is a function of the pressure ratio r_p of the compressor only, or

$$\eta_{\text{th,Brayton}} = 1 - (r_p)^{(1-k)/k} \tag{7.31}$$

The thermal efficiency of a Brayton cycle can often be increased by employing the regeneration process, which uses the turbine-exhaust gas to preheat the working fluid as it leaves the compressor. The efficiency of the regenerator is defined as

$$\eta_{\text{regen}} \equiv \frac{\text{actual heat-transfer rate in regenerator}}{\text{maximum heat-transfer rate that could occur in regenerator}}$$

With the cold air-standard assumptions the thermal efficiency of an ideal Brayton cycle modified with ideal regeneration is

$$\eta_{\text{th}} = 1 - \left(\frac{T_1}{T_4}\right)(r_p)^{(k-1)/k} \tag{7.36}$$

where T_1 is the temperature of the working fluid entering the compressor and T_4 is the temperature of the working fluid entering the turbine. Regeneration becomes more attractive when it is coupled with reheating of the turbine-exhaust gases and intercooling of the air between the compressor stages.

Modified Brayton-cycle engines are used extensively for aircraft power plants. Modifications to the basic turbojet engine include a propjet engine, a fan-jet engine, and a turbojet engine with an afterburner.

The processes that make up the Carnot, Otto, Diesel, and Brayton cycles are summarized in Table 7.8.

TABLE 7.8 SUMMARY OF PROCESSES FOR TYPICAL, IDEAL GAS POWER CYCLES

Cycle	Application	Compression process (compressor)	Heating process (combustion)	Expansion process (turbine)	Cooling process (exhaust)
Carnot power cycle	Standard of comparison	Totally reversible, adiabatic	Totally reversible, isothermal	Totally reversible, adiabatic	Totally reversible, isothermal
Ideal Otto cycle	Spark-ignition IC engine	Internally reversible, adiabatic	Internally reversible, constant volume	Internally reversible, adiabatic	Internally reversible, constant volume
Ideal diesel cycle	Compression-ignition IC engine	Internally reversible, adiabatic	Internally reversible, constant pressure	Internally reversible, adiabatic	Internally reversible, constant volume
Ideal Brayton cycle	Gas-turbine engines	Internally reversible, adiabatic	Internally reversible, constant pressure	Internally reversible, adiabatic	Internally reversible, constant pressure

TABLE 7.9 SUMMARY OF PROCESSES FOR TYPICAL, IDEAL GAS REFRIGERATION CYCLES

Cycle	Application	Compression process (compressor)	Cooling process	Expansion process (turbine)	Heating process
Carnot refrigeration cycle	Standard of comparison	Totally reversible, adiabatic	Totally reversible, isothermal	Totally reversible, adiabatic	Totally reversible, isothermal
Reversed Brayton cycle	Refrigerator or heat pump	Internally reversible, adiabatic	Internally reversible, constant pressure	Internally reversible, adiabatic	Internally reversible, constant pressure

The standard of comparison for all refrigeration cycles is the Carnot refrigeration cycle, which, when operating between energy reservoirs with temperatures T_H and T_L, has a COP of

$$\beta_{R,\text{Car}} = \frac{1}{(T_H/T_L) - 1} \tag{7.6}$$

when used for the purpose of cooling (refrigeration), and a COP of

$$\beta_{H,\text{Car}} = \frac{1}{1 - (T_L/T_H)} \tag{7.7}$$

when used for the purpose of heating, that is, as a heat pump.

An example of a gas refrigeration cycle is the ideal, cold air-standard Brayton refrigeration cycle, which has a coefficient of performance of

$$\beta_R = \frac{1}{[(T_2 - T_3)/(T_1 - T_4)] - 1} \tag{7.47}$$

when used as a refrigerator, and a COP of

$$\beta_H = \frac{1}{1 + [(T_1 - T_4)/(T_3 - T_2)]} \tag{7.50}$$

when used as a heat pump.

The Carnot refrigeration cycle and reversed Brayton-cycle processes are summarized in Table 7.9.

PROBLEMS

7.1 A refrigeration unit has a refrigeration capacity of 2 kW, and the nameplate of the compressor lists the power requirement of 1 kW. The ambient-air temperature for this refrigeration unit is 20°C, and the unit is capable of maintaining the refrigerated space at −10°C. Does the nameplate information violate the second law of thermodynamics?

7.2 Soil that has an average temperature of 50°F is used as a low-temperature thermal-energy reservoir for a heat pump. The heat pump has a power requirement of 10 hp and heats a home to an average temperature of 75°F. Calculate the maximum heat-transfer rate that can be provided to the home.

7.3 A power plant has a thermal efficiency of 35 percent. The temperature of the fuel in the boiler is 1000°C, and there is heat transfer from the plant to a lake that has a temperature of 15°C. Does the power-plant cycle violate the second law of thermodynamics?

7.4 An automobile engine burns gasoline at a rate of 3.2 gal/h while providing a power output of 25 hp. The heat content of gasoline is 19,400 Btu/lb_m. When the gasoline burns, it provides a maximum temperature of 1650°F, and the ambient temperature that the engine operates in is 60°F. Determine the thermal efficiency of the engine and show that the engine does not violate the second law of thermodynamics for these conditions. The density of the fuel is 7.0 lb_m/gal.

7.5 A power cycle operates as shown in the accompanying schematic diagram. Determine the following:

(a) the thermal efficiency of the cycle
(b) the thermal efficiency of a Carnot cycle operating between the same reservoirs
(c) the power output of the actual cycle
(d) the irreversibility of the cycle ($T_0 = T_L$).

Problem 7.5

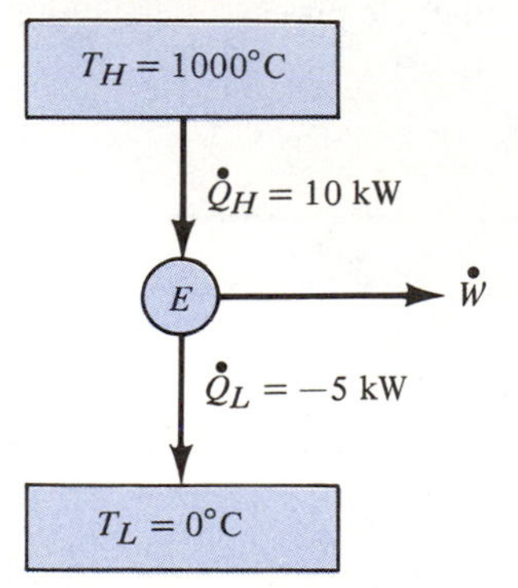

7.6 A refrigeration cycle operates as shown in the accompanying schematic diagram. Determine the following:

(a) the coefficient of performance of the cycle
(b) the COP of a Carnot refrigeration cycle operating between the same reservoirs
(c) the power required to operate the actual cycle
(d) the irreversibility of the cycle ($T_0 = T_H$)

Problem 7.6

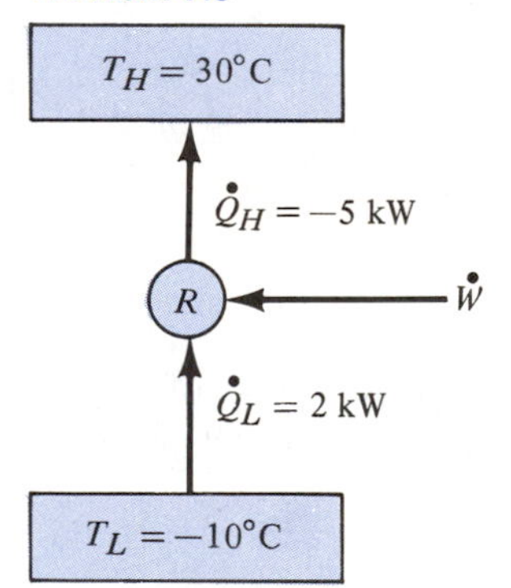

7.7 A Carnot refrigeration cycle operates between reservoirs at temperatures of 12°F and 95°F. The heat transfer with the high-temperature reservoir is 9500 Btu. Determine the heat transfer with the low-temperature reservoir and the work input to the cycle

7.8 The net power produced by a Carnot heat engine is 15 hp. The temperatures of the two reservoirs are 100°F and 1000°F. Determine the heat-transfer rates with the two reservoirs.

7.9 **[ICD211]** A closed system executes an air-standard power cycle consisting of the following three processes. Step (1-2): Isothermal compression from 130 kPa, 295 K to 800 kPa. Step (2-3): Constant-pressure heat addition from a source at 900 K. Step (3-1): Isentropic, reversible expansion. Assume that air is an ideal gas with c_p = 1.005 kJ/kg·K. Determine the following:

(a) the heat transfer and work for each process in kJ/kg

(b) the cycle efficiency

7.10 A Carnot cycle utilizes air as a working fluid in a closed system. The maximum and minimum temperatures that occur during the cycle are 500 and 0°C. The maximum and minimum pressures that occur during the cycle are 1 MPa and 100 kPa. Calculate the following quantities:

(a) the thermal efficiency of the cycle

(b) the net work provided by the cycle per unit mass of air

(c) the heat transfer per unit mass of air to the cycle from the high-temperature reservoir

(d) the irreversibility of the cycle per unit mass of air

7.11 A closed system executes a cold air-standard Carnot-heat-engine cycle that is heated by energy from a reservoir at 740°F. Heat transfer from the heat engine is to a reservoir at a temperature of 40°F. The heat transfer to the heat engine from the high-temperature reservoir is 120 Btu. The pressure at the beginning of the isentropic-compression process is 20 psia, and the pressure at the end of the heat-addition process is 380 psia. Calculate the heat transfer with the low-temperature reservoir, the mass of the working fluid, the net work provided by the cycle, the thermal efficiency of the cycle, and the state of the air at the beginning of each of the four processes in the cycle.

7.12 A Carnot cycle uses air as a working fluid. The temperatures of the high- and low-temperature reservoirs are 800 and 100°F. The pressures of the air at the beginning and the end of the heat-addition process are 300 and 50 psia, respectively. Draw both the *T-s* and *P-v* diagrams for the cycle. Determine the following:

(a) the thermal efficiency of the cycle

(b) the heat transfer from the high-temperature reservoir per unit mass of air

(c) the heat transfer to the low-temperature reservoir per unit mass of air

(d) the net work provided by the cycle per unit mass of air

(e) the irreversibility of the cycle per unit mass of air

7.13 A Carnot power cycle utilizes water as a working fluid. The temperature limits of the cycle are 100 and 800°C. The water is a saturated vapor prior to the isentropic-compression process, and the pressure at the end of the isentropic-expansion process is 10 kPa. Draw the *T-s* diagram for this cycle and calculate the following quantities:

(a) the thermal efficiency of the cycle

(b) the heat transfer for both reservoirs per unit mass of water

(c) the irreversibility of the cycle per unit mass of water

(d) the net work output of the cycle per unit mass of water

7.14 A Carnot refrigeration cycle utilizes refrigerant-12 as a working fluid in a closed system. The temperature limits of the cycle are 40 and 180°F. The maximum and minimum pressures that occur during the cycle are 150 and 15 psia. Draw the *P-v* and *T-s* diagrams for the cycle, and calculate the heat transfer with the two reservoirs per unit mass of R-12, the net work required by the cycle per unit mass of R-12, and the COP of the cycle.

7.15 An ideal Stirling cycle uses 1 kg of nitrogen as a working fluid. The N_2 begins its isothermal-expansion process at 1000 K and 500 kPa, and during the expansion process the volume of the nitrogen increases by a factor of 5. The temperature of the low-temperature reservoir is 20°C. Draw the P-v process diagram for the cycle. Calculate the following:

(a) the heat transfer with the regenerator per cycle
(b) the net work per cycle
(c) the heat transfer with both reservoirs per cycle
(d) the thermal efficiency of the cycle
(e) the irreversibility of the cycle ($T_L = T_0 = 20°C$)

7.16 An ideal Ericsson cycle uses 10 lb_m of helium as a working fluid. At the start of the compression process the state of the helium is 20 psia, 1000°F. The expansion process occurs at a pressure of 650 psia, and the temperature of the low-temperature reservoir is 77°F. Determine the following:

(a) the thermal efficiency of the cycle
(b) the heat transfer with the regenerator
(c) the net work of the cycle
(d) the heat transfer with both reservoirs

7.17 [ICC111] A Stirling cycle using air as the working fluid has a maximum temperature of 2460°R. The minimum temperature and pressure are 540°R and 14.7 psia. The compression ratio is 10. Assume air to be an ideal gas with $c_p = 0.24$ Btu/lb_m·°R, independent of temperature. The environment (state 0) is at 540°R and 14.7 psia. Determine the following:

(a) the heat transfer and work per unit mass of air for each process of this cycle
(b) the cycle efficiency with an ideal regenerator and with no regenerator

7.18 An ideal Ericsson cycle operates between temperature reservoirs at 1000 and 0°C. The maximum and minimum pressures during the cycle are 3 MPa and 200 kPa. Using the air-standard assumptions, determine the following:

(a) the thermal efficiency of the cycle
(b) the heat transfer with the regenerator per unit mass of working fluid
(c) the net work per unit mass of working fluid
(d) the heat transfer with both reservoirs per unit mass of working fluid
(e) the irreversibility per unit mass of working fluid for each process of the cycle ($T_L = T_0 = 0°C$)

7.19 An ideal Otto cycle operates with a compression ratio of 8.5. The minimum and maximum temperatures that occur during the cycle are 100 and 800°C. The pressure at the beginning of the compression process is 100 kPa, and there is 600 kJ of heat transfer to the air from the fuel during each cycle. Apply the air-standard assumptions.

(a) Draw the P-v process diagram for the cycle.
(b) Determine the work required for the compression process.
(c) Determine the work done during the expansion process.
(d) Determine the heat transfer to the low-temperature reservoir.
(e) Determine the thermal efficiency of the cycle.
(f) Determine the irreversibility of each process of the cycle if $T_H = 1000°C$ and $T_L = T_0 = 30°C$.
(g) Why is this "ideal" cycle irreversible?

7.20 An ideal Stirling cycle operates between temperature reservoirs at 1200 and 100°F. The upper and lower pressure limits during the cycle are 600 and 150 psia. Assume that the cycle uses a constant mass of 2 lb_m of CO_2 as a working fluid. Determine the following:

(a) the thermal efficiency of the cycle
(b) the heat transfer with the regenerator per cycle
(c) the net work per cycle
(d) the heat transfer with both reservoirs per cycle

(e) the irreversibility of each process of the cycle ($T_L = T_0 = 100°C$)

7.21 Determine the thermal efficiency of an ideal Otto cycle that has a compression ratio of 10. Use the cold air-standard assumptions.

7.22 An ideal Otto cycle operates between pressure and temperature limits of 100 kPa and 1.4 MPa, and 300 and 1200 K, respectively. Apply the air-standard assumptions and calculate the heat transfer with the two thermal reservoirs and the net work of the cycle per unit mass of the working fluid. Calculate the thermal efficiency of the cycle, the mean effective pressure, and the irreversibility of the cycle per unit mass if $T_H = 1300$ K and $T_L = T_0 = 280$ K. Explain the source(s) of the irreversibility.

7.23 Work Problem 7.22 by using the cold air-standard assumptions.

7.24 The compression ratio of an ideal air-standard Otto cycle is 10.0. The state of the air at the beginning of the compression process is 300 K, 100 kPa. The heat transfer in the combustion process is 750 kJ/kg. Determine the work output per unit mass of the Otto cycle, the thermal efficiency of the cycle, and the heat transfer per unit mass to the low-temperature reservoir.

7.25 An engine operates on an ideal Otto cycle and utilizes carbon dioxide as a working fluid. The minimum pressure and temperature that occur during the cycle are 15 psia and 530°R. The maximum pressure and temperature that occur during the cycle are 500 psia and 2500°R. Determine the thermal efficiency of the cycle, the net work output per unit mass of working fluid, and the heat transfer during the combustion process per unit mass of working fluid. Calculate the irreversibility per unit mass for the four processes that make up the cycle if $T_H = 2800°R$ and $T_L = T_0 = 530°R$.

7.26 An ideal, air-standard Otto cycle has a compression ratio of 8.0 and has 150 kJ/kg of heat transfer from the air to the environment. The state of the air at the end of the expansion process is 0.6 MPa, 500 K. Determine the heat transfer per unit mass during the combustion process, the work output per unit mass, and the irreversibility per unit mass of the four processes that make up the cycle if $T_H = 1300$ K and $T_L = T_0 = 300$ K. Which process has the largest irreversibility? Why?

7.27 An ideal, cold air-standard Otto cycle has a compression ratio of 9.5. The conditions at the beginning of the compression stroke are 100 kPa and 300 K, and the temperature at the end of the combustion process is 1300 K. Determine thermal efficiency of the cycle and compare it with the thermal efficiency of a Carnot cycle operating between reservoirs with temperatures of 300 and 1300 K. Calculate the net work output and the heat transfer from the high-temperature reservoir pet unit mass of working fluid. Determine the mean effective pressure of the cycle. Determine the irreversibility per unit mass of each process in the cycle if $T_H = 2500$ K and $T_L = T_0 = 270$ K.

7.28 An ideal, air-standard Otto cycle has a compression ratio of 9.0. The heat transfer per unit mass during the combustion process is 800 kJ/kg. The temperature and pressure of the working fluid prior to the compression process are 300 K and 100 kPa. Assuming that the working fluid has variable specific heats, calculate the thermal efficiency of the cycle, the maximum temperature during the cycle, the net work output of the cycle per unit mass, and the mean effective pressure of the cycle.

7.29 Work Problem 7.28 by using the cold air-standard assumptions.

7.30 The autoignition temperature of gasoline (the temperature at which gasoline will ignite without the presence of a spark) is approximately 730°F. Suppose that a gasoline-air mixture is compressed isentropically from 80°F in an Otto-cycle engine. Calculate the maximum compression ratio of the engine that will prevent ignition of the mixture at the end of the compression process.

7.31 The *Atkinson cycle* is identical to an ideal Otto cycle except that the heat rejection occurs at constant pressure instead of constant volume. Sketch the *P-v* and *T-s* diagrams for the Atkinson cycle. Referring to the states shown in Figure 7.5, show that the thermal efficiency of a cold air-standard Atkinson cycle is given by

$$\eta_{\text{th}} = 1 - \left[\frac{k}{(V_1/V_2)^{k-1}}\right]\left[\frac{(P_3/P_2)^{1/k} - 1}{(P_3/P_2) - 1}\right]$$

7.32 **[ICB111]** Consider an ideal, cold air-standard Otto cycle that has heat transfer of 2800 kJ/kg from a source at 5000 K, a compression ratio of 8, and a pressure and temperature at the beginning of the compression process of 100 kPa, 290 K (the conditions of state 1). Assume that the working fluid has the properties of pure air with c_p = 1.005 kJ/kg·K. Determine the following:

(a) the maximum pressure for this cycle in kPa
(b) the maximum temperature in K
(c) the thermal efficiency
(d) the mean effective pressure

7.33 Over 15 million model T Ford engines were built between 1909 and 1927. These four-cylinder engines had a displacement of 2892 cm^3 with a bore of 95 mm and a stroke of 102 mm. The engine developed 22 hp at 200 rpm. A 6-cyclinder Rolls Royce engine produced in 1925 was rated at 43 hp at 1000 rpm. The cylinder diameter was 4.25 in and the stroke was 5.5 in, providing a displacement of 468 in^3. Calculate the mean effective pressures for each of these engines.

7.34 A certain 8-cylinder spark-ignition engine develops 200 hp at 4500 rpm with a compression ratio of 9. Use a cold air-standard analysis to determine the temperature and pressure at the end states of each process of an ideal Otto cycle that corresponds to these performance specifications. Assume that the temperature and pressure at the beginning of the compression process are 70°F and 14.7 psia, respectively. Calculate the heat transfer per unit mass, the work per unit mass, and the irreversibility per unit mass for each process and for the cycle. Determine the thermal efficiency of the cycle. Use k = 1.4 and assume that T_H = 4500°R and $T_0 = T_L$ = 530°R. (Engine displacement equals 348 in^3.)

7.35 An ideal, air-standard Otto cycle operates with a compression ratio of 9.0. At the end of the cooling process the air is at 140°F and 20 psia. During the combustion process the heat transfer to the air is 80 Btu/lb$_m$. Draw the *P-v* diagram for the cycle, and calculate the pressure and temperature of the air at the end of each process that makes up the cycle. Calculate the mean effective pressure, the net work of the cycle per unit mass of air, and the heat transfer to the low-temperature reservoir per unit mass of air.

7.36 Work Problem 7.35 by using the cold air-standard assumptions.

7.37 An ideal diesel cycle has a compression ratio of 18.2 and a cutoff ratio of 2.4. The state of air at the beginning of the compression process is 200 kPa, 37°C. Use the air-standard assumptions and calculate the following:

(a) the net work during the cycle per unit mass of working fluid
(b) the heat transfer with both reservoirs per unit mass of working fluid

(c) the mean effective pressure of the cycle
(d) the thermal efficiency of the cycle
(e) the irreversibility per unit mass of working fluid for each process during the cycle if $T_H = 2500$ K and $T_L = 270$ K
(f) which process has the largest irreversibility? Why?

7.38 An ideal, air-standard diesel cycle has a compression ratio of 19.0, and the heat transfer to the working fluid during the combustion process is 600 Btu/lb$_m$. The state of the air at the beginning of the compression process is 70°F, 14 psia. The total displacement volume of the engine at BDC is 250 in^3, and the engine runs at 2000 rev/min. Determine the power output of the engine, the fuel consumption rate (in gal/h), the thermal efficiency of the engine, and the irreversibility of the cycle if the energy content of the fuel is 20,000 Btu/lb$_m$. Assume that $T_H = 3200$°F and $T_L = T_0 = 40$°F.

7.39 The working fluid begins the compression process of an ideal, air-standard diesel cycle at 20°C and 90 kPa. The maximum temperature during the cycle is 1200°C. Calculate the mean effective pressure, the net work output of the cycle per unit mass of the working fluid, the thermal efficiency, and the irreversibility per unit mass for the cycle if $T_H = 1400$°C and $T_L = T_0 = 5$°C. Assume that the compression ratio is (a) 17 and (b) 20. Which compression ratio yields superior thermodynamic performance? Explain.

7.40 A *dual cycle* is a close approximation to the actual performance of a compression-ignition engine, and it consists of a two-segment heating process. The first segment is a constant-volume process, and the second segment is a constant-pressure process. The compression, expansion, and heat-rejection processes are identical to those in the ideal diesel and Otto cycles. Draw the *P-v* and *T-s* diagrams for an ideal dual cycle. Show that the expression for the thermal efficiency for an ideal dual cycle is

$$\eta_{th} = 1 - \frac{T_4 - T_1}{(T_x - T_2) + k(T_3 - T_x)}$$

when the cold air-standard assumptions are used. State 2 represents the end of the compression process, state 3 is the beginning of the expansion process, and state 4 is the beginning of the cooling process; x represents the state at the end of the constant-volume heating process.

7.41 The state of the working fluid at the beginning of the compression process of an ideal dual cycle is given by $P = 15$ psia and $T = 500$°R. The compression ratio of the cycle is 13.0, and the pressure ratio during the constant-volume heating process is 1.5. The volume ratio (final volume divided by the initial volume) of the constant-pressure segment of the heating process is 1.75. Using the cold air-standard assumptions, calculate the pressures and temperatures at the five states around the cycle, the net work provided by the cycle per unit mass of working fluid, and the thermal efficiency of the cycle. A description of the dual cycle is given in Problem 7.40.

7.42 **[ICA111]** An air-standard, ideal diesel cycle has a compression ratio of 15 and the heat transfer to the air during the combustion process is 1800 kJ/kg. At the beginning of the compression process (state 1), the temperature is 15°C and the pressure is 100 kPa. Assume that the working fluid is air and use the air tables. Determine the following:
(a) the temperature in °C and pressure in kPa at the end of the compression process (state 2)

(b) the work in kJ/kg during the constant-pressure heat-transfer process (2-3)
(c) the thermal efficiency of the cycle

7.43 [ICA121] An air-standard, ideal diesel cycle has a compression ratio of 15. The pressure at the beginning of the compression process (state 1) is 14.7 psia, and the temperature is 60°F. The maximum temperature during the cycle is 3790°F. Assume that the working fluid has the properties of pure air and use the air tables. Determine the following:
(a) the temperature and pressure at the end of the compression process (state 2)
(b) the thermal efficiency
(c) the mean effective pressure for the cycle

7.44 [ICA211] An air-standard, ideal diesel cycle has a compression ratio of 15 and the heat transfer to the air during the combustion process is 1900 kJ/kg. At the beginning of the compression process (state 1), the temperature is 290 K and the pressure is 100 kPa. Assume that the working fluid has the properties of pure air which behaves as an ideal gas with c_p = 1.003 kJ/kg·K. Determine the following:
(a) the temperature in K and pressure in kPa at the end of the compression process (state 2)
(b) the work in kJ/kg during the constant-pressure heat-transfer process (2-3)
(c) the thermal efficiency of the cycle

7.45 Write a computer program that will calculate the net work per unit mass of working fluid, the heat transfer to the cycle, the heat transfer from the cycle, and the cycle efficiency for an ideal diesel cycle. The state of the air at the beginning of the compression process is 30 psia and 98°F. Use the cold air-standard assumption and k = 1.4. Calculate your results for a range of compression ratios between 15 and 25 in steps of 2.5 and for a range of cutoff ratios between 1.5 and 3.0 in steps of 0.5.

7.46 An ideal diesel cycle has a compression ratio of 18.0 and a cutoff ratio of 2.4. Using the cold air-standard assumptions, calculate the thermal efficiency of the cycle.

7.47 Suppose that inlet temperatures to the compressor and the turbine are fixed for an ideal, cold air-standard Brayton cycle. Show that the maximum net work for the cycle results when

$$T_2 = (T_1 T_3)^{1/2}$$

where 1 is the state at the inlet to the compressor, 2 is the state at the inlet to the combustion chamber, and 3 is the state at the beginning of the heat-rejection process.

7.48 An ideal diesel cycle has a compression ratio of 17.5. The state of the working fluid at the beginning of the compression process is 18 psia, 70°F, and at the end of the combustion process the temperature is 2200°F. Apply the air-standard assumptions and calculate the thermal efficiency of the cycle and the mean effective pressure. Calculate the net work produced by the cycle, the heat transfer with the low-temperature reservoir, and the irreversibility of the cycle for 1 lb_m of working fluid if T_H = 2600°F and $T_L = T_0$ = 50°F. Why is this "ideal" cycle irreversible?

7.49 Work Problem 7.48 by assuming that the cold air-standard assumptions apply.

7.50 The heat transfer to the working fluid in an ideal Brayton cycle is 1700 kJ/kg. The pressure ratio of the cycle is 9.0, and the working fluid enters the compressor at 25°C and 100 kPa. Determine the thermal efficiency of the cycle, assuming that air-standard conditions apply.

7.51 At the beginning of the cooling process of an air-standard, ideal Brayton cycle, the pressure and temperature of the air are 20 psia and 1240°F. The state of the air leaving the compressor is 200 psia, 1440°F. Determine the thermal efficiency of the cycle.

7.52 Air in an ideal, air-standard Brayton cycle enters the compressor at 100 kPa and 300 K, and it enters the turbine at 1300 K and 1 MPa. Calculate the net work output per unit mass and the thermal efficiency of the cycle. Determine the irreversibility per unit mass for each process and for the cycle if $T_H = 1400$ K and $T_L = T_0 = 280$ K. Which process has the largest irreversibility?

7.53 A simple ideal Brayton cycle has a pressure ratio of 8.0. The working fluid enters the compressor at a pressure of 15 psia and a temperature of 100°F. The temperature at the end of the combustion process is 2300°R. Sketch the *T-s* and *P-v* diagrams for the cycle; calculate the compressor work and turbine work on a per-unit-mass basis. Determine the thermal efficiency of the cycle. Use the air-standard assumptions.

7.54 Work Problem 7.53 by using the cold air-standard assumption. Compute the change in thermal efficiency of the cycle due to the application of the cold air-standard assumption.

7.55 The pressure ratio of both the turbine and the compressor in an ideal, air-standard gas turbine cycle is 8.5. The mass flow rate of air through the cycle is 5 lb_m/s. The air enters the compressor at 13 psia and 520°R, and it enters the turbine section at 1600°R. Determine the power output of the gas turbine and the net power output of the cycle.

7.56 Suppose the maximum and minimum temperatures T_{max} and T_{min} of an ideal, cold air-standard Brayton cycle are fixed. Show that an expression for the optimum net work output (that is, the greatest amount of work output for the given temperatures) is

$$w_{opt} = c_p T_{min} \sqrt{\frac{T_{max}}{T_{min}} - 1}$$

Show that an expression for the pressure ratio that produces the optimum work is

$$r_{opt} = \left(\frac{T_{max}}{T_{min}}\right)^{k/(2k-2)}$$

7.57 An ideal Brayton cycle has a minimum pressure and temperature of 120 kPa and 27°C. The maximum pressure and temperature during the cycle are 1.2 MPa and 1400 K. Determine the net work and the heat transfer with the high-temperature reservoir for a unit mass of working fluid. Calculate the thermal efficiency and the irreversibility per unit mass for the cycle if $T_H = 1500$ K and $T_L = T_0 = 250$ K.

7.58 A gas turbine is used as a standby power unit. The minimum pressure and temperature that occur during the cycle are 15 psia and 80°F. The maximum pressure and temperature that occur during the cycle are 120 psia and 1540°F. Determine the minimum flow rate of working fluid necessary to develop 50 hp of power. Use the air-standard assumptions and use an ideal Brayton cycle as a model for the gas-turbine-cycle operation. Calculate the cycle thermal efficiency and the irreversibility of each process of the cycle if $T_H = 2200$°R and $T_L = T_0 = 500$°R.

7.59 [GTA111] In an open-cycle gas turbine, air enters the compressor at 290 K and 100 kPa, and leaves at a pressure of 500 kPa. The air is heated in a combustor to a temperature of 1140 K from a source at 1400 K and exhausts from the turbine at 100 kPa. Assume that there is no pressure drop across the combustor and that the compressor and turbine operate reversibly. Using air tables for properties determine:

(a) the turbine work output per unit mass
(b) the compressor work input per unit mass
(c) the cycle efficiency

7.60 [GTA211] In an open-cycle gas turbine, air enters the compressor at 530°R, 14.7 psia and exits at 50 psia. The air is heated in the combustor to 1460°R from a source at 2000°R and exhausts from the turbine at 14.7 psia. The power output desired from the turbine is 25,000 Btu/h. Assume that the compressor and turbine operate reversibly and adiabatically and that there is no pressure drop across the combustor. Neglect the effects of the added fuel. Determine:

(a) the required air-mass flow rate in lb_m/h

(b) the heat-transfer rate to the combustor in Btu/h

(c) the cycle efficiency

7.61 The maximum and minimum temperatures in an ideal, cold air-standard Brayton cycle are T_{max} and T_{min}. Write a computer program to calculate the net work output per unit mass, the heat transfer in the combustion process per unit mass, the heat rejected during the cycle per unit mass, the cycle thermal efficiency, the turbine inlet pressure, the compressor exit temperature, and the turbine exit temperature. Calculate these quantities for pressure ratios from 5 to 35 in steps of 2, T_{max} = 2000 K, T_{min} = 300 K, k = 1.4 and a compressor inlet pressure of 100 kPa. Use your results to verify the values of w_{opt} and r_{opt} given in Problem 7.56.

7.62 Consider an ideal, cold air-standard Brayton cycle operating with the air entering the compressor at 85°F and entering the turbine at 2700°F. Write a computer program to calculate the net work per unit mass, the heat transfer per unit mass in the combustion chamber, the cycle thermal efficiency, and the pressure ratio. Calculate these quantities for values of temperature leaving the compressor between 900 and 1800°R in increments of 50°R and k = 1.4. Use the results to verify the expression for temperature T_2 in Problem 7.47 which will produce a maximum amount of work.

7.63 [GTA411] In an open-cycle gas turbine, air enters the compressor at 293 K, 100 kPa and exits at 350 kPa. The air is heated in the combustor to 800 K from a source at 1100 K. The power output desired from the turbine is 75 kW. Assume that the compressor and turbine operate reversibly and adiabatically, and that there is no pressure drop across the combustor. Use the cold air-standard assumption with c_v = 0.7175 kJ/kg·K. Neglect the effects of the added fuel. Determine the following:

(a) the required air-mass flow rate in kg/s

(b) the heat-transfer rate to the combustor in kW

(c) the cycle thermal efficiency

7.64 Suppose that an air-standard cycle is designed to operate with the following limits: maximum cycle pressure of 3 MPa, minimum cycle pressure of 100 kPa, maximum cycle temperature of 1000°C, minimum cycle temperature of 25°C. Determine the net work of the cycle per unit mass of the working fluid and the thermal efficiency, assuming that the cycle is:

(a) an ideal Otto cycle

(b) an ideal diesel cycle

(c) an ideal Brayton cycle

7.65 An air-standard, ideal Brayton cycle has a pressure ratio of 16.0. The minimum and maximum temperatures of the cycle are 540 and 1800°R. The pressure at the inlet to the compressor is 20 psia. Determine the pressure and temperature of the air at the beginning of the four processes that make up the cycle. Calculate the net work and the heat transfer to the air during the combustion process, both for a unit mass of working fluid. Determine the irreversibility per unit mass of each process if T_H = 2000°R and T_L = T_0 = 500°R. Why is this "ideal" cycle irreversible?

7.66 An ideal, air-standard Brayton cycle operates with a regenerator that has an efficiency of 100 percent. The air enters the compressor at 100 kPa and 300 K, and it enters the turbine at 1300 K and 1 MPa. Calculate the work output per unit mass, the heat transfer in the regenerator per unit mass, and the thermal efficiency of the cycle. Determine the irreversibility per unit mass for each process and for the cycle if T_H = 1400 K and $T_L = T_0$ = 280 K. Which process has the largest irreversibility?

7.67 The pressure ratio for an ideal Brayton cycle with regeneration is 15.8. The air enters the compressor at 80°F, and it enters the turbine at 2240°F. Using the cold air-standard assumptions, calculate the net work output per unit mass and the thermal efficiency of the cycle. The efficiency of the regenerator is 70 percent. Compare the efficiency and the irreversibility per unit mass for this cycle with the efficiency and the irreversibility that would result if the regenerator were not present. Which cycle would have the best thermodynamic performance? Explain.

7.68 Work Problem 7.58 if a regenerator with an efficiency of 75 percent is inserted in the cycle.

7.69 The pressure ratio of both the compressor and the turbine in a closed, ideal Brayton cycle with regeneration is 10.0. Air enters the turbine at 847°C and enters the compressor at 27°C and 100 kPa. The regenerator in the cycle has an efficiency of 85 percent. Determine the thermal efficiency of the cycle and the mass flow rate of air necessary to produce 25 kW. Use the air-standard assumptions. Determine which process contributes the greatest irreversibility of the cycle if T_H = 1400 K and $T_L = T_0$ = 290 K.

7.70 Air enters the compressor of an ideal, air-standard Brayton cycle with regeneration at 300 K and 130 kPa. The pressure ratio of the compressor is 14.5, and the regenerator in the cycle has an efficiency of 45 percent. The temperature of the air entering the turbine is 1600 K. Calculate the following:

- **(a)** the net work output per unit mass of working fluid
- **(b)** the heat transfer in the regenerator per unit mass of working fluid
- **(c)** the thermal efficiency of the cycle
- **(d)** the irreversibility per unit mass for each process and for the cycle if T_H = 1800 K and $T_L = T_0$ = 250 K.

7.71 An ideal, air-standard Brayton cycle with regeneration is attached to an electric generator. The air enters the compressor with a temperature of 500°R and a pressure of 14 psia. The regenerator in the cycle has an efficiency of 55 percent. The inlet state to the turbine is 300 psia, 2800°R. Determine the mass flow rate of air necessary to generate 5 MW of electric energy, assuming that the generator has an efficiency of 93 percent.

7.72 A turbojet engine is attached to a test stand and operated to determine its performance characteristics. During a test the measured temperature and pressure of the air at the inlet of the engine are 27°C and 100 kPa, respectively. The temperature of the air entering the turbine is 727°C, and the pressure ratio of the compressor is 7.0. The inlet area of the engine is 0.4 m^2 and the exit area is 0.2 m^2. The mass flow rate of air into the engine during the test is 50 kg/s, and the pressure in the exit plane of the engine is 100 kPa. Apply the air-standard assumptions and assume all components in the engine operate ideally. Calculate the thrust developed by the engine, the heat-transfer rate in the combustion section, the velocity of the gas at the inlet and exit areas of the engine, and the temperature of the gas as it leaves the engine.

7.73 A jet aircraft flies with a speed of 700 ft/s into air at a pressure of 8 psia and a temperature of −40°F. The pressure ratio of the compressor is 7.0, and the temperature of the gas entering the turbine section is 1800°R. The inlet and exit areas of the engine are 5 ft^2 and 3.5 ft^2, respectively. Apply the air-standard assumptions and assume all components of the jet engine are ideal. Neglect the presence of the diffuser and nozzle sections. Calculate the velocity of the gas leaving the engine, the thrust produced by the engine, and the propulsive efficiency of the cycle.

7.74 A constant-area afterburner section is attached to the engine described in Problem 7.72. The conditions given in that problem remain the same, except that fuel with a mass flow rate of 0.5 kg/s is injected into the afterburner section and burned. The heat content of the fuel is 35,000 kJ/kg. Neglect pressure changes in the afterburner section. Determine the thrust developed by the engine and the temperature and velocity of the gas as it leaves the afterburner section.

7.75 Air enters the compressor of a stationary jet engine at 480°R and 12 psia. The compressor has an adiabatic efficiency of 100 percent and a pressure ratio of 9.5. The heat-transfer rate in the combustion process is 43 × 10^6 Btu/h, and the temperature of the gases as they enter the turbine is 2100°R. The pressure of the gases at the exit of the turbine is 12 psia, and the turbine has an adiabatic efficiency of 100 percent. The cross-sectional area of the compressor inlet is 2 ft^2 and at the turbine exit the cross-sectional area is 1 ft^2. Determine the thrust produced by the engine.

7.76 Work Problem 7.50 assuming that the single-stage compression process is replaced by two-stage compression separated by intercooling. Both compression stages have a pressure ratio of 3.0, and the air leaving the intercooler has a temperature of 25°C.

7.77 Show that the work per unit mass required to steadily compress an ideal gas along a polytropic process (polytropic exponent equal to n) in a two-stage process separated by intercooling is given by the expression

$$w = \frac{nRT_1}{1-n}\left[\left(\frac{P_x}{P_1}\right)^{(n-1)/n} - 1 + \left(\frac{P_2}{P_x}\right)^{(n-1)/n} - 1\right]$$

where P_1 is the pressure entering the first-stage compressor, P_2 is the pressure leaving the second-stage compressor, and P_x is the intercooler pressure. Assume that the temperature of the gas entering both compression stages is T_1. Show that the minimum amount of work required for this two-stage compression process will result if the intercooler pressure is given by the expression

$$P_x = (P_1 P_2)^{1/2}$$

Show that for this value of P_x the work required during each compression stage is identical.

7.78 In an ideal Brayton cycle with ideal intercooling and reheating, air enters the first-stage compressor at 37°C and 100 kPa. Both turbine stages and both compressor stages have pressure ratios of 3.2, and the air enters the first-stage turbine with a temperature of 827°C. The mass flow rate of air through the cycle is 0.45 kg/s. Use the air-standard assumptions. Determine the heat-transfer rates in the intercooler and the reheater. Calculate the total power developed by both stages of the turbine, the total power required by both compressor stages, and the thermal efficiency of the cycle.

7.79 Work Problem 7.78 if a regenerator with an efficiency of 80 percent is added to the cycle.

7.80 **(a)** Air at 60°F and 15 psia enters a compressor, where it is steadily compressed in a reversible adiabatic process. The compressor has a pressure ratio of 9. Calculate the work per lb_m required for the compression process and the temperature of the air at the outlet of the compressor.

(b) Given the same inlet conditions as in part a, the compression process is achieved in a two-stage compressor separated by intercooling. Each compressor stage has a pressure ratio of 3, and the air is cooled to 60°F in the intercooler before it enters the second stage of the compressor. Determine the heat transfer per lb_m in the intercooler, the temperature of the air at the outlet of the second stage of the compressor, and the percent reduction in energy required to compress the air compared with that of the compression process described in part a.

7.81 An air-standard, ideal gas refrigeration cycle operates with a compressor that has a pressure ratio of 3.5 and the turbine has a pressure ratio of 3.5. The air enters the compressor at 300 K and 150 kPa, and it enters the turbine at 350 K. Calculate the COP of the cycle.

7.82 An ideal Brayton refrigeration cycle has a compression device and an expansion device, each having a pressure ratio of 6.0. Air enters the compressor at 60°F and 45 psia. The temperature of the air entering the expansion device is 120°F. Determine the minimum flow rate of air if the refrigeration unit is to supply a cooling capacity of 40,000 Btu/h. Determine the minimum power required to operate the cycle and the COP of the cycle. Calculate the irreversibility of each process of the cycle if $T_H = T_0 = 100°F$ and $T_L = 70°F$.

7.83 [ICD111] A closed system using nitrogen as the working fluid executes a heat-pump cycle consisting of the following three internally reversible processes. Step (1-2): Adiabatic compression from 100 kPa, 310 K to 420 kPa. Step (2-3): Constant-volume process with heat transfer to the surroundings at 310 K. Step (3-1): Constant-temperature process. Assume that nitrogen is an ideal gas with $c_p = 1.038$ kJ/kg·K. The environment (state 1) is at 310 K and 100 kPa. Determine the following:

(a) the heat transfer and work for each process in kJ/kg

(b) the COP if this cycle operates as a refrigerator

7.84 Work Problem 7.65 assuming that the compressor has an adiabatic efficiency of 80 percent and that the adiabatic efficiency of the turbine is 84 percent.

7.85 Work Problem 7.71 assuming that the compressor and turbine have adiabatic efficiencies of 90 and 92 percent, respectively. Determine the percent increase in mass flow rate necessary to produce the 5.0 MW of electric energy.

7.86 In an air-standard gas refrigeration cycle (Brayton refrigeration cycle), the air enters the compressor at 40°F and 10 psia and it enters the turbine at 130°F and 45 psia. The mass flow rate of air through the cycle is 40 lb_m/min, and the heat-exchange processes occur at constant pressure. The adiabatic efficiencies of the compressor and the turbine are 85 and 90 percent, respectively. Determine the following:

(a) the heat-transfer rate with the low-temperature reservoir

(b) the heat-transfer rate with the high-temperature reservoir

(c) the net power required to operate the cycle

(d) the COP of the cycle

7.87 Work Problem 7.81 assuming that the compressor has an adiabatic efficiency of 87 percent and the turbine has an adiabatic efficiency of 82 percent.

7.88 Work Problem 7.55 assuming that the compressor has an adiabatic efficiency of

85 percent and the turbine has an adiabatic efficiency of 90 percent.

7.89 The states for an air-standard Brayton cycle with regeneration illustrated in the accompanying sketch are given in the table below. For this cycle, assume that heat transfer to the cycle is from a reservoir at a temperature of 1900 K and that heat transfer from the cycle is to the environment at $T_0 = 250$ K.

State	T, K	P, kPa
1	1600.0	2000
2s	761.5	100
2	832.7	100
3	791.4	100
4	310.0	100
5s	714.6	2000
5	757.2	2000
6	798.9	2000

Determine the adiabatic efficiency of the compressor and the turbine, the regeneration efficiency, the thermal efficiency of the cycle, and the irreversibility per unit mass for each of the following: turbine, combustion chamber, regenerator, overall cycle. What mass flow rate of air would be required to produce 5 MW of power with this cycle?

7.90 Work Problem 7.75 assuming that the turbine and compressor have adiabatic efficiencies of 91 percent and 87 percent, respectively. What is the percent change in thrust that results from nonisentropic expansion and compression of the air?

7.91 **[GTB111]** In an open-cycle gas turbine, air enters the compressor at 290 K, 100 kPa, with a flow rate of 0.6 kg/s. The air leaving the compressor is at a pressure of 500 kPa. The air is heated in the combustor to a temperature of 1200 K from a source at 1500 K. Assume a compressor efficiency of 0.8, a turbine efficiency of 0.85, and a pressure drop between the compressor and turbine of 15 kPa. Using the air tables for property data, determine the following:
(a) the turbine power output in kW
(b) the compressor power input in kW
(c) the cycle efficiency

7.92 **[GTB211]** In an open-cycle gas turbine, air enters the compressor at 520°R, 14.7 psia, with a mass flow rate of 50 lb_m/h. The air leaving the compressor is at a pressure of 70 psia. The air is heated in the combustor to a temperature of 2060°R from a source at 2460°R. Assume a compressor effi-

Problem 7.89

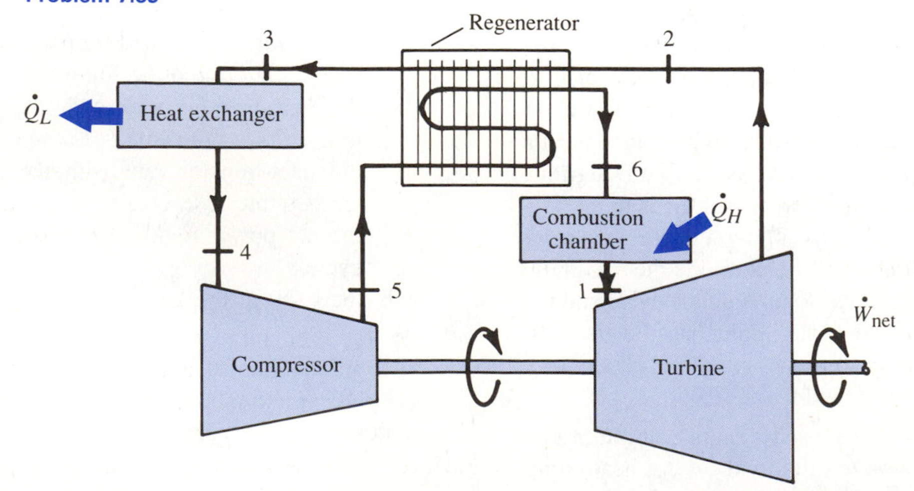

ciency of 0.8, a turbine efficiency of 0.85, and a pressure drop between the compressor and turbine of 2 psia. Use the cold air-standard assumption with $c_v = 0.171$ Btu/lb$_m$·°R. Determine the following:

(a) the turbine power output in Btu/h
(b) the compressor power input in Btu/h
(c) the cycle efficiency

7.93 [GTC111] In an open-cycle gas turbine system with a regenerator, air enters the compressor at a rate of 0.5 kg/s with atmospheric conditions of 290 K and 100 kPa. The air is compressed to 600 kPa by the compressor and then heated to 1000 K. Part of the energy for heating comes from the regenerator whose effectiveness is 0.7. The air leaves the turbine at 711 K, passes through the regenerator, and is then exhausted to the atmosphere. Assume a compressor efficiency of 0.8. Neglect pressure losses and use air tables for property values. Determine:

(a) the temperature entering the combustor in K
(b) the heat-transfer rate to the combustor in kW
(c) the turbine efficiency
(d) the cycle efficiency

7.94 [GTC211] In an open-cycle gas-turbine system with a regenerator, air enters the compressor at a rate of 300 lb$_m$/h with atmospheric conditions of 530°R and 14.7 psia. The air is compressed to 73.5 psia by the compressor and then heated to 1800°R. Part of the energy for heating comes from the regenerator whose effectiveness is 0.7. The air leaves the turbine at 1322°R, passes through the regenerator, and is then exhausted to the atmosphere. Assume a compressor efficiency of 0.8 and neglect pressure losses. Use the cold air-standard assumption with $c_p = 0.24$ Btu/lb$_m$·°R. Determine the following:

(a) the temperature entering the combustor in °R
(b) the heat-transfer rate to the combustor in Btu/h
(c) the turbine efficiency
(d) the cycle efficiency

7.95 [GTE111] The airspeed of a jet aircraft is 240 m/s in still air at 55 kPa and 255 K. The mass flow rate of air through the engine is 7.5 kg/s. The pressure ratio across the compressor is 11 and the maximum cycle temperature is 1280 K. The efficiencies of the compressor, turbine, and nozzle are 0.84, 0.865, and 0.94, respectively. Assume that the power output of the turbine is equal in magnitude to the power input to the compressor. Using the air tables for property data, determine:

(a) the compressor power in kW
(b) the pressure at the turbine outlet in kPa
(c) the exit jet velocity in m/s
(d) the thrust in kN

7.96 [GTD111] In an open-cycle, multistage gas-turbine system, air enters the compressor at a rate of 200 lb$_m$/h and atmospheric conditions of 530°R, 14.7 psia, where it is compressed to 100 psia. This process requires 20,000 Btu/h of power input to the compressor. The air is then heated in the combustor to a temperature of 1960°R from a source at 2460°R, before entering a high-pressure turbine. This turbine expands the air to 40 psia and provides 20,000 Btu/h of power. The air is then reheated to 1960°R and enters the low-pressure turbine which produces 20,000 Btu/h of power. Use the air tables and determine the following:

(a) the compressor efficiency
(b) the low-pressure turbine efficiency
(c) the high-pressure turbine efficiency
(d) the cycle efficiency

7.97 [GTD121] In an open-cycle, multistage gas-turbine system, air enters the compressor at a rate of 200 lb$_m$/h and atmospheric conditions of 530°R, 14.7 psia, where it is compressed to 100 psia. The air is then heated in the combustor to a temperature of 1960°R from a source at 2460°R, before entering a high-pressure turbine. This turbine expands the air to 40 psia. The air is then reheated to 1960°R and enters the low-pressure turbine. Assume a compressor efficiency of 0.926, and turbine efficiencies of 0.909 and 0.841, respectively. Using the air tables for property data, determine:

(a) the low-pressure turbine power output in Btu/h

(b) the high-pressure turbine power output in Btu/h

(c) the cycle efficiency

7.98 [GTD211] In an open-cycle, multistage gas-turbine system, air enters the compressor at a rate of 0.5 kg/s and atmospheric conditions of 298 K, 100 kPa, where it is compressed to 690 kPa. This requires 119 kW of power input to the compressor. The air is then heated in the combustor to a temperature of 1100 K from a source at 1400 K, before entering a high-pressure turbine. This turbine expands the air to 275 kPa and provides 117.8 kW of power. The air is then reheated to 1100 K and enters the low-pressure turbine which produces 118.8 kW of power. Use the cold air-standard assumption with c_p = 1.005 kJ/kg·K. Determine the following:

(a) the compressor efficiency

(b) the low-pressure turbine efficiency

(c) the high-pressure turbine efficiency

(d) the cycle efficiency

7.99 An actual internal-combustion engine is designed on the Otto cycle. The air prior to the compression process is at a temperature of 100°F and a pressure of 15 psia. In an effort to prevent preignition of the fuel, the temperature of the fuel-air mixture is limited to 760°F at the end of the compression process. Assume that the properties of the working fluid are k = 1.35 and c_p = 0.26 Btu/lb_m·°R and that the compression process is internally reversible and follows the path $Pv^n = C$. Calculate the maximum compression ratio that will prevent preignition of the fuel under the following conditions:

(a) The compression process is adiabatic and internally reversible.

(b) The compression process occurs such that there is 10 Btu/lb_m of heat transfer from the working fluid during the compression process.

(c) The compression process occurs such that there is 20 Btu/lb_m of heat transfer from the working fluid during the compression process.

7.100 The mass flow rate of air through an actual Brayton cycle is 1.2 kg/s. The pressure ratios of the compressor and turbine are both 8.5, and the temperature of the air as it enters the compressor and the turbine is 27 and 527°C, respectively. The adiabatic efficiency of the compressor and the turbine is 82 and 85 percent, respectively. The pressure of the air entering the compressor is 100 kPa. Determine the net power output of the cycle, the thermal efficiency of the cycle, and the irreversibility for the cycle if T_H = 800°C and $T_L = T_0$ = 15°C.

8 Vapor Cycles

8.1 INTRODUCTION

This chapter considers vapor power cycles and vapor refrigeration cycles. The working fluid for these cycles is alternately condensed and vaporized, as opposed to the gas cycles considered in Chapter 7 for which the working fluid always remains within the gaseous phase. When a working fluid remains in the saturation region at constant pressure, its temperature is also constant. Thus the condensation or the evaporation of a fluid in a heat exchanger is a process that closely approximates the isothermal heat-transfer processes of the Carnot cycle. In a gas cycle a heat-transfer process occurring at a constant pressure deviates significantly from a constant-temperature process. Owing to this fact, vapor cycles more closely approximate the behavior of the Carnot cycle than do gas cycles and, in general, they tend to perform more efficiently than the gas cycles.

The vapor power cycle employing water as a working fluid is used extensively to generate electricity. Steam power plants provide most of the electric power for the world and are therefore important to engineers interested in energy-conversion systems. Similarly, most refrigeration systems in use today operate on a vapor refrigeration cycle that is described in this chapter.

Various modifications to the basic vapor cycles are examined for the purpose of improving the thermodynamic performance of the cycle. Considering the extent of the electric-power-generation network and the abundant use of refrigeration systems throughout the world, even seemingly small increases in the performance of vapor cycles can have a surprisingly large impact on world energy resources.

8.2 IDEAL RANKINE CYCLE

The ideal *Rankine cycle* is the model for the steam power plant and, in its most basic form, consists of four components: a pump, a boiler, a turbine, and a condenser, as shown schematically in Figure 8.1. The corresponding *T-s* diagram for the ideal Rankine cycle is also shown in Figure 8.1.

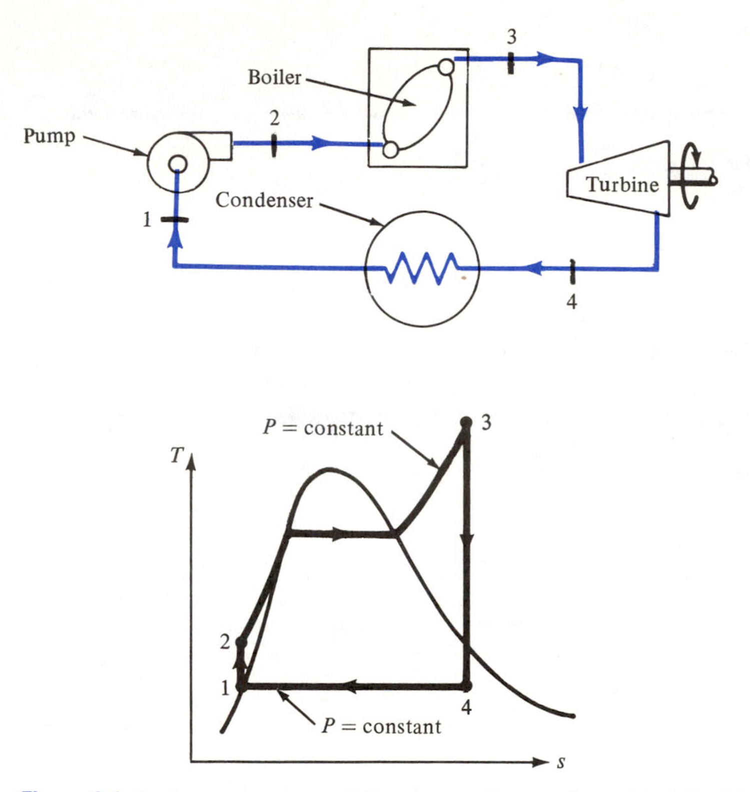

Figure 8.1 Basic components and T-s process diagram for an ideal Rankine cycle.

Water is the most common working fluid in the Rankine cycle. As indicated in Figure 8.1, water enters the pump at state 1 in the liquid phase and is compressed to a high pressure at state 2. In the ideal Rankine cycle the pumping process 1-2 is assumed to be internally reversible and adiabatic or isentropic, and it is represented by a vertical line on the T-s diagram.

Even though the water is in a liquid phase in the pump, the pressure change is large enough that the specific volume of the water changes slightly and the water emerges from the pump at state 2 at a temperature higher than the inlet temperature. States 1 and 2 on the T-s diagram in Figure 8.1 are very close together, and the vertical distance between them has been exaggerated for purposes of clarity.

After leaving the pump at state 2 as a compressed liquid, the water enters the boiler, where it is heated by the high-temperature combustion gases. The water enters the boiler as a liquid at state 2 and leaves as a superheated vapor at state 3. The boiler is basically a large heat exchanger and is assumed to be an internally reversible, constant-pressure device in the ideal cycle.

The superheated vapor leaves the boiler at state 3 indicated in Figure 8.1 and enters the turbine, where it expands and produces work by rotating the shaft of an electric generator. In the turbine-expansion process the water is assumed to follow an internally reversible adiabatic process, leaving at state 4. The ideal-expansion process in the turbine is therefore isentropic and, as such, is represented by a vertical line on the T-s diagram of Figure 8.1. The water leaves the turbine at state 4 and enters a condenser, where heat transfer to the surroundings occurs. Like the boiler, the condenser is merely a heat exchanger. In the ideal condenser, therefore, the water is assumed to undergo an internally reversible, constant-pressure process. The water that enters the condenser is usually a mixture of water vapor and liquid droplets with a high quality. This mixture is condensed so that the exit state is very close to saturated-liquid conditions at the operating pressure of the condenser. The pressure throughout the condenser is assumed to be constant, and if the water in the condenser remains entirely in the saturation region, the process is also one of constant temperature, as shown in Figure 8.1. The liquid water leaves the condenser at state 1 and reenters the pump to start the cycle once again.

Most power plants operate on the Rankine cycle, although actual power plants differ from the ideal Rankine cycle in that the individual components are not internally reversible. The four basic components that make up a Rankine cycle are common to all power plants regardless of whether the fuel used to produce the steam is a nuclear material, oil, natural gas, or coal. Similarly, the basic cycle is unchanged even though the heat transfer from the water in the condenser might be to a cooling tower, to the environmental air, or to water in a nearby river or lake.

Power-plant equipment varies greatly in complexity depending upon the size of the plant, the environmental regulations, and the number of modifications made to the cycle in the quest to achieve higher cycle efficiencies. Simple boilers are merely enclosures in which the water tubes line the enclosure and absorb energy from the high-temperature combustion gases that circulate through the central portion of the boiler. Large modern boilers like the one shown in Figure 8.2 are designed for efficient heat transfer from the products of combustion to the water in the boiler tubes. They often consist of several subcomponents such as superheaters, reheaters, economizers, and air preheaters.

Expressions that relate the work and the heat transfer to the properties of the water at the four states shown in Figure 8.1 can be obtained by applying the conservation-of-energy equation to each of the four components in the Rankine cycle. In the analysis of the ideal cycle the kinetic- and potential-energy changes of the water are usually assumed to be much smaller than the work and heat-transfer terms, and they are therefore usually neglected in the analysis.

Since the pumping process in an ideal Rankine cycle is adiabatic and internally reversible, the conservation-of-energy equation on a unit-mass basis applied to the pump reduces to

$$w_{12} = h_1 - h_2 \tag{8.1}$$

The enthalpy change in Equation 8.1 can be evaluated from the second $T\,ds$ equation, Equation 5.20:

$$h_1 - h_2 = \int_2^1 dh = \int_2^1 T\,ds + \int_2^1 v\,dP \tag{8.2}$$

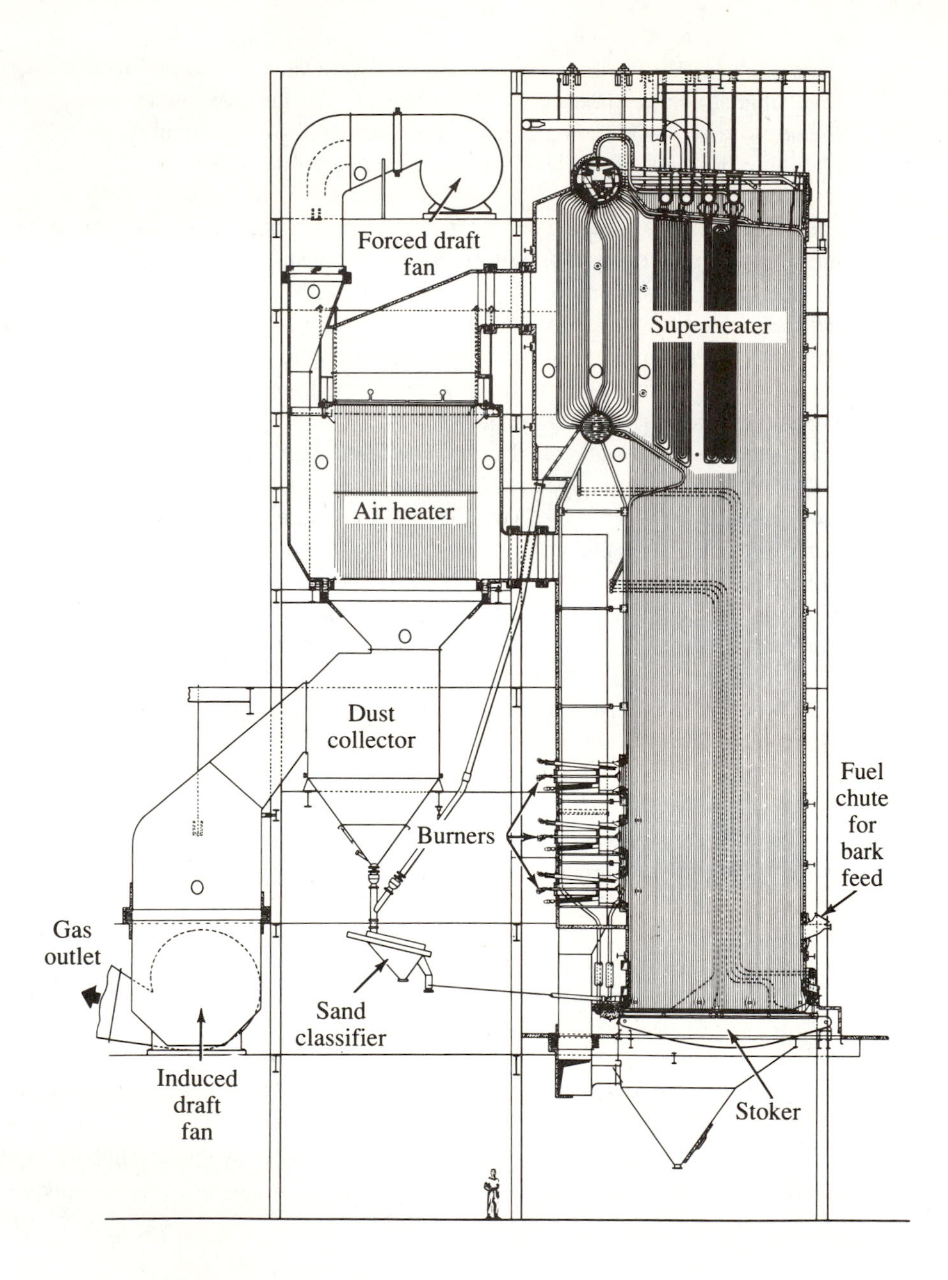

Figure 8.2 This schematic shows a typical industrial boiler that is adaptable to coal, oil, or even tree bark that has a considerable energy content even though it is often discarded during pulping operations. The performance of this boiler is enhanced by forcing the combustion air into the boiler by means of a forced-draft fan and by drawing the products of combustion out of the boiler through an induced-draft fan. The combustion air is also heated in an air heater before it enters the boiler to further increase boiler efficiency. Inlet air temperatures up to 450°F are common for this type of boiler. (Courtesy of Babcock and Wilcox.)

The water in the pump is in the liquid phase and can be assumed to be incompressible. Since the process is isentropic ($ds = 0$) and the specific volume is essentially constant, Equation 8.2 reduces to

$$h_1 - h_2 = v \int_2^1 dP = v(P_1 - P_2) \tag{8.3}$$

Thus the work input required by the isentropic pump per unit mass of water is

$$w_{12} = v(P_1 - P_2) \tag{8.4}$$

The turbine in an ideal Rankine cycle is also internally reversible and adiabatic, so that the work per unit mass developed by the turbine is

$$w_{34} = h_3 - h_4 \tag{8.5}$$

Since work interactions are absent in the boiler and the condenser processes, the heat transfer per unit mass in the boiler is

$$q_{23} = h_3 - h_2 \tag{8.6}$$

and the heat transfer per unit mass in the condenser is

$$q_{41} = h_1 - h_4 \tag{8.7}$$

The thermal efficiency of the cycle is the ratio of the net work produced during the cycle to the heat transfer to the water in the boiler, or

$$\eta_{th} = \frac{w_{34} + w_{12}}{q_{23}} \tag{8.8}$$

This equation can be written in terms of the enthalpy of the working fluid at the four states shown in Figure 8.1 by substituting the results of Equations 8.1, 8.5, and 8.6:

$$\eta_{th} = \frac{(h_3 - h_4) + (h_1 - h_2)}{h_3 - h_2} \tag{8.9}$$

These principles are illustrated in the following example.

EXAMPLE 8.1

An ideal Rankine cycle uses water as a working fluid, which circulates at a rate of 80 kg/s. The boiler pressure is 6 MPa, and the condenser pressure is 10 kPa. The water enters the turbine at 600°C and leaves the condenser as a saturated liquid. Assume that heat transfer to the working fluid in the boiler is from a reservoir at 1400 K and that the fluid in the condenser is cooled by heat transfer to the surroundings at 25°C. Calculate the following quantities:

(a) The power required to operate the pump
(b) The heat-transfer rate to the water in the boiler
(c) The power developed by the turbine
(d) The heat-transfer rate in the condenser
(e) The thermal efficiency of the cycle
(f) The irreversibility associated with each process and the total cycle

Solution. For the components in the simple power-plant cycle, the potential-energy and kinetic-energy changes are usually relatively small when compared with heat-transfer and work terms and can be neglected. The states of the working fluid for this problem are those shown in Figure 8.1.

(a) The specific volume of the water in the pump may be approximated by using saturated-liquid properties at the inlet conditions. From Table B.2 at 10 kPa,

$$v = 1.01 \times 10^{-3}\ \text{m}^3/\text{kg}$$

and the work per unit mass required by the pump can be determined by using Equation 8.4:

$$w_{12} = (1.01 \times 10^{-3}\text{m}^3/\text{kg})(10^4\ \text{Pa} - 6 \times 10^6\ \text{Pa})(1\ \text{N/m}^2\cdot\text{Pa})$$
$$= -6050\ \text{N}\cdot\text{m/kg} = -6.05\ \text{kJ/kg}$$

The power required by the pump is

$$\dot{W}_{12} = \dot{m}w_{12} = (80\ \text{kg/s})(-6.05\ \text{kJ/kg}) = \underline{\underline{-484\ \text{kW}}}$$

(b) From Equation 8.6 the heat-transfer rate in the boiler is

$$\dot{Q}_{23} = \dot{m}(h_3 - h_2)$$

The enthalpy at state 2 can be determined by using Equation 8.1:

$$h_2 = h_1 - w_{12} = h_f - w_{12}$$
$$= 191.8\ \text{kJ/kg} + 6.05\ \text{kJ/kg} = 197.9\ \text{kJ/kg}$$

and

$$\dot{Q}_{23} = (80\ \text{kg/s})(3658.1 - 197.9)\text{kJ/kg} = \underline{\underline{277{,}000\ \text{kW}}} \text{ or } \underline{\underline{277\ \text{MW}}}$$

(c) The enthalpy at state 3 can be determined from the given temperature and pressure and the enthalpy at state 4 can be evaluated by recognizing that process 3-4 is isentropic. Therefore, $s_4 = s_3$, and the state of the water leaving the turbine is in the saturation region. The quality at state 4 is determined as follows:

$$s_4 = s_3 = s_f + x_4 s_{fg} = 7.1673\ \text{kJ/kg}\cdot\text{K}$$

$$x_4 = \frac{s_4 - s_f}{s_{fg}} = \frac{(7.1673 - 0.6493)\text{kJ/kg}\cdot\text{K}}{7.4990\ \text{kJ/kg}\cdot\text{K}} = 0.869$$

The enthalpy at state 4 is

$$h_4 = h_f + x_4 h_{fg}$$
$$= 191.8\ \text{kJ/kg} + 0.869(2391.9\ \text{kJ/kg}) = 2270\ \text{kJ/kg}$$

Thus the power output of the turbine is

$$\dot{W}_{34} = \dot{m}(h_3 - h_4)$$
$$= (80\ \text{kg/s})(3658.1 - 2270)\text{kJ/kg} = \underline{\underline{111\ \text{MW}}}$$

(d) The heat-transfer rate in the condenser is

$$\dot{Q}_{41} = \dot{m}(h_1 - h_4) = (80\ \text{kg/s})(191.8 - 2270)\text{kJ/kg} = \underline{\underline{-166\ \text{MW}}}$$

(e) The thermal efficiency of the cycle, determined by Equation 8.8 written on a rate basis, is

$$\eta_{\text{th}} = \frac{\dot{W}_{\text{net}}}{\dot{Q}_{23}} = \frac{\dot{W}_{34} + \dot{W}_{12}}{\dot{Q}_{23}}$$

$$= \frac{(111{,}000 - 484)\text{kW}}{277{,}000\ \text{kW}} = 0.399 = \underline{\underline{39.9\ \text{percent}}}$$

While a thermal efficiency of less than 40 percent may appear to be quite low, it is the highest efficiency that could be achieved with a Rankine cycle operating between the stated temperature and pressure limits because the cycle has been assumed to be an ideal one. The components in an actual Rankine power cycle would necessarily have irreversibilities such as fluid and mechanical friction that are not considered in an ideal cycle. These irreversibilities would cause the thermal efficiency of an actual cycle to be less than that of the ideal Rankine cycle. Even though the efficiency calculated in this problem represents the maximum efficiency of any Rankine cycle operating with the given temperature limits, it does not represent the maximum efficiency of any cycle for those limits. Because of the irreversibilities associated with the heat-transfer processes in the boiler and condenser, the ideal-Rankine-cycle efficiency is less than the efficiency that could be achieved with a Carnot cycle operating between these same two temperature limits.

Notice that approximately 60 percent of the energy transferred to the water in the boiler is eventually rejected to the environment during the condensation process. The condenser heat transfer occurs at a temperature of 45.8°C, and the energy is therefore of such low ''quality'' that it has little practical use. In the past few years several ideas have been proposed for deriving some benefit from this low-grade energy. Some of the proposals have suggested using the steam from the turbine exhaust to heat industrial plants or to heat agricultural areas in order to extend the seasonal growing period.

Notice also that only about 0.44 percent of the work produced by the turbine is required to operate the pump. By contrast, in the Brayton cycle discussed in Section 7.9 approximately 40 percent (see Example 7.5) of the work output of the turbine was used to power the compressor. This trend can be explained by comparing the work required to compress a gas with that required to pump a liquid. From Equations 8.1 and 8.2 the work required to steadily and isentropically change the pressure of a substance with negligible changes in kinetic and potential energies is

$$w_{12} = h_1 - h_2 = -\int_1^2 v\,dP$$

That is, the magnitude of the compression work is equal to the area between the process curve and the P axis on a P-v diagram, as shown in Figure 8.3. The paths followed during the compression of a liquid and of a gas are indicated in

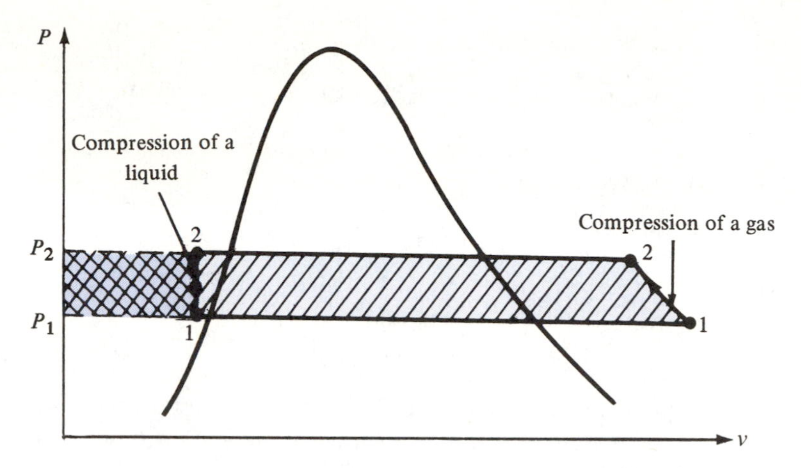

Figure 8.3 Comparison of work required to compress a gas and a liquid.

the figure, which graphically illustrates that compressing a gas through a given pressure difference requires a greater work input than compressing a liquid through the same pressure difference.

(f) The component and cycle irreversibilities are determined from a second-law analysis. The irreversibility for each process can be calculated by using Equation 6.47, which reduces to

$$i = T_0 \left[(s_e - s_i) + \frac{q_k}{T_k} \right]$$

and the change in availability for each process is, from Equation 6.32,

$$\psi_e - \psi_i = (h_e - h_i) - T_0(s_e - s_i)$$

The properties at each state of the cycle are summarized as follows:

State	T, °C	P, MPa	h, kJ/kg	s, kJ/kg·K	Note
1	45.82	0.01	191.8	0.6493	Saturated liquid
2		6	197.9	0.6493	Compressed liquid
3	600	6	3658.1	7.1673	
4	45.82	0.01	2270.0	7.1673	$x_4 = 0.869$

These properties were used to calculate the data for each process summarized in Table 8.1.

The turbine and pump processes in the ideal cycle are each isentropic and adiabatic and therefore totally reversible. Thus the irreversibility for each of these processes is zero. The heat-transfer processes, although internally reversible, are not totally reversible because the heat transfer occurs through a finite

TABLE 8.1 IDEAL RANKINE CYCLE
$T_H = 1400$ K, $T_L = T_0 = 298$ K

Process	q, kJ/kg	w, kJ/kg	$h_e - h_i$, kJ/kg	$s_e - s_i$, kJ/kg·K	$\psi_e - \psi_i$, kJ/kg	i, kJ/kg
1-2	0.0	−6.1	6.1	0.0	6.1	0.0
2-3	3460.2	0.0	3460.2	6.518	1517.8	1205.8
3-4	0.0	1388.1	−1388.1	0.0	−1388.1	0.0
4-1	−2078.2	0.0	−2078.2	−6.518	−135.8	135.8
Cycle	1382.0	1382.0	0.0	0.0	0.0	1341.6

temperature difference between the system and its surroundings. The irreversibility is dominated by the irreversibility of the boiler process, because of the large temperature difference between the water in the boiler and the high-temperature reservoir.

Notice that the change in the availability for each of the reversible, adiabatic, steady-flow processes is equal to the work for the process. Therefore, the second-law efficiency, or effectiveness, of these devices is 100 percent. From the definition of effectiveness in Example 6.4,

$$\epsilon_T = \frac{w_T}{\psi_3 - \psi_4} = 1.0$$

and

$$\epsilon_p = \frac{\psi_1 - \psi_2}{w_p} = 1.0$$

This result is not true for irreversible adiabatic processes, and this point is illustrated further in the analysis of an actual cycle in Example 8.6.

The net irreversibility for the cycle could be calculated more directly if desired. For example, from Equation 6.38,

$$\begin{aligned} i_{cyc} &= T_0 \oint ds_{sys} + \sum_k T_0 \oint \frac{\delta q_k}{T_k} \\ &= 0 + T_0\left(\frac{q_H}{T_H} + \frac{q_L}{T_L}\right) = -T_0\left(\frac{q_{23}}{T_H} + \frac{q_{41}}{T_0}\right) \\ &= (-298 \text{ K})\left(\frac{3460.2 \text{ kJ/kg}}{1400 \text{ K}} + \frac{-2078.2 \text{ kJ/kg}}{298 \text{ K}}\right) = 1341.6 \text{ kJ/kg} \end{aligned}$$

Another means of calculating the cycle irreversibility is provided by Equation 7.4,

$$i_{cyc} = q_H(\eta_{Car} - \eta_{act})$$

Since the Carnot cycle efficiency is

$$\eta_{\text{Car}} = 1 - \frac{T_L}{T_H} = 1 - \frac{298 \text{ K}}{1400 \text{ K}} = 0.787$$

and

$$\eta_{\text{act}} = \eta_{\text{th}} = 0.399$$

then

$$i_{\text{cyc}} = (3460.2 \text{ kJ/kg})(0.787 - 0.399) = 1342.6 \text{ kJ/kg}$$

This last method more clearly illustrates the fact that the irreversibility represents the difference between the reversible work (work done by a totally reversible process or cycle) and the work produced by an actual cycle.

Any modification to a power cycle that increases the average temperature of the working fluid during the heating process or decreases the average temperature of the working fluid during the cooling process will produce an increase in cycle thermal efficiency. The influence of these two modifications on the operation and thermal efficiency of a Rankine cycle can be illustrated with the aid of Figures 8.4 and 8.5. First, consider the changes to the cycle caused by increasing the average temperature of the water in the boiler. Suppose that the heating process in the boiler is extended so that the water reaches state 3′ instead of state 3, as shown in Figure 8.4. This modification increases the average temperature in the boiler and therefore increases the thermal efficiency of the cycle. The temperature of the steam that enters the turbine is limited, however, by metallurgical considerations, and it cannot simply be increased without first considering the effects of the high temperatures on the strength and integrity of the turbine blades.

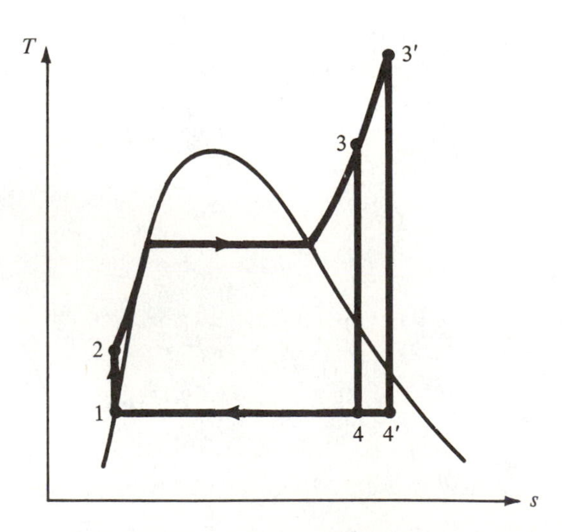

Figure 8.4 Changes in the ideal Rankine cycle caused by an increase in boiler temperature.

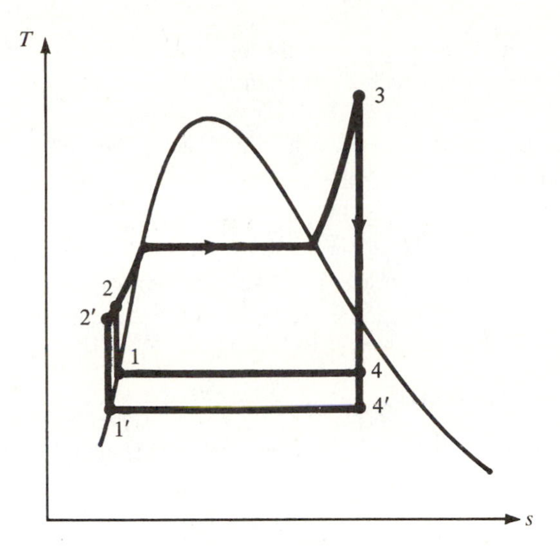

Figure 8.5 Changes in the ideal Rankine cycle caused by a decrease in condenser temperature.

The increase in average boiler temperature results in another desirable change to the cycle. The quality of the steam leaving the turbine after the isentropic expansion to state 4′ is higher than it is at state 4. This increase is important since the quality of the steam in the turbine exhaust should be maintained above about 90 percent to prevent excessive erosion of the turbine blades caused by water droplets in the low-pressure sections of the turbine.

Another modification that produces an increase in the thermal efficiency of a basic Rankine cycle is a reduction in the temperature of the fluid in the condenser, as depicted in Figure 8.5. Rather than assume that the steam is expanded in the turbine to state 4, assume that the condenser temperature is reduced to a lower value such that the water leaves the turbine and enters the condenser at state 4′. The condenser now operates between states 4′ and 1′ such that the steam leaving the condenser is completely condensed to a saturated liquid at state 1′. The pump now compresses the liquid water along an isentropic path between states 1′ and 2′. The net work produced by this modified Rankine cycle is increased, as can be seen by the increase in area within the *T-s* cycle shown in Figure 8.5. However, reducing the condenser temperature does have an undesirable side effect. The quality at state 4′ is lower than that at state 4 when the steam leaves the turbine at a higher condenser temperature. This effect would cause increased wear on the low-pressure turbine blades.

EXAMPLE 8.2

The ideal Rankine cycle described in Example 8.1 is modified so that the temperature of the water entering the turbine is increased to 700°C. Calculate the thermal efficiency of this modified cycle. Compare the irreversibilities of the cycle in Example 8.1 with the irreversibilities of this modified cycle. For the purpose of comparison assume that the heat-transfer rate to the boiler in each cycle is the same.

Solution. The states for this cycle are shown in Figure 8.4. The enthalpies at states 1 and 2, determined in Example 8.1, are

$$h_1 = 191.8 \text{ kJ/kg} \qquad h_2 = 197.9 \text{ kJ/kg}$$

Increasing the turbine-inlet temperature does not affect the work per unit mass required for the pumping process. The enthalpy and entropy for the water entering the turbine at 6 MPa and 700°C are

$$h_{3'} = 3893.6 \text{ kJ/kg} \qquad s_{3'} = 7.4227 \text{ kJ/kg·K}$$

The turbine process is isentropic, so $s_{4'} = s_{3'}$ and the quality at state 4′ can be determined from

$$x_{4'} = \frac{s_{4'} - s_f}{s_{fg}} = \frac{(7.4227 - 0.6493)\text{kJ/kg·K}}{7.4990 \text{ kJ/kg·K}} = 0.9032$$

Increasing the turbine-inlet temperature by 100°C has increased the quality of the water leaving the turbine by approximately 3 percent. The enthalpy at state 4′ is then

$$\begin{aligned} h_{4'} &= h_f + x_{4'}h_{fg} = 191.8 \text{ kJ/kg} + 0.9032(2392 \text{ kJ/kg}) \\ &= 2352.3 \text{ kJ/kg} \end{aligned}$$

The thermal efficiency, determined by Equation 8.9, is

$$\begin{aligned} \eta_{\text{th}} &= \frac{h_{3'} - h_{4'} + h_1 - h_2}{h_{3'} - h_2} \\ &= \frac{(3893.6 - 2352.3 + 191.8 - 197.9)\text{kJ/kg}}{(3893.6 - 197.9)\text{kJ/kg}} \\ &= 0.415 = \underline{\underline{41.5 \text{ percent}}} \end{aligned}$$

Increasing the turbine-inlet temperature by 100°C has increased the thermal efficiency of the cycle by 1.6 percent.

As in Example 8.1, the irreversibility for each process is calculated from Equation 6.47:

$$i = T_0\left[(s_e - s_i) + \frac{q_k}{T_k}\right]$$

and the change in availability is determined from Equation 6.32:

$$\psi_e - \psi_i = (h_e - h_i) - T_0(s_e - s_i)$$

The properties at each state of this modified cycle are summarized in the following table; notice that the modification affects only the properties at states 3′ and 4′.

State	T, °C	P, MPa	h, kJ/kg	s, kJ/kg·K	Note
1	45.82	0.01	191.8	0.6493	Saturated liquid
2		6	197.9	0.6493	Compressed liquid
3′	700	6	3893.6	7.4227	
4′	45.82	0.01	2352.3	7.4227	$x_{4'} = 0.9032$

The cycle characteristics obtained with these properties are summarized in Table 8.2. These data should be compared with the corresponding table entries presented in Example 8.1, Table 8.1.

For a more realistic comparison of the two cycles, we can examine the performance of each on the basis of the same heat-transfer rate to the boiler. From Example 8.1,

$$\dot{Q}_{23} = \dot{Q}_{23'} = 277 \text{ MW}$$

If the two cycles have the same heat-transfer rate but different amounts of heat transfer per unit mass, then the mass flow rate for this modified cycle must be different, or

$$\dot{Q}_{23'} = \dot{m}(h_{3'} - h_2)$$

or

$$\dot{m} = \frac{\dot{Q}_{23'}}{(h_{3'} - h_2)} = \frac{277{,}000 \text{ kW}}{3695.7 \text{ kJ/kg}} = 75.0 \text{ kg/s}$$

In other words, the mass flow rate required for the modified cycle is about 5 kg/s less than that of the previous cycle (a 6 percent decrease).

The net power produced by each cycle is

$$\dot{W}_{\text{net,unmodified}} = (80 \text{ kg/s})(1382.0 \text{ kJ/kg}) = 110{,}600 \text{ kW}$$

and

$$\dot{W}_{\text{net,modified}} = (75.0 \text{ kg/s})(1535.2 \text{ kJ/kg}) = 115{,}100 \text{ kW}$$

TABLE 8.2 IDEAL RANKINE CYCLE
$T_H = 1400$ K, $T_L = T_0 = 298$ K

Process	q, kJ/kg	w, kJ/kg	$h_e - h_i$, kJ/kg	$s_e - s_i$, kJ/kg·K	$\psi_e - \psi_i$, kJ/kg	i, kJ/kg
1-2	0.0	−6.1	6.1	0.0	6.1	0.0
2-3′	3695.7	0.0	3695.7	6.7734	1677.2	1231.8
3′-4′	0.0	1541.3	−1541.3	0.0	−1541.3	0.0
4′-1	−2160.5	0.0	−2160.5	−6.7734	−142.0	142.0
Cycle	1535.2	1535.2	0.0	0.0	0.0	1373.8

Thus the power output of the modified cycle is 4 percent higher than that of the unmodified cycle for the same heat-transfer rate in the boiler, even though the mass flow rate for the modified cycle is 6 percent lower.

The irreversibility rate is also improved for the modified cycle:

$$\dot{I}_{\text{unmodified}} = (80\ \text{kg/s})(1341.6\ \text{kJ/kg}) = \underline{\underline{107{,}300\ \text{kW}}}$$

and

$$\dot{I}_{\text{modified}} = (75.0\ \text{kg/s})(1373.8\ \text{kJ/kg}) = \underline{\underline{103{,}000\ \text{kW}}}$$

Even though the irreversibility per unit mass for the modified cycle is higher, the irreversibility rate is 4 percent less than that for the unmodified cycle.

Increasing the maximum temperature can be seen to have definite advantages from the viewpoint of both an energy analysis and a second-law analysis when a suitable basis for comparison is chosen. For a given heat-transfer rate to the boiler, the required mass flow rate is reduced, the net power output is increased, and the irreversibility rate is decreased.

The irreversibilities for both cycles result from the heat transfer across finite temperature differences in the condenser and the boiler. Since the area under the *T-s* curve is equal to the heat transfer for an internally reversible process, the changes in heat transfer that occur in the boiler and condenser due to increases in the turbine-inlet temperature can be determined by examining Figure 8.4. Increasing the turbine-inlet temperature from 3 to 3′ increases the heat transfer per unit mass in the boiler (the area under the line segment 1-2-3′ is greater than the area under the curve 1-2-3) and increases the heat transfer in the condenser per unit mass (the area under 1-4′ is greater than the area under 1-4). Since increasing the turbine-inlet temperature increases the heat transfer per unit mass in both the boiler and the condenser, it also increases the irreversibility of the cycle per unit mass for each of these processes. ■

8.3 IDEAL RANKINE CYCLE MODIFIED WITH REHEAT

As with all cycles, modifications to the ideal Rankine cycle that would improve the thermal efficiency should be explored. Two modifications that achieve this objective are reheat and regeneration. The ideal Rankine cycle modified with the *reheat process* is illustrated schematically along with the accompanying *T-s* diagram in Figure 8.6. The *reheat cycle* is characterized by a two-stage expansion of the steam in the turbine. The steam is first expanded in a high-pressure section of the turbine and then routed back to the boiler, where it is reheated before returning to the low-pressure section of the turbine. The steam in the reheat cycle therefore expands in a two-stage process separated by reheat in the boiler. Like the process that occurs in the boiler, the ideal reheat process is also assumed to be an internally reversible, constant-pressure process, shown as path 4-5 in Figure 8.6. The isentropic expansion in the high-pressure turbine is process 3-4, and the isentropic expansion in the low-pressure turbine is process 5-6. The steam leaves the low-pressure turbine at state 6, and it is condensed to a saturated liquid at state 1. The liquid enters the pump at that state, and the cycle is repeated.

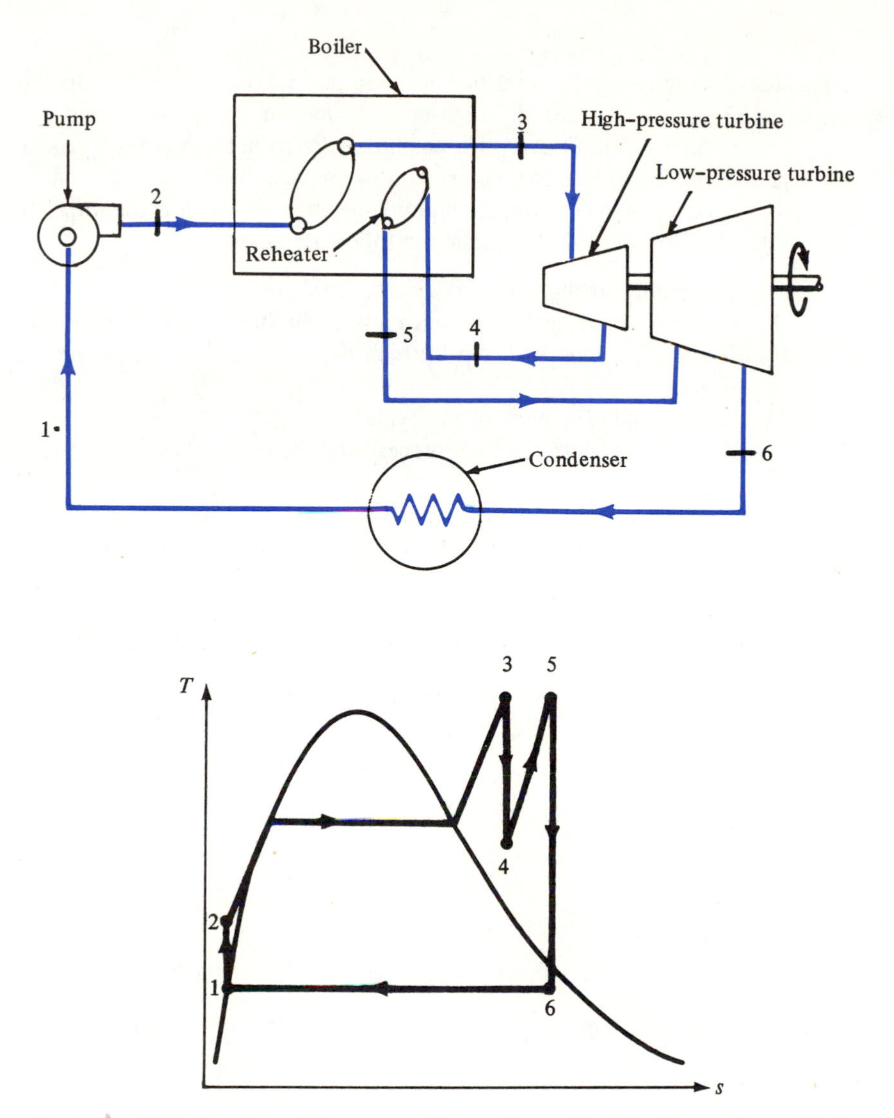

Figure 8.6 Components and T-s process diagram for an ideal Rankine cycle modified with the reheat process.

As can be seen from the T-s diagram in Figure 8.6, the average temperature of the water in the boiler is not significantly influenced by the reheat process; therefore, the Rankine cycle modified by reheat does not produce a noticeable change in the thermal efficiency of the cycle. However, the presence of reheat does have a desirable influence on the operation of the cycle since it causes an increase in the quality of the exhaust steam in the low-pressure turbine, and it thereby reduces the probability of damage to the turbine blades.

■ EXAMPLE 8.3

The ideal Rankine cycle described in Example 8.1 is modified to include the reheat process. The water leaves the high-pressure turbine at 0.5 MPa and enters the low-pressure turbine at 500°C. All other conditions stated in Example 8.1 remain unchanged. To have a common basis for comparison of the two cycles, assume that the total heat-transfer rate to each cycle from the high-temperature reservoir is the same (from Example 8.1, $\dot{Q}_H = 277{,}000$ kW). Calculate the following quantities:

(a) The quality of the steam leaving the low-pressure turbine
(b) The total heat transfer per unit mass in the boiler and the reheater
(c) The total power produced by both the high- and low-pressure sections of the turbine
(d) The thermal efficiency of the cycle
(e) The irreversibility for each process and for the cycle

Solution. The presence of the reheat process does not affect the states 1, 2, or 3 shown in Figure 8.1. The enthalpy values for these states taken from Example 8.1 are

$$h_1 = 191.8 \text{ kJ/kg} \qquad h_2 = 197.9 \text{ kJ/kg} \qquad h_3 = 3658.1 \text{ kJ/kg}$$

The expansion process in the turbine is assumed to be isentropic, so that

$$s_3 = s_4 = 7.1673 \text{ kJ/kg·K}$$

The second known property at state 4 is the given pressure:

$$P_4 = 0.5 \text{ MPa}$$

Therefore, from Table B.3

$$h_4 = 2909.0 \text{ kJ/kg}$$

The reheat process is one of constant pressure, or

$$P_5 = P_4 = 0.5 \text{ MPa} \qquad T_5 = 500°\text{C}$$

From Table B.3 at these two conditions, the enthalpy and entropy at state 5 are

$$h_5 = 3483.9 \text{ kJ/kg} \qquad s_5 = 8.0873 \text{ kJ/kg·K}$$

(a) The expansion process in the low-pressure turbine is isentropic, so

$$s_5 = s_6 = 8.0873 \text{ kJ/kg·K}$$

$$x_6 = \frac{s_6 - s_f}{s_{fg}} = \frac{(8.0873 - 0.6493)\text{kJ/kg·K}}{7.4990 \text{ kJ/kg·K}}$$

$$= 0.992 = \underline{\underline{99.2 \text{ percent}}}$$

The enthalpy of the water leaving the low-pressure turbine is

$$h_6 = h_f + x_6 h_{fg} = 191.8 \text{kJ/kg} + 0.992(2392)\text{kJ/kg}$$
$$= 2564.7 \text{ kJ/kg}$$

The properties at each state are summarized in the following table:

State	T, °C	P, MPa	h, kJ/kg	s, kJ/kg·K	Note
1	45.82	0.01	191.8	0.6493	Saturated liquid
2		6	197.9	0.6493	Compressed liquid
3	600	6	3658.1	7.1673	
4	225.7	0.5	2909.0	7.1673	
5	500	0.5	3483.9	8.0873	
6	45.82	0.01	2564.7	8.0873	$x_6 = 0.992$

(b) The total heat transfer per unit mass in the boiler and reheater is

$$q_{\text{in}} = q_{23} + q_{45} = (h_3 - h_2) + (h_5 - h_4)$$
$$= (3658.1 - 197.9)\text{kJ/kg} + (3483.9 - 2909.0)\text{kJ/kg}$$
$$= 3460.2 \text{ kJ/kg} + 574.9 \text{ kJ/kg} = \underline{\underline{4035 \text{ kJ/kg}}}$$

Since the total heat-transfer rate to the cycle is assumed to be the same as that for the simple cycle in Example 8.1, we can calculate the mass flow rate for this reheat cycle as

$$\dot{Q}_H = \dot{Q}_{\text{in}} = 277{,}000 \text{ kW} = \dot{m}q_{\text{in}}$$

or

$$\dot{m} = \frac{277{,}000 \text{ kW}}{4035 \text{ kJ/kg}} = 68.6 \text{ kg/s}$$

(c) The power produced by the high-pressure turbine is

$$\dot{W}_{34} = \dot{m}(h_3 - h_4) = (68.6 \text{ kg/s})(3658.1 - 2909.0)\text{kJ/kg} = 51{,}390 \text{ kW}$$

The power produced by the low-pressure turbine is

$$\dot{W}_{56} = \dot{m}(h_5 - h_6) = (68.6 \text{ kg/s})(3483.9 - 2564.7)\text{kJ/kg} = 63{,}060 \text{ kW}$$

The total turbine power is therefore

$$\dot{W}_{\text{tot}} = \dot{W}_{34} + \dot{W}_{56} = 51{,}390 \text{ kW} + 63{,}060 \text{ kW} = \underline{\underline{114{,}450 \text{ kW}}}$$

(d) The power input to the pump is

$$\dot{W}_{12} = \dot{m}(h_1 - h_2)$$
$$= (68.6 \text{ kg/s})(191.8 - 197.9)\text{kJ/kg} = -418 \text{ kW}$$

Thus the thermal efficiency of the cycle is

$$\eta_{\text{th}} = \frac{\dot{W}_{\text{net}}}{\dot{Q}_{\text{in}}} = \frac{\dot{W}_{\text{tot}} + \dot{W}_{12}}{\dot{Q}_{\text{in}}}$$
$$= \frac{114{,}450 \text{ kW} - 418 \text{ kW}}{277{,}000 \text{ kW}} = 0.412 = \underline{\underline{41.2 \text{ percent}}}$$

In this example the thermal efficiency of the Rankine cycle modified with the reheat process is 1.3 percentage points greater than that without reheat.

(e) For each process the irreversibility is determined from Equation 6.47:

$$i = T_0\left[(s_e - s_i) + \frac{q_k}{T_k}\right]$$

and, in addition, the change in availability for each process, from Equation 6.32, is

$$\psi_e - \psi_i = (h_e - h_i) - T_0(s_e - s_i)$$

The results of these and other calculations for the ideal reheat cycle are summarized in Table 8.3.

Comparing the entries in this table with those presented in Example 8.1, Table 8.1, for the simple cycle, we find that the irreversibility per unit mass of the heat-rejection process is higher even though the temperature is unchanged. This increase is attributed to the fact that the heat transfer per unit mass is higher. However, the irreversibility *rate* for the heat-rejection process in the reheat cycle is less than that for the simple cycle because the mass flow rate is substantially smaller. Note also that the reheat process adds another source of irreversibility into the cycle—another process involving heat transfer through a finite temperature difference. But since the required mass flow rate for this cycle is smaller than that of the simple cycle, the irreversibility rate for this cycle is less. The calculation to support these conclusions follows. We use the subscript "simple" to denote the simple ideal Rankine cycle discussed in Example 8.1 and the subscript "reheat" to denote the ideal reheat cycle.

$$\dot{Q}_{\text{in,simple}} = \dot{Q}_{\text{in,reheat}} = 277{,}000 \text{ kW}$$

From previous calculations,

$$\dot{m}_{\text{simple}} = 80 \text{ kg/s} \quad \text{and} \quad \dot{m}_{\text{reheat}} = 68.6 \text{ kg/s}$$

TABLE 8.3 IDEAL RANKINE CYCLE
$T_H = 1400$ K, $T_L = T_0 = 298$ K

Process	q, kJ/kg	w, kJ/kg	$h_e - h_i$, kJ/kg	$s_e - s_i$, kJ/kg·K	$\psi_e - \psi_i$, kJ/kg	i, kJ/kg
1-2	0.0	−6.1	6.1	0.0	6.1	0.0
2-3	3460.2	0.0	3460.2	6.518	1517.8	1205.8
3-4	0.0	749.1	−749.1	0.0	−749.1	0.0
4-5	574.9	0.0	574.9	0.920	300.7	151.8
5-6	0.0	919.2	−919.2	0.0	−919.2	0.0
6-1	−2372.9	0.0	−2372.9	−7.438	−156.4	156.4
Cycle	1662.2	1662.2	0.0	0.0	−0.1	1514.0

The net power output for each cycle is

$$\dot{W}_{\text{net,simple}} = 111{,}000 \text{ kW}$$

from Example 8.1, and

$$\dot{W}_{\text{net,reheat}} = (68.6 \text{ kg/s})(1662.2 \text{ kJ/kg}) = 114{,}000 \text{ kW}$$

The irreversibility rate for each cycle is

$$\dot{I}_{\text{simple}} = 107{,}300 \text{ kW}$$

and

$$\dot{I}_{\text{reheat}} = \dot{m}_{\text{reheat}} i_{\text{cyc}} = (68.6 \text{ kg/s})(1514.0 \text{ kJ/kg}) = \underline{\underline{103{,}900 \text{ kW}}}$$

Thus for the same total heat-transfer rate to each cycle, reheat has the following advantages: The mass flow rate is decreased by 14 percent, the net power output is increased by 3.2 percent, and the irreversibility rate is decreased by 3.2 percent.

The primary sources of irreversibility are the heat-addition processes. The decrease in the irreversibility rate for the cycle can be attributed to the fact that the average temperature of the working fluid during the overall heat-addition process is increased by including the reheat section. ■

8.4 IDEAL RANKINE CYCLE MODIFIED WITH REGENERATION

Regeneration is one common means of increasing a cycle's thermal efficiency by increasing the average temperature of the working fluid during the heating process. The term *regeneration* was used in the previous chapter in connection with the Brayton cycle, whereby the thermal efficiency of the gas-turbine cycle was increased by extracting energy from the gas leaving the turbine exhaust. In a somewhat similar fashion, a portion of the steam that enters the turbine in the Rankine cycle can be extracted before it completely expands to the condenser pressure, and it can be used to preheat the liquid water before it enters the boiler. In this manner the temperature in the boiler is increased, and the thermal efficiency of the cycle is likewise increased. The regeneration process takes place in a device called a *feedwater heater*. The T-s diagram and system schematic for a Rankine cycle modified with regeneration are shown in Figure 8.7.

Steam leaves the boiler at state 5 as a superheated vapor and enters the high-pressure stage of the turbine. In the ideal cycle the steam expands isentropically within the turbine. However, some of the steam is extracted at an intermediate pressure at state 6 and is routed to an open feedwater heater. The remaining steam continues the isentropic expansion through the turbine and leaves the low-pressure stages of the turbine at state 7, where it enters the condenser. This steam condenses along a constant-pressure path in the condenser, enters the isentropic pump at state 1, and leaves the pump at state 2.

After leaving the pump, the condensed water enters the feedwater heater at state 2 and mixes with the steam extracted from the turbine at state 6. The proportion of water

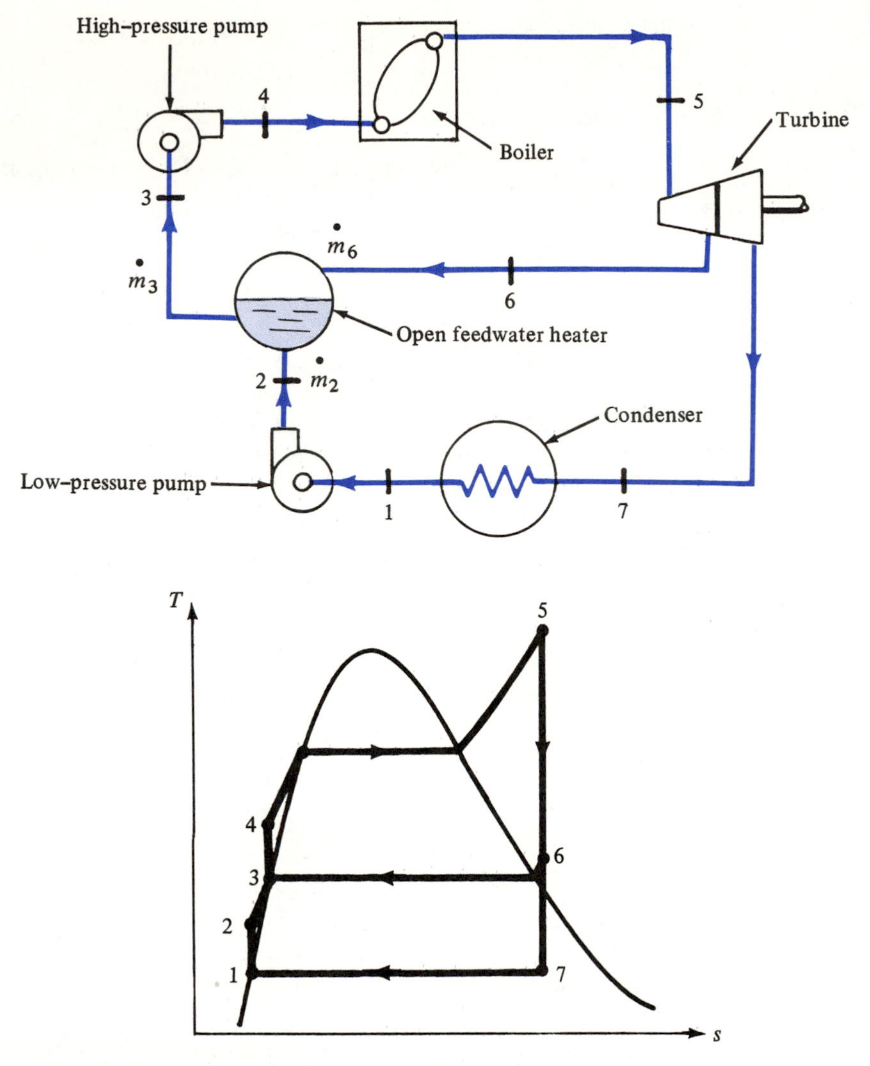

Figure 8.7 Components and T-s process diagram for an ideal Rankine cycle modified with a regeneration process.

to steam entering the feedwater heater is regulated so that the state of the water leaving the feedwater heater is a saturated liquid at state 3. The steam extracted from the turbine at state 6 condenses at constant pressure along path 6-3, and the temperature of the liquid water leaving the pump increases along constant-pressure path 2-3.

The feedwater heater considered here is called an *open feedwater heater* because the steam and water are mixed in an open process where there is direct contact between and mixing of the two fluids. In an open feedwater heater the pressures of the liquid and vapor streams are equal, as shown in Figure 8.7 where $P_2 = P_3 = P_6$. This mixing process is not internally reversible, and therefore the open feedwater heater is an additional source of irreversibility in the cycle. However, the regeneration process increases the average temperature at which heat transfer to the cycle occurs, and it therefore results in an overall decrease in the irreversibility of the cycle. In an ideal cycle the feedwater heater is assumed to be perfectly insulated. Therefore, the energy transferred from the extracted steam in the feedwater heater is equal to the amount of energy added to the liquid water arriving from the low-pressure pump. After leaving the feedwater heater, the water is a saturated liquid at state 3. The pressure of the water is below the boiler pressure, so a second high-pressure pump is necessary to increase the pressure of the water to the boiler pressure at state 4. In the ideal cycle the high-pressure pump is assumed to operate isentropically along path 3-4. The water then enters the boiler, and heat transfer to the water from the high-temperature combustion gases takes place such that the water follows the constant pressure path 4-5. The superheated vapor leaves the boiler at state 5, and the cycle is repeated.

Several salient features of the Rankine cycle modified with regeneration should be recognized. First, the thermal efficiency of the cycle is increased because the average temperature of the water in the boiler is increased, as is evident in Figure 8.7. Second, more equipment is necessary when regeneration is used; the cycle requires a feedwater heater and one additional pump, neither of which is required for the unmodified Rankine cycle.

In most power plants several feedwater heaters are staged, each feedwater heater operating with steam extracted at different locations and at different pressures within the turbine. The selection of the number of feedwater heaters used in any particular plant is made after a complete thermodynamic and economic analysis of the cycle has been performed. Even though the inclusion of more feedwater heaters will increase plant efficiency, the increase in efficiency is not unlimited. Furthermore, the marginal increases in plant efficiency must be balanced against the increased costs of the additional equipment.

An alternative to the open feedwater heater is a *closed feedwater heater* in which the liquid water and extracted steam are kept separate so that they never come into direct contact with each other. In the closed feedwater heater the tubes separating the feedwater and steam represent a thermal resistance to heat transfer and therefore the energy-transfer process is somewhat less efficient than it is in the open feedwater heater. Furthermore, the open feedwater heater is less complex and less expensive because the internal piping necessary in a closed feedwater heater is not required. The closed feedwater heater, however, does permit the water and steam to be maintained at different pressures so that separate pumps for each feedwater heater are not required. The principles of a Rankine cycle with regeneration are illustrated in the following example.

EXAMPLE 8.4

The ideal Rankine cycle described in Example 8.1 is modified to include the regeneration process. The steam leaving the turbine, which enters the open feedwater heater, has a pressure of 0.5 MPa. All other conditions stated in Example 8.1 remain unchanged. Assume that the heat-transfer rate to the boiler is the same as that in Example 8.1. Calculate the following quantities:

(a) The work per unit mass required to operate both pumps
(b) The mass flow rate of water in the boiler

(c) The power developed by the turbine
(d) The heat-transfer rate in the condenser
(e) The thermal efficiency of the cycle
(f) The irreversibility for each process and for the cycle

Solution. The given properties at the states shown in Figure 8.7 are

$$P_1 = P_7 = 10 \text{ kPa} \qquad P_4 = P_5 = 6 \text{ MPa}$$

$$T_5 = 600°\text{C} \qquad P_2 = P_3 = P_6 = 0.5 \text{ MPa}$$

Property values from Example 8.1 are

$$h_5 = 3658.1 \text{ kJ/kg} \qquad s_7 = s_6 = s_5 = 7.1673 \text{ kJ/kg·K}$$

$$h_7 = 2270 \text{ kJ/kg} \qquad v_1 = 1.010 \times 10^{-3}\text{m}^3/\text{kg}$$

$$h_1 = 191.8 \text{ kJ/kg} \qquad s_1 = s_2 = 0.6493 \text{ kJ/kg·K}$$

(a) From Equation 8.4 the work per unit mass required by the low-pressure pump is

$$w_{12} = h_1 - h_2 = v_1(P_1 - P_2)$$
$$= (1.01 \times 10^{-3}\text{m}^3/\text{kg})(10 - 500)\text{kN/m}^2 = \underline{\underline{-0.495 \text{ kJ/kg}}}$$

Thus the enthalpy at state 2 is

$$h_2 = h_1 - w_{12} = (191.8 + 0.5)\text{kJ/kg} = 192.3 \text{ kJ/kg}$$

State 3 is a saturated liquid at a pressure of 0.5 MPa, so that

$$h_3 = 640.4 \text{ kJ/kg} \qquad v_3 = 1.093 \times 10^{-3}\text{m}^3/\text{kg}$$

$$s_3 = 1.8610 \text{ kJ/kg·K}$$

From Equation 8.4 the work per unit mass required by the high-pressure pump is

$$w_{34} = h_3 - h_4 = v_3(P_3 - P_4)$$
$$= (1.093 \times 10^{-3}\text{m}^3/\text{kg})(500 - 6000)\text{kN/m}^2 = \underline{\underline{-6.0 \text{ kJ/kg}}}$$

Therefore, the enthalpy at state 4 is

$$h_4 = h_3 - w_{34} = (640.4 + 6.0)\text{kJ/kg} = 646.4 \text{ kJ/kg}$$

The enthalpy at state 6 can be determined by interpolating in Table B.3 at the known pressure and entropy:

$$h_6 = 2909.0 \text{ kJ/kg}$$

The properties for the seven states identified for this cycle are summarized in the following table.

State	T, °C	P, MPa	h, kJ/kg	s, kJ/kg·K	Note
1	45.82	0.01	191.8	0.6493	Saturated liquid
2		0.5	192.3	0.6493	Compressed liquid
3	151.87	0.5	640.4	1.8610	Saturated liquid
4		6.0	646.4	1.8610	Compressed liquid
5	600	6.0	3658.1	7.1673	
6	225.7	0.5	2909.0	7.1673	
7	45.82	0.01	2270.0	7.1673	$x_7 = 0.869$

(b) The mass flow rate of water in the boiler can be determined from the conservation-of-energy equation applied to the boiler and the fact that the heat-transfer rate is the same as that calculated in Example 8.1:

$$\dot{Q}_{45} = \dot{m}_4(h_5 - h_4)$$

or

$$\dot{m}_4 = \frac{\dot{Q}_{45}}{(h_5 - h_4)} = \frac{277{,}000\,\text{kW}}{(3658.1 - 646.4)\text{kJ/kg}} = \underline{\underline{91.975\,\text{kg/s}}}$$

(The number of significant figures in this calculation and those that follow is more than is usually retained, in an attempt to minimize the influence of round-off error.)

The mass flow rate of water extracted from the turbine and circulated through the open feedwater heater can be determined by applying the conservation-of-mass and -energy equations to the feedwater heater. The conservation of mass results in the equations

$$\dot{m}_2 + \dot{m}_6 = \dot{m}_3$$

and

$$\dot{m}_3 = \dot{m}_4 = 91.975\text{ kg/s}$$

The feedwater heater is adiabatic and performs no work, so the conservation-of-energy equation can be written as

$$\dot{m}_6 h_6 + \dot{m}_2 h_2 - \dot{m}_3 h_3 = 0$$

Combining these equations to eliminate $\dot{m}_6$ gives

$$(\dot{m}_3 - \dot{m}_2)h_6 + \dot{m}_2 h_2 - \dot{m}_3 h_3 = 0$$

or

$$\dot{m}_2 = \dot{m}_3\left(\frac{h_6 - h_3}{h_6 - h_2}\right)$$

$$= (91.975\text{ kg/s})\left[\frac{(2909.0 - 640.4)\text{kJ/kg}}{(2909.0 - 192.3)\text{kJ/kg}}\right] = 76.804\text{ kg/s}$$

and the mass flow rate of steam extracted from the turbine is

$$\dot{m}_6 = \dot{m}_3 - \dot{m}_2 = (91.975 - 76.804)\text{kg/s} = 15.171 \text{ kg/s}$$

(c) The total power output of the turbine, from the conservation-of-energy equation, is

$$-\dot{W}_T + \dot{m}_5 h_5 - \dot{m}_7 h_7 - \dot{m}_6 h_6 = 0$$

But

$$\dot{m}_7 = \dot{m}_2 \qquad \dot{m}_5 = \dot{m}_3 \qquad \dot{m}_6 = \dot{m}_3 - \dot{m}_2$$

Therefore

$$\begin{aligned} \dot{W}_T &= \dot{m}_3(h_5 - h_6) + \dot{m}_2(h_6 - h_7) \\ &= (91.975 \text{ kg/s})(3658.1 - 2909.0)\text{kJ/kg} \\ &\quad + (76.804 \text{ kg/s})(2909.0 - 2270)\text{kJ/kg} \\ &= \underline{\underline{117{,}976 \text{ kW}}} \end{aligned}$$

The total power required by the pumps is

$$\begin{aligned} \dot{W}_p &= \dot{W}_{12} + \dot{W}_{34} = \dot{m}_2 w_{12} + \dot{m}_3 w_{34} \\ &= (76.804 \text{ kg/s})(-0.5 \text{ kJ/kg}) + (91.975 \text{ kg/s})(-6.0 \text{ kJ/kg}) \\ &= -590.3 \text{ kW} \end{aligned}$$

(d) The heat-transfer rate in the condenser is

$$\begin{aligned} \dot{Q}_{71} &= \dot{m}_2(h_1 - h_7) = (76.804 \text{ kg/s})(191.8 - 2270)\text{kJ/kg} \\ &= \underline{\underline{-159{,}614 \text{ kW}}} \end{aligned}$$

(e) The thermal efficiency of the cycle is

$$\eta_{\text{th}} = \frac{\dot{W}_{\text{net}}}{\dot{Q}_{45}} = \frac{\dot{W}_T + \dot{W}_p}{\dot{Q}_{45}} = \frac{(117{,}976 - 590.3)\text{kW}}{277{,}000 \text{ kW}} = 0.424 = \underline{\underline{42.4 \text{ percent}}}$$

The presence of the regenerator produces an increase in the thermal efficiency of 6.2 percent over the simple ideal cycle's efficiency.

(f) For each process the irreversibility rate is determined from Equation 6.47:

$$\dot{I} = T_0 \left(\sum_{\text{exit}} \dot{m}s - \sum_{\text{inlet}} \dot{m}s + \frac{\dot{Q}_k}{T_k} \right)$$

and the rate of change of availability is determined by using Equation 6.32 in the form

$$\sum_{\text{exit}} \dot{m}\psi - \sum_{\text{inlet}} \dot{m}\psi = \sum_{\text{exit}} \dot{m}(h - T_0 s) - \sum_{\text{inlet}} \dot{m}(h - T_0 s)$$

TABLE 8.4 RANKINE CYCLE WITH REGENERATION
$T_H = 1400$ K, $T_L = T_0 = 298$ K

Process	$\dot{Q}$, kW	$\dot{W}$, kW	$\sum_e \dot{m}h - \sum_i \dot{m}h$, kW	$\sum_e \dot{m}s - \sum_i \dot{m}s$, kW/K	$\dot{I}$, kW
1-2	0.0	−38.4	38.4	0.0	0.0
3-4	0.0	−551.9	551.9	0.0	0.0
4-5	277,000	0.0	277,000	488.0	86,463
5-6	0.0	68,898	−68,898	0.0	0.0
FWH	0.0	0.0	0.0	12.6	3,755
6-7	0.0	49,078	−49,078	0.0	0.0
7-1	−159,614	0.0	−159,614	−500.6	10,435
Cycle	117,386	117,386	0.3	0.0	100,653

The characteristics of this cycle are summarized in Table 8.4 (the abbreviation FWH represents feedwater heater). Since the mass flow rates are different for various components in the cycle, the quantities are not presented on a unit-mass basis, as was done in the preceding examples.

Comparing the results in Table 8.4 with those for the ideal simple cycle in Example 8.1, we find that the cycle modified with reheat with same heat-transfer rate to the boiler has the following features:

1. A 15 percent higher mass flow rate in the boiler
2. A 6.2 percent higher net power output
3. A 6.2 percent lower irreversibility rate

Even though the feedwater heater adds another source of irreversibility to the cycle, the irreversibility rate of the boiler process is significantly reduced because the average temperature of the fluid in the boiler is increased. Therefore, there is a net decrease in the irreversibility rate of the cycle for the same heat-transfer rate as occurs in the simple ideal cycle. The irreversibility for an open feedwater heater is due to the mixing of two fluid streams—an irreversible process. ■

8.5 IDEAL VAPOR-COMPRESSION REFRIGERATION CYCLE

The most widely used cycle for air-conditioning systems and refrigeration plants is the *vapor-compression refrigeration cycle*. In this cycle the refrigerant in the vapor phase is compressed in a compressor, causing the temperature to exceed that of the high-temperature reservoir, usually the environment. The hot, high-pressure refrigerant is then circulated through a heat exchanger, called a *condenser*, where it is cooled by heat transfer to the environment. As a result of the heat transfer in the condenser, the refrigerant condenses from a vapor to a liquid. After leaving the condenser, the refrigerant passes through a throttling device, where the pressure and temperature of the refrigerant both decrease. The cold refrigerant leaves the throttling device and enters a second heat exchanger, called an *evaporator*, located in the refrigerated space. Heat transfer in the evaporator causes the refrigerant to evaporate or change from a saturated mixture of liquid and vapor into a superheated vapor. The vapor leaving the evaporator is then routed back into the compressor, and the cycle is repeated. The throttling valve is irreversible because it relies on fluid friction for its operation. Nonetheless, the vapor-compression refrigeration cycle is still referred to as an ideal cycle.

Air-conditioning units come in many capacities and sizes ranging from small residential air conditioners that are capable of cooling single rooms up to integrated units that can provide comfort cooling for large commercial and industrial buildings. Most small units use reciprocating compressors (see Figure 8.8), and heat rejection from these units normally occurs to the outdoor air. Large units typically employ water-cooled centrifugal compressors (see Figure 4.24). The water that is heated in the condenser is pumped outdoors, often to cooling towers that reject energy to outdoor air. The chilled water that is produced in the evaporator is pumped to a remote location where the conditioned air is needed.

A schematic of the four basic components of a vapor-compression refrigeration cycle is shown in Figure 8.9 along with the *T-s* diagram for the ideal cycle. The refrigerant

Figure 8.8 A two-cylinder reciprocating refrigeration compressor. The electric drive motor on the left and the compressor section are enclosed to minimize leakage of the refrigerant. (Courtesy of The Trane Company.)

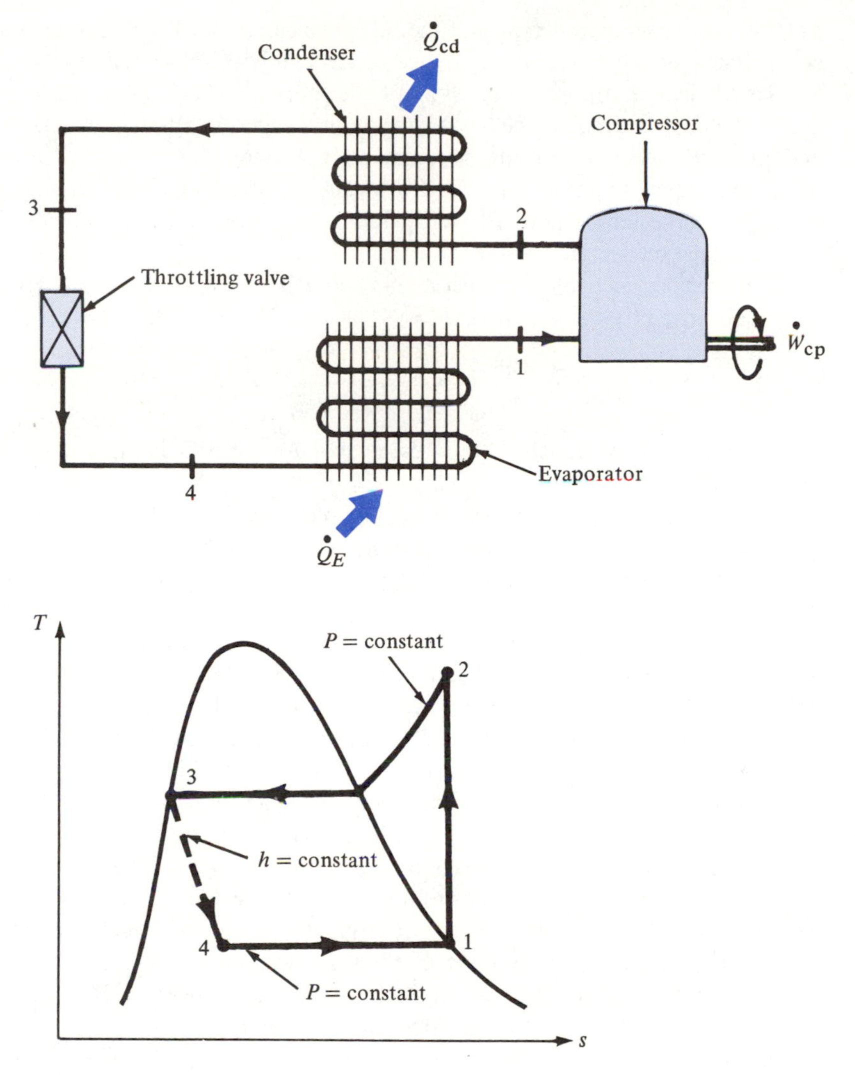

Figure 8.9 Schematic and T-s process diagram of an ideal vapor-compression refrigeration cycle.

enters the compressor as a vapor at state 1 and is compressed isentropically to state 2 in the superheat region. The temperature of the refrigerant at state 2 exceeds that of the environment, so that when the refrigerant enters the condenser, heat transfer to the environment along the constant-pressure path 2-3 will occur. As the refrigerant cools in the condenser and the temperature drops, the refrigerant condenses, forming a liquid at state 3. At state 3 the pressure of the liquid refrigerant is still equal to the discharge pressure of the compressor ($P_3 = P_2$), but owing to the heat transfer in the condenser, the refrigerant has a temperature less than the temperature of the compressor discharge. In the throttling valve the refrigerant follows a constant-enthalpy path from state 3 to 4 as the pressure and temperature of the refrigerant both decrease. The throttling valve is occasionally an adjustable *needle valve* or frequently a long *capillary tube*, which provides sufficient restriction to the flow of the refrigerant to create a large pressure drop across the device. Finally, the low-temperature refrigerant at state 4 enters the evaporator and proceeds along a constant-pressure path, leaving the evaporator at state 1. The refrigerant enters the evaporator as a low-quality mixture and leaves as a saturated (or slightly superheated) vapor. The refrigerant changes phase because of the heat transfer to it from the refrigerated space.

The performance of a refrigeration system is measured by the coefficient of performance, given by the equation

$$\beta_R = -\frac{\dot{Q}_e}{\dot{W}_{cp}} \tag{8.10}$$

where $\dot{Q}_e$ is the heat-transfer rate in the evaporator and $\dot{W}_{cp}$ is the power input required by the compressor.

A major source of irreversibility in the refrigeration cycle can be attributed to the evaporator and condenser heat-transfer processes, caused by finite temperature differences between the system and its surroundings. In fact, the only sources of irreversibility in the ideal vapor-compression refrigeration cycle are the heat-transfer processes (external irreversibilities) and the throttling process. In an attempt to reduce the irreversibilities associated with the heat transfer and to increase the cycle COP, modern designs call for the heat exchangers to have large surface areas. The larger surface areas of the heat exchangers produce smaller temperature differences between the refrigerant and the thermal reservoir, thereby reducing the source of the irreversibility. With good design practices the COPs of most air-conditioning and refrigeration systems fall in the range of 2 to 5.

One measure of the refrigeration capacity of refrigeration systems that is widely used, especially for large refrigeration systems, is the ton, which is defined as 12,000 Btu/h or 200 Btu/min. This term was originally used to represent the amount of heat transfer required to produce a ton of ice from approximately 1 ton of liquid water at 32°F. The refrigeration systems in an ice plant usually operated 24 h each day and the current definition of the ton corresponds approximately to the heat-transfer rate required to produce 1 ton of ice in 24 h.

EXAMPLE 8.5

An ideal vapor-compression refrigeration cycle uses refrigerant-12 as a working fluid in an air-conditioning system. The refrigerant enters the compressor as a saturated vapor at 40°F and leaves the condenser as a saturated liquid at 130°F. The mass flow rate of the refrigerant is 1.5 lb_m/s. Heat transfer to the refrigerant in the evaporator is from a reservoir at 60°F (the cool space), and the refrigerant in the condenser is cooled by heat transfer to the environment, which is at a temperature of 115°F. Calculate the following quantities:

(a) The heat-transfer rate in the condenser
(b) The heat-transfer rate in the evaporator
(c) The coefficient of performance of the cycle
(d) The irreversibility for each component and for the cycle

Solution. Properties of refrigerant-12 are obtained from Tables C.1E and C.3E. Changes in kinetic and potential energy of the refrigerant are small compared with enthalpy changes for a refrigeration cycle, and they are neglected in the following calculations.

(a) From Tables C.1E and C.3E the properties of the refrigerant for the states shown in Figure 8.9 are

$$h_1 = h_g(40°\text{F}) = 81.97 \text{ Btu/lb}_\text{m}$$

$$P_1 = P_4 = P_\text{sat}(40°\text{F}) = 51.71 \text{ psia}$$

$$s_1 = s_2 = s_g(40°\text{F}) = 0.1669 \text{ Btu/lb}_\text{m}\cdot°\text{R}$$

$$P_2 = P_3 = P_\text{sat}(130°\text{F}) = 195.2 \text{ psia}$$

$$h_4 = h_3 = h_f(130°\text{F}) = 39.0 \text{ Btu/lb}_\text{m}$$

$$s_3 = s_f(130°\text{F}) = 0.0768 \text{ Btu/lb}_\text{m}\cdot°\text{R}$$

The value for h_2 is obtained at the known values for s_2 and P_2. Interpolating twice in Table C.3E gives

$$h_2 = 92.16 \text{ Btu/lb}_\text{m}$$

Furthermore, s_4 is determined from Table C.1E at the known pressure and enthalpy of state 4:

$$x_4 = \frac{h_4 - h_f}{h_{fg}} = \frac{(39.0 - 17.35)\text{Btu/lb}_\text{m}}{64.62 \text{ Btu/lb}_\text{m}} = 0.335$$

$$s_4 = s_f + x_4 s_{fg}$$

$$= 0.0376 \text{ Btu/lb}_\text{m}\cdot°\text{R} + 0.335(0.1293 \text{ Btu/lb}_\text{m}\cdot°\text{R}) = 0.0809 \text{ Btu/lb}_\text{m}°\text{R}$$

These properties are summarized in the following table:

State	T, °F	P, psia	h, Btu/lb$_\text{m}$	s, Btu/lb$_\text{m}\cdot$°R	Note
1	40	51.71	81.97	0.1669	Saturated vapor
2	140.3	195.20	92.16	0.1669	
3	130	195.20	39.0	0.0768	Saturated liquid
4	40	51.71	39.0	0.0809	$x_4 = 0.335$

The heat-transfer rate in the condenser is

$$\dot{Q}_{cd} = \dot{m}(h_3 - h_2) = (1.5 \text{ lb}_\text{m}/\text{s})(39.0 - 92.16)\text{Btu/lb}_\text{m} = \underline{\underline{-79.7 \text{ Btu/s}}}$$

(b) The heat-transfer rate in the evaporator is

$$\dot{Q}_e = \dot{m}(h_1 - h_4) = (1.5 \text{ lb}_\text{m}/\text{s})(81.97 - 39.0)\text{Btu/lb}_\text{m} = \underline{\underline{64.5 \text{ Btu/s}}}$$

(c) The coefficient of performance of the cycle is

$$\beta = -\frac{\dot{Q}_e}{\dot{W}_{cp}} = -\frac{h_1 - h_4}{h_1 - h_2} = -\frac{(81.97 - 39.0)\text{Btu/lb}_\text{m}}{(81.97 - 92.16)\text{Btu/lb}_\text{m}} = \underline{\underline{4.22}}$$

(d) For each process the irreversibility per unit mass is determined from Equation 6.47:

$$i = T_0 \left[(s_e - s_i) + \frac{q_k}{T_k} \right]$$

Notice that T_0 is not equal to the temperature of the low-temperature reservoir. Instead, it is the lowest naturally occurring temperature, that is, the environment. Thus $T_0 = T_H$.

The results of the irreversibility calculations are included in Table 8.5. Notice that the cycle irreversibility is about evenly distributed among the evaporator, condenser, and throttling processes in this example. The irreversibility of the throttling process arises from the frictional losses, which cause the pressure decrease, and is due entirely to the increase in entropy of the fluid. The irreversibility in the evaporator and in the condenser is due to heat transfer through a finite temperature difference between the system and its surroundings. For reduction of the irreversibility the average temperature difference must be reduced. Thus the average temperature of the fluid in the condenser should be decreased and the temperature of the fluid in the evaporator should be increased. These changes would require better heat-exchange equipment from a thermodynamic standpoint (for example, larger heat-transfer surface area).

The cycle irreversibility could also be determined from Equation 6.38:

$$i_\text{cyc} = T_0 \oint ds_\text{sys} + \sum_k T_0 \oint \frac{\delta q_k}{T_k}$$

$$= 0 + T_0 \left(\frac{q_{H,\text{res}}}{T_H} + \frac{q_{L,\text{res}}}{T_L} \right) = -T_0 \left(\frac{q_{23}}{T_0} + \frac{q_{41}}{T_L} \right)$$

$$= -(575^\circ\text{R}) \left(\frac{-53.16\ \text{Btu/lb}_\text{m}}{575^\circ\text{R}} + \frac{42.97\ \text{Btu/lb}_\text{m}}{520^\circ\text{R}} \right) = 5.65\ \text{Btu/lb}_\text{m}$$

TABLE 8.5 IDEAL VAPOR-COMPRESSION REFRIGERATION CYCLE $T_H = T_0 = 115°\text{F} = 575°\text{R}$, $T_L = 60°\text{F} = 520°\text{R}$

Process	q, Btu/lb$_\text{m}$	w, Btu/lb$_\text{m}$	$h_e - h_i$, Btu/lb$_\text{m}$	$s_e - s_i$, Btu/lb$_\text{m}$·°R	i, Btu/lb$_\text{m}$
1-2	0.0	−10.19	10.19	0.0	0.0
2-3	−53.16	0.0	−53.16	−0.0901	1.35
3-4	0.0	0.0	0.0	0.0041	2.36
4-1	42.97	0.0	42.97	0.0860	1.94
Cycle	−10.19	−10.19	0.0	0.0	5.65

An even simpler means of determining the irreversibility of the cycle is to use the expression

$$i = q_L \left(\frac{1}{\beta_{\mathrm{R,Car}}} - \frac{1}{\beta_{R,\mathrm{act}}} \right)$$

The derivation of this equation is presented as an exercise for the reader at the end of the chapter. The coefficient of performance for a Carnot refrigerator operating between the same two reservoirs is, from Equation 5.10,

$$\beta_{\mathrm{R,Car}} = \frac{1}{(T_H/T_L) - 1} = \frac{1}{(575°\mathrm{R}/520°\mathrm{R}) - 1} = 9.45$$

Therefore, since $q_L = -q_{41}$,

$$i = (-42.97\ \mathrm{Btu/lb_m}) \left(\frac{1}{9.45} - \frac{1}{4.22} \right) = 5.64\ \mathrm{Btu/lb_m}$$

8.6 ACTUAL VAPOR CYCLES

The operation of actual vapor cycles can deviate from the performance of the ideal cycles described in the preceding sections primarily because of irreversibilities associated with mechanical and fluid friction as well as additional heat transfer. Figure 8.10 illustrates typical deviations that can occur between the ideal Rankine cycle and the actual cycle owing to the inevitable irreversibilities that accompany the performance of real components. The actual pumping process, which includes the effects of fluid friction, follows the path 1-2a, during which the entropy of the water increases. The ideal, isentropic pumping process is shown for comparison as path 1-2s. As a result of irreversibilities present in the pump, the actual pump will require a greater work input than an isentropic pump.

Fluid friction present in the boiler is another possible source of irreversibility in the actual cycle. Motion of the water through the boiler produces a pressure drop, as shown in the figure. Rather than following the constant-pressure path assumed for the ideal cycle, which would result in an inlet state to the turbine of $3p$ shown in Figure 8.10, the actual boiler process is accompanied by a pressure drop such that the actual inlet state to the turbine is $3a$. The availability of the steam at state $3a$ is less than that at state $3p$, and therefore the pressure drop in the boiler has a detrimental effect on the cycle performance. Heat transfer, which takes place from the water in the turbine to the surroundings, is another source of irreversibility. The actual process in the turbine occurs between states $3a$ and $4a$, while the ideal process is shown as the isentropic path $3p$-$4s$. As a result of unavoidable irreversibilities that occur in the turbine, an actual turbine is capable of providing less power than an isentropic turbine. Pressure drops that result from fluid friction in both the boiler and the condenser also have a detrimental effect on the performance of a Rankine cycle. The pressure drop in the condenser produces a further decrease in the thermal efficiency of the actual cycle.

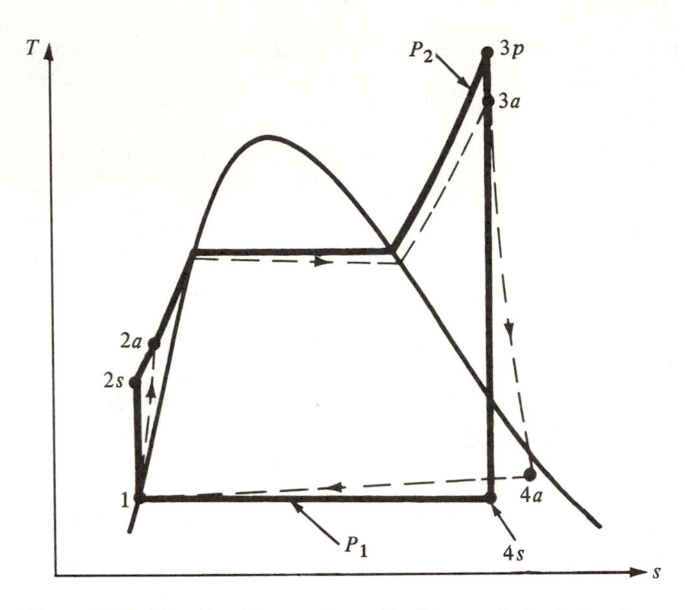

Figure 8.10 The *T-s* diagram for a Rankine cycle including irreversibilities in all components.

EXAMPLE 8.6

A simple power plant operates with an inlet boiler pressure of 6 MPa and an inlet condenser pressure of 10 kPa. Owing to frictional effects the pressure of the water drops 0.4 MPa in the boiler and 1 kPa in the condenser. The temperature of the steam as it enters the turbine is 600°C. The adiabatic efficiency of the pump and of the turbine are 82 percent and 87 percent, respectively. The pump and the turbine are adiabatic, and the water enters the pump as a saturated liquid.

Determine the thermal efficiency of the cycle and compare it with the thermal efficiency of the ideal rankine cycle in Example 8.1. (Example 8.1 uses the same boiler and condenser pressure and the same inlet turbine temperature.) Calculate the irreversibility of each process and the cycle irreversibility.

Assume that heat transfer to the water in the boiler is from a reservoir at 1400 K and that the environment is at a temperature of 25°C.

Solution. The states that exist throughout the actual cycle and ideal cycle are shown in Figure 8.10. The thermal efficiency of the actual cycle is

$$\eta_{th} = \frac{w_{net}}{q_H} = \frac{w_{3a-4a} + w_{1-2a}}{q_{2a-3a}}$$

The heat transfer to the water in the boiler per unit mass is

$$q_{2a-3a} = h_{3a} - h_{2a}$$

The work required to operate the adiabatic pump is

$$w_{1-2a} = h_1 - h_{2a}$$

and the work output from the steam in the turbine is

$$w_{3a-4a} = h_{3a} - h_{4a}$$

Substituting these last three expressions into the equation for the thermal efficiency of the cycle gives the result

$$\eta_{th} = \frac{h_{3a} - h_{4a} + h_1 - h_{2a}}{h_{3a} - h_{2a}}$$

The task now is to determine the enthalpy values at states 1, $2a$, $3a$, and $4a$. The property values at states 1 and $3a$, from Tables B.1 and B.2, are

$$v_1 = 1.009 \times 10^{-3}\ \text{m}^3/\text{kg} \qquad h_{3a} = 3661.3\ \text{kJ/kg}$$

$$h_1 = 183.3\ \text{kJ/kg} \qquad s_{3a} = 7.2020\ \text{kJ/kg·K}$$

$$s_1 = 0.6223\ \text{kJ/kg·K}$$

The enthalpy at state $2a$ can be determined by using Equation 8.3 applied to the isentropic path 1-$2s$:

$$v_1(P_2 - P_1) = h_{2s} - h_1$$

and the expression for the adiabatic pump efficiency,

$$\eta_p = \frac{h_{2s} - h_1}{h_{2a} - h_1} = \frac{v_1(P_2 - P_1)}{h_{2a} - h_1}$$

can be solved for h_{2a}:

$$h_{2a} = \frac{v_1(P_2 - P_1)}{\eta_p} + h_1$$

$$= \left[\frac{(1.009 \times 10^{-3}\ \text{m}^3/\text{kg})(6 \times 10^3 - 9)\text{kPa}}{0.82}\right]$$

$$\times\ (1\ \text{kN/m}^2\text{·kPa})(1\ \text{kJ/kN·m}) + 183.3\ \text{kJ/kg}$$

$$= 190.7\ \text{kJ/kg}$$

Then the entropy at state $2a$ can be determined from Table B.4 at the known pressure and enthalpy:

$$s_{2a} = 0.6253\ \text{kJ/kg·K}$$

States $4a$ and $4s$ are in the saturation region, so the quality must be calculated before properties at either state can be determined. State $4s$ is on an isentropic path connecting $3a$, so the quality at state $4s$ is given by

$$x_{4s} = \frac{s_{4s} - s_f}{s_{fg}} = \frac{(7.2020 - 0.6493)\text{kJ/kg·K}}{7.499\ \text{kJ/kg·K}} = 0.874$$

and the enthalpy at state $4s$ is

$$h_{4s} = h_f + x_{4s}h_{fg} = 191.8\ \text{kJ/kg} + 0.874(2392)\text{kJ/kg} = 2282\ \text{kJ/kg}$$

The enthalpy at state $4a$ can be determined by applying the conservation-of-energy equation to the turbine and using the definition of the adiabatic-turbine efficiency. Thus

$$-w_{3a-4a} + (h_{3a} - h_{4a}) = 0$$

or

$$h_{4a} = h_{3a} - w_{3a-4a}$$

Also,

$$\eta_T = \frac{w_a}{w_s} = \frac{w_{3a-4a}}{h_{3a} - h_{4s}} = \frac{h_{3a} - h_{4a}}{h_{3a} - h_{4s}}$$

Therefore

$$\begin{aligned} h_{4a} &= h_{3a} - \eta_T(h_{3a} - h_{4s}) \\ &= 3661.3 \text{ kJ/kg} - 0.87(3661.3 - 2282)\text{kJ/kg} \\ &= 2461.3 \text{ kJ/kg} \end{aligned}$$

The quality and entropy at state $4a$ can now be determined. The result is

$$x_{4a} = 0.949 \quad \text{and} \quad s_{4a} = 7.7659 \text{ kJ/kg·K}$$

The properties of the various states of the cycle are summarized in the following table:

State	T, °C	P, MPa	h, kJ/kg	s, kJ/kg·K	Note
1	43.6	0.009	183.3	0.6223	Saturated liquid
$2a$	44.3	6	190.7	0.6253	
$3a$	600	5.6	3661.3	7.2020	
$4a$	45.8	0.01	2461.3	7.7659	$x_{4a} = 0.949$

The thermal efficiency of the cycle can now be calculated:

$$\begin{aligned} \eta_{\text{th}} &= \frac{(3661.3 - 2461.3 + 183.3 - 190.7)\text{kJ/kg}}{(3661.3 - 190.7)\text{kJ/kg}} \\ &= 0.344 = \underline{\underline{34.4 \text{ percent}}} \end{aligned}$$

The thermal efficiency of the ideal Rankine cycle in Example 8.1 was 39.9 percent. The efficiency of the ideal cycle is therefore 5.5 percentage points higher than that of the actual cycle considered here, because the ideal cycle consists of an isentropic pump and turbine and a frictionless boiler and condenser. The actual cycle includes irreversibilities in each of the four components in the cycle.

For each process the irreversibility per unit mass is determined from Equation 6.47:

$$i = T_0\left[(s_e - s_i) + \frac{q_k}{T_k}\right]$$

and the results of the calculations are summarized in Table 8.6.

TABLE 8.6 ACTUAL RANKINE CYCLE
T_H = 1400 K, T_L = T_0 = 298 K

Process	q, kJ/kg	w, kJ/kg	$h_e - h_i$, kJ/kg	$s_e - s_i$, kJ/kg·K	i, kJ/kg
1-2*a*	0.0	−7.4	7.4	0.0030	0.9
2*a*-3*a*	3470.6	0.0	3470.6	6.5767	1221.1
3*a*-4*a*	0.0	1215.0	−1215.0	0.5190	169.7
4*a*-1	−2263.0	0.0	−2263.0	−7.0987	147.6
Cycle	1207.6	1207.6	0.0	0.0	1539.3

Notice that each of the actual processes is irreversible, and that the largest component irreversibility is associated with the heat-transfer process in the boiler.

If we compare these results with those obtained in Example 8.1 for the ideal Rankine cycle, we find the following, for approximately the same heat input per unit mass.

1. The net work output for the actual cycle is about 14 percent lower than that of the ideal cycle.
2. The irreversibility for the actual cycle is about 15 percent greater than that of the ideal cycle. ■

The irreversibilities that occur in a refrigeration cycle produce detrimental effects similar to those discussed for the Rankine cycle. The compressor and condenser in a typical refrigeration or air-conditioning system are placed outdoors, but the throttling valve and evaporator are placed in the refrigerated space. The physical separation of the components dictates that long lengths of tubing frequently must be used to route the refrigerant through the compressor, evaporator, and condenser. Therefore, frictional effects can often be significant in air-conditioning and refrigeration systems, and their influence on the coefficient of performance should be assessed. In order of importance, the major areas of fluid friction in a refrigeration system occur in the suction line leading to the compressor from the evaporator followed by the discharge line that connects the compressor and the condenser. The line between the throttling valve and the evaporator is usually so short that the pressure drop in it is too small to be a significant factor in the irreversibility of the cycle. As noted in Equation 8.4, the work required to adiabatically compress a substance is proportional to the specific volume of the substance. A pressure drop in the suction line leading to the compressor will increase the specific volume of the refrigerant and cause an increase in the energy required by the compressor. Therefore, the pressure drop caused by fluid friction in the suction line has the effect of increasing the work required by the compressor, which translates to a reduction in the cycle COP.

A typical *T*-*s* diagram for an actual vapor-compression refrigeration cycle is shown in Figure 8.11. The irreversible process that occurs in the compressor from state 1 to 2 can produce either a decrease or an increase in the entropy. Heat transfer to the surroundings from the compressor tends to produce a decrease in entropy, while frictional effects that are present in the compressor cause an increase in entropy. Which factor will override the other can only be determined by an analysis of the actual compression

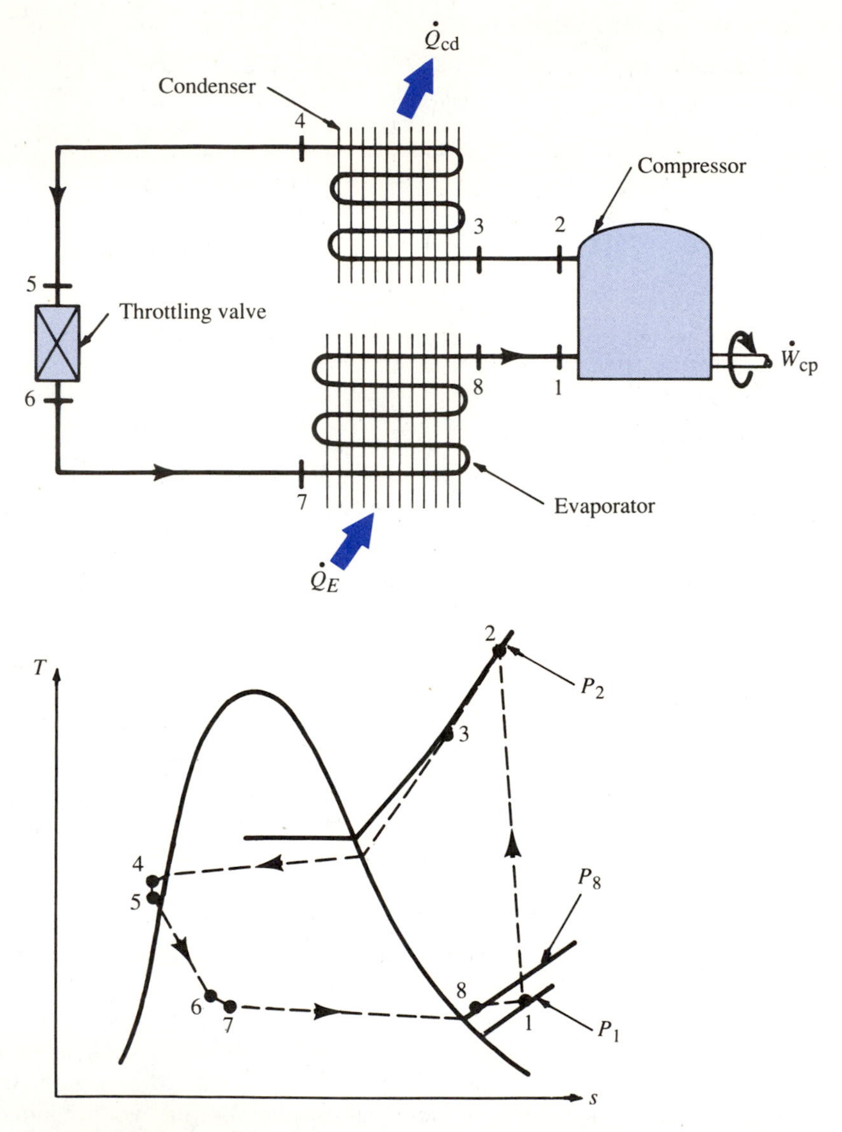

Figure 8.11 The T-s diagram for a vapor-compression refrigeration cycle including irreversibilities in all components.

process. Friction in the discharge line leaving the compressor results in a pressure drop, so that the refrigerant enters the condenser at state 3. The entropy of the refrigerant decreases during the process from 2 to 3 owing to heat transfer with the surroundings.

In the condenser the refrigerant condenses from a superheated vapor at state 3 to a compressed liquid at state 4, and fluid friction causes the pressure to decrease slightly during the process. A further pressure drop occurs in the line that connects the condenser to the throttling valve. The throttling process occurs between states 5 and 6, during which the refrigerant follows a constant-enthalpy path from the compressed-liquid region into the saturation region. A small pressure drop occurs in the line leading to the evaporator such that the refrigerant enters the evaporator at state 7. A further drop in pressure caused by fluid friction in the evaporator results in the refrigerant leaving the evaporator at state 8. Fluid friction and heat transfer to the refrigerant in the suction line increases the entropy and decreases the pressure of the refrigerant such that it enters the compressor at state 1, and the cycle can then be repeated.

EXAMPLE 8.7

An air-conditioning unit using refrigerant-12 is shown in Figure 8.11. Assume that the throttling valve is a constant-enthalpy device. Heat transfer from the condenser is to the environment, which has a temperature of 105°F. Heat transfer to the evaporator is from the cooled region, which has a temperature of 77°F. The heat transfer per lb_m of refrigerant from the compressor to the environment is 2.6 Btu/lb_m. Properties at the various states are as follows:

$$P_1 = 15 \text{ psia} \qquad P_2 = 180 \text{ psia} \qquad P_5 = 165 \text{ psia}$$

$$T_1 = 40°\text{F} \qquad T_2 = 200°\text{F} \qquad T_5 = 110°\text{F}$$

States 2 and 3 are identical, states 4 and 5 are identical, states 6 and 7 are identical, and the properties at state 8 are

$$P_8 = 20 \text{ psia} \qquad T_8 = 35°\text{F}$$

Assume that $P_7 = P_8$, and that during process 8-1 the enthalpy of the refrigerant is increased owing to heat transfer from the environment to the refrigerant in the suction line.

Determine the COP of the air-conditioning unit and compare it with the COP of a Carnot refrigeration unit operating between reservoirs with temperatures of 77 and 105°F. Calculate the irreversibility per unit mass for each process and for the cycle.

Solution. With reference to the states shown in Figure 8.11, the coefficient of performance of the cycle is

$$\beta_R = -\frac{\dot{Q}_{78}}{\dot{W}_{12}} = -\frac{q_{78}}{w_{12}}$$

where the heat transfer to the evaporator is

$$q_{78} = h_8 - h_7$$

and the work required to operate the compressor is

$$w_{12} = q_{12} + h_1 - h_2$$

Substituting these two expressions into the equation for the COP gives

$$\beta_R = -\left(\frac{h_8 - h_7}{q_{12} + h_1 - h_2}\right)$$

The property values at states 1, 2, and 8 can be determined from Table C.3E:

$$h_1 = 83.88\ \text{Btu/lb}_\text{m} \qquad s_1 = 0.1900\ \text{Btu/lb}_\text{m}\cdot{}^\circ\text{R}$$

$$h_2 = 103.64\ \text{Btu/lb}_\text{m} \qquad s_2 = 0.1863\ \text{Btu/lb}_\text{m}\cdot{}^\circ\text{R}$$

$$h_8 = 82.92\ \text{Btu/lb}_\text{m} \qquad s_8 = 0.1835\ \text{Btu/lb}_\text{m}\cdot{}^\circ\text{R}$$

The throttling valve is a constant-enthalpy device, so that

$$h_5 = h_6 = h_7$$

and the enthalpy of the refrigerant at state 5 in the compressed-liquid region is

$$h_7 = h_5 \simeq h_f(110^\circ\text{F}) = 33.94\ \text{Btu/lb}_\text{m}$$

Furthermore,

$$s_5 \simeq s_f(110^\circ\text{F}) = 0.0682\ \text{Btu/lb}_\text{m}\cdot{}^\circ\text{R}$$

and the quality of state 7, corresponding to the known values of h_7 and P_7, can be determined from Table C.2E:

$$x_7 = 0.388$$

The entropy at state 7 is therefore

$$s_7 = 0.0756\ \text{Btu/lb}_\text{m}\cdot{}^\circ\text{R}$$

The properties at the various states of the cycle are summarized below:

State	T, °F	P, psia	h, Btu/lb$_\text{m}$	s, Btu/lb$_\text{m}$·°R	Note
1	40	15	83.88	0.1900	
2	200	180	103.64	0.1863	
5	110	165	33.94	0.0682	Approximately saturated liquid
7	−8.2	20	33.94	0.0756	$x_7 = 0.388$
8	35	20	82.92	0.1835	

The value for the cycle COP can now be calculated:

$$\beta_R = -\frac{(82.92 - 33.94)\text{Btu/lb}_\text{m}}{(-2.6 + 83.88 - 103.64)\text{Btu/lb}_\text{m}} = \underline{\underline{2.19}}$$

The COP of a Carnot refrigerator operating between 537°R and 565°R is

$$\beta_{R,\text{Car}} = \frac{1}{(T_H/T_L) - 1} = \frac{1}{(565^\circ\text{R}/537^\circ\text{R}) - 1} = \underline{\underline{19.2}}$$

The actual refrigeration cycle, which includes the inevitable irreversibilities, requires a work input of nearly ten times the amount required to operate the totally reversible Carnot refrigerator while providing the same amount of cooling.

TABLE 8.7 ACTUAL VAPOR-COMPRESSION REFRIGERATION CYCLE
$T_H = T_0 = 105°F = 565°R$, $T_L = 77°F = 537°R$

Process	q, Btu/lb$_m$	w, Btu/lb$_m$	$h_e - h_i$, Btu/lb$_m$	$s_e - s_i$ Btu/lb$_m$·°R	i, Btu/lb$_m$
1-2	−2.6	−22.36	19.76	−0.0037	0.510
2-5	−69.7	0.0	−69.7	−0.1181	2.974
5-7	0.0	0.0	0.0	0.0074	4.181
7-8	48.98	0.0	48.98	0.1079	9.440
8-1	0.96	0.0	0.96	0.0065	2.662
Cycle	−22.36	−22.36	0.0	0.0	19.8

For each process the irreversibility per unit mass is determined from Equation 6.47:

$$i = T_0 \left[(s_e - s_i) + \frac{q_k}{T_k} \right]$$

The lowest naturally occurring reservoir temperature, T_0, in this instance is the environment temperature of 565°R(105°F). The results of the irreversibility calculations are included in Table 8.7.

Each of the processes in the actual cycle is irreversible. The largest irreversibilities occur in the throttling process (from frictional losses) and the heat-transfer process in the evaporator. The evaporator irreversibility results from a combination of losses due to heat transfer through a finite temperature difference and fluid friction.

We can check the result for the cycle irreversibility by using the expression presented in Example 8.5:

$$i = q_L \left(\frac{1}{\beta_{R,\text{Car}}} - \frac{1}{\beta_{R,\text{act}}} \right) = -q_{78} \left(\frac{1}{\beta_{R,\text{Car}}} - \frac{1}{\beta_{R,\text{act}}} \right)$$

$$= -48.98 \text{ Btu/lb}_m \left(\frac{1}{19.2} - \frac{1}{2.19} \right) = 19.8 \text{ Btu/lb}_m$$

8.7 SUMMARY

The basic processes that make up vapor power and refrigeration cycles are summarized in Tables 8.8 and 8.9. The Rankine cycle is the model for the design of a steam power plant that is used to generate the vast majority of the electric energy throughout the world. The ideal Rankine cycle consists of four internally reversible processes: an isentropic process in the pump, a constant-pressure heat-transfer process in the boiler, an isentropic expansion in the turbine, and a constant-pressure heat-transfer process in the condenser. The ratio of pump work to turbine work is significantly less for the Rankine cycle than the ratio of compressor work to turbine work for a gas cycle such as the Brayton cycle.

TABLE 8.8 SUMMARY OF PROCESSES FOR TYPICAL IDEAL VAPOR POWER CYCLES

Cycle	Application	Compression process	Heating process	Expansion process	Cooling process
Carnot cycle	Standard of comparison	Totally reversible, adiabatic	Totally reversible, isothermal	Totally reversible, adiabatic	Totally reversible, isothermal
Ideal Rankine cycle	Power-plant cycle	Internally reversible, adiabatic	Internally reversible, constant pressure	Internally reversible, adiabatic	Internally reversible, constant pressure

TABLE 8.9 SUMMARY OF PROCESSES FOR A TYPICAL IDEAL REFRIGERATION CYCLE

Cycle	Application	Compression process	Cooling process	Expansion process	Heating process
Ideal vapor-compression refrigeration cycle	Air conditioning, refrigeration, or heat pump	Internally reversible, adiabatic	Internally reversible, constant pressure	Throttling process (internally irreversible)	Internally reversible, constant pressure

Turbine-blade erosion can be reduced in the low-pressure stages of the turbine by modifying the Rankine cycle to include the reheat process. In the reheat cycle steam is extracted after partial expansion in the high-pressure stages of the turbine, and it is sent back to the boiler to be reheated at constant pressure. The steam is then returned to the turbine, where it is expanded to the condenser pressure in the low-pressure region of the turbine. The reheat process does not greatly affect the thermal efficiency of the Rankine cycle, but it does reduce the moisture content of the steam in the low-pressure stages of the turbine.

Significant increases in the thermal efficiency of the Rankine cycle can be achieved by a process called regeneration. In the regeneration process a portion of the steam in the turbine is bled off and routed through a feedwater heater. In an open feedwater heater the steam is allowed to mix with condensate from the condenser, and it heats the liquid water that enters the feedwater heater from a low-pressure pump. In a closed feedwater heater the extracted steam condenses on the tubes of the feedwater heater, and it heats the liquid water as it is pumped into the boiler.

The vapor-compression refrigeration cycle more closely approaches the behavior of a Carnot refrigeration cycle than does the gas refrigeration cycle discussed in Chapter 7. Evaporation caused by heat transfer to the refrigerant from the low-temperature region and condensation caused by heat transfer from the refrigerant to the high-temperature region closely approximate the desirable constant-temperature heat-transfer process. As a result, the coefficient of performance of a vapor-compression refrigeration cycle will exceed that of a comparable gas refrigeration cycle.

PROBLEMS

8.1 A Carnot power cycle uses water as a working fluid. Heat transfer to the cycle is from a reservoir at 240°F in an amount such that the water changes from a saturated liquid to a saturated vapor. The low-temperature reservoir is at a temperature of 80°F. Draw the cycle on a *T-s* diagram and include the saturation lines. Determine the following:

(a) the thermal efficiency of the cycle
(b) the heat transfer to the working fluid from the high-temperature reservoir per unit mass of the working fluid
(c) the net work provided by the cycle per unit mass of the working fluid
(d) the irreversibility of the cycle

8.2 A Carnot power cycle takes the form of a piston and cylinder, and refrigerant-12 is used as the working fluid. At the beginning of the heating process the refrigerant has a quality of 0.4 and a temperature of 80°C. At the end of the expansion process the temperature and pressure are 10°C and 0.20 MPa, respectively. Draw the cycle on a *T-s* diagram and include the saturation lines. Determine the following:

(a) the thermal efficiency of the cycle
(b) the maximum and minimum pressures during the cycle
(c) the net work provided by the cycle per unit mass of the refrigerant
(d) the heat transfer with the high-temperature reservoir per unit mass of refrigerant
(e) the compression ratio of the engine
(f) the irreversibility of the cycle

8.3 A Carnot power cycle operates with water as the working fluid. The cycle has a thermal efficiency of 50 percent and a mass flow rate of 0.1 lb_m/s. The temperature of the high temperature reservoir is 540°F and during the heat rejection process the water changes from a saturated vapor to a saturated liquid. Calculate the following information:

(a) The heat transfer rate with the low temperature reservoir
(b) The heat transfer rate with the high temperature reservoir
(c) The net power output of the cycle

8.4 Suppose a Carnot refrigeration cycle using R-12 is designed to transfer heat from steam. The cooling of the steam takes place in a heat exchanger which is the evaporator of the refrigeration cycle. The steam condenses steadily from a saturated vapor to a saturated liquid in the heat exchanger at 50 kPa. The heat transfer rate in the heat exchanger is 15kW and the heat exchanger process is totally reversible. During the heat rejection process of the refrigeration cycle, the R-12 changes from a saturated vapor to a saturated liquid at 100°C. Determine the following information:

(a) The thermal efficiency of the Carnot cycle
(b) The mass flow rate of steam through the heat exchanger
(c) The mass flow rate of the refrigerant through the Carnot cycle
(d) The net power required to drive the refrigeration cycle.

8.5 A Carnot refrigeration cycle utilizes refrigerant-12 as a working fluid in a closed system and operates between reservoirs that have temperatures of -15 and 50°C. The quality of the refrigerant prior to the isentropic-compression process is 1.0. The refrigerant is a saturated liquid as it enters the heat-addition portion of the cycle. Calculate the following quantities:

(a) the coefficient of performance of the cycle.
(b) the heat transfer to the high-temperature reservoir and the heat transfer from the low-temperature reservoir per unit mass of refrigerant.

(c) the net work input per unit mass of refrigerant required to operate the cycle.

8.6 Show that the irreversibility of an actual power cycle on a unit-mass basis is given by

$$i = q_H \left[\eta_{\text{Car}} - \eta_{\text{act}}\right]$$

where q_H is the heat transfer to the power cycle from the high temperature reservoir.

8.7 Show that the irreversibility of an actual refrigeration cycle on a unit-mass basis is given by

$$i = q_L \left(\frac{1}{\beta_{R,\text{act}}} - \frac{1}{\beta_{R,\text{Car}}}\right)$$

where q_L is the heat transfer to the refrigeration cycle from the low-temperature reservoir.

8.8 A refrigeration device based on the Carnot cycle provides 2 kW of cooling and uses refrigerant-12 as a working fluid. At the beginning of the isentropic-expansion process, the refrigerant is a saturated liquid. At the end of the isentropic-compression process, the pressure of the refrigerant is 0.5 MPa. Assume T_H = 50°C and T_L = 0°C. Draw the P-v and T-s diagrams for the cycle. Calculate the COP of the cycle, the mass flow rate of refrigerant, the quality at the end of the isentropic-expansion process, and the irreversibility of the cycle.

8.9 A Carnot refrigeration cycle operates between temperature reservoirs with T_H = 130°F and T_L = 70°F. The working fluid is refrigerant-12. The maximum and minimum pressures during the cycle are 160 psia and 20 psia. Determine the heat transfer with both reservoirs and the net work of the cycle, all on a per-unit-mass basis of refrigerant.

8.10 A Carnot refrigeration cycle uses refrigerant-12 as a working fluid. The temperatures of the two reservoirs are 50 and 10°C. During the isothermal heat-transfer process in the condenser, the refrigerant changes from a saturated vapor to a saturated liquid. Determine the heat transfer with both reservoirs and the net work of the cycle, all on a per-unit-mass basis.

8.11 Water enters the boiler of a simple, ideal Rankine cycle at 600 psia and 160°F. Steam enters the turbine at 1200°F, and the condenser pressure is 10 psia. Water leaves the condenser as a saturated liquid. Calculate the work of the turbine, the work of the pump, the heat transfer to the water in the boiler, and the heat transfer from the water in the condenser, all on a unit-mass basis. Determine the thermal efficiency of the cycle and the irreversibility of each process of the cycle on a per-unit-mass basis if T_H = 1500°F and $T_0 = T_L$ = 80°F. Why is this "ideal" cycle irreversible?

8.12 Water enters the turbine of a simple, ideal Rankine cycle at 700°C, and it leaves the condenser as a saturated liquid. The boiler pressure is 3.5 MPa, and the condenser pressure is 20 kPa. The mass flow rate of water through the cycle is 2.0 kg/s. Determine the power generated by the turbine, the heat-transfer rate for the boiler, the quality of the steam leaving the turbine, and the thermal efficiency of the cycle. Determine the irreversibility rate for each process of the cycle if T_H = 1000 K and $T_0 = T_L$ = 300 K. Explain the source of the irreversibility for each process.

8.13 The generator of a single-unit power plant is designed to provide 2 MW of electric energy. The efficiency (ratio of electric-power output to shaft-power input) of the generator is 93 percent, and the thermal efficiency (ratio of heat-transfer rate to water to heat-transfer rate from the fuel) of the boiler is 82 percent. For design purposes suppose that the power plant is assumed to operate as a simple, ideal Rankine cycle. The boiler pressure is 2.5 MPa, and the condenser pressure is 20 kPa. The steam enters the turbine at 700°C and leaves the condenser as a saturated liquid. The heat transfer from the fuel is 40,000 kJ per kg of fuel. Determine the mass rate of flow of fuel that must be provided in the boiler, the mass flow rate of steam through the cycle, the power output of the turbine, and the thermal efficiency of the cycle.

8.14 An ideal Rankine cycle operates such that the water leaves the condenser as a saturated liquid at a pressure of 8 psia. The pressure of the water leaving the pump is 500 psia, and the temperature of the steam entering the turbine is 1200°F. Sketch the cycle on a *T-s* diagram. Calculate the work per unit mass of the turbine and the cycle thermal efficiency.

8.15 Work Problem 8.14 for a boiler pressure of 600 psia. What is the percent increase in thermal efficiency of the cycle caused by the increase in boiler pressure? Why does increasing the boiler pressure cause an increase in the thermal efficiency?

8.16 Work Problem 8.14 for a turbine-inlet temperature of 1400°F. What is the percent increase in thermal efficiency of the cycle caused by the increase in turbine-inlet temperature? Why does increasing the turbine-inlet temperature cause an increase in the thermal efficiency?

8.17 Work Problem 8.14 for condenser pressure of 4 psia. What is the percent increase in thermal efficiency of the cycle caused by the decrease in condenser pressure? Why does decreasing the condenser pressure cause an increase in the thermal efficiency?

8.18 The *Mollier diagram* (plot of *h* as a function of *s*, see Figures B.5 and B.5E in the Appendix) is a useful tool when working problems that involve water as a working fluid. Work Problem 8.11 by using the Mollier diagram rather than using properties from the steam tables.

8.19 Work Problem 8.12 by determining water properties from the Mollier diagram (Figure B.5) rather than using properties from the steam tables.

8.20 [VPA111] In a Rankine cycle, steam is heated in the boiler by a source at 1400 K, and enters the turbine at 400°C and 3.5 MPa. The condenser pressure is 15 kPa. Assume that the pump and turbine efficiencies are unity, and that there are no pressure losses in the boiler or condenser. Also assume that water enters the pump as saturated liquid. Determine the following:

(a) the turbine power output in kJ/kg
(b) the pump power input in kJ/kg
(c) the heat transfer in the boiler in kJ/kg
(d) the cycle efficiency

8.21 [VPA121] In a Rankine cycle, steam is heated in the boiler by a source at 1400 K, and enters the turbine at 3.5 MPa. Steam exits the turbine at 15 kPa and a quality of 0.847. Assume that the pump and turbine efficiencies are unity, and that there are no pressure losses in the boiler or condenser. Also assume that water enters the pump as saturated liquid. The dead state conditions (state 0) are 21°C and 101.3 kPa. Determine the following:

(a) the turbine power output in kJ/kg
(b) the inlet temperature to the turbine in °C
(c) the irreversibility of the boiler process in kJ/kg
(d) the cycle efficiency

8.22 A nuclear power plant is designed to operate on an ideal Rankine cycle. Before the power plant is built and the equipment is selected, several design trade-offs are to be considered. The condenser pressure is fixed at 10 psia and the steam leaves the condenser as saturated liquid. The inlet temperature of the steam into the turbine is limited to 1200°F. Plot curves of cycle efficiency, quality of the steam leaving the turbine, and turbine work per lb_m of steam for turbine inlet pressures between 1000 and 2500 psia.

8.23 An ideal Rankine cycle has a turbine power output of 3.5 MW and a boiler pressure of 1200 psia. At the turbine exhaust the pressure is 3 psia and the steam has a quality of 96 percent. The liquid leaving the condenser is saturated. Determine the mass flow rate of steam, the heat-transfer rate in the boiler, and the thermal efficiency of the cycle. If the reservoir temperatures T_H and T_L corresponded to the maximum and minimum cycle temperatures, respectively, then determine the maximum thermal efficiency for these temperature limits and the irreversibility rate for the ideal cycle. What is the source of the irreversibility? Assume $T_0 = T_L$.

8.24 The turbine power output in an ideal Rankine cycle is 2 MW. Saturated liquid at 20 kPa leaves the condenser, and the vapor at the turbine exhaust has a quality of 95 percent. The boiler pressure is 1.4 MPa. Determine the mass flow rate of steam, the heat-transfer rate in the boiler, the thermal efficiency of the cycle, and the irreversibility rate for this cycle. Assume T_H = 900°C and $T_L = T_0$ = 30°C.

8.25 The turbine in an ideal Rankine cycle produces a power output of 5 MW. The temperature and pressure of the steam entering the turbine are 1000°F and 500 psia, respectively. The quality of the steam at the turbine exhaust is 100 percent, and saturated liquid leaves the condenser. Determine the mass flow rate of steam, the heat-transfer rate in the boiler, and the thermal efficiency of the cycle.

8.26 Two ideal Rankine cycles each produce 1 MW of turbine power output. In each cycle, the boiler pressure is 250 psia, the condenser pressure is 5 psia, and the condensate leaves the condenser as a saturated liquid. However, the steam is a saturated vapor at the exhaust from the turbine in cycle *A*, while the steam is superheated by 40°F at the exhaust from the turbine in cycle *B*. Compare the steam mass flow rates, the boiler heat-transfer rates, the thermal efficiencies, and the cycle irreversibility rates for these two cycles. Comment on the significance of your results. Assume T_H = 2000°F and $T_L = T_0$ = 90°F.

8.27 An ideal Rankine cycle operates such that the water leaves the condenser as a saturated liquid at a pressure of 40 kPa. The pressure of the water leaving the pump is 3.0 MPa, and the temperature of the steam entering the turbine is 600°C. Assume that T_H = 800°C and $T_L = T_0$ = 25°C. Determine the work per unit mass of the turbine and the irreversibility per unit mass for the condenser process.

8.28 Water enters the boiler of an ideal Rankine cycle with reheat at 1000 psia and 100°F. Steam enters the turbine at 1400°F, and the condenser pressure is 8 psia. Water leaves the condenser as a saturated liquid. Steam is extracted from the high-pressure turbine at 300 psia, and it enters the low-pressure turbine with a temperature of 1200°F. Determine the thermal efficiency of the cycle, the heat transfer in the boiler, and the work output of the turbine, both on a unit-mass basis. Calculate the irreversibility per unit mass for each process of the cycle if T_H = 1600°F and $T_0 = T_L$ = 80°F.

8.29 The Rankine cycle described in Problem 8.12 is modified to include a reheat process. The reheater pressure is 0.6 MPa, and the temperature of the water as it leaves the reheater is 600°C. Work Problem 8.12 for these new conditions.

8.30 Consider an ideal Rankine cycle modified with the reheat process. The boiler pressure is 2000 psia, the condenser pressure is 6 psia, and the reheat pressure is 500 psia. The temperature of the water entering both the low-pressure and the high-pressure turbines is 2000°F. The mass flow rate of water through the cycle is 10 lb_m/s. The water leaving the condenser is a saturated liquid.

Sketch the *T-s* diagram for the cycle, and calculate the following information:

(a) the power generated by the turbine
(b) the power required by the pump
(c) the heat-transfer rate in the boiler, excluding the reheater
(d) the heat-transfer rate in the reheater
(e) the thermal efficiency of the cycle
(f) the irreversibility rate for each process in the cycle if $T_H = 2400°F$ and $T_0 = T_L = 70°F$.

8.31 An ideal Rankine cycle with reheat has a boiler pressure of 4.0 MPa and a condenser pressure of 10 kPa. Steam enters both the high- and low-pressure turbines at 700°C, and it leaves the condenser as a saturated liquid. The quality of the steam leaving the low-pressure turbine is 0.95. Calculate the following:

(a) the pressure in the reheater
(b) the work per kilogram provided by both the high- and the low-pressure turbines
(c) the heat transfer per kilogram in the condenser and the reheater
(d) the irreversibility per kilogram for each process in the cycle if $T_H = 900°C$ and $T_0 = T_L = 20°C$.

8.32 Work Problem 8.28 by using the Mollier diagram (Figure B.5E) rather than obtaining properties from the steam tables.

8.33 Work Problem 8.31 by using the Mollier diagram (Figure B.5) rather than obtaining properties from the steam tables.

8.34 [VPC111] In a Rankine cycle, steam is heated in a boiler from a source at 1400 K, and enters the high-pressure turbine at 350°C and 3.5 MPa. It expands to 1.2 MPa and is then reheated to 350°C in a reheater before entering the low-pressure turbine. The condenser pressure is 15 kPa. Assume that the pump and turbine operate ideally and that there are no pressure losses in the boiler, reheater, or condenser. Also assume that water enters the pump as saturated liquid. Determine the following:

(a) the total turbine work output in kJ/kg
(b) the cycle efficiency

8.35 Consider an ideal Rankine cycle modified with the regeneration process. The boiler pressure is 1750 psia, and the condenser pressure is 8 psia. The temperature of the steam as it enters the turbine is 1400°F. The open feedwater heater operates at 80 psia, and the water leaving the feedwater heater and the condenser is a saturated liquid. The mass flow rate of steam through the boiler is 7.0 lb_m/s. Determine the following information:

(a) the power required to operate both pumps
(b) the mass flow rate of steam extracted from the turbine that enters the feedwater heater
(c) the quality of steam as it enters the condenser
(d) the heat-transfer rate of the boiler
(e) the net power output of the turbine
(f) the thermal efficiency of the cycle
(g) the irreversibility rate of each process of the cycle if $T_H = 1600°F$ and $T_0 = T_L = 70°F$.

8.36 An ideal Rankine cycle is modified to include regeneration with a single, open feedwater heater. The boiler pressure is 6.0 MPa, the condenser pressure is 20 kPa, and the feedwater-heater pressure is 1 MPa. The water enters both pumps as a saturated liquid. The temperature of the steam entering the turbine is 900°C. The cycle is designed to provide a net power output of 2.5 MW. Sketch the cycle on a *T-s* diagram. Determine the following information:

(a) the rate of pump work to turbine work (often called the *backwork ratio*)
(b) the mass flow rate of steam through the boiler necessary to provide the 2.5 MW of shaft-power output
(c) the mass flow rate of water entering the feedwater heater

(d) the mass flow rate of water through the condenser
(e) the thermal efficiency of the cycle
(f) the process that causes the greatest irreversibility if $T_H = 1000°C$ and $T_0 = T_L = 15°C$.

8.37 Consider an ideal Rankine cycle that includes regeneration with two open feedwater heaters and three pumps. The boiler pressure is 1250 psia, the condenser pressure is 5 psia, and the feedwater heaters operate at 250 psia and 120 psia. The mass flow rate of water through the boiler is 9.0 lb_m/s. Water leaves both the feedwater heaters and the condenser as a saturated liquid. The steam enters the turbine at 1600°F. Sketch the cycle on a *T-s* diagram, and calculate the following information:
(a) the mass flow rate of water through each of the three pumps
(b) the total power required to operate the three pumps
(c) the total power provided by the turbine
(d) the heat-transfer rate in the condenser and the boiler
(e) the thermal efficiency of the cycle
(f) the irreversibility rate of the cycle if $T_H = 1800°F$ and $T_0 = T_L = 60°F$

8.38 Work Problem 8.37 for the same boiler and condenser pressures but with only one open feedwater heater and two pumps. The feedwater heater operates at 200 psia, and the water enters the turbine at 1600°F. Determine the increase in thermal efficiency due to the two feedwater heaters as opposed to only a single feedwater heater.

8.39 Work Problem 8.37 by using property values from the Mollier chart (Figure B.5E) instead of properties from the steam tables.

8.40 Suppose that an ideal Rankine cycle with reheat and regeneration operates as shown in the sketch that accompanies Problem 8.62. The pumps and turbines are assumed to be isentropic devices, and the boiler has negligible pressure drops ($P_4 = P_5 = 3.0$ MPa). All other conditions are identical to those given in Problem 8.62. Determine the information asked for in Problem 8.62.

8.41 An ideal Rankine cycle is modified to include both the reheat and regeneration processes. Steam enters the high-pressure section of the turbine at 1600°F and 1000 psia, and a portion of the steam is extracted from the turbine at 200 psia and routed to an open feedwater heater. The remainder of the steam is sent to the reheater, where it is heated to 1400°F at 200 psia. The reheated steam then enters the low-pressure turbine, where it expands to the condenser pressure of 5 psia. The mass flow rate of steam through the high-pressure turbine is 20 lb_m/s. Water leaves both the condenser and the feedwater heater as a saturated liquid. Sketch a *T-s* diagram of the cycle. Calculate the following information:
(a) the work output of the low-pressure turbine per lb_m
(b) the work output of the high-pressure turbine per lb_m
(c) the heat transfer in the boiler per lb_m
(d) the heat transfer in the reheater per lb_m
(e) the heat transfer in the condenser per lb_m
(f) the percent of steam extracted from the high-pressure turbine that is routed to the feedwater heater
(g) the thermal efficiency of the cycle

8.42 [VPD111] In a Rankine cycle with regeneration, steam is heated in a boiler by a source at 1367 K, and enters the high-pressure turbine at 400°C and 3.5 MPa. Steam is extracted at 800 kPa for the purpose of heating the feedwater in an open feedwater heater. The condenser pressure is 15 kPa. Assume that the pump and turbine operate ideally and that there are no pressure losses in the boilers or condenser. Also assume that water enters both pumps as saturated liquid.

For a unit mass flow through the boiler, determine:

(a) the heat transfer to the system

(b) the cycle efficiency

8.43 You wish to design a Rankine cycle with the ultimate goal of generating 5 MW of electric power output. The generator is known to have an efficiency (ratio of electric power output to mechanical power input) of 96 percent. The following design alternatives are to be considered:

(a) an ideal Rankine cycle with no regeneration

(b) an ideal regenerative Rankine cycle with one open feedwater heater operating at 200 psia

(c) an ideal regenerative Rankine cycle with two open feedwater heaters, one operating at 600 and the other operating at 100 psia

The water enters the pump(s) as a saturated liquid. The state of the water at the inlet to the high-pressure section of the turbine is 1000°F, 2000 psia and the condenser pressure is 10 psia. For the three design alternatives calculate the cycle efficiency and the steam mass flow rate through the boiler required to produce the required power output.

8.44 [VRA111] A vapor-compression refrigeration cycle uses R-12 as the refrigerant. The cycle has a maximum pressure of 1 MPa and a minimum pressure of 100 kPa. The R-12 is saturated liquid leaving the condenser and saturated vapor leaving the evaporator. The heat-transfer processes occur at constant pressure and the compressor is assumed to be reversible and adiabatic. Determine the following:

(a) the maximum room temperature in °C

(b) the minimum refrigerated space temperature in °C

(c) the coefficient of performance (COP) of the cycle

8.45 Refrigerant-12 is used in an ideal vapor-compression refrigeration cycle. The condenser pressure is 190 psia, and the evaporator pressure is 40 psia. The refrigerant enters the compressor as a saturated vapor and leaves the condenser as a saturated liquid. The mass flow rate of the refrigerant is 12 lb_m/min. Determine the cooling capacity of the cycle and the coefficient of performance of the cycle. Calculate the irreversibility rate for each process of the cycle when $T_0 = T_H = 110°F$ and $T_L = 70°F$. Why is this ''ideal'' cycle irreversible?

8.46 A meat-processing plant requires a refrigeration capacity of 180 kW. Suppose that you wish to estimate the power requirements necessary to provide for this cooling capacity by assuming that the refrigeration system operates on an ideal vapor-compression refrigeration cycle. You select a refrigerant-12 unit with a condenser pressure of 1.4 MPa and an evaporator pressure of 320 kPa. Assume that the R-12 enters the compressor as a saturated vapor and leaves the condenser as a saturated liquid. Determine the power requirements of the cycle, the cycle COP, and mass flow rate of the refrigerant.

8.47 A vapor-compression refrigeration cycle uses refrigerant-12 as a working fluid in a facility that produces ice. The refrigeration capacity of the plant is 5×10^5 Btu/h. The refrigerant enters the compressor at $-10°F$ as a saturated vapor, and it leaves the condenser as a saturated liquid. The pressure ratio of the compressor is 5.0. Assuming that the cycle is ideal, calculate the mass flow rate of the refrigerant, the heat-transfer rate from the refrigerant to the environment, and the power required by the refrigeration system. Which process of the cycle produces the greatest irreversibility if $T_H = T_0 = 100°F$ and $T_L = 10°F$?

8.48 An ideal vapor-compression refrigeration cycle is used as a residential heat pump. The heat requirements of the home are 18 kW. Refrigerant-12 enters the compressor at 300 kPa and 0°C. The condenser pressure is 900 kPa, and the refrigerant leaves the condenser as a saturated liquid. Determine the mass flow rate of the refrigerant required to satisfy the heating requirements of the home. Determine the heat-transfer rate with the environment, the power requirement for the heat pump, and the COP of the cycle. Calculate the irreversibility rate for each process of the cycle if $T_H = T_0 = 30°C$ and $T_L = 5°C$.

8.49 An ideal vapor-compression refrigeration cycle uses refrigerant-12 as working fluid. The evaporator pressure is 30 psia, and the condenser pressure is 240 psia. The refrigerant leaves the condenser as a saturated liquid and the evaporator as a saturated vapor. The mass flow rate is 5.4 lb_m/min. Determine the cooling capacity of the refrigeration cycle, in Btu/h, the power requirements of the cycle, and the COP of the cycle.

8.50 [VRA111] A vapor-compression refrigeration cycle uses R-12 as the refrigerant. The cycle has a maximum pressure of 120 psia and a minimum pressure of 15 psia. The R-12 is saturated liquid leaving the condenser and saturated vapor leaving the evaporator. The heat-transfer processes occur at constant pressure and the compressor is assumed to be reversible and adiabatic. Determine the following:

(a) the maximum room temperature in °F
(b) the minimum refrigerated space temperature in °F
(c) the coefficient of performance (COP) of the cycle.

8.51 [VRA121] In an ideal vapor-compression refrigeration cycle, R-12 enters the evaporator with a quality of 0.4081 and leaves as saturated vapor at a pressure of 100 kPa. The condenser pressure is 1 MPa. The condenser rejects heat to the surroundings at temperature 294 K, and the evaporator heat transfer is from a space maintained at 250 K. Assume that the heat-transfer processes occur at constant pressure and that the compressor is reversible and adiabatic. Determine the following:

(a) the entropy change of the R-12 as it passes through the valve in kJ/kg·K
(b) the coefficient of performance (COP) of the cycle
(c) the irreversibility of the cycle in kJ/kg

8.52 The performance of a simple refrigeration cycle can be improved by using a multistage compression process as shown in the figure. The cycle consists of two compressors and two separate refrigerant loops, one at high pressure and a second one at lower pressure. The refrigerants in both loops mix in an open heat exchanger which is a condenser for the low-pressure loop and an evaporator for the high-pressure loop. This type of arrangement is called *cascading* and it can be shown to improve cycle coefficient of performance over that of a simple cycle that has only a single compressor. The optimum pressure for the heat exchanger is given by

$$P_{\text{opt}} = \sqrt{P_H P_L}$$

where the pressures P_H and P_L are shown in the figure. Assume that both cycles are ideal and use R-12 as a working fluid. Both cycles have the refrigerant enter the compressor as a saturated vapor and $P_H = 150$ psia, $P_L = 20$ psia. The heat exchanger operates at the optimum design pressure. The refrigerant enters the two expansion values as a saturated liquid. If the refrigeration cycle is designed to have a cooling capacity of 5×10^5 Btu/h, calculate the following information:

(a) the power required by both compressors
(b) the mass flow rate of refrigerant through both the high- and low-pressure loops
(c) the cycle COP
(d) the heat-transfer rate in the condenser

Problem 8.52

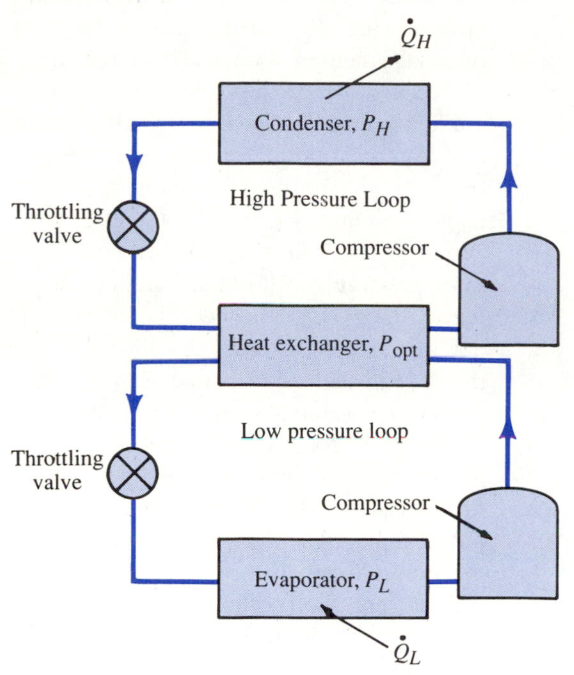

8.53 Work Problem 8.52 assuming a single compression process replaces the dual compression process. The single compressor operates between pressures of 20 and 150 psia and the heat exchanger is eliminated, producing a simple, ideal vapor-compression refrigeration cycle. Show that the simple refrigeration cycle requires more power to produce the required 5×10^5 Btu/h of cooling. Draw a T-s diagram for both cycles and explain why the cycle with multistage compression is more efficient than the cycle with only a single compression process.

8.54 An ideal Rankine cycle is used to power the compressor of an ideal refrigeration cycle as shown in the figure. The refrigeration cycle is designed to provide cooling in the evaporator at a rate of 1.5×10^6 Btu/h. The Rankine cycle utilizes steam and the turbine inlet state is 500°F, 400 psia. The refrigeration cycle utilizes R-12 and the inlet state to the compressor is saturated vapor at 20 psia. The steam and R-12 are not mixed in the condenser and they therefore exist at two different condenser pressures. The work output of the turbine is regulated to exactly match the work requirements of the compressor. The water entering the pump and the R-12 entering the throttling valve are both saturated liquids. The pressure of the refrigerant in the condenser is 10 psia. Calculate the following quantities:

(a) the mass flow rate of refrigerant
(b) the mass flow rate of water
(c) the thermal efficiency of the Rankine cycle
(d) the coefficient of performance of the refrigeration cycle
(e) the heat-transfer rate in the condenser of the refrigeration cycle and the boiler of the power cycle

Problem 8.54

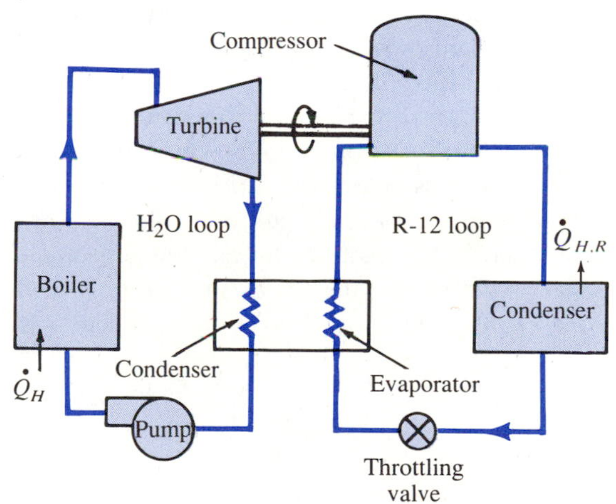

8.55 A low-temperature refrigeration system uses refrigerant-12 as the working fluid. The saturation temperatures in the evaporator and condenser are −40°C and 40°C, respectively. The refrigerant leaves the condenser as a saturated liquid and leaves the evaporator as a saturated vapor. Assuming ideal vapor-compression refrigeration, determine the compression ratio of the compressor, the refrigeration capacity per unit mass of refrigerant, and the coefficient of performance.

8.56 The refrigeration capacity of an ideal vapor-compression refrigeration system is 5000 Btu/min. The working fluid is refrigerant-12, the evaporator pressure is 30 psia, and the condenser pressure is 160 psia. The refrigerant leaves the condenser as a saturated liquid and leaves the evaporator as a saturated vapor. Determine the mass flow rate of refrigerant, the power input to the compressor, the coefficient of performance, and the irreversibility rate for the cycle. Assume $T_H = T_0 = 90°F$ and $T_L = 30°F$.

8.57 An ideal vapor-compression refrigeration system that uses refrigerant-12 as the working fluid is modified to include a liquid-to-suction heat exchanger as shown in the accompanying sketch. Saturated liquid at 35°C leaves the condenser and enters the heat exchanger where the temperature of the liquid is reduced to 25°C. The cooling in the heat exchanger is accomplished by routing saturated vapor at −10°C leaving the evaporator to the heat exchanger. Compare the coefficient of performance, the power input to the compressor, and the refrigerant mass flow rate for this modified cycle with the corresponding quantities for an unmodified ideal cycle having the same evaporator and condenser temperatures. Comment on your results.

Problem 8.57

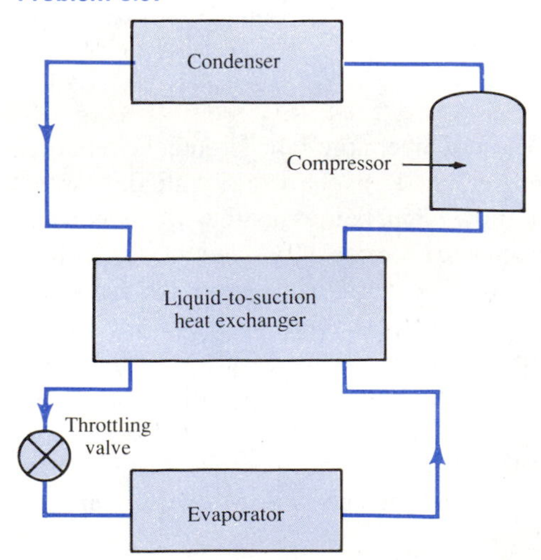

8.58 An ideal vapor-compression refrigeration system using refrigerant-12 is modified with two-stage compression and an open intercooler as shown in the sketch. Saturated liquid leaving the condenser at 900 kPa is throttled to a pressure of 300 kPa, and then this low-pressure refrigerant enters the intercooler. Saturated liquid leaving the intercooler at 300 kPa is throttled to the evaporator pressure of 100 kPa. Saturated vapor leaves the evaporator and is compressed to 300 kPa by the low-pressure compressor and the superheated vapor is routed to the intercooler. The saturated vapor leaves the intercooler to be compressed to 900 kPa by the high-pressure compressor. For a refrigeration capacity of 120 kW, determine the power input to the low-pressure compressor, the power input to the high-pressure compressor, and the coefficient of performance of the cycle. Compare the coefficient of performance with that for a simple ideal cycle having the same evaporator and condenser pressures and the same refrigeration capacity.

Problem 8.58

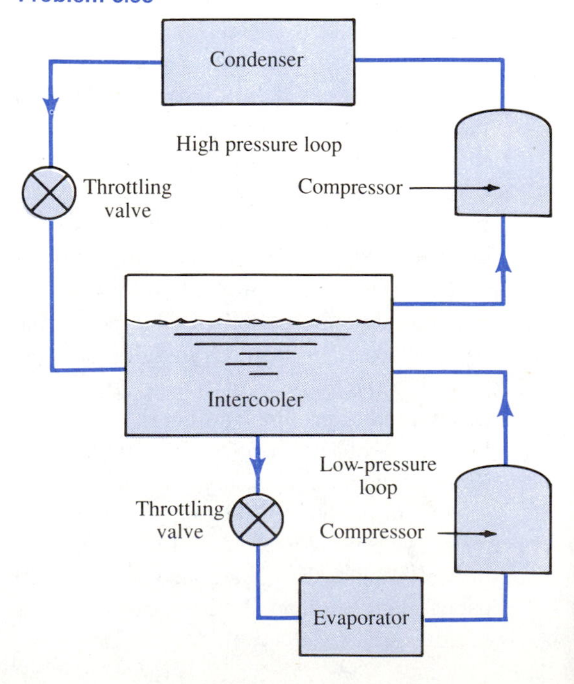

8.59 An ideal vapor-compression refrigeration system using refrigerant-12 as the working fluid has a condenser pressure of 150 psia and a compressor pressure ratio of 5. The liquid leaving the condenser is subcooled to 11°F below its saturation temperature. Compare the coefficient of performance, compressor work per unit mass, and irreversibility per unit mass for this cycle with those of an ideal cycle having the same refrigeration capacity but no subcooling. Assume $T_H = T_0 = 90°\text{F}$ and $T_L = 25°\text{F}$.

8.60 The refrigerant-12 vapor leaving the evaporator in an ideal vapor-compression refrigeration cycle is superheated 8°C above the saturation temperature corresponding to the evaporator pressure of 600 kPa. The compressor pressure ratio is 2 and the R-12 leaving the condenser is a saturated liquid. Compare the coefficient of performance, the compressor work per unit mass, the refrigerant temperature at the compressor exit, and the irreversibility per unit mass for this cycle with those of an ideal cycle having the same refrigeration capacity but no evaporator superheat. Assume $T_H = T_0 = 40°\text{C}$ and $T_L = 10°\text{C}$.

8.61 The evaporator pressure in an ideal vapor compression refrigeration cycle is 25 psia, and the refrigerant temperature at the compressor exit is 120°F. The working fluid, which is refrigerant-12, leaves the condenser as a saturated liquid and leaves the evaporator as a saturated vapor. The refrigerant mass flow rate is 0.75 $\text{lb}_\text{m}/\text{s}$. Determine the evaporator and condenser heat transfer rates, the coefficient of performance for this cycle operating as a refrigerator, and the coefficient of performance for this cycle operating as a heat pump.

8.62 A Rankine cycle with reheat and regeneration is illustrated in the accompanying figure. The states of the water throughout the cycle are as follows:

$P_1 = P_8 = 15\,\text{kPa}$ $\quad x_3 = 0.0$ $\quad P_5 = 2.9\,\text{MPa}$

$x_1 = 0.0$ $\quad P_4 = 3.0\,\text{MPa}$ $\quad T_5 = 600°\text{C}$

$P_2 = P_3 = P_6 = P_7 = 350\,\text{kPa}$

$T_7 = 500°\text{C}$

Problem 8.62

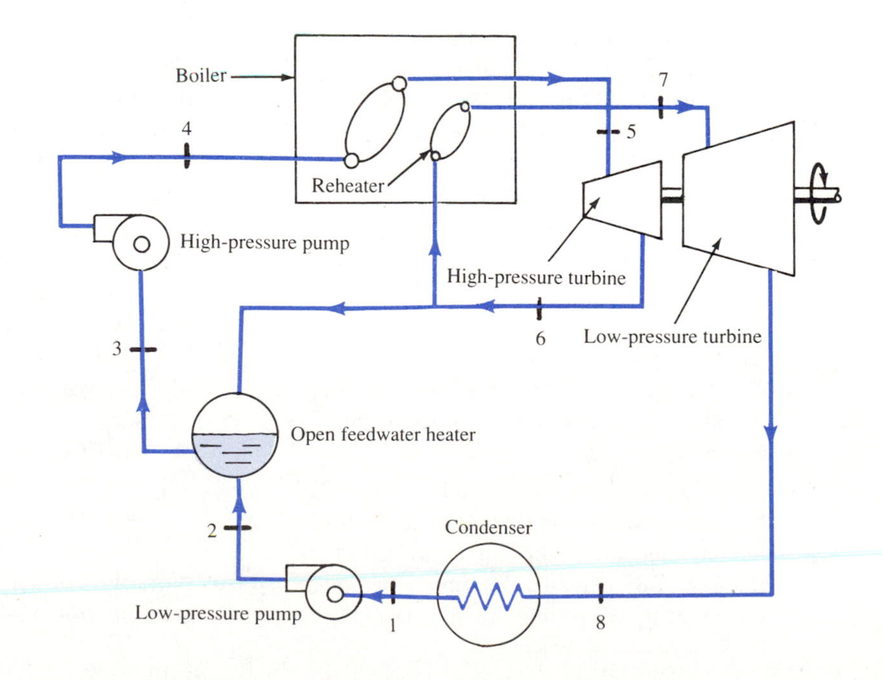

The adiabatic efficiency of each pump is 88 percent, and the adiabatic efficiency of each turbine is 91 percent. Neglect heat transfer in the turbines and pumps. Sketch the cycle on a T-s diagram, and determine the following information:

(a) the work output per kilogram for the high- and low-pressure turbines
(b) the work input per kilogram for the high- and low-pressure pumps
(c) the heat transfer per kilogram in the boiler and the reheater
(d) the heat transfer per kilogram in the condenser
(e) the ratio of the mass flow rate of water through the high-pressure turbine to the mass flow rate of water through the low-pressure turbines
(f) the thermal efficiency of the cycle

8.63 A simple Rankine cycle operates like the one shown in Figure 8.1 except that the cycle is not ideal. Properties at various states throughout the cycle are as follows:

$$P_1 = 25\text{ kPa} \quad P_2 = 2.0\text{ MPa} \quad T_3 = 600°\text{C}$$

$$x_1 = 0.0 \quad P_3 = 1.9\text{ MPa} \quad P_4 = 30\text{ kPa}$$

The mass flow rate of water through the cycle is 1.2 kg/s. The adiabatic efficiency of the pump is 87 percent, and the adiabatic efficiency of the turbine is 90 percent. Determine the following:

(a) the power developed by the turbine
(b) the heat-transfer rate in the boiler
(c) the heat-transfer rate in the condenser
(d) the power required for the pumping process
(e) the thermal efficiency of the cycle
(f) the irreversibility rate for each process and for the cycle if T_H = 850°C and $T_0 = T_L$ = 10°C.

8.64 Given the conditions stated in Problem 8.12, the turbine has an adiabatic efficiency of 86 percent, and the pump has an adiabatic efficiency of 83 percent. Determine the power generated by the turbine, the power required by the pump, the heat-transfer rate in the boiler, and the thermal efficiency of the cycle.

8.65 Given the conditions stated in Problem 8.14, a turbine having an adiabatic efficiency of 92 percent replaces the isentropic turbine. The pump remains isentropic. Determine the work per lb_m of the turbine and the thermal efficiency of the cycle. What is the reduction in turbine work and thermal efficiency due to irreversibilities in the turbine?

8.66 Work Problem 8.49 assuming that the isentropic compressor is replaced by one that has an adiabatic efficiency of 85 percent.

8.67 Work Problem 8.48 assuming that the isentropic compressor is replaced by one that has an adiabatic efficiency of 88 percent.

8.68 Refer to the states shown in Figure 8.9. The properties in a refrigeration cycle that uses refrigerant-12 as a working fluid are

$$P_1 = 300\text{ kPa} \quad P_2 = P_3 = 1.6\text{ MPa} \quad x_5 = 0.0$$

$$x_1 = 1.0 \quad T_2 = T_3 \quad P_4 = P_5 = 1.55\text{ MPa}$$

$$P_6 = P_7 = 320\text{ kPa} \quad P_8 = P_1 \quad T_4 = T_5$$

$$T_6 = T_7 \quad x_8 = 0.98$$

The compressor is adiabatic and has an adiabatic efficiency of 90 percent. The expansion device is a constant-enthalpy device. Calculate the following information:

(a) the heat transfer per kilogram in the condenser
(b) the heat transfer per kilogram in the evaporator
(c) the heat transfer per kilogram in the suction line to the compressor
(d) the work per kilogram in the compressor
(e) the COP of the cycle
(f) the irreversibility per kilogram for each process if $T_0 = T_H$ = 35°C and T_L = 20°C.

8.69 Work Problem 8.45 assuming that the isentropic compressor is replaced by a compressor that has an adiabatic efficiency of 88 percent.

8.70 [VPC211] In a Rankine cycle, steam is heated in a boiler from a source at 1400 K, and enters the high-pressure turbine at 350°C and 3.5 MPa. It expands to 1.2 MPa and is then reheated to 350°C in a reheater before entering the low-pressure turbine. The condenser pressure is 15 kPa. Assume that the pump and turbine efficiencies are 0.8 and 0.9, respectively, and that there are no pressure losses in the boiler, reheater, or condenser. Also assume that water enters the pump as saturated liquid. Determine the following:

- **(a)** the total turbine power output in kJ/kg
- **(b)** the cycle efficiency

8.71 [VPE111] In a Rankine cycle, steam is heated in a boiler by a source at 2460°R, and enters the high-pressure turbine at 700°F and 500 psia. Steam is extracted at 120 psia and is used to supply a feedwater heater. The remaining steam is then reheated to 700°F in a reheater before entering the low-pressure turbine. The condenser pressure is 2 psia. Assume that the pump and turbine efficiencies are 0.8 and 0.9, respectively, and that there are no pressure losses in the boilers or condenser. Also assume that water enters the pumps as saturated liquid. Determine for a unit mass flow of water passing through the boiler:

- **(a)** the fraction, F, of the steam that is extracted for the feedwater heater
- **(b)** the heat transfer to the system
- **(c)** the cycle efficiency

8.72 [VPB111] In a Rankine cycle, steam is heated in the boiler by a source at 1400 K, and enters the turbine at 350°C and 4 MPa. The condenser pressure is 15 kPa. Assume that the pump and turbine efficiencies are 0.8 and 0.9, respectively, and that there are no pressure losses in the boiler or condenser. Also assume that water enters the pump as saturated liquid. Determine the following:

- **(a)** the turbine power output in kJ/kg
- **(b)** the pump power input in kJ/kg
- **(c)** the heat transfer in the boiler in kJ/kg
- **(d)** the cycle efficiency

8.73 [VPB221] In a Rankine cycle, water leaves the condenser as saturated liquid with a pressure of 2 psia. Steam leaves the boiler at 500 psia and 700°F. Before entering the turbine, the steam is throttled to 400 psia. Assume a pump efficiency of 0.8 and a turbine efficiency of 0.9. If the environmental conditions (state 0) are 70°F and 14.7 psia, determine the following:

- **(a)** the turbine work per unit mass flow in Btu/lb_m
- **(b)** the irreversibility for the valve process in Btu/lb_m
- **(c)** the cycle efficiency

8.74 [VPB211] In a Rankine cycle, water leaves the condenser as saturated liquid with a pressure of 15 kPa. Steam is heated in the boiler by a source at 1400 K and leaves at 4 MPa and 350°C. Before entering the turbine, the steam is throttled to 3 MPa. Assume a pump efficiency of 0.8 and a turbine efficiency of 0.9. Also assume that there are no pressure losses in the boiler and condenser. Determine the following:

- **(a)** the turbine work output in kJ/kg
- **(b)** the pump work input in kJ/kg
- **(c)** the heat transfer in the boiler in kJ/kg
- **(d)** the cycle efficiency

8.75 [VRB111] A refrigerated cooler is to be maintained at 475°R in a 530°R environment. The condensing temperature is 85°F and the evaporation temperature is 0°F. Refrigerant-12 enters the compressor at a temperature of 10°F and a pressure of 23.9 psia. The compressor efficiency is 0.85 and the mass flow rate of the refrigerant is 19.2 lb_m/h. Refrigerant leaves the condenser as saturated liquid. Neglect the pressure drops in the condenser and evaporator. Determine the following:

(a) the compressor power input
(b) the evaporator heat-transfer rate
(c) the coefficient of performance

8.76 **[VRB121]** A refrigerated cooler is to be maintained at 264 K in a 294 K environment. Refrigerant-12 of 0.5 kg/s enters the compressor at −10°C and 150 kPa, and exits at 60.7°C and 800 kPa. The R-12 leaves the condenser as saturated liquid. Neglect the pressure drops in the condenser and evaporator. Determine the following:
(a) the compressor power input in kW
(b) the compressor efficiency
(c) the coefficient of performance

8.77 **[VRC111]** A heat pump maintains a house at 530°R when the outdoor temperature is 460°R. The condenser temperature is 85°F and the evaporator temperature is 0°F. Refrigerant-12 enters the compressor at a temperature of 10°F and a pressure of 23.9 psia. The compressor efficiency is 0.85 and the mass flow rate of the refrigerant is 19.2 lb_m/h. Refrigerant leaves the condenser as saturated liquid. Neglect the pressure drops in the condenser and evaporator. Determine the following:
(a) the compressor power input in Btu/h
(b) the rate of heat transfer to the house in Btu/h
(c) the coefficient of performance for heating

8.78 **[VRC121]** A heat pump maintains a house at 294 K when the outdoor temperature is 258 K. Refrigerant-12 enters the evaporator with a quality of 0.3085 and pressure 150 kPa, and leaves at a temperature of −10°C. The condenser pressure is 800 kPa. The compressor efficiency is 0.85 and the mass flow rate of the refrigerant is 2.5 kg/s. Neglect the pressure drops in the condenser and evaporator. Determine the following:
(a) the irreversibility of the valve process in kW
(b) the coefficient of performance for heating
(c) the total irreversibility in kW

8.79 **[VRD111]** A refrigerated cooler is to be maintained at 480°R in a 530°R environment. An unconventional refrigeration cycle has been proposed which uses refrigerant-12 as the working fluid. It leaves the evaporator as saturated vapor at −10°F and enters the cool end of a heat exchanger. The heat exchanger raises the temperature of the fluid to 80°F before the compressor increases its pressure to 200 psia. The fluid leaves the condenser at 100°F and enters the hot end of the heat exchanger. Assume a compressor efficiency of 0.9 and neglect the pressure drop across the heat exchanger, evaporator, and condenser. Determine the following:
(a) the compressor work per unit mass flow
(b) the coefficient of performance of the cycle

Problem 8.79

Condenser
Heat exchanger
Compressor
Throttling valve
Evaporator

8.80 **[VRD211]** A cascade system is commonly utilized for refrigeration cycles that operate over a wide temperature range. The system is comprised of a high-temperature cycle using R-12, and a low-temperature cycle using R-13. (The properties of R-12 and R-13 at specific states in this cycle are in the accompanying table.) The refrigerated space is maintained at 205 K, and heat transfer from the cycle is to the surroundings at 295 K. The conditions at each state are given in the table below. Determine the following:

(a) the ratio of the mass flow of R-13 to R-12

(b) the irreversibility of the heat-exchanger process in kJ/kg

(c) the coefficient of performance (COP) of the system

(d) the total irreversibility in kJ/kg

Problem 8.80

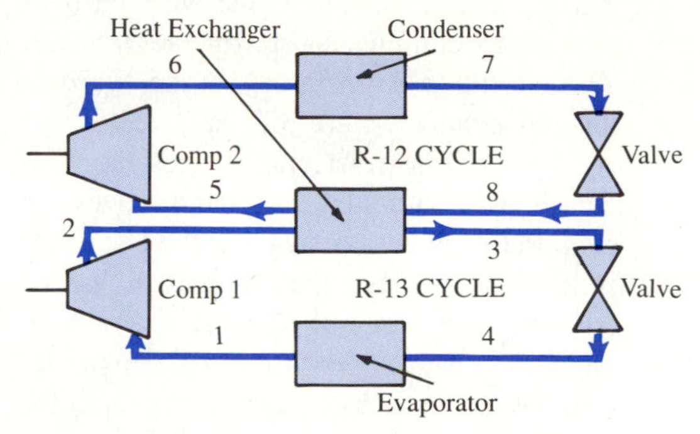

State	1	2	3	4	5	6	7	8
T (°C)	−73.0	24.0	−13.0	−73.0	−15.0	50.0	35.0	−14.0
P (kPa)	155	1400	1395	160	183	850	848	190
h (kJ/kg)	188	234	103	103	181	212	69.5	69.5
s (kJ/kg·K)	0.968	1.000	0.501	0.550	0.705	0.719	0.256	0.273
x	1.000	Superheat	0	0.410	1.000	Superheat	0	0.292

8.81 **[VRD311]** A vapor-compression refrigeration cycle is powered by a Rankine cycle to form a dual-loop cycle as shown in the figure. The working fluid that is used in both the refrigeration and Rankine cycles is R-12. The refrigeration cycle maintains a space at 265 K. The high-temperature reservoir for the Rankine cycle is at 500 K, and heat transfer from the cycle is to the surroundings at 293 K. The conditions at each state are given in the table below. Determine the following:

(a) the ratio of the mass flow in the power loop to that in the refrigeration loop

(b) the heat transfer in the mixer in kJ/kg

(c) the irreversibility of the mixer in kJ/kg

(d) the coefficient of performance (COP) of the system

(e) the total irreversibility in kJ/kg

Problem 8.81

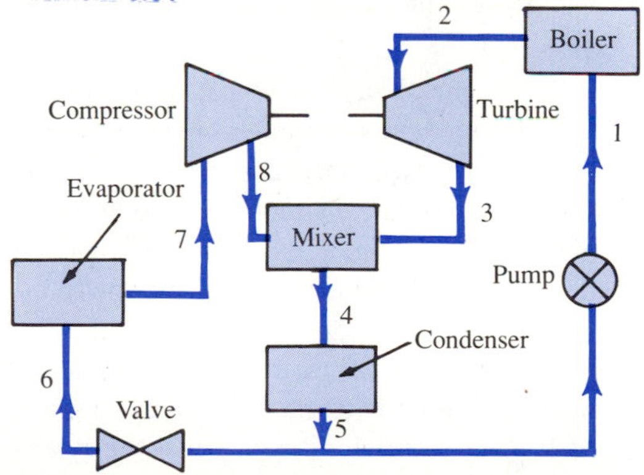

State	1	2	3	4	5	6	7	8
T (°C)	45.0	110	41.6	41.6	41.6	−15.4	−15.4	61.2
P (kPa)	4000	4000	1000	1000	1000	180	180	1000
h (kJ/kg)	82.0	212	200	204	76.3	76.3	181	219
s (kJ/kg·K)	0.287	0.644	0.670	0.682	0.277	0.300	0.705	0.728
x	Superheat	1.000	0.970	1.000	0	0.342	1.000	Superheat

8.82 Refrigerant-12 is used as the working fluid in a vapor-compression refrigeration cycle. The saturation temperatures in the evaporator and condensor are 30°F and 130°F, respectively. The refrigerant leaves the condenser as a saturated liquid and leaves the evaporator as a saturated vapor. The cycle is ideal except that the compressor has an adiabatic efficiency of 87 percent. For a refrigeration capacity of 1.2×10^5 Btu/h, determine the mass flow rate of refrigerant, the power input to the compressor, and the coefficient of performance of the cycle.

8.83 A vapor-compression refrigeration system has a refrigeration capacity of 80 kW. The working fluid is refrigerant-12, the condenser pressure is 1 MPa, and the evaporator pressure is 200 kPa. The refrigerant leaves the evaporator as a saturated vapor and leaves the condenser as a saturated liquid. The cycle is ideal except that the compressor has an adiabatic efficiency of 85 percent. Determine the mass flow rate of refrigerant, the power input to the compressor, the coefficient of performance, and the irreversibility rate for the cycle. Assume $T_H = T_0 =$ 35°C and $T_L =$ 0°C.

8.84 Refer to the sketch accompanying Problem 8.58 for the vapor-compression refrigeration cycles with two-stage compression and an open intercooler. Saturated liquid refrigerant-12 leaving the condenser at 125 psia is throttled to 70 psia, and then this low-pressure refrigerant enters the intercooler. Saturated liquid leaving the intercooler at 70 psia is throttled to the evaporator pressure of 15 psia. Saturated vapor leaves the evaporator and is compressed to 70 psia in the low-pressure compressor. The superheated vapor is then routed to the intercooler, and saturated vapor leaves the intercooler to be compressed to 125 psia by the high-pressure compressor. Both compressors have adiabatic compressor efficiencies of 88 percent. For a refrigeration capacity of 6000 Btu/min, determine the power input to each compressor and the coefficient of performance for the cycle.

8.85 Refer to the sketch accompanying Problem 8.57 for the vapor-compression refrigeration cycle with a liquid-to-suction heat exchanger. Saturated liquid refrigerant-12 at 90°F leaves the condenser and enters the heat exchanger where the temperature of the refrigerant is reduced to 70°F. The cooling in the heat exchanger is accomplished by routing saturated vapor at 20°F leaving the evaporator to the heat exchanger, and the compressor has an adiabatic efficiency of 86 percent. For a power input to the compressor of 3.5 hp, determine the cooling capacity and the coefficient of performance of the cycle, the irreversibility rate for the condenser and for the cycle. Assume $T_H = T_0$ = 80°F and $T_L =$ 30°F.

8.86 The compressor in a refrigerant-12 vapor-compression refrigeration cycle has an adiabatic efficiency of 87 percent and a pressure ratio of 4. The condenser pressure is 1.2 MPa, the liquid leaving the condenser is subcooled by 5°C, and the vapor leaving the evaporator is superheated by 5°C. Neglect pressure drops in the evaporator and condenser, and determine the refrigeration capacity per unit mass, the coefficient of performance, and the irreversibility per unit mass for each component and for the cycle. What impact do condenser subcooling and evaporator superheating have on the performance of this cycle?

8.87 A refrigerant-12 vapor-compression refrigeration system requires a compressor power input of 26.7 kW when the evaporator and condenser pressures are 250 kPa and 1 MPa, respectively. The liquid is saturated at the condenser exit, the vapor is saturated at the evaporator exit, and the refrigerant temperature at the compressor exit is 60°C. Determine the refrigeration capacity, the adiabatic efficiency of the compressor, the coefficient of performance, and the irreversibility rate of the compressor. Assume $T_0 =$ 25°C.

9 Thermodynamic Relationships

9.1 INTRODUCTION

Only a very few of the many thermodynamic properties can be measured experimentally, and the others must be evaluated in terms of those properties that can be measured. Experimental measurements of the pressure, specific volume, temperature, and specific heats are the most accessible, and the objective of the study of thermodynamic relations is to develop expressions that relate these properties to properties such as internal energy, enthalpy, and entropy. The state postulate provides the basis for beginning this development, and the requisite mathematical foundation is presented in the following section. The remainder of this chapter is devoted to the derivation of several important thermodynamic relations and a discussion of their application.

9.2 MATHEMATICAL PRELIMINARIES

The state postulate presented in Chapter 2 was used to conclude that two independent, intensive thermodynamic properties are sufficient to completely specify the state of a simple, compressible, homogeneous substance. That the other properties associated with the state are thus fixed can be expressed in the mathematical form

$$x = x(y, z) \tag{9.1}$$

Here y and z are independent variables or properties and x represents any of the other properties associated with the state specified by y and z.

The property x is a continuously differentiable function of the two variables y and z if the substance is homogeneous, and from principles established in calculus the differential of x can be written as

$$dx = \left(\frac{\partial x}{\partial y}\right)_z dy + \left(\frac{\partial x}{\partial z}\right)_y dz \tag{9.2}$$

The differential dx is an exact differential, as defined in Chapter 1, and has the characteristic that its mixed second partial derivatives are equal,

$$\frac{\partial}{\partial z}\left[\left(\frac{\partial x}{\partial y}\right)_z\right]_y = \frac{\partial}{\partial y}\left[\left(\frac{\partial x}{\partial z}\right)_y\right]_z \tag{9.3}$$

In other words, the order of differentiation is immaterial since x is a property (i.e., a continuously differentiable function of y and z). In fact, the condition expressed by Equation 9.3 can be used as a *test for exactness* for a differential dx. That is, a differential dx of the form

$$dx = M\,dy + N\,dz$$

is exact if and only if

$$\frac{\partial M}{\partial z} = \frac{\partial N}{\partial y} \tag{9.4}$$

If Equation 9.4 is satisfied, then

$$M = \left(\frac{\partial x}{\partial y}\right)_z \quad \text{and} \quad N = \left(\frac{\partial x}{\partial z}\right)_y$$

The relation expressed by Equation 9.1 could also be written as $y = y(x, z)$ if x and z were instead considered to be independent variables, and the differential of y would then be given by

$$dy = \left(\frac{\partial y}{\partial x}\right)_z dx + \left(\frac{\partial y}{\partial z}\right)_x dz \tag{9.5}$$

Equations 9.2 and 9.5 can be combined to eliminate dy,

$$dx = \left[\left(\frac{\partial x}{\partial y}\right)_z\left(\frac{\partial y}{\partial x}\right)_z dx + \left(\frac{\partial x}{\partial y}\right)_z\left(\frac{\partial y}{\partial z}\right)_x dz\right] + \left(\frac{\partial x}{\partial z}\right)_y dz$$

Upon rearrangement this equation can be expressed as

$$\left[1 - \left(\frac{\partial x}{\partial y}\right)_z\left(\frac{\partial y}{\partial x}\right)_z\right] dx = \left[\left(\frac{\partial x}{\partial y}\right)_z\left(\frac{\partial y}{\partial z}\right)_x + \left(\frac{\partial x}{\partial z}\right)_y\right] dz \tag{9.6}$$

Since x and z can be considered to be independent variables, the terms in brackets in Equation 9.6 must each be equal to zero. Two additional relations result from this observation.

The value of x can vary independently of z, so if dx is nonzero while dz vanishes, then

$$\left(\frac{\partial x}{\partial y}\right)_z \left(\frac{\partial y}{\partial x}\right)_z = 1$$

or

$$\left(\frac{\partial x}{\partial y}\right)_z = \frac{1}{(\partial y/\partial x)_z} \tag{9.7}$$

This result establishes the fact that the partial derivative can be inverted to yield the reciprocal of the derivative. Furthermore, if dz is nonzero while dx equals zero, then from Equation 9.6

$$\left(\frac{\partial x}{\partial y}\right)_z \left(\frac{\partial y}{\partial z}\right)_x = -\left(\frac{\partial x}{\partial z}\right)_y$$

and from Equation 9.7 this result can be expressed as

$$\left(\frac{\partial x}{\partial y}\right)_z \left(\frac{\partial y}{\partial z}\right)_x \left(\frac{\partial z}{\partial x}\right)_y = -1 \tag{9.8}$$

This last equation is called the *cyclic relation* and finds frequent application in the development of thermodynamic relations.

EXAMPLE 9.1

The equation of state of an ideal gas is given by $Pv = RT$. Verify that the differential dv of the property v is an exact differential, and show that the cyclic relation, Equation 9.8, holds.

Solution. Since $v = v(P, T)$ is a property of a simple, compressible, homogeneous substance, the results of this section assure us that dv is an exact differential and that the cyclic relation holds. Therefore, verification of these facts is not necessary, although illustrating their meaning with a specific example is worthwhile.

The equation of state is first rearranged so that the specific volume is written as an explicit function of the two independent properties temperature and pressure; that is, $v = v(T, P)$,

$$v = v(T, P) = \frac{RT}{P}$$

Since v is a property, dv is given by the form of Equation 9.2,

$$dv = \left(\frac{\partial v}{\partial T}\right)_P dT + \left(\frac{\partial v}{\partial P}\right)_T dP$$

and the partial derivatives can be evaluated as

$$\left(\frac{\partial v}{\partial T}\right)_P = \frac{R}{P} \quad \text{and} \quad \left(\frac{\partial v}{\partial P}\right)_T = -\frac{RT}{P^2}$$

The test for exactness, Equation 9.3, can be used to show that dv is exact:

$$\frac{\partial}{\partial P}\left[\left(\frac{\partial v}{\partial T}\right)_P\right]_T = -\frac{R}{P^2}$$

and

$$\frac{\partial}{\partial T}\left[\left(\frac{\partial v}{\partial P}\right)_T\right]_P = -\frac{R}{P^2}$$

Since the mixed partial derivatives are equal, dv is an exact differential.

The cyclic relation, Equation 9.8, can be written in terms of the variables v, T, and P as

$$\left(\frac{\partial v}{\partial T}\right)_P \left(\frac{\partial T}{\partial P}\right)_v \left(\frac{\partial P}{\partial v}\right)_T = -1$$

Our task is to verify that the product of the three partial derivatives does indeed equal -1. The first derivative in this expression has been determined above. The second can be obtained by writing the equation of state in the form $T = T(v, P)$ and taking the indicated derivative, with the result

$$\left(\frac{\partial T}{\partial P}\right)_v = \frac{v}{R}$$

and the third partial derivative follows from Equation 9.7 and results obtained above:

$$\left(\frac{\partial P}{\partial v}\right)_T = \frac{1}{(\partial v/\partial P)_T} = -\frac{P^2}{RT}$$

Substituting these results into the cyclic relation above gives

$$\left(\frac{R}{P}\right)\left(\frac{v}{R}\right)\left(-\frac{P^2}{RT}\right) = -\frac{Pv}{RT} = -1$$

■

9.3 THE GIBBS EQUATIONS AND THE MAXWELL RELATIONS

The mathematical expressions developed in the preceding section can be applied to the Gibbs equations to yield a set of important thermodynamic relations between entropy and the measurable properties pressure, temperature, and specific volume. The first two of the Gibbs, or $T\,ds$, equations for simple compressible systems were derived in Chapter

5 and can be written in the following forms:

$$du = T\,ds - P\,dv \tag{9.9}$$

and

$$dh = T\,ds + v\,dP \tag{9.10}$$

Two other Gibbs equations can be developed with the aid of two thermodynamic properties that are usually associated with the study of chemical equilibrium. These properties are the *Helmholtz function*, denoted by the symbol a and defined as

$$a \equiv u - Ts \tag{9.11}$$

and the *Gibbs function*, denoted by the symbol g and defined as

$$g \equiv h - Ts \tag{9.12}$$

The third Gibbs equation is derived by first writing the differential of the Helmholtz function,

$$da = du - T\,ds - s\,dT$$

and then substituting Equation 9.9 into this result to obtain

$$da = -s\,dT - P\,dv \tag{9.13}$$

The fourth Gibbs equation is derived in a similar manner, by writing the differential of the Gibbs function,

$$dg = dh - T\,ds - s\,dT$$

and substituting Equation 9.10 to obtain

$$dg = -s\,dT + v\,dP \tag{9.14}$$

Each of the four Gibbs equations,

$$du = T\,ds - P\,dv \tag{9.9}$$

$$dh = T\,ds + v\,dP \tag{9.10}$$

$$da = -s\,dT - P\,dv \tag{9.13}$$

$$dg = -s\,dT + v\,dP \tag{9.14}$$

is a first-order differential form and represents an exact differential of the properties u, h, a, or g. Because of this fact, the mixed second partial derivatives must be equal as expressed in Equation 9.3. For example, if u is expressed as a function of s and v, then $u = u(s, v)$, and

$$du = \left(\frac{\partial u}{\partial s}\right)_v ds + \left(\frac{\partial u}{\partial v}\right)_s dv \tag{9.15}$$

Since du is an exact differential, then

$$\frac{\partial}{\partial v}\left[\left(\frac{\partial u}{\partial s}\right)_v\right]_s = \frac{\partial}{\partial s}\left[\left(\frac{\partial u}{\partial v}\right)_s\right]_v$$

The differential of a function of two variables is unique, and a comparison of the coefficients of ds and dv in Equations 9.9 and 9.15 for the differential du reveals that

$$\left(\frac{\partial u}{\partial s}\right)_v = T \quad \text{and} \quad \left(\frac{\partial u}{\partial v}\right)_s = -P$$

Therefore, the requirement that the mixed second partial derivatives be equal yields the relation

$$\left(\frac{\partial T}{\partial v}\right)_s = -\left(\frac{\partial P}{\partial s}\right)_v \tag{9.16}$$

Invoking this condition, in turn, on Equations 9.10, 9.13, and 9.14 results in the following additional relations:

$$\left(\frac{\partial T}{\partial P}\right)_s = \left(\frac{\partial v}{\partial s}\right)_P \tag{9.17}$$

$$\left(\frac{\partial s}{\partial v}\right)_T = \left(\frac{\partial P}{\partial T}\right)_v \tag{9.18}$$

$$\left(\frac{\partial s}{\partial P}\right)_T = -\left(\frac{\partial v}{\partial T}\right)_P \tag{9.19}$$

The relations expressed in Equations 9.16 through 9.19 are called the *Maxwell relations*.

As an example of the usefulness of the Maxwell relations, consider the application of Equation 9.19 in determining the change in entropy that results from a pressure change during an isothermal process. According to Equation 9.19, an indirect measurement of the entropy change during an isothermal process can be obtained by recording the change in specific volume due to a change in temperature during a constant-pressure process. Thus the pressure dependence of entropy can be determined from experimental measurements of the P-v-T behavior of a pure substance. This result is illustrated in the following example.

■ EXAMPLE 9.2

Estimate the entropy change of refrigerant-12 during a constant-temperature process at 40°C during which the pressure of the refrigerant increases from 200 to 300 kPa. Evaluate the accuracy of this estimate by comparison with results obtained using tabulated values of entropy.

Solution. If the entropy is considered to be a function of temperature and pressure, $s = s(T, P)$, the differential of s is written as

$$ds = \left(\frac{\partial s}{\partial T}\right)_P dT + \left(\frac{\partial s}{\partial P}\right)_T dP$$

For an isothermal process $dT = 0$, and the entropy change can be reduced to

$$ds_T = \left(\frac{\partial s}{\partial P}\right)_T dP$$

From the fourth Maxwell relation,

$$\left(\frac{\partial s}{\partial P}\right)_T = -\left(\frac{\partial v}{\partial T}\right)_P$$

Thus

$$ds_T = -\left(\frac{\partial v}{\partial T}\right)_P dP$$

and integrating the expression between states 1 and 2 along an isothermal path yields

$$(s_2 - s_1)_T = -\int_1^2 \left(\frac{\partial v}{\partial T}\right)_P dP$$

An estimate of the entropy change can be obtained by assuming that $(\partial v/\partial T)_P$ remains relatively constant over the range of pressures from 200 to 300 kPa and by evaluating this derivative at the average pressure of 250 kPa:

$$(s_2 - s_1)_T \simeq -\left(\frac{\partial v}{\partial T}\right)_{P_{\text{avg}}} (P_2 - P_1)$$

To estimate the value of the derivative, we use a linear approximation at 250 kPa and 40°C and property values from Table C.3:

$$\left(\frac{\partial v}{\partial T}\right)_{P=250\text{ kPa}} \simeq \left(\frac{v_{50^\circ\text{C}} - v_{30^\circ\text{C}}}{323\text{ K} - 303\text{ K}}\right)_{P=250\text{ kPa}}$$

$$= \frac{(0.08560 - 0.07956)\text{m}^3/\text{kg}}{20\text{ K}}$$

$$= 3.02 \times 10^{-4}\ \text{m}^3/\text{kg}\cdot\text{K}$$

Therefore

$$(s_2 - s_1)_T \simeq -(3.02 \times 10^{-4}\ \text{m}^3/\text{kg}\cdot\text{K})[(300 - 200)\text{kPa}](1\ \text{kJ}/\text{m}^3\cdot\text{kPa})$$

$$= \underline{\underline{-0.0302\ \text{kJ/kg}\cdot\text{K}}}$$

From tabular values the actual entropy change is found to be

$$(s_2 - s_1)_T = (0.7901 - 0.8207)\text{kJ/kg}\cdot\text{K} = \underline{\underline{-0.0306\ \text{kJ/kg}\cdot\text{K}}}$$

In this example the entropy change is underestimated by only about 1 percent. The procedure outlined above illustrates how *P-v-T* data can be used to develop tables of properties such as entropy, which cannot be measured directly. ■

9.4 GENERAL EQUATIONS FOR *du, dh,* AND *ds*

The internal energy, enthalpy, and entropy of a substance cannot be measured directly. However, changes in these properties can be related to the specific heats and the P-v-T behavior of the substance. If descriptions of the P-v-T behavior and the specific heats are available, from either theoretical models or experimental data, changes in internal energy, enthalpy, and entropy can ultimately be evaluated. So that this objective can be achieved, this section is devoted to the development of general equations for changes in internal energy, enthalpy, and entropy of simple compressible substances written entirely in terms of pressure, specific volume, temperature, and specific heats.

9.4.1 Internal Energy

The general equation for the change in internal energy can be derived in a straightforward manner, beginning with the first $T\,ds$ equation, Equation 9.9,

$$du = T\,ds - P\,dv \tag{9.20}$$

If the entropy is considered to be a function of temperature and specific volume, $s = s(T, v)$, the differential ds can be written as

$$ds = \left(\frac{\partial s}{\partial T}\right)_v dT + \left(\frac{\partial s}{\partial v}\right)_T dv \tag{9.21}$$

Substituting this expression for ds into Equation 9.20 and rearranging yields the result

$$du = T\left(\frac{\partial s}{\partial T}\right)_v dT + \left[T\left(\frac{\partial s}{\partial v}\right)_T - P\right] dv \tag{9.22}$$

Recall that in Chapter 2 when the internal energy was written as a function of T and v, $u = u(T, v)$, the change in internal energy for a simple compressible substance was found to be (Equation 2.18)

$$du = c_v\,dT + \left(\frac{\partial u}{\partial v}\right)_T dv \tag{9.23}$$

For a simple compressible substance, if $u = u(T, v)$, then du is unique, and therefore the coefficients of dT and dv, respectively, in Equations 9.22 and 9.23 must be identical. Thus

$$\left(\frac{\partial s}{\partial T}\right)_v = \frac{c_v}{T} \tag{9.24}$$

and

$$\left(\frac{\partial u}{\partial v}\right)_T = T\left(\frac{\partial s}{\partial v}\right)_T - P \tag{9.25}$$

Furthermore, with the third Maxwell relation from Equation 9.18, Equation 9.25 can be written as

$$\left(\frac{\partial u}{\partial v}\right)_T = T\left(\frac{\partial P}{\partial T}\right)_v - P \tag{9.26}$$

Use of Equation 9.26 in Equation 9.23 yields the general equation for the change in internal energy for a simple compressible substance:

$$du = c_v\, dT + \left[T\left(\frac{\partial P}{\partial T}\right)_v - P\right] dv \tag{9.27}$$

Notice that if the P-v-T behavior and the constant-volume specific heat of a homogeneous substance are adequately characterized, then Equation 9.27 is sufficient to determine the change in internal energy.

9.4.2 Enthalpy

The general equation for the change in enthalpy is derived in a manner similar to that used for the internal energy. From the second $T\,ds$ equation, Equation 9.10,

$$dh = T\, ds + v\, dP \tag{9.28}$$

In this instance the entropy is considered to be a function of temperature and pressure, $s = s(T, P)$, and the differential ds is

$$ds = \left(\frac{\partial s}{\partial T}\right)_P dT + \left(\frac{\partial s}{\partial P}\right)_T dP \tag{9.29}$$

Substituting this expression for ds into Equation 9.28 and rearranging yields

$$dh = T\left(\frac{\partial s}{\partial T}\right)_P dT + \left[T\left(\frac{\partial s}{\partial P}\right)_T + v\right] dP \tag{9.30}$$

From Chapter 2, with enthalpy written as a function of T and P, the change in enthalpy for a simple compressible substance was found to be (Equation 2.22)

$$dh = c_p\, dT + \left(\frac{\partial h}{\partial P}\right)_T dP \tag{9.31}$$

A comparison of the coefficients of dT and dP in Equations 9.30 and 9.31 reveals that

$$\left(\frac{\partial s}{\partial T}\right)_P = \frac{c_p}{T} \tag{9.32}$$

and

$$\left(\frac{\partial h}{\partial P}\right)_T = \left[T\left(\frac{\partial s}{\partial P}\right)_T + v\right] \tag{9.33}$$

With the fourth Maxwell relation, Equation 9.19, Equation 9.33 can be written as

$$\left(\frac{\partial h}{\partial P}\right)_T = \left[v - T\left(\frac{\partial v}{\partial T}\right)_P\right] \tag{9.34}$$

Combining Equation 9.34 with Equation 9.31 results in a general equation for the change in enthalpy for a simple compressible substance:

$$dh = c_p\,dT + \left[v - T\left(\frac{\partial v}{\partial T}\right)_P\right] dP \tag{9.35}$$

9.4.3 Entropy

General equations for the entropy change are easily obtained from results of the preceding paragraphs. From Equation 9.21 the differential ds for $s = s(T, v)$ is given by

$$ds = \left(\frac{\partial s}{\partial T}\right)_v dT + \left(\frac{\partial s}{\partial v}\right)_T dv \tag{9.36}$$

In Equation 9.24 the first partial derivative above has been identified as

$$\left(\frac{\partial s}{\partial T}\right)_v = \frac{c_v}{T}$$

The second partial derivative in Equation 9.36 can be expressed in terms of P, v, and T by using the third Maxwell relation, Equation 9.18. With these results Equation 9.36 can be written as

$$ds = \frac{c_v}{T}\,dT + \left(\frac{\partial P}{\partial T}\right)_v dv \tag{9.37}$$

Similarly, from Equation 9.29 the differential ds for $s = s(T, P)$ is given by

$$ds = \left(\frac{\partial s}{\partial T}\right)_P dT + \left(\frac{\partial s}{\partial P}\right)_T dP \tag{9.38}$$

In Equation 9.32 the first partial derivative above was identified as

$$\left(\frac{\partial s}{\partial T}\right)_P = \frac{c_p}{T}$$

The other partial derivative in Equation 9.38 can be expressed in terms of P, v, and T by using the fourth Maxwell relation, Equation 9.19. Therefore, Equation 9.38 can be written as

$$ds = \frac{c_p}{T}\,dT - \left(\frac{\partial v}{\partial T}\right)_P dP \tag{9.39}$$

■ EXAMPLE 9.3

Experimental measurements of pressure, specific volume, and temperature of a certain gas indicate that the P-v-T behavior of the gas can be described by the equation

$$Pv = RT + aP$$

where a is a constant, for a limited range of pressure and temperature. Use the general equations for du, dh, and ds to derive expressions for the changes in internal energy, enthalpy, and entropy of the gas. Assume that the specific heats of the gas are constant. Compare these results with the corresponding equations for an ideal gas developed in Chapter 2.

Solution. The general equation for du, Equation 9.27, is

$$du = c_v\, dT + \left[T\left(\frac{\partial P}{\partial T}\right)_v - P\right] dv$$

The equation of state can be written in a form that is explicit in P as

$$P = \frac{RT}{v - a}$$

and the partial derivative indicated above can be evaluated as

$$\left(\frac{\partial P}{\partial T}\right)_v = \frac{R}{v - a}$$

Thus for this gas

$$du = c_v\, dT + \left(\frac{RT}{v - a} - P\right) dv$$

or

$$du = c_v\, dT$$

and

$$u_2 - u_1 = \int_1^2 c_v\, dT = \underline{\underline{c_v(T_2 - T_1)}}$$

Notice that the internal-energy change for this gas is the same as that for an ideal gas, and it depends only on temperature.

From Equation 9.35 the general equation for dh is given by

$$dh = c_p\, dT + \left[v - T\left(\frac{\partial v}{\partial T}\right)_P\right] dP$$

and for this gas

$$v = \frac{RT}{P} + a$$

Therefore

$$\left(\frac{\partial v}{\partial T}\right)_P = \frac{R}{P}$$

and

$$dh = c_p\, dT + \left(v - \frac{RT}{P}\right) dP$$

or

$$dh = c_p\, dT + adP$$

and the enthalpy change is therefore

$$\underline{\underline{h_2 - h_1 = c_p(T_2 - T_1) + a(P_2 - P_1)}}$$

This expression for the enthalpy change differs from that derived for an ideal gas in that a term involving the pressure change appears. For this gas the enthalpy depends on pressure as well as temperature.

The entropy change of the gas can be evaluated by using Equation 9.37 or Equation 9.39. From Equation 9.39

$$ds = \frac{c_p}{T}\, dT - \left(\frac{\partial v}{\partial T}\right)_P dP$$

Thus for this gas ds can be written as

$$ds = \frac{c_p}{T}\, dT - \frac{R}{P} dP$$

so that

$$\underline{\underline{s_2 - s_1 = c_p \ln \frac{T_2}{T_1} - R \ln \frac{P_2}{P_1}}}$$

This expression is identical to the one derived in Chapter 2 for an ideal gas with constant specific heats.

If ideal-gas behavior were assumed for this gas, calculated values of enthalpy would be in error because the pressure dependence of h would be neglected. The magnitude of the error would depend on the value of the constant a as well as the pressure change involved. ■

■ EXAMPLE 9.4

Use the general equation for du and P-v-T data from Table B.3E to estimate the internal-energy change of water during an isothermal process at 800°F from 750 to 1000 psia. Compare this value with the internal-energy change calculated by using tabular values of u.

Solution. From Equation 9.27

$$du = c_v\, dT + \left[T\left(\frac{\partial P}{\partial T}\right)_v - P\right] dv$$

and for an isothermal process ($dT = 0$),

$$du_T = \left[T\left(\frac{\partial P}{\partial T}\right)_v - P\right]dv$$

To estimate the internal-energy change, we estimate the average value of the term in brackets and assume that this value remains constant during the change of state. Thus

$$(u_2 - u_1)_T \simeq \left[T\left(\frac{\partial P}{\partial T}\right)_v - P\right]_{\text{avg}} (v_2 - v_1)$$

For the average value of the bracketed term we can average the values obtained by evaluating this term at each of the end states. Thus at each state we must estimate $[T(\partial P/\partial T)_v - P]$. At state 1 where $v = 0.93908$ ft^3/lb$_m$,

$$\left(\frac{\partial P}{\partial T}\right)_{v_1} \simeq \left(\frac{(1000 - 500)\text{psia}}{(1163.5 - 472.8)°\text{F}}\right)_{v_1 = 0.93908\ \text{ft}^3/\text{lb}_m} = 0.724\ \text{psia}/°\text{R}$$

and at state 2 where $v = 0.68772$ ft^3/lb$_m$

$$\left(\frac{\partial P}{\partial T}\right)_{v_2} \simeq \left(\frac{(1250 - 750)\text{psia}}{(1057 - 568.1)°\text{F}}\right)_{v_2 = 0.68772\ \text{ft}^3/\text{lb}_m} = 1.023\ \text{psia}/°\text{R}$$

Therefore

$$\left[T\left(\frac{\partial P}{\partial T}\right)_v - P\right]_1 = (1260°\text{R})(0.724\ \text{psia}/°\text{R}) - 750\ \text{psia} = 162.2\ \text{psia}$$

and

$$\left[T\left(\frac{\partial P}{\partial T}\right)_v - P\right]_2 = (1260°\text{R})(1.023\ \text{psia}/°\text{R}) - 1000\ \text{psia} = 289\ \text{psia}$$

so that

$$\left[T\left(\frac{\partial P}{\partial T}\right)_v - P\right]_{\text{avg}} \simeq \frac{(162.2 + 289)\text{psia}}{2} = 225.6\ \text{psia}$$

and

$$(u_2 - u_1)_T = (225.6\ \text{psia})(0.68772 - 0.93908)\text{ft}^3/\text{lb}_m$$

$$= \frac{-\left(56.7\,\dfrac{\text{lb}_f\cdot\text{ft}^3}{\text{in}^2\cdot\text{lb}_m}\right)(144\ \text{in}^2/\text{ft}^2)}{778\ \text{ft}\cdot\text{lb}_f/\text{Btu}}$$

$$= \underline{\underline{-10.5\ \text{Btu}/\text{lb}_m}}$$

Using tabular values of internal energy, we find

$$(u_2 - u_1) = (1261.2 - 1270.3)\text{Btu}/\text{lb}_m = \underline{\underline{-9.1\ \text{Btu}/\text{lb}_m}}$$

The estimate of the internal-energy change obtained from P-v-T data is in error by about 13 percent. This error could be reduced substantially if property tables with smaller increments of pressure and temperature were used. ■

9.5 GENERAL EQUATIONS FOR SPECIFIC HEATS

In the preceding section thermodynamic relations for changes in internal energy, enthalpy, and entropy in terms of pressure, temperature, specific volume, and the specific heats were developed. In this section further relations are developed in order to describe the dependence of the specific heats on temperature and pressure or specific volume. These relations, together with a means of measuring the specific heat as a function of temperature, will complete the set of mathematical equations necessary to evaluate property changes for a simple, compressible, homogeneous substance from the P-v-T behavior of the substance.

The first specific-heat relation is obtained by equating Equations 9.37 and 9.39 for ds. After multiplying by the absolute temperature and rearranging, we obtain the result

$$(c_p - c_v)\, dT = T\left(\frac{\partial v}{\partial T}\right)_P dP + T\left(\frac{\partial P}{\partial T}\right)_v dv$$

For $P = P(T, v)$ the differential of P in this equation can be replaced by

$$dP = \left(\frac{\partial P}{\partial T}\right)_v dT + \left(\frac{\partial P}{\partial v}\right)_T dv$$

with the result

$$(c_p - c_v)\, dT = T\left(\frac{\partial v}{\partial T}\right)_P\left(\frac{\partial P}{\partial T}\right)_v dT + T\left[\left(\frac{\partial v}{\partial T}\right)_P\left(\frac{\partial P}{\partial v}\right)_T + \left(\frac{\partial P}{\partial T}\right)_v\right] dv$$

But the expression in brackets is zero since, by Equation 9.8, the cyclic relation requires that

$$\left(\frac{\partial v}{\partial T}\right)_P\left(\frac{\partial P}{\partial v}\right)_T = -\left(\frac{\partial P}{\partial T}\right)_v \tag{9.40}$$

Therefore, the difference between the specific heats can be expressed as

$$c_p - c_v = T\left(\frac{\partial v}{\partial T}\right)_P\left(\frac{\partial P}{\partial T}\right)_v \tag{9.41}$$

The partial derivative of the pressure in Equation 9.41 is seldom measured and can be eliminated by substituting Equation 9.40:

$$c_p - c_v = -T\left(\frac{\partial v}{\partial T}\right)_P^2\left(\frac{\partial P}{\partial v}\right)_T \tag{9.42}$$

The partial derivatives appearing in Equation 9.42 are related to two important thermodynamic properties. The first, called the *volume expansivity*, is denoted by the Greek symbol β and is defined as

$$\beta \equiv \frac{1}{v}\left(\frac{\partial v}{\partial T}\right)_P \tag{9.43}$$

The second, called the *isothermal compressibility*, is denoted by the Greek symbol κ and is defined as

$$\kappa \equiv -\frac{1}{v}\left(\frac{\partial v}{\partial P}\right)_T \tag{9.44}$$

Experimental evidence shows that the specific volume of a substance decreases as the pressure is increased isothermally, $(\partial v/\partial P)_T < 0$, and the negative sign is introduced into the definition in Equation 9.44 so that the isothermal compressibility is always positive. With these definitions Equation 9.42 can be written as

$$c_p - c_v = \frac{vT\beta^2}{\kappa} \tag{9.45}$$

Equation 9.45 is primarily used to study the temperature dependence of the constant-volume specific heat, and it is a particularly significant result in many respects. The isothermal compressibility κ is positive for all known substances, and all other terms in Equation 9.45 are positive; therefore, the constant-pressure specific heat must always be greater than or equal to the constant-volume specific heat, or

$$c_p \geq c_v$$

Furthermore, Equation 9.45 shows that the difference between c_p and c_v approaches zero as the absolute temperature approaches zero. And finally, for incompressible substances (v = constant or $dv = 0$) the volume expansivity is zero, and therefore the constant-pressure specific heat and the constant-volume specific heat of an incompressible substance are equal.

The influence exerted by the specific volume on the constant-volume specific heat can be examined by resorting to Equation 9.37 for the differential entropy change:

$$ds = \frac{c_v}{T}\,dT + \left(\frac{\partial P}{\partial T}\right)_v dv$$

Using Equation 9.3 as a test for exactness (i.e., that the mixed second partial derivatives must be equal) yields

$$\frac{\partial}{\partial v}\left(\frac{c_v}{T}\right)_T = \frac{\partial}{\partial T}\left[\left(\frac{\partial P}{\partial T}\right)_v\right]_v$$

or

$$\left(\frac{\partial c_v}{\partial v}\right)_T = T\left(\frac{\partial^2 P}{\partial T^2}\right)_v \tag{9.46}$$

Similarly, the influence of pressure on the constant-pressure specific heat can be determined from Equation 9.39 for the differential entropy change,

$$ds = \frac{c_p}{T}\,dT - \left(\frac{\partial v}{\partial T}\right)_P dP$$

and the test for exactness. This manipulation results in the expression

$$\left(\frac{\partial c_p}{\partial P}\right)_T = -T\left(\frac{\partial^2 v}{\partial T^2}\right)_P \tag{9.47}$$

With the temperature dependence of the constant-pressure specific heat established through experimental measurements at low pressure, the specific heat at other pressures can be obtained by integrating Equation 9.47 along an isothermal path. Denoting these low-pressure specific-heat values by $c_{p,0}$ (the zero-pressure constant-pressure specific heat), the integration yields

$$c_p = c_{p,0} - T\int_0^P \left(\frac{\partial^2 v}{\partial T^2}\right)_P dP \tag{9.48}$$

If an equation of state is available from experimental data or theoretical considerations that is explicit in v, Equation 9.48 can be evaluated. Should the equation of state be more conveniently expressed as an explicit function of pressure, evaluation of Equation 9.46 is usually preferred.

EXAMPLE 9.5

The constant-pressure specific heat of steam at 300°C and 1.6 MPa is approximately 2.25 kJ/kg·K. Use Equation 9.47 and P-v-T data from Table B.3 to estimate the specific heat of steam at 300°C and 2.0 MPa. Compare this estimate with the specific heat calculated by using the definition of the constant-pressure specific heat given in Equation 2.21.

Solution. This example illustrates how values of specific heat can be obtained by using P-v-T data and an experimental measurement of the specific heat at a single pressure. The specific heat at the indicated state can be obtained by integrating Equation 9.47 along an isothermal path (T = 300°C). Thus

$$c_p(P_2, T_1) - c_p(P_1, T_1) = -T_1\int_{P_1}^{P_2} \left(\frac{\partial^2 v}{\partial T^2}\right)_P dP$$

We estimate the value of the integral by evaluating the partial derivative at the average pressure of 1.8 MPa and assuming that the value of the derivative remains constant:

$$c_p(P_2, T_1) \simeq c_p(P_1, T_1) - T_1\left(\frac{\partial^2 v}{\partial T^2}\right)_{P_{\text{avg}}} (P_2 - P_1)$$

The derivative can be approximated by combining Taylor's series expansions of v at 250 and 350°C and a pressure equal to the average pressure of 1.8 MPa. This procedure gives

$$\left(\frac{\partial^2 v}{\partial T^2}\right)_{P_{avg}} \simeq \left[\frac{v_{350°C} - 2v_{300°C} + v_{250°C}}{(\Delta T)^2}\right]_{P=1.8\text{ MPa}}$$

Using P-v-T data from Table B.3 in this expression gives

$$\left(\frac{\partial^2 v}{\partial T^2}\right)_{P_{avg}} \simeq \frac{[0.15456 - 2(0.14019) + 0.12493]\text{m}^3/\text{kg}}{(50\text{ K})^2}$$

$$= (-3.56 \times 10^{-7}\text{ m}^3/\text{kg·K}^2)(10^3\text{kN/m}^2\text{·MPa})$$

$$= -3.56 \times 10^{-4}\text{ kJ/kg·K}^2\text{·MPa}$$

Therefore

$$c_p(P_2, T_1) \simeq 2.25\text{ kJ/kg·K} - (573\text{ K})$$

$$\times (-3.56 \times 10^{-4}\text{kJ/kg·K}^2\text{·MPa})(0.4\text{ MPa})$$

$$= \underline{\underline{2.331\text{ kJ/kg·K}}}$$

The accuracy of this value can be checked by using Equation 2.21,

$$c_p = \left(\frac{\partial h}{\partial T}\right)_P$$

evaluated at state 2,

$$c_p(P_2, T_2) \simeq \left(\frac{h_{350°C} - h_{250°C}}{350°C - 250°C}\right)_{P=2.0\text{ MPa}}$$

$$= \frac{(3136.6 - 2901.6)\text{kJ/kg}}{100\text{ K}} = \underline{\underline{2.35\text{ kJ/kg·K}}}$$

■

■ EXAMPLE 9.6

Calculate the volume expansivity and the isothermal compressibility of water at 150°F and 2000 psia. Use these values to estimate the difference between c_p and c_v of water at this state.

Solution. The volume expansivity is defined in Equation 9.43,

$$\beta = \frac{1}{v}\left(\frac{\partial v}{\partial T}\right)_P$$

and can be approximated by using data from Table B.4E as

$$\beta \simeq \frac{1}{v}\left(\frac{\Delta v}{\Delta T}\right)_P = \frac{1}{0.016242\text{ ft}^3/\text{lb}_m}\left[\frac{(0.016526 - 0.016034)\text{ft}^3/\text{lb}_m}{(200 - 100)°\text{F}}\right]_{P=2000\text{ psia}}$$

$$= \underline{\underline{3.03 \times 10^{-4}\,°\text{R}^{-1}}}$$

Similarly, the isothermal compressibility, Equation 9.44, is estimated as

$$\kappa = -\frac{1}{v}\left(\frac{\partial v}{\partial P}\right)_T \simeq -\frac{1}{v}\left(\frac{\Delta v}{\Delta P}\right)_T$$

$$= -\frac{1}{0.016242\ \text{ft}^3/\text{lb}_\text{m}}\left[\frac{(0.016195 - 0.016291)\text{ft}^3/\text{lb}_\text{m}}{(3000 - 1000)\text{psia}}\right]_{T=150°\text{F}}$$

$$= \underline{\underline{2.96 \times 10^{-6}\ \text{psia}^{-1}}}$$

From these values the difference between c_p and c_v for water at the given state, from Equation 9.45, is estimated to be

$$c_p - c_v = \frac{vT\beta^2}{\kappa} = \frac{(0.016242\ \text{ft}^3/\text{lb}_\text{m})(610°\text{R})(3.03 \times 10^{-4}°\text{R}^{-1})^2(144\ \text{in}^2/\text{ft}^2)}{2.96 \times 10^{-6}\text{in}^2/\text{lb}_\text{f}}$$

$$= 44.3\ \text{ft}\cdot\text{lb}_\text{f}/\text{lb}_\text{m}\cdot°\text{R} = \underline{\underline{0.057\ \text{Btu}/\text{lb}_\text{m}\cdot°\text{R}}}$$

9.6 OTHER THERMODYNAMIC RELATIONS

Other thermodynamic relations, in addition to those presented thus far, can be derived from the Gibbs equations and the Maxwell relations. Two of particular interest, the Clapeyron equation and the Joule-Thompson coefficient, are considered in this section. The Clapeyron equation provides a means of evaluating changes in enthalpy or entropy during phase changes, while the Joule-Thompson coefficient can be used to facilitate measurements of the constant-pressure specific heat of a fluid.

9.6.1 The Clapeyron Equation

During a change of phase that occurs at constant temperature, the pressure also remains constant. The phase change could be a solid-liquid, solid-vapor, or liquid-vapor transition. The entropy change during any isothermal process can be described in terms of the P-v-T behavior, using the third Maxwell relation, Equation 9.18,

$$\left(\frac{\partial s}{\partial v}\right)_T = \left(\frac{\partial P}{\partial T}\right)_v \tag{9.49}$$

During an isothermal change of phase the pressure is a unique function of temperature and is independent of the specific volume. Therefore, the term $(\partial P/\partial T)_v$ can be written as a total derivative,

$$\left(\frac{\partial P}{\partial T}\right)_v = \left(\frac{dP}{dT}\right)_{\text{sat}}$$

for the isothermal phase change. The subscript "sat" is used to indicate that the derivative represents the slope of the saturation P-T curve at a specific saturation state (see, for example, Figure 2.9). Since this slope is independent of specific volume, Equation 9.49

can be integrated to yield

$$s_2 - s_1 = \left(\frac{dP}{dT}\right)_{\text{sat}} (v_2 - v_1)$$

where the subscripts 1 and 2 denote saturation states corresponding to the specific phase change. For instance, the preceding equation applied to a transition from liquid to vapor can be written as

$$s_g - s_f = \left(\frac{dP}{dT}\right)_{\text{sat}} (v_g - v_f)$$

or

$$\left(\frac{dP}{dT}\right)_{\text{sat}} = \frac{s_{fg}}{v_{fg}} \tag{9.50}$$

Equation 9.50 is an example of a means of evaluating an entropy change associated with a phase change entirely from measurements of pressure, specific volume, and temperature. With use of the second $T\,ds$ equation, Equation 9.50 can alternatively be written in terms of the enthalpy of vaporization, h_{fg}. From Equation 9.10

$$dh = T\,ds + v\,dP$$

For the isothermal phase change dP is zero and T is constant. Therefore, integration of this expression gives

$$s_{fg} = \frac{h_{fg}}{T} \tag{9.51}$$

Finally, substituting Equation 9.51 into Equation 9.50 gives the desired relation,

$$\left(\frac{dP}{dT}\right)_{\text{sat}} = \frac{h_{fg}}{Tv_{fg}} \tag{9.52}$$

which is known as the *Clapeyron equation*. This important relation permits the determination of the enthalpy of vaporization h_{fg} entirely from P-v-T measurements.

The Clapeyron equation can be written in a general form applicable to a change of phase between saturation states 1 and 2 as

$$\left(\frac{dP}{dT}\right)_{\text{sat}} = \frac{h_{21}}{Tv_{21}} \tag{9.53}$$

■ EXAMPLE 9.7

Use the Clapeyron equation to estimate the value of the enthalpy of vaporization of refrigerant-12 at 40°C.

Solution. With the Clapeyron equation, Equation 9.52,

$$\left(\frac{dP}{dT}\right)_{\text{sat}} = \frac{h_{fg}}{Tv_{fg}}$$

the enthalpy of vaporization can be determined from experimental measurements of pressure, temperature, and specific volume of liquid and vapor saturation states:

$$h_{fg} = Tv_{fg}\left(\frac{dP}{dT}\right)_{\text{sat}}$$

The slope of the saturation P-T curve at 40°C can be estimated from P-T data as

$$\left(\frac{dP}{dT}\right)_{\text{sat}} \simeq \left(\frac{P_{45°\text{C}} - P_{35°\text{C}}}{45°\text{C} - 35°\text{C}}\right)_{\text{sat}}$$

Using tabular values from Table C.1 yields

$$\left(\frac{dP}{dT}\right)_{\text{sat}} \simeq \frac{(1.0825 - 0.84701)\text{MPa}}{10°\text{C}}$$

$$= 0.02355 \text{ MPa/°C} = 0.02355 \text{ MPa/K}$$

and at 40°C

$$v_{fg} = 0.017563 \text{ m}^3/\text{kg}$$

Therefore, the enthalpy of vaporization is approximately

$$h_{fg} \simeq (313 \text{ K})(0.017563 \text{ m}^3/\text{kg})(0.02355 \text{ MPa/K})(10^3 \text{ kN/m}^2\cdot\text{MPa})$$

$$= \underline{\underline{129.46 \text{ kJ/kg}}}$$

The tabulated value of h_{fg} at 40°C from Table C.1 is 129.38 kJ/kg. ■

9.6.2 The Joule-Thompson Coefficient

If the pressure of a flowing fluid is decreased by means of an adiabatic throttling process in which kinetic and potential energy changes are negligible, the enthalpies of the fluid at the inlet and the exit of the throttling device are equal. This result can be established by applying the conservation-of-energy equation to the steady-flow throttling process. Throttling the fluid to a lower pressure can alter the temperature of the fluid. In fact, the fluid temperature can increase, decrease, or even remain unchanged.

The *Joule-Thompson coefficient*, used to measure the temperature change of a fluid during a throttling process, is defined by the expression

$$\mu \equiv \left(\frac{\partial T}{\partial P}\right)_h \tag{9.54}$$

Positive values of the Joule-Thompson coefficient signify that the temperature decreases as a result of a pressure drop caused by throttling; whereas negative values signify that the temperature increases. If the coefficient is zero, throttling will not cause a change in the temperature of the fluid.

The Joule-Thompson coefficient has a convenient geometrical interpretation since it represents the slope of lines of constant enthalpy on a temperature-pressure diagram. Such a diagram can be constructed from experimental measurements of the temperature

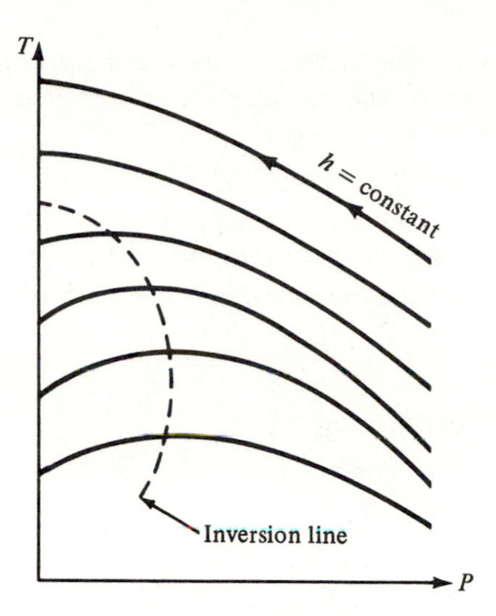

Figure 9.1 Typical T-P diagram from throttling-process measurements.

and pressure changes that occur during a throttling process. A typical T-P diagram for a gas is illustrated in Figure 9.1. Since throttling causes a decrease in pressure, throttling processes proceed from right to left along a line of constant h on this diagram. The *inversion line* in this figure is the locus of all points of zero slope or zero values of the Joule-Thompson coefficient, and the temperature at the intersection of the inversion line and a line of constant h is called the *inversion temperature*. Notice that not all lines of constant h have an inversion point. The uppermost line in Figure 9.1, for example, always has a negative slope. Thus throttling the gas from any initial state represented by this line will cause the temperature of the gas to increase. On the other hand, if the throttling process begins with the gas at a state that is to the left of the inversion line, throttling can result in a decrease in the temperature of the gas. This cooling effect can be used to advantage in systems designed to liquefy gases. However, if the Joule-Thompson coefficient is always negative at ordinary temperatures and pressures, as it is for hydrogen and helium, the gas must be cooled by other means until a state is reached at which the Joule-Thompson coefficient is positive or until the inversion temperature can be crossed before throttling can be used to cool the gas further.

The Joule-Thompson coefficient can be related to the constant-pressure specific heat in the following manner. With the cyclic relation, Equation 9.8, Equation 9.54 can be expressed as

$$\mu = -\frac{(\partial h/\partial P)_T}{(\partial h/\partial T)_P}$$

Substituting the definition of c_p from Equation 2.21 and using Equation 9.34 and rearranging gives

$$\mu c_p = T\left(\frac{\partial v}{\partial T}\right)_P - v \tag{9.55}$$

Equation 9.55 can be used with experimental measurements of P-v-T data and the Joule-Thompson coefficient to determine the specific heat of gases at high pressure.

■ EXAMPLE 9.8

Calculate the Joule-Thompson coefficient for steam at 700°F and 1000 psia. Estimate the temperature that would result from throttling steam at 700°F and 1000 psia to 300 psia. First assume ideal-gas behavior for the steam and then use property values from Table B.3E.

Solution. If ideal-gas behavior is assumed, then

$$Pv = RT$$

and

$$\left(\frac{\partial v}{\partial T}\right)_P = \frac{R}{P}$$

Substituting this result into Equation 9.55, we find that

$$\mu_{\text{ideal}} = \frac{1}{c_p}\left(\frac{RT}{P} - v\right) = \underline{\underline{0}}$$

The Joule-Thompson coefficient is identically zero for an ideal gas. Therefore, if the behavior of a gas is ideal, changing its temperature with a throttling process is not possible.

Steam is not an ideal gas, however, and we should expect that throttling would produce a temperature change. The Joule-Thompson coefficient of the steam at 700°F and 1000 psia can be estimated by using property values from the steam tables and Equation 9.54,

$$\mu = \left(\frac{\partial T}{\partial P}\right)_h \simeq \left(\frac{\Delta T}{\Delta P}\right)_h$$

or by interpolation

$$\mu \simeq \left(\frac{726.1°\text{F} - 674.7°\text{F}}{1250\ \text{psia} - 750\ \text{psia}}\right)_{h=1324.3\ \text{Btu/lb}_\text{m}} = \underline{\underline{0.103°\text{F/psia}}}$$

Notice that the Joule-Thompson coefficient is positive, indicating that the temperature of the steam will decrease as the pressure is decreased during the throttling process. If we assume that the Joule-Thompson coefficient is approximately constant, then we should expect that the temperature of the steam would decrease by about 72°F if it is throttled from 1000 to 300 psia. By interpolating in Table B.3E at 300 psia and an enthalpy of 1324.3 Btu/lb$_\text{m}$, we find that the corresponding temperature is 619°F, a decrease of about 81°F. This result compares favorably with the estimate obtained from the Joule-Thompson coefficient. More accurate results could be obtained if property tables having finer subdivisions were used. ■

9.7 SOME APPLICATIONS OF THERMODYNAMIC RELATIONS

In this section the application of the thermodynamic relations is illustrated for ideal gases, incompressible substances, and the pure substances. Given an equation of state and measurements of the constant-pressure specific heat as a function of temperature at a single pressure, we can determine other properties of interest such as internal energy, enthalpy, and entropy.

A procedure that could be used to develop tables of properties is outlined for pure substances. These same concepts are also used in the next chapter to establish a means of determining properties of real gases.

9.7.1 Ideal Gases

In Chapter 2 the ideal gas was defined as a gas that obeys the equation of state

$$Pv = RT$$

and whose internal energy is a function only of the absolute temperature, or

$$u = u(T)$$

Now that the thermodynamic relations of the preceding sections have been established, the fact that the internal energy of an ideal gas is a function of temperature alone can be verified.

With the P-v-T behavior described by the ideal-gas equation of state and the assumption that the constant-pressure specific heat as a function of temperature at low pressure is known, evaluation of other properties of the gas is possible, as illustrated in the following analysis.

From Equation 9.27 the general equation for internal energy is

$$du = c_v\, dT + \left[T\left(\frac{\partial P}{\partial T}\right)_v - P\right] dv$$

For an ideal gas, however, the bracketed term is

$$T\left(\frac{\partial P}{\partial T}\right)_v - P = T\left(\frac{R}{v}\right) - P = 0$$

so that for an ideal gas

$$du = c_v\, dT \tag{9.56}$$

To verify that u is a function of temperature alone, then, we must establish the fact that c_v is independent of specific volume. From Equation 9.46

$$\left(\frac{\partial c_v}{\partial v}\right)_T = T\left(\frac{\partial^2 P}{\partial T^2}\right)_v$$

and for the ideal gas this expression becomes

$$\left(\frac{\partial c_v}{\partial v}\right)_T = T\frac{\partial}{\partial T}\left(\frac{R}{v}\right)_v = 0$$

Thus the constant-volume specific heat of an ideal gas depends, at most, on temperature, and therefore the internal energy of an ideal gas is, at most, a function of temperature.

The fact that the enthalpy and the constant-pressure specific heat of an ideal gas depend on temperature alone can be established as follows. From

$$dh = c_p\, dT \tag{9.57}$$

and Equation 9.42, the relationship between c_p and c_v can be determined:

$$c_p - c_v = -T\left(\frac{\partial v}{\partial T}\right)_P^2\left(\frac{\partial P}{\partial v}\right)_T$$

For an ideal gas this equation becomes

$$c_p - c_v = -T\left(\frac{R}{P}\right)^2\left(-\frac{RT}{v^2}\right)$$

or

$$c_p - c_v = R \tag{9.58}$$

Since c_v is a function of temperature alone and R is a constant, the constant-pressure specific heat of an ideal gas is a function of temperature alone. Furthermore, Equation 9.57 can be used to conclude that the enthalpy of an ideal gas depends, at most, on temperature.

The entropy change can be determined from Equation 9.39:

$$ds = \frac{c_p}{T}\, dT - \left(\frac{\partial v}{\partial T}\right)_P dP$$

For an ideal gas this equation can be written as

$$ds = \frac{c_p}{T}\, dT - \frac{R}{P} dP \tag{9.59}$$

Changes in internal energy, enthalpy, and entropy of an ideal gas can now be evaluated by using the following procedures:

1. With $c_p(T)$ given, Equation 9.57 could be integrated to obtain the change in enthalpy:

$$h_2 - h_1 = \int_1^2 c_p\, dT \tag{9.60}$$

2. The internal-energy change could then be determined from the definition of h and the equation of state:

$$h = u + Pv = u + RT \tag{9.61}$$

Alternatively, internal-energy changes could be obtained by using Equation 9.58 to determine the specific heat at constant volume and integrating Equation 9.56.

3. Finally, the entropy change could be evaluated by integrating Equation 9.59:

$$s_2 - s_1 = \int_1^2 \frac{c_p}{T}\,dT - R \ln \frac{P_2}{P_1} \tag{9.62}$$

These are the same results obtained in Chapter 2 for the ideal gas. However, in Chapter 2 the fact that the internal energy of an ideal gas depends only on temperature was assumed; whereas this fact has now been established by using thermodynamic relations.

9.7.2 Incompressible Substances

An incompressible substance is one for which the specific volume remains constant regardless of changes in temperature, pressure, and other properties. For such a substance dv is zero, and Equation 9.27 for the internal-energy changes reduces to

$$du = c_v\,dT \tag{9.63}$$

Furthermore, all partial derivatives of the specific volume must be zero for an incompressible substance, and Equation 9.35 for the change in enthalpy can be simplified to

$$dh = c_p\,dT + v\,dP \tag{9.64}$$

For the same reason Equation 9.41 can be used to conclude that the constant-pressure specific heat and the constant-volume specific heat of an incompressible substance are equal:

$$c_p = c_v \tag{9.65}$$

Because of this fact, dropping the subscripts on the specific heats and simply using c to denote the specific heat of an incompressible substance is customary practice, or

$$c = c_p = c_v \tag{9.66}$$

The fact that the specific heat of an incompressible substance is independent of pressure and therefore depends, at most, on temperature should be evident from Equation 9.47,

$$c = c(T) \tag{9.67}$$

Finally, the entropy change for an incompressible substance, from Equation 9.37, is simply

$$ds = \frac{c}{T}\,dT \tag{9.68}$$

Notice that since the specific heat of an incompressible substance depends only on temperature, the expression for the internal-energy change is identical to Equation 9.56 for an ideal gas, although the reason for the simplifications that led to Equations 9.56 and 9.63 are distinctly different. Unlike the enthalpy change of the ideal gas, however, the enthalpy change of an incompressible substance is both temperature- and pressure-dependent. Also, the entropy change of an incompressible substance depends only on temperature.

The following steps could then be used to determine changes in internal energy, enthalpy, and entropy for an incompressible substance:

1. With $c(T)$ given, Equation 9.63 can be integrated to obtain the change in internal energy:

$$u_2 - u_1 = \int_1^2 c\, dT \tag{9.69}$$

2. The enthalpy change can then be determined from $h = u + Pv$ or by integrating Equation 9.64:

$$h_2 - h_1 = u_2 - u_1 + v(P_2 - P_1) \tag{9.70}$$

3. The entropy change can be determined by integrating Equation 9.68:

$$s_2 - s_1 = \int_1^2 \frac{c}{T} dT \tag{9.71}$$

These are the same results obtained previously in Chapter 2 for an incompressible substance. Since the specific heats of most liquids and solids are approximately linear functions of temperature for rather large ranges of temperature, the use of an average value of the specific heat in the above equations is usually sufficient, and the equations can be integrated directly.

9.7.3 Pure Substances

The thermodynamic properties of pure substances, such as those presented in Appendixes B and C for water and refrigerant-12, are derived from experimental measurements of pressure, specific volume, temperature, and the specific heats with the aid of thermodynamic relations such as those presented earlier in this chapter. The general equations for *du*, *dh*, and *ds* are used in single-phase regions, and relations such as the Clapeyron equation are used to evaluate changes in enthalpy and entropy during phase changes.

Consider, for example, the problem of determining the change in enthalpy of a real gas as it changes state from (T_1, P_1) to (T_2, P_2). The general equation for *dh* (Equation 9.35) must then be integrated, but the choice of the path of integration is arbitrary since enthalpy is a property. A convenient choice is the path 1-1′-2′-2 illustrated in Figure 9.2, which consists of two isothermal processes and one constant-pressure process.

The first term in Equation 9.35 vanishes for the isothermal processes 1-1′ and 2-2′, and the second term vanishes for the constant-pressure process 1′-2′. Therefore

$$h_2 - h_1 = \int_1^{1'} \left[v - T\left(\frac{\partial v}{\partial T}\right)_P \right] dP + \int_{1'}^{2'} c_p\, dT + \int_{2'}^{2} \left[v - T\left(\frac{\partial v}{\partial T}\right)_P \right] dP$$

For evaluation of these integrals only information regarding the temperature dependence of c_p at a single pressure and either experimental *P-v-T* data or a suitable equation of state for the gas are necessary. For the evaluation of properties of real gases, a very low pressure is usually chosen for the process 1′-2′, so that ideal-gas specific-heat data can be used. In the next chapter the procedure used to evaluate the properties of real gases is examined in more detail.

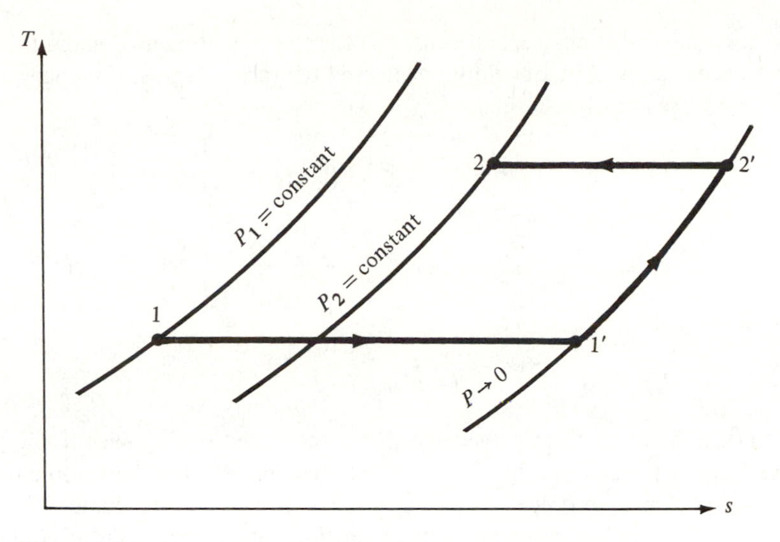

Figure 9.2 Paths of integration for enthalpy change of a gas.

9.8 SUMMARY

The purpose of the study of thermodynamic relations is to relate properties that cannot be measured directly to those that are much easier to measure or evaluate experimentally. Thermodynamic relations have been developed from a knowledge of the behavior of properties and the mathematical characteristics attributable to them. Perhaps the most important of these relations are the Gibbs equations,

$$du = T\,ds - P\,dv \tag{9.9}$$

$$dh = T\,ds + v\,dP \tag{9.10}$$

$$da = -s\,dT - P\,dv \tag{9.13}$$

$$dg = -s\,dT + v\,dP \tag{9.14}$$

and the Maxwell relations,

$$\left(\frac{\partial T}{\partial v}\right)_s = -\left(\frac{\partial P}{\partial s}\right)_v \tag{9.16}$$

$$\left(\frac{\partial T}{\partial P}\right)_s = \left(\frac{\partial v}{\partial s}\right)_P \tag{9.17}$$

$$\left(\frac{\partial s}{\partial v}\right)_T = \left(\frac{\partial P}{\partial T}\right)_v \tag{9.18}$$

$$\left(\frac{\partial s}{\partial P}\right)_T = -\left(\frac{\partial v}{\partial T}\right)_P \tag{9.19}$$

From these relations general equations for changes in internal energy, enthalpy, and entropy were derived that could be expressed entirely in terms of the specific heats and the P-v-T behavior of a substance:

$$du = c_v\, dT + \left[T\left(\frac{\partial P}{\partial T}\right)_v - P\right] dv \quad (9.27)$$

$$dh = c_p\, dT + \left[v - T\left(\frac{\partial v}{\partial T}\right)_P\right] dP \quad (9.35)$$

$$ds = \frac{c_p}{T}\, dT - \left(\frac{\partial v}{\partial T}\right)_P dP \quad (9.39)$$

Further relations were considered that can be used to characterize the pressure and temperature dependence of the specific heats. In addition, the Clapeyron equation, which deals with changes in properties during phase changes, and the Joule-Thompson coefficient, which describes the temperature change resulting from a throttling process, were introduced.

The chapter concluded with illustrations of the use of the thermodynamic relations and general equations in determining property changes for ideal gases, incompressible substances, and pure substances.

PROBLEMS

9.1 Calculate the value of $(\partial s/\partial v)_T$ for air at 100°F and 30 psia.

9.2 Use the P-v-T data for water in Table B.3 to estimate the value of $(\partial s/\partial P)_T$ at 30 MPa and 500°C. From this result, estimate the change in entropy of water as a result of an isothermal process at 500°C during which the pressure decreases from 35 to 25 MPa. Evaluate the accuracy of this estimate by comparing the approximate value with the entropy change determined from Table B.3.

9.3 Using data from Table B.3E for water at 150 psia and 1000°F, approximate the derivatives appearing in the fourth Maxwell relation (Equation 9.19). Do these values satisfy Equation 9.19? Explain.

9.4 Estimate the entropy change of refrigerant-12 when it undergoes a constant-temperature process at 110°C while the pressure increases from 1.6 MPa to 2.0 MPa. Use only P-v-T data from Table C.3.

9.5 Using the cyclic relation and the first Maxwell relation, derive the other Maxwell relations.

9.6 Derive an expression for $(\partial u/\partial P)_T$ and for $(\partial h/\partial v)_T$ that involves only P, v, and T.

9.7 Use Equation 9.26 to estimate $(\partial u/\partial v)_T$ for refrigerant-12 at 400 psia and 300°F. From this result alone, would you expect ideal-gas behavior for refrigerant-12 at this state to be an appropriate assumption? Explain.

9.8 Use Equation 9.34 to estimate $(\partial h/\partial P)_T$ for water at (a) 1.2 MPa and 250°C and (b) 1.4 MPa and 900°C. From these results alone, would you expect ideal-gas behavior for water at either or both of these states to be an appropriate assumption? Explain.

9.9 A certain gas obeys the equation of state $P(v - a/T) = RT$ where a is a constant. Derive an expression for (a) an isothermal change in enthalpy between pressures P_1 and

P_2 at temperature T_1 and (b) $c_p - c_v$ in terms of R, P, and T for this gas.

9.10 Evaluate $(\partial h/\partial s)_T$ for a gas that obeys the ideal-gas equation of state. Discuss the significance of this result after examining the slopes of isotherms in Figure B.5 in the Appendix.

9.11 Show that lines of constant pressure on a T-s diagram for a gas diverge. Discuss the implications of this result by considering, for example, an isentropic gas turbine that operates between fixed pressure limits.

9.12 Show that the slope of a constant-pressure line on a T-s diagram in the superheat region is less than the slope of a constant-specific-volume line through the same point. Sketch several lines of constant pressure and constant specific volume in the superheat region.

9.13 Derive the relation

$$\left(\frac{\partial \beta}{\partial P}\right)_T = -\left(\frac{\partial \kappa}{\partial T}\right)_P$$

9.14 For a gas that obeys the equation of state $P(v\text{-}a) = RT$, where a is a constant, and whose constant-volume specific heat is constant, show the following:

(a) The constant-pressure specific heat of the gas is constant.

(b) An internally reversible adiabatic process for this gas can be represented by the equation

$$P(v - a)^k = \text{constant}$$

9.15 Derive the relation

$$\left(\frac{\partial v}{\partial T}\right)_s = -\frac{c_v \kappa}{\beta T}$$

9.16 For an ideal gas, show that

$$\left(\frac{\partial h}{\partial v}\right)_T \quad \left(\frac{\partial c_p}{\partial P}\right)_T \quad \left(\frac{\partial c_v}{\partial P}\right)_T \quad \left(\frac{\partial c_p}{\partial v}\right)_T$$

are all zero.

9.17 Describe an experiment or experiments that could theoretically be used to determine the constant-pressure specific heat of a gas from P-v-T measurements only.

9.18 Derive the relation

$$c_v = -T\left(\frac{\partial v}{\partial T}\right)_s\left(\frac{\partial P}{\partial T}\right)_v$$

9.19 Derive the relation

$$c_p = T\left(\frac{\partial P}{\partial T}\right)_s\left(\frac{\partial v}{\partial T}\right)_P$$

9.20 Use the expression in Problem 9.18 to estimate the constant-volume specific heat of water at 7000 psia and 1500°F. Then estimate the value by using the definition of the constant-volume specific heat, and compare the two values.

9.21 Use the expression in Problem 9.19 to estimate the constant-pressure specific heat of refrigerant-12 at 0.8 MPa and 100°C. Then estimate the value by using the definition of the constant-pressure specific heat, and compare the two values.

9.22 Use Equation 9.47 to estimate the constant-pressure specific heat of refrigerant-12 at 200 psia and 200°F. The constant-pressure specific heat at 180 psia, 200°F is approximately 0.175 $\text{Btu/lb}_\text{m}\cdot$°R.

9.23 Use Equation 9.46 to estimate the constant-volume specific heat of water at 7 MPa and 600°C. The constant-volume specific heat at 0.06525 $\text{m}^3\text{/kg}$, 600°C is approximately 1.805 kJ/kg·K.

9.24 Evaluate the volume expansivity and the isothermal compressibility of nitrogen at 180°F and 50 psia.

9.25 Derive the following expression for the change in specific volume:

$$\frac{dv}{v} = \beta\, dT - \kappa\, dP$$

9.26 The values of v, β, and κ for copper at 300 K are 111.1×10^{-6} $\text{m}^3\text{/kg}$, 49.2×10^{-6}

K^{-1}, and 77.6×10^{-9} cm^2/N, respectively. Calculate the value of $c_p - c_v$ for copper at this temperature. Use the value of c_p from Table H.3, and calculate c_v for copper. Determine the percent error associated with assuming that $c_v = c_p$.

9.27 Use the expression in Problem 9.25 and the values for β and κ given in Problem 9.26 to estimate the specific-volume change for copper for the following changes of state:

(a) a temperature increase from 300 to 500 K at constant pressure

(b) a pressure increase from 1 kPa to 1 MPa at constant temperature

(c) a change from 300 K and 1 kPa to 500 K and 1 MPa

Assume that the volume expansivity and the isothermal compressibility remain relatively constant.

9.28 Estimate the values of the volume expansivity and the isothermal compressibility for water at 0.3 MPa and 400°C by using data from Table B.3. Compare these values with those obtained by assuming ideal-gas behavior for water at this state.

9.29 Use the values of h_{fg}, v_{fg}, and P_{sat} at 40°F to estimate the saturation pressure of refrigerant-12 at 80°F. Evaluate the accuracy of your estimate.

9.30 Use P-v-T data from Table B.1 to estimate the enthalpy of vaporization of water at 65°C.

9.31 Use P-v-T data from Table C.1 to estimate the change in internal energy of refrigerant-12 as it changes from a saturated liquid to a saturated vapor at a constant temperature of 20°C.

9.32 Use the Clapeyron equation to estimate the enthalpy of vaporization of water at (a) 180°F and (b) 750 psia. Compare your results with the tabulated values.

9.33 Use the Clapeyron equation to estimate the enthalpy of vaporization of refrigerant-12 at (a) −40°C and (b) 40°C. Compare your results with the tabulated results.

9.34 Derive an expression for the Joule-Thompson coefficient for a gas that obeys the equation of state $P(v\text{-}a) = RT$, where a is a positive constant. Is it possible to cool this gas in a throttling process? Explain.

9.35 Derive an expression for the inversion temperature for a gas that obeys the equation of state

$$\left(P + \frac{a}{v^2}\right)(v - b) = RT$$

9.36 Estimate the Joule-Thompson coefficient for refrigerant-12 at 50 psia and 150°F. Using this value, estimate the temperature of the refrigerant if it is throttled from 50 psia and 150°F to 10 psia. Evaluate the accuracy of your estimate.

9.37 Estimate the Joule-Thompson coefficient for steam at 30 MPa and 550°C. Use this value to estimate the pressure drop required to cause a 50°C change in temperature of the steam if it is throttled from 30 MPa and 550°C. Evaluate the accuracy of your estimate.

10 Thermodynamic Behavior of Real Gases

10.1 INTRODUCTION

The concept of an ideal gas, first introduced in Chapter 2, pertains to a gas whose P-v-T behavior is described by the equation of state,

$$Pv = RT \tag{10.1}$$

This equation predicts the behavior of a real gas if the pressure of the gas is significantly lower than its critical pressure and the temperature of the gas is significantly higher than its critical temperature. A question that naturally arises is, which equation of state should be used when a gas is at a state that cannot adequately be described by the ideal-gas equation of state?

This chapter explores equations of state that can accurately predict P-v-T data for real gases whose behavior is not adequately described by Equation 10.1. Several real-gas equations of state are discussed, and these equations are used to evaluate changes in internal energy, enthalpy, and entropy of real gases.

10.2 DEPARTURE FROM IDEAL-GAS BEHAVIOR

The departure of the ideal-gas equation of state from the actual P-v-T behavior of a real gas can become significant as the pressure of the gas increases and the temperature decreases. In other words, the ideal-gas equation becomes less accurate as the state of the gas moves closer to the saturation region.

The ideal-gas equation of state accurately describes the behavior of a gas when the forces between individual molecules of the gas are negligible. As the pressure on the gas is reduced or the temperature is increased, the mean free path between adjacent molecules becomes greater and the assumption of ideal-gas behavior becomes more accurate. As the pressure increases and the temperature decreases, the molecules are more tightly packed, and the attractive forces between the molecules are more significant. Under these conditions the actual density of the gas will exceed that predicted by the ideal-gas equation of state.

EXAMPLE 10.1

Determine the specific volume of water vapor at the following two states:

(a) $P = 5000$ psia, $T = 1000°\text{F}$
(b) $P = 10{,}000$ psia, $T = 800°\text{F}$

Use the ideal-gas equation of state and compare the predicted values of the specific volume with values from Table B.3E.

Solution.

(a) Using the ideal-gas equation of state

$$v = \frac{RT}{P} = \frac{(10.73\ \text{psia}\cdot\text{ft}^3/\text{lb}_\text{m}\cdot\text{mol}\cdot°\text{R})(1460°\text{R})}{(18.01\ \text{lb}_\text{m}/\text{lb}_\text{m}\cdot\text{mol})(5000\ \text{psia})}$$

$$= \underline{\underline{0.174\ \text{ft}^3/\text{lb}_\text{m}}}$$

The tabulated value for the specific volume can be obtained from Table B.3E:

$$v = \underline{\underline{0.13105\ \text{ft}^3/\text{lb}_\text{m}}}$$

The percent error in the specific volume as a result of using the ideal-gas equation of state is

$$\text{Percent error} = (100)\left(\frac{v_{\text{ideal}} - v_{\text{tables}}}{v_{\text{tables}}}\right)$$

$$= (100)\left[\frac{(0.174 - 0.13105)\text{ft}^3/\text{lb}_\text{m}}{0.13105\ \text{ft}^3/\text{lb}_\text{m}}\right] = 32.8\ \text{percent}$$

The rather large error results from the fact that the state is relatively close to the saturation region.

(b) For this state the pressure is greater and the temperature is less than the values given in part a, so the state is closer to the saturation region. We should expect that the error in approximating the specific volume by the ideal-gas equation of state will exceed the error determined for the state specified in part a.

The specific volume approximated by using the ideal-gas equation of state is

$$v = \frac{RT}{P} = \frac{(10.73\,\text{psia}\cdot\text{ft}^3/\text{lb}_\text{m}\cdot\text{mol}\cdot{}^\circ\text{R})(1260^\circ\text{R})}{(18.01\ \text{lb}_\text{m}/\text{lb}_\text{m}\cdot\text{mol})(10{,}000\,\text{psia})}$$

$$= \underline{\underline{0.07507\ \text{ft}^3/\text{lb}_\text{m}}}$$

The value from Table B.3E is

$$v = \underline{\underline{0.02765\ \text{ft}^3/\text{lb}_\text{m}}}$$

and the error that results from using the ideal-gas assumption is

$$\text{Percent error} = 100 \cdot \left[\frac{(0.07507 - 0.02765)\text{ft}^3/\text{lb}_\text{m}}{0.02765\,\text{ft}^3/\text{lb}_\text{m}}\right] = 172\,\text{percent}$$

At this state the ideal-gas assumption predicts a specific volume that is nearly three times the actual value. Clearly, this error is unacceptable, and this example illustrates the magnitude of the error that can result when the ideal-gas equation of state is used to determine properties near the saturation region. ■

■ EXAMPLE 10.2

Refrigerant-12 (dichlorodifluoromethane) has a molecular weight of 120.91. Determine the percent error in the specific volume calculated from the ideal-gas equation of state for a pressure of 1.0 MPa and a temperature of 60°C.

Solution. The specific volume predicted by the ideal-gas assumption is

$$v = \frac{RT}{P} = \frac{(8.314\,\text{kPa}\cdot\text{m}^3/\text{kg}\cdot\text{mol}\cdot\text{K})(333\ \text{K})}{(120.91\ \text{kg}/\text{kg}\cdot\text{mol})(1.0\ \text{MPa})}\left(\frac{1\,\text{MPa}}{10^3\ \text{kPa}}\right)$$

$$= 0.0229\ \text{m}^3/\text{kg}$$

The value from Table C.3 at this state is

$$v = 0.0195\ \text{m}^3/\text{kg}$$

The percent error in the ideal-gas prediction is

$$\text{Percent error} = 100 \cdot \left[\frac{(0.0229 - 0.0195)\text{m}^3/\text{kg}}{0.0195\,\text{m}^3/\text{kg}}\right] = \underline{\underline{17.4\,\text{percent}}}$$ ■

10.3 COMPRESSIBILITY FACTOR

As the previous examples illustrate, the errors that result from using the ideal-gas equation of state can be sizable. For this reason other equations of state that are more accurate over larger ranges of pressure and temperature should be examined. A particularly convenient equation of state is

$$Pv = ZRT \tag{10.2}$$

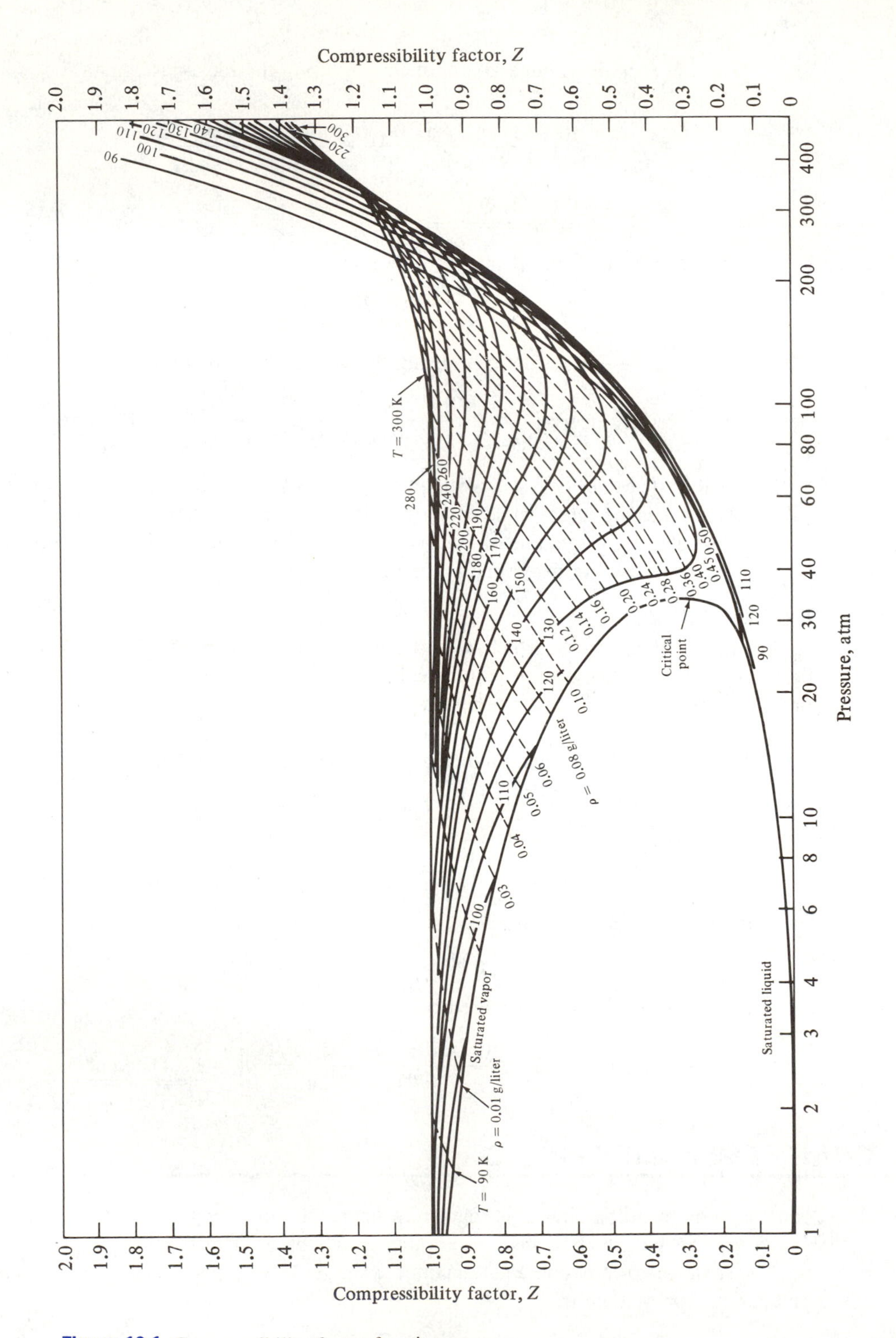

Figure 10.1 Compressibility factor for nitrogen.

which defines the *compressibility factor* Z. Comparison of this equation of state with the ideal-gas equation of state reveals that the compressibility factor can be visualized as a measure of how closely a gas approximates ideal-gas behavior. When Z is equal to 1.0, the gas behaves ideally. At those states for which Z is not equal to 1.0, the gas is not strictly ideal, and its behavior is more nonideal as the value for Z deviates farther from 1.0.

Values for the compressibility factor for a particular gas can be determined from Equation 10.2 and experimental measurements of the equilibrium values for P, v, and T. Once the compressibility factor has been calculated for a wide range of states, it can be conveniently presented in graphical form. Compressibility factors for nitrogen obtained in this fashion are presented in Figure 10.1 as a function of pressure and temperature. Compressibility-factor data are available for many other gases, and graphs similar to Figure 10.1 can be found in referencces listed in the Thermodynamic Properties section of the Bibliography.

From Figure 10.1 the fact that the compressibility factor approaches 1.0 as the pressure on the gas is reduced to zero regardless of the gas temperature should be apparent. In fact, experimental evidence indicates that all gases approach ideal-gas behavior as the pressure is reduced to zero. The compressibility factor for nitrogen is less than 1.0 for temperatures up to about 300 K and pressures up to about 100 atm. For pressures above 100 atm the compressibility factor for nitrogen can exceed 1.0. Notice also that the magnitude of the error that can result from using the ideal-gas assumption near the critical point and at pressures in excess of 300 atm is substantial.

EXAMPLE 10.3

Calculate the specific volume of nitrogen for a pressure of 450 psia and a temperature of 300°R.

(a) Assume ideal-gas behavior.
(b) Use the compressibility factor and Equation 10.2.

Solution.

(a) Assuming ideal-gas behavior for nitrogen yields

$$v_{\text{ideal}} = \frac{RT}{P} = \frac{(10.73\ \text{psia}\cdot\text{ft}^3/\text{lb}_\text{m}\cdot\text{mol}\cdot°\text{R})(300°\text{R})}{(28.01\ \text{lb}_\text{m}/\text{lb}_\text{m}\cdot\text{mol})(450\ \text{psia})} = \underline{\underline{0.2554\ \text{ft}^3/\text{lb}_\text{m}}}$$

(b) The pressure of the nitrogen, in atmospheres, is

$$P = \frac{450\ \text{psia}}{14.7\ \text{psia/atm}} = 30.6\ \text{atm}$$

From Figure 10.1 the compressibility factor at the given pressure and a temperature of 300°R (167 K) is

$$Z \simeq 0.88$$

and the specific volume is

$$v = \frac{ZRT}{P} = Zv_{\text{ideal}} = 0.88(0.2554\ \text{ft}^3/\text{lb}_\text{m})$$

$$= \underline{\underline{0.255\ \text{ft}^3/\text{lb}_\text{m}}}$$

The specific volume calculated from the ideal-gas equation of state is 14 percent greater than the value calculated from Equation 10.2. The specific volume could also be determined by using the density value plotted in Figure 10.1.

$$\rho \simeq 0.07\ \text{g/cm}^3$$

or

$$v = \frac{1}{\rho} = \left(\frac{1}{0.07\ \text{g/cm}^3}\right)\left(\frac{1\ \text{m}^3}{10^6\text{cm}^3}\right)\left(\frac{10^3\text{g}}{\text{kg}}\right)\frac{(3.2808\ \text{ft/m})^3}{(2.2046\ \text{lb}_\text{m}/\text{kg})} = 0.229\ \text{ft}^3/\text{lb}_\text{m}$$

The difference between this value and the one calculated from Equation 10.2 can be attributed to the limited accuracy obtainable from reading values from Figure 10.1. ■

■ EXAMPLE 10.4

Nitrogen has a density of 0.14 g/cm^3 and a temperature of 150 K. Estimate the compressibility factor and the pressure of the nitrogen.

Solution. From Figure 10.1 the value for Z is

$$Z \simeq \underline{\underline{0.71}}$$

and the pressure, from Figure 10.1, is approximately

$$P = (44\ \text{atm})(0.1013\ \text{MPa/atm}) = \underline{\underline{4.46\ \text{MPa}}}$$

The pressure could also be calculated from Equation 10.2:

$$P = \frac{ZRT}{v} = \rho ZRT$$

$$= (0.14\ \text{g/cm}^3)(0.71)\left(\frac{8.314 \times 10^{-3}\ \text{MPa}\cdot\text{m}^3/\text{kg}\cdot\text{mol}\cdot\text{K}}{28.013\ \text{kg/kg}\cdot\text{mol}}\right)(150\ \text{K})$$

$$\times \left(\frac{1\ \text{kg}}{1000\ \text{g}}\right)(10^6\ \text{cm}^3/\text{m}^3)$$

$$= \underline{\underline{4.43\ \text{MPa}}}$$

■

10.4 PRINCIPLE OF CORRESPONDING STATES

Compressibility-factor charts for all gases exhibit similar trends and can be consolidated into a single chart with suitable choices for the chart coordinates. If the pressure and temperature values are normalized with respect to the pressure and temperature at the critical state, the compressibility charts for practically all gases are approximately identical. This observation is the basis for the *principle of corresponding states*, which states that the compressibility factors for practically all gases are identical when the gases have the same reduced pressure and reduced temperature.

The ratio of a property to the value of the property at the critical state is called a *reduced property*. The *reduced pressure* is defined as

$$P_R \equiv \frac{P}{P_c} \tag{10.3}$$

The *reduced temperature*, defined in a similar fashion, is

$$T_R \equiv \frac{T}{T_c} \tag{10.4}$$

and the *reduced specific volume* is

$$v_R \equiv \frac{v}{v_c} \tag{10.5}$$

The reduced specific volume defined in Equation 10.5 is not often used in calculations, and it is usually replaced by a *pseudo-reduced specific volume*, which is defined by the equation

$$v_R' = \frac{v}{RT_c/P_c} \tag{10.6}$$

According to the principle of corresponding states, the compressibility factor for any gas is a function only of P_R and T_R,

$$Z = Z(P_R, T_R) \tag{10.7}$$

and can be represented on a single chart called the *generalized compressibility chart*. Generalized compressibility charts are included in Appendix E. The data are separated for purposes of clarity and accuracy into three figures, each applying to a different range of reduced pressures. Values for properties at the critical state for a number of common gases are listed in Table H.1 and Table H.1E in the Appendix.

■ EXAMPLE 10.5

Approximate the specific volume of methane (CH_4) at a pressure of 4710 psia and a temperature of 228°F.

Solution. The critical constants and molecular weight of methane from Table H.1E are

$$T_c = 344°\text{R} \qquad P_c = 672.9 \text{ psia} \qquad M = 16.043\,\text{lb}_\text{m}/\text{lb}_\text{m}\cdot\text{mol}$$

The reduced properties are

$$T_R = \frac{T}{T_c} = \frac{688°\text{R}}{344°\text{R}} = 2.0$$

$$P_R = \frac{P}{P_c} = \frac{4710 \text{ psia}}{672.9 \text{ psia}} = 7.0$$

From the generalized compressibility chart in Figure E.3E, the pseudo-reduced specific volume is

$$v'_R = 0.30$$

and the specific volume calculated by using Equation 10.6 is

$$v = \frac{v'_R RT_c}{P_c} = \frac{(0.30)(10.73 \text{ psia·ft}^3/\text{lb}_\text{m}\text{·mol·°R})(344°\text{R})}{(672.9 \text{ psia})(16.043 \text{ lb}_\text{m}/\text{lb}_\text{m}\text{·mol})}$$

$$= \underline{\underline{0.103 \text{ ft}^3/\text{lb}_\text{m}}}$$

If the methane had been assumed to be an ideal gas, the predicted value of the specific volume would have been

$$v = \frac{RT}{P} = \left(\frac{10.73 \text{ psia·ft}^3/\text{lb}_\text{m}\text{·mol·°R}}{16.043 \text{ lb}_\text{m}/\text{lb}_\text{m}\text{·mol}}\right)\left(\frac{688°\text{R}}{4710 \text{ psia}}\right)$$

$$= \underline{\underline{0.0977 \text{ ft}^3/\text{lb}_\text{m}}}$$

which is in error by approximately 4.5 percent. ■

■ EXAMPLE 10.6

Approximate the temperature of carbon dioxide at a pressure of 22.17 MPa and a specific volume of $3.11 \times 10^{-3}\text{m}^3/\text{kg}$.

Solution. The critical constants of carbon dioxide from Table H.1 are

$$P_c = 7.39 \text{ MPa} \qquad T_c = 304.2 \text{ K}$$

The reduced pressure of the CO_2 is

$$P_R = \frac{P}{P_c} = \frac{22.17 \text{ MPa}}{7.39 \text{ MPa}} = 3.0$$

and the pseudo-reduced specific volume is

$$v'_R = \frac{v}{RT_c/P_c} = \frac{(3.11 \times 10^{-3}\text{m}^3/\text{kg})(7.39 \text{ MPa})(44.01 \text{ kg/kg·mol})}{(8.314 \times 10^{-3} \text{ MPa·m}^3/\text{kg·mol·K})(304.2 \text{ K})}$$

$$= 0.4$$

From Figure E.3 at the known values of P_R and v'_R,

$$T_R = 1.5$$

and therefore

$$T = T_R T_c = 1.5(304.2 \text{ K}) = \underline{\underline{456 \text{ K}}}$$

If the carbon dioxide had been assumed to be an ideal gas, the predicted temperature would have been

$$T = \frac{Pv}{R} = \frac{(22.17 \text{ MPa})(3.11 \times 10^{-3}\text{m}^3/\text{kg})(44.01 \text{ kg/kg·mol})}{(8.314 \times 10^{-3}\text{ MPa·m}^3/\text{kg·mol·K})} = \underline{\underline{365 \text{ K}}}$$

which is in error by approximately 20 percent. ■

■ EXAMPLE 10.7

Calculate the specific volume of refrigerant-12 at a pressure of 160 psia and a temperature of 140°F using the generalized compressibility chart, and compare the answer with the tabulated value.

Solution. From the critical properties of Table H.1E, the reduced properties are calculated.

$$P_R = \frac{P}{P_c} = \frac{160 \text{ psia}}{598.3 \text{ psia}} = 0.267$$

$$T_R = \frac{T}{T_c} = \frac{600°\text{R}}{693°\text{R}} = 0.87$$

The value of Z from Figure E.2 at this reduced state is

$$Z \simeq 0.86$$

and the specific volume is

$$v = \frac{ZRT}{P} = \frac{(0.86)(10.73 \text{ psia·ft}^3/\text{lb}_\text{m}\text{·mol·°R})(600°\text{R})}{(160 \text{ psia})(120.91 \text{ lb}_\text{m}/\text{lb}_\text{m}\text{·mol})}$$

$$= \underline{\underline{0.286 \text{ ft}^3/\text{lb}_\text{m}}}$$

The specific volume for R-12 at the given state from Table C.3E is $v = 0.277 \text{ ft}^3/\text{lb}_\text{m}$. A comparison of this result shows that the error between this value and the tabulated specific volume is 3.2 percent. ■

10.5 OTHER EQUATIONS OF STATE

A number of other equations of state have been developed in an attempt to describe the P-v-T behavior of substances in an analytical form. While no single equation of state presently available is adequate for the entire range of temperatures and pressures, the accuracy of predictions provided by the ideal-gas equation of state can be improved. In this section several equations of state are discussed that are representative of the many available in the literature. They can be classified as being either generalized, empirical, or theoretical equations.

The *generalized equations of state*, usually two-constant equations, are the least complicated mathematically. Because of their relative simplicity, their range of applicability is limited to states for which the density is much less than the density at the critical state. The two constants that appear in the equations are evaluated on the basis of the generalized behavior of a substance. Since the critical isotherm on a P-v diagram has a point of inflection at the critical state, the slope (or first partial derivative) and the second partial derivative of pressure with respect to specific volume are zero at the critical state. These two conditions are sufficient to evaluate two arbitrary constants.

The earliest of the generalized equations of state is the *van der Waals equation*,

$$\left(P + \frac{a}{\bar{v}^2}\right)(\bar{v} - b) = \bar{R}T \tag{10.8}$$

The van der Waals equation represents an attempt to improve upon the ideal-gas equation of state on the basis of physical reasoning. For example, the constant b is intended to account somewhat for the volume occupied by the gas molecules, and the term involving the constant a is included to correct for the intermolecular forces between molecules. Both of these terms are neglected in the ideal-gas model.

The derivatives at the critical state used to evaluate the constants a and b can be expressed as

$$\left(\frac{\partial P}{\partial \bar{v}}\right)_{T_c, P_c} = 0 \quad \text{and} \quad \left(\frac{\partial^2 P}{\partial \bar{v}^2}\right)_{T_c, P_c} = 0 \tag{10.9}$$

From the van der Waals equation of state and Equation 10.9, the constants a and b can be determined as

$$a = \left(\frac{27}{64}\right)\left(\frac{\bar{R}^2 T_c^2}{P_c}\right) \tag{10.10}$$

and

$$b = \frac{\bar{R}T_c}{8P_c} \tag{10.11}$$

With these values of a and b the compressibility factor at the critical state predicted by the van der Waals equation is given by

$$Z_c = \frac{P_c \bar{v}_c}{\bar{R}T_c} = \frac{3}{8} \tag{10.12}$$

This value is higher than the critical-state compressibility factor observed experimentally for any known substance. Generally, the compressibility factor at the critical state is less than about 0.3.

The accuracy of the predictions provided by the van der Waals equation can be improved if a and b are, instead, evaluated from experimental P-v-T measurements. Then the parameters a and b are found to be temperature-dependent rather than constant values.

Among the other two-constant, generalized equations of state are the *Berthelot equation*,

$$\left(P + \frac{a}{T\bar{v}^2}\right)(\bar{v} - b) = \bar{R}T \tag{10.13}$$

and the *Redlich-Kwong equation*,

$$P = \frac{\bar{R}T}{\bar{v} - b} - \frac{a}{\bar{v}(\bar{v} + b)T^{1/2}} \tag{10.14}$$

The constants a and b in Equations 10.13 and 10.14 are different from the van der Waals constants, but they too can be determined from the generalized behavior described by Equation 10.9. The evaluation of these constants is left as an exercise in the problems at the end of the chapter. The Berthelot equation is more accurate than the van der Waals equation, but its application is limited to low pressures. The Redlich-Kwong equation, on the other hand, is found to have reasonable accuracy at high pressures.

Examples of the *empirical equations of state* are the *Beattie-Bridgeman equation*, which has five constants,

$$P = \frac{\bar{R}T}{\bar{v}^2}\left(1 - \frac{c}{\bar{v}T^3}\right)(\bar{v} + B) - \frac{A}{\bar{v}^2} \tag{10.15}$$

where

$$A = A_0\left(1 - \frac{a}{\bar{v}}\right) \quad \text{and} \quad B = B_0\left(1 - \frac{b}{\bar{v}}\right) \tag{10.16}$$

and the *Benedict-Webb-Rubin equation*, which has eight constants,

$$P = \frac{\bar{R}T}{\bar{v}} + \left(B_0\bar{R}T - A_0 - \frac{C_0}{T^2}\right)\frac{1}{\bar{v}^2} + \frac{b\bar{R}T - a}{\bar{v}^3} + \frac{a\alpha}{\bar{v}^6} + \frac{c}{\bar{v}^3T^2}\left(1 + \frac{\gamma}{\bar{v}^2}\right)e^{-\gamma/\bar{v}^2} \tag{10.17}$$

These equations differ from the generalized equations in that they are much more complex mathematically and that the constants are determined from experimental data rather than generalized behavior. Although the additional complexity is a disadvantage for some applications, the predictions provided by these equations are generally much better than those of the two-constant generalized equations. The Beattie-Bridgeman equation is a fairly popular equation of state, with good accuracy for states that have densities lower than about 0.8 ρ_c. The more complex Benedict-Webb-Rubin equation is applicable for higher densities up to about 2.5 ρ_c. Values for the constants that appear in Equations 10.15 through 10.17 are presented in Tables H.2 and H.2E for several common gases.

A well-known *theoretical equation of state* is the *virial equation*, which is developed from principles of kinetic theory. In this equation $P\bar{v}$ is expressed as an infinite series of the form

$$P\bar{v} = \bar{R}T\left(1 + \frac{B}{\bar{v}} + \frac{C}{\bar{v}^2} + \frac{D}{\bar{v}^3} + \cdots\right) \tag{10.18}$$

The coefficients B, C, D, . . . in this equation are not constants but are, rather, functions of temperature and are called *virial coefficients*. The first virial coefficient is given by the product RT. Expressions for the other virial coefficients could be developed from kinetic theory or statistical mechanics and experimental measurements, although the evaluation of the coefficients is quite complex. The advantage of the virial equation of state is that it is accurate for a much wider range of densities than the generalized equations and the empirical equations discussed previously.

10.6 GENERALIZED CHARTS FOR PROPERTIES OF REAL GASES

In this section the evaluation of the internal energy, enthalpy, and entropy of real gases is considered. This evaluation can be accomplished with the aid of the general equations for enthalpy and entropy developed in Chapter 9 and an equation of state that describes the P-v-T behavior of the real gas.

The behavior of all real gases approaches ideal-gas behavior at low pressure, and this fact can be exploited in evaluating real-gas properties. As indicated in Section 9.7.3, one procedure that can be used to determine the enthalpy change of a real gas is to imagine that the gas undergoes an isothermal process from the actual initial state to a state at a very low pressure P_0, where the behavior of the gas is ideal. This process would be followed by a constant-pressure process during which the temperature of the gas attains the actual final temperature, and finally the gas would reach the final state by means of a second isothermal process. This technique is illustrated in Figure 10.2. The asterisk superscript used in Figure 10.2 denotes an ideal-gas state and the zero subscript denotes states at very low or zero pressure.

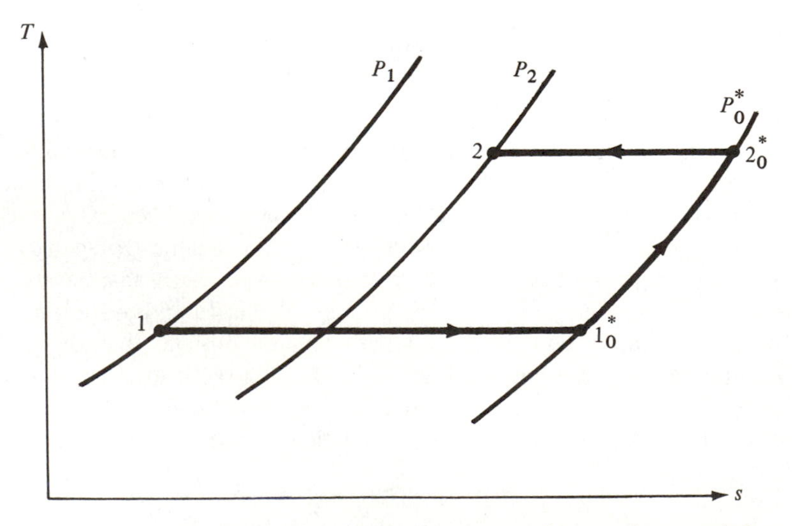

Figure 10.2 Illustration of processes used to evaluate real-gas enthalpy change.

Recall the general equation for the change in enthalpy of a simple compressible substance, Equation 9.35, written here on a molar basis,

$$d\bar{h} = \bar{c}_p\, dT + \left[\bar{v} - T\left(\frac{\partial \bar{v}}{\partial T}\right)_P\right] dP \tag{10.19}$$

Notice that for the series of processes depicted in Figure 10.2, the first term on the right side of Equation 10.19 vanishes along the isothermal paths and need only be evaluated for the constant-pressure process at zero pressure, where the behavior of the gas is ideal. The specific heat of the ideal gas is $\bar{c}_{p,0}$, and it depends only on temperature. Furthermore, the second term on the right side of Equation 10.19 vanishes along the constant-pressure path. In other words, the enthalpy change of the real gas can be separated into three terms consisting of the enthalpy change of the gas under ideal conditions, $d\bar{h}^*$, and the enthalpy changes that occur as a result of pressure changes during an isothermal process, $d\bar{h}_T$,

$$\bar{h}_2 - \bar{h}_1 = \int_{P_1}^{0} d\bar{h}_T + \int_{T_1}^{T_2} d\bar{h}^* + \int_0^{P_2} d\bar{h}_T \tag{10.20}$$

where

$$d\bar{h}_T = \left[\bar{v} - T\left(\frac{\partial \bar{v}}{\partial T}\right)_P\right] dP \tag{10.21}$$

and

$$d\bar{h}^* = \bar{c}_{p,0}\, dT$$

The integrals involving $d\bar{h}_T$ can be interpreted as being corrections that must be applied to the ideal-gas enthalpy change in order to obtain the enthalpy change of the real gas. These corrections are necessary since the enthalpy of the real gas, unlike that of an ideal gas, depends on pressure as well as temperature. In principle, all that is required for the evaluation of $d\bar{h}_T$ is an equation of state that adequately describes the P-v-T behavior of the real gas. This idea is illustrated in the following example.

EXAMPLE 10.8

Derive an expression for the change in enthalpy of a gas during an isothermal process from state 1 to state 2 assuming that the P-v-T behavior of the gas is given by the following equations of state:

(a) $P(\bar{v} - b) = \bar{R}T$

(b) The Berthelot equation of state (Equation 10.13)

Solution. The isothermal change in enthalpy is given by Equation 10.21,

$$d\bar{h}_T = \left[\bar{v} - T\left(\frac{\partial \bar{v}}{\partial T}\right)_P\right] dP$$

(a) The equation in this form is well-suited for equations of state that can be written explicitly in terms of specific volume such as the equation proposed in part a.

Solving the first equation of state for $\bar{v}$ and evaluating the derivative with respect to temperature, we obtain

$$\left(\frac{\partial \bar{v}}{\partial T}\right)_P = \frac{\bar{R}}{P}$$

Substituting into Equation 10.21 yields

$$d\bar{h}_T = \left(\bar{v} - \frac{\bar{R}T}{P}\right) dP$$

and eliminating $\bar{v}$ by using the equation of state gives

$$d\bar{h}_T = b\,dP$$

Therefore

$$\int_1^2 d\bar{h}_T = \int_{P_1}^{P_2} b\,dP = \underline{\underline{b(P_2 - P_1)}}$$

(b) If the P-v-T behavior is described by the Berthelot equation of state, Equation 10.21 is not particularly convenient because the Berthelot equation cannot easily be expressed in a form that is explicit in $\bar{v}$. It can, however, be expressed in a pressure-explicit form, $P = P(\bar{v}, T)$. Therefore, we express dP in Equation 10.21 in terms of $\bar{v}$ and T, using Equation 9.2:

$$dP = \left(\frac{\partial P}{\partial T}\right)_{\bar{v}} dT + \left(\frac{\partial P}{\partial \bar{v}}\right)_T d\bar{v}$$

For an isothermal process this equation reduces to

$$dP = \left(\frac{\partial P}{\partial \bar{v}}\right)_T d\bar{v}$$

so that Equation 10.21 can be written as

$$d\bar{h}_T = \left[\bar{v}\left(\frac{\partial P}{\partial \bar{v}}\right)_T - T\left(\frac{\partial \bar{v}}{\partial T}\right)_P\left(\frac{\partial P}{\partial \bar{v}}\right)_T\right] d\bar{v}$$

Using the cyclic relation, Equation 9.8, to simplify the second term in the brackets gives

$$d\bar{h}_T = \left[\bar{v}\left(\frac{\partial P}{\partial \bar{v}}\right)_T + T\left(\frac{\partial P}{\partial T}\right)_{\bar{v}}\right] d\bar{v}$$

For the Berthelot equation of state the derivatives become

$$\left(\frac{\partial P}{\partial \bar{v}}\right)_T = -\frac{\bar{R}T}{(\bar{v} - b)^2} + \frac{2a}{T\bar{v}^3}$$

and

$$\left(\frac{\partial P}{\partial T}\right)_{\bar{v}} = \frac{\bar{R}}{(\bar{v} - b)} + \frac{a}{T^2\bar{v}^2}$$

Substituting into the expression for $d\bar{h}_T$ above and simplifying yields

$$d\bar{h}_T = \left[\frac{3a}{T\bar{v}^2} - \frac{\bar{R}Tb}{(\bar{v} - b)^2}\right] d\bar{v}$$

Thus

$$\int_1^2 d\bar{h}_T = \int_{\bar{v}_1}^{\bar{v}_2} \left[\frac{3a}{T\bar{v}^2} - \frac{\bar{R}Tb}{(\bar{v} - b)^2}\right] d\bar{v}$$

$$= \frac{3a}{T}\left(\frac{1}{\bar{v}_1} - \frac{1}{\bar{v}_2}\right) + \bar{R}Tb\left(\frac{1}{\bar{v}_2 - b} - \frac{1}{\bar{v}_1 - b}\right)$$

■

Equation 10.21 can be cast in a more general form if the equation of state is written in terms of a compressibility factor. The compressibility factor could be obtained from any of the equations of state that adequately describe the P-v-T behavior of the gas. Some of these equations were discussed in the preceding section. Thus

$$\bar{v} = \frac{Z\bar{R}T}{P}$$

and

$$\left(\frac{\partial \bar{v}}{\partial T}\right)_P = \frac{Z\bar{R}}{P} + \frac{\bar{R}T}{P}\left(\frac{\partial Z}{\partial T}\right)_P$$

Substituting these expressions into Equation 10.21 and simplifying yields

$$d\bar{h}_T = -\frac{\bar{R}T^2}{P}\left(\frac{\partial Z}{\partial T}\right)_P dP \tag{10.22}$$

In order to generalize the results that follow, we perform a transformation to the reduced coordinates T_R and P_R. Since $P = P_cP_R$ and $T = T_cT_R$, and since P_c and T_c are constants, Equation 10.22 can be written as

$$d\bar{h}_T = -\frac{\bar{R}(T_cT_R)^2}{(P_cP_R)}\left[\frac{\partial Z}{\partial (T_cT_R)}\right]_{P_R} d(P_cP_R)$$

or

$$\frac{d\bar{h}_T}{\bar{R}T_c} = -T_R^2\left(\frac{\partial Z}{\partial T_R}\right)_{P_R} d(\ln P_R)$$

This expression can be integrated along the isothermal path connecting a real-gas state at a reduced pressure P_R to the ideal-gas state at zero pressure to obtain

$$\frac{(\bar{h}^* - \bar{h})_T}{\bar{R}T_c} = -\int_{P_R}^{0} T_R^2 \left(\frac{\partial Z}{\partial T_R}\right)_{P_R} d(\ln P_R)$$

or

$$\frac{(\bar{h}^* - \bar{h})_T}{\bar{R}T_c} = T_R^2 \int_{0}^{P_R} \left(\frac{\partial Z}{\partial T_R}\right)_{P_R} d(\ln P_R) \tag{10.23}$$

Equation 10.23 represents the normalized (or dimensionless) change in enthalpy of a real gas due to an isothermal pressure change from zero pressure to a pressure corresponding to P_R. With this result the change in enthalpy of a real gas that undergoes a change of state from state 1 to 2, as in Figure 10.2, can, from Equation 10.20, be expressed as

$$\bar{h}_2 - \bar{h}_1 = (\bar{h}_2 - \bar{h}_2^*)_{T_2} + (\bar{h}_2^* - \bar{h}_1^*) + (\bar{h}_1^* - \bar{h}_1)_{T_1}$$

or

$$\bar{h}_2 - \bar{h}_1 = \bar{R}T_c \left[\frac{(\bar{h}_1^* - \bar{h}_1)_{T_1}}{\bar{R}T_c} - \frac{(\bar{h}_2^* - \bar{h}_2)_{T_2}}{\bar{R}T_c}\right] + (\bar{h}_2^* - \bar{h}_1^*) \tag{10.24}$$

The internal-energy change can be determined from the enthalpy change and the equation of state for the real gas. Since

$$\bar{h} = \bar{u} + P\bar{v} \qquad \text{and} \qquad P\bar{v} = Z\bar{R}T$$

then

$$\begin{aligned} \bar{u}_2 - \bar{u}_1 &= (\bar{h}_2 - \bar{h}_1) - (P_2\bar{v}_2 - P_1\bar{v}_1) \\ &= (\bar{h}_2 - \bar{h}_1) - \bar{R}(Z_2T_2 - Z_1T_1) \end{aligned} \tag{10.25}$$

A closed-form solution for the integral in Equation 10.23 is seldom possible because the expressions for the compressibility factor are rarely simple functions of T_R and P_R. For this reason a numerical evaluation of the integral is usually necessary.

The data from the generalized compressibility chart can be employed to evaluate the integral in Equation 10.23 numerically and to construct a generalized chart for enthalpy. The necessary procedure can be visualized by imagining that the compressibility factor is first plotted as a function of T_R with P_R as a parameter (instead of plotting Z as a function of P_R with T_R as a parameter, as is done in Figure E.1). On such a figure $(\partial Z/\partial T_R)_{P_R}$ is determined from the slopes of lines of constant P_R. Next, the product $T_R^2(\partial Z/\partial T_R)_{P_R}$ could be plotted as a function of $\ln P_R$. The value of the integral in Equation 10.23 would then be given by the area under the curve from zero reduced pressure to the desired value of P_R. The generalized chart for enthalpy presented in Figure E.4 was developed in an analogous manner and can be used with Equation 10.24 to determine enthalpy changes for real gases.

■ EXAMPLE 10.9

Determine the change in enthalpy and change in internal energy of a real gas that undergoes a change of state from state 1 to 2 if the equation of state of the gas is given by

$$P\bar{v} = Z\bar{R}T$$

where the compressibility factor in terms of reduced coordinates is given by

$$Z = 1 + \frac{AP_R}{T_R}$$

Solution. In this example we illustrate the use of the general equation for dh in terms of reduced coordinates, Equation 10.24,

$$\bar{h}_2 - \bar{h}_1 = \bar{R}T_c\left[\frac{(\bar{h}_1^* - \bar{h}_1)_{T_1}}{\bar{R}T_c} - \frac{(\bar{h}_2^* - \bar{h}_2)_{T_2}}{\bar{R}T_c}\right] + (\bar{h}_2^* - \bar{h}_1^*)$$

where the terms in brackets are defined in Equation 10.23 as

$$\frac{(\bar{h}^* - \bar{h})_T}{\bar{R}T_c} = T_R^2 \int_0^{P_R} \left(\frac{\partial Z}{\partial T_R}\right)_{P_R} d(\ln P_R)$$

For the given compressibility factor this integral can be written as

$$\frac{(\bar{h}^* - \bar{h})_T}{\bar{R}T_c} = T_R^2 \int_0^{P_R} \left(-\frac{AP_R}{T_R^2}\right)\frac{dP_R}{P_R} = -A\int_0^{P_R} dP_R = -AP_R$$

Thus the enthalpy change of the gas is given by

$$\bar{h}_2 - \bar{h}_1 = \bar{R}T_c(-AP_{R_1} + AP_{R_2}) + (\bar{h}_2^* - \bar{h}_1^*)$$

or

$$\underline{\underline{\bar{h}_2 - \bar{h}_1 = (\bar{h}_2^* - \bar{h}_1^*) + A\bar{R}T_c(P_{R_2} - P_{R_1})}}$$

The enthalpy difference $(\bar{h}_2^* - \bar{h}_1^*)$ is simply the enthalpy change that would occur for the same change of temperature if the behavior of the gas was ideal. This term can therefore be evaluated by using Equation 2.32 for an ideal gas,

$$\bar{h}_2^* - \bar{h}_1^* = \int_{T_1}^{T_2} \bar{c}_{p,0}\, dT$$

Thus the term $A\bar{R}T_c(P_{R_2} - P_{R_1})$ can be interpreted as being a correction that must be added to the ideal-gas enthalpy change in order to obtain the enthalpy change of the real gas.

The internal-energy change can be determined from Equation 10.25,

$$\bar{u}_2 - \bar{u}_1 = \bar{h}_2 - \bar{h}_1 - \bar{R}(Z_2T_2 - Z_1T_1)$$

For this gas the internal-energy changes becomes

$$\underline{\underline{\bar{u}_2 - \bar{u}_1 = \bar{h}_2 - \bar{h}_1 - \bar{R}\left[(T_2 - T_1) + A\left(\frac{P_{R_2}T_2}{T_{R_2}} - \frac{P_{R_1}T_1}{T_{R_1}}\right)\right]}}$$ ■

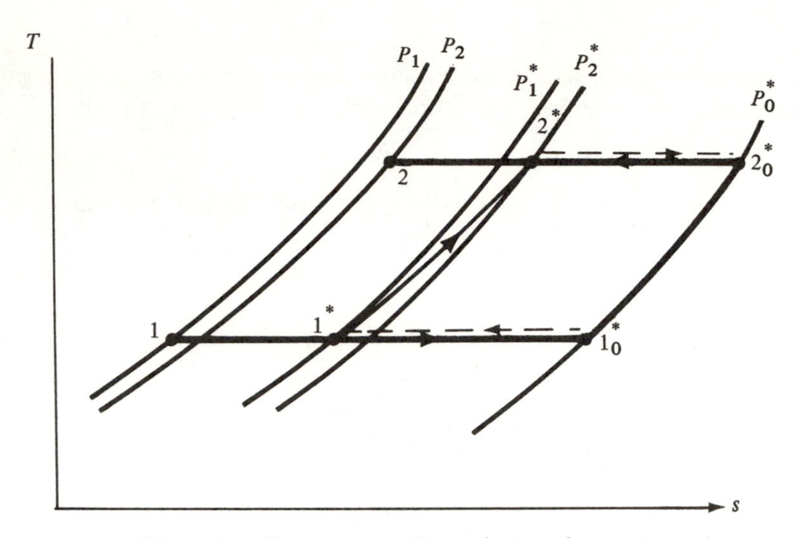

Figure 10.3 Illustration of processes used to evaluate real-gas entropy change.

The entropy change for a real gas is determined in much the same manner as that for the enthalpy change. That is, an expression for the entropy change is related to the entropy change of an ideal gas by imagining that the real-gas states are connected to the ideal-gas states by isothermal paths. The most significant difference arises from the fact that the entropy of an ideal gas depends on the pressure as well as the temperature; whereas the enthalpy of an ideal gas depends only on temperature.

The real gas is assumed to undergo an isothermal process from the actual initial state to an ideal-gas state at zero pressure. Then under conditions of ideal-gas behavior, the temperature is increased to the same temperature as exists in the actual final state. Finally, an isothermal process from this ideal-gas state to the actual final state completes the imaginary path connecting the real-gas states. This process is illustrated in Figure 10.3. The states indicated by 1_0^* and 2_0^* in this figure represent the ideal-gas states at zero pressure, which have the same temperature as the corresponding real-gas states 1 and 2. A difficulty arises with this series of processes since the entropy approaches an infinite value as the pressure is reduced to zero. This difficulty can be avoided, however, by imagining isothermal processes during which the pressure of the gas increases from zero pressure to a pressure that is the same as the pressure at the real-gas state, while the behavior of the gas remains ideal (dashed lines in Figure 10.3). These states are designated by 1* and 2* in Figure 10.3, and the complete series of paths assumed to connect the real-gas states is also indicated in the figure.

Recall the general equation for the change in entropy of a simple compressible substance, Equation 9.39, on a molar basis:

$$d\bar{s} = \frac{\bar{c}_p}{T}\,dT - \left(\frac{\partial \bar{v}}{\partial T}\right)_P dP \tag{10.26}$$

For the series of processes depicted in Figure 10.3, the first term vanishes along the four isothermal processes, and the entropy change corresponding to the change of state from

1* to 2* can be determined from Equation 5.25 since ideal-gas behavior is assumed during this change of state. Therefore, the entropy change can be expressed as

$$\bar{s}_2 - \bar{s}_1 = \int_{P_1}^{0} d\bar{s}_T + \int_{0}^{P_1} d\bar{s}_T^* + \int_{1*}^{2*} d\bar{s}^* + \int_{P_2}^{0} d\bar{s}_T^* + \int_{0}^{P_2} d\bar{s}_T \tag{10.27}$$

where

$$d\bar{s}_T = -\left(\frac{\partial \bar{v}}{\partial T}\right)_P dP$$

and $d\bar{s}^*$ for process 1*-2* can be evaluated from Equation 5.25 for ideal-gas behavior. Notice that the temperatures and pressures at ideal-gas states 1* and 2* are the same as those at the corresponding real-gas states 1 and 2. Therefore, integrating $d\bar{s}^*$ will yield the entropy change of the gas that would occur if the behavior of the gas were ideal, and the other terms in Equation 10.27 can be interpreted as being corrections that must be applied as a result of nonideal behavior.

If the change in entropy between a real-gas state at T and P and an ideal-gas state at the same values of T and P is denoted by $(\bar{s}_P^* - \bar{s}_P)_T$, then

$$(\bar{s}_P^* - \bar{s}_P)_T = \int_{P}^{0} d\bar{s}_T + \int_{0}^{P} d\bar{s}_T^* = \int_{0}^{P} (d\bar{s}_T^* - d\bar{s}_T) \tag{10.28}$$

For ideal-gas behavior $P\bar{v} = \bar{R}T$, and from Equation 10.26,

$$d\bar{s}_T^* = -\left(\frac{\partial \bar{v}}{\partial T}\right)_P dP = -\left(\frac{\bar{R}}{P}\right) dP$$

For real-gas behavior a compressibility factor can be used, $P\bar{v} = Z\bar{R}T$, so that

$$d\bar{s}_T = -\left(\frac{\partial \bar{v}}{\partial T}\right)_P dP = -\left[\frac{Z\bar{R}}{P} + \frac{\bar{R}T}{P}\left(\frac{\partial Z}{\partial T}\right)_P\right] dP$$

Substituting these expressions into Equation 10.28 yields

$$(\bar{s}_P^* - \bar{s}_P)_T = \int_{0}^{P} \left[\frac{(Z-1)\bar{R}}{P} + \frac{\bar{R}T}{P}\left(\frac{\partial Z}{\partial T}\right)_P\right] dP \tag{10.29}$$

This integral can be generalized by transforming to the reduced coordinates T_R and P_R, with the result

$$\frac{(\bar{s}_P^* - \bar{s}_P)_T}{\bar{R}} = \int_{0}^{P_R} \left[(Z-1) + T_R\left(\frac{\partial Z}{\partial T_R}\right)_{P_R}\right] d(\ln P_R) \tag{10.30}$$

Equation 10.27 can then be written in the form

$$\bar{s}_2 - \bar{s}_1 = (\bar{s}_{P_1}^* - \bar{s}_{P_1})_{T_1} + (\bar{s}_2^* - \bar{s}_1^*) - (\bar{s}_{P_2}^* - \bar{s}_{P_2})_{T_2}$$

or

$$\bar{s}_2 - \bar{s}_1 = \bar{R}\left[\frac{(\bar{s}_{P_1}^* - \bar{s}_{P_1})_{T_1}}{\bar{R}} - \frac{(\bar{s}_{P_2}^* - \bar{s}_{P_2})_{T_2}}{\bar{R}}\right] + (\bar{s}_2^* - \bar{s}_1^*) \tag{10.31}$$

The data from the generalized compressibility chart can be used, as it was for enthalpy changes, to numerically evaluate the integral in Equation 10.30 and to construct a generalized chart for entropy such as the one presented in Figure E.5.

EXAMPLE 10.10

Use the generalized charts to evaluate the enthalpy change and the entropy change per unit mole of carbon dioxide due to a change of state from 80°C and 7.5 MPa to 135°C and 15 MPa.

Solution. The critical temperature and pressure for carbon dioxide from Table H.1 are 304.2 K and 7.39 MPa, respectively. Therefore, the reduced temperatures and pressures at states 1 and 2 are

$$T_{R_1} = \frac{T_1}{T_c} = \frac{353 \text{ K}}{304.2 \text{ K}} = 1.16$$

$$T_{R_2} = \frac{T_2}{T_c} = \frac{408 \text{ K}}{304.2 \text{ K}} = 1.34$$

$$P_{R_1} = \frac{P_1}{P_c} = \frac{7.5 \text{ MPa}}{7.39 \text{ MPa}} = 1.01$$

$$P_{R_2} = \frac{P_2}{P_c} = \frac{15 \text{ MPa}}{7.39 \text{ MPa}} = 2.03$$

and the compressibility factors for these states, from Figure E.3, are approximately

$$Z_1 \simeq 0.76 \quad \text{and} \quad Z_2 \simeq 0.74$$

The behavior of the gas is therefore far from ideal, and Equations 10.24 and 10.31 should be used to determine the enthalpy and entropy changes. From the generalized chart for enthalpy, Figure E.4, we obtain

$$\frac{(\bar{h}_1^* - \bar{h}_1)_{T_1}}{\bar{R}T_c} \simeq 1.0 \qquad \frac{(\bar{h}_2^* - \bar{h}_2)_{T_2}}{\bar{R}T_c} \simeq 1.3$$

From the ideal-gas properties for carbon dioxide, Table D.2,

$$\bar{h}_2^* - \bar{h}_1^* = [(311.3 - 260.6)\text{kJ/kg}](44 \text{ kg/kg·mol}) = 2231 \text{ kJ/kg·mol}$$

From Equation 10.24 the enthalpy change for this change of state is

$$\bar{h}_2 - \bar{h}_1 = \bar{R}T_c\left[\frac{(\bar{h}_1^* - \bar{h}_1)_{T_1}}{\bar{R}T_c} - \frac{(\bar{h}_2^* - \bar{h}_2)_{T_2}}{\bar{R}T_c}\right] + (\bar{h}_2^* - \bar{h}_1^*)$$

$$= (8.314 \text{ kJ/kg·mol·K})(304.2 \text{ K})(1.0 - 1.3) + 2231 \text{ kJ/kg·mol}$$

$$= \underline{\underline{1472 \text{ kJ/kg·mol}}}$$

From the generalized chart for entropy, Figure E.5, we obtain

$$\frac{(\bar{s}^*_{P_1} - \bar{s}_{P_1})_{T_1}}{\bar{R}} \simeq 0.70 \qquad \frac{(\bar{s}^*_{P_2} - \bar{s}_{P_2})_{T_2}}{\bar{R}} \simeq 0.81$$

From the ideal-gas properties for carbon dioxide, Table D.2,

$$\begin{aligned}\bar{s}^*_2 - \bar{s}^*_1 &= (\bar{s}^\circ_2 - \bar{s}^\circ_1) - \bar{R}\ln\frac{P_2}{P_1} \\ &= [(5.138 - 5.005)\text{kJ/kg}\cdot\text{K}](44\,\text{kg/kg}\cdot\text{mol}) \\ &\quad - (8.314\ \text{kJ/kg}\cdot\text{mol}\cdot\text{K})\ln\left(\frac{15\ \text{MPa}}{7.5\ \text{MPa}}\right) \\ &= 0.089\ \text{kJ/kg}\cdot\text{mol}\cdot\text{K}\end{aligned}$$

From Equation 10.31 the entropy change is

$$\begin{aligned}\bar{s}_2 - \bar{s}_1 &= \bar{R}\left[\frac{(\bar{s}^*_{P_1} - \bar{s}_{P_1})_{T_1}}{\bar{R}} - \frac{(\bar{s}^*_{P_2} - \bar{s}_{P_2})_{T_2}}{\bar{R}}\right] + (\bar{s}^*_2 - \bar{s}^*_1) \\ &= (8.314\,\text{kJ/kg}\cdot\text{mol}\cdot\text{K})(0.70 - 0.81) + 0.089\ \text{kJ/kg}\cdot\text{mol}\cdot\text{K} \\ &= \underline{\underline{-0.826\ \text{kJ/kg}\cdot\text{mol}\cdot\text{K}}}\end{aligned}$$

10.7 SUMMARY

Ideal-gas behavior is an appropriate assumption when the pressure of a gas is much less than the critical pressure and the temperature of the gas is greater than the critical temperature. Other equations of state are used to describe the P-v-T behavior of gases when the ideal-gas assumption cannot be justified. A particularly convenient equation of state is

$$Pv = ZRT \tag{10.2}$$

which defines the compressibility factor Z.

A generalized compressibility chart can be developed on the basis of the principle of corresponding states, using the reduced pressure

$$P_R \equiv \frac{P}{P_c} \tag{10.3}$$

and the reduced temperature

$$T_R \equiv \frac{T}{T_c} \tag{10.4}$$

Other equations of state have been developed in an attempt to describe the P-v-T behavior of substances in analytical form. These include the van der Waals equation, the Berthelot equation, the Redlich-Kwong equation, the Beattie-Bridgeman equation, the Benedict-Webb-Rubin equation, and the virial equation of state.

Generalized charts are also used to predict the enthalpy and entropy changes of real gases. These charts are developed with the aid of the reduced temperature and reduced pressure. Values obtained from the generalized enthalpy and generalized entropy charts are used to correct the enthalpy and entropy changes based on ideal-gas behavior in order to predict the changes in these properties for real gases. The enthalpy change of a real gas is determined from

$$\bar{h}_2 - \bar{h}_1 = \bar{R}T_c\left[\frac{(\bar{h}_1^* - \bar{h}_1)_{T_1}}{\bar{R}T_c} - \frac{(\bar{h}_2^* - \bar{h}_2)_{T_2}}{\bar{R}T_c}\right] + (\bar{h}_2^* - \bar{h}_1^*) \tag{10.24}$$

and the entropy change is determined from

$$\bar{s}_2 - \bar{s}_1 = \bar{R}\left[\frac{(\bar{s}_{P_1}^* - \bar{s}_{P_1})_{T_1}}{\bar{R}} - \frac{(\bar{s}_{P_2}^* - \bar{s}_{P_2})_{T_2}}{\bar{R}}\right] + (\bar{s}_2^* - \bar{s}_1^*) \tag{10.31}$$

PROBLEMS

10.1 [CDD111] A fixed mass (m = 1.794 kg) of carbon dioxide at 40°C and 7.3 MPa is confined in a spherical vessel having a diameter of 0.25 m. At these conditions, carbon dioxide does not strictly obey the ideal-gas equation of state. Calculate the following:

- **(a)** the specific volume of the carbon dioxide in m^3/kg
- **(b)** the volume if the ideal-gas law were obeyed
- **(c)** the compressibility factor at these conditions

10.2 Use the generalized compressibility chart to estimate the volume occupied by 2 lb_m of carbon dioxide at 1350 psia and 390°F. Compare this value with the value predicted by the ideal-gas equation of state.

10.3 Using the compressibility chart, estimate the mass of oxygen at 2 MPa and −100°C contained in a vessel having a volume of 0.65 m^3. Estimate the error involved in assuming ideal-gas behavior for oxygen at this state.

10.4 Ten lb_m of ethylene occupies a volume of 1.1 ft^3 at 460°F. Estimate the pressure of the ethylene by using the compressibility chart. Estimate the error involved in assuming ideal-gas behavior for ethylene at this state.

10.5 Use the compressibility chart to estimate the temperature of 5 lb_m of nitrous oxide in a 0.40 ft^3 container at 6500 psia.

10.6 Use the compressibility chart to estimate the pressure of steam at 500°C with a specific volume of 0.003892 m^3/kg. Compare this result with the tabulated value as well as the estimate based on ideal-gas behavior.

10.7 Use the compressibility chart to estimate the temperature of refrigerant-12 (dichlorodifluoromethane) at 500 psia with a specific volume of 0.10833 ft^3/lb_m. Compare this result with the tabulated value as well as the estimate based on ideal-gas behavior.

10.8 [CDD211] Carbon dioxide at 380.2 K and 11.05 MPa is placed in a cylinder having a volume of 0.015 m^3. At these conditions, carbon dioxide does not obey the ideal-gas equation of state. Estimate the following:

(a) the compressibility factor at these conditions

(b) the mass of carbon dioxide in the cylinder in kg

10.9 Begin with the equation of state written in terms of the compressibility factor and derive an expression for the Joule-Thompson coefficient. Use this result and Figure E.3 to estimate the reduced temperature corresponding to the inversion temperature of a gas at low pressure.

10.10 Calculate the temperature and pressure of steam and refrigerant-12 at a reduced temperature of 1.2 and a reduced pressure of 0.9. Use these results and interpolated values of specific volume from the superheat tables in the Appendix to calculate the compressibility factors for these two substances at the specific states. Do the results confirm the principle of corresponding states? Explain.

10.11 Derive an expression for the change in internal energy and the change in entropy of a gas that obeys the van der Waals equation of state.

10.12 Derive an expression for the change in internal energy of a gas that obeys the Berthelot equation of state.

10.13 Derive an expression for the difference between $\bar{c}_p$ and $\bar{c}_v$ for a gas that obeys the van der Waals equation of state.

10.14 Verify Equations 10.10 and 10.11 for a van der Waals gas.

10.15 Evaluate the constants appearing in the Berthelot equation of state in terms of critical-state properties, using Equation 10.9.

10.16 Evaluate the constants appearing in the Redlich-Kwong equation of state in terms of critical-state properties, using Equation 10.9.

10.17 Derive expressions for the volume expansivity and the isothermal compressibility for a van der Waals gas.

10.18 Nitrogen expands isothermally from 340°F and 1200 psia to 700 psia. Determine the change in internal energy, enthalpy, and entropy of the gas, assuming that the nitrogen obeys the van der Waals equation of state. Compare your results with the values obtained assuming ideal-gas behavior.

10.19 Estimate the pressure of 5 kg of carbon dioxide which occupies a volume of 0.07 m^3 at 75°C using the Beattie-Bridgeman equation of state. Compare this result with the value obtained using the generalized compressibility chart.

10.20 Steam at 1200°F has a specific volume of 0.0963 ft^3/lb_m. Estimate the pressure of steam at this state by using (a) the ideal-gas equation of state, (b) the van der Waals equation, (c) the Redlich-Kwong equation, and (d) the generalized compressibility chart. Evaluate the accuracy of each method.

10.21 Refrigerant-12 at 800 kPa and 70°C has a specific volume of 0.02638 m^3/kg. Estimate the temperature of refrigerant-12 at this state by using (a) the ideal-gas equation of state, (b) the van der Waals equation, (c) the Redlich-Kwong equation, and (d) the generalized compressibility chart. Evaluate the accuracy of each method.

10.22 Estimate the pressure of nitrogen at -100°F and a specific volume of 0.073 ft^3/lb_m by using (a) the Benedict-Webb-Rubin equation, (b) the Beattie-Bridgeman equation, and (c) the ideal-gas equation of state.

10.23 Estimate the specific volume of nitrogen at 7 MPa and -110°C by using the Benedict-Webb-Rubin equation of state. Compare this result with the value obtained using the generalized compressibility chart.

10.24 Develop an equation for the compressibility factor in terms of reduced temperature and reduced pressure by using the van der Waals equation of state.

10.25 Construct a compressibility chart similar to Figure E.3 for a van der Waals gas based on the results of Problem 10.24. Plot the compressibility factor as a function of P_R for $P_R < 5.0$ and the following values of reduced temperature: $T_R = 1.0, 1.3, 2.0,$ and 3.5. Discuss the qualitative differences between your figure and Figure E.3.

10.26 Develop an equation for the compressibility factor in terms of reduced temperature and reduced pressure by using the Redlich-Kwong equation of state.

10.27 Construct a compressibility chart similar to Figure E.3 based on the Redlich-Kwong equation of state and the result obtained in Problem 10.26. Plot the compressibility factor as a function of P_R for $P_R < 5.0$ and the following values of reduced temperature: $T_R = 1.0, 1.3, 2.0,$ and 3.5. Discuss the qualitative differences between your figure and Figure E.3.

10.28 Determine the change in volume, the change in internal energy, and the change in entropy of 1.6 lb_m of propane due to a change in state from 1200 psia and 240°F to 600 psia and 420°F.

10.29 Ethane is compressed isothermally in a frictionless-piston-cylinder assembly from 5 MPa and 90°C to 10 MPa and the heat transfer from the ethane is to the environment at 25°C and 100 kPa. Determine the work, heat transfer, and irreversibility per unit mass of ethane.

10.30 A rigid storage tank containing 8 ft^3 of ethylene at 1500 psia and 300°F is allowed to cool until the temperature of the gas reaches 100°F. The cooling is due to heat transfer to the environment at 75°F and 14.7 psia. Determine the final pressure of the ethylene, the heat transfer, and the irreversibility.

10.31 Carbon dioxide undergoes an adiabatic expansion in a frictionless-piston-cylinder assembly from an initial state of 210°C, 22 MPa to a final pressure of 11 MPa. Determine the work of expansion for this process. How much error would be introduced if ideal-gas behavior were assumed for the carbon dioxide?

10.32 A paddle wheel is used to agitate the contents of a rigid, insulated tank having a volume of 7 ft^3. The tank contains propane initially at 260°F, and 1200 psia. When the paddle wheel is stopped, the temperature of the propane is 320°F. Determine the amount of work done on the gas by the paddle wheel and the irreversibility of the process. (Assume $T_0 = 77°F$ and $P_0 =$ 14.7 psia.)

10.33 An argon storage tank is, for some time, exposed to the environment, which is at 25°C and 100 kPa. The argon is initially at −115°C and 6 MPa, and heat transfer to the gas from the environment causes the temperature of the gas to reach −45°C. The tank contains 1.85 kg of argon. Determine the heat transfer and the irreversibility of the process.

10.34 Oxygen reportedly expands isothermally in a piston-cylinder assembly from 250°F and 4500 psia to 1000 psia. During the process the heat transfer to the oxygen from a reservoir at a temperature of 600°F is 100 Btu/lb_m. Determine whether this process is internally reversible, internally irreversible, or impossible. (Assume $T_0 = 77°F$, and $P_0 = 14.7$ psia.)

10.35 An adiabatic throttling valve is used to throttle nitrogen from 15 MPa and −60°C to 2.3 MPa. Determine the exit temperature and the irreversibility per unit mass of nitrogen. (Assume $T_0 = 25°C$.)

10.36 Ethane is compressed in an adiabatic compressor from 40 psia and 130°F to a pressure of 1200 psia. The mass flow rate of ethane is 0.7 lb_m/s and the adiabatic-compressor efficiency is 83 percent. Determine the power required to operate the compressor.

10.37 Propane enters an adiabatic turbine at 3 MPa and 175°C, and it exhausts at 300 kPa and 130°C. Determine the work output per unit mass of propane and the adiabatic efficiency of the turbine.

10.38 Determine the maximum exit velocity that can be achieved by carbon dioxide if it enters an adiabatic nozzle at 660°F and 1300 psia with negligible velocity and is expanded to 60 psia.

10.39 Determine the work per unit mass required to steadily compress methane isentropically from −60°C and 6 MPa to 18 MPa, and calculate the ratio of the exit specific volume to the inlet specific volume.

10.40 Determine the work output per unit mass of an isentropic turbine that expands argon from 100°F and 2000 psia to a pressure of 500 psia.

10.41 Oxygen is throttled from −40°C and 18 MPa to 3.5 MPa in a steady-flow, adiabatic throttling valve. Determine the exit temperature of the oxygen. Compare your result with that obtained assuming ideal-gas behavior for oxygen. What conclusion can be reached regarding the Joule-Thompson coefficient of the oxygen during the process described above?

10.42 An insulated storage vessel with a volume of 6 ft^3 is initially evacuated. A valve on the vessel is opened, and carbon dioxide gas flows into the vessel from a supply line, where the condition of the gas is maintained at 1000 psia and 200°F. The valve is closed when the pressure in the vessel reaches 500 psia. Determine the final temperature and mass of the carbon dioxide in the tank.

10.43 A tank containing 2.4 m^3 of ethane initially at 95°C, 5 MPa is fitted with a valve. The valve is opened and ethane is fed into the tank from a supply line where the temperature is 95°C and the pressure is 12 MPa. The valve is closed when the pressure in the tank reaches 10 MPa. During this process heat transfer from the contents of the tank to the environment occurs in order to maintain the temperature of the ethane constant. Calculate the heat transfer for this process.

10.44 [FLE111] Propane flows through a pipe where the inlet conditions are 600 psia, 700°R and the propane has a velocity of 60 ft/s. At the exit, the pressure is 150 psia and the temperature is 620°R. Assume that the pipe has a constant diameter. Determine the following:

(a) the exit velocity in ft/s

(b) the heat transfer to the gas in $Btu/lb_m{\cdot}mol$

10.45 [FLE211] Ethane is heated in a frictionless piston cylinder at constant pressure of 5370 kPa from 325 K to a final state where the compressibility factor is 0.895. Determine the following:

(a) the final temperature in K

(b) the work output in kJ/kg·mol

(c) the required heat transfer in kJ/kg·mol

10.46 [SLF211] Ethylene enters a compressor at a temperature of 660°R and a pressure of 300 psia. The gas exits at 760°R and 440 psia. Assume real-gas behavior at both the inlet and outlet. Determine the following:

(a) the work required in $Btu/lb_m{\cdot}mol$

(b) the increase in entropy in $Btu/lb_m{\cdot}mol{\cdot}°R$

10.47 [SLF111] Methane in a tank of volume 8 m^3 is initially at 210 K and 5.8 MPa. Heat transfer from a source at 500 K occurs until the gas is at a temperature of 344 K. At these two states the gas is not ideal; thus the compressibility charts must be used. Determine the following:

(a) the mass of the methane in kg·mol

(b) the required heat transfer in kJ

(c) the total entropy change in kJ/K (T_0 = 300 K)

11 Nonreacting-Gas Mixtures

11.1 INTRODUCTION

Mixtures of substances are commonly encountered in systems and processes of interest to the engineer. A mixture consisting of two phases of the same substance, such as a mixture of liquid water and water vapor, has already been considered, and in this chapter mixtures of different gases are discussed. While the basic thermodynamic principles presented thus far are valid for nonreacting-gas mixtures, the principles cannot be applied properly unless the thermodynamic behavior and properties of such mixtures can be evaluated.

Since the number of possible mixture compositions is infinite, to attempt to tabulate the thermodynamic properties of mixtures would be impractical. The alternative is to develop procedures for predicting mixture properties on the basis of the composition of the mixture and the thermodynamic properties of the individual components of the mixture. In this chapter nonreacting-gas mixtures are discussed in detail, beginning with descriptions of mixture compositions. This discussion if followed by a presentation of the procedures used to predict the properties of gas mixtures, and the chapter concludes with the study of a mixture of considerable engineering significance, the air-water-vapor mixture.

11.2 MIXTURE CHARACTERISTICS

The composition of a mixture of gases is usually specified either by the number of moles of each component in the mixture (a *molar analysis*) or by the mass of each component in the mixture (a *gravimetric analysis*). If the number of moles of component i in a gas mixture composed of k components is designated by N_i, then the total number of moles of gas in the mixture, N_m,* is given by the sum of the number of moles of each of the k component gases in the mixture,

$$N_m = \sum_{i=1}^{k} N_i \tag{11.1}$$

The *mole fraction* of component i, denoted by y_i, is defined as the ratio of the number of moles of component i to the total number of moles present in the mixture,

$$y_i \equiv \frac{N_i}{N_m} \tag{11.2}$$

The fact that the sum of the mole fractions is equal to unity becomes evident if Equation 11.1 is divided by N_m,

$$\sum_{i=1}^{k} y_i = 1 \tag{11.3}$$

Similarly, if the mass of component i is designated by m_i, then the total mass of the mixture, m_m, is determined by summing the mass of each of the k component gases in the mixture,

$$m_m = \sum_{i=1}^{k} m_i \tag{11.4}$$

The *mass fraction* of component i, denoted by mf_i, is defined as the ratio of the mass of component i to the total mass of the mixture,

$$mf_i \equiv \frac{m_i}{m_m} \tag{11.5}$$

The sum of the mass fractions is also equal to unity,

$$\sum_{i=1}^{k} mf_i = 1 \tag{11.6}$$

The mass of a component is related to the number of moles and the molecular weight through the relationship

$$m_i = N_i M_i \tag{11.7}$$

*In this chapter, the subscript m denotes a property of a mixture and the subscript i denotes a property of an individual component of a mixture.

Similarly, the *apparent molecular weight* of a mixture is defined as the total mass of the mixture divided by the total number of moles in the mixture,

$$M_m \equiv \frac{m_m}{N_m} \tag{11.8}$$

Substituting Equations 11.4 and 11.7 into Equation 11.8 yields

$$M_m = \frac{1}{N_m} \sum_{i=1}^{k} N_i M_i = \sum_{i=1}^{k} y_i M_i \tag{11.9}$$

In other words, the apparent molecular weight of a gas mixture can be determined from the sum of the products of the mole fraction and molecular weight of each of the components in the mixture. Furthermore, the *apparent gas constant* for the mixture, based on the apparent molecular weight of the mixture, can be defined as

$$R_m \equiv \frac{\overline{R}}{M_m} \tag{11.10}$$

EXAMPLE 11.1

A gas mixture consists of 3 mol O_2, 5.5 mol N_2, and 1.8 mol CO_2. Determine the mole fraction and mass fraction of each component, the apparent molecular weight of the mixture, and the apparent gas constant for the mixture.

Solution. The mass of each component can be determined from the number of moles of the component present in the mixture and the component molecular weight, using Equation 11.7,

$$m_i = N_i M_i$$

Therefore

$$m_{O_2} = (3 \text{ kg·mol})(32 \text{ kg/kg·mol}) = 96 \text{ kg}$$

$$m_{N_2} = (5.5 \text{ kg·mol})(28 \text{ kg/kg·mol}) = 154 \text{ kg}$$

$$m_{CO_2} = (1.8 \text{ kg·mol})(44 \text{ kg/kg·mol}) = 79.2 \text{ kg}$$

and the total mass present in the mixture is

$$m_m = m_{O_2} + m_{N_2} + m_{CO_2} = 329.2 \text{ kg}$$

The mass fractions of the components, from Equation 11.5, are

$$mf_{O_2} = \frac{m_{O_2}}{m_m} = \frac{96 \text{ kg}}{329.2 \text{ kg}} = \underline{\underline{0.292}}$$

$$mf_{N_2} = \frac{154 \text{ kg}}{329.2 \text{ kg}} = \underline{\underline{0.468}}$$

$$mf_{CO_2} = \frac{79.2 \text{ kg}}{329.2 \text{ kg}} = \underline{\underline{0.241}}$$

Notice that the sum of the mass fractions is 1.001 rather than exactly 1.0 due to round-off in the preceding calculations.

The total number of moles in the mixture is 10.3, and the mole fractions can be determined from Equation 11.2 as

$$y_{O_2} = \frac{N_{O_2}}{N_m} = \frac{3\ \text{kg·mol}}{10.3\ \text{kg·mol}} = \underline{\underline{0.291}}$$

$$y_{N_2} = \frac{5.5\ \text{kg·mol}}{10.3\ \text{kg·mol}} = \underline{\underline{0.534}}$$

$$y_{CO_2} = \frac{1.8\ \text{kg·mol}}{10.3\ \text{kg·mol}} = \underline{\underline{0.175}}$$

From Equation 11.8 the apparent molecular weight of the mixture is

$$M_m = \frac{m_m}{N_m} = \frac{329.2\ \text{kg}}{10.3\ \text{kg·mol}} = \underline{\underline{31.96\ \text{kg/kg·mol}}}$$

and the apparent gas constant for the mixture (Equation 11.10) is

$$R_m = \frac{\overline{R}}{M_m} = \frac{8.314\ \text{kJ/kg·mol·K}}{31.96\ \text{kg/kg·mol}} = \underline{\underline{0.26\ \text{kJ/kg·K}}}$$ ■

11.3 *P-v-T* BEHAVIOR OF IDEAL- AND REAL-GAS MIXTURES

The *P-v-T* behavior of nonreacting-gas mixtures is usually described on the basis of either Dalton's law of additive pressures or Amagat's law of additive volumes. Each of these laws is discussed in the following paragraphs.

Dalton's law of additive pressures states that the total pressure exerted by a mixture of gases is equal to the sum of the pressures exerted by each component of the mixture if each component existed separately at the same temperature and volume as the mixture. When this *component pressure* is denoted by p_i, Dalton's law can be expressed as

$$P_m = \sum_{i=1}^{k} p_i \tag{11.11}$$

where $p_i = p_i(T, \mathbf{V}_m)$.

The equation of state that adequately describes the *P-v-T* behavior of component i (refer to Chapter 10) should be used to calculate the component pressure p_i appearing in Equation 11.11. In many instances the generalized equation of state,

$$p_i = \frac{Z_{i,D} N_i \overline{R} T}{\mathbf{V}_m} \tag{11.12}$$

is preferred, where $Z_{i,D}$ denotes that the compressibility factor is to be evaluated at the temperature and volume of the mixture as required by Dalton's law. That is,

$$Z_{i,D} = Z_i(T, \mathbf{V}_m, N_i) \tag{11.13}$$

For ideal-gas behavior the compressibility factor has a value of unity, so that

$$p_i = \frac{N_i \bar{R} T}{\mathbf{V}_m}$$

for ideal gases.

An *average compressibility factor* for a mixture based on Dalton's law can be defined in terms of the component compressibility factors by substituting Equation 11.12 into Equation 11.11:

$$P_m = \frac{Z_{m,D} N_m \bar{R} T}{\mathbf{V}_m} = \sum_{i=1}^{k} \frac{Z_{i,D} N_i \bar{R} T}{\mathbf{V}_m} \tag{11.14}$$

Thus

$$Z_{m,D} N_m = \sum_{i=1}^{k} N_i Z_{i,D}$$

or

$$Z_{m,D} = \sum_{i=1}^{k} y_i Z_{i,D} \tag{11.15}$$

That is, the average compressibility factor of the mixture is given by the sum of the products of the mole fraction and the component compressibility factor for each of the components. Furthermore, since the component compressibility factor is unity for an ideal gas, the compressibility factor for a mixture consisting entirely of ideal gases is also unity.

Since Dalton's law assumes that each component behaves as if it were alone at the temperature and volume of the mixture, the presence of the molecules of other gas components is neglected. Thus Dalton's law should be expected to best predict the behavior of gas mixtures at low pressure, where the interaction between molecules is less significant.

The product $y_i P_m$ is called the *partial pressure*. Combining Equations 11.12 and 11.14 establishes a relationship between the component pressure and the partial pressure:

$$p_i = \left(\frac{Z_{i,D} N_i}{Z_{m,D} N_m}\right) P_m = \left(\frac{Z_{i,D}}{Z_{m,D}}\right) y_i P_m \tag{11.16}$$

Since the compressibility factors are unity for ideal gases, Equation 11.16 shows that the component pressure and the partial pressure are identical for ideal gases. Thus

$$y_i = \frac{p_i}{P_m} \tag{11.17}$$

for ideal-gas mixtures. For real-gas mixtures, the component pressure and the partial pressure differ by a factor equal to the ratio of the component compressibility factor to the average compressibility factor for the mixture.

Another widely used method of predicting the *P-v-T* behavior of gas mixtures is *Amagat's law of additive volumes*, which states that the total volume of a mixture of gases is equal to the sum of the volumes each component would occupy if each component existed separately at the same temperature and pressure as the mixture. When this *component volume* is denoted by $\mathbf{V}_i$, Amagat's law can be expressed as

$$\mathbf{V}_m = \sum_{i=1}^{k} \mathbf{V}_i \tag{11.18}$$

where $\mathbf{V}_i = \mathbf{V}_i(T, P_m)$.

As with Dalton's law, the equation of state that adequately describes the *P-v-T* behavior of component *i* should be used to calculate the component volume $\mathbf{V}_i$, appearing in Equation 11.18. If compressibility factors and the generalized equation of state are used, then

$$\mathbf{V}_i = \frac{Z_{i,A} N_i \overline{R} T}{P_m} \tag{11.19}$$

In this instance $Z_{i,A}$ denotes the component compressibility factor evaluated at the temperature and pressure of the mixture in accordance with Amagat's law,

$$Z_{i,A} = Z_i(T, P_m, N_i) \tag{11.20}$$

A compressibility factor for the mixture based on Amagat's law can also be defined in terms of the component compressibility factors. Substituting Equation 11.19 into Equation 11.18 gives

$$\mathbf{V}_m = \frac{Z_{m,A} N_m \overline{R} T}{P_m} = \sum_{i=1}^{k} \frac{Z_{i,A} N_i \overline{R} T}{P_m} \tag{11.21}$$

Thus

$$Z_{m,A} = \sum_{i=1}^{k} y_i Z_{i,A} \tag{11.22}$$

Although Equation 11.15 and Equation 11.22 are similar in form, they generally produce different results, because the method of computing the component compressibility factors differs from one equation to the other. By accounting for the volume occupied by the other gas components, Amagat's law includes the effect of intermolecular interactions to some extent. Thus the use of Amagat's law should be expected to produce better predictions at higher pressures than could be obtained with Dalton's law.

If Equation 11.22 is combined with Equation 11.19, a relationship between the component volume and mixture volume results,

$$\frac{\mathbf{V}_i}{\mathbf{V}_m} = \frac{Z_{i,A} N_i}{Z_{m,A} N_m} = \left(\frac{Z_{i,A}}{Z_{m,A}}\right) y_i \tag{11.23}$$

Thus for ideal-gas mixtures the *volume fraction* $\mathbf{V}_i/\mathbf{V}_m$ of component i is equal to the mole fraction y_i. Combining this result with Equation 11.17 yields

$$y_i = \frac{p_i}{P_m} = \frac{\mathbf{V}_i}{\mathbf{V}_m} = \frac{N_i}{N_m} \tag{11.24}$$

for ideal-gas mixtures only. Equation 11.24 shows that a molar analysis for a mixture of ideal gases is identical to the volumetric analysis of the mixture.

EXAMPLE 11.2

A mixture of nitrogen and methane (CH_4) at 35°C and a total pressure of 17 MPa has a mole fraction of nitrogen equal to 0.4. Estimate the volume occupied by 1 kg of this mixture, using the following:

(a) Ideal-gas behavior for both components
(b) Compressibility factors based on Amagat's law
(c) Compressibility factors based on Dalton's law

Solution.

(a) If each gas is assumed to be ideal, the volume of the mixture can be calculated from

$$\mathbf{V}_{m,\text{ideal}} = \frac{N_m \bar{R} T}{P_m}$$

This result can be verified, for instance, from either Equation 11.14 (Dalton's law) or Equation 11.21 (Amagat's law) since component compressibility factors as well as the mixture compressibility factor are unity for ideal gases and ideal-gas mixtures.

The number of moles in the mixture can be determined by combining Equations 11.8 and 11.9:

$$N_m = \frac{m_m}{M_m} = \frac{m_m}{\sum_{i=1}^{k} y_i M_i}$$

$$N_m = \frac{1\text{ kg}}{(0.4)(28\text{ kg/kg}\cdot\text{mol}) + (0.6)(16\text{ kg/kg}\cdot\text{mol})} = 0.048\text{ kg}\cdot\text{mol}$$

Therefore, if ideal-gas behavior is assumed, the mixture volume is estimated to be

$$\mathbf{V}_{m,\text{ideal}} = \frac{N_m \bar{R} T}{P_m} = \frac{(0.048\text{ kg}\cdot\text{mol})(8.314\text{ kPa}\cdot\text{m}^3/\text{kg}\cdot\text{mol}\cdot\text{K})(308\text{ K})}{17 \times 10^3\text{ kPa}}$$

$$= \underline{\underline{7.23 \times 10^{-3}\text{m}^3}}$$

(b) If Amagat's law is used and the compressibility factor is chosen to describe the P-v-T behavior of the gases, then the mixture volume is determined from Equation 11.21:

$$\mathbf{V}_{m,A} = \frac{Z_{m,A} N_m \overline{R} T}{P_m}$$

or

$$\mathbf{V}_{m,A} = Z_{m,A} \mathbf{V}_{m,\text{ideal}}$$

where the mixture compressibility factor based on Amagat's law is given by Equation 11.22,

$$Z_{m,A} = \sum_{i=1}^{k} y_i Z_{i,A}$$

and the individual-component compressibility factors are based on the temperature and pressure of the mixture in accordance with Amagat's law,

$$Z_{i,A} = Z_i(T, P_m, N_i)$$

The critical constants, from Table H.1, are used to evaluate the reduced temperatures and pressures:

$$T_{R,N_2} = \frac{T}{T_{c,N_2}} = \frac{308\text{ K}}{126.2\text{ K}} = 2.44$$

$$P_{R,N_2} = \frac{P_m}{P_{c,N_2}} = \frac{17\text{ MPa}}{3.39\text{ MPa}} = 5.01$$

$$T_{R,CH_4} = \frac{T}{T_{c,CH_4}} = \frac{308\text{ K}}{191.1\text{ K}} = 1.61$$

$$P_{R,CH_4} = \frac{P_m}{P_{c,CH_4}} = \frac{17\text{ MPa}}{4.64\text{ MPa}} = 3.66$$

The component compressibility factors are determined from Figure E.3 at the corresponding reduced properties:

$$Z_{N_2,A} \simeq 1.04 \qquad Z_{CH_4,A} \simeq 0.86$$

Thus the approximate value of the mixture compressibility factor based on Amagat's law is

$$Z_{m,A} = y_{N_2} Z_{N_2,A} + y_{CH_4} Z_{CH_4,A}$$

$$= 0.4(1.04) + 0.6(0.86) = 0.932$$

and the mixture volume is estimated to be

$$\mathbf{V}_{m,A} = Z_{m,A} \mathbf{V}_{m,\text{ideal}}$$

$$= 0.932(7.23 \times 10^{-3}\text{ m}^3) = \underline{\underline{6.74 \times 10^{-3}\text{ m}^3}}$$

The volume predicted from Amagat's law and the generalized compressibility factors is approximately 7 percent lower than that predicted assuming ideal-gas behavior.

(c) If compressibility factors based on Dalton's law are used, the mixture volume is determined from Equation 11.14:

$$\mathbf{V}_{m,D} = \frac{Z_{m,D} N_m \overline{R} T}{P_m}$$

or

$$\mathbf{V}_{m,D} = Z_{m,D} \mathbf{V}_{m,\text{ideal}}$$

However, with Dalton's law the mixture compressibility factor is determined from

$$Z_{m,D} = \sum_{i=1}^{k} y_i Z_{i,D}$$

where the component compressibility factors are based on the temperature and volume of the mixture,

$$Z_{i,D} = Z_i(T, \mathbf{V}_m, N_i)$$

In this instance, therefore, the solution is not straightforward because to determine the mixture volume, we need the component compressibility factors, but the component compressibility factors depend on the mixture volume, which is unknown. An iterative solution is therefore required, and the mixture volume will be assumed equal to $\mathbf{V}_{m,\text{ideal}}$ to begin the iterative-solution procedure.

The reduced-temperature values are the same as those calculated in part b, and the pseudo-reduced specific volume is determined from Equation 10.6:

$$v'_{R,i} = \frac{\overline{v}_i P_{c,i}}{\overline{R} T_{c,i}} = \frac{\mathbf{V}_m P_{c,i}}{N_i \overline{R} T_{c,i}} = \frac{\mathbf{V}_m P_{c,i}}{y_i N_m \overline{R} T_{c,i}}$$

Thus for the first iteration

$$v'_{R,N_2} = \frac{(7.23 \times 10^{-3}\ \text{m}^3)(3.39 \times 10^3\ \text{kPa})}{(0.4)(0.048\ \text{kg}\cdot\text{mol})(8.314\ \text{kPa}\cdot\text{m}^3/\text{kg}\cdot\text{mol}\cdot\text{K})(126.2\ \text{K})}$$

$$= 1.217$$

$$v'_{R,CH_4} = \frac{(7.23 \times 10^{-3}\ \text{m}^3)(4.64 \times 10^3\ \text{kPa})}{(0.6)(0.048\ \text{kg}\cdot\text{mol})(8.314\ \text{kPa}\cdot\text{m}^3/\text{kg}\cdot\text{mol}\cdot\text{K})(191.1\ \text{K})}$$

$$= 0.733$$

and the component compressibility factors based on Dalton's law, from Figure E.3, are approximately

$$Z_{N_2,D} \simeq 0.99 \qquad Z_{CH_4,D} \simeq 0.89$$

The estimate of the mixture compressibility factor in this instance is

$$Z_{m,D} = y_{N_2} Z_{N_2,D} + y_{CH_4} Z_{CH_4,D}$$

$$= 0.4(0.99) + 0.6(0.89) = 0.93$$

and the volume of the mixture is approximately

$$\begin{aligned} V_{m,D} &= Z_{m,D} V_{m,\text{ideal}} \\ &= 0.93(7.23 \times 10^{-3}\ \text{m}^3) = 6.72 \times 10^{-3}\ \text{m}^3 \end{aligned}$$

At the end of the first iteration we find that the calculated value of V_m is about 7 percent lower than the assumed value. If greater accuracy is desired, more iterations can be performed following the same procedure used above. The most recent calculated value of V_m could be used for the assumed mixture volume to begin the next iteration. After two iterations, however, linear interpolation (or extrapolation) will usually yield a fairly accurate estimate of the mixture volume.

In this example the calculated value of V_m after the second iteration is

$$V_m = \underline{\underline{6.68 \times 10^{-3}\ \text{m}^3}}$$

and a third iteration gives approximately the same result. ■

In practice, Dalton's law generally predicts mixture compressibility factors that are too high at low pressures and too low at high pressures. One means of improving the predictions at low pressures is to use a modified form of Dalton's law, which is called *Bartlett's rule of additive pressures*. The modification consists of evaluating the component compressibility factors by using the temperature and molar specific volume of the mixture rather than the molar specific volume of the component. That is,

$$p_i = \frac{Z_{i,B} N_i \overline{R} T}{V_m} \tag{11.25}$$

where $Z_{i,B} = Z_i(T, V_m, N_m)$.

Thus the average compressibility factor for the mixture based on Barlett's rule, obtained by combining Equations 11.11 and 11.25, is developed as

$$P_{m,B} = \frac{Z_{m,B} N_m \overline{R} T}{V_m} = \sum_{i=1}^{k} \frac{Z_{i,B} N_i \overline{R} T}{V_m} \tag{11.26}$$

or

$$Z_{m,B} = \sum_{i=1}^{k} y_i Z_{i,B} \tag{11.27}$$

An approximate method of determining the compressibility factor for a gas mixture is also available. In this method, known as *Kay's rule*, a *pseudo-critical temperature* and a *pseudo-critical pressure* for the mixture are defined on the basis of the critical temperatures and pressures of the components. The pseudo-critical values are, in turn, used to evaluate a mixture compressibility factor. These pseudo-critical values are defined as

$$T_c' \equiv \sum_{i=1}^{k} y_i T_{c_i} \tag{11.28}$$

and

$$P'_c \equiv \sum_{i=1}^{k} y_i P_{c_i} \tag{11.29}$$

Thus the compressibility factor obtained with Kay's rule can be expressed as

$$Z_{m,K} = Z(T'_R, P'_R) \tag{11.30}$$

where the *pseudo-reduced temperature and pressure* for the mixture are given by

$$T'_R = \frac{T}{T'_c} \qquad P'_R = \frac{P_m}{P'_c}$$

EXAMPLE 11.3

Use compressibility factors based on (a) Bartlett's rule and (b) Kay's rule to estimate the total pressure exerted by a mixture of carbon dioxide and ethane (C_2H_6) that occupies a volume of 4.8 ft^3 at a temperature of 720°R. The mole fraction of carbon dioxide in the mixture is 0.3, and the mixture consists of 2 lb_m·mol of gas.

Solution.

(a) Bartlett's rule is a modified form of Dalton's law of additive pressures in which the component pressures are determined from component compressibility factors evaluated at the temperature and the molar specific volume of the mixture:

$$P_{m,B} = \frac{Z_{m,B} N_m \overline{R} T}{\mathbf{V}_m}$$

where

$$Z_{m,B} = \sum_{i=1}^{k} y_i Z_{i,B}$$

and

$$Z_{i,B} = Z_i(T, \mathbf{V}_m, N_m)$$

With critical property values from Table H.1E, the reduced temperature, and the pseudo-reduced specific volume can be calculated for each component:

$$T_{R,CO_2} = \frac{T}{T_{c,CO_2}} = \frac{720°\text{R}}{548°\text{R}} = 1.31$$

$$v'_{R,CO_2} = \frac{\overline{v}_m P_{c,CO_2}}{\overline{R} T_{c,CO_2}}$$

$$= \frac{(4.8\ \text{ft}^3)(1072\ \text{psia})}{(2\ \text{lb}_\text{m}\cdot\text{mol})(10.73\ \text{psia}\cdot\text{ft}^3/\text{lb}_\text{m}\cdot\text{mol}\cdot°\text{R})(548°\text{R})} = 0.438$$

$$T_{R,C_2H_6} = \frac{720°\text{R}}{550°\text{R}} = 1.31$$

$$v'_{R,C_2H_6} = \frac{(4.8\ \text{ft}^3)(707.8\ \text{psia})}{(2\ \text{lb}_\text{m}\cdot\text{mol})(10.73\ \text{psia}\cdot\text{ft}^3/\text{lb}_\text{m}\cdot\text{mol}\cdot°\text{R})(550°\text{R})} = 0.288$$

From Figure E.3, therefore, the component compressibility factors are approximately

$$Z_{CO_2,B} \simeq 0.70 \qquad Z_{C_2H_6,B} \simeq 0.66$$

and the mixture compressibility factor based on Bartlett's rule is

$$\begin{aligned} Z_{m,B} &= y_{CO_2} Z_{CO_2,B} + y_{C_2H_6} Z_{C_2H_6,B} \\ &= 0.3(0.70) + 0.7(0.66) = 0.67 \end{aligned}$$

Thus the estimate of the total pressure of the mixture is

$$\begin{aligned} P_{m,B} &= \frac{Z_{m,B} N_m \overline{R} T}{\mathbf{V}_m} \\ &= \frac{(0.67)(2\ \text{lb}_m\cdot\text{mol})(10.73\ \text{psia}\cdot\text{ft}^3/\text{lb}_m\cdot\text{mol}\cdot{}^\circ\text{R})(720^\circ\text{R})}{(4.8\ \text{ft}^3)} \\ &= \underline{\underline{2160\ \text{psia}}} \end{aligned}$$

(b) To use Kay's rule, we first must calculate the pseudo-critical temperature and pressure by using Equations 11.28 and 11.29:

$$\begin{aligned} T'_c &= y_{CO_2} T_{c,CO_2} + y_{C_2H_6} T_{c,C_2H_6} \\ &= 0.3(548^\circ\text{R}) + 0.7(550^\circ\text{R}) = 549.4^\circ\text{R} \\ P'_c &= y_{CO_2} P_{c,CO_2} + y_{C_2H_6} P_{c,C_2H_6} \\ &= 0.3(1072\ \text{psia}) + 0.7(707.8\ \text{psia}) = 817\ \text{psia} \end{aligned}$$

The pseudo-reduced temperature and specific volume are

$$T'_R = \frac{T}{T'_c} = \frac{720^\circ\text{R}}{549.4^\circ\text{R}} = 1.31$$

$$\begin{aligned} v'_R &= \frac{\overline{v} P'_c}{\overline{R} T'_c} = \frac{(4.8\ \text{ft}^3)(817\ \text{psia})}{(2\ \text{lb}_m\cdot\text{mol})(10.73\ \text{psia}\cdot\text{ft}^3/\text{lb}_m\cdot\text{mol}\cdot{}^\circ\text{R})(549.4^\circ\text{R})} \\ &= 0.333 \end{aligned}$$

(The pseudo-reduced specific volume for Kay's rule is used here instead of P'_R since the mixture pressure is unknown.) With these pseudo-reduced properties the mixture compressibility factor can be determined from Figure E.3:

$$Z_{m,K} \simeq 0.67$$

Therefore, the total pressure exerted by the mixture is estimated to be

$$P_{m,K} = \frac{Z_{m,K}N_m\overline{R}T}{V_m}$$

$$= \frac{(0.67)(2\ \text{lb}_\text{m}\cdot\text{mol})(10.73\ \text{psia}\cdot\text{ft}^3/\text{lb}_\text{m}\cdot\text{mol}\cdot{}^\circ\text{R})(720^\circ\text{R})}{(4.8\ \text{ft}^3)}$$

$$= \underline{\underline{2160\ \text{psia}}}$$

Notice that in this example Bartlett's rule and Kay's rule predict approximately the same value for the total pressure. With Kay's rule, however, the estimate is obtained with much less effort, and the accuracy of Kay's rule is often acceptable for engineering calculations. ■

11.4 PROPERTIES OF IDEAL- AND REAL-GAS MIXTURES

All extensive properties of a nonreacting-gas mixture can be obtained by summing the contributions of the individual components of the mixture. For example, the internal energy of a mixture can be expressed as

$$U_m = \sum_{i=1}^{k} U_i \tag{11.31}$$

Therefore, the specific internal energy of a mixture on a mass basis is given by

$$u_m = \frac{1}{m_m}\sum_{i=1}^{k} m_i u_i = \sum_{i=1}^{k} mf_i u_i \tag{11.32}$$

and on a molar basis

$$\overline{u}_m = \frac{1}{N_m}\sum_{i=1}^{k} N_i\overline{u}_i = \sum_{i=1}^{k} y_i\overline{u}_i \tag{11.33}$$

Likewise, the enthalpy of a mixture can be expressed as

$$H_m = \sum_{i=1}^{k} H_i \tag{11.34}$$

Thus the specific enthalpy of a mixture on a mass basis is

$$h_m = \sum_{i=1}^{k} mf_i h_i \tag{11.35}$$

and on a molar basis is

$$\overline{h}_m = \sum_{i=1}^{k} y_i\overline{h}_i \tag{11.36}$$

Expressions for the constant-volume specific heat of a mixture can be obtained by differentiating Equation 11.32,

$$c_{v,m} = \left(\frac{\partial u_m}{\partial T}\right)_{\bar{v}} = \sum_{i=1}^{k} mf_i \left(\frac{\partial u_i}{\partial T}\right)_{\bar{v}}$$

or

$$c_{v,m} = \sum_{i=1}^{k} mf_i c_{v,i} \tag{11.37}$$

Similarly, on a molar basis, differentiating Equation 11.33 gives

$$\bar{c}_{v,m} = \left(\frac{\partial \bar{u}_m}{\partial T}\right)_v = \sum_{i=1}^{k} y_i \left(\frac{\partial \bar{u}_i}{\partial T}\right)_v$$

or

$$\bar{c}_{v,m} = \sum_{i=1}^{k} y_i \bar{c}_{v,i} \tag{11.38}$$

In like manner, expressions for the constant-pressure specific heat are obtained by differentiating Equations 11.35 and 11.36, with the results

$$c_{p,m} = \sum_{i=1}^{k} mf_i c_{p,i} \tag{11.39}$$

and on a molar basis

$$\bar{c}_{p,m} = \sum_{i=1}^{k} y_i \bar{c}_{p,i} \tag{11.40}$$

Finally, the entropy of the gas mixture is evaluated as

$$S_m = \sum_{i=1}^{k} S_i \tag{11.41}$$

and the specific entropies of a mixture on a mass basis and a molar basis, respectively, are given by

$$s_m = \sum_{i=1}^{k} mf_i s_i \tag{11.42}$$

and

$$\bar{s}_m = \sum_{i=1}^{k} y_i \bar{s}_i \tag{11.43}$$

A question naturally arises concerning the evaluation of the properties of the individual components in the preceding equations—namely, which temperature and pressure or which temperature and volume should be used to determine the component properties? This question is addressed in the succeeding paragraphs.

11.4.1 Ideal-Gas Mixtures

For ideal-gas mixtures an extension of Dalton's law is used for the evaluation of component properties. That is, each component is assumed to behave as an ideal gas at the temperature and volume of the mixture. Since the internal energy and enthalpy of an ideal gas depend on temperature alone, these properties for each component of the mixture are evaluated at the temperature of the mixture. This same conclusion holds for the constant-volume specific heat and the constant-pressure specific heat of each component of an ideal-gas mixture.

The entropy of an ideal gas, however, depends not only on temperature but also on pressure. According to Dalton's law, each component behaves as if it were alone at the temperature and volume of the mixture, and therefore the pressure of the component is the corresponding component pressure, p_i. The entropy change of each component of the ideal-gas mixture can thus be expressed, using Equation 5.28, as

$$s_{i,2} - s_{i,1} = s^\circ_{i,2} - s^\circ_{i,1} - R_i \ln \frac{p_{i,2}}{p_{i,1}} \tag{11.44}$$

on a mass basis, and on a molar basis as

$$\bar{s}_{i,2} - \bar{s}_{i,1} = \bar{s}^\circ_{i,2} - \bar{s}^\circ_{i,1} - \bar{R} \ln \frac{p_{i,2}}{p_{i,1}} \tag{11.45}$$

These expressions, together with either Equation 11.42 or Equation 11.43 as appropriate, can be used to determine the entropy change of an ideal-gas mixture. Thus

$$s_{m,2} - s_{m,1} = \sum_{i=1}^{k} mf_i \left(s^\circ_{i,2} - s^\circ_{i,1} - R_i \ln \frac{p_{i,2}}{p_{i,1}} \right) \tag{11.46}$$

and on a molar basis

$$\bar{s}_{m,2} - \bar{s}_{m,1} = \sum_{i=1}^{k} y_i \left(\bar{s}^\circ_{i,2} - \bar{s}^\circ_{i,1} - \bar{R} \ln \frac{p_{i,2}}{p_{i,1}} \right) \tag{11.47}$$

The application of these results to an ideal-gas mixture is illustrated in the following example.

EXAMPLE 11.4

Dry atmospheric air is a mixture of oxygen, nitrogen, and argon with trace amounts of other gases. The volumetric composition of air is frequently taken to be 21 percent O_2, 78 percent N_2, and 1 percent Ar; the presence of the other gases is neglected. For this composition for air, determine the apparent molecular weight. Calculate the enthalpy change and the entropy change of the mixture on a molar basis due to a change from an initial state of 27°C and 100 kPa to a final state of 227°C and 200 kPa. Compare these values with those obtained by using the properties for air from Table D.1.

Solution. Examining the critical constants for these gases in Table H.1, we find that for the specified states the temperatures are higher than T_c and the pressures are much lower than P_c for each gas. Therefore, the behavior of each is essentially ideal. For ideal-gas mixtures the volume fraction of each component is equal to the mole fraction (Equation 11.24);

$$y_i = \frac{V_i}{V_m}$$

Therefore

$$y_{O_2} = 0.21 \qquad y_{N_2} = 0.78 \qquad y_{Ar} = 0.01$$

and the apparent molecular weight of the mixture, from Equation 11.9, is

$$M_m = y_{O_2}M_{O_2} + y_{N_2}M_{N_2} + y_{Ar}M_{Ar}$$
$$= 0.21(31.999) + 0.78(28.013) + 0.01(39.948) = \underline{\underline{28.97}}$$

The enthalpy change for a mixture of nonreacting ideal gases is determined by using Equation 11.36 evaluated at the end states:

$$\bar{h}_{m,2} - \bar{h}_{m,1} = \left(\sum_{i=1}^{k} y_i\bar{h}_i\right)_2 - \left(\sum_{i=1}^{k} y_i\bar{h}_i\right)_1$$

Since the mixture composition remains unchanged, this equation can be written as

$$\bar{h}_{m,2} - \bar{h}_{m,1} = \sum_{i=1}^{k} y_i(\bar{h}_{i,2} - \bar{h}_{i,1})$$

Each component is ideal, and therefore the enthalpy depends on temperature alone. Tables D.5 and D.6 can be used to evaluate the enthalpy changes of the nitrogen and oxygen, respectively:

$$\bar{h}_{N_2,2} - \bar{h}_{N_2,1} = (520.5 - 311.4)\text{kJ/kg}(28\ \text{kg/kg·mol}) = 5855\ \text{kJ/kg·mol}$$
$$\bar{h}_{O_2,2} - \bar{h}_{O_2,1} = (461.5 - 273.0)\text{kJ/kg}(32\ \text{kg/kg·mol}) = 6032\ \text{kJ/kg·mol}$$

Argon is a monatomic gas whose constant-pressure specific heat is essentially independent of temperature. From Equation 2.38 the specific heat is given by

$$\bar{c}_{p,Ar} = \frac{5}{2}\bar{R}$$

Therefore

$$\bar{h}_{Ar,2} - \bar{h}_{Ar,1} = \bar{c}_{p,Ar}(T_2 - T_1) = \frac{5}{2}\bar{R}(T_2 - T_1)$$
$$= \left(\frac{5}{2}\right)(8.314\ \text{kJ/kg·mol·K})(500 - 300)\text{K} = 4157\ \text{kJ/kg·mol}$$

The enthalpy change of the mixture is therefore

$$\bar{h}_{m,2} - \bar{h}_{m,1} = [0.21(6032) + 0.78(5855) + 0.01(4157)]\text{kJ/kg·mol}$$
$$= \underline{\underline{5875\ \text{kJ/kg·mol}}}$$

If Table D.1 is used for the properties of air, the enthalpy change is

$$\bar{h}_{m,2} - \bar{h}_{m,1} = (503.3 - 300.4)\text{kJ/kg}(28.97\ \text{kg/kg·mol}) = \underline{\underline{5878\ \text{kJ/kg·mol}}}$$

The entropy change for a mixture of ideal gases is determined by using Equation 11.47 evaluated at the end states:

$$\bar{s}_{m,2} - \bar{s}_{m,1} = \sum_{i=1}^{k} y_i \left(\bar{s}^{\circ}_{i,2} - \bar{s}^{\circ}_{i,1} - \bar{R} \ln \frac{p_{i,2}}{p_{i,1}} \right)$$

For ideal gases the component pressures are equal to the partial pressures (Equation 11.17); therefore

$$\frac{p_{i,2}}{p_{i,1}} = \frac{y_i P_2}{y_i P_1} = \frac{P_2}{P_1} = \frac{200\ \text{kPa}}{100\ \text{kPa}} = 2$$

since the composition remains constant.

The temperature-dependent part of the entropy change for oxygen and nitrogen can be evaluated by using Table D.6 and Table D.5, respectively:

$$\bar{s}^{\circ}_{O_2,2} - \bar{s}^{\circ}_{O_2,1} = (6.8969 - 6.4168)\text{kJ/kg·K}(32\ \text{kg/kg·mol}) = 15.36\ \text{kJ/kg·mol·K}$$

$$\bar{s}^{\circ}_{N_2,2} - \bar{s}^{\circ}_{N_2,1} = (7.380 - 6.8463)\text{kJ/kg·K}(28\ \text{kg/kg·mol}) = 14.94\ \text{kJ/kg·mol·K}$$

The corresponding term for argon, which has constant specific heats, can be evaluated from the definition of s° (Equation 5.27);

$$\begin{aligned}
\bar{s}^{\circ}_{\text{Ar},2} - \bar{s}^{\circ}_{\text{Ar},1} &= \int_0^{T_2} \bar{c}_{p,\text{Ar}} \frac{dT}{T} - \int_0^{T_1} \bar{c}_{p,\text{Ar}} \frac{dT}{T} \\
&= \int_{T_1}^{T_2} \bar{c}_{p,\text{Ar}} \frac{dT}{T} = \bar{c}_{p,\text{Ar}} \ln \frac{T_2}{T_1} \\
&= \frac{5}{2}\bar{R} \ln \frac{T_2}{T_1} = \left(\frac{5}{2}\right) (8.314\ \text{kJ/kg·mol·K}) \ln \left(\frac{500\ \text{K}}{300\ \text{K}}\right) \\
&= 10.62\ \text{kJ/kg·mol·K}
\end{aligned}$$

Since the pressure-dependent part of the entropy change is the same for each component in this example, it can be evaluated as

$$\bar{R} \ln \frac{P_2}{P_1} = (8.314\ \text{kJ/kg·mol·K}) \ln 2 = 5.763\ \text{kJ/kg·mol·K}$$

Therefore, the entropy change for the mixture is

$$\begin{aligned}
\bar{s}_{m,2} - \bar{s}_{m,1} &= [0.21(15.36 - 5.763) + 0.78(14.94 - 5.763) \\
&\quad + 0.01(10.62 - 5.763)]\text{kJ/kg·mol·K} \\
&= \underline{\underline{9.22\ \text{kJ/kg·mol·K}}}
\end{aligned}$$

If Table D.1 is used for the properties of air, the entropy change is calculated as

$$\bar{s}_{m,2} - \bar{s}_{m,1} = (\bar{s}_2^\circ - \bar{s}_1^\circ)_m - \bar{R}\ln\frac{P_2}{P_1} = (s_2^\circ - s_1^\circ)_m M_m - \bar{R}\ln\frac{P_2}{P_1}$$

$$= (6.2193 - 5.7016)\text{kJ/kg·K}(28.97\text{ kg/kg·mol})$$

$$- (8.314\text{ kJ/kg·mol·K})\ln\left(\frac{200\text{ kPa}}{100\text{ kPa}}\right)$$

$$= \underline{\underline{9.23\text{ kJ/kg·mol·K}}}$$

The properties of air tabulated in Table D.1 are developed from the mixture relationships discussed in this chapter, so the fact that the comparisons in this example are quite good is not surprising. Remember, however, that air is a mixture of several gases rather than a single gas. As such, it does not have a true critical temperature or critical pressure, although pseudo-critical properties for air can be defined as is done in Kay's rule. ■

11.4.2 Real-Gas Mixtures

The method used to evaluate the internal energy, enthalpy, and entropy of a mixture of real gases is similar to that used to determine the entropy of a mixture of ideal gases. The similarity exists because each of the properties of a real gas depends on pressure (or specific volume) as well as temperature. However, a number of alternatives are available for describing the P-v-T behavior of the gases, and the method employed for estimating the enthalpy, internal energy, and entropy should be consistent with the description of the P-v-T behavior.

So that the nonideal behavior of the components in the mixture is accounted for, the generalized equations (or generalized charts) for enthalpy and entropy can be used to determine the enthalpy and entropy changes of each component. Since a mixture is involved, some additional consideration of the development of the generalized equations and charts is necessary. Recall that the development of the generalized equation for an enthalpy change (Chapter 9) began with Equation 9.28 written on a molar basis:

$$d\bar{h} = T\,d\bar{s} + \bar{v}\,dP$$

If this equation is written for a mixture, then

$$d\bar{h}_m = T_m d\bar{s}_m + \bar{v}_m dP_m$$

Substituting Equations 11.36 and 11.43, we can write this equation as

$$d\left(\sum_{i=1}^{k} y_i\bar{h}_i\right) = T_m d\left(\sum_{i=1}^{k} y_i\bar{s}_i\right) + \left(\sum_{i=1}^{k} y_i\bar{v}_i\right) dP_m$$

or

$$\sum_{i=1}^{k} y_i(d\bar{h}_i - T_m d\bar{s}_i - \bar{v}_i dP_m) = 0$$

and therefore

$$d\bar{h}_i = T_m d\bar{s}_i + \bar{v}_i dP_m$$

This result is the form of the equation for each component of a mixture that would be used to develop the generalized equation (or generalized chart) for each component in a mixture of real gases. The important point to recognize here is that the pressure in this equation is the pressure of the mixture rather than the component pressure. Therefore, the results obtained in Chapters 9 and 10 relating to the generalized equations or charts for enthalpy and entropy can be used provided that the pressures are interpreted as being the mixture pressures and that the reduced pressure for each component is based on the pressure of the mixture. This procedure corresponds to using Amagat's law to describe the *P-v-T* behavior of the component.

If the temperature and pressure of the mixture are given, Amagat's law can be used to describe the *P-v-T* behavior of the mixture, and the temperature and pressure of the mixture are used directly to compute the reduced pressure and temperature of each component for use with the generalized enthalpy and entropy charts. On the other hand, if the temperature and volume of the mixture are given, Dalton's law is preferred for the *P-v-T* behavior of the mixture. However, the temperature and pressure of the mixture should be used to compute reduced temperatures and pressures for use with the generalized charts for enthalpy and entropy. The mixture pressure in this instance would be determined from the sum of the component pressures in accordance with Dalton's law. Another alternative, which is reasonably accurate and easier to implement, is to use the generalized charts with pseudo-critical properties based on Kay's rule. The determination of the enthalpy change of a real-gas mixture is illustrated in the following example.

EXAMPLE 11.5

A mixture of nitrogen and oxygen at 315°R and a total pressure of 1500 psia has a mole fraction of nitrogen equal to 0.4. Estimate the molar specific volume of the mixture and the enthalpy change of the mixture if the pressure is increased to 4500 psia during an isothermal process.

Solution. At the stated temperature and pressures, the behavior of the gases is not ideal, and the enthalpy change of each component should be determined by using the generalized chart for enthalpy and the methods discussed in Chapter 10. The enthalpy change of the mixture is determined by using Equation 11.36,

$$\bar{h}_{m,2} - \bar{h}_{m,1} = y_{N_2}(\bar{h}_2 - \bar{h}_1)_{N_2} + y_{O_2}(\bar{h}_2 - \bar{h}_1)_{O_2}$$

From Equation 10.24 the enthalpy change of component i is given by

$$(\bar{h}_2 - \bar{h}_1)_i = \bar{R}T_{c,i}\left[\frac{(\bar{h}_1^* - \bar{h}_1)_{T_1}}{\bar{R}T_{c,i}} - \frac{(\bar{h}_2^* - \bar{h}_2)_{T_2}}{\bar{R}T_{c,i}}\right]_i + (\bar{h}_2^* - \bar{h}_1^*)_i$$

The last term on the right side represents the ideal-gas enthalpy change and is zero for this example because the process is isothermal. The terms in brackets represent the additional change in enthalpy due to nonideal behavior and are evaluated by using the generalized chart for enthalpy. For this purpose the reduced temperature and reduced pressure of each component are required. The reduced pressure, however, should be based on the mixture pressure.

The reduced pressures and temperatures are computed by using data from Table H.1E:

$$T_{R,N_2} = \frac{T}{T_{c,N_2}} = \frac{315°R}{227°R} = 1.39$$

$$T_{R,O_2} = \frac{T}{T_{c,O_2}} = \frac{315°R}{279°R} = 1.13$$

$$P_{R_1,N_2} = \frac{P_1}{P_{c,N_2}} = \frac{1500 \text{ psia}}{491.7 \text{ psia}} = 3.05$$

$$P_{R_2,N_2} = \frac{P_2}{P_{c,N_2}} = \frac{4500 \text{ psia}}{491.7 \text{ psia}} = 9.15$$

$$P_{R_1,O_2} = \frac{P_1}{P_{c,O_2}} = \frac{1500 \text{ psia}}{736.8 \text{ psia}} = 2.04$$

$$P_{R_2,O_2} = \frac{P_2}{P_{c,O_2}} = \frac{4500 \text{ psia}}{736.8 \text{ psia}} = 6.11$$

The corresponding compressibility factors are found from Figure E.3E:

$$Z_{1,N_2} = 0.73 \qquad Z_{2,N_2} = 1.1$$

$$Z_{1,O_2} = 0.48 \qquad Z_{2,O_2} = 0.8$$

The mixture compressibility factors at the end states are, therefore, approximately

$$Z_{m,1} = y_{N_2}Z_{1,N_2} + y_{O_2}Z_{1,O_2}$$

$$= 0.4(0.73) + 0.6(0.48) = 0.58$$

$$Z_{m,2} = 0.4(1.1) + 0.6(0.8) = 0.92$$

From Amagat's law, Equation 11.21, the molar specific volumes at the end states are

$$\bar{v}_{m,1} = \frac{Z_{m,1}\bar{R}T}{P_{m,1}} = \frac{(0.58)(10.73 \text{ psia}\cdot\text{ft}^3/\text{lb}_m\cdot\text{mol}\cdot°\text{R})(315°\text{R})}{1500 \text{ psia}}$$

$$= \underline{\underline{1.31 \text{ ft}^3/\text{lb}_m\cdot\text{mol}}}$$

$$\bar{v}_{m,2} = \frac{Z_{m,2}\bar{R}T}{P_{m,2}} = \frac{(0.92)(10.73 \text{ psia}\cdot\text{ft}^3/\text{lb}_m\cdot\text{mol}\cdot°\text{R})(315°\text{R})}{4500 \text{ psia}}$$

$$= \underline{\underline{0.69 \text{ ft}^3/\text{lb}_m\cdot\text{mol}}}$$

From the generalized enthalpy chart, Figure E.4E, we obtain

$$\left(\frac{\bar{h}_1^* - \bar{h}_1}{\bar{R}T_c}\right)_{N_2} = 1.7 \qquad \left(\frac{\bar{h}_1^* - \bar{h}_1}{\bar{R}T_c}\right)_{O_2} = 2.4$$

$$\left(\frac{\bar{h}_2^* - \bar{h}_2}{\bar{R}T_c}\right)_{N_2} = 2.5 \qquad \left(\frac{\bar{h}_2^* - \bar{h}_2}{\bar{R}T_c}\right)_{O_2} = 3.6$$

Therefore

$$(\bar{h}_2 - \bar{h}_1)_{N_2} = \bar{R}T_{c,N_2}(1.7 - 2.5)$$

$$= (1.986\ \text{Btu/lb}_\text{m}\cdot\text{mol}\cdot{}^\circ\text{R})(227^\circ\text{R})(1.7 - 2.5)$$

$$= -361\ \text{Btu/lb}_\text{m}\cdot\text{mol}$$

$$(\bar{h}_2 - \bar{h}_1)_{O_2} = (1.986\ \text{Btu/lb}_\text{m}\cdot\text{mol}\cdot{}^\circ\text{R})(279^\circ\text{R})(2.4 - 3.6)$$

$$= -665\ \text{Btu/lb}_\text{m}\cdot\text{mol}$$

and the change in enthalpy of the mixture is

$$\bar{h}_{m,2} - \bar{h}_{m,1} = 0.4(-361\ \text{Btu/lb}_\text{m}\cdot\text{mol}) + 0.6(-665\ \text{Btu/lb}_\text{m}\cdot\text{mol})$$

$$= \underline{\underline{-543\ \text{Btu/lb}_\text{m}\cdot\text{mol}}}$$

We could also estimate the enthalpy change by using Kay's rule. The pseudo-critical temperature and pressure are calculated first:

$$P_c' = y_{N_2}P_{c,N_2} + y_{O_2}P_{c,O_2}$$

$$= 0.4(491.7\ \text{psia}) + 0.6(736.8\ \text{psia}) = 638.8\ \text{psia}$$

$$T_c' = y_{N_2}T_{c,N_2} + y_{O_2}T_{c,O_2}$$

$$= 0.4(227^\circ\text{R}) + 0.6(279^\circ\text{R}) = 258.2^\circ\text{R}$$

The corresponding pseudo-reduced temperatures and pressures are

$$T_R' = \frac{T}{T_c'} = \frac{315^\circ\text{R}}{258.2^\circ\text{R}} = 1.22$$

$$P_{R,1}' = \frac{P_1}{P_c'} = \frac{1500\ \text{psia}}{638.8\ \text{psia}} = 2.35$$

$$P_{R,2}' = \frac{P_2}{P_c'} = \frac{4500\ \text{psia}}{638.8\ \text{psia}} = 7.04$$

From Figure E.4E we find

$$\left(\frac{\bar{h}_1^* - \bar{h}_1}{\bar{R}T_c}\right)_m = 2.1 \qquad \left(\frac{\bar{h}_2^* - \bar{h}_2}{\bar{R}T_c}\right)_m = 3.2$$

and therefore the change in the enthalpy of the mixture is approximately

$$\bar{h}_{m,2} - \bar{h}_{m,1} = \bar{R}T'_c(2.1 - 3.2)$$

$$= (1.986\ \text{Btu/lb}_\text{m}\cdot\text{mol}\cdot{}^\circ\text{R})(258.2^\circ\text{R})(2.1 - 3.2)$$

$$= \underline{\underline{-564\ \text{Btu/lb}_\text{m}\cdot\text{mol}}}$$

This value is within 4 percent of the value obtained above. ■

11.5 MIXTURES OF AIR AND WATER VAPOR

Psychrometrics, or *psychrometry*, is the name given to the study of air-water-vapor mixtures. Several simplifying assumptions are commonly used in psychrometrics. The gas mixture of interest is assumed to be composed of dry air and water vapor, each of which is considered to behave as an ideal gas. The specific heats of the air are assumed to be constant, and the enthalpy of the water vapor depends only on temperature since ideal-gas behavior is assumed.

These assumptions are justified on the basis that in typical engineering applications of psychrometry, the temperatures of interest generally range from slightly below 0°C (32°F) up to about 40 or 50°C (i.e., about 100 to 120°F), while the total pressure of the mixture is generally about one standard atmosphere. The validity of assuming that the enthalpy of the water vapor depends only on temperature for the range of temperatures of interest can best be confirmed by examining the charts for water given in Figure B.5 or B.6. Note that lines of constant enthalpy and lines of constant temperature are nearly identical in the superheated-vapor region at low temperatures, and therefore the enthalpy of the water vapor is taken to be equal to the value of the enthalpy of the saturated vapor at the temperature of the mixture.

The total pressure of a mixture of dry air and water vapor, according to Dalton's law, is the sum of the component pressures:

$$P = p_a + p_v \tag{11.48}$$

The subscript a is used to denote the component pressure of the dry air, while the subscript v is used to denote the component pressure of the water vapor (also called the vapor pressure). Furthermore, since the mixture is composed of ideal gases, the component pressure and partial pressure of a component are identical (see Equation 11.17).

The analysis of a mixture such as air and water vapor involves other considerations in addition to those associated with ordinary gas mixtures because of the possibility of vapor condensation. Vapor condensation will occur if the component pressure of the water vapor reaches the saturation pressure corresponding to the temperature of the mixture.

The temperature of the mixture is called the *dry bulb temperature*. This value is the temperature that would be measured by an ordinary thermometer with a dry sensing element. If the partial pressure of the water vapor, p_v, reaches the saturation pressure corresponding to the dry bulb temperature of the mixture, some of the vapor will begin to condense. The condensation process can be visualized by referring to the T-s diagram

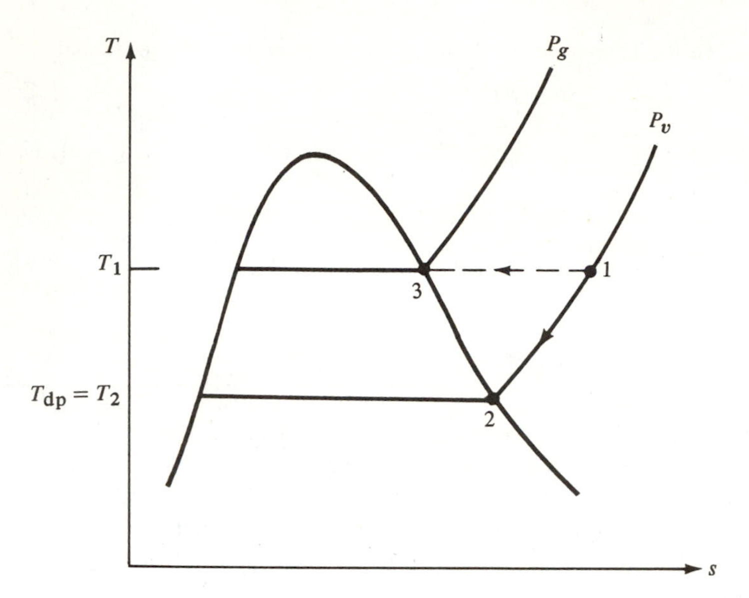

Figure 11.1 A T-s diagram illustrating dew-point temperature and saturation pressure.

shown in Figure 11.1. In this figure the mixture, initially at a dry bulb temperature T_1, is cooled at constant total pressure. If the mole fractions of the components are fixed, then the partial pressures are also fixed (Equation 11.24), and the water vapor follows the path 1-2, where p_v is constant. The dry bulb temperature of the mixture decreases during the process until a temperature is reached, at state 2, where the partial pressure of the water vapor is equal to the saturation pressure corresponding to the temperature of the mixture. Any further cooling will result in condensation of the water vapor. The temperature T_2 is called the *dew-point temperature*. This value (T_{dp} in Figure 11.1) is the temperature at which a moist-air mixture becomes saturated as a result of cooling at constant pressure.

If the mixture undergoes the constant-temperature process indicated by the path 1-3 in Figure 11.1, the partial pressure of the water vapor in the mixture increases continuously until state 3 is reached. At this state the partial pressure of the water vapor in the mixture is equal to the saturation pressure corresponding to the dry bulb temperature T_1 of the mixture. This pressure is denoted by p_g.

The amount of water vapor present in a moist-air mixture can be specified in a number of ways. As with other gas mixtures, the mole fractions of the components could be used. However, two additional measures of moisture content are also commonly used. Perhaps the more frequently used of the two is the *relative humidity*, denoted by the symbol ϕ. The relative humidity is defined as the ratio of the partial pressure of the water vapor to the saturation pressure corresponding to the dry bulb temperature of the mixture,

$$\phi \equiv \frac{p_v}{p_g} \tag{11.49}$$

Alternative expressions for the relative humidity can be obtained with the aid of the ideal-gas equation of state. Thus

$$\phi = \frac{p_v}{p_g} = \frac{R_v T_v / v_v}{R_g T_g / v_g} = \frac{v_g}{v_v} \tag{11.50}$$

since T_v and T_g are equal to the dry bulb temperature of the mixture and $R_v = R_g$. Equation 11.50 can also be expressed as

$$\phi = \frac{m_v}{m_g} \tag{11.51}$$

As can be seen from Equation 11.49, the relative humidity has a maximum value of unity or 100 percent at saturation when $p_v = p_g$. The relative humidity, however, does not give a direct indication of the relative mass of water vapor and dry air present in the mixture, as does a quantity called the humidity ratio.

The *humidity ratio*, also referred to as the *specific humidity*, is defined as the ratio of the mass of water vapor to the mass of dry air present in a moist-air mixture and is denoted by the symbol ω,

$$\omega \equiv \frac{m_v}{m_a} \tag{11.52}$$

The humidity ratio is zero for dry air and increases as the relative amount of water vapor in the mixture increases. The humidity ratio can be expressed in terms of the partial pressures of the components since ideal-gas behavior is assumed. On the basis of Dalton's law,

$$\omega = \frac{m_v}{m_a} = \frac{p_v V M_v / \bar{R}T}{p_a V M_a / \bar{R}T} = \frac{M_v}{M_a}\left(\frac{p_v}{p_a}\right) = 0.622\left(\frac{p_v}{p_a}\right) \tag{11.53}$$

The expression for the humidity ratio can be written in terms of the total pressure and the partial pressure of the water vapor if Equation 11.48 is used to eliminate p_a:

$$\omega = 0.622\left(\frac{p_v}{P - p_v}\right) \tag{11.54}$$

Furthermore, the humidity ratio and the relative humidity can be related by combining Equations 11.49 and 11.53:

$$\phi = \frac{p_v}{p_g} = \frac{\omega p_a}{0.622 p_g} \tag{11.55}$$

EXAMPLE 11.6

A moist-air mixture has a dry bulb temperature of 85°F and a relative humidity of 60 percent. The total pressure of the mixture is 14.7 psia. Determine the partial pressure of the water vapor, the humidity ratio, and the dew-point temperature.

Solution. Since the relative humidity is known, the partial pressure of the water vapor can be determined from Equation 11.49:

$$p_v = \phi p_g$$

and p_g is the saturation pressure of water corresponding to a dry bulb temperature of 85°F. From Table B.1E

$$p_g = 0.6 \text{ psia}$$

and therefore

$$p_v = 0.6(0.6 \text{ psia}) = \underline{\underline{0.36 \text{ psia}}}$$

The humidity ratio can next be calculated from Equation 11.54:

$$\omega = 0.622\left(\frac{p_v}{P - p_v}\right)$$

$$= (0.622)\left[\frac{0.36 \text{ psia}}{(14.7 - 0.36)\text{psia}}\right] = 1.56 \times 10^{-2}$$

The humidity ratio represents the ratio of the mass of water vapor to the mass of dry air in the mixture. Therefore

$$\omega = \underline{\underline{1.56 \times 10^{-2} \text{lb}_\text{m} \text{ H}_2\text{O/lb}_\text{m} \text{ dry air}}}$$

The dew-point temperature is the temperature at which the vapor pressure of the water is equal to the saturation pressure (see Figure 11.1). The dew-point temperature is, therefore, the saturation temperature corresponding to p_v and can be found by interpolation from Table B.1E:

$$T_{dp} \simeq \underline{\underline{70°\text{F}}}$$

If this moist air is cooled at constant total pressure to a temperature of 70°F, the water vapor in the mixture will begin to condense. ■

Another property of interest in the study of moist-air mixtures is the enthalpy of the mixture. Since the extensive enthalpy is the sum of the enthalpies of the components, then

$$H_m = H_a + H_v$$

or

$$H_m = m_a h_a + m_v h_v$$

Because many engineering applications of psychrometrics involve processes in which the mass of dry air present in a mixture of dry air and water vapor does not change, the specific enthalpy and all other intensive properties of the mixture are expressed on the basis of a unit mass of dry air rather than a unit mass of the mixture. Thus the *specific enthalpy of the mixture*, h_m, is defined as the enthalpy of the mixture per unit mass of dry air, or

$$h_m \equiv \frac{H_m}{m_a} = h_a + \left(\frac{m_v}{m_a}\right) h_v = h_a + \omega h_v \tag{11.56}$$

where h_a is the enthalpy of dry air at the dry bulb temperature and h_v is the enthalpy of saturated vapor at the dry bulb temperature.

For convenience in psychrometric calculations, the air and the water vapor in the moist air are assumed to behave as ideal gases having constant specific heats. The reference enthalpy of the dry air is chosen to be zero at the reference temperature of 0°C or 0°F, so that

$$h_a = c_{p,a}T$$

where T is the dry bulb temperature. In SI units, the dry bulb temperature is measured in Celsius degrees and $c_{p,a} = 1.0$ kJ/kg·°C. In English units, the dry bulb temperature is in Fahrenheit degrees and $c_{p,a} = 0.24$ Btu/lb$_m$·°F.

The reference enthalpy of the water vapor is chosen to be equal to the enthalpy of saturated vapor at a reference temperature of 0.01°C or 32.018°F. Thus

$$h_v = c_{p,v}(T - T_{v,\text{ref}}) + h_{v,\text{ref}}$$

The enthalpy of saturated water vapor may be approximated, with reasonable accuracy, as

$$h_v = c_{p,v}T + \Delta h_{v,\text{ref}} \tag{11.57}$$

where T is the dry bulb temperature and $\Delta h_{v,\text{ref}}$ is a shorthand notation for the quantity $(h_{v,\text{ref}} - c_{p,v}T_{v,\text{ref}})$. In SI units, the dry bulb temperature is measured in Celsius degrees, $c_{p,v} = 1.82$ kJ/kg·°C and $\Delta h_{v,\text{ref}} = 2500.5$ kJ/kg. In English units, the dry bulb temperature is in Fahrenheit degrees, $c_{p,v} = 0.435$ Btu/lb$_m$·°F and $\Delta h_{v,\text{ref}} = 1061$ Btu/lb$_m$. Thus h_m may be expressed as

$$h_m = c_{p,a}T + \omega h_v \tag{11.58}$$

where h_m and h_v are the enthalpy of the mixture and the enthalpy of saturated water vapor, respectively, evaluated at the dry bulb temperature of the mixture.

After substitution of Equations 11.57 and 11.58 into Equation 11.56, we may express the enthalpy of the mixture per unit mass of dry air as

$$h_m = c_{p,a}T + \omega(c_{p,v}T + \Delta h_{v,\text{ref}}) \tag{11.59}$$

Thus, the mixture enthalpy depends only on the dry bulb temperature and the humidity ratio.

The entropy of a moist-air mixture can be calculated in a similar, but slightly different, manner. The extensive entropy is given by

$$S_m = S_a + S_v$$

or

$$S_m = m_a s_a + m_v s_v \tag{11.60}$$

and the *specific entropy of the mixture*, s_m, defined as the entropy of the mixture per unit mass of dry air, is

$$s_m \equiv \frac{S_m}{m_a} = s_a + \left(\frac{m_v}{m_a}\right) s_v = s_a + \omega s_v \tag{11.61}$$

Since the air is assumed to be an ideal gas with constant specific heats,

$$s_a - s_{a,\text{ref}} = c_{p,a} \ln\left(\frac{T}{T_{a,\text{ref}}}\right) - R_a \ln\left(\frac{p_a}{P_{a,\text{ref}}}\right) \tag{11.62}$$

where T is the dry bulb temperature in absolute temperature units and p_a is the partial pressure of the air. The reference entropy of the dry air is chosen to be zero at the reference temperature of 273.15 K (0°C) or 459.67°R (0°F) and a reference pressure of one atmosphere (101.325 kPa or 14.696 psia).

For the water vapor, also assumed ideal,

$$s_v - s_{v,\text{ref}} = c_{p,v} \ln\left(\frac{T}{T_{v,\text{ref}}}\right) - R_v \ln\left(\frac{p_v}{P_{v,\text{ref}}}\right) \tag{11.63}$$

where T is the dry bulb temperature in absolute temperature units and p_v is the partial pressure of the water vapor. The reference entropy is the entropy of saturated vapor at the reference temperature of 273.16 K (0.01°C) in SI units or 491.69°R (32.018°F) in English units. The reference pressure is the saturation pressure corresponding to the reference temperature, 0.61172 kPa in SI units or 0.088724 psia in English units.

After substitution of Equations 11.62 and 11.63 into Equation 11.61, the entropy of the mixture per unit mass of dry air may be expressed as

$$\begin{aligned} s_m = {} & c_{p,a} \ln\left(\frac{T}{T_{a,\text{ref}}}\right) - R_a \ln\left(\frac{p_a}{P_{a,\text{ref}}}\right) \\ & + \omega \left[c_{p,v} \ln\left(\frac{T}{T_{v,\text{ref}}}\right) - R_v \ln\left(\frac{p_v}{P_{v,\text{ref}}}\right) + s_{v,\text{ref}} \right] \end{aligned} \tag{11.64}$$

The various reference quantity and constants used to evaluate the enthalpy and entropy per unit mass of dry air for air-water-vapor mixtures are summarized in Table 11.1.

Another quantity frequently used in psychrometric applications is the volume of the mixture per unit mass of dry air,

$$v = \frac{V}{m_a}$$

With the ideal-gas equation of state this quantity can be expressed as

$$v = \frac{m_m \bar{R} T}{M_m P m_a}$$

Noting that the partial pressure of the air is related to the mole fraction by Equation 11.17,

$$p_a = y_a P = \left(\frac{m_a}{M_a}\right)\left(\frac{M_m}{m_m}\right) P$$

the specific volume of the mixture on the basis of a unit mass of dry air is found to be

$$v = \frac{\bar{R} T}{M_a p_a} = v_a \tag{11.65}$$

TABLE 11.1 SUMMARY OF CONSTANTS USED TO EVALUATE ENTHALPY* AND ENTROPY† OF AIR-WATER-VAPOR MIXTURES

Constant	SI units	English units
$c_{p,a}$	1.0 kJ/kg·°C	0.24 Btu/lb$_m$·°F
R_a	0.287 kJ/kg·K	0.06855 Btu/lb$_m$·°R
$c_{p,v}$	1.82 kJ/kg·°C	0.435 Btu/lb$_m$·°F
R_v	0.4615 kJ/kg·K	0.1102 Btu/lb$_m$·°R
$T_{a,\text{ref}}$	0°C (273.15 K)	0°F (459.67°R)
$P_{a,\text{ref}}$	101.325 kPa	14.696 psia
$T_{v,\text{ref}}$	0.01°C (273.16 K)	32.018°F (491.69°R)
$P_{v,\text{ref}}$	0.61173 kPa	0.088724 psia
$\Delta h_{v,\text{ref}}$	2500.5 kJ/kg	1061 Btu/lb$_m$
$s_{v,\text{ref}}$	9.1541 kJ/kg·K	2.1864 Btu/lb$_m$·°R

* $h_m = c_{p,a}T + \omega(c_{p,v}T + \Delta h_{v,\text{ref}})$; T in °C or °F, as appropriate (11.59)

† $$s_m = c_{p,a} \ln\left(\frac{T}{T_{a,\text{ref}}}\right) - R_a \ln\left(\frac{p_a}{P_{a,\text{ref}}}\right) + \omega\left[c_{p,v} \ln\left(\frac{T}{T_{v,\text{ref}}}\right) - R_v \ln\left(\frac{p_v}{P_{v,\text{ref}}}\right) + s_{v,\text{ref}}\right] \quad (11.64)$$ T in K or °R, as appropriate

In other words, the volume of the mixture per unit mass of dry air is equal to the specific volume of the dry air. This conclusion is a result of using Dalton's law to describe the P-v-T behavior of the mixture.

EXAMPLE 11.7

One kilogram of moist air initially at a total pressure of 1 atm has a dry bulb temperature of 20°C and a relative humidity of 60 percent and is contained in a closed, rigid vessel. Determine the heat transfer to the moist air required to increase the dry bulb temperature to 50°C. Calculate the final pressure and final relative humidity of the mixture. Suppose that the heat transfer to the system is from a heat source that has a temperature of 100°C. Determine the total entropy change associated with this process.

Solution. The system is chosen to be the moist air (assumed ideal) in the container, and a sketch of the system as well as a P-v diagram for the process is shown in Figure 11.2.

Since the mixture is assumed ideal, the final pressure can be determined from the ideal-gas equation of state:

$$P_m \mathbf{V}_m = m_m R_m T$$

The system is closed, and the container is rigid. Using P_1 and P_2 to denote the mixture pressure of the initial and final states, respectively, we find that

$$\frac{P_2}{P_1} = \frac{T_2}{T_1}$$

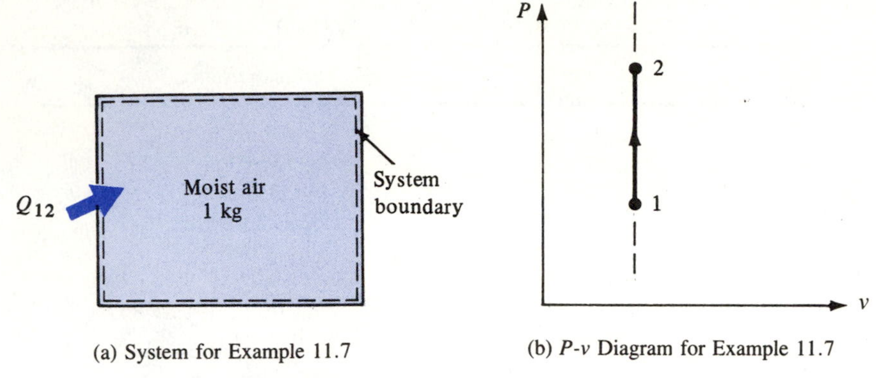

(a) System for Example 11.7

(b) P-v Diagram for Example 11.7

Figure 11.2 Sketches for Example 11.7.

or

$$P_2 = P_1\left(\frac{T_2}{T_1}\right) = (1\ \text{atm})\left(\frac{323\ \text{K}}{293\ \text{K}}\right) = \underline{\underline{1.102\ \text{atm}}}$$

Applying the conservation-of-mass equation to the dry air in the closed system, we find that

$$m_{a1} = m_{a2} = m_a$$

Next, the conservation-of-mass equation applied to the water vapor yields

$$m_{v1} = m_{v2}$$

or

$$\omega_1 m_{a1} = \omega_2 m_{a2}$$

Thus

$$\omega_1 = \omega_2 = \omega$$

since the mass of dry air is constant. The fact that the humidity ratio remains constant can be used to determine the relative humidity of the mixture at the final state. First, however, the vapor pressure of the initial state must be determined from Equation 11.49:

$$p_{v1} = \phi p_{g1} = 0.6(2.339\ \text{kPa}) = 1.403\ \text{kPa}$$

The humidity ratio can now be calculated from Equation 11.54:

$$\omega = \omega_1 = 0.622\left(\frac{p_{v1}}{P_1 - p_{v1}}\right)$$

$$= (0.622)\left[\frac{1.403\ \text{kPa}}{(101.3 - 1.403)\text{kPa}}\right]$$

$$= 8.736 \times 10^{-3}$$

Next, Equation 11.54 can be used to evaluate the vapor pressure at state 2:

$$p_{v2} = \frac{\omega P_2}{0.622 + \omega}$$

$$= \frac{(8.736 \times 10^{-3})(1.102 \text{ atm})(101.3 \text{ kPa/atm})}{0.622 + 8.736 \times 10^{-3}}$$

$$= 1.546 \text{ kPa}$$

and the relative humidity can be calculated from Equation 11.49:

$$\phi_2 = \frac{p_{v2}}{p_{g2}} = \frac{1.546 \text{ kPa}}{12.344 \text{ kPa}} = \underline{\underline{0.125 \text{ or } 12.5 \text{ percent}}}$$

The heat transfer to the system can be determined by applying the conservation-of-energy equation to the closed system and noting that no work is done since the volume does not change:

$$Q_{12} = U_2 - U_1$$

With the definition of the enthalpy this equation can be written as

$$Q_{12} = (H_2 - H_1) - \mathbf{V}(P_2 - P_1)$$

since the volume of the system is constant. If the enthalpy and specific volume of the mixture per unit mass of dry air are used, then this equation becomes

$$Q_{12} = m_a[(h_2 - h_1) - v(P_2 - P_1)]$$

Since the humidity ratio is constant, the enthalpy change of the mixture per unit mass of dry air, from Equation 11.59, is

$$h_2 - h_1 = c_{p,a}(T_2 - T_1) + \omega c_{p,v}(T_2 - T_1)$$

Substituting the appropriate values, we obtain

$$h_2 - h_1 = (1 \text{ kJ/kg dry air}\cdot{}^\circ\text{C})(50 - 20)^\circ\text{C}$$

$$+ (8.736 \times 10^{-3} \text{kg H}_2\text{O/kg dry air})$$

$$\times (1.82 \text{ kJ/kg H}_2\text{O}\cdot{}^\circ\text{C})(50 - 20)^\circ\text{C}$$

$$= 30.48 \text{ kJ/kg dry air}$$

The volume of the mixture per unit mass of dry air, from Equation 11.65, is

$$v = v_a = \frac{\overline{R}T_a}{M_a p_a} = \frac{\overline{R}T_a}{M_a(P - p_v)}$$

and can be calculated from the conditions at state 1:

$$v = \frac{(8.314 \text{ kPa}\cdot\text{m}^3/\text{kg}\cdot\text{mol}\cdot\text{K})(293 \text{ K})}{(28.97 \text{ kg dry air/kg}\cdot\text{mol})(101.3 - 1.403)\text{kPa}}$$

$$= 0.842 \text{ m}^3/\text{kg dry air}$$

The mass of moist air is given, and the mass of dry air must be calculated in order to determine the heat transfer. The total mass is the sum of the mass of dry air and the mass of water vapor:

$$m_m = m_a + m_v = m_a\left(1 + \frac{m_v}{m_a}\right) = m_a(1 + \omega)$$

Therefore

$$m_a = \frac{m_m}{1 + \omega} = \frac{1 \text{ kg}}{(1 + 8.736 \times 10^{-3})\text{kg/kg dry air}}$$
$$= 0.991 \text{ kg dry air}$$

and the heat transfer to the system during the process is

$$Q_{12} = (0.991 \text{ kg dry air})\{30.48 \text{ kJ/kg dry air} - (0.842 \text{ m}^3\text{/kg dry air})$$
$$\times [(1.102 - 1)\text{atm}](101.3 \text{ kPa/atm})(1 \text{ kJ/kPa}\cdot\text{m}^3)\}$$
$$= \underline{\underline{+21.6 \text{ kJ}}}$$

From Equation 6.2 the total entropy change for a closed system is given by

$$\Delta S_{\text{tot}} = \Delta S_{\text{sys}} + \Delta S_{\text{surr}}$$

where the surroundings consist only of the thermal-energy reservoir. The entropy change of the system can be expressed as

$$\Delta S_{\text{sys}} = m_a(s_{m,2} - s_{m,1})$$

and using Equation 11.64 and noting that the humidity ratio does not change during the process, we can write this equation as

$$\Delta S_{\text{sys}} = m_a\left[(c_{p,a} + \omega c_{p,v}) \ln\frac{T_2}{T_1} - (R_a + \omega R_v) \ln\frac{P_2}{P_1}\right]$$

Thus

$$\Delta S_{\text{sys}} = (0.991 \text{ kg dry air})$$
$$\times \Bigg\{[1.0 \text{ kJ/kg dry air}\cdot\text{K} + (8.736 \times 10^{-3} \text{ kg H}_2\text{O/kg dry air})$$
$$\times (1.82 \text{ kJ/kg H}_2\text{O}\cdot\text{K})] \ln\left(\frac{323 \text{ K}}{293 \text{ K}}\right)$$
$$- [0.287 \text{ kJ/kg}\cdot\text{K} + (8.736 \times 10^{-3} \text{ kg H}_2\text{O/kg dry air})$$
$$\times (0.4615 \text{ kJ/kg}\cdot\text{K})] \ln\left(\frac{1.102 \text{ atm}}{1 \text{ atm}}\right)\Bigg\}$$
$$= 7.013 \times 10^{-2} \text{ kJ/K}$$

The entropy change of the surroundings is given by

$$\Delta S_{surr} = \frac{Q_{surr}}{T_{surr}}$$

since the heat source is isothermal and undergoes an internally reversible process. Furthermore, the heat transfer from the surroundings is equal in magnitude but opposite in direction to the heat transfer to the system. Thus

$$\Delta S_{surr} = \frac{Q_{surr}}{T_{surr}} = -\frac{Q_{12}}{T_{surr}}$$

$$= -\frac{21.6 \text{ kJ}}{373 \text{ K}} = -5.79 \times 10^{-2} \text{ kJ/K}$$

and the total entropy change is

$$\Delta S_{tot} = \Delta S_{sys} + \Delta S_{surr}$$

$$= (7.013 \times 10^{-2} - 5.79 \times 10^{-2})\text{kJ/K} = \underline{\underline{1.22 \times 10^{-2} \text{ kJ/K}}}$$

The total entropy change is positive, indicating that the process is irreversible. The irreversibility arises due to heat transfer between the system and its surroundings, which occurs through a finite temperature difference. ■

There is an inherent difficulty associated with using either the relative humidity or the humidity ratio to specify the composition of the moist-air mixture because neither can be measured directly. An indirect-measurement scheme must therefore be devised. One method that is conceptually simple is called an *adiabatic-saturation process*. In this process, shown schematically in Figure 11.3(a) and diagrammatically in Figure 11.3(b), an unsaturated airstream is passed steadily through a long insulated channel that contains a pool of liquid water. If the channel is sufficiently long, the airstream will exit at saturated conditions since moisture will be added to the airstream as a result of evaporation. The temperature at the exit, T_2 in Figure 11.3(b), is called the *adiabatic-saturation temperature*. In addition, the saturated liquid water is assumed to be added to the channel at a rate sufficient to make up for the water being evaporated, and its temperature is assumed to be T_2.

This steady-flow process can be analyzed on the basis of the conservation of mass and of energy to develop a relationship between the inlet-air conditions and the adiabatic-saturation temperature. First, the conservation-of-mass equation is applied to the dry air and the water vapor separately, with the results

$$\dot{m}_{a1} = \dot{m}_{a2} = \dot{m}_a$$

and

$$\dot{m}_{v1} + \dot{m}_f = \dot{m}_{v2}$$

The first of these equations states that the mass flow rate of dry air in the mixture does not change as the airstream passes through the channel. The second equation states that the mass flow rate of vapor at the exit (state 2) is higher than that at the inlet (state 1)

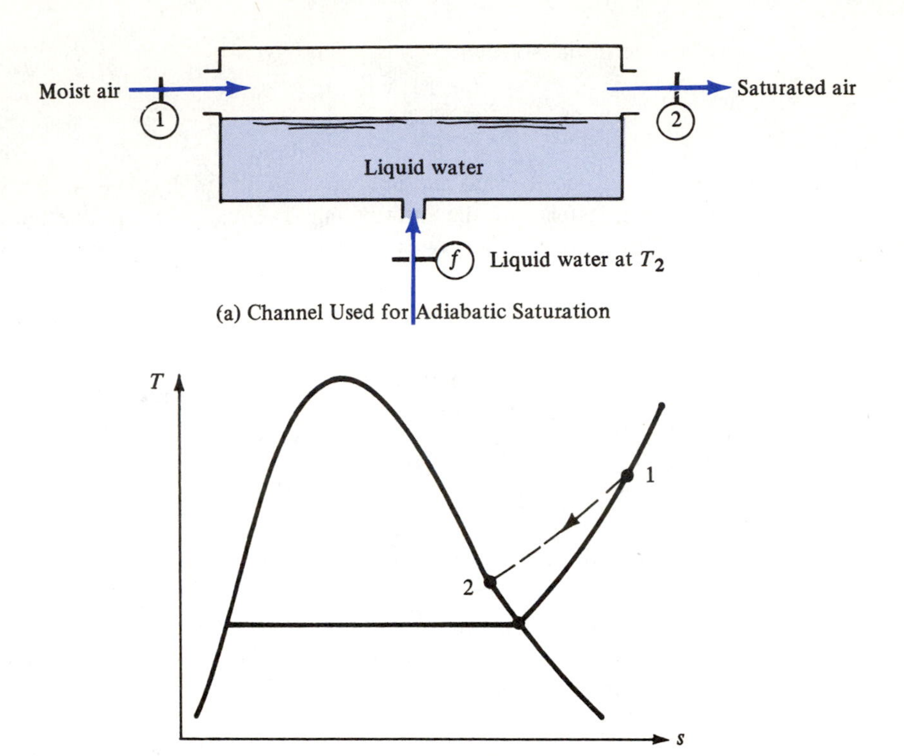

Figure 11.3 Adiabatic-saturation process.

by an amount equal to the rate of evaporation along the length of the channel (the amount of liquid added is equal to the amount of water evaporated). This last equation can be expressed in terms of the humidity ratios since, as a result of Equation 11.52,

$$\dot{m}_v = \omega \dot{m}_a$$

Therefore

$$\omega_1 \dot{m}_{a1} + \dot{m}_f = \omega_2 \dot{m}_{a2}$$

or

$$\dot{m}_f = (\omega_2 - \omega_1)\dot{m}_a \tag{11.66}$$

Applying the conservation-of-energy equation to the system indicated in Figure 11.3 and noting that heat transfer and work are absent, we have

$$0 = \dot{m}_{a2}h_2 - \dot{m}_f h_f - \dot{m}_{a1}h_1$$

In this equation the mixture enthalpies of the airstream are expressed on the basis of a unit mass of dry air, as indicated in Equation 11.56. Substituting the result from the conservation of mass for the dry air and Equation 11.66 gives

$$0 = \dot{m}_a(h_2 - h_1) - \dot{m}_a(\omega_2 - \omega_1)h_f$$

With Equation 11.58 this expression can be written as

$$0 = c_{p,a}(T_2 - T_1) + \omega_2 h_{v2} - \omega_1 h_{v1} - (\omega_2 - \omega_1)h_f$$

or

$$\omega_1 = \frac{c_{p,a}(T_2 - T_1) + \omega_2(h_{v2} - h_f)}{h_{v1} - h_f} \tag{11.67}$$

The humidity ratio ω_2 corresponds to the humidity ratio of a saturated mixture at the temperature T_2. Examination of Equation 11.67 reveals that all quantities on the right side can be determined if the dry bulb temperature T_1 and the adiabatic-saturation temperature T_2 can be measured. In other words, the humidity ratio of the entering airstream could be determined indirectly with two temperature measurements.

While the adiabatic-saturation process is conceptually simple, it is not a practical means of determining the adiabatic-saturation temperature because a very long channel would be required in order to achieve saturated conditions at the exit. An alternative solution is to measure a quantity called the *wet bulb temperature*. The wet bulb temperature can be measured by placing a wetted wick over the bulb of a thermometer and passing the moist-air stream over the wick. The wet bulb temperature is often measured by attaching the wicked thermometer to a shaft affixed to a handle so that the thermometer can be rotated rapidly (a *sling psychrometer*). The movement of the unsaturated air through the wick causes evaporation and a decrease in the temperature of the thermometer fluid. The wet bulb temperature measured in this way is found to be a reasonably accurate approximation to the adiabatic-saturation temperature at ordinary temperatures and pressures. The temperature T_2 in Equation 11.67 can therefore be taken to be equal to the wet bulb temperature, and the humidity ratio can be determined by measuring the wet bulb and dry bulb temperatures of a moist-air mixture.

EXAMPLE 11.8

The wet bulb and dry bulb temperatures of a moist-air mixture at a total pressure of 1 atm are measured with a sling psychrometer and are found to be 50 and 70°F, respectively. Determine the humidity ratio, the relative humidity, and the enthalpy and volume of the mixture per unit mass of dry air.

Solution. The humidity ratio is determined from Equation 11.67,

$$\omega_1 = \frac{c_{p,a}(T_2 - T_1) + \omega_2(h_{v2} - h_f)}{h_{v1} - h_f}$$

where T_2 corresponds to the wet bulb temperature and T_1 corresponds to the dry bulb temperature. Values for h_{v2} and h_{v1} can be found in Table B.1E:

$$h_{v1} = 1091.7 \text{ Btu/lb}_m \text{ H}_2\text{O} \qquad h_{v2} = 1082.9 \text{ Btu/lb}_m \text{ H}_2\text{O}$$

and h_f is evaluated at the wet-bulb temperature:

$$h_f = h_{f2} = 18.05 \text{ Btu/lb}_m \text{ H}_2\text{O}$$

The humidity ratio ω_2 corresponds to a state of saturated moist air at the wet bulb temperature. At temperature T_2,

$$p_{v2} = P_{sat} = 0.17813 \text{ psia}$$

and from Equation 11.54

$$\omega_2 = 0.622\left(\frac{p_{v2}}{P - p_{v2}}\right) = 0.622\left[\frac{0.17813 \text{ psia}}{(14.7 - 0.17813)\text{psia}}\right]$$

$$= 7.63 \times 10^{-3}\,\text{lb}_m\ H_2O/\text{lb}_m \text{ dry air}$$

Therefore

$$\omega_1 = \frac{(0.24 \text{ Btu/lb}_m \text{ dry air}\cdot{}^\circ\text{F})(50 - 70)^\circ\text{F}}{(1091.7 - 18.05)\text{Btu/lb}_m\ H_2O}$$

$$+ \frac{(7.63 \times 10^{-3}\text{lb}_m\ H_2O/\text{lb}_m \text{ dry air})(1082.9 - 18.05)\text{Btu/lb}_m H_2O}{(1091.7 - 18.05)\text{Btu/lb}_m\ H_2O}$$

$$= \underline{\underline{3.1 \times 10^{-3}\ \text{lb}_m\ H_2O/\text{lb}_m \text{ dry air}}}$$

The partial pressure of the water vapor in the mixture can be determined from Equation 11.54:

$$\omega_1 = 0.622\left(\frac{p_{v1}}{P - p_{v1}}\right)$$

or

$$p_{v1} = \frac{\omega_1 P}{0.622 + \omega_1}$$

$$= \frac{(3.1 \times 10^{-3})(14.7 \text{ psia})}{0.622 + 3.1 \times 10^{-3}} = 0.0729 \text{ psia}$$

Next, the relative humidity is calculated from Equation 11.49

$$\phi_1 = \frac{p_{v1}}{p_{g1}}$$

where p_{g1} is the saturation water pressure corresponding to the dry bulb temperature of the air.

$$\phi_1 = \frac{0.0729 \text{ psia}}{0.36328 \text{ psia}} = \underline{\underline{0.20 \text{ or } 20 \text{ percent}}}$$

The volume of the mixture per unit mass of dry air is the same as the specific volume of the dry air in the mixture (Equation 11.65):

$$v = \frac{\overline{R}T}{M_a p_a}$$

and $p_a = P - p_v$. So

$$v = \frac{(10.73 \text{ psia}\cdot\text{ft}^3/\text{lb}_\text{m}\cdot\text{mol}\cdot°\text{R})(530°\text{R})}{(28.97 \text{ lb}_\text{m}/\text{lb}_\text{m}\cdot\text{mol})(14.7 - 0.0729)\text{psia}}$$

$$= \underline{\underline{13.4 \text{ ft}^3/\text{lb}_\text{m} \text{ dry air}}}$$

Finally, the enthalpy of the mixture per unit mass of dry air can be calculated from Equation 11.59:

$$h_{m1} = c_{p,a}T_1 + \omega_1(c_{p,v}T_1 + \Delta h_{v,\text{ref}})$$

where T_1 is in degrees Fahrenheit. Thus

$$h_{m1} = (0.24 \text{ Btu/lb}_\text{m}\cdot°\text{F})(70°\text{F}) + (3.1 \times 10^{-3}\text{lb}_\text{m} \text{ H}_2\text{O/lb}_\text{m} \text{ dry air})$$
$$\times [(0.435 \text{ Btu/lb}_\text{m} \text{ H}_2\text{O}\cdot°\text{F})(70°\text{F}) + 1061 \text{ Btu/lb}_\text{m} \text{ H}_2\text{O}]$$
$$= \underline{\underline{20.2 \text{ Btu/lb}_\text{m} \text{ dry air}}}$$

11.6 THE PSYCHROMETRIC CHART

For engineering applications the properties of air–water-vapor mixtures are often presented in a graphical form on a *psychrometric chart*. This chart eliminates the need to perform the actual calculations required to determine the properties. Notice that in the preceding set of equations, any two properties of the air–water-vapor mixture, together with the total pressure, completely specify the thermodynamic state of the mixture. Thus the properties of the mixture can be presented on a two-dimensional figure such as the psychrometric chart illustrated in the sketch in Figure 11.4 for a given total pressure.

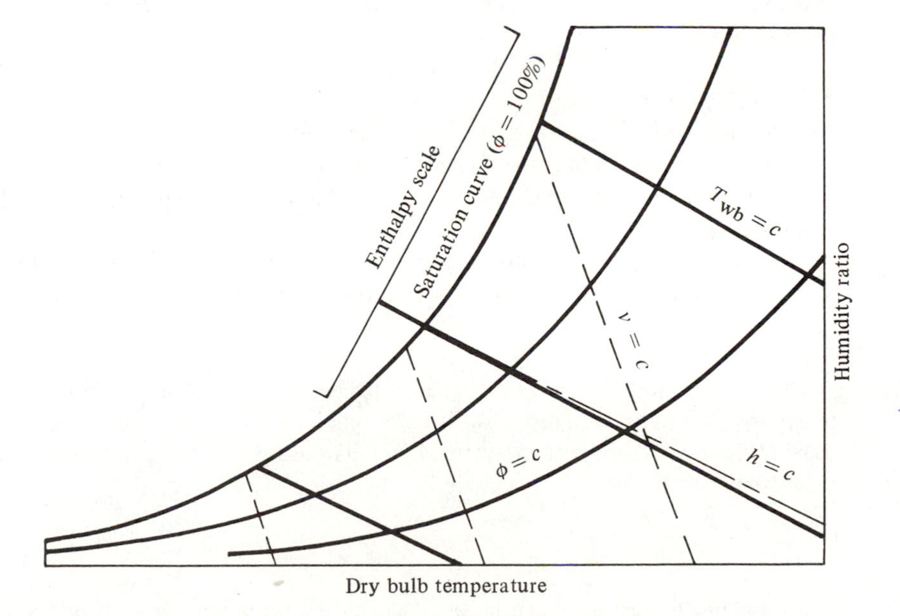

Figure 11.4 Sketch of a typical psychrometric chart.

Psychrometric charts are usually constructed for a total pressure of 1 atm. Corrections for total pressures other than 1 atm can easily be made, and some psychrometric charts have correction tables appended to them for this purpose. A psychrometric chart for a total pressure of 1 atm is presented in Figure F.1 for SI units and in Figure F.1E for English units in the Appendix.

While the specific format can vary somewhat, the salient features, such as those shown in Figure 11.4, are predominantly the same for all psychrometric charts. The horizontal axis is a dry bulb temperature scale, and the vertical axis on the right side of the chart typically has scales for both the humidity ratio and the partial pressure of the water vapor (or vapor pressure). The solid lines that slant up and to the left in this figure are lines of constant wet bulb temperature. The dashed lines that slant up and to the left indicate lines of constant specific volume of the mixture on the basis of a unit mass of dry air. The solid lines that curve up and to the right on the psychrometric chart are lines of constant relative humidity. The topmost of these lines, which denotes a relative humidity of 100 percent, is also called the *saturation curve*. Above the saturation curve is a scale for the enthalpy of the mixture per unit mass of dry air. Notice that lines of constant enthalpy are very nearly parallel to constant wet bulb lines. Since the deviation between the slopes of these lines is quite small, constant-enthalpy lines are often assumed to be parallel to constant wet bulb lines.

For a saturated mixture the wet bulb temperature and the dew-point temperature are equal. Therefore, the dew-point temperature can be determined by evaluating the wet bulb temperature on the saturation curve. Recall that the dew-point temperature is reached by cooling an unsaturated mixture at constant total pressure. Thus the vapor pressure (and therefore the humidity ratio) remains constant, and the path on the psychrometric chart would be a horizontal line from right to left that intersects the saturation curve. The dew-point temperature, then, would be indicated by that intersection with the saturation curve.

■ EXAMPLE 11.9

Use the psychrometric chart to determine the humidity ratio, the wet bulb temperature, the dew-point temperature, the vapor pressure, and the enthalpy and specific volume of moist air at 1 atm total pressure with a dry bulb temperature of 40°C and a relative humidity of 30 percent.

If this air is cooled at constant total pressure until the dry bulb temperature reaches 30°C, what effect, if any, does this cooling have on the relative humidity and the humidity ratio?

Solution. The initial state of the air is indicated on the sketch of a psychrometric chart in Figure 11.5. The humidity ratio is determined by the intersection of a horizontal line passing through state 1 and the vertical axis on the right of the chart. From the psychrometric chart in Figure F.1 this value is

$$\omega_1 \simeq \underline{\underline{0.0139 \text{ kg } H_2O/\text{kg dry air}}}$$

The wet bulb temperature can be obtained by interpolating between two adjacent lines of constant wet bulb temperature or, alternatively, by constructing a line parallel to a

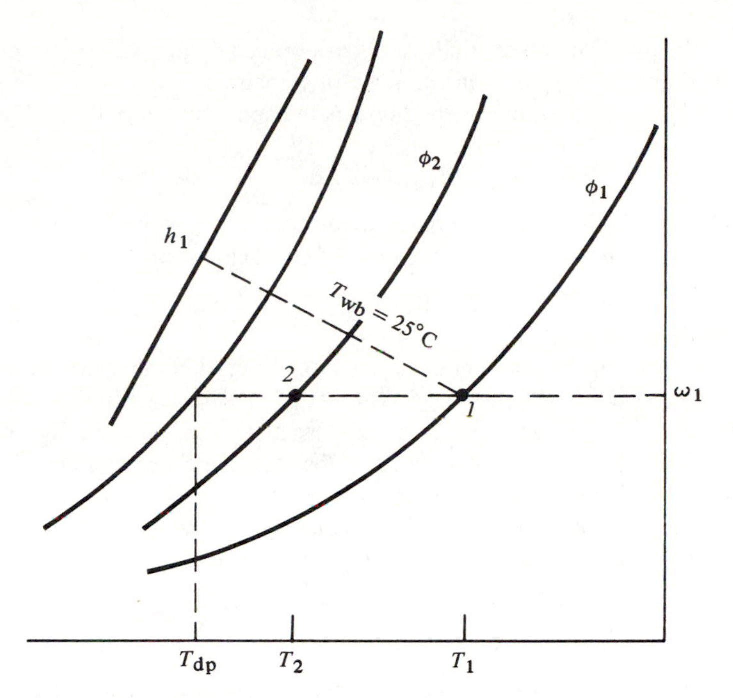

Figure 11.5 Sketch of psychrometric chart for Example 11.9.

constant wet bulb line so that it intersects the saturation curve, as shown in Figure 11.5. The wet bulb temperature is the value of the saturation temperature corresponding to this point of intersection:

$$T_{wb} \simeq \underline{\underline{25°C}}$$

The dew-point temperature is reached if the air is cooled at constant total pressure until saturation conditions are achieved. For this process the vapor pressure would remain constant, and therefore the mole fractions of dry air and water vapor would not change (Equation 11.24). Thus the humidity ratio ω would be constant, and the dew-point temperature can be determined by intersecting the saturation curve with a horizontal line that passes through state 1:

$$T_{dp} \simeq \underline{\underline{19°C}}$$

A scale for vapor pressure is not provided on the psychrometric chart in Figure F.1, but the vapor pressure can be calculated with the aid of Equation 11.54:

$$\omega_1 = 0.622 \left(\frac{p_{v1}}{P - p_{v1}} \right)$$

or

$$p_{v1} = \frac{\omega_1 P}{0.622 + \omega_1} = \frac{(0.0139)(101.3\,\text{kPa})}{0.622 + 0.0139} = \underline{\underline{2.21\ \text{kPa}}}$$

The enthalpy per unit mass of dry air is determined by intersecting the enthalpy scale with a line drawn parallel to a line of constant enthalpy and that passes through state 1. Notice that this line is essentially the same as the constant wet bulb line for state 1:

$$h_1 \simeq \underline{\underline{76 \text{ kJ/kg dry air}}}$$

Finally, the specific volume of the air based on a unit mass of dry air can be estimated by interpolating between two adjacent lines of constant specific volume in Figure F.1:

$$v_1 \simeq \underline{\underline{0.908 \text{ m}^3\text{/kg dry air}}}$$

If the air is cooled at constant total pressure to a dry bulb temperature of 30°C, state 2 is reached, as shown in Figure 11.5. This state has the same humidity ratio as state 1 because cooling at constant total pressure does not alter the partial pressures of the components unless the temperature reaches the dew-point temperature. Therefore, the mole fractions and mass fractions of the components (assumed ideal) do not change, and ω_1 and ω_2 must be equal. Notice, however, that the relative humidity of the air at state 2 is higher than that at state 1. Recall that the relative humidity is defined as

$$\phi = \frac{p_v}{p_g}$$

Even though the vapor pressure does not change, p_g does change because p_g is the saturation pressure corresponding to the dry bulb temperature of the mixture. As the dry bulb temperature is decreased, the corresponding saturation pressure of water also decreases. Therefore, the relative humidity is higher at state 2 even though the vapor pressure is identical to the vapor pressure of state 1. ■

11.7 AIR-CONDITIONING APPLICATIONS

In the broadest sense air-conditioning processes are used to change the state or composition of a mass or stream of air. These processes include simple heating or cooling wherein the moisture content (or humidity ratio) of the air remains constant, heating with humidification in order to increase both the temperature and the moisture content of the air, cooling with dehumidification, evaporative cooling, adiabatic mixing of two or more airstreams, and cooling-tower processes. *Humidification* is a process that results in an increase in the amount of moisture in an air–water-vapor mixture. *Dehumidification*, on the other hand, results in a decrease in the moisture content of the mixture. These processes are used to condition the air in buildings in order to provide comfortable surroundings for the occupants, to maintain the environmental control necessary for equipment such as precision-measurement devices or electronic computers, and for a variety of other commercial and industrial processes.

The basic processes mentioned above are described in detail in this section. Not only are those processes important individually, but more complex air-conditioning processes can be devised by combining these basic processes. The emphasis in this section is on understanding the analysis of the basic steady-flow processes through application of the conservation-of-mass and -energy principles in addition to visualizing the features of the processes with the aid of the psychrometric chart. In each of these applications the conservation-of-mass equation is applied separately to the dry air and to the water vapor present in the moist-air streams.

11.7.1 Heating and Cooling

Simple heating and cooling processes are used to increase or decrease the dry bulb temperature of moist air without adding or subtracting moisture. In other words, the humidity ratio of the air remains unchanged during the process, and the process can therefore be represented by a horizontal line on the psychrometric chart. In Figure 11.6(b) a simple heating process is indicated by the path 1-2 and a simple cooling process is indicated by the path 3-4.

These processes are typical of those that occur in many residential heating and cooling systems. Their main disadvantage is that they provide no means of controlling

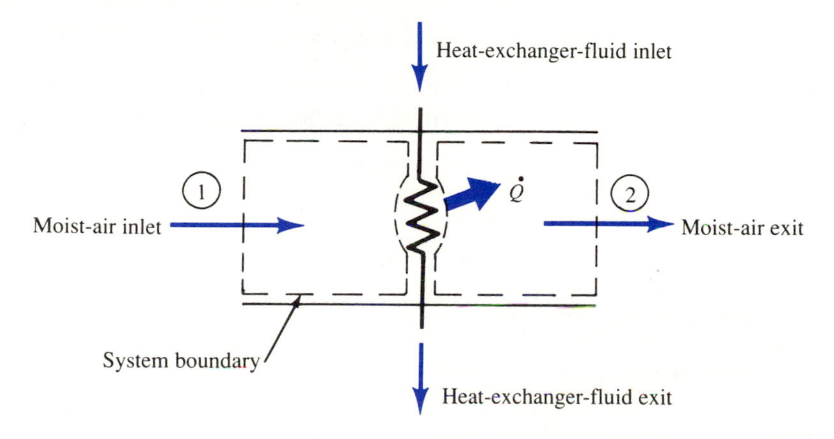

(a) Schematic of Heating or Cooling Section

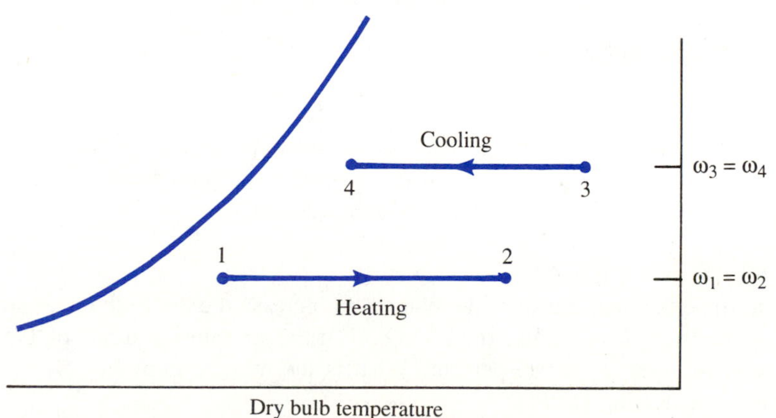

(b) Process Representation for Simple Heating or Cooling

Figure 11.6 Simple heating and cooling processes.

the amount of moisture in the air, and less-than-comfortable environments can result. For example, if relatively dry, cold winter air is heated to, say, 25°C, the relative humidity of the air will decrease somewhat [see Figure 11.6(b)], and the final relative humidity may be well below a comfortable level. This condition can cause dry skin, respiratory difficulties, and an increase in static electricity. On the other hand, in the summer the outdoor air in many locations is hot and quite humid. If this air is simply cooled to lower temperatures, the relative humidity of the air will increase [see Figure 11.6(b)], resulting in uncomfortable surroundings.

In heating and cooling applications, moist air is passed through a heat exchanger such as the one depicted schematically in Figure 11.6(a). For cooling purposes a cold refrigerant circulates in the heat-exchanger coils to cool the moist-air stream, and for heating purposes a hot fluid is circulated to warm the airstream. As shown in the figure, the thermodynamic system is chosen to exclude the heat-exchanger fluid, and heat transfer occurs between the moist-air stream and the heat-exchanger surfaces. Applying the conservation-of-mass equation (Equation 3.16) for steady flow to the dry air first gives

$$\dot{m}_{a1} = \dot{m}_{a2} = \dot{m}_a$$

In other words, the mass flow rate of dry air does not change during the process. Next, applying the conservation-of-mass equation to the water vapor yields

$$\omega_1 \dot{m}_{a1} = \omega_2 \dot{m}_{a2}$$

or

$$\omega_1 = \omega_2$$

This equation simply states that the humidity ratio remains unchanged; that is, the amount of water vapor in the air is constant since the air is neither humidified nor dehumidified.

Finally, applying the conservation-of-energy equation (Equation 4.31), noting that no work is done, and assuming negligible changes in the kinetic and potential energies of the airstream leads to the result

$$\dot{Q} = \dot{m}_{a2} h_2 - \dot{m}_{a1} h_1 = \dot{m}_a (h_2 - h_1)$$

or on the basis of a unit mass of dry air,

$$q = h_2 - h_1$$

Thus the heat transfer per unit mass of dry air required for simple heating or cooling is equal to the change in enthalpy of the airstream (per unit mass of dry air) between the exit and the inlet of the heating or cooling section.

EXAMPLE 11.10

Determine the heat-transfer rate required to increase the dry bulb temperature of a steady flow of moist air at 1 atm from 50 to 85°F if the air enters at a rate of 1750 ft^3/min with a relative humidity of 80 percent. What is the relative humidity of the airstream as it leaves the heating section?

Solution. The heating section and the process diagram are as shown in Figure 11.6. The conservation-of-mass equation applied to the dry air as it passes through the heating section gives

$$\dot{m}_{a1} = \dot{m}_{a2} = \dot{m}_a$$

Similarly, the result for the water vapor is

$$\omega_1 \dot{m}_{a1} = \omega_2 \dot{m}_{a2}$$

or

$$\omega_1 = \omega_2 = \omega$$

Next, the conservation-of-energy equation for steady flow with negligible changes in kinetic and potential energies yields

$$\dot{Q} = \dot{m}_{a2} h_2 - \dot{m}_{a1} h_1 = \dot{m}_a (h_2 - h_1)$$

since no work is done during the process.

The mass flow rate of dry air can be determined from the volumetric flow rate at the inlet to the heating section:

$$\dot{m}_a = \frac{A_1 V_1}{v_1}$$

The specific volume v_1 is found by interpolation from the psychrometric chart:

$$v_1 \simeq 12.95 \text{ ft}^3/\text{lb}_\text{m} \text{ dry air}$$

and

$$\dot{m}_a = \frac{1750 \text{ ft}^3/\text{min}}{12.95 \text{ ft}^3/\text{lb}_\text{m} \text{ dry air}} = 135.1 \text{ lb}_\text{m} \text{ dry air/min}$$

Since the mass of water vapor in the air does not change, state 2 is located at the point of intersection of the 85°F dry bulb line and the horizontal line that passes through state 1 on the psychrometric chart. Therefore, the enthalpy values at states 1 and 2 can be obtained from the psychrometric chart, and

$$\begin{aligned} \dot{Q} &= \dot{m}_a (h_2 - h_1) \\ &= (135.1 \text{ lb}_\text{m} \text{ dry air/min})(27.0 - 18.6)\text{Btu/lb}_\text{m} \text{ dry air} \\ &= \underline{\underline{1135 \text{ Btu/min}}} \end{aligned}$$

From the psychrometric chart the relative humidity of the moist-air stream leaving the heating section is

$$\phi_2 \simeq \underline{\underline{0.23 \text{ or } 23 \text{ percent}}}$$

An important fact to recognize is that the volumetric flow rate at the exit is different from that at the inlet. The mass flow rate of dry air is constant, but the specific volume increases during the process. Therefore, the volumetric flow rate at the exit is higher than the volumetric flow rate at the inlet. ■

11.7.2 Cooling with Dehumidification

If the simple cooling process results in unacceptably high relative humidities, some means of removing moisture from the air must be provided. This removal is easily accomplished by dehumidification, that is, cooling the air to its dew-point temperature so that some of the water vapor in the air will condense. Cooling with dehumidification, therefore, can be acheived with basically the same equipment used for simple cooling provided that the temperature of the refrigerant in the heat exchanger is low enough to cool the airstream below its dew-point temperature.

A schematic of the cooling and dehumidifying section is shown in Figure 11.7(a), and the process, as it would appear on a psychrometric chart, is shown in Figure 11.7(b). The first part of the process is a simple cooling process during which the humidity ratio of the air remains unchanged [1-2 in Figure 11.7(b)]. When the dew-point temperature

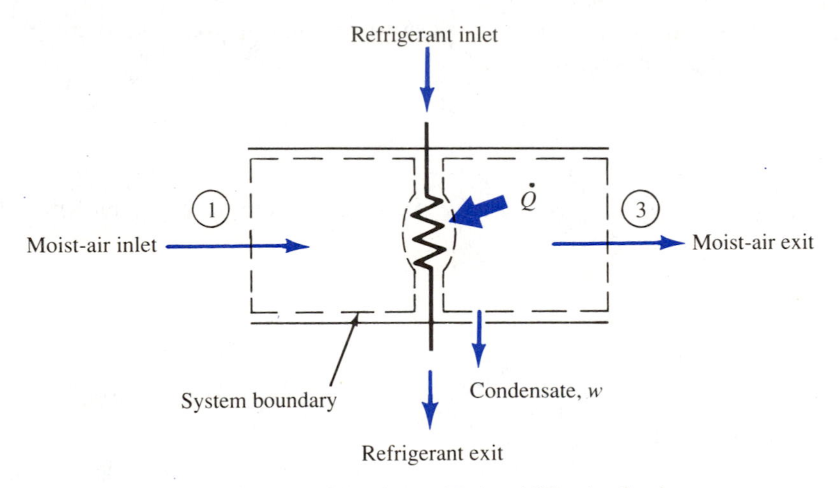

(a) Schematic of Cooling and Dehumidification Section

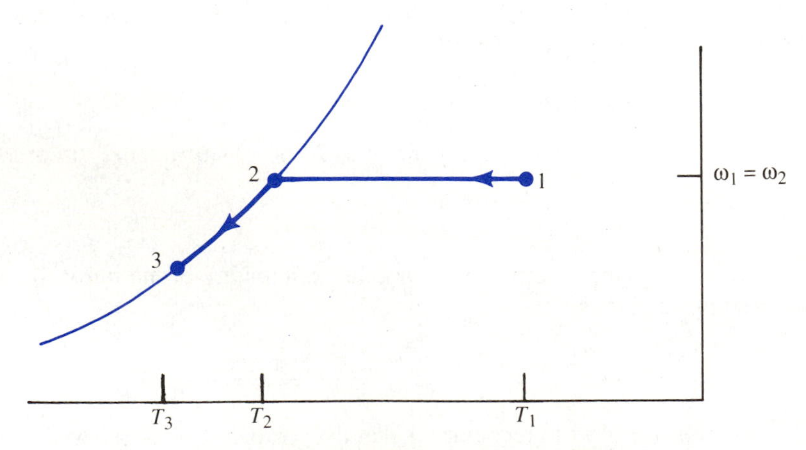

(b) Process Representation for Cooling and Dehumidification

Figure 11.7 Cooling with dehumidification.

corresponding to the initial state of the air is reached at state 2, vapor begins to condense. Further cooling results in additional condensation, while the air remains at saturated conditions, as indicated by the path 2-3. The condensate that leaves the cooling section is assumed to be cooled to the final dew-point temperature (T_3) before leaving. While this assumption is only approximately true, it introduces negligible error since the enthalpy of the liquid water is significantly smaller than the enthalpy of the airstream.

The analysis of this process is similar to that described in the previous section. The conservation-of-mass equation for steady flow is first applied to the dry air between the inlet and exit of the cooling and dehumidifying section:

$$\dot{m}_{a3} = \dot{m}_{a1} = \dot{m}_a$$

Thus the mass flow rate of dry air is not affected by this process.

Next the conservation-of-mass equation is applied to the water, with the result

$$\omega_3\dot{m}_{a3} + \dot{m}_{w2} = \omega_1\dot{m}_{a1}$$

or

$$\dot{m}_{w2} = (\omega_1 - \omega_3)\dot{m}_a$$

The mass flow rate of condensate ($\dot{m}_{w2}$) is therefore equal to the product of the mass flow rate of dry air and the change in the humidity ratio that occurs during the process. Applying the conservation-of-energy equation for negligible changes in kinetic and potential energies yields

$$\dot{Q} = \dot{m}_{a3}h_3 + \dot{m}_{w2}h_{w2} - \dot{m}_{a1}h_1$$

since no work is done during the process. Eliminating $\dot{m}_{w2}$, using the results from the conservation-of-mass equation, gives

$$\dot{Q} = \dot{m}_a(h_3 - h_1) + \dot{m}_a(\omega_1 - \omega_3)h_{w2}$$

The last term in this expression represents the amount of energy removed from the device due to the flow of condensate. This term is usually much smaller in magnitude than the first term on the right side and is therefore often neglected in engineering calculations.

The temperature of the air leaving the cooling and dehumidifying section is low, and the humidity ratio is relatively low. This low-temperature moist air is generally not used directly for comfort-cooling applications. The air can be passed through a heating section to reheat it to a more desirable temperature, or it can be mixed with a warm-air stream to obtain the design conditions required for specific applications.

■ EXAMPLE 11.11

A steady-flow cooling and dehumidification process with a reheat section is employed to deliver moist air at a dry bulb temperature of 25°C and a relative humidity of 50 percent. The air enters the cooling section at a dry bulb temperature of 40°C with a relative humidity of 40 percent at a rate of 75 m^3/min. Determine the heat-transfer rate required in the cooling and dehumidification section and the heat-transfer rate required in the reheat section. The total pressure is 1 atm.

Solution. A sketch of the system and a process diagram are shown in Figure 11.8. The reheat section is used to simply heat the air to a more comfortable temperature after it has been cooled and dehumidified.

We first analyze the cooling and dehumidification process. The conservation-of-mass equation for the dry air yields

$$\dot{m}_{a1} = \dot{m}_{a2} = \dot{m}_a$$

and for the water gives

$$\omega_1 \dot{m}_{a1} = \omega_2 \dot{m}_{a2} + \dot{m}_w$$

or

$$\dot{m}_w = \dot{m}_a(\omega_1 - \omega_2)$$

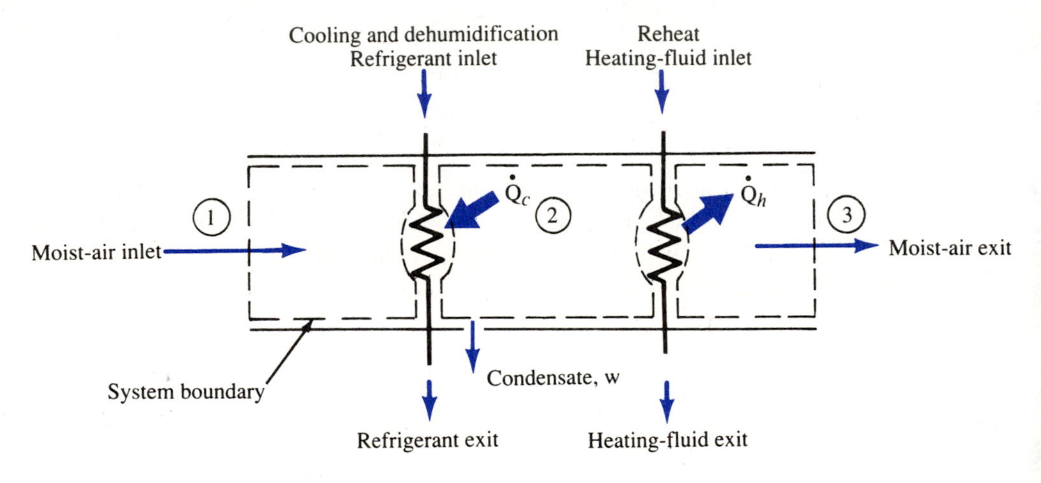

(a) Cooling and Dehumidification with Reheat

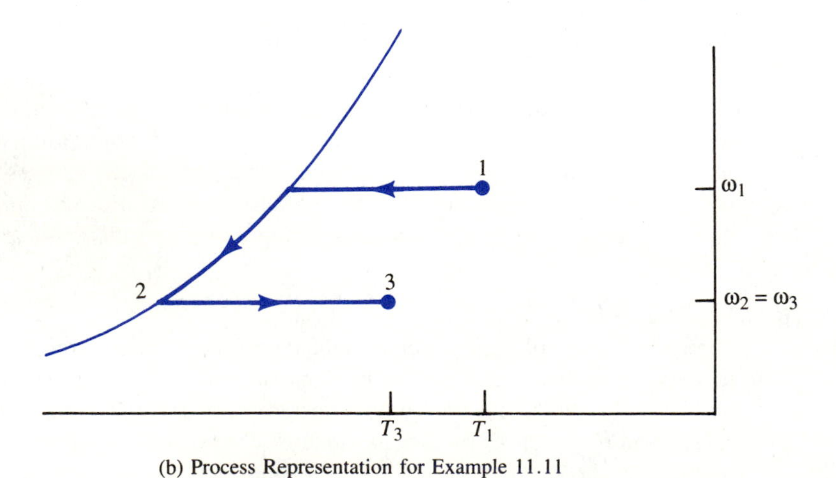

(b) Process Representation for Example 11.11

Figure 11.8 Sketches for Example 11.11.

where $\dot{m}_w$ is the mass flow rate of condensate leaving the cooling and dehumidification section.

Applying the conservation-of-energy equation, we obtain

$$\dot{Q}_c = \dot{m}_{a2}h_2 + \dot{m}_w h_w - \dot{m}_{a1}h_1$$
$$= \dot{m}_a[(h_2 - h_1) + (\omega_1 - \omega_2)h_w]$$

where kinetic- and potential-energy changes have been neglected and the results from the conservation-of-mass equations have been substituted. The mixture enthalpies and the humidity ratios are obtained from the psychrometric chart, and h_w is the enthalpy of saturated liquid water at the temperature corresponding to state 2 (14°C). Note that state 2 is fixed by the fact that it has the same humidity ratio as state 3 (process 2-3 is simple heating), and it lies on the saturation curve since it is part of the dehumidification process.

The mass flow rate of dry air is determined from the volumetric flow rate and the specific volume, which is obtained from the psychrometric chart, Figure F.1:

$$\dot{m}_{a1} = \frac{A_1 V_1}{v_1} = \frac{75 \text{ m}^3/\text{min}}{0.91 \text{ m}^3/\text{kg dry air}} = 82.4 \text{ kg dry air/min}$$

The heat-transfer rate in the cooling section can therefore be calculated as follows:

$$\dot{Q}_c = (82.4 \text{ kg dry air/min})[(39 - 88.1) \text{ kJ/kg dry air}$$
$$+ (0.0187 - 0.0099) \text{ kg H}_2\text{O/kg dry air } (58.8 \text{ kJ/kg H}_2\text{O})]$$
$$= \underline{\underline{-4000 \text{ kJ/min}}}$$

For the heating section, application of the conservation-of-mass equation to the dry air and water vapor, in turn, yields

$$\dot{m}_{a2} = \dot{m}_{a3} = \dot{m}_a$$

and

$$\omega_2 \dot{m}_{a2} = \omega_3 \dot{m}_{a3}$$

or

$$\omega_2 = \omega_3$$

In other words, neither the mass flow rate of dry air nor the amount of moisture changes as the airstream passes through the heating section. Next, the conservation-of-energy equation for negligible changes in kinetic and potential energies gives

$$\dot{Q}_h = \dot{m}_{a3}h_3 - \dot{m}_{a2}h_2 = \dot{m}_a(h_3 - h_2)$$

The mass flow rate of dry air is the same as that calculated previously for the cooling section, and h_3 can be obtained directly from the psychrometric chart. Therefore, the heat-transfer rate in the heating section is

$$\dot{Q}_h = (82.4 \text{ kg dry air/min})(50.2 - 39) \text{ kJ/kg dry air}$$
$$= \underline{\underline{923 \text{ kJ/min}}}$$

The magnitude of the heat-transfer rate for the cooling and dehumidification process is significantly larger than that for the heating section because dehumidification requires additional energy to condense the water vapor. ■

11.7.3 Heating with Humidification

As mentioned previously, heating cold, dry air can produce warm air with a very low relative humidity, and therefore addition of moisture to the air in order to achieve the desired combination of dry bulb temperature and relative humidity is often necessary. Adding moisture to an airstream is called *humidification* and can be achieved in a variety of ways. One example of humidification is the *adiabatic-saturation* process, where air is passed over a pool of liquid water so that evaporation causes an increase in the amount of moisture in the air. Other humidification processes can also be produced by spraying water or steam into an airstream.

A schematic of a heating and humidification section is shown in Figure 11.9(a). The air is first passed over a heat-exchanger surface so that simple heating occurs. This section is followed by a humidifier section, indicated schematically by the spray apparatus in the figure. The humidifier is considered to be adiabatic, and since the spray can range from relatively low temperature liquid-water droplets to higher-temperature steam, the dry bulb temperature of the air can decrease or increase, depending on the condition of the water being added to the air. The path for a typical heating and humidification process, as it would appear on a psychrometric chart, is shown in Figure 11.9(b). The heating process, 1-2 in the figure, is represented by a horizontal line because the humidity ratio of the air is unchanged. The humdification-process curve, 2-3 in this figure, slopes up and to the right, indicating that the dry bulb temperature as well as the humidity ratio is increased during the process. The slope of the line for path 2-3 is controlled by the amount and condition of the water added in the humidifier section. Depending on the amount of water added and its state, line 2-3 could range from one that slopes up and to the left to one that is nearly horizontal in an extreme case. In many applications of heating and humidifying, the temperature of the water added does not differ significantly from the dry bulb temperature of the airstream.

The analysis of the heating and humidification process is quite similar to that for cooling and dehumidification. The conservation-of-mass equation for steady flow applied to the dry air passing through the heating and humidifying sections yields

$$\dot{m}_{a3} = \dot{m}_{a1} = \dot{m}_a$$

and applied to the water passing through these same two sections results in

$$\omega_3 \dot{m}_{a3} = \omega_1 \dot{m}_{a3} + \dot{m}_s$$

Thus

$$\dot{m}_s = (\omega_3 - \omega_1)\dot{m}_a$$

In other words, the mass flow rate of water added in the humidifier section, $\dot{m}_s$, is equal to the product of the mass flow rate of dry air and the increase in the humidity ratio of the air.

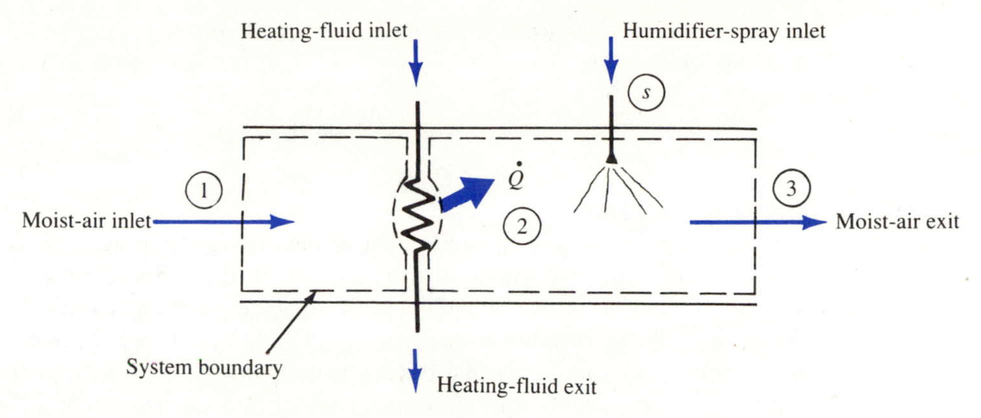

(a) Schematic of Heating and Humidification Section

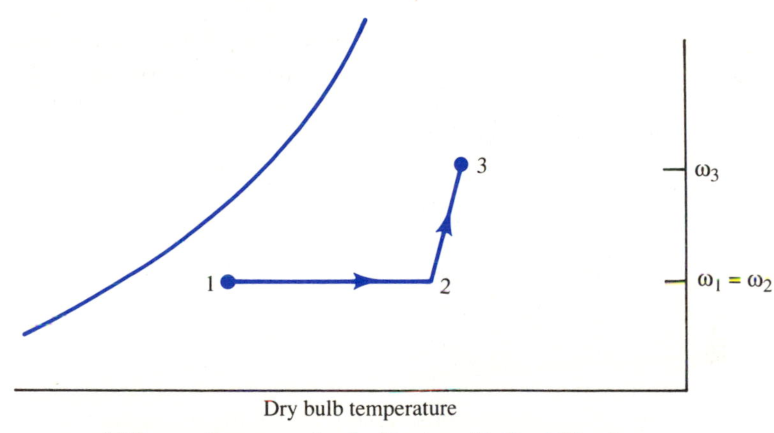

(b) Process Representation for Heating with Humidification

Figure 11.9 Heating and humidification.

Applying the conservation-of-energy equation for steady flow with negligible changes in kinetic and potential energies to the heating section alone gives

$$\dot{Q} = \dot{m}_{a2}h_2 - \dot{m}_{a1}h_1 = \dot{m}_a(h_2 - h_1)$$

That is, the heat-transfer rate to the air in the heating section is equal to the product of the mass flow rate of dry air and the increase in the enthalpy of the airstream.

Next, application of the conservation of energy to the adiabatic-humidifier section yields

$$\dot{m}_s h_s + \dot{m}_{a2}h_2 - \dot{m}_{a3}h_3 = 0$$

$$\dot{m}_s h_s + \dot{m}_a(h_2 - h_3) = 0$$

This expression can be used to determine the enthalpy of the water added to the air in the humidifier section,

$$h_s = \frac{\dot{m}_a}{\dot{m}_s}(h_3 - h_2)$$

EXAMPLE 11.12

A steady-flow heating and humidification process is used to provide moist air at a dry bulb temperature of 77°F with a relative humidity of 45 percent. Outdoor air at 40°F dry bulb and 90 percent relative humidity enters the heating section at a rate of 2100 ft^3/min and its dry bulb temperature is increased to 75°F. Steam is then sprayed into the air in the humidifier section so that the desired exit conditions are achieved. Determine the heat-transfer rate required in the heating section. Suppose the steam added in the humidifier section is at a pressure of 1 atm. Determine the temperature of the steam. Assume that the processes occur at a constant total pressure of 1 atm.

Solution. The system and the process diagram for this example are similar to those shown in Figure 11.9. We begin the analysis by first considering the air in the heating section to be the system and applying the conservation-of-mass equation to the dry air and the water vapor, in turn:

$$\dot{m}_{a1} = \dot{m}_{a2} = \dot{m}_a$$

and

$$\omega_1 \dot{m}_{a1} = \omega_2 \dot{m}_{a2}$$

or

$$\omega_1 = \omega_2$$

Also, the conservation-of-energy equation, for negligible changes in kinetic and potential energies, reduces to

$$\dot{Q}_h = \dot{m}_{a2}h_2 - \dot{m}_{a1}h_1 = \dot{m}_a(h_2 - h_1)$$

The mass flow rate of dry air is determined from the volumetric flow rate at state 1,

$$\dot{m}_a = \dot{m}_{a1} = \frac{A_1 V_1}{v_1}$$

with v_1 obtained from the psychrometric chart:

$$\dot{m}_a = \frac{2100 \text{ ft}^3/\text{min}}{12.67 \text{ ft}^3/\text{lb}_\text{m} \text{ dry air}} = 165.7 \text{ lb}_\text{m} \text{ dry air/min}$$

State 2 is fixed by the given dry bulb temperature along with the fact that ω_2 is the same as ω_1. Using enthalpy values for states 1 and 2 from the psychrometric chart, we find

$$\dot{Q}_h = (165.7 \text{ lb}_\text{m} \text{ dry air/min})(23.0 - 14.7)\text{Btu/lb}_\text{m} \text{ dry air}$$

$$= \underline{\underline{1375 \text{ Btu/min}}}$$

If we now choose the air passing through the humidifier section as the system and apply the conservation-of-mass equation to the dry air and the water, we obtain

$$\dot{m}_{a2} = \dot{m}_{a3} = \dot{m}_a$$

and

$$\omega_2 \dot{m}_{a2} + \dot{m}_s = \omega_3 \dot{m}_{a3}$$

or

$$\dot{m}_s = (\omega_3 - \omega_2)\dot{m}_a$$

Thus with values for the humidity ratios from the psychrometric chart, we can calculate the steam mass flow rate into the humidifier:

$$\begin{aligned}\dot{m}_s &= (8.9 \times 10^{-3} - 4.7 \times 10^{-3})\text{lb}_\text{m}\,\text{H}_2\text{O/lb}_\text{m}\text{ dry air }(165.7\text{ lb}_\text{m}\text{ dry air/min})\\ &= 0.696\text{ lb}_\text{m}\,\text{H}_2\text{O/min}\end{aligned}$$

Applying the conservation-of-energy equation to the adiabatic humidifier, we obtain

$$\dot{m}_s h_s + \dot{m}_{a2} h_2 - \dot{m}_{a3} h_3 = 0$$

or

$$h_s = \frac{\dot{m}_a}{\dot{m}_s}(h_3 - h_2)$$

Therefore, the enthalpy of the steam added to the humidifier is

$$\begin{aligned}h_s &= \left(\frac{165.7\text{ lb}_\text{m}\text{ dry air/min}}{0.696\text{ lb}_\text{m}\,\text{H}_2\text{O/min}}\right)(28.3 - 23.1)\text{Btu/lb}_\text{m}\text{ dry air}\\ &= 1238\text{ Btu/lb}_\text{m}\end{aligned}$$

The temperature of the steam can be found from the superheated-steam tables at a pressure of 1 atm and an enthalpy of 1238 Btu/lb$_\text{m}$. By interpolation we find that for this process the steam temperature would be

$$T_s \simeq \underline{\underline{396°\text{F}}}$$

■

11.7.4 Evaporative Cooling

Since energy is required to evaporate water, a cooling process based on evaporation, called *evaporative cooling*, can be used to reduce the dry bulb temperature of air without the need for a traditional air-conditioning system. However, evaporative cooling causes an increase in the relative humidity of an airstream, so that it is most effective when the air to be cooled is warm and has a low relative humidity. These conditions are often encountered in locations such as the southwestern United States where evaporative cooling is both popular and economical.

The evaporative-cooling process is shown schematically in Figure 11.10(a), and a process diagram, as it would appear on a psychrometric chart, is depicted in Figure 11.10(b). Warm, dry air enters the evaporative cooler, where it passes through a spray

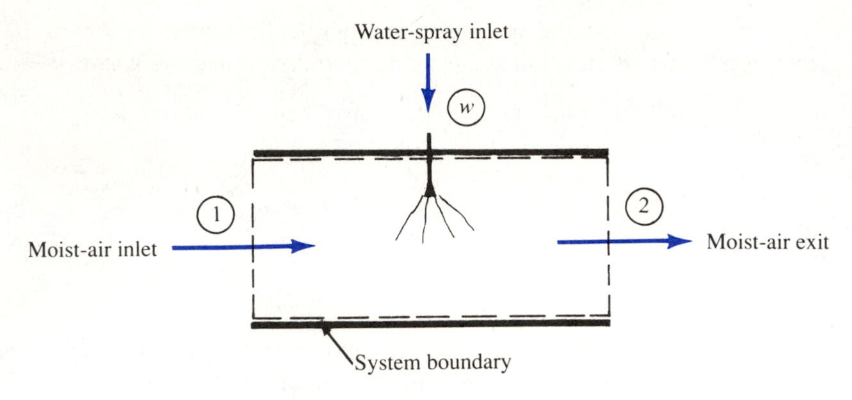

(a) Schematic of Evaporative-Cooling Section

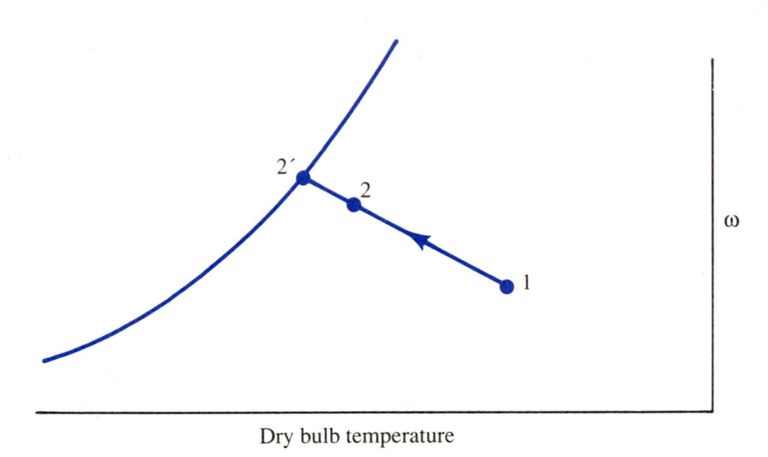

(b) Process Representation for Evaporative Cooling

Figure 11.10 Evaporative cooling.

of liquid water. Energy extracted from the air to evaporate the water causes the dry bulb temperature of the air to decrease. As shown in the process diagram, the humidity ratio of the air increases as the water evaporates and mixes with the airstream. The evaporative-cooling section is considered to be adiabatic because the heat transfer to the surroundings is usually negligible.

In the analysis of the evaporative-cooling process, the conservation-of-mass equation for steady flow is first applied to the dry air,

$$\dot{m}_{a1} = \dot{m}_{a2} = \dot{m}_a$$

and then to the water,

$$\omega_1 \dot{m}_{a1} + \dot{m}_w = \omega_2 \dot{m}_{a2}$$

or

$$\dot{m}_w = \dot{m}_a(\omega_2 - \omega_1)$$

This last expression is identical to that obtained for a humidifier. That is, the mass flow rate of water added to the evaporative cooler is equal to the product of the mass flow rate of the dry air and the increase in the humidity ratio of the airstream.

Similarly, applying the conservation-of-energy equation to the steady-flow, adiabatic evaporative cooler gives

$$\dot{m}_{a1}h_1 + \dot{m}_w h_w - \dot{m}_{a2}h_2 = 0$$

or

$$\dot{m}_w h_w = \dot{m}_a(h_2 - h_1)$$

This expression is also identical in form to the one obtained for a humidifier. An important difference, however, is that the water added to the evaporative cooler is a liquid with a significantly lower enthalpy than the enthalpy of the steam sprayed into the humidifier. For this reason the product of $\dot{m}_w$ and h_w in the evaporative cooling process is quite small, and therefore the enthalpy of the entering airstream and the enthalpy of the leaving airstream differ only by a small amount. Because of this fact, the evaporative-cooling process is usually considered to be one for which the enthalpy of the moist airstream remains unchanged. This fact indicates that the wet bulb temperature is approximately constant throughout the process. The assumption that the evaporative-cooling process is one of constant enthalpy or constant wet bulb temperature is reasonably accurate and quite acceptable for most engineering calculations.

■ EXAMPLE 11.13

Air at a total pressure of 1 atm has a dry bulb temperature of 35°C and a relative humidity of 15 percent and is passed through an evaporative cooler. Determine the minimum dry bulb temperature that could be attained with this process. Suppose that the air leaves the evaporative cooler with a dry bulb temperature of 20°C. Determine the relative humidity of the air. Determine the percent change in the velocity of the airstream assuming that the cross-sectional area of the evaporative cooler is constant.

Solution. The evaporative cooler and the corresponding process diagram are illustrated in Figure 11.10. Applying the conservation-of-mass equation for steady flow to the dry air and then to the water in the evaporative cooler, we obtain

$$\dot{m}_{a1} = \dot{m}_{a2} = \dot{m}_a$$

and

$$\omega_1 \dot{m}_{a1} + \dot{m}_w = \omega_2 \dot{m}_{a2}$$

or

$$\dot{m}_w = \dot{m}_a(\omega_2 - \omega_1)$$

Since the magnitude of $(\omega_2 - \omega_1)$ is very small and less than unity, we can conclude that

$$\dot{m}_w << \dot{m}_a$$

Next we apply the conservation-of-energy equation to the adiabatic evaporative cooler to obtain

$$\dot{m}_{a1}h_1 + \dot{m}_w h_w - \dot{m}_{a2}h_2 = 0$$

Since the enthalpy of the liquid water, h_w, is of the same order of magnitude as h_1, and since $\dot{m}_w << \dot{m}_{a1}$ then

$$\dot{m}_w h_w << \dot{m}_{a1}h_1$$

Thus we conclude from the conservation-of-energy equation that h_1 and h_2 are approximately equal:

$$h_1 \simeq h_2$$

This result indicates that the evaporative-cooling process is one for which the enthalpy of the moist-air stream is essentially constant.

The minimum dry bulb temperature for evaporative cooling can be found from the intersection of the saturation curve and a constant-enthalpy line passing through state 1. This state is state 2′ in Figure 11.10(b):

$$T_{2'} = T_{db,\min} \simeq \underline{\underline{17°\text{C}}}$$

If the evaporative-cooling process continues until state 2′ is reached, the air will be saturated and the corresponding dry bulb temperature of the air will be approximately equal to the wet bulb temperature of the initial state.

If the process continues only until the dry bulb temperature reaches 20°C, the corresponding relative humidity, found from the intersection of the 20°C dry bulb line and the constant-enthalpy line, is

$$\phi_2 \simeq \underline{\underline{0.76 \text{ or } 76 \text{ percent}}}$$

The relationship between the velocities at the inlet and exit can be developed from the conservation of mass for the dry air:

$$\dot{m}_{a1} = \dot{m}_{a2}$$

Therefore

$$\frac{A_1 V_1}{v_1} = \frac{A_2 V_2}{v_2}$$

If the cross-section area does not change, then

$$\frac{V_2}{V_1} = \frac{v_2}{v_1}$$

Since ideal-gas behavior is assumed and since $v = v_a$, then

$$\frac{V_2}{V_1} = \frac{v_2}{v_1} = \frac{v_{a2}}{v_{a1}}$$

The specific volumes are determined from the psychrometric chart:

$$v_{a1} \simeq 0.88 \text{ m}^3/\text{kg dry air}$$

$$v_{a2} \simeq 0.845 \text{ m}^3/\text{kg dry air}$$

Therefore

$$\frac{V_2}{V_1} = \frac{0.845}{0.88} = 0.96$$

We conclude that the velocity decreases by 4 percent as a result of the evaporative-cooling process. ■

11.7.5 Adiabatic Mixing

A number of air-conditioning applications require that two airstreams be mixed. Mixing processes are particularly important when design standards dictate that outdoor air be mixed with recirculated and reconditioned air to meet minimum fresh-air requirements for the occupants. The mixing is usually achieved by simply merging two airstreams into a single stream, as indicated by the duct system shown schematically in Figure 11.11(a). The airstreams are assumed to have different dry bulb temperatures and humidity ratios, as indicated in the figure. The heat transfer to the surroundings during the process is small, so that the process is termed *adiabatic mixing*. The mixture airstream leaves with a dry bulb temperature and a humidity ratio that are between the corresponding values for the entering airstreams.

The conditions of the mixture airstream can be determined by applying the conservation-of-mass and -energy equations to the adiabatic-mixing section. The conservation-of-mass equation for the dry air requires that

$$\dot{m}_{a1} + \dot{m}_{a2} = \dot{m}_{a3}$$

and for the water

$$\omega_1 \dot{m}_{a1} + \omega_2 \dot{m}_{a2} = \omega_3 \dot{m}_{a3}$$

Combining these two equations to eliminate $\dot{m}_{a3}$ gives

$$\omega_3 = \frac{\omega_1 \dot{m}_{a1} + \omega_2 \dot{m}_{a2}}{\dot{m}_{a1} + \dot{m}_{a2}}$$

An alternative relationship that has a convenient geometrical interpretation can be obtained by combining the two conservation-of-mass equations to eliminate $\dot{m}_{a2}$ (or $\dot{m}_{a1}$) instead,

$$\frac{\dot{m}_{a1}}{\dot{m}_{a3}} = \frac{\omega_3 - \omega_2}{\omega_1 - \omega_2}$$

This equation shows that the ratio of the vertical distance between ω_3 and ω_2 on the psychrometric chart and the vertical distance between ω_1 and ω_2 is the same as the ratio of $\dot{m}_{a1}$ to $\dot{m}_{a3}$. Thus state 3 must lie on the dashed horizontal line indicated in Figure 11.11(b). The location of this line is determined from

$$\omega_3 - \omega_2 = (\omega_1 - \omega_2)\frac{\dot{m}_{a1}}{\dot{m}_{a3}}$$

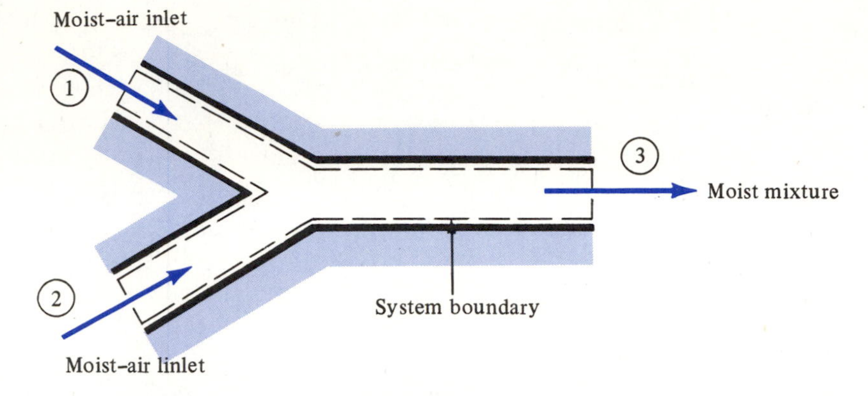

(a) Schematic of Adiabatic–Mixing Section

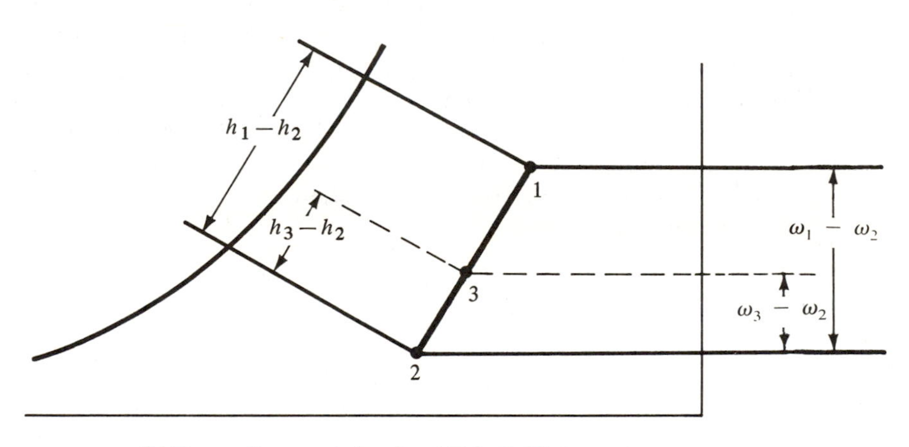

(b) Process Representation for Adiabatic Mixing

Figure 11.11 Adiabatic mixing.

The location of state 3 can be fixed precisely after the conservation-of-energy equation is applied to the mixing process to give

$$\dot{m}_{a1}h_1 + \dot{m}_{a2}h_2 - \dot{m}_{a3}h_3 = 0$$

Substituting the results from the conservation of mass for the dry air to eliminate $\dot{m}_{a2}$ gives, after rearrangement,

$$\frac{\dot{m}_{a1}}{\dot{m}_{a3}} = \frac{h_3 - h_2}{h_1 - h_2}$$

This equation shows that the ratio of the distance between h_3 and h_2 along the enthalpy scale on the psychrometric chart and the distance between h_1 and h_2 along the same scale is equal to the ratio of $\dot{m}_{a1}$ to $\dot{m}_{a3}$. Thus state 3 must also lie on the slanted dashed line in Figure 11.11(b). The intersection of the two dashed lines in this figure locates state 3 on the psychrometric chart. As can be verified from further geometrical considerations, the results of this analysis lead to the conclusion that the mixture state (state 3) lies on a straight line connecting states 1 and 2 on the psychrometric chart, and that the ratio of the distance between state 3 and state 2 on this line and the distance between state 1 and state 2 is the same as the ratio of $\dot{m}_{a1}$ to $\dot{m}_{a3}$. Thus, for example, if $\dot{m}_{a1}$ is much smaller than $\dot{m}_{a2}$, then state 3 will be very nearly the same as state 2 (i.e., the mixture is composed almost entirely of air that was initially at state 2).

Under certain conditions condensation can occur when two airstreams are mixed adiabatically. For example, if states 1 and 2 in Figure 11.11(b) are close to the saturation curve, the analysis presented above would predict that state 3 would lie on a straight line connecting states 1 and 2 and would be to the left of the saturation curve. This result means that condensation occurs as a result of mixing, and the analysis should be modified accordingly. With steady-flow conditions assumed, the condensate is assumed to leave at temperature T_3, and state 3 is assumed to be saturated. Since the energy leaving the system as a result of the condensate flow is small, the enthalpy at state 3 is calculated as described above, and the mass flow rate of condensate can be calculated from the conservation-of-mass equation applied to the water.

■ EXAMPLE 11.14

Moist air with a dry bulb temperature of 60°F and a relative humidity of 20 percent is mixed adiabatically at a rate of 1600 ft³/min with a second airstream flowing at a rate of 2450 ft³/min with a dry bulb temperature of 77°F and a relative humidity of 40 percent. Determine the volumetric flow rate, the dry bulb temperature, and the relative humidity of the mixture airstream, and calculate the irreversibility for this process. Assume that the process occurs at a total pressure of 1 atm and that the surroundings are at a pressure of 1 atm and a temperature of 77°F.

Solution. The system and the process diagram are shown in Figure 11.11. The properties at state 3 can be determined by applying the conservation-of-mass and -energy equations for steady flow. For the air

$$\dot{m}_{a1} + \dot{m}_{a2} = \dot{m}_{a3}$$

and for the water vapor

$$\dot{m}_{v1} + \dot{m}_{v2} = \dot{m}_{v3}$$

or

$$\omega_1 \dot{m}_{a1} + \omega_2 \dot{m}_{a2} = \omega_3 \dot{m}_{a3}$$

Combining these two equations, we find

$$\omega_3 = \frac{\omega_1 \dot{m}_{a1} + \omega_2 \dot{m}_{a2}}{\dot{m}_{a3}}$$

The humidity ratios and the specific volumes at states 1 and 2 can be determined from the psychrometric chart, Figure F.1E:

$$\omega_1 = 2.2 \times 10^{-3}\,\mathrm{lb_m\,H_2O/lb_m\ dry\ air}$$

$$\omega_2 = 7.9 \times 10^{-3}\,\mathrm{lb_m\,H_2O/lb_m\ dry\ air}$$

$$v_1 = 13.1\ \mathrm{ft^3/lb_m\,dry\ air} \qquad v_2 = 13.7\ \mathrm{ft^3/lb_m\ dry\ air}$$

The mass flow rates are calculated next:

$$\dot{m}_a = \frac{AV}{v}$$

Thus

$$\dot{m}_{a1} = \frac{(AV)_1}{v_1} = \frac{1600\,\mathrm{ft^3/min}}{13.1\ \mathrm{ft^3/lb_m\ dry\ air}} = 122\ \mathrm{lb_m\ dry\,air/min}$$

and

$$\dot{m}_{a2} = \frac{(AV)_2}{v_2} = \frac{2450\,\mathrm{ft^3/min}}{13.7\ \mathrm{ft^3/lb_m\ dry\ air}} = 179\ \mathrm{lb_m\ dry\,air/min}$$

and the mass flow rate of the mixture is

$$\dot{m}_{a3} = \dot{m}_{a1} + \dot{m}_{a2} = (122 + 179)\mathrm{lb_m\ dry\ air/min}$$
$$= 301\ \mathrm{lb_m\ dry\ air/min}$$

The humidity ratio at state 3 is then

$$\omega_3 = \frac{(2.2 \times 10^{-3}\ \mathrm{lb_m\ H_2O/lb_m\ dry\ air})(122\,\mathrm{lb_m\ dry\ air/min})}{(301\ \mathrm{lb_m\ dry\ air/min})}$$
$$+ \frac{(7.9 \times 10^{-3}\ \mathrm{lb_m\ H_2O/lb_m\ dry\ air})(179\ \mathrm{lb_m\ dry\,air/min})}{(301\ \mathrm{lb_m\,dry\ air/min})}$$
$$= 5.6 \times 10^{-3}\ \mathrm{lb_m\ H_2O/lb_m\ dry\,air}$$

The enthalpy at state 3 is determined from the conservation-of-energy equation for steady flow. Neglecting kinetic- and potential-energy changes and noting that work and heat transfer are absent, we obtain

$$\dot{m}_{a1}h_1 + \dot{m}_{a2}h_2 - \dot{m}_{a3}h_3 = 0$$

or

$$h_3 = \frac{\dot{m}_{a1}h_1 + \dot{m}_{a2}h_2}{\dot{m}_{a3}}$$

Using enthalpy values from Equation F.1E, we have

$$h_3 = \frac{(122 \text{ lb}_m \text{ dry air/min})(16.8 \text{ Btu/lb}_m \text{ dry air})}{(301 \text{ lb}_m \text{ dry air/min})}$$
$$+ \frac{(179 \text{ lb}_m \text{ dry air/min})(27.2 \text{ Btu/lb}_m \text{ dry air})}{(301 \text{ lb}_m \text{ dry air/min})}$$
$$= 23.0 \text{ Btu/lb}_m \text{ dry air}$$

The two properties, ω_3 and h_3, are sufficient to locate state 3 on the psychrometric chart. At this state the following properties are determined from the chart:

$$v_3 = 13.5 \text{ ft}^3/\text{lb}_m \text{ dry air}$$
$$T_{3,db} = \underline{\underline{70.5°\text{F}}}$$
$$\phi_3 = \underline{\underline{36 \text{ percent}}}$$

The volumetric flow rate at state 3 is therefore

$$(AV)_3 = \dot{m}_{a3}v_3 = (301 \text{ lb}_m \text{ dry air/min})(13.5 \text{ ft}^3/\text{lb}_m \text{ dry air})$$
$$= \underline{\underline{4064 \text{ ft}^3/\text{min}}}$$

The irreversibility rate for this process can be evaluated by using Equation 6.47 for steady flow:

$$\dot{I} = T_0 \left(\frac{dS}{dt}\right)_{\text{tot}} = T_0 \left(\dot{m}_{a3}s_3 - \dot{m}_{a2}s_2 - \dot{m}_{a1}s_1 + \sum_k \frac{\dot{Q}_k}{T_k}\right)$$

Since the process is adiabatic, the last term in this expression vanishes (the net rate of change of entropy of the surroundings is zero). Thus

$$\dot{I} = T_0 \left(\dot{m}_{a3}s_3 - \dot{m}_{a2}s_2 - \dot{m}_{a1}s_1\right)$$

The entropy values can be calculated from Equation 11.64:

$$s_m = c_{p,a} \ln\left(\frac{T}{T_{a,\text{ref}}}\right) - R_a \ln\left(\frac{p_a}{P_{a,\text{ref}}}\right)$$
$$+ \omega\left[c_{p,v} \ln\left(\frac{T}{T_{v,\text{ref}}}\right) - R_v \ln\left(\frac{p_v}{P_{v,\text{ref}}}\right) + s_{v,\text{ref}}\right]$$

The partial pressures are determined from Equations 11.48 and 11.49

$$p_{v1} = \phi_1 p_{g1} = 0.2(0.2564 \text{ psia}) = 0.05128 \text{ psia}$$
$$p_{a1} = (14.7 - 0.05)\text{psia} = 14.65 \text{ psia}$$
$$p_{v2} = 0.4(0.464 \text{ psia}) = 0.186 \text{ psia}$$
$$p_{a2} = (14.7 - 0.19)\text{psia} = 14.5 \text{ psia}$$
$$p_{v3} = 0.36(0.37 \text{ psia}) = 0.133 \text{ psia}$$
$$p_{a3} = (14.7 - 0.13)\text{psia} = 14.57 \text{ psia}$$

Thus the entropy of the mixture at each state is

$$s_1 = \left(0.24 \frac{\text{Btu}}{\text{lb}_\text{m}\text{ dry air}\cdot°\text{R}}\right) \ln\left(\frac{519.67°\text{R}}{459.67°\text{R}}\right) - \left(0.0686 \frac{\text{Btu}}{\text{lb}_\text{m}\text{ dry air}\cdot°\text{R}}\right)$$

$$\times \ln\left(\frac{14.65\text{ psia}}{14.696\text{ psia}}\right) + \left(2.2 \times 10^{-3} \frac{\text{lb}_\text{m}\text{ H}_2\text{O}}{\text{lb}_\text{m}\text{ dry air}}\right)$$

$$\times \left[\left(0.435 \frac{\text{Btu}}{\text{lb}_\text{m}\text{ H}_2\text{O}\cdot°\text{R}}\right) \ln\left(\frac{519.67°\text{R}}{491.69°\text{R}}\right) - \left(0.1102 \frac{\text{Btu}}{\text{lb}_\text{m}\text{ dry air}\cdot°\text{R}}\right)\right.$$

$$\left.\times \ln\left(\frac{0.05128\text{ psia}}{0.08872\text{ psia}}\right) + 2.1864 \frac{\text{Btu}}{\text{lb}_\text{m}\text{ dry air}\cdot°\text{R}}\right)\right]$$

$$= 0.0347\text{ Btu/lb}_\text{m}\text{ dry air}\cdot°\text{R}$$

$$s_2 = (0.24) \ln\left(\frac{536.67}{459.67}\right) - (0.0686) \ln\left(\frac{14.52}{14.696}\right)$$

$$+ (7.9 \times 10^{-3}) \left[(0.435) \ln\left(\frac{536.67}{491.69}\right) - (0.1102) \ln\left(\frac{0.186}{0.08872}\right) + 2.1864\right]$$

$$= 0.0549\text{ Btu/lb}_\text{m}\text{ dry air}\cdot°\text{R}$$

$$s_3 = (0.24) \ln\left(\frac{530.17}{459.67}\right) - (0.0686) \ln\left(\frac{14.57}{14.696}\right)$$

$$+ (5.6 \times 10^{-3}) \left[(0.435) \ln\left(\frac{530.17}{491.69}\right) - (0.1102) \ln\left(\frac{0.133}{0.08872}\right) + 2.1864\right]$$

$$= 0.047\text{ Btu/lb}_\text{m}\text{ dry air}\cdot°\text{R}$$

The irreversibility rate for the process is therefore

$$\dot{I} = T_0(\dot{m}_{a3}s_3 - \dot{m}_{a2}s_2 - \dot{m}_{a1}s_1)$$

$$= (537°\text{R})[(301\text{ lb}_\text{m}\text{ dry air/min})(0.047\text{ Btu/lb}_\text{m}\text{ dry air}\cdot°\text{R})$$

$$-(179)(0.0549) - (122)(0.0347)]$$

$$= \underline{\underline{46.5\text{ Btu/min}}}$$

The irreversibility rate is positive because mixing processes are irreversible. ■

11.7.6 Wet-Cooling Tower

Water is often used as a cooling fluid in power-plant condensers as well as in a variety of other industrial applications. In many instances this water must be recirculated and reused because the water supply is scarce or because thermal pollution might result if the warm water is directed to nearby water sources. However, the water to be recirculated must be cooled before it can be reused. A wet-cooling tower is one means of providing the necessary cooling.

A schematic of a wet-cooling tower is shown in Figure 11.12. The principle of operation is essentially the same as that described previously for the evaporative cooling process. Warm water, from a condenser, is sprayed into the top of the tower so that water droplets come into contact with unsaturated atmospheric air. The housing of the cooling tower is open at the top and the bottom. Air is drawn into the bottom and leaves at the top either by natural draft or by forced draft (a fan).

The falling water droplets result in a large surface area of water being exposed to the air. The water is cooled by means of evaporation, which causes the temperature and moisture content of the air to increase. The air is assumed to leave the cooling tower in a saturated state, and makeup water must be provided to replace the water that is evaporated and carried off by the airstream. The cooled water is collected in a reservoir at the bottom of the tower and is subsequently pumped back to the condenser. Two typical cooling tower designs are shown in Figure 11.13.

For the purpose of analysis, a system boundary is drawn around the cooling tower, which operates in steady flow, as shown in Figure 11.12. From the conservation of mass applied to the dry air,

$$\dot{m}_{a1} = \dot{m}_{a2} = \dot{m}_a$$

and for the water

$$\omega_1 \dot{m}_{a1} + \dot{m}_3 + \dot{m}_5 = \omega_2 \dot{m}_{a2} + \dot{m}_4$$

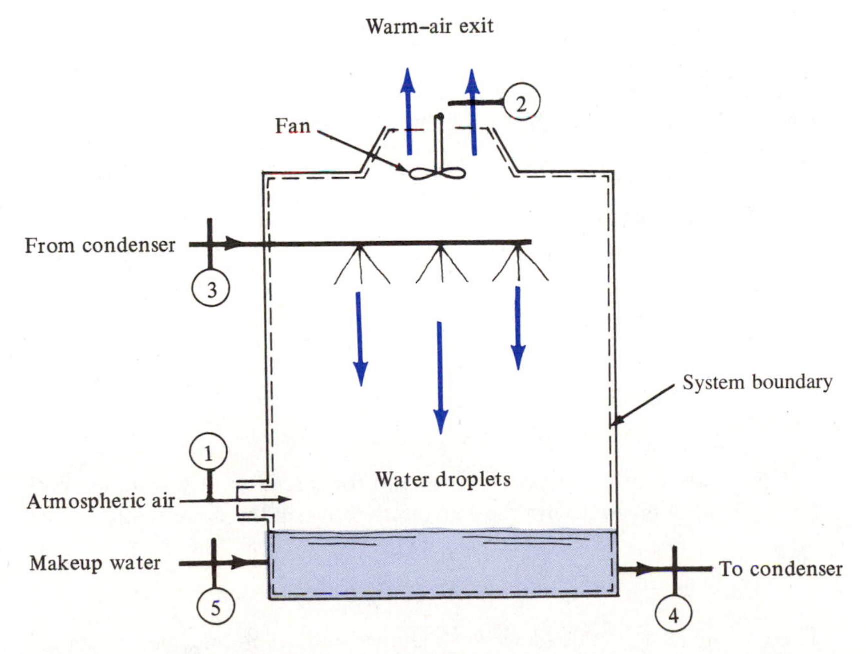

Figure 11.12 Schematic of a wet-cooling tower.

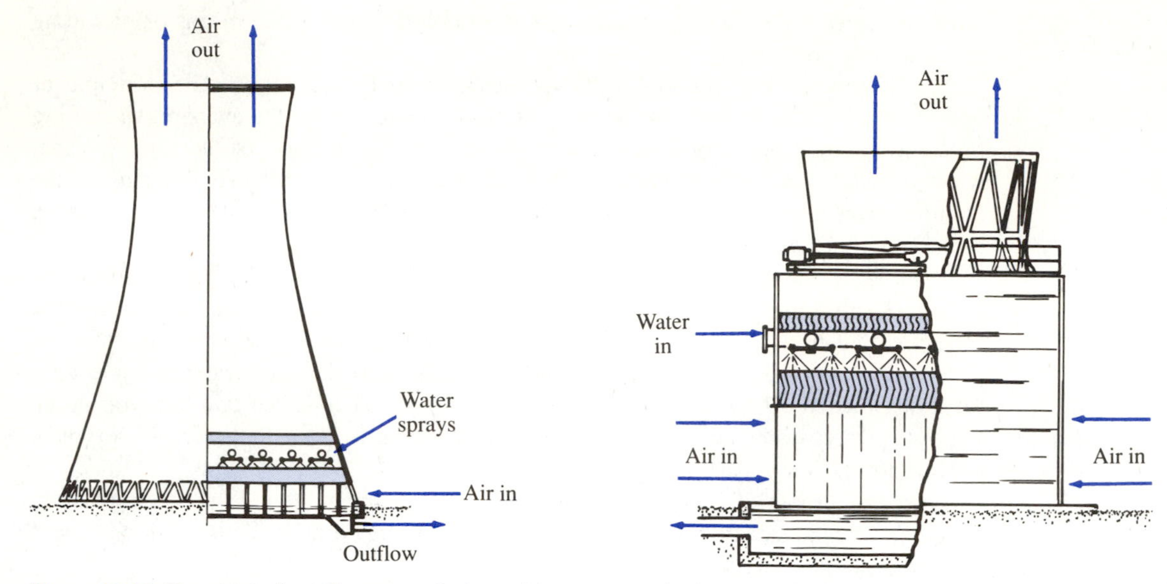

Figure 11.13 Two typical cooling tower designs. Air movement in the atmospheric, hyperbolic natural-draft tower shown on the left arises from density differences that exist between the hot air inside the tower and the more dense ambient air. These large cooling towers are often used as condensers in power plants where large concentrated heat-rejection loads are common. A fan is used to draw the air through the mechanical, induced-draft, counterflow tower on the right. (Source: *Cooling Tower Fundamentals*, The Marley Cooling Tower Co., 1985.)

Usually, however, the mass flow rate of the condenser water is the same entering and leaving the cooling tower,

$$\dot{m}_3 = \dot{m}_4 = \dot{m}_w$$

so that the conservation-of-mass equation for water can be used to relate the mass flow rate of makeup water to the change in the humidity ratio of the air as follows:

$$\dot{m}_5 = (\omega_2 - \omega_1)\dot{m}_a$$

The conservation-of-energy equation is applied next with the assumption that heat transfer to the surroundings is negligible and changes in kinetic and potential energies are also negligible:

$$-\dot{W} + \dot{m}_{a1}h_1 + \dot{m}_3h_3 + \dot{m}_5h_5 - \dot{m}_{a2}h_2 - \dot{m}_4h_4 = 0$$

or

$$-\dot{W} + \dot{m}_a(h_1 - h_2) + \dot{m}_w(h_3 - h_4) + \dot{m}_5h_5 = 0$$

For natural-draft towers $\dot{W} = 0$, and for forced-draft towers the power input to the fan is small compared with the remaining terms in the conservation-of-energy equation. Thus

$$\dot{m}_w(h_3 - h_4) + \dot{m}_a(h_1 - h_2) + \dot{m}_5h_5 = 0$$

The enthalpies of the condensate streams as well as the makeup-water streams are taken to be equal to the enthalpy of saturated liquid at the appropriate stream temperature.

EXAMPLE 11.15

Water from a power-plant condenser enters a forced-draft, wet-cooling tower at 45°C at a rate of 130 kg/s and is cooled to 30°C. Atmospheric air at 101.3 kPa with a dry bulb temperature of 25°C and a relative humidity of 50 percent enters the tower, and the air leaves in a saturated state at 32°C. The makeup water enters at 30°C. Determine the volumetric flow rate of air entering the cooling tower and the mass flow rate of makeup water. Neglect the power input to the fan.

Solution. The system is shown schematically in Figure 11.12. From the conservation of mass for the dry air,

$$\dot{m}_{a1} = \dot{m}_{a2} = \dot{m}_a$$

and for the water

$$\omega_1 \dot{m}_{a1} + \dot{m}_3 + \dot{m}_5 = \omega_2 \dot{m}_{a2} + \dot{m}_4$$

If we assume that the mass flow rate of condenser water is the same at the inlet and exit ($\dot{m}_3 = \dot{m}_4 = \dot{m}_w$), then

$$\dot{m}_5 = (\omega_2 - \omega_1)\dot{m}_a$$

The mass flow rate of makeup water ($\dot{m}_5$) and the mass flow rate of dry air are both unknown. Thus another equation is needed to relate these two quantities. This relationship can be developed by applying the conservation-of-energy equation:

$$\dot{m}_{a1}h_1 + \dot{m}_3 h_3 + \dot{m}_5 h_5 - \dot{m}_{a2}h_2 - \dot{m}_4 h_4 = 0$$

In this equation we have assumed negligible heat transfer to the surroundings and negligible changes in kinetic and potential energies, and we have neglected the power input to the fan.

Rearranging this last equation, we have

$$\dot{m}_w(h_3 - h_4) + \dot{m}_a(h_1 - h_2) + \dot{m}_5 h_5 = 0$$

and substituting the previous expression for the makeup water in order to eliminate $\dot{m}_5$ gives

$$\dot{m}_w(h_3 - h_4) + \dot{m}_a[h_1 - h_2 + (\omega_2 - \omega_1)h_5] = 0$$

Thus

$$\dot{m}_a = -\dot{m}_w \left[\frac{h_3 - h_4}{h_1 - h_2 + (\omega_2 - \omega_1)h_5} \right]$$

From the psychrometric chart, Figure F.1,

$$h_1 = 50.2 \text{ kJ/kg dry air} \qquad h_2 = 111.2 \text{ kJ/kg dry air}$$

$$\omega_1 = 0.00989 \text{ kg H}_2\text{O/kg dry air} \qquad \omega_2 = 0.0309 \text{ kg H}_2\text{O/kg dry air}$$

and from Table B.1

$$h_3 = 188.4 \text{ kJ/kg H}_2\text{O} \qquad h_4 = h_5 = 125.7 \text{ kJ/kg H}_2\text{O}$$

Therefore, the mass flow rate of dry air is

$$\dot{m}_a = -130\ \frac{\text{kg H}_2\text{O}}{s} \times \left[\frac{(188.4 - 125.7)\ \dfrac{\text{kJ}}{\text{kg H}_2\text{O}}}{(50.2 - 111.2)\ \dfrac{\text{kJ}}{\text{kg dry air}} + \left[(0.0309 - 0.00989)\ \dfrac{\text{kg H}_2\text{O}}{\text{kg dry air}}\right]\left(125.7\ \dfrac{\text{kJ}}{\text{kg H}_2\text{O}}\right)} \right]$$

$$= 139.7 \text{ kg dry air/s}$$

The volumetric flow rate of air entering is given by

$$(AV)_1 = \dot{m}_a v_1$$

The volume per unit mass of dry air, from Figure F.1, is

$$v_1 \simeq 0.86 \text{ m}^3/\text{kg dry air}$$

Thus

$$(AV)_1 = (139.7 \text{ kg dry air/s})(0.86 \text{ m}^3/\text{kg dry air}) = \underline{\underline{120 \text{ m}^3/\text{s}}}$$

The mass flow rate of makeup water can now be determined:

$$\dot{m}_5 = (\omega_2 - \omega_1)\dot{m}_a$$
$$= [(0.0309 - 0.00989)\text{kg H}_2\text{O/kg dry air}](139.7 \text{ kg dry air/s})$$
$$= \underline{\underline{2.94 \text{ kg H}_2\text{O/s}}}$$

■

11.8 SUMMARY

This chapter was devoted to the study of nonreacting-gas mixtures. The study began with a description of the mixture composition, which can be specified by the mole fraction

$$y_i = \frac{N_i}{N_m} \tag{11.2}$$

or the mass fraction

$$mf_i = \frac{m_i}{m_m} \tag{11.5}$$

of each of the i components in a mixture.

A number of relationships were discussed that are used to describe the P-v-T behavior of mixtures. Dalton's law of additive pressures, Amagat's law of additive volumes, Bartlett's rule, and Kay's rule are examples of these relationships. Each can be used in conjunction with the principle of corresponding states to develop component compressibility factors as well as mixture compressibility factors.

The methods used to determine the internal energy, enthalpy, entropy, and specific heats of mixtures were discussed in detail. For ideal-gas mixtures the enthalpy of the mixture is determined from

$$\bar{h}_m = \sum_{i=1}^{k} y_i \bar{h}_i \tag{11.36}$$

where h_i depends on temperature alone; and the entropy change of a mixture is determined from

$$\bar{s}_{m,2} - \bar{s}_{m,1} = \sum_{i=1}^{k} y_i \left(\bar{s}^\circ_{i,2} - \bar{s}^\circ_{i,1} - \bar{R} \ln \frac{p_{i,2}}{p_{i,1}} \right) \tag{11.47}$$

The enthalpy and entropy changes for real-gas mixtures and individual components of the mixtures can be evaluated by using the generalized charts for enthalpy and entropy provided that the reduced-temperature and reduced-pressure values are based on the temperature and pressure of the mixture.

A mixture of considerable importance in engineering applications is the air–water-vapor mixture. The study of this mixture is called psychrometrics. Ideal-gas behavior is assumed for the air and the water vapor, and the mixture composition is specified by using the relative humidity:

$$\phi = \frac{p_v}{p_g} \tag{11.49}$$

or the humidity ratio,

$$\omega = \frac{m_v}{m_a} \tag{11.52}$$

The enthalpy of the mixture per unit mass of dry air is determined from

$$h_m = c_{p,a} T + \omega (c_{p,v} T + \Delta h_{v,\mathrm{ref}}) \tag{11.59}$$

where T is the dry bulb temperature of the mixture in degrees Celsius for SI units and degrees Fahrenheit for English units. The entropy of the mixture per unit mass of dry air is determined from

$$\begin{aligned} s_m = {} & c_{p,a} \ln \left(\frac{T}{T_{a,\mathrm{ref}}} \right) - R_a \ln \left(\frac{p_a}{P_{a,\mathrm{ref}}} \right) \\ & + \omega \left[c_{p,v} \ln \left(\frac{T}{T_{v,\mathrm{ref}}} \right) - R_v \ln \left(\frac{p_v}{P_{v,\mathrm{ref}}} \right) + s_{v,\mathrm{ref}} \right] \end{aligned} \tag{11.64}$$

The reference values and constants appearing in Equations 11.59 and 11.64 are summarized in Table 11.1.

A number of important air-conditioning processes were examined, and analyses of these processes were performed on the basis of the conservation-of-mass and -energy principles. The second-law analysis of several of these processes was also illustrated. Common air-conditioning processes include simple heating and cooling, during which the humidity ratio remains constant; cooling with dehumidification; heating with humidification; evaporative cooling; and adiabatic mixing.

The psychrometric chart is used to evaluate the mixture properties and to aid in visualizing the changes that occur during the air-conditioning processes.

PROBLEMS

11.1 A gaseous mixture has the following composition on a molar basis: O_2, 15 percent; N_2, 60 percent; CO_2, 25 percent. Determine the following:
- **(a)** the composition of the mixture on a mass basis
- **(b)** the partial pressure of each component if the total pressure is 150 kPa and the temperature is 30°C
- **(c)** the apparent molecular weight of the mixture

11.2 A gaseous mixture has the following composition on a mass basis: CO_2, 35 percent; CO, 10 percent; N_2, 55 percent. Determine the following:
- **(a)** the composition of the mixture on a molar basis
- **(b)** the partial pressure of each component if the total pressure is 80 psia and the temperature is 450°F
- **(c)** the apparent gas constant for the mixture

11.3 A gas mixture has the following molar analysis: N_2, 35 percent; Ar, 20 percent; CO, 45 percent. The gas is expanded from 950 kPa and 100°C to 225 kPa and 40°C in a closed system containing 2.35 kg of the mixture. Determine the composition of the mixture on a mass basis, the initial partial pressure of each component, the apparent molecular weight of the mixture, the change in volume caused by the expansion process, and the pseudo-critical temperature and pressure of the mixture.

11.4 A rigid tank contains 3 lb_m of a mixture of neon and nitrogen at 100 psia and 220°F. Sixty percent of the mixture on a molar basis is nitrogen. Determine the partial pressure of each component and the volume of the tank.

11.5 A mixture of oxygen and hydrogen contains 70 percent oxygen by mass. The mixture is confined in a storage tank with a volume of 0.1 m^3 initially at 600 kPa and 75°C. A valve on the tank is opened, and hydrogen is added to the tank until the pressure reaches 800 kPa while the temperature is maintained constant. Determine the partial pressure of each component in the initial and the final states and the mass of hydrogen added to the tank during the process.

11.6 Calculate the psuedo-critical temperature and pressure of air, which is assumed to have the following composition on a molar basis: O_2, 21 percent; N_2, 78 percent; Ar, 1 percent. Use the pseudo-critical temperature and pressure and the compressibility chart to estimate the compressibility factor for air at 675 kPa and 190°C. Compare this value with that obtained assuming ideal-gas behavior for each component.

11.7 A gaseous mixture of 75 percent propane and 25 percent ethane on a molar basis occupies a volume of 0.25 ft^3 at 250°F. Calculate the pressure of 3.2 lb_m of the mixture, using (a) ideal-gas behavior, (b) Dalton's law and compressibility factors, and (c) Kay's rule and compressibility factors.

11.8 A 2-kg mixture of 30 percent methane and 70 percent oxygen on a molar basis is contained in a rigid tank at 20 MPa and 5°C. Determine the volume of the tank, using

(a) the compressibility chart and Amagat's law; (b) the compressibility chart and Dalton's law, and (c) the compressibility chart and Kay's rule.

11.9 A closed, frictionless-piston-cylinder assembly contains a mixture having the following composition on a mass basis: CO_2, 40 percent; O_2, 25 percent; Ne, 35 percent. The cylinder contains 1.8 lb_m of the mixture at 500°F and 65 psia. Determine the magnitude and direction of the work if the mixture undergoes a constant-pressure process to 200°F.

11.10 A closed, frictionless-piston-cylinder assembly contains 70 percent nitrous oxide and 30 percent oxygen by volume initially at 200 psia and 180°F. The mixture is expanded isothermally until the pressure reaches 20 psia. Determine the magnitude and direction of the work per unit mass of mixture for this process and the partial pressure of each component in the final state.

11.11 Determine the work per unit mass of mixture required to compress a mixture of ethane and methane isothermally in a frictionless-piston-cylinder assembly from 90°F and 725 psia to 2200 psia. The mixture contains 40 percent ethane on a mass basis. Use (a) the generalized chart and Amagat's law and (b) the generalized chart and Kay's rule.

11.12 A frictionless-piston-cylinder assembly contains 3 kg of a mixture of 67 percent ethylene and 33 percent carbon dioxide on a molar basis initially at 35°C in an initial volume of 0.02 m^3. The mixture undergoes a constant-pressure process until the volume is increased by 50 percent. Estimate the final temperature of the mixture. Determine the work for this process, using (a) the compressibility chart and Dalton's law and (b) the compressibility chart and Kay's rule.

11.13 A gaseous mixture contains 0.15 lb_m N_2, 0.5 lb_m CO_2, and 0.2 lb_m Ar at 100°F and 35 psia. Determine the change in volume, the change in internal energy, the change in enthalpy, and the change in entropy of the mixture if the state of the mixture is changed to 200°F and 15 psia.

11.14 The mixture in Problem 11.1 is contained in a rigid vessel and is heated to a temperature of 120°C by heat transfer from a thermal-energy reservoir having a temperature of 300°C. The volume of the vessel is 0.4 m^3, and the environment is at 25°C and 100 kPa. Determine the heat transfer required and the irreversibility for the process.

11.15 The mixture in Problem 11.2 (at 80 psia, 450°F) is contained in a frictionless-piston-cylinder assembly and is cooled to 180°F during a constant-pressure process by heat transfer to the environment (T_0 = 77°F, P_0 = 14.7 psia). The cylinder holds 5.2 lb_m of the mixture. Determine the work, heat transfer, and irreversibility associated with this process.

11.16 A mixture of 60 percent nitrogen and 40 percent oxygen on a molar basis is contained in a frictionless-piston-cylinder assembly at 200 kPa and 175°C. The cylinder has an initial volume of 0.8 m^3, and the mixture is compressed adiabatically until the pressure reaches 450 kPa. Determine the final temperature of the mixture and the work for this process.

11.17 Two well-insulated, rigid storage tanks are connected by a valve. One tank has a volume of 15 ft^3 and contains nitrogen at 120°F and 40 psia. The other tank contains carbon monoxide at 80°F and 65 psia and has a volume of 8 ft^3. The valve is opened, the gases mix, and the mixture is allowed to reach equilibrium. Determine the final temperature and pressure of the mixture and the irreversibility of the process. (Assume T_0 = 77°F, P_0 = 14.7 psia.)

11.18 A gaseous mixture has the following composition on a volumetric basis: N_2, 60 percent; CO_2, 40 percent. The mixture, flowing at a rate of 2 kg/s, enters a steady-flow adiabatic compressor at 30°C and 150 kPa and is compressed to 450 kPa. The compressor has an adiabatic efficiency of 85 percent. Neglect changes in kinetic and potential energy, and determine the exit temperature of the mixture and the irreversibility of the process. (Assume T_0 = 25°C.)

11.19 A gaseous mixture of 60 percent carbon monoxide and 40 percent argon on a mass basis is expanded in an adiabatic nozzle from 460°F and 90 psia to a pressure of 40 psia. The adiabatic-nozzle efficiency is 95 percent, and the inlet velocity is negligible. Determine the exit temperature and the irreversibility per unit mass of mixture. (Assume T_0 = 77°F.)

11.20 A mixture of 55 percent argon and 45 percent oxygen on a molar basis is contained in a frictionless-piston-cylinder assembly initially at −45°C and 16 MPa. The pressure of the mixture is reduced to 8 MPa as a result of an isothermal process during which there is heat transfer to the mixture from the environment, which is at 25°C and 100 kPa. Determine the work, the heat transfer, and the irreversibility for this process per unit mass of mixture.

11.21 A mixture contains the following composition on a mass basis: CO_2, 50 percent; O_2, 30 percent; N_2, 20 percent. Seven lb_m of the mixture is contained in a rigid vessel initially at −60°F and 725 psia. Heat transfer to the mixture from a reservoir at a temperature of 570°F occurs until the temperature of the mixture reaches 300°F. Determine the heat transfer for this process. Use Kay's rule.

11.22 A mixture containing 60 percent nitrogen and 40 percent carbon dioxide on a mass basis is expanded in an adiabatic turbine from 200°C and 20 MPa to 95°C and 7 MPa. Determine the work output and the irreversibility per unit mass of mixture. (Assume T_0 = 25°C.)

11.23 An equimolar mixture of nitrogen and oxygen is throttled from 400°F and 2900 psia to 210°F and 290 psia. Determine the heat transfer per unit mass of the mixture.

11.24 [PSA111] A coal gasification process yields a gas mixture consisting of hydrogen (H_2), carbon monoxide (CO), and carbon dioxide (CO_2). The composition of the mixture and other data for the pure gases appear in the table below.

Gas	Volume %	Mol. wt.	c_p	c_v
H_2	35	2	14.21 kJ/kg·K	10.08 kJ/kg·K
CO	50	28	1.041 kJ/kg·K	0.7445 kJ/kg·K
CO_2	15	44	0.8418 kJ/kg·K	0.6529 kJ/kg·K

A sample of this gas is stored in a 0.35-m^3 cylinder at 25°C and 100 kPa. Assume ideal-gas behavior. Calculate the following:

(a) the apparent molecular weight of the gas mixture

(b) the mass of gas in the cylinder in kg

(c) the heat transfer in kJ required to heat the gas mixture at constant volume from 25°C to 80°C.

11.25 [PSA211] Hydrogen (H_2) and nitrogen (N_2) are mixed in an adiabatic steady-flow process in the ratio of 0.1 lb_m/h of hydrogen to 1 lb_m/h of nitrogen. The hydrogen enters the mixing chamber at 540°R, 75 psia (state 1) and the nitrogen enters at 900°R, 75 psia (state 2). The gas mixture exits at 70 psia. Assume hydrogen and nitrogen behave as ideal gases with constant-pressure specific heats of 6.86 Btu/lb_m·mol·°R and 6.94 Btu/lb_m·mol·°R, respectively. Calculate the following:

(a) the temperature of the exiting gas mixture in °R

(b) the mole fractions of H_2 and N_2 in the gas mixture

(c) the entropy change of the mixture per unit mol of the mixture

11.26 Determine the mass of dry air and the mass of water vapor contained in a room 7.5 m by 10 m by 3 m at 101.3 kPa if the dry bulb temperature and relative humidity are 26°C and 50 percent, respectively.

11.27 For a day when the barometric pressure is 30 inHg, the temperature in a room is 72°F dry bulb, and the wet bulb temperature is 48°F, determine the relative humidity, the humidity ratio, and the partial pressure of the water vapor.

11.28 At one section of a circular duct (0.3-m diameter), a stream of moist air with a dry bulb temperature of 30°C and a relative humidity of 40 percent has a velocity of 11 m/s. Determine the dew-point temperature, the enthalpy per unit mass of dry air, the humidity ratio, the mass flow rate, and the volumetric flow rate if the total pressure is 1 atm.

11.29 Repeat Problem 11.28 for a total pressure of 1.5 atm.

11.30 A moist-air stream flowing at a rate of 1.3 ft^3/s has a wet bulb temperature of 80°F and a dry bulb temperature of 95°F. Determine the relative humidity, humidity ratio, dew-point temperature, and mass flow rate if the total pressure is 1 atm.

11.31 Repeat Problem 11.30 for a total pressure of 1.25 atm.

11.32 Air enters an air-conditioning unit with a dry bulb temperature of 34°C, a relative humidity of 70 percent, and a total pressure of 101.3 kPa at a rate of 0.6 m^3/s. Determine the humidity ratio, the dew-point temperature, the wet bulb temperature, and the mass flow rate of the air.

11.33 Two lb_m of moist air with a total pressure of 18 psia and a dry bulb temperature of 110°F is contained in a tank with a volume of 24 ft^3. Determine the relative humidity of the air, the mass of water vapor in the tank, and the internal energy of the air (in Btu).

11.34 A rigid tank with a volume of 2.4 m^3 contains a moist-air mixture having 2.7 kg of dry air at a dry bulb temperature of 36°C and a total pressure of 1 atm. Determine the relative humidity of the air, the mass of water vapor in the tank, and the internal energy of the air (in kilojoules).

11.35 Repeat Problem 11.34 if the total pressure is 105 kPa.

11.36 How much heat transfer is required to increase the dry bulb temperature of 2 lb_m of moist air in a closed, rigid container by 40°F if the initial dry-bulb temperature is 60°F, the initial relative humidity is 70 percent, and the initial pressure is 20 psia? Compare this value with the heat transfer required for the same temperature change if the container held dry air instead of moist air.

11.37 A moist-air mixture with a dry bulb temperature of 48°C, a relative humidity of 40 percent, and a total pressure of 160 kPa is contained in a rigid vessel having a volume of 1.65 m^3. The air is cooled until the dry bulb temperature reaches 23°C. Determine the heat transfer, the final pressure of the air, and the amount of water vapor condensed during the process.

11.38 Air with a dry bulb temperature of 100°F and a relative humidity of 30 percent enters a reversible adiabatic nozzle at a pressure of 70 psia with a negligible velocity. Determine the pressure at which condensation will occur. At this pressure, determine the temperature and velocity of the air.

11.39 Air with a dry bulb temperature of 22°C and a relative humidity of 60 percent is contained in a frictionless-piston-cylinder assembly that has an initial volume of 0.5 m^3. The air is expanded at a constant pressure of 1 atm until the dry bulb temperature reaches 35°C. During this process heat transfer is to the air from a reservoir having a temperature of 80°C. Determine the heat transfer to the air during the process (in kilojoules), the work, and the irreversibility of the process. (Assume T_0 = 25°C, P_0 = 1 atm.)

11.40 A rigid pressure vessel contains 4.5 lb_m of air with initial dry bulb and wet bulb temperatures of 90 and 75°F, respectively, and an initial pressure of 45 psia. The air is cooled to a dry bulb temperature of 60°F by heat transfer to the surroundings (T_0 = 50°F, P_0 = 14.7 psia). Determine the initial and final relative humidities of the air, the heat transfer (in Btu) required for this process, and the irreversibility.

11.41 A moist-air mixture with a dry bulb temperature of 34°C, a relative humidity of 20 percent, and a total pressure of 150 kPa is contained in a frictionless-piston-cylinder assembly that has an initial volume of 0.6 m^3. The air undergoes an isothermal (constant dry bulb temperature) process until the total pressure has doubled. During the process heat transfer from the air is to the surroundings (T_0 = 25°C, P_0 = 100 kPa). Determine the relative humidity of the air in the final state, the heat transfer and work (in kilojoules), and the irreversibility of the process.

11.42 [PSE211] An air-water vapor mixture at 100°F, 14.7 psia, and 70 percent relative humidity is contained in a 1000-ft^3 closed tank. The mixture is cooled until water droplets just begin to form. Assume ideal-gas behavior for the mixture. The constant-pressure specific heat for air is 0.24 Btu/lb_m·°R. Determine the following:

(a) the final temperature and pressure
(b) the heat transfer for this process

11.43 [PSD112] An air–water vapor mixture at 27°C, 101.3 kPa, and 50 percent relative humidity is steadily compressed to 55°C, 290 kPa at a dry air flow rate of 0.3 kg/s. It is then cooled in a condenser at constant pressure to 27°C. Calculate the following:

(a) the condensation temperature in °C at the beginning of the compression process
(b) the condensation temperature in °C at the end of the compression process
(c) the relative humidity at the compressor outlet
(d) the rate of water removal in the condenser in kg/s

11.44 A cloud with a height of 3000 ft containing moist air with a temperature of 80°F and a relative humidity of 90 percent has its temperature decreased to 35°F. Estimate the depth of rainfall produced by the cloud.

11.45 A moist-air stream at a total pressure of 1 atm enters a cooling section and is cooled to its dew-point temperature of 50°F by heat transfer in the amount of 10 Btu/lb_m dry air from the air stream. Determine the dry bulb temperature, the wet bulb temperature, and the relative humidity of the air as it enters the cooling section.

11.46 Atmospheric air with a dry bulb temperature of 31°C and a wet bulb temperature of 22°C is to be cooled to a dry bulb temperature of 20°C. The air enters a cooling section at the rate of 0.8 kg/s. Determine the heat-transfer rate in the cooling section and the relative humidity of the cooled air.

11.47 A home-heating system is designed to have a heating capacity of 1.2×10^5 Btu/h. The system was designed to heat atmospheric air at a dry bulb temperature of 40°F and a relative humidity of 30 percent to 72°F. Atmospheric air enters instead at a dry bulb temperature of 35°F and a relative humidity of 20 percent. Determine the relative humidity of the heated air at design conditions and the dry bulb temperature and relative humidity of the heated air at the off-design condition. Assume that the mass flow rate of dry air is the same for all operating conditions.

11.48 Atmospheric air with a dry bulb temperature of 30°C and a relative humidity of 40 percent enters the cooling section of an air-conditioning system at a rate of 10 kg/s and is cooled to 18°C. Determine the volumetric flow rate of the air leaving the cooling section, the heat-transfer rate required for this process, and the entropy change of the air per unit mass of dry air. If the velocity of the air stream is not to exceed 15 m/s in the cooling section, what should be the dimensions of the cooling section if a rectangular duct with an aspect ratio of 1.5:1 is used?

11.49 A heat-transfer rate of 80,000 Btu/h is required to heat a moist-air stream, which enters a heating section at a dry bulb temperature of 40°F and leaves with a relative humidity of 10 percent, with a dry bulb temperature of 95°F. Determine the relative humidity of the air entering the heating section and the mass flow rate of dry air. The total pressure is 1 atm.

11.50 A stream of atmospheric air at a dry bulb temperature of 10°C with a relative humidity of 60 percent is heated to 25°C by being passed over a heat-exchanger surface. Determine the relative humidity of the heated air, the heat transfer required per unit mass of dry air in the air stream, and the entropy change of the air stream per unit mass of dry air.

11.51 Atmospheric air with a dry bulb temperature of 50°F and a relative humidity of 40 percent enters a heating section at the rate of 2000 ft^3/min. Heating is accomplished by passing the air over the surface of a heat exchanger in which hot combustion gases are flowing. The combustion gases enter at 500°F and leave at 395°F and flow at the rate of 1.5 lb_m/s. Assume that the combustion gases are ideal, that they have the same properties as dry air, and that there is negligible heat loss to the surroundings (T_0 = 50°F). Determine the dry bulb temperature and relative humidity of the moist air leaving the heating section and the irreversibility rate of this process.

11.52 The heat exchanger in the cooling section of a certain air-conditioning system uses refrigerant-12 to cool a moist-air stream. The refrigerant enters the heat exchanger at 5°C with a quality of 10 percent and leaves as a saturated vapor at the same temperature. Atmospheric air with a dry bulb temperature of 32°C and a relative humidity of 30 percent enters the cooling section at a rate of 0.9 m^3/s and is cooled to a dry bulb temperature of 14°C. Determine the relative humidity of the air leaving the cooling section, the mass flow rate of refrigerant-12 in the heat exchanger, the heat-transfer rate from the air, and the irreversibility rate for this process. Assume negligible heat loss to the surroundings, which are at a temperature T_0 = 32°C.

11.53 Atmospheric air with a dry bulb temperature of 85°F and a relative humidity of 65 percent is cooled to 55°F dry bulb in a steady-flow process. Determine the heat transfer from the air per unit mass of dry air and the amount of water vapor condensed per unit mass of dry air.

11.54 A combination cooling and reheat process is used to deliver air at a dry bulb temperature of 20°C and a relative humidity of 40 percent. The air enters at a dry bulb temperature of 29°C with a relative humidity of 70 percent and a volumetric flow rate of 45 m^3/min. Determine the heat-transfer rate in the cooling section, the heat-transfer rate in the heating section, and the mass flow rate of condensate from the cooling section.

11.55 Air is delivered to a computer room at a dry bulb temperature of 60°F and a relative humidity of 40 percent. This condition is achieved by cooling and reheating 1000 ft^3/min of atmospheric air having a dry bulb temperature of 85°F and a relative humidity of 30 percent. Determine the required heat-transfer rate in the cooling section and the heat-transfer rate in the reheat section.

11.56 A common method for dehumidifying air employs silica gel to adsorb some of the moisture in the air. Suppose that 2 kg/s of moist air with a dry bulb temperature of 35°C and a relative humidity of 50 percent enters the dehumidifier and leaves with a humidity ratio of 0.0011 kg H_2O/kg dry air. Determine the rate of water adsorption in the dehumidifier and the dry bulb temperature of the air leaving the dehumidifier. Assume that the dehumidifier process is adiabatic.

11.57 **[PSD211]** Atmospheric air at 30°C, 101.3 kPa, and 60 percent relative humidity enters a cooling coil condenser unit at a volumetric flow rate of 940 liter/s. The liquid water that condenses is removed at 16°C. Subsequent heating results in conditioned air at a temperature of 24°C and a relative humidity of 50 percent. Calculate the following:

(a) the heat-transfer rate in the cooling coil condenser in kW

(b) the rate of condensate removal in kg/s

(c) the heat-transfer rate in the heater in kW

11.58 Outdoor air with a dry bulb temperature of 35°F and a relative humidity of 90 percent is to be heated and humidified so that it can be delivered to a heated space at a dry bulb temperature of 74°F and a relative humidity of 45 percent. Saturated steam at 240°F is used in the humidifier section. Determine the mass flow rate of steam required per unit mass of dry air and the heat transfer per unit mass of dry air required in the heating section.

11.59 So that air at 25°C dry bulb with a relative humidity of 50 percent can be delivered, atmospheric air at 12°C dry bulb and 15 percent relative humidity is to be heated and then humidified. Steam at 200 kPa and 200°C is sprayed into the air in the humidifier section. Determine the mass flow rate of steam required per unit mass of dry air, and the heat transfer per unit mass of dry air required in the heating section. Determine the irreversibility associated with the process that occurs in the humidifier section, and explain the source of the irreversibility. (Assume $T_0 = 12°C$.)

11.60 Air at 14.7 psia, 70°F dry bulb and 60 percent relative humidity enters an industrial dryer and is discharged with a dry bulb temperature of 85°F and a relative humidity of 90 percent. Determine the volumetric flow rate of air entering the dryer if 12 lb_m/min of water must be removed from the product in the dryer.

11.61 In a certain industrial dryer 4 kg/h of water is to be removed from a moist material. The drying is accomplished by passing air over this material. Air with a dry bulb temperature of 19°C and a relative humidity of 20 percent is preheated to 50°C and passed over the moist material. It leaves with a dry bulb temperature of 40°C and a relative humidity of 80 percent. Determine the mass flow rate of dry air required for this process.

11.62 **[PSE111]** In a ventilating system, fresh air is taken in at 40°F dry bulb and 35°F wet bulb and a dry air mass flow rate of 44,100 lb_m/h. The air is then split into two streams. One stream is heated to 65°F and humidified to a relative humidity of 100 percent at a mass flow rate of 30,900 lb_m/h. Water is supplied to the humidifier at a temperature of 40°F. The other stream is simply heated to 85°F. The two streams are then mixed to produce the conditioned ventilation air. Determine the following:

(a) temperature (in °F) and the relative humidity of the conditioned air
(b) volumetric flow rate in ft^3/min of the conditioned air
(c) the heat-transfer rate and the water-supply rate

11.63 [PSC311] An air–water vapor mixture steadily enters a heater-humidifier at point 1 at 12°C, 101.3 kPa, 50 percent relative humidity, and a volumetric flow rate of 150 liters/s. Water is sprayed into the air and heat transfer occurs so that the air-vapor mixture leaves the unit at point 2 at 32°C, 101.3 kPa, with a dew-point temperature of 20°C. Determine the following:
(a) the relative humidity and wet-bulb temperature at point 2
(b) the mass flow rate of dry air through the unit in kg/s
(c) the rate at which water is supplied to the humidifier in kg/s
(d) the heat-transfer rate to the air-vapor mixture flowing through the unit in kW

11.64 Desert air with a dry bulb temperature of 102°F and a relative humidity of 10 percent enters an evaporative cooler at a rate of 1400 ft^3/min and leaves with a dry bulb temperature of 75°F. Determine the relative humidity of the air leaving the evaporative cooler and the mass flow rate of water that must be added.

11.65 Atmospheric air with a dry bulb temperature of 35°C and a relative humidity of 10 percent is passed through an evaporative cooler in order to achieve a relative humidity of 55 percent. Determine the final dry bulb temperature of the air and the amount of water added per unit mass of dry air flowing through the evaporative cooler.

11.66 A stream of atmospheric air at a dry bulb temperature of 45°F with a relative humidity of 40 percent is heated to 90°F before being passed through an evaporative cooler. The air leaving the evaporative cooler is at 70°F. Determine the heat transfer per unit mass of dry air required in the heating section, the relative humidity of the air leaving the evaporative cooler, and the entropy change of the air stream per unit mass of dry air.

11.67 Atmospheric air at 16°C dry bulb with a relative humidity of 20 percent is to be conditioned so that it achieves a state of 29°C dry bulb and 40 percent relative humidity. Consider the following two processes, which achieve the desired result, and determine which is preferred from a thermodynamic viewpoint: In the first process the air is heated and then passed through an evaporative cooler. In the second the air is heated and then passed through a humidifier section that employs a steam spray, with steam entering as a saturated vapor at 125 kPa. In both processes assume that heat transfer to the air in the heating section is from a thermal-energy reservoir that has a temperature of 250°C. Assume T_0 = 16°C and that water for evaporative cooling is added at the wet bulb temperature of the air entering the evaporative cooler.

11.68 Atmospheric air with a dry bulb temperature of 80°F and a relative humidity of 20 percent is to be conditioned by a combination of heating and evaporative cooling so that a relative humidity of 40 percent is achieved at a dry bulb temperature of 80°F. Two alternatives are considered. In the first the air is heated and then passed through an evaporative cooler. In the second the evaporative cooling precedes heating. In both cases heat transfer to the air in the heating section is from a thermal-energy reservoir having a temperature of 360°F. Determine which of these alternatives is preferred from a thermodynamic viewpoint. Assume T_0 = 80°F and that water for evaporative cooling is added at the wet bulb temperature of the air entering the evaporative cooler.

11.69 Atmospheric air enters an evaporative cooler with a dry bulb temperature of 30°C and a relative humidity of 20 percent. What is the lowest exit temperature attainable using evaporative cooling in this instance? Suppose that the relative humidity of the air leaving the evaporative cooler is 80 percent. Determine the exit-air temperature.

11.70 Atmospheric air with a dry bulb temperature of 95°F enters an evaporative cooler and leaves with a dry bulb temperature of 78°F and a relative humidity of 70 percent. Determine the relative humidity of the air entering the evaporative cooler and the irreversibility per unit mass of dry air flowing through the cooler. Assume $T_0 = 95°F$ and that water for evaporative cooling is added at the wet bulb temperature of the air entering the evaporative cooler.

11.71 [PSC112] Atmospheric air at 10°C, 101.3 kPa, and 40 percent relative humidity is steadily heated to 38°C and then passed through an evaporative cooler from which it emerges at a temperature of 25°C. The entire process occurs at constant pressure. The volumetric air flow rate at the inlet state is 150 liters/s. Calculate the following:

- **(a)** the heat-transfer rate to the air in kW
- **(b)** the relative humidity at the outlet of the heater
- **(c)** the mass flow rate of water to the evaporative cooler in kg/s
- **(d)** the relative humidity at the outlet of the evaporative cooler

11.72 [PSC212] Atmospheric air at 100°F, 14.7 psia, and 15 percent relative humidity steadily enters an evaporative cooler at a volumetric flow rate of 1400 ft^3/min. Water is supplied to the cooler at 60°F. The air leaving the cooler has a humidity ratio of 0.01 lb_m/lb_m. Calculate the following:

- **(a)** the dry bulb temperature of the conditioned air
- **(b)** the dew-point temperature of the conditioned air
- **(c)** the rate at which water is supplied to the cooler in lb_m/h
- **(d)** the lowest temperature to which the conditioned air could be cooled in an evaporative cooling process

11.73 A stream of air with a dry bulb temperature of 18°C and a relative humidity of 30 percent is mixed adiabatically with 40 m^3/min of air with a dry bulb temperature of 35°C and a relative humidity of 50 percent to produce a mixture having a dry bulb temperature of 32°C. Determine the volumetric flow rate of the first air stream, the mass flow rate of the mixture, and the relative humidity of the mixture. The total pressure is 1 atm.

11.74 A stream of air with a dry bulb temperature of 90°F and a relative humidity of 30 percent and flowing at 800 ft^3/min is mixed adiabatically with a second air stream flowing at 600 ft^3/min that has a dry bulb temperature of 70°F and a relative humidity of 40 percent. Determine the dry bulb temperature, the relative humidity, and the mass flow rate of the mixture. The total pressure is 1 atm.

11.75 A stream of air with a dry bulb temperature of 42°C and a relative humidity of 30 percent flows at a rate of 0.5 kg/s. This stream is mixed adiabatically with a second air stream that also has a mass flow rate of 0.5 kg/s to produce a mixture having a dry bulb temperature of 29°C and a relative humidity of 40 percent. Determine the dry bulb temperature and the relative humidity of the second air stream. The total pressure is 1 atm.

11.76 Atmospheric air with a dry bulb temperature of 87°F and a relative humidity of 20 percent is cooled to 53°F and then mixed adiabatically with a stream of recirculated air having a dry bulb temperature of 80°F and a relative humidity of 50 percent to produce a mixture with a dry bulb temperature of 70°F that flows at the rate of 2600 ft^3/min. Determine the relative humidity of the mixture, the mass flow rate of the recirculated air, and the heat-transfer rate in the cooling section.

11.77 A natural-draft cooling tower is used to cool 110 kg/s of water from 42 to 28°C. Atmospheric air enters the cooling tower at a dry bulb temperature of 37°C with a relative humidity of 30 percent and leaves saturated at 39°C. Makeup water is available at 24°C. Determine the mass flow rate of air entering the tower and the mass flow rate of makeup water required.

11.78 A wet-cooling tower is used to remove 7.5 MW of energy from power-plant-condenser water by reducing the temperature of the liquid water from 110 to 80°F. The outdoor air enters the tower with a dry bulb temperature of 68°F and a relative humidity of 35 percent and leaves with a dry bulb temperature of 92°F and a relative humidity of 90 percent. Makeup water is available at 70°F. Determine the volumetric flow rate of air entering the tower and the mass flow rate of makeup water.

11.79 A cooling tower located at a high altitude where the atmospheric pressure is 90 kPa is used to cool 110 kg/s of water from 40 to 28°C. Air enters the tower at a dry bulb temperature of 35°C with a relative humidity of 70 percent and leaves saturated at 37°C. Makeup water enters at 28°C. Determine the mass flow rate of air entering the tower and the mass flow rate of makeup water required.

11.80 A wet-cooling tower is used to cool 2400 lb_m/min of water from 100 to 75°F. Atmospheric air enters the tower at a dry bulb temperature of 83°F and a relative humidity of 35 percent. The air leaves saturated at 90°F. Makeup water is available at 65°F. Determine the mass flow rate of makeup water and the irreversibility of this process. (Assume T_0 = 70°F.)

11.81 A wet-cooling tower cools 800 kg/min of water from 38 to 24°C. Atmospheric air enters the tower with a dry bulb temperature of 20°C and a relative humidity of 40 percent at a rate of 500 m^3/min. This process requires 15 kg/min of 22°C makeup water. Determine the relative humidity and dry bulb temperature of the air leaving the cooling tower and the irreversibility of the process. (Assume T_0 = 20°C.)

11.82 [PSB111] Air enters a cooling tower at a volumetric flow rate of 13,500 ft^3/min at 95°F, 14.7 psia , and 78°F wet bulb temperature. The air leaves the tower at 90°F, 14.7 psia, and 95 percent relative humidity. Water enters the tower at 100°F at a flow rate of 45,000 lb_m/h. Calculate the following:

(a) the relative humidity of the entering air

(b) the temperature of the water leaving the tower in °F

(c) the mass flow rate of water leaving the tower in lb_m/h

11.83 [PSB121] It is desired to cool water from 40°C to 34.72°C by use of a cooling tower. Atmospheric air enters the tower at a volumetric flow rate of 6400 liters/s with dry bulb and wet bulb temperatures of 35°C and 25°C, respectively. The rate of water flow to the tower is 6 kg/s and the rate of water returning from the tower is 5.93 kg/s. Determine the temperature and relative humidity of the air leaving the cooling tower.

12 Thermodynamics of Chemical Reactions

12.1 INTRODUCTION

In previous chapters the study of thermodynamics has been limited to situations for which the composition of the working fluid remains chemically unchanged during the thermodynamic process. For example, the water that enters a turbine was assumed to be chemically identical to the water that leaves the turbine, and the air entering a compressor was assumed to remain chemically unchanged during the compression process. Even during mixing processes, the individual components have been assumed to be chemically nonreacting, and therefore the composition of the fluid leaving the device was simply a homogeneous mixture of the entering components. In this chapter, problems involving a change in chemical composition will be considered from a thermodynamic viewpoint.

Combustion processes occur quite frequently in engineering problems, and they represent an important segment of an engineer's education. While the examples and problems in this chapter are largely focused upon combustion applications, the principles developed apply equally well to any chemical reaction.

The topics covered in this chapter build on the fundamentals and principles established in an introductory chemistry course and on the material covered in the previous chapters. The principles of conservation of mass and conservation of energy are applied to systems that experience a change in chemical composition.

The conservation equations derived in previous chapters have been applied thus far only to nonreacting systems. These same equations, however, apply equally well to systems that involve chemically reacting systems, but the methods used to evaluate the thermodynamic properties in the conservation equations must be modified.

Finally, in this chapter the second law of thermodynamics is applied to a chemically reacting system. The application of the second law permits the determination of how far the chemical reaction will proceed before the system reaches chemical equilibrium. Even though the second law cannot provide a measure of the rate of the reaction, it plays an important role in predicting the equilibrium state of a reaction, a subject that is discussed in the next chapter.

12.2 CONSERVATION OF MASS

In a chemical reaction the components that exist prior to the reaction are called *reactants*, and those that exist after the reaction are called *products*. The relationship between these terms can be expressed in equation form as

$$\text{Reactants} \rightarrow \text{products} \tag{12.1}$$

In other words, the reactants form products as a result of the reaction.

The principle of conservation of mass developed in Chapter 3 states that the mass is a conserved quantity that can be neither created nor destroyed. Applying this principle to a process that involves a chemical reaction, such as the one expressed by Equation 12.1, means that the mass of the reactants and the mass of the products must be equal. For example, consider the chemical reaction in which hydrogen and oxygen react completely to form water. For this particular reaction the properly balanced chemical equation from a conservation-of-mass standpoint is

$$H_2 + \frac{1}{2}O_2 \rightarrow H_2O \tag{12.2}$$

Equation 12.2 implies that 1 mol of hydrogen will combine with 1/2 mol of oxygen to form 1 mol of water, or 2 kg of hydrogen will combine with 16 kg of oxygen to form 18 kg of water.

The total mass of the reactants and the total mass of the products are the same, ensuring that the total mass has been conserved during the reaction. Furthermore, notice that the total mass of each chemical element has also been conserved, even though the elements exist in different chemical compounds in the reactants and in the products. However, the total number of moles of products and of reactants are not necessarily equal, because the molecular weights of the products and the reactants are not usually the same. In other words, mass is the conserved quantity and not the number of moles. The properly balanced chemical equation expressed by Equation 12.2 is simply one that ensures that the mass of each chemical element is conserved during the reaction.

■ EXAMPLE 12.1

Reactants consisting of 1 mol of hydrogen, 1 mol of carbon monoxide, and 2 mol of oxygen combine chemically to form products that consist only of carbon dioxide, water, and oxygen. Determine the mass of each product that is formed.

Solution. The chemical equation is

$$H_2 + CO + 2O_2 \rightarrow xCO_2 + yH_2O + zO_2$$

where x, y, and z represent the unknown numbers of moles of CO_2, H_2O, and O_2 in the products, respectively. The conservation of mass can be applied to each element in turn. Thus for the hydrogen

$$2 = 2y$$

or

$$y = 1$$

and for the carbon

$$x = 1$$

and finally, for the oxygen

$$1 + 4 = 2x + y + 2z$$

or

$$z = 1$$

The properly balanced chemical equation for the reaction is therefore

$$H_2 + CO + 2O_2 \rightarrow CO_2 + H_2O + O_2$$

The mass of the products is determined as follows:

$$1 \text{ mol } CO_2 = \underline{\underline{44 \text{ lb}_m \text{ } CO_2}}$$

$$1 \text{ mol } H_2O = \underline{\underline{18 \text{ lb}_m \text{ } H_2O}}$$

$$1 \text{ mol } O_2 = \underline{\underline{32 \text{ lb}_m \text{ } O_2}}$$

Notice that 94 lb_m of reactants forms 94 lb_m of products, because the total mass is conserved during the reaction. ■

12.3 THEORETICAL COMBUSTION PROCESSES

A few terms will be defined at this point to facilitate the discussion of chemical reactions. *Combustion* is a general term referring to a chemical reaction involving a fuel and an oxidizer that results in a release of energy. The *fuel* is simply a combustible material that may come in many different forms, although the emphasis here will be on *hydrocarbon fuels*, that is, fuels that contain the elements hydrogen and carbon. Hydrocarbon fuels can exist in either the vapor, liquid, or solid phases. Typical examples of vapor-hydrocarbon fuels are butane, propane, and methane—the major component in natural gas. Examples of liquid-hydrocarbon fuels are alcohol, kerosene, and methanol. Coal is an example of a solid-hydrocarbon fuel.

Gasoline and *diesel oil* are used so frequently as energy sources that they deserve special attention. These liquid fuels consist of complex blends of numerous hydrocarbon compounds and additives. The particular composition of gasoline may vary depending on the source of the crude oil, the refiner, and the ultimate market for the gasoline. While the actual properties of gasoline may vary, the analysis of problems involving gasoline can be simplified by assuming that the properties are those of the single compound *octane*, C_8H_{18}. Likewise, the properties of diesel fuel are assumed to be those of *dodecane*, $C_{12}H_{26}$, a less refined fuel than gasoline. The chemical compositions of other common liquid- and gaseous-hydrocarbon fuels are provided in Tables G.1 and G.1E.

Whenever a fuel burns, an *oxidizer* must be present, and the oxidizer most frequently available for the combustion process is air. The composition of dry air on a volume basis (or a molar basis, if air is assumed to be an ideal gas) is 20.9 percent oxygen, 78.1 percent nitrogen, 0.9 percent argon, and lesser amounts of carbon dioxide, neon, helium, and hydrogen. For this composition, air has an apparent molecular weight of 28.97.

If the presence of all elements in dry air except oxygen and nitrogen is neglected, then the composition of dry air is approximately 79 percent nitrogen and 21 percent oxygen by volume. Therefore, for 1 mol of air, the composition of dry air is approximately

$$0.21 \text{ mol } O_2 + 0.79 \text{ mol } N_2 \rightarrow 1 \text{ mol of air} \tag{12.3}$$

or for 1 mol of oxygen, the composition of dry air is

$$1 \text{ mol } O_2 + 3.76 \text{ mol } N_2 \rightarrow 4.76 \text{ mol air} \tag{12.4}$$

When air is used as the oxidizer in the combustion of a fuel, each mole of oxygen is accompanied by 3.76 mol of nitrogen.

Nitrogen, the major component of air, is a relatively stable element and does not readily react chemically with other compounds. Nitrogen is therefore assumed to be completely inert at relatively low temperatures, although at elevated temperatures it can dissociate into monatomic nitrogen or form compounds such as nitric oxide and nitrogen dioxide.

For hydrocarbon fuels the term *complete combustion* implies that the oxidizer is available in sufficient quantities such that no unoxidized carbon in the form of C or CO and no unoxidized hydrogen in the form of H_2 or OH can exist in the products. On the other hand, the term *incomplete combustion* implies a deficiency in the amount of oxidizer so that C, CO, OH, H_2, and some unburned fuel are all possible chemical components in the products.

Even with insufficient oxygen to completely burn or oxidize all of the fuel, the hydrogen in the fuel usually completely reacts with the available oxygen to form water, because the oxidation of hydrogen proceeds to completion much more readily than the oxidation of carbon to form carbon dioxide. Therefore when the amount of oxidizer is less than sufficient to completely burn the fuel (i.e., incomplete combustion occurs), the formation of water occurs to completion, leaving no free hydrogen, while some carbon monoxide or unburned fuel may remain in the products.

So that a quantitative measure can be placed on the completeness of a combustion process, the *stoichiometric* or *theoretical proportions* of the oxidizer are defined. The stoichiometric proportions of an oxidizer (or air, if air is the oxidizer in the combustion process) are defined as the minimum amount of oxidizer theoretically necessary to provide sufficient oxygen for complete combustion of all carbon, hydrogen, and any other combustible elements in the fuel. In other words, the stoichiometric proportion of air is the theoretical amount of air necessary to oxidize all carbon in the fuel to carbon dioxide and all hydrogen in the hydrocarbon fuel to water. Theoretically, when stoichiometric proportions of air are provided, there will be no uncombined oxygen in the products. The term theoretical proportions of air is often used instead of stoichiometric proportions, and in this text they are synonymous.

To illustrate the concept of stoichiometric proportions, suppose that the stoichiometric amount of air required to burn 1 mol of propane (C_3H_8) is desired. The theoretically correct chemical reaction is

$$C_3H_8 + x(O_2 + 3.76N_2) \rightarrow yCO_2 + zH_2O + wN_2 \tag{12.5}$$

where the symbols x, y, z, and w represent the unknown numbers of moles of air, carbon dioxide, water, and nitrogen in the balanced chemical equation. Since a stoichiometric reaction is assumed, no unburned fuel, uncombined oxygen, hydrogen, or carbon monoxide can appear in the products. Applying the conservation-of-mass principle to the carbon, hydrogen, oxygen, and nitrogen results in four equations,

$$\begin{aligned} y &= 3 \\ 2z &= 8 \\ 2x &= 2y + z \\ 3.76x &= w \end{aligned}$$

which can be solved for the values of x, y, z, and w. Thus, the balanced chemical equation for stoichiometric proportions of propane burned in dry air is

$$C_3H_8 + 5(O_2 + 3.76\,N_2) \rightarrow 3CO_2 + 4H_2O + 18.8N_2 \tag{12.6}$$

Another term that is frequently used to quantify the amount of fuel and oxidizer in a combustion process is the air-fuel ratio. The *air-fuel ratio*, designated by the symbol AFR, is simply the ratio of the amount of air in a reaction to the amount of fuel. The air-fuel ratio can be written on either a molar basis, as the ratio of the number of moles of air to the number of moles of fuel, or on a mass basis, as the ratio of the mass of air to the mass of fuel. For example, the molar air-fuel ratio for stoichiometric proportions of air and propane given by chemical equation 12.6 is

$$\text{AFR} = \frac{5 \text{ mol } O_2 + 18.8 \text{ mol } N_2}{1 \text{ mol propane}} = 23.8 \text{ mol air/mol fuel} \tag{12.7}$$

This same air-fuel ratio on a mass basis is

$$\text{AFR} = \frac{5(32)\text{ kg}O_2 + 18.8(28)\text{ kg}N_2}{44 \text{ kg propane}} = 15.6\,\text{kg air/kg fuel} \tag{12.8}$$

Any air in excess of the stoichiometric proportions is referred to as *excess air*. Therefore, 50 percent excess air is equivalent to 150 percent stoichiometric proportions. Also, 100 percent excess air is the same as 200 percent stoichiometric proportions.

The concepts of conservation of mass, stoichiometric proportions, and air-fuel ratio for combustion processes are illustrated in the next three examples.

■ EXAMPLE 12.2

Methane (CH_4) is burned in stoichiometric proportions of pure oxygen. Assuming complete combustion, balance the chemical equation and calculate the air-fuel ratio on a molar basis and on a mass basis.

Solution. The balance chemical equation for this combustion reaction is

$$CH_4 + 2O_2 \rightarrow CO_2 + 2H_2O$$

No carbon monoxide, hydrogen, or oxygen can appear in the products, because the combustion process is assumed to be complete.

Even though the oxidizer is oxygen in this problem, the terminology *air-fuel ratio* has been retained rather than the more appropriate term *oxygen-fuel ratio*. The air-fuel ratio on a molar basis is

$$\text{AFR} = \frac{2 \text{ mol } O_2}{1 \text{ mol } CH_4} = \underline{\underline{2 \text{ mol } O_2/\text{mol fuel}}}$$

and on a mass basis it is

$$\text{AFR} = \frac{64 \text{ kg } O_2}{16 \text{ kg } CH_4} = \underline{\underline{4 \text{ kg } O_2/\text{kg fuel}}}$$

■

■ EXAMPLE 12.3

Octane (C_8H_{18}) is burned in 80 percent excess air. Assuming complete combustion, write the balanced chemical equation for this reaction and calculate the air-fuel ratio on a molar basis.

Solution. Before determining the actual chemical equation, we must first determine the stoichiometric proportions. The balanced stoichiometric equation is

$$C_8H_{18} + 12.5(O_2 + 3.76 \ N_2) \rightarrow 8CO_2 + 9H_2O + 47N_2$$

For 80 percent excess air the number of moles of air is increased above the stoichiometric proportions by 80 percent, and the balanced chemical equation becomes

$$C_8H_{18} + (1.8)(12.5)(O_2 + 3.76N_2) \rightarrow 8CO_2 + 9H_2O + 10.0O_2 + 84.6N_2$$

The air-fuel ratio is therefore

$$\text{AFR} = \frac{1.8(12.5 \text{ mol } O_2) + 1.8(12.5)(3.76 \text{ mol } N_2)}{1 \text{ mol octane}}$$

$$= \underline{\underline{107.1 \text{ mol air/mol fuel}}}$$

■

■ EXAMPLE 12.4

Butane (C_4H_{10}) is mixed with 80 percent excess air and burned. Prior to being mixed with the fuel, the air has a temperature of 20°C and a relative humidity of 75 percent. Write the balanced chemical equation for this reaction and determine the dew-point temperature of the products. Assume complete combustion and ideal-gas behavior for the reactants and products. The total pressure of the products and reactants remains constant at 1 atm.

Solution. In this problem the air contains water in the vapor phase, and the amount of water in the air can be determined from the given value of relative humidity. Since the reactants are assumed to be ideal gases, the mole fraction of the water vapor is equal to the ratio of the partial pressure of the water vapor to the total pressure of the mixture of dry air and water vapor (see Equation 11.17). The saturation pressure of the water vapor at 20°C from Table B.1 is

$$p_g = P_{\text{sat}} = 2.339 \text{ kPa}$$

The partial pressure of the water vapor in the air is

$$p_v = \phi p_g = 0.75(2.339 \text{ kPa}) = 1.754 \text{ kPa}$$

The mole fraction of the water vapor is therefore

$$y_v = \frac{N_v}{N} = \frac{p_v}{P} = \frac{1.754 \text{ kPa}}{101.3 \text{ kPa}} = 0.01731$$

The ratio of the number of moles of water vapor (designated by the subscript v) in each mole of dry air (designated by the subscript a) can be determined from:

$$y_v = \frac{N_v}{N} = \frac{N_v}{N_a + N_v} = \frac{N_v/N_a}{1 + (N_v/N_a)} = 0.01731$$

Thus

$$\frac{N_v}{N_a} = 0.0176$$

With this result the stoichiometric equation is

$$C_4H_{10} + x(O_2 + 3.76N_2) + 0.0176(4.76x)H_2O \rightarrow yCO_2 + zH_2O + wN_2$$

The mass balances on the carbon, nitrogen, hydrogen, and oxygen result in the four equations for the unknowns x, y, z, and w,

$$y = 4$$

$$3.76x = w$$

$$10 + 2(4.76)(0.0176)x = 2z$$

$$2x + 0.0176(4.76)x = 2y + z$$

which, when solved simultaneously, result in the values

$$x = 6.5$$

$$y = 4$$

$$z = 5.545$$

$$w = 24.44$$

The balanced stoichiometric equation, using moist air as an oxidizer, is therefore

$$C_4H_{10} + 6.5(O_2 + 3.76\ N_2) + 0.545H_2O \rightarrow 4CO_2 + 5.545H_2O + 24.44N_2$$

The chemical equation for 80 percent excess air is

$$C_4H_{10} + 1.8(6.5)(O_2 + 3.76\ N_2) + 1.8(0.545)H_2O \rightarrow 4CO_2 + xH_2O + 1.8(24.44)N_2 + yO_2$$

When mass balances are applied to the hydrogen and oxygen in the reaction, values for x and y are

$$x = 5.981$$

$$y = 5.20$$

resulting in the following balanced chemical equation for 80 percent excess moist air:

$$C_4H_{10} + 11.7(O_2 + 3.76N_2) + 0.981H_2O \rightarrow 4CO_2 + 5.981H_2O + 43.992N_2 + 5.2O_2$$

The total number of moles of products is

$$N_{prod} = 4 + 5.981 + 43.992 + 5.2 = 59.173\ \text{mol}$$

giving a mole fraction of the water vapor in the products of

$$\frac{N_{H_2O}}{N_{prod}} = \frac{5.981}{59.173} = 0.1011$$

The partial pressure of the water vapor in the products is then

$$p_v = P_{prod}\left(\frac{N_{H_2O}}{N_{prod}}\right) = (101.3\ \text{kPa})(0.1011) = 10.24\ \text{kPa}$$

which corresponds to a dew-point temperature (see discussion relating to Figure 11.1), from Table B.1, of

$$T_{dp} = T_{sat} = \underline{\underline{46.2°\text{C}}}$$

To determine the effect that the moisture in the combustion air has on the dew point of the products, we could repeat the calculations, using dry air as the oxidizer. The details are left to the reader. The balanced chemical equation for stoichiometric proportions of dry air is

$$C_4H_{10} + 6.5(O_2 + 3.76N_2) \rightarrow 4CO_2 + 5H_2O + 24.44\ N_2$$

and for 80 percent excess dry air the balanced equation is

$$C_4H_{10} + 11.7(O_2 + 3.76N_2) \rightarrow 4CO_2 + 5H_2O + 5.2O_2 + 43.99N_2$$

The mole fraction of the water vapor in the products is

$$\frac{N_{H_2O}}{N_{prod}} = \frac{5}{58.19} = 0.0859$$

producing a partial pressure of water vapor of

$$p_v = P_{prod}\left(\frac{N_{H_2O}}{N_{prod}}\right) = (101.3\,\text{kPa})(0.0859) = 8.70\ \text{kPa}$$

and a dew-point temperature of

$$T_{dp} = T_{sat} = 43.0°\text{C}$$

The water vapor in the combustion air increases the dew-point temperature of the combustion products by slightly more than 3°C. ■

12.4 ACTUAL COMBUSTION PROCESSES

In the preceding section the combustion processes were assumed to proceed to completion. If a fuel is given and the combustion is assumed to be complete, then the composition of the products can be determined simply by applying the conservation of mass to the *combustion equation* (the balanced chemical equation for the combustion process). In an *actual combustion process*, however, the combustion of the fuel is often incomplete, resulting in unoxidized compounds and unburned fuel in the products.

In actual combustion processes, even though the air-fuel ratio may exceed stoichiometric proportions, some carbon monoxide and uncombined oxygen often exist in the products of combustion because of incomplete combustion. The amount of carbon monoxide and oxygen in the products depends primarily on the degree to which the products are mixed prior to and during the combustion process. Usually, however, no uncombined hydrogen exists in the products, even when a deficiency of oxidizer exists, because the affinity of hydrogen for oxygen to form water is so great.

With incomplete combustion the composition of the products can be determined only by actually measuring the presence of the individual components in the products. A common device used to analyze the composition of the exhaust gases of a combustion process is an *Orsat gas analyzer*. The use of this device involves collecting a representative sample of the combustion gases, cooling it to room temperature, and measuring its volume. The gas sample is then brought into contact with chemicals that react with and successively absorb the carbon dioxide, oxygen, and carbon monoxide. After each of the compounds is absorbed, the reduction in the original gas volume is measured and recorded. The reduction in volume divided by the original gas volume is equal to the volume fraction of each absorbed constituent. If the combustion gases are ideal gases, the volume fractions are also equal to the mole fractions of the constituents. Therefore, measuring the volume of the combustion-gas sample after each component is absorbed allows the Orsat analyzer to be used to determine the composition of the products of combustion.

After the carbon dioxide, oxygen, and carbon monoxide are absorbed and measured, the remainder of the gas volume is assumed to be nitrogen. The Orsat analysis is assumed to be carried out on a dry sample of gas. Strictly speaking, the gas sample is saturated with water vapor, but since the test is conducted at constant temperature, the amount of water vapor in the gas sample remains constant during the absorption process.

Another device that is often used to analyze a gas sample is a *gas chromatograph*, which is capable of separating the various components in a volatile gas mixture. The central component of the gas chromatograph, shown schematically in Figure 12.1, is a column consisting of a long tube packed with a permeable absorbent material. An inert carrier gas is metered into the column at a controlled rate, and a sample injector is used to introduce the gas to be analyzed into the column. The carrier gas and sample gas mix together and enter the column, where each chemical component in the mixture is partially absorbed in the permeable material that fills the column.

As equilibrium is established between the gas vapors that occupy the pore spaces in the absorbant and the gases that are absorbed in the permeable material, the carrier gas continually sweeps each component down the column in a compact zone at a well-defined velocity. The speed at which the individual gas zones propagate through the column depends on the flow rate of the carrier gas and the extent to which each gas is absorbed in the permeable material that is packed in the column. Since gases differ in their absorption characteristics, the presence of each gas can be detected from their different speeds of progress through the column. Each gas will exit the column over a discrete time interval, and the presence of each gas can be detected as a peak in the continuous-output curve produced by the detector.

The detector can assume one of several different forms. Two of the most popular detectors are a thermal-conductivity cell and a flame-ionization detector. The thermal-conductivity cell employs a filament heated at a constant rate and placed in the gas stream, where it is cooled by the gas sample as it leaves the column. As the composition

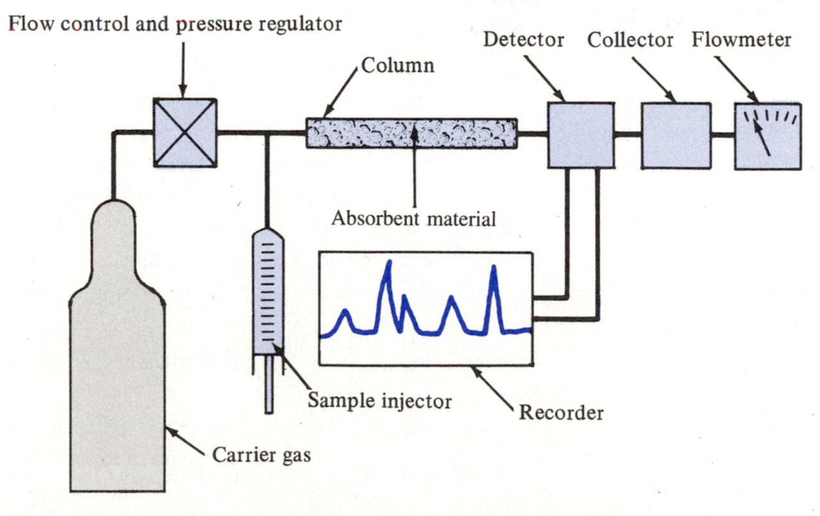

Figure 12.1 Basic components of a simple gas chromatograph.

of the gas sample changes, the rate of cooling experienced by the filament changes, resulting in a change in the resistance of the filament which can be measured with a bridge network and displayed by a recorder.

The flame-ionization detector consists of a collector electrode charged with a dc potential and placed above a small hydrogen flame. As the gas sample passes through the flame and burns, the electrical conductivity of the mixture changes. The current picked up by the electrode changes, and the current can be amplified and recorded.

The information provided by a gas chromatograph includes a measure of the purity of the gas sample, a qualitative identification of the individual chemical components, as well as a quantitative analysis of the gas mixture. If the record provided by the detector contains only a single peak, the sample consists of only a single, pure substance. A quantitative measure of the presence of each chemical component can be obtained by measuring the distance between the various peaks recorded on the output of the detector. The area under each of the peaks of the curves is proportional to the concentration of each gas component.

EXAMPLE 12.5

Hexane gas is burned in dry air. A 100-ml sample of dry combustion products is collected in an Orsat gas analyzer, and the volumes of each component are determined to be 8.5 ml CO_2, 3.0 ml O_2, 5.2 ml CO, and 83.3 ml N_2. Determine the air-fuel ratio on a molar basis for this combustion process and the dew-point temperature of the products, assuming that the products of combustion have a total pressure of 1 atm.

Solution. The fuel is C_6H_{14} and the products of combustion will be CO, CO_2, O_2, N_2, and H_2O. If the products are assumed to be ideal gases, then the volume fractions given for the products are equivalent to mole fractions of the constituents (see Equation 11.24). The balanced chemical equation for the reaction is therefore

$$xC_6H_{14} + y(O_2 + 3.76N_2) \rightarrow 5.2CO + 8.5CO_2 + 3O_2 + 83.3N_2 + zH_2O$$

where the symbols x, y, and z represent the unknown numbers of moles of fuel and air in the reactants and the water in the products, respectively. The number of moles of fuel is unknown because the equation has been partially balanced by assuming 100 ml of dry products. The number of moles of water in the products is unknown because the Orsat gas analysis does not provide a measure of the amount of water in the products.

Values for x, y, and z can be determined from mass balances on the carbon, hydrogen, oxygen, and nitrogen. These balances produce the following equations:

$$6x = 5.2 + 8.5 \qquad \text{(balance on C)}$$

$$14x = 2z \qquad \text{(balance on H)}$$

$$2y = 5.2 + 17 + 6 + z \qquad \text{(balance on O)}$$

$$3.76y = 83.3 \qquad \text{(balance on } N_2\text{)}$$

Solving these equations simultaneously gives

$$x = 2.28$$

$$y = 22.15$$

$$z = 15.98$$

Thus the balanced chemical equation is

$$2.28C_6H_{14} + 22.15(O_2 + 3.76N_2) \rightarrow 5.2CO + 8.5CO_2 + 3O_2 + 83.3N_2 + 15.98H_2O$$

The air-fuel ratio on a molar basis is

$$\text{AFR} = \frac{(4.76)(22.15)\text{mol air}}{2.28 \text{ mol fuel}} = \underline{\underline{46.2 \text{ mol air/mol fuel}}}$$

The ratio of the partial pressure of the water vapor to the total pressure of the products is

$$\frac{p_v}{P_{\text{prod}}} = \frac{N_{H_2O}}{N_{\text{prod}}} = \frac{15.98 \text{ mol } H_2O}{(5.2 + 8.5 + 3 + 83.3 + 15.98)\text{mol products}} = 0.138$$

Thus, the partial pressure of the water vapor in the products is

$$p_v = 0.138(101.3 \text{ kPa}) = 14 \text{ kPa}$$

and the dew-point temperature is the saturation temperature of water corresponding to this pressure:

$$T_{dp} = \underline{\underline{52.4°\text{C}}}$$

■

12.5 ENTHALPY OF FORMATION, ENTHALPY OF COMBUSTION, AND HEATING VALUES

For nonreacting substances the changes in properties that the substance experiences during a process can be determined solely from a knowledge of the thermodynamic states at the beginning and at the end of the process. For chemically reacting substances, on the other hand, the change in chemical composition as well as the change in thermodynamic state must be known before changes in properties can be determined. For example, consider a simple reaction that forms water in a steady-flow process by burning hydrogen in stoichiometric proportions of oxygen such that the products and reactants are at the same pressure and temperature. The chemical reaction is

$$H_2 + \frac{1}{2} O_2 \rightarrow H_2O \tag{12.9}$$

Obviously, the products are chemically different from the reactants and there must be heat transfer from the system during the reaction in order for the products to return to the identical state of the reactants. If kinetic- and potential-energy changes are neglected, then the conservation of energy requires that the heat transfer from the combustion chamber during the reaction is

$$\bar{q} = \sum_P N_i\bar{h}_i - \sum_R N_i\bar{h}_i \tag{12.10}$$

where the subscripts P and R refer to the properties of the products and reactants, respectively, and the bar superscript indicates molar enthalpy values. Equation 12.10 has been integrated with respect to time so that the result is no longer a rate equation. This form of the conservation-of-energy equation is frequently preferred when dealing with chemical-reaction problems because it is expressed in terms of the number of moles of reactants and products rather than the mass of each chemical component. If the chemical equation is balanced assuming one mole of fuel, then the resulting heat transfer determined from Equation 12.10 will be the heat transfer per mole of fuel as indicated by the bar superscript on the symbol q.

The amount of heat transfer during a combustion process cannot be evaluated solely by using tabular enthalpy values of the products and the reactants at the known states because the tables used thus far do not account for changes in properties due to a change in the chemical composition of the substance. Instead, those tabulated property values account only for changes that result from a change in thermodynamic state. To illustrate this point, suppose that a combustion process occurs so that the state of the products and reactants is the same as the reference state where the enthalpy is defined to be zero. Applying Equation 12.10 to a problem such as this would suggest that when hydrogen is mixed with oxygen and burned, the heat transfer is zero. However, this cannot be correct, because burning hydrogen in the presence of oxygen liberates a significant amount of energy. The fact that has not been considered is the change in properties (enthalpy in this case) that results from a change in chemical composition of the system. This omission in the analysis will now be corrected.

To establish property changes associated with changes in chemical composition, we must first select a reference state at which properties are defined to be zero. The usual convention for a reference state is as follows:

All stable elements are assigned zero values of enthalpy at 1 atm pressure and a temperature of 25°C or 77°F.

To indicate that a property is evaluated at the standard reference state, we use a zero superscript. This choice of a standard reference state is completely arbitrary, but it is a logical choice because it coincides with ordinary room pressure and temperature. Note that only stable elements are assigned a value of zero enthalpy at the standard reference state, and *stable* simply means that the particular element is chemically stable at 1 atm and 25°C or 77°F. For example, the stable form of hydrogen at the standard reference state is H_2 gas and not monatomic hydrogen H. The stable form of oxygen is O_2, and the stable form of nitrogen is N_2 at the standard reference state.

Once a reference state has been selected, changes in properties caused by changes in chemical composition can be determined. For this purpose the *enthalpy of formation* is introduced and defined as follows:

The enthalpy of formation of a compound is the difference between the enthalpy of 1 mol of the compound at the standard reference state and the enthalpy of the stable elements that form the compound evaluated at the standard reference state.

The enthalpy of formation is assigned the symbol $\Delta\bar{h}_f^0$ and its definition can be expressed mathematically as

$$\Delta\bar{h}_f^0 \equiv \bar{h}^0_{\text{compound}} - \sum (N_i\bar{h}_i^0)_{\text{element}} \tag{12.11}$$

where the zero superscripts signify that the enthalpies are evaluated at the standard reference state. Since the enthalpy of all stable elements is defined to be zero at the standard reference state, Equation 12.11 can be reduced to

$$\Delta\bar{h}_f^0 = \bar{h}^0_{\text{compound}} \tag{12.12}$$

This equation states that the enthalpy of a compound at the standard reference state is numerically equal to the enthalpy of formation of that compound at the standard reference state.

Values for the enthalpy of formation can be measured experimentally or determined from other properties by means of statistical thermodynamics, and they have been determined for many compounds. The enthalpy of formation in SI units for common compounds is provided in Table G.1 in the Appendix. The corresponding values in the English system of units appear in Table G.1E. The sign associated with the values appearing in Tables G.1 and G.1E corresponds to the sign convention established for heat transfer. That is, if heat transfer occurs from the compound during the reaction when the compound is formed from its elements, then the enthalpy of formation is negative. If there is heat transfer to the compound to bring its state back to the standard reference state when it is formed from its stable elements, then the enthalpy of formation is positive. In other words, all compounds that result from *exothermic reactions* (those reactions that liberate energy when the reaction occurs at constant temperature) have a negative value for $\Delta\bar{h}_f^0$, and all compounds that are formed when *endothermic reactions* take place (those reactions that absorb energy when the reaction occurs at constant temperature) have positive values for $\Delta\bar{h}_f^0$.

The *enthalpy of combustion* is another term that is often used when a fuel is burned in air. The enthalpy of combustion is designated by the symbol $\Delta\bar{h}_c$, and it is defined as follows:

> **The enthalpy of combustion of 1 mol of fuel is the difference between the enthalpy of the products at a given reference state and the enthalpy of the reactants at the same reference state when the reaction consists of complete combustion of 1 mol of the fuel in stoichiometric proportions of air.**

This definition for the enthalpy of combustion is very useful because it facilitates the determination of the heat transfer during the steady-flow combustion of a fuel with stoichiometric proportions of air. Equation 12.10 shows that the heat transfer during a steady-flow reaction when potential- and kinetic-energy changes are neglected is equal to the difference between the enthalpy of the products and the reactants. If the reaction is combustion of a fuel in stoichiometric proportions of air and the products and reactants are both assumed to be at the same state, then the heat transfer during the combustion process is simply equal to the enthalpy of combustion at the given state. Thus, the enthalpy of combustion at the standard reference state of 25°C (77°F) and 1 atm pressure can be expressed as

$$\Delta \bar{h}_c^0 \equiv \sum_P N_i \bar{h}_i^0 - \sum_R N_i \bar{h}_i^0 \tag{12.13}$$

The zero superscripts are used in Equation 12.13 to indicate that the enthalpies of the products and the reactants are evaluated at the standard reference state. Table G.2 lists enthalpy of combustion values for several common hydrocarbon fuels at 25°C and 1 atm in SI units. The same values in English units are listed in Table G.2E for a reference state of 1 atm and 77°F.

The enthalpy of combustion at the standard reference state can also be expressed in terms of the enthalpy of formation of each of the compounds that exist in the products and reactants of the combustion equation by substituting the results of Equation 12.12 into Equation 12.13:

$$\Delta \bar{h}_c^0 = \sum_P N_i \Delta \bar{h}_{fi}^0 - \sum_R N_i \Delta \bar{h}_{fi}^0 \tag{12.14}$$

Another term widely used in the analysis of combustion processes is the *heating value* of the fuel, defined as follows:

The heating value of a fuel is the amount of heat transfer from the products when the fuel is burned in stoichiometric proportions of air in a steady-flow process such that the reactants are at the standard reference state and the products are returned to the standard reference state.

This definition is similar to the one for the enthalpy of combustion. In fact, the heating value is numerically equal to the enthalpy of combustion, but it has the opposite sign, or

$$\Delta \bar{h}_c = -(\text{heating value})$$

The *higher heating value* (HHV) is the heating value of the fuel when the water in the products is in the liquid phase. The *lower heating value* (LHV) is the heating value of the fuel when the water in the products is in the vapor phase. The higher heating value is therefore greater than the lower heating value by an amount equal to the energy released when the water vapor in the products condenses to a liquid. Thus the difference between the higher heating value and the lower heating value is equal to the product of the number of moles of water formed in the products and the heat of vaporization of water on a molar basis:

$$\text{HHV} = \text{LHV} + (N\bar{h}_{fg})_{H_2O} \tag{12.15}$$

The heat of vaporization of water at the standard temperature is 44,000 kJ/kg·mol or 18,900 Btu/lb_m·mol. The latent heat of vaporization of several substances in SI units is given in Table G.2, and the English unit values appear in Table G.2E in the Appendix.

The enthalpy of combustion and the heating values of a fuel are terms reserved for the combustion of a fuel in stoichiometric proportions of air. A term that is not so restricted is the *heat* or *enthalpy of reaction*. The enthalpy of reaction is the difference between the enthalpy of the products and the enthalpy of the reactants when the reaction occurs at constant temperature. Not being limited to a specific reaction, the enthalpy of reaction can be determined for reactions other than stoichiometric-combustion chemical reactions.

EXAMPLE 12.6

Determine the enthalpy of formation of methane and of acetylene at 77°F and 1 atm.

Solution. The value for the enthalpy of formation, from Table G.1E, for methane (CH_4) is

$$\Delta \bar{h}_f^0 = \underline{\underline{-32{,}181\ \text{Btu/lb}_\text{m}\cdot\text{mol}}}$$

and for acetylene (C_2H_2) is

$$\Delta \bar{h}_f^0 = \underline{\underline{97{,}481\ \text{Btu/lb}_\text{m}\cdot\text{mol}}}$$

The value for methane is negative, indicating that methane releases energy when it is formed from carbon and hydrogen at 77°F and 1 atm pressure. Acetylene, on the other hand, has a positive value for the heat of formation, which implies that the reaction is endothermic. ■

EXAMPLE 12.7

Calculate the enthalpy of combustion at 25°C and 1 atm for butane gas, using tabular values for the enthalpy of formation. Check the computed value for $\Delta \bar{h}_c^0$ with the one listed in Table G.2.

(a) Assume that the water in the products is in the vapor phase.
(b) Assume that the water in the products is in the liquid phase.

Solution. The balanced chemical equation for stoichiometric combustion of butane (C_4H_{10}) in air is

$$C_4H_{10} + 6.5(O_2 + 3.76N_2) \rightarrow 4CO_2 + 5H_2O + 24.44N_2$$

From Equation 12.14 the enthalpy of combustion for this fuel is, then,

$$\Delta \bar{h}_c^0 = 4(\Delta \bar{h}_f^0)_{CO_2} + 5(\Delta \bar{h}_f^0)_{H_2O} - (\Delta \bar{h}_f^0)_{C_4H_{10}}$$

since the enthalpy of formation for stable elements at the standard reference state is zero.

(a) Assuming that the water in the products is in the vapor phase and using values for $\Delta \bar{h}_f^0$ from Table G.1 results in

$$\begin{aligned}\Delta \bar{h}_c^0 &= (4\ \text{kg}\cdot\text{mol CO}_2/\text{kg}\cdot\text{mol C}_4\text{H}_{10})(-393{,}522\ \text{kJ/kg}\cdot\text{mol CO}_2)\\ &\quad + (5\ \text{kg}\cdot\text{mol H}_2\text{O}/\text{kg}\cdot\text{mol C}_4\text{H}_{10})(-241{,}826\ \text{kJ/kg}\cdot\text{mol H}_2\text{O})\\ &\quad - (-126{,}150\ \text{kJ/kg}\cdot\text{mol C}_4\text{H}_{10})\\ &= \underline{\underline{-2.657 \times 10^6\ \text{kJ/kg}\cdot\text{mol C}_4\text{H}_{10}}}\end{aligned}$$

This value is also equal in magnitude to the lower heating value because the reaction is a stoichiometric combustion of the fuel, the products and reactants are maintained at the standard reference state, and the water in the products is in the vapor phase.

(b) When the water in the products is in the liquid phase, the value for $\Delta\bar{h}_c^0$ is

$$\begin{aligned}\Delta\bar{h}_c^0 &= (4\ \text{kg·mol CO}_2/\text{kg·mol C}_4\text{H}_{10})(-393{,}522\ \text{kJ/kg·mol CO}_2)\\ &\quad + (5\ \text{kg·mol H}_2\text{O/kg·mol C}_4\text{H}_{10})(-285{,}826\ \text{kJ/kg·mol H}_2\text{O})\\ &\quad - (-126{,}150\ \text{kJ/kg·mol C}_4\text{H}_{10})\\ &= \underline{\underline{-2.877 \times 10^6\ \text{kJ/kg·mol C}_4\text{H}_{10}}}\end{aligned}$$

This value is equal in magnitude to the higher heating value, and it checks closely with the value given in Table G.2.

The higher heating value can also be calculated by using Equation 12.15:

$$\begin{aligned}\text{HHV} &= \text{LHV} + 5(\bar{h}_{fg})_{\text{H}_2\text{O}}\\ &= 2.657 \times 10^6\ \text{kJ/kg·mol C}_4\text{H}_{10}\\ &\quad + (5\ \text{kg·mol H}_2\text{O/kg·mol C}_4\text{H}_{10})(44{,}000\ \text{kJ/kg·mol H}_2\text{O})\\ &= 2.877 \times 10^6\ \text{kJ/kg·mol C}_4\text{H}_{10}\end{aligned}$$

■

12.6 CONSERVATION OF ENERGY FOR CHEMICALLY REACTING SYSTEMS

The conservation of energy applied to an open, steady-flow system containing a chemically reacting substance that experiences negligible changes in kinetic and potential energy is

$$\dot{Q} - \dot{W} + \sum_R \dot{N}_i\bar{h}_i - \sum_P \dot{N}_i\bar{h}_i = 0$$

A more convenient form of this equation can be written by dividing all terms by the molar flow rate of the fuel, $\dot{N}_f$. The form of the conservation equation that emerges is one in which each term is expressed on the basis of a unit mole of fuel. The result is

$$\bar{q} - \bar{w} + \sum_R N_i\bar{h}_i - \sum_P N_i\bar{h}_i = 0 \qquad (12.16)$$

The heat transfer and work terms in Equation 12.16 are written in terms of a single mole of fuel, and this fact is designated by the bar superscript. The symbols N_i in Equation 12.16 represent the number of moles of products and reactants per mole of fuel, and these values can be determined by simply applying the conservation of mass to the combustion equation assuming a single mole of fuel. The symbols $\bar{h}_i$ represent the enthalpy of component i per unit mole of that component. The heat-transfer rate that occurs from the system during the chemical reaction can be determined from

$$\dot{Q} = \bar{q}\dot{N}_f$$

The enthalpy of the products and reactants in Equation 12.16 must account for changes in state as well as changes in chemical composition that occur during the process. Assuming that both products and reactants are ideal gases, the enthalpies of the products and reactants are solely functions of the temperature. The total enthalpy of an ideal gas can therefore be written as the sum of the enthalpy changes caused by changes in temperature, the *sensible-enthalpy change*, and the enthalpy changes that result from changes in the chemical composition. If these enthalpy changes are defined relative to the standard reference state, the enthalpy of the products at a temperature of T_P is

$$\sum_P N_i\bar{h}_i = \sum_P N_i[\Delta\bar{h}^0_{f_i} + \bar{h}_i(T_P) - \bar{h}^0_i] \tag{12.17}$$

The enthalpy of formation at the standard reference state, $\Delta\bar{h}^0_f$, accounts for changes in chemical composition, while the remaining two terms on the right side of Equation 12.17 represent the sensible-enthalpy change resulting from a change in temperature.

Similarly, if the reactants are ideal gases, then the enthalpy of the reactants at a temperature of T_R is

$$\sum_R N_i\bar{h}_i = \sum_R N_i[\Delta\bar{h}^0_{f_i} + \bar{h}_i(T_R) - \bar{h}^0_i] \tag{12.18}$$

Substituting Equations 12.17 and 12.18 into the conservation-of-energy equation for an open, steady-flow system (Equation 12.16) yields

$$\bar{q} - \bar{w} + \sum_R N_i[\Delta\bar{h}^0_{f_i} + \bar{h}_i(T_R) - \bar{h}^0_i] - \sum_P N_i[\Delta\bar{h}^0_{f_i} + \bar{h}_i(T_P) - \bar{h}^0_i] = 0 \tag{12.19}$$

This equation is applicable to a steady-flow, open system consisting of ideal gases with negligible changes in kinetic and potential energy. The values for the enthalpy of formation can be determined from Tables G.1 and G.1E, and the sensible-enthalpy changes for a number of common gases are listed in Tables G.3 through G.8 for SI units and in Tables G.3E through G.8E for English units.

For a closed, stationary system containing an ideal gas, the conservation of energy for a chemically reacting system can be written in the form

$$Q - W + \sum_R N_i\bar{u}_i - \sum_P N_i\bar{u}_i = 0 \tag{12.20}$$

The internal energy of the products and reactants can be written in terms of the enthalpy by using the definition of the enthalpy written on a molar basis, $\bar{h} = \bar{u} + P\bar{v}$,

$$\sum_R N_i\bar{u}_i - \sum_P N_i\bar{u}_i = \sum_R N_i\bar{h}_i - \sum_P N_i\bar{h}_i - \left[\sum_R (NP\bar{v})_i - \sum_P (NP\bar{v})_i\right] \tag{12.21}$$

and substituting the ideal equation of state results in

$$\sum_R N_i\bar{u}_i - \sum_P N_i\bar{u}_i = \sum_R N_i\bar{h}_i - \sum_P N_i\bar{h}_i - \left[\sum_R N_i\bar{R}T_i - \sum_P N_i\bar{R}T_i\right] \tag{12.22}$$

Finally, if Equations 12.17, 12.18, and 12.22 are substituted into Equation 12.20 and the result is divided by the number of moles of fuel, the conservation-of-energy equation for a stationary, closed system consisting of chemically reacting ideal gases becomes

$$\overline{q} - \overline{w} + \sum_R N_i[\Delta \overline{h}^0_{f_i} + \overline{h}_i(T_R) - \overline{h}^0_i - \overline{R}T_R] - \sum_P N_i[\Delta \overline{h}^0_{f_i} + \overline{h}_i(T_P) - \overline{h}^0_i - \overline{R}T_P] = 0 \quad (12.23)$$

This equation, like Equation 12.19, is written on the basis of a unit mole of fuel. Therefore, the symbols N_i represent the number of moles of products and reactants taken from the balanced chemical equation, assuming a single mole of fuel. Even though the work term has been retained in both Equations 12.19 and 12.23, most practical steady-flow combustion processes occur with the absence of work, and therefore in most cases the work can be eliminated from the conservation-of-energy equation.

EXAMPLE 12.8

Determine the enthalpy for the following:

(a) For 1 lb_m·mol of nitrogen at 1200°R, 1 atm pressure relative to a reference state of 77°F, 1 atm

(b) For 1 lb_m·mol of carbon dioxide gas at 1600°R, 2 atm pressure relative to a reference state of 77°F, 1 atm

Solution.

(a) Nitrogen at 1200°R and 1 atm behaves as an ideal gas, and the enthalpy is only a function of the temperature. The enthalpy of formation for nitrogen at the standard reference state is zero because nitrogen is a stable element at this state. Therefore, the enthalpy of nitrogen at 1200°R and 1 atm consists only of the sensible-enthalpy term. Using the sensible-enthalpy value from Table G.5E at 1200°R yields

$$\begin{aligned} H_{N_2} &= N_{N_2}[\Delta \overline{h}^0_f + \overline{h}_{N_2}(1200°\text{R}) - \overline{h}^0_{N_2}] = N_{N_2}[\overline{h}_{N_2}(1200°\text{R}) - \overline{h}^0_{N_2}] \\ &= (1\ \text{lb}_\text{m}\cdot\text{mol})(4693\ \text{Btu/lb}_\text{m}\cdot\text{mol}) = \underline{\underline{4693\ \text{Btu}}} \end{aligned}$$

(b) The carbon dioxide at 1600°R behaves as an ideal gas, and the enthalpy is not a function of the pressure. With the enthalpy of formation and the sensible-enthalpy values from Table G.3E, the enthalpy of the carbon dioxide is

$$\begin{aligned} H_{CO_2} &= N_{CO_2}[\Delta \overline{h}^0_f + \overline{h}_{CO_2}(1600°\text{R}) - \overline{h}^0_{CO_2}] \\ &= (1\ \text{lb}_\text{m}\cdot\text{mol})(-169{,}184\ \text{Btu/lb}_\text{m}\cdot\text{mol} + 11{,}798\ \text{Btu/lb}_\text{m}\cdot\text{mol}) \\ &= \underline{\underline{-157{,}386\ \text{Btu}}} \end{aligned}$$

The value of the enthalpy is negative because the formation of CO_2 from its elements is an endothermic reaction. ■

EXAMPLE 12.9

Propane gas at 77°F enters a steady-flow combustion chamber with a molar flow rate of 2 $lb_m \cdot mol/s$ where it is mixed with 100 percent excess air at 77°F. The mixture burns and the products leave the combustion chamber at 1000°R. The combustion process occurs at a constant pressure of 1 atm. Assuming complete combustion and that water in the products is in the vapor phase, calculate the heat-transfer rate from the combustion chamber.

Solution. The stoichiometric reaction assuming a single mole of fuel is

$$C_3H_8 + 5(O_2 + 3.76\ N_2) \rightarrow 3CO_2 + 4H_2O + 18.8\,N_2$$

and for 100 percent excess air the combustion reaction is

$$C_3H_8 + 10(O_2 + 3.76\ N_2) \rightarrow 3CO_2 + 4H_2O + 5O_2 + 37.6N_2$$

If the gases inside the combustion chamber are assumed to be the thermodynamic system shown in Figure 12.2, then no work crosses the boundary of the system. The products and reactants are ideal gases at the given states, and their enthalpies are functions only of temperature. Applying Equation 12.19 results in

$$\bar{q} = \sum_P N_i[\Delta\bar{h}^0_{f_i} + \bar{h}_i(T_P) - \bar{h}^0_i] - \sum_R N_i[\Delta\bar{h}^0_{f_i} + \bar{h}_i(T_R) - \bar{h}^0_i]$$

The enthalpy of formation at the standard reference state is zero for all stable elements. Also, the sensible-enthalpy changes of the reactants are equal to zero, because the temperature of the reactants is equal to the reference temperature of 77°F. So

$$\bar{q} = 3[\Delta\bar{h}^0_{fCO_2} + \bar{h}_{CO_2}(1000°R) - \bar{h}^0_{CO_2}] + 4[\Delta\bar{h}^0_{fH_2O} + \bar{h}_{H_2O}(1000°R) - \bar{h}^0_{H_2O}]$$
$$+ 5[\bar{h}_{O_2}(1000°R) - \bar{h}^0_{O_2}] + 37.6[\bar{h}_{N_2}(1000°R) - \bar{h}^0_{N_2}] - 1(\Delta\bar{h}^0_{fC_3H_8})$$

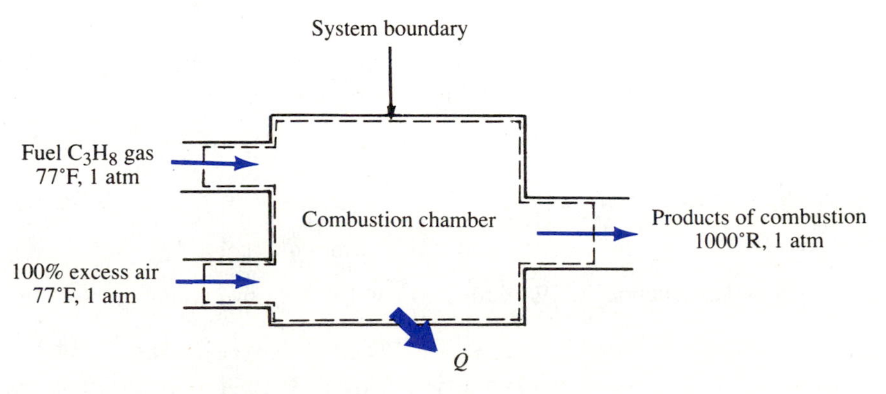

Figure 12.2 Sketch for Example 12.9.

Substituting tabular values results in

$$\begin{aligned}\overline{q} &= (3\ \text{lb}_m\cdot\text{mol } CO_2/\text{lb}_m\cdot\text{mol } C_3H_8)(-169{,}184 + 4655)\text{Btu}/\text{lb}_m\cdot\text{mol } CO_2 \\ &\quad + (4\ \text{lb}_m\cdot\text{mol } H_2O/\text{lb}_m\cdot\text{mol } C_3H_8)(-103{,}966 + 3826)\text{Btu}/\text{lb}_m\cdot\text{mol } H_2O \\ &\quad + (5\ \text{lb}_m\cdot\text{mol } O_2/\text{lb}_m\cdot\text{mol } C_3H_8)(3365\ \text{Btu}/\text{lb}_m\cdot\text{mol } O_2) \\ &\quad + (37.6\ \text{lb}_m\cdot\text{mol } N_2/\text{lb}_m\cdot\text{mol } C_3H_8)(3251\ \text{Btu}/\text{lb}_m\cdot\text{mol } N_2) \\ &\quad - (-44{,}649\ \text{Btu}/\text{lb}_m\cdot\text{mol } C_3H_8) \\ &= -7.104 \times 10^5\ \text{Btu}/\text{lb}_m\cdot\text{mol } C_3H_8\end{aligned}$$

The negative sign indicates that there is heat transfer from the system during the combustion process. This value of heat transfer is 26 percent less than the higher heating value for propane listed in Table G.2E. The heat transfer is less than the higher heating value because the reaction is not one of stoichiometric combustion of the fuel, the products do not leave the combustion chamber at the standard reference state, and the water in the products is not in the liquid phase.

The heat-transfer rate from the combustion chamber can be determined by multiplying the molar flow rate of the fuel and the heat transfer based on a unit mole of fuel, or

$$\begin{aligned}\dot{Q} = \overline{q}\dot{N}_f &= \left(-7.104 \times 10^5 \frac{\text{Btu}}{\text{lb}_m\cdot\text{mol } C_3H_8}\right)\left(\frac{2\ \text{lb}_m\cdot\text{mol } C_3H_8}{\text{s}}\right) \\ &= \underline{\underline{-1.42 \times 10^6\ \text{Btu/s}}}\end{aligned}$$

■

■ EXAMPLE 12.10

A steady-flow industrial furnace is designed to operate on gaseous methane at a constant pressure of 1 atm. Methane gas enters the furnace at 25°C and mixes with 40 percent excess air at 400 K. The products of combustion leave the furnace at 700 K. The furnace is designed to have a steady heat-transfer rate of 200 kW. Assuming complete combustion and that the water in the products is in the vapor phase, calculate the mass flow rate of methane and air necessary to provide the required heat-transfer rate from the furnace.

Solution. A schematic diagram of the furnace is shown in Figure 12.3. The chemical equation for combustion of CH_4 in 40 percent excess air is

$$CH_4 + 2.8(O_2 + 3.76N_2) \rightarrow CO_2 + 2H_2O + 0.8O_2 + 10.53\ N_2$$

The air-fuel ratio is

$$\text{AFR} = 13.33\ \text{mol air/mol } CH_4$$

Applying Equation 12.19 on the basis of 1 mol of fuel results in

$$\begin{aligned}\overline{q} &= 1[\Delta\overline{h}^0_{fCO_2} + \overline{h}_{CO_2}(700\text{ K}) - \overline{h}^0_{CO_2}] + 2[\Delta\overline{h}^0_{fH_2O} + \overline{h}_{H_2O}(700\text{ K}) - \overline{h}^0_{H_2O}] \\ &\quad + 0.8[\overline{h}_{O_2}(700\text{ K}) - \overline{h}^0_{O_2}] + 10.53[\overline{h}_{N_2}(700\text{ K}) - \overline{h}^0_{N_2}] \\ &\quad - 1(\Delta\overline{h}^0_{fCH_4}) - 2.8[\overline{h}_{O_2}(400\text{ K}) - \overline{h}^0_{O_2}] - 10.53[\overline{h}_{N_2}(400\text{ K}) - \overline{h}^0_{N_2}]\end{aligned}$$

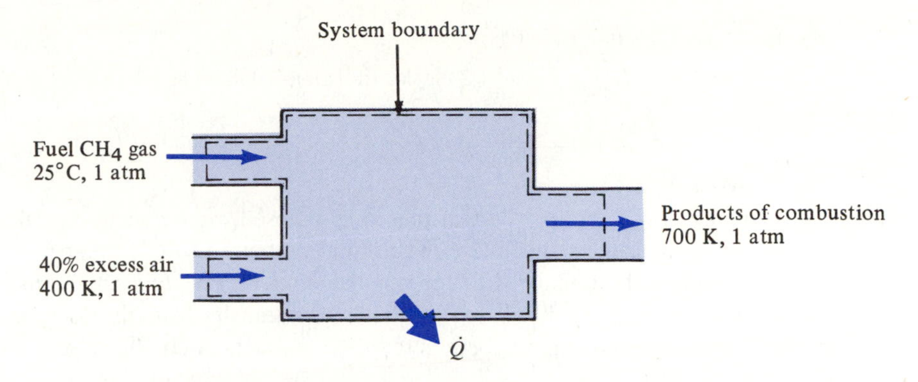

Figure 12.3 Sketch for Example 12.10.

Substituting tabular values for the enthalpy of formation and sensible-enthalpy changes results in the heat transfer from the furnace for each mole of CH_4 burned:

$$\begin{aligned}\bar{q} &= (1\ \text{kg·mol CO}_2/\text{kg·mol CH}_4)(-393{,}522 + 17{,}754)\text{kJ/kg·mol CO}_2\\ &\quad + (2\ \text{kg·mol H}_2\text{O/kg·mol CH}_4)(-241{,}826 + 14{,}192)\text{kJ/kg·mol H}_2\text{O}\\ &\quad + (0.8\ \text{kg·mol O}_2/\text{kg·mol CH}_4)(12{,}499\ \text{kJ/kg·mol O}_2)\\ &\quad + (10.53\ \text{kg·mol N}_2/\text{kg·mol CH}_4)(11{,}937\ \text{kJ/kg·mol N}_2)\\ &\quad - (-74{,}850\ \text{kJ/kg·mol CH}_4)\\ &\quad - (2.8\ \text{kg·mol O}_2/\text{kg·mol CH}_4)(3025\ \text{kJ/kg·mol O}_2)\\ &\quad - (10.53\ \text{kg·mol N}_2/\text{kg·mol CH}_4)(2971\ \text{kJ/kg·mol N}_2)\\ &= -660{,}245\ \text{kJ/kg·mol CH}_4\end{aligned}$$

The molar flow rate of fuel to produce a heat-transfer rate of 200 kW is

$$\dot{N}_{CH_4} = \frac{\dot{Q}}{\bar{q}} = \frac{-200\ \text{kW}}{-660{,}245\ \text{kJ/kg·mol CH}_4} = 3.029 \times 10^{-4}\ \text{kg·mol CH}_4/\text{s}$$

The mass flow rate of methane is

$$\begin{aligned}\dot{m}_{CH_4} = M_{CH_4}\dot{N}_{CH_4} &= (16\ \text{kg CH}_4/\text{kg·mol CH}_4)(3.029 \times 10^{-4}\ \text{kg·mol CH}_4/\text{s})\\ &= \underline{\underline{4.85 \times 10^{-3}\ \text{kg/s}}}\end{aligned}$$

The molar flow rate of air is

$$\begin{aligned}\dot{N}_a = (\text{AFR})\dot{N}_{CH_4} &= (13.33\ \text{kg·mol air/kg·mol CH}_4)\\ &\quad \times (3.029 \times 10^{-4}\ \text{kg·mol CH}_4/\text{s})\\ &= 4.038 \times 10^{-3}\ \text{kg·mol air/s}\end{aligned}$$

and the mass flow rate of air is

$$\dot{m}_a = M_a \dot{N}_a = (28.97 \text{ kg air/kg·mol air})(4.038 \times 10^{-3} \text{ kg·mol air/s})$$
$$= \underline{\underline{0.117 \text{ kg air/s}}}$$

EXAMPLE 12.11

One mole of octane vapor at 25°C is placed in a closed, rigid container with 200 percent excess air at the same temperature. The initial pressure in the container is 1 atm. The octane is mixed thoroughly with the air and ignited. The mixture burns completely, resulting in products at 1200 K. Calculate the heat transfer from the container during the process, and determine the final pressure of the products of combustion in the container.

Solution. The balanced chemical equation for a single mole of fuel and 200 percent excess air is

$$C_8H_{18} + 37.5(O_2 + 3.76\ N_2) \rightarrow 8CO_2 + 9H_2O + 25O_2 + 141N_2$$

Assuming that both products and reactants are ideal gases, the conservation of energy for a closed, rigid system is Equation 12.23, which reduces to

$$\begin{aligned}\bar{q} = {} & 8[\Delta\bar{h}^0_{fCO_2} + \bar{h}_{CO_2}(1200\text{ K}) - \bar{h}^0_{CO_2} - \bar{R}T_{CO_2}]_{TP} \\ & + 9[\Delta\bar{h}^0_{fH_2O} + \bar{h}_{H_2O}(1200\text{ K}) - \bar{h}^0_{H_2O} - \bar{R}T_{H_2O}]_{TP} \\ & + 25[\bar{h}_{O_2}(1200\text{ K}) - \bar{h}^0_{O_2} - \bar{R}T_{O_2}]_{TP} + 141[\bar{h}_{N_2}(1200\text{ K}) - \bar{h}^0_{N_2} - \bar{R}T_{N_2}]_{TP} \\ & - 1(\Delta\bar{h}^0_{fC_8H_{18}} - \bar{R}T_{C_8H_{18}})_{TR} + 37.5(\bar{R}T_{O_2}) + 141(\bar{R}T_{N_2})_{TR}\end{aligned}$$

Substituting the tabular values results in the heat transfer from the container:

$$\begin{aligned}\bar{q} = {} & \left(8\ \frac{\text{kg·mol } CO_2}{\text{kg·mol } C_8H_{18}}\right)[-393{,}522 + 44{,}473 - 8.314(1200)]\ \frac{\text{kJ}}{\text{kg·mol } CO_2} \\ & + \left(9\ \frac{\text{kg·mol } H_2O}{\text{kg·mol } C_8H_{18}}\right)[-241{,}826 + 34{,}506 - 8.314(1200)]\ \frac{\text{kJ}}{\text{kg·mol } H_2O} \\ & + \left(25\ \frac{\text{kg·mol } O_2}{\text{kg·mol } C_8H_{18}}\right)[29{,}761 - 8.314(1200)]\ \frac{\text{kJ}}{\text{kg·mol } O_2} \\ & + \left(141\ \frac{\text{kg·mol } N_2}{\text{kg·mol } C_8H_{18}}\right)[28{,}109 - 8.314(1200)]\frac{\text{kJ}}{\text{kg·mol } N_2} \\ & - 1[-208{,}450 - 8.314(298)]\frac{\text{kJ}}{\text{kg·mol } C_8H_{18}} \\ & + \left(37.5\ \frac{\text{kg·mol } O_2}{\text{kg·mol } C_8H_{18}}\right)[(8.314)298]\ \frac{\text{kJ}}{\text{kg·mol } O_2} \\ & + \left(141\ \frac{\text{kg·mol } N_2}{\text{kg·mol } C_8H_{18}}\right)[(8.314)298]\ \frac{\text{kJ}}{\text{kg·mol } N_2} \\ = {} & \underline{\underline{-1.12 \times 10^6 \text{ kJ/kg·mol}}}\end{aligned}$$

Taking the ratio of the ideal-gas equation of state between initial and final states results in the value for the final pressure:

$$\frac{P_2}{P_1} = \frac{N_2 T_2}{N_1 T_1}$$

$$P_2 = P_1 \left(\frac{N_2}{N_1}\right)\left(\frac{T_2}{T_1}\right) = (1 \text{ atm}) \left(\frac{183 \text{ mol}}{179.5 \text{ mol}}\right)\left(\frac{1200 \text{ K}}{298 \text{ K}}\right) = \underline{\underline{4.11 \text{ atm}}}$$

■

12.7 ADIABATIC FLAME TEMPERATURE

Estimating the maximum temperature that products can reach during a combustion process is often desirable, particularly in the design of combustion equipment. The maximum temperature that the products could achieve would occur if the combustion process was adiabatic, because any heat transfer during the combustion process would reduce the total energy of the products, and the temperature of the products would be likewise reduced.

The maximum temperature of the products for a given reaction is called the *adiabatic flame temperature*. For a steady-flow, adiabatic-combustion process with no work and negligible changes in kinetic and potential energy, the conservation of energy (Equation 12.16) reduces to

$$\sum_R N_i \bar{h}_i = \sum_P N_i \bar{h}_i \tag{12.24}$$

and when both reactants and products are ideal gases, this equation can be expressed as

$$\sum_R N_i[\Delta \bar{h}^0_{f_i} + \bar{h}_i(T_R) - \bar{h}^0_i] = \sum_P N_i[\Delta \bar{h}^0_{f_i} + \bar{h}_i(T_P) - \bar{h}^0_i] \tag{12.25}$$

The left side of Equation 12.25 can be evaluated once the reactants and their states are specified. The right side of Equation 12.25 can be used to determine the adiabatic flame temperature that corresponds to the temperature of the products T_P. However, an iterative procedure must be used, because the sensible-enthalpy terms for the products are unknown until the value for the adiabatic flame temperature has been determined. The iterative procedure used to calculate the adiabatic flame temperature is illustrated in the following example.

■ EXAMPLE 12.12

Liquid propane at 25°C is burned in air at 25°C in a steady-flow process. Assume complete combustion and that the gaseous products and reactants are ideal gases. Determine the adiabatic flame temperature for the following:

(a) When the propane is burned in stoichiometric proportions of air
(b) When the propane is burned in 200 percent excess air

Solution.

(a) The stoichiometric reaction of propane when burned in air is

$$C_3H_8 + 5(O_2 + 3.76N_2) \rightarrow 3CO_2 + 4H_2O + 18.8N_2$$

The conservation of energy applied to the adiabatic reaction is Equation 12.25:

$$\sum_R N_i[\Delta \bar{h}^0_{f_i} + \bar{h}_i(25°C) - \bar{h}^0_i] = \sum_P N_i[\Delta \bar{h}^0_{f_i} + \bar{h}_i(T_p) - \bar{h}^0_i]$$

where T_P represents the adiabatic flame temperature. The sensible-enthalpy terms for the reactants are zero, because the reactants are at the standard reference temperature. The enthalpy of vaporization must be included in the evaluation of the enthalpy of the fuel because the enthalpy values in Table G.1 are for propane in the vapor phase. Substitution of the values from tables in Appendix G results in

$$(-15{,}060 - 103{,}850)\text{kJ/kg·mol } C_3H_8$$

$$= \left(3\,\frac{\text{kg·mol } CO_2}{\text{kg·mol } C_3H_8}\right)[-393{,}522 + \bar{h}(T_P) - \bar{h}^0]_{CO_2}\,\frac{\text{kJ}}{\text{kg·mol } CO_2}$$

$$+ \left(4\,\frac{\text{kg·mol } H_2O}{\text{kg·mol } C_3H_8}\right)[-241{,}826 + \bar{h}(T_P) - \bar{h}^0]_{H_2O}\,\frac{\text{kJ}}{\text{kg·mol } H_2O}$$

$$+ \left(18.8\,\frac{\text{kg·mol } N_2}{\text{kg·mol } C_3H_8}\right)[0 + \bar{h}(T_P) - \bar{h}^0]_{N_2}\,\frac{\text{kJ}}{\text{kg·mol } N_2}$$

or

$$2.029 \times 10^6 \text{ kJ/kg·mol } C_3H_8$$

$$= \left(3\,\frac{\text{kg·mol } CO_2}{\text{kg·mol } C_3H_8}\right)[\bar{h}(T_P) - \bar{h}^0]_{CO_2}\,\frac{\text{kJ}}{\text{kg·mol } CO_2}$$

$$+ \left(4\,\frac{\text{kg·mol } H_2O}{\text{kg·mol } C_3H_8}\right)[\bar{h}(T_P) - \bar{h}^0]_{H_2O}\,\frac{\text{kJ}}{\text{kg·mol } H_2O}$$

$$+ \left(18.8\,\frac{\text{kg·mol } N_2}{\text{kg·mol } C_3H_8}\right)[\bar{h}(T_P) - \bar{h}^0]_{N_2}\,\frac{\text{kJ}}{\text{kg·mol } N_2}$$

The necessity for an iterative procedure for determining the value for T_P (the adiabatic flame temperature) is now evident, because this equation contains three values for the sensible enthalpy, each of which depends on T_P. One way to minimize the effort in determining T_P is to initially assume that all of the products are nitrogen. Notice that this assumption is not such an unreasonable one because 18.8 mol of the 25.8 mol of products is nitrogen.

If the products are assumed to be exclusively nitrogen, the conservation-of-energy equation reduces to

$$2.029 \times 10^6 \text{ kJ/kg·mol } C_3H_8 = (25.8 \text{ kg·mol } N_2/\text{kg·mol } C_3H_8)[\bar{h}(T_P) - \bar{h}^0]_{N_2} \text{ kJ/kg·mol } N_2$$

or

$$[\bar{h}(T_P) - \bar{h}^0]_{N_2} \text{ kJ/kg·mol } N_2 = 78{,}640 \text{ kJ/kg·mol } N_2$$

With values from Table G.5 the adiabatic flame temperature is approximately

$$T_P \simeq 2600 \text{ K}$$

Using this approximate value for T_P in the right side of the conservation-of-energy equation above results in

$$\begin{aligned}
&(3 \text{ kg·mol } CO_2/\text{kg·mol } C_3H_8)(128{,}073 \text{ kJ/kg·mol } CO_2) \\
&+ (4 \text{ kg·mol } H_2O/\text{kg·mol } C_3H_8)(104{,}520 \text{ kJ/kg·mol } H_2O) \\
&+ (18.8 \text{ kg·mol } N_2/\text{kg·mol } C_3H_8)(77{,}963 \text{ kJ/kg·mol } N_2) \\
&= 2.268 \times 10^6 \text{ kJ/kg·mol } C_3H_8
\end{aligned}$$

This value for the enthalpy of the products is greater than the enthalpy value of the reactants. Therefore, the assumed adiabatic flame temperature is too high. Assuming that T_P is 2400 K and repeating the process gives

$$\begin{aligned}
&(3 \text{ kg·mol } CO_2/\text{kg·mol } C_3H_8)(115{,}779 \text{ kJ/kg·mol } CO_2) \\
&+ (4 \text{ kg·mol } H_2O/\text{kg·mol } C_3H_8)(93{,}741 \text{ kJ/kg·mol } H_2O) \\
&+ (18.8 \text{ kg·mol } N_2/\text{kg·mol } C_3H_8)(70{,}640 \text{ kJ/kg·mol } N_2) \\
&= 2.050 \times 10^6 \text{ kJ/kg·mol } C_3H_8
\end{aligned}$$

Further iterations show that the adiabatic flame temperature is approximately

$$T_P \simeq \underline{\underline{2380 \text{ K}}}$$

(b) For 200 percent excess air the balanced chemical equation is

$$C_3H_8 + 15(O_2 + 3.76N_2) \rightarrow 3CO_2 + 4H_2O + 10\,O_2 + 56.4N_2$$

The adiabatic flame temperature is determined by application of the conservation-of-energy equation,

$$\begin{aligned}
&(-15{,}060 - 103{,}850)\text{kJ/kg·mol } C_3H_8 \\
&= \left(3 \frac{\text{kg·mol } CO_2}{\text{kg·mol } C_3H_8}\right)[-393{,}522 + \bar{h}(T_P) - \bar{h}^0]_{CO_2} \frac{\text{kJ}}{\text{kg·mol } CO_2} \\
&+ \left(4 \frac{\text{kg·mol } H_2O}{\text{kg·mol } C_3H_8}\right)[-241{,}826 + \bar{h}(T_P) - \bar{h}^0]_{H_2O} \frac{\text{kJ}}{\text{kg·mol } H_2O} \\
&+ \left(10 \frac{\text{kg·mol } O_2}{\text{kg·mol } C_3H_8}\right)[\bar{h}(T_P) - \bar{h}^0]_{O_2} \frac{\text{kJ}}{\text{kg·mol } O_2} \\
&+ \left(56.4 \frac{\text{kg·mol } N_2}{\text{kg·mol } C_3H_8}\right)[\bar{h}(T_P) - \bar{h}^0]_{N_2} \frac{\text{kJ}}{\text{kg·mol } N_2}
\end{aligned}$$

or

$$
\begin{aligned}
&2.029 \times 10^6 \text{ kJ/kg}\cdot\text{mol } C_3H_8 \\
&\quad = \left(\frac{3 \text{ kg}\cdot\text{mol } CO_2}{\text{kg}\cdot\text{mol } C_3H_8}\right) [\bar{h}(T_P) - \bar{h}^0]_{CO_2} \frac{\text{kJ}}{\text{kg}\cdot\text{mol } CO_2} \\
&\quad + \left(\frac{4 \text{ kg}\cdot\text{mol } H_2O}{\text{kg}\cdot\text{mol } C_3H_8}\right) [\bar{h}(T_P) - \bar{h}^0]_{H_2O} \frac{\text{kJ}}{\text{kg}\cdot\text{mol } H_2O} \\
&\quad + \left(\frac{10 \text{ kg}\cdot\text{mol } O_2}{\text{kg}\cdot\text{mol } C_3H_8}\right) [\bar{h}(T_p) - \bar{h}^0]_{O_2} \frac{\text{kJ}}{\text{kg}\cdot\text{mol } O_2} \\
&\quad + \left(\frac{56.4 \text{ kg}\cdot\text{mol } N_2}{\text{kg}\cdot\text{mol } C_3H_8}\right) [\bar{h}(T_p) - \bar{h}^0]_{N_2} \frac{\text{kJ}}{\text{kg}\cdot\text{mol } N_2}
\end{aligned}
$$

Assuming that all of the products are nitrogen to obtain an estimate for T_P, we have

$$
[\bar{h}(T_P) - \bar{h}^0]_{N_2} \text{ kJ/kg}\cdot\text{mol } N_2 = 27{,}643 \text{ kJ/kg}\cdot\text{mol } N_2
$$

$$
T_P \simeq 1200 \text{ K}
$$

Using this estimated temperature and values of properties from Table G in the Appendixes results in the conservation of energy in the following form:

$$
\begin{aligned}
&(3 \text{ kg}\cdot\text{mol } CO_2/\text{kg}\cdot\text{mol } C_3H_8)(44{,}473 \text{ kJ/kg}\cdot\text{mol } CO_2) \\
&+ (4 \text{ kg}\cdot\text{mol } H_2O/\text{kg}\cdot\text{mol } C_3H_8)(34{,}506 \text{ kJ/kg}\cdot\text{mol } H_2O) \\
&+ (10 \text{ kg}\cdot\text{mol } O_2/\text{kg}\cdot\text{mol } C_3H_8)(29{,}761 \text{ kJ/kg}\cdot\text{mol } O_2) \\
&+ (56.4 \text{ kg}\cdot\text{mol } N_2/\text{kg}\cdot\text{mol } C_3H_8)(28{,}109 \text{ kJ/kg}\cdot\text{mol } N_2) \\
&\quad = 2.15 \times 10^6 \text{ kJ/kg}\cdot\text{mol } C_3H_8
\end{aligned}
$$

Further iterations show that

$$
T_P \simeq \underline{\underline{1150 \text{ K}}}
$$

The adiabatic flame temperature of the products for the reaction when 200 percent excess air is present is significantly less than that for stoichiometric combustion. When excess air is present in the combustion process, the adiabatic flame temperature is reduced, because the additional products absorb some of the energy that results from the combustion of the fuel.

The maximum adiabatic flame temperature results when a fuel is burned in stoichiometric proportions. Incomplete combustion or combustion when excess air is present reduces the adiabatic flame temperature. This conclusion suggests that the temperature of the products can be controlled by simply regulating the air-fuel ratio. ■

12.8 SECOND-LAW ANALYSIS FOR CHEMICALLY REACTING SYSTEMS

A second-law analysis and the determination of entropy changes of chemically reacting mixtures are important in the study of chemical equilibrium. In this section the principles needed to evaluate the entropy changes of chemically reacting systems are briefly considered, and the subject of chemical equilibrium will be pursued in more detail in the next chapter.

Suppose that a mixture of reactants is placed in a combustion chamber where a reaction occurs, products begin to form, and the temperature of the surroundings is T_0. As the reaction continues, the mass of the reactants is reduced as more products are formed. The second law of thermodynamics can be applied to the reaction to determine to what extent the reaction continues toward completion. In this way the second law can be used to establish the equilibrium composition that exists between the products and reactants.

In Chapter 5 we found that the total entropy change of a system and its surroundings (i.e., the entropy change of an isolated system) always increases (Equation 5.47), or

$$(dS)_{tot} \geq 0$$

If the isolated system consists of a closed combustion chamber and its surroundings, this equation can be written as

$$dS_{sys} + dS_{surr} \geq 0 \tag{12.26}$$

Since the surroundings essentially proceed along an internally reversible path while maintaining a constant temperature T_0, then the entropy change of the surroundings is

$$dS_{surr} = \frac{\delta Q_{surr}}{T_0} \tag{12.27}$$

Furthermore, the heat transfer to the surroundings is equal in magnitude but opposite in sign to the heat transfer from the system,

$$\delta Q_{surr} = -\delta Q_{sys} \tag{12.28}$$

Substitution of Equations 12.27 and 12.28 into Equation 12.26 results in the second law of thermodynamics for a closed, chemically reacting system,

$$dS_{sys} \geq \frac{\delta Q_{sys}}{T_0} \tag{12.29}$$

Equation 12.29 can be used to formulate a criterion for chemical equilibrium. That is, as long as the entropy change experienced by the reaction within the system exceeds or equals the value for the heat transfer for the system divided by the absolute temperature of the surroundings, then the reaction can continue without violating the second law. A reaction for which the entropy change is less than Q_{sys}/T_0 cannot occur because it would

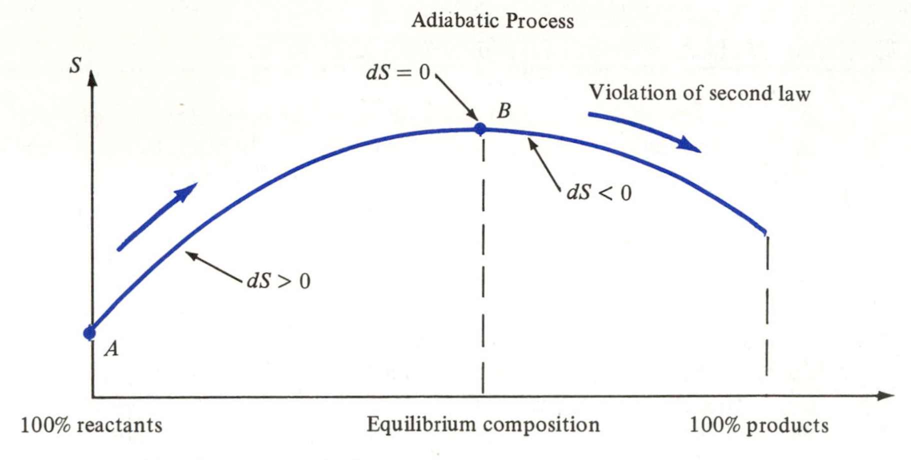

Figure 12.4 Equilibrium criteria for an adiabatic, chemical reaction in a closed system.

violate the second law. If the reaction takes place in a closed, adiabatic-combustion chamber, then Equation 12.29 reduces to

$$dS_{sys} \geq 0 \tag{12.30}$$

Suppose that reactants are mixed together in a closed, adiabatic system and a chemical reaction begins at point A in Figure 12.4. The second law in the form of Equation 12.30 states that the reactants can chemically change into products as the reaction continues as long as the entropy increases. The reaction can, therefore, continue until point B is reached, which is the point of maximum entropy for the reaction. Any further continuation in the chemical reaction would produce a decrease in the entropy, which would be a violation of the second law. Therefore, the reaction stops at point B, and this state represents a condition of chemical equilibrium. Thus the criterion for chemical equilibrium of a closed, adiabatic system is

$$(dS)_{\text{closed,adiabatic}} = 0 \tag{12.31}$$

Application of the criterion for chemical equilibrium expressed by either Equation 12.29 or 12.31 requires a logical and systematic method of determining the entropy change for reacting substances. Similar to the procedure used to determine values for the enthalpy of reacting compounds, the method used to determine values for entropy changes involves establishing a reference state at which the entropy is defined to be zero.

Through the extensive work of W. Nernst, M. Planck, and T. Richardson in the early part of the twentieth century, the reference state for entropy has been established. The result is called the *Nernst theorem*, or, more frequently, the *third law of thermodynamics*. The third law of thermodynamics states the following:

> **The entropy of all pure substances in thermodynamic equilibrium approaches zero as the temperature of the substance approaches absolute zero.**

The third law of thermodynamics establishes a common base for the value of the entropy for all pure substances.

In the remainder of this section attention is focused on combustion processes in which all of the reactants and products are ideal gases. Determining the entropy of ideal gases is slightly more complicated than determining the enthalpy and internal energy because the enthalpy and internal energy of ideal gases are functions only of temperature, whereas the entropy is a function of both pressure and temperature. Furthermore, the entropy of component i in a mixture of ideal gases, according to Dalton's law (Equation 11.11), must be evaluated at the temperature of the mixture and at the partial pressure of component i:

$$S_i = S_i(T, p_i) \tag{12.32}$$

The difference in entropy between the products P and the reactants R in a chemical reaction is

$$\Delta S = S_P - S_R = \sum_P N_i \bar{s}_i(T_P, p_i) - \sum_R N_i \bar{s}_i(T_R, p_i) \tag{12.33}$$

To evaluate the entropy for each ideal-gas component in the products and reactants, we can use Equation 5.25 written on a molar basis:

$$d\bar{s} = \bar{c}_p \frac{dT}{T} - \frac{\bar{R}\, dP}{P} \tag{12.34}$$

If a reference state specified by $P_0 = 1$ atm and $T_0 = 0$ K or 0°R is established, then Equation 12.34 can be integrated between the reference state (T_0, P_0) and the arbitrary state specified by (T, P), with the result

$$\int_{T_0, P_0}^{T,P} d\bar{s} = \int_{T_0}^{T} \bar{c}_p \frac{dT}{T} - \bar{R} \int_{P_0}^{P} \frac{dP}{P} \tag{12.35}$$

The first integral on the right side of the equation has been defined in Equation 5.27 as $\bar{s}°(T)$, and the other term can easily be integrated. Thus

$$\bar{s}(T, P) - \bar{s}(T_0, P_0) = \bar{s}°(T) - \bar{R} \ln \frac{P}{P_0} \tag{12.36}$$

According to the third law of thermodynamics, the value for the entropy at the reference state (T_0, P_0) is zero. If Equation 12.36 is evaluated at the partial pressure p_i, then Equation 12.36 becomes the entropy of component i in a reacting mixture, or

$$\bar{s}_i(T, p_i) = s_i°(T) - \bar{R} \ln \frac{p_i}{P_0} \tag{12.37}$$

Values for $\bar{s}°(T)$ in SI units for common gases are listed in Tables G.3 through G.8 in the Appendix. Also, values for $\bar{s}°(T)$ evaluated at 25°C are provided in Table G.1 for many common fuels. The corresponding values for $\bar{s}°(T)$ in English units appear in Tables G.3E through G.8E and English values of $\bar{s}°$ at 77°F for common fuels are in Table G.1E.

Upon combining Equations 12.37 and 12.33, the entropy change that occurs during a chemical reaction can be written as

$$\Delta S = \sum_P N_i \left[\bar{s}_i^\circ(T_P) - \bar{R} \ln \frac{p_i}{P_0} \right] - \sum_R N_i \left[\bar{s}_i^\circ(T_R) - \bar{R} \ln \frac{p_i}{P_0} \right] \qquad (12.38)$$

This equation can also be written in terms of the mole fractions of the components. For a mixture of ideal gases Equation 11.17 states that the ratio of the partial pressure of a gas to the total pressure of the mixture is equal to the mole fraction of the gas, or

$$p_i = y_i P_T \qquad (12.39)$$

where P_T is the total pressure of the mixture of gases and y_i is the mole fraction. If the total pressure of the reactants and products is measured in *atmospheres*, then $P_0 = 1$ atm and Equation 12.38 reduces to

$$\Delta S = \sum_P N_i[\bar{s}_i^\circ(T_P) - \bar{R} \ln (y_i P_T)] - \sum_R N_i[\bar{s}_i^\circ(T_R) - \bar{R} \ln (y_i P_T)] \qquad (12.40)$$

The procedure used to calculate the entropy change during a combustion process is illustrated in the following example.

■ EXAMPLE 12.13

Ethane gas is burned in stoichiometric proportions of air in a closed, flexible container. The reactants are initally at 1 atm and 537°R, and the temperature of the products is 1800°R. During the combustion process there is heat transfer to the surroundings, which have a temperature of 537°R. Assuming complete combustion and that the pressure in the container remains constant, determine the entropy change during the process on the basis of 1 mol of ethane. Determine whether the combustion process violates the second law of thermodynamics, and calculate the irreversibility of the process.

Solution. The balanced chemical equation for this stoichiometric reaction is

$$C_2H_6 + 3.5(O_2 + 3.76N_2) \rightarrow 2CO_2 + 3H_2O + 13.16N_2$$

The total number of moles of reactants is

$$\sum_R N_i = 1 + 3.5 + 13.16 = 17.66$$

The mole fractions of the reactants are

$$y_{C_2H_6} = \frac{1}{17.66} = 0.057$$

$$y_{O_2} = \frac{3.5}{17.66} = 0.198$$

$$y_{N_2} = \frac{13.16}{17.66} = 0.745$$

With Equation 12.40 and the values for $\bar{s}^\circ(T)$ from the tables in the Appendix, the absolute entropy of each of the reactants is

$$\bar{s}_{C_2H_6} = \bar{s}^\circ - \bar{R} \ln (y_{C_2H_6} P_T) = 54.82 \text{ Btu/lb}_m\text{·mol·°R}$$
$$- (1.986 \text{ Btu/lb}_m\text{·mol·°R}) \ln [(0.057)(1.0)]$$
$$= 60.51 \text{ Btu/lb}_m\text{·mol·°R}$$

$$\bar{s}_{O_2} = \bar{s}^\circ - \bar{R} \ln (y_{O_2} P_T) = 49.00 \text{ Btu/lb}_m\text{·mol·°R}$$
$$- (1.986 \text{ Btu/lb}_m\text{·mol·°R}) \ln [(0.198)(1.0)]$$
$$= 52.22 \text{ Btu/lb}_m\text{·mol·°R}$$

$$\bar{s}_{N_2} = \bar{s}^\circ - \bar{R} \ln (y_{N_2} P_T) = 45.77 \text{ Btu/lb}_m\text{·mol·°R}$$
$$- (1.986 \text{ Btu/lb}_m\text{·mol·°R}) \ln [(0.745)(1.0)]$$
$$= 46.35 \text{ Btu/lb}_m\text{·mol·°R}$$

$$S_R = \sum_R N_i \bar{s}_i = [1(60.51) + 3.5(52.22) + 13.16(46.35)] \text{Btu/°R}$$
$$= 853 \text{ Btu/°R}$$

The total number of moles of products is

$$\sum_P N_i = 2 + 3 + 13.16 = 18.16$$

The mole fractions of the products are

$$y_{CO_2} = \frac{2}{18.16} = 0.110$$

$$y_{H_2O} = \frac{3}{18.16} = 0.165$$

$$y_{N_2} = \frac{13.16}{18.16} = 0.725$$

The absolute entropies of the products are

$$\bar{s}_{CO_2} = 64.32 \text{ Btu/lb}_m\text{·mol·°R} - (1.986 \text{ Btu/lb}_m\text{·mol·°R}) \ln [(0.110)(1.0)]$$
$$= 68.70 \text{ Btu/lb}_m\text{·mol·°R}$$

$$\bar{s}_{H_2O} = 55.59 \text{ Btu/lb}_m\text{·mol·°R} - (1.986 \text{ Btu/lb}_m\text{·mol·°R}) \ln [(0.165)(1.0)]$$
$$= 59.17 \text{ Btu/lb}_m\text{·mol·°R}$$

$$\bar{s}_{N_2} = 54.50 \text{ Btu/lb}_m\text{·mol·°R} - (1.986 \text{ Btu/lb}_m\text{·mol·°R}) \ln [(0.725)(1.0)]$$
$$= 55.14 \text{ Btu/lb}_m\text{·mol·°R}$$

$$S_P = \sum_P N_i \bar{s}_i = [2(68.70) + 3(59.17) + 13.16(55.14)] \text{Btu/°R} = 1041 \text{ Btu/°R}$$

The entropy change during the combustion process for each mole of C_2H_6 is

$$\Delta S = S_P - S_R = (1041 - 853)\text{Btu/°R} = 188\ \text{Btu/°R}$$

The heat transfer to the surroundings during a constant-pressure process occurring in a closed system is

$$\begin{aligned}\bar{q}_{sys} &= \sum_P N_i\bar{h}_i - \sum_R N_i\bar{h}_i \\ &= \sum_P N_i[\Delta\bar{h}^0_{f_i} + \bar{h}_i(T_P) - \bar{h}^0_i] - \sum_R N_i[\Delta\bar{h}^0_{f_i} + \bar{h}_i(T_R) - \bar{h}^0_i] \\ &= (2\ \text{lb}_m\cdot\text{mol CO}_2/\text{lb}_m\cdot\text{mol C}_2\text{H}_6)(-169{,}184 + 14{,}350)\text{Btu/lb}_m\cdot\text{mol CO}_2 \\ &\quad + (3\ \text{lb}_m\cdot\text{mol H}_2\text{O/lb}_m\cdot\text{mol C}_2\text{H}_6)(-103{,}966 + 11{,}178)\text{Btu/lb}_m\cdot\text{mol H}_2\text{O} \\ &\quad + (13.16\ \text{lb}_m\cdot\text{mol N}_2/\text{lb}_m\cdot\text{mol C}_2\text{H}_6)(9227)\text{Btu/lb}_m\cdot\text{mol N}_2 \\ &\quad - (-36{,}407\ \text{Btu/lb}_m\cdot\text{mol C}_2\text{H}_6) \\ &= -4.302 \times 10^5\ \text{Btu/lb}_m\cdot\text{mol C}_2\text{H}_6\end{aligned}$$

The entropy change of the surroundings for each mole of C_2H_6 is

$$\Delta S_{surr} = \frac{Q_{surr}}{T_0} = -\frac{Q_{sys}}{T_0} = \frac{4.302 \times 10^5\text{Btu}}{537\text{°R}} = 801\ \text{Btu/°R}$$

The total entropy change during the process for each mole of C_2H_6 is

$$\Delta S_{tot} = \Delta S_{sys} + \Delta S_{surr} = 188\ \text{Btu/°R} + 801\ \text{Btu/°R} = \underline{\underline{989\ \text{Btu/°R}}}$$

Since the total entropy increases during the process, the second law is not violated.

The irreversibility of the process is given by Equation 6.23:

$$I = T_0(\Delta S_{tot}) = (537\text{°R})(989\ \text{Btu/°R}) = \underline{\underline{5.31 \times 10^5\ \text{Btu}}}$$

■

12.9 SUMMARY

The principle of conservation of mass, when applied to a chemical reaction, requires that the mass of each chemical element be conserved during the reaction. That is, the mass of each element in the reactants must equal the mass of that element in the products.

Combustion is a chemical reaction involving a fuel and an oxidizer that results in a transfer of energy. The fuel is usually a compound consisting of hydrogen and carbon and the oxidizer is often air. The composition of dry air is assumed to consist of nitrogen and oxygen in the following proportions:

$$4.76\ \text{mol air} = 1\ \text{mol oxygen} + 3.76\ \text{mol nitrogen}$$

The amount of air necessary to provide for complete combustion such that all carbon in the fuel forms carbon dioxide, and all hydrogen in the fuel forms water in the products, is called the stoichiometric or theoretical proportion of air. Any amount of air above stoichiometric proportions is called excess air.

The air-fuel ratio is the ratio of the amount of air to the amount of fuel in a combustion reaction. The air-fuel ratio can be expressed in terms of either a mass or a molar ratio.

The amount of the individual products that exists in an actual combustion process can be determined with a number of instruments. Two of the most common devices are the Orsat gas analyzer and the gas chromatograph.

Whenever the conservation of energy is applied to a system in which a chemical reaction occurs, the changes in properties due to both changes in state and changes in chemical composition must be determined. To simplify matters, a standard reference state is selected. The enthalpy of all stable elements is set equal to zero at the standard reference state of 1 atm pressure and 25°C or 77°F.

For determination of changes in properties resulting from changes in chemical composition, the enthalpy of formation is introduced. The enthalpy of formation of a compound is defined as the difference between the enthalpy of a compound at the standard reference state and the enthalpy of the elements that form the compound at the same reference state:

$$\Delta \bar{h}_f^0 \equiv \bar{h}_{\text{compound}}^0 - \sum (N_i \bar{h}_i^0)_{\text{element}} \tag{12.11}$$

So that enthalpy changes due to changes in the thermodynamic state may be considered, sensible-enthalpy changes are provided in the tables in Appendix G for several common ideal gases. The total enthalpy of an ideal gas relative to the reference state can be written as the sum of the enthalpy of formation and the sensible-enthalpy change measured relative to the standard reference state, or

$$\bar{h}_i(T) = [\Delta \bar{h}_{f_i}^0 + \bar{h}_i(T) - \bar{h}_i^0] \tag{12.17}$$

The conservation of energy applied to an open, steady-flow system consisting of ideal gases with negligible changes in kinetic and potential energy results in

$$\bar{q} - \bar{w} + \sum_R N_i[\Delta \bar{h}_{f_i}^0 + \bar{h}_i(T_R) - \bar{h}_i^0] - \sum_P N_i[\Delta \bar{h}_{f_i}^0 + \bar{h}_i(T_P) - \bar{h}_i^0] = 0 \tag{12.19}$$

where the symbols N_i are the numbers of moles of products and reactants based on a single mole of fuel.

If a combustion process involves only ideal gases and it takes place in a closed system, the conservation of energy takes the form

$$\bar{q} - \bar{w} + \sum_R N_i[\Delta \bar{h}_{f_i}^0 + \bar{h}_i(T_R) - \bar{h}_i^0 - \bar{R}T_R] - \sum_P N_i[\Delta \bar{h}_{f_i}^0 + \bar{h}_i(T_P) - \bar{h}_i^0 - \bar{R}T_P] = 0 \tag{12.23}$$

The enthalpy of combustion of a fuel is the difference between the enthalpy of the products at the standard reference state and the enthalpy of the reactants at the same state when the reaction is one of stoichiometric combustion. The heating value of a fuel is equal in magnitude but opposite in sign to the enthalpy of combustion of the fuel,

$$\Delta H_c = -(\text{heating value})$$

The higher heating value of a fuel is the amount of heat transfer during a steady-flow process when the fuel is burned in stoichiometric proportions of air such that the initial temperature of the reactants and the final temperature of the products are equal to the standard reference temperature and the water in the products is in the liquid phase. The definition of the lower heating value is identical to the one for the higher heating value, except that the water in the products is in the vapor phase. The difference between the higher heating value and the lower heating value is equal to the product of the number of moles of water formed and the molar heat of vaporization for water at the standard reference temperature, or

$$\text{HHV} = \text{LHV} + (N\bar{h}_{fg})_{H_2O} \tag{12.15}$$

The adiabatic flame temperature is the temperature that the products of combustion would reach if a fuel was burned in an adiabatic, steady-flow process. The adiabatic flame temperature can be determined by applying the conservation of energy to an adiabatic-combustion process, which results in

$$\sum_R N_i[\Delta\bar{h}^0_{f_i} + \bar{h}_i(T_R) - \bar{h}^0_i] = \sum_P N_i[\Delta\bar{h}^0_{f_i} + \bar{h}_i(T_P) - \bar{h}^0_i] \tag{12.25}$$

The maximum adiabatic flame temperature results from burning a fuel in stoichiometric proportions of air.

Calculation of the entropy change that occurs during a chemical reaction was initiated by considering the third law of thermodynamics. The third law states that the entropy of all pure substances in equilibrium approaches zero as the temperature approaches absolute zero. This law establishes a common base for which the entropy of all pure substances can be measured. The entropy change that occurs during a chemical reaction involving ideal gases is

$$\Delta S = \sum_P N_i[\bar{s}^\circ_i(T_P) - \bar{R}\ln(y_iP_T)] - \sum_R N_i[\bar{s}^\circ_i(T_R) - \bar{R}\ln(y_iP_T)] \tag{12.40}$$

where y_i represents the mole fractions of the various chemical components and P_T is the total pressure, measured in atmospheres.

If a chemical reaction takes place inside a closed system and the process is adiabatic, then the reaction can continue as long as the entropy of the system increases. Once the entropy of the system reaches a maximum value, the reaction will have reached a condition of chemical equilibrium. Any adiabatic process for which the entropy of this system would decrease would violate the second law of thermodynamics. Criteria for chemical equilbrium are discussed in more detail in the next chapter.

PROBLEMS

12.1 Determine the molar air-fuel ratio when hydrogen is burned in stoichiometric proportions of air.

12.2 Calculate the molar and mass air-fuel ratios for propane burned in stoichiometric proportions of air.

12.3 Calculate the air-fuel ratio on a mass basis when pentane is burned in stoichiometric proportions of pure oxygen.

12.4 Determine the molar air-fuel ratio for ethane burned in 30 percent excess air.

12.5 Determine the mole fractions of all the products of combustion when acetylene is burned in 50 percent excess air.

12.6 Calculate the air-fuel ratio on a mass and molar basis when octane is burned in 50 percent excess air. Determine the dew-point temperature of the products if the pressure of the products is 1 atm.

12.7 Propylene is burned in 20 percent excess air. Calculate the air-fuel ratio on a mass and molar basis.

12.8 Hydrogen is burned in 20 percent excess air. Determine the air-fuel ratio on a mass basis and the mole fractions of the products of combustion. Determine the dew-point temperature of the products if the pressure of the products is 1 atm.

12.9 A mixture of oxygen and methane is placed in a container with a volume of 1 m^3. The mixture has a total pressure of 200 kPa and a temperature of 35°C. The partial pressure of the methane is 50 kPa. The mixture is ignited and the combustion process is complete. Determine the air-fuel ratio on a mass basis and the mole fractions of the products of combustion.

12.10 The combustible range of gasoline is between 1.3 and 6 percent fuel vapor by volume. Assume that gasoline has the same properties as octane (C_8H_{18}). Determine the air-fuel ratio and percent excess (or deficiency) air when combustion of gasoline occurs at the low and high ends of its combustible range.

12.11 The mass flow rate of air through the carburetor of an automobile is 1.3 lb_m/min. The gasoline is burned in 20 percent excess air. Calculate the fuel-consumption rate, in gal/min, assuming that the properties of gasoline are the same as those of octane.

12.12 One kilogram of hexane is burned in 20 kg of air. Compute the percent excess air used in the combustion process.

12.13 The ratio of mass of air to butane used in the burning of butane is 12 to 1. Calculate the percentage of theoretical air used in the combustion process.

12.14 One gal of gasoline (assume properties of C_8H_{18}) is burned in 30 percent excess air. Determine the mass of air required, assuming complete combustion. The density of gasoline is 52.4 lb_m/ft^3.

12.15 Determine the dew-point temperature when octane is burned in 150 percent theoretical dry air. The pressure of the products is (a) 100 kPa, (b) 500 kPa, and (c) 800 kPa.

12.16 Calculate the dew-point temperature when butane is burned in 20 percent excess dry air and the pressure of the products is 1 atm.

12.17 Pentane gas is burned in 60 percent excess dry air and the pressure of the products is 45 psia. Determine the dew-point temperature of the products.

12.18 An automobile engine operates with 20 percent excess dry air. Assume that the fuel has properties of octane. The exhaust temperature as it leaves the tail pipe on a cold day immediately after engine start-up is 20°C. Will the water in the combustion gases condense under these conditions? After the engine has warmed up, the temperature of the exhaust gases has increased to 80°C. Under these conditions will the water in the combustion gases condense?

12.19 Pentane is burned in 80 percent excess dry air. The products of combustion are cooled to 20°C and 1 atm pressure. Calculate the mass of water that condenses per mol of pentane burned.

12.20 Ethane is burned in 30 percent excess air at 1 atm pressure. The air has a relative humidity of 90 percent and a temperature of 90°F. Assuming complete combustion, calculate the dew-point temperature of the products.

12.21 Propane is burned in 50 percent excess air. The products and reactants are at 1 atm pressure, and the air has a temperature of 40°C. Assuming complete combustion, calculate the dew-point temperature when (a) the air is dry and (b) the air has a relative humidity of 60 percent.

12.22 One mol of butane is mixed with 50 percent excess air and placed in a closed, rigid container at 77°F and 1 atm pressure. The mixture is ignited and the combustion process is complete. Determine the pressure in the container if the final temperature of the products is 1440°R. Calculate the volume of the container necessary to hold the reactants.

12.23 **[CRA111]** Ethanol (C_2H_5OH) is burned with 50 percent excess dry air. The combustion products are cooled to 30°C. The total pressure is 100 kPa. Assume that air is an ideal gas consisting of 79 percent nitrogen and 21 percent oxygen by volume. Assuming complete combustion, determine:

(a) the air-fuel ratio in units of kg/kg
(b) The dew point of the combustion products in °C
(c) The ratio of condensate to ethanol combusted in units of kg/kg

12.24 An analysis of the flue gas from a furnace reveals the following volumetric percentages on a dry basis: CO, 1.70 percent; CO_2, 8.00 percent; O_2, 4.32 percent; N_2, 85.98 percent. The fuel is CH_4 and the combustion occurs at a constant pressure of 1 atm. Calculate the molar air-fuel ratio, the percent excess air, and the dew-point temperature of the products.

12.25 Pentane is burned in air, and a volumetric analysis of the products on a dry basis is as follows: CO_2, 6.20 percent; CO, 1.40 percent; O_2, 10.36 percent; N_2, 82.04 percent. Determine the air-fuel ratio on a mass basis and the percent excess air in the combustion process.

12.26 One mol of a hypothetical hydrocarbon fuel, C_xH_y, is burned in air at 1 atm pressure. Analysis of the combustion products on a dry basis shows that the percentages of individual gases on a volume basis are as follows: CO, 2 percent; CO_2, 7 percent; O_2, 6 percent; N_2, 85 percent. Determine the composition of the fuel (values for x and y) that will produce the given analysis of the products and the dew-point temperature of the products.

12.27 **[CRA211]** An unknown hydrocarbon fuel is combusted with air (assumed to be 21 percent oxygen and 79 percent nitrogen on a volumetric basis), and a volumetric analysis of the combustion products yields the following percentages on a dry basis. CO_2 = 10.5 percent; O_2 = 5.3 percent; and N_2 = 84.2 percent. Determine the following:

(a) the mass fraction of carbon in the fuel
(b) the air-fuel ratio on a mass basis for this reaction
(c) the percent theoretical air

12.28 A mixture of 2 mol of methane and 5 mol of propane is burned in stoichiometric air. Calculate the air-fuel ratio on a mass basis. Determine the higher heating value for the mixture at 77°F.

12.29 Determine the heat of combustion for gaseous hydrogen at 77°F. Water in the products is in the liquid phase.

12.30 Calculate the heat of combustion for liquid methyl alcohol at 25°C. Water in the products is in the liquid phase.

12.31 Calculate the heat of combustion for liquid octane at 77°F. Water in the products is in the liquid phase.

12.32 Determine the higher and lower heating values at 25°C for gaseous butane.

12.33 Determine the higher and lower heating values at 77°F for hexane in (a) the liquid phase and (b) the gaseous phase.

12.34 Determine the enthalpy of combustion at 25°C when liquid butane is burned in (a) pure oxygen and (b) air. Assume water in the products is in the vapor phase.

12.35 Calculate the enthalpy of combustion at 77°F when propane is burned in (a) air and

(b) pure oxygen. Assume water in the products is in the vapor phase.

12.36 Octane vapor is burned in air. The products and reactants are both maintained at 1 atm pressure and 25°C. Assume water in the products is in the vapor phase. Calculate the heat of reaction for (a) stoichiometric proportions of air, (b) 40 percent excess air, and (c) 70 percent excess air. Compare the answers with the lower and higher heating values.

12.37 Calculate the higher and lower heating values at 77°F for methane by using the values for enthalpy of formation and compare the higher heating value with the value in Table G.2E.

12.38 Compute the higher and lower heating values at 25°C for methane by using the values for enthalpy of formation and compare the higher heating value with the value in Table G.2.

12.39 Compute the higher heating value at 77°F for propylene by using the values for enthalpy of formation and compare it with the value given in Table G.2E.

12.40 Determine the higher and lower heating values at 25°C for toluene in (a) the liquid phase and (b) the gaseous phase.

12.41 Methane is burned in 40 percent excess dry air in an open, steady-flow system. Assuming complete combustion, calculate the heat transfer from the combustion chamber based on 1 mol of fuel if the products and reactants are maintained at 1 atm and 77°F.

12.42 Hydrogen is burned in 20 percent excess air. The state of the reactants is 1 atm, 25°C. After complete combustion the products are returned to the original state. Determine the heat-transfer rate per mole of hydrogen. Assume water in the products remains in the vapor phase.

12.43 A closed, flexible combustion chamber is designed to maintain a constant pressure during the combustion process. Initially, the combustion chamber has a volume of 100 ft^3 and contains acetylene gas and 50 percent excess air at 30 psia and 77°F. The mixture is ignited and burns to completion. The products are cooled to 77°F. Determine the mass of acetylene and air in the chamber prior to ignition. Calculate the heat transfer from the chamber during the process.

12.44 Ethane gas is burned steadily in an open combustion chamber in stoichiometric proportions of air. The ethane and air enter the combustion chamber at 100 kPa and 25°C, and the products are cooled to 25°C. The mass flow rate of the ethane entering the combustion chamber is 2.1 kg/h. Determine the mass flow rate of air into the chamber and the heat-transfer rate for the process.

12.45 Ethyl alcohol vapor mixes with 30 percent excess air and enters a steady-flow combustion chamber at 1 atm and 25°C. The mass flow rate of the alcohol is 2 kg/min. The combustion process is complete, and the products leave the combustion chamber at 1 atm and 400 K. Determine the following:

(a) the air-fuel ratio on a mass basis
(b) the dew-point temperature of the products
(c) the mass flow rate of the air
(d) the heat-transfer rate from the combustion chamber

12.46 Calculate the mass flow rate of vaporized diesel fuel and air necessary to produce a heating rate of 1.5×10^4 Btu/h if the diesel fuel is burned in a steady-flow combustion chamber with 60 percent excess air. The reactants enter at 1 atm and 537°R, and the products leave at 1 atm and 700°R. Assume that combustion is complete and that the products of diesel fuel are the same as those of $C_{12}H_{26}$.

12.47 Octane enters a steady-flow combustion chamber at 77°F and 1 atm, where it mixes with 60 percent excess air at the same pressure and temperature. The products leave the combustion chamber at 1000°R and 1 atm. Assume that combustion is complete and that all gases are ideal. Calculate the heat transfer from the combustion chamber for each mole of octane burned if the octane is (a) a liquid or (b) a vapor.

12.48 Hydrogen is mixed with 40 percent excess air and burned steadily in a combustion chamber. The air and hydrogen enter the combustion chamber at 25°C and 1 atm. The heat transfer from the combustion chamber is 200,000 kJ/kg·mol hydrogen burned. Calculate the temperature of the products as they leave the combustion chamber.

12.49 Benzene is mixed with 100 percent excess air and burned completely in a closed, rigid tank. The reactants are originally at 1 atm and 77°F, and the products reach a temperature of 1200°R. Calculate the heat transfer per lb_m·mol of benzene during the process and the final pressure in the tank.

12.50 In a power-plant furnace 20 percent excess air is heated to 500 K before it enters a steady-flow combustion chamber, where it mixes with methane at 25°C. The products of combustion leave the furnace at 1200 K. Calculate the heat transfer in the furnace for each mole of methane that enters the combustion chamber.

12.51 [CRB311] A gaseous mixture of methane (CH_4) and hydrogen (H_2) has the following composition, expressed as a volumetric fraction: $CH_4 = 0.85$; $H_2 = 0.15$. The gas mixture is burned with a stoichiometric amount of air at 537°R. Combustion is complete and products exit at 720 °R.

(a) Calculate the temperature in °R used in determining the enthalpy of the combustion products for the calculation of the HHV.

(b) What is the higher heating value (HHV) of this gas mixture in units of Btu/lb_m·mol of mixture?

(c) If the gas mixture were combusted with pure oxygen instead of air, would the HHV be higher, lower, or the same?

12.52 [CRB211] A mixture of carbon monoxide and 50 percent excess air, initially at 298 K and 140 kPa, are combusted in a constant-volume cylinder. A maximum temperature of 2500 K is measured. Assume complete combustion. Determine the following:

(a) the maximum observed pressure in units of kPa

(b) the heat transfer in units of kg/kg·mol

12.53 [CRB111] Butane (C_4H_{10}) is burned with air and a volumetric analysis of the combustion products on a dry basis yields the following composition: CO_2 = 7.8 percent; CO = 1.1 percent; O_2 = 8.2 percent; and N_2 = 82.9 percent. The butane and air are initially at 537°R and 14.7 psia. The combustion gases exit at 720°R and 14.7 psia. Determine the following:

(a) the percent theoretical air

(b) the heat transfer per unit mass of butane

12.54 A rigid bomb calorimeter is charged with 1 mol of ethane vapor and 30 percent excess air at 1 atm and 77°F. The reactants ignite and burn to completion. Determine the temperature of the products if (a) the reaction occurs adiabatically and (b) the heat transfer from the mixture during the process is 5×10^5 Btu.

12.55 One mole of hydrogen is placed in a rigid, closed container with 30 percent excess oxygen. The oxygen-hydrogen mixture is maintained at 1 atm and 25°C. The mixture is ignited and the reaction continues to completion. Assuming that the reaction occurs adiabatically, determine the volume of the container and the final temperature and pressure of the products.

12.56 Hydrogen at 77°F is mixed with oxygen at 77°F and burned in a steady-flow adiabatic process. Calculate the temperatures of the products for the following proportions of oxygen: stoichiometric proportions, 20 percent excess oxygen, and 40 percent excess oxygen.

12.57 Propane with a temperature of 25°C is mixed with air at the same temperature and burned in a steady-flow, adiabatic process. Determine the air-fuel ratio required for the temperature of the products to be 1500 K.

12.58 Carbon monoxide at 77°F and 1 atm is mixed with an oxidizer at the same pressure and temperature. Assuming complete combustion, calculate the adiabatic flame temperature for 100 percent excess oxidizer if the oxidizer is (a) air and (b) oxygen. Why is the adiabatic flame temperature higher when oxygen is used?

12.59 Determine the adiabatic flame temperature when acetylene is burned in 40 percent excess air at 1 atm pressure and 77°F. Assume complete combustion.

12.60 Ethyl alcohol vapor is burned in 20 percent excess air. The reactants enter the combustion chamber at 25°C and 1 atm. Assuming complete combustion, calculate the adiabatic flame temperature for this reaction.

12.61 Hydrogen at 77°F and 1 atm is mixed with air at the same temperature and pressure and burns completely. Determine the adiabatic flame temperature for the following proportions of air and fuel: (a) stoichiometric proportions, (b) 100 percent excess air, and (c) 200 percent excess air. Explain why the adiabatic flame temperature decreases as the percent excess air increases.

12.62 Determine the adiabatic flame temperature when diesel fuel vapor (assume properties of dodecane, $C_{12}H_{26}$) is burned with 100 percent excess air. Assume that complete combustion occurs and that the reactants are at a pressure of 1 atm and a temperature of 25°C.

12.63 A system consists of a mixture of ethane gas and stoichiometric proportions of air. The gas mixture burns to completion in a steady process. The reactants are at 77°F and the products are at 2000°R. Assuming that the reaction occurs at a constant pressure of 1 atm, determine the entropy change of the system per unit mole of ethane and the irreversibility of the process if $T_0 = 77°F$. What is the source of the irreversibility?

12.64 One mole of methane is burned in 50 percent excess air in an insulated, closed rigid container. The reactants originally have a temperature of 25°C and a total pressure of 1 atm. Calculate the temperature and pressure of the products, the entropy change of the system per unit mole of ethane, and the irreversibility of the process if $T_0 = 25°C$. What is the source of the irreversibility?

12.65 One mole of methane is burned in 50 percent excess air in an insulated, closed rigid container. The reactants originally have a temperature of 77°F and a total pressure of 1 atm. Calculate the temperature and pressure of the products, the entropy change that occurs during the process, and the irreversibility of the process if $T_0 = 77°F$.

12.66 Methane at 298 K, 1 atm is mixed with 50 percent excess air with a temperature of 400 K and burned to completion. The products leave the combustion chamber at 700 K, 1 atm. Assuming that the mass flow rate of the methane is 0.1 kg/s, calculate the heat-transfer rate from the combustion chamber, the entropy change of the process, and the irreversibility of the process if $T_0 = 25°C$.

12.67 Hydrogen peroxide vapor enters a heat exchanger at 150 psia and 77°F, where it is heated until it decomposes into water vapor and oxygen. The H_2O and O_2 leave the heat exchanger at 150 psia and 1100°R and enter a well-insulated turbine, where they expand to a pressure of 1 atm and a temperature of 500°R. Determine the heat transfer per mole of H_2O_2 in the heat exchanger, the work output of the turbine per mole of H_2O_2, and the entropy change in the heat exchanger per mole of H_2O_2.

12.68 Liquid ethyl alcohol is burned in 30 percent excess air. The reactants enter the steady-flow combustion chamber at 1 atm and 25°C, and the products leave at 1 atm and 500 K. The mass flow rate of the alcohol is 0.25 kg/min. Calculate the heat-transfer rate from the combustion chamber, the mass flow rate of air required for combustion, and the entropy change that occurs within the combustion chamber per mole of alcohol.

12.69 Propane at 1 atm and 77°F is mixed with 40 percent excess air at 1 atm and 600°R and burned in a steady-flow process. The products leave the combustion chamber at 1 atm and 1200°R. Calculate the entropy change and irreversibility during the process on a mole basis of the propane. Assume that $T_0 = 77°F$.

12.70 [CRC211] Hydrogen at 298 K and 1 atm is steadily burned with a stoichiometric amount of pure oxygen at 298 K and 1 atm in an adiabatic reactor. Determine the following assuming that no dissociation reactions occur:

- **(a)** the adiabatic flame temperature in K
- **(b)** the total entropy change per unit mole in kJ/kg·mol·K of hydrogen assuming that the products are at the adiabatic flame temperature
- **(c)** If the hydrogen were combusted with 50 percent excess oxygen, would the adiabatic flame temperature be higher, lower, or the same?

12.71 [CRC111] A mixture of methanol (CH_3OH) vapor and 120 percent theoretical air is burned in a steady-flow process. The mixture enters the combustion chamber at 298 K and 140 kPa. The combustion products emerge from the combustion chamber at 400 K and 140 kPa. Heat transfer from the mixture is to an environment at 375 K. The enthalpy of formation of methanol at 298 K is −200,670 kJ/kg·mol. Calculate:

- **(a)** the heat transfer in units of kJ/kg·mol
- **(b)** the absolute entropy of the reactants per mole of methanol
- **(c)** the absolute entropy of the products per mole of methanol
- **(d)** the total entropy change per mole of methanol for this process

13 Chemical Equilibrium

13.1 INTRODUCTION

In the application of the conservation of energy to chemically reacting systems in the previous chapter, the assumption was tacitly made that the reactions proceeded to completion. Because many reactions reach a condition of equilibrium before the reaction reaches completion, the equilibrium composition of a reaction must be first established.

In this chapter criteria for chemical equilibrium will be determined by applying the second law of thermodynamics to chemically reacting systems. One such condition for chemical equilibrium was established in the previous chapter. For an isolated system the condition for chemical equilibrium was shown to be the composition that results in a maximum value of entropy of the system. While this condition for chemical equilibrium is valid, it has some serious limitations, primarily because chemical reactions that occur within an isolated system are seldom encountered. So that this restriction is eased, other conditions for chemical equilibrium are established in this chapter. These criteria permit the determination of the component properties when the system is in chemical equilibrium.

The criteria for chemical equilibrium established in this chapter also have some limitations. No conclusions concerning the rate at which the chemical reaction will take place can be drawn from the equilibrium criteria. The question of how rapidly a chemical reaction occurs can only be answered after the chemical kinetics of the reaction have been studied, a subject that is beyond the scope of this text.

This chapter begins with a general discussion of equilibrium, and several criteria for chemical equilibrium are then established. Finally, the equilibrium constant is introduced and shown to be a useful quantity for the determination of chemical equilibrium for a constant-temperature, constant-pressure reaction involving ideal gases.

13.2 EQUILIBRIUM

A system is said to be in a state of equilibrium if no changes in properties occur when the system is isolated from its surroundings. If no changes in the pressure occur within the system when it is isolated, the system is in *mechanical equilibrium*; if the temperature does not change when the system is isolated, the system is in *thermal equilibrium*. However, even though a system may be in both mechanical and thermal equilibrium, it is not necessarily in chemical equilibrium. For example, consider two chemical components that exist as a mixture in a system. If the system is isolated from the surroundings and the pressure and temperature remain unchanged, a chemical reaction may still occur, as evidenced by a change in the chemical composition of the mixture. If the chemical composition of the mixture does not change with time when the system is isolated from its surroundings, the system is said to be in *chemical equilibrium*. When in chemical equilibrium, a system is chemically homogeneous throughout and the chemical composition does not vary with time. When a system is in a state of mechanical, thermal, and chemical equilibrium, it is often said to be in *complete equilibrium*.

The general problem of determining the composition of a system at a state of chemical equilibrium is as follows: Suppose several chemically different species or components are mixed together in a system and they begin to react chemically. As the reaction continues, the chemical composition of the system changes. In such circumstances, how is it possible to determine when the system reaches a condition of chemical equilibrium? In the next section several criteria are derived that give property restrictions that exist at chemical equilibrium. These criteria will aid in determining how far a chemical reaction will proceed before reaching a condition of chemical equilibrium.

13.3 EQUILIBRIUM CRITERIA

In Chapter 12 the second law of thermodynamics was applied to an isolated system containing a chemical reaction to arrive at a condition for equilibrium including chemical equilibrium. In this section, the analysis is extended in order to develop alternative criteria for equilibrium.

Consider a closed, stationary system containing a simple compressible substance that approaches an equilibrium state from an initial condition that is not an equilibrium state. The conservation-of-energy equation for this system is

$$\delta Q - \delta W = dU \tag{13.1}$$

If irreversible work modes are absent, then

$$\delta W = P\, d\mathbf{V} \tag{13.2}$$

and Equation 13.1 may be expressed as

$$\delta Q = dU + P\, d\mathbf{V} \tag{13.3}$$

Furthermore, the second law of thermodynamics for this system (Equation 5.17) is

$$T\, dS \geq \delta Q \tag{13.4}$$

Even though irreversible work modes are not present, the process under consideration is not internally reversible unless the system is infinitesimally close to an equilibrium state throughout the process. Since the present process involves attaining an equilibrium state from an initial condition not at equilibrium, the process is not internally reversible and Equation 13.4 must be an inequality.

If Equation 13.3 is substituted into Equation 13.4, then

$$T\,dS \geq dU + P\,d\mathbf{V} \tag{13.5}$$

or

$$T\,dS - dU - P\,d\mathbf{V} \geq 0 \tag{13.6}$$

Equation 13.6 must be satisfied as the system approaches equilibrium, and it may be used to formulate alternative criteria for equilibrium. For example, if the closed system is isolated, then Equation 13.1 reduces to $dU = 0$ since δQ and δW are zero, and $d\mathbf{V} = 0$ because $\delta W = P\,d\mathbf{V} = 0$. Thus, for an isolated system U and $\mathbf{V}$ are constant and Equation 13.6 reduces to

$$(dS)_{U,\mathbf{V}} \geq 0 \tag{13.7}$$

where the subscripts U and $\mathbf{V}$ indicate that the process occurs at constant internal energy and constant volume. This equation reveals that the approach to a state of equilibrium by an isolated system proceeds in the direction of increasing entropy and that the entropy is a maximum at the equilibrium state. Thus the equilibrium criterion for an isolated system is expressed as

$$(dS)_{U,\mathbf{V}} = 0 \tag{13.8}$$

In reacting systems, a reaction can continue as long as the entropy increases. When the entropy reaches its maximum value, chemical equilibrium exists and the reaction will cease. Reactions that would result in a decrease in the entropy of an isolated system cannot occur, because they would violate the second law of thermodynamics. These conclusions are summarized graphically in Figure 13.1.

An alternative equilibrium criterion can be developed from Equation 13.6 with the aid of the Helmholtz function introduced in Equation 9.11:

$$A \equiv U - TS \tag{13.9}$$

Differentiating this expression gives

$$dA = dU - T\,dS - S\,dT \tag{13.10}$$

and combining Equation 13.10 and Equation 13.6 yields

$$-dA - S\,dT - P\,d\mathbf{V} \geq 0 \tag{13.11}$$

or

$$dA + S\,dT + P\,d\mathbf{V} \leq 0 \tag{13.12}$$

Thus, if a substance reacts chemically and the reaction occurs at constant temperature and constant volume,

$$(dA)_{T,\mathbf{V}} \leq 0 \tag{13.13}$$

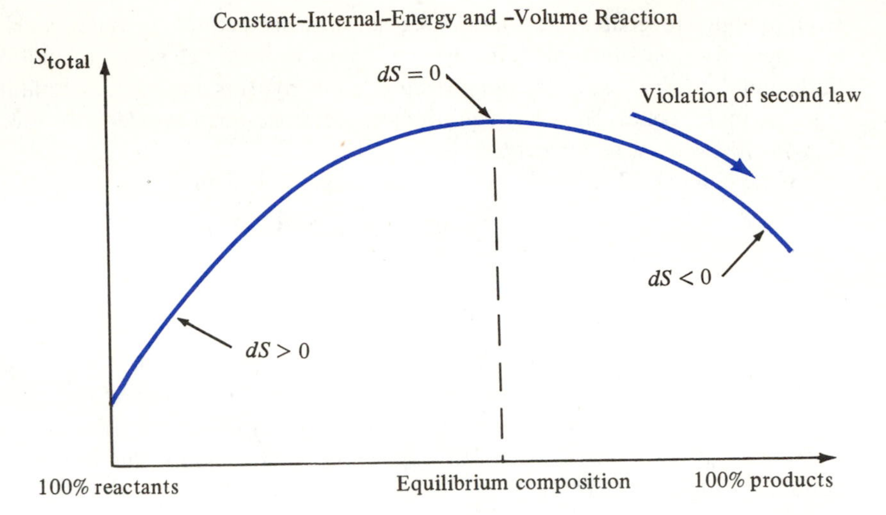

Figure 13.1 Equilibrium criteria for constant-internal-energy and -volume reaction.

Therefore, the equilibrium criterion for a chemical reaction that occurs in a closed system containing a simple compressible substance at constant volume and temperature is that the value for the Helmholtz function is a minimum, or

$$(dA)_{T,V} = 0 \tag{13.14}$$

That is, the reaction can continue as long as the Helmholtz function continues to decrease. Once the Helmholtz function reaches a minimum value, the equilibrium composition will be attained, and the reaction will cease. Reactions that would cause the Helmholtz function to increase cannot occur, because a violation of the second law of thermo-

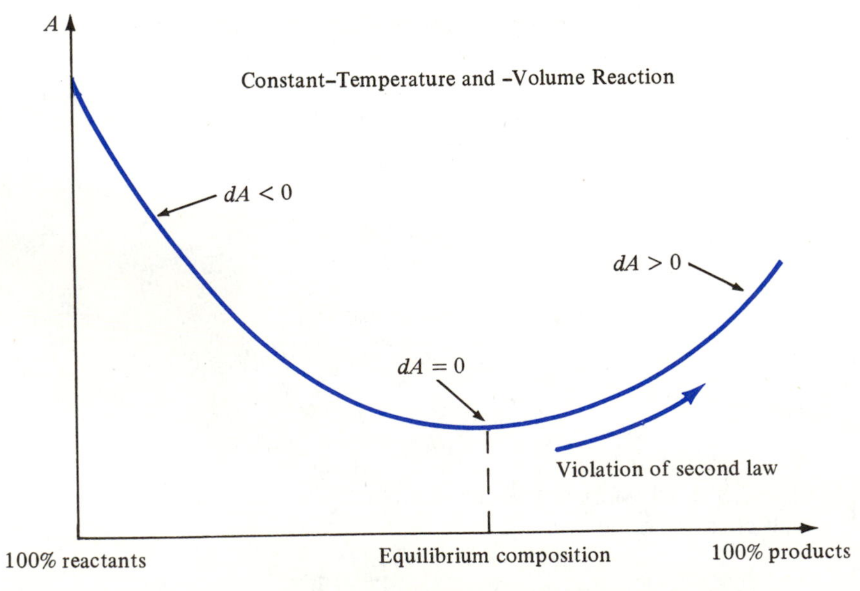

Figure 13.2 Equilibrium criteria for constant-temperature and -volume reaction.

dynamics would occur. A graphical interpretation of these results is shown in Figure 13.2.

Yet another alternative equilibrium criterion can be developed with the aid of the Gibbs function, defined in Equation 9.12:

$$G \equiv H - TS \tag{13.15}$$

Upon substituting $H = U + PV$ in Equation 13.15 and differentiating, we obtain

$$dG = dU + P\,dV + V\,dP - T\,dS - S\,dT$$

Combining this last equation and Equation 13.6 gives

$$-dG - S\,dT + V\,dP \geq 0$$

or

$$dG + S\,dT - V\,dP \leq 0 \tag{13.16}$$

From this equation we find that if a substance reacts chemically and the reaction occurs at constant temperature and constant pressure, then

$$(dG)_{T,P} \leq 0 \tag{13.17}$$

Therefore, a criterion for chemical equilibrium for a reaction that occurs in a closed system at constant temperature and pressure is that the value for the Gibbs function is a minimum, or

$$(dG)_{T,P} = 0 \tag{13.18}$$

The reaction can continue only as long as the Gibbs function continues to decrease. The reaction cannot proceed if the Gibbs function increases, because a violation of the second law of thermodynamics would occur. These results are summarized in Figure 13.3.

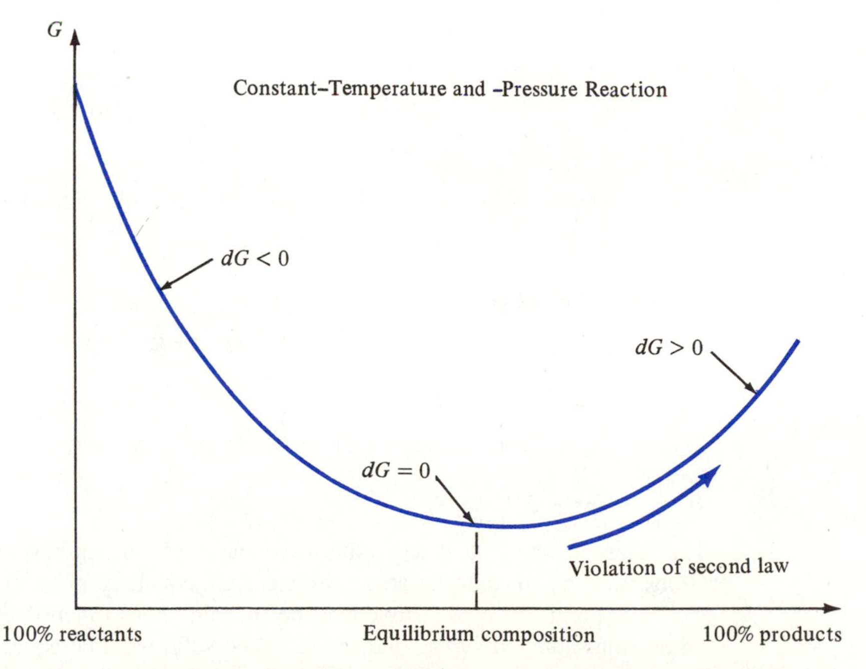

Figure 13.3 Equilibrium criteria for constant-temperature and -pressure reaction.

■ EXAMPLE 13.1

Show that the Gibbs functions of a saturated liquid and of a saturated vapor are equal when the two phases are in equilibrium. Demonstrate this equality for the following:

(a) Water at 100°C
(b) Refrigerant-12 at 0°C

Solution. If the two phases are in equilibrium, then they have the same temperature and the same pressure. The Gibbs function for the mixture consisting of saturated liquid and saturated vapor is

$$G = m_f g_f + m_g g_g$$

If some of the liquid evaporates at constant pressure and temperature, the change in the Gibbs function for the mixture is

$$(dG)_{T,P} = g_f\, dm_f + g_g\, dm_g$$

because g_g and g_f are constant when the pressure and temperature remain constant. Because of the conservation of mass,

$$-dm_g = dm_f$$

The condition for equilibrium expressed by Equation 13.18 for the two phases now becomes

$$(dG)_{T,P} = 0 = (g_f - g_g)dm_f$$

Therefore, for equilibrium

$$g_f = g_g$$

(a) For saturated water at 100°C,

$$g_f = h_f - Ts_f = 419.06 \text{ kJ/kg} - (373.15 \text{ K})(1.3069 \text{ kJ/kg·K})$$
$$= -68.6 \text{ kJ/kg}$$
$$g_g = h_g - Ts_g = 2675.7 \text{ kJ/kg} - (373.15 \text{ K})(7.3545 \text{ kJ/kg·K})$$
$$= -68.6 \text{ kJ/kg}$$

(b) For refrigerant-12 at 0°C,

$$g_f = h_f - Ts_f = 36.136 \text{ kJ/kg} - (273.15 \text{ K})(0.14225 \text{ kJ/kg·K})$$
$$= -2.72 \text{ kJ/kg}$$
$$g_g = h_g - Ts_g = 188.58 \text{ kJ/kg} - (273.15 \text{ K})(0.70034 \text{ kJ/kg·K})$$
$$= -2.72 \text{ kJ/kg}$$

This example shows that when any two phases are in equilibrium, the specific Gibbs functions of the saturated substances are equal. If this result is applied to the triple point, we could prove that the specific Gibbs function of the saturated liquid, the saturated vapor, and the saturated solid must be equal.

This result provides an additional valuable relationship between properties at saturated conditions. Since

$$g_f = g_g$$

then

$$h_f - Ts_f = h_g - Ts_g$$

or

$$h_{fg} = Ts_{fg}$$

Therefore, the tabulated values for s_{fg} can be calculated from the saturation temperature and the latent heat of vaporization of the substance. ■

13.4 THE EQUILIBRIUM CONSTANT

Three criteria for chemical equilibrium have been developed thus far. They are as follows:

1. For a constant-internal-energy and constant-volume process, the value of the entropy is a maximum at equilibrium (Equation 13.8).
2. For a constant-temperature and constant-volume process, the value of the Helmholtz function is a minimum at equilibrium (Equation 13.14).
3. For a constant-temperature and constant-pressure process, the value of the Gibbs function is a minimum at equilibrium (Equation 13.18).

The third of these equilibrium conditions is perhaps the most useful criterion for equilibrium, because many reactions occur at constant temperature and constant pressure. Also, this criterion takes on added significance because it can be shown that if a chemical reaction does not occur for the condition of constant temperature and constant pressure, then the reaction cannot occur under any condition.

The equilibrium criterion expressed in terms of the Gibbs function (Equation 13.18) can be used to determine the equilibrium composition. First, the specific Gibbs function, defined as

$$g \equiv h - Ts \tag{13.19}$$

can be differentiated to yield

$$dg = dh - T\,ds - s\,dT \tag{13.20}$$

Substituting one of the $T\,ds$ equations (Equation 5.20) gives an alternative expression for the differential change in the Gibbs function,

$$dg = v\,dP - s\,dT \tag{13.21}$$

Suppose now that a reaction involving ideal gases occurs at constant temperature and constant total pressure. If Equation 13.21 is used to evaluate the Gibbs function of each chemical component in a mixture, the appropriate temperature and pressure must be applied at each state. Equation 13.21 can be integrated over a given change in state, and the result can be used as a condition for chemical equilibrium.

Dalton's law (Equation 11.11) requires that the Gibbs function be evaluated at the temperature of the mixture and at the partial pressure of the individual species. Therefore, even though the total pressure of the mixture remains constant during the reaction, the partial pressure of each species changes as the reaction proceeds. During the chemical reaction the partial pressures of the products being formed increase, and the partial pressures of the reactants decrease as the reactants are transformed into products.

For a chemical species designated by a subscript i in the mixture evaluated at a temperature T and a partial pressure p_i, the value for the Gibbs function for each species measured relative to a reference state can be determined by integrating Equation 13.21. The integration is between a state given by the temperature T and the partial pressure p_i and a reference state at the same temperature but at a pressure P_0. The result is

$$\int_{T,P_0}^{T,p_i} dg_i = \int_{P_0}^{p_i} v\, dP = RT \int_{P_0}^{p_i} \frac{dP}{P} = RT \ln \frac{p_i}{P_0} \tag{13.22}$$

or

$$g_i(T, p_i) - g_i(T, P_0) = RT \ln \frac{p_i}{P_0} \tag{13.23}$$

The Gibbs function evaluated at the state (T, P_0) is given the symbol $g_i^0(T)$ and the reference pressure P_0 is assigned a value of 1 atm. Therefore, the Gibbs function for species i in a mixture of ideal gases is

$$g_i(T, p_i) = g_i^0(T) + RT \ln \frac{p_i}{P_0} \tag{13.24}$$

Equation 13.24 is the Gibbs function of an ideal gas at the temperature T and the partial pressure p_i, and it is often called the *chemical potential* of an ideal gas. As will be shown later in this section, the chemical potential is useful in predicting the equilibrium compositions of reacting-ideal-gas mixtures. On a molar basis the chemical potential of an ideal gas is

$$\bar{g}_i(T, p_i) = \bar{g}_i^0(T) + \bar{R}T \ln \frac{p_i}{P_0} \tag{13.25}$$

If the partial pressures are measured in atmospheres, then Equation 13.25 can be simplified to

$$\bar{g}_i(T, p_i) = \bar{g}_i^0(T) + \bar{R}T \ln p_i \tag{13.26}$$

The restrictions that apply to Equation 13.26 are the following:

1. All chemical species that make up the reacting mixture must be ideal gases.
2. The reaction must occur at constant temperature and constant total pressure.
3. The partial pressure of each component that makes up the reacting mixture is measured in atmospheres.

The condition for chemical equilibrium for a simple compressible substance proceeding along a constant-temperature and constant-pressure reaction expressed by Equation 13.18 is

$$(dG)_{T,P} = 0 \tag{13.27}$$

The equilibrium criterion expressed by Equation 13.27 can now be applied to a specific mixture of chemical components that exists in equilibrium. Suppose four chemical species represented by the symbols A, B, C, and D exist at equilibrium in a system at a given total pressure and temperature. Let the number of moles of each component be N_A, N_B, N_C, and N_D, respectively. Therefore, the mixture at equilibrium consists of the four components A, B, C, and D in the following proportions:

$$N_A A + N_B B + N_C C + N_D D \tag{13.28}$$

Corresponding to the set of chemical components, A, B, C, and D, there is a corresponding stoichiometric reaction

$$\nu_A A + \nu_B B \rightarrow \nu_C C + \nu_D D \tag{13.29}$$

which represents a complete reaction consisting of reactants A and B producing products C and D. The number of moles of the reactants in the stoichiometric equation are ν_A and ν_B, while the number of moles of products are ν_C and ν_D.

The correlation between the *actual* mixture of components at equilibrium and the *stoichiometric* reaction needs to be emphasized. The actual mixture of four components at equilibrium consists of the four chemical species A, B, C, and D. The corresponding stoichiometric reaction expressed by Equation 13.29 consists of the identical four components A, B, C, and D. The stoichiometric proportions ν_A through ν_D in Equation 13.29 can be evaluated once the reaction is specified; however, the number of moles N_A through N_D at equilibrium in the actual reaction are unknown and must be determined in order to fix the composition of the reaction at equilibrium. Once the chemical components that exist at equilibrium are specified, the corresponding stoichiometric reaction can be selected as the one that involves all of the possible components as either products or reactants.

Suppose that the mixture of ideal gases expressed by Equation 13.28 exists in a system that is maintained at constant pressure and constant temperature. The reaction occurs to an infinitesimal extent such that a small amount of the reactants A and B in the mixture are transformed into the products C and D. The condition for chemical equilibrium expressed by Equation 13.27 for this situation is

$$(dG)_{T,P} = \overline{g}_C dN_C + \overline{g}_D dN_D + \overline{g}_A dN_A + \overline{g}_B dN_B = 0 \tag{13.30}$$

The change in the number of moles of the components A, B, C, and D in Equation 13.30 is related to the number of moles ν that satisfy the corresponding stoichiometric equation, Equation 13.29. We define the *degree of the reaction* ϵ by the relationships

$$\begin{aligned} dN_A &= -\nu_A d\epsilon \qquad & dN_C &= \nu_C d\epsilon \\ dN_B &= -\nu_B d\epsilon \qquad & dN_D &= \nu_D d\epsilon \end{aligned} \tag{13.31}$$

These equations state that the change in the number of moles is proportional to the number of moles in the stoichiometric reaction and the degree of the reaction. The minus sign appears in the first two equations because the number of moles of the reactants N_A and N_B decreases as the reaction progresses, while the number of moles of the products N_C and N_D increases.

When the results of Equation 13.31 are substituted into Equation 13.30, the condition for chemical equilibrium becomes

$$\nu_C \overline{g}_C + \nu_D \overline{g}_D - \nu_A \overline{g}_A - \nu_B \overline{g}_B = 0 \tag{13.32}$$

Substitution of the expression for the Gibbs function for each component given in Equation 13.26 results in

$$\begin{aligned} &\nu_C[\overline{g}_C^0(T) + \overline{R}T \ln p_C] + \nu_D[\overline{g}_D^0(T) + \overline{R}T \ln p_D] \\ &- \nu_A[\overline{g}_A^0(T) + \overline{R}T \ln p_A] - \nu_B[\overline{g}_B^0(T) + \overline{R}T \ln p_B] = 0 \end{aligned} \tag{13.33}$$

The discussion that follows applies to any stoichiometric reaction regardless of the number of reactants or products, but the analysis is simplified somewhat if two reactants A and B and two products C and D are considered.

For convenience, the grouping of the $\overline{g}^0(T)$ terms in Equation 13.33 is defined as the *standard-state Gibbs function change*:

$$\Delta G^0(T) \equiv \nu_C \overline{g}_C^0(T) + \nu_D \overline{g}_D^0(T) - \nu_A \overline{g}_A^0(T) - \nu_B \overline{g}_B^0(T) \tag{13.34}$$

Substituting this definition into Equation 13.33 results in

$$\begin{aligned} \Delta G^0(T) &= -\overline{R}T[\ln (p_C)^{\nu_C} + \ln (p_D)^{\nu_D} - \ln (p_A)^{\nu_A} - \ln (p_B)^{\nu_B}] \\ &= -\overline{R}T \left[\ln \left(\frac{p_C^{\nu_C} p_D^{\nu_D}}{p_A^{\nu_A} p_B^{\nu_B}} \right) \right] \end{aligned} \tag{13.35}$$

Next, the *equilibrium constant*, K_P, is defined as

$$K_P \equiv \frac{p_C^{\nu_C} p_D^{\nu_D}}{p_A^{\nu_A} p_B^{\nu_B}} \tag{13.36}$$

With this definition of K_P, Equation 13.35, written in terms of the equilibrium constant, is

$$\Delta G^0(T) + \overline{R}T \ln K_P = 0 \tag{13.37}$$

Equation 13.37 represents the condition of chemical equilibrium as specified by the second law of thermodynamics for conditions of an ideal-gas reaction occurring at constant temperature and constant pressure.

The definition of the standard-state Gibbs function change given by Equation 13.34 indicates that $\Delta G^0(T)$ is only a function of temperature and its value is independent of the total pressure. Therefore, Equation 13.37 requires that the value for K_P also be independent of pressure. Thus it is a function of temperature, and values for the natural logarithm of K_P for several stoichiometric reactions appear in Table G.9.

The magnitude of the equilibrium constant is indicative of how easy or how difficult it is for a particular chemical reaction to take place. With reference to the definition of K_P given in Equation 13.36, those reactions that occur largely to completion result in greater values for the partial pressures of the products (p_C and p_D) and smaller values for the partial pressures of the reactants (p_A and p_B). Therefore, any factor that drives the reaction more toward completion will increase the value of K_P. A reaction that does not proceed toward completion to any great extent will result in large values for the partial pressures of the reactants (p_A and p_B) and small values for the partial pressures of the products (p_C and p_D) at equilibrium, resulting in a small value for the equilibrium constant.

The value for the equilibrium constant is always positive, but since the information provided in Table G.9 is ln K_P, both positive and negative values for ln K_P appear in the table. Negative values for ln K_P correspond to K_P values less than 1.0, and positive values of ln K_P are for K_P values greater than 1.0.

As a general rule, if the value of ln K_P listed in Table G.9 is less than -7, the components are so stable that the reaction will simply not occur. On the other hand, if ln K_P is greater than $+7$, the reaction occurs so readily that the reaction can be assumed to proceed to completion. Those reactions for which $-7 < \ln K_P < +7$ occur to varying degrees, so the equilibrium composition is somewhere between negligible completion and total completion, depending on the value for K_P.

To illustrate these conclusions, refer to the values of ln K_P listed in Table G.9. Consider the reaction $H_2 \rightarrow 2H$. At temperatures less than 2400 K, the reaction does not occur to any appreciable degree. That is, the equilibrium composition consists only of H_2, and no monatomic hydrogen appears in the equilibrium composition at temperatures less than 2400 K. Similarly, at temperatures above approximately 6000 K, the reaction occurs so readily that the equilibrium composition consists primarily of all H with negligible amounts of H_2. All reactions listed in Table G.9 are ones for which an increase in temperature encourages the reaction to proceed further to completion, because all of the reactions have values of K_P that increase with increasing temperature.

EXAMPLE 13.2

Calculate the equilibrium constant for the reaction

$$CO_2 \rightarrow CO + \frac{1}{2}O_2$$

at 77°F by using Equation 13.37. Compare the calculated value of K_P with the one in Table G.9E.

Solution. Substituting Equation 13.34 into Equation 13.37 and solving for K_P gives

$$\ln K_P = -\frac{1}{\overline{R}T}[\overline{g}^0_{CO}(537°R) + \frac{1}{2}\overline{g}^0_{O_2}(537°R) - \overline{g}^0_{CO_2}(537°R)]$$

Values for the Gibbs function for all stable elements are assigned a value of zero at the standard reference state of 298 K (537°R) and 1 atm, and values for the Gibbs function of formation for several compounds at 537°R are given in Table G.1E. Gibbs functions for states other than the standard reference state can be determined by using the definition of the Gibbs function and values of enthalpy and entropy from Tables G.3E through G.8E.

Substituting the values from Table G.1E gives

$$\ln K_p = -\frac{1}{(1.986\ \text{Btu/lb}_\text{m}\cdot\text{mol}\cdot°\text{R})(537°\text{R})}[-58{,}966\ \text{Btu/lb}_\text{m}\cdot\text{mol} - (-169{,}552\ \text{Btu/lb}_\text{m}\cdot\text{mol})]$$

$$= -103.7$$

$$K_P = \underline{\underline{9.21 \times 10^{-46}}}$$

The value for $\ln K_P$ checks with the one listed in Table G.9E. Since $\ln K_P < -7$, carbon dioxide will not dissociate into carbon monoxide and oxygen to any appreciable extent at 77°F. ■

■ EXAMPLE 13.3

Determine the equilibrium constant at 2000 K for the reaction.

$$CO + \frac{1}{2}O_2 \rightarrow CO_2$$

Solution. In Table G.9 the reaction that is closest to this one is

$$CO_2 \rightarrow CO + \frac{1}{2}O_2$$

which is the reaction that occurs in the opposite direction. For this reaction at 2000 K

$$\ln K_P = -6.635$$

$$K_P = \frac{p_{CO}\,p_{O_2}^{1/2}}{p_{CO_2}} = 1.3136 \times 10^{-3}$$

For the reaction

$$CO + \frac{1}{2}O_2 \rightarrow CO_2$$

the equilibrium constant is

$$K_P = \frac{p_{CO_2}}{p_{CO}\,p_{O_2}^{1/2}} = \frac{1}{1.3136 \times 10^{-3}} = \underline{\underline{761.3}}$$

at 2000 K. This example illustrates the general rule that reversing the direction of the reaction inverts the value of the equilibrium constant or changes the sign on the values of $\ln K_P$ in Table G.9. The magnitude of K_P for this reaction indicates that the combination of carbon monoxide with oxygen to form carbon dioxide at 2000 K occurs to completion. ■

■ EXAMPLE 13.4

Determine the value for K_P at 5400°R for the reaction

$$2H_2O \rightarrow 2H_2 + O_2$$

Solution. The reaction closest to this one that is listed in Table G.9E is the stoichiometric reaction

$$H_2O \rightarrow H_2 + \frac{1}{2} O_2$$

For this reaction the value for K_P at 5400°R is

$$\ln K_P = -3.086$$

$$K_P = \frac{p_{H_2} p_{O_2}^{1/2}}{p_{H_2O}} = 4.5684 \times 10^{-2}$$

For the reaction involving twice the number of moles of reactants and products,

$$2H_2O \rightarrow 2H_2 + O_2$$

the definition of K_P is

$$K_P = \frac{p_{H_2}^2 p_{O_2}}{p_{H_2O}^2} = (4.5684 \times 10^{-2})^2 = \underline{\underline{2.087 \times 10^{-3}}}$$

This example shows that the expression for K_P in Table G.9E can be used for reactions that involve multiples of the number of moles of reactants and products. ■

The products of combustion from mobile sources such as internal-combustion engines and stationary sources such as chemical plants and electric power plants produce pollutants that can have a major impact on the health of individuals. If all combustion processes of hydrocarbon fuels were to occur under theoretical conditions, the fuel could be completely oxidized to provide heat transfer and the by-products of the combustion process would be carbon dioxide and water vapor. Unfortunately, theoretical combustion rarely, if ever, takes place when a fuel burns under actual conditions. Impurities in the fuel, poor control of the air-fuel ratio, incomplete combustion, and variations in the combustion temperature help to promote pollutants such as sulfur oxides, nitrogen oxides, carbon monoxide, and particulate matter.

The presence of carbon monoxide is a major factor in air pollution. It is a colorless and odorless gas that can cause physiological changes in body tissue, and in sufficient quantities it can cause death. Carbon monoxide is an intermediate product resulting from incomplete combustion of the hydrocarbon fuel in oxygen or air. When the air-fuel ratio is below stoichiometric proportions (i.e., a ''rich'' mixture of fuel is burned), there is insufficient oxygen to completely oxidize all of the carbon in the fuel to produce carbon dioxide, and some carbon monoxide will result in the products of combustion. Even when the air-fuel ratio exceeds stoichiometric proportions (i.e., a ''lean'' mixture of fuel is burned), carbon monoxide can still be formed in the products because of poor mixing of the fuel and air in the combustion zone. Furthermore, combustion temperatures may be sufficiently high so that the chemical-equilibrium considerations will dictate that some carbon dioxide will dissociate into carbon monoxide. The values for K_P for this dissociation reaction given in Tables G.9 and G.9E indicate that the formation of CO from CO_2 will be negligible at temperatures below approximately 2000 K (3600°R). However, as the combustion-zone temperatures exceed approximately 2000 K, the dissociation of carbon dioxide can no longer be neglected. Once carbon monoxide is produced and emitted into the atmosphere, it is a very stable compound and it has an average lifetime of several months before it is oxidized to carbon dioxide.

Combustion of a hydrocarbon fuel with air can produce several oxides of nitrogen including nitric oxide (NO), nitrogen dioxide (NO_2), nitrous oxide (N_2O), nitrogen trioxide (N_2O_3), and nitrogen pentoxide (N_2O_5). A mixture of these compounds produced by a combustion process is usually called NO_x (or NOX) to imply an unknown mixture of oxides of nitrogen. Of the several possible oxides of nitrogen, only NO, NO_2, and N_2O are produced in sufficient amounts to be considered as pollutants. Nitric oxide (NO) is a colorless gas, and at the concentrations present in the atmosphere it is usually not considered to be a health hazard. However, NO is known to be an active ingredient in the photochemical formation of smog, and it can react to form NO_2, which is a reddish-brown gas that in high concentrations is known to affect the respiratory system in humans.

For temperatures up to about 1000 K (1800°R) the equilibrium constant indicates that formation of nitric oxide from diatomic nitrogen and oxygen is not likely. At elevated temperatures typical of conventional combustion processes, the formation of NO is possible, and the dominant component of NO_x in combustion exhaust gases is NO. When the combustion exhaust gases are released in the environment, the gas temperature drops and the nitric oxide becomes thermodynamically unstable. However, the reaction rates of NO with O_2 to form NO_2 and the decomposition of NO into N_2 and O_2 are restricted at lower temperatures. This being the case, the concentration of NO in the exhaust gases is essentially frozen at the levels existing after the high-temperature combustion process.

Since the equilibrium composition of chemical reactions that take place during combustion is governed by the temperatures that occur during the combustion process, one logical way to regulate the composition of exhaust gases is to simply regulate the combustion temperature by changing the air-fuel ratio. The stoichiometric air-fuel ratio produces the maximum combustion temperature and encourages the formation of NO_x. The maximum equilibrium NO_x concentration in actual combustion processes occurs with a small percentage of excess air. The formation of carbon monoxide is also influenced by the air-fuel ratio. Carbon monoxide occurs primarily because sufficient oxygen is not present during combustion to completely oxidize all of the carbon in the fuel. Consequently, lean mixtures (air-fuel ratios in excess of stoichiometric proportions) reduce the presence of CO in the exhaust gases. A lean mixture also helps reduce the amount of unburned fuel in the exhaust gases. Unfortunately, when the air-fuel ratio is adjusted to minimize the amount of unburned fuel and CO in the exhaust, the conditions are such that the amount of NO in the exhaust tends to reach a maximum.

Other factors influence the presence of pollutants in the combustion products. The formation of NO and the presence of unburned hydrocarbons can be reduced by retarding the timing of the spark in an internal-combustion engine. Retarding the spark does have limitations, however, because when the spark advance is reduced below a certain amount, the engine will begin to lose power. The ratio of the surface area of the cylinder to the volume of the combustion chamber also has an effect on the amount of the pollutants formed during combustion. Once the combustion is initiated by the spark, the flame front travels through the air-fuel mixture until it eventually reaches the walls of the combustion chamber. The walls are cooled by either a water jacket or air, and the fuel immediately adjacent to the cylinder wall can be cooled to such an extent that the flame front is "quenched," leaving a layer of unburned fuel in the cylinder. This layer of unburned fuel is removed during the exhaust stroke along with the other products of combustion. Reducing the ratio of combustion-chamber surface area to volume reduces the amount of fuel quenching, and the proportion of unburned fuel in the exhaust gases is thereby reduced. Therefore, modifications such as changing the shape of the combustion chamber (as in a stratified-charge engine) decrease the amount of pollutants. These modifications do have severe limitations, however, because they can have a significant influence on fuel consumption, engine power, and engine roughness.

13.5 CALCULATION OF EQUILIBRIUM COMPOSITIONS

At this stage of the treatment of chemical equilibrium, the equilibrium constant will be used rather than the Gibbs function to determine the equilibrium compositions of ideal gases. For the purpose of this study the Gibbs function is merely a convenient vehicle to establish the definition of K_P.

The definition of K_P in Equation 13.36 can be rewritten in terms of the total pressure of the reaction. Since all chemical species in the reaction are assumed to be ideal gases, the mole fraction of each component is also equal to the ratio of the component partial pressure to total pressure (see Equation 11.17), or

$$p_i = y_i P = \left(\frac{N_i}{N}\right) P \tag{13.38}$$

Substituting this result into the definition of K_P yields

$$K_P = \frac{(N_C P/N)^{\nu_C}(N_D P/N)^{\nu_D}}{(N_A P/N)^{\nu_A}(N_B P/N)^{\nu_B}} \tag{13.39}$$

or

$$K_P = \left[\frac{N_C^{\nu_C} N_D^{\nu_D}}{N_A^{\nu_A} N_B^{\nu_B}}\right]\left(\frac{P}{N}\right)^{\nu_C + \nu_D - \nu_A - \nu_B} \tag{13.40}$$

where the total number of moles is

$$N = N_A + N_B + N_C + N_D \tag{13.41}$$

The fact that the equilibrium composition is a function of both the temperature and the total pressure at which the reaction is carried out is clearly illustrated by Equation 13.40. The value for K_P is a function of temperature only, and the influence of pressure can be seen in Equation 13.40 as P raised to the $(\nu_C + \nu_D - \nu_A - \nu_B)$ power. Note that if the number of moles of products equals the number of moles of reactants in the *stoichiometric* reaction (that is, $\nu_C + \nu_D = \nu_A + \nu_B$), the pressure has no effect on the equilibrium composition. If the number of moles of the products in the stoichiometric reaction exceeds the number of moles of reactants ($\nu_C + \nu_D > \nu_A + \nu_B$), then an increase in pressure will increase the number of moles of reactants and decrease the number of moles of products listed in the stoichiometric reaction. If the number of moles of products is less than the number of moles of reactants in the stoichiometric reaction ($\nu_C + \nu_D < \nu_A + \nu_B$), then an increase in the pressure will have the opposite effect.

The scheme used to determine the equilibrium composition at a given temperature and pressure for a known initial mixture of several chemical components is now reasonably straightforward. Once the value for K_P for a given stoichiometric reaction is determined from Table G.9 or G.9E for a given temperature, Equation 13.40 represents a single expression for the unknown values for the number of moles at equilibrium: N_A, N_B, N_C, and N_D. The total pressure is a known value, and the total number of moles at equilibrium, N, is related to the number of moles of each chemical species by Equation 13.41.

The coefficients of the stoichiometric equation, ν_A, ν_B, ν_C, and ν_D, are known once the stoichiometric equation that corresponds to the given components at equilibrium has been selected. Equation 13.40 therefore represents one equation with four unknowns, N_A, N_B, N_C, and N_D. However, other equations that result from the application of the conservation of mass applied to each chemical species can be formulated that will relate these quantities. The problem then consists of solving simultaneously the equations that result from application of the conservation of mass and the equation relating K_P to the number of moles of each chemical component at equilibrium.

Finally, one caution should be stated. The values for K_P listed in Tables G.9 and G.9E are defined in terms of a *stoichiometric* reaction, and values of ν_A, ν_B, ν_C, and ν_D are known once the stoichiometric reaction is specified. However, the reaction for which the equilibrium composition is to be determined is not the stoichiometric reaction, but one that involves all of the chemical species that are possible at equilibrium. For example, the equilibrium composition that results when oxygen and carbon monoxide are mixed and maintained at a constant temperature and pressure could possibly contain CO, CO_2, and O_2. Thus the correct stoichiometric reaction would be

$$CO + \frac{1}{2}O_2 \rightarrow CO_2 \tag{13.42}$$

because it contains all of the possible chemical species that could exist at equilibrium. Once the stoichiometric reaction is selected, the values for ν_A, ν_B, ν_C, and ν_D are known. For this particular reaction they are

$$\nu_A = 1 \qquad \nu_B = \frac{1}{2} \qquad \nu_C = 1 \qquad \nu_D = 0$$

The use of the equilibrium constant in determining the equilibrium composition is illustrated in the following examples.

■ EXAMPLE 13.5

Nitrogen is placed in a container and maintained at 1 atm pressure. Determine the amount of N_2 that dissociates into monatomic nitrogen if the nitrogen is allowed to reach equilibrium at (a) 3000 K and (b) 6000 K.

Solution. The chemical reaction for an initial composition of pure nitrogen is

$$N_2 \rightarrow xN_2 + yN$$

where x and y represent the number of moles of diatomic and monatomic nitrogen that exist at equilibrium. Applying the conservation of mass to the nitrogen in this reaction gives a relationship between x and y:

$$y = 2 - 2x$$

The reaction that will satisfy the conservation of mass is therefore

$$N_2 \rightarrow xN_2 + (2 - 2x)N$$

The total number of moles at equilibrium is

$$N = x + 2 - 2x = 2 - x$$

and the partial pressures of N_2 and N, measured in atmospheres at equilibrium for a total pressure of 1 atm, are

$$p_{N_2} = \frac{x}{2 - x} \qquad p_N = \frac{2 - 2x}{2 - x}$$

The reaction listed in Table G.9 that involves all of the components in the equilibrium mixture (N_2 and N) for this particular problem is

$$N_2 \rightarrow 2N$$

and the definition of K_P for this reaction is

$$K_P = \frac{p_N^2}{p_{N_2}}$$

Substituting the partial pressures for the equilibrium mixture into this expression yields

$$K_P = \left(\frac{2 - 2x}{2 - x}\right)^2 \left(\frac{2 - x}{x}\right)$$

or

$$(4 + K_P)x^2 - (8 + 2K_P)x + 4 = 0$$

Once the value for K_P is determined, this expression can be used to determine the number of moles of N_2 and N in the equilibrium mixture of products.

(a) At 3000 K the value for $\ln K_P$, from the Table G.9, is

$$\ln K_P = -22.359$$

Thus

$$K_P = 1.948 \times 10^{-10}$$

Substituting this value of K_P into the above polynomial relating x and K_P and solving for x gives

$$x = 1.0 \qquad y = 0$$

Therefore, the equilibrium composition at 3000 K is

$$N_2 \rightarrow N_2$$

That is, diatomic nitrogen does not dissociate to any measurable extent at 3000 K. This result should be expected, because nitrogen is known to be very stable and practically inert even at relatively high temperatures.

(b) At 6000 K the value for ln K_P, from Table G.9, is

$$\ln K_P = -2.865$$

and

$$K_P = 5.698 \times 10^{-2}$$

Substituting into the polynomial for x gives

$$x = 0.881 \qquad y = 0.238$$

Therefore, the equilibrium composition at 6000 K is

$$N_2 \rightarrow 0.881\ N_2 + 0.238\ N$$

which indicates that the dissociation reaction is appreciable at this temperature, because approximately 12 percent of the mass of the diatomic nitrogen has dissociated to monatomic nitrogen at 6000 K. ■

■ EXAMPLE 13.6

Initially, a mixture of 1 mol of carbon monoxide and 1/2 mol of diatomic oxygen is placed in a container and maintained at 4680°R and a constant total pressure. Determine the equilibrium composition of this mixture at (a) 1 atm and (b) 10 atm.

Solution. The reaction for this example is

$$CO + \frac{1}{2} O_2 \rightarrow xCO + yCO_2 + zO_2$$

where the number of moles of CO, CO_2, and O_2 are unknown values at equilibrium. Monatomic oxygen is not considered as a possible component at equilibrium because the value of K_P at 4680°R for the reaction $O_2 \rightarrow 2O$ indicates that the dissociation of O_2 at this temperature is not significant. Applying the conservation of mass to both the carbon and the oxygen in this reaction results in two mass-balance equations,

$$1 = x + y$$

$$2 = x + 2y + 2z$$

Solving these two equations in terms of the single unknown x gives

$$y = 1 - x$$

$$z = \frac{x}{2}$$

The chemical reaction in terms of the single unknown x is now

$$CO + \frac{1}{2} O_2 \rightarrow xCO + (1 - x)CO_2 + \left(\frac{x}{2}\right) O_2$$

where the right side of the chemical equation represents the composition at chemical equilibrium. The total number of moles in the mixture at equilibrium is then

$$N = x + 1 - x + \frac{x}{2} = 1 + \frac{x}{2}$$

The partial pressures of the three components that make up the mixture at equilibrium in terms of the total pressure P are

$$p_{CO} = \frac{xP}{1 + (x/2)} = \frac{2xP}{2 + x}$$

$$p_{CO_2} = \frac{(1 - x)P}{1 + (x/2)} = \frac{2(1 - x)P}{2 + x}$$

$$p_{O_2} = \frac{(x/2)P}{1 + (x/2)} = \frac{xP}{2 + x}$$

The stoichiometric reaction listed in Table G.9E that contains all of the chemical species that exist in the equilibrium mixture for this problem is

$$CO_2 \rightarrow CO + \frac{1}{2} O_2$$

The value for $\ln K_P$ at 4680°R for this reaction is

$$\ln K_P = -2.801$$

or

$$K_P = \frac{p_{CO} p_{O_2}^{1/2}}{p_{CO_2}} = 0.06075$$

Substituting values for each component partial pressure in the equilibrium mixture results in an equation for the value of x in terms of the total pressure of the equilibrium mixture:

$$0.06075 = \frac{[2xP/(2 + x)][xP/(2 + x)]^{1/2}}{[2(1 - x)P/(2 + x)]}$$

or

$$x^3P - 0.00369x^3 + 0.01107x - 0.00738 = 0$$

(a) At a total pressure of 1 atm, $P = 1$ and the above algebraic equation for x can be solved to give

$$x = 0.176$$

$$y = 1 - x = 0.824$$

$$z = \frac{x}{2} = 0.088$$

The equilibrium composition at 4680°R and 1 atm is therefore

$$CO + \frac{1}{2}O_2 \rightarrow 0.176\ CO + 0.824\ CO_2 + 0.088\ O_2$$

(b) At a total pressure of 10 atm, $P = 10$ and the expression for x can be solved to give

$$x = 0.086$$

$$y = 1 - x = 0.914$$

$$z = \frac{x}{2} = 0.043$$

The equilibrium composition at 4680°R and 10 atm is therefore

$$CO + \frac{1}{2}O_2 \rightarrow 0.086\ CO + 0.914\ CO_2 + 0.043\ O_2$$

The results illustrate that increasing the pressure for this particular reaction produces less CO and O_2 and more CO_2 at equilibrium. This behavior should have been expected because the number of moles of reactants is less than the number of moles of products in the stoichiometric reaction. ■

13.6 EFFECT OF INERT GASES ON EQUILIBRIUM

The presence of inert substances in a chemical reaction can have an effect on the equilibrium composition. As an example, consider the situation of the stoichiometric reaction of oxidizing carbon monoxide with oxygen to form carbon dioxide. The reaction is

$$CO + \frac{1}{2}O_2 \rightarrow CO_2 \tag{13.43}$$

The definition of the equilibrium constant for this reaction is

$$K_P = \frac{p_{CO_2}}{p_{CO}p_{O_2}^{1/2}}$$

Now suppose that the reaction is modified by including the presence of 1 mol of inert nitrogen. This reaction becomes

$$CO + \frac{1}{2}O_2 + N_2 \rightarrow CO_2 + N_2$$

The equilibrium constant for this reaction is

$$K_P = \frac{p_{CO_2}p_{N_2}}{p_{CO}p_{O_2}^{1/2}p_{N_2}} = \frac{p_{CO_2}}{p_{CO}p_{O_2}^{1/2}}$$

which is identical to the equilibrium constant for the reaction without the inert gas. The presence of the inert gas, therefore, will not change the value of K_P, but it will influence the equilibrium composition. If the expression for K_P for the chemical reaction given in Equation 13.43 is written in terms of the mole fractions of the components by using Equation 13.40, the result is

$$K_P = \left(\frac{N_{CO_2}}{N_{CO}N_{O_2}^{1/2}}\right)\left(\frac{P}{N}\right)^{-1/2} \tag{13.44}$$

where P is the total pressure of the reaction and N is the total number of moles of the gas mixture at equilibrium. The value of N is changed when an inert gas is present in the reaction, and therefore Equation 13.44 shows that the presence of an inert gas will change the equilibrium composition even though it does not alter the value of K_P.

These conclusions are illustrated in the following example.

EXAMPLE 13.7

A mixture of 1 mol carbon monoxide, 1/2 mol of diatomic oxygen, and 1 mol of nitrogen is placed in a container and maintained at 4680°R and a pressure of 1 atm. Determine the equilibrium composition.

Solution. This problem is the same as part a of Example 13.6 except that the reaction includes the presence of nitrogen. At the given temperature of 4680°R, nitrogen will not dissociate, although oxides of nitrogen may appear. If the nitrogen is assumed to be completely inert, the form of the nitrogen in the products will be diatomic nitrogen. The reaction is therefore

$$CO + \frac{1}{2}O_2 + N_2 \rightarrow xCO + yCO_2 + zO_2 + N_2$$

Mass balances for the carbon and oxygen result in the same relationships as given in Example 13.6:

$$y = 1 - x$$

$$z = \frac{x}{2}$$

The total number of moles of gases at equilibrium is

$$N = x + y + z + 1 = x + 1 - x + \frac{x}{2} + 1 = \frac{x}{2} + 2$$

With the value for K_P from Table G.9E for the reaction $CO_2 \rightarrow CO + 1/2O_2$ at 4680°R, the reaction for equilibrium written in terms of the single unknown x for a reaction pressure of 1 atm is

$$0.06075 = \frac{p_{CO}p_{O_2}^{1/2}}{p_{CO_2}} = \frac{\{x/[(x/2) + 2]\}\{(x/2)/[(x/2) + 2]\}^{1/2}}{(1 - x)/[(x/2) + 2]}$$

The value for x that satisfies this equation is

$$x = 0.212$$

and the number of moles of carbon dioxide and oxygen at equilibrium are

$$y = 0.788 \qquad z = 0.106$$

The equilibrium reaction at 4680°R and 1 atm is therefore

$$CO + \frac{1}{2} O_2 + N_2 \rightarrow 0.212\, CO + 0.788\, CO_2 + 0.106\, O_2 + N_2$$

Comparing this equilibrium composition with the one obtained for part a of Example 13.6 shows that the presence of inert nitrogen retards the reaction, because less of the oxygen has reacted to form CO_2 in the equilibrium mixture when the nitrogen is present. ■

13.7 EQUILIBRIUM FOR SIMULTANEOUS REACTIONS

In the previous discussion of chemical equilibrium, the reactions were selected so that the equilibrium composition could be determined by using the equilibrium constant for only a single stoichiometric reaction. Most practical chemical reactions are more complex than this, because they usually involve two or more reactions that occur simultaneously. Whenever a chemical species in the reaction can participate in two or more simultaneous reactions, the final equilibrium composition can be determined by applying the equilibrium constants of each possible simultaneous reaction.

Consider as an example an initial mixture of H_2O that is maintained at a constant pressure and temperature sufficient to produce an equilibrium composition of H_2, O_2, H_2O, and OH. No single stoichiometric reaction listed in Table G.9 contains all of these products. However, there are two stoichiometric reactions that can occur simultaneously and include all possible components at equilibrium. These stoichiometric reactions are

$$H_2O \rightarrow H_2 + \frac{1}{2} O_2 \qquad \text{(reaction 1)}$$

and

$$H_2O \rightarrow \frac{1}{2} H_2 + OH \qquad \text{(reaction 2)}$$

Following the identical procedure used in Section 13.4 to derive the value of K_P in terms of the Gibbs function, we can show that the values for the equilibrium constant for these two reactions are

$$\ln K_{P1} = -\frac{\Delta G_1^0}{\overline{R}T} = \frac{p_{H_2} p_{O_2}^{1/2}}{p_{H_2O}}$$

and

$$\ln K_{P2} = -\frac{\Delta G_2^0}{\overline{R}T} = \frac{p_{H_2}^{1/2} p_{OH}}{p_{H_2O}}$$

Once the values for the two equilibrium constants have been determined, these equations can be solved simultaneously for a relationship among the number of moles of each component at equilibrium. Additional relationships for the equilibrium composition can be obtained by applying the conservation of mass to each chemical species involved in the reaction.

The following example illustrates the procedure to be used in determining the equilibrium composition when simultaneous reactions can occur.

■ EXAMPLE 13.8

One mol of carbon dioxide and 1/2 mol of oxygen are mixed and heated to a temperature of 3600 K and a pressure of 1 atm. The products consist of CO_2, CO, O_2, and O. Determine the equilibrium composition of the products.

Solution. The chemical reaction is

$$CO_2 + \frac{1}{2} O_2 \rightarrow xCO_2 + yCO + zO_2 + wO$$

The conservation of mass for the carbon and oxygen will provide two independent equations relating the unknown number of moles x, y, z, and w. Conservation of mass for the carbon yields

$$1 = x + y$$

and conservation of mass applied to the oxygen gives

$$3 = 2x + y + 2z + w$$

Substituting the value of y in terms of x and rearranging gives

$$z = 1 - \frac{x}{2} - \frac{w}{2}$$

Two additional relations among the quantities x, y, z, and w must be determined before the equilibrium composition can be established.

Two independent stoichiometric reactions that involve all of the components that exist at equilibrium will provide two additional equations for the equilibrium composition. They are

$$CO_2 \rightarrow CO + \frac{1}{2} O_2 \qquad \text{(reaction 1)}$$

or

$$O_2 \rightarrow 2O \qquad \text{(reaction 2)}$$

The equilibrium constants for these two reactions at 3600 K are

$$\ln K_{P1} = 0.701 \qquad K_{P1} = 2.016$$

$$\ln K_{P2} = -0.926 \qquad K_{P2} = 0.396$$

The total number of moles that exist at equilibrium is

$$N = x + y + z + w = x + (1 - x) + \left(1 - \frac{x}{2} - \frac{w}{2}\right) + w = 2 - \frac{x}{2} + \frac{w}{2}$$

The definitions of the equilibrium constants for the two reactions are

$$K_{P1} = \frac{p_{CO}p_{O_2}^{1/2}}{p_{CO_2}} = \left(\frac{y}{N}\right)\left(\frac{z}{N}\right)^{1/2}\left(\frac{N}{x}\right) = \left(\frac{y}{x}\right)\left(\frac{z}{N}\right)^{1/2}$$

$$K_{P2} = \frac{p_O^2}{p_{O_2}} = \left(\frac{w}{N}\right)^2\left(\frac{N}{z}\right) = \frac{w^2}{Nz}$$

Substituting values for K_{P1}, K_{P2}, and the expressions that stem from the application of the conservation of mass results in

$$2.016 = \left(\frac{1 - x}{x}\right)\left[\frac{1 - (x/2) - (w/2)}{2 - (x/2) + (w/2)}\right]^{1/2}$$

and

$$0.396 = \frac{w^2}{[2 - (x/2) + (w/2)][1 - (x/2) - (w/2)]}$$

Solving these two equations simultaneously gives

$$x = 0.20$$

$$w = 0.70$$

and

$$y = 1 - x = 0.80$$

$$z = 1 - \frac{x}{2} - \frac{w}{2} = 0.55$$

The equilibrium composition at 1 atm and 3600 K is therefore

$$0.20\ CO_2 + 0.80\ CO + 0.55\ O_2 + 0.70\ O$$

■

13.8 SUMMARY

If the composition of a mixture of different chemical species does not change with time when the system is isolated from its surroundings, then the system is in chemical equilibrium.

Several criteria can be used to determine the equilibrium composition of a mixture of different chemical species. All of the equilibrium criteria result from an application of the second law of thermodynamics. They are as follows:

1. For a constant-internal-energy and constant-volume process, the value of the entropy of the system is a maximum at equilibrium, or

$$(dS)_{U,V} = 0 \tag{13.8}$$

2. For a constant-temperature and constant-volume process, the value of the Helmholtz function of the system is a minimum at equilibrium, or

$$(dA)_{T,V} = 0 \tag{13.13}$$

3. For a constant-temperature and constant-pressure process, the value of the Gibbs function of the system is a minimum at equilibrium, or

$$(dG)_{T,P} = 0 \tag{13.18}$$

If the equilibrium criterion expressed by Equation 13.18 is applied to a reaction that consists entirely of ideal gases, the second law of thermodynamics requires that at chemical equilibrium

$$\Delta G^0(T) + \overline{R}T \ln K_P = 0 \tag{13.37}$$

where the standard-state Gibbs function change is defined as

$$\Delta G^0(T) \equiv \nu_C \overline{g}_C^0(T) + \nu_D \overline{g}_D^0(T) - \nu_A \overline{g}_A^0(T) - \nu_B \overline{g}_B^0(T) \tag{13.34}$$

and the equilibrium constant is defined as

$$K_P = \frac{p_C^{\nu_C} p_D^{\nu_D}}{p_A^{\nu_A} p_B^{\nu_B}} \tag{13.36}$$

The partial pressures that appear in Equation 13.36 are measured in atmospheres, and the ν_i's are the number of moles of products and reactants in the stoichiometric reaction

$$\nu_A A + \nu_B B \rightarrow \nu_C C + \nu_D D \tag{13.29}$$

The equilibrium constant is a function only of the temperature, and values for K_P for a number of common stoichiometric reactions are listed in Tables G.9 and G.9E. The value of K_P does not depend on the amount of the various components present before the chemical reaction occurs. Equation 13.36 can be used to relate the number of moles of each chemical species that exists at equilibrium. Other equations relating the number of moles result from the conservation of mass applied to each chemical species. Applying both the conservation of mass and the second law of thermodynamics in the form of the equilibrium constant allows us to determine the extent to which the chemical reaction can proceed at a given temperature and total pressure.

PROBLEMS

13.1 Calculate the equilibrium constant at 5400°R for the reaction $NO \rightarrow \frac{1}{2}O_2 + \frac{1}{2}N_2$.

13.2 Determine the equilibrium constant for the reaction $H_2 + \frac{1}{2}O_2 \rightarrow H_2O$ at 2600 K.

13.3 Calculate the equilibrium constant at 7200°R for the reaction $H_2 + 2OH \rightarrow 2H_2O$.

13.4 Determine the equilibrium constant for the reaction $O_2 + 2CO \rightarrow 2CO_2$ if the reaction occurs at (a) 500 K and (b) 5000 K.

13.5 Does the reaction $O_2 + N_2 \rightarrow 2NO$ occur more readily at low or high temperatures?

13.6 Does the reaction $\frac{1}{2}H_2 + OH \rightarrow H_2O$ occur more readily with increasing or decreasing temperature?

13.7 Does the reaction $CO_2 \rightarrow CO + \frac{1}{2}O_2$ react further to completion as temperature increases or decreases?

13.8 Which of the following two reactions is more likely to take place at 7200°R: $CO_2 \rightarrow CO + \frac{1}{2}O_2$ or $N_2 + O_2 \rightarrow 2NO$?

13.9 At a temperature of 4000 K, which of the two elements, N_2 or O_2, is more likely to dissociate to its monatomic form?

13.10 Determine the equilibrium constant at 77°F for the reaction $CO + \frac{1}{2}O_2 \rightarrow CO_2$ by using the Gibbs function of formation values from Table G.1E. Compare your value of K_P with the one that appears in Table G.9E.

13.11 Use the Gibbs function of formation values in Table G.1 and calculate the equilibrium constant for the reaction $H_2 + \frac{1}{2}O_2 \rightarrow H_2O$ at 25°C. Compare the value for K_P with the one in Table G.9.

13.12 Determine the value for the equilibrium constant for the reaction $H_2 + 2OH \rightarrow 2H_2O$ at 77°F by using the Gibbs function of formation values from Table G.1E. Compare your value of K_P with the one that appears in Table G.9E.

13.13 Calculate the equilibrium constant for the reaction $CO_2 \rightarrow CO + \frac{1}{2}O_2$ at 2000 K by using the properties in the Appendix. Compare your answer with the K_P value in Table G.9.

13.14 Initially, a mixture of 2 mol H_2 and 1 mol OH is placed in a container at 1 atm pressure and a temperature of 9000°R. Determine the equilibrium composition.

13.15 A mixture of 3 mol of carbon dioxide and 0.5 mol of carbon monoxide is maintained at 1 atm and 2600 K. Determine the equilibrium composition.

13.16 Determine the equilibrium composition of 2 mol N_2 and 1 mol O_2 at 1 atm and 5400°R.

13.17 One mol H_2 is maintained at 3800 K and 1 atm pressure. Determine the percentage of mass of H_2 that dissociates to H under these conditions.

13.18 Ten mol of diatomic oxygen is initially maintained at a temperature of 7200°R and a pressure of 2 atm. Determine the number of moles of monatomic oxygen that exist at equilibrium.

13.19 Six mol CO_2 is placed in a container and maintained at a fixed temperature and pressure. Determine the number of moles of CO and O_2 that exist at equilibrium if the pressure and temperature are (a) 2000 K, 1 atm, (b) 4000 K, 1 atm, and (c) 4000 K, 5 atm. Neglect any O in the products.

13.20 A mixture of 30 percent CO_2, 60 percent CO, and 10 percent H_2 by volume is placed in a container and maintained at 4680°R and 1 atm. Determine the composition of the mixture at equilibrium.

13.21 One mol O_2, 5 mol CO, and 3 mol CO_2 are mixed and heated to 2400 K, 1 atm. Determine the equilibrium composition.

13.22 One mol CO_2 is heated to a temperature of 5400°R and a pressure of 2 atm. Assuming that the equilibrium composition consists only of CO_2, CO, and O_2, determine the number of moles of CO_2 at equilibrium.

13.23 Determine the molar percent of diatomic oxygen that dissociates at a temperature of 3800 K and a pressure of 1 atm.

13.24 Determine the molar percent of diatomic nitrogen that dissociates at a temperature of 10,800°R and a pressure of 1 atm.

13.25 [CRD111] Diatomic oxygen (O_2) is heated at a constant pressure of 0.1 atm from 537°R to 5400°R. During this process, the oxygen dissociates to its equilibrium condition according to the reaction

$$O_2 \leftrightharpoons 2O$$

Calculate the following:

(a) the standard-state Gibbs' free energy change in Btu/lb_m·mol for this reaction at equilibrium

(b) the equilibrium constant

(c) the mole fraction of monatomic oxygen at equilibrium

(d) the heat transfer (per mole of O_2) to the reaction vessel

13.26 [CRD211] A gas mixture at 2820 K and 10 atm consists of 0.1 kg·mol of carbon monoxide (CO), 0.6 kg·mol of carbon dioxide (CO_2), 0.3 kg·mol of oxygen (O_2), and 2 kg·mol of nitrogen (N_2). The mixture is at equilibrium for the reaction:

$$2\,CO + O_2 \leftrightharpoons 2\,CO_2$$

Assume ideal-gas behavior and assume that the nitrogen is inert. Determine the following:

(a) the mole fraction of carbon monoxide at equilibrium

(b) the equilibrium constant at 10 atm and 2820 K

(c) the Gibbs' free energy change in kJ/kg·mol of oxygen for this reaction at 2820 K and a standard state of 1 atm

(d) If the pressure is reduced to 5 atm, calculate the new value of the equilibrium constant.

(e) If the pressure is reduced to 5 atm, determine if the equilibrium amount of the CO_2 is higher, lower, or the same as it was at 10 atm.

13.27 Three mol CO, 2 mol O_2, and 2 mol N_2 are placed in a container at 2 atm pressure and 3000 K. Determine the equilibrium composition assuming that the nitrogen is inert.

13.28 One mol H_2, 2 mol O_2, and 2 mol N_2 are placed in a container at 2 atm pressure and 5400°R. Determine the equilibrium composition assuming that the nitrogen is inert.

13.29 One mol CO_2, 2 mol H_2O, and 1 mol Ar are mixed in a container and heated to 3000 K and 1 atm. Assuming that the argon is inert, determine the equilibrium composition.

13.30 Three mol O_2, 2 mol N_2, and 6 mol Ar reach equilibrium at 6300°R and 50 atm pressure. Determine the equilibrium composition if the argon is assumed to be inert.

13.31 A mixture of 2 mol H_2, 1 mol CO_2, and 1 mol N_2 is heated to 1800 K and 6 atm. Determine the equilibrium composition if the nitrogen is inert.

13.32 One mol H_2, 1 mol O_2, and 1 mol Ar are maintained at 1 atm and 6480°R. Assuming that the products consist only of H_2, O_2, H_2O, and Ar, determine the equilibrium composition.

13.33 Two mol O_2, 1 mol CO, and 3 mol N_2 are placed in a container and maintained at 3000 K and 3 atm pressure. Determine the equilibrium composition at this state if the nitrogen is assumed to be inert.

13.34 Two mol H_2O are heated to 3600 K and 2 atm pressure. The substances at equilibrium are H_2O, OH, H_2, and O_2. Determine the equilibrium composition.

13.35 Two mol NO and 1 mol H_2O are heated to 6120°R and 1 atm pressure. Determine the equilibrium composition assuming that the equilibrium mixture consists only of NO, H_2O, O_2, H_2, and N_2.

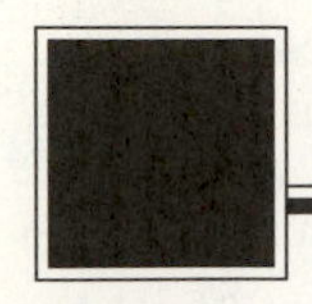

Bibliography

General References, Classical Thermodynamics

Babits, G.: *Applied Thermodynamics,* Allyn & Bacon, Boston, 1968.

Badger, Parker H.: *Equilibrium Thermodynamics,* Allen & Bacon, Boston, 1967.

Balzhiser, R. E., and M. R. Samuels: *Engineering Thermodynamics*, Prentice-Hall, Englewood Cliffs, NJ, 1977.

Bejan, A.: *Advanced Engineering Thermodynamics,* Wiley, New York, 1988.

Burghardt, M. D.: *Engineering Thermodynamics with Applications,* 3d ed., Harper & Row, New York, 1986.

Callen, H. B.: *Thermodynamics,* Wiley, New York, 1960.

Crawford, Franzo H., and William D. Van Vorst: *Thermodynamics for Engineers,* Harcourt, Brace and World, New York, 1968.

Dixon, J. R.: *Thermodynamics I: An Introduction to Energy,* Prentice-Hall, Englewood Cliffs, NJ, 1976.

Doolittle, J. S., and F. J. Hale: *Thermodynamics for Engineers,* Wiley, New York, 1983.

Hall, N. A., and W. E. Ibele: *Engineering Thermodynamics,* Prentice-Hall, Englewood Cliffs, NJ, 1960.

Hatsopoulos, G. N., and J. H. Keenan: *Principles of General Thermodynamics,* Wiley, New York, 1965.

Holman, J. P.: *Thermodynamics,* 3d ed., McGraw-Hill, New York, 1980.

Howell, J. R., and R. O. Buckius: *Fundamentals of Engineering Thermodynamics,* McGraw-Hill, New York, 1987.

Hsieh, Jui Sheng: *Principles of Thermodynamics,* Scripta Book Co., Washington, DC, 1975.

Huang, F. F.: *Engineering Thermodynamics Fundamentals and Applications,* Macmillan, New York, 1976.

Jones, J. B., and G. A. Hawkins: *Engineering Thermodynamics,* 2d ed., Wiley, New York, 1986.

Karlekar, B. V.: *Thermodynamics for Engineers,* Prentice-Hall, Englewood Cliffs, NJ, 1983.

Keenan, J. H.: *Thermodynamics,* MIT Press, Cambridge, MA, 1970.

Kestin, J.: *A Course in Thermodynamics,* vol. 1, Blaisdell, Waltham, MA, 1966.

Lay, Joachim E.: *Thermodynamics,* 2d ed., Charles E. Merrill Books, Columbus OH, 1963.

Lewis, G. N., and M. Randall: *Thermodynamics,* 2d ed. (revised by K. S. Pitzer and L. Brewer), McGraw-Hill, New York, 1961.

Look, D. C., Jr., and H. S. Sauer, Jr.: *Engineering Thermodynamics,* PWS Publishers, Boston, MA, 1986.

Moran, M. J., and H. N. Shapiro: *Fundamentals of Engineering Thermodynamics,* Wiley, New York, 1988.

Myers, Glen E.: *Engineering Thermodynamics,* Prentice-Hall, Englewood Cliffs, NJ, 1989.

Obert, E. F.: *Concepts of Thermodynamics,* McGraw-Hill, New York, 1960.

Obert, E. F., and R. A. Gaggioli: *Thermodynamics,* McGraw-Hill, New York, 1963.

Penner, S. S.: *Thermodynamics for Scientists and Engineers,* Addison-Wesley, Reading, MA, 1968.

Reynolds, W. C.: *Thermodynamics,* 2d ed., McGraw-Hill, New York, 1968.

Reynolds, W. C., and H. C. Perkins: *Engineering Thermodynamics,* 2d ed., McGraw-Hill, New York, 1977.

Rolle, K. C.: *Introduction to Thermodynamics,* 2d ed., Charles E. Merrill, Columbus, OH, 1980.

Saad, M. A.: *Thermodynamics for Engineers,* Prentice-Hall, Englewood Cliffs, NJ, 1966.

Sears, F. W.: *Thermodynamics,* 2d ed., Addison-Wesley, Reading, MA, 1953.

Silver, H. F., and J. E. Nydahl: *Introduction to Engineering Thermodynamics,* West, St. Paul, 1977.

Spalding, D. B., and E. H. Cole: *Engineering Thermodynamics an Introductory Text,* McGraw-Hill, New York, 1959.

Sweigert, Ray L., and Mario J. Goglia: *Thermodynamics,* Ronald Press, New York, 1955.

Tribus, M.: *Thermostatics and Thermodynamics,* Van Nostrand, New York, 1961.

Vanderslice, J. T., H. W. Schamp, Jr., and E. A. Mason: *Thermodynamics,* Prentice-Hall, Englewood Cliffs, NJ, 1966.

Van Wylen, G. J.: *Thermodynamics,* Wiley, New York, 1959.

Van Wylen, G. J., and R. E. Sonntag: *Fundamentals of Classical Thermodynamics,* 2d ed., Wiley, New York, 1973.

Wark, K. *Thermodynamics,* 4th ed., McGraw-Hill, New York, 1983.

Zemansky, M. W., and H. C. Van Ness: *Basic Engineering Thermodynamics,* McGraw-Hill, 1966.

General References, Statistical Thermodynamics

Allis, W. P., and M. A. Herlin: *Thermodynamics and Statistical Mechanics,* McGraw-Hill, New York, 1952.

Crawford, F. H.: *Heat, Thermodynamics and Statistical Physics,* Harcourt, Brace Jovanovich, New York, 1963.

El-Saden, M. R.: *Engineering Thermodynamics,* Van Nostrand, New York, 1965.

Fay, J. A.: *Molecular Thermodynamics,* Addison-Wesley, Reading, MA, 1961.

Fay, J. A.: *Molecular Thermodynamics,* Addison-Wesley, Reading, MA, 1965.

Hill, L.: *Statistical Mechanics,* McGraw-Hill, New York, 1956.

Incropera, F. P.: *Introduction to Molecular Structure and Thermodynamics,* Wiley, New York, 1974.

Jeans, J. H.: *The Dynamical Theory of Gases,* Dover, New York, 1954.

Kennard, E., II: *Kinetic Theory of Gases,* McGraw-Hill, New York, 1938.

Kestin, J.: *A Course in Thermodynamics,* vol. 2, Blaisdell, Waltham, MA, 1968.

Knuth, E. S.: *Introduction to Statistical Thermodynamics,* McGraw-Hill, New York, 1966.

Lee, J. F., F. W. Sears, and D. L. Turcotte: *Statistical Thermodynamics,* Addison-Wesley, Reading, MA, 1963.

Mayer, J. F., and M. G. Mayer: *Statistical Mechanics,* Wiley, New York, 1940.

Pierce, F. J.: *Microscopic Thermodynamics,* Wiley, New York, 1968.

Present, R. D.: *Kinetic Theory of Gases,* McGraw-Hill, New York, 1958.

Schrödinger, E.: *Statistical Thermodynamics,* Cambridge University Press, New York, 1952.

Sears, F. W.: *An Introduction to Thermodynamics, the Kinetic Theory of Gases, and Statistical Mechanics,* Addison-Wesley, Reading, MA, 1950.

Sommerfeld, A.: *Thermodynamics and Statistical Mechanics,* Academic Press, New York, 1956.

Sonntag, R. E., and G. J. Van Wylen: *Fundamentals of Statistical Thermodynamics,* Wiley, New York, 1966.

Tolman, C.: *The Principles of Statistical Mechanics,* Oxford University Press, New York, 1938.

Thermodynamic Properties

American Society of Mechanical Engineers: *Thermodynamic and Transport Properties of Gases, Liquids and Solids,* McGraw-Hill, New York, 1959.

American Society of Mechanical Engineers: *Thermodynamic and Transport Properties of Steam,* ASME, New York, 1967.

Beattie, J. A., and O. C. Bridgeman: "A New Equation of State for Fluids," in *Proceedings of the American Academy of Arts and Sciences,* vol. 63, 1928, p. 229.

Benedict, M., G. Webb, and L. Rubin: "An Empirical Equation for the Thermodynamic Properties of Light Hydrocarbons and Their Mixtures," *J. Chem. Phys.* 8:334 (1940).

Combustion Engineering, Steam Tables: *Properties of Saturated and Superheated Steam,* Combustion Engineering, New York, 1967.

DuPont Freon Refrigerants, *Thermodynamic Properties of "Freon" 12 Refrigerants,* Technical Bulletin T-12-SI, DuPont De Nemours International S.A., Geneva, Switzerland.

Haar, L., J. S. Gallagher, and G. S. Kell: *NBS/NRC Steam Tables,* Hemisphere Publishing Corp., New York, 1984.

Hilsenrath, J., et al.: *Tables of Thermal Properties of Gases,* National Bureau of Standards Circular 564, Government Printing Office, Washington, DC, 1955.

JANAF Thermochemical Tables, 3d ed., *Journal of Physical and Chemical Reference Data,* American Chemical Society and American Institute of Physics, 1986.

Keenan, J. H., J. Chao, and J. Kaye: *Gas Tables,* International Version, Wiley, New York, 1983.

Keenan, J. H., and F. G. Keyes: *Thermodynamic Properties of Steam,* Wiley, New York, 1936.

Nelson, L. C., and E. F. Obert: "Generalized *p-v-T* Properties of Gases," *Trans. ASME,* October 1954, p. 1057.

Selected Values of Chemical Thermodynamic Properties, NBS Technical Notes, 270–1 and 270–2, 1955.

Selected Values of Physical and Thermodynamic Properties of Hydrocarbons and Related Compounds, API Research Project 44, Carnegie Press, Pittsburgh.

Stewart, R. B., R. T. Jacobsen, and S. G. Penoncello: *ASHRAE Thermodynamic Properties of Refrigerants,* ASHRAE, Atlanta, GA 1986.

Sweigert, R. L., and M. W. Beardsley: Bulletin no. 2, Georgia School of Technology, 1938.

Cycles and Applications

American Society of Heating, Refrigerating and Air Conditioning Engineers, *Handbooks and Product Directories: Applications,* 1978; *Fundamentals,* 1977; *Systems,* 1976; *Equipment,* 1975; New York, ASHRAE.

Barron, R.: *Cryogenic Systems,* McGraw-Hill, New York, 1966.

Culp, A. W.: *Principles of Energy Conversion,* McGraw-Hill, New York, 1979.

Clifford, George E.: *Heating, Ventilating and Air Conditioning,* Reston Publishing, Reston, VA, 1986.

Duffie, John A., and William A. Beckman: *Solar Engineering of Thermal Processes,* Wiley, New York, 1980.

Durham, F. P.: *Aircraft Jet Power Plants,* Prentice-Hall, Englewood Cliffs, NJ, 1951.

El-Wakil: *Powerplant Technology,* McGraw-Hill, New York, 1984.

Gaffert, G. A.: *Steam Power Stations,* 4th ed., McGraw-Hill, New York, 1952.

Hill, P. G., and C. R. Peterson: *Mechanics and Thermodynamics of Propulsion,* Addison-Wesley, Reading, MA, 1965.

Morgan, H. E.: *Turbojet Fundamentals,* McGraw-Hill, New York, 1958.

Moran, M. J.: *Availability Analysis: A Guide to Efficient Energy Use,* Prentice-Hall, Englewood Cliffs, NJ, 1982.

Obert, E. F.: *Internal Combustion Engines and Air Pollution,* Harper & Row, New York, 1973.

Potter, P. J.: *Power Plant Theory and Design,* 2d ed., Wiley, New York, 1959.

Sorenson, Harry A.: *Energy Conversion Systems,* Wiley, New York, 1983.

Stoecker, W. F.: *Refrigeration and Air Conditioning,* McGraw-Hill, New York, 1958.

Stoecker, W. F.: *Design of Thermal Systems,* 2d ed., McGraw-Hill, New York, 1980.

Stoecker, W. F., and J. W. Jones: *Refrigeration and Air Conditioning,* McGraw-Hill, New York, 1982.

Wood, B. D.: *Applications of Thermodynamics,* 2d ed., Addison-Wesley, Reading, MA, 1982.

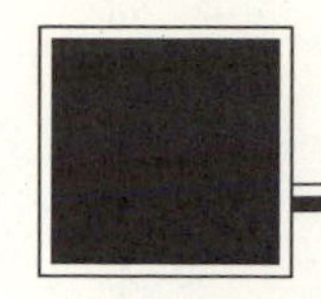

Appendixes
Tables, Figures, and Charts

SI Units

English Units*

*Tables in this Appendix are labeled identically to the SI property tables except that an E has been added to the table designation to indicate English units were used throughout.

Dimensions and Units

TABLE A.1 SI UNIT PREFIXES

Multiples	Prefix	Symbol
10^{18}	Exa	E
10^{15}	Peta	P
10^{12}	Tera	T
10^{9}	Giga	G
10^{6}	Mega	M
10^{3}	Kilo	k
10^{2}	Hecto	h
10	Deka	da
10^{-1}	Deci	d
10^{-2}	Centi	c
10^{-3}	Milli	m
10^{-6}	Micro	μ
10^{-9}	Nano	n
10^{-12}	Pico	p
10^{-15}	Femto	f
10^{-18}	Atto	a

TABLE A.2 SI PRIMARY UNITS

Dimension	Unit	Symbol
Length	Meter	m
Mass	Kilogram	kg
Time	Second	s
Electric current	Ampere	A
Thermodynamic temperature	Kelvin	K
Luminous intensity	Candela	cd
Amount of substance	Mole	mol

TABLE A.3 SI SECONDARY UNITS

Dimension	Unit	Symbol	Definition
Force	Newton	N	1 N = 1 kg·m/s^2
Energy	Joule	J	1 J = 1 N·m
Power	Watt	W	1 W = 1 J/s
Pressure	Pascal	Pa	1 Pa = 1 N/m^2

TABLE A.4 UNITS IN THE SI SYSTEM

Dimension	Unit
Energy, work, heat	J
Force	N
Length	m
Mass	kg
Power	W
Time	s

TABLE A.5 COMMONLY USED CONVERSION FACTORS

Dimension	Conversion factor	Conversion factor
Acceleration	1 m/s^2 = 3.2808 ft/s^2	1 ft/s^2 = 0.3048 m/s^2
Area	1 m^2 = 1550 in^2	1 in^2 = 6.4516×10^{-4} m^2
	1 m^2 = 10.764 ft^2	1 ft^2 = 9.2903×10^{-2} m^2
Density	1 g/cm^3 = 62.428 lb$_m$/ft^3	1 lb$_m$/ft^3 = 0.0160185 g/m^3
	1 kg/m^3 = 0.062428 lb$_m$/ft^3	1 lb$_m$/ft^3 = 16.0185 kg/m^3
Energy	1 J = 0.73756 ft·lb$_f$	1 ft·lb$_f$ = 1.35582 J
	1 J = 9.4782×10^{-4} Btu	1 Btu = 1.055056 kJ
		1 Btu = 778.169 ft·lb$_f$
Force	1 N = 0.22481 lb$_f$	1 lb$_f$ = 4.4482 N
Length	1 m = 39.370 in	1 in = 0.0254 m
	1 m = 3.2808 ft	1 ft = 0.3048 m
	1 m = 6.2137×10^{-4} mi	1 mi = 1.6093 km
Mass	1 g = 2.2046×10^{-3} lb$_m$	1 lb$_m$ = 453.59 g
	1 kg = 2.2046 lb$_m$	1 lb$_m$ = 0.45359 kg
Mass flow rate	1 kg/s = 2.2046 lb$_m$/s	1 lb$_m$/s = 0.45359 kg/s
Power	1 W = 0.73756 ft·lb$_f$/s	1 ft·lb$_f$/s = 1.3558 W
	1 W = 3.4121 Btu/h	1 Btu/h = 0.29307 W
	1 W = 1.3410×10^{-3} hp	1 hp = 745.7 W
		1 hp = 550 ft·lb$_f$/s
Pressure	1 Pa = 1.4504×10^{-4} lb$_f$/in^2	1 lb$_f$/in^2 = 6.8948×10^3 Pa
	1 Pa = 0.020886 lb$_f$/ft^2	1 lb$_f$/ft^2 = 47.880 Pa
	1 Pa = 4.015×10^{-3} in H_2O (4°C)	1 in H_2O = 249.08 Pa (39.2°F)
	1 Pa = 2.953×10^{-4} in Hg (0°C)	1 in Hg = 3.3864 kPa (32°F)
	1 Pa = 9.869×10^{-6} atm	1 atm = 101.325 kPa
	1 bar = 10^5 Pa	1 atm = 1.01325 bar
Specific energy	1 kJ/kg = 0.42992 Btu/lb$_m$	1 Btu/lb$_m$ = 2.326 kJ/kg
	1 kJ/kg = 334.56 ft·lb$_f$/lb$_m$	1 ft·lb$_f$/lb$_m$ = 2.989 J/kg
Specific heat	1 kJ/kg·K = 0.23885 Btu/lb$_m$·°R	1 Btu/lb$_m$·°R = 4.1868 kJ/kg·K
Temperature	K = (5/9)(°R)	°R = 1.8(K)
	K = °C + 273.15	°R = °F + 459.67
	°C = (5/9)(°F − 32)	°F = 1.8(°C) + 32
Temperature difference	1 K = 1°C	1°R = 1°F
Velocity	1 m/s = 3.2808 ft/s	1 ft/s = 0.3048 m/s
	1 m/s = 2.237 mi/h	1 mi/h = 0.44704 m/s
Volume	1 m^3 = 6.1024×10^4 in^3	1 in^3 = 1.6387×10^{-5} m^3
	1 m^3 = 35.315 ft^3	1 ft^3 = 2.8317×10^{-2} m^3
	1 m^3 = 264.17 gal (U.S. liquid)	1 gal = 3.7854×10^{-3} m^3
		1 ft^3 = 7.48 gal
Volume flow rate	1 m^3/s = 35.315 ft^3/s	1 ft^3/s = 2.8317×10^{-2} m^3/s
	1 m^3/s = 2.1189×10^3 ft^3/min	1 ft^3/min = 4.7195×10^{-4} m^3/s
	1 m^3/s = 1.585×10^4 gal/min	1 gal/min = 6.309×10^{-5} m^3/s

Note: To convert a dimension from SI units to English units, multiply by the conversion factor in the middle column or divide by the conversion factor in the last column. Using the dimension of power as an example, we will convert 3500 W to horsepower. From the table the applicable conversion factor is

$$1\text{ W} = 1.3410 \times 10^{-3}\text{ hp}$$

Thus

$$(3500\text{ W}) \cdot (1.3410 \times 10^{-3}\text{ hp/W}) = 4.694\text{ hp}$$

TABLE A.6 SI PHYSICAL CONSTANTS

Dimension	Conversion factor
Acceleration of gravity at sea level	$g = 9.807$ m/s^2
Avogadro's number	$N_A = 6.022169 \times 10^{26}$ molecules/kg·mol
Boltzmann's constant	$k = 1.380622 \times 10^{-23}$ J/K
Planck's constant	$h = 6.626196 \times 10^{-34}$ J·s
Speed of light in a vacuum	$c = 2.997925 \times 10^{8}$ m/s
Universal gas constant	$\overline{R} = 8.31434$ kPa·m^3/kg·mol·K
	$= 8.31434$ kJ/kg·mol·K
	$= 8.31434$ N·m/kg·mol·K
	$= 0.08206$ liter·atm/g·mol·K

Properties of Water

TABLE B.1 SATURATED WATER—TEMPERATURE TABLE: SI UNITS

Temp., °C, T	Press., kPa, P	Specific volume, m³/kg			Internal energy, kJ/kg			Enthalpy, kJ/kg			Entropy, kJ/kg·K		
		Sat. liquid, v_f	Evap., v_{fg}	Sat. vapor, v_g	Sat. liquid, u_f	Evap., u_{fg}	Sat. vapor, u_g	Sat. liquid, h_f	Evap., h_{fg}	Sat. vapor, h_g	Sat. liquid, s_f	Evap., s_{fg}	Sat. vapor, s_g
0.01	0.61173	0.0010002	205.99	205.99	0.00	2374.5	2374.5	0.00	2500.5	2500.5	0.0000	9.1541	9.1541
1.00	0.65716	0.0010002	192.44	192.44	4.18	2371.7	2375.9	4.18	2498.2	2502.4	0.0153	9.1124	9.1277
5	0.87260	0.0010001	147.02	147.02	21.02	2360.4	2381.4	21.02	2488.7	2509.7	0.0763	8.9473	9.0236
10	1.2281	0.0010003	106.32	106.32	41.99	2346.3	2388.3	41.99	2476.9	2518.9	0.1510	8.7477	8.8986
15	1.7056	0.0010009	77.896	77.897	62.92	2332.3	2395.2	62.92	2465.1	2528.0	0.2242	8.5550	8.7792
20	2.3388	0.0010018	57.777	57.778	83.83	2318.2	2402.0	83.84	2453.3	2537.2	0.2962	8.3689	8.6651
25	3.1690	0.0010030	43.360	43.361	104.75	2304.1	2408.9	104.75	2441.5	2546.3	0.3670	8.1889	8.5559
30	4.2455	0.0010044	32.897	32.898	125.67	2290.0	2415.7	125.67	2429.7	2555.3	0.4365	8.0148	8.4513
35	5.6267	0.0010060	25.221	25.222	146.58	2275.9	2422.5	146.59	2417.8	2564.4	0.5050	7.8461	8.3511
40	7.3814	0.0010079	19.528	19.529	167.50	2261.7	2429.2	167.50	2405.9	2573.4	0.5723	7.6828	8.2550
45	9.5898	0.0010099	15.262	15.263	188.41	2247.5	2435.9	188.42	2393.9	2582.3	0.6385	7.5244	8.1629
50	12.344	0.0010122	12.036	12.037	209.31	2233.3	2442.6	209.33	2381.9	2591.2	0.7037	7.3708	8.0745
55	15.752	0.0010146	9.571	9.572	230.22	2219.0	2449.2	230.24	2369.8	2600.0	0.7679	7.2216	7.9896
60	19.932	0.0010171	7.673	7.674	251.13	2204.7	2455.8	251.15	2357.6	2608.8	0.8312	7.0768	7.9080
65	25.022	0.0010199	6.198	6.199	272.05	2190.3	2462.4	272.08	2345.4	2617.5	0.8935	6.9360	7.8295
70	31.176	0.0010228	5.043	5.044	292.98	2175.9	2468.8	293.01	2333.1	2626.1	0.9549	6.7990	7.7540
75	38.563	0.0010258	4.132	4.133	313.92	2161.3	2475.2	313.96	2320.7	2634.6	1.0155	6.6657	7.6812
80	47.373	0.0010290	3.408	3.409	334.88	2146.7	2481.6	334.93	2308.1	2643.1	1.0753	6.5359	7.6111
85	57.815	0.0010324	2.828	2.829	355.86	2132.0	2487.9	355.92	2295.5	2651.4	1.1343	6.4093	7.5436
90	70.117	0.0010359	2.361	2.362	376.86	2117.2	2494.0	376.93	2282.7	2659.6	1.1925	6.2858	7.4783
95	84.529	0.0010396	1.982	1.983	397.89	2102.2	2500.1	397.98	2269.8	2667.7	1.2501	6.1653	7.4154
100	101.32	0.0010434	1.673	1.674	418.96	2087.2	2506.1	419.06	2256.7	2675.7	1.3069	6.0476	7.3545
105	120.79	0.0010474	1.419	1.420	440.05	2072.0	2512.1	440.18	2243.4	2683.6	1.3630	5.9325	7.2956
110	143.24	0.0010515	1.210	1.211	461.19	2056.7	2517.9	461.34	2229.9	2691.3	1.4186	5.8200	7.2386
115	169.02	0.0010558	1.036	1.037	482.36	2041.2	2523.5	482.54	2216.3	2698.8	1.4735	5.7098	7.1833
120	198.48	0.0010603	0.8911	0.8922	503.57	2025.5	2529.1	503.78	2202.4	2706.2	1.5278	5.6020	7.1297
125	232.01	0.0010649	0.7698	0.7709	524.82	2009.7	2534.5	525.07	2188.3	2713.4	1.5815	5.4962	7.0777
130	270.02	0.0010697	0.6676	0.6687	546.12	1993.7	2539.8	546.41	2174.0	2720.4	1.6346	5.3925	7.0272
135	312.93	0.0010746	0.5813	0.5823	567.46	1977.5	2545.0	567.80	2159.4	2727.2	1.6873	5.2908	6.9780
140	361.19	0.0010797	0.5079	0.5090	588.85	1961.1	2550.0	589.24	2144.6	2733.8	1.7394	5.1908	6.9302
145	415.29	0.0010850	0.4453	0.4464	610.30	1944.5	2554.8	610.75	2129.5	2740.2	1.7910	5.0926	6.8836
150	475.72	0.0010904	0.3918	0.3929	631.80	1927.7	2559.5	632.32	2114.1	2746.4	1.8421	4.9960	6.8381
155	542.99	0.0010961	0.3457	0.3468	653.35	1910.6	2564.0	653.95	2098.3	2752.3	1.8927	4.9010	6.7937
160	617.66	0.0011019	0.3060	0.3071	674.97	1893.3	2568.3	675.65	2082.3	2758.0	1.9429	4.8073	6.7503
165	700.29	0.0011080	0.2716	0.2727	696.65	1875.7	2572.4	697.43	2065.9	2763.3	1.9927	4.7151	6.7078
170	791.47	0.0011142	0.2417	0.2428	718.40	1857.9	2576.3	719.28	2049.2	2768.5	2.0421	4.6241	6.6662
175	891.80	0.0011206	0.2157	0.2168	740.22	1839.7	2579.9	741.22	2032.0	2773.3	2.0910	4.5343	6.6254

TABLE B.1 SATURATED WATER—TEMPERATURE TABLE: SI UNITS *(Continued)*

Temp., °C, T	Press., MPa, P	Specific volume, m³/kg: Sat. liquid, v_f	Evap., v_{fg}	Sat. vapor, v_g	Internal energy, kJ/kg: Sat. liquid, u_f	Evap., u_{fg}	Sat. vapor, u_g	Enthalpy, kJ/kg: Sat. liquid, h_f	Evap., h_{fg}	Sat. vapor, h_g	Entropy, kJ/kg·K: Sat. liquid, s_f	Evap., s_{fg}	Sat. vapor, s_g
180	1.0019	0.0011273	0.1929	0.1940	762.12	1821.3	2583.4	763.25	2014.5	2777.8	2.1397	4.4456	6.5853
185	1.1225	0.0011342	0.1729	0.1741	784.10	1802.5	2586.6	785.37	1996.6	2782.0	2.1879	4.3580	6.5459
190	1.2542	0.0011414	0.1554	0.1565	806.17	1783.4	2589.6	807.60	1978.2	2785.8	2.2358	4.2713	6.5071
195	1.3976	0.0011488	0.1399	0.1410	828.33	1763.9	2592.3	829.93	1959.4	2789.4	2.2834	4.1855	6.4689
200	1.5536	0.0011564	0.1262	0.1273	850.58	1744.1	2594.7	852.38	1940.1	2792.5	2.3308	4.1005	6.4312
205	1.7229	0.0011644	0.1140	0.1152	872.95	1723.9	2596.9	874.96	1920.4	2795.3	2.3778	4.0162	6.3940
210	1.9062	0.0011726	0.1032	0.1044	895.43	1703.3	2598.7	897.66	1900.0	2797.7	2.4246	3.9326	6.3572
215	2.1042	0.0011811	0.09357	0.09475	918.02	1682.3	2600.3	920.51	1879.2	2799.7	2.4712	3.8496	6.3208
220	2.3178	0.0011900	0.08497	0.08616	940.75	1660.8	2601.6	943.51	1857.8	2801.3	2.5175	3.7671	6.2847
225	2.5479	0.0011992	0.07726	0.07846	963.61	1638.9	2602.5	966.67	1835.7	2802.4	2.5637	3.6851	6.2488
230	2.7951	0.0012088	0.07035	0.07155	986.62	1616.5	2603.1	990.00	1813.1	2803.1	2.6097	3.6034	6.2132
235	3.0604	0.0012188	0.06412	0.06534	1009.79	1593.5	2603.3	1013.52	1789.7	2803.3	2.6556	3.5221	6.1777
240	3.3447	0.0012292	0.05851	0.05974	1033.12	1570.0	2603.1	1037.24	1765.7	2803.0	2.7013	3.4410	6.1423
245	3.6488	0.0012401	0.05345	0.05469	1056.64	1546.0	2602.6	1061.16	1741.0	2802.1	2.7470	3.3600	6.1070
250	3.9736	0.0012515	0.04886	0.05011	1080.35	1521.3	2601.6	1085.32	1715.4	2800.7	2.7926	3.2791	6.0717
255	4.3202	0.0012633	0.04470	0.04596	1104.26	1496.0	2600.2	1109.72	1689.1	2798.8	2.8382	3.1981	6.0363
260	4.6894	0.0012758	0.04091	0.04219	1128.40	1470.0	2598.4	1134.38	1661.9	2796.2	2.8838	3.1171	6.0009
265	5.0823	0.0012889	0.03747	0.03876	1152.77	1443.3	2596.0	1159.32	1633.7	2793.0	2.9294	3.0358	5.9652
270	5.4999	0.0013026	0.03434	0.03564	1177.41	1415.8	2593.2	1184.57	1604.5	2789.1	2.9751	2.9542	5.9293
275	5.9431	0.0013171	0.03146	0.03278	1202.32	1387.4	2589.7	1210.15	1574.4	2784.5	3.0209	2.8722	5.8931
280	6.4132	0.0013324	0.02883	0.03016	1227.53	1358.2	2585.7	1236.08	1543.1	2779.2	3.0669	2.7896	5.8565
285	6.9111	0.0013486	0.02642	0.02777	1253.08	1328.0	2581.1	1262.40	1510.6	2773.0	3.1131	2.7064	5.8195
290	7.4380	0.0013658	0.02419	0.02556	1278.98	1296.7	2575.7	1289.14	1476.8	2765.9	3.1595	2.6223	5.7818
295	7.9952	0.0013841	0.02215	0.02354	1305.28	1264.4	2569.7	1316.34	1441.5	2757.8	3.2062	2.5372	5.7434
300	8.5838	0.0014037	0.02027	0.02167	1332.01	1230.8	2562.8	1344.05	1404.7	2748.7	3.2534	2.4508	5.7042
305	9.2051	0.0014247	0.01852	0.01994	1359.22	1195.8	2555.0	1372.33	1366.2	2738.5	3.3010	2.3630	5.6640
310	9.8605	0.0014473	0.01690	0.01834	1386.96	1159.3	2546.2	1401.23	1325.8	2727.0	3.3491	2.2735	5.6226
315	10.551	0.0014718	0.01539	0.01686	1415.32	1121.0	2536.3	1430.84	1283.4	2714.2	3.3979	2.1820	5.5799
320	11.279	0.0014984	0.01398	0.01548	1444.36	1080.9	2525.2	1461.26	1238.4	2699.7	3.4476	2.0880	5.5356
325	12.046	0.0015277	0.01266	0.01419	1474.18	1038.4	2512.6	1492.58	1190.9	2683.5	3.4982	1.9911	5.4893
330	12.852	0.0015601	0.01143	0.01299	1504.9	993.5	2498.4	1525.0	1140.3	2665.3	3.5501	1.8906	5.4407
335	13.701	0.0015963	0.01025	0.01185	1536.8	945.5	2482.3	1558.6	1086.1	2644.7	3.6035	1.7858	5.3893
340	14.594	0.0016373	0.00915	0.01079	1569.9	894.0	2463.9	1593.8	1027.5	2621.3	3.6587	1.6758	5.3345
345	15.533	0.0016845	0.00809	0.00978	1604.7	838.0	2442.7	1630.9	963.6	2594.5	3.7164	1.5589	5.2753
350	16.521	0.0017401	0.00707	0.00881	1641.7	776.2	2417.9	1670.4	893.1	2563.5	3.7774	1.4331	5.2105
355	17.561	0.0018076	0.00607	0.00788	1681.5	706.9	2388.4	1713.3	813.4	2526.7	3.8429	1.2950	5.1379
360	18.655	0.0018936	0.00507	0.00696	1725.6	626.6	2352.2	1761.0	721.0	2482.0	3.9153	1.1389	5.0542
365	19.809	0.0020120	0.00402	0.00603	1776.8	528.4	2305.2	1816.7	607.9	2424.6	3.9994	0.9526	4.9520
370	21.030	0.0022068	0.00279	0.00499	1843.3	391.9	2235.2	1889.7	450.2	2340.2	4.1094	0.7004	4.8098
373.976	22.055	0.003106	0	0.003106	2017	0	2017	2086	0	2086	4.409	0	4.409

Values generated from property formulation given in *NBS/NRC Steam Tables: Thermodynamic and Transport Properties and Computer Programs for Vapor and Liquid States of Water in SI* [illegible] Publishing Corp., Washington, 1984.

TABLE B.2 SATURATED WATER—PRESSURE TABLE: SI UNITS

Press., kPa, P	Temp., °C, T	Specific volume, m^3/kg Sat. liquid, v_f	Evap., v_{fg}	Sat. vapor, v_g	Internal energy, kJ/kg Sat. liquid, u_f	Evap., u_{fg}	Sat. vapor, u_g	Enthalpy, kJ/kg Sat. liquid, h_f	Evap., h_{fg}	Sat. vapor, h_g	Entropy, kJ/kg·K Sat. liquid, s_f	Evap., s_{fg}	Sat. vapor, s_g
0.61173	0.010	0.0010002	205.99	205.99	0.00	2374.5	2374.5	0.00	2500.5	2500.5	0.0000	9.1541	9.1541
1.0	6.970	0.0010001	129.19	129.19	29.29	2354.8	2384.1	29.29	2484.0	2513.3	0.1059	8.8678	8.9737
1.5	13.021	0.0010006	87.970	87.971	54.63	2337.8	2392.5	54.64	2469.8	2524.4	0.1954	8.6305	8.8258
2.0	17.497	0.0010013	66.997	66.998	73.36	2325.2	2398.6	73.37	2459.2	2532.6	0.2603	8.4612	8.7216
2.5	21.080	0.0010020	54.248	54.249	88.35	2315.2	2403.5	88.36	2450.8	2539.1	0.3116	8.3295	8.6411
3.0	24.083	0.0010028	45.660	45.661	100.92	2306.7	2407.6	100.92	2443.7	2544.6	0.3541	8.2214	8.5755
4.0	28.966	0.0010041	34.797	34.798	121.35	2292.9	2414.3	121.35	2432.1	2553.5	0.4223	8.0503	8.4725
5.0	32.881	0.0010053	28.190	28.191	137.72	2281.9	2419.6	137.73	2422.8	2560.5	0.4761	7.9169	8.3930
7.5	40.299	0.0010080	19.236	19.237	168.75	2260.9	2429.6	168.76	2405.1	2573.9	0.5763	7.6731	8.2494
10	45.817	0.0010103	14.673	14.674	191.82	2245.2	2437.0	191.83	2391.9	2583.8	0.6493	7.4990	8.1482
15	53.983	0.0010141	10.022	10.023	225.97	2221.9	2447.9	225.98	2372.2	2598.2	0.7550	7.2516	8.0066
20	60.073	0.0010172	7.649	7.650	251.44	2204.5	2455.9	251.46	2357.5	2608.9	0.8321	7.0747	7.9068
25	64.980	0.0010198	6.204	6.205	271.96	2190.4	2462.3	271.99	2345.5	2617.5	0.8932	6.9366	7.8298
40	75.877	0.0010264	3.993	3.994	317.59	2158.8	2476.4	317.63	2318.5	2636.1	1.0261	6.6427	7.6688
50	81.339	0.0010299	3.240	3.241	340.49	2142.8	2483.3	340.54	2304.8	2645.3	1.0912	6.5017	7.5928
75	91.783	0.0010372	2.216	2.218	384.35	2111.9	2496.2	384.43	2278.1	2662.5	1.2131	6.2425	7.4556
100	99.632	0.0010431	1.693	1.694	417.41	2088.3	2505.7	417.51	2257.6	2675.1	1.3027	6.0562	7.3589
125	105.993	0.0010482	1.374	1.375	444.25	2069.0	2513.2	444.38	2240.7	2685.1	1.3741	5.9100	7.2841
150	111.378	0.0010527	1.158	1.160	467.02	2052.4	2519.4	467.18	2226.2	2693.4	1.4338	5.7894	7.2232
175	116.070	0.0010568	1.003	1.004	486.89	2037.9	2524.7	487.07	2213.3	2700.4	1.4851	5.6866	7.1717
200	120.241	0.0010605	0.8848	0.8859	504.59	2024.8	2529.4	504.80	2201.7	2706.5	1.5304	5.5968	7.1272
225	124.005	0.0010640	0.7923	0.7934	520.59	2012.9	2533.5	520.83	2191.2	2712.0	1.5708	5.5171	7.0879
250	127.443	0.0010672	0.7177	0.7188	535.22	2001.9	2537.2	535.49	2181.4	2716.8	1.6075	5.4453	7.0528
275	130.612	0.0010702	0.6563	0.6574	548.73	1991.8	2540.5	549.02	2172.2	2721.3	1.6411	5.3800	7.0211
300	133.555	0.0010731	0.6048	0.6059	561.29	1982.2	2543.5	561.61	2163.7	2725.3	1.6721	5.3200	6.9921
325	136.306	0.0010759	0.5610	0.5620	573.04	1973.3	2546.3	573.39	2155.6	2729.0	1.7009	5.2645	6.9654
350	138.891	0.0010786	0.5232	0.5243	584.10	1964.8	2548.9	584.48	2147.9	2732.4	1.7278	5.2129	6.9407
375	141.330	0.0010811	0.4903	0.4914	594.56	1956.7	2551.3	594.96	2140.6	2735.6	1.7531	5.1645	6.9177
450	147.938	0.0010882	0.4129	0.4140	622.93	1934.7	2557.6	623.42	2120.5	2743.9	1.8211	5.0357	6.8567
500	151.866	0.0010925	0.3738	0.3749	639.84	1921.3	2561.2	640.38	2108.2	2748.6	1.8610	4.9604	6.8214
550	155.492	0.0010967	0.3415	0.3426	655.48	1908.9	2564.4	656.08	2096.8	2752.9	1.8977	4.8917	6.7894
600	158.863	0.0011006	0.3145	0.3156	670.05	1897.3	2567.3	670.71	2086.0	2756.7	1.9315	4.8285	6.7601
650	162.017	0.0011043	0.2915	0.2926	683.71	1886.2	2569.9	684.43	2075.7	2760.2	1.9631	4.7700	6.7330
700	164.983	0.0011079	0.2717	0.2728	696.58	1875.8	2572.4	697.35	2066.0	2763.3	1.9925	4.7154	6.7079
750	167.786	0.0011114	0.2544	0.2555	708.76	1865.8	2574.6	709.59	2056.6	2766.2	2.0203	4.6642	6.6845
800	170.444	0.0011148	0.2393	0.2404	720.33	1856.3	2576.6	721.23	2047.7	2768.9	2.0464	4.6161	6.6625
850	172.974	0.0011180	0.2258	0.2269	731.37	1847.1	2578.5	732.32	2039.0	2771.4	2.0712	4.5706	6.6418
900	175.388	0.0011212	0.2138	0.2149	741.92	1838.3	2580.2	742.93	2030.7	2773.6	2.0948	4.5274	6.6222

H_2O

TABLE B.2 SATURATED WATER—PRESSURE TABLE: SI UNITS (*Continued*)

		Specific volume, m^3/kg			Internal energy, kJ/kg			Enthalpy, kJ/kg			Entropy, kJ/kg·K		
Press., MPa, P	Temp., °C, T	Sat. liquid, v_f	Evap., v_{fg}	Sat. vapor, v_g	Sat. liquid, u_f	Evap., u_{fg}	Sat. vapor, u_g	Sat. liquid, h_f	Evap., h_{fg}	Sat. vapor, h_g	Sat. liquid, s_f	Evap., s_{fg}	Sat. vapor, s_g
1.00	179.916	0.0011272	0.1933	0.1944	761.75	1821.6	2583.3	762.88	2014.8	2777.7	2.1388	4.4471	6.5859
1.10	184.100	0.0011330	0.1763	0.1775	780.14	1805.9	2586.0	781.39	1999.9	2781.2	2.1793	4.3737	6.5529
1.20	187.996	0.0011385	0.1621	0.1633	797.31	1791.1	2588.4	798.68	1985.7	2784.3	2.2167	4.3059	6.5226
1.30	191.644	0.0011438	0.1501	0.1512	813.44	1777.0	2590.5	814.93	1972.1	2787.0	2.2515	4.2430	6.4945
1.40	195.079	0.0011489	0.1396	0.1408	828.67	1763.6	2592.3	830.28	1959.1	2789.4	2.2842	4.1841	6.4683
1.75	205.764	0.0011656	0.1123	0.1134	876.38	1720.8	2597.2	878.42	1917.3	2795.7	2.3850	4.0034	6.3884
2.00	212.417	0.0011767	0.09841	0.09959	906.33	1693.2	2599.5	908.69	1890.0	2798.7	2.4471	3.8924	6.3396
2.25	218.452	0.0011872	0.08753	0.08872	933.70	1667.5	2601.2	936.37	1864.5	2800.8	2.5032	3.7926	6.2958
2.50	223.989	0.0011973	0.07875	0.07995	958.98	1643.3	2602.3	961.97	1840.2	2802.2	2.5544	3.7016	6.2560
3.0	233.892	0.0012166	0.06545	0.06666	1004.64	1598.6	2603.3	1008.29	1795.0	2803.3	2.6454	3.5401	6.1855
3.5	242.595	0.0012348	0.05582	0.05705	1045.31	1557.6	2602.9	1049.63	1753.0	2802.6	2.7251	3.3989	6.1240
4	250.392	0.0012524	0.04852	0.04977	1082.22	1519.3	2601.5	1087.23	1713.4	2800.6	2.7962	3.2727	6.0689
5	263.977	0.0012861	0.03815	0.03944	1147.77	1448.8	2596.5	1154.20	1639.5	2793.7	2.9201	3.0524	5.9725
6	275.621	0.0013190	0.03112	0.03244	1205.44	1383.8	2589.3	1213.35	1570.6	2783.9	3.0266	2.8619	5.8886
7	285.864	0.0013515	0.02602	0.02737	1257.54	1322.7	2580.2	1267.00	1504.8	2771.8	3.1211	2.6919	5.8130
8	295.042	0.0013843	0.02214	0.02352	1305.51	1264.1	2569.6	1316.59	1441.2	2757.8	3.2066	2.5365	5.7431
9	303.379	0.0014177	0.01907	0.02049	1350.36	1207.3	2557.6	1363.11	1378.9	2742.0	3.2855	2.3916	5.6771
10	311.031	0.0014522	0.01657	0.01803	1392.78	1151.5	2544.3	1407.30	1317.2	2724.5	3.3591	2.2548	5.6139
11	318.112	0.0014881	0.01450	0.01599	1433.33	1096.2	2529.5	1449.70	1255.7	2705.4	3.4287	2.1238	5.5525
12	324.709	0.0015259	0.01273	0.01426	1472.45	1041.0	2513.4	1490.76	1193.7	2684.5	3.4953	1.9968	5.4921
13	330.888	0.0015662	0.01121	0.01278	1510.5	985.2	2495.7	1530.9	1130.9	2661.8	3.5595	1.8723	5.4318
14	336.701	0.0016096	0.00988	0.01149	1547.9	928.4	2476.3	1570.4	1066.7	2637.1	3.6220	1.7491	5.3711
15	342.192	0.0016571	0.00868	0.01034	1585.0	870.0	2455.0	1609.8	1000.2	2610.1	3.6837	1.6255	5.3092
16	347.394	0.0017099	0.007601	0.009311	1622.1	809.2	2431.3	1649.5	930.8	2580.3	3.7452	1.5000	5.2451
17	352.335	0.0017698	0.006603	0.008373	1659.9	744.9	2404.8	1690.0	857.2	2547.1	3.8073	1.3704	5.1777
18	357.038	0.0018399	0.005665	0.007505	1698.9	675.7	2374.6	1732.0	777.7	2509.7	3.8714	1.2340	5.1054
19	361.522	0.0019251	0.004756	0.006681	1740.3	599.0	2339.3	1776.8	689.4	2466.2	3.9393	1.0862	5.0255
20	365.800	0.0020359	0.003838	0.005874	1786.0	510.1	2296.1	1826.7	586.9	2413.5	4.0146	0.9184	4.9330
21	369.881	0.002200	0.002812	0.005020	1841.4	396.0	2237.4	1887.6	455.2	2342.8	4.1062	0.7079	4.8141
22.055	373.976	0.003106	0	0.003106	2017	0	2017	2086	0	2086	4.409	0	4.409

Values generated from property formulation given in *NBS/NRC Steam Tables: Thermodynamic and Transport Properties and Computer Programs for Vapor and Liquid States of Water in SI Units*, Lester Haar, John S. Gallagher and George S. Kell, Hemisphere Publishing Corp., Washington, 1984.

TABLE B.3 SUPERHEATED WATER: SI UNITS

Temp., °C	v, m³/kg	u, kJ/kg	h, kJ/kg	s, kJ/kg·K	Temp., °C	v, m³/kg	u, kJ/kg	h, kJ/kg	s, kJ/kg·K
	P = 0.010 MPa (45.817°C)					P = 0.050 MPa (81.339°C)			
Sat.	14.674	2437.0	2583.8	8.1482	Sat.	3.241	2483.3	2645.3	7.5928
50	14.869	2443.1	2591.8	8.1731					
100	17.196	2515.0	2687.0	8.4471	100	3.419	2511.2	2682.1	7.6941
150	19.513	2587.4	2782.5	8.6873	150	3.890	2585.2	2779.7	7.9394
200	21.826	2660.8	2879.0	8.9030	200	4.356	2659.4	2877.2	8.1572
250	24.136	2735.5	2976.9	9.0995	250	4.821	2734.5	2975.6	8.3548
300	26.446	2811.7	3076.2	9.2808	300	5.284	2811.0	3075.2	8.5367
400	31.063	2968.8	3279.4	9.6075	400	6.209	2968.3	3278.8	8.8640
500	35.680	3132.4	3489.2	9.8979	500	7.134	3132.0	3488.7	9.1547
600	40.296	3302.8	3705.7	10.1612	600	8.058	3302.5	3705.4	9.4182
700	44.912	3480.2	3929.4	10.4036	700	8.981	3480.1	3929.1	9.6606
800	49.527	3664.8	4160.1	10.6292	800	9.905	3664.7	4159.9	9.8863
900	54.143	3856.4	4397.8	10.8410	900	10.828	3856.3	4397.7	10.0981
1000	58.758	4054.7	4642.3	11.0409	1000	11.751	4054.6	4642.2	10.2981
1100	63.374	4259.5	4893.2	11.2307	1100	12.675	4259.4	4893.1	10.4878
1200	67.989	4470.3	5150.2	11.4113	1200	13.598	4470.2	5150.1	10.6684
1300	72.604	4686.8	5412.9	11.5838	1300	14.521	4686.8	5412.8	10.8409
1400	77.220	4908.6	5680.8	11.7488	1400	15.444	4908.5	5680.7	11.0060
1500	81.835	5135.1	5953.5	11.9071	1500	16.367	5135.0	5953.4	11.1643
	P = 0.100 MPa (99.632°C)					P = 0.2 MPa (120.241°C)			
Sat.	1.6943	2505.7	2675.1	7.3589	Sat.	0.8859	2529.4	2706.5	7.1272
100	1.6961	2506.3	2675.9	7.3609					
150	1.9364	2582.4	2776.1	7.6129	150	0.9597	2576.7	2768.6	7.2793
200	2.172	2657.6	2874.8	7.8335	200	1.0803	2653.9	2870.0	7.5059
250	2.406	2733.3	2973.9	8.0325	250	1.1988	2730.8	2970.5	7.7078
300	2.639	2810.1	3073.9	8.2152	300	1.3162	2808.2	3071.4	7.8920
400	3.103	2967.7	3278.0	8.5432	400	1.5493	2966.6	3276.4	8.2216
500	3.566	3131.6	3488.2	8.8342	500	1.7814	3130.8	3487.1	8.5133
600	4.028	3302.3	3705.0	9.0979	600	2.013	3301.7	3704.3	8.7773
700	4.490	3479.8	3928.8	9.3405	700	2.244	3479.4	3928.3	9.0201
800	4.952	3664.5	4159.7	9.5662	800	2.475	3664.1	4159.2	9.2460
900	5.414	3856.1	4397.5	9.7781	900	2.707	3855.8	4397.1	9.4579
1000	5.875	4054.5	4642.0	9.9781	1000	2.938	4054.2	4641.7	9.6580
1100	6.337	4259.3	4893.0	10.1678	1100	3.168	4259.1	4892.8	9.8477
1200	6.799	4470.1	5150.0	10.3485	1200	3.399	4470.0	5149.8	10.0284
1300	7.260	4686.7	5412.7	10.5210	1300	3.630	4686.5	5412.5	10.2010
1400	7.722	4908.4	5680.6	10.6861	1400	3.861	4908.3	5680.5	10.3661
1500	8.184	5135.0	5953.3	10.8444	1500	4.092	5134.8	5953.2	10.5244

TABLE B.3 SUPERHEATED WATER: SI UNITS (*Continued*)

H₂O

Temp., °C	v, m³/kg	u, kJ/kg	h, kJ/kg	s, kJ/kg·K	Temp., °C	v, m³/kg	u, kJ/kg	h, kJ/kg	s, kJ/kg·K
	P = 0.3 MPa (133.555°C)					P = 0.4 MPa (143.643°C)			
Sat.	0.6059	2543.5	2725.3	6.9921	Sat.	0.4625	2553.5	2738.5	6.8961
150	0.6339	2570.7	2760.9	7.0779	150	0.4708	2564.4	2752.8	6.9300
200	0.7163	2650.2	2865.1	7.3108	200	0.5342	2646.4	2860.1	7.1699
250	0.7963	2728.2	2967.1	7.5157	250	0.5951	2725.6	2963.6	7.3779
300	0.8753	2806.3	3068.9	7.7015	300	0.6548	2804.4	3066.3	7.5654
350	0.9536	2885.3	3171.3	7.8729	350	0.7139	2883.8	3169.4	7.7378
400	1.0315	2965.4	3274.9	8.0327	400	0.7726	2964.3	3273.3	7.8982
500	1.1867	3130.1	3486.1	8.3252	500	0.8894	3129.3	3485.0	8.1913
600	1.3414	3301.1	3703.5	8.5895	600	1.0056	3300.5	3702.7	8.4561
700	1.4958	3478.9	3927.7	8.8325	700	1.1215	3478.5	3927.1	8.6993
800	1.6500	3663.8	4158.8	9.0585	800	1.2373	3663.4	4158.3	8.9254
900	1.8042	3855.5	4396.7	9.2705	900	1.3530	3855.2	4396.4	9.1375
1000	1.9582	4054.0	4641.4	9.4706	1000	1.4686	4053.7	4641.1	9.3377
1100	2.112	4258.8	4892.5	9.6605	1100	1.5841	4258.6	4892.3	9.5275
1200	2.266	4469.8	5149.6	9.8412	1200	1.6997	4469.6	5149.4	9.7083
1300	2.420	4686.3	5412.4	10.0137	1300	1.8152	4686.1	5412.2	9.8808
1400	2.574	4908.1	5680.3	10.1788	1400	1.9307	4907.9	5680.2	10.0460
1500	2.728	5134.7	5953.1	10.3372	1500	2.046	5134.6	5953.0	10.2043
	P = 0.5 MPa (151.866°C)					P = 0.6 MPa (158.863°C)			
Sat.	0.3749	2561.2	2748.6	6.8214	Sat.	0.3156	2567.3	2756.7	6.7601
200	0.4249	2642.5	2854.9	7.0585	200	0.3520	2638.5	2849.7	6.9658
250	0.4743	2723.0	2960.1	7.2699	250	0.3938	2720.3	2956.6	7.1806
300	0.5225	2802.5	3063.7	7.4591	300	0.4344	2800.5	3061.2	7.3716
350	0.5701	2882.3	3167.4	7.6324	350	0.4742	2880.9	3165.4	7.5459
400	0.6173	2963.1	3271.7	7.7935	400	0.5137	2961.9	3270.2	7.7076
450	0.6642	3045.1	3377.2	7.9446	450	0.5529	3044.1	3375.9	7.8591
500	0.7109	3128.5	3483.9	8.0873	500	0.5920	3127.7	3482.9	8.0021
600	0.8041	3299.9	3701.9	8.3524	600	0.6697	3299.3	3701.2	8.2676
700	0.8969	3478.0	3926.5	8.5958	700	0.7472	3477.6	3925.9	8.5112
800	0.9896	3663.0	4157.8	8.8221	800	0.8246	3662.7	4157.4	8.7376
900	1.0822	3854.9	4396.0	9.0342	900	0.9018	3854.6	4395.6	8.9498
1000	1.1748	4053.4	4640.8	9.2345	1000	0.9789	4053.2	4640.5	9.1501
1100	1.2673	4258.4	4892.0	9.4244	1100	1.0560	4258.2	4891.8	9.3401
1200	1.3597	4469.4	5149.2	9.6051	1200	1.1331	4469.2	5149.0	9.5209
1300	1.4521	4686.0	5412.0	9.7777	1300	1.2101	4685.8	5411.9	9.6935
1400	1.5446	4907.8	5680.1	9.9429	1400	1.2872	4907.6	5679.9	9.8587
1500	1.6369	5134.4	5952.9	10.1013	1500	1.3642	5134.3	5952.8	10.0170

TABLE B.3 SUPERHEATED WATER: SI UNITS (*Continued*)

Temp., °C	*v*, m³/kg	*u*, kJ/kg	*h*, kJ/kg	*s*, kJ/kg·K	Temp., °C	*v*, m³/kg	*u*, kJ/kg	*h*, kJ/kg	*s*, kJ/kg·K
	P = 0.8 MPa (170.444°C)					*P* = 1.0 MPa (179.916°C)			
Sat.	0.2404	2576.6	2768.9	6.6625	Sat.	0.19438	2583.3	2777.7	6.5859
200	0.2607	2630.2	2838.8	6.8151	200	0.2059	2621.5	2827.4	6.6932
250	0.2931	2714.8	2949.3	7.0373	250	0.2326	2709.2	2941.9	6.9234
300	0.3241	2796.6	3055.9	7.2319	300	0.2579	2792.7	3050.6	7.1219
350	0.3544	2877.9	3161.4	7.4084	350	0.2825	2874.9	3157.3	7.3005
400	0.3843	2959.6	3267.0	7.5713	400	0.3066	2957.2	3263.8	7.4648
450	0.4139	3042.2	3373.3	7.7237	450	0.3304	3040.3	3370.7	7.6180
500	0.4433	3126.1	3480.7	7.8673	500	0.3541	3124.5	3478.6	7.7622
600	0.5018	3298.1	3699.6	8.1335	600	0.4011	3297.0	3698.1	8.0292
700	0.5601	3476.7	3924.7	8.3775	700	0.4478	3475.7	3923.6	8.2736
800	0.6182	3661.9	4156.5	8.6041	800	0.4944	3661.2	4155.5	8.5005
900	0.6762	3854.0	4394.9	8.8165	900	0.5408	3853.4	4394.2	8.7130
1000	0.7341	4052.7	4639.9	9.0169	1000	0.5872	4052.1	4639.3	8.9135
1100	0.7920	4257.7	4891.3	9.2070	1100	0.6335	4257.3	4890.8	9.1037
1200	0.8498	4468.8	5148.6	9.3878	1200	0.6798	4468.4	5148.2	9.2846
1300	0.9076	4685.4	5411.5	9.5605	1300	0.7261	4685.1	5411.2	9.4573
1400	0.9654	4907.3	5679.6	9.7257	1400	0.7724	4907.0	5679.4	9.6225
1500	1.0232	5134.0	5952.5	9.8841	1500	0.8186	5133.7	5952.3	9.7810
	P = 1.2 MPa (187.996°C)					*P* = 1.4 MPa (195.079°C)			
Sat.	0.16328	2588.4	2784.3	6.5226	Sat.	0.14079	2592.3	2789.4	6.4683
200	0.16925	2612.3	2815.4	6.5890	200	0.14297	2602.6	2802.7	6.4966
250	0.19231	2703.5	2934.3	6.8281	250	0.16347	2697.6	2926.4	6.7454
300	0.2138	2788.6	3045.2	7.0307	300	0.18226	2784.5	3039.7	6.9523
350	0.2345	2871.8	3153.3	7.2115	350	0.2003	2868.8	3149.1	7.1354
400	0.2548	2954.8	3260.6	7.3771	400	0.2178	2952.4	3257.3	7.3024
450	0.2748	3038.3	3368.1	7.5312	450	0.2351	3036.4	3365.5	7.4573
500	0.2946	3122.8	3476.4	7.6760	500	0.2521	3121.2	3474.2	7.6027
600	0.3339	3295.8	3696.5	7.9437	600	0.2860	3294.6	3694.9	7.8712
700	0.3730	3474.8	3922.4	8.1885	700	0.3195	3473.9	3921.2	8.1164
800	0.4118	3660.4	4154.6	8.4156	800	0.3529	3659.7	4153.7	8.3438
900	0.4506	3852.7	4393.4	8.6284	900	0.3861	3852.1	4392.7	8.5567
1000	0.4893	4051.6	4638.7	8.8290	1000	0.4193	4051.1	4638.1	8.7574
1100	0.5279	4256.8	4890.3	9.0192	1100	0.4525	4256.4	4889.8	8.9477
1200	0.5665	4468.0	5147.8	9.2002	1200	0.4856	4467.6	5147.4	9.1288
1300	0.6051	4684.7	5410.9	9.3729	1300	0.5187	4684.4	5410.5	9.3016
1400	0.6437	4906.7	5679.1	9.5382	1400	0.5517	4906.4	5678.8	9.4669
1500	0.6822	5133.4	5952.1	9.6967	1500	0.5848	5133.1	5951.8	9.6254

TABLE B.3 SUPERHEATED WATER: SI UNITS *(Continued)*

Temp., °C	*v*, m³/kg	*u*, kJ/kg	*h*, kJ/kg	*s*, kJ/kg·K	Temp., °C	*v*, m³/kg	*u*, kJ/kg	*h*, kJ/kg	*s*, kJ/kg·K
	P = 1.6 MPa (201.410°C)					*P* = 1.8 MPa (207.151°C)			
Sat.	0.12375	2595.3	2793.3	6.4207	Sat.	0.11037	2597.7	2796.4	6.3781
225	0.13283	2644.0	2856.5	6.5506	225	0.11669	2635.9	2846.0	6.4795
250	0.14181	2691.5	2918.4	6.6718	250	0.12493	2685.2	2910.1	6.6052
300	0.15860	2780.4	3034.1	6.8833	300	0.14019	2776.1	3028.5	6.8214
350	0.17456	2865.7	3145.0	7.0688	350	0.15456	2862.6	3140.8	7.0093
400	0.19005	2950.0	3254.1	7.2372	400	0.16846	2947.6	3250.8	7.1792
450	0.2053	3034.4	3362.8	7.3930	450	0.18207	3032.5	3360.2	7.3359
500	0.2203	3119.6	3472.0	7.5390	500	0.19549	3118.0	3469.9	7.4825
600	0.2500	3293.4	3693.4	7.8082	600	0.2220	3292.2	3691.8	7.7524
700	0.2794	3473.0	3920.0	8.0539	700	0.2482	3472.1	3918.8	7.9986
800	0.3086	3659.0	4152.8	8.2815	800	0.2742	3658.2	4151.9	8.2264
900	0.3378	3851.5	4392.0	8.4946	900	0.3002	3850.9	4391.2	8.4397
1000	0.3669	4050.6	4637.6	8.6954	1000	0.3261	4050.1	4637.0	8.6406
1100	0.3959	4255.9	4889.3	8.8858	1100	0.3519	4255.5	4888.8	8.8311
1200	0.4249	4467.2	5147.0	9.0669	1200	0.3777	4466.8	5146.6	9.0122
1300	0.4538	4684.0	5410.2	9.2397	1300	0.4034	4683.7	5409.9	9.1851
1400	0.4828	4906.1	5678.5	9.4051	1400	0.4292	4905.8	5678.2	9.3505
1500	0.5117	5132.9	5951.6	9.5636	1500	0.4549	5132.6	5951.4	9.5090
	P = 2.0 MPa (212.417°C)					*P* = 2.2 MPa (217.288°C)			
Sat.	0.09959	2599.5	2798.7	6.3396	Sat.	0.09070	2600.9	2800.5	6.3042
225	0.10374	2627.5	2835.0	6.4133	225	0.09309	2618.8	2823.6	6.3509
250	0.11141	2678.8	2901.6	6.5438	250	0.10032	2672.1	2892.8	6.4866
300	0.12545	2771.8	3022.7	6.7651	300	0.11338	2767.5	3016.9	6.7133
350	0.13856	2859.4	3136.6	6.9556	350	0.12547	2856.2	3132.3	6.9064
400	0.15119	2945.1	3247.5	7.1269	400	0.13706	2942.7	3244.2	7.0792
450	0.16352	3030.5	3357.5	7.2845	450	0.14835	3028.5	3354.9	7.2378
500	0.17567	3116.3	3467.7	7.4318	500	0.15944	3114.7	3465.5	7.3857
600	0.19960	3291.0	3690.2	7.7024	600	0.18128	3289.8	3688.6	7.6571
700	0.2232	3471.1	3917.6	7.9490	700	0.2028	3470.2	3916.4	7.9040
800	0.2467	3657.5	4150.9	8.1771	800	0.2242	3656.7	4150.0	8.1325
900	0.2701	3850.3	4390.5	8.3905	900	0.2455	3849.7	4389.8	8.3460
1000	0.2934	4049.5	4636.4	8.5916	1000	0.2667	4049.0	4635.8	8.5472
1100	0.3167	4255.0	4888.4	8.7821	1100	0.2879	4254.6	4887.9	8.7378
1200	0.3399	4466.4	5146.2	8.9633	1200	0.3090	4466.0	5145.8	8.9191
1300	0.3631	4683.3	5409.5	9.1363	1300	0.3301	4683.0	5409.2	9.0921
1400	0.3863	4905.4	5678.0	9.3017	1400	0.3512	4905.1	5677.7	9.2575
1500	0.4094	5132.3	5951.1	9.4603	1500	0.3722	5132.0	5950.9	9.4161

TABLE B.3 SUPERHEATED WATER: SI UNITS (*Continued*)

Temp., °C	v, m³/kg	u, kJ/kg	h, kJ/kg	s, kJ/kg·K	Temp., °C	v, m³/kg	u, kJ/kg	h, kJ/kg	s, kJ/kg·K
	P = 2.5 MPa (223.989°C)					P = 3.0 MPa (233.892°C)			
Sat.	0.07995	2602.3	2802.2	6.2560	Sat.	0.06666	2603.3	2803.3	6.1855
250	0.08698	2661.7	2879.1	6.4069	250	0.07056	2643.1	2854.8	6.2857
300	0.09888	2760.8	3008.0	6.6424	300	0.08113	2749.2	2992.6	6.5375
350	0.10975	2851.4	3125.8	6.8395	350	0.09052	2843.2	3114.8	6.7420
400	0.12009	2938.9	3239.2	7.0146	400	0.09935	2932.7	3230.7	6.9210
450	0.13013	3025.5	3350.9	7.1746	450	0.10787	3020.5	3344.1	7.0835
500	0.13997	3112.2	3462.2	7.3235	500	0.11618	3108.1	3456.6	7.2339
550	0.14968	3199.6	3573.8	7.4634	550	0.12435	3196.1	3569.1	7.3750
600	0.15930	3288.0	3686.3	7.5960	600	0.13243	3285.0	3682.3	7.5084
700	0.17833	3468.8	3914.7	7.8436	700	0.14839	3466.5	3911.7	7.7571
800	0.19720	3655.6	4148.6	8.0724	800	0.16418	3653.8	4146.3	7.9865
900	0.2160	3848.8	4388.7	8.2862	900	0.17987	3847.2	4386.8	8.2008
1000	0.2347	4048.2	4634.9	8.4876	1000	0.19548	4046.9	4633.4	8.4024
1100	0.2533	4253.9	4887.1	8.6783	1100	0.2110	4252.8	4885.9	8.5934
1200	0.2719	4465.4	5145.2	8.8597	1200	0.2266	4464.4	5144.2	8.7749
1300	0.2905	4682.5	5408.7	9.0327	1300	0.2421	4681.6	5407.9	8.9480
1400	0.3090	4904.7	5677.3	9.1982	1400	0.2576	4903.9	5676.6	9.1136
1500	0.3276	5131.6	5950.6	9.3569	1500	0.2730	5130.9	5950.0	9.2723
	P = 4.0 MPa (250.392°C)					P = 5.0 MPa (263.977°C)			
Sat.	0.04977	2601.5	2800.6	6.0689	Sat.	0.03944	2596.5	2793.7	5.9725
275	0.05455	2667.0	2885.2	6.2269	275	0.04140	2630.5	2837.5	6.0532
300	0.05882	2724.4	2959.7	6.3598	300	0.04530	2697.0	2923.5	6.2067
350	0.06644	2826.1	3091.8	6.5811	350	0.05193	2808.0	3067.7	6.4482
400	0.07340	2919.8	3213.4	6.7688	400	0.05781	2906.5	3195.5	6.6456
450	0.08002	3010.3	3330.4	6.9364	450	0.06330	2999.8	3316.3	6.8187
500	0.08642	3099.7	3445.4	7.0902	500	0.06856	3091.1	3433.9	6.9760
550	0.09268	3189.0	3559.7	7.2335	550	0.07367	3181.8	3550.2	7.1218
600	0.09884	3278.9	3674.3	7.3687	600	0.07869	3272.8	3666.2	7.2586
700	0.11096	3461.8	3905.7	7.6195	700	0.08850	3457.1	3899.7	7.5117
800	0.12291	3650.0	4141.7	7.8503	800	0.09815	3646.3	4137.0	7.7438
900	0.13475	3844.1	4383.1	8.0654	900	0.10768	3841.0	4379.4	7.9598
1000	0.14652	4044.3	4630.4	8.2676	1000	0.11715	4041.7	4627.4	8.1626
1100	0.15824	4250.5	4883.5	8.4590	1100	0.12656	4248.3	4881.1	8.3543
1200	0.16993	4462.5	5142.2	8.6408	1200	0.13594	4460.5	5140.2	8.5365
1300	0.18159	4679.9	5406.2	8.8142	1300	0.14529	4678.1	5404.5	8.7101
1400	0.19322	4902.3	5675.2	8.9799	1400	0.15462	4900.8	5673.8	8.8760
1500	0.2048	5129.5	5948.9	9.1388	1500	0.16393	5128.1	5947.7	9.0350

TABLE B.3 SUPERHEATED WATER: SI UNITS (*Continued*)

Temp., °C	*v*, m³/kg	*u*, kJ/kg	*h*, kJ/kg	*s*, kJ/kg·K	Temp., °C	*v*, m³/kg	*u*, kJ/kg	*h*, kJ/kg	*s*, kJ/kg·K
	P = 6.0 MPa (275.621°C)					*P* = 7.0 MPa (285.864°C)			
Sat.	0.03244	2589.3	2783.9	5.8886	Sat.	0.02737	2580.2	2771.8	5.8130
300	0.03615	2666.3	2883.2	6.0659	300	0.02946	2631.4	2837.6	5.9293
350	0.04222	2788.9	3042.2	6.3322	350	0.03523	2768.5	3015.1	6.2269
400	0.04739	2892.7	3177.0	6.5404	400	0.03993	2878.4	3157.9	6.4474
450	0.05214	2989.1	3301.9	6.7195	450	0.04416	2978.1	3287.3	6.6329
500	0.05665	3082.4	3422.3	6.8805	500	0.04813	3073.6	3410.5	6.7978
550	0.06100	3174.6	3540.6	7.0287	550	0.05194	3167.2	3530.8	6.9486
600	0.06525	3266.6	3658.1	7.1673	600	0.05565	3260.3	3649.8	7.0889
650	0.06942	3359.1	3775.6	7.2982	650	0.05927	3353.6	3768.5	7.2211
700	0.07353	3452.4	3893.6	7.4227	700	0.06284	3447.6	3887.5	7.3466
800	0.08164	3642.5	4132.3	7.6561	800	0.06985	3638.7	4127.6	7.5815
900	0.08964	3837.9	4375.7	7.8730	900	0.07675	3834.8	4372.0	7.7992
1000	0.09756	4039.1	4624.5	8.0764	1000	0.08358	4036.4	4621.5	8.0032
1100	0.10544	4246.0	4878.6	8.2686	1100	0.09035	4243.7	4876.2	8.1958
1200	0.11328	4458.5	5138.2	8.4510	1200	0.09709	4456.6	5136.2	8.3785
1300	0.12109	4676.4	5402.9	8.6248	1300	0.10381	4674.6	5401.3	8.5526
1400	0.12888	4899.2	5672.5	8.7910	1400	0.11050	4897.6	5671.1	8.7189
1500	0.13666	5126.7	5946.6	8.9501	1500	0.11718	5125.3	5945.5	8.8782
	P = 8.0 MPa (295.042°C)					*P* = 9.0 MPa (303.379°C)			
Sat.	0.02352	2569.6	2757.8	5.7431	Sat.	0.02049	2557.6	2742.0	5.6772
325	0.02738	2677.2	2896.2	5.9809	325	0.02326	2645.6	2854.9	5.8696
350	0.02995	2746.7	2986.3	6.1286	350	0.02579	2723.4	2955.5	6.0345
400	0.03431	2863.5	3138.0	6.3630	400	0.02993	2848.1	3117.5	6.2848
450	0.03816	2966.9	3272.2	6.5554	450	0.03350	2955.4	3256.9	6.4847
500	0.04174	3064.6	3398.5	6.7243	500	0.03677	3055.5	3386.4	6.6579
550	0.04515	3159.8	3521.0	6.8778	550	0.03986	3152.2	3511.0	6.8141
600	0.04845	3254.0	3641.5	7.0200	600	0.04284	3247.6	3633.2	6.9583
650	0.05166	3348.1	3761.4	7.1535	650	0.04574	3342.6	3754.3	7.0932
700	0.05482	3442.8	3881.4	7.2800	700	0.04858	3438.0	3875.2	7.2207
800	0.06101	3634.9	4122.9	7.5163	800	0.05413	3631.0	4118.2	7.4584
900	0.06708	3831.6	4368.3	7.7349	900	0.05957	3828.5	4364.6	7.6779
1000	0.07309	4033.8	4618.5	7.9395	1000	0.06493	4031.2	4615.5	7.8831
1100	0.07904	4241.5	4873.8	8.1325	1100	0.07024	4239.2	4871.4	8.0765
1200	0.08496	4454.6	5134.2	8.3156	1200	0.07552	4452.6	5132.2	8.2599
1300	0.09084	4672.9	5399.6	8.4899	1300	0.08076	4671.1	5398.0	8.4344
1400	0.09671	4896.1	5669.8	8.6564	1400	0.08599	4894.5	5668.5	8.6011
1500	0.10257	5123.9	5944.4	8.8158	1500	0.09120	5122.5	5943.3	8.7606

TABLE B.3 SUPERHEATED WATER: SI UNITS (*Continued*)

Temp., °C	v, m³/kg	u, kJ/kg	h, kJ/kg	s, kJ/kg·K	Temp., °C	v, m³/kg	u, kJ/kg	h, kJ/kg	s, kJ/kg·K
	P = 10.0 MPa (311.031°C)					P = 12.5 MPa (327.847°C)			
Sat.	0.018026	2544.3	2724.5	5.6140	Sat.	0.013495	2504.8	2673.5	5.4621
350	0.02242	2698.1	2922.2	5.9425	350	0.016121	2623.3	2824.8	5.7097
400	0.02641	2832.0	3096.1	6.2114	400	0.02001	2788.7	3038.8	6.0409
450	0.02975	2943.6	3241.1	6.4193	450	0.02299	2912.8	3200.2	6.2724
500	0.03278	3046.2	3374.0	6.5971	500	0.02560	3022.2	3342.2	6.4625
550	0.03563	3144.6	3500.9	6.7561	550	0.02801	3125.0	3475.2	6.6291
600	0.03836	3241.1	3624.7	6.9022	600	0.03029	3224.7	3603.3	6.7802
650	0.04101	3337.1	3747.1	7.0385	650	0.03248	3322.9	3728.9	6.9202
700	0.04359	3433.1	3869.0	7.1671	700	0.03461	3420.8	3853.4	7.0514
750	0.04613	3529.7	3991.0	7.2893	750	0.03669	3518.8	3977.4	7.1757
800	0.04863	3627.2	4113.5	7.4062	800	0.03873	3617.4	4101.5	7.2942
900	0.05355	3825.3	4360.9	7.6266	900	0.04273	3817.4	4351.6	7.5169
1000	0.05840	4028.5	4612.5	7.8324	1000	0.04666	4021.9	4605.1	7.7243
1100	0.06320	4236.9	4869.0	8.0263	1100	0.05054	4231.3	4863.0	7.9192
1200	0.06796	4450.6	5130.3	8.2100	1200	0.05437	4445.7	5125.4	8.1037
1300	0.07270	4669.4	5396.4	8.3847	1300	0.05819	4665.1	5392.4	8.2790
1400	0.07741	4893.0	5667.1	8.5516	1400	0.06198	4889.1	5663.8	8.4463
1500	0.08211	5121.1	5942.2	8.7112	1500	0.06575	5117.6	5939.5	8.6063
	P = 15.0 MPa (342.192°C)					P = 17.5 MPa (354.715°C)			
Sat.	0.010340	2455.0	2610.1	5.3092	Sat.	0.007931	2390.2	2529.0	5.1422
350	0.011469	2519.3	2691.3	5.4404					
400	0.015652	2739.9	2974.7	5.8799	400	0.012452	2684.0	2901.9	5.7198
450	0.018450	2879.9	3156.6	6.1410	450	0.015180	2844.7	3110.3	6.0191
500	0.02080	2997.3	3309.3	6.3452	500	0.017359	2971.2	3275.0	6.2395
550	0.02292	3104.9	3448.8	6.5201	550	0.019282	3084.2	3421.7	6.4234
600	0.02490	3207.9	3581.5	6.6767	600	0.02106	3190.8	3559.3	6.5857
650	0.02679	3308.6	3710.5	6.8204	650	0.02273	3294.0	3691.8	6.7334
700	0.02862	3408.3	3837.6	6.9544	700	0.02434	3395.6	3821.6	6.8703
750	0.03039	3507.8	3963.7	7.0808	750	0.02590	3496.7	3949.9	6.9989
800	0.03213	3607.6	4089.6	7.2009	800	0.02742	3597.7	4077.5	7.1206
900	0.03552	3809.4	4342.2	7.4260	900	0.03037	3801.4	4332.9	7.3481
1000	0.03883	4015.2	4597.7	7.6350	1000	0.03324	4008.5	4590.3	7.5587
1100	0.04209	4225.6	4857.0	7.8310	1100	0.03606	4219.9	4851.0	7.7558
1200	0.04532	4440.8	5120.5	8.0163	1200	0.03885	4435.8	5115.7	7.9418
1300	0.04851	4660.7	5388.4	8.1922	1300	0.04160	4656.4	5384.5	8.1184
1400	0.05169	4885.3	5660.6	8.3599	1400	0.04434	4881.4	5657.3	8.2865
1500	0.05485	5114.1	5936.8	8.5203	1500	0.04706	5110.7	5934.2	8.4472

TABLE B.3 SUPERHEATED WATER: SI UNITS (*Continued*)

Temp., °C	v, m³/kg	u, kJ/kg	h, kJ/kg	s, kJ/kg·K	Temp., °C	v, m³/kg	u, kJ/kg	h, kJ/kg	s, kJ/kg·K
	P = 20.0 MPa (365.800°C)					P = 25.0 MPa			
Sat.	0.005870	2295.8	2413.2	4.9323					
375	0.007668	2447.6	2601.0	5.2246	375	0.0019794	1799.8	1849.3	4.0340
400	0.009946	2617.9	2816.9	5.5521	400	0.006001	2428.1	2578.1	5.1388
450	0.012701	2806.8	3060.8	5.9026	450	0.009167	2721.5	2950.6	5.6755
500	0.014769	2944.1	3239.4	6.1417	500	0.011123	2886.1	3164.2	5.9616
550	0.016549	3063.0	3393.9	6.3355	550	0.012716	3018.6	3336.5	6.1778
600	0.018169	3173.3	3536.7	6.5039	600	0.014126	3137.3	3490.4	6.3593
650	0.019687	3279.2	3672.9	6.6556	650	0.015423	3249.0	3634.5	6.5198
700	0.02113	3382.8	3805.5	6.7955	700	0.016644	3356.8	3773.0	6.6659
750	0.02253	3485.4	3936.0	6.9263	750	0.017812	3462.7	3908.0	6.8012
800	0.02388	3587.8	4065.4	7.0498	800	0.018938	3567.6	4041.1	6.9282
900	0.02651	3793.3	4323.5	7.2797	900	0.02110	3777.1	4304.7	7.1631
1000	0.02905	4001.8	4582.8	7.4919	1000	0.02319	3988.3	4568.0	7.3785
1100	0.03154	4214.2	4845.0	7.6901	1100	0.02522	4202.8	4833.2	7.5790
1200	0.03400	4430.9	5110.9	7.8769	1200	0.02721	4421.1	5101.3	7.7674
1300	0.03642	4652.1	5380.5	8.0540	1300	0.02917	4643.5	5372.8	7.9457
1400	0.03883	4877.6	5654.2	8.2227	1400	0.03112	4870.0	5647.9	8.1152
1500	0.04122	5107.3	5931.6	8.3837	1500	0.03304	5100.4	5926.5	8.2770
	P = 30.0 MPa					P = 35.0 MPa			
375	0.0017913	1737.7	1791.4	3.9303	375	0.0017007	1702.4	1761.9	3.8714
400	0.002793	2066.9	2150.7	4.4723	400	0.002106	1914.6	1988.3	4.2136
450	0.006736	2620.2	2822.3	5.4435	450	0.004960	2499.4	2672.9	5.1969
500	0.008676	2823.2	3083.5	5.7936	500	0.006924	2755.0	2997.3	5.6320
550	0.010158	2972.0	3276.8	6.0361	550	0.008333	2923.3	3215.0	5.9052
600	0.011431	3100.1	3443.1	6.2324	600	0.009511	3061.9	3394.7	6.1174
650	0.012583	3218.0	3595.5	6.4022	650	0.010558	3186.5	3556.0	6.2971
700	0.013655	3330.4	3740.1	6.5547	700	0.011523	3303.6	3706.9	6.4563
750	0.014671	3439.7	3879.8	6.6948	750	0.012431	3416.5	3851.6	6.6012
800	0.015645	3547.3	4016.7	6.8254	800	0.013295	3526.9	3992.2	6.7355
900	0.017504	3760.8	4285.9	7.0653	900	0.014935	3744.4	4267.2	6.9804
1000	0.019281	3974.8	4553.3	7.2840	1000	0.016491	3961.3	4538.5	7.2025
1100	0.02100	4191.4	4821.4	7.4867	1100	0.017993	4180.0	4809.7	7.4075
1200	0.02269	4411.2	5091.8	7.6768	1200	0.019456	4401.5	5082.4	7.5992
1300	0.02434	4634.9	5365.1	7.8563	1300	0.02089	4626.4	5357.6	7.7799
1400	0.02598	4862.4	5641.7	8.0267	1400	0.02231	4854.9	5635.6	7.9512
1500	0.02759	5093.7	5921.5	8.1891	1500	0.02370	5087.0	5916.6	8.1143

TABLE B.3 SUPERHEATED WATER: SI UNITS (*Continued*)

Temp., °C	v, m³/kg	u, kJ/kg	h, kJ/kg	s, kJ/kg·K	Temp., °C	v, m³/kg	u, kJ/kg	h, kJ/kg	s, kJ/kg·K
	P = 40.0 MPa					P = 50.0 MPa			
375	0.0016405	1676.5	1742.2	3.8280	375	0.0015593	1638.2	1716.1	3.7632
400	0.0019096	1854.5	1930.8	4.1134	400	0.0017301	1787.6	1874.1	4.0022
450	0.003691	2365.5	2513.1	4.9463	450	0.002486	2160.4	2284.7	4.5894
500	0.005619	2681.9	2906.7	5.4745	500	0.003892	2529.6	2724.2	5.1780
550	0.006973	2872.6	3151.6	5.7819	550	0.005111	2767.7	3023.3	5.5535
600	0.008077	3022.7	3345.8	6.0111	600	0.006098	2942.7	3247.7	5.8184
650	0.009046	3154.5	3516.3	6.2011	650	0.006949	3089.7	3437.2	6.0296
700	0.009930	3276.6	3673.8	6.3673	700	0.007713	3222.1	3607.8	6.2097
750	0.010754	3393.1	3823.3	6.5171	750	0.008420	3346.2	3767.2	6.3695
800	0.011536	3506.4	3967.8	6.6551	800	0.009083	3465.3	3919.5	6.5148
900	0.013010	3728.1	4248.5	6.9052	900	0.010322	3695.4	4211.5	6.7751
1000	0.014401	3947.9	4523.9	7.1305	1000	0.011479	3921.1	4495.0	7.0070
1100	0.015737	4168.6	4798.1	7.3378	1100	0.012582	4146.1	4775.2	7.2188
1200	0.017035	4391.7	5073.1	7.5311	1200	0.013647	4372.4	5054.7	7.4154
1300	0.018304	4617.9	5350.1	7.7130	1300	0.014684	4601.1	5335.3	7.5996
1400	0.019553	4847.4	5629.5	7.8852	1400	0.015702	4832.6	5617.7	7.7736
1500	0.02079	5080.3	5911.8	8.0491	1500	0.016703	5067.2	5902.3	7.9389
	P = 60.0 MPa					P = 70.0 MPa			
375	0.0015034	1609.3	1699.5	3.7139	375	0.0014608	1585.8	1688.0	3.6734
400	0.0016328	1745.1	1843.0	3.9312	400	0.0015667	1713.1	1822.8	3.8774
450	0.002084	2055.0	2180.0	4.4134	450	0.0018917	1991.2	2123.6	4.3081
500	0.002955	2394.6	2571.9	4.9373	500	0.002464	2294.4	2466.9	4.7669
550	0.003955	2663.7	2901.0	5.3505	550	0.003227	2569.1	2795.0	5.1784
600	0.004826	2862.7	3152.3	5.6471	600	0.003972	2785.7	3063.8	5.4957
650	0.005582	3025.0	3359.9	5.8786	650	0.004642	2961.9	3286.8	5.7442
700	0.006259	3167.9	3543.5	6.0723	700	0.005245	3114.7	3481.9	5.9502
750	0.006880	3299.7	3712.5	6.2417	750	0.005800	3253.9	3659.9	6.1286
800	0.007461	3424.6	3872.3	6.3942	800	0.006317	3384.5	3826.7	6.2878
900	0.008538	3663.1	4175.4	6.6643	900	0.007272	3631.3	4140.3	6.5674
1000	0.009536	3894.5	4466.7	6.9027	1000	0.008153	3868.4	4439.1	6.8118
1100	0.010482	4123.7	4752.6	7.1189	1100	0.008985	4101.8	4730.7	7.0323
1200	0.011391	4353.3	5036.7	7.3186	1200	0.009781	4334.5	5019.1	7.2351
1300	0.012273	4584.5	5320.8	7.5052	1300	0.010551	4568.1	5306.7	7.4240
1400	0.013135	4818.0	5606.1	7.6810	1400	0.011302	4803.6	5594.8	7.6015
1500	0.013981	5054.2	5893.0	7.8476	1500	0.012037	5041.4	5884.0	7.7694

Values generated from property formulation given in *NBS/NRC Steam Tables: Thermodynamic and Transport Properties and Computer Programs for Vapor and Liquid States of Water in SI Units,* Lester Haar, John S. Gallagher and George S. Kell, Hemisphere Publishing Corp., Washington, 1984.

TABLE B.4 COMPRESSED LIQUID WATER: SI UNITS

P = 5.0 MPa (263.977°C)

Temp., °C	*v*, m³/kg	*u*, kJ/kg	*h*, kJ/kg	*s*, kJ/kg·K
0	0.0009977	0.06	5.05	0.0002
20	0.0009995	83.53	88.52	0.2951
40	0.0010057	166.89	171.92	0.5703
60	0.0010149	250.26	255.34	0.8285
80	0.0010267	333.74	338.87	1.0720
100	0.0010409	417.54	422.75	1.3031
120	0.0010576	501.87	507.16	1.5234
140	0.0010768	586.88	592.26	1.7346
160	0.0010988	672.73	678.22	1.9377
180	0.0011240	759.66	765.28	2.1342
200	0.0011529	848.03	853.79	2.3253
220	0.0011867	938.32	944.25	2.5126
240	0.0012266	1031.3	1037.4	2.6977
260	0.0012751	1128.0	1134.3	2.8830
Sat.	0.0012861	1147.8	1154.2	2.9201

P = 10.0 MPa (311.031°C)

Temp., °C	*v*, m³/kg	*u*, kJ/kg	*h*, kJ/kg	*s*, kJ/kg·K
0	0.0009952	0.15	10.10	0.0004
20	0.0009973	83.22	93.20	0.2940
40	0.0010035	166.30	176.33	0.5684
60	0.0010127	249.40	259.53	0.8259
80	0.0010244	332.61	342.85	1.0688
100	0.0010384	416.13	426.52	1.2992
120	0.0010549	500.15	510.70	1.5190
140	0.0010737	584.80	595.53	1.7295
160	0.0010953	670.24	681.19	1.9319
180	0.0011199	756.68	767.88	2.1276
200	0.0011481	844.43	855.91	2.3177
220	0.0011807	933.95	945.75	2.5036
240	0.0012190	1025.8	1038.0	2.6870
260	0.0012650	1121.0	1133.7	2.8699
280	0.0013222	1221.0	1234.2	3.0550
300	0.0013975	1328.4	1342.4	3.2470
Sat.	0.0014522	1392.8	1407.3	3.3592

P = 15.0 MPa (342.192°C)

Temp., °C	*v*, m³/kg	*u*, kJ/kg	*h*, kJ/kg	*s*, kJ/kg·K
0	0.0009928	0.22	15.11	0.0006
20	0.0009951	82.92	97.85	0.2929
40	0.0010014	165.72	180.74	0.5664
60	0.0010105	248.56	263.72	0.8233
80	0.0010221	331.50	346.84	1.0656
100	0.0010360	414.75	430.29	1.2955
120	0.0010522	498.46	514.24	1.5146
140	0.0010708	582.77	598.83	1.7245
160	0.0010919	667.81	684.19	1.9262
180	0.0011159	753.79	770.53	2.1211
200	0.0011434	840.97	858.12	2.3102
220	0.0011750	929.75	947.38	2.4950
240	0.0012118	1020.7	1038.9	2.6768
260	0.0012556	1114.6	1133.4	2.8575
280	0.0013092	1212.6	1232.2	3.0394
300	0.0013777	1316.7	1337.4	3.2261
320	0.0014725	1431.0	1453.0	3.4244
340	0.0016307	1567.1	1591.6	3.6540
Sat.	0.0016571	1585.0	1609.8	3.6837

P = 20.0 MPa (365.800°C)

Temp., °C	*v*, m³/kg	*u*, kJ/kg	*h*, kJ/kg	*s*, kJ/kg·K
0	0.0009904	0.28	20.08	0.0007
20	0.0009929	82.62	102.48	0.2918
40	0.0009992	165.14	185.13	0.5645
60	0.0010084	247.73	267.90	0.8207
80	0.0010199	330.42	350.82	1.0624
100	0.0010336	413.40	434.07	1.2917
120	0.0010496	496.81	517.81	1.5103
140	0.0010678	580.79	602.15	1.7196
160	0.0010886	665.46	687.23	1.9207
180	0.0011121	750.98	773.22	2.1147
200	0.0011389	837.62	860.40	2.3030
220	0.0011695	925.72	949.12	2.4866
240	0.0012051	1015.8	1039.9	2.6670
260	0.0012469	1108.5	1133.4	2.8458
280	0.0012974	1204.8	1230.7	3.0250
300	0.0013605	1306.2	1333.4	3.2073
320	0.0014442	1415.6	1444.5	3.3978
340	0.0015685	1539.4	1570.7	3.6070
360	0.0018248	1703.2	1739.7	3.8778
Sat.	0.002035	1785.8	1826.5	4.0142

TABLE B.4 COMPRESSED LIQUID WATER: SI UNITS (*Continued*)

Temp., °C	*v*, m³/kg	*u*, kJ/kg	*h*, kJ/kg	*s*, kJ/kg·K	Temp., °C	*v*, m³/kg	*u*, kJ/kg	*h*, kJ/kg	*s*, kJ/kg·K
		P = 30.0 MPa					*P* = 50.0 MPa		
0	0.0009857	0.35	29.92	0.0005	0	0.0009767	0.36	49.20	−0.0008
20	0.0009887	82.02	111.68	0.2893	20	0.0009805	80.83	129.85	0.2842
40	0.0009951	164.02	193.87	0.5606	40	0.0009872	161.87	211.23	0.5527
60	0.0010042	246.12	276.25	0.8156	60	0.0009962	243.07	292.88	0.8054
80	0.0010155	328.32	358.79	1.0562	80	0.0010072	324.35	374.71	1.0440
100	0.0010290	410.78	441.64	1.2844	100	0.0010201	405.84	456.84	1.2702
120	0.0010445	493.63	524.96	1.5019	120	0.0010349	487.65	539.40	1.4857
140	0.0010622	576.98	608.85	1.7100	140	0.0010517	569.86	622.44	1.6917
160	0.0010822	660.93	693.39	1.9098	160	0.0010704	652.51	706.04	1.8893
180	0.0011048	745.61	778.76	2.1025	180	0.0010914	735.71	790.28	2.0795
200	0.0011303	831.24	865.15	2.2890	200	0.0011148	819.60	875.34	2.2631
220	0.0011593	918.11	952.89	2.4707	220	0.0011411	904.38	961.44	2.4413
240	0.0011925	1006.6	1042.4	2.6485	240	0.0011706	990.3	1048.9	2.6151
260	0.0012311	1097.3	1134.2	2.8240	260	0.0012042	1077.7	1137.9	2.7854
280	0.0012766	1190.7	1229.0	2.9986	280	0.0012428	1167.0	1229.2	2.9534
300	0.0013317	1288.0	1328.0	3.1743	300	0.0012876	1258.7	1323.1	3.1201
320	0.0014008	1390.7	1432.8	3.3540	320	0.0013406	1353.4	1420.5	3.2871
340	0.0014925	1501.5	1546.3	3.5422	340	0.0014046	1452.1	1522.3	3.4559
360	0.0016269	1626.1	1674.9	3.7486	360	0.0014845	1555.8	1630.1	3.6289

Values generated from property formulation given in *NBS/NRC Steam Tables: Thermodynamic and Transport Properties and Computer Programs for Vapor and Liquid States of Water in SI Units,* Lester Haar, John S. Gallagher and George S. Kell, Hemisphere Publishing Corp., Washington, 1984.

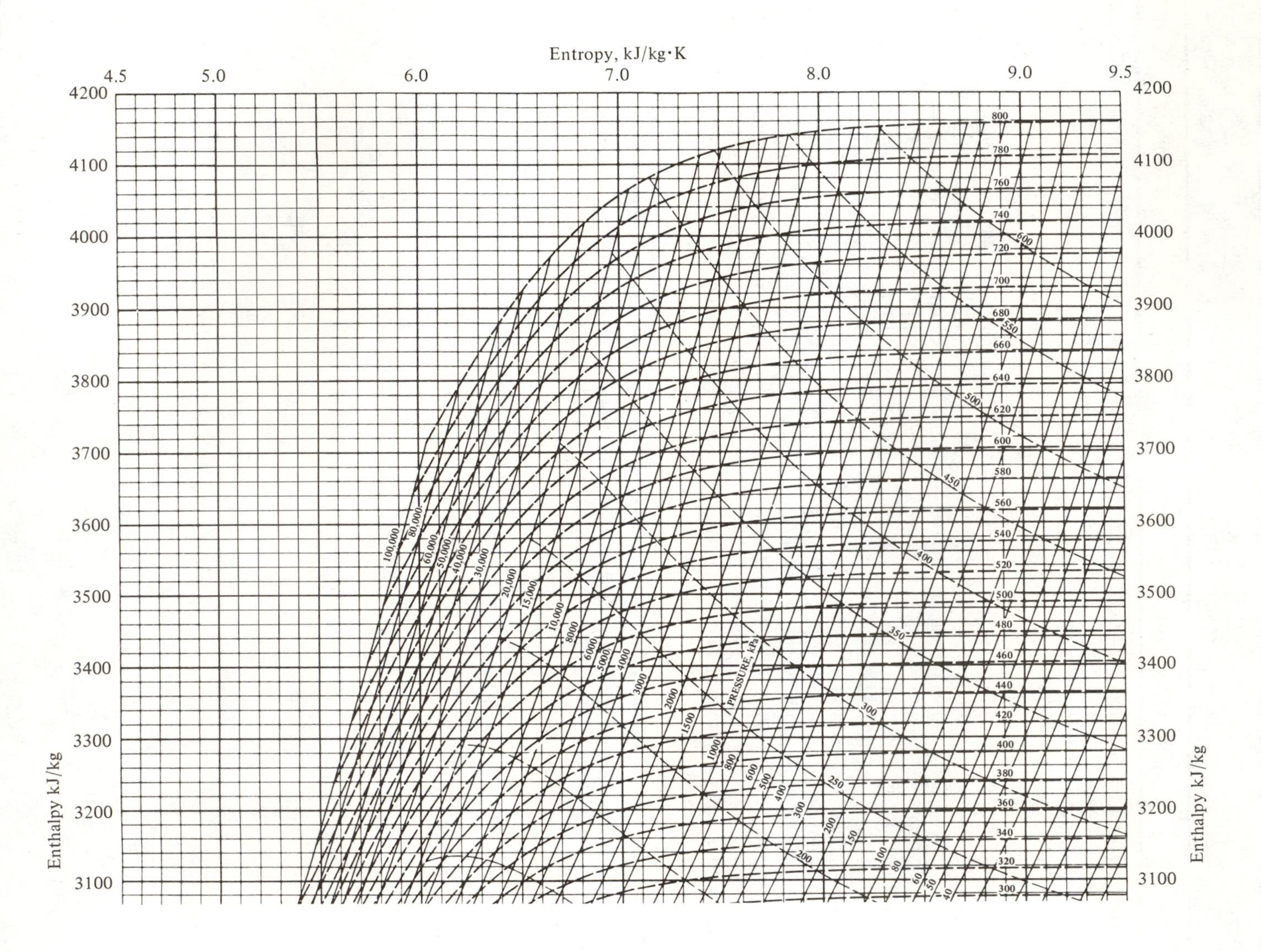
Entropy, kJ/kg·K
Enthalpy kJ/kg
Enthalpy kJ/kg
PRESSURE, kPa

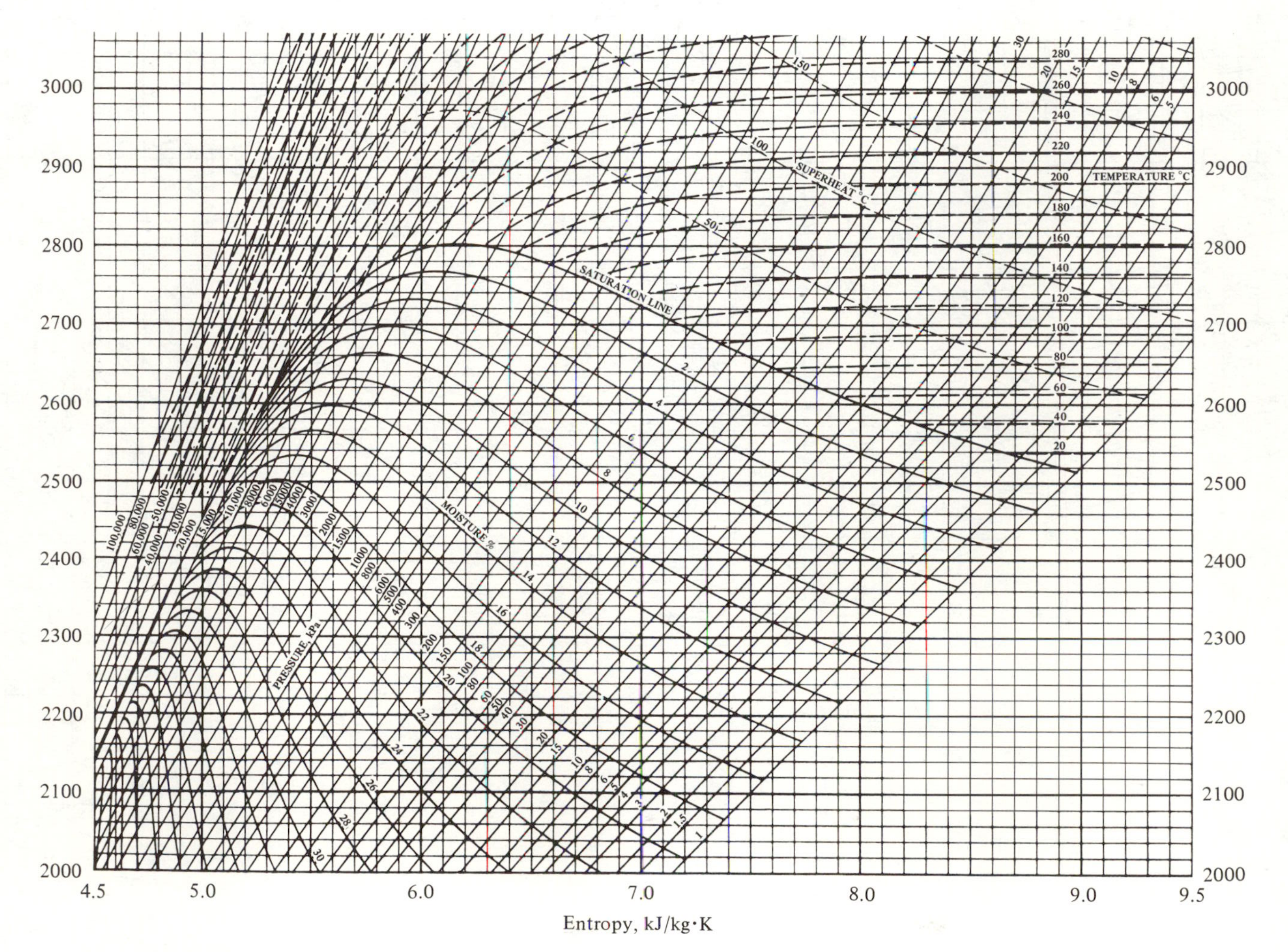

Figure B.5 Mollier diagram for water. (Source: ASME Steam Tables, American Society of Mechanical Engineers, New York, 1967.)

H_2O

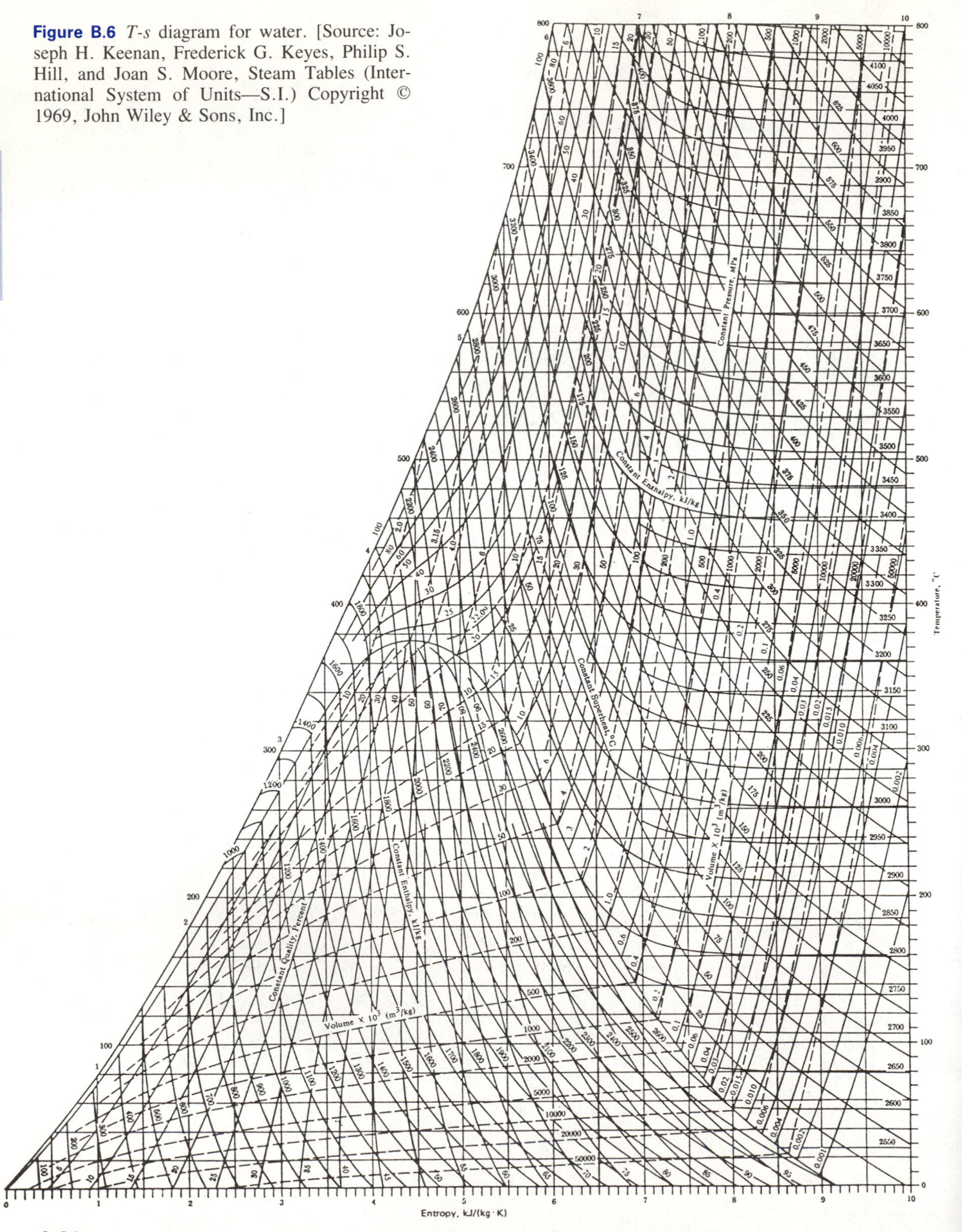

Figure B.6 *T-s* diagram for water. [Source: Joseph H. Keenan, Frederick G. Keyes, Philip S. Hill, and Joan S. Moore, Steam Tables (International System of Units—S.I.) Copyright © 1969, John Wiley & Sons, Inc.]

Properties of Refrigerant-12

TABLE C.1 SATURATED REFRIGERANT-12—TEMPERATURE TABLE: SI UNITS

		Specific volume, m³/kg			Internal energy, kJ/kg			Enthalpy, kJ/kg			Entropy, kJ/kg·K		
Temp., °C, T	Press., MPa, P	Sat. liquid, v_f	Evap., v_{fg}	Sat. vapor, v_g	Sat. liquid, u_f	Evap., u_{fg}	Sat. vapor, u_g	Sat. liquid, h_f	Evap., h_{fg}	Sat. vapor, h_g	Sat. liquid, s_f	Evap., s_{fg}	Sat. vapor, s_g
−90	0.002836	0.00060537	4.4258	4.4264	−42.726	177.21	134.49	−42.724	189.76	147.04	−0.20593	1.03611	0.83018
−85	0.004230	0.00061023	3.0443	3.0450	−38.504	174.92	136.42	−38.501	187.80	149.30	−0.18318	0.99814	0.81495
−80	0.006160	0.00061521	2.1431	2.1438	−34.278	172.65	138.37	−34.275	185.85	151.58	−0.16102	0.96222	0.80120
−75	0.008774	0.00062030	1.5409	1.5416	−30.048	170.40	140.35	−30.042	183.92	153.88	−0.13940	0.92818	0.78878
−70	0.012246	0.00062551	1.1295	1.1302	−25.808	168.16	142.35	−25.801	181.99	156.19	−0.11827	0.89584	0.77757
−65	0.016776	0.00063085	0.84269	0.84332	−21.558	165.92	144.37	−21.547	180.06	158.51	−0.097597	0.86505	0.76745
−60	0.022591	0.00063633	0.63893	0.63956	−17.294	163.69	146.40	−17.279	178.13	160.85	−0.077352	0.83569	0.75833
−55	0.029944	0.00064195	0.49166	0.49230	−13.013	161.46	148.45	−12.993	176.18	163.19	−0.057499	0.80762	0.75012
−50	0.039115	0.00064772	0.38351	0.38415	−8.712	159.22	150.51	−8.686	174.22	165.53	−0.038008	0.78073	0.74272
−45	0.050408	0.00065365	0.30290	0.30355	−4.389	156.97	152.58	−4.356	172.24	167.88	−0.018850	0.75492	0.73607
−40	0.064152	0.00065975	0.24198	0.24264	−0.042	154.70	154.66	0.000	170.22	170.22	0.000000	0.73010	0.73010
−35	0.080701	0.00066604	0.19536	0.19603	4.332	152.41	156.74	4.386	168.18	172.56	0.018566	0.70618	0.72475
−30	0.10043	0.00067252	0.15926	0.15993	8.736	150.10	158.84	8.804	166.09	174.90	0.036871	0.68309	0.71996
−25	0.12373	0.00067921	0.13098	0.13166	13.173	147.76	160.93	13.257	163.96	177.22	0.054935	0.66074	0.71568
−20	0.15101	0.00068613	0.10861	0.10929	17.644	145.38	163.03	17.748	161.78	179.53	0.072777	0.63908	0.71185
−15	0.18272	0.00069330	0.090731	0.091424	22.152	142.97	165.12	22.279	159.55	181.82	0.090415	0.61804	0.70845
−10	0.21928	0.00070074	0.076318	0.077019	26.698	140.51	167.21	26.852	157.25	184.10	0.10786	0.59756	0.70542
−5	0.26117	0.00070847	0.064599	0.065308	31.285	138.01	169.30	31.470	154.88	186.35	0.12514	0.57759	0.70273
0	0.30885	0.00071651	0.054993	0.055709	35.915	135.46	171.37	36.136	152.44	188.58	0.14225	0.55809	0.70034
5	0.36282	0.00072491	0.047059	0.047783	40.588	132.85	173.43	40.851	149.92	190.77	0.15921	0.53899	0.69821
10	0.42356	0.00073370	0.040459	0.041192	45.307	130.18	175.48	45.618	147.31	192.93	0.17604	0.52026	0.69630
15	0.49158	0.00074292	0.034931	0.035674	50.072	127.44	177.51	50.438	144.61	195.05	0.19274	0.50186	0.69460
20	0.56740	0.00075261	0.030272	0.031025	54.887	124.63	179.52	55.314	141.81	197.12	0.20932	0.48373	0.69305
25	0.65155	0.00076285	0.026321	0.027084	59.751	121.74	181.49	60.248	138.89	199.14	0.22579	0.46584	0.69164

TABLE C.1 SATURATED REFRIGERANT-12—TEMPERATURE TABLE: SI UNITS (*Continued*)

		Specific volume, m^3/kg			Internal energy, kJ/kg			Enthalpy, kJ/kg			Entropy, kJ/kg·K		
Temp., °C, T	Press., MPa, P	Sat. liquid, v_f	Evap., v_{fg}	Sat. vapor, v_g	Sat. liquid, u_f	Evap., u_{fg}	Sat. vapor, u_g	Sat. liquid, h_f	Evap., h_{fg}	Sat. vapor, h_g	Sat. liquid, s_f	Evap., s_{fg}	Sat. vapor, s_g
30	0.74457	0.00077368	0.022950	0.023723	64.667	118.77	183.43	65.243	135.86	201.10	0.24217	0.44815	0.69032
35	0.84701	0.00078520	0.020058	0.020843	69.636	115.70	185.34	70.301	132.69	202.99	0.25846	0.43060	0.68906
40	0.95944	0.00079750	0.017563	0.018360	74.662	112.53	187.19	75.427	129.38	204.80	0.27467	0.41315	0.68782
45	1.0825	0.00081069	0.015399	0.016209	79.747	109.24	188.99	80.625	125.91	206.53	0.29083	0.39575	0.68657
50	1.2167	0.00082492	0.013512	0.014337	84.896	105.82	190.72	85.900	122.26	208.16	0.30694	0.37833	0.68527
55	1.3627	0.00084035	0.011858	0.012698	90.115	102.25	192.37	91.260	118.41	209.67	0.32302	0.36084	0.68386
60	1.5212	0.00085721	0.010400	0.011257	95.411	98.51	193.92	96.715	114.33	211.05	0.33911	0.34319	0.68230
65	1.6929	0.00087577	0.0091076	0.0099834	100.80	94.57	195.37	102.28	109.99	212.27	0.35525	0.32528	0.68052
70	1.8786	0.00089640	0.0079553	0.0088517	106.29	90.40	196.68	107.97	105.34	213.31	0.37147	0.30699	0.67845
75	2.0791	0.00091956	0.0069208	0.0078404	111.90	85.93	197.83	113.81	100.32	214.13	0.38784	0.28815	0.67599
80	2.2953	0.00094593	0.0059847	0.0069307	117.67	81.10	198.77	119.84	94.84	214.68	0.40446	0.26855	0.67301
85	2.5282	0.00097646	0.0051294	0.0061059	123.64	75.81	199.45	126.11	88.78	214.89	0.42146	0.24789	0.66934
90	2.7790	0.0010126	0.0043379	0.0053505	129.88	69.91	199.79	132.69	81.96	214.66	0.43902	0.22570	0.66472
95	3.0490	0.0010566	0.0035918	0.0046484	136.50	63.14	199.63	139.72	74.09	213.81	0.45747	0.20124	0.65871
100	3.3399	0.0011129	0.0028671	0.0039800	143.68	55.04	198.73	147.40	64.62	212.02	0.47734	0.17317	0.65051
105	3.6538	0.0011918	0.0021195	0.0033113	151.86	44.59	196.45	156.22	52.33	208.55	0.49985	0.13839	0.63823
110	3.9943	0.0013394	0.0011776	0.0025170	162.69	27.41	190.10	168.04	32.11	200.16	0.52978	0.08381	0.61359
111.8	4.125	0.001792	0.0	0.001792	164.75	0.0	164.75	172.14	0.0	172.14	0.54025	0.0	0.54025

Values generated from property formulation given in *ASHRAE Thermodynamic Properties of Refrigerants*, R.B. Stewart, R.T. Jacobsen and S.G. Penoncello, ASHRAE Inc., 1986.

TABLE C.2 SATURATED REFRIGERANT-12—PRESSURE TABLE: SI UNITS

		Specific volume, m^3/kg			Internal energy, kJ/kg			Enthalpy, kJ/kg			Entropy, kJ/kg·K		
Press., MPa, P	Temp., °C, T	Sat. liquid, v_f	Evap., v_{fg}	Sat. vapor, v_g	Sat. liquid, u_f	Evap., u_{fg}	Sat. vapor, u_g	Sat. liquid, h_f	Evap., h_{fg}	Sat. vapor, h_g	Sat. liquid, s_f	Evap., s_{fg}	Sat. vapor, s_g
0.002	−94.73	0.00060086	6.1180	6.1186	−46.046	178.73	132.68	−46.044	190.96	144.92	−0.22791	1.07029	0.84238
0.004	−86.21	0.00060905	3.1995	3.2001	−39.010	174.95	135.94	−39.008	187.75	148.74	−0.18851	1.00433	0.81582
0.006	−80.35	0.00061486	2.1967	2.1973	−34.586	172.82	138.24	−34.582	186.00	151.42	−0.16254	0.96472	0.80218
0.008	−76.25	0.00061901	1.6802	1.6808	−31.190	171.04	139.85	−31.185	184.49	153.30	−0.14480	0.93697	0.79217
0.010	−73.00	0.00062236	1.3646	1.3652	−28.425	169.57	141.15	−28.419	183.22	154.80	−0.13093	0.91542	0.78449
0.015	−66.78	0.00062894	0.93531	0.93594	−23.102	166.75	143.65	−23.092	180.78	157.69	−0.10492	0.87599	0.77107
0.020	−62.07	0.00063405	0.71563	0.71626	−19.071	164.63	145.56	−19.058	178.94	159.88	−0.085679	0.84772	0.76205
0.030	−54.96	0.00064199	0.49082	0.49146	−12.984	161.45	148.46	−12.965	176.17	163.21	−0.057357	0.80743	0.75007
0.040	−49.57	0.00064822	0.37561	0.37626	−8.3406	159.03	150.68	−8.3147	174.05	165.73	−0.036343	0.77846	0.74212
0.050	−45.16	0.00065345	0.30519	0.30585	−4.5319	157.04	152.51	−4.4992	172.30	167.80	−0.019474	0.75575	0.73628
0.060	−41.41	0.00065801	0.25755	0.25820	−1.2734	155.34	154.07	−1.2339	170.79	169.56	−0.005297	0.73702	0.73172
0.070	−38.13	0.00066208	0.22309	0.22375	1.5922	153.85	155.44	1.6385	169.46	171.10	0.006983	0.72104	0.72803
0.080	−35.19	0.00066579	0.19696	0.19763	4.1615	152.50	156.66	4.2148	168.26	172.47	0.017849	0.70710	0.72495
0.090	−32.54	0.00066921	0.17644	0.17711	6.4987	151.28	157.77	6.5589	167.16	173.71	0.027617	0.69470	0.72232
0.100	−30.10	0.00067239	0.15989	0.16057	8.6484	150.15	158.79	8.7156	166.13	174.85	0.036508	0.68354	0.72005
0.125	−24.75	0.00067955	0.12973	0.13041	13.397	147.64	161.04	13.482	163.85	177.34	0.055837	0.65964	0.71547
0.150	−20.17	0.00068589	0.10930	0.10998	17.489	145.46	162.95	17.592	161.86	179.45	0.072164	0.63981	0.71198
0.175	−16.15	0.00069163	0.094509	0.095200	21.110	143.53	164.64	21.231	160.07	181.30	0.086370	0.62283	0.70920
0.200	−12.55	0.00069691	0.083288	0.083985	24.375	141.77	166.15	24.514	158.43	182.94	0.098988	0.60793	0.70692
0.250	−6.27	0.00070647	0.067356	0.068062	30.116	138.65	168.77	30.293	155.49	185.78	0.12077	0.58262	0.70338
0.300	−0.88	0.00071507	0.056555	0.057270	35.095	135.91	171.01	35.309	152.88	188.19	0.13924	0.56150	0.70074
0.350	3.87	0.00072297	0.048731	0.049454	39.524	133.44	172.97	39.777	150.50	190.28	0.15538	0.54329	0.69867
0.400	8.13	0.00073036	0.042791	0.043521	43.535	131.18	174.72	43.827	148.30	192.13	0.16976	0.52723	0.69699
0.500	15.58	0.00074402	0.034347	0.035091	50.631	127.12	177.75	51.003	144.29	195.29	0.19468	0.49973	0.69441

TABLE C.2 SATURATED REFRIGERANT-12—PRESSURE TABLE: SI UNITS (*Continued*)

		Specific volume, m^3/kg			Internal energy, kJ/kg			Enthalpy, kJ/kg			Entropy, kJ/kg·K		
Press., MPa, P	Temp., °C, T	Sat. liquid, v_f	Evap., v_{fg}	Sat. vapor, v_g	Sat. liquid, u_f	Evap., u_{fg}	Sat. vapor, u_g	Sat. liquid, h_f	Evap., h_{fg}	Sat. vapor, h_g	Sat. liquid, s_f	Evap., s_{fg}	Sat. vapor, s_g
0.600	22.00	0.00075663	0.028616	0.029373	56.824	123.49	180.31	57.278	140.65	197.93	0.21592	0.47656	0.69247
0.700	27.67	0.00076854	0.024458	0.025227	62.365	120.17	182.53	62.903	137.29	200.19	0.23454	0.45639	0.69092
0.800	32.76	0.00077996	0.021297	0.022077	67.407	117.08	184.49	68.031	134.12	202.15	0.25118	0.43843	0.68961
0.900	37.41	0.00079103	0.018808	0.019599	72.055	114.18	186.24	72.767	131.11	203.88	0.26629	0.42217	0.68846
1.00	41.70	0.00080187	0.016794	0.017595	76.382	111.42	187.81	77.184	128.22	205.40	0.28016	0.40724	0.68740
1.20	49.40	0.00082315	0.013724	0.014547	84.277	106.24	190.51	85.265	122.70	207.97	0.30501	0.38042	0.68543
1.40	56.21	0.00084430	0.011487	0.012331	91.393	101.36	192.75	92.575	117.44	210.02	0.32693	0.35657	0.68350
1.60	62.34	0.00086568	0.009775	0.010641	97.923	96.69	194.62	99.308	112.33	211.64	0.34667	0.33483	0.68150
1.80	67.93	0.00088758	0.0084168	0.0093044	104.00	92.16	196.16	105.60	107.31	212.91	0.36474	0.31461	0.67935
2.00	73.07	0.00091029	0.0073070	0.0082173	109.72	87.69	197.41	111.54	102.30	213.84	0.38151	0.29549	0.67699
2.25	78.98	0.00094027	0.0061680	0.0071082	116.48	82.12	198.60	118.60	96.00	214.59	0.40106	0.27261	0.67367
2.50	84.41	0.00097263	0.0052260	0.0061986	122.93	76.46	199.39	125.36	89.53	214.89	0.41944	0.25037	0.66982
2.75	89.44	0.0010082	0.0044238	0.0054320	129.17	70.60	199.77	131.94	82.77	214.71	0.43702	0.22827	0.66530
3.00	94.12	0.0010481	0.0037207	0.0047688	135.30	64.41	199.70	138.44	75.57	214.01	0.45414	0.20576	0.65990
3.25	98.50	0.0010943	0.0030849	0.0041791	141.44	57.66	199.10	145.00	67.69	212.69	0.47116	0.18213	0.65329
3.50	102.60	0.0011499	0.0024860	0.0036359	147.76	50.02	197.79	151.79	58.72	210.51	0.48858	0.15628	0.64486
3.75	106.46	0.0012227	0.0018831	0.0031058	154.57	40.73	195.30	159.16	47.79	206.95	0.50730	0.12590	0.63321
4.00	110.08	0.0013436	0.0011565	0.0025001	162.93	26.97	189.90	168.30	31.59	199.90	0.53044	0.08244	0.61288
4.125	111.90	0.001792	0.0	0.001792	164.75	0.0	164.75	172.14	0.0	172.14	0.54025	0.0	0.54025

Values generated from property formulation given in *ASHRAE Thermodynamic Properties of Refrigerants*, R.B. Stewart, R.T. Jacobsen and S.G. Penoncello, ASHRAE, Inc. 1986.

TABLE C.3 SUPERHEATED REFRIGERANT-12: SI UNITS

P = 0.05 MPa (−45.164°C)

Temp., °C	v, m³/kg	u, kJ/kg	h, kJ/kg	s, kJ/kg·K
Sat.	0.30585	152.51	167.80	0.73628
−20	0.34239	164.62	181.74	0.79425
−10	0.35670	169.60	187.43	0.81630
0	0.37092	174.67	193.21	0.83787
10	0.38507	179.83	199.09	0.85898
20	0.39917	185.09	205.05	0.87968
30	0.41322	190.44	211.10	0.89997
40	0.42724	195.87	217.24	0.91989
50	0.44122	201.40	223.46	0.93944
60	0.45519	207.00	229.76	0.95864
70	0.46913	212.69	236.14	0.97752
80	0.48305	218.45	242.60	0.99607
90	0.49696	224.29	249.14	1.01432

P = 0.10 MPa (−30.100°C)

Temp., °C	v, m³/kg	u, kJ/kg	h, kJ/kg	s, kJ/kg·K
Sat.	0.16057	158.79	174.85	0.72005
−20	0.16821	163.85	180.67	0.74350
−10	0.17567	168.92	186.49	0.76605
0	0.18303	174.07	192.38	0.78801
10	0.19032	179.31	198.34	0.80944
20	0.19755	184.62	204.37	0.83039
30	0.20473	190.01	210.48	0.85089
40	0.21188	195.48	216.67	0.87097
50	0.21900	201.04	222.94	0.89066
60	0.22608	206.67	229.28	0.90999
70	0.23315	212.38	235.69	0.92896
80	0.24020	218.16	242.18	0.94760
90	0.24723	224.02	248.74	0.96591

P = 0.15 MPa (−20.173°C)

Temp., °C	v, m³/kg	u, kJ/kg	h, kJ/kg	s, kJ/kg·K
Sat.	0.10998	162.95	179.45	0.71198
−20	0.11007	163.04	179.55	0.71239
−10	0.11526	168.22	185.51	0.73547
0	0.12035	173.46	191.51	0.75786
10	0.12537	178.76	197.57	0.77962
20	0.13032	184.13	203.68	0.80083
30	0.13522	189.57	209.86	0.82155
40	0.14008	195.09	216.10	0.84181
50	0.14490	200.67	222.41	0.86165
60	0.14970	206.33	228.79	0.88109
70	0.15448	212.07	235.24	0.90017
80	0.15924	217.87	241.76	0.91889
90	0.16399	223.75	248.34	0.93728

P = 0.20 MPa (−12.550°C)

Temp., °C	v, m³/kg	u, kJ/kg	h, kJ/kg	s, kJ/kg·K
Sat.	0.083985	166.15	182.94	0.70692
−10	0.085103	167.50	184.50	0.71287
0	0.088978	172.33	190.62	0.73571
10	0.092860	178.20	196.78	0.75783
20	0.096676	183.63	202.97	0.77933
30	0.10044	189.13	209.21	0.80027
40	0.10416	194.68	215.51	0.82072
50	0.10785	200.30	221.87	0.84071
60	0.11151	205.99	228.30	0.86028
70	0.11514	211.75	234.78	0.87946
80	0.11876	217.58	242.33	0.89827
90	0.12236	223.47	247.94	0.91673

P = 0.25 MPa (−6.270°C)

Temp., °C	v, m³/kg	u, kJ/kg	h, kJ/kg	s, kJ/kg·K
Sat.	0.068062	168.77	185.78	0.70338
0	0.070120	172.17	189.70	0.71791
10	0.073332	177.63	195.96	0.74041
20	0.076473	183.12	202.24	0.76221
30	0.079558	188.67	208.56	0.78339
40	0.082599	194.27	214.92	0.80403
50	0.085604	199.93	221.33	0.82418
60	0.088579	205.65	227.79	0.84388
70	0.091530	211.43	234.31	0.86317
80	0.094461	217.28	240.89	0.88207
90	0.097375	223.19	247.53	0.90061
100	0.10027	229.17	254.23	0.91881
110	0.10316	235.20	260.99	0.93668

P = 0.30 MPa (−0.882°C)

Temp., °C	v, m³/kg	u, kJ/kg	h, kJ/kg	s, kJ/kg·K
Sat.	0.057270	171.01	188.19	0.70074
0	0.057519	171.49	188.75	0.70280
10	0.060291	177.04	195.12	0.72572
20	0.062987	182.60	201.50	0.74784
30	0.065624	188.20	207.89	0.76928
40	0.068214	193.85	214.31	0.79013
50	0.070767	199.54	220.77	0.81044
60	0.073288	205.30	227.28	0.83028
70	0.075784	211.11	233.84	0.84968
80	0.078260	216.98	240.46	0.86867
90	0.080717	222.91	247.12	0.88729
100	0.083160	228.90	253.85	0.90556
110	0.085591	234.95	260.63	0.92349

TABLE C.3 SUPERHEATED REFRIGERANT-12: SI UNITS (*Continued*)

Temp., °C	v, m³/kg	u, kJ/kg	h, kJ/kg	s, kJ/kg·K	Temp., °C	v, m³/kg	u, kJ/kg	h, kJ/kg	s, kJ/kg·K
	P = 0.40 MPa (8.129°C)					P = 0.50 MPa (15.582°C)			
Sat.	0.043521	174.72	192.13	0.69699	Sat.	0.035091	177.75	195.29	0.69441
20	0.046088	181.51	199.94	0.72421	20	0.035897	180.34	198.29	0.70473
30	0.048174	187.23	206.50	0.74621	30	0.037666	186.21	205.04	0.72736
40	0.050208	192.98	213.06	0.76749	40	0.039375	192.07	211.75	0.74915
50	0.052200	198.76	219.64	0.78817	50	0.041038	197.94	218.46	0.77021
60	0.054158	204.58	226.24	0.80829	60	0.042663	203.83	225.16	0.79066
70	0.056089	210.45	232.88	0.82793	70	0.044258	209.76	231.89	0.81055
80	0.057998	216.36	239.56	0.84713	80	0.045829	215.73	238.65	0.82996
90	0.059887	222.34	246.29	0.86591	90	0.047380	221.75	245.44	0.84892
100	0.061761	228.36	253.07	0.88432	100	0.048914	227.81	252.27	0.86748
110	0.063621	234.44	259.89	0.90237	110	0.050434	233.92	259.14	0.88565
120	0.065470	240.58	266.77	0.92008	120	0.051942	240.09	266.06	0.90348
130	0.067310	246.77	273.70	0.93748	130	0.053439	246.30	273.02	0.92097
	P = 0.60 MPa (21.998°C)					P = 0.70 MPa (27.665°C)			
Sat.	0.029373	180.31	197.93	0.69247	Sat.	0.025227	182.53	200.19	0.69092
30	0.030624	185.13	203.50	0.71109	30	0.025556	183.97	201.86	0.69646
40	0.032126	191.11	210.39	0.73343	40	0.026921	190.10	208.95	0.71946
50	0.033575	197.08	217.23	0.75494	50	0.028224	196.19	215.95	0.74145
60	0.034983	203.06	224.05	0.77573	60	0.029482	202.26	222.90	0.76263
70	0.036358	209.06	230.87	0.79591	70	0.030703	208.33	229.82	0.78312
80	0.037706	215.08	237.71	0.81554	80	0.031895	214.42	236.74	0.80300
90	0.039033	221.15	244.57	0.83470	90	0.033065	220.53	243.68	0.82236
100	0.040343	227.25	251.46	0.85341	100	0.034214	226.68	250.63	0.84124
110	0.041637	233.40	258.38	0.87172	110	0.035348	232.86	257.60	0.85969
120	0.042918	239.59	265.34	0.88966	120	0.036469	239.08	264.61	0.87775
130	0.044189	245.83	272.34	0.90725	130	0.037578	245.35	271.66	0.89544
	P = 0.80 MPa (32.764°C)					P = 0.90 MPa (37.414°C)			
Sat.	0.022077	184.49	202.15	0.68961	Sat.	0.019599	186.24	203.88	0.68846
40	0.022990	189.04	207.43	0.70666	40	0.019904	187.90	205.82	0.69468
50	0.024192	195.25	214.61	0.72922	50	0.021035	194.27	213.20	0.71789
60	0.025341	201.42	221.70	0.75083	60	0.022106	200.55	220.45	0.73998
70	0.026451	207.58	228.74	0.77165	70	0.023132	206.79	227.61	0.76117
80	0.027528	213.73	235.75	0.79181	80	0.024124	213.02	234.73	0.78162
90	0.028581	219.90	242.76	0.81138	90	0.025087	219.25	241.83	0.80143
100	0.029613	226.09	249.78	0.83044	100	0.026029	225.49	248.91	0.82069
110	0.030628	232.31	256.81	0.84904	110	0.026953	231.75	256.01	0.83945
120	0.031629	238.57	263.87	0.86723	120	0.027861	238.04	263.12	0.85776
130	0.032618	244.86	270.96	0.88502	130	0.028757	244.37	270.25	0.87567
140	0.033597	251.20	278.08	0.90247	140	0.029642	250.73	277.41	0.89321
150	0.034567	257.58	285.23	0.91958	150	0.030519	257.13	284.59	0.91040

TABLE C.3 SUPERHEATED REFRIGERANT-12: SI UNITS (*Continued*)

R-12

Temp., °C	v, m³/kg	u, kJ/kg	h, kJ/kg	s, kJ/kg·K	Temp., °C	v, m³/kg	u, kJ/kg	h, kJ/kg	s, kJ/kg·K
	P = 1.00 MPa (41.698°C)					*P* = 1.20 MPa (49.402°C)			
Sat.	0.017595	187.81	205.40	0.68740	Sat.	0.014547	190.51	207.97	0.68543
50	0.018489	193.22	211.71	0.70719	50	0.014607	190.92	208.45	0.68692
60	0.019504	199.64	219.14	0.72983	60	0.015556	197.66	216.33	0.71094
70	0.020467	205.98	226.45	0.75144	70	0.016438	204.25	223.97	0.73355
80	0.021392	212.29	233.68	0.77221	80	0.017271	210.74	231.47	0.75507
90	0.022286	218.58	240.87	0.79228	90	0.018068	217.18	238.86	0.77572
100	0.023157	224.87	248.03	0.81174	100	0.018836	223.59	246.20	0.79564
110	0.024009	231.18	255.19	0.83067	110	0.019583	230.00	253.50	0.81495
120	0.024844	237.51	262.35	0.84913	120	0.020311	236.41	260.78	0.83371
130	0.025666	243.86	269.53	0.86715	130	0.021024	242.83	268.06	0.85199
140	0.026477	250.25	276.73	0.88479	140	0.021725	249.28	275.34	0.86985
150	0.027279	256.67	283.95	0.90207	150	0.022415	255.75	282.65	0.88731
160	0.028072	263.13	291.21	0.91901	160	0.023097	262.25	289.97	0.90441
	P = 1.40 MPa (56.214°C)					*P* = 1.60 MPa (62.344°C)			
Sat.	0.012331	192.75	210.02	0.68350	Sat.	0.010641	194.62	211.64	0.68150
60	0.012673	195.44	213.18	0.69305					
70	0.013517	202.34	221.27	0.71697	70	0.011278	200.21	218.26	0.70100
80	0.014298	209.07	229.09	0.73943	80	0.012035	207.24	226.50	0.72468
90	0.015033	215.69	236.73	0.76078	90	0.012733	214.08	234.46	0.74690
100	0.015734	222.24	244.27	0.78125	100	0.013390	220.81	242.23	0.76801
110	0.016409	228.76	251.73	0.80099	110	0.014016	227.46	249.88	0.78825
120	0.017063	235.26	259.15	0.82010	120	0.014618	234.07	257.45	0.80776
130	0.017700	241.76	266.54	0.83867	130	0.015201	240.65	264.97	0.82665
140	0.018324	248.27	273.93	0.85676	140	0.015768	247.24	272.47	0.84500
150	0.018936	254.80	281.31	0.87442	150	0.016323	253.82	279.94	0.86288
160	0.019539	261.35	288.70	0.89169	160	0.016868	260.43	287.41	0.88034
170	0.02013	267.92	296.11	0.9086	170	0.017403	267.05	294.89	0.89740
	P = 1.80 MPa (67.930°C)					*P* = 2.00 MPa (73.073°C)			
Sat.	0.009304	196.16	212.91	0.67935	Sat.	0.008217	197.41	213.84	0.67699
70	0.009476	197.76	214.82	0.68495					
80	0.010238	205.22	223.65	0.71032	80	0.008757	202.93	220.44	0.69587
90	0.010921	212.35	232.00	0.73364	90	0.009443	210.44	229.33	0.72068
100	0.011550	219.28	240.07	0.75555	100	0.010061	217.63	237.75	0.74356
110	0.012143	226.08	247.94	0.77637	110	0.010632	224.63	245.89	0.76509
120	0.012707	232.82	255.69	0.79633	120	0.011170	231.50	253.84	0.78558
130	0.013250	239.50	263.35	0.81559	130	0.011682	238.30	261.67	0.80524
140	0.013775	246.17	270.96	0.83423	140	0.012176	245.06	269.41	0.82421
150	0.014287	252.82	278.54	0.85235	150	0.012654	251.79	277.10	0.84260
160	0.014787	259.48	286.10	0.87001	160	0.013120	258.51	284.75	0.86047
170	0.015277	266.15	293.65	0.88724	170	0.013575	265.23	292.38	0.87789
180	0.015760	272.83	301.20	0.90410	180	0.014021	271.96	300.01	0.89490

TABLE C.3 SUPERHEATED REFRIGERANT-12: SI UNITS (*Continued*)

Temp., °C	v, m³/kg	u, kJ/kg	h, kJ/kg	s, kJ/kg·K	Temp., °C	v, m³/kg	u, kJ/kg	h, kJ/kg	s, kJ/kg·K
	P = 2.25 MPa (78.983°C)					P = 2.50 MPa (84.414°C)			
Sat.	0.007108	198.60	214.59	0.67367	Sat.	0.006199	199.39	214.89	0.66982
90	0.007921	207.74	225.56	0.70435	90	0.006637	204.52	221.11	0.68709
100	0.008543	215.37	234.59	0.72888	100	0.007293	212.82	231.05	0.71409
110	0.009102	222.67	243.15	0.75151	110	0.007856	220.52	240.16	0.73820
120	0.009619	229.76	251.41	0.77279	120	0.008364	227.89	248.80	0.76047
130	0.010106	236.73	259.47	0.79305	130	0.008835	235.07	257.15	0.78144
140	0.010570	243.62	267.40	0.81249	140	0.009278	242.11	265.31	0.80142
150	0.011016	250.46	275.25	0.83124	150	0.009701	249.07	273.32	0.82060
160	0.011448	257.27	283.03	0.84942	160	0.010108	255.98	281.25	0.83911
170	0.011869	264.06	290.77	0.86708	170	0.010502	262.86	289.11	0.85705
180	0.012280	270.85	298.48	0.88430	180	0.010885	269.72	296.93	0.87450
190	0.012683	277.65	306.19	0.90112	190	0.011260	276.57	304.72	0.89151
200	0.013079	284.46	313.89	0.91757	200	0.011627	283.43	312.50	0.90812
	P = 2.75 MPa (89.441°C)					P = 3.00 MPa (94.120°C)			
Sat.	0.005432	199.77	214.71	0.66530	Sat.	0.004769	199.70	214.01	0.65990
100	0.006224	209.84	226.96	0.69861	100	0.005267	206.20	222.00	0.68148
110	0.006811	218.13	236.86	0.72481	110	0.005909	215.41	233.14	0.71096
120	0.007321	225.87	246.00	0.74836	120	0.006434	223.65	242.95	0.73624
130	0.007784	233.29	254.70	0.77021	130	0.006897	231.39	252.08	0.75917
140	0.008214	240.52	263.11	0.79082	140	0.007319	238.84	260.80	0.78054
150	0.008619	247.63	271.33	0.81048	150	0.007713	246.12	269.25	0.80076
160	0.009007	254.65	279.42	0.82937	160	0.008086	253.27	277.53	0.82008
170	0.009380	261.62	287.41	0.84762	170	0.008443	260.34	285.67	0.83867
180	0.009742	268.55	295.34	0.86532	180	0.008787	267.36	293.72	0.85664
190	0.010094	275.47	303.23	0.88254	190	0.009121	274.35	301.71	0.87408
200	0.010438	282.39	311.09	0.89933	200	0.009446	281.32	309.66	0.89106
210	0.010775	289.30	318.93	0.9157	210	0.009764	288.29	317.58	0.90762
	P = 3.25 MPa (98.495°C)					P = 3.50 MPa (102.599°C)			
Sat.	0.004179	199.10	212.69	0.65329	Sat.	0.003636	197.79	210.51	0.64486
110	0.005105	212.22	228.82	0.69608	110	0.004357	208.27	223.52	0.67916
120	0.005664	221.18	239.59	0.72384	120	0.004978	218.38	235.80	0.71083
130	0.006134	229.33	249.27	0.74817	130	0.005468	227.09	246.23	0.73703
140	0.006555	237.06	258.37	0.77045	140	0.005891	235.16	255.78	0.76044
150	0.006941	244.53	267.09	0.79132	150	0.006274	242.87	264.83	0.78207
160	0.007303	251.83	275.57	0.81113	160	0.006629	250.34	273.54	0.80243
170	0.007647	259.02	283.88	0.83009	170	0.006963	257.66	282.03	0.82181
180	0.007978	266.14	292.06	0.84836	180	0.007282	264.88	290.37	0.84041
190	0.008296	273.20	300.17	0.86605	190	0.007589	272.03	298.59	0.85837
200	0.008606	280.24	308.21	0.88323	200	0.007885	279.14	306.74	0.87576
210	0.008908	287.26	316.21	0.89996	210	0.008173	286.22	314.82	0.89268
220	0.009203	294.27	324.18	0.91630	220	0.008455	293.28	322.87	0.90916

Values generated from property formulation given in *ASHRAE Thermodynamic Properties of Refrigerants*, R.B. Stewart, R.T. Jacobsen and S.G. Penoncello, ASHRAE, Inc. 1986.

Properties of Ideal Gases

TABLE D.1 IDEAL-GAS PROPERTIES OF AIR—MASS BASIS: SI UNITS
Mol. Wt.: 28.967

T, K	h, kJ/kg	u, kJ/kg	s^0, kJ/kg·K	c_p, kJ/kg·K	P_r	v_r
0	0.0	0.0	0.0			
100	99.93	71.23	4.6004	1.0000	0.02977	964.30
120	119.97	85.53	4.7805	1.0000	0.05574	616.28
140	140.01	99.83	4.9366	1.0000	0.09603	418.07
160	160.05	114.12	5.0721	1.0000	0.15396	298.50
180	180.09	128.42	5.1904	1.0000	0.23252	222.42
200	200.13	142.72	5.2950	1.0000	0.33472	171.50
220	220.17	157.03	5.3908	1.0000	0.46740	135.15
240	240.22	171.34	5.4779	1.0000	0.63299	108.85
260	260.28	185.65	5.5586	1.0000	0.83848	89.009
280	280.34	199.97	5.6335	1.0000	1.0887	73.826
300	300.43	214.32	5.7016	1.0000	1.3800	62.396
320	320.54	228.69	5.7654	1.0023	1.7238	53.283
340	340.66	243.07	5.8270	1.0073	2.1361	45.688
360	360.81	257.48	5.8851	1.0112	2.6157	39.505
380	381.01	271.94	5.9397	1.0133	3.1639	34.476
400	401.25	286.44	5.9916	1.0150	3.7903	30.291
420	421.54	300.98	6.0411	1.0167	4.5039	26.767
440	441.88	315.58	6.0884	1.0188	5.3108	23.781
460	462.28	330.25	6.1338	1.0217	6.2200	21.228
480	482.75	344.98	6.1774	1.0256	7.2403	19.029
500	503.30	359.78	6.2193	1.0300	8.3792	17.128
520	523.93	374.67	6.2598	1.0340	9.6489	15.470
540	544.63	389.63	6.2989	1.0380	11.057	14.020
560	565.42	404.68	6.3367	1.0420	12.612	12.746
580	586.30	419.82	6.3732	1.0460	14.326	11.621
600	607.26	435.04	6.4087	1.0500	16.210	10.624

TABLE D.1 IDEAL-GAS PROPERTIES OF AIR—MASS BASIS: SI UNITS (*Continued*)
Mol. Wt.: 28.967

T, K	h, kJ/kg	u, kJ/kg	s^0, kJ/kg·K	c_p, kJ/kg·K	P_r	v_r
620	628.32	450.36	6.4433	1.0537	18.286	9.7327
640	649.48	465.78	6.4769	1.0574	20.558	8.9364
660	670.73	481.29	6.5096	1.0614	23.039	8.2232
680	692.07	496.89	6.5415	1.0655	25.742	7.5824
700	713.51	512.59	6.5725	1.0700	28.682	7.0051
720	735.05	528.39	6.6029	1.0758	31.882	6.4822
740	756.68	544.28	6.6325	1.0819	35.351	6.0087
760	778.41	560.27	6.6615	1.0881	39.105	5.5787
780	800.23	576.35	6.6898	1.0942	43.160	5.1875
800	822.15	592.52	6.7175	1.1000	47.534	4.8307
820	844.16	608.80	6.7447	1.1045	52.259	4.5039
840	866.26	625.16	6.7713	1.1086	57.343	4.2048
860	888.46	641.61	6.7975	1.1126	62.805	3.9305
880	910.73	658.15	6.8231	1.1163	68.667	3.6785
900	933.10	674.77	6.8482	1.1200	74.949	3.4467
920	955.55	691.48	6.8729	1.1240	81.679	3.2330
940	978.08	708.27	6.8971	1.1280	88.875	3.0359
960	1000.7	725.15	6.9209	1.1320	96.560	2.8537
980	1023.4	742.10	6.9443	1.1360	104.76	2.6852
1000	1046.2	759.13	6.9673	1.1400	113.49	2.5291
1050	1103.4	802.03	7.0231	1.1506	137.87	2.1860
1100	1161.1	845.38	7.0768	1.1600	166.21	1.8997
1150	1219.2	889.14	7.1285	1.1653	199.00	1.6587
1200	1277.7	933.29	7.1783	1.1700	236.71	1.4551
1250	1336.6	977.81	7.2264	1.1772	279.87	1.2820
1300	1395.8	1022.7	7.2728	1.1850	329.01	1.1341
1350	1455.3	1067.9	7.3177	1.1928	384.73	1.0072
1400	1515.2	1113.3	7.3612	1.2000	447.67	0.89763
1450	1575.3	1159.1	7.4034	1.2053	518.62	0.80251
1500	1635.7	1205.1	7.4444	1.2100	598.20	0.71974
1550	1696.3	1251.4	7.4842	1.2150	687.08	0.64752
1600	1757.2	1297.9	7.5228	1.2200	786.09	0.58422
1650	1818.3	1344.7	7.5604	1.2253	896.11	0.52851
1700	1879.6	1391.6	7.5970	1.2300	1018.0	0.47934
1750	1941.1	1438.8	7.6326	1.2328	1152.6	0.43580
1800	2002.8	1486.1	7.6674	1.2350	1300.9	0.39714
1850	2064.7	1533.6	7.7013	1.2372	1464.1	0.36270
1900	2126.7	1581.3	7.7344	1.2400	1643.0	0.33193
1950	2188.9	1629.2	7.7667	1.2450	1838.8	0.30439
2000	2251.3	1677.2	7.7983	1.2500	2052.7	0.27967
2050	2313.8	1725.4	7.8292	1.2528	2285.9	0.25741
2100	2376.5	1773.7	7.8594	1.2550	2539.6	0.23735
2150	2439.3	1822.1	7.8890	1.2575	2815.1	0.21922
2200	2502.2	1870.7	7.9179	1.2600	3113.7	0.20280
2250	2565.3	1919.4	7.9462	1.2625	3436.8	0.18791

TABLE D.1 IDEAL-GAS PROPERTIES OF AIR—MASS BASIS: SI UNITS (*Continued*)
Mol. Wt.: 28.967

T, K	h, kJ/kg	u, kJ/kg	s^0, kJ/kg·K	c_p, kJ/kg·K	P_r	v_r
2300	2628.4	1968.3	7.9740	1.2650	3785.8	0.17438
2350	2691.8	2017.2	8.0012	1.2676	4162.3	0.16206
2400	2755.2	2066.3	8.0279	1.2700	4567.9	0.15081
2450	2818.7	2115.5	8.0541	1.2716	5004.2	0.14053
2500	2882.3	2164.8	8.0798	1.2730	5473.2	0.13111
2550	2946.1	2214.1	8.1051	1.2749	5977.1	0.12245
2600	3009.9	2263.6	8.1299	1.2770	6517.0	0.11451
2650	3073.8	2313.2	8.1542	1.2796	7093.0	0.10724
2700	3137.9	2362.9	8.1781	1.2820	7708.6	0.10054
2750	3202.0	2412.7	8.2016	1.2836	8367.4	0.09434
2800	3266.2	2462.5	8.2248	1.2850	9070.6	0.08860
2850	3330.5	2512.5	8.2476	1.2865	9820.1	0.08330
2900	3394.9	2562.5	8.2700	1.2880	10618	0.07840
2950	3459.3	2612.6	8.2920	1.2895	11464	0.07386
3000	3523.9	2662.8	8.3137	1.2910	12364	0.06965
3050	3588.5	2713.0	8.3351	1.2924	13319	0.06573
3100	3653.2	2763.4	8.3561	1.2940	14332	0.06209
3150	3717.9	2813.8	8.3768	1.2961	15404	0.05870
3200	3782.8	2864.2	8.3972	1.2980	16538	0.05554
3250	3847.6	2914.8	8.4173	1.2991	17740	0.05258
3300	3912.6	2965.4	8.4372	1.3000	19011	0.04982
3350	3977.6	3016.1	8.4567	1.3009	20350	0.04725
3400	4042.7	3066.8	8.4760	1.3020	21763	0.04484
3450	4107.9	3117.6	8.4950	1.3034	23254	0.04258
3500	4173.0	3168.4	8.5138	1.3050	24826	0.04047
3550	4238.3	3219.3	8.5323	1.3070	26482	0.03848
3595	4297.1	3265.2	8.5488	1.3090	28046	0.03679

Adapted from *Gas Tables International Version,* Thermodynamic Properties of Air Products of Combustion and Component Gases Compressible Flow Functions, Joseph Keenan, Jing Chao and Joseph Kaye, John Wiley and Sons, N.Y., 1983.

TABLE D.2 IDEAL-GAS PROPERTIES OF CARBON DIOXIDE (CO_2)—MASS BASIS: SI UNITS
Mol. Wt.: 44.01

T, K	h, kJ/kg	u, kJ/kg	s^0, kJ/kg·K	c_p, kJ/kg·K	P_r	v_r
0	0.0	0.0	0.0			
100	66.08	47.18	4.0675	0.6637	0.00045	42,377
150	99.52	71.18	4.3331	0.6914	0.00182	15,435
200	135.20	97.41	4.5439	0.7353	0.00555	6808.0
250	173.38	126.15	4.7154	0.7888	0.01376	3440.3
300	214.34	157.66	4.8631	0.8457	0.03007	1884.6
350	257.92	191.80	4.9974	0.8966	0.06120	1080.7
400	303.73	228.16	5.1196	0.9390	0.11691	646.39
450	351.65	266.64	5.2327	0.9785	0.21268	399.89
500	401.48	307.02	5.3375	1.0140	0.37035	255.06
550	452.99	349.08	5.4357	1.0461	0.62294	166.83
600	506.05	392.69	5.5279	1.0752	1.0149	111.69
650	560.47	437.68	5.6151	1.1019	1.6099	76.283
700	616.18	483.94	5.6976	1.1262	2.4915	53.078
750	673.04	531.35	5.7761	1.1484	3.7748	37.538
800	730.97	579.84	5.8508	1.1687	5.6070	26.955
850	789.90	629.32	5.9223	1.1873	8.1835	19.623
900	849.67	679.64	5.9906	1.2043	11.750	14.471
950	910.05	730.57	6.0561	1.2198	16.623	10.797
1000	971.22	782.29	6.1191	1.2340	23.193	8.1457
1050	1033.4	835.05	6.1796	1.2470	31.953	6.2082
1100	1096.3	888.49	6.2379	1.2590	43.500	4.7773
1150	1159.6	942.30	6.2941	1.2700	58.575	3.7091
1200	1223.3	996.59	6.3484	1.2802	78.064	2.9041
1250	1287.5	1051.4	6.4008	1.2896	103.05	2.2917
1300	1352.2	1106.7	6.4516	1.2983	134.80	1.8219
1350	1417.4	1162.3	6.5007	1.3061	174.85	1.4586
1400	1482.9	1218.4	6.5483	1.3134	224.99	1.1756
1450	1548.7	1274.8	6.5945	1.3202	287.33	0.95338
1500	1614.8	1331.5	6.6394	1.3265	364.35	0.77779
1550	1681.3	1388.5	6.6830	1.3325	458.91	0.63810
1600	1748.1	1445.8	6.7254	1.3380	574.34	0.52630
1650	1815.1	1503.4	6.7666	1.3431	714.48	0.43629
1700	1882.4	1561.2	6.8068	1.3478	883.73	0.36342
1750	1949.9	1619.3	6.8459	1.3523	1087.2	0.30410
1800	2017.6	1677.6	6.8841	1.3565	1330.5	0.25558
1850	2085.6	1736.1	6.9213	1.3606	1620.3	0.21571
1900	2153.7	1794.7	6.9577	1.3644	1963.8	0.18278
1950	2222.0	1853.6	6.9931	1.3680	2369.7	0.15546
2000	2290.5	1912.6	7.0278	1.3713	2847.1	0.13271
2050	2359.1	1971.8	7.0617	1.3745	3406.6	0.11369
2100	2427.9	2031.2	7.0949	1.3775	4060.2	0.09771
2150	2496.9	2090.7	7.1273	1.3803	4820.9	0.08425
2200	2565.9	2150.3	7.1591	1.3830	5703.7	0.07287
2250	2635.1	2210.1	7.1902	1.3856	6724.9	0.06321
2300	2704.5	2270.0	7.2207	1.3880	7902.5	0.05499
2350	2774.0	2330.0	7.2506	1.3903	9256.4	0.04796
2400	2843.5	2390.1	7.2799	1.3926	10,809	0.04195
2450	2913.2	2450.4	7.3086	1.3947	12,584	0.03678
2500	2983.0	2510.7	7.3368	1.3968	14,609	0.03233

TABLE D.2 IDEAL-GAS PROPERTIES OF CARBON DIOXIDE (CO_2)—MASS BASIS: SI UNITS Mol. Wt.: 44.01 *(Continued)*

T, K	h, kJ/kg	u, kJ/kg	s^0, kJ/kg·K	c_p, kJ/kg·K	P_r	v_r
2550	3052.9	2571.1	7.3645	1.3988	16,915	0.02848
2600	3122.9	2631.7	7.3916	1.4008	19,533	0.02515
2650	3193.0	2692.3	7.4183	1.4026	22,497	0.02225
2700	3263.1	2753.1	7.4446	1.4043	25,848	0.01973
2750	3333.4	2813.9	7.4704	1.4060	29,629	0.01753
2800	3403.7	2874.7	7.4957	1.4077	33,885	0.01561
2850	3474.2	2935.7	7.5206	1.4093	38,663	0.01393
2900	3544.7	2996.8	7.5452	1.4109	44,021	0.01245
2950	3615.2	3057.9	7.5693	1.4125	50,021	0.01114
3000	3685.9	3119.1	7.5931	1.4140	56,725	0.00999
3050	3756.6	3180.4	7.6164	1.4153	64,197	0.00898
3100	3827.4	3241.8	7.6395	1.4167	72,516	0.00808
3150	3898.3	3303.2	7.6621	1.4180	81,763	0.00728
3200	3969.2	3364.7	7.6845	1.4193	92,026	0.00657
3250	4040.2	3426.2	7.7065	1.4205	103,400	0.00594
3300	4111.3	3487.9	7.7282	1.4218	115,990	0.00538
3350	4182.4	3549.5	7.7496	1.4230	129,890	0.00487
3400	4253.6	3611.3	7.7707	1.4243	145,240	0.00442
3450	4324.8	3673.1	7.7915	1.4254	162,140	0.00402
3500	4396.1	3734.9	7.8120	1.4266	180,740	0.00366
3550	4467.5	3796.8	7.8322	1.4278	201,180	0.00333
3600	4538.9	3858.8	7.8522	1.4289	223,610	0.00304
3650	4610.4	3920.8	7.8719	1.4300	248,230	0.00278
3700	4681.9	3982.9	7.8914	1.4310	275,170	0.00254
3750	4753.5	4045.0	7.9106	1.4321	304,620	0.00233
3800	4825.1	4107.2	7.9296	1.4332	336,790	0.00213
3850	4896.8	4169.5	7.9483	1.4342	371,920	0.00196
3900	4968.6	4231.8	7.9668	1.4353	410,220	0.00180
3950	5040.3	4294.1	7.9851	1.4363	451,920	0.00165
4000	5112.2	4356.5	8.0032	1.4373	497,270	0.00152
4050	5184.1	4418.9	8.0211	1.4383	546,580	0.00140
4100	5256.0	4481.4	8.0387	1.4392	600,120	0.00129
4150	5328.0	4544.0	8.0562	1.4402	658,220	0.00119
4200	5400.0	4606.6	8.0734	1.4412	721,190	0.00110
4250	5472.1	4669.2	8.0905	1.4421	789,340	0.00102
4300	5544.2	4731.9	8.1074	1.4431	863,050	0.00094
4350	5616.4	4794.6	8.1240	1.4440	942,750	0.00087
4400	5688.6	4857.4	8.1406	1.4449	1,028,800	0.00081
4450	5760.9	4920.2	8.1569	1.4458	1,121,700	0.00075
4500	5833.2	4983.1	8.1730	1.4467	1,221,900	0.00070
4550	5905.6	5046.0	8.1890	1.4475	1,329,800	0.00065
4600	5978.0	5108.9	8.2049	1.4484	1,446,000	0.00060
4650	6050.4	5171.9	8.2205	1.4493	1,571,000	0.00056
4700	6122.9	5235.0	8.2360	1.4502	1,705,400	0.00052
4750	6143.5	5246.1	8.2514	1.4510	1,849,800	0.00049
4800	6268.0	5361.2	8.2666	1.4518	2,004,600	0.00045
4850	6807.8	5891.6	8.2816	1.4526	2,170,800	0.00042
4900	7243.9	6318.2	8.2965	1.4535	2,349,000	0.00039
4950	6953.2	6018.0	8.3113	1.4544	2,540,000	0.00037
5000	6558.7	5614.1	8.3259	1.4553	2,744,600	0.00034

Adapted from *JANAF Thermochemical Tables* Third Edition, M. W. Chase et al., American Chemical Society and the American Institute of Physics for the National Bureau of Standards, 1985.

TABLE D.3 IDEAL-GAS PROPERTIES OF CARBON MONOXIDE (CO)—MASS BASIS: SI UNITS
Mol. Wt.: 28.01

T, K	h, kJ/kg	u, kJ/kg	s^0, kJ/kg·K	c_p, kJ/kg·K	P_r	v_r
0	0.0	0.0	0.0			
100	103.60	73.92	5.9210	1.0390	0.00134	22,203
150	155.61	111.08	6.3324	1.0386	0.00535	8250.1
200	207.53	148.16	6.6413	1.0392	0.01513	3923.1
250	259.46	185.25	6.8754	1.0400	0.03331	2232.7
300	311.49	222.44	7.0628	1.0404	0.06262	1422.1
350	363.59	259.70	7.2236	1.0424	0.10761	965.68
400	415.81	297.08	7.3629	1.0475	0.17208	690.01
450	468.36	334.78	7.4871	1.0546	0.26148	511.04
500	521.31	372.89	7.5983	1.0637	0.38027	390.29
550	574.77	411.51	7.7004	1.0746	0.53636	304.43
600	628.80	450.70	7.7942	1.0868	0.73580	242.05
650	683.48	490.54	7.8818	1.0997	0.98833	195.23
700	738.80	531.01	7.9637	1.1128	1.3023	159.55
750	794.78	572.15	8.0410	1.1259	1.6896	131.77
800	851.40	613.93	8.1140	1.1388	2.1611	109.88
850	908.65	656.34	8.1835	1.1512	2.7307	92.399
900	966.50	699.35	8.2496	1.1630	3.4119	78.298
950	1024.9	742.95	8.3128	1.1742	4.2214	66.801
1000	1083.9	787.09	8.3732	1.1847	5.1753	57.355
1050	1143.4	831.72	8.4313	1.1944	6.2932	49.527
1100	1203.3	876.82	8.4871	1.2035	7.5938	42.998
1150	1263.7	922.38	8.5407	1.2121	9.0993	37.515
1200	1324.5	968.34	8.5925	1.2201	10.832	32.884
1250	1385.7	1014.7	8.6425	1.2274	12.818	28.947
1300	1447.3	1061.4	8.6907	1.2343	15.082	25.586
1350	1509.2	1108.4	8.7374	1.2407	17.652	22.702
1400	1571.3	1155.8	8.7827	1.2467	20.557	20.215
1450	1633.8	1203.4	8.8265	1.2522	23.830	18.062
1500	1696.5	1251.3	8.8691	1.2573	27.502	16.190
1550	1759.5	1299.4	8.9104	1.2621	31.608	14.556
1600	1822.8	1347.8	8.9505	1.2667	36.183	13.126
1650	1886.2	1396.4	8.9895	1.2709	41.270	11.868
1700	1949.8	1445.2	9.0275	1.2749	46.906	10.758
1750	2013.7	1494.2	9.0646	1.2786	53.135	9.7762
1800	2077.7	1543.4	9.1006	1.2821	60.000	8.9049
1850	2141.9	1592.7	9.1358	1.2854	67.551	8.1293
1900	2206.2	1642.2	9.1701	1.2885	75.832	7.4372
1950	2270.7	1691.9	9.2036	1.2914	84.893	6.8183
2000	2335.4	1741.7	9.2364	1.2942	94.787	6.2632
2050	2400.2	1791.7	9.2683	1.2968	105.57	5.7639
2100	2465.0	1841.7	9.2996	1.2992	117.30	5.3140
2150	2530.0	1891.9	9.3302	1.3015	130.04	4.9077
2200	2595.2	1942.2	9.3602	1.3037	143.85	4.5398
2250	2660.6	1992.7	9.3895	1.3068	158.80	4.2056
2300	2725.9	2043.2	9.4183	1.3079	174.94	3.9027
2350	2790.9	2093.4	9.4462	1.3014	192.22	3.6290
2400	2855.9	2143.5	9.4736	1.2967	210.79	3.3797
2450	2921.1	2193.9	9.5005	1.3050	230.80	3.1510
2500	2986.6	2244.5	9.5270	1.3151	252.31	2.9412

TABLE D.3 IDEAL-GAS PROPERTIES OF CARBON MONOXIDE (CO)—MASS BASIS: SI UNITS
Mol. Wt.: 28.01 *(Continued)*

T, K	h, kJ/kg	u, kJ/kg	s^0, kJ/kg·K	c_p, kJ/kg·K	P_r	v_r
2550	3052.4	2295.4	9.5530	1.3176	275.44	2.7481
2600	3118.3	2346.5	9.5786	1.3182	300.24	2.5705
2650	3184.2	2397.6	9.6037	1.3197	326.75	2.4074
2700	3250.2	2448.8	9.6284	1.3210	355.08	2.2571
2750	3316.3	2500.0	9.6526	1.3225	385.31	2.1185
2800	3382.5	2551.4	9.6765	1.3239	417.53	1.9906
2850	3448.7	2602.8	9.6999	1.3251	451.86	1.8722
2900	3515.0	2654.2	9.7230	1.3263	488.37	1.7626
2950	3581.4	2705.7	9.7457	1.3275	527.13	1.6612
3000	3647.8	2757.3	9.7680	1.3287	568.28	1.5670
3050	3714.2	2808.9	9.7900	1.3298	611.97	1.4794
3100	3780.7	2860.6	9.8116	1.3309	658.26	1.3979
3150	3847.3	2912.3	9.8329	1.3320	707.21	1.3221
3200	3913.9	2964.1	9.8539	1.3330	758.99	1.2515
3250	3980.6	3015.9	9.8746	1.3340	813.75	1.1855
3300	4047.3	3067.8	9.8949	1.3349	871.58	1.1239
3350	4114.1	3119.7	9.9150	1.3359	932.59	1.0663
3400	4180.9	3171.7	9.9348	1.3368	996.90	1.0124
3450	4247.8	3223.7	9.9543	1.3376	1064.7	0.96187
3500	4314.7	3275.8	9.9736	1.3385	1136.0	0.91453
3550	4381.6	3327.9	9.9926	1.3394	1211.1	0.87010
3600	4448.6	3380.0	10.011	1.3403	1290.0	0.82837
3650	4515.7	3432.2	10.030	1.3412	1372.9	0.78916
3700	4582.7	3484.4	10.048	1.3420	1460.0	0.75227
3750	4649.8	3536.7	10.066	1.3427	1551.3	0.71753
3800	4717.0	3589.0	10.084	1.3435	1647.1	0.68480
3850	4784.2	3641.4	10.101	1.3442	1747.5	0.65395
3900	4851.4	3693.8	10.119	1.3450	1852.7	0.62483
3950	4918.7	3746.2	10.136	1.3457	1963.0	0.59731
4000	4986.0	3798.7	10.153	1.3465	2078.2	0.57131
4050	5053.3	3851.2	10.170	1.3472	2198.6	0.54678
4100	5120.7	3903.7	10.186	1.3479	2324.5	0.52356
4150	5188.1	3956.3	10.202	1.3486	2456.2	0.50154
4200	5255.6	4008.9	10.219	1.3493	2593.7	0.48067
4250	5323.1	4061.5	10.235	1.3500	2736.9	0.46093
4300	5390.6	4114.2	10.250	1.3506	2886.4	0.44221
4350	5458.1	4166.9	10.266	1.3513	3042.4	0.42440
4400	5525.7	4219.6	10.281	1.3520	3205.1	0.40749
4450	5593.3	4272.4	10.297	1.3526	3374.4	0.39145
4500	5661.0	4325.2	10.312	1.3532	3550.5	0.37621
4550	5728.6	4378.1	10.327	1.3539	3734.2	0.36169
4600	5796.3	4430.9	10.342	1.3545	3925.2	0.34786
4650	5864.1	4483.8	10.356	1.3551	4123.7	0.33472
4700	5931.9	4536.8	10.371	1.3557	4329.9	0.32220
4750	5999.7	4589.7	10.385	1.3563	4544.4	0.31026
4800	6067.5	4642.7	10.399	1.3569	4767.3	0.29887
4850	6135.3	4695.7	10.413	1.3575	4998.6	0.28801
4900	6203.2	4748.8	10.427	1.3581	5238.7	0.27764
4950	6271.2	4801.8	10.441	1.3587	5487.9	0.26774
5000	6339.1	4854.9	10.455	1.3593	5746.3	0.25828

Adapted from *JANAF Thermochemical Tables* Third Edition, M. W. Chase et al., American Chemical Society and the American Institute of Physics for the National Bureau of Standards, 1985.

TABLE D.4 IDEAL-GAS PROPERTIES OF HYDROGEN (H_2)—MASS BASIS: SI UNITS
Mol. Wt.: 2.016

T, K	h, kJ/kg	u, kJ/kg	s^0 kJ/kg·K	c_p, kJ/kg·K	P_r	v_r
0	0.0	0.0	0.0			
100	1,487.7	1,075.2	49.967	13.966	0.23263	1,773.0
150	2,148.1	1,529.5	55.249	13.430	0.83734	733.45
200	2,824.1	1,999.2	59.236	13.615	2.2012	374.75
250	3,516.6	2,485.5	62.325	14.060	4.6555	221.48
300	4,226.4	2,989.1	64.914	14.311	8.7202	141.89
350	4,945.2	3,501.7	67.129	14.426	14.923	96.734
400	5,668.0	4,018.2	69.060	14.476	23.829	69.235
450	6,392.7	4,536.7	70.766	14.499	36.040	51.499
500	7,118.0	5,055.8	72.294	14.515	52.206	39.502
550	7,844.0	5,575.5	73.680	14.530	73.045	31.058
600	8,570.9	6,096.3	74.943	14.548	99.231	24.939
650	9,298.9	6,618.0	76.110	14.573	131.67	20.362
700	10,028	7,141.3	77.190	14.605	171.09	16.875
750	10,760	7,666.4	78.200	14.645	218.53	14.156
800	11,493	8,193.7	79.146	14.695	274.87	12.004
850	12,229	8,723.5	80.039	14.755	341.31	10.272
900	12,969	9,256.5	80.883	14.823	418.89	8.8616
950	13,712	9,793.4	81.687	14.899	509.02	7.6977
1000	14,459	10,334	82.453	14.984	612.94	6.7290
1050	15,210	10,879	83.187	15.074	732.22	5.9145
1100	15,966	11,429	83.890	15.170	868.33	5.2249
1150	16,727	11,984	84.566	15.270	1023.1	4.6360
1200	17,493	12,544	85.218	15.374	1198.3	4.1303
1250	18,265	13,109	85.848	15.480	1396.0	3.6930
1300	19,041	13,680	86.458	15.588	1618.3	3.3133
1350	19,823	14,255	87.048	15.696	1867.3	2.9818
1400	20,611	14,837	87.621	15.805	2145.5	2.6913
1450	21,404	15,423	88.177	15.914	2455.5	2.4355
1500	22,202	16,016	88.719	16.022	2799.8	2.2097
1550	23,006	16,613	89.246	16.128	3181.5	2.0094
1600	23,815	17,216	89.759	16.234	3603.4	1.8313
1650	24,629	17,824	90.260	16.337	4068.9	1.6725
1700	25,449	18,437	90.749	16.439	4581.2	1.5305
1750	26,273	19,056	91.228	16.539	5144.2	1.4031
1800	27,103	19,679	91.695	16.636	5761.5	1.2886
1850	27,937	20,307	92.152	16.732	6436.6	1.1854
1900	28,776	20,939	92.599	16.825	7173.9	1.0924
1950	29,619	21,576	93.037	16.916	7978.1	1.0081
2000	30,467	22,218	93.467	17.005	8853.5	0.93171
2050	31,320	22,865	93.888	17.091	9804.8	0.86234
2100	32,177	23,515	94.301	17.176	10,837	0.79922
2150	33,037	24,170	94.706	17.258	11,956	0.74167
2200	33,902	24,829	95.104	17.338	13,167	0.68913
2250	34,771	25,491	95.494	17.416	14,474	0.64114
2300	35,644	26,158	95.878	17.493	15,884	0.59721
2350	36,520	26,828	96.255	17.567	17,405	0.55689
2400	37,401	27,502	96.625	17.639	19,041	0.51986
2450	38,284	28,179	96.990	17.710	20,801	0.48580
2500	39,171	28,860	97.349	17.780	22,691	0.45443

H_2

TABLE D.4 IDEAL-GAS PROPERTIES OF HYDROGEN (H_2)—MASS BASIS: SI UNITS Mol. Wt.: 2.016 (*Continued*)

T, K	h, kJ/kg	u, kJ/kg	s^0 kJ/kg·K	c_p, kJ/kg·K	P_r	v_r
2550	40,062	29,545	97.701	17.847	24,717	0.42551
2600	40,956	30,233	98.048	17.913	26,887	0.39883
2650	41,854	30,924	98.390	17.978	29,209	0.37419
2700	42,754	31,618	98.727	18.042	31,692	0.35138
2750	43,658	32,315	99.058	18.104	34,347	0.33023
2800	44,564	33,016	99.385	18.165	37,181	0.31060
2850	45,474	33,719	99.707	18.224	40,200	0.29241
2900	46,387	34,426	100.02	18.283	43,416	0.27550
2950	47,302	35,135	100.34	18.341	46,838	0.25977
3000	48,221	35,847	100.65	18.397	50,477	0.24513
3050	49,142	36,562	100.95	18.453	54,346	0.23147
3100	50,066	37,280	101.25	18.509	58,455	0.21873
3150	50,993	38,001	101.55	18.563	62,813	0.20684
3200	51,922	38,724	101.84	18.616	67,433	0.19572
3250	52,855	39,450	102.13	18.669	72,328	0.18533
3300	53,789	40,179	102.42	18.721	77,511	0.17560
3350	54,727	40,910	102.70	18.773	82,995	0.16648
3400	55,667	41,643	102.98	18.824	88,795	0.15793
3450	56,609	42,380	103.25	18.874	94,922	0.14991
3500	57,554	43,118	103.52	18.924	101,390	0.14238
3550	58,502	43,860	103.79	18.974	108,220	0.13530
3600	59,451	44,603	104.06	19.023	115,410	0.12865
3650	60,404	45,349	104.32	19.072	123,010	0.12238
3700	61,358	46,098	104.58	19.120	131,010	0.11648
3750	62,316	46,849	104.84	19.168	139,430	0.11093
3800	63,276	47,603	105.09	19.216	148,290	0.10569
3850	64,238	48,358	105.34	19.264	157,620	0.10075
3900	65,202	49,116	105.59	19.311	167,420	0.09608
3950	66,168	49,877	105.84	19.357	177,730	0.09167
4000	67,137	50,640	106.08	19.404	188,550	0.08750
4050	68,109	51,405	106.32	19.450	199,910	0.08356
4100	69,082	52,172	106.56	19.496	211,830	0.07983
4150	70,058	52,942	106.80	19.541	224,330	0.07630
4200	71,036	53,714	107.03	19.586	237,440	0.07296
4250	72,017	54,488	107.26	19.632	251,180	0.06979
4300	73,000	55,265	107.49	19.676	265,570	0.06678
4350	73,985	56,043	107.72	19.720	280,650	0.06393
4400	74,972	56,824	107.95	19.764	296,430	0.06122
4450	75,961	57,607	108.17	19.808	312,960	0.05865
4500	76,952	58,392	108.39	19.851	330,240	0.05620
4550	77,946	59,180	108.61	19.894	348,300	0.05388
4600	78,942	59,969	108.83	19.936	367,160	0.05167
4650	79,940	60,761	109.05	19.977	386,870	0.04957
4700	80,940	61,555	109.26	20.019	407,460	0.04757
4750	81,942	62,350	109.47	20.059	428,950	0.04567
4800	82,945	63,148	109.68	20.099	451,370	0.04386
4850	83,951	63,948	109.89	20.139	474,770	0.04213
4900	84,959	64,750	110.10	20.178	499,180	0.04049
4950	85,969	65,553	110.30	20.216	524,640	0.03891
5000	86,981	66,359	110.51	20.254	551,190	0.03741

Adapted from *JANAF Thermochemical Tables* Third Edition, M. W. Chase et al., American Chemical Society and the American Institute of Physics for the National Bureau of Standards, 1985.

TABLE D.5 IDEAL-GAS PROPERTIES OF NITROGEN (N_2)—MASS BASIS: SI UNITS
Mol. Wt.: 28.013

T, K	h, kJ/kg	u, kJ/kg	s^0, kJ/kg·K	c_p, kJ/kg·K	P_r	v_r
0	0.0	0.0	0.0			
100	103.59	73.91	5.7048	1.0389	0.00647	4,588.3
150	155.57	111.05	6.1184	1.0390	0.02606	1,695.6
200	207.51	148.15	6.4250	1.0390	0.07321	810.81
250	259.45	185.25	6.6569	1.0392	0.15992	464.00
300	311.42	222.38	6.8463	1.0397	0.30279	294.06
350	363.43	259.55	7.0067	1.0411	0.51974	199.87
400	415.55	296.83	7.1459	1.0441	0.83080	142.90
450	467.88	334.32	7.2691	1.0490	1.2584	106.14
500	520.50	372.10	7.3800	1.0559	1.8283	81.169
550	573.51	410.27	7.4811	1.0647	2.5706	63.508
600	626.99	448.90	7.5741	1.0748	3.5159	50.650
650	681.02	488.10	7.6607	1.0860	4.7068	40.991
700	735.61	527.85	7.7415	1.0978	6.1804	33.616
750	790.80	568.20	7.8177	1.1099	7.9897	27.862
800	846.59	609.15	7.8897	1.1221	10.182	23.319
850	903.01	650.73	7.9581	1.1340	12.821	19.677
900	960.00	692.88	8.0232	1.1455	15.966	16.731
950	1017.6	735.61	8.0854	1.1566	19.692	14.319
1000	1075.7	778.86	8.1450	1.1672	24.070	12.331
1050	1134.3	822.63	8.2022	1.1772	29.187	10.678
1100	1193.4	866.88	8.2572	1.1866	35.127	9.2943
1150	1252.9	911.59	8.3102	1.1955	41.987	8.1293
1200	1312.9	956.75	8.3612	1.2038	49.865	7.1424
1250	1373.3	1002.3	8.4105	1.2116	58.879	6.3011
1300	1434.1	1048.2	8.4582	1.2190	69.138	5.5807
1350	1495.2	1094.5	8.5043	1.2258	80.764	4.9612
1400	1556.6	1141.1	8.5490	1.2322	93.886	4.4258
1450	1618.4	1188.0	8.5924	1.2382	108.65	3.9610
1500	1680.4	1235.2	8.6344	1.2438	125.20	3.5560
1550	1742.8	1282.7	8.6753	1.2491	143.68	3.2018
1600	1805.4	1330.5	8.7150	1.2540	164.26	2.8910
1650	1868.2	1378.4	8.7537	1.2586	187.11	2.6173
1700	1931.2	1426.6	8.7913	1.2629	212.41	2.3755
1750	1994.4	1475.0	8.8280	1.2670	240.34	2.1611
1800	2057.9	1523.6	8.8638	1.2708	271.11	1.9706
1850	2121.5	1572.4	8.8986	1.2744	304.91	1.8008
1900	2185.3	1621.4	8.9327	1.2778	341.95	1.6491
1950	2249.3	1670.5	8.9659	1.2810	382.45	1.5133
2000	2313.4	1719.8	8.9983	1.2841	426.65	1.3913
2050	2377.7	1769.3	9.0301	1.2869	474.81	1.2814
2100	2442.1	1818.8	9.0611	1.2896	527.17	1.1823
2150	2506.6	1868.5	9.0915	1.2922	583.98	1.0927
2200	2571.3	1918.3	9.1212	1.2947	645.52	1.0115
2250	2636.1	1968.3	9.1504	1.2970	712.07	0.93784
2300	2701.0	2018.4	9.1789	1.2992	783.92	0.87081
2350	2766.0	2068.6	9.2069	1.3013	861.40	0.80971
2400	2831.1	2118.8	9.2343	1.3033	944.79	0.75395
2450	2896.4	2169.2	9.2612	1.3053	1034.4	0.70300
2500	2961.7	2219.7	9.2876	1.3071	1130.5	0.65635
2550	3027.1	2270.2	9.3135	1.3089	1,233.6	0.61353
2600	3092.6	2320.9	9.3389	1.3106	1,343.9	0.57419

TABLE D.5 IDEAL-GAS PROPERTIES OF NITROGEN (N_2)—MASS BASIS: SI UNITS
Mol. Wt.: 28.013 (*Continued*)

T, K	h, kJ/kg	u, kJ/kg	s^0, kJ/kg·K	c_p, kJ/kg·K	P_r	v_r
2650	3158.1	2371.6	9.3639	1.3122	1,461.9	0.53800
2700	3223.8	2422.4	9.3884	1.3137	1,587.9	0.50466
2750	3289.5	2473.3	9.4125	1.3152	1,722.4	0.47389
2800	3355.3	2524.2	9.4362	1.3166	1,865.6	0.44545
2850	3421.1	2575.3	9.4596	1.3180	2,018.1	0.41914
2900	3487.1	2626.4	9.4825	1.3193	2,180.3	0.39477
2950	3553.1	2677.5	9.5051	1.3206	2,352.5	0.37218
3000	3619.2	2728.8	9.5273	1.3219	2,535.3	0.35121
3050	3685.3	2780.0	9.5491	1.3231	2,729.0	0.33172
3100	3751.5	2831.4	9.5706	1.3242	2,934.2	0.31358
3150	3817.7	2882.8	9.5918	1.3253	3,151.4	0.29667
3200	3884.0	2934.2	9.6127	1.3264	3,381.2	0.28090
3250	3950.3	2985.7	9.6333	1.3275	3,623.8	0.26618
3300	4016.7	3037.3	9.6536	1.3285	3,879.9	0.25244
3350	4083.2	3088.9	9.6735	1.3295	4,150.2	0.23957
3400	4149.7	3140.6	9.6933	1.3305	4,435.2	0.22753
3450	4216.2	3192.3	9.7127	1.3314	4,735.2	0.21625
3500	4282.8	3244.0	9.7318	1.3323	5,051.0	0.20566
3550	4349.4	3295.8	9.7508	1.3332	5,383.4	0.19572
3600	4416.1	3347.7	9.7694	1.3341	5,732.9	0.18638
3650	4482.9	3399.6	9.7878	1.3350	6,099.5	0.17761
3700	4549.6	3451.5	9.8060	1.3358	6,484.3	0.16936
3750	4616.4	3503.4	9.8239	1.3366	6,888.4	0.16158
3800	4683.3	3555.5	9.8416	1.3374	7,312.2	0.15424
3850	4750.2	3607.5	9.8591	1.3382	7,755.6	0.14734
3900	4817.1	3659.6	9.8764	1.3389	8,219.9	0.14082
3950	4884.1	3711.7	9.8935	1.3397	8,706.8	0.13465
4000	4951.1	3763.9	9.9103	1.3404	9,216.1	0.12882
4050	5018.1	3816.1	9.9270	1.3412	9,747.6	0.12332
4100	5085.2	3868.3	9.9434	1.3419	10,303	0.11811
4150	5152.3	3920.6	9.9597	1.3426	10,884	0.11317
4200	5219.5	3972.9	9.9758	1.3433	11,491	0.10849
4250	5286.6	4025.2	9.9917	1.3439	12,122	0.10406
4300	5353.8	4077.6	10.007	1.3446	12,781	0.09985
4350	5421.1	4130.0	10.023	1.3452	13,469	0.09586
4400	5488.4	4182.5	10.038	1.3459	14,186	0.09206
4450	5555.7	4234.9	10.054	1.3465	14,932	0.08845
4500	5623.0	4287.4	10.069	1.3471	15,709	0.08502
4550	5690.4	4339.9	10.083	1.3478	16,516	0.08176
4600	5757.8	4392.5	10.098	1.3484	17,356	0.07866
4650	5825.2	4445.1	10.113	1.3490	18,230	0.07571
4700	5892.7	4497.8	10.127	1.3496	19,139	0.07289
4750	5960.2	4550.4	10.142	1.3503	20,082	0.07020
4800	6027.7	4603.1	10.156	1.3509	21,062	0.06764
4850	6095.3	4655.8	10.170	1.3515	22,080	0.06520
4900	6162.9	4708.6	10.184	1.3521	23,136	0.06286
4950	6230.5	4761.3	10.197	1.3527	24,232	0.06063
5000	6298.2	4814.2	10.211	1.3534	25,369	0.05850

Adapted from *JANAF Thermochemical Tables* Third Edition, M. W. Chase et al., American Chemical Society and the American Institute of Physics for the National Bureau of Standards, 1985.

TABLE D.6 IDEAL-GAS PROPERTIES OF OXYGEN (O_2)—MASS BASIS: SI UNITS
Mol. Wt.: 31.999

T, K	h, kJ/kg	u, kJ/kg	s^0, kJ/kg·K	c_p, kJ/kg·K	P_r	v_r
0	0.0	0.0	0.0			
100	90.75	64.77	5.4160	0.9096	0.00530	4906.4
150	136.25	97.27	5.7783	0.9097	0.02135	1812.3
200	181.73	129.76	6.0466	0.9102	0.05997	866.60
250	227.29	162.33	6.2499	0.9126	0.13113	495.39
300	273.04	195.09	6.4168	0.9183	0.24921	312.79
350	319.20	228.26	6.5590	0.9280	0.43081	211.10
400	365.89	261.95	6.6837	0.9408	0.69623	149.28
450	413.33	296.40	6.7954	0.9558	1.0701	109.26
500	461.49	331.57	6.8969	0.9716	1.5816	82.144
550	510.47	367.56	6.9904	0.9875	2.2660	63.072
600	560.24	404.34	7.0769	1.0029	3.1612	49.317
650	610.76	441.86	7.1578	1.0173	4.3160	39.135
700	661.96	480.08	7.2336	1.0307	5.7784	31.477
750	713.79	518.92	7.3052	1.0430	7.6110	25.606
800	766.22	558.35	7.3728	1.0542	9.8744	21.051
850	819.19	598.33	7.4370	1.0644	12.644	17.468
900	872.66	638.81	7.4981	0.0736	15.994	14.621
950	926.55	679.71	7.5564	1.0821	20.017	12.332
1000	980.85	721.01	7.6121	1.0897	24.801	10.477
1050	1035.5	762.68	7.6654	1.0967	30.453	8.9592
1100	1090.5	804.69	7.7166	1.1032	37.080	7.7082
1150	1145.8	847.01	7.7658	1.1091	44.806	6.6690
1200	1201.4	889.62	7.8131	1.1146	53.757	5.8002
1250	1257.3	932.49	7.8587	1.1198	64.073	5.0691
1300	1313.4	975.61	7.9027	1.1247	75.900	4.4504
1350	1369.7	1019.0	7.9453	1.1293	89.401	3.9236
1400	1426.3	1062.5	7.9864	1.1337	104.74	3.4730
1450	1483.1	1106.4	8.0263	1.1379	122.11	3.0855
1500	1540.1	1150.4	8.0649	1.1420	141.69	2.7508
1550	1597.3	1194.6	8.1024	1.1460	163.69	2.4604
1600	1654.7	1239.0	8.1389	1.1499	188.33	2.2075
1650	1712.3	1283.6	8.1743	1.1538	215.85	1.9862
1700	1770.1	1328.4	8.2088	1.1575	246.50	1.7920
1750	1828.1	1373.3	8.2424	1.1613	280.54	1.6208
1800	1886.2	1418.5	8.2752	1.1649	318.25	1.4696
1850	1944.5	1463.9	8.3072	1.1686	359.92	1.3356
1900	2003.1	1509.4	8.3384	1.1722	405.87	1.2164
1950	2061.8	1555.1	8.3689	1.1759	456.41	1.1101
2000	2120.6	1601.0	8.3987	1.1795	511.91	1.0152
2050	2179.7	1647.1	8.4279	1.1830	572.73	0.93004
2100	2239.0	1693.3	8.4564	1.1866	639.25	0.85359
2150	2298.4	1739.7	8.4844	1.1901	711.86	0.78477
2200	2358.0	1786.3	8.5118	1.1936	791.00	0.72268
2250	2417.7	1833.1	8.5386	1.1972	877.16	0.66650
2300	2477.7	1880.1	8.5650	1.2006	970.80	0.61559
2350	2537.8	1927.2	8.5908	1.2041	1072.4	0.56938
2400	2598.1	1974.5	8.6162	1.2075	1182.5	0.52737
2450	2658.6	2022.0	8.6412	1.2109	1301.5	0.48912
2500	2719.2	2069.6	8.6656	1.2143	1430.1	0.45422

TABLE D.6 IDEAL-GAS PROPERTIES OF OXYGEN (O_2)—MASS BASIS: SI UNITS
Mol. Wt.: 31.999 (*Continued*)

T, K	h, kJ/kg	u, kJ/kg	s^0, kJ/kg·K	c_p, kJ/kg·K	P_r	v_r
2550	2780.0	2117.4	8.6897	1.2176	1,569.1	0.42227
2600	2841.0	2165.4	8.7134	1.2209	1,718.9	0.39303
2650	2902.1	2213.5	8.7367	1.2242	1,880.0	0.36626
2700	2963.4	2261.8	8.7596	1.2274	2,053.3	0.34168
2750	3024.8	2310.3	8.7822	1.2306	2,239.4	0.31908
2800	3086.4	2358.9	8.8044	1.2337	2,439.2	0.29827
2850	3148.2	2407.7	8.8262	1.2368	2,653.3	0.27909
2900	3210.1	2456.6	8.8478	1.2399	2,882.7	0.26140
2950	3272.2	2505.6	8.8690	1.2429	3,128.0	0.24505
3000	3334.4	2554.9	8.8899	1.2458	3,390.1	0.22994
3050	3396.7	2604.3	8.9105	1.2487	3,670.0	0.21594
3100	3459.3	2653.8	8.9308	1.2515	3,968.6	0.20297
3150	3521.9	2703.4	8.9509	1.2543	4,286.8	0.19093
3200	3584.7	2753.2	8.9706	1.2571	4,625.7	0.17975
3250	3647.6	2803.2	8.9902	1.2598	4,986.6	0.16935
3300	3710.7	2853.2	9.0094	1.2624	5,370.3	0.15967
3350	3773.9	2903.4	9.0284	1.2650	5,777.7	0.15066
3400	3837.2	2953.7	9.0472	1.2675	6,210.1	0.14226
3450	3900.6	3004.2	9.0657	1.2700	6,669.0	0.13442
3500	3964.1	3054.7	9.0840	1.2724	7,155.4	0.12710
3550	4027.8	3105.4	9.1021	1.2748	7,670.5	0.12025
3600	4091.7	3156.3	9.1199	1.2772	8,215.8	0.11385
3650	4155.6	3207.2	9.1375	1.2795	8,792.7	0.10786
3700	4219.6	3258.2	9.1550	1.2817	9,402.7	0.10225
3750	4283.7	3309.4	9.1722	1.2839	10,047	0.09698
3800	4348.0	3360.6	9.1892	1.2861	10,728	0.09204
3850	4412.3	3412.0	9.2060	1.2882	11,446	0.08740
3900	4476.8	3463.4	9.2227	1.2903	12,202	0.08305
3950	4541.4	3515.0	9.2391	1.2924	13,000	0.07895
4000	4606.0	3566.7	9.2554	1.2945	13,840	0.07510
4050	4670.8	3618.5	9.2715	1.2965	14,724	0.07147
4100	4735.7	3670.4	9.2874	1.2985	15,654	0.06805
4150	4800.7	3722.4	9.3032	1.3004	16,632	0.06483
4200	4865.7	3774.4	9.3188	1.3024	17,661	0.06179
4250	4930.9	3826.6	9.3342	1.3043	18,741	0.05892
4300	4996.2	3878.9	9.3494	1.3062	19,875	0.05622
4350	5061.5	3931.2	9.3646	1.3081	21,065	0.05366
4400	5127.0	3983.7	9.3795	1.3100	22,313	0.05124
4450	5192.5	4036.3	9.3943	1.3120	23,621	0.04895
4500	5258.2	4088.9	9.4090	1.3139	24,993	0.04678
4550	5323.9	4141.7	9.4235	1.3158	26,431	0.04473
4600	5389.8	4194.5	9.4379	1.3177	27,937	0.04278
4650	5455.7	4247.5	9.4522	1.3196	29,513	0.04094
4700	5521.7	4300.5	9.4663	1.3215	31,161	0.03919
4750	5587.8	4353.6	9.4803	1.3235	32,884	0.03753
4800	5654.1	4406.9	9.4941	1.3255	34,686	0.03596
4850	5720.4	4460.2	9.5079	1.3275	36,570	0.03446
4900	5786.8	4513.6	9.5215	1.3295	38,540	0.03304
4950	5853.3	4567.2	9.5350	1.3315	40,597	0.03168
5000	5920.0	4620.8	9.5484	1.3336	42,745	0.03039

Adapted from *JANAF Thermochemical Tables* Third Edition, M. W. Chase et al., American Chemical Society and the American Institute of Physics for the National Bureau of Standards, 1985.

TABLE D.7 IDEAL-GAS PROPERTIES OF WATER VAPOR (H_2O)—MASS BASIS: SI UNITS
Mol. Wt.: 18.015

T, K	h, kJ/kg	u, kJ/kg	s^0, kJ/kg·K	c_p, kJ/kg·K	P_r	v_r
0	0.0	0.0	0.0			
100	182.57	136.42	8.4588	1.8484	0.00080	57,990
150	275.02	205.79	9.1914	1.8482	0.00389	17,620
200	367.58	275.27	9.7409	1.8512	0.01280	7,210.0
250	460.29	344.91	10.158	1.8568	0.03161	3,657.8
300	553.20	414.74	10.493	1.8649	0.06538	2,117.8
350	646.74	485.21	10.783	1.8794	0.12234	1,320.7
400	741.37	556.76	11.034	1.9018	0.21110	874.50
450	837.12	629.43	11.261	1.9271	0.34461	602.88
500	934.15	703.39	11.464	1.9553	0.53591	430.59
550	1,032.6	778.81	11.652	1.9851	0.80547	315.19
600	1,132.6	855.74	11.826	2.0163	1.1737	235.93
650	1,234.3	934.29	11.989	2.0484	1.6702	179.63
700	1,337.5	1,014.5	12.142	2.0813	2.3260	138.89
750	1,442.4	1,096.3	12.287	2.1150	3.1833	108.74
800	1,549.0	1,179.8	12.424	2.1493	4.2882	86.101
850	1,657.4	1,265.1	12.556	2.1843	5.7008	68.816
900	1,767.5	1,352.1	12.681	2.2196	7.4873	55.476
950	1,879.4	1,440.9	12.802	2.2552	9.7313	45.056
1000	1,993.0	1,531.5	12.919	2.2907	12.527	36.843
1050	2,108.4	1,623.8	13.032	2.3261	15.989	30.308
1100	2,225.6	1,717.9	13.141	2.3611	20.249	25.071
1150	2,344.5	1,813.8	13.246	2.3956	25.462	20.845
1200	2,465.1	1,911.3	13.349	2.4295	31.805	17.413
1250	2,587.4	2,010.5	13.449	2.4626	39.487	14.610
1300	2,711.4	2,111.4	13.546	2.4948	48.744	12.309
1350	2,836.9	2,213.9	13.641	2.5261	59.853	10.410
1400	2,964.0	2,317.9	13.733	2.5564	73.123	8.8362
1450	3,092.5	2,423.3	13.823	2.5857	88.911	7.5267
1500	3,222.5	2,530.3	13.912	2.6139	107.62	6.4328
1550	3,353.9	2,638.6	13.998	2.6411	129.70	5.5153
1600	3,486.6	2,748.2	14.082	2.6672	155.68	4.7432
1650	3,620.6	2,859.1	14.164	2.6922	186.14	4.0911
1700	3,755.8	2,971.2	14.245	2.7163	221.72	3.5386
1750	3,892.2	3,084.6	14.324	2.7394	263.17	3.0690
1800	4,029.7	3,199.0	14.402	2.7615	311.29	2.6687
1850	4,168.4	3,314.6	14.478	2.7827	366.97	2.3266
1900	4,308.0	3,431.1	14.552	2.8030	431.24	2.0334
1950	4,448.6	3,548.7	14.625	2.8223	505.21	1.7814
2000	4,590.2	3,667.2	14.697	2.8409	590.12	1.5642
2050	4,732.7	3,786.6	14.767	2.8591	687.34	1.3765
2100	4,876.1	3,907.0	14.836	2.8766	798.37	1.2140
2150	5,020.4	4,028.1	14.904	2.8932	924.85	1.0729
2200	5,165.4	4,150.1	14.971	2.9091	1,068.6	0.95016
2250	5,311.3	4,272.9	15.037	2.9244	1,231.7	0.84308
2300	5,457.9	4,396.4	15.101	2.9390	1,416.3	0.74949
2350	5,605.2	4,520.6	15.164	2.9531	1,624.7	0.66755
2400	5,753.2	4,645.5	15.227	2.9666	1,859.6	0.59565
2450	5,901.8	4,771.1	15.288	2.9796	2,123.7	0.53243
2500	6,051.1	4,897.3	15.348	2.9921	2,420.2	0.47674

TABLE D.7 IDEAL-GAS PROPERTIES OF WATER VAPOR (H_2O)—MASS BASIS: SI UNITS
Mol. Wt.: 18.015 (*Continued*)

T, K	h, kJ/kg	u, kJ/kg	s^0, kJ/kg·K	c_p, kJ/kg·K	P_r	v_r
2550	6,201.0	5,024.1	15.408	3.0042	2,752.4	0.42759
2600	6,351.5	5,151.6	15.466	3.0157	3,123.9	0.38412
2650	6,502.6	5,279.5	15.524	3.0269	3,538.8	0.34561
2700	6,654.2	5,408.1	15.580	3.0376	4,001.2	0.31143
2750	6,806.3	5,537.2	15.636	3.0479	4,515.8	0.28105
2800	6,959.0	5,666.7	15.691	3.0579	5,087.5	0.25401
2850	7,112.1	5,796.8	15.745	3.0675	5,721.5	0.22989
2900	7,265.7	5,927.3	15.799	3.0768	6,423.7	0.20836
2950	7,419.8	6,058.3	15.851	3.0858	7,200.5	0.18908
3000	7,574.3	6,189.7	15.903	3.0945	8,058.3	0.17182
3050	7,729.2	6,321.6	15.955	3.1028	9,004.3	0.15633
3100	7,884.6	6,453.9	16.005	3.1109	10,046	0.14242
3150	8,040.3	6,586.5	16.055	3.1188	11,191	0.12991
3200	8,196.4	6,719.6	16.104	3.1264	12,449	0.11864
3250	8,353.0	6,853.0	16.153	3.1337	13,830	0.10846
3300	8,509.8	6,986.8	16.201	3.1408	15,343	0.09927
3350	8,667.0	7,120.9	16.248	3.1477	16,997	0.09096
3400	8,824.6	7,255.4	16.295	3.1544	18,806	0.08344
3450	8,982.5	7,390.2	16.341	3.1609	20,781	0.07662
3500	9,140.7	7,525.4	16.386	3.1672	22,936	0.07043
3550	9,299.2	7,660.8	16.431	3.1733	25,285	0.06480
3600	9,458.0	7,796.6	16.476	3.1793	27,840	0.05968
3650	9,617.1	7,932.6	16.519	3.1850	30,618	0.05502
3700	9,776.5	8,068.9	16.563	3.1906	33,634	0.05077
3750	9,936.1	8,205.5	16.606	3.1961	36,907	0.04689
3800	10,096	8,342.3	16.648	3.2014	40,454	0.04335
3850	10,256	8,479.5	16.690	3.2066	44,298	0.04011
3900	10,417	8,616.8	16.731	3.2117	48,458	0.03714
3950	10,577	8,754.5	16.772	3.2166	52,954	0.03443
4000	10,738	8,892.4	16.813	3.2213	57,809	0.03193
4050	10,900	9,030.5	16.853	3.2260	63,049	0.02965
4100	11,061	9,168.8	16.892	3.2305	68,699	0.02754
4150	11,223	9,307.3	16.932	3.2350	74,785	0.02561
4200	11,385	9,446.1	16.970	3.2393	81,337	0.02383
4250	11,547	9,585.1	17.009	3.2435	88,386	0.02219
4300	11,709	9,724.3	17.047	3.2476	95,964	0.02068
4350	11,871	9,863.7	17.084	3.2516	104,100	0.01928
4400	12,034	10,003	17.121	3.2556	112,840	0.01800
4450	12,197	10,143	17.158	3.2594	122,200	0.01681
4500	12,360	10,283	17.195	3.2632	132,240	0.01571
4550	12,523	10,423	17.231	3.2668	143,000	0.01469
4600	12,687	10,564	17.267	3.2704	154,510	0.01374
4650	12,850	10,704	17.302	3.2740	166,810	0.01287
4700	13,014	10,845	17.337	3.2774	179,960	0.01205
4750	13,178	10,986	17.372	3.2808	194,020	0.01130
4800	13,342	11,127	17.406	3.2841	209,030	0.01060
4850	13,506	11,268	17.440	3.2872	225,040	0.00995
4900	13,671	11,409	17.474	3.2903	242,100	0.00934
4950	13,835	11,551	17.507	3.2934	260,270	0.00878
5000	14,000	11,693	17.540	3.2966	279,620	0.00825

Adapted from *JANAF Thermochemical Tables* Third Edition, M. W. Chase et al., American Chemical Society and the American Institute of Physics for the National Bureau of Standards, 1985.

TABLE D.8 ZERO-PRESSURE SPECIFIC HEATS FOR COMMON IDEAL GASES AS A FUNCTION OF TEMPERATURE

Temp., K	c_{p0}, kJ/kg·K	c_{v0}, kJ/kg·K	k	c_{p0}, kJ/kg·K	c_{v0}, kJ/kg·K	k	c_{p0}, kJ/kg·K	c_{v0}, kJ/kg·K	k	Temp., K
	Air			Carbon dioxide, CO_2			Carbon monoxide, CO			
250	1.000	0.713	1.403	0.789	0.600	1.315	1.040	0.743	1.400	250
300	1.000	0.713	1.403	0.846	0.657	1.288	1.040	0.743	1.399	300
350	1.010	0.723	1.397	0.897	0.708	1.268	1.042	0.745	1.398	350
400	1.015	0.728	1.394	0.939	0.750	1.252	1.047	0.750	1.396	400
450	1.020	0.733	1.391	0.979	0.790	1.239	1.055	0.758	1.392	450
500	1.030	0.743	1.386	1.014	0.825	1.229	1.064	0.767	1.387	500
600	1.050	0.763	1.376	1.075	0.886	1.213	1.087	0.790	1.376	600
700	1.070	0.783	1.367	1.126	0.937	1.202	1.113	0.816	1.364	700
800	1.100	0.813	1.353	1.169	0.980	1.193	1.139	0.842	1.353	800
	Hydrogen, H_2			Nitrogen, N_2			Oxygen, O_2			
250	14.060	9.936	1.415	1.039	0.742	1.400	0.913	0.653	1.398	250
300	14.311	10.187	1.405	1.040	0.743	1.400	0.918	0.658	1.395	300
350	14.426	10.302	1.400	1.041	0.744	1.399	0.928	0.668	1.389	350
400	14.476	10.352	1.398	1.044	0.747	1.397	0.941	0.681	1.382	400
450	14.499	10.375	1.398	1.049	0.752	1.395	0.956	0.696	1.373	450
500	14.515	10.391	1.397	1.056	0.759	1.391	0.972	0.712	1.365	500
600	14.548	10.424	1.396	1.075	0.778	1.382	1.003	0.743	1.350	600
700	14.605	10.480	1.394	1.098	0.801	1.371	1.031	0.771	1.337	700
800	14.695	10.570	1.390	1.122	0.825	1.360	1.054	0.794	1.327	800

At low pressures the constant-pressure and constant-volume specific heats for all monatomic gases are independent of pressure and temperature. Also, the values of c_p and c_v on a molar basis are the same for all monatomic gases. Therefore, for monatomic gases such as argon, helium, and neon, the specific heats can be assumed constant at the following values:

$$\bar{c}_p = 20.8 \text{ kJ/kg·mol·K} \qquad \bar{c}_v = 12.5 \text{ kJ/kg·mol·K}$$

at states where the gases may be assumed to be ideal gases.

SOURCE: Kenneth Wark, *Thermodynamics*, 3d ed., McGraw-Hill, 1977, p. 804, table A-4M. Originally published in "Tables of Thermal Properties of Gases," *NBS Circular* 564, 1955.

TABLE D.9 ZERO-PRESSURE SPECIFIC HEATS AND k AT 27°C FOR COMMON GASES

Gas	Chemical formula	Molecular weight	R, kJ/kg·K	c_{p0}, kJ/kg·K	c_{v0}, kJ/kg·K	k
Air	—	28.97	0.287 00	1.000	0.713	1.403
Argon	Ar	39.948	0.208 13	0.5203	0.3122	1.667
Butane	C_4H_{10}	58.124	0.143 04	1.7164	1.5734	1.091
Carbon dioxide	CO_2	44.010	0.188 92	0.846	0.657	1.289
Carbon monoxide	CO	28.010	0.296 83	1.040	0.744	1.400
Ethane	C_2H_6	30.070	0.276 50	1.7662	1.4897	1.186
Ethylene	C_2H_4	28.054	0.296 37	1.5482	1.2518	1.237
Helium	He	4.003	2.077 03	5.1926	3.1156	1.667
Hydrogen	H_2	2.016	4.124 18	14.311	10.187	1.405
Methane	CH_4	16.043	0.518 35	2.2537	1.7354	1.299
Neon	Ne	20.183	0.411 95	1.0299	0.6179	1.667
Nitrogen	N_2	28.013	0.296 80	1.039	0.743	1.400
Octane	C_8H_{18}	114.23	0.072 79	1.7113	1.6385	1.044
Oxygen	O_2	31.999	0.259 83	0.918	0.658	1.395
Propane	C_3H_8	44.097	0.188 55	1.6794	1.4909	1.126
Water vapor	H_2O	18.015	0.461 52	1.865	1.4035	1.329

SOURCE: Adapted from Gordon J. Van Wylen and Richard E. Sonntag, *Fundamentals of Classical Thermodynamics*, SI Version, 2d ed., Revised Printing, Copyright © 1976, 1978 by John Wiley & Sons, Inc., Table A-8, p. 683.

TABLE D.10 MOLECULAR WEIGHT AND GAS CONSTANT FOR COMMON GASES

Gas	Molecular weight, M, kg/kg·mol	Gas constant, R, N·m/kg·K or Pa·m^3/kg·K
Air	28.97	287.0
Argon	39.948	208.1
Carbon dioxide	44.010	188.9
Carbon monoxide	28.010	296.8
Helium	4.003	2077
Hydrogen	2.016	4124
Nitrogen	28.013	296.8
Oxygen	31.999	259.8
Water vapor	18.015	461.5

TABLE D.11 EXPRESSIONS FOR CONSTANT-PRESSURE SPECIFIC HEATS FOR COMMON GASES AT ZERO PRESSURE

$$\bar{c}_{p0}/\bar{R} = a + bT + cT^2 + dT^3 + eT^4$$

T is in degrees kelvin, equations valid from 300 to 1000K

Gas	a	$b \times 10^3$	$c \times 10^6$	$d \times 10^9$	$e \times 10^{12}$
CO	3.710	−1.619	3.692	−2.032	0.240
CO_2	2.401	8.735	−6.607	2.002	
H_2	3.057	2.677	−5.180	5.521	−1.812
H_2O	4.070	−1.108	4.152	−2.964	0.807
O_2	3.626	−1.878	7.056	−6.764	2.156
N_2	3.675	−1.208	2.324	−0.632	−0.226
Air (dry)	3.640	−1.101	2.466	−0.942	
NH_3	3.591	0.494	8.345	−8.383	2.730
NO	4.046	−3.418	7.982	−6.114	1.592
NO_2	3.459	2.065	6.687	−9.556	3.620
SO_2	3.267	5.324	0.684	−5.281	2.559
SO_3	2.578	14.556	−9.176	−0.792	1.971
CH_4	3.826	−3.979	24.558	−22.733	6.963
C_2H_2	1.410	19.057	−24.501	16.391	−4.135
C_2H_4	1.426	11.383	7.989	−16.254	6.749

SOURCE: Kenneth Wark, *Thermodynamics,* 4th ed., McGraw-Hill, 1983; original source NASA SP-273, Government Printing Office, Washington, DC, 1971.

Generalized Charts for Gases

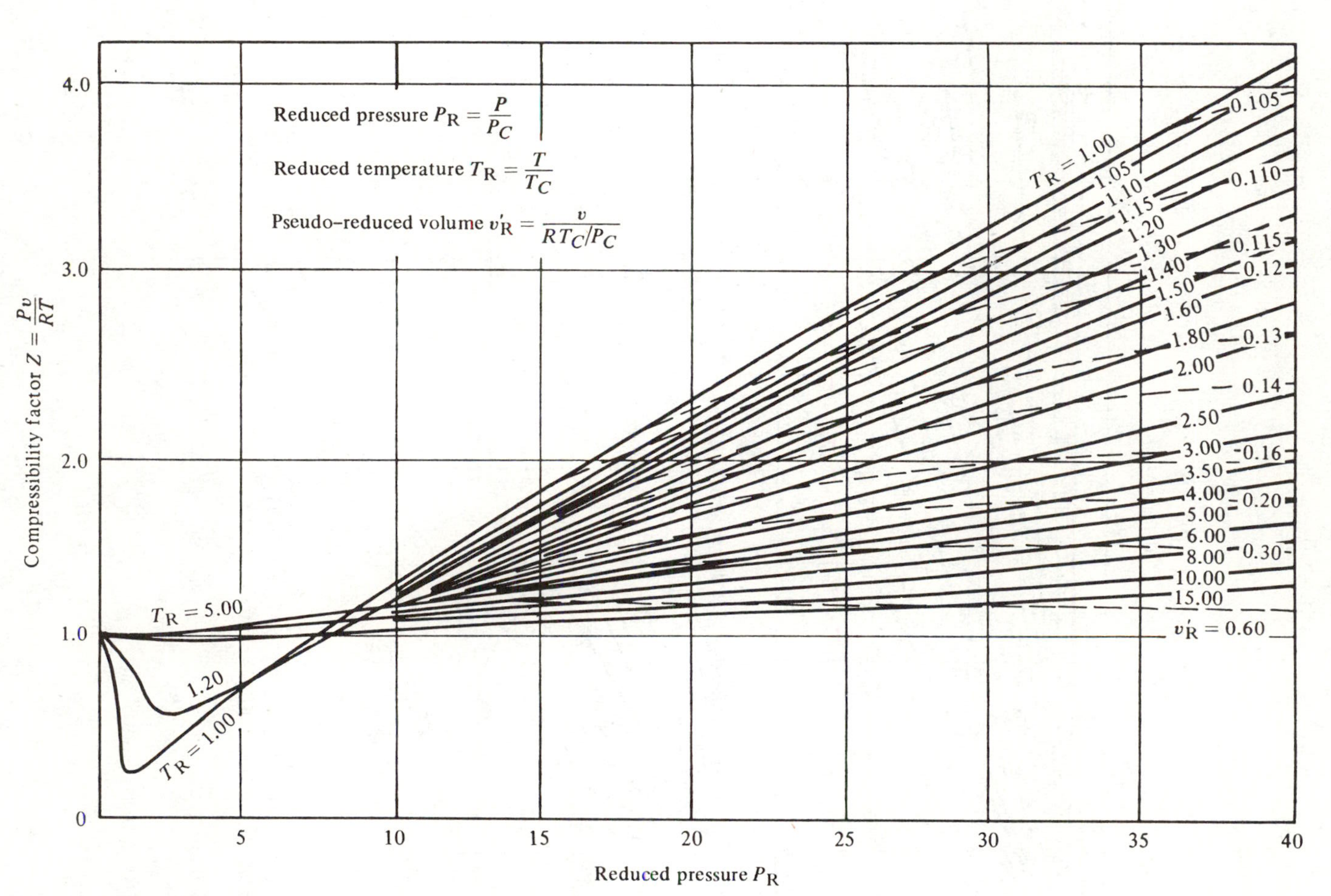

Figure E.1 Compressibility chart, high-pressure range. (Source: J.P. Holman, *Thermodynamics*, 3d ed., McGraw-Hill, New York, 1960.)

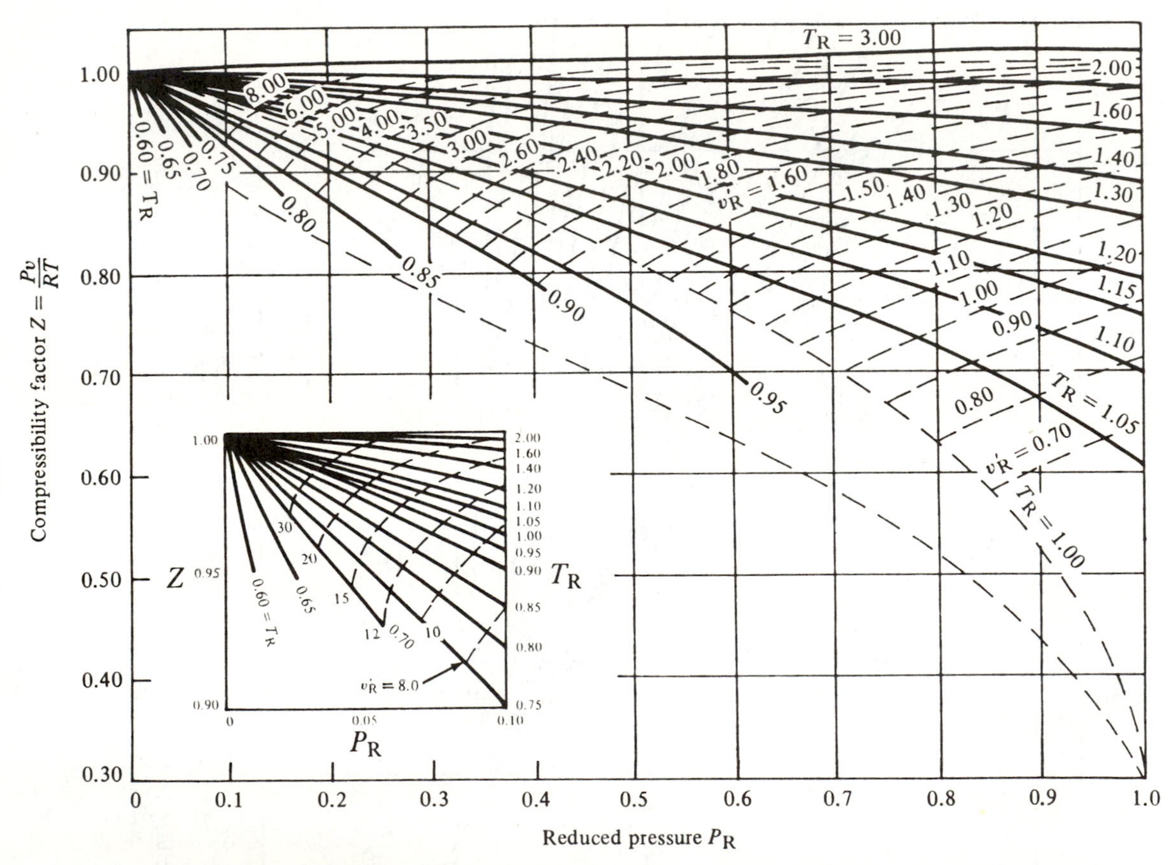

Figure E.2 Compressibility chart, low-pressure range. (Source: J.P. Holman, *Thermodynamics,* 3d ed., McGraw-Hill, 1980, p. 287, Figure 7-6. Originally published in E.F. Obert, *Concepts of Thermodynamics,* McGraw-Hill, New York, 1960.)

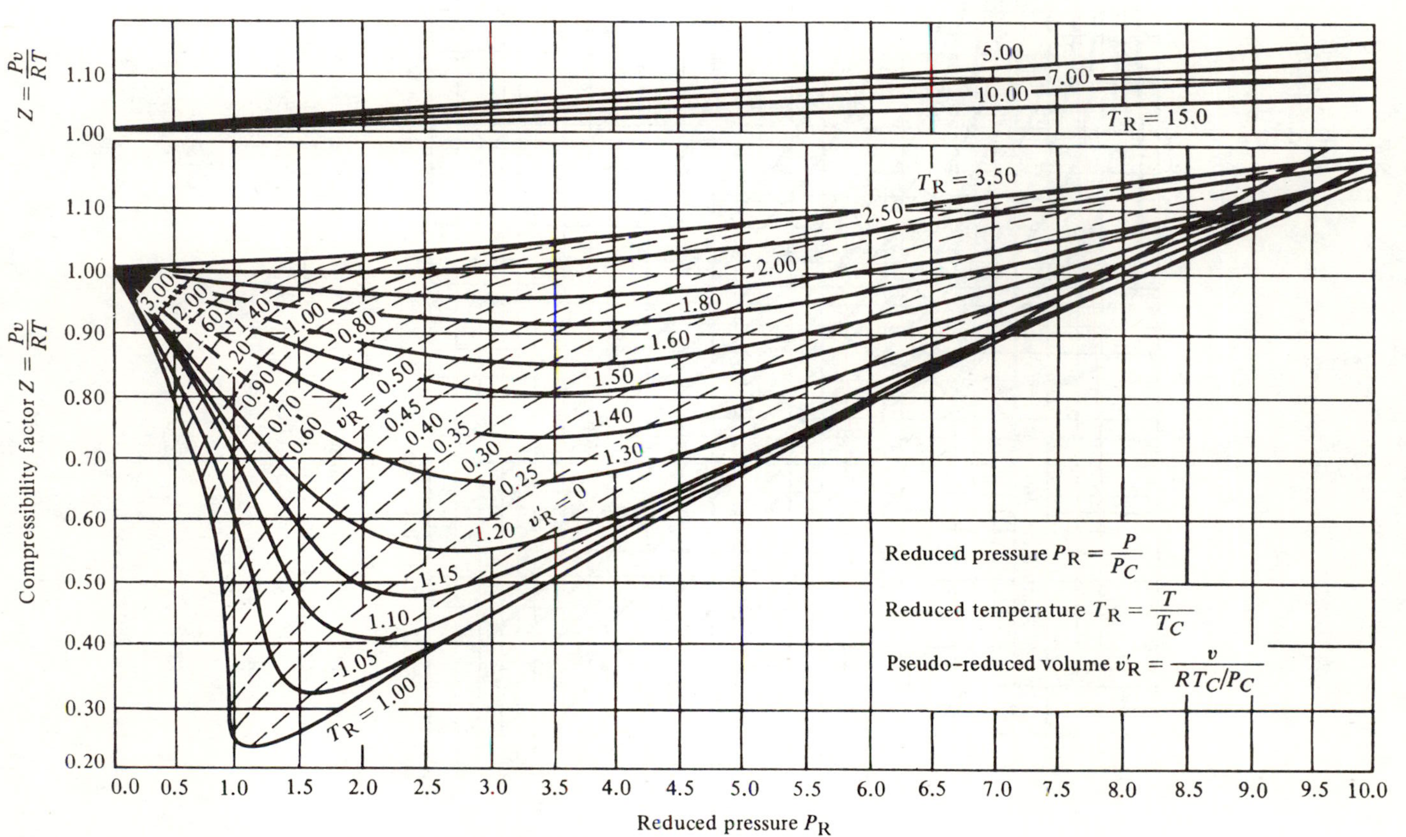

Figure E.3 Compressibility chart, moderate-pressure range. (Source: J.P. Holman, *Thermodynamics,* 3d ed., McGraw-Hill, 1980, p. 285, Figure 7-4. Originally published in E.F. Obert, *Concepts of Thermodynamics,* McGraw-Hill, New York, 1960.)

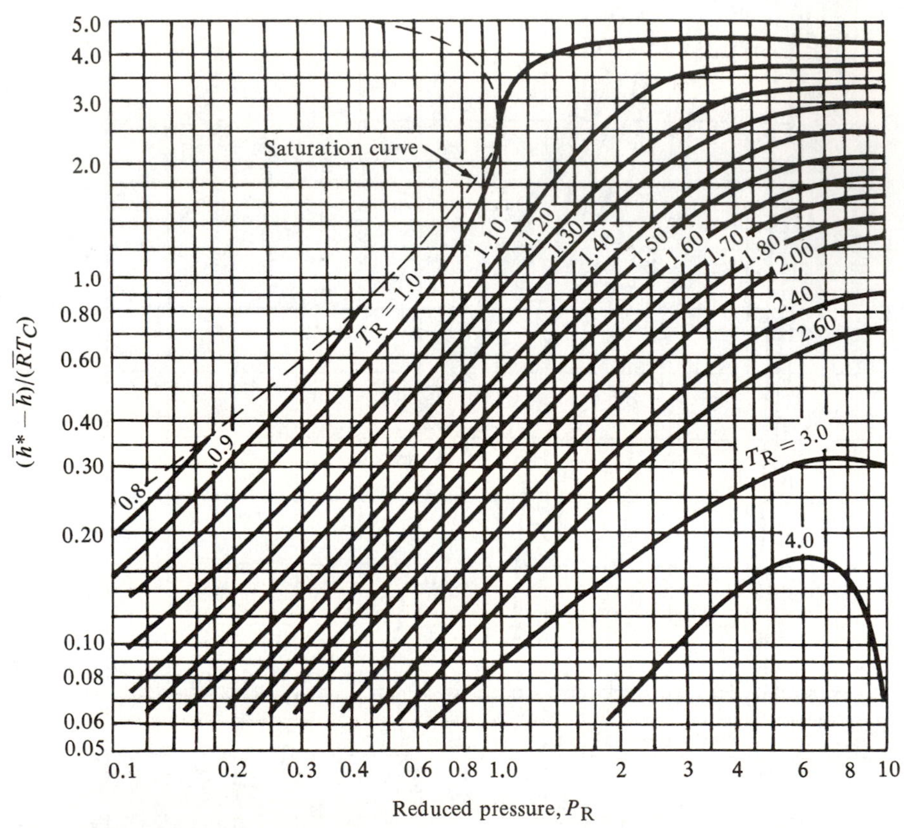

Figure E.4 Generalized enthalpy chart. (Source: Kenneth Wark, *Thermodynamics,* 4th ed., McGraw-Hill, 1983, p. 821, Figure A-27M. Originally published by A.L. Lydersen, R.A. Greenkorn, and O.A. Hougen, ''Engineering Experiment Station Report No. 4,'' University of Wisconsin, 1955.)

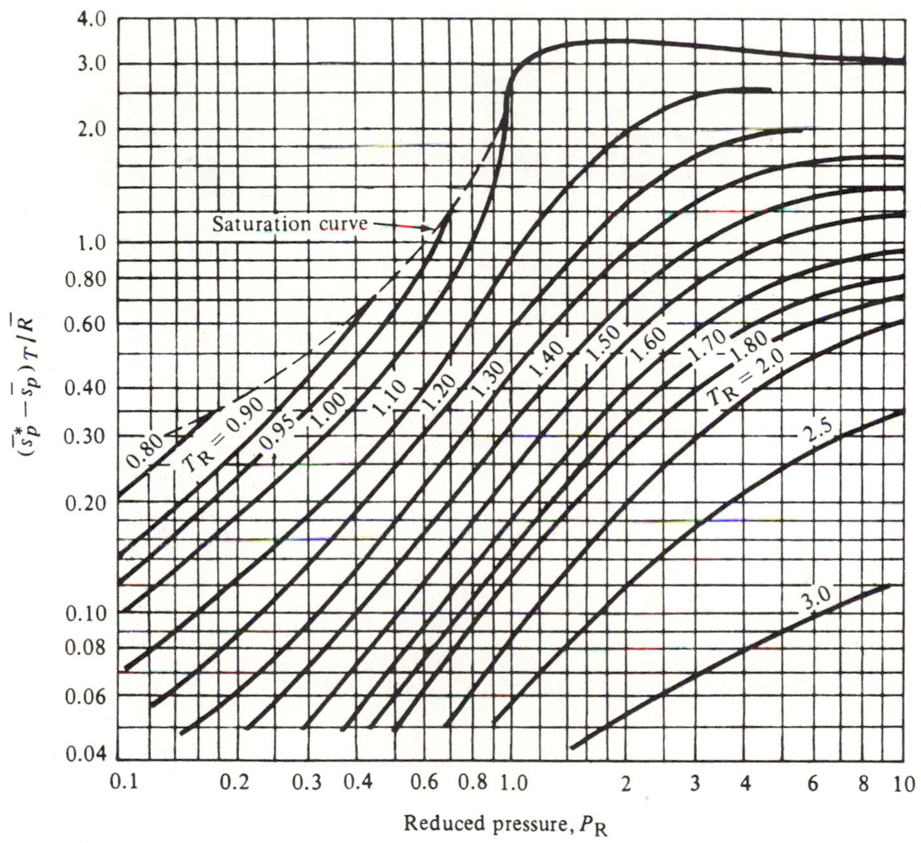

Figure E.5 Generalized entropy chart. (Source: Kenneth Wark, *Thermodynamics,* 4th ed., McGraw-Hill, 1983, p. 822, Figure A-28M. Originally published by A.L. Lydersen, R.A. Greenkorn, and O.A. Hougen, ''Engineering Experiment Station Report No. 4,'' University of Wisconsin, 1955.)

Properties of Air-Water Mixtures

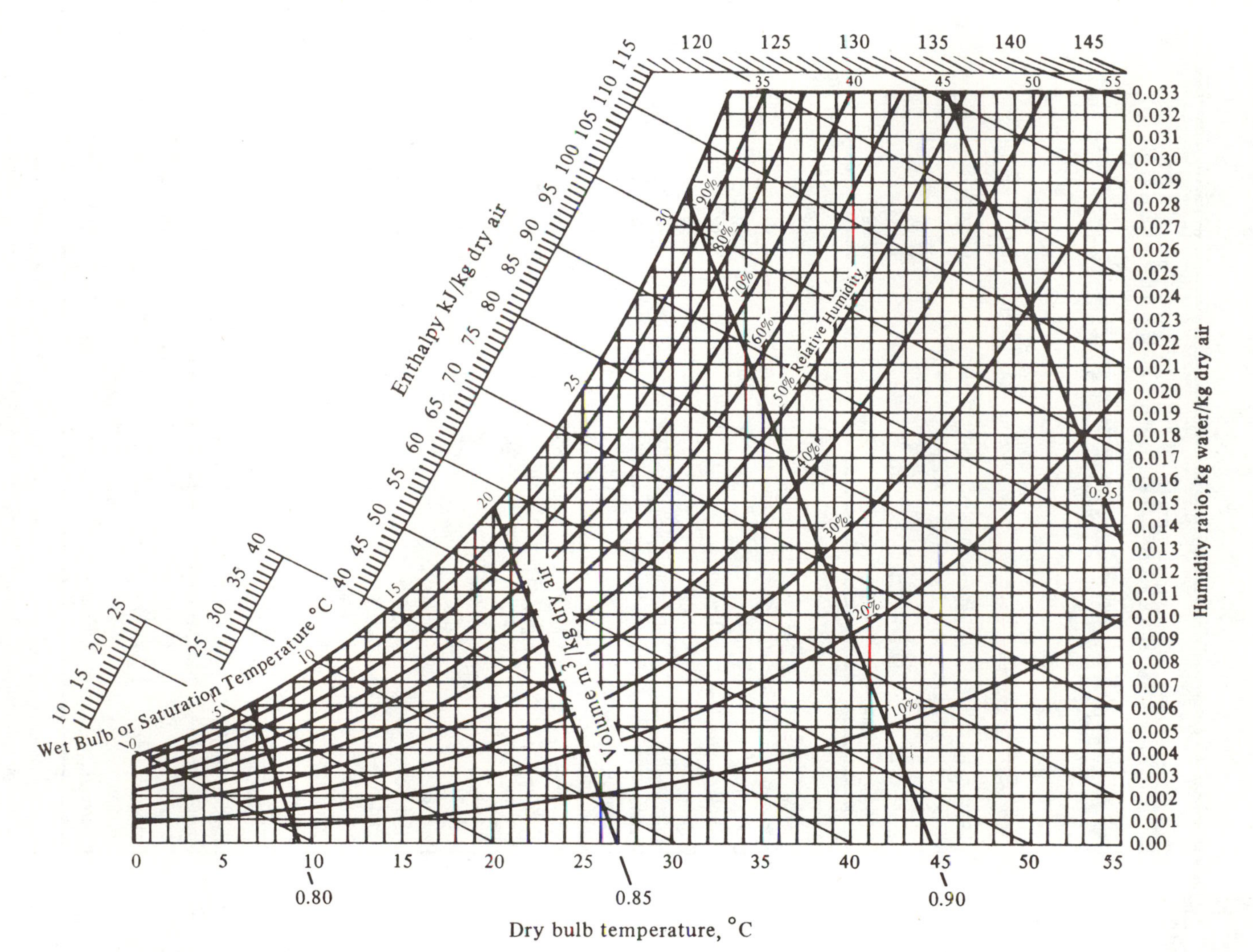

Figure F.1 Psychrometric chart for 1 atm total pressure (101.3 kPa).

Properties of Chemically Reacting Substances

TABLE G.1 ENTHALPY OF FORMATION, GIBBS FUNCTION OF FORMATION, AND ABSOLUTE ENTROPY AT 25°C, 1 atm

Substance	Formula	$\Delta\bar{h}_f^\circ$, kJ/kg·mol	$\Delta\bar{g}_f^\circ$, kJ/kg·mol	$\bar{s}^\circ$, kJ/kg·mol·K
Carbon	$C(s)$	0	0	5.74
Hydrogen	$H_2(g)$	0	0	130.68
Nitrogen	$N_2(g)$	0	0	191.61
Oxygen	$O_2(g)$	0	0	205.14
Carbon monoxide	$CO(g)$	−110,530	−137,150	197.65
Carbon dioxide	$CO_2(g)$	−393,522	−394,360	213.80
Water	$H_2O(g)$	−241,826	−228,590	188.83
Water	$H_2O(l)$	−285,826	−237,180	69.92
Hydrogen peroxide	$H_2O_2(g)$	−136,310	−105,600	232.63
Ammonia	$NH_3(g)$	−46,190	−16,590	192.33
Methane	$CH_4(g)$	−74,850	−50,790	186.16
Acetylene	$C_2H_2(g)$	+226,730	+209,170	200.85
Ethylene	$C_2H_4(g)$	+52,280	+68,120	219.83
Ethane	$C_2H_6(g)$	−84,680	−32,890	229.49
Propylene	$C_3H_6(g)$	+20,410	+62,720	266.94
Propane	$C_3H_8(g)$	−103,850	−23,490	269.91
n-Butane	$C_4H_{10}(g)$	−126,150	−15,710	310.12
n-Octane	$C_8H_{18}(g)$	−208,450	+16,530	466.73
n-Octane	$C_8H_{18}(l)$	−249,950	+6,610	360.79
n-Dodecane	$C_{12}H_{26}$	−291,010	+50,150	622.83
Benzene	$C_6H_6(g)$	+82,930	+129,660	269.20
Methyl alcohol	$CH_3OH(g)$	−200,670	−162,000	239.70
Methyl alcohol	$CH_3OH(l)$	−238,660	−166,360	126.80
Ethyl alcohol	$C_2H_5OH(g)$	−235,310	−168,570	282.59
Ethyl alcohol	$C_2H_5OH(l)$	−277,690	−174,890	160.70
Oxygen	$O(g)$	+249,170	+231,770	161.06
Hydrogen	$H(g)$	+217,999	+203,290	114.72
Nitrogen	$N(g)$	+472,680	+455,510	153.30
Hydroxyl	$OH(g)$	+38,987	+34,280	183.70

SOURCES: From the JANAF Thermochemical Tables, Dow Chemical Co., 1971; *Selected Values of Chemical Thermodynamic Properties,* NBS Tech. Note 270-3, 1968; and *API Research Project 44,* Carnegie Press, 1953.

TABLE G.2 ENTHALPY OF COMBUSTION AND ENTHALPY OF VAPORIZATION AT 25°C, 1 atm
Water appears as a liquid in the product of combustion

Substance	Formula	$\Delta \bar{h}_c^\circ = -\text{HHV}$, kJ/kg·mol	$\bar{h}_{fg}$, kJ/kg·mol
Hydrogen	$H_2(g)$	−285,840	
Carbon	$C(s)$	−393,520	
Carbon monoxide	$CO(g)$	−282,990	
Methane	$CH_4(g)$	−890,360	
Acetylene	$C_2H_2(g)$	−1,299,600	
Ethylene	$C_2H_4(g)$	−1,410,970	
Ethane	$C_2H_6(g)$	−1,559,900	
Propylene	$C_3H_6(g)$	−2,058,500	
Propane	$C_3H_8(g)$	−2,220,000	15,060
n-Butane	$C_4H_{10}(g)$	−2,877,100	21,060
n-Pentane	$C_5H_{12}(g)$	−3,536,100	26,410
n-Hexane	$C_6H_{14}(g)$	−4,194,800	31,530
n-Heptane	$C_7H_{16}(g)$	−4,853,500	36,520
n-Octane	$C_8H_{18}(g)$	−5,512,200	41,460
Benzene	$C_6H_6(g)$	−3,301,500	33,830
Toluene	$C_7H_8(g)$	−3,947,900	39,920
Methyl alcohol	$CH_3OH(g)$	−764,540	37,900
Ethyl alcohol	$C_2H_5OH(g)$	−1,409,300	42,340

SOURCE: Kenneth Wark, *Thermodynamics*, 3d ed., McGraw-Hill, 1977, pp. 834–835, table A-23M.

TABLE G.3 IDEAL-GAS ENTHALPY AND ABSOLUTE ENTROPY FOR CO_2 AND CO AT 1 atm: SI UNITS

	Carbon dioxide (CO_2) $\Delta\bar{h}_f^\circ = -393{,}522$ kJ/kg·mol, $M = 44.010$			Carbon monoxide (CO) $\Delta\bar{h}_f^\circ = -110{,}530$ kJ/kg·mol, $M = 28.010$	
Temp., K	$\bar{h}_T - \bar{h}^0$, kJ/kg·mol	$\bar{s}^0$, kJ/kg·mol·K	Temp., K	$\bar{h}_T - \bar{h}^0$, kJ/kg·mol	$\bar{s}^0$, kJ/kg·mol·K
0	−9 364	0.0	0	−8 671	0.0
100	−6 456	179.009	100	−5 769	165.850
200	−3 414	199.975	200	−2 858	186.025
298.15	0	213.795	298.15	0	197.653
300	69	214.025	300	54	197.833
400	4 003	225.314	400	2 976	206.238
500	8 305	234.901	500	5 931	212.831
600	12 907	243.283	600	8 942	218.319
700	17 754	250.750	700	12 023	223.066
800	22 806	257.494	800	15 177	227.277
900	28 030	263.645	900	18 401	231.074
1000	33 379	269.299	1000	21 690	234.538
1100	38 884	274.528	1100	25 035	237.726
1200	44 473	279.390	1200	28 430	240.679
1300	50 148	283.932	1300	31 868	243.431
1400	55 896	288.191	1400	35 343	246.006
1500	61 705	292.199	1500	38 850	248.426
1600	67 569	295.983	1600	42 385	250.707
1700	73 480	299.566	1700	45 945	252.865
1800	79 431	302.968	1800	49 526	254.912
1900	85 419	306.205	1900	53 126	256.859
2000	91 439	309.293	2000	56 744	258.714
2100	97 488	312.244	2100	60 376	260.486
2200	103 562	315.070	2200	64 021	262.182
2300	109 660	317.781	2300	67 683	263.809
2400	115 779	320.385	2400	71 324	265.359
2500	121 917	322.890	2500	74 985	266.854
2600	128 073	325.305	2600	78 673	268.300
2700	134 246	327.634	2700	82 369	269.695
2800	140 433	329.885	2800	86 074	271.042
2900	146 636	332.061	2900	89 786	272.345
3000	152 852	334.169	3000	93 504	273.605
3200	165 321	338.192	3200	100 960	276.011
3400	177 836	341.986	3400	108 438	278.278
3600	190 393	345.574	3600	115 937	280.421
3800	202 989	348.979	3800	123 454	282.453
4000	215 622	352.219	4000	130 989	284.386
4200	228 290	355.310	4200	138 540	286.228
4400	240 991	358.264	4400	146 106	287.988
4600	253 725	361.094	4600	153 687	289.673
4800	266 489	363.810	4800	161 282	291.289
5000	279 283	366.422	5000	168 890	292.842
5200	292 109	368.937	5200	176 511	294.336
5400	304 971	371.364	5400	184 146	295.777
5600	317 870	373.709	5600	191 775	297.164
5800	330 806	375.979	5800	199 434	298.508
6000	343 779	378.178	6000	207 106	299.808

Adapted from *JANAF Thermochemical Tables* Third Edition, M. W. Chase et al., American Chemical Society and the American Institute of Physics for the National Bureau of Standards, 1985.

TABLE G.4 IDEAL-GAS ENTHALPY AND ABSOLUTE ENTROPY FOR H_2 AND H AT 1 atm: SI UNITS

	Hydrogen, diatomic (H_2)			Hydrogen, monatomic (H)	
	$\Delta\bar{h}_f^\circ$ = 0 kJ/kg·mol, M = 2.016			$\Delta\bar{h}_f^\circ$ = 217,999 kJ/kg·mol, M = 1.008	
Temp., K	$\bar{h}_T - \bar{h}^0$, kJ/kg·mol	$\bar{s}^0$, kJ/kg·mol·K	Temp., K	$\bar{h}_T - \bar{h}^0$, kJ/kg·mol	$\bar{s}^0$, kJ/kg·mol·K
0	−8 467	0.0	0	−6 197	0.0
100	−5 468	100.727	100	−4 119	92.009
200	−2 774	119.412	200	−2 040	106.417
298.15	0	130.680	298.15	0	114.716
300	53	130.858	300	38	114.845
400	2 959	139.216	400	2 117	120.825
500	5 882	145.737	500	4 196	125.463
600	8 811	151.077	600	6 274	129.253
700	11 749	155.606	700	8 353	132.457
800	14 702	159.548	800	10 431	135.232
900	17 676	163.051	900	12 510	137.681
1000	20 680	166.216	1000	14 589	139.871
1100	23 719	169.112	1100	16 667	141.852
1200	26 797	171.790	1200	18 746	143.660
1300	29 918	174.288	1300	20 824	145.324
1400	33 082	176.633	1400	22 903	146.865
1500	36 290	178.846	1500	24 982	148.299
1600	39 541	180.944	1600	27 060	149.640
1700	42 835	182.940	1700	29 139	150.900
1800	46 169	184.846	1800	31 217	152.088
1900	49 541	186.669	1900	33 296	153.212
2000	52 951	188.418	2000	35 375	154.278
2100	56 397	190.099	2100	37 453	155.293
2200	59 876	191.718	2200	39 532	156.260
2300	63 387	193.278	2300	41 610	157.184
2400	66 928	194.785	2400	43 689	158.068
2500	70 498	196.243	2500	45 768	158.917
2600	74 096	197.654	2600	47 846	159.732
2700	77 720	199.021	2700	49 925	160.516
2800	81 369	200.349	2800	52 004	161.272
2900	85 043	201.638	2900	54 082	162.002
3000	88 740	202.891	3000	56 161	162.706
3200	96 202	205.299	3200	60 318	164.048
3400	103 750	207.587	3400	64 475	165.308
3600	111 380	209.767	3600	68 632	166.496
3800	119 089	211.851	3800	72 790	167.620
4000	126 874	213.848	4000	76 947	168.686
4200	134 734	215.765	4200	81 104	169.700
4400	142 667	217.610	4400	85 261	170.667
4600	150 670	219.389	4600	89 418	171.591
4800	158 741	221.106	4800	93 576	172.476
5000	166 876	222.767	5000	97 733	173.325
5200	175 071	224.374	5200	101 890	174.140
5400	183 322	225.931	5400	106 047	174.924
5600	191 621	227.440	5600	110 204	175.680
5800	199 963	228.903	5800	114 362	176.410
6000	208 341	230.323	6000	118 519	177.114

Adapted from *JANAF Thermochemical Tables* Third Edition, M. W. Chase et al., American Chemical Society and the American Institute of Physics for the National Bureau of Standards, 1985.

TABLE G.5 IDEAL-GAS ENTHALPY AND ABSOLUTE ENTROPY FOR N_2 AND N AT 1 atm: SI UNITS

	Nitrogen, diatomic (N_2) $\Delta\bar{h}_f^\circ = 0$ kJ/kg·mol, $M = 28.013$			Nitrogen, monatomic (N) $\Delta\bar{h}_f^\circ = 472{,}680$ kJ/kg·mol, $M = 14.007$	
Temp., K	$\bar{h}_T - \bar{h}^0$, kJ/kg·mol	$\bar{s}^0$, kJ/kg·mol·K	Temp., K	$\bar{h}_T - \bar{h}^0$, kJ/kg·mol	$\bar{s}^0$, kJ/kg·mol·K
0	−8 670	0.0	0	−6 197	0.0
100	−5 768	159.811	100	−4 119	130.593
200	−2 857	179.985	200	−2 040	145.001
298.15	0	191.609	298.15	0	153.300
300	54	191.789	300	38	153.429
400	2 971	200.181	400	2 117	159.408
500	5 911	206.739	500	4 196	164.047
600	8 894	212.176	600	6 274	167.836
700	11 937	216.866	700	8 353	171.041
800	15 046	221.017	800	10 431	173.816
900	18 223	224.757	900	12 510	176.264
1000	21 463	228.170	1000	14 589	178.454
1100	24 760	231.313	1100	16 667	180.436
1200	28 109	234.226	1200	18 746	182.244
1300	31 503	236.943	1300	20 824	183.908
1400	34 936	239.487	1400	22 903	185.448
1500	38 405	241.880	1500	24 982	186.882
1600	41 904	244.138	1600	27 060	188.224
1700	45 429	246.275	1700	29 139	189.484
1800	48 978	248.304	1800	31 218	190.672
1900	52 548	250.234	1900	33 296	191.796
2000	56 137	252.074	2000	35 375	192.863
2100	59 742	253.833	2100	37 454	193.877
2200	63 361	255.517	2200	39 534	194.844
2300	66 995	257.132	2300	41 614	195.769
2400	70 640	258.684	2400	43 695	196.655
2500	74 296	260.176	2500	45 777	197.504
2600	77 963	261.614	2600	47 860	198.322
2700	81 639	263.001	2700	49 945	199.109
2800	85 323	264.341	2800	52 033	199.868
2900	89 015	265.637	2900	54 124	200.601
3000	92 715	266.891	3000	56 218	201.311
3200	100 134	269.285	3200	60 420	202.667
3400	107 577	271.541	3400	64 646	203.948
3600	115 041	273.675	3600	68 902	205.164
3800	122 525	275.698	3800	73 194	206.325
4000	130 027	277.622	4000	77 532	207.437
4200	137 545	279.456	4200	81 920	208.508
4400	145 078	281.208	4400	86 367	209.542
4600	152 625	282.885	4600	90 877	210.544
4800	160 187	284.494	4800	95 457	211.519
5000	167 763	286.041	5000	100 111	212.469
5200	175 352	287.529	5200	104 843	213.397
5400	182 955	288.964	5400	109 655	214.305
5600	190 571	290.348	5600	114 550	215.195
5800	198 201	291.687	5800	119 528	216.068
6000	205 848	292.984	6000	124 590	216.926

Adapted from *JANAF Thermochemical Tables* Third Edition, M. W. Chase et al., American Chemical Society and the American Institute of Physics for the National Bureau of Standards, 1985.

TABLE G.6 IDEAL-GAS ENTHALPY AND ABSOLUTE ENTROPY FOR NO AND NO_2 AT 1 atm: SI UNITS

	Nitric oxide (NO) $\Delta\bar{h}_f^\circ$ = 90,291 kJ/kg·mol, M = 30.006			Nitrogen dioxide (NO_2) $\Delta\bar{h}_f^\circ$ = 33,100 kJ/kg·mol, M = 46.006	
Temp., K	$\bar{h}_T - \bar{h}^0$, kJ/kg·mol	$\bar{s}^0$, kJ/kg·mol·K	Temp., K	$\bar{h}_T - \bar{h}^0$, kJ/kg·mol	$\bar{s}^0$, kJ/kg·mol·K
0	−9 192	0.0	0	−10 186	0.0
100	−6 073	177.031	100	−6 861	202.563
200	−2 951	198.747	200	−3 495	225.852
298.15	0	210.758	298.15	0	240.034
300	55	210.943	300	68	240.262
400	3 040	219.529	400	3 927	251.342
500	6 059	226.263	500	8 099	260.638
600	9 144	231.886	600	12 555	268.755
700	12 307	236.761	700	17 250	275.988
800	15 548	241.087	800	22 138	282.512
900	18 858	244.985	900	27 179	288.449
1000	22 229	248.536	1000	32 344	293.889
1100	25 653	251.799	1100	37 605	298.903
1200	29 120	254.816	1200	42 946	303.550
1300	32 626	257.621	1300	48 351	307.876
1400	36 164	260.243	1400	53 808	311.920
1500	39 729	262.703	1500	59 309	315.715
1600	43 319	265.019	1600	64 846	319.288
1700	46 929	267.208	1700	70 414	322.663
1800	50 557	269.282	1800	76 007	325.861
1900	54 201	271.252	1900	81 624	328.897
2000	57 859	273.128	2000	87 259	331.788
2100	61 530	274.919	2100	92 911	334.545
2200	65 212	276.632	2200	98 577	337.181
2300	68 904	278.273	2300	104 257	339.706
2400	72 606	279.849	2400	109 947	342.128
2500	76 316	281.363	2500	115 648	344.455
2600	80 034	282.822	2600	121 357	346.694
2700	83 759	284.227	2700	127 075	348.852
2800	87 491	285.585	2800	132 799	350.934
2900	91 229	286.896	2900	138 530	352.945
3000	94 973	288.165	3000	144 267	354.889
3200	102 477	290.587	3200	155 756	358.597
3400	110 000	292.867	3400	167 262	362.084
3600	117 541	295.022	3600	178 783	365.377
3800	125 098	297.065	3800	190 316	368.495
4000	132 671	299.008	4000	201 859	371.455
4200	140 257	300.858	4200	213 412	374.274
4400	147 857	302.626	4400	224 973	376.963
4600	155 469	304.318	4600	236 540	379.534
4800	163 094	305.940	4800	248 114	381.996
5000	170 730	307.499	5000	259 692	384.360
5200	178 377	308.998	5200	271 276	386.631
5400	186 034	310.443	5400	282 863	388.818
5600	193 703	311.838	5600	294 455	390.926
5800	201 381	313.185	5800	306 049	392.960
6000	209 070	314.488	6000	317 647	394.926

Adapted from *JANAF Thermochemical Tables* Third Edition, M. W. Chase et al., American Chemical Society and the American Institute of Physics for the National Bureau of Standards, 1985.

TABLE G.7 IDEAL-GAS ENTHALPY AND ABSOLUTE ENTROPY FOR O_2 AND O AT 1 atm: SI UNITS

	Oxygen, diatomic (O_2) $\Delta\bar{h}_f^\circ = 0$ kJ/kg·mol, $M = 31.999$			Oxygen, monatomic (O) $\Delta\bar{h}_f^\circ = 249{,}170$ kJ/kg·mol, $M = 15.999$	
Temp., K	$\bar{h}_T - \bar{h}^0$, kJ/kg·mol	$\bar{s}^0$, kJ/kg·mol·K	Temp., K	$\bar{h}_T - \bar{h}^0$, kJ/kg·mol	$\bar{s}^0$, kJ/kg·mol·K
0	−8 683	0.0	0	−6 725	0.0
100	−5 779	173.307	100	−4 518	135.947
200	−2 868	193.485	200	−2 186	152.153
298.15	0	205.147	298.15	0	161.058
300	54	205.329	300	41	161.194
400	3 025	213.871	400	2 207	167.430
500	6 084	220.693	500	4 343	172.197
600	9 244	226.451	600	6 462	176.060
700	12 499	231.466	700	8 570	179.310
800	15 835	235.921	800	10 671	182.116
900	19 241	239.931	900	12 767	184.585
1000	22 703	243.578	1000	14 860	186.790
1100	26 212	246.922	1100	16 950	188.782
1200	29 761	250.010	1200	19 039	190.599
1300	33 344	252.878	1300	21 126	192.270
1400	36 957	255.556	1400	23 212	193.816
1500	40 599	258.068	1500	25 296	195.254
1600	44 266	260.434	1600	27 381	196.599
1700	47 958	262.672	1700	29 464	197.862
1800	51 673	264.796	1800	31 547	199.053
1900	55 413	266.818	1900	33 630	200.179
2000	59 175	268.748	2000	35 713	201.247
2100	62 691	270.595	2100	37 796	202.263
2200	66 769	272.366	2200	39 878	203.232
2300	70 600	274.069	2300	41 962	204.158
2400	74 453	275.709	2400	44 045	205.045
2500	78 328	277.290	2500	46 130	205.896
2600	82 224	278.819	2600	48 216	206.714
2700	86 141	280.297	2700	50 303	207.502
2800	90 079	281.729	2800	52 391	208.261
2900	94 036	283.118	2900	54 481	208.995
3000	98 013	284.466	3000	56 574	209.704
3200	106 023	287.050	3200	60 767	211.057
3400	114 102	289.499	3400	64 971	212.332
3600	122 245	291.826	3600	69 190	213.537
3800	130 447	294.044	3800	73 424	214.682
4000	138 705	296.162	4000	77 675	215.772
4200	147 015	298.189	4200	81 945	216.814
4400	155 374	300.133	4400	86 234	217.812
4600	163 783	302.002	4600	90 543	218.769
4800	172 240	303.801	4800	94 872	219.690
5000	180 749	305.538	5000	99 222	220.578
5200	189 311	307.217	5200	103 592	221.435
5400	197 933	308.844	5400	107 982	222.264
5600	206 618	310.424	5600	112 391	223.065
5800	215 375	311.960	5800	116 818	223.842
6000	224 210	313.457	6000	121 264	224.596

Adapted from *JANAF Thermochemical Tables* Third Edition, M. W. Chase et al., American Chemical Society and the American Institute of Physics for the National Bureau of Standards, 1985.

TABLE G.8 IDEAL-GAS ENTHALPY AND ABSOLUTE ENTROPY FOR H_2O AND OH AT 1 atm: SI UNITS

	Water (H_2O)			Hydroxyl (OH)	
	$\Delta\bar{h}_f^\circ = -241{,}826$ kJ/kg·mol, $M = 18.015$			$\Delta\bar{h}_f^\circ = 38{,}987$ kJ/kg·mol, $M = 17.007$	
Temp., K	$\bar{h}_T - \bar{h}^0$, kJ/kg·mol	$\bar{s}^0$, kJ/kg·mol·K	Temp., K	$\bar{h}_T - \bar{h}^0$, kJ/kg·mol	$\bar{s}^0$, kJ/kg·mol·K
0	−9 904	0.0	0	−9 172	0.0
100	−6 615	152.388	100	−6 139	149.590
200	−3 282	175.485	200	−2 976	171.592
298.15	0	188.834	298.15	0	183.708
300	62	189.042	300	55	183.894
400	3 452	198.788	400	3 035	192.466
500	6 925	206.534	500	5 992	199.066
600	10 501	213.052	600	8 943	204.447
700	14 192	218.739	700	11 902	209.007
800	18 002	223.825	800	14 880	212.983
900	21 938	228.459	900	17 888	216.526
1000	26 000	232.738	1000	20 935	219.736
1100	30 191	236.731	1100	24 024	222.680
1200	34 506	240.485	1200	27 160	225.408
1300	38 942	244.035	1300	30 342	227.955
1400	43 493	247.407	1400	33 569	230.346
1500	48 151	250.620	1500	36 839	232.602
1600	52 908	253.690	1600	40 151	234.740
1700	57 758	256.630	1700	43 502	236.771
1800	62 693	259.451	1800	46 889	238.707
1900	67 706	262.161	1900	50 310	240.557
2000	72 790	264.769	2000	53 762	242.327
2100	77 941	267.282	2100	57 243	244.026
2200	83 153	269.706	2200	60 752	245.658
2300	88 421	272.048	2300	64 285	247.228
2400	93 741	274.312	2400	67 841	248.741
2500	99 108	276.503	2500	71 419	250.202
2600	104 520	278.625	2600	75 017	251.613
2700	109 973	280.683	2700	78 633	252.978
2800	115 464	282.680	2800	82 267	254.300
2900	120 990	284.619	2900	85 918	255.581
3000	126 549	286.504	3000	89 584	256.824
3200	137 757	290.120	3200	96 960	259.203
3400	149 073	293.550	3400	104 387	261.455
3600	160 485	296.812	3600	111 863	263.591
3800	171 980	299.919	3800	119 381	265.624
4000	183 552	302.887	4000	126 939	267.562
4200	195 191	305.726	4200	134 534	269.415
4400	206 892	308.448	4400	142 164	271.189
4600	218 650	311.061	4600	149 827	272.893
4800	230 458	313.574	4800	157 521	274.530
5000	242 313	315.993	5000	165 246	276.107
5200	254 215	318.327	5200	173 001	277.627
5400	266 164	320.582	5400	180 784	279.096
5600	278 161	322.764	5600	188 597	280.517
5800	290 204	324.877	5800	196 438	281.892
6000	302 295	326.926	6000	204 308	283.226

Adapted from *JANAF Thermochemical Tables* Third Edition, M. W. Chase et al., American Chemical Society and the American Institute of Physics for the National Bureau of Standards, 1985.

TABLE G.9 LOGARITHMS TO THE BASE e OF THE EQUILIBRIUM CONSTANTS, ln K_p

For the reaction $\nu_A A + \nu_B B \rightleftharpoons \nu_C C + \nu_D D$, the equilibrium constant K_p is defined as: $K_p \equiv \dfrac{p_C^{\nu_C} p_D^{\nu_D}}{p_A^{\nu_A} p_B^{\nu_B}}$

Temp., K	ln K_p						
	$H_2 \rightleftharpoons 2H$	$O_2 \rightleftharpoons 2O$	$N_2 \rightleftharpoons 2N$	$H_2O \rightleftharpoons H_2 + \frac{1}{2}O_2$	$H_2O \rightleftharpoons \frac{1}{2}H_2 + OH$	$CO_2 \rightleftharpoons CO + \frac{1}{2}O_2$	$\frac{1}{2}N_2 + \frac{1}{2}O_2 \rightleftharpoons NO$
298	−164.005	−186.975	−367.480	−92.208	−106.208	−103.762	−35.052
500	−92.827	−105.630	−213.372	−52.691	−60.281	−57.616	−20.295
1000	−39.803	−45.150	−99.127	−23.163	−26.034	−23.529	−9.388
1200	−30.874	−35.005	−80.011	−18.182	−20.283	−17.871	−7.569
1400	−24.463	−27.742	−66.329	−14.609	−16.099	−13.842	−6.270
1600	−19.637	−22.285	−56.055	−11.921	−13.066	−10.830	−5.294
1800	−15.866	−18.030	−48.051	−9.826	−10.657	−8.497	−4.536
2000	−12.840	−14.622	−41.645	−8.145	−8.728	−6.635	−3.931
2200	−10.353	−11.827	−36.391	−6.768	−7.148	−5.120	−3.433
2400	−8.276	−9.497	−32.011	−5.619	−5.832	−3.860	−3.019
2600	−6.517	−7.521	−28.304	−4.648	−4.719	−2.801	−2.671
2800	−5.002	−5.826	−25.117	−3.812	−3.763	−1.894	−2.372
3000	−3.685	−4.357	−22.359	−3.086	−2.937	−1.111	−2.114
3200	−2.534	−3.072	−19.937	−2.451	−2.212	−0.429	−1.888
3400	−1.516	−1.935	−17.800	−1.891	−1.576	0.169	−1.690
3600	−0.609	−0.926	−15.898	−1.392	−1.088	0.701	−1.513
3800	0.202	−0.019	−14.199	−0.945	−0.501	1.176	−1.356
4000	0.934	0.796	−12.660	−0.542	−0.044	1.599	−1.216
4500	2.486	2.513	−9.414	0.312	0.920	2.490	−0.921
5000	3.725	3.895	−6.807	0.996	1.689	3.197	−0.686
5500	4.743	5.023	−4.666	1.560	2.318	3.771	−0.497
6000	5.590	5.963	−2.865	2.032	2.843	4.245	−0.341

SOURCE: Gordon J. Van Wylen and Richard E. Sonntag, *Fundamentals of Classical Thermodynamics*, 2d ed., SI Version 2d ed., John Wiley & Sons, Inc., 1976, p.697, table A.12. Based on thermodynamic data given in the JANAF Thermochemical Tables, Thermal Research Laboratory, The Dow Chemical Company, Midland, MI.

Miscellaneous Properties

TABLE H.1 CRITICAL CONSTANTS

Substance	Formula	Molecular weight	Temp., K	Pressure, MPa	Volume, $m^3/kg \cdot mol$
Ammonia	NH_3	17.03	405.5	11.28	0.0724
Argon	Ar	39.948	151	4.86	0.0749
Benzene	C_6H_6	78.115	562	4.92	0.2603
Bromine	Br_2	159.808	584	10.34	0.1355
n-Butane	C_4H_{10}	58.124	425.2	3.80	0.2547
Carbon dioxide	CO_2	44.01	304.2	7.39	0.0943
Carbon monoxide	CO	28.01	133	3.50	0.0930
Carbon tetrachloride	CCl_4	153.82	556.4	4.56	0.2759
Chlorine	Cl_2	70.906	417	7.71	0.1242
Chloroform	$CHCl_3$	119.38	536.6	5.47	0.2403
Deuterium (normal)	D_2	4.00	38.4	1.66	
Dichlorodifluoromethane (R-12)	CCl_2F_2	120.91	384.9	4.13	0.2167
Dichlorofluoromethane (R-22)	$CHCl_2F$	102.92	451.7	5.17	0.1973
Ethane	C_2H_6	30.070	305.5	4.88	0.1480
Ethyl alcohol	C_2H_5OH	46.07	516	6.38	0.1673
Ethylene	C_2H_4	28.054	282.4	5.12	0.1242
Helium	He	4.003	5.3	0.23	0.0578
Helium^3	He	3.00	3.3	0.12	
n-Hexane	C_6H_{14}	86.178	507.9	3.03	0.3677
Hydrogen (normal)	H_2	2.016	33.3	1.30	0.0649
Krypton	Kr	83.80	209.4	5.50	0.0924
Methane	CH_4	16.043	191.1	4.64	0.0993
Methyl alcohol	CH_3OH	32.042	513.2	5.95	0.1180
Methyl chloride	CH_3Cl	50.488	416.3	6.68	0.1430
Neon	Ne	20.183	44.5	2.73	0.0417
Nitrogen	N_2	28.013	126.2	3.39	0.0899
Nitrous oxide	N_2O	44.013	309.7	7.27	0.0961
Oxygen	O_2	31.999	154.8	5.08	0.0780
Propane	C_3H_8	44.097	370	4.26	0.1998
Propene	C_3H_6	42.081	365	4.62	0.1810
Propyne	C_3H_4	40.065	401	5.35	
Sulfur dioxide	SO_2	64.063	430.7	7.88	0.1217
Trichlorofluoromethane	CCl_3F	137.37	471.2	4.38	0.2478
Water	H_2O	18.015	647.1	22.06	0.0560
Xenon	Xe	131.30	289.8	5.88	0.1186

Pseudo-critical properties for air: Air is a mixture of gases and therefore does not have true critical properties. However, pseudo-critical properties can be defined using Kay's rule for mixtures. Values for the pseudo-critical temperature and pressure obtained in this manner for air ($M = 28.97$) are:

$$Tc = 133\,\text{K}$$

$$Pc = 3.76\,\text{MPa}$$

SOURCE: Reprinted with permission from K.A. Robe and R.E. Lynn, Jr., *Chemical Review*, 52 (1953): 117–236. Copyright 1982, American Chemical Society.

TABLE H.2 CONSTANTS FOR BEATTIE-BRIDGEMAN AND THE BENEDICT-WEBB-RUBIN EQUATIONS OF STATE

The Beattie-Bridgeman equation of state is

$$P = \frac{\bar{R}T}{\bar{v}^2}\left(1 - \frac{c}{\bar{v}T^3}\right)(\bar{v} + B) - \frac{A}{\bar{v}^2}$$

where

$$A = A_0\left(1 - \frac{a}{\bar{v}}\right) \qquad B = B_0\left(1 - \frac{b}{\bar{v}}\right)$$

When P is in kilopascals, $\bar{v}$ is in m³/kg·mol, T is in degrees kelvin, and $\bar{R} = 8.314$ kPa·m³/kg·mol·K, then the constants in the Beattie-Bridgeman equation are as given in the table.

Gas	A_0	a	B_0	b	$c \times 10^{-4}$
Air	131.8441	0.01931	0.04611	−0.001101	4.34
Argon	130.7802	0.02328	0.03931	0.0	5.99
Carbon dioxide	507.2836	0.07132	0.10476	0.07235	66.00
Helium	2.1886	0.05984	0.01400	0.0	0.0040
Hydrogen	20.0117	−0.00506	0.02096	−0.04359	0.0504
Nitrogen	136.2315	0.02617	0.05046	−0.00691	4.20
Oxygen	151.0857	0.02562	0.04624	0.004208	4.80

The Benedict-Webb-Rubin equation of state is

$$P = \frac{\bar{R}T}{\bar{v}} + \left(B_0\bar{R}T - A_0 - \frac{C_0}{T^2}\right)\frac{1}{\bar{v}^2} + \frac{b\bar{R}T - a}{\bar{v}^3} + \frac{a\alpha}{\bar{v}^6} + \frac{c}{\bar{v}^3T^2}\left(1 + \frac{\gamma}{\bar{v}^2}\right)e^{-\gamma/\bar{v}^2}$$

When P is in kilopascals, $\bar{v}$ is in m³/kg·mol, T is in degrees kelvin, and $\bar{R} = 8.314$ kPa·m³/kg·mol·K, then the constants in the Benedict-Webb-Rubin equation are as given in the table.

Gas	a	A_0	b	B_0	c	C_0	α	γ
n-Butane, C_4H_{10}	190.68	1021.6	0.039998	0.12436	3.205×10^7	1.006×10^8	1.101×10^{-3}	0.0340
Carbon dioxide, CO_2	13.86	277.30	0.007210	0.04991	1.511×10^6	1.404×10^7	8.470×10^{-5}	0.00539
Carbon monoxide, CO	3.71	135.87	0.002632	0.05454	1.054×10^5	8.673×10^5	1.350×10^{-4}	0.0060
Methane, CH_4	5.00	187.91	0.003380	0.04260	2.578×10^5	2.286×10^6	1.244×10^{-4}	0.0060
Nitrogen, N_2	2.54	106.73	0.002328	0.04074	7.379×10^4	8.164×10^5	1.272×10^{-4}	0.0053

SOURCE: Adapted from: Gordon J. Van Wylen and Richard E. Sonntag, *Fundamentals of Classical Thermodynamics,* SI Version, 2d ed., Revised Printing, © 1976, 1978 by John Wiley & Sons, Inc., Table 3-3 p. 47, and Kenneth Wark, *Thermodynamics,* 4th ed., McGraw-Hill Book Company, 1983, table A-21M, p. 815.

TABLE H.3 SPECIFIC HEATS OF COMMON LIQUIDS AND SOLIDS

Liquids

Substance	State	c_p, kJ/kg·K
Ammonia	sat., −20°C	4.52
	sat., 50°C	5.10
Benzene	1 atm, 15°C	1.80
	1 atm, 65°C	1.92
Bismuth	1 atm, 425°C	0.144
	1 atm, 760°C	0.164
Ethyl alcohol	1 atm, 25°C	2.43
Glycerin	1 atm, 10°C	2.32
	1 atm, 50°C	2.58
Mercury	1 atm, 10°C	0.138
	1 atm, 315°C	0.134
Propane	1 atm, 0°C	2.41
Refrigerant-12	sat., −40°C	0.883
	sat., −20°C	0.908
	sat., 50°C	1.02
Sodium	1 atm, 95°C	1.38
	1 atm, 540°C	1.26
Water	1 atm, 0°C	4.213
	1 atm, 25°C	4.177
	1 atm, 50°C	4.178
	1 atm, 100°C	4.213

Solids at 1 atm

Substance	T, °C	c_p, kJ/kg·K
Aluminum	−250	0.0163
	−200	0.318
	−100	0.699
	0	0.870
	100	0.941
	300	1.04
Carbon (diamond)	25	0.519
Carbon (graphite)	25	0.711
Chromium	25	0.448
Copper	−223	0.0967
	−173	0.252
	−100	0.328
	−50	0.361
	0	0.381
	27	0.385
	100	0.393
	200	0.403
Gold	25	0.129
Ice	−200	0.678
	−140	1.096
	−60	1.640
	−11	2.033
	−2.2	1.682
Iron	20	0.448
Lead	−270	0.0033
	−259	0.0305
	−100	0.118
	0	0.124
	100	0.134
	300	0.149
Nickel	25	0.444
Silver	20	0.233
	500	0.243
Sodium	25	1.226
Tungsten	25	0.134
Zinc	25	0.385

SOURCE: Adapted from: Kenneth Wark, *Thermodynamics*, 4th ed., 1983, McGraw-Hill Book Company, table A-19M, p. 813.

TABLE H.4 TRIPLE-POINT PROPERTIES FOR COMMON SUBSTANCES

Substance	*T*, K	*P*, kPa
Ammonia (NH_3)	195.4	6.18
Carbon dioxide (CO_2)	216.6	516.6
Helium 4 (λ point)	2.17	5.07
Hydrogen (H_2)	13.84	7.09
Nitrogen (N_2)	63.18	12.56
Oxygen (O_2)	54.36	0.152
Water (H_2O)	273.16	0.6113

Dimensions and Units

TABLE A.1E ENGLISH UNIT PREFIXES

Multiples	Prefix	Symbol
10^{18}	Exa	E
10^{15}	Peta	P
10^{12}	Tera	T
10^{9}	Giga	G
10^{6}	Mega	M
10^{3}	Kilo	k
10^{2}	Hecto	h
10	Deka	da
10^{-1}	Deci	d
10^{-2}	Centi	c
10^{-3}	Milli	m
10^{-6}	Micro	μ
10^{-9}	Nano	n
10^{-12}	Pico	p
10^{-15}	Femto	f
10^{-18}	Atto	a

TABLE A.2E ENGLISH PRIMARY UNITS

Dimension	Unit	Symbol
Length	Foot	ft
Mass	Pound mass	lb_m
Force	Pound force	lb_f
Time	Second	s
Electric current	Ampere	A
Thermodynamic temperature	Rankine	°R

TABLE A.3E ENGLISH SECONDARY UNITS

Dimension	Unit	Symbol	Definition
Energy	Foot·pound force	$ft \cdot lb_f$	$1\ ft \cdot lb_f = 1\ ft \times 1\ lb_f$
	British thermal unit	Btu	$1\ Btu = 778.169\ ft \cdot lb_f$
Power	Horsepower	hp	$1\ hp = 550\ ft \cdot lb_f/s$
Pressure	Pound per square inch	psi or lb_f/in^2	$1\ lb_f/in^2 = 1\ lb_f/1\ in^2$

TABLE A.4E UNITS IN THE ENGLISH SYSTEM

Dimension	Unit
Energy, work, heat	Btu
Force	lb_f
Length	ft
Mass	lb_m
Power	hp
Time	s

TABLE A.5E COMMONLY USED CONVERSION FACTORS

Dimension	Conversion factor	Conversion factor
Acceleration	1 m/s^2 = 3.2808 ft/s^2	1 ft/s^2 = 0.3048 m/s^2
Area	1 m^2 = 1550 in^2	1 in^2 = 6.4516 × 10^{-4} m^2
	1 m^2 = 10.764 ft^2	1 ft^2 = 9.2903 × 10^{-2} m^2
Density	1 g/cm^3 = 62.428 lb$_m$/ft^3	1 lb$_m$/ft^3 = 0.0160185 g/m^3
	1 kg/m^3 = 0.062428 lb$_m$/ft^3	1 lb$_m$/ft^3 = 16.0185 kg/m^3
Energy	1 J = 0.73756 ft·lb$_f$	1 ft·lb$_f$ = 1.35582 J
	1 J = 9.4782 × 10^{-4} Btu	1 Btu = 1.055056 kJ
		1 Btu = 778.169 ft·lb$_f$
Force	1 N = 0.22481 lb$_f$	1 lb$_f$ = 4.4482 N
Length	1 m = 39.370 in	1 in = 0.0254 m
	1 m = 3.2808 ft	1 ft = 0.3048 m
	1 m = 6.2137 × 10^{-4} mi	1 mi = 1.6093 km
Mass	1 g = 2.2046 × 10^{-3} lb$_m$	1 lb$_m$ = 453.59 g
	1 kg = 2.2046 lb$_m$	1 lb$_m$ = 0.45359 kg
Mass flow rate	1 kg/s = 2.2046 lb$_m$/s	1 lb$_m$/s = 0.45359 kg/s
Power	1 W = 0.73756 ft·lb$_f$/s	1 ft·lb$_f$/s = 1.3558 W
	1 W = 3.4121 Btu/h	1 Btu/h = 0.29307 W
	1 W = 1.3410 × 10^{-3} hp	1 hp = 745.7 W
		1 hp = 550 ft·lb$_f$/s
Pressure	1 Pa = 1.4504 × 10^{-4} lb$_f$/in^2	1 lb$_f$/in^2 = 6.8948 × 10^3 Pa
	1 Pa = 0.020886 lb$_f$/ft^2	1 lb$_f$/ft^2 = 47.880 Pa
	1 Pa = 4.015 × 10^{-3} in H_2O (4°C)	1 in H_2O = 249.08 Pa (39.2°F)
	1 Pa = 2.953 × 10^{-4} in Hg (0°C)	1 in Hg = 3.3864 kPa (32°F)
	1 Pa = 9.869 × 10^{-6} atm	1 atm = 101.325 kPa
	1 bar = 10^5 Pa	1 atm = 1.01325 bar
Specific energy	1 kJ/kg = 0.42992 Btu/lb$_m$	1 Btu/lb$_m$ = 2.326 kJ/kg
	1 kJ/kg = 334.56 ft·lb$_f$/lb$_m$	1 ft·lb$_f$/lb$_m$ = 2.989 J/kg
Specific heat	1 kJ/kg·K = 0.23885 Btu/lb$_m$·°R	1 Btu/lb$_m$·°R = 4.1868 kJ/kg·K
Temperature	K = (5/9)(°R)	°R = 1.8(K)
	K = °C + 273.15	°R = °F + 459.67
	°C = (5/9)(°F − 32)	°F = 1.8(°C) + 32
Temperature difference	1 K = 1°C	1°R = 1°F
Velocity	1 m/s = 3.2808 ft/s	1 ft/s = 0.3048 m/s
	1 m/s = 2.237 mi/h	1 mi/h = 0.44704 m/s
Volume	1 m^3 = 6.1024 × 10^4 in^3	1 in^3 = 1.6387 × 10^{-5} m^3
	1 m^3 = 35.315 ft^3	1 ft^3 = 2.8317 × 10^{-2} m^3
	1 m^3 = 264.17 gal (U.S. liquid)	1 gal = 3.7854 × 10^{-3} m^3
		1 ft^3 = 7.48 gal
Volume flow rate	1 m^3/s = 35.315 ft^3/s	1 ft^3/s = 2.8317 × 10^{-2} m^3/s
	1 m^3/s = 2.1189 × 10^3 ft^3/min	1 ft^3/min = 4.7195 × 10^{-4} m^3/s
	1 m^3/s = 1.585 × 10^4 gal/min	1 gal/min = 6.309 × 10^{-5} m^3/s

Note: To convert a dimension from English units to SI units, multiply by the conversion factor in the last column or divide by the conversion factor in the middle column. Using the dimension of power as an example, we will convert 5 hp to W. From the table the applicable conversion factor is

$$1 \text{ hp} = 745.7 \text{ W}$$

Thus

$$(5 \text{ hp})(745.7 \text{ W/hp}) = 3729 \text{ W}$$

TABLE A.6E ENGLISH UNIT PHYSICAL CONSTANTS

Dimension	Conversion factor
Acceleration of gravity at sea level	$g = 32.174\ \mathrm{ft/s^2}$
Avogadro's number	$N_A = 2.7316 \times 10^{26}$ molecules/$\mathrm{lb_m \cdot mol}$
Boltzmann's constant	$k = 7.2699 \times 10^{-27}\ \mathrm{Btu/°R}$
Planck's constant	$h = 6.2804 \times 10^{-37}\ \mathrm{Btu \cdot s}$
Speed of light in a vacuum	$c = 9.836 \times 10^{8}\ \mathrm{ft/s}$
Universal gas consant	$\overline{R} = 1545\ \mathrm{ft \cdot lb_f/lb_m \cdot mol \cdot °R}$
	$= 1.986\ \mathrm{Btu/lb_m \cdot mol \cdot °R}$
	$= 10.73\ \mathrm{psia \cdot ft^3/lb_m \cdot mol \cdot °R}$

Properties of Water

TABLE B.1E SATURATED WATER—TEMPERATURE TABLE: ENGLISH UNITS

		Specific volume, ft^3/lb_m			Internal energy, Btu/lb_m			Enthalpy, Btu/lb_m			Entropy, $Btu/lb_m \cdot {}^{\circ}R$		
Temp., °F, T	Press., psia, P	Sat. liquid, v_f	Evap., v_{fg}	Sat. vapor, v_g	Sat. liquid, u_f	Evap., u_{fg}	Sat. vapor, u_g	Sat. liquid, h_f	Evap., h_{fg}	Sat. vapor, h_g	Sat. liquid, s_f	Evap., s_{fg}	Sat. vapor, s_g
32.018	0.088725	0.016022	3299.5	3299.5	0.00	1020.9	1020.9	0.00	1075.0	1075.0	0.00000	2.1864	2.1864
40	0.12174	0.016019	2443.4	2443.4	8.03	1015.5	1023.5	8.03	1070.5	1078.5	0.01621	2.1424	2.1586
50	0.17813	0.016023	1703.1	1703.1	18.05	1008.7	1026.8	18.05	1064.9	1082.9	0.03606	2.0893	2.1254
60	0.25636	0.016034	1206.3	1206.3	28.05	1002.0	1030.1	28.05	1059.2	1087.3	0.05548	2.0383	2.0938
70	0.36328	0.016051	867.40	867.42	38.04	995.3	1033.3	38.04	1053.6	1091.7	0.07453	1.9892	2.0637
80	0.50732	0.016073	632.71	632.72	48.03	988.6	1036.6	48.03	1048.0	1096.0	0.09321	1.9419	2.0351
90	0.69878	0.016100	467.66	467.68	58.02	981.8	1039.9	58.03	1042.3	1100.3	0.11156	1.8962	2.0078
100	0.95008	0.016131	350.05	350.06	68.01	975.1	1043.1	68.02	1036.6	1104.6	0.12957	1.8522	1.9818
110	1.2760	0.016166	265.13	265.15	78.00	968.3	1046.3	78.01	1030.9	1108.9	0.14726	1.8097	1.9569
120	1.6940	0.016205	203.08	203.10	87.99	961.5	1049.5	88.00	1025.2	1113.2	0.16464	1.7685	1.9332
130	2.2245	0.016247	157.20	157.22	97.98	954.7	1052.7	97.99	1019.4	1117.4	0.18173	1.7288	1.9105
140	2.8909	0.016292	122.91	122.92	107.97	947.8	1055.8	107.98	1013.6	1121.6	0.19853	1.6903	1.8888
150	3.7203	0.016341	97.002	97.018	117.96	941.0	1058.9	117.97	1007.8	1125.7	0.21505	1.6530	1.8680
160	4.7434	0.016393	77.240	77.257	127.96	934.1	1062.0	127.97	1001.9	1129.8	0.23132	1.6168	1.8481
170	5.9947	0.016448	62.024	62.041	137.96	927.1	1065.1	137.98	995.9	1133.9	0.24733	1.5816	1.8290
180	7.5129	0.016507	50.203	50.220	147.98	920.1	1068.1	148.00	989.9	1137.9	0.26311	1.5475	1.8106
190	9.3411	0.016568	40.942	40.958	158.01	913.1	1071.1	158.04	983.8	1141.9	0.27867	1.5144	1.7930
200	11.527	0.016632	33.628	33.644	168.05	905.9	1074.0	168.08	977.7	1145.8	0.29401	1.4821	1.7761
210	14.122	0.016700	27.807	27.824	178.11	898.8	1076.9	178.15	971.4	1149.6	0.30914	1.4506	1.7598
212	14.696	0.016713	26.791	26.807	180.12	897.3	1077.5	180.16	970.2	1150.4	0.31214	1.4444	1.7566
220	17.185	0.016770	23.141	23.158	188.18	891.5	1079.7	188.23	965.1	1153.4	0.32408	1.4200	1.7441
230	20.775	0.016843	19.375	19.392	198.27	884.2	1082.5	198.34	958.7	1157.0	0.33882	1.3901	1.7289
240	24.962	0.016920	16.314	16.331	208.39	876.8	1085.2	208.47	952.2	1160.6	0.35338	1.3609	1.7143
250	29.815	0.017000	13.812	13.829	218.52	869.3	1087.8	218.62	945.5	1164.1	0.36776	1.3324	1.7001
260	35.413	0.017082	11.754	11.771	228.68	861.7	1090.4	228.79	938.8	1167.6	0.38198	1.3044	1.6864
270	41.838	0.017169	10.051	10.068	238.86	854.1	1092.9	239.00	931.9	1170.9	0.39603	1.2771	1.6732
280	49.176	0.017258	8.6347	8.6519	249.07	846.3	1095.4	249.23	924.9	1174.1	0.40992	1.2504	1.6603
290	57.519	0.017351	7.4502	7.4676	259.31	838.4	1097.7	259.49	917.7	1177.2	0.42367	1.2241	1.6478
300	66.966	0.017447	6.4549	6.4723	269.57	830.4	1099.9	269.78	910.4	1180.2	0.43727	1.1984	1.6356
310	77.619	0.017547	5.6143	5.6319	279.86	822.2	1102.1	280.11	902.9	1183.0	0.45073	1.1731	1.6238
320	89.585	0.017651	4.9013	4.9190	290.18	814.0	1104.2	290.48	895.2	1185.7	0.46406	1.1482	1.6123
330	102.98	0.017758	4.2938	4.3116	300.54	805.6	1106.1	300.88	887.4	1188.3	0.47726	1.1237	1.6010
340	117.91	0.017870	3.7741	3.7920	310.94	797.0	1108.0	311.33	879.4	1190.7	0.49035	1.0997	1.5900
350	134.51	0.017986	3.3276	3.3456	321.37	788.3	1109.7	321.82	871.1	1193.0	0.50332	1.0759	1.5792
360	152.90	0.018106	2.9425	2.9607	331.85	779.4	1111.3	332.36	862.7	1195.1	0.51618	1.0525	1.5687
370	173.21	0.018231	2.6093	2.6275	342.37	770.4	1112.8	342.95	854.0	1197.0	0.52894	1.0294	1.5583

TABLE B.1E SATURATED WATER—TEMPERATURE TABLE: ENGLISH UNITS (*Continued*)

		Specific volume, ft³/lbm			Internal energy, Btu/lbm			Enthalpy, Btu/lbm			Entropy, Btu/lbm·°R		
Temp., °F, T	Press., psia, P	Sat. liquid, v_f	Evap., v_{fg}	Sat. vapor, v_g	Sat. liquid, u_f	Evap., u_{fg}	Sat. vapor, u_g	Sat. liquid, h_f	Evap., h_{fg}	Sat. vapor, h_g	Sat. liquid, s_f	Evap., s_{fg}	Sat. vapor, s_g
380	195.58	0.018361	2.3198	2.3381	352.94	761.2	1114.1	353.60	845.1	1198.7	0.54161	1.0065	1.5481
390	220.15	0.018496	2.0674	2.0859	363.56	751.7	1115.3	364.31	836.0	1200.3	0.55419	0.9839	1.5381
400	247.06	0.018636	1.8467	1.8654	374.23	742.1	1116.4	375.08	826.6	1201.6	0.56668	0.9615	1.5282
410	276.47	0.018782	1.6531	1.6719	384.96	732.3	1117.3	385.92	816.9	1202.8	0.57911	0.9393	1.5184
420	308.52	0.018935	1.4828	1.5017	395.76	722.2	1118.0	396.84	806.9	1203.7	0.59146	0.9173	1.5087
430	343.38	0.019094	1.3324	1.3515	406.63	711.9	1118.6	407.84	796.6	1204.5	0.60375	0.8954	1.4992
440	381.21	0.019260	1.1993	1.2186	417.57	701.4	1119.0	418.93	786.0	1204.9	0.61600	0.8737	1.4897
450	422.17	0.019433	1.0812	1.1006	428.59	690.6	1119.2	430.11	775.1	1205.2	0.62819	0.8520	1.4802
460	466.43	0.019615	0.97605	0.99566	439.70	679.5	1119.2	441.39	763.8	1205.1	0.64035	0.8305	1.4708
470	514.18	0.019805	0.88225	0.90205	450.89	668.1	1119.0	452.78	752.1	1204.8	0.65248	0.8090	1.4614
480	565.59	0.020005	0.79834	0.81834	462.19	656.4	1118.6	464.29	740.0	1204.3	0.66459	0.7875	1.4521
490	620.85	0.020215	0.72310	0.74329	473.60	644.4	1118.0	475.92	727.5	1203.4	0.67669	0.7660	1.4427
500	680.15	0.020436	0.65542	0.67586	485.12	632.0	1117.1	487.70	714.5	1202.2	0.68879	0.7445	1.4333
510	743.68	0.020669	0.59445	0.61512	496.77	619.2	1116.0	499.62	701.0	1200.6	0.70090	0.7229	1.4238
520	811.65	0.020916	0.53937	0.56029	508.56	605.9	1114.5	511.70	687.0	1198.7	0.71303	0.7013	1.4143
530	884.26	0.021178	0.48951	0.51069	520.50	592.3	1112.8	523.97	672.4	1196.4	0.72520	0.6795	1.4047
540	961.74	0.021456	0.44423	0.46569	532.61	578.2	1110.8	536.43	657.3	1193.7	0.73741	0.6575	1.3949
550	1044.30	0.021753	0.40304	0.42479	544.89	563.5	1108.4	549.10	641.4	1190.5	0.74970	0.6353	1.3850
560	1132.17	0.022071	0.36546	0.38753	557.38	548.3	1105.7	562.00	624.9	1186.9	0.76206	0.6128	1.3749
570	1225.60	0.022413	0.33110	0.35351	570.09	532.4	1102.5	575.17	607.5	1182.7	0.77454	0.5900	1.3645
580	1324.84	0.022782	0.29958	0.32236	583.05	515.8	1098.8	588.63	589.3	1177.9	0.78715	0.5668	1.3539
590	1430.15	0.023183	0.27060	0.29378	596.29	498.4	1094.7	602.42	570.0	1172.4	0.79992	0.5430	1.3429
600	1541.82	0.023620	0.24385	0.26747	609.85	480.1	1089.9	616.59	549.7	1166.3	0.81289	0.5187	1.3316
610	1660.14	0.024102	0.21909	0.24319	623.78	460.7	1084.5	631.18	528.0	1159.2	0.82611	0.4936	1.3197
620	1785.43	0.024637	0.19607	0.22071	638.14	440.2	1078.3	646.28	505.0	1151.2	0.83964	0.4677	1.3073
630	1918.04	0.025239	0.17457	0.19981	653.02	418.2	1071.2	661.98	480.1	1142.1	0.85354	0.4406	1.2941
640	2058.33	0.025923	0.15436	0.18028	668.53	394.4	1062.9	678.41	453.1	1131.6	0.86794	0.4122	1.2801
650	2206.74	0.026717	0.13521	0.16193	684.84	368.5	1053.3	695.75	423.8	1119.5	0.88299	0.3818	1.2648
660	2363.72	0.027661	0.11688	0.14454	702.18	339.9	1042.0	714.28	390.9	1105.2	0.89890	0.3492	1.2481
670	2529.83	0.028822	0.09903	0.12785	720.95	307.4	1028.3	734.45	353.8	1088.2	0.91606	0.3131	1.2292
680	2705.73	0.030331	0.08118	0.11151	741.89	269.3	1011.2	757.08	310.0	1067.1	0.93516	0.2720	1.2072
690	2892.30	0.032492	0.062354	0.094846	766.63	221.8	988.3	784.02	255.1	1039.1	0.95774	0.2219	1.1796
700	3090.93	0.036470	0.039113	0.075580	801.01	150.07	951.08	821.87	172.44	994.31	0.98943	0.1487	1.1381
705.157	3198.8	0.049748	0.0	0.049748	867.28	0.0	867.28	896.72	0.0	896.72	1.05308	0.0	1.05308

Values generated from property formulation given in *NBS/NRC Steam Tables: Thermodynamic and Transport Properties and Computer Programs for Vapor and Liquid States of Water in SI Units*, Lester Haar, John S. Gallagher and George S. Kell. Hemisphere Publishing Corp., Washington, 1984.

TABLE B.2E SATURATED WATER—PRESSURE TABLE: ENGLISH UNITS

		Specific volume, ft³/lbₘ			Internal energy, Btu/lbₘ			Enthalpy, Btu/lbₘ			Entropy, Btu/lbₘ·°R		
Press., psia, P	Temp., F, T	Sat. liquid, v_f	Evap., v_{fg}	Sat. vapor, v_g	Sat. liquid, u_f	Evap., u_{fg}	Sat. vapor, u_g	Sat. liquid, h_f	Evap., h_{fg}	Sat. vapor, h_g	Sat. liquid, s_f	Evap., s_{fg}	Sat. vapor, s_g
0.08875	32.018	0.016022	3299.5	3299.5	0.00	1020.9	1020.9	0.00	1075.0	1075.0	0.00000	2.1864	2.1864
0.2	53.134	0.016026	1526.0	1526.0	21.18	1006.6	1027.8	21.19	1063.1	1084.3	0.04219	2.0731	2.1153
0.3	64.456	0.016041	1039.5	1039.5	32.50	999.0	1031.5	32.50	1056.7	1089.2	0.06401	2.0162	2.0802
0.4	72.840	0.016057	791.93	791.94	40.88	993.4	1034.3	40.88	1052.0	1092.9	0.07987	1.9756	2.0554
0.5	79.556	0.016072	641.39	641.41	47.59	988.9	1036.5	47.59	1048.2	1095.8	0.09240	1.9439	2.0363
0.6	85.188	0.016086	539.96	539.98	53.22	985.1	1038.3	53.22	1045.0	1098.2	0.10278	1.9180	2.0208
0.7	90.056	0.016100	466.86	466.88	58.08	981.8	1039.9	58.08	1042.3	1100.3	0.11167	1.8960	2.0076
0.8	94.353	0.016113	411.62	411.64	62.37	978.9	1041.3	62.38	1039.8	1102.2	0.11944	1.8769	1.9963
0.9	98.207	0.016125	368.36	368.38	66.22	976.3	1042.5	66.23	1037.6	1103.9	0.12637	1.8600	1.9863
1.0	101.707	0.016137	333.55	333.56	69.72	973.9	1043.6	69.72	1035.6	1105.4	0.13262	1.8448	1.9774
1.5	115.658	0.016187	227.70	227.72	83.66	964.5	1048.1	83.66	1027.7	1111.3	0.15714	1.7862	1.9434
2.0	126.048	0.016230	173.73	173.75	94.03	957.4	1051.4	94.04	1021.7	1115.7	0.17501	1.7443	1.9193
3.0	141.445	0.016299	118.71	118.73	109.41	946.9	1056.3	109.42	1012.8	1122.2	0.20093	1.6848	1.8857
4.0	152.945	0.016356	90.630	90.646	120.90	938.9	1059.8	120.91	1006.0	1126.9	0.21986	1.6422	1.8621
5.0	162.219	0.016405	73.521	73.538	130.18	932.5	1062.7	130.19	1000.6	1130.7	0.23489	1.6089	1.8438
6.0	170.039	0.016449	61.975	61.991	138.00	927.1	1065.1	138.02	995.9	1133.9	0.24739	1.5815	1.8289
7.0	176.829	0.016488	53.642	53.659	144.80	922.3	1067.1	144.82	991.8	1136.6	0.25813	1.5583	1.8164
8.0	182.848	0.016524	47.337	47.354	150.83	918.1	1068.9	150.86	988.2	1139.0	0.26756	1.5380	1.8056
9.0	188.267	0.016557	42.394	42.411	156.27	914.3	1070.6	156.29	984.9	1141.2	0.27599	1.5200	1.7960
10	193.204	0.016588	38.413	38.429	161.22	910.8	1072.0	161.25	981.9	1143.1	0.28360	1.5039	1.7875
14.696	212.000	0.016713	26.791	26.807	180.12	897.3	1077.5	180.17	970.2	1150.4	0.31214	1.4444	1.7566
15	213.036	0.016721	26.282	26.299	181.16	896.6	1077.7	181.21	969.5	1150.7	0.31369	1.4413	1.7549
20	227.969	0.016828	20.078	20.094	196.22	885.7	1081.9	196.28	960.0	1156.3	0.33584	1.3961	1.7319
25	240.085	0.016921	16.291	16.308	208.47	876.7	1085.2	208.55	952.1	1160.7	0.35350	1.3606	1.7141
30	250.353	0.017002	13.733	13.750	218.88	869.1	1087.9	218.98	945.3	1164.3	0.36827	1.3314	1.6996
35	259.308	0.017077	11.884	11.901	227.98	862.3	1090.2	228.09	939.2	1167.3	0.38100	1.3064	1.6874
40	267.275	0.017145	10.484	10.501	236.09	856.2	1092.3	236.21	933.8	1170.0	0.39221	1.2845	1.6767
45	274.471	0.017208	9.3861	9.4033	243.42	850.6	1094.0	243.57	928.8	1172.3	0.40226	1.2651	1.6674
50	281.047	0.017267	8.5007	8.5180	250.14	845.5	1095.6	250.30	924.1	1174.4	0.41137	1.2476	1.6590
55	287.111	0.017324	7.7713	7.7886	256.35	840.7	1097.0	256.52	919.8	1176.3	0.41971	1.2317	1.6514
60	292.747	0.017377	7.1596	7.1770	262.12	836.2	1098.3	262.31	915.7	1178.0	0.42742	1.2170	1.6444
65	298.016	0.017428	6.6390	6.6565	267.53	832.0	1099.5	267.74	911.8	1179.6	0.43458	1.2034	1.6380
70	302.970	0.017476	6.1904	6.2079	272.62	828.0	1100.6	272.85	908.2	1181.0	0.44128	1.1908	1.6321
75	307.648	0.017523	5.7996	5.8171	277.44	824.2	1101.6	277.68	904.7	1182.3	0.44758	1.1790	1.6266
80	312.083	0.017568	5.4561	5.4737	282.01	820.5	1102.5	282.27	901.3	1183.6	0.45352	1.1679	1.6214

TABLE B.2E SATURATED WATER—PRESSURE TABLE: ENGLISH UNITS *(Continued)*

Press., psia, P	Temp., F, T	Specific volume, ft³/lbm — Sat. liquid, v_f	Evap., v_{fg}	Sat. vapor, v_g	Internal energy, Btu/lbm — Sat. liquid, u_f	Evap., u_{fg}	Sat. vapor, u_g	Enthalpy, Btu/lbm — Sat. liquid, h_f	Evap., h_{fg}	Sat. vapor, h_g	Entropy, Btu/lbm·°R — Sat. liquid, s_f	Evap., s_{fg}	Sat. vapor, s_g
85	316.302	0.017612	5.1516	5.1693	286.36	817.0	1103.4	286.64	898.1	1184.7	0.45915	1.1574	1.6165
90	320.327	0.017654	4.8799	4.8975	290.52	813.7	1104.2	290.82	895.0	1185.8	0.46449	1.1474	1.6119
95	324.178	0.017695	4.6358	4.6535	294.51	810.5	1105.0	294.82	892.0	1186.8	0.46959	1.1379	1.6075
100	327.871	0.017735	4.4153	4.4330	298.34	807.4	1105.7	298.66	889.1	1187.7	0.47446	1.1289	1.6034
110	334.838	0.017812	4.0324	4.0502	305.57	801.4	1107.0	305.93	883.5	1189.5	0.48361	1.1120	1.5957
120	341.319	0.017885	3.7113	3.7292	312.31	795.9	1108.2	312.71	878.3	1191.0	0.49206	1.0965	1.5886
130	347.385	0.017955	3.4380	3.4559	318.64	790.6	1109.2	319.07	873.3	1192.4	0.49994	1.0821	1.5820
140	353.094	0.018022	3.2024	3.2204	324.61	785.6	1110.2	325.08	868.5	1193.6	0.50731	1.0686	1.5759
150	358.490	0.018088	2.9972	3.0152	330.26	780.8	1111.0	330.77	864.0	1194.7	0.51424	1.0560	1.5702
160	363.610	0.018151	2.8167	2.8349	335.64	776.2	1111.8	336.18	859.6	1195.8	0.52080	1.0441	1.5649
170	368.484	0.018212	2.6568	2.6750	340.77	771.8	1112.5	341.34	855.4	1196.7	0.52701	1.0328	1.5599
180	373.138	0.018271	2.5140	2.5323	345.68	767.5	1113.2	346.29	851.3	1197.5	0.53293	1.0222	1.5551
190	377.594	0.018329	2.3858	2.4041	350.39	763.4	1113.8	351.03	847.3	1198.3	0.53857	1.0120	1.5505
200	381.869	0.018386	2.2699	2.2883	354.92	759.4	1114.3	355.60	843.4	1199.0	0.54397	1.0022	1.5462
250	401.040	0.018651	1.8254	1.8441	375.34	741.1	1116.5	376.21	825.6	1201.8	0.56798	0.9592	1.5271
300	417.423	0.018895	1.5246	1.5435	392.97	724.9	1117.8	394.02	809.5	1203.5	0.58828	0.9229	1.5112
350	431.810	0.019123	1.3071	1.3262	408.60	710.1	1118.7	409.84	794.7	1204.6	0.60598	0.8915	1.4974
400	444.686	0.019340	1.1422	1.1616	422.72	696.4	1119.1	424.15	781.0	1205.1	0.62172	0.8635	1.4852
450	456.376	0.019548	1.0128	1.0323	435.66	683.6	1119.2	437.29	767.9	1205.2	0.63595	0.8383	1.4742
500	467.107	0.019749	0.90831	0.92806	447.64	671.5	1119.1	449.47	755.5	1205.0	0.64897	0.8151	1.4641
550	477.043	0.019945	0.82222	0.84213	458.84	659.9	1118.7	460.87	743.6	1204.5	0.66101	0.7938	1.4548
600	486.310	0.020136	0.74992	0.77005	469.38	648.9	1118.2	471.61	732.1	1203.7	0.67222	0.7740	1.4462
700	503.199	0.020509	0.63523	0.65574	488.84	627.9	1116.8	491.49	710.2	1201.7	0.69266	0.7376	1.4303
800	518.333	0.020874	0.54818	0.56905	506.59	608.2	1114.8	509.68	689.4	1199.0	0.71100	0.7049	1.4159
900	532.084	0.021234	0.47971	0.50094	523.01	589.4	1112.4	526.55	669.4	1195.9	0.72774	0.6749	1.4026
1000	544.713	0.021594	0.42434	0.44593	538.37	571.3	1109.7	542.37	649.9	1192.3	0.74319	0.6470	1.3902
1200	567.320	0.022319	0.34001	0.36233	566.66	536.7	1103.4	571.61	612.2	1183.8	0.77118	0.5961	1.3673
1400	587.196	0.023067	0.27848	0.30155	592.54	503.4	1095.9	598.52	575.5	1174.0	0.79632	0.5498	1.3461
1600	604.989	0.023855	0.23127	0.25512	616.75	470.6	1087.3	623.81	539.0	1162.8	0.81945	0.5063	1.3257
1800	621.127	0.024702	0.19360	0.21830	639.79	437.7	1077.5	648.02	502.2	1150.2	0.84118	0.4647	1.3059
2000	635.910	0.025632	0.16248	0.18811	662.11	404.3	1066.4	671.59	464.5	1136.1	0.86200	0.4239	1.2859
2200	649.558	0.026679	0.13604	0.16272	684.10	369.7	1053.8	694.96	425.1	1120.1	0.88230	0.3832	1.2655
2400	662.233	0.027898	0.11286	0.14076	706.23	333.0	1039.2	718.62	383.1	1101.7	0.90261	0.3415	1.2441
2600	674.059	0.029382	0.09182	0.12120	729.13	292.8	1021.9	743.27	336.9	1080.2	0.92352	0.2972	1.2207
2800	685.128	0.031324	0.07176	0.10308	753.93	246.6	1000.5	770.16	283.7	1053.9	0.94614	0.2479	1.1940
3000	695.504	0.034261	0.05067	0.084929	783.40	187.6	971.0	802.42	215.8	1018.2	0.97314	0.1868	1.1599
3198.8	705.156	0.049748	0.0	0.049748	867.28	0.0	867.28	896.72	0.0	896.72	1.05308	0.0	1.05308

H_2O

TABLE B.3E SUPERHEATED WATER: ENGLISH UNITS

H_2O

Temp., °F	v, ft³/lbm	u, Btu/lbm	h, Btu/lbm	s, Btu/lbm·°R	Temp., °F	v, ft³/lbm	u, Btu/lbm	h, Btu/lbm	s, Btu/lbm·°R
	P = 1.0 psia (101.707°F)					P = 5.0 psia (162.219°F)			
Sat.	333.56	1043.6	1105.4	1.9774	Sat.	73.538	1062.7	1130.7	1.8438
150	362.58	1060.2	1127.3	2.0148					
200	392.52	1077.3	1149.9	2.0505	200	78.154	1076.0	1148.4	1.8713
300	452.25	1111.7	1195.4	2.1148	300	90.240	1111.1	1194.6	1.9365
400	511.91	1146.8	1241.5	2.1718	400	102.24	1146.4	1241.0	1.9939
500	571.53	1182.6	1288.4	2.2233	500	114.20	1182.3	1288.0	2.0456
600	631.13	1219.2	1336.0	2.2705	600	126.15	1219.0	1335.7	2.0929
700	690.71	1256.6	1384.4	2.3142	700	138.08	1256.4	1384.2	2.1366
800	750.30	1294.8	1433.7	2.3549	800	150.01	1294.7	1433.5	2.1774
900	809.88	1334.0	1483.8	2.3933	900	161.94	1333.9	1483.7	2.2158
1000	869.45	1374.0	1534.9	2.4295	1000	173.86	1373.9	1534.8	2.2520
1200	988.60	1456.9	1639.8	2.4968	1200	197.70	1456.8	1639.7	2.3194
1400	1107.7	1543.5	1748.5	2.5586	1400	221.54	1543.4	1748.4	2.3812
1600	1226.9	1633.9	1860.9	2.6160	1600	245.37	1633.8	1860.8	2.4386
1800	1346.0	1727.8	1976.9	2.6698	1800	269.20	1727.8	1976.9	2.4923
2000	1465.2	1825.3	2096.4	2.7204	2000	293.03	1825.3	2096.4	2.5430
2200	1584.3	1926.0	2219.2	2.7684	2200	316.86	1926.0	2219.1	2.5910
2400	1703.4	2029.7	2344.9	2.8140	2400	340.69	2029.6	2344.9	2.6365
2600	1822.6	2136.1	2473.4	2.8574	2600	364.52	2136.1	2473.3	2.6800
	P = 10.0 psia (193.204°F)					P = 14.696 psia (212.000°F)			
Sat.	38.429	1072.0	1143.1	1.7875	Sat.	26.807	1077.5	1150.4	1.7566
200	38.853	1074.5	1146.4	1.7925					
300	44.987	1110.3	1193.5	1.8590	300	30.525	1109.5	1192.5	1.8156
400	51.031	1145.9	1240.3	1.9169	400	34.667	1145.4	1239.7	1.8739
500	57.038	1182.0	1287.5	1.9689	500	38.771	1181.6	1287.1	1.9261
600	63.027	1218.7	1335.4	2.0163	600	42.857	1218.5	1335.0	1.9736
700	69.005	1256.2	1383.9	2.0601	700	46.932	1256.0	1383.7	2.0175
800	74.978	1294.6	1433.3	2.1009	800	51.001	1294.4	1433.1	2.0583
900	80.946	1333.7	1483.5	2.1393	900	55.065	1333.6	1483.4	2.0967
1000	86.911	1373.8	1534.7	2.1755	1000	59.127	1373.7	1534.5	2.1330
1100	92.875	1414.8	1586.7	2.2100	1100	63.188	1414.7	1586.6	2.1675
1200	98.837	1456.7	1639.6	2.2429	1200	67.246	1456.7	1639.5	2.2004
1400	110.76	1543.4	1748.3	2.3047	1400	75.361	1543.3	1748.3	2.2623
1600	122.68	1633.8	1860.8	2.3622	1600	83.474	1633.7	1860.7	2.3197
1800	134.60	1727.8	1976.9	2.4159	1800	91.585	1727.7	1976.8	2.3735
2000	146.51	1825.2	2096.4	2.4666	2000	99.695	1825.2	2096.3	2.4241
2200	158.43	1925.9	2219.1	2.5146	2200	107.80	1925.9	2219.1	2.4721
2400	170.34	2029.6	2344.8	2.5601	2400	115.91	2029.6	2344.8	2.5177
2600	182.26	2136.0	2473.3	2.6036	2600	124.02	2136.0	2473.3	2.5611

Temp., °F	v, ft³/lbm	u, Btu/lbm	h, Btu/lbm	s, Btu/lbm·°R	Temp., °F	v, ft³/lbm	u, Btu/lbm	h, Btu/lbm	s, Btu/lbm·°R
	P = 20.0 psia (227.969°F)					P = 40.0 psia (267.275°F)			
Sat.	20.094	1081.9	1156.3	1.7319	Sat.	10.501	1092.3	1170.0	1.6767
250	20.794	1090.2	1167.1	1.7475					
300	22.357	1108.6	1191.3	1.7804	300	11.038	1105.1	1186.8	1.6993
400	25.426	1144.9	1239.0	1.8393	400	12.622	1142.8	1236.2	1.7604
500	28.455	1181.3	1286.6	1.8917	500	14.163	1179.9	1284.7	1.8138
600	31.466	1218.2	1334.7	1.9394	600	15.685	1217.2	1333.3	1.8620
700	34.466	1255.8	1383.4	1.9833	700	17.196	1255.1	1382.3	1.9062
800	37.460	1294.2	1432.9	2.0242	800	18.701	1293.6	1432.1	1.9473
900	40.450	1333.5	1483.2	2.0627	900	20.202	1333.0	1482.5	1.9859
1000	43.437	1373.6	1534.4	2.0990	1000	21.700	1373.2	1533.8	2.0223
1100	46.422	1414.6	1586.4	2.1335	1100	23.196	1414.3	1586.0	2.0569
1200	49.406	1456.6	1639.4	2.1664	1200	24.690	1456.3	1639.0	2.0898
1400	55.371	1543.3	1748.2	2.2283	1400	27.677	1543.0	1747.9	2.1517
1600	61.334	1633.7	1860.7	2.2857	1600	30.661	1633.5	1860.4	2.2092
1800	67.295	1727.7	1976.8	2.3395	1800	33.644	1727.5	1976.6	2.2630
2000	73.255	1825.2	2096.3	2.3902	2000	36.626	1825.0	2096.1	2.3137
2200	79.214	1925.9	2219.1	2.4381	2200	39.607	1925.8	2218.9	2.3617
2400	85.173	2029.6	2344.8	2.4837	2400	42.587	2029.5	2344.7	2.4073
2600	91.132	2136.0	2473.3	2.5271	2600	45.567	2135.9	2473.2	2.4507
	P = 60.0 psia (292.747°F)					P = 80.0 psia (312.083°F)			
Sat.	7.1770	1098.3	1178.0	1.6444	Sat.	5.4737	1102.5	1183.6	1.6214
325	7.5396	1111.4	1195.1	1.6667	325	5.5866	1108.0	1190.7	1.6306
400	8.3520	1140.6	1233.3	1.7132	400	6.2160	1138.3	1230.4	1.6788
500	9.3986	1178.4	1282.8	1.7676	500	7.0159	1177.0	1280.8	1.7344
600	10.424	1216.1	1331.9	1.8163	600	7.7941	1215.1	1330.5	1.7836
700	11.439	1254.3	1381.3	1.8609	700	8.5609	1253.5	1380.2	1.8285
800	12.448	1293.0	1431.2	1.9022	800	9.3213	1292.4	1430.4	1.8700
900	13.452	1332.5	1481.9	1.9408	900	10.077	1332.0	1481.2	1.9088
1000	14.454	1372.8	1533.3	1.9773	1000	10.831	1372.4	1532.7	1.9453
1100	15.453	1413.9	1585.5	2.0119	1100	11.582	1413.6	1585.0	1.9800
1200	16.452	1456.0	1638.6	2.0449	1200	12.332	1455.7	1638.2	2.0130
1300	17.449	1498.9	1692.7	2.0766	1300	13.082	1498.7	1692.3	2.0447
1400	18.446	1542.8	1747.6	2.1069	1400	13.830	1542.6	1747.3	2.0751
1600	20.437	1633.3	1860.2	2.1644	1600	15.325	1633.1	1860.0	2.1326
1800	22.427	1727.4	1976.4	2.2182	1800	16.819	1727.1	1976.2	2.1865
2000	24.416	1824.9	2096.0	2.2689	2000	18.311	1824.8	2095.9	2.2372
2200	26.404	1925.6	2218.8	2.3169	2200	19.803	1925.5	2218.7	2.2852
2400	28.392	2029.4	2344.6	2.3625	2400	21.294	2029.3	2344.5	2.3308
2600	30.379	2135.8	2473.1	2.4060	2600	22.785	2135.7	2473.0	2.3742

TABLE B.3E SUPERHEATED WATER: ENGLISH UNITS (*Continued*)

Temp., °F	*v*, ft³/lbm	*u*, Btu/lbm	*h*, Btu/lbm	*s*, Btu/lbm·°R	Temp., °F	*v*, ft³/lbm	*u*, Btu/lbm	*h*, Btu/lbm	*s*, Btu/lbm·°R
	P = 100 psia (327.871°F)					*P* = 125 psia (344.400°F)			
Sat.	4.4330	1105.7	1187.7	1.6034	Sat.	3.5873	1108.7	1191.7	1.5852
350	4.5907	1115.3	1200.3	1.6191	350	3.6205	1111.3	1195.0	1.5893
400	4.9335	1136.0	1227.3	1.6515	400	3.9063	1133.0	1223.4	1.6233
500	5.5859	1175.5	1278.8	1.7082	500	4.4415	1173.6	1276.3	1.6816
600	6.2156	1214.1	1329.1	1.7580	600	4.9528	1212.7	1327.3	1.7322
700	6.8338	1252.7	1379.2	1.8032	700	5.4520	1251.7	1377.8	1.7777
800	7.4453	1291.8	1429.5	1.8449	800	5.9445	1291.0	1428.5	1.8196
900	8.0526	1331.5	1480.5	1.8838	900	6.4327	1330.9	1479.6	1.8587
1000	8.6571	1371.9	1532.1	1.9204	1000	6.9181	1371.4	1531.4	1.8955
1100	9.2597	1413.2	1584.6	1.9552	1100	7.4016	1412.8	1584.0	1.9303
1200	9.8609	1455.4	1637.8	1.9883	1200	7.8837	1455.0	1637.3	1.9634
1300	10.461	1498.4	1692.0	2.0199	1300	8.3647	1498.0	1691.5	1.9951
1400	11.061	1542.3	1747.0	2.0503	1400	8.8451	1542.0	1746.6	2.0256
1600	12.258	1632.9	1859.7	2.1079	1600	9.8040	1632.7	1859.5	2.0832
1800	13.453	1727.1	1976.0	2.1618	1800	10.761	1726.9	1975.8	2.1371
2000	14.648	1824.6	2095.7	2.2125	2000	11.718	1824.5	2095.5	2.1879
2200	15.842	1925.4	2218.6	2.2605	2200	12.674	1925.3	2218.4	2.2359
2400	17.036	2029.2	2344.4	2.3062	2400	13.629	2029.0	2344.3	2.2815
2600	18.229	2135.6	2473.0	2.3496	2600	14.584	2135.5	2472.9	2.3250
	P = 150 psia (358.490°F)					*P* = 175 psia (370.837°F)			
Sat.	3.0152	1111.0	1194.7	1.5702	Sat.	2.6017	1112.9	1197.1	1.5574
375	3.0984	1118.7	1204.7	1.5823	375	2.6203	1114.9	1199.8	1.5606
400	3.2203	1129.9	1219.3	1.5995	400	2.7293	1126.6	1215.0	1.5786
500	3.6782	1171.6	1273.7	1.6595	500	3.1327	1169.7	1271.1	1.6404
600	4.1107	1211.4	1325.5	1.7108	600	3.5091	1210.0	1323.7	1.6925
700	4.5307	1250.7	1376.5	1.7568	700	3.8726	1249.7	1375.1	1.7389
800	4.9439	1290.2	1427.4	1.7989	800	4.2291	1289.4	1426.4	1.7813
900	5.3527	1330.2	1478.8	1.8382	900	4.5813	1329.6	1477.9	1.8207
1000	5.7587	1370.9	1530.7	1.8750	1000	4.9306	1370.4	1530.0	1.8577
1100	6.1628	1412.3	1583.4	1.9099	1100	5.2780	1411.9	1582.8	1.8926
1200	6.5655	1454.6	1636.8	1.9431	1200	5.6239	1454.2	1636.3	1.9259
1300	6.9672	1497.7	1691.1	1.9749	1300	5.9689	1497.4	1690.7	1.9577
1400	7.3680	1541.7	1746.3	2.0053	1400	6.3130	1541.4	1745.9	1.9882
1600	8.1681	1632.4	1859.2	2.0630	1600	6.9997	1632.2	1858.9	2.0459
1800	8.9667	1726.7	1975.6	2.1169	1800	7.6848	1726.5	1975.3	2.0998
2000	9.7643	1824.3	2095.3	2.1677	2000	8.3689	1824.1	2095.2	2.1506
2200	10.561	1925.1	2218.3	2.2157	2200	9.0523	1925.0	2218.1	2.1987
2400	11.358	2028.9	2344.2	2.2614	2400	9.7352	2028.8	2344.0	2.2443
2600	12.154	2135.4	2472.8	2.3048	2600	10.418	2135.3	2472.7	2.2878

TABLE B.3E SUPERHEATED WATER: ENGLISH UNITS (*Continued*)

Temp., °F	v, ft³/lbm	u, Btu/lbm	h, Btu/lbm	s, Btu/lbm·°R	Temp., °F	v, ft³/lbm	u, Btu/lbm	h, Btu/lbm	s, Btu/lbm·°R
	P = 200 psia (381.869°F)					P = 250 psia (401.040°F)			
Sat.	2.2883	1114.3	1199.0	1.5462	Sat.	1.8441	1116.5	1201.8	1.5271
400	2.3601	1123.2	1210.6	1.5598					
500	2.7232	1167.6	1268.4	1.6235	500	2.1492	1163.5	1262.9	1.5945
600	3.0577	1208.6	1321.8	1.6765	600	2.4255	1205.8	1318.1	1.6492
700	3.3790	1248.7	1373.7	1.7233	700	2.6877	1246.6	1371.0	1.6969
800	3.6930	1288.6	1425.3	1.7659	800	2.9424	1287.0	1423.2	1.7401
900	4.0027	1328.9	1477.1	1.8055	900	3.1926	1327.7	1475.4	1.7800
1000	4.3095	1369.8	1529.3	1.8426	1000	3.4399	1368.8	1527.9	1.8172
1100	4.6143	1411.4	1582.2	1.8776	1100	3.6852	1410.5	1581.0	1.8524
1200	4.9178	1453.8	1635.8	1.9109	1200	3.9291	1453.0	1634.8	1.8859
1300	5.2202	1497.0	1690.2	1.9428	1300	4.1720	1496.4	1689.4	1.9178
1400	5.5218	1541.1	1745.5	1.9733	1400	4.4140	1540.5	1744.7	1.9484
1500	5.8228	1586.1	1801.6	2.0027	1500	4.6555	1585.6	1801.0	1.9778
1600	6.1233	1632.0	1858.6	2.0310	1600	4.8965	1631.5	1858.0	2.0062
1800	6.7234	1726.3	1975.1	2.0850	1800	5.3774	1725.9	1974.7	2.0603
2000	7.3224	1824.0	2095.0	2.1358	2000	5.8572	1823.6	2094.6	2.1111
2200	7.9207	1924.8	2218.0	2.1839	2200	6.3364	1924.5	2217.7	2.1592
2400	8.5185	2028.6	2343.9	2.2296	2400	6.8151	2028.4	2343.7	2.2049
2600	9.1159	2135.2	2472.6	2.2730	2600	7.2934	2135.0	2472.4	2.2484
	P = 300 psia (417.423°F)					P = 350 psia (431.810°F)			
Sat.	1.5435	1117.8	1203.5	1.5112	Sat.	1.3262	1118.7	1204.6	1.4974
450	1.6355	1135.0	1225.8	1.5362	450	1.3728	1128.9	1217.8	1.5122
500	1.7657	1159.1	1257.1	1.5697	500	1.4908	1154.5	1251.1	1.5478
600	2.0038	1203.0	1314.2	1.6263	600	1.7022	1200.0	1310.3	1.6065
700	2.2268	1244.5	1368.1	1.6750	700	1.8974	1242.4	1365.3	1.6561
800	2.4420	1285.4	1421.0	1.7187	800	2.0844	1283.8	1418.8	1.7004
900	2.6526	1326.4	1473.6	1.7589	900	2.2668	1325.1	1471.9	1.7409
1000	2.8602	1367.7	1526.5	1.7964	1000	2.4461	1366.6	1525.1	1.7787
1100	3.0658	1409.6	1579.8	1.8318	1100	2.6234	1408.7	1578.6	1.8142
1200	3.2700	1452.3	1633.8	1.8653	1200	2.7992	1451.5	1632.8	1.8478
1300	3.4732	1495.7	1688.5	1.8973	1300	2.9740	1495.0	1687.6	1.8799
1400	3.6755	1539.9	1744.0	1.9280	1400	3.1480	1539.3	1743.2	1.9107
1500	3.8773	1585.0	1800.3	1.9575	1500	3.3214	1584.5	1799.6	1.9402
1600	4.0786	1631.0	1857.4	1.9859	1600	3.4943	1630.5	1856.8	1.9687
1800	4.4800	1725.5	1974.2	2.0400	1800	3.8391	1725.1	1973.7	2.0228
2000	4.8805	1823.3	2094.2	2.0909	2000	4.1828	1823.0	2093.9	2.0737
2200	5.2802	1924.2	2217.4	2.1390	2200	4.5258	1924.0	2217.1	2.1219
2400	5.6795	2028.1	2343.4	2.1847	2400	4.8683	2027.9	2343.2	2.1676
2600	6.0783	2134.7	2472.2	2.2282	2600	5.2105	2134.5	2472.0	2.2111

TABLE B.3E SUPERHEATED WATER: ENGLISH UNITS (*Continued*)

H_2O

Temp., °F	v, ft³/lbm	u, Btu/lbm	h, Btu/lbm	s, Btu/lbm·°R	Temp., °F	v, ft³/lbm	u, Btu/lbm	h, Btu/lbm	s, Btu/lbm·°R
	P = 400 psia (444.686°F)					P = 450 psia (456.376°F)			
Sat.	1.1616	1119.1	1205.1	1.4852	Sat.	1.0323	1119.2	1205.2	1.4742
500	1.2839	1149.7	1244.8	1.5279	500	1.1222	1144.7	1238.1	1.5094
600	1.4757	1197.0	1306.2	1.5889	600	1.2993	1193.9	1302.1	1.5729
700	1.6502	1240.3	1362.4	1.6395	700	1.4578	1238.1	1359.5	1.6246
800	1.8162	1282.2	1416.6	1.6844	800	1.6075	1280.5	1414.4	1.6701
900	1.9774	1323.8	1470.1	1.7253	900	1.7523	1322.4	1468.4	1.7113
1000	2.1355	1365.6	1523.6	1.7632	1000	1.8939	1364.5	1522.2	1.7495
1100	2.2915	1407.8	1577.4	1.7989	1100	2.0334	1406.9	1576.2	1.7853
1200	2.4461	1450.7	1631.8	1.8327	1200	2.1715	1449.9	1630.8	1.8192
1300	2.5997	1494.3	1686.8	1.8648	1300	2.3085	1493.7	1685.9	1.8514
1400	2.7524	1538.7	1742.5	1.8956	1400	2.4447	1538.1	1741.7	1.8823
1500	2.9045	1584.0	1799.0	1.9252	1500	2.5803	1583.4	1798.3	1.9119
1600	3.0562	1630.0	1856.3	1.9537	1600	2.7154	1629.6	1855.7	1.9405
1700	3.2074	1676.9	1914.4	1.9813	1700	2.8501	1676.5	1913.9	1.9681
1800	3.3584	1724.7	1973.3	2.0079	1800	2.9845	1724.3	1972.8	1.9948
2000	3.6596	1822.6	2093.5	2.0589	2000	3.2526	1822.3	2093.1	2.0458
2200	3.9600	1923.7	2216.8	2.1071	2200	3.5200	1923.4	2216.5	2.0940
2400	4.2600	2027.6	2343.0	2.1528	2400	3.7868	2027.4	2342.7	2.1397
2600	4.5596	2134.3	2471.8	2.1963	2600	4.0533	2134.1	2471.6	2.1833
	P = 500 psia (467.107°F)					P = 750 psia (510.958°F)			
Sat.	0.92806	1119.1	1205.0	1.4642	Sat.	0.60963	1115.8	1200.4	1.4229
500	0.99207	1139.3	1231.1	1.4919					
600	1.1580	1190.7	1297.9	1.5582	600	0.73127	1173.5	1275.0	1.4965
700	1.3039	1235.9	1356.5	1.6111	700	0.84085	1224.3	1341.0	1.5562
800	1.4405	1278.8	1412.1	1.6571	800	0.93908	1270.3	1400.6	1.6055
900	1.5721	1321.1	1466.6	1.6987	900	1.0315	1314.4	1457.5	1.6490
1000	1.7006	1363.4	1520.7	1.7371	1000	1.1206	1357.9	1513.4	1.6886
1100	1.8269	1406.0	1575.0	1.7731	1100	1.2074	1401.4	1568.9	1.7254
1200	1.9518	1449.1	1629.7	1.8071	1200	1.2926	1445.2	1624.6	1.7600
1300	2.0756	1493.0	1685.0	1.8394	1300	1.3768	1489.5	1680.6	1.7928
1400	2.1985	1537.5	1741.0	1.8704	1400	1.4601	1534.5	1737.2	1.8240
1500	2.3209	1582.9	1797.6	1.9001	1500	1.5427	1580.2	1794.3	1.8540
1600	2.4428	1629.1	1855.1	1.9286	1600	1.6249	1626.7	1852.2	1.8828
1700	2.5642	1676.1	1913.3	1.9563	1700	1.7067	1673.9	1910.8	1.9106
1800	2.6854	1723.9	1972.4	1.9830	1800	1.7882	1721.9	1970.1	1.9374
2000	2.9270	1821.9	2092.8	2.0340	2000	1.9504	1820.3	2091.0	1.9886
2200	3.1679	1923.1	2216.2	2.0823	2200	2.1118	1921.6	2214.7	2.0370
2400	3.4083	2027.1	2342.5	2.1280	2400	2.2728	2025.9	2341.3	2.0829
2600	3.6483	2133.8	2471.4	2.1716	2600	2.4334	2132.7	2470.4	2.1265

TABLE B.3E SUPERHEATED WATER: ENGLISH UNITS (*Continued*)

P = 1000 psia (544.713°F)

Temp., °F	v, ft^3/lb$_m$	u, Btu/lb$_m$	h, Btu/lb$_m$	s, Btu/lb$_m$·°R
Sat.	0.44594	1109.7	1192.3	1.3902
575	0.48546	1135.3	1225.1	1.4225
600	0.51376	1153.2	1248.3	1.4447
700	0.60793	1211.8	1324.3	1.5133
800	0.68772	1261.2	1388.5	1.5665
900	0.76092	1307.4	1448.2	1.6121
1000	0.83040	1352.2	1505.9	1.6530
1100	0.89751	1396.6	1562.7	1.6907
1200	0.96301	1441.2	1619.4	1.7259
1300	1.0274	1486.0	1676.2	1.7591
1400	1.0909	1531.4	1733.3	1.7907
1500	1.1537	1577.5	1791.0	1.8209
1600	1.2160	1624.2	1849.2	1.8499
1700	1.2780	1671.7	1908.2	1.8778
1800	1.3396	1719.9	1967.8	1.9048
2000	1.4621	1818.6	2089.1	1.9563
2200	1.5838	1920.2	2213.3	2.0048
2400	1.7051	2024.6	2340.1	2.0507
2600	1.8259	2131.6	2469.5	2.0945

P = 1250 psia (572.515°F)

Temp., °F	v, ft^3/lb$_m$	u, Btu/lb$_m$	h, Btu/lb$_m$	s, Btu/lb$_m$·°R
Sat.	0.34543	1101.6	1181.5	1.3619
575	0.34873	1104.4	1185.0	1.3653
600	0.37850	1128.7	1216.2	1.3951
700	0.46688	1198.1	1306.1	1.4764
800	0.53638	1251.8	1375.9	1.5341
900	0.59830	1300.2	1438.6	1.5821
1000	0.65617	1346.4	1498.2	1.6244
1100	0.71153	1391.8	1556.4	1.6630
1200	0.76522	1437.1	1614.1	1.6988
1300	0.81772	1482.5	1671.7	1.7325
1400	0.86934	1528.3	1729.4	1.7644
1500	0.92030	1574.7	1787.6	1.7949
1600	0.97074	1621.8	1846.3	1.8241
1700	1.0208	1669.5	1905.6	1.8522
1800	1.0705	1717.9	1965.5	1.8793
2000	1.1691	1816.9	2087.3	1.9310
2200	1.2671	1918.7	2211.8	1.9796
2400	1.3645	2023.3	2338.9	2.0257
2600	1.4615	2130.5	2468.5	2.0695

P = 1500 psia (596.323°F)

Temp., °F	v, ft^3/lb$_m$	u, Btu/lb$_m$	h, Btu/lb$_m$	s, Btu/lb$_m$·°R
Sat.	0.27691	1091.7	1168.6	1.3358
650	0.33273	1146.5	1238.8	1.4008
700	0.37157	1183.1	1286.2	1.4426
800	0.43502	1241.8	1362.6	1.5059
900	0.48968	1292.7	1428.6	1.5564
1000	0.53991	1340.5	1490.4	1.6002
1100	0.58750	1386.9	1550.0	1.6397
1200	0.63334	1432.9	1608.7	1.6762
1300	0.67795	1478.9	1667.1	1.7104
1400	0.72167	1525.2	1725.5	1.7427
1500	0.76471	1572.0	1784.2	1.7734
1600	0.80723	1619.3	1843.4	1.8028
1700	0.84933	1667.2	1903.0	1.8311
1800	0.89110	1715.9	1963.2	1.8583
1900	0.93260	1765.2	2024.0	1.8847
2000	0.97388	1815.2	2085.5	1.9102
2200	1.0559	1917.3	2210.4	1.9590
2400	1.1374	2022.1	2337.8	2.0052
2600	1.2186	2129.3	2467.6	2.0491

P = 1750 psia (617.229°F)

Temp., °F	v, ft^3/lb$_m$	u, Btu/lb$_m$	h, Btu/lb$_m$	s, Btu/lb$_m$·°R
Sat.	0.22679	1080.1	1153.5	1.3108
650	0.26263	1122.0	1207.1	1.3599
700	0.30218	1166.2	1264.1	1.4102
800	0.36220	1231.3	1348.6	1.4802
900	0.41191	1285.0	1418.4	1.5336
1000	0.45679	1334.4	1482.4	1.5790
1100	0.49887	1382.0	1543.5	1.6195
1200	0.53913	1428.7	1603.3	1.6567
1300	0.57812	1475.3	1662.5	1.6913
1400	0.61620	1522.1	1721.6	1.7240
1500	0.65359	1569.2	1780.8	1.7550
1600	0.69045	1616.8	1840.4	1.7846
1700	0.72689	1665.0	1900.4	1.8131
1800	0.76299	1713.8	1960.9	1.8405
1900	0.79882	1763.3	2022.0	1.8669
2000	0.83443	1813.5	2083.7	1.8925
2200	0.90512	1915.8	2208.9	1.9415
2400	0.97528	2020.8	2336.6	1.9878
2600	1.0451	2128.2	2466.7	2.0317

TABLE B.3E SUPERHEATED WATER: ENGLISH UNITS (*Continued*)

Temp., °F	*v*, ft³/lbm	*u*, Btu/lbm	*h*, Btu/lbm	*s*, Btu/lbm·°R	Temp., °F	*v*, ft³/lbm	*u*, Btu/lbm	*h*, Btu/lbm	*s*, Btu/lbm·°R
	P = 2000 psia (635.910°F)					*P* = 2500 psia (668.246°F)			
Sat.	0.18813	1066.5	1136.1	1.2860	Sat.	0.13073	1030.9	1091.4	1.2327
675	0.22944	1122.4	1207.3	1.3499	675	0.14161	1051.1	1116.6	1.2549
700	0.24868	1147.2	1239.2	1.3778	700	0.16838	1098.0	1175.9	1.3067
800	0.30720	1220.1	1333.8	1.4562	800	0.22916	1195.6	1301.7	1.4112
900	0.35344	1277.1	1407.9	1.5128	900	0.27121	1260.3	1385.8	1.4755
1000	0.39438	1328.2	1474.2	1.5599	1000	0.30687	1315.4	1457.4	1.5264
1100	0.43236	1376.9	1536.9	1.6015	1100	0.33920	1366.6	1523.5	1.5702
1200	0.46846	1424.5	1597.9	1.6394	1200	0.36950	1415.9	1586.8	1.6096
1300	0.50325	1471.7	1657.9	1.6745	1300	0.39844	1464.3	1648.6	1.6457
1400	0.53711	1518.9	1717.7	1.7076	1400	0.42640	1512.5	1709.7	1.6795
1500	0.57026	1566.4	1777.4	1.7389	1500	0.45363	1560.7	1770.6	1.7114
1600	0.60288	1614.3	1837.4	1.7687	1600	0.48032	1609.3	1831.5	1.7417
1700	0.63507	1662.7	1897.8	1.7973	1700	0.50656	1658.2	1892.6	1.7707
1800	0.66692	1711.8	1958.6	1.8249	1800	0.53247	1707.7	1954.0	1.7985
1900	0.69850	1761.5	2020.0	1.8514	1900	0.55809	1757.7	2015.9	1.8253
2000	0.72986	1811.8	2081.9	1.8771	2000	0.58349	1808.4	2078.3	1.8512
2200	0.79204	1914.4	2207.5	1.9262	2200	0.63375	1911.5	2204.7	1.9005
2400	0.85368	2019.5	2335.5	1.9726	2400	0.68348	2017.0	2333.2	1.9471
2600	0.91494	2127.1	2465.8	2.0166	2600	0.73281	2124.9	2463.9	1.9913
	P = 3000 psia (695.504°F)					*P* = 3500 psia			
Sat.	0.08484	970.8	1017.9	1.1596					
700	0.09794	1004.0	1058.3	1.1946	700	0.03066	759.9	779.8	0.95099
800	0.17582	1167.6	1265.2	1.3675	800	0.13633	1134.6	1222.9	1.3225
900	0.21602	1242.3	1362.3	1.4418	900	0.17628	1223.1	1337.2	1.4101
1000	0.24838	1302.1	1440.0	1.4970	1000	0.20652	1288.2	1421.9	1.4703
1100	0.27705	1355.9	1509.7	1.5432	1100	0.23263	1345.0	1495.7	1.5192
1200	0.30353	1407.0	1575.5	1.5841	1200	0.25641	1398.0	1564.1	1.5617
1300	0.32858	1456.8	1639.2	1.6214	1300	0.27870	1449.2	1629.7	1.6001
1400	0.35262	1506.0	1701.7	1.6559	1400	0.29994	1499.4	1693.7	1.6354
1500	0.37591	1555.0	1763.7	1.6884	1500	0.32042	1549.3	1756.8	1.6685
1600	0.39864	1604.2	1825.5	1.7192	1600	0.34033	1599.1	1819.5	1.6997
1700	0.42093	1653.7	1887.4	1.7485	1700	0.35979	1649.1	1882.2	1.7294
1800	0.44287	1703.6	1949.5	1.7766	1800	0.37889	1699.5	1944.9	1.7578
1900	0.46452	1754.0	2011.9	1.8036	1900	0.39771	1750.3	2007.9	1.7851
2000	0.48595	1804.9	2074.7	1.8297	2000	0.41630	1801.5	2071.2	1.8114
2100	0.50719	1856.5	2138.0	1.8549	2100	0.43470	1853.3	2134.9	1.8367
2200	0.52827	1908.6	2201.8	1.8794	2200	0.45294	1905.7	2199.0	1.8613
2400	0.57004	2014.5	2331.0	1.9262	2400	0.48903	2012.0	2328.7	1.9083
2600	0.61142	2122.7	2462.1	1.9705	2600	0.52473	2120.5	2460.4	1.9528

TABLE B.3E SUPERHEATED WATER: ENGLISH UNITS (*Continued*)

Temp., °F	v, ft³/lbm	u, Btu/lbm	h, Btu/lbm	s, Btu/lbm·°R	Temp., °F	v, ft³/lbm	u, Btu/lbm	h, Btu/lbm	s, Btu/lbm·°R
		P = 4000 psia					P = 4500 psia		
700	0.02871	742.2	763.4	0.93453	700	0.02757	730.5	753.4	0.92372
800	0.10524	1094.8	1172.7	1.2738	800	0.07966	1045.3	1111.6	1.2186
900	0.14624	1202.3	1310.6	1.3796	900	0.12271	1180.1	1282.2	1.3496
1000	0.17507	1273.7	1403.3	1.4454	1000	0.15060	1258.7	1384.1	1.4220
1100	0.19932	1333.8	1481.3	1.4972	1100	0.17343	1322.4	1466.8	1.4768
1200	0.22109	1388.9	1552.6	1.5415	1200	0.19365	1379.6	1540.9	1.5229
1300	0.24131	1441.5	1620.1	1.5810	1300	0.21226	1433.7	1610.5	1.5636
1400	0.26045	1492.8	1685.6	1.6172	1400	0.22977	1486.1	1677.5	1.6007
1500	0.27882	1543.5	1749.9	1.6509	1500	0.24650	1537.7	1742.9	1.6350
1600	0.29661	1594.0	1813.5	1.6826	1600	0.26264	1588.8	1807.5	1.6671
1700	0.31395	1644.5	1876.9	1.7126	1700	0.27832	1640.0	1871.7	1.6975
1800	0.33093	1695.4	1940.3	1.7413	1800	0.29365	1691.2	1935.8	1.7265
1900	0.34763	1746.5	2003.8	1.7688	1900	0.30869	1742.8	1999.8	1.7543
2000	0.36409	1798.1	2067.6	1.7953	2000	0.32349	1794.7	2064.1	1.7809
2100	0.38036	1850.2	2131.7	1.8208	2100	0.33810	1847.0	2128.6	1.8066
2200	0.39646	1902.8	2196.2	1.8455	2200	0.35255	1899.9	2193.5	1.8315
2400	0.42830	2009.5	2326.5	1.8928	2400	0.38107	2007.0	2324.3	1.8789
2600	0.45973	2118.3	2458.6	1.9374	2600	0.40919	2116.2	2456.9	1.9238
		P = 5000 psia					P = 6000 psia		
700	0.02677	721.5	746.3	0.91540	700	0.025625	707.9	736.3	0.90258
800	0.05927	986.7	1041.5	1.1578	800	0.039460	896.9	940.7	1.0708
900	0.10382	1156.2	1252.2	1.3198	900	0.075851	1104.2	1188.4	1.2608
1000	0.13105	1243.1	1364.4	1.3996	1000	0.10192	1210.6	1323.8	1.3571
1100	0.15276	1310.7	1452.0	1.4577	1100	0.12192	1286.7	1422.1	1.4223
1200	0.17173	1370.2	1529.1	1.5057	1200	0.13899	1351.2	1505.5	1.4742
1300	0.18905	1425.9	1600.8	1.5476	1300	0.15433	1410.1	1581.4	1.5186
1400	0.20525	1479.4	1669.3	1.5855	1400	0.16855	1465.9	1653.1	1.5583
1500	0.22066	1531.8	1736.0	1.6204	1500	0.18197	1520.1	1722.1	1.5944
1600	0.23547	1583.7	1801.6	1.6530	1600	0.19478	1573.4	1789.6	1.6280
1700	0.24983	1635.4	1866.5	1.6838	1700	0.20715	1626.2	1856.2	1.6596
1800	0.26383	1687.1	1931.2	1.7131	1800	0.21915	1678.8	1922.2	1.6895
1900	0.27755	1739.0	1995.8	1.7411	1900	0.23087	1731.6	1987.9	1.7179
2000	0.29102	1791.3	2060.6	1.7680	2000	0.24235	1784.5	2053.6	1.7452
2100	0.30431	1843.9	2125.5	1.7938	2100	0.25364	1837.7	2119.3	1.7714
2200	0.31743	1897.0	2190.7	1.8188	2200	0.26477	1891.3	2185.2	1.7966
2400	0.34329	2004.5	2322.2	1.8665	2400	0.28664	1999.6	2317.9	1.8447
2600	0.36876	2114.0	2455.2	1.9115	2600	0.30812	2109.7	2451.8	1.8900

TABLE B.3E SUPERHEATED WATER: ENGLISH UNITS (*Continued*)

Temp., °F	v, ft³/lbₘ	u, Btu/lbₘ	h, Btu/lbₘ	s, Btu/lbₘ·°R	Temp., °F	v, ft³/lbₘ	u, Btu/lbₘ	h, Btu/lbₘ	s, Btu/lbₘ·°R
		P = 7000 psia					P = 8000 psia		
700	0.024814	697.3	729.4	0.89261	700	0.024183	688.5	724.3	0.88432
800	0.033425	855.1	898.4	1.0320	800	0.030600	830.8	876.1	1.0096
900	0.057589	1051.0	1125.6	1.2056	900	0.046553	1004.9	1073.8	1.1605
1000	0.081628	1176.9	1282.6	1.3174	1000	0.067182	1143.2	1242.6	1.2806
1100	0.10019	1262.2	1392.0	1.3899	1100	0.084281	1237.6	1362.4	1.3600
1200	0.11581	1331.9	1481.9	1.4459	1200	0.098674	1312.6	1458.6	1.4199
1300	0.12970	1394.1	1562.2	1.4928	1300	0.11140	1378.2	1543.2	1.4694
1400	0.14246	1452.4	1636.9	1.5342	1400	0.12302	1438.9	1621.0	1.5125
1500	0.15442	1508.4	1708.4	1.5716	1500	0.13386	1496.7	1694.9	1.5512
1600	0.16579	1563.1	1777.8	1.6062	1600	0.14412	1552.8	1766.2	1.5866
1700	0.17671	1617.0	1845.9	1.6385	1700	0.15395	1607.9	1835.8	1.6197
1800	0.18728	1670.6	1913.2	1.6689	1800	0.16342	1662.5	1904.4	1.6507
1900	0.19756	1724.1	1980.1	1.6979	1900	0.17262	1716.8	1972.3	1.6801
2000	0.20761	1777.7	2046.7	1.7255	2000	0.18159	1771.0	2039.8	1.7081
2100	0.21747	1831.5	2113.2	1.7520	2100	0.19037	1825.4	2107.2	1.7350
2200	0.22718	1885.6	2179.8	1.7776	2200	0.19900	1879.9	2174.5	1.7608
2400	0.24620	1994.7	2313.7	1.8261	2400	0.21587	1989.9	2309.5	1.8097
2600	0.26483	2105.5	2448.5	1.8717	2600	0.23236	2101.3	2445.3	1.8556
		P = 9000 psia					P = 10000 psia		
700	0.023668	681.0	720.5	0.87715	700	0.023232	674.4	717.4	0.87080
800	0.028866	813.6	861.7	0.99379	800	0.027645	800.3	851.4	0.98152
900	0.040115	969.7	1036.5	1.1272	900	0.036169	943.3	1010.3	1.1027
1000	0.056879	1111.1	1205.8	1.2476	1000	0.049548	1082.1	1173.8	1.2189
1100	0.072359	1213.3	1333.8	1.3325	1100	0.063292	1189.9	1307.0	1.3073
1200	0.085624	1293.4	1436.0	1.3961	1200	0.075471	1274.6	1414.3	1.3740
1300	0.097352	1362.5	1524.6	1.4479	1300	0.086312	1347.0	1506.7	1.4281
1400	0.10803	1425.6	1605.5	1.4926	1400	0.096188	1412.4	1590.4	1.4744
1500	0.11797	1485.2	1681.6	1.5325	1500	0.10536	1473.8	1668.7	1.5154
1600	0.12735	1542.7	1754.8	1.5689	1600	0.11401	1532.6	1743.6	1.5527
1700	0.13630	1598.9	1825.9	1.6027	1700	0.12224	1590.0	1816.2	1.5871
1800	0.14491	1654.4	1895.7	1.6343	1800	0.13015	1646.4	1887.2	1.6193
1900	0.15326	1709.4	1964.7	1.6641	1900	0.13780	1702.2	1957.2	1.6496
2000	0.16138	1764.4	2033.1	1.6925	2000	0.14523	1757.8	2026.5	1.6783
2100	0.16931	1819.3	2101.3	1.7197	2100	0.15249	1813.2	2095.4	1.7058
2200	0.17710	1874.3	2169.3	1.7458	2200	0.15960	1868.8	2164.1	1.7321
2400	0.19230	1985.1	2305.4	1.7951	2400	0.17345	1980.4	2301.3	1.7819
2600	0.20712	2097.1	2442.1	1.8413	2600	0.18693	2093.0	2438.9	1.8284

Values generated from property formulation given in *NBS/NRC Steam Tables: Thermodynamic and Transport Properties and Computer Programs for Vapor and Liquid States of Water in SI Units*, Lester Haar, John S. Gallagher and George S. Kell, Hemisphere Publishing Corp., Washington, 1984.

TABLE B.4E COMPRESSED LIQUID WATER: ENGLISH UNITS

Temp., °F	v, ft³/lbm	u, Btu/lbm	h, Btu/lbm	s, Btu/lbm·°R	Temp., °F	v, ft³/lbm	u, Btu/lbm	h, Btu/lbm	s, Btu/lbm·°R
	P = 500 psia (467.107°F)					P = 1000 psia (544.713°F)			
32	0.015994	0.01	1.49	0.00002	32	0.015966	0.04	3.00	0.00007
50	0.015997	18.01	19.49	0.03598	50	0.015971	17.98	20.93	0.03590
100	0.016106	67.84	69.33	0.12927	100	0.016082	67.68	70.65	0.12896
150	0.016316	117.68	119.19	0.21459	150	0.016291	117.40	120.42	0.21412
200	0.016605	167.66	169.19	0.29342	200	0.016579	167.26	170.33	0.29281
250	0.016970	218.02	219.59	0.36706	250	0.016940	217.50	220.64	0.36631
300	0.017415	268.97	270.58	0.43648	300	0.017379	268.30	271.51	0.43558
350	0.017953	320.72	322.38	0.50251	350	0.017908	319.85	323.17	0.50143
400	0.018607	373.65	375.37	0.56600	400	0.018549	372.51	375.95	0.56468
450	0.019421	428.35	430.15	0.62793	450	0.019342	426.84	430.42	0.62626
Sat	0.019749	447.65	449.47	0.64898	500	0.020362	483.78	487.55	0.68738
					Sat	0.021594	538.38	542.37	0.74320
	P = 2000 psia (635.910°F)					P = 3000 psia (695.504°F)			
32	0.015912	0.09	5.98	0.00014	32	0.015858	0.12	8.93	0.00016
50	0.015920	17.90	23.79	0.03572	50	0.015870	17.81	26.62	0.03551
100	0.016034	67.35	73.28	0.12835	100	0.015988	67.02	75.90	0.12773
150	0.016242	116.85	122.87	0.21321	150	0.016195	116.32	125.31	0.21230
200	0.016526	166.49	172.61	0.29162	200	0.016474	165.75	174.89	0.29045
250	0.016880	216.48	222.73	0.36485	250	0.016822	215.49	224.83	0.36342
300	0.017309	266.98	273.39	0.43383	300	0.017241	265.71	275.28	0.43211
350	0.017822	318.17	324.76	0.49932	350	0.017740	316.55	326.40	0.49727
400	0.018440	370.34	377.16	0.56211	400	0.018336	368.27	378.45	0.55964
450	0.019195	423.97	431.08	0.62306	450	0.019058	421.28	431.86	0.62002
500	0.020148	479.84	487.29	0.68321	500	0.019954	476.21	487.29	0.67934
550	0.021418	539.30	547.22	0.74406	550	0.021118	534.13	545.85	0.73880
600	0.023305	605.34	613.97	0.80854	600	0.022747	597.01	609.64	0.80044
Sat	0.025633	662.12	671.60	0.86200	650	0.025451	670.20	684.33	0.86925
					Sat	0.034243	783.27	802.29	0.97300

TABLE B.4E COMPRESSED LIQUID WATER: ENGLISH UNITS (*Continued*)

Temp., °F	v, ft³/lbm	u, Btu/lbm	h, Btu/lbm	s, Btu/lbm·°R	Temp., °F	v, ft³/lbm	u, Btu/lbm	h, Btu/lbm	s, Btu/lbm·°R
		P = 4000 psia					P = 5000 psia		
32	0.015807	0.15	11.85	0.00014	32	0.015756	0.16	14.74	0.00008
50	0.015821	17.73	29.44	0.03528	50	0.015773	17.64	32.23	0.03502
100	0.015942	66.70	78.50	0.12711	100	0.015897	66.39	81.10	0.12649
150	0.016148	115.80	127.76	0.21140	150	0.016102	115.30	130.20	0.21050
200	0.016424	165.02	177.18	0.28930	200	0.016375	164.31	179.46	0.28817
250	0.016765	214.54	226.95	0.36202	250	0.016710	213.61	229.07	0.36065
300	0.017175	264.48	277.20	0.43044	300	0.017111	263.30	279.13	0.42881
350	0.017661	314.99	328.07	0.49529	350	0.017585	313.50	329.77	0.49336
400	0.018237	366.29	379.79	0.55727	400	0.018143	364.40	381.19	0.55498
450	0.018929	418.73	432.75	0.61714	450	0.018808	416.33	433.73	0.61438
500	0.019778	472.86	487.50	0.67572	500	0.019615	469.72	487.87	0.67232
550	0.020855	529.49	544.93	0.73404	550	0.020622	525.27	544.35	0.72967
600	0.022304	590.05	606.56	0.79361	600	0.021934	584.02	604.32	0.78763
650	0.024471	657.58	675.70	0.85732	650	0.023777	647.93	669.93	0.84810

Values generated from property formulation given in *NBS/NRC Steam Tables: Thermodynamic and Transport Properties and Computer Programs for Vapor and Liquid States of Water in SI Units*, Lester Haar, John S. Gallagher and George S. Kell, Hemisphere Publishing Corp., Washington, 1984.

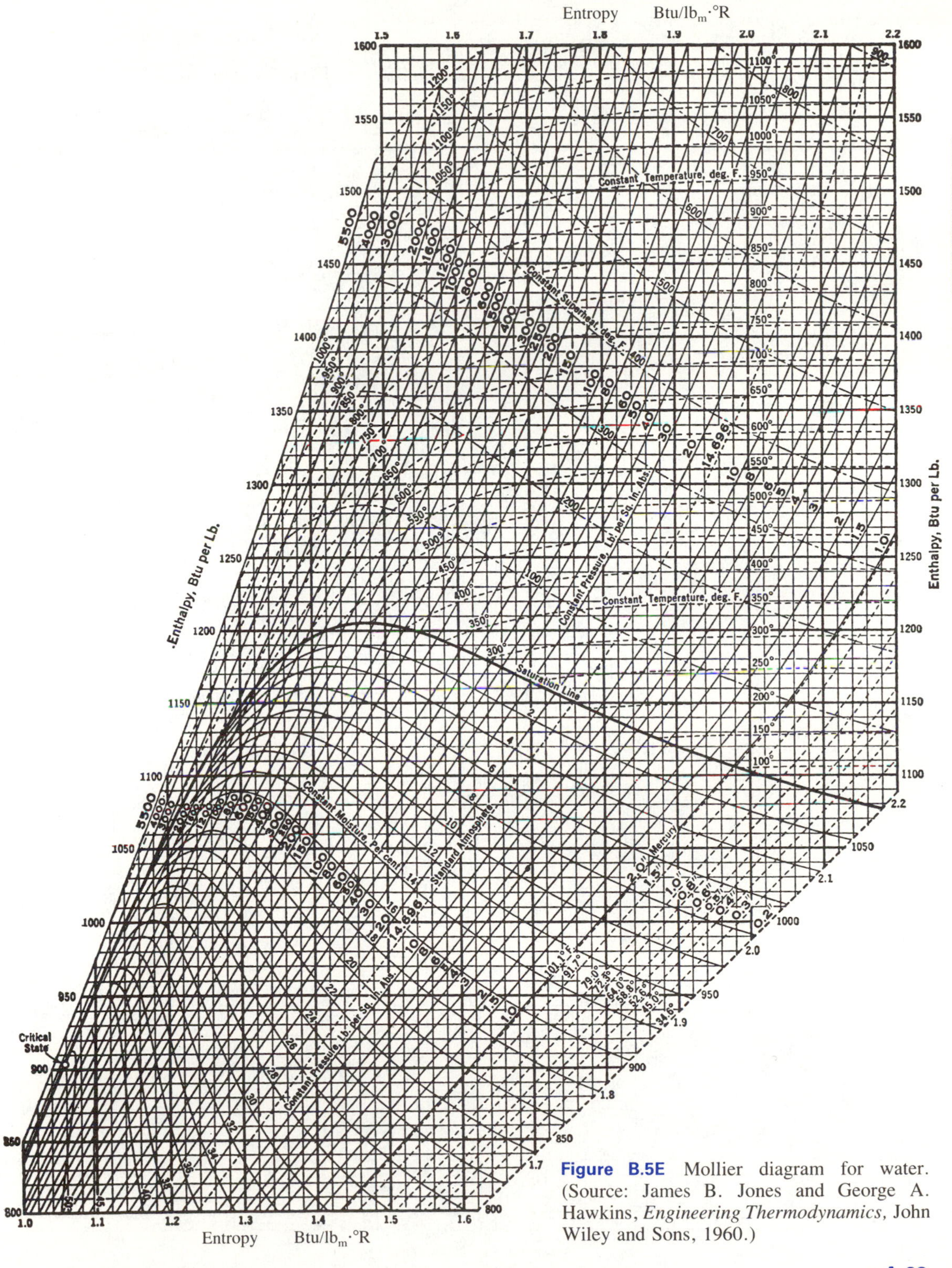

H₂O

Figure B.5E Mollier diagram for water. (Source: James B. Jones and George A. Hawkins, *Engineering Thermodynamics,* John Wiley and Sons, 1960.)

Figure B.6E T-s diagram for water. (Source: Joseph H. Keenan and Frederick G. Keyes, *Thermodynamic Properties of Steam,* John Wiley and Sons, 1936.)

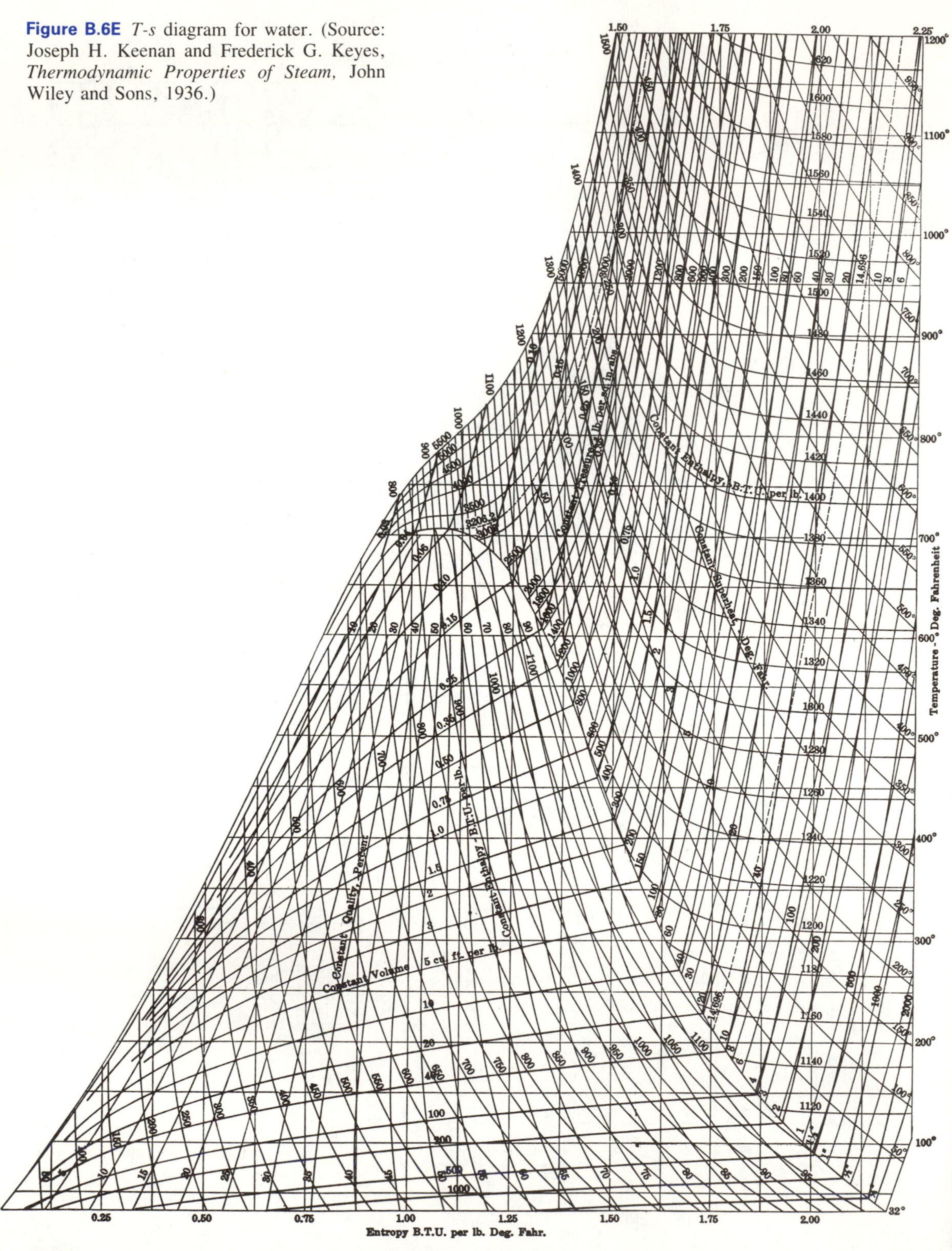

Properties of Refrigerant-12

TABLE C.1E SATURATED REFRIGERANT-12—TEMPERATURE TABLE: ENGLISH UNITS

		Specific volume, ft³/lbm			Internal energy, Btu/lbm			Enthalpy, Btu/lbm			Entropy, Btu/lbm·°R		
Temp., °F, T	Press., psia, P	Sat. liquid, v_f	Evap., v_{fg}	Sat. vapor, v_g	Sat. liquid, u_f	Evap., u_{fg}	Sat. vapor, u_g	Sat. liquid, h_f	Evap., h_{fg}	Sat. vapor, h_g	Sat. liquid, s_f	Evap., s_{fg}	Sat. vapor, s_g
−130	0.41131	0.0096971	70.895	70.904	−18.381	76.239	57.858	−18.380	81.638	63.258	−0.049218	0.24764	0.19842
−120	0.64047	0.0097837	46.849	46.858	−16.363	75.144	58.781	−16.362	80.700	64.338	−0.043187	0.23758	0.19440
−110	0.96829	0.0098726	31.847	31.857	−14.343	74.060	59.718	−14.341	79.771	65.430	−0.037326	0.22813	0.19081
−100	1.4252	0.0099638	22.210	22.220	−12.319	72.985	60.666	−12.317	78.847	66.530	−0.031620	0.21922	0.18760
−90	2.0473	0.010057	15.851	15.861	−10.291	71.916	61.625	−10.287	77.925	67.638	−0.026058	0.21080	0.18474
−80	2.8765	0.010154	11.552	11.563	−8.2561	70.849	62.593	−8.2506	77.002	68.751	−0.020626	0.20281	0.18219
−70	3.9604	0.010253	8.5809	8.5912	−6.2129	69.782	63.569	−6.2054	76.075	69.869	−0.015314	0.19523	0.17991
−60	5.3526	0.010355	6.4846	6.4950	−4.1600	68.712	64.552	−4.1497	75.139	70.990	−0.010112	0.18800	0.17789
−50	7.1124	0.010460	4.9779	4.9884	−2.0956	67.637	65.541	−2.0818	74.193	72.111	−0.005010	0.18110	0.17609
−40	9.3045	0.010568	3.8762	3.8868	−0.0182	66.553	66.535	0.0000	73.232	73.232	0.000000	0.17450	0.17450
−30	11.999	0.010680	3.0577	3.0684	2.0737	65.459	67.533	2.0974	72.253	74.350	0.004927	0.16816	0.17309
−20	15.270	0.010796	2.4408	2.4516	4.1815	64.351	68.533	4.2121	71.253	75.465	0.009777	0.16206	0.17184
−10	19.197	0.010916	1.9694	1.9803	6.3068	63.227	69.534	6.3456	70.228	76.574	0.014557	0.15618	0.17073
0	23.863	0.011041	1.6047	1.6157	8.4507	62.085	70.536	8.4995	69.176	77.675	0.019274	0.15049	0.16976
10	29.356	0.011171	1.3191	1.3303	10.615	60.922	71.536	10.675	68.092	78.768	0.023932	0.14498	0.16891
20	35.765	0.011307	1.0932	1.1045	12.800	59.734	72.534	12.874	66.974	79.849	0.028537	0.13963	0.16816
30	43.182	0.011448	0.91256	0.92401	15.007	58.521	73.528	15.098	65.818	80.916	0.033093	0.13441	0.16751
40	51.705	0.011597	0.76679	0.77838	17.237	57.278	74.515	17.348	64.620	81.968	0.037604	0.12932	0.16693
50	61.432	0.011753	0.64809	0.65984	19.491	56.003	75.494	19.625	63.375	83.000	0.042075	0.12435	0.16642
60	72.462	0.011917	0.55062	0.56254	21.771	54.693	76.464	21.931	62.081	84.012	0.046508	0.11946	0.16597
70	84.900	0.012091	0.46995	0.48204	24.076	53.344	77.420	24.266	60.732	84.998	0.050906	0.11466	0.16557
80	98.850	0.012276	0.40267	0.41495	26.408	51.952	78.360	26.633	59.323	85.956	0.055273	0.10992	0.16520
90	114.42	0.012474	0.34617	0.35864	28.768	50.514	79.281	29.032	57.848	86.880	0.059613	0.10524	0.16485
100	131.72	0.012685	0.29838	0.31106	31.156	49.023	80.179	31.466	56.301	87.766	0.063928	0.10060	0.16452

TABLE C.1E SATURATED REFRIGERANT-12—TEMPERATURE TABLE: ENGLISH UNITS (*Continued*)

		Specific volume, ft^3/lb_m			Internal energy, Btu/lb_m			Enthalpy, Btu/lb_m			Entropy, $Btu/lb_m \cdot °R$		
Temp., °F T	Press., psia, P	Sat. liquid, v_f	Evap., v_{fg}	Sat. vapor, v_g	Sat. liquid, u_f	Evap., u_{fg}	Sat. vapor, u_g	Sat. liquid, h_f	Evap., h_{fg}	Sat. vapor, h_g	Sat. liquid, s_f	Evap., s_{fg}	Sat. vapor, s_g
110	150.87	0.012914	0.25768	0.27060	33.576	47.473	81.049	33.937	54.672	88.609	0.068224	0.09597	0.16420
120	171.99	0.013162	0.22281	0.23597	36.028	45.857	81.885	36.448	52.953	89.400	0.072504	0.09135	0.16385
130	195.20	0.013433	0.19272	0.20615	38.517	44.164	82.681	39.003	51.130	90.133	0.076777	0.08671	0.16349
140	220.63	0.013731	0.16659	0.18032	41.047	42.381	83.428	41.608	49.187	90.795	0.081050	0.08202	0.16307
150	248.43	0.014063	0.14374	0.15780	43.624	40.493	84.116	44.271	47.105	91.375	0.085335	0.07726	0.16260
160	278.74	0.014438	0.12360	0.13804	46.257	38.474	84.731	47.002	44.854	91.856	0.089649	0.07238	0.16203
170	311.74	0.014864	0.10570	0.12057	48.960	36.295	85.254	49.818	42.396	92.214	0.094013	0.06733	0.16134
180	347.61	0.015360	0.08963	0.10499	51.752	33.907	85.660	52.741	39.677	92.418	0.098461	0.06203	0.16049
190	386.56	0.015950	0.075014	0.090963	54.666	31.242	85.908	55.808	36.612	92.419	0.10304	0.05635	0.15940
200	428.84	0.016672	0.061459	0.078132	57.751	28.184	85.935	59.075	33.064	92.139	0.10784	0.05012	0.15796
210	474.75	0.017604	0.048504	0.066108	61.099	24.519	85.618	62.646	28.784	91.430	0.11299	0.04298	0.15597
220	524.70	0.018921	0.035342	0.054263	64.910	19.760	84.670	66.748	23.194	89.942	0.11882	0.03413	0.15295
230	579.33	0.021456	0.018863	0.040319	69.992	11.792	81.784	72.294	13.815	86.110	0.12662	0.02003	0.14665
233.2	598.3	0.02871	0.0	0.02871	76.22	0.0	76.22	79.40	0.0	79.4	0.1368	0.0	0.1368

Values generated from property formulation given in *ASHRAE Thermodynamic Properties of Refrigerants*. R.B. Stewart, R.T. Jacobsen and S.G. Penoncello, ASHRAE, Inc. 1986.

TABLE C.2E SATURATED REFRIGERANT-12—PRESSURE TABLE: ENGLISH UNITS

		Specific volume, ft^3/lb_m			Internal energy, Btu/lb_m			Enthalpy, Btu/lb_m			Entropy, $Btu/lb_m \cdot °R$		
Press., psia, P	Temp., °F, T	Sat. liquid, v_f	Evap., v_{fg}	Sat. vapor, v_g	Sat. liquid, u_f	Evap., u_{fg}	Sat. vapor, u_g	Sat. liquid, h_f	Evap., h_{fg}	Sat. vapor, h_g	Sat. liquid, s_f	Evap., s_{fg}	Sat. vapor, s_g
0.4	−132.65	0.0096744	72.312	72.322	−18.389	75.999	57.610	−18.389	81.355	62.967	−0.050790	0.24878	0.19799
0.6	−122.29	0.0097637	49.680	49.690	−16.632	75.197	58.566	−16.631	80.717	64.086	−0.044528	0.23925	0.19472
0.8	−114.78	0.0098298	38.050	38.060	−15.290	74.557	59.268	−15.288	80.194	64.906	−0.040107	0.23252	0.19242
1.0	−109.07	0.0098810	30.915	30.925	−14.184	73.990	59.806	−14.182	79.714	65.532	−0.036795	0.22737	0.19057
1.5	−98.51	0.0099776	21.184	21.194	−12.045	72.854	60.809	−12.042	78.738	66.696	−0.030787	0.21801	0.18723
2.0	−90.60	0.010052	16.203	16.213	−10.428	71.996	61.567	−10.424	77.996	67.572	−0.026390	0.21133	0.18494
3.0	−78.70	0.010167	11.110	11.120	−7.9954	70.715	62.719	−7.9897	76.887	68.897	−0.019929	0.20182	0.18189
4.0	−69.67	0.010256	8.5020	8.5123	−6.1473	69.748	63.601	−6.1397	76.046	69.906	−0.015142	0.19499	0.17985
5.0	−62.31	0.010331	6.9089	6.9192	−4.6360	68.960	64.324	−4.6265	75.357	70.730	−0.011306	0.18964	0.17834
7.5	−48.07	0.010480	4.7380	4.7485	−1.6954	67.428	65.733	−1.6808	74.008	72.328	−0.004036	0.17981	0.17577
10.0	−37.22	0.010599	3.6243	3.6349	0.5618	66.250	66.812	0.5815	72.962	73.543	0.001378	0.17271	0.17409
12.5	−28.34	0.010699	2.9431	2.9538	2.4226	65.276	67.699	2.4474	72.088	74.536	0.005737	0.16713	0.17287
15.0	−20.76	0.010787	2.4819	2.4927	4.0212	64.436	68.457	4.0511	71.329	75.381	0.009412	0.16251	0.17193
17.5	−14.11	0.010867	2.1481	2.1590	5.4319	63.691	69.123	5.4672	70.652	76.119	0.012602	0.15857	0.17117
20	−8.15	0.010939	1.8950	1.9059	6.7009	63.018	69.719	6.7414	70.036	76.778	0.015432	0.15511	0.17055
25	2.20	0.011070	1.5357	1.5468	8.9259	61.831	70.757	8.9771	68.940	77.917	0.020305	0.14926	0.16957
30	11.08	0.011186	1.2922	1.3034	10.849	60.795	71.644	10.911	67.974	78.885	0.024430	0.14440	0.16883
40	25.88	0.011389	0.98218	0.99357	14.095	59.024	73.119	14.180	66.299	80.478	0.031223	0.13654	0.16777
50	38.10	0.011568	0.79214	0.80371	16.812	57.516	74.328	16.919	64.850	81.770	0.036752	0.13028	0.16703
60	48.61	0.011731	0.66324	0.67497	19.176	56.183	75.358	19.306	63.552	82.858	0.041454	0.12503	0.16649
70	57.87	0.011882	0.56983	0.58171	21.284	54.975	76.258	21.438	62.361	83.798	0.045567	0.12049	0.16606
80	66.20	0.012024	0.49887	0.51089	23.197	53.861	77.058	23.375	61.251	84.626	0.049238	0.11648	0.16571
90	73.79	0.012160	0.44305	0.45521	24.956	52.822	77.778	25.159	60.206	85.364	0.052564	0.11286	0.16542
100	80.78	0.012291	0.39793	0.41022	26.590	51.842	78.432	26.818	59.211	86.029	0.055611	0.10956	0.16517
125	96.24	0.012604	0.31545	0.32806	30.254	49.590	79.844	30.545	56.892	87.438	0.062306	0.10234	0.16465
150	109.56	0.012904	0.25933	0.27223	33.470	47.542	81.012	33.828	54.745	88.573	0.068037	0.09617	0.16421

TABLE C.2E SATURATED REFRIGERANT-12—PRESSURE TABLE: ENGLISH UNITS (*Continued*)

Press., psia, P	Temp., °F, T	Specific volume, ft^3/lb_m Sat. liquid, v_f	Evap., v_{fg}	Sat. vapor, v_g	Internal energy, Btu/lb_m Sat. liquid, u_f	Evap., u_{fg}	Sat. vapor, u_g	Enthalpy, Btu/lb_m Sat. liquid, h_f	Evap., h_{fg}	Sat. vapor, h_g	Entropy, $Btu/lb_m \cdot °R$ Sat. liquid, s_f	Evap., s_{fg}	Sat. vapor, s_g
175	121.35	0.013197	0.21848	0.23168	36.362	45.633	81.995	36.790	52.713	89.503	0.073082	0.09072	0.16381
200	131.96	0.013489	0.18731	0.20080	39.009	43.822	82.832	39.509	50.759	90.268	0.077613	0.08580	0.16341
225	141.63	0.013783	0.16265	0.17644	41.464	42.081	83.545	42.038	48.858	90.896	0.081748	0.08125	0.16300
250	150.54	0.014083	0.14258	0.15667	43.764	40.387	84.151	44.416	46.988	91.404	0.085567	0.07700	0.16257
275	158.81	0.014391	0.12587	0.14026	45.940	38.722	84.662	46.673	45.131	91.805	0.089134	0.07297	0.16211
300	166.54	0.014710	0.11167	0.12638	48.015	37.070	85.085	48.832	43.274	92.106	0.092495	0.06910	0.16160
325	173.79	0.015043	0.099410	0.11445	50.007	35.416	85.424	50.913	41.399	92.312	0.095689	0.06535	0.16104
350	180.64	0.015395	0.088660	0.10406	51.935	33.746	85.681	52.932	39.492	92.425	0.098749	0.06168	0.16043
400	193.27	0.016169	0.070485	0.086653	55.653	30.293	85.946	56.850	35.514	92.364	0.10458	0.05439	0.15897
450	204.71	0.017078	0.055315	0.072394	59.288	26.552	85.840	60.711	31.161	91.872	0.11021	0.04690	0.15711
500	215.16	0.018215	0.041819	0.060034	62.986	22.255	85.241	64.673	26.127	90.799	0.11588	0.03872	0.15460
550	224.75	0.019829	0.028447	0.048276	67.032	16.729	83.762	69.052	19.627	88.678	0.12206	0.02868	0.15074
598.3	233.2	0.02871	0.0	0.02871	76.22	0.0	76.22	79.40	0.0	79.40	0.1368	0.0	0.1368

Values generated from property formulation given in *ASHRAE Thermodynamic Properties of Refrigerants*. R.B. Stewart, R.T. Jacobsen and S.G. Penoncello, ASHRAE, Inc. 1986.

TABLE C.3E SUPERHEATED REFRIGERANT-12: ENGLISH UNITS

P = 10 psia (−37.220°F)

Temp., °F	v, ft³/lbm	u, Btu/lbm	h, Btu/lbm	s, Btu/lbm·°R
Sat.	3.6349	66.812	73.543	0.17409
−20	3.7999	68.810	75.847	0.17943
0	3.9891	71.173	78.559	0.18547
20	4.1761	73.581	81.314	0.19133
40	4.3615	76.036	84.112	0.19705
60	4.5457	78.538	86.955	0.20263
80	4.7289	81.086	89.843	0.20808
100	4.9113	83.680	92.774	0.21341
120	5.0931	86.318	95.749	0.21863
140	5.2744	89.000	98.767	0.22375
160	5.4552	91.72	101.83	0.22877
180	5.6357	94.49	104.92	0.23369
200	5.8159	97.29	108.06	0.23852

P = 15 psia (−20.758°F)

Temp., °F	v, ft³/lbm	u, Btu/lbm	h, Btu/lbm	s, Btu/lbm·°R
Sat.	2.4927	68.457	75.381	0.17193
−20	2.4977	68.547	75.485	0.17216
0	2.6280	70.949	78.248	0.17831
20	2.7562	73.388	81.043	0.18426
40	2.8827	75.868	83.875	0.19004
60	3.0078	78.390	86.744	0.19567
80	3.1320	80.954	89.653	0.20117
100	3.2553	83.561	92.603	0.20653
120	3.3780	86.211	95.593	0.21178
140	3.5001	88.902	98.624	0.21692
160	3.6218	91.63	101.69	0.22196
180	3.7432	94.40	104.80	0.22689
200	3.8642	97.21	107.95	0.23174

P = 20 psia (−8.155°F)

Temp., °F	v, ft³/lbm	u, Btu/lbm	h, Btu/lbm	s, Btu/lbm·°R
Sat.	1.9059	69.719	76.778	0.17055
0	1.9469	70.718	77.929	0.17307
20	2.0458	73.190	80.767	0.17912
40	2.1429	75.696	83.632	0.18497
60	2.2386	78.239	86.529	0.19065
80	2.3333	80.820	89.461	0.19619
100	2.4271	83.441	92.430	0.20159
120	2.5202	86.102	95.436	0.20687
140	2.6129	88.802	98.479	0.21203
160	2.7050	91.54	101.56	0.21708
180	2.7968	94.32	104.68	0.22203
200	2.8883	97.13	107.83	0.22689
220	2.9796	99.99	111.02	0.23165

P = 30 psia (11.076°F)

Temp., °F	v, ft³/lbm	u, Btu/lbm	h, Btu/lbm	s, Btu/lbm·°R
Sat.	1.3034	71.644	78.885	0.16883
20	1.3344	72.780	80.193	0.17158
40	1.4024	75.342	83.132	0.17758
60	1.4688	77.929	86.089	0.18338
80	1.5342	80.546	89.069	0.18901
100	1.5986	83.196	92.076	0.19448
120	1.6623	85.881	95.115	0.19982
140	1.7254	88.601	98.186	0.20502
160	1.7881	91.36	101.29	0.21012
180	1.8504	94.15	104.43	0.21510
200	1.9123	96.97	107.60	0.21998
220	1.9740	99.84	110.80	0.22476

P = 40 psia (25.883°F)

Temp., °F	v, ft³/lbm	u, Btu/lbm	h, Btu/lbm	s, Btu/lbm·°R
Sat.	0.99357	73.119	80.478	0.16777
40	1.0313	74.972	82.610	0.17209
60	1.0834	77.607	85.631	0.17802
80	1.1342	80.263	88.663	0.18375
100	1.1840	82.944	91.713	0.18930
120	1.2330	85.654	94.787	0.19469
140	1.2815	88.394	97.886	0.19995
160	1.3294	91.17	101.01	0.20508
180	1.3770	93.97	104.17	0.21010
200	1.4242	96.81	107.36	0.21500
220	1.4712	99.68	110.58	0.21981
240	1.5179	102.59	113.83	0.22452
260	1.5644	105.52	117.11	0.22915

P = 50 psia (38.104°F)

Temp., °F	v, ft³/lbm	u, Btu/lbm	h, Btu/lbm	s, Btu/lbm·°R
Sat.	0.80371	74.328	81.770	0.16703
40	0.80793	74.583	82.064	0.16762
60	0.85154	77.272	85.156	0.17369
80	0.89376	79.970	88.245	0.17952
100	0.93492	82.684	91.340	0.18515
120	0.97526	85.421	94.451	0.19061
140	1.0149	88.184	97.581	0.19592
160	1.0541	90.97	100.73	0.20110
180	1.0929	93.80	103.91	0.20615
200	1.1313	96.65	107.12	0.21108
220	1.1694	99.53	110.36	0.21591
240	1.2073	102.44	113.62	0.22065
260	1.2450	105.38	116.91	0.22529

TABLE C.3E SUPERHEATED REFRIGERANT-12: ENGLISH UNITS (*Continued*)

Temp., °F	v, ft^3/lb$_m$	u, Btu/lb$_m$	h, Btu/lb$_m$	s, Btu/lb$_m$·°R
		P = 60 psia (48.606°F)		
Sat.	0.67497	75.358	82.858	0.16649
60	0.69650	76.923	84.661	0.17000
80	0.73313	79.666	87.811	0.17594
100	0.76861	82.416	90.956	0.18167
120	0.80320	85.182	94.106	0.18720
140	0.83710	87.968	97.269	0.19256
160	0.87045	90.78	100.45	0.19778
180	0.90335	93.61	103.65	0.20286
200	0.93590	96.48	106.88	0.20783
220	0.96815	99.37	110.13	0.21268
240	1.0002	102.29	113.41	0.21744
260	1.0319	105.25	116.71	0.22209
280	1.0636	108.23	120.04	0.22666

Temp., °F	v, ft^3/lb$_m$	u, Btu/lb$_m$	h, Btu/lb$_m$	s, Btu/lb$_m$·°R
		P = 70 psia (57.871°F)		
Sat.	0.58171	76.258	83.798	0.16606
60	0.58529	76.556	84.143	0.16673
80	0.61806	79.350	87.362	0.17280
100	0.64957	82.140	90.560	0.17862
120	0.68011	84.936	93.752	0.18423
140	0.70992	87.747	96.949	0.18965
160	0.73914	90.58	100.16	0.19491
180	0.76790	93.43	103.38	0.20004
200	0.79628	96.31	106.63	0.20503
220	0.82435	99.21	109.90	0.20991
240	0.85216	102.15	113.19	0.21469
260	0.87976	105.10	116.51	0.21936
280	0.90717	108.09	119.85	0.22394

Temp., °F	v, ft^3/lb$_m$	u, Btu/lb$_m$	h, Btu/lb$_m$	s, Btu/lb$_m$·°R
		P = 80 psia (66.199°F)		
Sat.	0.51089	77.058	84.626	0.16571
80	0.53143	79.021	86.894	0.16997
100	0.56005	81.854	90.150	0.17590
120	0.58762	84.684	93.389	0.18158
140	0.61440	87.521	96.623	0.18707
160	0.64056	90.373	99.862	0.19238
180	0.66623	93.24	103.11	0.19754
200	0.69150	96.14	106.38	0.20257
220	0.71645	99.05	109.67	0.20748
240	0.74113	101.99	112.97	0.21228
260	0.76558	104.96	116.30	0.21697
280	0.78985	107.96	119.66	0.22157
300	0.81396	110.98	123.04	0.22607

Temp., °F	v, ft^3/lb$_m$	u, Btu/lb$_m$	h, Btu/lb$_m$	s, Btu/lb$_m$·°R
		P = 90 psia (73.788°F)		
Sat.	0.45521	77.778	85.364	0.16542
80	0.46372	78.678	86.406	0.16736
100	0.49019	81.557	89.726	0.17341
120	0.51552	84.423	93.014	0.17918
140	0.53999	87.289	96.288	0.18473
160	0.56379	90.163	99.559	0.19010
180	0.58708	93.05	102.84	0.19530
200	0.60995	95.96	106.13	0.20036
220	0.63248	98.89	109.43	0.20530
240	0.65473	101.84	112.75	0.21012
260	0.67675	104.82	116.10	0.21483
280	0.69858	107.82	119.46	0.21944
300	0.72025	110.85	122.85	0.22396

Temp., °F	v, ft^3/lb$_m$	u, Btu/lb$_m$	h, Btu/lb$_m$	s, Btu/lb$_m$·°R
		P = 100 psia (80.777°F)		
Sat.	0.41022	78.432	86.029	0.16517
100	0.43408	81.249	89.287	0.17110
120	0.45767	84.154	92.629	0.17696
140	0.48033	87.051	95.945	0.18259
160	0.50229	89.949	99.250	0.18801
180	0.52369	92.86	102.56	0.19326
200	0.54465	95.78	105.87	0.19836
220	0.56526	98.72	109.19	0.20332
240	0.58558	101.69	112.53	0.20816
260	0.60566	104.67	115.89	0.21289
280	0.62555	107.68	119.27	0.21752
300	0.64526	110.72	122.67	0.22206
320	0.66483	113.78	126.09	0.22650

Temp., °F	v, ft^3/lb$_m$	u, Btu/lb$_m$	h, Btu/lb$_m$	s, Btu/lb$_m$·°R
		P = 120 psia (93.339°F)		
Sat.	0.34187	79.584	87.180	0.16474
100	0.34925	80.593	88.353	0.16685
120	0.37044	83.588	91.820	0.17294
140	0.39052	86.553	95.231	0.17872
160	0.40979	89.506	98.611	0.18427
180	0.42843	92.46	101.98	0.18961
200	0.44657	95.42	105.34	0.19479
220	0.46433	98.39	108.70	0.19981
240	0.48178	101.37	112.08	0.20471
260	0.49897	104.38	115.47	0.20948
280	0.51594	107.40	118.87	0.21415
300	0.53273	110.45	122.29	0.21871
320	0.54937	113.52	125.73	0.22318

R-12

TABLE C.3E SUPERHEATED REFRIGERANT-12: ENGLISH UNITS (*Continued*)

P = 140 psia (104.444°F)

Temp., °F	v, ft³/lbm	u, Btu/lbm	h, Btu/lbm	s, Btu/lbm·°R
Sat.	0.29228	80.569	88.147	0.16438
120	0.30752	82.979	90.951	0.16928
140	0.32595	86.025	94.475	0.17526
160	0.34341	89.039	97.941	0.18095
180	0.36015	92.04	101.37	0.18640
200	0.37635	95.03	104.79	0.19166
220	0.39211	98.04	108.20	0.19675
240	0.40753	101.05	111.61	0.20170
260	0.42268	104.08	115.03	0.20652
280	0.43759	107.12	118.46	0.21122
300	0.45231	110.18	121.91	0.21582
320	0.46687	113.27	125.37	0.22031
340	0.48130	116.37	128.85	0.22472

P = 160 psia (114.440°F)

Temp., °F	v, ft³/lbm	u, Btu/lbm	h, Btu/lbm	s, Btu/lbm·°R
Sat.	0.25457	81.425	88.967	0.16405
120	0.25968	82.315	90.009	0.16585
140	0.27708	85.460	93.669	0.17206
160	0.29332	88.546	97.236	0.17791
180	0.30872	91.60	100.75	0.18349
200	0.32351	94.64	104.23	0.18884
220	0.33782	97.68	107.69	0.19401
240	0.35176	100.72	111.14	0.19902
260	0.36539	103.77	114.59	0.20389
280	0.37877	106.83	118.05	0.20863
300	0.39195	109.91	121.52	0.21326
320	0.40496	113.01	125.01	0.21778
340	0.41783	116.12	128.50	0.22221

P = 180 psia (123.557°F)

Temp., °F	v, ft³/lbm	u, Btu/lbm	h, Btu/lbm	s, Btu/lbm·°R
Sat.	0.22486	82.173	89.668	0.16373
140	0.23860	84.851	92.803	0.16903
160	0.25404	88.023	96.490	0.17508
180	0.26850	91.14	100.09	0.18080
200	0.28226	94.23	103.64	0.18626
220	0.29548	97.30	107.15	0.19151
240	0.30829	100.38	110.65	0.19658
260	0.32076	103.45	114.14	0.20150
280	0.33297	106.53	117.63	0.20628
300	0.34497	109.63	121.13	0.21095
320	0.35678	112.74	124.64	0.21550
340	0.36844	115.87	128.15	0.21996
360	0.37997	119.02	131.68	0.22432

P = 200 psia (131.958°F)

Temp., °F	v, ft³/lbm	u, Btu/lbm	h, Btu/lbm	s, Btu/lbm·°R
Sat.	0.20080	82.832	90.268	0.16341
140	0.20729	84.187	91.864	0.16609
160	0.22229	87.464	95.696	0.17238
180	0.23610	90.656	99.400	0.17826
200	0.24910	93.80	103.03	0.18384
220	0.26149	96.92	106.60	0.18918
240	0.27342	100.02	110.15	0.19433
260	0.28500	103.13	113.68	0.19930
280	0.29629	106.23	117.21	0.20413
300	0.30735	109.35	120.73	0.20884
320	0.31821	112.48	124.26	0.21342
340	0.32891	115.62	127.80	0.21790
360	0.33948	118.78	131.35	0.22229

P = 250 psia (150.540°F)

Temp., °F	v, ft³/lbm	u, Btu/lbm	h, Btu/lbm	s, Btu/lbm·°R
Sat.	0.15666	84.151	91.404	0.16257
160	0.16358	85.857	93.430	0.16586
180	0.17680	89.310	97.495	0.17232
200	0.18875	92.63	101.37	0.17829
220	0.19985	95.88	105.14	0.18391
240	0.21034	99.09	108.83	0.18927
260	0.22038	102.28	112.48	0.19441
280	0.23007	105.45	116.10	0.19937
300	0.23949	108.62	119.70	0.20418
320	0.24868	111.79	123.30	0.20885
340	0.25768	114.97	126.90	0.21341
360	0.26653	118.16	130.49	0.21785
380	0.27525	121.36	134.10	0.22220

P = 300 psia (166.538°F)

Temp., °F	v, ft³/lbm	u, Btu/lbm	h, Btu/lbm	s, Btu/lbm·°R
Sat.	0.12638	85.085	92.106	0.16160
180	0.13561	87.684	95.217	0.16652
200	0.14750	91.292	99.486	0.17309
220	0.15809	94.73	103.51	0.17910
240	0.16783	98.08	107.40	0.18474
260	0.17698	101.36	111.19	0.19008
280	0.18569	104.61	114.93	0.19521
300	0.19408	107.85	118.63	0.20014
320	0.20219	111.07	122.30	0.20491
340	0.21010	114.29	125.96	0.20955
360	0.21783	117.52	129.62	0.21407
380	0.22542	120.75	133.27	0.21847

TABLE C.3E SUPERHEATED REFRIGERANT-12: ENGLISH UNITS (*Continued*)

Temp., °F	v, ft³/lbm	u, Btu/lbm	h, Btu/lbm	s, Btu/lbm·°R	Temp., °F	v, ft³/lbm	u, Btu/lbm	h, Btu/lbm	s, Btu/lbm·°R
	P = 350 psia (180.639°F)					P = 400 psia (193.270°F)			
Sat.	0.10405	85.681	92.425	0.16043	Sat.	0.086653	85.946	92.364	0.15897
200	0.11674	89.683	97.249	0.16785	200	0.091667	87.595	94.384	0.16205
220	0.12749	93.42	101.68	0.17447	220	0.10358	91.861	99.534	0.16974
240	0.13697	96.95	105.83	0.18048	240	0.11327	95.68	104.07	0.17633
260	0.14565	100.37	109.81	0.18610	260	0.12180	99.29	108.31	0.18230
280	0.15376	103.72	113.69	0.19141	280	0.12959	102.77	112.37	0.18786
300	0.16147	107.03	117.50	0.19650	300	0.13686	106.17	116.31	0.19312
320	0.16887	110.32	121.26	0.20139	320	0.14377	109.53	120.18	0.19815
340	0.17602	113.59	125.00	0.20612	340	0.15039	112.87	124.00	0.20299
360	0.18298	116.86	128.72	0.21071	360	0.15678	116.18	127.79	0.20767
380	0.18977	120.13	132.43	0.21518	380	0.16300	119.49	131.56	0.21221
400	0.19643	123.40	136.13	0.21954	400	0.16906	122.79	135.32	0.21663
420	0.20297	126.68	139.83	0.22380	420	0.17500	126.10	139.06	0.22094
	P = 450 psia (204.712°F)					P = 500 psia (215.162°F)			
Sat.	0.072394	85.840	91.872	0.15711	Sat.	0.060034	85.241	90.799	0.15460
220	0.083594	89.908	96.874	0.16456	220	0.064948	87.077	93.090	0.15798
240	0.094170	94.21	102.06	0.17208	240	0.077996	92.436	99.657	0.16751
260	0.10286	98.08	106.65	0.17855	260	0.087275	96.71	104.79	0.17475
280	0.11054	101.73	110.94	0.18444	280	0.095050	100.60	109.40	0.18106
300	0.11757	105.26	115.05	0.18992	300	0.10197	104.28	113.72	0.18683
320	0.12414	108.71	119.05	0.19512	320	0.10833	107.84	117.87	0.19222
340	0.13038	112.11	122.97	0.20008	340	0.11430	111.32	121.90	0.19733
360	0.13636	115.48	126.84	0.20486	360	0.11998	114.75	125.86	0.20222
380	0.14214	118.83	130.68	0.20948	380	0.12542	118.16	129.77	0.20693
400	0.14775	122.17	134.49	0.21397	400	0.13068	121.54	133.64	0.21149
420	0.15323	125.51	138.28	0.21833	420	0.13580	124.91	137.49	0.21591
440	0.15860	128.85	142.07	0.22259	440	0.14079	128.28	141.32	0.22022
	P = 550 psia (224.748°F)					P = 600 psia			
Sat.	0.048276	83.762	88.678	0.15074					
240	0.063384	90.116	96.571	0.16217	240	0.048281	86.432	91.796	0.15460
260	0.073984	95.13	102.66	0.17075	260	0.062212	93.20	100.12	0.16634
280	0.082099	99.34	107.70	0.17767	280	0.071001	97.93	105.82	0.17416
300	0.089050	103.22	112.29	0.18379	300	0.078120	102.07	110.75	0.18075
320	0.095303	106.91	116.62	0.18941	320	0.084349	105.93	115.30	0.18666
340	0.10109	110.49	120.79	0.19469	340	0.090015	109.63	119.63	0.19214
360	0.10653	114.00	124.85	0.19971	360	0.095289	113.22	123.81	0.19730
380	0.11172	117.46	128.84	0.20452	380	0.10027	116.75	127.89	0.20222
400	0.11670	120.89	132.78	0.20916	400	0.10504	120.23	131.90	0.20694
420	0.12153	124.31	136.68	0.21364	420	0.10963	123.68	135.86	0.21150
440	0.12622	127.71	140.56	0.21800	440	0.11408	127.12	139.79	0.21592
460	0.13081	131.10	144.42	0.22225	460	0.11841	130.54	143.70	0.22021

Values generated from property formulation given in *ASHRAE Thermodynamic Properties of Refrigerants*. R.B. Stewart, R.T. Jacobsen and S.G. Penoncello, ASHRAE, Inc. 1986.

R-12

Properties of Ideal Gases

TABLE D.1E IDEAL-GAS PROPERTIES OF AIR—MASS BASIS: ENGLISH UNITS
Mol. Wt.: 28.967

T, °R	h, Btu/lb$_m$	u, Btu/lb$_m$	s^0, Btu/lb$_m$·°R	c_p, Btu/lb$_m$·°R	P_r	v_r
0	0.0	0.0	0.0			
200	47.75	34.04	1.1234	0.23885	0.04265	746.07
220	52.54	37.45	1.1462	0.23885	0.05945	588.63
240	57.32	40.87	1.1672	0.23885	0.08079	472.91
260	62.11	44.28	1.1867	0.23885	0.10726	386.32
280	66.89	47.70	1.2046	0.23885	0.13941	320.42
300	71.68	51.11	1.2213	0.23885	0.17772	269.45
320	76.47	54.53	1.2367	0.23885	0.22269	229.39
340	81.25	57.94	1.2512	0.23885	0.27482	197.42
360	86.04	61.36	1.2647	0.23885	0.33472	171.50
380	90.83	64.78	1.2777	0.23885	0.40460	149.81
400	95.62	68.19	1.2900	0.23885	0.48410	131.81
420	100.40	71.61	1.3016	0.23885	0.57388	116.74
440	105.19	75.03	1.3127	0.23885	0.67468	104.01
460	109.98	78.45	1.3234	0.23885	0.78871	93.007
480	114.77	81.87	1.3338	0.23885	0.91712	83.464
500	119.56	85.29	1.3436	0.23885	1.0588	75.308
520	124.36	88.71	1.3530	0.23885	1.2133	68.341
540	129.16	92.14	1.3618	0.23885	1.3800	62.396
560	133.96	95.57	1.3703	0.23904	1.5629	57.136
580	138.77	99.01	1.3787	0.23951	1.7662	52.367
600	143.57	102.44	1.3869	0.24015	1.9908	48.062
620	148.38	105.88	1.3949	0.24087	2.2374	44.189
640	153.19	109.32	1.4026	0.24140	2.5039	40.759
660	158.01	112.77	1.4101	0.24171	2.7902	37.722
680	162.84	116.22	1.4173	0.24197	3.0992	34.989
700	167.67	119.68	1.4243	0.24221	3.4322	32.523
720	172.51	123.15	1.4311	0.24243	3.7903	30.291
740	177.35	126.62	1.4377	0.24265	4.1757	28.260
760	182.20	130.10	1.4442	0.24289	4.5888	26.411

TABLE D.1E IDEAL-GAS PROPERTIES OF AIR—MASS BASIS: ENGLISH UNITS (*Continued*)
Mol. Wt.: 28.967

T, °R	h, Btu/lb$_m$	u, Btu/lb$_m$	s^0, Btu/lb$_m$·°R	c_p, Btu/lb$_m$·°R	P_r	v_r
780	187.05	133.58	1.4505	0.24315	5.0310	24.724
800	191.92	137.07	1.4566	0.24345	5.5035	23.180
820	196.79	140.58	1.4627	0.24385	6.0085	21.763
840	201.68	144.09	1.4685	0.24433	6.5474	20.459
860	206.57	147.61	1.4743	0.24486	7.1212	19.258
880	211.47	151.14	1.4799	0.24542	7.7313	18.151
900	216.38	154.68	1.4855	0.24601	8.3792	17.128
920	221.30	158.23	1.4909	0.24654	9.0681	16.179
940	226.23	161.79	1.4962	0.24707	9.7983	15.299
960	231.18	165.36	1.5014	0.24760	10.571	14.482
980	236.13	168.95	1.5065	0.24813	11.389	13.722
1000	241.10	172.54	1.5115	0.24867	12.253	13.015
1050	253.56	181.58	1.5237	0.24999	14.628	11.447
1100	266.10	190.69	1.5353	0.25128	17.340	10.116
1150	278.72	199.88	1.5465	0.25251	20.426	8.9783
1200	291.42	209.15	1.5574	0.25383	23.915	8.0020
1250	304.19	218.49	1.5678	0.25526	27.841	7.1596
1300	317.05	227.92	1.5779	0.25712	32.254	6.4273
1350	329.98	237.43	1.5876	0.25915	37.191	5.7886
1400	342.99	247.02	1.5971	0.26118	42.694	5.2292
1450	356.09	256.68	1.6063	0.26304	48.812	4.7370
1500	369.25	266.42	1.6152	0.26447	55.607	4.3016
1550	382.50	276.24	1.6239	0.26578	63.120	3.9159
1600	395.81	286.12	1.6323	0.26702	71.406	3.5731
1650	409.20	296.08	1.6406	0.26830	80.526	3.2675
1700	422.66	306.11	1.6486	0.26963	90.540	2.9941
1750	436.18	316.21	1.6565	0.27096	101.51	2.7492
1800	449.77	326.37	1.6641	0.27228	113.49	2.5291
1850	463.42	336.59	1.6716	0.27371	126.57	2.3307
1900	477.13	346.87	1.6789	0.27509	140.82	2.1516
1950	490.90	357.21	1.6860	0.27637	156.29	1.9895
2000	504.72	367.61	1.6930	0.27738	173.09	1.8425
2100	532.54	388.57	1.7066	0.27870	211.00	1.5871
2200	560.55	409.73	1.7197	0.28018	255.19	1.3747
2300	588.76	431.08	1.7322	0.28219	306.39	1.1970
2400	617.14	452.60	1.7443	0.28429	365.38	1.0474
2500	645.68	474.29	1.7559	0.28625	433.02	0.92063
2600	674.38	496.13	1.7672	0.28775	510.32	0.81243
2700	703.22	518.11	1.7781	0.28900	598.20	0.71974
2800	732.19	540.23	1.7886	0.29033	697.57	0.64007
2900	761.28	562.47	1.7988	0.29168	809.56	0.57123
3000	790.49	584.83	1.8087	0.29306	935.37	0.51144

TABLE D.1E IDEAL-GAS PROPERTIES OF AIR—MASS BASIS: ENGLISH UNITS (*Continued*)
Mol. Wt.: 28.967

T, °R	h, Btu/lb$_m$	u, Btu/lb$_m$	s^0, Btu/lb$_m$·°R	c_p, Btu/lb$_m$·°R	P_r	v_r
3100	819.82	607.29	1.8183	0.29411	1076.2	0.45934
3200	849.24	629.86	1.8277	0.29475	1233.3	0.41377
3300	878.77	652.53	1.8367	0.29531	1408.0	0.37374
3400	908.38	675.29	1.8456	0.29600	1601.8	0.33848
3500	938.09	698.14	1.8542	0.29722	1816.2	0.30731
3600	967.88	721.07	1.8626	0.29856	2052.7	0.27967
3700	997.74	744.09	1.8708	0.29929	2313.0	0.25508
3800	1027.7	767.17	1.8788	0.29988	2598.9	0.23316
3900	1057.7	790.34	1.8866	0.30055	2912.0	0.21356
4000	1087.8	813.57	1.8942	0.30121	3254.2	0.19601
4100	1118.0	836.87	1.9016	0.30187	3627.4	0.18024
4200	1148.2	860.23	1.9089	0.30256	4033.7	0.16604
4300	1178.4	883.65	1.9160	0.30322	4475.1	0.15322
4400	1208.8	907.14	1.9230	0.30367	4954.2	0.14162
4500	1239.2	930.68	1.9298	0.30405	5473.2	0.13111
4600	1269.6	954.27	1.9365	0.30455	6035.3	0.12154
4700	1300.1	977.92	1.9431	0.30514	6641.7	0.11284
4800	1330.7	1001.6	1.9495	0.30582	7293.7	0.10494
4900	1361.3	1025.4	1.9558	0.30638	7996.0	0.09772
5000	1391.9	1049.2	1.9620	0.30677	8752.5	0.09110
5100	1422.6	1073.0	1.9681	0.30716	9565.0	0.08502
5200	1453.4	1096.9	1.9741	0.30755	10436	0.07946
5300	1484.2	1120.8	1.9799	0.30795	11368	0.07435
5400	1515.0	1144.8	1.9857	0.30835	12364	0.06965
5500	1545.9	1168.8	1.9914	0.30873	13429	0.06531
5600	1576.8	1192.8	1.9969	0.30917	14565	0.06131
5700	1607.7	1216.9	2.0024	0.30972	15775	0.05762
5800	1638.7	1241.1	2.0078	0.31015	17064	0.05420
5900	1669.7	1265.2	2.0131	0.31041	18438	0.05103
6000	1700.7	1289.4	2.0183	0.31065	19896	0.04809

Adapted from *Gas Tables International Version*, Thermodynamic Properties of Air Products of Combustion and Component Gases Compressible Flow Functions, Joseph Keenan, Jing Chao and Joseph Kaye, John Wiley and Sons, N.Y., 1983.

TABLE D.2E IDEAL-GAS PROPERTIES OF CARBON DIOXIDE (CO_2)—MASS BASIS: ENGLISH UNITS Mol. Wt.: 44.010

T, °R	h, Btu/lb$_m$	u, Btu/lb$_m$	s^0, Btu/lb$_m$·°R	c_p, Btu/lb$_m$·°R	P_r	v_r
0	0.0	0.0	0.0			
200	31.53	22.50	0.98692	0.15958	0.00063	33,208
250	39.51	28.23	1.0221	0.16328	0.00137	18,965
300	47.79	34.25	1.0530	0.16828	0.00271	11,534
350	56.37	40.58	1.0802	0.17431	0.00496	7,398.9
400	65.28	47.23	1.1044	0.18111	0.00848	4,959.2
450	74.54	54.24	1.1262	0.18841	0.01376	3,440.3
500	84.17	61.61	1.1463	0.19596	0.02146	2,446.8
550	94.18	69.36	1.1653	0.20348	0.03266	1,767.6
600	104.53	77.46	1.1833	0.21036	0.04866	1,294.3
650	115.18	85.85	1.2003	0.21653	0.07102	960.82
700	126.12	94.54	1.2165	0.22215	0.10174	722.24
750	137.35	103.51	1.2320	0.22753	0.14348	548.76
800	148.85	112.75	1.2469	0.23271	0.19945	421.13
850	160.61	122.25	1.2612	0.23759	0.27349	326.30
900	172.60	131.99	1.2748	0.24219	0.37035	255.06
950	184.82	141.95	1.2881	0.24655	0.49648	200.86
1000	197.25	152.13	1.3008	0.25067	0.65864	159.38
1100	222.71	173.08	1.3251	0.25828	1.1271	102.44
1200	248.89	194.74	1.3478	0.26518	1.8670	67.464
1300	275.72	217.06	1.3693	0.27141	3.0045	45.414
1400	303.14	239.97	1.3896	0.27704	4.7138	31.173
1500	331.11	263.42	1.4089	0.28214	7.2281	21.781
1600	359.56	287.36	1.4273	0.28676	10.857	15.468
1700	388.35	311.64	1.4448	0.29094	16.005	11.148
1800	417.55	336.33	1.4615	0.29474	23.193	8.1457
1900	447.28	361.55	1.4775	0.29818	33.086	6.0273
2000	477.35	387.10	1.4929	0.30132	46.517	4.5127
2100	507.63	412.87	1.5077	0.30418	64.532	3.4155
2200	538.17	438.90	1.5219	0.30680	88.419	2.6115
2300	568.97	465.19	1.5356	0.30919	119.76	2.0157
2400	600.00	491.71	1.5488	0.31136	160.48	1.5697
2500	631.24	518.43	1.5615	0.31332	212.87	1.2326
2600	662.66	545.34	1.5739	0.31514	279.73	0.97554
2700	694.26	572.43	1.5858	0.31683	364.35	0.77779
2800	726.02	599.68	1.5973	0.31841	470.64	0.62443
2900	757.94	627.08	1.6085	0.31986	603.20	0.50460
3000	789.99	654.62	1.6194	0.32118	767.43	0.41029
3100	822.17	682.29	1.6300	0.32240	969.61	0.33556
3200	854.47	710.07	1.6402	0.32356	1,217.0	0.27597
3300	886.88	737.97	1.6502	0.32466	1,518.1	0.22815
3400	919.40	765.98	1.6599	0.32569	1,882.4	0.18957
3500	952.02	794.08	1.6694	0.32664	2,321.2	0.15826
3600	984.72	822.28	1.6786	0.32753	2,847.1	0.13271
3700	1017.5	850.56	1.6876	0.32837	3,474.4	0.11177
3800	1050.4	878.93	1.6963	0.32916	4,219.5	0.09452
3900	1083.3	907.37	1.7049	0.32990	5,100.9	0.08025
4000	1116.4	935.88	1.7132	0.33060	6,139.4	0.06838
4100	1149.5	964.46	1.7214	0.33126	7,358.6	0.05848

TABLE D.2E IDEAL-GAS PROPERTIES OF CARBON DIOXIDE (CO_2)—MASS BASIS: ENGLISH UNITS Mol. Wt.: 44.010 (*Continued*)

T, °R	h, Btu/lb$_m$	u, Btu/lb$_m$	s^0, Btu/lb$_m$·°R	c_p, Btu/lb$_m$·°R	P_r	v_r
4200	1182.6	993.11	1.7294	0.33190	8,784.2	0.05018
4300	1215.8	1021.8	1.7372	0.33249	10,445	0.04321
4400	1249.1	1050.6	1.7449	0.33306	12,375	0.03732
4500	1282.5	1079.4	1.7524	0.33361	14,609	0.03233
4600	1315.8	1108.3	1.7597	0.33415	17,190	0.02809
4700	1349.3	1137.2	1.7669	0.33466	20,161	0.02447
4800	1382.8	1166.2	1.7739	0.33513	23,569	0.02137
4900	1416.3	1195.2	1.7809	0.33559	27,473	0.01872
5000	1449.9	1224.3	1.7876	0.33604	31,931	0.01643
5100	1483.5	1253.4	1.7943	0.33648	37,009	0.01446
5200	1517.2	1282.6	1.8008	0.33691	42,778	0.01276
5300	1550.9	1311.8	1.8073	0.33733	49,321	0.01128
5400	1584.7	1341.0	1.8136	0.33772	56,725	0.00999
5500	1618.4	1370.3	1.8198	0.33809	65,079	0.00887
5600	1652.3	1399.6	1.8259	0.33843	74,488	0.00789
5700	1686.1	1428.9	1.8319	0.33878	85,066	0.00703
5800	1720.0	1458.3	1.8378	0.33912	96,938	0.00628
5900	1754.0	1487.7	1.8436	0.33946	110,240	0.00562
6000	1787.9	1517.2	1.8493	0.33979	125,100	0.00503
6100	1821.9	1546.7	1.8549	0.34011	141,700	0.00452
6200	1855.9	1576.2	1.8604	0.34043	160,180	0.00406
6300	1890.0	1605.7	1.8659	0.34074	180,740	0.00366
6400	1924.1	1635.3	1.8712	0.34104	203,570	0.00330
6500	1958.2	1664.9	1.8765	0.34134	228,890	0.00298
6600	1992.4	1694.5	1.8817	0.34163	256,940	0.00270
6700	2026.5	1724.2	1.8869	0.34191	287,940	0.00244
6800	2060.7	1753.9	1.8919	0.34220	322,140	0.00222
6900	2095.0	1783.6	1.8969	0.34248	359,870	0.00201
7000	2129.2	1813.4	1.9019	0.34275	401,430	0.00183
7100	2163.5	1843.2	1.9067	0.34302	447,110	0.00167
7200	2197.8	1873.0	1.9115	0.34329	497,270	0.00152
7300	2232.2	1902.8	1.9163	0.34355	552,310	0.00139
7400	2266.5	1932.6	1.9209	0.34381	612,630	0.00127
7500	2300.9	1962.5	1.9256	0.34407	678,650	0.00116
7600	2335.4	1992.4	1.9301	0.34432	750,810	0.00106
7700	2369.8	2022.4	1.9346	0.34457	829,580	0.00097
7800	2404.3	2052.3	1.9391	0.34481	915,500	0.00089
7900	2438.8	2082.3	1.9435	0.34505	1,009,100	0.00082
8000	2473.3	2112.3	1.9478	0.34529	1,111,100	0.00076
8100	2507.8	2142.3	1.9521	0.34553	1,221,900	0.00070
8200	2542.4	2172.4	1.9563	0.34576	1,342,300	0.00064
8300	2577.0	2202.5	1.9605	0.34600	1,473,000	0.00059
8400	2611.6	2232.6	1.9647	0.34624	1,614,800	0.00055
8500	2633.7	2250.1	1.9688	0.34646	1,768,300	0.00050
8600	2662.6	2274.5	1.9728	0.34667	1,934,500	0.00047
8700	2847.8	2455.3	1.9768	0.34689	2,114,100	0.00043
8800	3083.4	2686.3	1.9808	0.34711	2,308,300	0.00040
8900	3007.9	2606.3	1.9847	0.34735	2,518,100	0.00037
9000	2819.7	2413.6	1.9886	0.34758	2,744,600	0.00034

Adapted from *JANAF Thermochemical Tables* Third Edition, M. W. Chase et al., American Chemical Society and the American Institute of Physics for the National Bureau of Standards, 1985.

TABLE D.3E IDEAL-GAS PROPERTIES OF CARBON MONOXIDE (CO)—MASS BASIS: ENGLISH UNITS Mol. Wt.: 28.010

T, °R	h, Btu/lb$_m$	u, Btu/lb$_m$	s^0, Btu/lb$_m$·°R	c_p, Btu/lb$_m$·°R	P_r	v_r
0	0.0	0.0	0.0			
200	49.52	35.34	1.4385	0.24812	0.00188	17,383
250	61.93	44.21	1.4930	0.24807	0.00406	10,040
300	74.34	53.07	1.5395	0.24810	0.00782	6,286.1
350	86.74	61.93	1.5790	0.24818	0.01367	4,218.1
400	99.14	70.78	1.6129	0.24829	0.02204	2,998.6
450	111.55	79.64	1.6422	0.24840	0.03331	2,232.7
500	123.97	88.52	1.6680	0.24847	0.04794	1,721.2
550	136.41	97.41	1.6915	0.24850	0.06677	1,358.3
600	148.85	106.31	1.7132	0.24871	0.09064	1,091.8
650	161.30	115.22	1.7331	0.24918	0.12012	892.64
700	173.77	124.14	1.7516	0.24987	0.15586	740.73
750	186.28	133.11	1.7689	0.25071	0.19893	621.87
800	198.84	142.12	1.7851	0.25168	0.25019	527.50
850	211.45	151.19	1.8004	0.25280	0.31032	451.83
900	224.12	160.31	1.8148	0.25405	0.38027	390.29
950	236.86	169.51	1.8286	0.25547	0.46206	339.09
1000	249.68	178.78	1.8418	0.25699	0.55624	296.52
1100	275.54	197.55	1.8664	0.26026	0.78709	230.47
1200	301.74	216.66	1.8892	0.26369	1.0857	182.29
1300	328.29	236.12	1.9104	0.26719	1.4648	146.36
1400	355.18	255.93	1.9304	0.27065	1.9404	118.99
1500	382.41	276.07	1.9492	0.27399	2.5291	97.806
1600	409.97	296.53	1.9669	0.27717	3.2501	81.182
1700	437.84	317.32	1.9838	0.28016	4.1247	67.968
1800	466.00	338.39	1.9999	0.28295	5.1753	57.355
1900	494.42	359.72	2.0153	0.28552	6.4283	48.742
2000	523.10	381.30	2.0300	0.28791	7.9099	41.696
2100	552.00	403.12	2.0441	0.29015	9.6504	35.885
2200	581.12	425.15	2.0576	0.29221	11.682	31.056
2300	610.44	447.38	2.0707	0.29409	14.040	27.016
2400	639.94	469.79	2.0832	0.29583	16.759	23.615
2500	669.60	492.36	2.0953	0.29746	19.881	20.737
2600	699.42	515.09	2.1070	0.29893	23.447	18.286
2700	729.38	537.96	2.1183	0.30030	27.502	16.190
2800	759.48	560.96	2.1293	0.30158	32.092	14.388
2900	789.70	584.09	2.1399	0.30277	37.268	12.832
3000	820.03	607.34	2.1502	0.30388	43.086	11.482
3100	850.47	630.69	2.1601	0.30490	49.599	10.307
3200	881.01	654.14	2.1698	0.30585	56.868	9.2795
3300	911.63	677.67	2.1793	0.30675	64.955	8.3780
3400	942.35	701.30	2.1884	0.30759	73.926	7.5844
3500	973.15	725.01	2.1974	0.30837	83.846	6.8837
3600	1,004.0	748.80	2.2061	0.30911	94.787	6.2632
3700	1,035.0	772.66	2.2145	0.30980	106.83	5.7116
3800	1,066.0	796.58	2.2228	0.31044	120.04	5.2202
3900	1,097.1	820.55	2.2309	0.31104	134.52	4.7810
4000	1,128.2	844.63	2.2388	0.31175	150.35	4.3873
4100	1,159.4	868.77	2.2465	0.31236	167.62	4.0336

TABLE D.3E IDEAL-GAS PROPERTIES OF CARBON MONOXIDE (CO)—MASS BASIS: ENGLISH UNITS Mol. Wt.: 28.010 (*Continued*)

T, °R	h, Btu/lb$_m$	u, Btu/lb$_m$	s^0, Btu/lb$_m$·°R	c_p, Btu/lb$_m$·°R	P_r	v_r
4200	1190.6	892.80	2.2540	0.31137	186.32	3.7173
4300	1221.6	916.75	2.2613	0.30986	206.54	3.4332
4400	1252.7	940.80	2.2684	0.31143	228.50	3.1755
4500	1284.0	964.97	2.2755	0.31410	252.31	2.9412
4600	1315.4	989.30	2.2824	0.31474	278.11	2.7276
4700	1346.9	1013.7	2.2891	0.31493	305.98	2.5331
4800	1378.4	1038.1	2.2958	0.31530	335.99	2.3559
4900	1410.0	1062.6	2.3023	0.31568	368.28	2.1941
5000	1441.6	1087.1	2.3087	0.31606	402.96	2.0462
5100	1473.2	1111.6	2.3149	0.31641	440.18	1.9107
5200	1504.8	1136.2	2.3211	0.31672	480.06	1.7863
5300	1536.5	1160.8	2.3271	0.31704	522.71	1.6721
5400	1568.3	1185.4	2.3330	0.31735	568.28	1.5670
5500	1600.0	1210.1	2.3389	0.31765	616.98	1.4700
5600	1631.8	1234.8	2.3446	0.31794	668.90	1.3806
5700	1663.6	1259.5	2.3502	0.31822	724.15	1.2980
5800	1695.4	1284.2	2.3558	0.31849	782.96	1.2216
5900	1727.3	1309.0	2.3612	0.31874	845.49	1.1508
6000	1759.2	1333.8	2.3666	0.31899	911.89	1.0850
6100	1791.1	1358.6	2.3718	0.31923	982.32	1.0240
6200	1823.0	1383.5	2.3770	0.31947	1057.0	0.96732
6300	1855.0	1408.3	2.3821	0.31970	1136.0	0.91453
6400	1887.0	1433.2	2.3872	0.31994	1219.7	0.86533
6500	1919.0	1458.1	2.3921	0.32018	1308.1	0.81945
6600	1951.0	1483.1	2.3970	0.32040	1401.5	0.77661
6700	1983.0	1508.0	2.4019	0.32060	1500.0	0.73657
6800	2015.1	1533.0	2.4066	0.32080	1604.0	0.69911
6900	2047.2	1558.0	2.4113	0.32100	1713.6	0.66403
7000	2079.3	1583.0	2.4159	0.32120	1828.9	0.63116
7100	2111.4	1608.1	2.4205	0.32140	1950.5	0.60029
7200	2143.6	1633.1	2.4250	0.32160	2078.2	0.57131
7300	2175.8	1658.2	2.4294	0.32179	2212.4	0.54414
7400	2207.9	1683.3	2.4338	0.32198	2353.2	0.51857
7500	2240.2	1708.4	2.4381	0.32216	2501.3	0.49446
7600	2272.4	1733.6	2.4424	0.32234	2656.6	0.47177
7700	2304.6	1758.7	2.4466	0.32252	2819.2	0.45041
7800	2336.9	1783.9	2.4508	0.32270	2989.7	0.43024
7900	2369.2	1809.1	2.4549	0.32288	3168.4	0.41118
8000	2401.5	1834.3	2.4589	0.32304	3355.2	0.39320
8100	2433.8	1859.5	2.4629	0.32320	3550.5	0.37621
8200	2466.1	1884.7	2.4669	0.32338	3755.0	0.36012
8300	2498.5	1910.0	2.4708	0.32356	3968.6	0.34489
8400	2530.8	1935.3	2.4747	0.32371	4191.5	0.33048
8500	2563.2	1960.6	2.4785	0.32387	4424.3	0.31682
8600	2595.6	1985.9	2.4823	0.32402	4667.2	0.30387
8700	2628.0	2011.2	2.4861	0.32418	4920.5	0.29157
8800	2660.4	2036.5	2.4898	0.32435	5184.5	0.27991
8900	2692.9	2061.9	2.4934	0.32450	5459.7	0.26882
9000	2725.3	2087.3	2.4971	0.32466	5746.3	0.25828

Adapted from *JANAF Thermochemical Tables* Third Edition, M. W. Chase et al., American Chemical Society and the American Institute of Physics for the National Bureau of Standards, 1985.

TABLE D.4E IDEAL-GAS PROPERTIES OF HYDROGEN (H_2)—MASS BASIS: ENGLISH UNITS
Mol. Wt.: 2.016

T, °R	h, Btu/lb$_m$	u, Btu/lb$_m$	s^0, Btu/lb$_m$·°R	c_p, Btu/lb$_m$·°R	P_r	v_r
0	0.0	0.0	0.0			
200	702.14	505.12	12.246	3.2868	0.31932	1,426.0
250	859.88	613.60	12.946	3.2175	0.64949	874.24
300	1,019.6	724.11	13.543	3.2088	1.1910	574.62
350	1,181.5	836.71	14.055	3.2420	2.0019	400.40
400	1,345.5	951.50	14.497	3.2981	3.1362	292.49
450	1,511.9	1,068.6	14.886	3.3583	4.6555	221.48
500	1,681.2	1,188.6	15.242	3.3964	6.6844	171.41
550	1,851.1	1,309.3	15.567	3.4225	9.2929	135.62
600	2,022.5	1,431.4	15.866	3.4390	12.582	109.27
650	2,195.0	1,554.7	16.141	3.4492	16.649	89.460
700	2,367.6	1,678.1	16.397	3.4557	21.587	74.303
750	2,540.6	1,801.8	16.636	3.4598	27.503	62.487
800	2,713.7	1,925.7	16.859	3.4627	34.501	53.132
850	2,886.9	2,049.6	17.069	3.4649	42.700	45.614
900	3,060.2	2,173.6	17.267	3.4668	52.206	39.502
950	3,233.5	2,297.7	17.455	3.4688	63.165	34.465
1000	3,407.0	2,421.9	17.633	3.4709	75.678	30.281
1100	3,754.3	2,670.7	17.964	3.4759	105.88	23.806
1200	4,102.3	2,920.1	18.267	3.4830	143.99	19.097
1300	4,451.1	3,170.4	18.546	3.4923	191.14	15.585
1400	4,800.9	3,421.8	18.805	3.5044	248.67	12.901
1500	5,152.0	3,674.4	19.047	3.5191	317.98	10.809
1600	5,504.7	3,928.5	19.275	3.5366	400.62	9.1514
1700	5,859.3	4,184.7	19.490	3.5566	498.35	7.8165
1800	6,216.1	4,442.9	19.694	3.5788	612.94	6.7290
1900	6,575.2	4,703.5	19.888	3.6029	746.48	5.8322
2000	6,936.8	4,966.5	20.073	3.6285	901.06	5.0859
2100	7,300.9	5,232.2	20.251	3.6554	1,079.2	4.4590
2200	7,667.8	5,500.6	20.422	3.6832	1,283.3	3.9282
2300	8,037.6	5,771.8	20.586	3.7116	1,516.3	3.4756
2400	8,410.2	6,045.9	20.745	3.7403	1,781.2	3.0874
2500	8,785.6	6,322.9	20.898	3.7692	2,081.1	2.7526
2600	9,164.0	6,602.7	21.046	3.7981	2,419.4	2.4624
2700	9,545.2	6,885.4	21.190	3.8267	2,799.8	2.2097
2800	9,929.3	7,171.0	21.330	3.8550	3,226.3	1.9886
2900	10,316	7,459.4	21.465	3.8829	3,703.0	1.7945
3000	10,706	7,750.6	21.598	3.9102	4,234.2	1.6235
3100	11,098	8,044.5	21.726	3.9370	4,825.0	1.4722
3200	11,493	8,341.0	21.852	3.9632	5,480.2	1.3380
3300	11,891	8,640.1	21.974	3.9888	6,204.9	1.2186
3400	12,291	8,941.6	22.093	4.0137	7,004.5	1.1122
3500	12,693	9,245.6	22.210	4.0379	7,885.3	1.0171
3600	13,098	9,552.1	22.324	4.0616	8,853.5	0.93171
3700	13,506	9,861.0	22.436	4.0845	9,915.5	0.85503
3800	13,916	10,172	22.545	4.1067	11,078	0.78597
3900	14,327	10,485	22.652	4.1285	12,349	0.72362
4000	14,741	10,801	22.757	4.1495	13,736	0.66728
4100	15,157	11,118	22.859	4.1700	15,245	0.61626

TABLE D.4E IDEAL-GAS PROPERTIES OF HYDROGEN (H_2)—MASS BASIS: ENGLISH UNITS Mol. Wt.: 2.016 (*Continued*)

T, °R	h, Btu/lb$_m$	u, Btu/lb$_m$	s^0, Btu/lb$_m$·°R	c_p, Btu/lb$_m$·°R	P_r	v_r
4200	15,575	11,438	22.960	4.1899	16,885	0.56995
4300	15,995	11,759	23.059	4.2093	18,667	0.52783
4400	16,417	12,082	23.156	4.2282	20,599	0.48944
4500	16,841	12,408	23.251	4.2466	22,691	0.45443
4600	17,266	12,735	23.345	4.2645	24,951	0.42244
4700	17,694	13,064	23.437	4.2820	27,390	0.39319
4800	18,123	13,394	23.527	4.2991	30,018	0.36639
4900	18,553	13,726	23.616	4.3158	32,850	0.34178
5000	18,986	14,060	23.703	4.3322	35,899	0.31914
5100	19,420	14,396	23.789	4.3481	39,172	0.29832
5200	19,855	14,733	23.874	4.3637	42,684	0.27915
5300	20,293	15,071	23.957	4.3790	46,447	0.26146
5400	20,731	15,412	24.039	4.3941	50,477	0.24513
5500	21,171	15,753	24.120	4.4090	54,791	0.23001
5600	21,613	16,096	24.199	4.4236	59,401	0.21602
5700	22,056	16,441	24.278	4.4379	64,323	0.20305
5800	22,501	16,787	24.355	4.4520	69,574	0.19102
5900	22,947	17,134	24.431	4.4660	75,171	0.17984
6000	23,394	17,483	24.506	4.4797	81,133	0.16945
6100	23,842	17,833	24.581	4.4932	87,478	0.15978
6200	24,292	18,185	24.654	4.5067	94,225	0.15077
6300	24,744	18,538	24.726	4.5200	101,390	0.14238
6400	25,196	18,892	24.797	4.5331	109,000	0.13454
6500	25,650	19,247	24.868	4.5462	117,070	0.12723
6600	26,106	19,604	24.937	4.5591	125,630	0.12038
6700	26,562	19,962	25.006	4.5719	134,700	0.11397
6800	27,020	20,321	25.074	4.5847	144,300	0.10798
6900	27,479	20,682	25.141	4.5973	154,450	0.10236
7000	27,940	21,044	25.207	4.6098	165,200	0.09709
7100	28,401	21,407	25.272	4.6222	176,560	0.09214
7200	28,864	21,771	25.337	4.6346	188,550	0.08750
7300	29,328	22,137	25.401	4.6468	201,210	0.08313
7400	29,793	22,503	25.464	4.6589	214,560	0.07903
7500	30,260	22,871	25.527	4.6710	228,630	0.07517
7600	30,727	23,241	25.589	4.6829	243,470	0.07153
7700	31,196	23,611	25.650	4.6949	259,090	0.06810
7800	31,666	23,983	25.711	4.7066	275,550	0.06486
7900	32,138	24,355	25.771	4.7183	292,860	0.06181
8000	32,610	24,729	25.830	4.7298	311,090	0.05893
8100	33,084	25,104	25.889	4.7413	330,240	0.05620
8200	33,558	25,480	25.948	4.7526	350,350	0.05363
8300	34,034	25,858	26.005	4.7638	371,460	0.05120
8400	34,511	26,236	26.062	4.7748	393,640	0.04890
8500	34,989	26,616	26.119	4.7857	416,900	0.04672
8600	35,468	26,996	26.175	4.7964	441,290	0.04465
8700	35,948	27,378	26.230	4.8070	466,860	0.04270
8800	36,430	27,761	26.285	4.8173	493,670	0.04085
8900	36,912	28,144	26.340	4.8275	521,760	0.03909
9000	37,395	28,529	26.394	4.8375	551,190	0.03741

Adapted from *JANAF Thermochemical Tables* Third Edition, M. W. Chase et al., American Chemical Society and the American Institute of Physics for the National Bureau of Standards, 1985.

TABLE D.5E IDEAL-GAS PROPERTIES OF NITROGEN (N_2)—MASS BASIS: ENGLISH UNITS
Mol. Wt.: 28.013

T, °R	h, Btu/lb$_m$	u, Btu/lb$_m$	s^0, Btu/lb$_m$·°R	c_p, Btu/lb$_m$·°R	P_r	v_r
0	0.0	0.0	0.0			
200	49.50	35.33	1.3871	0.24815	0.00915	3,582.4
250	61.92	44.20	1.4419	0.24817	0.01980	2,063.2
300	74.33	53.06	1.4882	0.24817	0.03807	1,293.5
350	86.73	61.92	1.5275	0.24817	0.06622	871.06
400	99.14	70.78	1.5609	0.24818	0.10618	621.68
450	111.54	79.64	1.5900	0.24820	0.15992	464.00
500	123.95	88.50	1.6161	0.24825	0.23132	356.43
550	136.38	97.39	1.6398	0.24835	0.32289	280.87
600	148.80	106.26	1.6614	0.24852	0.43802	225.87
650	161.22	115.14	1.6813	0.24879	0.58004	184.78
700	173.67	124.05	1.6998	0.24918	0.75251	153.39
750	186.14	132.98	1.7170	0.24972	0.95926	128.92
800	198.65	141.94	1.7331	0.25040	1.2044	109.53
850	211.19	150.93	1.7483	0.25123	1.4926	93.904
900	223.78	159.97	1.7627	0.25220	1.8283	81.169
950	236.41	169.07	1.7764	0.25331	2.2176	70.643
1000	249.11	178.22	1.7894	0.25455	2.6650	61.877
1100	274.70	196.72	1.8138	0.25730	3.7583	48.263
1200	300.58	215.51	1.8363	0.26032	5.1644	38.316
1300	326.77	234.61	1.8573	0.26349	6.9406	30.885
1400	353.28	254.03	1.8769	0.26671	9.1569	25.211
1500	380.11	273.78	1.8954	0.26990	11.889	20.805
1600	407.26	293.84	1.9129	0.27300	15.220	17.334
1700	434.71	314.20	1.9296	0.27596	19.247	14.564
1800	462.45	334.85	1.9454	0.27878	24.070	12.331
1900	490.46	355.77	1.9606	0.28142	29.805	10.512
2000	518.72	376.95	1.9751	0.28390	36.569	9.0181
2100	547.23	398.37	1.9890	0.28621	44.494	7.7823
2200	575.97	420.01	2.0023	0.28837	53.725	6.7521
2300	604.91	441.86	2.0152	0.29038	64.418	5.8873
2400	634.03	463.90	2.0276	0.29224	76.729	5.1576
2500	663.34	486.12	2.0396	0.29397	90.834	4.5382
2600	692.82	508.51	2.0511	0.29558	106.93	4.0095
2700	722.46	531.06	2.0623	0.29708	125.20	3.5560
2800	752.24	553.75	2.0731	0.29847	145.86	3.1652
2900	782.15	576.57	2.0836	0.29976	169.14	2.8271
3000	812.18	599.51	2.0938	0.30096	195.26	2.5334
3100	842.33	622.57	2.1037	0.30208	224.49	2.2770
3200	872.59	645.75	2.1133	0.30313	257.08	2.0525
3300	902.95	669.02	2.1226	0.30411	293.30	1.8553
3400	933.41	692.39	2.1317	0.30503	333.43	1.6814
3500	963.96	715.85	2.1406	0.30589	377.77	1.5277
3600	994.60	739.39	2.1492	0.30669	426.65	1.3913
3700	1025.3	763.01	2.1576	0.30744	480.42	1.2699
3800	1056.1	786.70	2.1658	0.30816	539.40	1.1616
3900	1086.9	810.45	2.1739	0.30884	603.96	1.0648
4000	1117.8	834.28	2.1817	0.30948	674.46	0.97790
4100	1148.8	858.18	2.1893	0.31008	751.31	0.89983

TABLE D.5E IDEAL-GAS PROPERTIES OF NITROGEN (N_2)—MASS BASIS: ENGLISH UNITS
Mol. Wt.: 28.013 *(Continued)*

T, °R	h, Btu/lb$_m$	u, Btu/lb$_m$	s^0, Btu/lb$_m$·°R	c_p, Btu/lb$_m$·°R	P_r	v_r
4200	1179.9	882.12	2.1968	0.31065	834.93	0.82946
4300	1210.9	906.12	2.2041	0.31119	925.73	0.76590
4400	1242.1	930.17	2.2113	0.31171	1,024.1	0.70844
4500	1273.3	954.28	2.2183	0.31219	1,130.5	0.65635
4600	1304.5	978.44	2.2252	0.31266	1,245.5	0.60900
4700	1335.8	1002.6	2.2319	0.31311	1,369.5	0.56589
4800	1367.2	1026.9	2.2385	0.31353	1,503.0	0.52659
4900	1398.5	1051.2	2.2450	0.31393	1,646.6	0.49068
5000	1429.9	1075.5	2.2513	0.31432	1,800.8	0.45781
5100	1461.4	1099.8	2.2575	0.31469	1,966.3	0.42769
5200	1492.9	1124.2	2.2636	0.31505	2,143.4	0.40003
5300	1524.4	1148.7	2.2697	0.31539	2,332.9	0.37461
5400	1556.0	1173.2	2.2755	0.31572	2,535.3	0.35121
5500	1587.5	1197.7	2.2813	0.31604	2,751.2	0.32964
5600	1619.2	1222.2	2.2870	0.31635	2,981.4	0.30971
5700	1650.8	1246.7	2.2926	0.31664	3,226.6	0.29129
5800	1682.5	1271.3	2.2982	0.31693	3,487.4	0.27423
5900	1714.2	1295.9	2.3036	0.31720	3,764.4	0.25843
6000	1745.9	1320.6	2.3089	0.31747	4,058.5	0.24377
6100	1777.7	1345.3	2.3142	0.31773	4,370.5	0.23014
6200	1809.5	1370.0	2.3193	0.31798	4,701.1	0.21747
6300	1841.3	1394.7	2.3244	0.31822	5,051.0	0.20566
6400	1873.1	1419.4	2.3294	0.31846	5,421.4	0.19465
6500	1905.0	1444.2	2.3344	0.31869	5,812.9	0.18438
6600	1936.9	1469.0	2.3392	0.31892	6,225.7	0.17480
6700	1968.8	1493.8	2.3440	0.31913	6,661.5	0.16584
6800	2000.7	1518.6	2.3488	0.31935	7,121.4	0.15745
6900	2032.6	1543.5	2.3534	0.31956	7,605.5	0.14959
7000	2064.6	1568.4	2.3580	0.31976	8,114.9	0.14224
7100	2096.6	1593.3	2.3626	0.31996	8,651.6	0.13532
7200	2128.6	1618.2	2.3670	0.32016	9,216.1	0.12882
7300	2160.6	1643.1	2.3715	0.32035	9,808.1	0.12272
7400	2192.7	1668.1	2.3758	0.32053	10,430	0.11699
7500	2224.7	1693.0	2.3801	0.32072	11,083	0.11158
7600	2256.8	1718.0	2.3844	0.32090	11,768	0.10649
7700	2288.9	1743.0	2.3886	0.32108	12,485	0.10170
7800	2321.0	1768.1	2.3927	0.32125	13,236	0.09717
7900	2353.1	1793.1	2.3968	0.32142	14,024	0.09289
8000	2385.3	1818.2	2.4008	0.32159	14,848	0.08884
8100	2417.5	1843.3	2.4048	0.32176	15,709	0.08502
8200	2449.6	1868.3	2.4088	0.32193	16,608	0.08141
8300	2481.8	1893.5	2.4127	0.32209	17,547	0.07799
8400	2514.1	1918.6	2.4166	0.32226	18,529	0.07475
8500	2546.3	1943.7	2.4204	0.32242	19,554	0.07168
8600	2578.6	1968.9	2.4241	0.32259	20,622	0.06876
8700	2610.8	1994.1	2.4279	0.32275	21,736	0.06600
8800	2643.1	2019.3	2.4316	0.32292	22,898	0.06337
8900	2675.4	2044.5	2.4352	0.32308	24,108	0.06087
9000	2707.7	2069.7	2.4388	0.32324	25,369	0.05850

Adapted from *JANAF Thermochemical Tables* Third Edition, M. W. Chase et al., American Chemical Society and the American Institute of Physics for the National Bureau of Standards, 1985.

TABLE D.6E IDEAL-GAS PROPERTIES OF OXYGEN (O_2)—MASS BASIS: ENGLISH UNITS
Mol. Wt.: 31.999

T, °R	h, Btu/lbm	u, Btu/lbm	s^0, Btu/lbm·°R	c_p, Btu/lbm·°R	P_r	v_r
0	0.0	0.0	0.0			
200	43.37	30.95	1.3151	0.21727	0.00749	3829.9
250	54.23	38.72	1.3631	0.21729	0.01622	2205.2
300	65.09	46.47	1.4036	0.21730	0.03118	1382.5
350	75.95	54.23	1.4380	0.21737	0.05424	931.02
400	86.83	62.00	1.4673	0.21757	0.08699	664.29
450	97.72	69.79	1.4928	0.21796	0.13113	495.39
500	108.65	77.62	1.5158	0.21861	0.18995	379.99
550	119.57	85.44	1.5367	0.21955	0.26594	298.54
600	130.57	93.34	1.5558	0.22078	0.36213	239.18
650	141.67	101.33	1.5735	0.22227	0.48181	194.75
700	152.82	109.38	1.5901	0.22398	0.62888	160.68
750	164.07	117.52	1.6056	0.22587	0.80754	134.07
800	175.42	125.77	1.6202	0.22787	1.0224	112.95
850	186.86	134.11	1.6341	0.22995	1.2787	95.963
900	198.40	142.55	1.6473	0.23207	1.5816	82.144
950	210.06	151.10	1.6599	0.23418	1.9382	70.759
1000	221.82	159.76	1.6720	0.23627	2.3543	61.319
1100	245.66	177.39	1.6947	0.24031	3.3939	46.788
1200	269.88	195.41	1.7158	0.24406	4.7666	36.343
1300	294.46	213.79	1.7354	0.24751	6.5439	28.678
1400	319.37	232.48	1.7539	0.25063	8.8109	22.938
1500	344.57	251.48	1.7713	0.25344	11.660	18.571
1600	370.05	270.75	1.7877	0.25596	15.195	15.200
1700	395.76	290.26	1.8033	0.25823	19.534	12.563
1800	421.69	309.98	1.8181	0.26028	24.801	10.477
1900	447.81	329.89	1.8322	0.26213	31.139	8.8081
2000	474.11	349.99	1.8457	0.26381	38.697	7.4606
2100	500.57	370.24	1.8586	0.26536	47.647	6.3623
2200	527.18	390.64	1.8710	0.26678	58.165	5.4599
2300	553.92	411.18	1.8829	0.26811	70.447	4.7129
2400	580.79	431.85	1.8943	0.26936	84.705	4.0901
2500	607.79	452.64	1.9054	0.27055	101.17	3.5672
2600	634.90	473.55	1.9160	0.27168	120.07	3.1257
2700	662.13	494.57	1.9263	0.27277	141.69	2.7508
2800	689.46	515.69	1.9362	0.27383	166.29	2.4306
2900	716.90	536.92	1.9458	0.27486	194.19	2.1557
3000	744.43	558.25	1.9552	0.27587	225.71	1.9187
3100	772.07	579.68	1.9642	0.27687	261.20	1.7132
3200	799.80	601.20	1.9730	0.27785	301.02	1.5346
3300	827.63	622.83	1.9816	0.27883	345.57	1.3785
3400	855.57	644.56	1.9899	0.27979	395.27	1.2417
3500	883.59	666.38	1.9981	0.28075	450.56	1.1214
3600	911.71	688.29	2.0060	0.28171	511.91	1.0152
3700	939.93	710.31	2.0137	0.28265	579.83	0.92114
3800	968.25	732.42	2.0213	0.28360	654.84	0.83767
3900	996.65	754.62	2.0287	0.28454	737.49	0.76337
4000	1025.2	776.91	2.0359	0.28547	828.40	0.69702
4100	1053.7	799.30	2.0429	0.28640	928.23	0.63761

TABLE D.6E IDEAL-GAS PROPERTIES OF OXYGEN (O_2)—MASS BASIS: ENGLISH UNITS
Mol. Wt.: 31.999 (*Continued*)

T, °R	h, Btu/lb$_m$	u, Btu/lb$_m$	s^0, Btu/lb$_m$·°R	c_p, Btu/lb$_m$·°R	P_r	v_r
4200	1082.4	821.78	2.0498	0.28732	1,037.6	0.58430
4300	1111.2	844.35	2.0566	0.28823	1,157.3	0.53636
4400	1140.1	867.02	2.0632	0.28913	1,287.8	0.49320
4500	1169.0	889.77	2.0698	0.29003	1,430.1	0.45422
4600	1198.1	912.61	2.0761	0.29091	1,585.2	0.41890
4700	1227.2	935.54	2.0824	0.29179	1,753.7	0.38688
4800	1256.4	958.55	2.0886	0.29265	1,936.4	0.35783
4900	1285.8	981.65	2.0946	0.29350	2,134.4	0.33140
5000	1315.1	1004.8	2.1005	0.29434	2,348.7	0.30731
5100	1344.6	1028.1	2.1064	0.29516	2,580.3	0.28531
5200	1374.2	1051.5	2.1121	0.29597	2,830.4	0.26521
5300	1403.8	1074.9	2.1178	0.29677	3,099.9	0.24681
5400	1433.5	1098.4	2.1233	0.29755	3,390.1	0.22994
5500	1463.3	1122.0	2.1288	0.29832	3,702.2	0.21445
5600	1493.2	1145.7	2.1342	0.29908	4,037.5	0.20021
5700	1523.1	1169.4	2.1395	0.29981	4,397.4	0.18711
5800	1553.2	1193.2	2.1447	0.30053	4,783.4	0.17503
5900	1583.2	1217.1	2.1498	0.30124	5,196.9	0.16388
6000	1613.4	1241.0	2.1549	0.30193	5,639.2	0.15359
6100	1643.6	1265.1	2.1599	0.30261	6,111.8	0.14407
6200	1673.9	1289.1	2.1648	0.30327	6,616.6	0.13526
6300	1704.3	1313.3	2.1697	0.30391	7,155.4	0.12710
6400	1734.7	1337.5	2.1745	0.30455	7,729.6	0.11952
6500	1765.2	1361.8	2.1792	0.30517	8,341.2	0.11249
6600	1795.7	1386.1	2.1839	0.30577	8,992.3	0.10595
6700	1826.3	1410.5	2.1885	0.30637	9,684.8	0.09986
6800	1857.0	1435.0	2.1930	0.30695	10,421	0.09420
6900	1887.7	1459.5	2.1975	0.30752	11,202	0.08892
7000	1918.5	1484.1	2.2019	0.30808	12,031	0.08399
7100	1949.3	1508.7	2.2063	0.30863	12,909	0.07939
7200	1980.2	1533.4	2.2106	0.30918	13,840	0.07510
7300	2011.2	1558.2	2.2149	0.30971	14,825	0.07108
7400	2042.2	1582.9	2.2191	0.31024	15,867	0.06732
7500	2073.2	1607.8	2.2233	0.31075	16,970	0.06380
7600	2104.3	1632.7	2.2274	0.31127	18,134	0.06050
7700	2135.5	1657.6	2.2315	0.31178	19,364	0.05740
7800	2166.7	1682.6	2.2355	0.31229	20,662	0.05449
7900	2197.9	1707.7	2.2395	0.31280	22,030	0.05176
8000	2229.2	1732.8	2.2434	0.31330	23,473	0.04920
8100	2260.6	1757.9	2.2473	0.31381	24,993	0.04678
8200	2292.0	1783.1	2.2512	0.31432	26,595	0.04451
8300	2323.5	1808.4	2.2550	0.31482	28,281	0.04236
8400	2355.0	1833.7	2.2587	0.31533	30,054	0.04035
8500	2386.5	1859.0	2.2625	0.31585	31,917	0.03844
8600	2418.1	1884.4	2.2662	0.31637	33,875	0.03665
8700	2449.8	1909.9	2.2698	0.31690	35,933	0.03495
8800	2481.5	1935.4	2.2735	0.31743	38,095	0.03335
8900	2513.3	1961.0	2.2770	0.31798	40,364	0.03183
9000	2545.1	1986.6	2.2806	0.31854	42,745	0.03039

Adapted from *JANAF Thermochemical Tables* Third Edition, M. W. Chase et al., American Chemical Society and the American Institute of Physics for the National Bureau of Standards, 1985.

TABLE D.7E IDEAL-GAS PROPERTIES OF WATER VAPOR (H_2O)—MASS BASIS: ENGLISH UNITS
Mol. Wt.: 18.015

T, °R	h, Btu/lb$_m$	u, Btu/lb$_m$	s^0, Btu/lb$_m$·°R	c_p, Btu/lb$_m$·°R	P_r	v_r
0	0.0	0.0	0.0			
200	87.32	65.27	2.0636	0.44140	0.00118	43,189
250	109.40	81.84	2.1606	0.44137	0.00284	22,311
300	131.49	98.42	2.2433	0.44159	0.00602	12,707
350	153.61	115.02	2.3138	0.44203	0.01140	7,866.6
400	175.74	131.64	2.3740	0.44267	0.01969	5,218.7
450	197.89	148.28	2.4262	0.44350	0.03161	3,657.8
500	220.07	164.95	2.4723	0.44450	0.04804	2,670.5
550	242.28	181.65	2.5145	0.44568	0.07042	2,002.6
600	264.59	198.45	2.5535	0.44747	0.10028	1,534.4
650	287.05	215.40	2.5894	0.44994	0.13898	1,199.5
700	309.65	232.49	2.6228	0.45293	0.18811	954.23
750	332.40	249.72	2.6542	0.45616	0.25009	769.09
800	355.30	267.11	2.6838	0.45957	0.32720	627.13
850	378.36	284.67	2.7118	0.46320	0.42156	517.13
900	401.61	302.40	2.7382	0.46702	0.53591	430.59
950	425.06	320.34	2.7636	0.47093	0.67486	360.98
1000	448.70	338.47	2.7879	0.47495	0.84107	304.90
1100	496.60	375.35	2.8335	0.48328	1.2720	221.74
1200	545.36	413.08	2.8759	0.49184	1.8693	164.61
1300	594.99	451.69	2.9156	0.50066	2.6797	124.39
1400	645.50	491.17	2.9531	0.50970	3.7631	95.391
1500	696.93	531.59	2.9885	0.51891	5.1919	74.078
1600	749.30	572.93	3.0223	0.52827	7.0540	58.157
1700	802.60	615.20	3.0546	0.53769	9.4567	46.093
1800	856.82	658.41	3.0856	0.54713	12.527	36.843
1900	912.02	702.58	3.1155	0.55651	16.421	29.668
2000	968.14	747.67	3.1442	0.56578	21.319	24.053
2100	1025.2	793.68	3.1721	0.57490	27.441	19.622
2200	1083.1	840.58	3.1990	0.58381	35.041	16.098
2300	1141.9	888.38	3.2252	0.59248	44.419	13.276
2400	1201.6	937.03	3.2506	0.60088	55.926	11.003
2500	1262.1	986.51	3.2753	0.60900	69.970	9.1611
2600	1323.4	1036.8	3.2993	0.61681	87.020	7.6608
2700	1385.4	1087.8	3.3227	0.62432	107.62	6.4328
2800	1448.2	1139.6	3.3455	0.63151	132.39	5.4229
2900	1511.7	1192.0	3.3678	0.63840	162.04	4.5887
3000	1575.9	1245.2	3.3896	0.64497	197.40	3.8967
3100	1640.7	1299.0	3.4108	0.65126	239.37	3.3205
3200	1706.1	1353.4	3.4316	0.65725	289.02	2.8388
3300	1772.2	1408.4	3.4519	0.66297	347.51	2.4348
3400	1838.7	1463.9	3.4718	0.66842	416.16	2.0948
3500	1905.8	1520.0	3.4912	0.67359	496.47	1.8075
3600	1973.4	1576.6	3.5103	0.67854	590.12	1.5642
3700	2041.5	1633.7	3.5290	0.68336	698.97	1.3573
3800	2110.1	1691.2	3.5472	0.68796	825.08	1.1809
3900	2179.1	1749.2	3.5652	0.69231	970.75	1.0301
4000	2248.6	1807.6	3.5827	0.69646	1,138.6	0.90077
4100	2318.4	1866.5	3.6000	0.70043	1,331.5	0.78954

TABLE D.7E IDEAL-GAS PROPERTIES OF WATER VAPOR (H_2O)—MASS BASIS: ENGLISH UNITS Mol. Wt.: 18.015 (*Continued*)

T, °R	h, Btu/lb$_m$	u, Btu/lb$_m$	s^0, Btu/lb$_m$·°R	c_p, Btu/lb$_m$·°R	P_r	v_r
4200	2388.6	1925.7	3.6169	0.70422	1,552.4	0.69368
4300	2459.3	1985.3	3.6335	0.70785	1,804.9	0.61083
4400	2530.2	2045.2	3.6498	0.71133	2,092.8	0.53906
4500	2601.5	2105.5	3.6659	0.71466	2,420.2	0.47674
4600	2673.1	2166.1	3.6816	0.71784	2,791.6	0.42249
4700	2745.1	2227.0	3.6971	0.72089	3,212.2	0.37516
4800	2817.3	2288.2	3.7123	0.72382	3,687.4	0.33376
4900	2889.8	2349.7	3.7272	0.72662	4,223.2	0.29749
5000	2962.6	2411.5	3.7419	0.72932	4,826.0	0.26564
5100	3035.7	2473.5	3.7564	0.73191	5,502.9	0.23763
5200	3109.0	2535.8	3.7706	0.73440	6,261.4	0.21294
5300	3182.6	2598.3	3.7847	0.73680	7,110.3	0.19112
5400	3256.4	2661.1	3.7985	0.73910	8,058.3	0.17182
5500	3330.4	2724.1	3.8120	0.74132	9,115.2	0.15471
5600	3404.6	2787.3	3.8254	0.74345	10,291	0.13952
5700	3479.1	2850.7	3.8386	0.74552	11,597	0.12602
5800	3553.7	2914.4	3.8516	0.74751	13,047	0.11398
5900	3628.6	2978.2	3.8644	0.74943	14,653	0.10324
6000	3703.6	3042.2	3.8770	0.75128	16,429	0.09364
6100	3778.8	3106.4	3.8894	0.75307	18,390	0.08505
6200	3854.2	3170.8	3.9017	0.75480	20,553	0.07735
6300	3929.8	3235.3	3.9138	0.75647	22,936	0.07043
6400	4005.5	3300.0	3.9257	0.75810	25,558	0.06421
6500	4081.4	3364.9	3.9375	0.75967	28,438	0.05861
6600	4157.4	3429.9	3.9491	0.76118	31,596	0.05356
6700	4233.6	3495.1	3.9605	0.76265	35,056	0.04900
6800	4310.0	3560.4	3.9718	0.76409	38,842	0.04489
6900	4386.4	3625.8	3.9830	0.76548	42,983	0.04116
7000	4463.1	3691.4	3.9940	0.76683	47,505	0.03778
7100	4539.8	3757.2	4.0049	0.76813	52,437	0.03472
7200	4616.7	3823.0	4.0157	0.76940	57,809	0.03193
7300	4693.7	3889.0	4.0263	0.77064	63,656	0.02940
7400	4770.8	3955.1	4.0368	0.77184	70,013	0.02710
7500	4848.1	4021.3	4.0471	0.77301	76,916	0.02500
7600	4925.4	4087.6	4.0574	0.77415	84,407	0.02309
7700	5002.9	4154.1	4.0675	0.77525	92,529	0.02134
7800	5080.5	4220.6	4.0775	0.77633	101,330	0.01974
7900	5158.1	4287.3	4.0874	0.77737	110,850	0.01827
8000	5235.9	4354.1	4.0972	0.77840	121,130	0.01693
8100	5313.8	4420.9	4.1069	0.77940	132,240	0.01571
8200	5391.8	4487.9	4.1165	0.78037	144,240	0.01458
8300	5469.9	4555.0	4.1259	0.78132	157,170	0.01354
8400	5548.1	4622.1	4.1353	0.78225	171,100	0.01259
8500	5626.4	4689.4	4.1445	0.78316	186,100	0.01171
8600	5704.7	4756.7	4.1537	0.78405	202,240	0.01090
8700	5783.2	4824.1	4.1628	0.78489	219,590	0.01016
8800	5861.7	4891.6	4.1718	0.78570	238,220	0.00947
8900	5940.3	4959.2	4.1806	0.78654	258,190	0.00884
9000	6019.0	5026.9	4.1894	0.78739	279,620	0.00825

Adapted from *JANAF Thermochemical Tables* Third Edition, M. W. Chase et al., American Chemical Society and the American Institute of Physics for the National Bureau of Standards, 1985.

TABLE D.8E ZERO-PRESSURE SPECIFIC HEATS FOR COMMON IDEAL GASES AS A FUNCTION OF TEMPERATURE

Temp., °R	c_{p0}, Btu/lbm·°R	c_{v0}, Btu/lbm·°R	k	c_{p0}, Btu/lbm·°R	c_{v0}, Btu/lbm·°R	k	c_{p0}, Btu/lbm·°R	c_{v0}, Btu/lbm·°R	k	Temp., °R
	Air			Carbon dioxide, CO_2			Carbon monoxide, CO			
450	0.239	0.170	1.406	0.188	0.143	1.315	0.248	0.177	1.401	450
500	0.239	0.170	1.406	0.196	0.151	1.298	0.248	0.177	1.401	500
600	0.240	0.171	1.404	0.210	0.165	1.273	0.249	0.178	1.399	600
700	0.242	0.173	1.399	0.222	0.177	1.254	0.250	0.179	1.397	700
800	0.243	0.174	1.397	0.233	0.188	1.239	0.252	0.181	1.392	800
900	0.246	0.177	1.390	0.242	0.197	1.228	0.254	0.183	1.388	900
1000	0.248	0.179	1.385	0.251	0.206	1.218	0.257	0.186	1.382	1000
1100	0.251	0.182	1.379	0.258	0.213	1.211	0.260	0.189	1.376	1100
1200	0.254	0.185	1.373	0.265	0.220	1.205	0.264	0.193	1.368	1200
1300	0.257	0.188	1.367	0.271	0.226	1.199	0.267	0.196	1.362	1300
1400	0.261	0.192	1.359	0.277	0.232	1.194	0.271	0.200	1.355	1400
1500	0.264	0.195	1.354	0.282	0.237	1.190	0.274	0.203	1.350	1500
	Hydrogen, H_2			Nitrogen, N_2			Oxygen, O_2			
450	3.358	2.373	1.415	0.248	0.177	1.401	0.218	0.156	1.397	450
500	3.396	2.411	1.409	0.248	0.177	1.401	0.219	0.157	1.395	500
600	3.439	2.454	1.401	0.249	0.178	1.399	0.221	0.159	1.390	600
700	3.456	2.471	1.399	0.249	0.178	1.399	0.224	0.162	1.383	700
800	3.463	2.478	1.397	0.250	0.179	1.397	0.228	0.166	1.373	800
900	3.467	2.482	1.397	0.252	0.181	1.392	0.232	0.170	1.365	900
1000	3.471	2.486	1.396	0.255	0.184	1.386	0.236	0.174	1.356	1000
1100	3.476	2.491	1.395	0.257	0.186	1.382	0.240	0.178	1.348	1100
1200	3.483	2.498	1.394	0.260	0.189	1.376	0.244	0.182	1.341	1200
1300	3.492	2.507	1.393	0.263	0.192	1.370	0.248	0.186	1.333	1300
1400	3.504	2.519	1.391	0.267	0.196	1.362	0.251	0.189	1.328	1400
1500	3.519	2.534	1.389	0.270	0.199	1.357	0.253	0.191	1.325	1500

Specific heats for monatomic gases: At low pressures the constant-pressure and constant-volume specific heats for all monatomic gases are independent of pressure and temperature. Also, the values of c_p and c_v on a molar basis are the same for all monatomic gases. Therefore, for monatomic gases such as argon, helium, and neon, the specific heats can be assumed constant at the following values:

$$\bar{c}_p = 4.97 \text{ Btu/lb}_m\text{·mol·°R} \qquad \bar{c}_v = 2.98 \text{ Btu/lb}_m\text{·mol·°R}$$

at states where the gases may be assumed to be ideal gases.

SOURCE: Adapted from Kenneth Wark, *Thermodynamics*, 3d ed., McGraw-Hill, 1977, p. 804, table A-4M. Originally published in "Tables of Thermal Properties of Gases," *NBS Circular* 564, 1955.

lbm/lbm · mole.

TABLE D.9E ZERO-PRESSURE SPECIFIC HEATS AND k AT 80°F FOR COMMON GASES

Gas	Chemical formula	Molecular weight	R, Btu/lb$_m$·°R	c_{p0}, Btu/lb$_m$·°R	c_{v0}, Btu/lb$_m$·°R	k
Air		28.97	0.06855	0.2388	0.1703	1.402
Argon	Ar	39.948	0.04971	0.1243	0.07459	1.666
Butane	C_4H_{10}	58.124	0.03417	0.4050	0.3708	1.092
Carbon dioxide	CO_2	44.010	0.04513	0.2020	0.1569	1.287
Carbon monoxide	CO	28.010	0.07090	0.2485	0.1776	1.400
Ethane	C_2H_6	30.070	0.06605	0.4219	0.3559	1.185
Ethylene	C_2H_4	28.054	0.07079	0.3698	0.2990	1.237
Helium	He	4.003	0.4961	1.240	0.7439	1.667
Hydrogen	H_2	2.016	0.9851	3.4181	2.433	1.405
Methane	CH_4	16.043	0.1238	0.5383	0.4145	1.299
Neon	Ne	20.183	0.09840	0.2460	0.1476	1.667
Nitrogen	N_2	28.013	0.01090	0.2483	0.1774	1.400
Octane	C_8H_{18}	114.23	0.01739	0.4088	0.3914	1.044
Oxygen	O_2	31.999	0.06206	0.2193	0.1572	1.395
Propane	C_3H_8	44.097	0.04604	0.4011	0.3561	1.126
Water vapor	H_2O	18.015	0.1102	0.4454	0.3352	1.329

SOURCE: Gordon J. Van Wylen and Richard E. Sonntag, *Fundamentals of Classical Thermodynamics*, SI Version, 2d Revised Printing, Copyright 1976, 1978 by John Wiley & Sons, Inc., table A.8, p. 683.

TABLE D.10E MOLECULAR WEIGHT AND GAS CONSTANT FOR COMMON GASES

Gas	Molecular weight, M, lb$_m$/lb$_m$·mol	Gas constant, R	
		ft·lb$_f$/lb$_m$·°R	Btu/lb$_m$·°R
Air	28.97	53.34	0.06855
Argon	39.948	38.68	0.4971
Carbon dioxide	44.010	35.12	0.04513
Carbon monoxide	28.010	55.17	0.07090
Helium	4.003	386.0	0.4961
Hydrogen	2.016	766.5	0.9851
Nitrogen	28.013	55.17	0.07090
Oxygen	31.999	48.29	0.06206
Water vapor	18.015	85.75	0.1102

TABLE D.11E EXPRESSIONS FOR CONSTANT-PRESSURE SPECIFIC HEAT FOR COMMON GASES AT ZERO PRESSURE

$$\bar{c}_{p0}/\bar{R} = a + bT + cT^2 + dT^3 + eT^4$$

T is in °R, equations valid from 540 to 1800°R

Gas	a	$b \times 10^3$	$c \times 10^6$	$d \times 10^9$	$e \times 10^{12}$
CO	3.710	−1.009	1.140	−0.348	0.0229
CO_2	2.401	4.853	−2.039	0.343	
H_2	3.057	1.487	−1.793	0.947	−0.1726
H_2O	4.070	−0.616	1.281	−0.508	0.0769
O_2	3.626	−1.043	2.178	−1.160	0.2054
N_2	3.675	−0.671	0.717	−0.108	−0.0215
Air (dry)	3.640	−0.617	0.761	−0.162	
NH_3	3.591	0.274	2.576	−1.437	0.2601
NO	4.046	−1.899	2.464	−1.048	0.1517
NO_2	3.459	1.147	2.064	−1.639	0.3448
SO_2	3.267	2.958	0.211	−0.906	0.2438
SO_3	2.578	8.087	−2.832	−0.136	0.1878
CH_4	3.826	−2.211	7.580	−3.898	0.6633
C_2H_2	1.410	10.587	−7.562	2.811	−0.3939
C_2H_4	1.426	6.324	2.466	−2.787	0.6429

SOURCE: Kenneth Wark, *Thermodynamics*, 4th ed. McGraw-Hill, New York, 1983.

Generalized Charts for Gases

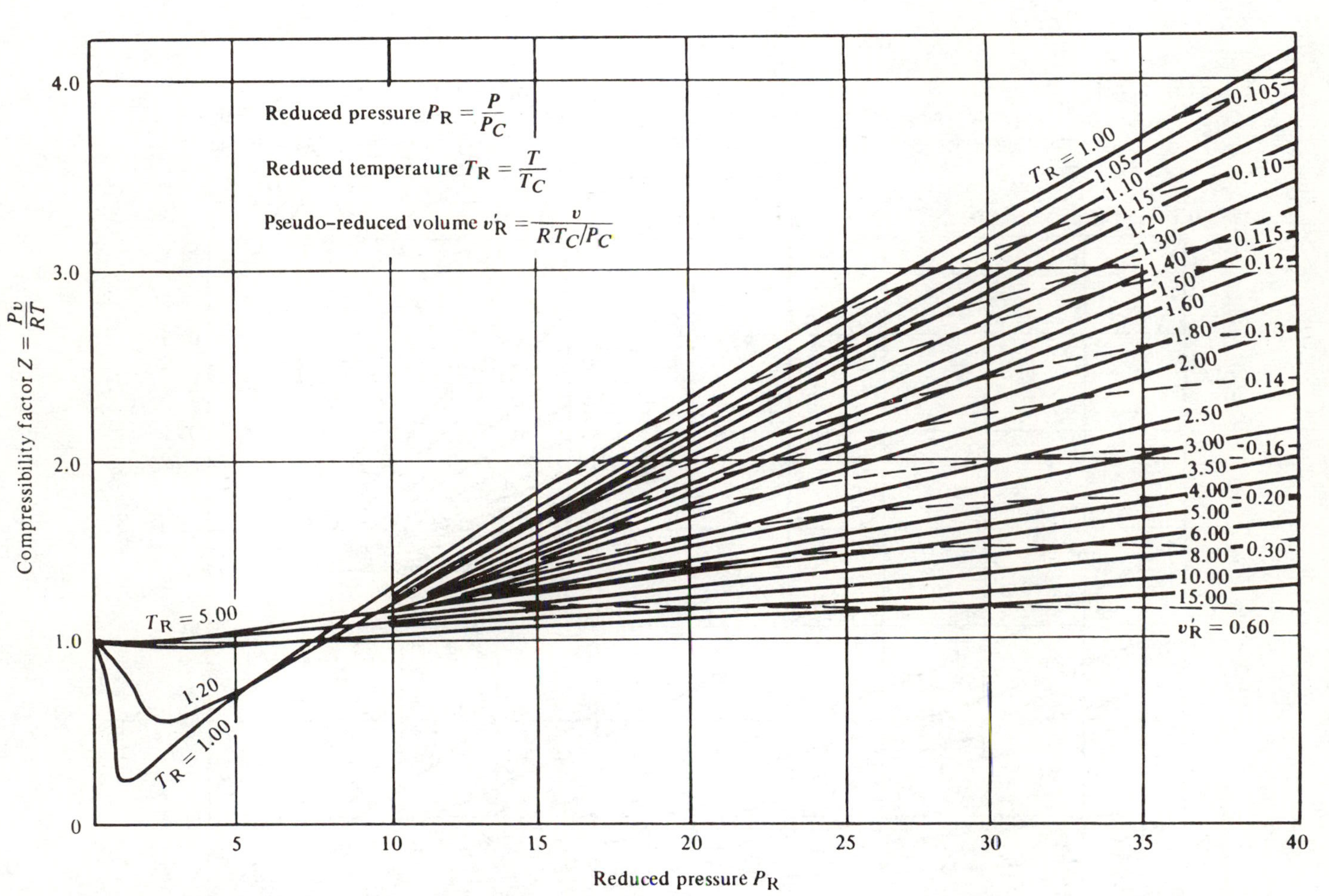

Figure E.1E Compressibility chart, high-pressure range. (Source: J.P. Holman, *Thermodynamics*, 3d ed., McGraw-Hill, New York, 1980, p. 286, Figure 7.5. Originally published in E. F. Obert, *Concepts of Thermodynamics*, McGraw-Hill, New York, 1960.)

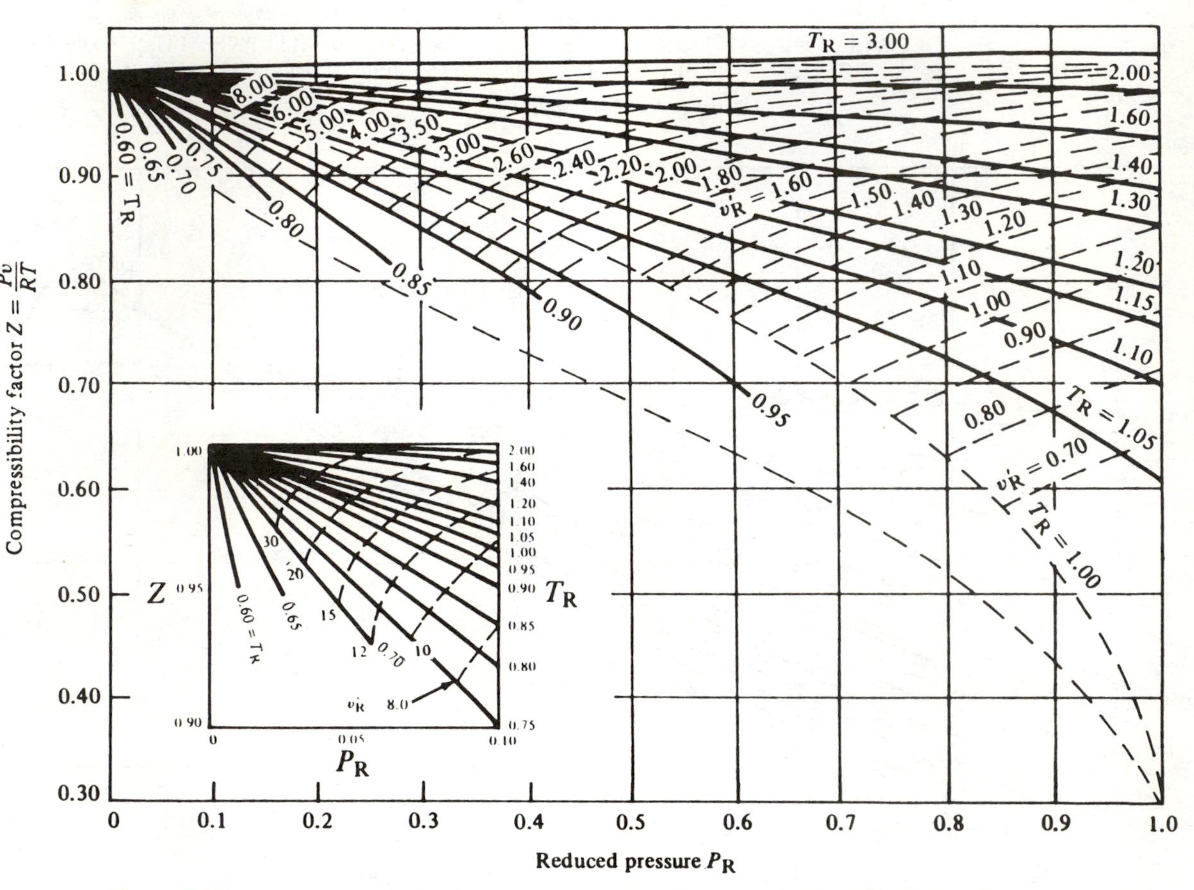

Figure E.2E Compressibility chart, low-pressure range. (Source: J.P. Holman, *Thermodynamics*, 3d ed., McGraw-Hill, New York, 1980, p. 287, Figure 7.6. Originally published in E. F. Obert, *Concepts of Thermodynamics*, McGraw-Hill, New York, 1960.)

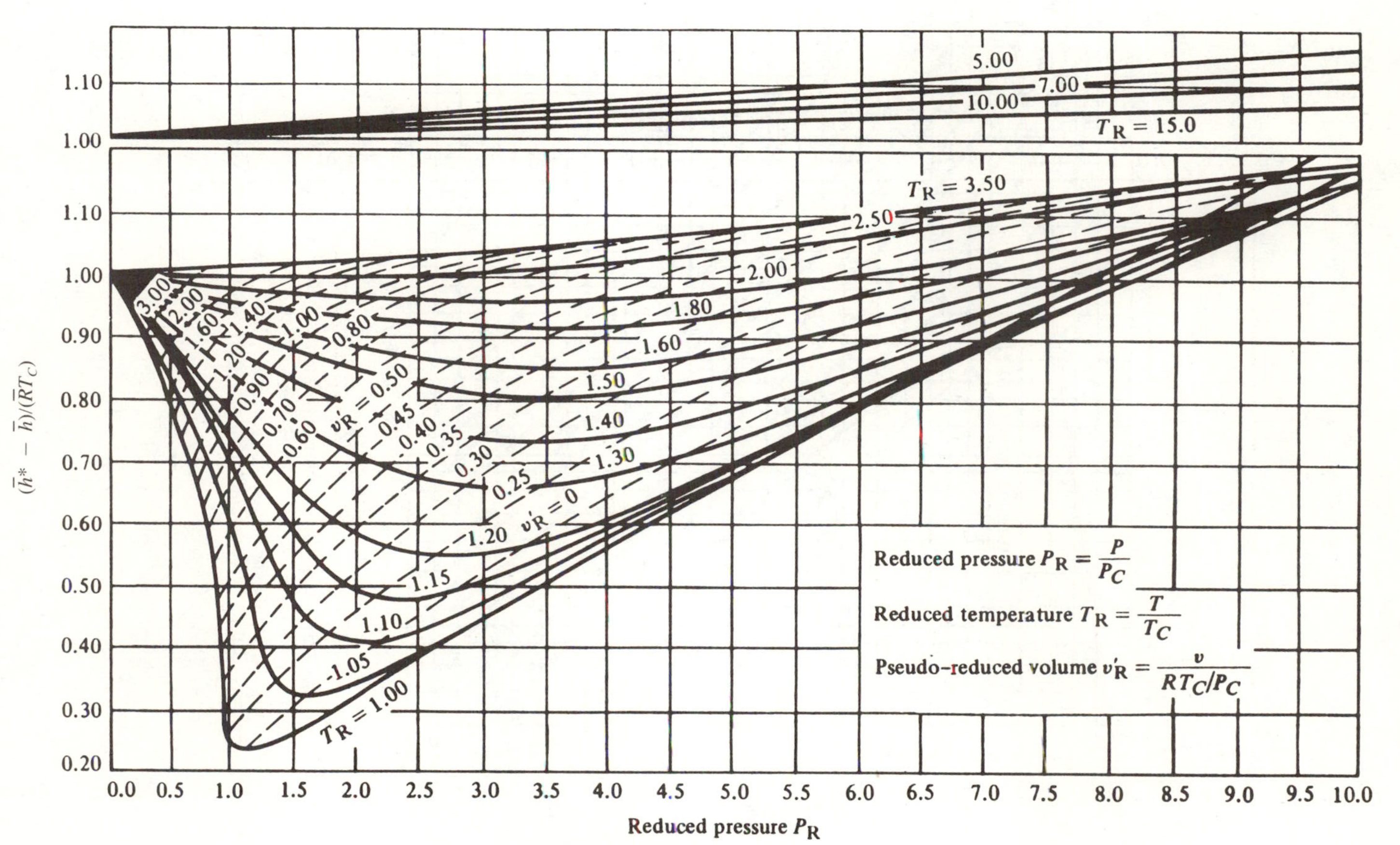

Figure E.3E Compressibility chart, moderate-pressure range. (Source: J.P. Holman, *Thermodynamics*, 3d ed., McGraw-Hill, New York, 1980, p. 285, Figure 7.4. Originally published in E. F. Obert, *Concepts of Thermodynamics*, McGraw-Hill, New York, 1960.)

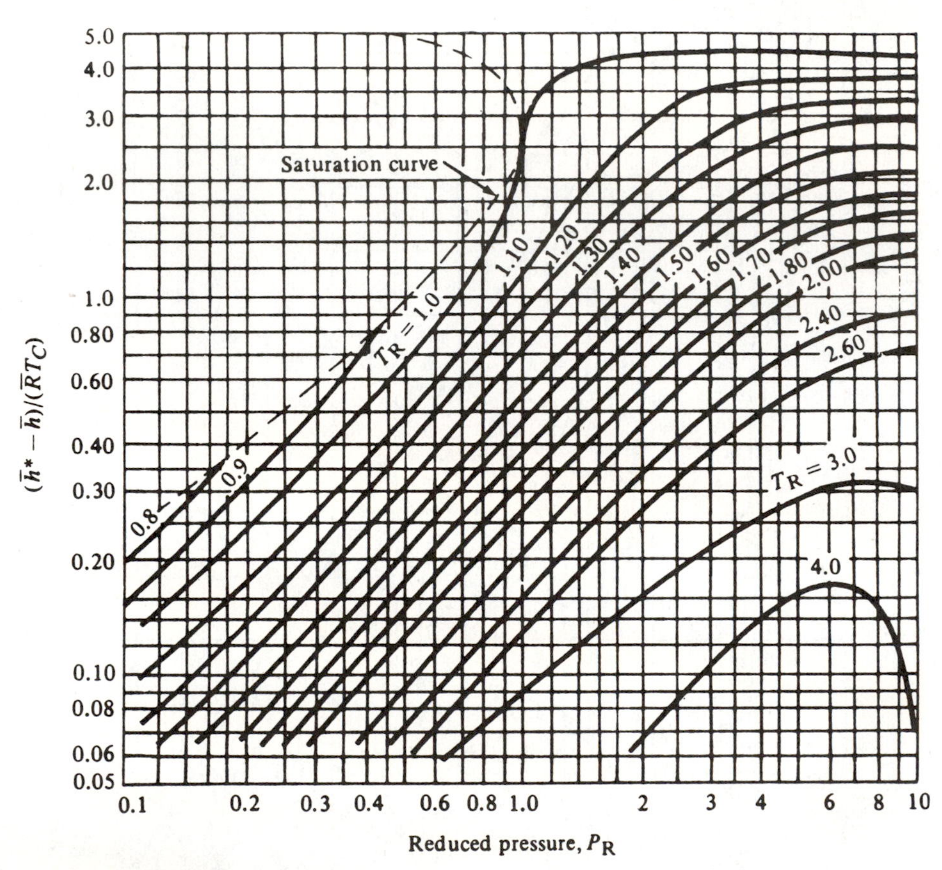

Figure E.4E Generalized enthalpy chart. (Source: Kenneth Wark, *Thermodynamics*, 4th ed., McGraw-Hill, New York, 1983, p. 821, Figure A.27M. Originally published by A. L. Lydersen, R.A. Greenkorn, and O.A. Hougen, "Engineering Experiment Station Report No. 4," University of Wisconsin, 1955.)

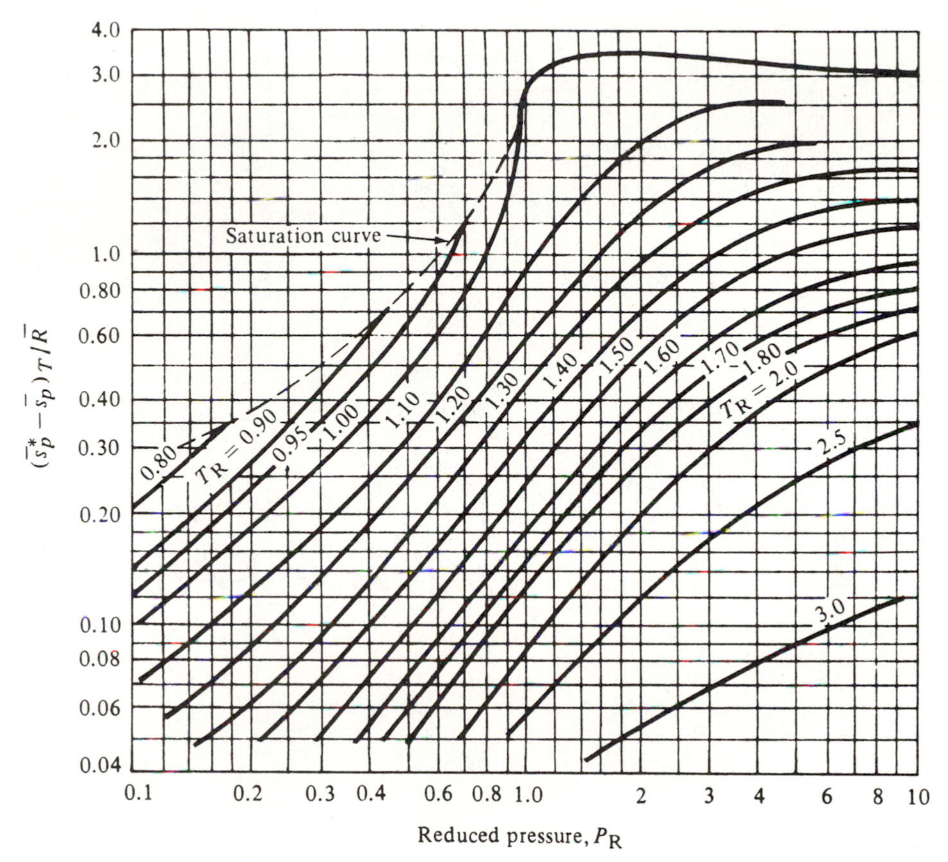

Figure E.5E Generalized entropy chart. (Source: Kenneth Wark, *Thermodynamics*, 4th ed., McGraw-Hill, New York, 1983, p. 822, Figure A.27M. Originally published by A. L. Lydersen, R.A. Greenkorn, and O.A. Hougen, ''Engineering Experiment Station Report No. 4,'' University of Wisconsin, 1955.)

Properties of Air-Water Mixtures

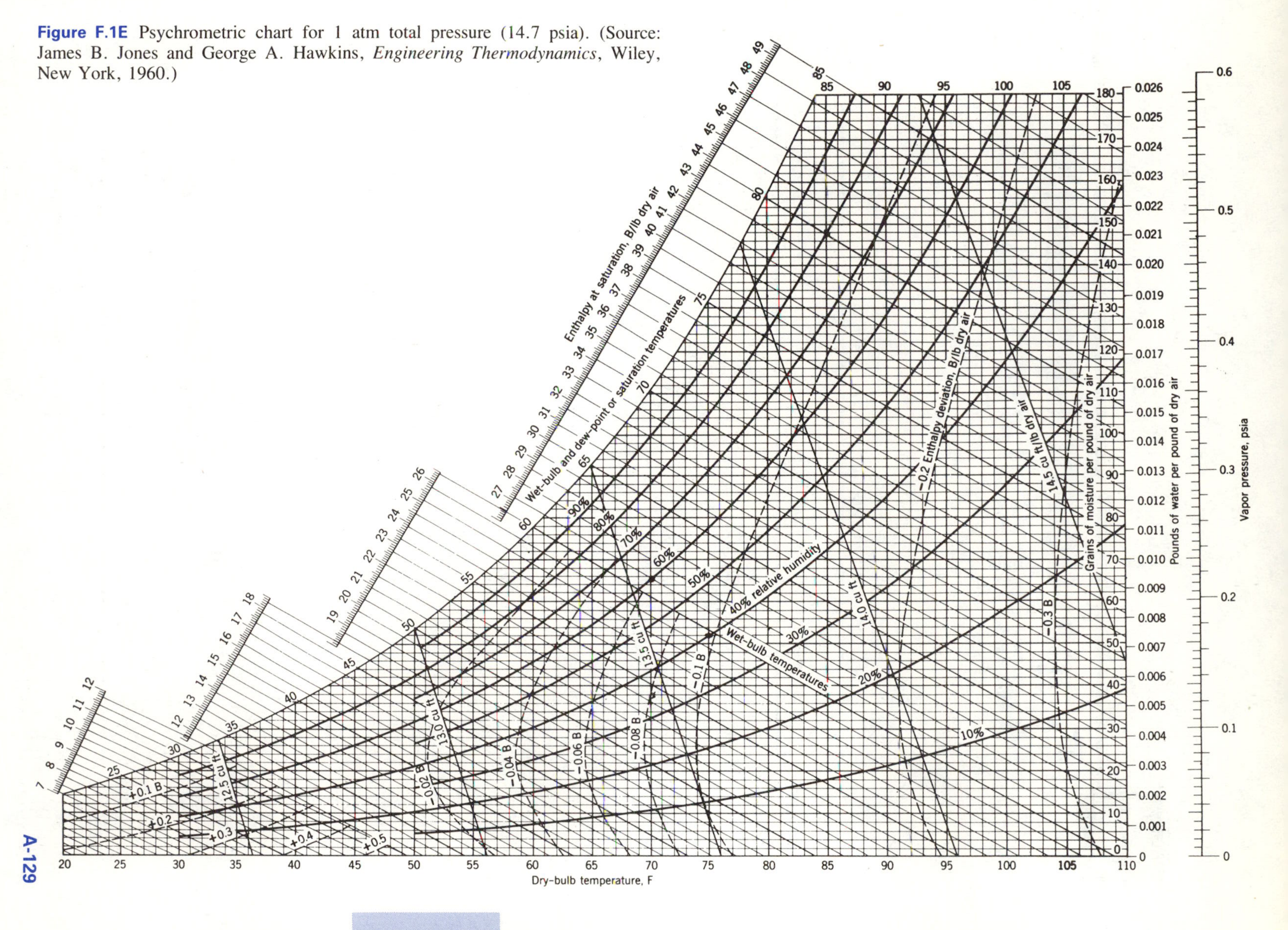

Figure F.1E Psychrometric chart for 1 atm total pressure (14.7 psia). (Source: James B. Jones and George A. Hawkins, *Engineering Thermodynamics*, Wiley, New York, 1960.)

Air, H_2O

Properties of Chemically Reacting Substances

TABLE G.1E ENTHALPY OF FORMATION, GIBBS FUNCTION OF FORMATION, AND ABSOLUTE ENTROPY AT 77°F, 1 atm

Substance	Formula	$\Delta\bar{h}_f^\circ$ Btu/lb$_m$·mol	$\Delta\bar{g}_f^\circ$ Btu/lb$_m$·mol	$\bar{s}^\circ$ Btu/lb$_m$·mol·°R
Carbon	$C(s)$	0	0	1.371
Hydrogen	$H_2(g)$	0	0	31.21
Nitrogen	$N_2(g)$	0	0	45.77
Oxygen	$O_2(g)$	0	0	49.00
Carbon monoxide	$CO(g)$	−47,519	−58,966	47.21
Carbon dioxide	$CO_2(g)$	−169,184	−169,552	51.07
Water	$H_2O(g)$	−103,966	−98,280	45.10
Water	$H_2O(l)$	−122,890	−101,973	16.70
Hydrogen peroxide	$H_2O_2(g)$	−58,605	−45,402	55.57
Ammonia	$NH_3(g)$	−19,859	−7,133	45.94
Methane	$CH_4(g)$	−32,181	−21,837	44.47
Acetylene	$C_2H_2(g)$	+97,481	+89,931	47.97
Ethylene	$C_2H_4(g)$	+22,477	+29,288	52.51
Ethane	$C_2H_6(g)$	−36,407	−14,141	54.82
Propylene	$C_3H_6(g)$	+8,775	+26,966	63.76
Propane	$C_3H_8(g)$	−44,649	−10,099	64.47
n-Butane	$C_4H_{10}(g)$	−54,237	−6,754	74.07
n-Octane	$C_8H_{18}(g)$	−89,621	+7,107	111.48
n-Octane	$C_8H_{18}(l)$	−107,464	+2,842	86.12
n-Dodecane	$C_{12}H_{26}$	−129,117	+21,562	148.77
Benzene	$C_6H_6(g)$	+35,655	+55,746	64.30
Methyl alcohol	$CH_3OH(g)$	−86,276	−69,650	57.25
Methyl alcohol	$CH_3OH(l)$	−102,610	−71,525	30.29
Ethyl alcohol	$C_2H_5OH(g)$	−101,169	−72,475	67.50
Ethyl alcohol	$C_2H_5OH(l)$	−119,390	−75,192	38.38
Oxygen	$O(g)$	+107,124	+99,647	38.47
Hydrogen	$H(g)$	+93,723	+87,403	27.40
Nitrogen	$N(g)$	+203,216	+195,842	36.62
Hydroxyl	$OH(g)$	+16,761	+14,738	43.88

SOURCES: Adapted from JANAF Thermochemical Tables, Dow Chemical Co., 1971; *Selected Values of Chemical Thermodynamic Properties*, NBS Tech. Note 270-3, 1968; and *API Research Project 44*, Carnegie Press, 1953.

TABLE G.2E ENTHALPY OF COMBUSTION AND ENTHALPY OF VAPORIZATION AT 77°F, 1 atm
Water appears as a liquid in the product of combustion

Substance	Formula	$\Delta \bar{h}_c^\circ = -\text{HHV}$ Btu/lb$_m$·mol	$\bar{h}_{fg}$ Btu/lb$_m$·mol
Hydrogen	$H_2(g)$	−122,894	
Carbon	$C(s)$	−169,190	
Carbon monoxide	$CO(g)$	−121,669	
Methane	$CH_4(g)$	−382,802	
Acetylene	$C_2H_2(g)$	−558,751	
Ethylene	$C_2H_4(g)$	−606,634	
Ethane	$C_2H_6(g)$	−670,665	
Propylene	$C_3H_6(g)$	−885,034	
Propane	$C_3H_8(g)$	−954,469	6,475
n-Butane	$C_4H_{10}(g)$	−1,236,984	9,055
n-Pentane	$C_5H_{10}(g)$	−1,520,315	11,355
n-Hexane	$C_6H_{14}(g)$	−1,803,517	13,556
n-Heptane	$C_7H_{16}(g)$	−2,086,719	15,701
n-Octane	$C_8H_{18}(g)$	−2,369,921	17,825
Benzene	$C_6H_6(g)$	−1,419,451	14,545
Toluene	$C_7H_8(g)$	−1,697,364	17,163
Methyl alcohol	$CH_3OH(g)$	−328,707	16,295
Ethyl alcohol	$C_2H_5OH(g)$	−605,916	18,204

SOURCE: Adapted from Kenneth Wark, *Thermodynamics*, 3d ed., McGraw-Hill, 1977, pp. 834–835, table A-23M.

TABLE G.3E IDEAL-GAS ENTHALPY AND ABSOLUTE ENTROPY FOR CO_2 AND CO AT 1 atm: ENGLISH UNITS

	Carbon dioxide (CO_2) $\Delta \bar{h}_f^\circ = -169{,}184$ Btu/lb$_m$·mol, $M = 44.010$			Carbon monoxide (CO) $\Delta \bar{h}_f^\circ = -47{,}519$ Btu/lb$_m$·mol, $M = 28.010$	
Temp., °R	$\bar{h}_T - \bar{h}^0$, Btu/lb$_m$·mol	$\bar{s}^0$, Btu/lb$_m$·mol·°R	Temp., °R	$\bar{h}_T - \bar{h}^0$, Btu/lb$_m$·mol	$\bar{s}^0$, Btu/lb$_m$·mol·°R
0	−4 026	0.0	0	−3 728	0.0
180	−2 776	42.756	180	−2 480	39.613
200	−2 638	43.434	200	−2 341	40.293
300	−1 923	46.342	300	−1 646	43.121
400	−1 153	48.606	400	−951	45.178
500	−322	50.449	500	−255	46.721
536.67	0	51.064	536.67	0	47.209
600	575	52.075	600	441	47.986
800	2 525	54.876	800	1 842	50.002
1000	4 655	57.248	1000	3 266	51.589
1200	6 928	59.318	1200	4 724	52.917
1400	9 315	61.157	1400	6 221	54.070
1600	11 798	62.814	1600	7 756	55.095
1800	14 350	64.321	1800	9 325	56.018
2000	16 982	65.703	2000	10 924	56.861
2200	19 659	66.979	2200	12 550	57.635
2400	22 380	68.162	2400	14 197	58.352
2600	25 138	69.266	2600	15 863	59.019
2800	27 926	70.299	2800	17 545	59.642
3000	30 742	71.270	3000	19 242	60.227
3200	33 579	72.186	3200	20 949	60.778
3400	36 437	73.052	3400	22 668	61.299
3600	39 312	73.873	3600	24 396	61.793
3800	42 202	74.655	3800	26 131	62.262
4000	45 105	75.399	4000	27 874	62.709
4200	48 021	76.111	4200	29 620	63.135
4400	50 948	76.791	4400	31 362	63.540
4600	53 884	77.444	4600	33 118	63.930
4800	56 830	78.071	4800	34 882	64.306
5000	59 784	78.674	5000	36 651	64.667
5200	62 746	79.254	5200	38 424	65.014
5400	65 715	79.815	5400	40 199	65.349
5600	68 690	80.356	5600	41 979	65.673
5800	71 672	80.879	5800	43 762	65.986
6000	74 660	81.386	6000	45 547	66.289
6200	77 654	81.876	6200	47 336	66.582
6400	80 653	82.352	6400	49 127	66.866
6600	83 657	82.815	6600	50 920	67.142
6800	86 667	83.264	6800	52 716	67.410
7000	89 682	83.701	7000	54 514	67.671
7200	92 701	84.126	7200	56 315	67.924
7400	95 724	84.540	7400	58 118	68.171
7600	98 753	84.944	7600	59 923	68.412
7800	101 786	85.338	7800	61 729	68.647
8000	104 823	85.723	8000	63 538	68.876
9000	120 070	87.518	9000	72 610	69.944
9900	133 885	88.981	9900	80 807	70.812

Adapted from *JANAF Thermochemical Tables* Third Edition, M. W. Chase et al., American Chemical Society and the American Institute of Physics for the National Bureau of Standards, 1985.

TABLE G.4E IDEAL-GAS ENTHALPY AND ABSOLUTE ENTROPY FOR H_2 AND H at 1 atm: ENGLISH UNITS

	Hydrogen, diatomic (H_2) $\Delta\bar{h}_f^\circ = 0$ Btu/lb$_m$·mol, $M = 2.016$			Hydrogen, monatomic (H) $\Delta\bar{h}_f^\circ = 93{,}723$ Btu/lb$_m$·mol, $M = 1.008$	
Temp., °R	$\bar{h}_T - \bar{h}^0$, Btu/lb$_m$·mol	$\bar{s}^0$, Btu/lb$_m$·mol·°R	Temp., °R	$\bar{h}_T - \bar{h}^0$, Btu/lb$_m$·mol	$\bar{s}^0$, Btu/lb$_m$·mol·°R
0	−3 640	0.0	0	−2 664	0.0
180	−2 351	24.058	180	−1 771	21.976
200	−2 225	24.687	200	−1 671	22.467
300	−1 585	27.301	300	−1 175	24.490
400	−928	29.224	400	−679	25.944
500	−251	30.727	500	−181	27.048
536.67	0	31.212	536.67	0	27.399
600	437	31.983	600	314	27.954
800	1 830	33.986	800	1 307	29.382
1000	3 228	35.546	1000	2 300	30.490
1200	4 630	36.823	1200	3 293	31.395
1400	6 038	37.908	1400	4 286	32.160
1600	7 457	38.856	1600	5 279	32.823
1800	8 891	39.700	1800	6 272	33.408
2000	10 344	40.465	2000	7 265	33.931
2200	11 817	41.167	2200	8 258	34.404
2400	13 314	41.818	2400	9 251	34.836
2600	14 833	42.427	2600	10 244	35.233
2800	16 376	42.998	2800	11 237	35.601
3000	17 942	43.538	3000	12 230	35.944
3200	19 529	44.050	3200	13 222	36.264
3400	21 137	44.538	3400	14 215	36.565
3600	22 765	45.003	3600	15 209	36.849
3800	24 412	45.448	3800	16 201	37.117
4000	26 076	45.875	4000	17 194	37.372
4200	27 758	46.285	4200	18 187	37.614
4400	29 454	46.680	4400	19 180	37.845
4600	31 167	47.060	4600	20 173	38.066
4800	32 893	47.428	4800	21 166	38.277
5000	34 633	47.783	5000	22 159	38.480
5200	36 386	48.127	5200	23 152	38.674
5400	38 151	48.460	5400	24 145	38.862
5600	39 929	48.783	5600	25 138	39.042
5800	41 718	49.097	5800	26 131	39.217
6000	43 519	49.402	6000	27 124	39.385
6200	45 330	49.699	6200	28 117	39.548
6400	47 153	49.988	6400	29 109	39.705
6600	48 986	50.270	6600	30 102	39.858
6800	50 829	50.546	6800	31 095	40.006
7000	52 683	50.814	7000	32 088	40.150
7200	54 546	51.077	7200	33 081	40.290
7400	56 420	51.333	7400	34 074	40.426
7600	58 303	51.584	7600	35 067	40.558
7800	60 196	51.830	7800	36 060	40.687
8000	62 098	52.071	8000	37 053	40.813
9000	71 744	53.207	9000	42 018	41.398
9900	80 595	54.144	9900	46 486	41.871

Adapted from *JANAF Thermochemical Tables* Third Edition, M. W. Chase et al., American Chemical Society and the American Institute of Physics for the National Bureau of Standards, 1985.

TABLE G.5E IDEAL-GAS ENTHALPY AND ABSOLUTE ENTROPY FOR N_2 AND N at 1 atm: ENGLISH UNITS

	Nitrogen, diatomic (N_2) $\Delta\bar{h}_f^\circ$ = 0 Btu/lb$_m$·mol, M = 28.013			Nitrogen, monatomic (N) $\Delta\bar{h}_f^\circ$ = 203,216 Btu/lb$_m$·mol, M = 14.007	
Temp., °R	$\bar{h}_T - \bar{h}^0$, Btu/lb$_m$·mol	$\bar{s}^0$, Btu/lb$_m$·mol·°R	Temp., °R	$\bar{h}_T - \bar{h}^0$, Btu/lb$_m$·mol	$\bar{s}^0$, Btu/lb$_m$·mol·°R
0	−3 727	0.0	0	−2 664	0.0
180	−2 480	38.170	180	−1 771	31.192
200	−2 341	38.858	200	−1 671	31.683
300	−1 645	41.690	300	−1 175	33.706
400	−950	43.727	400	−679	35.160
500	−255	45.273	500	−181	36.263
536.67	0	45.765	536.67	0	36.615
600	441	46.541	600	314	37.169
800	1 837	48.550	800	1 307	38.597
1000	3 251	50.127	1000	2 300	39.705
1200	4 693	51.441	1200	3 293	40.611
1400	6 169	52.578	1400	4 286	41.376
1600	7 681	53.587	1600	5 279	42.038
1800	9 227	54.497	1800	6 272	42.623
2000	10 804	55.328	2000	7 265	43.146
2200	12 407	56.092	2200	8 258	43.619
2400	14 034	56.800	2400	9 251	44.051
2600	15 681	57.459	2600	10 244	44.449
2800	17 345	58.075	2800	11 237	44.817
3000	19 025	58.655	3000	12 230	45.159
3200	20 717	59.201	3200	13 223	45.480
3400	22 421	59.717	3400	14 215	45.781
3600	24 135	60.207	3600	15 209	46.065
3800	25 857	60.673	3800	16 202	46.333
4000	27 587	61.116	4000	17 195	46.588
4200	29 324	61.540	4200	18 189	46.830
4400	31 068	61.946	4400	19 183	47.061
4600	32 817	62.334	4600	20 178	47.282
4800	34 571	62.708	4800	21 174	47.494
5000	36 330	63.067	5000	22 171	47.698
5200	38 093	63.412	5200	23 169	47.894
5400	39 860	63.746	5400	24 169	48.082
5600	41 631	64.068	5600	25 172	48.265
5800	43 405	64.379	5800	26 177	48.441
6000	45 182	64.680	6000	27 186	48.612
6200	46 962	64.972	6200	28 198	48.778
6400	48 745	65.255	6400	29 214	48.939
6600	50 530	65.530	6600	30 236	49.096
6800	52 318	65.797	6800	31 262	49.250
7000	54 109	66.056	7000	32 294	49.399
7200	55 902	66.309	7200	33 333	49.545
7400	57 696	66.555	7400	34 378	49.689
7600	59 493	66.794	7600	35 430	49.829
7800	61 292	67.028	7800	36 491	49.967
8000	63 093	67.256	8000	37 560	50.102
9000	72 125	68.320	9000	43 040	50.747
9900	80 293	69.185	9900	48 191	51.293

Adapted from *JANAF Thermochemical Tables* Third Edition, M. W. Chase et al., American Chemical Society and the American Institute of Physics for the National Bureau of Standards, 1985.

TABLE G.6E IDEAL-GAS ENTHALPY AND ABSOLUTE ENTROPY FOR NO AND NO_2 AT 1 atm: ENGLISH UNITS

	Nitric oxide (NO) $\Delta\bar{h}_f^\circ$ = 38,818 Btu/lb$_m$·mol, M = 30.006			Nitrogen dioxide (NO_2) $\Delta\bar{h}_f^\circ$ = 14,230 Btu/lb$_m$·mol, M = 46.006	
Temp., °R	$\bar{h}_T - \bar{h}^0$, Btu/lb$_m$·mol	$\bar{s}^0$, Btu/lb$_m$·mol·°R	Temp., °R	$\bar{h}_T - \bar{h}^0$, Btu/lb$_m$·mol	$\bar{s}^0$, Btu/lb$_m$·mol·°R
0	−3 952	0.0	0	−4 379	0.0
180	−2 611	42.283	180	−2 950	48.381
200	−2 458	43.037	200	−2 792	49.156
300	−1 708	46.097	300	−1 993	52.408
400	−979	48.240	400	−1 170	54.833
500	−261	49.834	500	−320	56.718
536.67	0	50.339	536.67	0	57.331
600	451	51.134	600	565	58.330
800	1 881	53.189	800	2 470	61.063
1000	3 338	54.815	1000	4 532	63.360
1200	4 834	56.178	1200	6 733	65.364
1400	6 372	57.363	1400	9 044	67.144
1600	7 948	58.415	1600	11 441	68.744
1800	9 557	59.362	1800	13 905	70.194
2000	11 194	60.224	2000	16 421	71.519
2200	12 853	61.015	2200	18 978	72.738
2400	14 532	61.745	2400	21 567	73.864
2600	16 228	62.424	2600	24 182	74.911
2800	17 937	63.057	2800	26 819	75.888
3000	19 658	63.650	3000	29 473	76.803
3200	21 388	64.209	3200	32 142	77.664
3400	23 128	64.736	3400	34 823	78.477
3600	24 875	65.235	3600	37 515	79.246
3800	26 629	65.710	3800	40 215	79.976
4000	28 388	66.161	4000	42 923	80.671
4200	30 153	66.592	4200	45 637	81.333
4400	31 923	67.003	4400	48 358	81.966
4600	33 698	67.398	4600	51 083	82.571
4800	35 476	67.776	4800	53 813	83.152
5000	37 258	68.140	5000	56 546	83.710
5200	39 043	68.490	5200	59 283	84.247
5400	40 831	68.827	5400	62 024	84.764
5600	42 622	69.153	5600	64 767	85.263
5800	44 416	69.468	5800	67 512	85.744
6000	46 212	69.772	6000	70 260	86.210
6200	48 011	70.067	6200	73 010	86.661
6400	49 812	70.353	6400	75 762	87.098
6600	51 616	70.630	6600	78 515	87.521
6800	53 421	70.900	6800	81 270	87.933
7000	55 229	71.162	7000	84 026	88.332
7200	57 038	71.417	7200	86 784	88.720
7400	58 849	71.665	7400	89 543	89.098
7600	60 662	71.907	7600	92 303	89.467
7800	62 478	72.142	7800	95 064	89.825
8000	64 294	72.372	8000	97 826	90.175
9000	73 401	73.445	9000	111 647	91.803
9900	81 628	74.316	9900	124 101	93.121

Adapted from *JANAF Thermochemical Tables* Third Edition, M. W. Chase et al., American Chemical Society and the American Institute of Physics for the National Bureau of Standards, 1985.

TABLE G.7E IDEAL-GAS ENTHALPY AND ABSOLUTE ENTROPY FOR O_2 AND O AT 1 atm: ENGLISH UNITS

	Oxygen, diatomic (O_2) $\Delta\bar{h}_f^\circ = 0$ Btu/lb$_m$·mol, $M = 31.999$			Oxygen, monatomic (O) $\Delta\bar{h}_f^\circ = 107{,}124$ Btu/lb$_m$·mol, $M = 15.999$	
Temp., °R	$\bar{h}_T - \bar{h}^0$, Btu/lb$_m$·mol	$\bar{s}^0$, Btu/lb$_m$·mol·°R	Temp., °R	$\bar{h}_T - \bar{h}^0$, Btu/lb$_m$·mol	$\bar{s}^0$, Btu/lb$_m$·mol·°R
0	−3 733	0.0	0	−2 891	0.0
180	−2 485	41.394	180	−1 942	32.470
200	−2 345	42.082	200	−1 828	33.033
300	−1 650	44.915	300	−1 268	35.317
400	−955	46.952	400	−724	36.915
500	−256	48.503	500	−193	38.096
536.67	0	48.999	536.67	0	38.468
600	445	49.784	600	331	39.050
800	1 880	51.845	800	1 358	40.529
1000	3 365	53.502	1000	2 374	41.663
1200	4 903	54.902	1200	3 383	42.582
1400	6 486	56.122	1400	4 387	43.357
1600	8 108	57.205	1600	5 389	44.025
1800	9 761	58.178	1800	6 389	44.614
2000	11 438	59.061	2000	7 387	45.140
2200	13 136	59.870	2200	8 385	45.615
2400	14 852	60.617	2400	9 382	46.049
2600	16 583	61.310	2600	10 378	46.448
2800	18 329	61.956	2800	11 373	46.817
3000	20 088	62.563	3000	12 369	47.160
3200	21 860	63.135	3200	13 364	47.481
3400	23 644	63.676	3400	14 359	47.783
3600	25 441	64.189	3600	15 354	48.067
3800	27 250	64.678	3800	16 349	48.336
4000	29 071	65.145	4000	17 344	48.591
4200	30 904	65.592	4200	18 339	48.834
4400	32 748	66.021	4400	19 334	49.066
4600	34 604	66.434	4600	20 331	49.287
4800	36 472	66.831	4800	21 327	49.499
5000	38 350	67.215	5000	22 325	49.703
5200	40 239	67.585	5200	23 323	49.898
5400	42 138	67.944	5400	24 322	50.087
5600	44 047	68.291	5600	25 323	50.269
5800	45 966	68.627	5800	26 326	50.445
6000	47 894	68.954	6000	27 329	50.615
6200	49 831	69.272	6200	28 335	50.780
6400	51 775	69.580	6400	29 343	50.940
6600	53 729	69.881	6600	30 352	51.095
6800	55 689	70.174	6800	31 364	51.246
7000	57 657	70.459	7000	32 378	51.393
7200	59 632	70.737	7200	33 394	51.536
7400	61 615	71.009	7400	34 413	51.676
7600	63 603	71.274	7600	35 434	51.812
7800	65 598	71.533	7800	36 458	51.945
8000	67 600	71.786	8000	37 485	52.075
9000	77 708	72.976	9000	42 658	52.684
9900	86 959	73.956	9900	47 371	53.183

Adapted from *JANAF Thermochemical Tables* Third Edition, M. W. Chase et al., American Chemical Society and the American Institute of Physics for the National Bureau of Standards, 1985.

TABLE G.8E IDEAL-GAS ENTHALPY AND ABSOLUTE ENTROPY FOR H_2O AND OH AT 1 atm: ENGLISH UNITS

	Water (H_2O) $\Delta\bar{h}_f^\circ = -103{,}966$ Btu/lb$_m$·mol, $M = 18.015$			Hydroxyl (OH) $\Delta\bar{h}_f^\circ = 16{,}761$ Btu/lb$_m$·mol, $M = 17.007$	
Temp., °R	$\bar{h}_T - \bar{h}^0$, Btu/lb$_m$·mol	$\bar{s}^0$, Btu/lb$_m$·mol·°R	Temp., °R	$\bar{h}_T - \bar{h}^0$, Btu/lb$_m$·mol	$\bar{s}^0$, Btu/lb$_m$·mol·°R
0	−4 258	0.0	0	−3 943	0.0
180	−2 844	36.397	180	−2 639	35.729
200	−2 685	37.177	200	−2 484	36.493
300	−1 889	40.414	300	−1 724	39.595
400	−1 092	42.769	400	−986	41.763
500	−293	44.540	500	−262	43.370
536.67	0	45.102	537.67	0	43.878
600	509	46.002	600	452	44.676
800	2 143	48.350	800	1 870	46.715
1000	3 826	50.225	1000	3 281	48.289
1200	5 567	51.811	1200	4 692	49.576
1400	7 371	53.200	1400	6 112	50.669
1600	9 241	54.448	1600	7 546	51.627
1800	11 178	55.589	1800	9 000	52.483
2000	13 183	56.645	2000	10 477	53.261
2200	15 254	57.631	2200	11 979	53.976
2400	17 389	58.560	2400	13 505	54.640
2600	19 583	59.438	2600	15 055	55.260
2800	21 832	60.271	2800	16 627	55.843
3000	24 132	61.064	3000	18 220	56.393
3200	26 479	61.821	3200	19 834	56.913
3400	28 867	62.545	3400	21 465	57.408
3600	31 294	63.239	3600	23 114	57.879
3800	33 756	63.905	3800	24 777	58.329
4000	36 251	64.544	4000	26 455	58.759
4200	38 774	65.160	4200	28 146	59.171
4400	41 324	65.753	4400	29 849	59.567
4600	43 899	66.325	4600	31 563	59.948
4800	46 497	66.878	4800	33 287	60.315
5000	49 115	67.412	5000	35 021	60.669
5200	51 752	67.929	5200	36 763	61.011
5400	54 406	68.430	5400	38 514	61.341
5600	57 077	68.916	5600	40 273	61.661
5800	59 763	69.387	5800	42 039	61.971
6000	62 464	69.845	6000	43 812	62.271
6200	65 177	70.290	6200	45 591	62.563
6400	67 903	70.722	6400	47 376	62.847
6600	70 639	71.144	6600	49 168	63.122
6800	73 387	71.554	6800	50 965	63.390
7000	76 145	71.953	7000	52 767	63.652
7200	78 913	72.343	7200	54 574	63.906
7400	81 690	72.724	7400	56 386	64.154
7600	84 475	73.095	7600	58 203	64.397
7800	87 268	73.458	7800	60 025	64.633
8000	90 069	73.812	8000	61 850	64.864
9000	104 176	75.474	9000	71 043	65.947
9900	117 006	76.832	9900	79 401	66.832

Adapted from *JANAF Thermochemical Tables* Third Edition, M. W. Chase et al., American Chemical Society and the American Institute of Physics for the National Bureau of Standards, 1985.

TABLE G.9E LOGARITHMS TO THE BASE e OF THE EQUILIBRIUM CONSTANTS, $\ln K_p$

For the reaction $\nu_A A + \nu_B B \rightleftharpoons \nu_C C + \nu_D D$, the equilibrium constant K_p is defined as: $K_p \equiv \dfrac{p_C^{\nu_C} p_D^{\nu_D}}{p_A^{\nu_A} p_B^{\nu_B}}$

Temp., °R	$\ln K_p$						
	$H_2 \rightleftharpoons 2H$	$O_2 \rightleftharpoons 2O$	$N_2 \rightleftharpoons 2N$	$H_2O \rightleftharpoons H_2 + \frac{1}{2}O_2$	$H_2O \rightleftharpoons \frac{1}{2}H_2 + OH$	$CO_2 \rightleftharpoons CO + \frac{1}{2}O_2$	$\frac{1}{2}N_2 + \frac{1}{2}O_2 \rightleftharpoons NO$
536	−164.005	−186.975	−367.480	−92.208	−106.208	−103.762	−35.052
900	−92.827	−105.630	−213.372	−52.691	−60.281	−57.616	−20.295
1800	−39.803	−45.150	−99.127	−23.163	−26.034	−23.529	−9.388
2160	−30.874	−35.005	−80.011	−18.182	−20.283	−17.871	−7.569
2520	−24.463	−27.742	−66.329	−14.609	−16.099	−13.842	−6.270
2880	−19.637	−22.285	−56.055	−11.921	−13.066	−10.830	−5.294
3240	−15.866	−18.030	−48.051	−9.826	−10.657	−8.497	−4.536
3600	−12.840	−14.622	−41.645	−8.145	−8.728	−6.635	−3.931
3960	−10.353	−11.827	−36.391	−6.768	−7.148	−5.120	−3.433
4320	−8.276	−9.497	−32.011	−5.619	−5.832	−3.860	−3.019
4680	−6.517	−7.521	−28.304	−4.648	−4.719	−2.801	−2.671
5040	−5.002	−5.826	−25.117	−3.812	−3.763	−1.894	−2.372
5400	−3.685	−4.357	−22.359	−3.086	−2.937	−1.111	−2.114
5760	−2.534	−3.072	−19.937	−2.451	−2.212	−0.429	−1.888
6120	−1.516	−1.935	−17.800	−1.891	−1.576	0.169	−1.690
6480	−0.609	−0.926	−15.898	−1.392	−1.088	0.701	−1.513
6840	0.202	−0.019	−14.199	−0.945	−0.501	1.176	−1.356
7200	0.934	0.796	−12.660	−0.542	−0.044	1.599	−1.216
8100	2.486	2.513	−9.414	0.312	0.920	2.490	−0.921
9000	3.725	3.895	−6.807	0.996	1.689	3.197	−0.686
9900	4.743	5.023	−4.666	1.560	2.318	3.771	−0.497
10800	5.590	5.963	−2.865	2.032	2.843	4.245	−0.341

SOURCE: Gordon J. Van Wylen and Richard E. Sonntag, *Fundamentals of Classical Thermodynamics*, 2d ed., SI Version 2d ed., John Wiley & Sons, Inc., 1976, p. 697, table A.12. Based on thermodynamic data given in the JANAF Thermochemical Tables, Thermal Research Laboratory, The Dow Chemical Company, Midland, MI.

Miscellaneous Properties

TABLE H.1E CRITICAL CONSTANTS

Substance	Formula	Molecular weight	Temp., °R	Pressure lb_f/in^2	Volume $ft^3/lb{\cdot}mol$
Ammonia	NH_3	17.03	730	1636	1.16
Argon	Ar	39.948	272	704.9	1.20
Benzene	C_6H_6	78.115	1012	713.6	4.17
Bromine	Br_2	159.808	1051	1500	2.17
n-Butane	C_4H_{10}	58.124	765	551.2	4.08
Carbon dioxide	CO_2	44.01	548	1072	1.51
Carbon monoxide	CO	28.01	239	507.6	1.49
Carbon tetrachloride	CCl_4	153.82	1002	661.4	4.42
Chlorine	Cl_2	70.906	751	1118	1.99
Chloroform	$CHCl_3$	119.38	966	793.4	3.85
Deuterium (normal)	D_2	4.00	69.1	240.8	
Dichlorodifluoromethane (R-12)	CCl_2F_2	120.91	692.9	598.3	3.47
Dichlorofluoromethane (R-22)	$CHCl_2F$	102.92	813	749.9	3.16
Ethane	C_2H_6	30.070	550	707.8	2.37
Ethyl alcohol	C_2H_5OH	46.07	929	925.4	2.68
Ethylene	C_2H_4	28.054	508	742.6	1.99
Helium	He	4.003	9.54	33.4	0.926
Helium[3]	He	3.00	5.94	17.4	
n-Hexane	C_6H_{14}	86.178	914	439.5	5.89
Hydrogen (normal)	H_2	2.016	59.9	188.6	1.04
Krypton	Kr	83.80	377	797.7	1.48
Methane	CH_4	16.043	344	672.9	1.59
Methyl alcohol	CH_3OH	32.042	924	1153	1.89
Methyl chloride	CH_3Cl	50.488	749	968.9	2.29
Neon	Ne	20.183	80.1	396.0	0.668
Nitrogen	N_2	28.013	227	491.7	1.44
Nitrous oxide	N_2O	44.013	557	1054	1.54
Oxygen	O_2	31.999	279	736.8	1.25
Propane	C_3H_8	44.097	666	617.9	3.20
Propene	C_3H_6	42.081	657	670.1	2.90
Propyne	C_3H_4	40.065	722	776.0	
Sulfur dioxide	SO_2	64.063	775	1143	1.95
Trichlorofluoromethane	CCl_3F	137.37	848	635.3	3.97
Water	H_2O	18.015	1165	3200	0.910
Xenon	Xe	131.30	522	852.8	1.90

Pseudo-critical properties for air: Air is a mixture of gases and therefore does not have true critical properties. However, pseudo-critical properties can be defined using Kay's rule for mixtures. Values for the pseudo-critical temperature and pressure obtained in this manner for air ($M = 28.97$) are:

$$T_c = 239°R$$

$$P_c = 545.4\ lb_f/in^2$$

SOURCE: Adapted from K. A. Robe and R. E. Lynn, Jr., *Chemical Review*, 52 (1953): 117–236. Copyright 1982, American Chemical Society.

TABLE H.2E CONSTANTS FOR BEATTIE-BRIDGEMAN AND THE BENEDICT-WEBB-RUBIN EQUATIONS OF STATE

The Beattie-Bridgeman equation of state is

$$P = \frac{\bar{R}T}{\bar{v}^2}\left(1 - \frac{c}{\bar{v}T^3}\right)(\bar{v} + B) - \frac{A}{\bar{v}^2}$$

where

$$A = A_0\left(1 - \frac{a}{\bar{v}}\right) \qquad B = B_0\left(1 - \frac{b}{\bar{v}}\right)$$

When P is in atm, $\bar{v}$ is in ft^3/lb$_m$·mol, T is in degrees Rankine, and $\bar{R}$ = 0.730 atm·ft^3/lb$_m$·mol·°R, then the constants in the Beattie-Bridgeman equation are as given in the table.

Gas	A_0	a	B_0	b	$c \times 10^{-4}$
Air	334	0.309	0.738	−0.0176	406
Argon	332	0.373	0.629	0	560
Carbon dioxide	1285	1.14	1.69	1.16	6170
Helium	5.6	0.958	0.224	0	0.37
Hydrogen	50.7	−0.081	0.336	−0.698	4.7
Nitrogen	345	0.419	0.808	−0.111	393
Oxygen	383	0.411	0.741	0.0674	449

The Benedict-Webb-Rubin equation of state is

$$P = \frac{\bar{R}T}{\bar{v}} + \left(B_0\bar{R}T - A_0 - \frac{C_0}{T^2}\right)\frac{1}{\bar{v}^2} + \frac{b\bar{R}T - a}{\bar{v}^3} + \frac{a\alpha}{\bar{v}^6} + \frac{c}{\bar{v}^3T^2}\left(1 + \frac{\gamma}{\bar{v}^2}\right)e^{-\gamma/\bar{v}^2}$$

When P is in atm, $\bar{v}$ is in ft^3/lb$_m$·mol, T is in degrees Rankine, and $\bar{R}$ = 0.730 atm·ft^3/lb$_m$·mol·°R, then the constants in the Benedict-Webb-Rubin equation are as given in the table.

Gas	a	A_0	b	B_0	$c \times 10^{-6}$	$C_0 \times 10^{-6}$	α	γ
n-Butane, C_4H_{10}	7747	2590	10.27	1.993	4219	826.3	4.531	8.732
Carbon dioxide, CO_2	563.1	703.0	1.852	0.7998	198.9	115.3	0.3486	1.384
Carbon monoxide, CO	150.7	344.5	0.676	0.8740	13.87	7.124	0.5556	1.541
Methane, CH_4	203.1	476.4	0.868	0.6827	33.93	18.78	0.5120	1.541
Nitrogen, N_2	103.2	270.6	0.598	0.6529	9.713	6.706	0.5235	1.361

SOURCE: Adapted from J. B. Jones and G. A. Hawkins, *Engineering Thermodynamics, an Introductory Textbook,* 2d ed., 1986, John Wiley & Sons, Inc., and Kenneth Wark, *Thermodynamics,* 3d ed., 1977, McGraw-Hill Book Company.

TABLE H.3E SPECIFIC HEATS OF COMMON LIQUIDS AND SOLIDS

Liquids		
Substance	State	c_p, Btu/lb$_m$·°R
Ammonia	sat., −5°F	1.08
	sat., 121°F	1.22
Benzene	1 atm, 58°F	0.430
	1 atm, 148°F	0.459
Bismuth	1 atm, 796°F	0.034
	1 atm, 1400°F	0.039
Ethyl alcohol	1 atm, 76°F	0.580
Glycerin	1 atm, 49°F	0.554
	1 atm, 121°F	0.616
Mercury	1 atm, 49°F	0.019
	1 atm, 598°F	0.032
Propane	1 atm, 32°F	0.576
Refrigerant-12	sat., −40°F	0.211
	sat., −5°F	0.217
	sat., 121°F	0.244
Sodium	1 atm, 202°F	0.330
	1 atm, 1003°F	0.301
Water	1 atm, 32°F	1.006
	1 atm, 76°F	0.998
	1 atm, 121°F	0.998
	1 atm, 212°F	1.006

Solids at 1 atm		
Substance	t, °F	c_p, Btu/lb$_m$·°R
Aluminum	−419	0.00389
	−329	0.076
	−150	0.167
	32	0.208
	212	0.225
	571	0.248
Carbon (diamond)	76	0.124
Carbon (graphite)	76	0.170
Chromium	76	0.107
Copper	−370	0.023
	−280	0.060
	−150	0.078
	−58	0.086
	32	0.091
	80	0.092
	212	0.094
	391	0.096
Gold	76	0.031
Ice	−329	0.162
	−221	0.262
	−77	0.392
	12	0.486
	28	0.402
Iron	67	0.107
Lead	−455	7.88×10^{-4}
	−435	7.28×10^{-3}
	−150	0.028
	32	0.030
	212	0.032
	571	0.036
Nickel	76	0.106
Silver	67	0.056
	391	0.058
Sodium	76	0.293
Tungsten	76	0.032
Zinc	76	0.092

SOURCE: Adapted from: Kenneth Wark, *Thermodynamics*, 4th ed., 1983, McGraw-Hill Book Company, table A-19M, p. 813.

TABLE H.4E TRIPLE-POINT PROPERTIES FOR COMMON SUBSTANCES

Substance	*T*, °R	*P*, lb_f/in^2
Ammonia (NH_3)	351.7	0.89
Carbon dioxide (CO_2)	389.9	74.92
Helium 4 (λ point)	3.91	0.735
Hydrogen (H_2)	24.9	1.03
Nitrogen (N_2)	113.7	1.82
Oxygen (O_2)	97.8	0.022
Water (H_2O)	491.7	0.0887

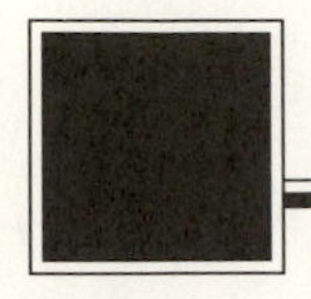

Index

D

G

H

I

Q

R

S

T